BEES AND BEEKEEPING

SCIENCE, PRACTICE AND WORLD RESOURCES

BEES AND BEEKEEPING

SCIENCE, PRACTICE AND WORLD RESOURCES

EVA CRANE OBE, DSc

formerly Director, International Bee Research Association

Comstock Publishing Associates

a division of Cornell University Press

Ithaca, New York

First published 1990 by Cornell University Press

Library of Congress Cataloging-in-Publication Data
Crane, Eva,
Bees and beekeeping: science, practice, and world resources / Eva Crane.
p. cm.
Includes bibliographical references.
ISBN 0-8014-2429-1
1. Bee culture. 2. Honeybee. I. Title.
SF523.C856 1990
638′.1—dc20 89-17477
CIP

Printed in Great Britain

This book is dedicated to the next generation of beekeepers, scientists and students throughout the world.

Contents

List of tables

Preface

Bees are kept in hives in almost every country of the world, and beekeepers have to operate in widely different conditions. This book presents the scientific principles underlying beekeeping management, and their practical application in different conditions. It also gives an account of honeybees as a world resource, both in producing honey and other hive products, and as pollinators that increase yields of seed and fruit crops. The book is the first attempt to provide an integrated picture of world beekeeping in the various continents, with a brief summary of its history.

During and since my time as Director of the International Bee Research Association (1949 to 1983), there have been many important developments in world beekeeping, and in the scientific research on which it is based. There has also been a greatly increased mechanization of apiary management and of the handling of hive products. Honey production in subtropical regions has expanded notably, as have beekeeping developments in the tropics. In selecting the contents for the book, I consulted some of the specialists named in Acknowledgements, and others, so that it would be as useful as possible to beekeepers, scientists and students throughout the world.

Different sections are written at somewhat different levels, according to how they are most likely to be used; for instance, handling bees is treated more simply than the composition of bee venom or propolis. Science and practice are interspersed in certain sections, where this seems useful. Some interesting topics for which the reader is unlikely to have a current reference source are given extra space. The book touches on many subjects that are currently under investigation, and will suggest lines for further research. I hope that it will also encourage readers to experiment with new ways in which beekeeping methods and equipment might be improved still further, and with the application of new technologies and materials – and to seek fresh conceptual approaches to beekeeping problems.

Overview of the book

Part I of the book describes the different bees that are used for beekeeping, and gives a short account of present-day knowledge about these bees, both as individuals and as members of their colony, leading a complex social life that endows the colony with some amazing capabilities.

Part II provides up-to-date information and instruction on modern beekeeping, based especially on expertise drawn from parts of the world where most advances are being made. Types of movable-frame hive and other equipment are described, together with methods of bee management in temperate, subtropical and tropical zones of the world, suitable for the different honeybees available. Attention is paid to ways of avoiding bee stings, and of dealing with their effects as necessary. Particular types of management have been devised, for bees used for crop pollination and for specialities such as queen rearing and package bee production.

In many countries the cost of hives and other beekeeping equipment is a major constraint to the spread of beekeeping. Part III presents well tried alternative equipment, including hives that are simpler and cheaper, but still fulfil the requirements of disease control legislation. The maintenance of bee health, and protection of bees against diseases, enemies, and injury from pesticides, form the subject of Part IV.

Part V covers the bees' plant resources on a world basis, and Appendix 1 lists 464 plants that are important honey sources, with their world distributions. Part V also deals with hive products, and tables provide world statistics of the production, export and import of honey and beeswax; tables showing numbers of hives in each country are in Part I. The tables show that the subtropics are the honey-exporting regions, and that most honey is imported by affluent countries of the north temperate zone. Other hive products have entered the commercial market during the past few decades: pollen, propolis, royal jelly, bee venom and bee brood. Production and trade figures quoted for them are more fragmentary, but they are probably the most complete yet published.

Part VI is concerned with the beekeepers themselves. One chapter is devoted to law as it affects beekeepers in different countries. Another deals with resources of many kinds that are available to beekeepers, including those with special needs and interests. Finally, the Gazetteer in Appendix 2 provides information needed by readers interested in beekeeping in any one of 177 individual countries and islands.

Cross-references are used in the text to link interrelated subjects in ways likely to be useful to the reader. The Bibliography lists some 2000 publications referred to in the text which give further information. Since the book is in English, publications listed for Further Reading and Reference at the end of each chapter are mostly in English. In the text, where reference could equally well be made to a publication in English or to one in another language, the English one is usually given; references in the Bibliography are to accessible journals in preference to those difficult for most readers to obtain. But where specific research is concerned, the original paper is usually cited whatever the language.

The book has 265 illustrations, and 90 tables which are listed on page x. SI units of measurements are used throughout the book, and conversions from non-metric units are on page xvii. Botanical names have been verified by the Royal Botanic Gardens, Kew.

When writing about a beekeeper and his actions, it has not been possible to eliminate the use of gender and still retain a fluent text. For ease of reading, references to 'the beekeeper' are 'he', 'him' or 'his', but the reader is asked to accept that in every case 'she' or 'her' is equally meant. Apart from cultural differences, which vary from one society to another, the intrinsic gender difference in beekeeping is that on average men are taller and physically stronger than women; they can lift heavier weights (see Section 5.41), and they can reach honey supers on taller hives. In some countries of south-east Asia, men tend to work with *Apis mellifera* and women with the smaller *A. cerana* bees.

Eva Crane

Acknowledgements

I have been helped enormously by colleagues throughout the world who have themselves advanced bee science and beekeeping. The following have actively contributed to the book by reading and improving substantial sections of the text, and in other ways:

Australia Stan Chambers, Bob Gulliford, Dr Francis Smith, Bruce White

Belgium Dr O. van Laere

Brazil Professor Warwick Kerr, Professor Paulo Nogueira-Neto

Canada John Corner, Professor Cameron Jay, Professor Reg Shuel, Professor Maurice Smith, Dr A. P. Tulloch

Egypt Professor S. E. Rashad

France Dr Jean Louveaux

German Federal Republic Dr Gudrun Koeniger

Greece Penelope Papadopoulo

India Professor L. R. Verma; also Central Bee Research Institute at Pune

Israel Professor A. Fahn

Italy Dr Franco Marletto; also Food and Agriculture Organization of the United Nations

Japan Institute of Honeybee Science at Tamagawa University

Kenya Peter Paterson

Netherlands Vincent Mulder, Dr Hayo Velthuis

New Zealand Trevor Bryant

Sweden Dr Ingemar Fries

Switzerland Rudy Kortbech-Olesen

Thailand Dr Pongthep Akratanakul

UK Vince Cook, Dr Hilary Fry, Dr Don Griffiths, Dr Bob Hider, Hans Kjaersgaard, David Lowe, Dr Harry Riches, Judge David Smith, John R. C. Walker; also CAB International Institute of Entomology in London and the Royal Botanic Gardens at Kew

USA Professor Michael Burgett, Dr Larry Connor, Dr Mercedes Delfinado-Baker, Dr Eric Erickson, Louis Hitchcock (Wake Island), Professor H. H. Laidlaw, Professor Charles Michener, Professor Roger Morse, Dr Tom Rinderer, Professor Nevin Weaver, Dr Jonathan White.

Many people in different countries have contributed information that is acknowledged in the text as a personal communication. Others have provided photographs, which are individually credited. Illustrations without an acknowledgement are my own.

I owe very much to my colleague Penelope Walker, who has done a great deal of work with me on the compilation and organization of the book, and has improved its structure and clarity. I also appreciate Sandra Braithwaite's work in word-processing all drafts of the text and tables.

The publisher Heinemann, for whom I edited *Honey: a comprehensive survey* in 1975, has again been most co-operative, allowing latitude in the content of the book and in the time scale for writing it. During this work I have been able to make extensive use of the Library of the International Bee Research Association, which was named the Eva Crane IBRA Library in 1987. Buckinghamshire County Library (Chalfont St Peter Branch) has also obtained many publications for me.

I have valued the support of the British Council, British Executive Service Overseas, and organizations in other countries, that contributed to the costs of travel during assignments for them, when I had opportunities to add to my own knowledge. I extend my thanks also to the many people who have given me hospitality, and taken me long distances through their own countries, to see the bees and beekeeping in different environments.

Finally I should like to put on record the debt this book owes to my husband James Crane, who died in 1978. His support and generosity of spirit made it possible for me to follow my wide interests in bees and beekeeping since 1942.

Conversion of non-metric units into SI units

SI = Système International d'Unités. More exact numerical equivalents will be found in standard works of reference.

Fractions and multiples

deci- (10^{-1})	d	deca- (10^{1})	da
centi- (10^{-2})	c	hecto- (10^{2})	h
milli- (10^{-3})	m	kilo- (10^{3})	k
micro- (10^{-6})	μ	mega- (10^{6})	M
nano- (10^{-9})	n	giga- (10^{9})	G
pico- (10^{-12})	p	tera- (10^{12})	T

Length

1 inch (in)	= 2.54 millimetre (mm)
1 foot (ft)	= 0.3048 metre (m)
1 yard (yd)	= 0.914 m
1 mile	= 1.61 kilometre (km)

Area

1 square inch	= 6.45 cm^2
1 square foot	= 0.093 m^2
1 square yard	= 0.836 m^2
1 square mile	= 2.59 km^2 = 259 hectare (ha)
1 acre	= 0.405 ha

Volume

1 imperial pint (pt) as used in UK	= 0.568 litre (dm^3)
as used in USA	= 0.473 litre
1 imperial gallon (gal) as used in UK	= 4.55 litre
as used in USA	= 3.79 litre
1 imperial fluid ounce (fl oz) as used in UK	= 28.4 ml (cm^3)
as used in USA	= 29.6 ml
1 cubic inch	= 16.39 cm^3 or 16.39 ml
1 cubic foot	= 0.028 m^3
1 cubic yard	= 0.765 m^3

Mass

1 ounce (oz) avoirdupois	= 28.4 gram (g)
1 pound (lb)	= 0.454 kilogram (kg)
1 cwt (long, UK) or 112 lb	= 50.8 kg
1 cwt (short, USA) or 100 lb	= 45.4 kg
1 zentner/centner/quintal	= 100 kg (in USSR = 50 kg)
1 ton (long, UK) or 20 cwt (long)	= 1.016 tonne
1 ton (short, USA) or 20 cwt (short)	= 0.907 tonne

Mass or volume per unit area

1 lb/acre	= 0.112 g/m^2
	= 1.12 kg/ha
1 (imperial, UK) gal/acre	= 11.2 litre/ha
1 (imperial, USA) gal/acre	= 9.36 litre/ha

Concentration of solutions

1 oz/(imperial, UK) gal	= 6.236 g/litre
1 oz/(imperial, USA) gal	= 7.49 g/litre
1 lb/cubic foot	= 16.02 g/litre

Speed

1 mile/hour (h)	= 1.61 km/h	= 0.44 m/second (s)
1 ft/second	= 1.097 km/h	= 0.305 m/s

Atmospheric pressure

1 lb/square inch	= 6.89 kN/m^2	= 0.0689 bar
	= 0.0703 kg/cm^2	= 51.7 mm mercury
	= 703 kg/m^2	
1 mm mercury	= 133 Newton $(N)/m^2$	
1 bar	= 10^5 N/m^2	

Temperature

$x°$ Fahrenheit $= \frac{5}{9}(x - 32)°$ Celsius

Power and energy

1 Watt	= 1 J/s
1 calorie	= 4.187 joule (J)
1 kilowatt-hour (kWh)	= 3.6 MJ

PART I

The Bees used in Beekeeping, and Background Information

1

The basis of beekeeping

1.1 INTRODUCTION

The bees of overriding world importance in beekeeping (the management of bees in hives), and the basis of the world's beekeeping industry, are races and strains of the honeybee *Apis mellifera* that are native to Europe. They have also been introduced into almost the whole of the New World (the Americas, Australia, New Zealand and Pacific Islands), where there are no native honeybees. In contrast to these temperate-zone bees, tropical African races of *Apis mellifera* are not so amenable to management in hives, and their introduction to Brazil in 1956 has substantially changed beekeeping in almost the whole of Latin America.

There are several other species of honeybee which are native to parts of Asia, especially in the tropics. *Apis cerana* can also be kept in hives. It is similar in many ways to *A. mellifera*, but is usually smaller and less productive, and for this reason *A. mellifera* has been introduced into some parts of *A. cerana* territory. In tropical Asia much honey is also collected from the large wild nests of *Apis dorsata*, and from the small but more accessible nests of *Apis florea*. In addition to these *Apis* species, many species of stingless bee (Meliponinae) are kept for their smaller yields of honey. All these bees are considered in Section 1.2.

Bees depend wholly on plants for their food, and both climate and soil determine what plants are able to grow and flower within the foraging range of bees from the colonies in a region. Colonies of native bees are likely to be large when the greatest flowering occurs. Many other plants have been introduced from one region to another by man, and they may flower at times of year when colonies are small; beekeepers must then carry out special 'colony management' if they are to get a honey harvest from such plants. Beekeepers may find it beneficial to grow special plants that provide their bees with forage, especially during periods of dearth. These subjects are dealt with in Chapter 12.

Bees foraging from a hive range over several square kilometres, but the hives themselves require a minimal amount of land. The bees' food resources – nectar, pollen and honeydew – have little or no economic use unless honeybees collect them. The economic yield of honey produced per hive varies greatly in different parts of the world. It depends on: the duration of the season when temperatures are high enough for the bees to fly, and for plants to grow and flower; the amounts of nectar, honeydew and pollen within flight range of the bees; the genetic honey-producing capability of the colonies; and the expertise and equipment of the beekeepers. Migratory beekeeping can greatly increase honey yields, and some beekeepers move their hives up to eight times a year to take advantage of honey flows from different plants.

Section 1.5 sets out the present pattern of world beekeeping, continent by continent. Table 1.5A shows the number of hives of bees, and honey yields per hive, in 147 individual countries, and Table 13.8A provides production and trade statistics for honey and beeswax. Section 1.4 gives a brief account of the history of beekeeping and hive products.

Bees pollinate fruit and seed crops, bringing a very much greater economic benefit to agriculture than the beekeepers' income from honey and other hive products. Other species besides honeybees are also reared for this purpose (Section 1.3).

1.2 SPECIES AND RACES OF BEES FROM WHICH HONEY IS HARVESTED

This book is mainly concerned with a few species of bees, all social, that are kept in hives by beekeepers.

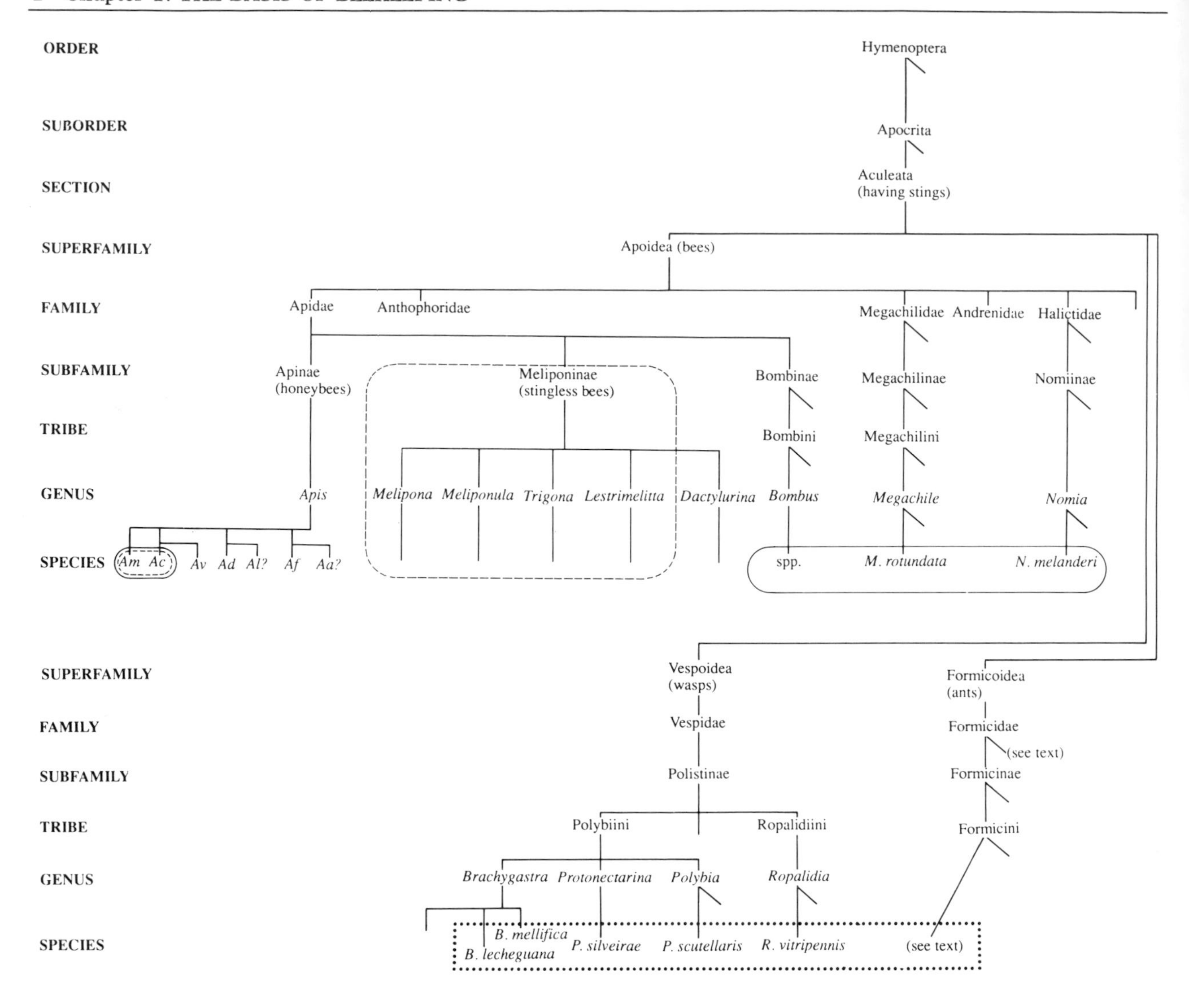

Figure 1.2a Diagram showing the taxonomy of the bees and other insects discussed in this book.
At each level, the line leading to the honeybees is on the left; in the centre are lines leading to other bees, and on the right are those leading to some other insects mentioned in this book. At any level, an unnamed branch indicates one or more members not discussed in the book. Species names within *Apis* follow Shao-wen et al. (1986), but *A. vechti* is added; see Section 1.23.

Aa	*andreniformis*	Af	*florea*	not included:
Ac	*cerana*	Al	*laboriosa*	*A. binghami*
Ad	*dorsata*	Am	*mellifera*	*A. breviligula*
		Av	*vechti*	

The two broken lines ----- enclose two *Apis* species, and genera of Meliponinae, that are widely managed for honey production. Honey is collected from wild colonies of all *Apis* spp. and many Meliponinae, and also from the other insects enclosed by the dotted line ······. Two solid lines ——— enclose species that are managed for crop pollination.

Other social species from which honey is harvested in the wild are also mentioned, and still other species (some social and some solitary) that are used for crop pollination are discussed. Figure 1.2a sets out the taxonomy of these bees.

Bees (Apoidea) are a superfamily of about 20 000 species, in the order Hymenoptera; other superfamilies in the Hymenoptera include the ants and wasps. The majority of bee species are 'solitary', not social: there is no worker caste, and each female makes her own nest and lays her eggs in it; she does not usually live there, and no males do so. But a minority of bee species are social, also some hundreds of species of wasps, and all 12 000 or more species of ants. So are all species of termites, the only social insects outside the Hymenoptera. In this entomological context 'social' means that the individual insect lives out its life in a social community, referred to as a colony. One characteristic of social bees is that workers produce enzymes (and pos-

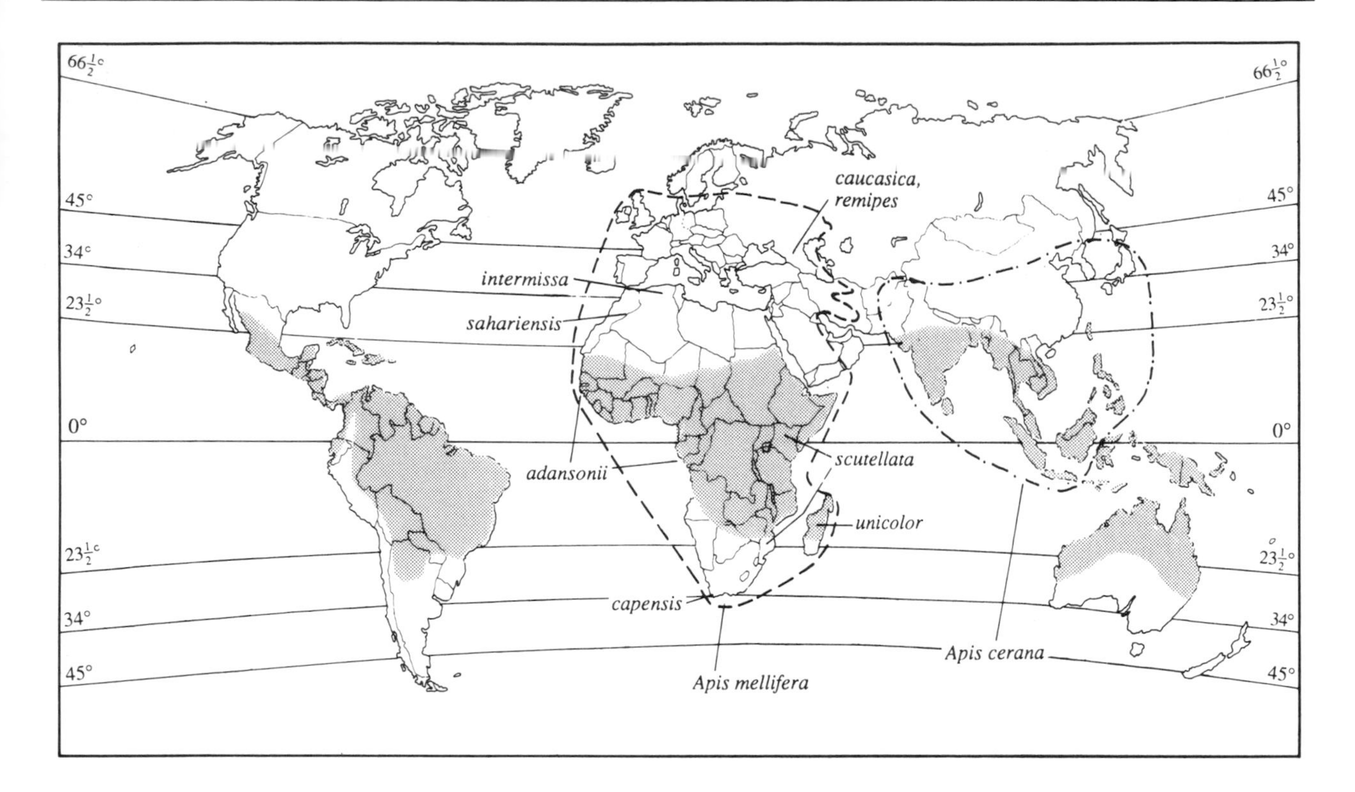

Figure 1.2b Natural distribution of species of honeybee (*Apis*) kept in hives and of stingless bees (Meliponinae).
---- *Apis mellifera* is native to Africa, and parts of Europe and the Middle East. Locations of some races outside Europe are indicated.
–·–·– *Apis cerana* is native to parts of Asia; distributions of other *Apis* species native to Asia are shown in Figure 1.23e.
Shaded areas: Meliponinae are native to many parts of the tropics.
Maps are based on information in Ruttner (1986, 1988), Sakagami (1982), Crane (1984*d*) and elsewhere. Transport by man has greatly reduced the separation between the distributions of *A. mellifera* and *A. cerana*.

sibly other secretions) that enable them to make honey, and thus to store a food safe from spoilage, for use in dearth periods. A beekeeper 'manages' a colony of bees so that it produces *more* honey than it needs, and he can then harvest the surplus.

Bees that produce enough honey to be worth harvesting belong to the two subfamilies of the family Apidae: Apinae (honeybees) and Meliponinae (stingless bees). Apinae has only one genus, *Apis*, of which the species *Apis mellifera* is of much greater economic importance than any other. Among the species of *Apis* and their races, those native to regions with cold winters (north temperate zone) in general store more honey than those native to subtropical and tropical regions where flight and foraging are not interrupted by a long cold period. Meliponinae, which has five genera and is confined to the tropics, is of less importance. Below, the *Apis* species are considered in Sections 1.21–1.23 and the Meliponinae in Section 1.24.

Referring to Figure 1.2a, all named and unnamed bees, and many of the other insects in the Aculeata, play a part in flower pollination. *Apis mellifera* and other species reared specifically for crop pollination are indicated in the figure.

Michener (1979) has summarized what is known of the relative abundance and diversity of bees in various parts of the world – both the superfamily Apoidea as a whole, and the families, subfamilies and tribes within it. His 1974 book is a comparative study of the social behaviour of the bees.

Since the late 1700s four species of honeybees have been recognized, and Figures 1.2b and 1.23e show their natural distributions. *Apis cerana* and *Apis mellifera* live in the Old World tropics, but during evolutionary times they succeeded in spreading into the north temperate zone of the Old World. Each builds a nest in a cavity, consisting of a number of parallel vertical combs, usually up to about ten (Figure 1.2c).

Apis cerana occupies tropical Asia, and in eastern Asia its range extends as far north as part of the USSR near the Pacific coast. The natural distribution of *Apis*

Figure 1.2c Large natural nest of *Apis cerana* inside a building, in Japan (photo: Tanji Inoue).
A nest of *A. mellifera* looks very similar.

*mellifera** occupies the continents of Europe and Africa except for the desert regions. It extends as far north as southern Scandinavia and the forests of Russia, and as far east as the Urals and, farther south, Iran. It has been widely introduced elsewhere and is now used for beekeeping in most of the world, and in this book statements refer to *Apis mellifera* except where otherwise indicated

Apis dorsata and *Apis florea* are confined to tropical Asia, and each species builds a nest in the open, consisting of a single vertical comb (see Figures 8.61b and 8.62a).

Ruttner's *Biogeography and taxonomy of honeybees* (1988) provides a detailed account of the evolution, geographical variability and taxonomy of the *Apis* species, and his *Geographical variability and classification* (1986) is also much referred to here.

1.21 Temperate-zone *Apis mellifera*

Although the genus *Apis* originated in the tropics, the course of history was such that European *Apis mellifera* was the bee first studied, and it still receives by far the most attention. It has been spread by man to many other parts of the world (Section 1.45), and to most readers it will be more familiar than any other honeybee. It is therefore discussed first.

During their evolution, many local populations of *A. mellifera* in Europe were in regions which became cut off from others, by the sea or by frozen mountain barriers. As a result, these populations became differentiated into rather distinct regional types, usually referred to as races. Figure 1.21a shows results of a multivariate analysis, based on measurements of the size of various parts of the body of the worker which increases in size from left to right. Three 'branches' appear: African, *A. m. carnica* and others in southern/eastern Europe, and *A. m. mellifera* and others in northern/western Europe. This analysis distinguishes the individual races within these branches.

Brother Adam's reports in *Bee World* from 1951 to 1965, republished as a book *In search of the best strains of bees* (1968*a*), provide many interesting sidelights on the various races in their native regions. His book

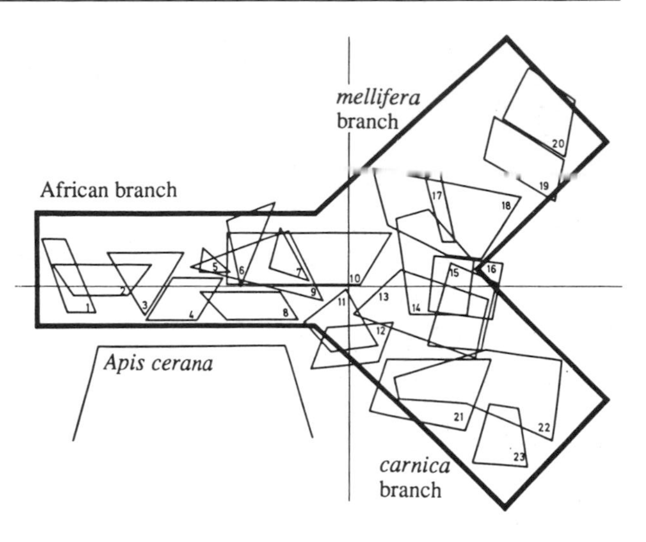

Figure 1.21a Separation of geographical races of *Apis mellifera* by multivariate analysis (principal components) (Ruttner, 1986).
In the text, the reference numbers in the diagram are inserted against the corresponding names:

1 *A. m. jemenitica*
2 *A. m. litorea*
3 *A. m. lamarckii*
4 *A. m. adansonii*
5 *A. m. capensis*
6 *A. m. unicolor*
7 *A. m. sahariensis*
8 *A. m. scutellata*
9 *A. m. syriaca*
10 *A. m. monticola*
11 *A. m. meda*
12 *A. m. cypria*
13 *A. m. iran*
14 *A. m. anatoliaca*
15 *A. m. sicula*
16 *A. m. adami*
17 *A. m. caucasica*
18 *A. m. intermissa*
19 *A. m. iberica*
20 *A. m. mellifera*
21 *A. m. ligustica*
22 *A. m. carnica*
23 *A. m. cecropia*

Breeding the honeybee (1987) evaluates different temperate-zone honeybees for 16 characters that are valuable in beekeeping.

1.21.1 Races of southern/eastern Europe

These are the bees that have been most extensively reared, bred, and exported to other parts of the world in recent times.

Apis mellifera ligustica (21)*, known as the Italian bee, or (Adam, 1968*a*) as the Ligurian bee from a narrow coastal strip near Genoa, at the head of the Italian peninsula. The bees were hemmed in on the inland side by the Alps, which were frozen and impenetrable during the Ice Age. Workers and queen have a conspicuous yellow colour on the abdomen. They are very adaptable to different environments, and in a season of good honey flows can produce large colonies and large honey yields; they are also comparatively efficient in preventing wax moth damage to their combs.

* The correct, official name of this bee is *Apis mellifera* L., which means the 'honey-bearing bee', not a very good descriptive name. It was the first name given to the bee by Linnaeus, and it appeared in the 10th edition of his *Systema Naturae* (1758). Later, he changed the name to *Apis mellifica*, the 'honey-making bee', which is much more appropriate. Unfortunately, this name has been invalidated by the International Rules of Nomenclature, in which 1758 is chosen as the date after which properly applied names cannot be changed. Linnaeus had his second thoughts too late from this point of view (Dade, 1962).

* This and subsequent numbers identify the races in Figure 1.21a.

A. m. carnica (22), the Carniolan bee, is named from Carniola (Krain), in the north of Yugoslavia, which is south of part of Carinthia (Kärnten) in Austria. Its natural distribution covers a wide area, stretching from the eastern Alps south into the Balkans; the bee in Greece is *A. m. cecropia* (23). Eastwards it reaches the Carpathians, where *A. m. carpathica* is regarded by some authors as a separate race, and as far as the Black Sea and into the Ukraine. Settlers took their bees from the Ukraine to the Russian Pacific seaboard, where the varroa mite transferred to them from the native honeybee *A. cerana* (see Section 11.42).

The Carniolan bee is rather large, with a dark abdomen, and is noted for its gentleness, its rapid colony growth early in spring, and its economical habits of food conservation and brood rearing. Because it has no yellow in its body colour, it was selected for large-scale introductions into several countries (e.g. Germany, Israel and Egypt; Sections 8.73, 8.23), where the less satisfactory native bees have some yellow in their coloration. Any unwanted crosses between Carniolan and local bees could then be identified by the presence of yellow in the colouring.

A. m. caucasica (17) and *A. m. remipes* are Caucasian bees, native to Georgia and the Caucasus mountains that stretch between the Black Sea and the Caspian Sea. The first is the larger, a dark or grey mountain bee; the second, lower down in Azerbaijan, is more yellow. Caucasian bees (especially the mountain bees) have the longest tongue of any *Apis mellifera*, and this makes them effective pollinators of crop plants whose flowers have a long corolla. They are noted for their extensive use of propolis. Honeybees of Central Anatolia (sometimes referred to as *A. m. anatoliaca*, 14) and some other parts of Turkey are thought to be related to Caucasians.

In addition Ruttner (1986) includes as separate races *A. m. syriaca* (9) in Syria and what was Palestine, and *A. m. meda* and *A. m. iran* (11, 13) in Iran; the former is discussed in detail by Ruttner et al. (1985).

1.21.2 Races of northern/western Europe

This region is the home of a group of dark bees, the most widespread of which is *A. m. mellifera* (20). Its natural area of distribution stretches from the coast and its offshore islands in the west to the Urals in the east, and from the Alps in the south to Scandinavia in the north. Hansson (1955) has suggested that its natural northern limit is about the same as that of hazel (*Corylus avellana*).

South of the Pyrenees is the Iberian peninsula, with the Iberian bee *A. m. iberica* (19), from which *A. m. mellifera* may be derived. *A. m. mellifera* and *A. m. iberica* were the bees used for beekeeping on the western seaboard of Europe, from which seafarers sailed to discover the New World, and settlers later took their hives of bees. In this way *A. m. mellifera* became the first honeybee in North America and the Caribbean islands, and *A. m. iberica* in Central and South America (Section 1.45). Other races were taken later.

1.21.3 Mediterranean island races

A. m. cypria (12), the Cyprian honeybee, is native to Cyprus which lies south of Turkey. It was the subject of much interest during the last century from the 1860s onwards, partly because beekeepers liked its reddish-yellow colour. However, it stings readily, and is not much used for beekeeping. Sicily, the island off the south-west tip of the Italian mainland, is the home of the Sicilian bee *A. m. siciliana* (15), for which the name *A. m. sicula* is sometimes incorrectly used. Both these bees may construct very large numbers of queen cells at swarming time.

The honeybee native to Crete (between Cyprus and Sicily) is intermediate in behaviour between European honeybees and those farther east. It was named *A. m. adami* (16) by Ruttner (1975*b*).

1.21.4 Races of North Africa

A. m. lamarckii (3), known earlier as *fasciata*, is the honeybee of the lower Nile valley at the eastern extremity of North Africa. It is small, with black and yellow abdominal bands, and it stings quite readily. It must have been the bee portrayed in the earliest records of beekeeping (Section 1.42), and in the countless representations of the bee as part of the hieroglyph of the titulary of the Pharoah of Ancient Egypt. Its brood-rearing cycle is shown in Figure 3.51b.

A. m. intermissa (18), known as the Punic or Tellian bee, is native to most of the coast of North Africa, from the Libyan desert to the Atlantic coastal belt. It is a black bee, which is easily alerted to sting and uses much propolis. It seems to be the only race that can live in the climatic extremes of this region. In a drought year as many as 80% of the colonies in an area may die, but in the next year with adequate rain so many swarms are produced that the colony population recovers. The bees are unsuited to more normal temperate-zone conditions, and those imported to Europe did not survive.

A. m. sahariensis (7) lives in oases in the Sahara desert that lie south of the Atlas Mountains, in Morocco and the western edge of Algeria. It is probably a relic of a much larger population that occupied the area when the climate and vegetation were more benign. The body colour is tan/yellow, and the bees are not very effective at defending their nest. Their most notable character is an ability to survive under very extreme conditions of heat and drought, with an annual foraging season of four months or less. Ruttner's morphometric analyses suggest that this bee is a link between honeybees of the western Mediterranean and of tropical Africa (Section 1.22).

In a small coastal area of the Rif Mountains in the north of Morocco is a local type of *A. mellifera* (named *major* by Ruttner, 1975*b*), notable for its large size.

1.22 Tropical African *Apis mellifera*

Bees described here are native to regions of Africa south of the Sahara. During and since evolutionary times, these regions were not isolated from each other by sea or frozen mountains, as parts of Europe were, and free hybridization occurred between adjoining groups of honeybees. So the groups are not sharply separated, and the term race is not very appropriate for them. Tropical forms of *Apis mellifera* are smaller than temperate-zone forms, and they have a more slender abdomen. Their colonies produce many more swarms; also, the whole colony may abscond as a result of damage or disturbance, or shortage of food (Section 3.34.5). The bees are easily alerted to sting and may attack *en masse*, and some bees may follow a person or animal a kilometre from the place of attack. This characteristic, common to most groups, allowed their survival in the African tropics where they were liable to be attacked by many 'enemies' (Section 11.5).

For further general information there is a Symposium report and a review (Fletcher, 1977, 1978), a bibliography (Crane, 1978*a*, No. 9), and a detailed ecological study on honeybees in Kenya (Kigatiira, 1984).

1.22.1 Races of West Africa

A. m. adansonii (4), first described from Senegal, has a natural distribution extending along the coastal lowlands of West Africa, from the westernmost point of Cape Verde in Senegal to Gabon just north of Angola.

Little is known about honeybees in the hinterland, or indeed about any honeybees west of the highlands of East Africa, 5000 km distant.

1.22.2 Races of East Africa

Perhaps because *A. m. adansonii* was named relatively early, by Latreille in 1804, the name came to be used for tropical African bees in general, including those of East Africa. In 1836 Lepeletier gave the name *A. m. scutellata* (8) to a bee 'de la Caffrerie', i.e. of the Kaffir, a Bantu people in Southern Africa. The bee inhabits open woodland, mainly at about 500–1500 m, including the miombo whose distribution is shown in Figure 12.7a, map 3. The name *A. m. adansonii* should not be used for bees in eastern and southern Africa; Ruttner regards them all as *A. m. scutellata*, except as noted below.

A. m. scutellata is of medium size for tropical African bees. A study of samples from countries stretching from Ethiopia to South Africa led Ruttner (1986) to conclude that it is 'a very well defined and relatively uniform race', although he also refers to considerable variability in its pronounced defensive behaviour. Many studies and development programmes have been carried out in the regions occupied by this bee, and its behaviour and other characteristics have been widely recorded, although it has often been referred to as *A. m. adansonii* (see e.g. F. G. Smith, 1953, 1960; Fletcher, 1977; Ruttner, 1986). Section 7.42 discusses bee management in relation to the bee's characteristics, of which swarming and absconding, and extreme readiness to sting, present major difficulties to the beekeeper. Ruttner quotes F. G. Smith's (1961*a*) suggestion that this may be the primary race of *A. mellifera* from which all others evolved in Africa, and also in Europe – reached by way of North Africa and the Iberian peninsula.

A different type of bee, and the only African one approaching European *A. mellifera* in size, exists at high altitudes in the same region. F. G. Smith (1961*a*) first recorded it on Mount Kilimanjaro in Tanzania and named it *A. m. monticola* (10); it has since been found also in Kenya, Burundi and Ethiopia. Larger than *A. m. scutellata*, it is dark, with longer hair on the abdomen, and is relatively gentle. Its reputed gentleness has led to consideration of its use for beekeeping at lower altitudes or even in temperate zones, but the bee's survival and performance there have been little studied. Dietz and Krell (1986) reported that most colonies sampled, especially those between 1500 and 2100 m, were 'aggressive to very aggressive'.

In contradistinction to *A. m. monticola*, F. G. Smith (1961*a*) gave the name *A. m. litorea* (2) to the bee on the coast of Tanzania, which was found later also in Kenya and Mozambique. It is very yellow, and small but with a relatively long tongue. Morphologically it is rather similar to *A. m. adansonii* on the west coast. In the area in which it lives there are nearly always plants in flower, and the colonies abscond from a dearth area to one with forage; see Section 3.34.5.

To the north of the East African highlands and plains, the belt of desert continues east from the Sahara, across the Nile valley and into the Arabian peninsula, whose native *A. mellifera* was named *jemenitica* (1) by Ruttner (1975*b*); see also Dutton et al. (1981). Ruttner found that honeybees from Chad and Sudan in Africa were morphologically of the same type, and (1986) described *A. m. jemenitica* as the

African bee of the dry thornbush [*Acacia*] savannah. Its survival through long drought periods may be compared with that of *A. m. sahariensis*.

1.22.3 Races of Southern Africa

The honeybees in Madagascar, off the east coast of Southern Africa, are considered to be indigenous (Tribe, 1987); they are uniformly black and were named *A. m. unicolor* (6). From Madagascar they were taken to Réunion and Mauritius nearby, which had no native honeybees. Both these islands later imported European bees, but *A. m. unicolor* continues to live wild as well as in hives, so it is looked upon as the native bee. In Madagascar, bees in the more temperate highlands show differences from those in the coastal regions, for instance they are gentle, and colonies do not abscond.

In the extreme south of Africa is the relic population of *A. m. capensis* (5), the Cape honeybee; its distribution area now extends little further than about 50 km from the Cape Peninsula. This bee is about the same size as *A. m. scutellata*, but has a dark abdomen. Its reproductive behaviour is most unusual, and is described briefly in Section 3.39. The development period of *A. m. capensis* is very short, the adult worker emerging only 9.6 days after the cell is sealed (see Table 2.41A and Section 11.42), and *Varroa* mites can hardly reproduce on its worker brood.

1.22.4 Africanized *Apis mellifera*

In 1956 more than a hundred queens were transported from Southern Africa to Brazil in the South American tropics, and 49 survived the journey (Kerr, 1957). In 1957 some colonies swarmed, and the bees hybridized with those of European origin that were used for beekeeping in South America. The spread of the resultant 'Africanized' bees through South and Central America and into Mexico, and its impact on beekeeping, are described in Section 7.43.

The Africanized bees show the influence of both African and European ancestry. They are intermediate in size, and can be distinguished from either by applying certain methods of statistical analysis to their morphological characteristics (Rinderer et al., 1986, 1987*a*; Sylvester & Rinderer, 1986; Buco et al., 1987); see also Section 7.43.2. Composition of nuclear DNA, desoxyribonucleic acid (Sheppard & Huettel, 1987) may be a distinguishing character. Other possible approaches include wing-beat frequency (Maxwell, 1987*b*), high-resolution supercritical fluid chromatography of beeswax (Garside, 1987), and assay of certain hydrocarbons extracted from whole bees (R.-K. Smith, 1988).

A number of new scientific findings on Africanized bees were published in the proceedings of an International Conference in Ohio (Needham et al., 1988).

1.23 Other species of *Apis*

Much less is known about other honeybees than about *Apis mellifera*. There are, however, a number of publications about them, although many of the early ones are anecdotal. Bibliographies are available (Crane, 1978*a*, Nos 11–13; 1987*b*, S32).

An *Apis cerana* colony builds a number of combs, in a cavity (like *A. mellifera*); see Figure 1.2c. *A. dorsata* and *A. florea* build a single comb in the open. Figure 1.23a shows the relative sizes of the workers, and Figure 1.23b indicates two measurements (forewing length and cubital index) that together can be used to distinguish these species. An effective tool in molecular systematics is electrophoretic investigation of isozymes, and in the course of studying isozymes in Apoidea, Shao-Wen et al. (1986) examined the esterase isozymes in species of *Apis*. Figure 1.23c shows the differentiation between the four species of *Apis* named above, and two further possible species; their status is not yet established, and as an interim measure they are referred to in this book as *Apis dorsata/laboriosa* and *Apis florea/andreniformis*. Figure 1.2a indicates the phylogenetic relationship suggested on the basis of zymograms.

In 1953, Maa put forward proposals to divide *Apis* into about 25 species. It seems unlikely that these proposals will ever be accepted in their entirety, but our knowledge about the honeybees in different parts of tropical Asia – still very incomplete – is now increasing; see for instance under *A. vechti* in Section 1.23.1. In this book, except in Section 1.2, only the species *mellifera*, *cerana*, *dorsata* and *florea* are referred to.

Michener (1974) regarded it as probable that an ancestral *Apis* species gave rise to two lines in the Asian tropics, one leading to *A. florea* and the other to the remaining species. Koeniger (1976*a*), on the other hand, proposed tropical Africa as being the ancestral homeland of *Apis*, and that cavity-nesting bees were the earlier species, which in turn gave rise to species nesting in the open. Ruttner (1988) discusses the evolution of the *Apis* species in detail, including a number of outstanding puzzles.

The development stages of immature bees of different *Apis* extend over different periods, as shown in Table 2.41A. This is an important factor in determining resistance to infestation by the mite *Varroa jacobsoni* (Section 11.42). Unlike *A. mellifera* (except *A. m. capensis*), *A. cerana* has only one caste (drone) whose developmental period is long enough to allow the mite to reproduce on it.

Figure 1.23a Relative sizes of workers of *Apis* species.
above 5 × natural size (Benton & Morse, 1968).
below About 2 × natural size (photo: Li Shao-wen).
Names follow Shao-wen et al. (1986), and are listed in full in Figure 1.2a.

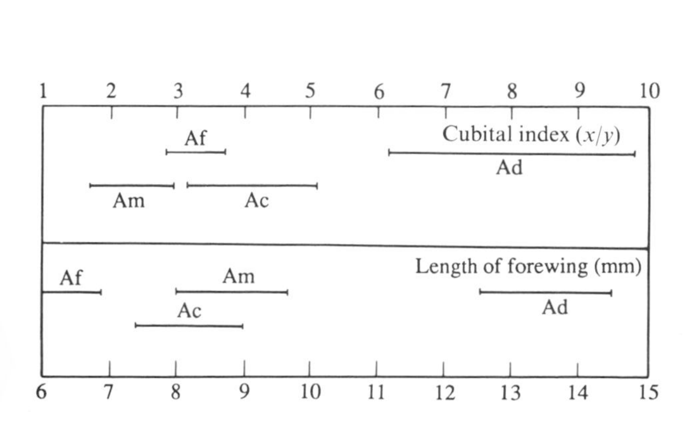

Figure 1.23b Two characters of the worker's wing that together distinguish four *Apis* species (data from Ruttner, 1986).
Cubital index and length of forewing are identified in Figure 2.22a.

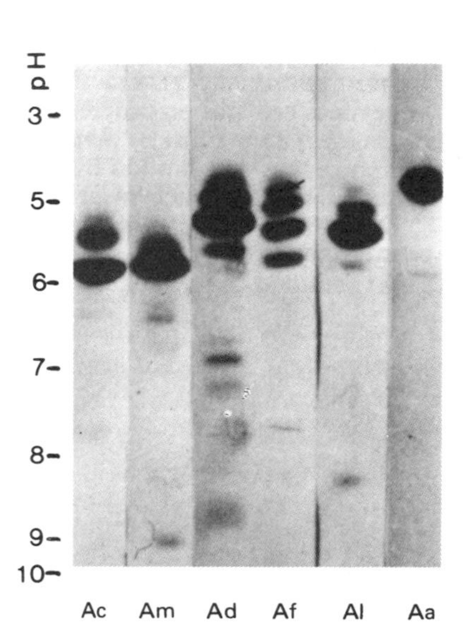

Figure 1.23c Esterase zymograms of six *Apis* species (Shao-wen et al., 1986).
Names as listed in Figure 1.2a.

The most significant morphological difference between the *Apis* – apart from size – is the shape of the drone's endophallus, which is shown in Figure 2.36b. This is a determining factor that prevents mating between a queen and drone of different species.

The various *Apis* show several behavioural differences in addition to the number of combs in their nest and the type of site chosen for it. Returning foragers of all *Apis* bees are able to remember distance of the forage, and its direction in relation to the sun's position, and they communicate both to other bees by performing a communication dance on the comb (Section 3.35). In *A. florea* (only) the forager usually dances on a horizontal surface, pointing directly towards the source of the forage, although dances on a vertical surface have also been observed (Gould et al., 1985). In *A. dorsata* the forager dances on a part of the vertical comb surface that is within view of the sky, transferring the sun's direction to 'upwards'. Foragers of *A. cerana* and *A. mellifera* dance as those of *A. dorsata* do, but in the dark cavity of their nest, out of sight of the sun. They dance on the comb, directing their dance at the angle to the (upward) vertical that the direction to the forage bore to the direction of the sun (Lindauer, 1957*a*).

A. cerana and *A. mellifera* both fan at the entrance of their nest or hive, either to ventilate it, or to attract other bees of the colony, in which case they fan with their scent gland exposed. But *A. cerana* bees face outwards, and *A. mellifera* face towards the entrance. There are variations between the honeys of different species, in the amounts of different sugars and in enzyme activities (Table 13.23B).

All the honeybees defend their nest and, in so doing, they may sting intruders. *A. florea* can have difficulty penetrating human skin, although it effectively stings attacking wasps or bees. Section 3.41 describes the sequence of events, and compares temperate-zone *A. mellifera* and Africanized bees (which are tropical *A. mellifera*). These latter bees, and *A. dorsata* in Asia, have the greatest reputation for stinging. Their alerting system is more effective: the bees are alerted by a smaller amount of the alarm pheromone (mainly isopentyl acetate, Table 2.35A). This substance is released when any honeybee stings, marking the target area and attracting other bees to it. The amount of isopentyl acetate produced per stinging bee was measured as about 40 μg in *A. dorsata*, 2 μg in *A. mellifera*, 1 μg in *A. cerana*, and 0.2 μg in *A. florea* (Morse et al., 1967). Table 14.51B compares the venoms of different *Apis*, including their toxicity and composition. *A. cerana* venom is about twice as toxic as that of *A. dorsata* or *A. mellifera*, and *A. florea* venom is least toxic of all.

Where different *Apis* species occur in the same area, they can interact with each other in various ways (Koeniger, 1982). Workers of the different species may rob each other's nests, and may compete for food or for nesting sites; drones may interfere during mating flights (see end of Section 8.23). Also, a parasite or disease of one species may transfer to another species which, not having evolved with it, is not resistant to it.

1.23.1 *Apis cerana* and *Apis vechti*

Apis cerana in Asia is the counterpart of *A. mellifera* in Africa and Europe, and Ruttner's (1988) book devotes 42 pages to it. The bees are in general smaller than *A. mellifera* (Figures 1.23a, 1.23b). However, the largest types in the two species have a similar size, and the same is true of the smallest (Ruttner, 1985). General morphological characteristics are quoted as: wing length 7–9 mm, tongue length 4.8–5.6 mm; body colour from dark grey to reddish-yellow. *A. cerana* is also differentiated from *A. mellifera* in that the median vein of the rear wing is forked, and tergite 6 of the abdomen has a tomentum (covering of hairs).

As with *A. mellifera*, body size increases in general at higher latitudes, and at higher altitudes. (The size of the worker affects frame spacing in hives, and cell size for comb foundation; the size of the queen determines the width of holes for queen excluders. Table 4.25A gives information needed for different areas.) In Kashmir (Shah & Shah, 1982) and parts of Himachal Pradesh (Verma, 1987), many characteristics of *A. cerana* approach those of *A. mellifera*. Table 7.41A gives some comparative data.

Apis cerana lives from sea level up to 2500 m, within the distribution area shown in Figure 1.2b. The species is separated from *A. mellifera* by desert area on either side of the border between Iran to the west and Afghanistan and Pakistan to the east (Nogge, 1974). Northward, *A. cerana* stretches to the Himalayas and also through eastern Asia to the temperate zone in northern China, Korea, and the Pacific seaboard of the USSR, and to Japan. Eastward, *A. cerana* extends across the whole of the Asian tropics reaching east to the Philippines, Borneo, and the Indonesian island chain as far as Bali, i.e. to the original Wallace Line, drawn west of Sulawesi, to indicate the boundary between the Oriental and Australasian faunal regions. *A. cerana* has been taken further east, most recently to Irian Jaya (in Indonesia) in 1985 or 1986 (Delfinado-Baker & Aggarwal, 1987*a*), and by 1987 it was in Vanimo, Papua New Guinea.

A. cerana is customarily divided into subspecies *A. c. cerana*, *A. c. indica* and *A. c. japonica*, in China, India and Japan, respectively, but quantitative studies are at a more preliminary stage than those on *A. mellifera*.

Ruttner (1985) published a preliminary separation of 40 characters (in 68 samples of *A. cerana* from a number of countries), obtained by multivariate analysis such as that used to produce Figure 1.21a for *A. mellifera*. Verma (1987) prepared a full report on an analysis of 55 characters of *A. cerana* in three areas of northern India, including Kashmir and Himachal Pradesh in the western Himalayas. The two sets of results lead to a provisional grouping of main ecotypes, roughly in order of increasing body size:

A. c. indica: Bali to Java and Sumatra, Malaysia, Thailand, Sri Lanka, S. India (both 'hill' and 'plains' varieties)
A. c. cerana: parts of N. India, N. Pakistan, Afghanistan, China, Far East (USSR); for Korea see below
A. c. japonica: Japan (including Tsushima between Honshu and Korea)
A. cerana in Kashmir (W. Himalayas, but separate from Himachal Pradesh), which are sufficiently large to be kept in Langstroth hives with *A. mellifera* foundation and comb-spacing.

A more detailed and (necessarily) more complex analysis is published by Ruttner (1988).

Further measurements on *A. cerana* in Korea and Japan might confirm whether those in Japan are morphologically separated from all those on mainland Asia. If so, then they are likely to be native to Japan; if not, the bees were probably taken by man from the mainland. *A. cerana* has been in the Japanese islands (except Hokkaido in the north) throughout historical times. Myeong and Seung (1986) studied *A. cerana* in Korea, but compared them only with imported *A. mellifera*.

Honey has been collected from wild colonies throughout their range, except possibly by Buddhists in Tibet, whose religion forbids any taking of life. Beekeeping was practised much less in Asia than in Africa or Europe. The keeping of *Apis cerana* in hives developed first in countries without *A. dorsata* (Crane, 1989*a*), and in some of these it has now been largely replaced by beekeeping with imported *A. mellifera* (Section 7.44). In regions with *A. dorsata*, there was little incentive to use *A. cerana*, and in some, *A. cerana* beekeeping did not start until *A. mellifera* was introduced in movable-frame hives. Areas of beekeeping with *A. cerana* and *A. mellifera* are often quite separate.

Beekeeping with *A. cerana* is dealt with in Section 7.41, and queen rearing is referred to in Section 8.28. One difficulty is that colonies of many tropical *A. cerana* ecotypes (like those of tropical *A. mellifera*) abscond if food becomes scarce, or as a response to disturbance of various kinds (Section 3.34.5). Verma (1986) found that *A. cerana* of Kashmir were much less prone to absconding than those of Himachal Pradesh. Table 7.41A compares many characters of *A. cerana* in Kashmir and other parts of India with *A. mellifera* in India.

Apis vechti/koschevnikovi

Some bees in Sabah (Malaysian NE part of Borneo, Figure 1.23e) rather like *Apis cerana* have been given the specific name *vechti* (Maa, 1953). By a comparison of the drone endophallus, Tingek et al. (1988) were able to show that *A. vechti* is indeed a separate species. In Sabah it inhabits some of the same area as *A. cerana*, but (Koeniger et al., 1988) the drones fly at a different time of day, *A. cerana* between 14 and 15 h and *A. vechti* between 17 and 18 h. Section 8.23 records a similar separation of mating times of three different *Apis* species in Sri Lanka. [Note in proof: *Apis koschevnikovi* was later published as the preferred name (Rinderer, 1988).]

1.23.2 *Apis dorsata* and *Apis dorsata/laboriosa*

Apis dorsata is larger than *A. mellifera*, *A. cerana* and *A. florea* (Figures 1.23a, 1.23b), and is known as the giant or rock honeybee. Ruttner (1988) summarizes much biological information. General morphological characteristics of the worker are: wing length 12.5–14.5 mm; tongue length about 6.7 mm; body colour yellow, with tergites 2 and 3 reddish-brown.

Section 8.61 describes the single-comb nest, fixed to a thick horizontal branch of a tree (Figure 8.61b), or under a roof or rock overhang that protects it against rain and direct sun during summer; see Figure 1.23d. In regions with good nectar and pollen resources there may be as many as 100 nests at one site. Worker and drone cells are of the same size; 4–6 queen cells may be built, at the lower edge of the comb.

Colonies commonly abscond or migrate regularly between the same two (or three) areas every year, building a nest and storing honey in each. This strategy enables the colony to enlarge the territory from which is obtains food, and the species to extend its range to regions where the forage cannot support colonies throughout the year. The distribution of *A. dorsata* is, however, more restricted than that of *A. cerana* or *A. florea* (Figure 1.23e). It does not extend north of the Himalayas as *A. cerana* does, or west to Iran and the Arabian peninsula like *A. florea*, but it reaches southern China. It has also reached further east than the others, not only to Borneo and Palawan, but also nearly 2000 km east of Bali, along the Indonesian island chain which (Sakagami et al., 1980) 'gave a good opportunity for transinsular dispersal to this strong flier', i.e. to the east of the Wallace Line. (Its most easterly record (Kei) is, however, still west of the amended Weber Line; see Good, 1974.)

Figure 1.23d A winter nesting site of *Apis dorsata/laboriosa*, viewed from below, Kaski District, Nepal, 1200 m (Underwood, 1986).
Two combs (arrowed) were still occupied in April, and about 10 were abandoned by colonies that had absconded as a result of human interference. A number of empty queen cells can be seen at the lower edge of the empty comb just left of centre.

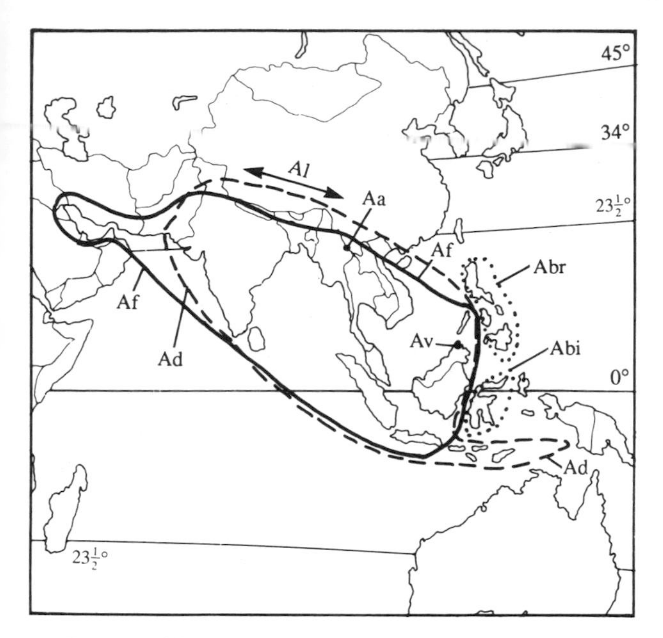

Figure 1.23e Some details of the natural distributions of *Apis dorsata*, *Apis florea* and related species (confirmed or possible), all in Asia.
Based on information from Sakagami et al. (1980), Ruttner (1985), Tingek et al. (1988), Wu & Kuang (1987).

Aa *andreniformis** Af *florea*
Abi *binghami* Al *laboriosa*
Abr *breviligula* Av *vechti**
Ad *dorsata*
* reported from one small area only

Names follow Shao-wen et al. (1986) and Tingek et al. (1988).

The very large '*Apis dorsata*', in parts of the high Himalayas that have a temperate climate and cold winters, has been proposed as a separate species *Apis laboriosa* F. Smith, on the basis of morphometric analysis and other evidence; workers are about 10% longer than those of *A. dorsata*. This reverts towards Maa's classification (1953); Roubik et al. (1985), Sakagami et al. (1980) and Underwood (1986) give further information. However, Ruttner (1988) points out that recognition or rejection of *A. laboriosa* as a species distinct from *A. dorsata* must await a detailed comparison of the endophallus of the drones: the structure of the endophallus varies considerably between *Apis* species (Figure 2.36b). Meanwhile, this very large bee is referred to below as *Apis dorsata/laboriosa*.

Evidence so far suggests that the bee exists in west Yunnan (China), in some parts of the Himalayas in Nepal, Bhutan, Tibet (Autonomous Region of China) and India, and probably in north Burma. Most of the studies have been made in Nepal, where its distribution does not overlap that of *A. dorsata* – which lives at the lower altitudes. At some sites *A. dorsata/laboriosa* establishes permanent nests on open rock faces, and Roubik et al. (1985) found it foraging up to 4100 m (*Apis dorsata* up to 1900 m). Underwood's (1986) observations were made from February to early May. He received reports of migratory behaviour in *A. dorsata/laboriosa* in later months, and nests had been found up to 3000–3500 m in June. Colonies at the winter sites (below 1700 m) apparently did not build comb, and bees taken from winter clusters were said to be 'full of honey'. Combs were built and brood was reared after spring started; Figure 1.23d shows a site in April.

Sakagami et al. (1980) present morphometric and other evidence, from which the existence of two other species of the *dorsata* group is considered likely: *A. breviligula* in the Philippines except Palawan, and *A. binghami* in Sulawesi (Figure 1.23e). Ruttner (1988) regards these as subspecies of *A. dorsata*.

Section 8.61 discusses the practical management of *A. dorsata* colonies, and colony behaviour that is especially relevant.

1.23.3 *Apis florea* and *Apis florea/andreniformis*

Apis florea, the dwarf or little honeybee, is smaller than *A. dorsata*, *A. mellifera* and *A. cerana* (Figure 1.23a). General morphological characteristics are: wing length 6.3–7 mm; tongue length 3.44 mm; body colour: scutellum black, abdominal segments 1 and 2 brick-red, segments 3–6 with bright white tomenta (hairy covering). Theses by Akratanakul (1977) and Whitcombe (1984) relate to this bee; see also Ruttner (1988).

The natural distribution of *A. florea* stretches from Oman through southern Iran and tropical mainland Asia, and to Palawan and the islands of Indonesia (Figure 1.23e). It lives mainly on the plains, and not above 500 m, although colonies may move up to (or even higher than) 1500 m; the Himalayas thus restrict its range to the north. In 1985, five colonies were found in Sudan in Africa, within 5 km of each other in irrigated gardens in suburbs of Khartoum, near the airport. It is most likely that they arrived – by accident or design – on a flight from western Asia, possibly Pakistan (Mogga & Ruttner, 1988). By the end of 1986 there were 21 colonies (Lord & Nagi, 1987), and by the end of 1988 about a thousand (Mogga, 1988), although *A. mellifera* does not thrive in the area.

A. florea nests in a wide variety of sites in the open, using either vegetation or a rock surface to support its single-comb nest (Figure 8.62a); unlike *A. dorsata*, it is not very selective. Its flight is rapid, and it successfully defends its nest against ants and also against large attacking insects by stinging them. Section 8.62 discusses the management of *A. florea* colonies and aspects of biology and behaviour that are especially relevant.

Wu and Kuang (1987) published diagnostic characters of queens, drones and workers of bees in Yunnan Province in China, which they refer to as distinct species under the names *Micrapis florea* Fabr. (at higher altitudes) and *M. andreniformis* Smith. The difference in esterase isozyme patterns can be seen in Figure 1.23c, and Ruttner (1988) seems inclined to regard them as distinct species. The names *Apis florea* and *A. florea/andreniformis* are meanwhile used here, and distributions are indicated in Figure 1.23e. The workers are shown in Figure 1.23a; *A. florea* has a larger body size and it builds a larger comb. Wu and Kuang (1987) report that *A. florea/andreniformis* in Yunnan is, unlike *A. florea*, easily alerted to sting; it might also attack *en masse*, and follow an intruder for 30–40 m.

A honeybee of the *A. florea* type, and showing similarities to Maa's (1953) *Apis iridipennis*, has been found in a tropical forest in Thailand. Farther west, Ruttner et al. (1985) found *A. florea* in Iran to be significantly larger than those in Oman, Pakistan or Sri Lanka.

1.24 Meliponinae (stingless bees)

Meliponinae are social bees in the family Apidae (Figure 1.2a); they live in permanent colonies that often contain many thousands of bees. They occur throughout most of the tropics, and Figure 1.2b shows their distribution. They extend from about 34°S in Argentina to 29°N in Mexico, and also occur in some islands of the Caribbean. Kerr and Maule estimated numbers of species known in different continents as follows; 120 more species have been studied and named since their paper (1964) which gave:

South America	183
Africa	32
Asia (+ Indonesia W of Wallace-Weber Line)	42
Australia, PNG, Solomon Islands	20

Some species can withstand cold periods, but stingless bees do not form a winter cluster as temperate-zone *Apis mellifera* and *A. cerana* do.

The taxonomy of stingless bees has been much studied and discussed in the last few decades. The following five genera are those recognized by Michener (1974):

Melipona, consisting of a number of tropical American species varying from somewhat smaller to somewhat larger than *Apis mellifera*, and in general larger than *Trigona*.

Meliponula, consisting of a single African species having the appearance of a small *Melipona*.

Trigona, a large genus of long-winged bees with representatives in all continents, whose size ranges from almost that of *A. mellifera* down to the shortest of all bees, 2 mm in length.

Dactylurina, a genus near to *Trigona*, represented by a single, slender-bodied African species.

Lestrimelitta, a genus of African and tropical American robber bees, dependent upon the products of nests of *Trigona* or, less commonly, *Melipona* or even *Apis*.

Stingless bees vary greatly in size, and their appearance varies widely: some have a slender body and some a broad one; some are shiny and some as hairy as bumble bees; their colours are various, and some are metallic. The sting is atrophied, the tip of the sting and the venom apparatus having 'dwindled away'. Schwarz (1948) discusses the implications of this, and describes their defensive mechanisms against man – which include biting, irritating by crawling into eyes, ears, etc., and ejecting a caustic fluid. Many stingless bees are good pollinators of various crops.

Stingless bees nest in cavities, which in different species may be underground, or in trees or other enclosed spaces such as termite nests. In any one area of Africa, Asia, Australasia or America, species are likely to be available that can be reared in simple nesting boxes or hives. Section 8.63, which discusses hives and management, gives information about some of the nests, and Table 8.63A indicates both the geographical and the species range of beekeeping with stingless bees. Probably all the species produce the enzymes necessary for making honey from nectar (see Table 13.23B), but only the larger bees store enough honey to make them worth keeping in hives, or their nests worth hunting for.

Schwarz (1948) and Nogueira-Neto (1951, 1953, 1970) give much information on stingless bees; Michener's book (1974) has a chapter on them and Crane (1978*b*, S33) provides many further references. Lindauer (1957*a*) and Lindauer and Kerr (1960) describe the bees' methods of communication, and Kerr and Maule (1964) discuss their distribution.

1.25 Other insects from which honey is harvested

Certain insects besides bees collect nectar and convert it to honey by enzymatic action (Section 13.23.3), and these are represented at the bottom of Figure 1.2a. Substantial amounts of honey (which is collected by native peoples for food) are stored by some polistine wasps: *Brachygastra* (formerly *Nectarina*) *mellifica*, *B. lecheguana* and *Polybia scutellaris* in tropical America and *Ropalidia vitripennis* in Madagascar (Michener, 1987). Some species build quite large nests, for instance a colony of *Brachygastra mellifica* may contain 10 000 to 15 000 adults.

Ants build no comb, but a colony of honey ants stores honey in the greatly distended crop of morphologically adapted workers known as repletes. The genus *Myrmecocystus* is native to more arid parts of Mexico and western USA; 27 species are known, including the following from which honey is (or was) harvested by Amerindians: *depilis, kennedyi, melliger, mendax, mexicanus, melanoticus, navajo, placodops, romainei, semirufus, testaceus* (Snelling, 1976, 1987). Harvesting consists of collecting honey-filled repletes; a nest of *M. melliger* may contain up to 1500 of them. *Leptomyrmex* (Dolichoderinae) is used in parts of Australia. In the Formicinae, Australian honey ants include *Melophorus* and *Camponotus*, various species of which were used by indigenous peoples, and Wilson (1971) refers to other Old World honey ants, including *Plagiolepis trimeni* in Natal.

1.3 BEES REARED FOR CROP POLLINATION

It has been estimated that a third of the total human diet in developed countries is derived directly or indirectly from insect-pollinated plants (Townsend, 1974*a*), and the proportion is also considerable in other countries. Techniques used in modern methods of crop production greatly reduce the number of wild pollinating insects, so it may be necessary to bring insects to the crop when it is in flower for the express purpose of pollinating it.

Bees are very effective as pollinators because any one bee is likely to visit only flowers of one species on a single foraging trip, as explained in Section 2.44. Almost all insects reared for crop pollination are bees. The most important are Apidae, of which the honeybee *Apis mellifera* is by far the most extensively used, and hives of them are shown in Figure 8.41a. Management for crop pollination is dealt with in Section 8.4, and the energetics of foraging in 2.43.1. Table 8.4A lists 177 important crop plants that are pollinated by honeybees and/or other insects, and gives information on the value of honeybees and the number of hives required per hectare. *A. cerana* is similarly useful (see Figure 8.42a), but its colonies are usually smaller, and less able to pollinate large areas of crop plantations. Management of *A. cerana* is described in Section 7.41.

Colonies of *A. dorsata* are not amenable to being kept in an enclosed hive, and cannot easily be moved from one place to another, although Ahmad et al. (1984*b*) have done so. This species is not reared for pollination, but it is an effective pollinator of certain crops in regions where it nests; the same is true for *A. florea*. Crane (1978*a*, Parts 12, 13) gives examples.

After the honeybees, the most widely important bees for pollination are the Meliponinae (stingless bees) which occur throughout the tropics. Some species are kept in hives, especially species of *Melipona* and *Trigona*; see Section 8.63. In addition to references given there, Rindfleisch (1980) presents the case for rearing stingless bees for crop pollination, but as far as is known this is not done commercially. Other (social) bees in the Apidae important for pollination are bumble bees (*Bombus*) in the temperate zones.

Table 8.5A lists 40 species of bees, together with crops which many of them have been reared to pollinate, either commercially or experimentally. Section 8.5 discusses the management of the most important for crop pollination, including the (solitary) species *Megachile rotundata* (Megachilidae) and *Nomia melanderi* (Halictidae).

1.4 THE HISTORY OF BEEKEEPING

It is likely that man hunted for wild nests of bees and took their honey during the whole of his existence. Beekeepers have used hives, and harvested honey from them, for at least 4500 years. Our knowledge of the history of both honey hunting and beekeeping has greatly increased in the past few decades, as a result of research by archaeologists, philologists, and scholars studying ancient records, and by travellers in some remote places of the world. Archaeological evidence has been presented elsewhere (Crane, 1983*a*), and a forthcoming book on the world history of beekeeping will deal more fully with other aspects of the subject. Here, a very brief sketch of some of the salient points must suffice.

It seems likely that honey was known by name very early in the development of human language, and the word *honey* is considerably older than *bee*. There are many references to honey in ancient records and literature, but most of them give no clue as to whether the honey was obtained by honey hunting or beekeeping. On their own, they must not be taken as evidence of beekeeping.

1.41 Honey collection from wild nests: the precursor of beekeeping

Early man probably took honey from bees' nests wherever he found them, and the collection of honey from wild nests continues to the present day except in some regions where it has been entirely superseded by beekeeping. The nests worth raiding were those of honeybees (*Apis*) and larger species of stingless bees (Meliponinae); Figure 1.2b shows the parts of the world where these are native.

Many mammals besides man raid bees' nests. Chimpanzees have been observed using their hands for the purpose, or a tool – a selected stick which is pushed into the nest, withdrawn, and the honey licked from it. Early man probably used similar methods, and they are mentioned in the Bible. Samson 'scraped the honey [from a nest inside a lion's carcass] into his hands and went on, eating as he went' (Judges 14: 8–9). Jonathan 'stretched out the stick that was in his hand, dipped the end of it in the honeycomb, put it into his mouth and was refreshed' (1 Samuel 14: 25–27).

These bees were *Apis mellifera*, and the earliest known evidence of honey hunting relates to the same species: a mesolithic rock painting in eastern Spain, dated to about 6000 BC (Figure 1.41a). At that time, not so very long after the end of the last major glaciation of the Ice Age around 9000 BC, the region probably enjoyed a climatic optimum. Several other mesolithic honey hunting scenes have been found in Spanish rock shelters (Dams, 1978), and many more in later paintings elsewhere (Crane, 1983*a*). In South Africa and Zimbabwe the bees were *A. mellifera*, and one painting in Zimbabwe shows smoke being used; it is likely that uncontrolled smoke or fire killed many colonies of bees when they were raided for their honey. In Asia the bees were *A. dorsata*, whose nests are depicted in a post-mesolithic Indian painting discovered in 1984 (Figure 1.41b).

As well as opportunistic raiding of nests, there has been, and still is, a more systematic harvesting from wild nests in known locations, especially nests of *A. dorsata*. There is an early written record of harvesting beeswax from such nests (Friedmann, 1955). In AD 983, Li Fang in China remarked that in many distant places bees lived 'on abrupt cliffs and rocks walls which are unclimbable', and that when they left there was an unlimited supply of wax [comb?], which was obtained 'by raising baskets to the top and lowering them to the bottom. In spring the bees all return . . .'. The nests were harvested after the colonies had absconded or migrated, and were unlikely to contain honey.

A. dorsata colonies have continued to yield much honey and wax, although the inclination of the bees to sting, and the inaccessibility of the nests, often make its collection a dangerous operation. In India alone some 14 000 tonnes of honey a year are collected from

Figure 1.41a Honey-collecting scene in a rock painting in La Araña shelter, Bicorp, eastern Spain, around 6000 BC, found in 1924; enlarged detail below (copy by E. Herñandez-Pacheco).

Figure 1.41b Rock painting showing honey collection from *Apis dorsata*, at Rajat Prapat, Central India; scale 1:3 (drawing: Mathpal, 1984).
On the right two men, each with a round pot tied from the waist, are climbing to reach bees' nests under a rock. On the left, two men (one carrying a pot) approach a nest, and another man approaches them with an axe.

A. dorsata. There is no reason to believe that methods in common use have changed much since prehistoric times. *A. dorsata* nests are widely visible (Figure 8.61a), although they may be high and inaccessible. On the other hand a nest of *A. mellifera* inside a forest tree can be quite difficult to find. The collection of honey from *A. mellifera* nests is well documented in Europe and the Mediterranean region, in tropical Africa (see e.g. Seyffert, 1930; Galton, 1971; Crane, 1983*a*), and in North America, where European *Apis mellifera* were taken in or about 1621. Methods for 'lining up' the bees were published by Dudley (1721) and many later authors, but they had been devised by Roman times, and were described by Columella in *De re rustica* IX.8: 7–13.

Robert Knox, writing in Sri Lanka in 1681 described honey collection from *A. cerana* bees, which are less inclined to sting. They 'build in hollow trees, or hollow holes in the ground . . . into which holes the men blow with their mouths, and Bees presently fly out. And then they put in their hands, and pull out the Combs, which they put in Pots or vessels, and carry away. They are not afraid of their stinging in the least, nor do they arm themselves with any cloths against them.'

Honey collection from wild *A. florea* nests is comparatively little recorded, perhaps because nests are often within easy reach from the ground, honey yields are small, and the bees are very gentle indeed. When their nest is disturbed, they are likely to reassemble a short distance away and start to build a new nest there.

The collection of honey from wild nests generally died away when and where modern beekeeping became more productive, but the tradition has continued until recent times even in parts of Europe, such as Hungary and Romania (Crane, 1983*a*). In tropical Asia, where consumption of sweet foods is traditionally low (Crane, 1975*e*), beekeeping did not advance much until about 1950, and a high proportion of the honey is still harvested from wild nests of *A. dorsata*.

Many species of stingless bees are native in different regions throughout the tropics (Figure 1.2b), and honey was also harvested from them, as it still is, although yields are generally small. A rock painting in Queensland, Australia, shows the process. Nests may be in the ground or in trees, and they are in general more accessible than most honeybee nests. The bees do not sting, and although they can annoy the honey hunter in various ways, they are not feared as some

Figure 1.41c Forest beekeepers at work in Germany (Krünitz, 1774).
One beekeeper is sitting in a sort of bosun's chair; he smokes the bees with a pipe, and uses a tool to cut out combs which are visible through the open doorway. The other stands below, with upstretched arms ready to receive combs dropped down to him. At the foot of the tree are two containers for the honey combs, and the door removed from the tree cavity.

tropical honeybees are. One day in 1984, in Johannesburg, I saw photographs both of nests, and of rock paintings of nests, of two common species, *Trigona gribodoi* and *T. denoiti*, and I also listened to a young African's description of locating and getting access to nests of the same two species, and of taking their honey.

A particular type of bee management was developed in the forests of northern Europe with European *A. mellifera*: the ownership and tending of colonies nesting wild in tree cavities. Galton (1971) provides many details of this 'forest beekeeping', and I have given a short account elsewhere (Crane, 1983*a*). It probably first developed around the middle course of the Volga river, finally reaching the western Slavs and the Germans. During the period between 1200 and 1900 forest beekeeping yielded large quantities of honey and wax. An upright door was cut in the side of the tree to give the beekeeper access to the combs of honey. Part of a tree trunk with such a door, dated to the first century BC, has been found in southern Poland, and Figure 1.41c shows a forest beekeeper harvesting honey 1800 years later. Finally, tree cavities in the forest beekeeping area became scarce, so separate hollowed logs were fixed high up in the trees (as we fix bird nesting boxes), and the bees that occupied them were looked after in the same way. Then individual upright log hives were stood in a group on the ground, and an *apiary* of the type used in modern beekeeping came into being.

Traces of tree beekeeping with *Trigona* bees by native peoples of the far north-west of Australia have been found (Dollin & Dollin, 1986), but there was no development into hive beekeeping, as there was in Europe.

1.42 Traditional beekeeping, using fixed-comb hives

1.42.1 Horizontal hives

By 2500 BC, before forest beekeeping is known to have existed, fully fledged beekeeping was being practised in Ancient Egypt. The hives used there, and in other Ancient civilizations, have been discussed in some detail by Crane and Graham (1985). The earliest representations of them are dated to 2450 BC, and appear in a honey-harvesting scene that formed part of a bas-relief in an Egyptian sun temple near the lower Nile. Figure 1.42a is from a later and better preserved painting that shows a very similar scene. Four such scenes are known, and all depict honey combs being removed from one end of hive(s) in a stack; in some of them smoke is used, and several also show honey being put into storage vessels. Stacks of several hundred hives, somewhat similar to those in the above scenes, constitute traditional Egyptian apiaries today, and honey harvesting is also similar; see Section 9.11.

The earliest written records that relate to the keeping of bees in hives are from about 1500 BC. They form part of a Hittite code of laws inscribed on clay tablets, found in the high treeless Anatolian plain 1000 km north of Egypt. They set out fines to be paid by a thief who stole either empty or occupied hives. The Hittites commonly used unbaked mud for building, and I think it likely that their hives were also made of mud. Traditional hives used in Anatolia today are still horizontal cylinders; those I have seen were woven.

Occasional records during the next thousand years throw sidelights on beekeeping in the Middle East. Rameses III lived from 1198 to 1167 BC, and one list of offerings he made to the Nile god amounted to about 15 tonnes of honey. This was almost certainly

Figure 1.42a Wall painting *c.* 1450 BC, in the tomb of Rekhmire, West Bank, Luxor, Egypt, showing (on the right) honey being harvested from mud hives – painted grey – and (left) being sealed in pottery containers – painted red (N. de Davies, 1944).

obtained from beekeeping, which already existed in Ancient Egypt; the only vegetated land there was cultivated. About 250 BC, beekeepers submitted an urgent petition to an official who had 'borrowed' their donkeys; they were desperate for the return of the animals so that they could move their hives away from a site that was due for irrigation flooding. This suggests that hives were migrated. Two Greeks in Egypt owned a thousand hives, and others owned several hundred each. In many places hives were taxed, and in Athens they were included among seizable securities.

We know much about one type of pottery hive that was used from about 400 BC to AD 600 in an area within 50 km of Athens. Since 1970, fragments of these hives have been found at 26 sites, and these finds have led to the identification of others excavated earlier. The hives were thimble-shaped pottery vessels used horizontally. The earliest were 40–50 cm long, and 40 cm or less in diameter at the mouth. Aristotle (384–322 BC), one of the most important authors of extant Greek writings on bees (see Section 1.43), said that an average (undescribed) hive produced 6–9 pints of honey, and a good one 12–15 or exceptionally 18 pints [5–7, 10–12, 14 kg]. Excavated hives of later dates were slightly longer, up to 60 cm or more, and similar hives 80 cm long are still used in a few islands of the Cyclades.

Such records and finds give indications of the extent and importance of beekeeping, but not of the management of bees. Two Greek books on beekeeping written about 330–320 BC would surely have done so, but they have not survived. Their authors were Philiscus of Thasos (an island which is still a great centre for the production of honeydew honey), and Aristomachus of Soli (near Mersin in modern Turkey).

In contrast to the paucity of extant Greek writings on beekeeping, many survive from Ancient Rome. They include clear descriptions of hives of nine different materials, of which the first three were not favoured: earthenware (pottery), bricks, dung; cork bark, logs, wooden boards, fennel stems, woven wicker; transparent mica or horn. The hives were used horizontally, and many are specified as about a foot in diameter (or in square cross-section) and 3 ft long, i.e. 30 × 30 × 90 cm.

Over an extended region of the world, traditional hives still have the shape of a horizontal cylinder or something similar: across North Africa and throughout most of Africa south of the Sahara; in southern Greece and many Mediterranean islands; in the Balkans and Asia Minor; in the Indus valley up to and including Kashmir, part of Afghanistan, and in other areas as distant as southern Thailand and Bali in Indonesia.

The use of such hives was convenient in many ways. By using end closures small enough to fit inside the cylinder, the space occupied by the bees could be reduced when desirable. It could be increased by adding an extension at one end; such extensions were found with hives excavated in the Athens region. Food for the bees could be placed inside the back end of the hive, and (Section 9.11) brood combs could, with care, be removed and replaced in the same or another hive. Honey combs could be harvested when required, by smoking the bees from the back of the hive to – and out of – the flight entrance at the front end; there was no need to kill the bees. Many of the hives were 'longer than the reach of a man's arm' so that some combs would be left intact. In many remote areas in the Old World, I have watched traditional beekeepers tend hives of this type, and harvest honey from them. Their knowledge about bees may be limited, but I am constantly impressed by their experience of the bees' needs, and their high standards of care and attention to them. The hives are often placed in stacks, sometimes under a shelter, or embedded in a specially constructed wall or in the wall of a house; or hung up out of reach of enemies. Unlike upright hives, or modern tiered hives, they are not usually placed on the ground, or on individual stands.

The history of beekeeping in Asia with *A. cerana* is dealt with elsewhere (Crane, 1989*a*). No archaeological finds relating to it are known.

In the New World, there was a rich tradition of beekeeping with the stingless bees of Central America (Schwarz, 1948); see Section 8.63. The Maya kept stingless bees, especially the species *Melipona beecheii*, in horizontal log hives. Excavations since 1970 have shown great similarities between the hives they used in past centuries and those used today (Figure 1.42b). The two ends were closed with discs of stone, circular or cut to fit the individual logs. No logs have survived from ancient times, but more than 400 stone end closures have been excavated, dating from the Late Preclassic period (*c.* 300 BC to AD 300) to the Late Postclassic which ended at the Spanish conquest in 1520 (Crane & Graham, 1985). The map in Figure 1.42c indicates the sites which have yielded these first archaeological finds relating to prehistoric American beekeeping.

Modern beekeeping with horizontal fixed-comb hives, its limitations, possibilities and applications, are discussed in Chapter 9.

1.42.2 Upright hives

North of the Alps and the Caucasus, and mountain ranges to the east, traditional hives were derived from the practice of forest beekeeping referred to at the end

Figure 1.42b End view of log hives (showing the end closures).
The hives contain stingless bees (*Melipona beecheii*) stacked on an A-frame rack under a thatched roof, Yucatán, Mexico, 1978; an empty hive lies in the foreground (photo: E. C. Weaver).

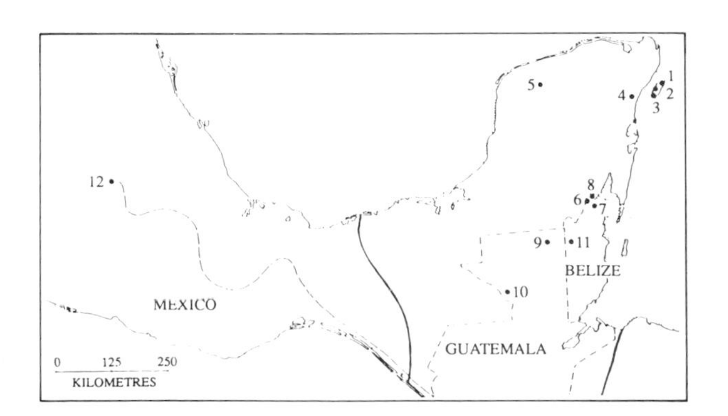

Figure 1.42c Map showing 12 sites at which stone discs resembling Maya hive closures have been found (Crane & Graham, 1985).
The two full lines indicate the approximate extent of Classic Maya culture. The broken line to the west shows an Olmec trade route.

of Section 1.41. They were upright logs – in some areas with a door in the side (Figure 1.42d). To the west of the forested areas, bees were kept in skeps: tallish baskets of woven wicker (Figure 1.42e), and later of coiled straw, placed mouth down on a base of some sort.

In general, bee management was more restricted with these upright hives than with the horizontal hives, from which honey could fairly easily be taken without killing the bees – although doubtless this was sometimes done. With upright hives that had no door at the side, I think the bees were more often killed. Skep beekeepers in England (Butler, 1609) – and probably elsewhere – were advised to overwinter skeps of medium weight, and to take the honey from the heaviest, and also from the lightest (whose bees would not survive the winter). The bees in the harvested hives were killed, for instance by drowning or by placing each skep in turn over a pit in the ground in which sulphur was burning; the fumes caused the bees to drop off the combs and into the fire. But by the 1500s or earlier, methods were being used that did not necessitate killing the bees, for instance by adding (above or below the skep) an extension in which the bees stored honey, or by 'driving' the bees at harvest time out of their own skep into another one.

Figure 1.42d Upright log hives fitted with doors, and one showing the door space, Poland, 1960 (photo: S. Kirkor).

Von Reussen. 1415

Reussen. Cap. lxxx.

DAs Lande Russia vnd Ruthenia, hat vor zeiten auch Roxolana geheissen. Es ligt hinder Poland/vnd stoßt gegen Mittag an die Moldaw/vnd Walachey. Es ist ein gantz fruchtbar Landt / vnnd hat auch vast viel Honig. So man die Acker ein wenig bawet mit dem Pflug/vñ Korn dareyn wirfft/ so geben sie 3. jar Frucht / doch daß man ein wenig Saamen im Acker laßt / wann man das Korn eynsammlet/so wechßt es ohn allen andern Ackerbaw. Die Binen lassen nit allein das Hönig in jren Körben vnd holen Bäumen/ sonder allenthalben auff dem Gestad der Flüß vnd in den Hülen der Felsen/ da tragen sie Honig zusammen/ darauß macht man köstlichen Meth vnd die grossen Wachsscheiben. In diesem Reussen seind die Gebiet/ Halicien/ Przemislien/ Sannock vnd Leopolis.

Figure 1.42e Wicker skeps depicted in Sebastian Münster's *Cosmographia*, Bern, 1545.

1.42.3 Immutability and change

One consistent characteristic of traditional hives and beekeeping methods has been their immutability, even through a period of millenia. In many remoter places of the Middle East, and elsewhere, one can see today apiaries and hives that follow exactly the descriptions written in Roman times, which themselves probably describe much earlier customs. I think that such knowledge as existed, about the habits of bees and how to 'manage' them, was passed down individually from father to son (early beekeeping, like honey hunting, was largely man's work). There was nothing to initiate improvements on what was being done – which probably seemed magical enough.

Over the centuries there seems to have been a shift in hive materials to what was more convenient in some way – less breakable, or less heavy – than what had been in use previously. Unbaked clay or mud used in Ancient Egypt was replaced by baked clay in Crete and Greece, and this by wooden or woven structures, possibly in Ancient Greece and certainly in Ancient Rome.

Such changes as occurred in a region were likely to occur if people migrated into it from elsewhere, bringing their own beekeeping customs with them. This would explain the existence of enclaves with one kind of beekeeping within an area where another type was used. For instance, the Arab occupation of much of Spain from AD 711 to 1492 (and the Roman influence earlier) led to a complex pattern of traditional beekeeping in that country, of which traces can be found today. In the north, upright hives of cork bark are used as in the south of France, but horizontal hives (of both woven wicker and pottery) still exist as far north as some of the foothills of the Pyrenees in Aragon, where the name for the apiary (*arnal*) is of Arabic origin, meaning *the bee* (Chevet & Chevet, 1987).

Before the end of the Middle Ages, the beekeeper – let alone the honey hunter – wore no special protective clothing against bee stings. In the Middle East and Europe, he probably sometimes covered his hair and most of his face with a cloth, and wrapped loose clothing round his body. Nonnus, a Roman writer living in Egypt around AD 450, wrote a mythological epic *Dionysiaca*, in a high-flown style. His description of the discovery of beekeeping by Aristaeus (V: 247-249), includes the sentence: 'He covered every limb from toenails to hair with a close-woven wrap of linen, to defend him from the formidable stings of the armed bees' [translation by A. J. Graham].

A number of mediaeval European illuminated manuscripts show beekeepers working at their hives. I know of none earlier than the 1400s which gives any indication of specially designed clothes for working with bees. From then onwards, complete protection for the head – and even for the body – is sometimes shown, and two types emerged. One was like a hood or fencing mask, with woven cane, cloth or metal in front of the eyes; examples are shown in Figures 1.42a and 15.21b. The other type, now widely used throughout the world, was a brimmed hat with a veil hanging from it. Modern descendants of both types are described in Section 5.11.

Much less clothing is worn in the tropics, and many honey hunters and traditional beekeepers customarily remove any clothing before harvesting honey, so that bees do not get trapped against the body, and sting. This practice is probably quite old.

1.43 Growth of knowledge about honeybees

All the beekeeping and the considerable honey and wax production in the Ancient World, and in mediaeval times, was done in ignorance of most of the facts that enlighten our dealings with bees today. The 1600s and 1700s saw a great flowering of many sciences, and Table 1.43A lists some of the significant advances in knowledge relevant to beekeeping that were made between 1568 and 1792. Some of the advances listed were applications of discoveries made in other fields of biology; in a few cases original observations on honeybees gave a new insight into much wider fields.

Aristotle (384–322 BC) knew that a colony of bees had a king or ruler bee, larger than the others. But he confessed that 'there is much difficulty about the generation of bees', saying:

> They must either bring the young brood from elsewhere . . . or generate them themselves, or bring some [the drones] and generate others. If they generate them it must be either with or without copulation: if the former, bees [workers] may be generated from the union of bees, drones from that of drones, and kings from that of kings, or all the others may be generated from one, as from what are called kings and leaders, or from the union of drones and bees [workers], for some say that the former are male, the latter female, while others say that the bees are male and the drones female . . .
> *De generatione animalium* X.759a,b.

One of Aristotle's observations on foraging by bees is, however, well attuned to modern behavioural studies, although the final sentence is a disappointment.

> On each expedition the bee does not fly from a flower of one kind to a flower of another, but flies from one violet, say, to another violet, and never

meddles with another flower until it has got back to the hive; on reaching the hive they throw off their load, and each bee on her return is followed by three or four companions. What it is that they gather is hard to see, and how they do it has not been observed.
Historia animalium IX.40

The teachings of the philosophers of the Ancient World were followed until well after the end of the Middle Ages, and beekeeping was carried on successfully in spite of much scientific ignorance. I think that in the mediaeval mind there was no difficulty in picturing the community in the hive as governed by a ruler – the large bee that stood out above the rest, and appeared to have authority over them. The ruler was referred to as king, sheik, pharaoh, or whatever other title was accorded to the human ruler of the society in which the beekeepers lived. There would have seemed no need to question the basis for this authority, let alone to seek a biological explanation for it. A few early writings refer to the ruler as female, for instance charms invoking a swarm to settle which date from AD 800–1000, but I know of no evidence that these were based on knowledge of the ruler's function in the colony. Luys Méndez de Torres in Spain first established the sex and function of the queen, and he stated in his book (1586) that the leader bee in the hive is a female that lays the eggs from which workers, drones and future queens develop.

By 1800, many of the basic facts about the behaviour, anatomy and physiology of European *Apis mellifera* were at last known (Table 1.43A). Also, the origin of substances produced by bees were differentiated from those the bees collected: brood and beeswax were produced in the hive; nectar, honeydew, pollen and propolis were produced by plants and collected from them by bees. Often, knowledge did not travel quickly outside the place in Europe where it was discovered, and much of it was discovered elsewhere at a later date.

In general the new knowledge had rather little

Table 1.43A Some significant advances before 1800 in knowledge relevant to beekeeping

	Year published	*Made by*	*Country*
a queenless colony with eggs or young larvae can rear a new queen	1568	Nickel Jakob	Germany
the queen is female and lays all the eggs in the colony	1586	Luys Méndez de Torres	Spain
drones are male	1609	Charles Butler	England
the honeybee was the first insect drawn under a microscope	1625	Prince Cesi	Italy
workers are female	1637	Richard Remnant	England
full anatomical drawings were made (with a commentary) of queen, drone and worker	1668–1673*	Jan Swammerdam	Netherlands
beeswax scales observed on the worker abdomen	1684	Martin John	Germany
nectar is produced by flowers in specific parts called '*mielliers*'	1717	S. Vaillant	France
1 described collection and use of propolis by honeybees	1739	R. A. F. de Réaumur	France
2 honeydew is produced by aphids	,,	,,	,,
beeswax comes 'from the body of the bee'	1744	H. C. Hornbostel	Germany
1 honeybees collect pollen (the 'male seed' of flowers, which fertilizes the ovum) from only one kind of flower on each flight	1750	Arthur Dobbs	Ireland
2 function of the queen's spermatheca described	,,	,,	,,
mating between queen and drone described	1771	Anton Janscha	Austria
bee dances described as a form of communication between foraging workers	1788	Ernst Spitzner	Germany
workers have (undeveloped) ovaries and can lay eggs	1792	François Huber	Switzerland

* Swammerdam died in 1680, and this work was not published until 1737–1738.

interaction with practical beekeeping, which was mostly continued on the same lines as during the centuries of scientific ignorance. The explosion of development in practical beekeeping was a separate phenomenon initiated around 1850 – and not in Europe but in North America.

1.44 The development of modern beekeeping, using movable-frame hives

The traditional hives described in Section 1.42 were fixed-comb hives: bees built their combs downwards from the top of the interior surface of the hive, and generally attached them to the sides as well. The beekeeper could remove a comb only by cutting it out. This was easier with horizontal hives opened at either end than with upright hives such as skeps, where access was only from the bottom and well away from where the comb was attached. But skep beekeeping could also be successful, as is evident from the well organized book published in England in 1609 by Charles Butler.

The stages of hive development that led to the movable-frame hive used today are set out in Table 1.44A, and discussed further elsewhere (Crane, 1983*a*). They had their origin in Sir George Wheler's (1682) description of a hive he found in use in Greece; Wheler's illustration is reproduced and the hive discussed, in Section 10.1. The hive was shaped like a giant flower pot, and the bees built their combs down from top-bars laid across the top but did not attach them to the inward-sloping sides. The combs could therefore be removed individually. The distance between the centre lines of adjacent top-bars had to be the same as the distance between the centre lines (midribs) of adjacent natural combs. This distance is about 35 mm for European bees, and slightly less for the smaller honeybees of the tropics (Table 4.25A).*

Wheler's account of the Greek top-bar hive had attracted beekeepers' attention in England and elsewhere, but progress was slow. In 1790 Abbot della Rocca published his *Traité complet sur les abeilles*, and it is clear from a passage in Volume 2 (pp. 467–469) that he understood the importance of the bee-space between the combs in top-bar hives he had seen on Crete, which were probably similar to those in Greece. François Huber (1792) understood the spacing too; he wrote of 'the equal distance uniformly preserved between the combs'. In England Robert Golding (1847) described an 'improved Grecian hive' with

* In 1933, hollow logs with top-bars were in use for *Apis cerana* in Tonkin and Annam (Vietnam), which enabled beekeepers to withdraw individual combs (Toumanoff, 1933; Toumanoff & Nanta, 1933). It seems possible that these movable-comb hives were developed independently of the Greek ones. A few are still in use (Mulder, 1988*a*).

Table 1.44A Hives that represent significant steps towards the development of the modern movable-frame hive

Adapted from Crane (1983*a*). All hives except the first two types were of wood; plastics are sometimes used for modern hives. Brackets indicate that combs/frames had to be cut out (but could be replaced) or had other deficiencies.

Hive type	*Where first developed*	*First known date*	*Author, or Figure number*	*Movable combs/ frames*	*With frames*	*Tiered boxes*	*Separate honey chambers*
horizontal mud/clay cylinders	Egypt	?	1.42a	(yes)	—	—	—
top-bar pottery/wicker	Crete/Greece	?	10.1a	yes	—	—	—
superimposed boxes	Europe	1649	W. Mew	—	—	yes	yes
collateral boxes	England/Europe	1756	T. Nutt	—	—	—	yes
leaf hive	Switzerland	1792	F. Huber	(yes)	yes	—	—
early box hives with frames	Europe	e.g. 1806	P. Prokopovich	(yes)	yes	yes	yes
tiered hive	England	1844	W. A. Munn	(yes)	(yes)	yes	yes
practical movable-frame hive	USA	1851	L. L. Langstroth 1.44a	yes	yes	yes	yes

appropriately spaced top-bars. All these hives had movable combs, and Chapter 10 describes more convenient types of them that are in wide use today.

Huber (1792) also developed a leaf hive in which four-sided frames were appropriately distanced, but they were not separately removable, being hinged along one upright side; the hive could be opened like a book – hence its name. Other frame hives were invented, for instance by Petr Prokopovich in Russia (*c.* 1806) and Johannes Dzierzon in Germany (1848). But their frames came close to the walls of the hive, and the bees attached the two surfaces together with comb or propolis, so it was necessary to cut the frames out. Both these hives were devised in the area of forest beekeeping, and had a door at the back like the upright log hives there.

In all the above developments, each hive was a single upright unit, but there had also been a different line of experimentation, using hives constructed of multiple units. In 1649 in England, William Mew had designed a wooden hive consisting of tiered boxes, without top-bars or frames. Bees tended to store honey in the upper boxes, and each was removed when it was full, complete with the combs fixed in it. A similar hive patented by Gedde (1675) had a framework inside each box to which the bees were supposed to attach their combs, and which was to be cut out complete with the combs full of honey. Other hives had fixed combs in tiered boxes; the Stewarton by Robin Ker in Scotland in 1819 had fixed top-bars. Augustus Munn in England advanced a stage further, and in 1844 described a hive which was tiered *and* had bar-frames of a sort.

The final, practical advance, incorporating movable combs in frames, in tiered boxes, was made in 1851 by the Rev. L. L. Langstroth in Philadelphia, USA. He owned many books on bees, including those by Huber, della Rocca and Golding (Johansson & Johansson, 1967, 1972*b*). He had two that described Munn's hive, which he fitted with four-sided frames suspended by lugs that were extensions of the top-bar. The lugs rested on grooves in the hive walls. He also distanced the sides of the frame at an appropriate 'bee-space' from the inner hive walls (Figure 4.2b), and the bees did not build comb across this gap. Langstroth also deepened the grooves on which the bars rested, in order to leave about 9 mm between the top-bars of the frames and the cover-board above. This facilitated the removal of the board; in earlier frame hives without this gap, the bees had fixed the frames to the cover. Langstroth wrote:

> Pondering . . . how I could get rid of the disagreeable necessity of cutting the attachments of the combs from the walls of the hives . . . the almost self-evident idea of using the same bee space as in the shallow chambers came into my mind, and in a moment the suspended movable frames, kept at a suitable distance from each other and the case containing them, came into being.

The bees 'respected' the bee-space left between the frames and the hive walls. The frames were, therefore, truly movable.

Langstroth's hive (Figure 1.44a) was of great practical utility, and its construction was relatively economical. The bees built their combs in rectangular wooden frames which could be lifted out individually, and the frames were suspended in tiered boxes that were easily separable. It was the prototype of movable-frame hives used today, which Section 4.2 describes in detail. The use of these hives spread rapidly, at first in North America and Europe, and then to beekeepers elsewhere who were ready for the improvements which the hives offered, and who learned about them from Langstroth's own book (1853) and other writings. Charles Dadant had emigrated from France to the USA in 1863, and he published in French (e.g. Dadant, 1874) as well as in English, spreading their use still further. In many countries beekeeping journals were started, and public-spirited beekeepers founded beekeeping societies, to give instruction on the new 'rational' beekeeping, and to discuss problems and new ideas.

A number of further advances followed the publication of Langstroth's 1853 book describing his hive and

Figure 1.44a Langstroth's movable-frame hive, illustrated in the second edition of his book (Langstroth, 1857).

Table 1.44B Some significant advances in the development of modern beekeeping

	Year published	Made by	Country
effective movable-frame hive	1851	L. L. Langstroth	USA
comb foundation	1857	J. Mehring	Germany
section honey production	1857	J. S. Harbison	USA
queens transported by post	1863	C. J. Robinson	USA
centrifugal honey extractor	1865	F. Hruschka	Austria
rollers for making comb foundation	1873	F. Weiss	USA
large-scale transportation of queens from Europe to America	1870	A. Grimm	USA
effective smoker	1875	M. Quinby/T. F. Bingham	USA
sale of bees by weight (leading to package bee industry)	1879	A. I. Root	USA
system for commercial queen rearing	1883	H. Alley	USA
queen rearing with artificial queen cells	1889	G. M. Doolittle	USA
effective bee-escape	1891	E. C. Porter	USA
effective system for swarm control	1892	G. W. Demaree	USA
hives of bees taken to pollinate crops	1895	M. B. Waite	USA
wired comb foundation	1920	H. Dadant	USA
successful instrumental insemination of queens	1926	L. R. Watson	USA

its use, and the most significant are set out in Table 1.44B. Many – like Langstroth's own advance – were made in the USA, and Pellett (1938) gives some details. In particular, the frame was fitted with an embossed wax sheet (comb foundation, 1857), and the framed comb built on this foundation could withstand the stress imposed on it when spun in a centrifugal honey extractor (1865). When the foundation was wired, the rate of spin could be increased.

Smoke had been used for quietening bees since the days of Ancient Egypt. But smokers had no mechanism to direct the smoke, and the type of smoker we use today (Section 5.21) was not developed until 1875, more than twenty years after the movable-frame hive.

1.45 The use of European honeybees in new regions

Until after 1600 all the honeybees (*Apis* species) were confined to the Old World where they had evolved. European honeybees (*Apis mellifera*) were taken in hives – almost certainly skeps – from England to North America from the 1600s onwards. The first may well have been those recorded, along with other livestock, in a letter dated 5 December 1621 sent from the Council of the Virginia Company in London to its Governor and Council in Virginia (D. A. Smith, 1977): 'We have by this Ship [the *Bona Vista* or the *Hopewell*] and the *Discovery* sent you [livestock listed] and Beehives . . .'

There are few records about beekeeping in North or Central America in the next two centuries. In 1721 Dudley said that 'all the Bees we have [in New England] in our Gardens, or in our Woods, . . . are the produce of such as were brought in Hives from England near a hundred Years ago.' Honeybees had been taken from Europe to the Leeward Islands early – to St Kitts in 1688 and Guadeloupe in 1689 (see Crane, 1989*b*). In 1784 they were taken from Florida to Cuba. They were fairly common throughout the eastern part of North America by 1800, and had been shipped to California – round South America – by 1853 (Pellett, 1938). They were transported from Spain and Portugal to South America, being first landed in Brazil in 1839, and reaching Chile and Peru perhaps in 1857.

Honeybees had been transported successfully by ship to Australia in 1810, and to New Zealand in 1839. In 1842 Cotton published explicit instructions for taking skeps of bees on the five-months' voyage to New Zealand. Apart from Hawaii (1857), many Pacific islands were without honeybees until after 1950.

In the present century European *Apis mellifera* has been imported into a number of countries where native honeybees were already being kept in hives, because they were better honey producers. For example European bees were taken to Egypt and Israel (Sections 8.23 and 8.73), where the native races were less productive and less manageable. European bees were also transported to Japan, China and Thailand that had native *A. cerana* (Section 7.44). Tropical *A. mellifera* was taken from Africa to South America in 1956, and the results of that venture are dealt with in Sections 1.22.4 and 7.43. In or about 1986 *A. florea* from Asia was found in Sudan, and *A. cerana* in Irian Jaya and Papua New Guinea.

The most bizarre attempt to introduce a new honeybee species was perhaps that made by Frank Benton in 1880: to take four colonies of the 'superior' *Apis dorsata* from south-east Asia to the USA (Pellett, 1938). The bees died on the journey.

Tables 1.5A and 7.1B include information on the honeybees now used for beekeeping in different countries, and Section 8.7 discusses considerations and problems associated with introducing honeybees into new areas.

1.46 Hive products through the ages

From the Ancient World onwards, honey has almost always been the most important product from hives and from wild colonies. In the Ancient World honey was used especially in medicine. It was by far the most popular 'drug' in Ancient Egypt, being mentioned 500 times in 900 remedies (Manjo, 1975), and it was a common ingredient of mediaeval medicines. It was often the only substance available that could make some of the other ingredients of medicines palatable.

Amphorae and other honey vessels have been excavated from Ancient World sites, and many records give evidence of international trade in honey in Ancient times, e.g. from Mediterranean ports in Spain, Sicily, Cyprus, Rhodes, Thasos, and mainland Greece and Turkey. It would seem that honeys from certain regions, and thus probably from certain plants, were especially prized, although it was not then understood that honey originated in nectar actually *produced* by flowers.

In Europe honey was used as a medicine and also as a food and sweetener: it was less expensive than sugar until the 1700s or later (Table 13.4A). The first book on honey was published in England in 1759: *The virtues of honey in preventing many of the worst disorders ...* by Sir John Hill. Honey was much valued as a food and sweetener in regions of the New World settled from northern Europe, but this has been much less so in Latin America.

By the late 1800s, the movable-frame hive and the centrifugal extractor were in common use in North America and Europe, and the amount of honey produced increased greatly. Also it was no longer most commonly sold as comb honey, but as a liquid in jars. There were many suspicions of adulteration – some justified – and this led to studies of the composition of honey, and to legislation to protect the honey industry (Section 15.71). Today, as throughout history, more honey is used as a food, eaten in the form in which it is produced and sold, than in any other way.

In traditional beekeeping, honey not eaten in the comb was separated from the wax by straining through a woven cloth or basket, and techniques for such weaving were developed very early in the Ancient World. The wax was melted and rendered into blocks or balls; dross sank to the bottom and was easily removed. With traditional hives, the beeswax yield was likely to be about 8–10% of the honey yield.

Beeswax is the easiest hive product to handle and transport, because it does not deteriorate, and it needs no container. In the Ancient World and in the Middle Ages it was used in payment of tribute and other dues, including fines, and it was an important article of commerce. It was needed for a wide range of purposes: for casting metal and modelling, for waterproofing, as an adhesive, as a component of ointments and cosmetics, and for candles and other sources of light. In Europe and neighbouring regions with a Christian tradition, the church needed large quantities of beeswax for candles, on the basis of the virginity of the bees. In the 1200s the Welsh Gwentian Code gave the reason thus: 'The origin of Bees is from Paradise, and on account of the sin of man they came hence, and God conferred his blessing upon them, and therefore the mass cannot be said without the wax.'

In the past hundred years the world market for beeswax has greatly decreased, other products having superseded it for many of the earlier uses. Also, with modern movable-frame hives and a centrifugal honey extractor, the beekeeper can use the same combs year after year, which reduces the amount of beeswax produced to 1½% to 2% of the honey yield, and enables him to get higher yields of honey, which is now a more profitable hive product. A considerable proportion of the beeswax produced goes back into the beekeeping industry as comb foundation (Table 13.7A). This and other modern uses are dealt with in Section 13.7.

The other contents of the combs – pollen, brood and brood food – are nutritious, but they are very perishable and could not be kept for future use, so when whole nests were harvested they were normally used straight away; see Section 14.7. Except in cultures where insects were not customarily eaten, brood was especially valued as a form of meat (Bodenheimer, 1951), and it is regarded as a delicacy by many peoples today, especially in eastern Asia; Table 14.61A gives its composition.

The commercialization of pollen, brood and royal jelly as separate hive products is dependent on modern methods of processing and storage, and it did not start until after 1950. Chapter 14 gives details of the present industries based on these materials, and on propolis which has antiseptic properties, and does not spoil.

1.5 THE PATTERN OF WORLD BEEKEEPING TODAY

Table 1.5A shows the present extent of beekeeping in different countries of the world, some of which are shown in Figure 1.5a. The table is presented in seven parts, for the continental regions: 1, Europe with all USSR; 2, North America; 3, Central America; 4, South America; 5, Oceania; 6, Asia; 7, Africa. Table 1.5B summarizes the above data, and includes both the honey production figures from Table 13.8A and some comparative figures for the 1970s.

Table 13.8A deals with honey and beeswax in individual countries, their production, exports and imports. Figures in all tables are more likely to be too low rather than too high, and the data available are not all on the same basis. For many countries in the tropics, statistics are not collected, and what is quoted is an estimate by the person best able to make one. Nevertheless many more countries can be documented now than in the 1970s (Crane, 1975*c*).

The three countries whose areas and populations are both very large – USSR, China and USA – also have the greatest number of hives of bees, and produce the most honey. But Europe is the continent with the greatest hive density (hives per km²), largely because it has a long and rich tradition of beekeeping, and a high human population density; the average honey yield per hive is low. Oceania has the lowest hive density (0.1/km²), and Africa the lowest average honey yield per hive, of any continent.

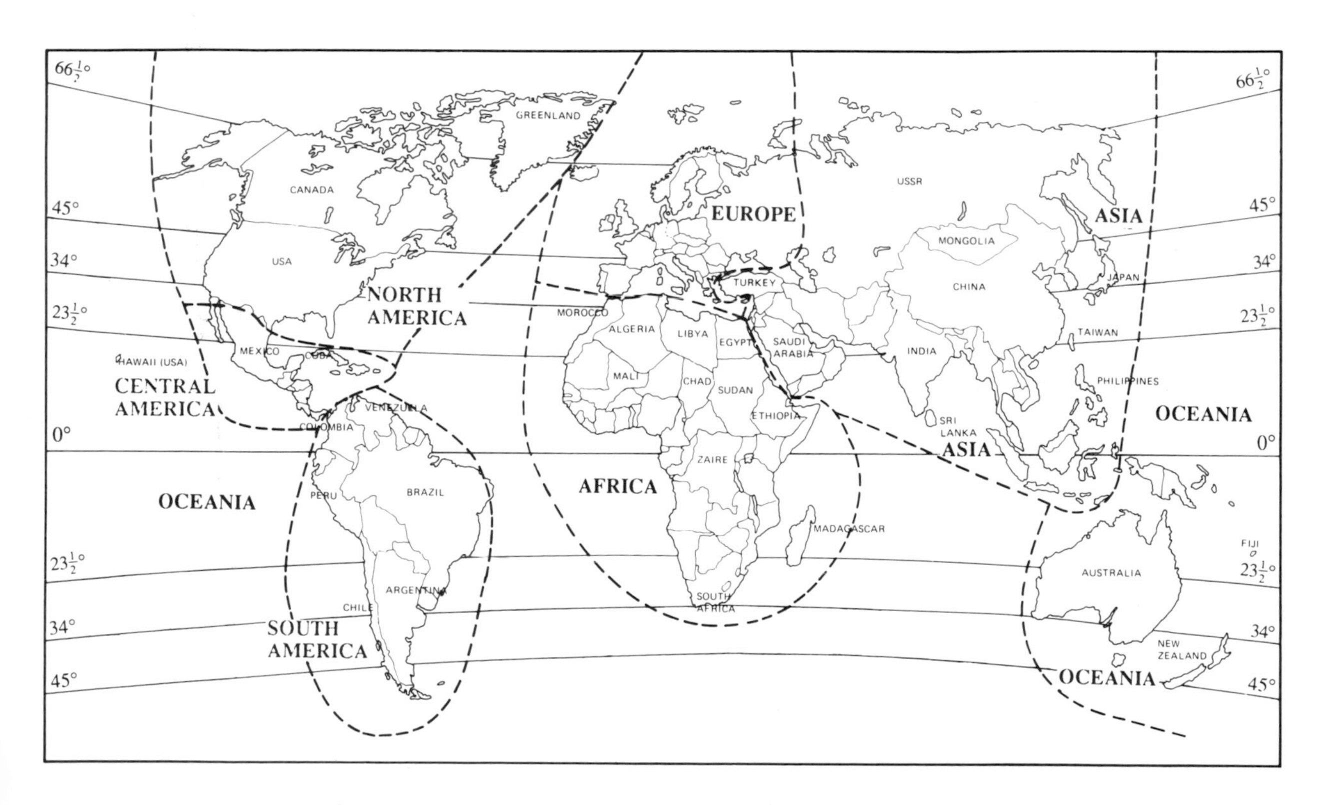

Figure 1.5a Map of the world showing continental regions and some of the larger countries.
In Sections 1.5 and 13.8, European and Asiatic parts of the USSR are treated as a single unit, with Europe.

The honey productivity of an area, as measured by the average honey yield per hive, influences the number of colonies of bees one beekeeper has. This number is not included in the tables here, but in the early 1970s the average number of colonies per beekeeper was twice as high in the New World (14) as in the Old World (7); see Crane (1975*c*). The profitability of beekeeping is governed by the honey yield per hive, together with labour costs, the selling price of honey and other hive products, and other factors; see Sections 6.7 and 7.6.

The following sections give a brief overview of beekeeping in continental regions represented in the seven parts of Tables 1.5A and 13.8A, with special reference to points of particular interest in each region. Photographs throughout the book illustrate beekeeping and bee forage in the different continents. Table 7.1A indicates general differences between the temperate and tropical regions, and the Gazetteer in Appendix 2 provides useful information relating to beekeeping in individual countries.

1.51 Europe and the USSR

Part 1 of Tables 1.5A and 13.8A refer to this region.

Europe excluding the USSR
The proportion of the land that can support vegetation is higher in Europe than in most continents. With its high human population and a long tradition of beekeeping, Europe has a greater concentration of beekeeping than any other continental region. The average is 6.1 hives/km² in eastern Europe (excluding the USSR), and 2.1 hives/km² in western Europe. The overall average (3.2) is over 40 times as high as in Australia.

Since the early 1970s, the total number of hives has increased from 13 to 15 million, and the total honey production from 122 to 165 thousand tonnes. Beekeeping is on the increase, and also its productivity. As everywhere, honey yields vary considerably from year to year; in Table 1.5A, the highest 1984 yields per hive in Europe were in Denmark (30 kg), Finland (27) and Hungary (26).

The average honey yield per hive, 11 kg a year, is low on a world scale. Widespread factors limiting honey yields include extensive industrial and residential use of land, large-scale mechanized farming, destruction of wild plants by herbicides, and – especially near the Atlantic seaboard – the weather. Some suburban areas are overstocked with hives, and this reduces individual hive yields.

The honey import–export balance is markedly different in eastern and western Europe. Figures below, calculated from Table 13.8A, are in thousand tonnes:

	Eastern Europe	*Western Europe*
honey production	79	86
net honey imports		127
net honey exports	29	

In eastern Europe all countries but one show a net export of honey, and in western Europe all but one show a net import. Section 13.81 discusses honey imports and exports on a world basis.

Beekeeping is not the chief means of livelihood for most European beekeepers, who have a few hives as a hobby, or up to about 100 as a sideline; the average holding is about 10 hives (Crane, 1975*c*). Nevertheless modern practices are used in most areas, and European beekeepers are probably more knowledgeable about bees than those in most other continents: they *need* such knowledge in order to succeed in their beekeeping. In regions where honey can be obtained more easily, there is less incentive to learn about the bees themselves.

All the bees kept in hives are *Apis mellifera*, and virtually no honey is collected from wild colonies. In most countries all beekeepers use movable-frame hives, although there are still appreciable numbers of traditional hives in certain parts of southern Europe.

The majority of countries in Europe have effective Beekeepers' Associations, and both these and Government extension services are important in providing instruction and advice. In any one country the percentage of beekeepers who belong to an Association varies from 95% or more down to less than 25%. Most countries publish at least one beekeeping journal.

European countries have legislation on many aspects of beekeeping, especially to prevent the spread of diseases and parasites (Table 15.4A). Legislation relating to honey is becoming more unified in the twelve countries that are members of the European Community (indicated in Table 1.5A, Part 1), and Section 15.7 gives some details.

The USSR
The Soviet Union includes areas in both Europe and Asia, and available statistics apply to the whole country; administratively, this is a single unit.

The State gives much encouragement to beekeeping in both public and private sectors, through education, research, supplies of hives (all movable-frame) and other equipment, queen rearing and distribution, bee breeding, and other channels. Legislation and standards cover many beekeeping supplies and products,

Table 1.5A Beekeeping in individual countries: number of hives of bees, hive density and honey yield per hive

Undated figures relate to 1984. Column 3 gives the total number of hives of bees, and (in brackets) the number of fixed-comb hives. In some tropical regions fewer than 50% of hives may be occupied at any one time.

The number of hives is quoted from printouts kindly prepared by FAO Statistics Division, or from another source:
[a] FAO 1984 figure from government or national institution
[b] FAO 1984 figure from some other source
FAO estimated (no symbol)
[n] information from another source, usually within the country

The honey yield per hive in a country is obtained from the number of hives given in this table, and the honey production in Table 13.8A. (The two figures are sometimes from different sources, and/or for different years; where they are widely disparate, the entry is ?.) In calculating continental (and world, Table 1.5B) mean yields, hive figures for countries with no honey production figure are excluded.

Part 1 Europe and USSR

EC indicates member of the European Community.
All honeybees are *Apis mellifera*, and virtually all hives have movable frames.

Country	*Area (1000 km²)*	*No. hives (1000s)*	*Year*	*No. hives/km²*	*Honey yield (kg/hive)*
Albania	29	115		4.0	6.2
Austria	84	500[n]	1986	6.0	9.0
Belgium-Luxembourg (EC)	33	100[n]	1976	3.0	10.0
Bulgaria	111	599[a]		5.4	16.7
Czechoslovakia	128	970[a]		7.6	9.9
Denmark (EC)	43	80[n]	1985	1.9	30.0
Finland	337	50[n]	1986	0.15	27.0
France (EC)	547	1200		2.2	16.8
German DR	108	524[a]		4.9	11.4
German FR (EC)	249	1147[b]		4.6	14.2
Greece (EC)	132	1250		9.5	9.6
Hungary	93	550		5.9	25.5
Irish Republic (EC)	70	28[n]	1986	0.40	6.4
Italy (EC)	301	810		2.7	7.4
Netherlands (EC)	37	70[n]	1976	1.9	7.1
Norway	324	68[n]	1982	0.21	17.8
Poland	313	2558		8.2	7.0
Portugal (EC)	92	200		2.2	15.5
Romania	238	1279[a]		5.4	12.0
Spain (EC)	505	1120 (see below)		2.2	9.0
Sweden	450	150[n]	1985	0.33	21.8
Switzerland	41	310[n]		7.6	7.5
UK (EC)	245	190		0.78	7.9
Yugoslavia	256	1200[n]	1983	4.7	4.7
TOTALS FOR EUROPE	4766	15 068	MEANS:	3.2	11.0
USSR (Europe + Asia)	22 402	8200		0.37	23.5
Earlier figures for Spain:		797 (232[n])	1980		

Iceland: no beekeeping; honey imported (Table 13.8A, Part 1)
Malta: has beekeeping; honey imported (Table 13.8A, Part 1)

(continued)

and disease control. Activities are organized through a centrally planned and controlled network of organizations throughout each of the 15 Republics of the Union, and both the total honey production and the honey yield per hive have probably doubled since the early 1970s, although the total number of hives (8 million) has not increased. The total production was 193 000 tonnes in 1984, and the honey yield per hive 24 kg (see Part 1 of Tables 13.8A and 1.5A). The country has a low average hive density – only 0.4/km^2 – because of its extensive cold region in the north where bees cannot live.

The public sector consists of bee farms that concentrate on different aspects of beekeeping: honey production, crop pollination, queen rearing and production of package bees. Some bee farms are sovkhozes (State farms) and others are kolkhozes (collective farms); in either type, the workers form a group which discusses and carries out policy that has been planned on a regional basis and distributed to them by the All-Union Ministry of Agriculture. Hives are systematically migrated for honey production, and also for crop pollination, in which case the rent paid for the hives by crop-growing sovkhozes and kolkhozes appears in the bee farm's accounts.

More attention is given now than formerly to aiding private beekeepers. Except in large towns, anyone may keep bees and sell the honey, either to a State organization or privately, e.g. from the door. Private beekeepers – who include a number of ex-servicemen – usually have 1–10 hives, and they form their own local Beekeepers' Associations, for which precise rules are laid down centrally. Retired people who take up beekeeping receive an addition to their pension. *Perestroika* has affected beekeeping, and Pchelovodstvo (1988*a*) prints in full a 1988 Decree of the Soviet of People's Deputies which now governs the formation and management of beekeeping co-operatives.

Apis mellifera is used for beekeeping throughout the USSR, which has a wide variety of native races and strains (see Section 1.21); *A. cerana* is native in the Far Eastern Province; see Figure 1.2b. Caucasians and Carniolans are often used for crop pollination, which is most important in the large arable regions, but also with greenhouse crops in the Arctic regions. Queen rearing, bee breeding and package bee production are mostly concentrated in the warmer south, for instance in the Ukrainian, Moldavian, Georgian, North Ossetian and Uzbek SSRs; queens are also reared in areas farther north that have especially valuable native races of bees, such as the Bashkir ASSR. Queens are distributed to other parts of the country on an area plan, each region receiving strains considered best for it. Package bees are referred to again in Section 8.32.

1.52 North America

Geographically Mexico is part of North America, but its beekeeping situation is more closely allied to that of Central America, so it is included in Section 1.53, and only the USA and Canada are considered here; Part 2 of Tables 1.5A and 13.8A refer to them. There are no honeybees native to the Americas; all in North America are descended from honeybees introduced from Europe. However, (tropical) Africanized honeybees were spreading north through Mexico by 1986; see Table 7.1B and Section 7.43.

Large-scale queen rearing, package bee production, and the use of both honeybees and other bees for crop pollination in huge monocultures, were all developed first in the USA; they are described in Sections 8.22, 8.31, 8.4 and 8.5, respectively.

The USA has 4.3 million hives, and except for an increase during the Second World War, the number has been fairly static for nearly a century – it was 4.1 million in 1899. The annual yield per hive has not increased much in recent decades, in spite of greatly improved hive management; in 1984 it was 17 kg. Reasons for the lack of increase include the large-scale mechanized agricultural methods which result in loss of bee forage, and pest control methods inimical to bees. Morse and Nowogrodzki (1983) estimate an average increase of 0.14 kg per hive per year between 1939 and 1981. In 1984 the USA produced 75 000 tonnes and imported a further 59 000 tonnes. A price support programme was operated for honey in the USA from 1950 to 1986; see Section 13.81.

In 1984 Canada produced 43 000 tonnes and exported 19 000 tonnes. It has the highest average honey yield per hive of any country in the world; 63 kg in 1984. Some regions of Canada share with Sweden, Finland, and parts of Siberia in the USSR, the benefit of a very long day-length in summer, and the associated rapid growth of plants and of honeybee colonies. Especially in the west, seed legumes are extensively grown, and honey yields per hive are highest in Manitoba, Saskatchewan and Alberta. Canada, like the USSR, has an extensive cold region, where bees cannot live, and the average hive density for the whole country is only 0.07/km^2.

In the past, Canadian beekeepers depended much on package bees from the USA, but for reasons explained in Section 8.31 they have developed methods for keeping colonies through the long cold winters (Section 6.54). Queen rearing within Canada has also been expanded, and queens and package bees are imported from New Zealand (Section 8.32); all imports of bees from the USA were prohibited from 1 January 1988 (Canadian Beekeeping, 1988).

Table 1.5A *(continued)*—explanatory notes are on page 31

Part 2 North America

All honeybees are *Apis mellifera*, and virtually all hives have movable frames.

Country	*Area (1000 km²)*	*No. hives (1000s)*		*No. hives/km²*	*Honey yield (kg/hive)*
Canada	9976	692		0.07	62.6
USA	9363	4300		0.46	17.4
TOTALS	19 339	4992	MEANS:	0.26	23.7

Greenland: no beekeeping; honey imported (Table 13.8A, Part 2)

Part 3 Central America

Africanized *Apis mellifera* are present in countries marked*; the year in which they were first reported is indicated in Table 7.1B. In many countries most, but not all, hives have movable frames; the number of others is included where it has been reported.

Country	*Area (1000 km²)*	*No. hives (1000s)*	*Year*	*No. hives/km²*	*Honey yield (kg/hive)*
Antigua/Barbuda	0.4	0.13[n]	1986	0.33	46
Bahamas	14	0.15[n] (0.05[n])	not stated	0.01	
Barbados	0.4	0.2[n] (some[n])	1985	0.5	5
*Belize	23	11 (few[n])		0.48	22.7
Bermuda	0.05	0.5[n] (0.03[n])	1986	10.0	
*Costa Rica	51	58†		0.59	30.0
Cuba	115	180		1.6	49.1
Dominica	0.8	0.2[n] (some[n])	1985	0.25	10
Dominican Republic	49	65†		1.3	20.0
*El Salvador	21	132†		6.3	19.7
Guadeloupe	2	0.8		0.4	5
*Guatemala	109	190 (most[n])		1.7	13.9
Haiti	28	19†		0.68	16.8
*Honduras	112	65[n] (30[n])	1982	0.58	18.5
Jamaica	11	70 (few[n])		6.4	14.3
Martinique	1	5		5.0	21.2
*Mexico	1973	2686[a]		1.4	25.0
Montserrat	0.1	0.15[n] (few[n])	1986	1.5	46.6
Netherlands Antilles	0.1	0.4[n] (0[n])	1977	4.0	
*Nicaragua	130	12[n] (0.9[n])	1986	0.09	12.5
*Panama	77	4[n] (few[n])	1986	0.05	
Puerto Rico	9	6		0.67	28.8
St Kitts/Nevis	0.4	0.24[n] (some[n])	1986	0.6	45.8
St Lucia	0.6	1.8[n] (0[n])	1986	3.0	38.3
*Trinidad	5	4[n]	1986	0.8	14.3
TOTALS	2732	3512	MEANS:	1.3	24.7

† *Earlier figures for:*

Costa Rica	30[n] (4[n])	1981
Dominican Republic	150[n] (100[n])	1972
El Salvador	130[n] (57[n])	1980
Haiti	15[n] (12[n])	1976

No figures are available for the following islands, but beekeeping is known to be practised in Cayman Islands, Grenada, St Vincent/Grenadines and Virgin Islands (UK and USA).

(continued)

North American beekeeping now faces several setbacks. The tracheal mite (*Acarapis woodi*) and the varroa mite (*Varroa jacobsoni*) reached the USA in or before 1984 and 1987, respectively; Sections 11.41 and 11.42 describe the effects of these mites. The likely arrival of Africanized bees has already been referred to. The beekeeping systems used in much of USA and Canada, with long-distance movement of package bees, and of colonies for honey production and crop pollination, make it difficult or impossible to restrict the spread of parasitic mites or of undesirable bees.

On a more positive note, North America has been the source of a great many technological advances in beekeeping, especially those that increase the effectiveness of large-scale operations with minimal labour requirements. Research and extension work have been at a high level; there are a number of active Beekeepers' Associations, and many books, bulletins and journals are published in the USA and Canada.

1.53 Central America

Here, Mexico is included in this region, which is referred to in Part 3 of Tables 1.5A and 13.8A. This is one of the beekeeping growth areas of the world, and by 1984 its honey production had increased to 87 thousand tonnes. The average hive density is 1.3/km^2, the highest for any continental region except Europe. All figures for Central America are dominated by those for Mexico, which produces 67 and exports 54 thousand tonnes of honey a year, and in 1984 was the world's largest exporter. The story of its rise to this position from a small start in the 1940s has been told by Willson (1975). In Central America most honey is produced for export, and the average domestic consumption was only 90 g/person in 1984; see Table 13.8D.

Much of the region is within the tropics. The mainland countries stretch from Mexico in the north to Panama in the south, and islands from the Bahamas, via Cuba, Hispaniola, Jamaica and a host of smaller Caribbean islands to Trinidad. Africanized bees are in all mainland countries (Figure 7.43a), and Table 7.1B shows their year of arrival in each. They are also in Trinidad, which is close to – and was part of – South America, but not Tobago, or any other Caribbean island. Parts of Central America have a fair number of box hives (see Table 1.5A, Part 3), and some top-bar hives (Section 10.21) are also used.

As well as the mainland, some of the islands that receive an adequate rainfall are especially favourable for beekeeping. Their relative isolation seems to have kept them free from bee diseases and parasites, and there are large stretches of uncultivated mountain land that yield nectar and pollen during much of the year. With modern hives and good management, honey yields can be high, although the need to import materials adds to the cost. Also, on a small island domestic sales are limited, and yet the total production is too small to interest the world market.

Stingless bees, especially *Melipona beecheii*, are kept for honey production in parts of Central America; see Section 8.63. In Panama, which has about 4000 hives of honeybees, Roubik (1986) estimates that honey is also harvested annually from up to a thousand wild nests of (Africanized) honeybees, and from 100–200 thousand wild colonies of stingless bees, which include *Melipona fuliginosa*, *M. fasciata*, *M. compressipes*; *Trigona pectoralis*, *T. capitata*, *T. frontalis* and *T. angustula*. Harvesting honey from wild nests of (European) honeybees is popular in many Caribbean islands, where it is known as honey-cutting.

1.54 South America

Hives are spread much more thinly in South America (Part 4 of Tables 1.5A and 13.8A); most of them are movable-frame hives. Between the early 1970s and 1984, honey production increased by over 50%, to 57 312 tonnes. Argentina, which produces well over half of this total, doubled its exports. Much of this country is in the more temperate south, where white clover and lucerne are among major honey sources, as well as the thistle-like cardoon, *Cynara carduncululs*. These all give light, mild honeys such as are favoured in many importing countries.

Argentina has been less affected by the change-over from European to Africanized bees than the more tropical parts of South America, where 'Africanization' has led many beekeepers with only a few hives to give up. Table 7.1B shows the year of arrival of Africanized bees in each country; see also Figure 7.43a. Sections 1.22.4 and 7.43 discuss these bees, and their management, respectively.

Brazil's honey production is increasing, but it is still very small on a world scale. Much of the beekeeping is concentrated in Santa Catarina Province in the south. Brazil has several bee research centres of world renown, and hosts a number of beekeepers' meetings – regional, national and international. Bees and beekeeping have also been intensively studied in Venezuela in the north, where the United States Department of Agriculture set up a laboratory for studying Africanized bees.

Table 1.5A (*continued*)—explanatory notes are on page 31

Part 4 South America

Africanized *Apis mellifera* are present in all countries except Chile; the year in which they were first reported is indicated in Table 7.1B. In many countries most hives have movable frames.

Country	*Area (1000 km²)*	*No. hives (1000s)*	*Year*	*No. hives/km²*	*Honey yield (kg/hive)*
Argentina	2767	1400[b]		0.51	25.0
Bolivia	1099	70		0.06	20.0
Brazil	8512	1700[n]		0.20	4.4
Chile	757	490		0.65	10.2
Colombia	1139	130[n] (120[n])	1980	0.11	16.9
Ecuador	284	66		0.23	17.0
French Guiana	91	(few[n])	(1975)		
Guyana	215	4		0.02	17.5
Paraguay	407	55 (few[n])		0.14	19.1
Peru	1285	84	(1961)	0.07	0.95
Surinam	163	4		0.02	18.0
Uruguay	176	155 (few[n])		0.88	22.6
Venezuela	912	14[n]		0.02	22.9
TOTALS	17 807	4088	MEANS:	0.23	14.0

Part 5 Oceania

All honeybees are *Apis mellifera*, and virtually all hives have movable frames.

Country	*Area (1000 km²)*	*No. hives (1000s)*	*Year*	*No. hives/km²*	*Honey yield (kg/hive)*
Australia	7687	537[n]		0.07	46.5
Cook Islands	0.2	0.2[n]	1980	1.0	
Fiji	18	0.5[n]	1980	0.03	10.0
French Polynesia	4	0.7		0.18	18.6
Hawaii	17	8[n]	1981	0.47	35.0
New Caledonia	19	0.5[n]	1980	0.03	
New Zealand	269	260		0.97	22.4
Niue	0.3	0.8		2.7	25.0
Pacific Trust Territory (incl. Marshall Is.)	2	few[n]	1980		
Papua New Guinea	462	2[n]	1980	0.004	
Samoa, America	0.2	0.2[n]	1980	1.0	
Samoa, Western	3	0.8[n] (0.2[n])	1980	0.27	
Solomon Islands	28	0.01[n]	1980	0.0004	
Tonga	0.7	1 (0[n])		1.4	18.0
Tuvalu	0.03	0.04		1.3	50.0
Vanuatu	16	0.1[n]	1980	0.006	
Wake Island	0.008	0.03[n]	1976	3.8	
Wallis/Futuna Is.	0.2	0.5		2.5	20.0
TOTALS	8527	812.4	MEANS:	0.10	38.5

No figures are available for the following islands, but beekeeping is known to be practised in Guam, and to have been considered in Kiribati, and Norfolk and Pitcairn Islands.

(*continued*)

1.55 Oceania

This region, referred to in Part 5 of Tables 1.5A and 13.8A, comprises Australia, New Zealand, the island of New Guinea, and Pacific islands. Like the Americas, it belongs to the New World, and has no native honeybees.

Australia is by far the largest land mass, and it has a great arid area which cannot support bees; as a consequence its average hive density is extremely low, only 0.07/km². The north of the country lies in the tropics, but most of the areas used for beekeeping are in the subtropics. Some give high yields to migratory beekeepers.

The bulk of the honey comes from indigenous species of *Eucalyptus* trees, some of which are referred to in Sections 12.4–12.8. Stands of different eucalypts may provide rich honey flows in succession, which a beekeeper can harvest by moving his hives from one flow to the next. Very large annual crops can thus be obtained, up to say 250 kg per hive. The price obtainable for honey sets a limit to the distances through which it is economically profitable for a beekeeper to take hives to harvest honey from an extra flow.

Until the 1950s the majority of beekeepers were professional, with 500 hives or more. Now they represent only 15% of the total; after the Second World War new settlers from Europe introduced small-scale beekeeping, and 40–45% of the total are 'semi-commercial' (with 50–300 hives), the remaining 40–45% having 6 hives or less.

In New Zealand the standards of beekeeping are in general high, and many innovations and useful practices have originated there, as also in Australia. There is detailed legislation, and an effective extension service. Much of the honey comes from white clover; it is both eaten at home and exported (Table 13.8A). The native flora yields some very interesting honeys, and a few that have objectionable properties. Queen rearing and package bee production have been developed, and an export trade established (Van Eaton, 1987); see also Section 8.32.

Since the early 1970s, the total honey production of Oceania has increased by 24%, from 14% more hives. Honeybees (*Apis mellifera*) have been introduced from Australia and New Zealand to additional Pacific islands, some of which now produce honey; Tonga has also reared some queens. The amounts of honey are too small to upset the dominance of Australia and New Zealand in this region. *Apis cerana* has been introduced from Asia into the island of New Guinea; see Section 1.23.1.

1.56 Asia

Part 6 of Tables 1.5A and 13.8A refer to Asia, an Old World continent and the only one with more than one species of native honeybee. (The Asiatic part of the USSR is covered in Part 1 of the Tables, and in Section 1.51.) The native hive bee is *Apis cerana*, except in the Mediterranean region in the west where it is *A. mellifera*; Figure 1.2b shows their distributions. European *A. mellifera* has been introduced successfully in some other parts of Asia which are temperate or subtropical (e.g. in Japan and China), or tropical but at higher altitudes and with large plantations of crops yielding nectar and pollen, such as longan (*Euphoria longan*) in northern Thailand. Section 7.44 discusses this subject.

Excluding the USSR, Asia has a quarter of the world's hives and produces a quarter of the world's honey. Both production and export are dominated by China, which has almost 7 million hives, over half the total in Asia. This country's very large honey export trade is based on beekeeping with introduced *A. mellifera*. In some areas hives are loaded on to special trains which take them to sites convenient for working a prolific honey flow, and collect them later. The countries with the greatest numbers of hives after China are Turkey and Iran, where *A. mellifera* is native.

Table 7.1B indicates which bees are kept in hives in individual countries, and Part 6 of Table 1.5A quotes the numbers of hives of *A. mellifera* and of *A. cerana* where this is known. The largest amount of honey probably comes from hives of *A. mellifera*, and where this bee does well, beekeeping can be very profitable, especially if labour costs are low. Except in Kashmir, *A. cerana* colonies give smaller yields – see Table 7.41A – and beekeepers may have to contend with absconding as well as swarming of colonies. There are at least a million modern hives of *A. cerana* in China, and in India, Nepal, and much of south-east Asia, beekeeping is based on *A. cerana* (Table 1.5A, Part 6). Some details are given in Section 7.41 and in Kevan (1989).

Eastern Asia is the world centre for the production of royal jelly (from *A. mellifera*), and probably also for its consumption. Section 14.43 quotes China's production as 400 tonnes a year, and Japan's imports in 1985 as 210 tonnes. Eastern Asia is probably the only part of the world where honeybee brood also has the status of a commercial product.

In many of the tropical countries of Asia where *A. cerana* is used for beekeeping, much more honey is harvested from wild *A. dorsata* nests (Sections 1.23.2 and 8.61) than from hives. Smaller amounts of honey

Table 1.5A (*continued*)—explanatory notes are on page 31
Part 6 Asia
Ac indicates *Apis cerana* and *Am Apis mellifera*; *Am*† indicates a few introduced colonies.
Most hives of *Am* have movable frames, and in some countries most hives of *Ac*. In other countries most hives of *Ac* have fixed combs.
The USSR, including its large area in Asia, is in Part 1.

Country	Bees	Area (1000 km²)		No. hives (1000s)	Year	No. hives/km²	Honey yield (kg/hive)
Afghanistan	Am Ac	648	Am	20^{n} (2^{n})		0.03	?
			Ac	(some^{n})			
Bangladesh	Ac	144		2^{n}	1987	0.014	1.0
Burma	Ac	677		2.5		0.004	2.0
China	Am Ac	9597	Am	5900	1988	0.72	23.3
			Ac	1000^{n} (500^{n})			
Cyprus	Am	9		65 (few^{n})		7.2	6.3
India	Ac Am†	3288	Ac	710^{n} (few^{n})	1980	0.22	6.3**
Indonesia	Am Ac	1904	Am + Ac	10^{n}	1987	0.01	10.0**
Iran	Am	1648		1260^{n} (340^{n})	1985	0.76	4.8
Iraq	Am	435		29^{n}			
				(some^{n})	1977	0.7	1.8
Israel	Am	21		80^{n}	1983	3.8	25.3
Japan	Am Ac	372	Am	291^{n}		0.78	23.4
			Ac	few^{n} (few^{n})			
Jordan	Am	98		40^{n} (33^{n})	1979	0.41	7.5
Korea, Republic	Am Ac	98	Am	300^{n}	1985	4.1	15.8
			Ac	100^{n}			
Lebanon	Am	10		105		10.5	2.9
Malaysia	Ac Am†	330	Ac	4^{n} (0.2^{n})	1986	0.012	
Mongolia	Am	1565		2		0.001	?
Nepal	Ac	141		75^{n}	1988	0.53	
Oman	Am	212		0.4^{n}	1986	0.005	
Pakistan	Ac Am	804	Ac	46^{n} (40^{n})		0.58	?
			Am	1^{n}			
Philippines	Am Ac	300	Am	1.5^{n}	1987	0.006	?
			Ac	0.3^{n}			
Singapore	Am	0.6		0.2^{n}	1951	0.33	
Sri Lanka	Ac	66		19^{n} (11^{n})		0.29	2.1
Syria	Am	185		147^{n}		0.79	4.6
Taiwan	Am Ac	36	Am	280^{n}	1988	7.8	11.5
			Ac	2^{n}			
Thailand	Ac Am	514	Am	36^{n}	1986	0.09	?
			Ac	8^{n} (most^{n})			
Turkey	Am	781		2500^{n} (1000^{n})	1985	3.2	14.2
Vietnam	Ac Am	330	Ac	54^{n} (most)	1983	0.27	3.6
			Am	35^{n}			
Yemen Arab Rep.	Am	195		21		0.11	14.3
Yemen, PDR	Am	333		4		0.01	17.5
TOTALS		24 742		13 151	MEANS:	0.53	17.5**

** excluding *Apis dorsata* honey listed for countries marked * in Table 13.8A, Part 6.

No figures are available for the following countries, but beekeeping is known to be practised in Bhutan, Brunei, Chagos Archipelago, Hong Kong, Kampuchea, Korea Democratic People's Republic and United Arab Emirates. Honey is also imported into Bahrain, Brunei, Kuwait, Macau, Qatar, Saudi Arabia and UAE; see Table 13.8A, Part 6.

(*continued*)

Table 1.5A (*continued*)—explanatory notes are on page 31

Part 7 Africa

All honeybees are *Apis mellifera*, and in every country some are kept in hives. The proportion of hives with movable frames varies greatly – from less than 1 in 1000 in some countries to more than 99% in others. The number of fixed-comb hives is included where it has been reported.

Country	*Area (1000 km²)*	*No. hives (1000s)*	*Year*	*No. hives/km²*	*Honey yield (kg/hive)*
Algeria	2382	207[n] (100[n])		0.09	7.7
Angola	1247	1000		0.80	15.0
Benin	113	4.3[n] (4.1[n])	1986	0.04	
Botswana	600	0.1[n] (0[n])	1983	0.0002	
Burundi	28	93		3.3	9.8
Cameroon	475	280		0.59	8.2
Central African Republic	623	650		1.0	10.5
Chad	1284	120		0.09	8.0
Egypt	1001	1100†		1.1	6.8
Ethiopia	1222	2520[n] (most[n])		2.1	8.3
Gabon	268	few[n] (few[n])			
Ghana	239	600[n] (most[n])	1986	2.5	
Guinea-Bissau	36	20[n]		0.56	15.0
Ivory Coast	322	3[n] (2.5[n])	1986	0.01	
Kenya	583	2100[n] (100[n])		3.6	5.7
Lesotho	30	0.03[n] (0[n])	1986	0.001	
Libya	1760	38†		0.02	14.2
Madagascar	587	300[n] (most[n])	1983	0.51	11.9
Malawi	118	80[n]	1979	0.68	
Mali	1240	501[n] (500[n])	1982	0.40	8.4
Mauritius	2	0.8[n]	1979		
Morocco	447	323[n] (273[n])	1980	0.72	9.6
Mozambique	783	28	1980	0.04	9.3
Nigeria	924	700?[n]	1950s	0.76	
Réunion	3	2.5[n] (1[n])	1981	0.83	21.2
Rwanda	26	4.4	1979	0.17	0.6
Senegal	197	20?[n]	1940	0.10	10.1
Seychelles	0.3	0.8[n] (0.4[n])		2.7	
Somalia	638	100[n] (most[n])	1986	0.16	
South Africa	1221	52		0.04	17.3
Sudan	2506	200?[n] (most[n])	1950s	0.08	
Tanzania	945	1500[n] (most[n])	1982	1.6	7.7
Tunisia	164	105[n] (65[n])		0.64	8.6
Uganda	236	43[a]		0.18	4.0
Zaire	2345	160?[n]	1960s	0.07	
Zambia	753	676[n] (674[n])	1986	0.9	
TOTALS	25 348	13 532	MEANS:	0.53	8.1

† *Earlier figures for:*			
	Egypt	900[n] (400[n])	1979
	Libya	15[n] (5[n])	1978

No figures are available for the following countries, but beekeeping is known to be practised in Burkina, Canary Islands, Gambia, Guinea, Liberia, Madeira, Namibia, St Helena, São Tomé and Principé, Sierra Leone, Swaziland, Togo and Zimbabwe. Honey is also imported into Cape Verde.

are collected from *A. florea*, and Section 8.62 describes a form of beekeeping with this bee.

1.57 Africa

Part 7 of Tables 1.5A and 13.8A refer to the African continent, which has several notable features from the beekeeping point of view. The earliest known evidence of beekeeping has been found there (Section 1.42.1), and the craft has been carried on for at least 4500 years, with a rich and varied tradition of hive making and bee management (Seyffert, 1930; Crane, 1983*a*, Nightingale, 1983).

The North African coastal area is part of the Mediterranean basin and has its own native honeybees (Section 1.21.4). The very much larger area south of the Sahara is the home of tropical *Apis mellifera* (Section 1.22); in the extreme south is the Cape bee (*A. m. capensis*), and on Madagascar is another race, *A. m. unicolor*; see Section 1.22.3. Tropical Africa has some very rich honey-producing country, including the miombo, a belt of *Brachystegia–Julbernardia* woodland which stretches from east to west coasts; see Figure 12.7a, map 3. Tropical Africa also has more, and more destructive, bee enemies and predators than any other world region, and for this reason traditional hives were generally hung in trees for protection.

Until after the Second World War, the primary commercial product from traditional hives was beeswax, and Africa still exports more than half the beeswax that reaches the world market; see Section 13.82 and Table 13.8E. In some areas the honey was used mostly for brewing beer. This situation is changing, and much more honey – of better quality – is now being produced and marketed, largely in the country of origin; see Table 13.8A.

Africa used to have a hive density higher than any other continental region except Europe, but its traditional beekeeping has declined. Modern beekeeping with tropical African bees presents various difficulties (Section 7.42), and a number of beekeeping development programmes have been organized. In Kenya a single-storey long top-bar hive was developed for use with tropical African bees; its full-width top-bars are especially convenient for handling these bees (Section 10.2). In spite of the many efforts in different countries, the overall average honey yield per hive shown in Table 1.5B is less than in any other continental region; this low figure may, however, be due partly to incompleteness of records – many Africans dislike counting things and consider it unlucky to do so.

Table 1.5B Beekeeping and honey production: world summary by continent

All figures are based on FAO data, except that in the absence of 1971 FAO figures for numbers of hives, those used are from Crane (1975*c*), with totals for Central and South America separated.
Areas, and numbers of hives/km² in 1984 or later, are for the countries tabulated in Parts 1–7 of Table 1.5A, and honey production figures are from Table 13.8B. In Asia, *Apis dorsata* honey is included in production figures, but not in yield/hive.

Continent	*Area (million km²)*	*No. hives (millions)*		*No. hives /km²*	*Honey production (1000 tonnes)*		*Honey yield (kg/hive)*
		c. 1971	1984	1984	1971	1984	1984
Europe	4.8	*13.0*	15.1	3.2	*122*	165	11.0
USSR	22.4	*10.0*	8.2	0.37	*220**	193*	23.5*
N. America	19.3	*4.8*	5.0	0.26	*158*	118	23.7
C. America	2.7	*1.7*	3.5	1.3		87	24.7
S. America	17.8	*2.3*	4.1	0.23	*37*	57	14.0
Oceania	8.5	*0.7*	0.8	0.10	*25*	31	38.5
Asia	24.6	*6.0*	13.2	0.53	*241**	246*	17.5*
Africa	25.3	*14.0*	13.5	0.53	*25*	95	8.1
WORLD TOTALS	125.4	*52.4*	63.4	0.51	*828*	992	15.3

* These and figures in Crane (1975*c*) may not all be calculated on the same basis.

1.6 FURTHER READING AND REFERENCE

Details of publications listed will be found in the Bibliography.

Species and races of bees from which honey is harvested

Michener, C. D. (1974) *The social behavior of the bees*
Rinderer, T. E. (ed.) (1986) *Bee genetics and breeding*
Ruttner, F. (1988) *Biogeography and taxonomy of honeybees*
Schwarz, H. F. (1948) *Stingless bees (Meliponidae) of the Western Hemisphere*

The history of beekeeping

Crane, E. (1983*a*) *The archaeology of beekeeping*
Fraser, H. M. (1951) *Beekeeping in antiquity*, 2nd ed.
Pellett, F. C. (1938) *History of American beekeeping*

The pattern of world beekeeping today

Crane, E. (1978*a*, *b*) *Bibliography of tropical apiculture* and *Satellite Bibliographies*
Crane, E. (1980*b*) *Apiculture*
Proceedings of the International Apicultural Congresses (listed in Section 16.82).

2

Honeybees as individual insects

2.1 INTRODUCTION

This chapter and the next describe the anatomy, life cycle, and behaviour of the honeybee. All applies to temperate-zone *Apis mellifera*, the bee used for most of the world's beekeeping, and much also to tropical *A. mellifera* and to *A. cerana*; where differences are known, they are indicated. Characteristics of bees as individual insects are dealt with here in Chapter 2, and those of bees as members of the colony in Chapter 3.

These characteristics of the honeybee constitute one of the foundations on which bee management must be planned and operated, and a more complete knowledge of them opens up new ways in which colony and apiary management can be made profitable for the beekeeper.

2.2 HONEYBEE ANATOMY

This Section provides an introduction to the anatomy of the adult honeybee (*Apis mellifera*). It will enable readers to locate the different structures of the bee's body that are referred to in the book, and to understand something of their function and operation. In the subject index, the first (bold) page number following an anatomical structure leads to a Figure in this chapter where the structure is pin-pointed. In Section 2.3 the internal systems (digestive, respiratory, reproductive, etc.) are described briefly, together with their physiology.

The following publications should be referred to for further information on subjects touched on here. Excellent books by Snodgrass (1925) and Dade (1962) give detailed descriptions of the anatomy of *Apis mellifera*; both books have stood the test of time and have been reprinted, in 1984 and 1985, respectively. Snodgrass's book is the more detailed, but the nonspecialist is likely to find Dade's book, and the chapter by Snodgrass in Dadant (1975), more useful.

Coloured anatomical drawings designed for teaching are available in *Bienenkundliche Lehrtafeln* (Jordan & Zecha, 1956), and in various wall charts (see IBRA, 1985/86). Erickson et al. (1986) published a comprehensive collection of scanning electron micrographs which show minute details of many structures of the honeybee. A few of them are included here.

In descriptions that follow:

- *dorsal* refers to the back part or surface (spinal for a vertebrate, upper for a flying bee);
- *ventral* refers to the opposite part or surface (front, or underside; the word floor is also used);
- *lateral* refers to the side part or surface, between dorsal and ventral.

Insects have an external skeleton (exoskeleton) consisting of layers of cuticle, which provides a hard protective casing that encloses the body, and the outer cuticular layer is waterproof. Much of the exoskeleton is covered with hairs; most are protective and provide thermal insulation, but some are important in the gathering of pollen, and in certain sense organs. The relatively thick rigid plates of the exoskeleton are joined together by membrane with only thin layers of cuticle, and in some places this membrane acts as a flexible joint. The cuticle has a very hard outer layer of sclerotin with some chitin, and an inner layer contains much more chitin.

The insect body is divided into three parts – head, thorax and abdomen, and each of these parts develops from segments that are formed during the embryonic

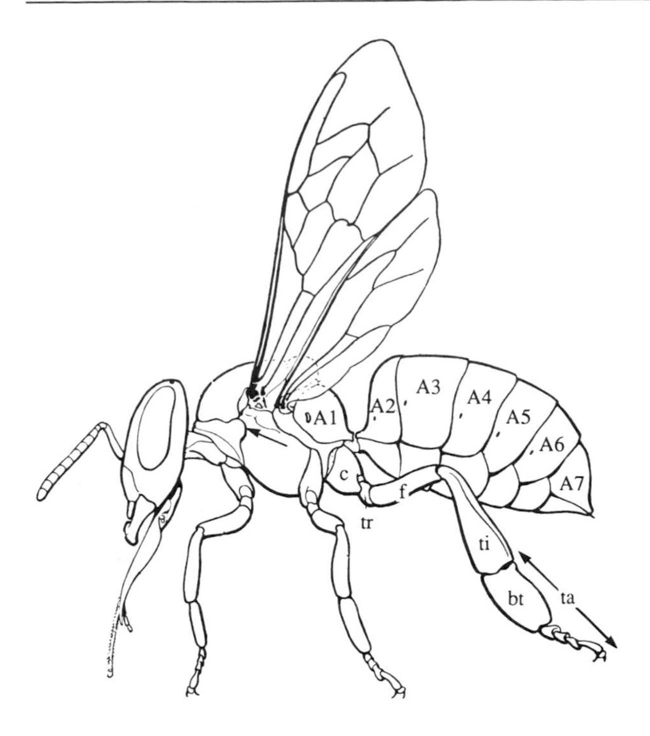

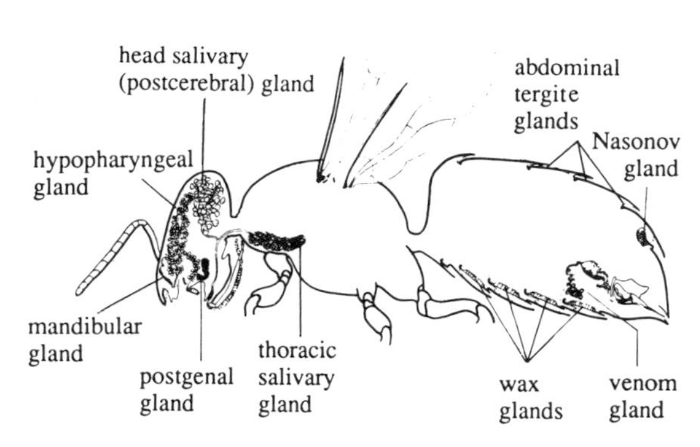

Figure 2.2a (*above*) External structure of the worker honeybee (*Apis mellifera*), with the hairy covering removed (adapted from Snodgrass, 1925): abdominal segments A1 to A7; first thoracic spiracle (arrow). For details of leg segments, see text (page 44) and also Figure 2.22a.
below Location of some important glands (adapted from Ribbands, 1953).

stage. Figure 2.2a shows an adult worker bee (imago) from the side. The head is shown on the left, and the thorax in the centre, with the two membranous wings and three legs that belong to the side of the body in view. On the right is the abdomen, and segments A2 to A7 are marked. The head has six, and the thorax three, not readily distinguishable segments. In the honeybee, the first abdominal segment A1 (the propodeum) is also incorporated into the thorax. Of the segments in the abdomen, A2 to A7 are visible; parts of A8 and A9 are modified to form the sting apparatus, and part of A10 to form the anus.

The head includes the mouthparts, and many organs of sensory perception especially in the five eyes and two antennae, which are co-ordinated by the brain. The thorax is the locomotor centre, and is almost entirely filled with muscles that operate the legs and wings. The abdomen, more spacious than the head or thorax, accommodates other organs – including those for reproduction, digestion, and stinging.

2.21 Head, mouthparts and antennae

Figure 2.21a shows similarities and differences between the heads of the queen, drone and worker.

The mouthparts of the worker are more specialized than those of queen or drone, and the proboscis – the part most easily seen – is longer. It consists of the postmentum and prementum and, joined to the latter, the labial palps, the long glossa (tongue) and two paraglossa. The glossa (salivary canal) is a hollow tube, and it ends in a small spoon-shaped labellum (or fla-

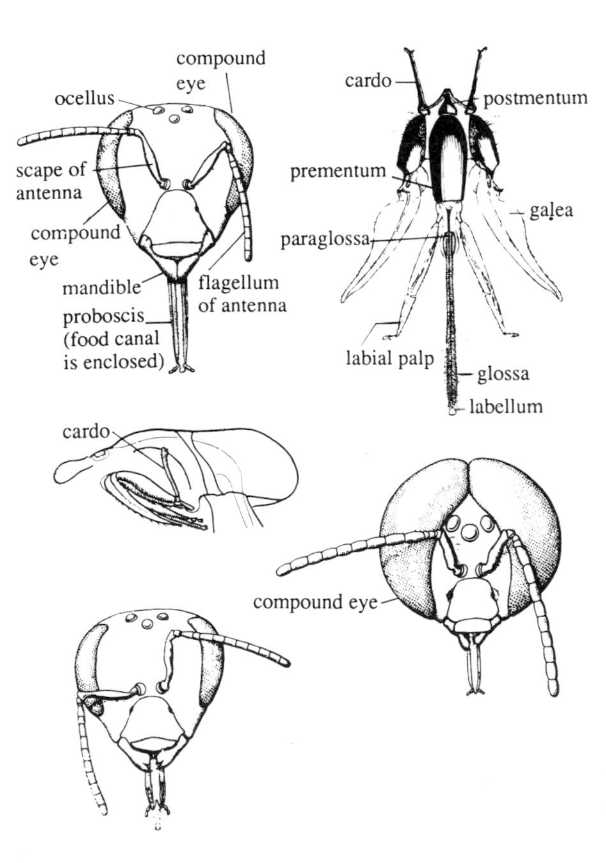

Figure 2.21a External anatomy of honeybee head (Dade, 1962).
above Head of worker on left; dissection of lower mouthparts, with proboscis, enlarged on right.
centre Proboscis folded, with cardines swung back, from side.
below Head of queen (*left*) and drone (*right*).

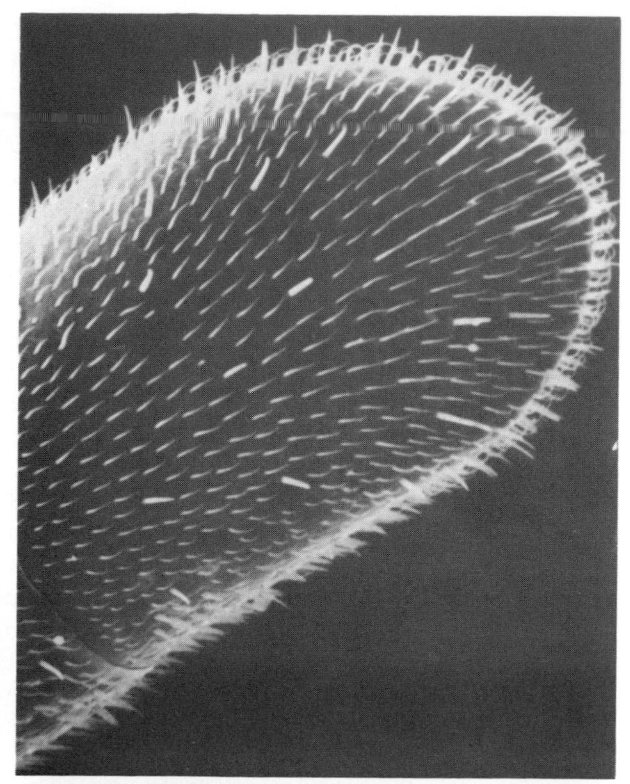

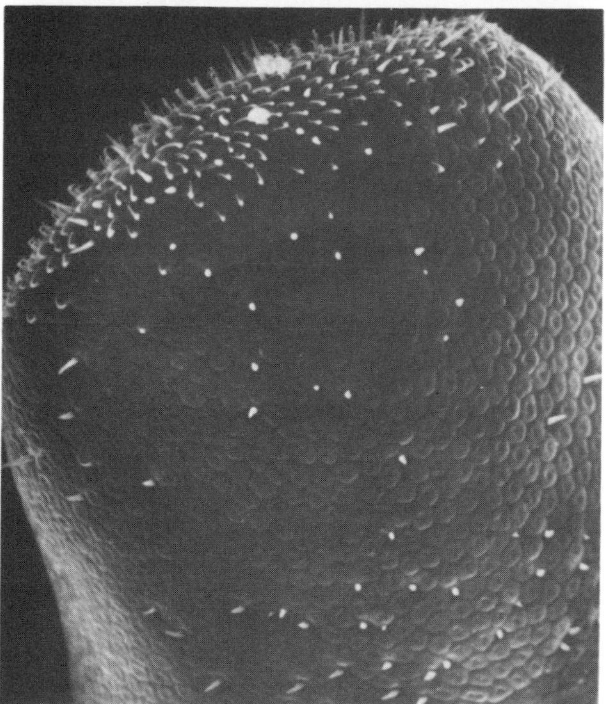

Figure 2.21b Tip of antenna showing sense organs, scanning electron microscope (SEM) photograph (Erickson et al., 1986, by publisher's permission).
above Ventral view, worker antenna (× 312), showing different forms of sensilla trichodea.
below Drone antenna (× 60) showing the packed, oval sensilla placodea (plate receptors), whiskery sensilla basiconica (short cone or peg organs) and sensilla chaetica (slender hairs).

bellum). When not in use, the parts of the proboscis in front of the prementum are folded back and tucked under the fossa (not shown). For action, they are swung forward on the cardines (singular = cardo), and then unfolded and extended so that they protrude well beyond the mouth; the proboscis is raised to mouth level and is grasped and steadied by the mandibles so that an airtight joint is formed. The galeae and labial palps are brought together to form a tube round the glossa, and this is the food canal. Liquid food is drawn (sucked) into the mouth through this tube. Solid food such as sugar cannot be imbibed in the same way; it is first moistened with saliva discharged along the interior of the glossa, and scrubbed with the bristly labellum, until some dissolves. The resulting solution is then lifted by the glossa and taken into the food canal. Solid particles with a diameter between 100 and 200 μm are caught between stylets in the mouthparts, and are not ingested. This was shown in a scanning electron micrograph by Peng and Marston (1986).

Each of the two antennae consists of a long segment, the scape, and the flagellum which has 11 shorter segments (12 in the drone); the segment next to the scape is the pedicel. There are many thousands of sensilla on the flagellum, including the bee's principal organs of taste and smell. Figure 2.21b shows examples; see also Section 2.34. The drone's antenna, which has 500 000 sensilla, is much longer and thicker than that of the worker or queen. The eyes of the honeybee – three simple (ocelli), and two compound, with unusual capabilities – are discussed in Section 2.34.

2.22 Thorax, wings and legs

The thorax (Figure 2.2a) consists of four segments, the true thoracic segments pro-, meso-, and metathorax, and A1 (the propodeum). Each segment consists of four plates, a tergite (dorsal), sternite (ventral) and two pleurites (lateral). Together, the plates of the three segments form a nearly spherical box. The boundaries between the plates, and between the segments, are not always distinct.

Figure 2.2a shows where the forewing and hind wing of the side of the body in view are jointed to the thorax, between the tergite and pleurite of the meso- and metathorax, respectively. The wings are formed quite late during development, in the prepupal stage; they grow out from the thorax as small pouches containing tracheae (breathing tubes, Section 2.33). The tracheae disappear as the wing becomes fully formed, but their positions are marked by 'veins' which stiffen the thin membrane of the wing in the adult insect (Figure 2.22a). The arrangement of these veins (venation) and quantitative data on them – particularly the cubital index – are used in taxonomic identifications; see Section 1.23.

The wings are moved by two systems of muscles in the thorax. The small direct flight muscles adjust the cant of the wings, and furl them. The large indirect flight muscles, which almost fill the interior of the thorax, operate the wings in flight. They move only the forewings: the hind wings are trailed passively. During flight a row of hamuli (wing hooks) on the leading edge of the hind wing (Figure 2.22a) engages automatically with a fold on the trailing edge of the forewing, so that the two wings move in unison. Characteristics of honeybee flight are discussed in Section 2.43. When the wings are 'disengaged' and folded

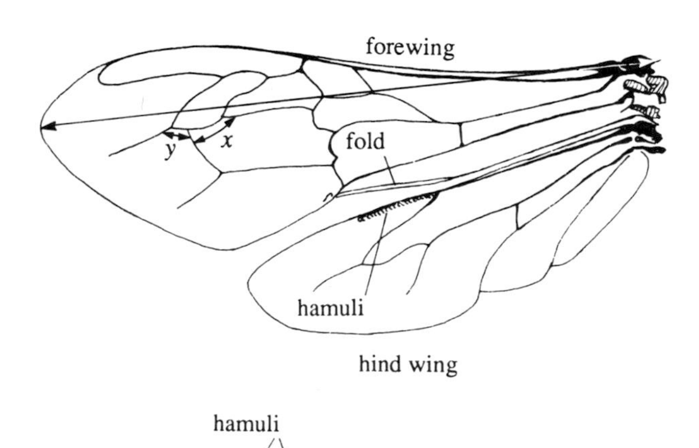

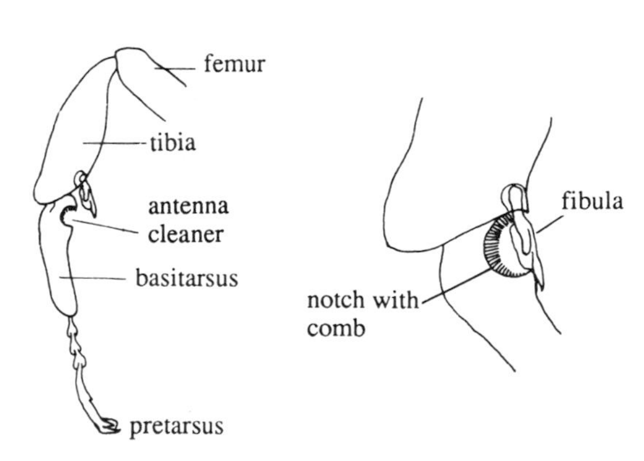

Figure 2.22a Wings and foreleg of worker (Dade, 1962).
above Wings, showing veins whose length ratio is the cubital index *x*/*y*; also arrowed line to indicate length of forewing.
centre Enlargement to show hamuli.
below Foreleg with antenna cleaner open, and (enlarged, on right) closed.
The hind leg is shown in Figure 2.43d.

back over the body, a bee is able to vibrate its wing muscles, thereby producing heat – or sometimes sounds such as those created by workers while dancing on the comb (Section 3.35).

A lobe on each side of the prothoracic tergite projects backward and partly covers the first spiracle (Section 2.33).

Each of the 6 legs has five segments, indicated in Figure 2.2a: coxa *c*, trochanter *tr*, femur *f*, tibia *ti* and tarsus *ta*. The tarsus has five subsegments, the first and largest being the basitarsus *bt*, and the final one the pretarsus (foot); see Figure 2.43d. Figure 2.2a indicates where the first segment (coxa) of the three legs is jointed to one of the thoracic pleurites.

The foreleg is close behind the head, and the bee uses hairs on its basitarsus to clean dust, pollen and other contaminants from the head. The antenna is cleaned with structures referred to as the antenna cleaner, shown in Figure 2.22a. The bee raises the foreleg and passes it over the antenna, which slips into the circular notch on the basitarsus. The tarsus is then flexed, and a jointed spur (fibula) on the tibia closes the notch, which is fitted with a comb of fine hairs. The antenna is brushed as it is drawn through the enclosing ring. Schönitzer and Renner (1984) have studied the action in detail.

The middle legs have no special tools, but hairs covering the inner side of the basitarsus of the worker are used for clearing pollen from the thorax and passing it to the hind legs. Meyer (1956*a*) published photographs of bees using a middle leg to manipulate propolis. But this leg is apparently not used in the manipulation of wax (Erickson et al., 1986), in spite of the presence of what has been referred to as a wax spur. The hind legs of the worker have highly specialized structures on the tibia and basitarsus that are used when a forager collects and packs pollen to carry back to the hive; see Section 2.43.3.

2.23 Abdomen

The abdominal segments A2 to A7 consist of only two plates, a dorsal plate (tergite) which overlaps a ventral plate (sternite), as shown in Figure 2.2a. Each tergite or sternite overlaps the one behind it, and adjacent tergites are joined together by a flexible intersegmental membrane which allows them to move easily in relation to each other; sternites are joined similarly. The intersegmental membranes are among the few soft parts of the integument that can be pierced by the sting of another bee, or by the mouthparts of the mite *Varroa jacobsoni*.

The abdomen contains important parts of several

internal systems that are referred to in Sections 2.31 to 2.36. The sting apparatus is described here, and also the wax glands which secrete the beeswax used by the bees to build comb; see Section 2.23.2. The scent gland or Nasonov gland, which lies under the tergite of A7, is discussed in Section 2.35; see also Figure 2.35a.

2.23.1 The sting apparatus, and venom production

The honeybee's sting apparatus is composed of greatly modified parts of abdominal segments A8 and A9, and lies in the sting chamber inside A7. Figure 2.23a shows a dissection of it. Venom is produced in the venom (or acid) gland, which widens to form a venom sac in which the venom is stored; its composition is described in Section 14.51. The venom sac is likely to have been filled by the time a worker is 14 days old, or 15–20 days according to E. Müller's findings in the 1930s (quoted by K. A. Forster, 1950), which also showed a sac content then of 0.3 mg venom; after the age of 18 days the venom sac degenerates and no more venom is produced. In most bees 12–19 days old Gałuska (1972) found some regeneration of venom after part of it had been released by electrical stimulation (see Section 14.52). Owen (1978*a*) found regeneration to be greatest at 2 weeks and absent at 40 days.

H. Inoue (1984) has shown how the composition of venom changes with a worker's age, and how the protein pattern varies through the season. H. Inoue et al. (1987) found that the content of different components was zero or at a very low level when the bee emerged; it then increased rapidly for about 14 days and thereafter reached a plateau, mostly at an age between 21 and 36 days in summer. Section 14.51 compares the venom of workers and queens, and of different honeybee species, and discusses the pharmacological use of venom in relation to its composition.

When a bee stings, the bulb of the sting (an inflatable organ that is an extension of the stylet) is filled with venom from the venom sac. Two umbrella valves inside the bulb force the venom into the venom canal inside the shaft of the sting, which consists of two barbed lancets and the stylet; the valves are attached to the lancets. As each lancet is protracted, the valve opens and drives the venom forward; as the lancet is retracted again the valve closes. The lancets are pushed forward alternately by a system of levers formed by the articulation of rigid plates in the sting chamber, operated by muscles. At each thrust, one of the two muscles shown protracts a lancet, pushing it more deeply into the tissue; simultaneously, the other muscle retracts the other lancet slightly. The final part

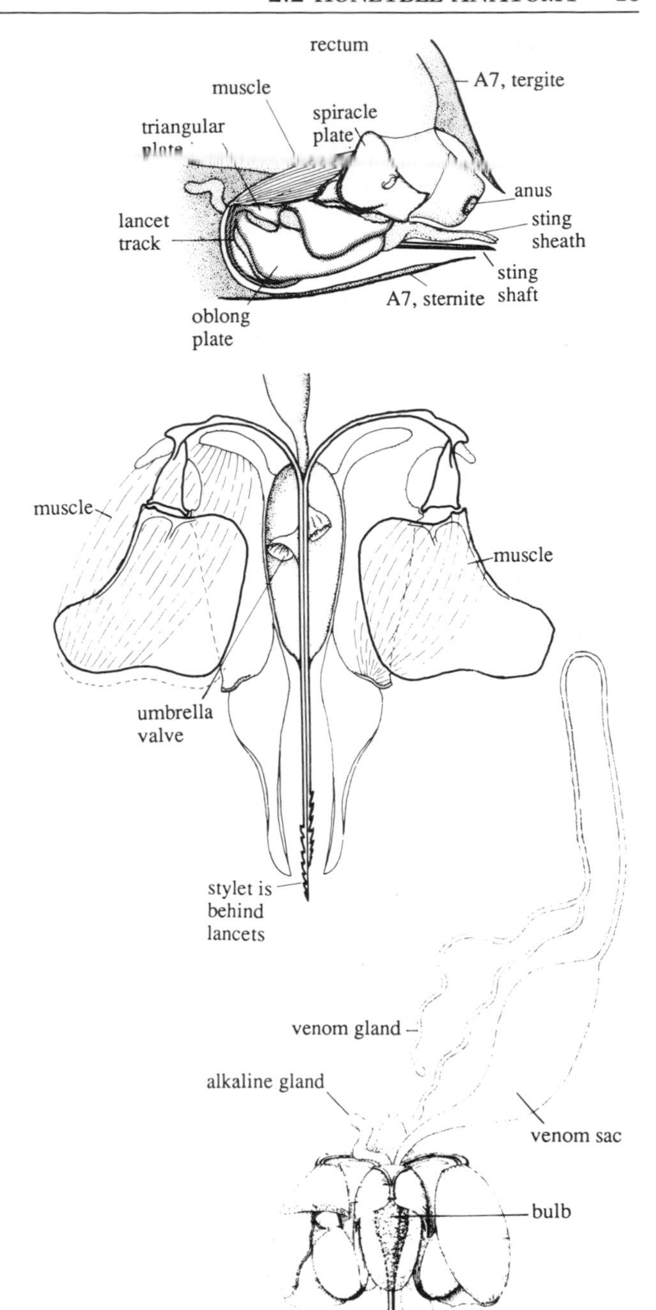

Figure 2.23a Dissection of worker sting apparatus (Dade, 1962).

above Lateral aspect of sting chamber.

centre Flattened-out diagram showing operation of sting, with moving parts in thick lines. Behind the two lancets is the stylet; these three rods together form the shaft and enclose the venom canal within it.

below Ventral view, apparatus lifted out and turned over.

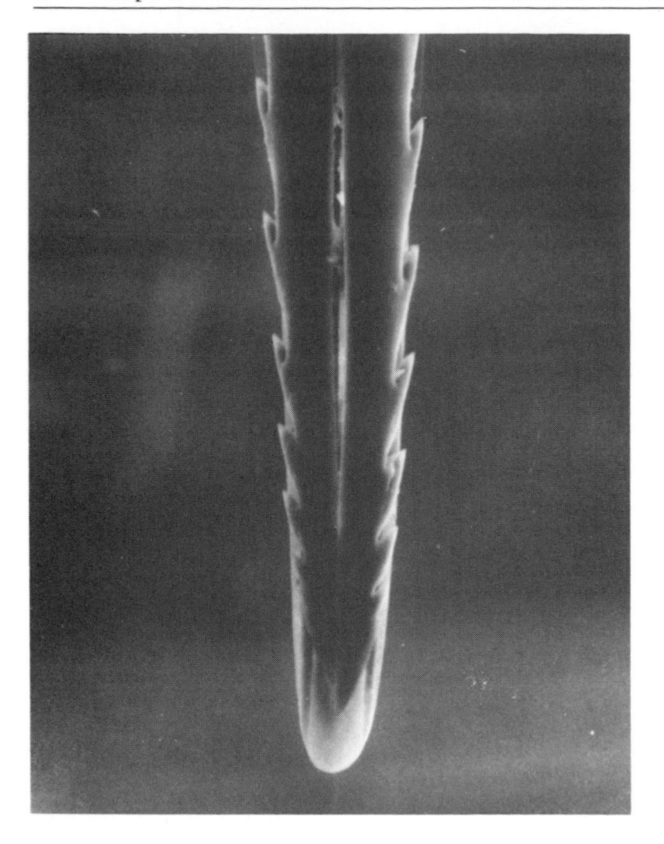

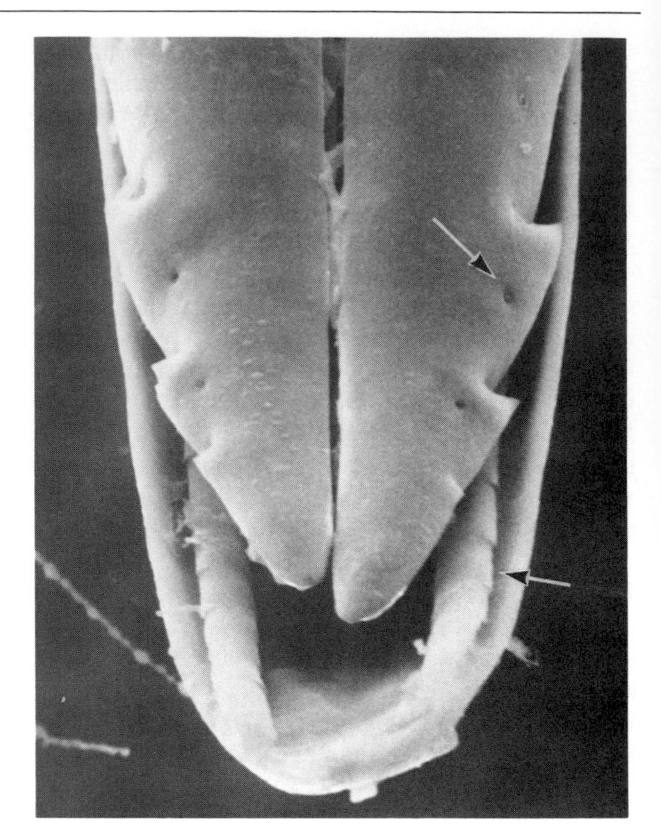

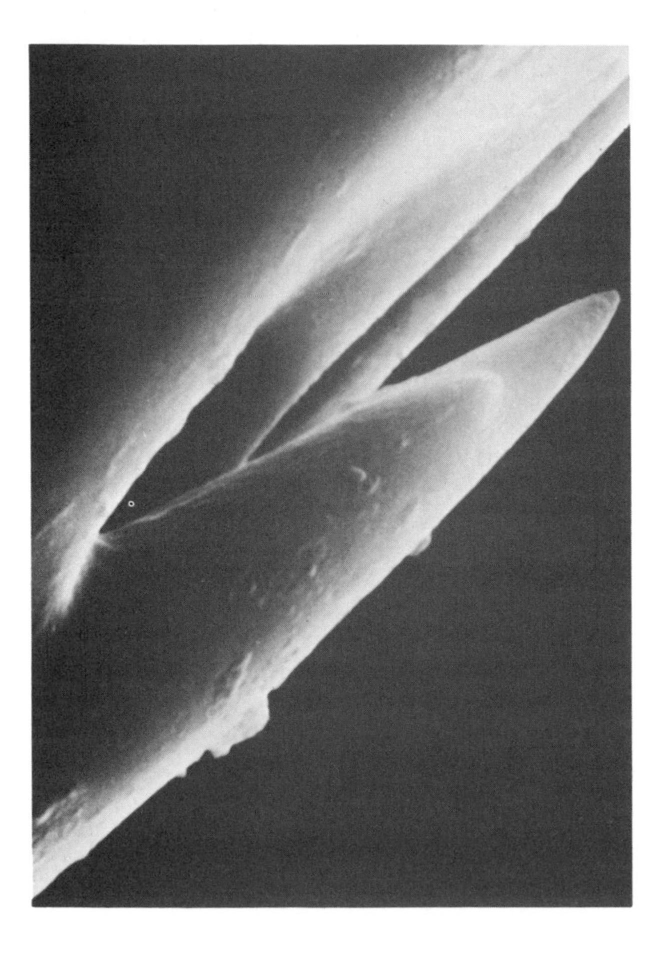

Figure 2.23b Lancets of sting of worker and queen, SEM photograph (Erickson et al., 1986, by publisher's permission).
above left Barbed lancets of worker, with tip of stylet extending just beyond them, × 360.
above right Lancets of queen sting (× 1040), with barbs much less acute; behind them, trough-like stylet with (*lower arrow*) serrated tracks on which the lancets slide back and forth. The indentation (*upper arrow*), and others like it, are probably sense organs. See text for other details.
below left Single barb, 10 times further enlarged (× 3600).

of the two tracks, with their serrations, are behind the lancets in Figure 2.23b, *above right*, at each side of the trough of the stylet which is behind them. The stylet and lancets that form the shaft are hollow rods with closed ends; together they completely enclose the venom canal along which the venom is driven from the bulb.

Figure 2.23b shows the two lancets of the sting, which are able to pierce the skin of a tough-skinned animal as well as the intersegmental membrane of another bee. The small indentation near each of the barbs, arrowed in Figure 2.23b, *above right*, may be a sensillum campaniformium, a receptor that monitors pressure – and thus the depth to which the sting has been inserted into the tissue (Erickson et al., 1986). The amount of venom released during the act of stinging is important for the harvesting of bee venom, and is discussed in Section 14.52.

An 'alkaline gland' produces a secretion that is poured directly into the sting chamber, which may possibly lubricate parts of the sting apparatus, perhaps in the queen only (Erickson et al., 1986); according to Pence (1981) it also produces a volatile fraction of the venom, referred to in Section 14.51.

When the bee tries to retract her sting from tough skin, the acutely pointed barbs ensure that the sting remains embedded; as a result, the weak membranes attaching it to the sting chamber and the spiracle plates are ripped; the bee is able to fly off, but she dies within a few days. A bee that has stung another bee can, however, retract her sting. (In the commercial production of bee venom, bees *en masse* are made to sting through a diaphragm so thin and smooth that retraction is possible; see Section 14.52.)

The queen has a longer sting than the worker, and Figure 2.23b, *above right*, shows that the few barbs on the lancets are much blunter (Erickson et al., 1986); also, the sting is firmly fixed in the sting chamber. She is therefore able to retract her sting after using it. Whereas the worker sting is straight, the queen's sting is curved downward, and Section 2.7 suggests a reason for this. The drone has no sting.

2.23.2 The wax glands

The four pairs of wax glands are situated on the front part of the inner side of the sternites of A4 to A7 (Figure 2.2a). They lie under oval plates known as mirrors; each pair is concealed by the overlying part of the preceding sternite. Hepburn (1986) provides the most detailed account of these glands and their secretory processes. The glands have a single layer of secretory epithelial cells – the wax cells – which increase in secretory activity, and in height from 15 μm to 50 μm, during about 9 days after the emergence of the adult worker. After the age of 17 days they start to regress, reaching 3 μm at 25 days, at which low level they remain. The wax cells may, however, be regenerated in older bees if their colony needs new comb. In general the need of the colony for comb plays a large part in determining the amount of wax secreted by bees capable of it; see Section 3.31.

In the USSR, Taranov (1959) measured the daily wax production per (young) bee, in different-sized colonies:

no. bees in colony	1100	1800	2400	3000	3700	5000
mg wax/day/ young bee	2.8	2.1	1.7	1.7	1.6	1.3

He calculated that during her lifetime – and usually during 1 or 2 weeks – a worker had a *potential* of producing about half her body weight in wax.

2.3 HONEYBEE PHYSIOLOGY AND THE BASIS OF BEHAVIOUR

The anatomy and physiology of internal systems of the honeybee are considered here, and also certain behaviour characteristics of the bees that result from the structure and operation of these systems.

2.31 The alimentary canal

The alimentary canal (Figure 2.31a) extends from the mouth where food is imbibed (Section 2.21), to the anus where waste material is voided. Food moves from the mouth through the oesophagus (which passes through the thorax to the abdomen), and in the worker it then enters the honey sac (crop, honey stomach), and can pass via the proventriculus to the ventriculus (midgut) where digestion occurs. Figure 2.31b shows a scanning electron micrograph of the proventriculus. This is a valve which remains closed while the forager collects nectar in her honey sac, carries it to the hive, and passes it back through her mouth to another bee in the hive. Only when the bee is consuming food is the valve opened, allowing the food to move through the valve to the ventriculus. The proventriculus also performs another function: it has closely set hairs (Figure 2.31b, *below*), which filter out pollen grains and other contaminants from the nectar in the honey sac, so that they collect in pouches behind the lips of the valve. When the pouches become full the pollen masses are passed through the valve into the ventriculus (Bailey, 1952). Peng and Marston (1986) showed that the proventricular hairs can filter out particles 0.5 to 100 μm in diameter. Such particles include not only pollen grains, but spores of *Nosema apis* and *Bacillus larvae*, and Torula yeast (*Candida utilis*).

Cells of the tissue that lines the ventriculus (epithelium) secrete digestive enzymes into the lumen of the ventriculus, and these cells can also proliferate, become detached, break, and mix with the food in the ventriculus. They contain enzymes that digest proteins in pollen, which seems to occur without fracture of the exine or intine walls of the pollen grain. Cruz-Landim (1985) suggests that the intine wall is permeable to digestive enzymes and to the products of digestion, but she found that only about 50% of ingested pollen is digested. (The epithelial cells are attacked by spores of the protozoan *Nosema apis*; see Section 11.31.) Most of the nutrients from digested food are absorbed through the walls of the ventriculus, while the undigested material is passed through the pyloric valve into a narrow tube that is the small intes-

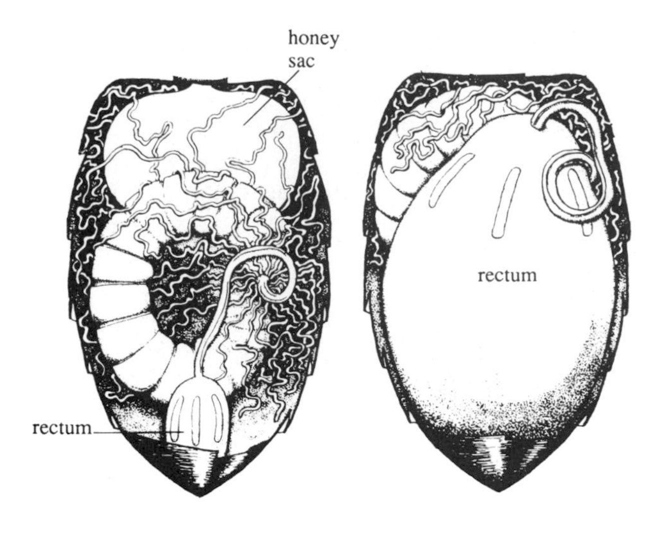

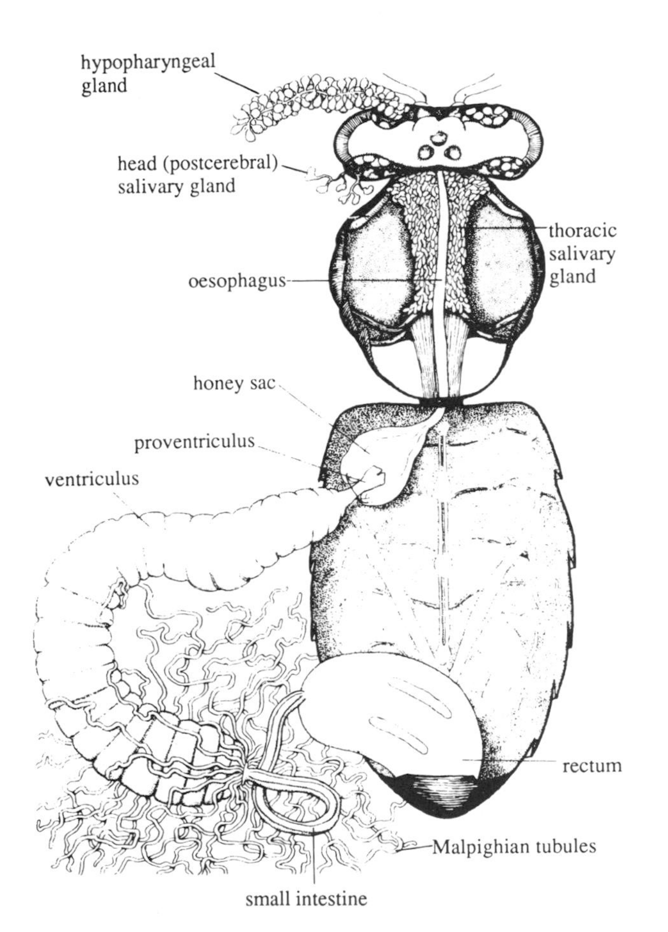

Figure 2.31a Dissections to show alimentary canal of worker (Dade, 1962).
above Dissection of abdomen.
left Forager returning to hive, honey sac full and rectum empty.
right Worker after confinement in hive, rectum full and much enlarged.
below Dissection with alimentary canal displayed, glands of head lifted out, and indirect flight muscles removed from thorax, to expose underlying organs.

tine of the hind gut (Weaver, 1988). Here, the remaining nutrients from the digested food are absorbed through the walls. Beyond the small intestine is the rectum (large intestine), where waste matter accumulates and water may be resorbed.

Figure 2.31a (*above*) shows how small the rectum is when empty, and how large it can become when distended, as after a period of winter confinement in the hive with unsuitable food; see Section 6.53.1. The rectal contents may even ferment. Such conditions cause dysentery, and they also make the bees 'fidget', with a resultant rise in temperature that may induce premature egg laying by the queen.

About a hundred narrow Malpighian tubules join the small intestine at its commencement; they carry waste matter which has passed from the blood into the tubules through their walls, whence it passes through the small intestine to the rectum. (The protozoan parasite *Malpighamoeba* attacks these tubules, giving rise to amoeba disease; see Section 11.32.)

Food reserves are stored in the worker's fat-body, a layer of cells concentrated mainly on the floor and roof of the abdomen. The fat-cells within it contain much fat, and also glycogen (a carbohydrate which serves as a reserve energy food) and – when the bees are not rearing brood – reserve protein. Also within the fat-body are oenocytes which probably have a function in wax production and secretion, and trophocytes containing hydrous iron oxides in granules, which may be important nutritionally, and in orientation as described in Section 2.34. Use of the food reserves in the bee's body is referred to in Sections 2.43 and 3.33.

Kapil (1959*a*) has made a detailed study of the alimentary canal of *A. cerana*.

2.32 Blood circulation

Honeybee blood (haemolymph) transports the dissolved nutrients absorbed from the small intestine to the tissues of the body, and transports dissolved waste material away from the tissues to the Malpighian tubules (see above). The blood contains white corpuscles (phagocytes) that ingest and destroy invading bacteria. Unlike vertebrates, insects have no red cor-

Opposite
Figure 2.31b Honey sac and proventricular valve (SEM photo: A. M. Millington-Ward).
above The proventriculus within the honey sac, as shown diagrammatically in Figure 2.31a; part of the oesophagus can be seen on the left, the Malpighian tubules on the right.
below The proventricular valve, with its hairs and four pouches.

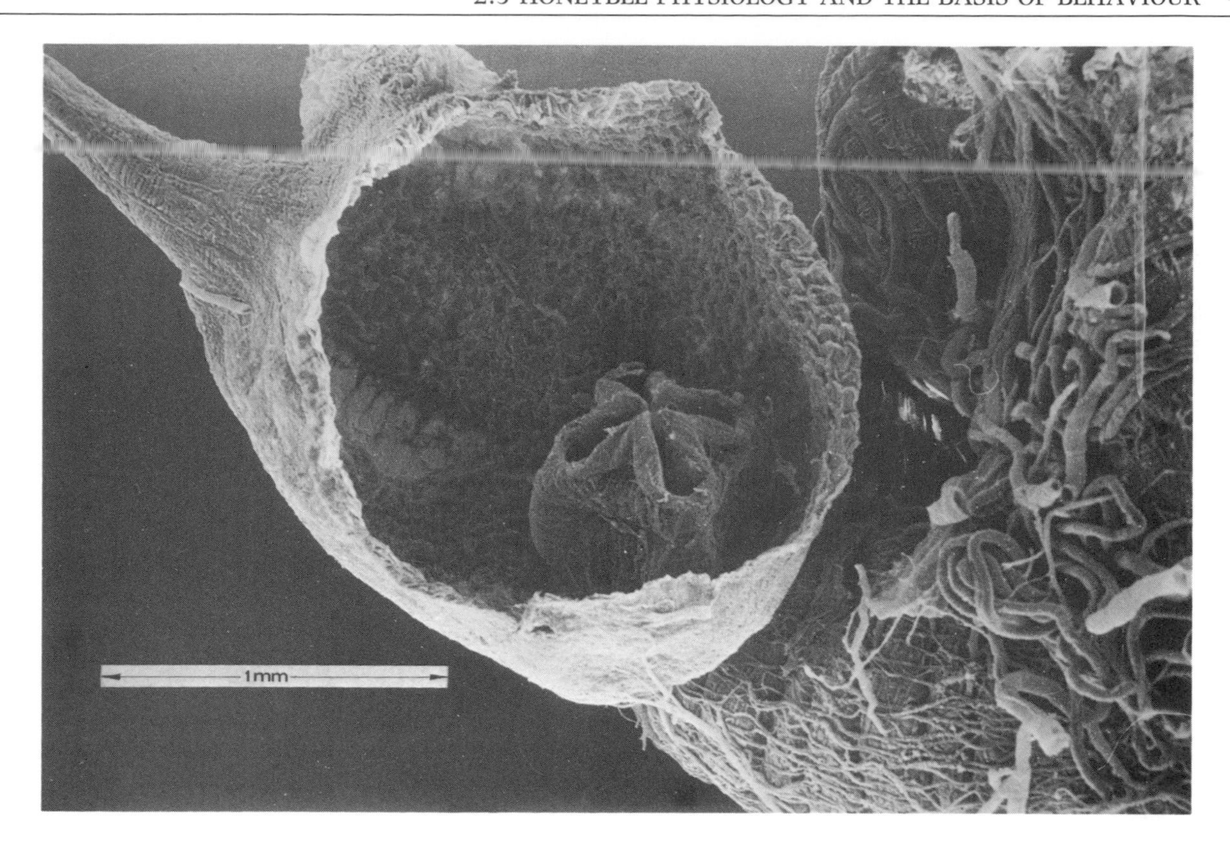
1mm

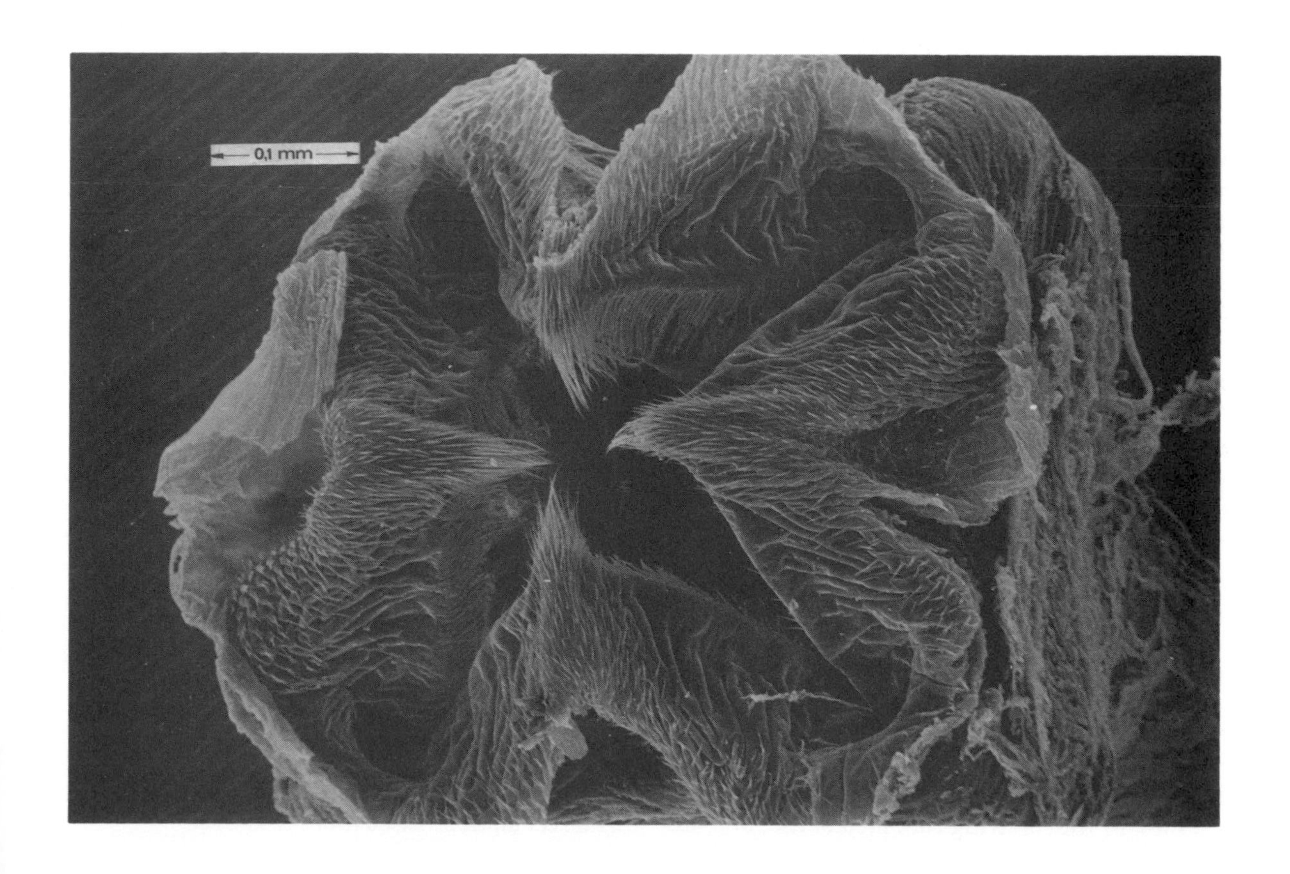
0,1 mm

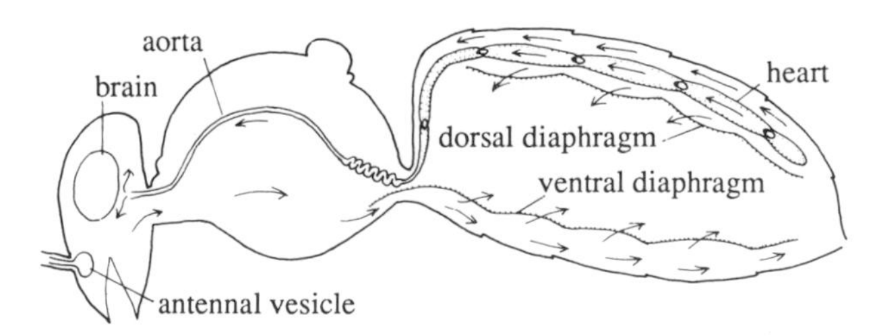

Figure 2.32a Circulatory system of honeybee (Dade, 1962).
Arrows indicate directions of flow of blood.

puscles (erythrocytes) and no lungs, and their blood plays only a localized part in respiration (Section 2.33).

In honeybees the elongated heart, which pulsates, lies just under the roof of the abdomen and is attached to the thin dorsal diaphragm (Figure 2.32a). The aorta runs from the heart through the thorax to the head, where its end opens below the brain; there is a pulsating antennal vesicle under the base of each antenna, and a small blood vessel runs along the antenna to its tip. Apart from these blood vessels, the blood is not confined to a circulatory system but fills the body cavity. Other organs besides the pulsating heart assist circulation of the blood. Pulsations of the dorsal and ventral diaphragms, brought about by a sequence of contractions of their muscles, drive the blood in the body cavity in the directions shown by the arrows in Figure 2.32a: forward in the dorsal part of the abdomen and backward in the ventral part. Blood drawn from the thorax is replaced by blood forced into the head by the aorta.

Septicaemia is a bacterial infection which spreads to the blood via the tracheae; see Section 11.33.

2.33 Respiratory system

In insects, as in most other animals, respiration involves the intake of air containing oxygen, and the release of carbon dioxide. The respiratory system of the honeybee is shown in Figure 2.33a. Air enters and leaves the body through ten pairs of spiracles (breathing holes). In the abdomen the spiracles are at the side margins of the tergites, and in the thorax they lie between adjacent segments. The first (prothoracic) pair, which are quite large, open between the pro- and mesothoracic segments. They cannot be completely closed, and in very young workers – perhaps before the guard hairs of these spiracles harden – tracheal mites (*Acarapis woodii*) enter the tracheae through them; see Section 11.41. The second – small – pair of spiracles are in the forewing root membrane, and are the only spiracles without a closing apparatus. The third – very large – pair are in the propodeum, and are called the first abdominal spiracles.

The spiracles are connected together internally by tracheal sacs along either side of the body, and transversely by narrower tracheae. (The term trachea is used because these tubes are maintained in a dilated state by spiral thickenings of cuticle in the wall, as are tracheae of terrestrial vertebrates.) The tracheal sacs serve as bellows; they contract under pressure from the surrounding blood when the abdomen is retracted and compressed, and expand when the abdomen is extended and dilated. The rapid rhythmic pulsations of a bee's abdomen, which can be seen when the bee is at rest, are thus respiratory movements.

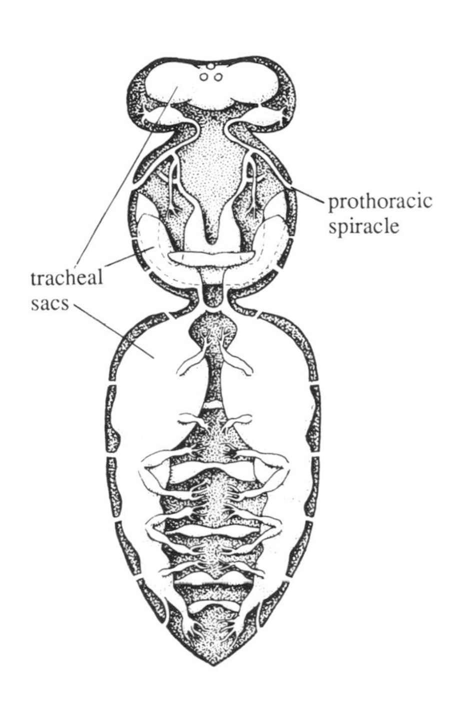

Figure 2.33a Respiratory system of honeybee: principal tracheae and tracheal sacs (Dade, 1962).

A network of smaller branches of the respiratory system extends to each part of the body and its organs. Tracheoles, tubes which are even smaller and have no spiral thickenings, are open at their outer extremities, which reach into the tissues and bring oxygen to the cells there; the oxygen has to diffuse in liquid through only a short distance. Carbon dioxide is removed simultaneously by the same means.

During flight, when much oxygen is needed, air is inhaled at the first spiracles and exhaled mainly at the third spiracles, the largest, which are adjacent to the indirect flight muscles (Bailey, 1954).

2.34 Sensory organs and the nervous system

Sense organs, mostly on the exterior of the bee's body, react to different forms of energy, as indicated in Table 2.34A. Sensory nerves lead from the receptive cells of the sense organs to a ganglion of the central nervous system. This system consists of the brain (cerebral ganglia) and the suboesophageal ganglion in the head, and the ventral nerve cord and segmental ganglia on the floor of the thorax and abdomen. (A ganglion is a knot of nervous tissue with nerve fibres radiating from it.) Motor nerves lead from cells in the central nervous system to the appropriate muscles and glands in the body. Association nerve fibres carry impulses from the ends of incoming sensory nerves to the roots of outgoing motor nerves, establishing a neural circuit: when an external stimulus sets up a nerve impulse, this activates the motor system, producing a certain movement, or causing a certain gland to secrete. Other association neurons carry impulses to the brain and other ganglia, thus allowing the co-ordination of behaviours. The neurobiology of honeybees in relation to their behaviour is the subject of a book by Menzel and Mercer (1987).

The brain, suboesophageal ganglion and ventral ganglia are masses of nerve cells and nerve fibres. Their organization depends mostly on intricate tracts of intercommunication formed by the association fibres. In comparison with most other insects, the worker honeybee has an unusually large brain in relation to its total body mass, and the association centres are especially prominent. In addition to its function in co-ordination and control, the brain receives nerve impulses directly from the sensory organs of the eyes and the antennae, and transmits nerve impulses to the motor centres of the ventral nerve cord. Without the brain, the insect can still act through its ventral ganglia, and a decapitated bee can therefore walk and even fly, and sting. But the brain inhibits and modifies these activities, so that the bee's behaviour can become goal-oriented.

Mechanical receptors

A mechanoreceptor reacts to mechanical deformation of some part of the receptor. It may respond to touch, including contact with a solid object or a current of air, or as a proprioreceptor registering deformation and stress, e.g. in the muscles and exoskeleton. Sensilla trichodea (trichoid sensilla) are sensory hairs, and some are shown in Figures 2.21b and 2.34a. They are often grouped together in a hair-plate (or bristle-field), as in gravity sense organs (below). Sensilla scolopophora (scolopophores or scolophores) are spindle-shaped. The dome-shaped sensilla campaniformia are more complex.

Sensilla trichodea at the tip of each antenna enable foragers to detect, learn and discriminate between microsculptured surfaces of flower petals (Kevan & Lane, 1985). The size of these sensilla is in the same range (*c.* 10 μm) as the features of the microsculpturing – which are sometimes different from one end of a petal to the other and may thus act as nectar guides (Section 12.21.1).

The honeybee's speed of motion through the air, and the air-current direction, are registered by sensory hairs which act as aerodynamic organs: on the compound eyes (Figure 2.34a), on other parts of the head, and on the wings. Perception of movement relative to the ground is visual; see below at the end of the discussion of light receptors.

In normal circumstances, the energy expended in flight gives the bee a measure of the distance flown, although with some error if the flight has been steeply up or down hill, or with or against a strong wind. She also uses information from pressure receptors on the wall of the honey sac (Lindauer, 1987); the pressure decreases as the honey sac empties. Section 3.35 explains how the forager's energy expenditure is linked with the tempo of her subsequent communication dance in the hive.

In other animals, sounds are perceived by an auditory organ that responds to air vibrations within a certain frequency range. Honeybees emit various sounds, and it has seemed strange that they have not been found to have any auditory organ to respond to these sounds – especially to sounds produced by workers performing communication dances (Section 3.35), and by virgin queens piping (3.34.4). Honeybees do, however, respond to vibrations of the substrate on which they stand (caused by the source of the sound), by means of the subgenual organ on the tibia of the foreleg. Michelson et al. (1986) suggest that these, and other, vibration signals produced in the above circumstances, are used for local communication within a restricted area of comb. [Note in proof: Towne and Kirchner (1989) show that honeybees can hear (perceive airborne sounds) by detecting air-particle oscillations (Table 2.34A); most vertebrates hear by detecting air-pressure oscillations.]

Gravity receptors

Gravity sense organs consist of sensilla trichodea in hair-plates or bristle-fields situated at certain joints; they respond to changes in the relative position of the two parts of the body that come together at the joint in question: head and thorax at the neck; thorax and abdomen at the petiole; possibly also coxa and trochanter on the leg. Details are given by Markl (1966),

Table 2.34A Examples of sensory organs of the honeybee that respond to different forms of energy
In column 2, s. = sensilla; text gives alternative names.

Type of stimulus	*Sensory organ(s) reported*	*Location*
Mechanical		
touch	s. trichodea	many sites
stress/strain	s. scolopophora	legs and head
	s. campaniformia	lancet of sting
gravitational	s. trichodea	hair-plates on each side of joint between: head and prothorax; thorax and abdomen (petiole); coxa and trochanter
movement	s. chaetica	antenna
flight speed (relative to ground)	interneurons	compound eye
flight speed (relative to air), air-current direction	long hairs	head (including compound eyes) and wing
energy expenditure (distance)	'pressure receptors'	wall of honey sac
oscillation of air particles	s. scolopophora	Johnston's organ in antenna
vibration of substrate	s. scolopophora	subgenual organ in tibia (foreleg only)
sound	by oscillation of air particles (above)	
heat/temperature	s. ampullaceae, s. coeloconica	last 5 antennal segments
Light		
intensity	ocelli	head
intensity, shape, colour, flicker	retinula cells	compound eye
polarization	UV-sensitive retinula cells only	compound eye
Chemical		
taste	s. chaetica, and probably s. trichodea, s. basiconica, and others	mouthparts
	s. chaetica	antenna
	s. chaetica	tarsus
carbon dioxide	s. ampullaceae, s. coeloconica	antenna
humidity	s. ampullaceae, s. coeloconica	antenna
olfaction	s. trichodea, s. basiconica, s. chaetica, s. coeloconica, s. placodea	antenna
Time	interneurons	corpora pedunculata
Electrical	certain parts of integument	see text
Magnetic	magnetite component of 'granules'	fat-body

and earlier papers he refers to. In Markl's training experiments, different species of bee showed one of two types of orientation to the direction of gravitational force. The more primitive type was shown by many non-*Apis* bees and by *Apis florea*, and the more highly developed type by *A. cerana*, *A. dorsata* and *A. mellifera*. In their dance communication, these latter bees, but not *A. florea*, can transfer an angle registered in a horizontal plane (relative to the sun's direction) to the same angle in a vertical plane (relative to the upward direction, i.e. against the gravitational force); see Jander and Jander (1970).

Reactions of 3400 *A. mellifera* workers, with a queen, to zero gravity were tested during 7 days on a NASA shuttle mission (Vandenberg et al., 1985). The bees flew within a small flight chamber, constructed 200 cm² of comb, and stored sugar syrup within it; mortality was low. The queen laid about 35 eggs, but these did not survive transfer to a hive on their return to earth. The geometry of the comb built in space was nearly normal.

Heat and temperature receptors

Sensilla ampullaceae (pit organs in long canals) and sensilla coeloconica (sunken peg organs or pit pegs), on terminal segments of worker and drone antennae, react to differentials of temperature, humidity, and carbon dioxide concentration (Lacher, 1964). Honeybees respond to a temperature decrease of only 0.25°; they are slightly less sensitive to a temperature increase (Ribbands, 1953).

Light receptors: the eyes

Each of the three simple eyes or ocelli of the honeybee (Figure 2.21a) has a single lens that refracts light to about 800 photoreceptor cells, but it cannot form an image on the retina of the eye. The ocelli are, however, able to monitor light intensity, period of exposure to light, and colour (Autrum & Metschl, 1963). They may regulate the daily start and finish of foraging and, by registering the brightness of the sky, help to maintain the bee in level flight. (Section 6.54 describes what can happen when a bee flies over snow that is brighter than the sky above.)

Insects have compound eyes consisting of many thousands of ommatidia, each with its own lens; unlike vertebrates, they receive an impression that is in the nature of a mosaic of dots of varying degrees of brightness. Figure 2.34a shows facets of honeybee ommatidia, and Figure 2.21a shows the position of the compound eyes in the head. These eyes are very much larger in the drone than in the queen or worker; Dade (1962) obtained the measurements:

	area (*mm*²)	*no. lenses*
worker	2.6	6900
queen	smaller	fewer
drone	9	8600

In the drone, the lenses are thus several times as large as those in the queen or worker; the eyes themselves are so large that the ocelli are displaced forward. Varela and Wiitanen (1970) have made a detailed study of the compound eye of the honeybee.

Section 2.43 discusses a honeybee's perception and memory of shape and colour.

In each ommatidium, behind the facet of the lens, are 9 retinula cells sensitive to different colours, one in the centre and 8 surrounding it (Figure 2.34b). These are surrounded by a layer of pigment cells. Von Frisch (1967) made a model of the cross-section of the 8 retinula cells, by fitting together 8 triangles of polaroid (polarizing screen) with their axes of polarization radiating from the centre. He was able to show that if retinulae let through only light that is polarized in a unique plane – like the polaroid triangles – the pattern registered by the compound eye would 'inform' the bee of the sun's position, even when the sun is obscured by clouds. In fact, only three retinula cells (1, 5 and 9 in Figure 2.34b) are sensitive to UV light and to its plane of polarization.

The response of honeybees to movement of a striped pattern below them, and in their visual field, is discussed by Ribbands (1953). Honeybees can react to a flicker fusion frequency of up to 54 per second, i.e. they can register the speed of separate stripes moving past them, up to 54 per second. This would provide a

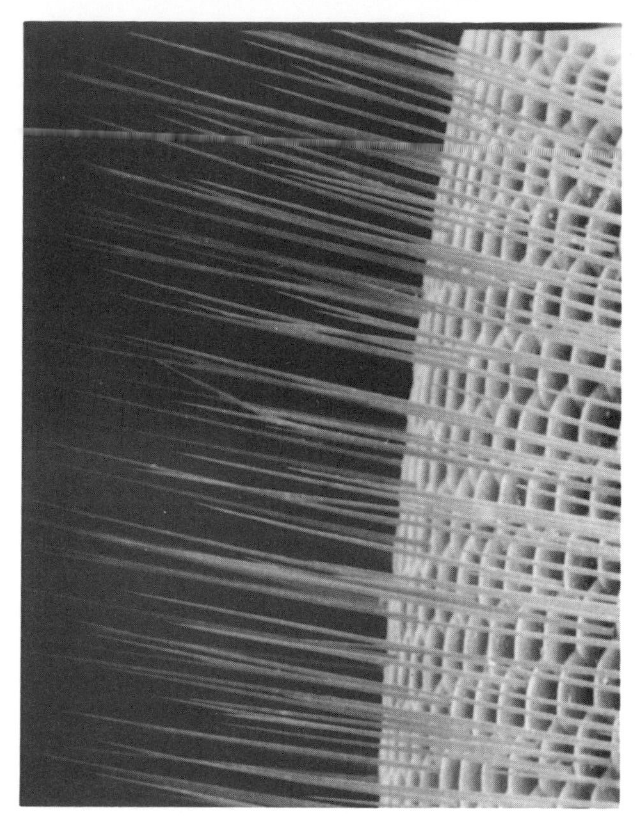

Figure 2.34a Part of compound eye of drone, SEM photograph × 550 (photo: E. H. Erickson).
The long hairs between the facets are mechanoreceptors that respond to flight speed through the air, and wind direction.

No.	3	7	8	4	2	6	2,6,4,8	5	1	9
SS	G	G	B	B	G	G	B	UV	UV	(UV)

Figure 2.34b Relative positions of the 9 retinula cells within an ommatidium of the honeybee compound eye, and their spectral sensitivities, SS (Ribi, 1987).
G = green; B = blue; UV = ultraviolet; the ommatidium shown is in the median eye region.

mechanism for monitoring speed over the ground. Because of the nature of the compound eye, bees orient very accurately to an object in relation to surrounding objects.

Response to an electric field
Sensitivity of honeybees to changes in electric fields has created interest, mainly in relation to the increase in the 'aggressiveness' of bees before thunderstorms, and also when their hives are placed under high-voltage transmission lines. Warnke (1976) has studied the subject in some detail, and he interprets the effects briefly as follows. The action of an external electric field influences the electrolytic (fluid) parts of the bee's body, and polarizes the outer sclerotin layer of cuticle as a dielectric. The differentiated capacitances and resistances, as well as the differentiated dielectric behaviours of the cuticle, produce great differences of potential between membranes and sclerotized layers, and sensory hairs possibly respond to these differences.

Response to a magnetic field
The worker honeybee is known to be able to detect magnetic fields. Trophocytes in the fat-body contain hydrous iron oxides in granules *c.* 0.5 μm in diameter scattered throughout the cytoplasm, about 1.7×10^{13} per cell. The accumulation of iron first appears during the larval or pupal stage, its source being ingested pollen. It is greatest (0.18% of the fat-body) at 9–12 days old when bees start to forage. These are the findings of Kuterbach and Walcott (1986) and Kuterbach (1987). The iron compounds possibly contain some magnetite (Fe_3O_4) and, if so, their alignment could produce a natural magnetic field in the horizontal plane of the bee at right angles to its body axis, as suggested by Gould (1980). This in turn could provide a mechanism enabling the worker to respond to the earth's magnetic field, thus aiding her orientation – perhaps especially in the tropics where the minimal daily variation in the sun's azimuth angle makes a 'solar compass' less useful (Section 2.43). The granules may, however, be paramagnetic and respond only to *fluctuations* in an external magnetic field.

Experiments by De Jong (1982) showed that if swarms at an early stage of comb building (within 5 days of swarming) were relocated in another hive, the workers tended to build their new combs in the same direction as the first ones. The bees thus seem to be able to use the earth's magnetic field as a directional reference in comb building, but their memory of the comb direction in relation to the field lasts only during the initial stages of comb building – in fact, while it is most useful to them.

No experiments are known to have been made on queens or drones.

Chemoreceptors
The honeybee has more than one type of chemoreceptor. Olfactory sense organs can perceive scent, and gustatory sense organs can perceive taste – especially sweetness. Sensilla ampullaceae and sensilla coeloconica on the antennae register both carbon dioxide concentration and humidity of the air (Lacher, 1964). There are taste organs (contact chemoreceptors) in various locations in the mouthparts, and also on the antennae, and the pretarsi and other parts of the legs. Some of the organs of smell respond with great specificity to one of the many volatile pheromones that are produced by different glands of other individual honeybees (Section 2.35). All the pheromones produced outside the hive, and some produced within the colony, are perceived by smell. Others, that are both produced and perceived within the colony and are relatively involatile, are eaten, or adsorbed on the body surface where they are perceived by contact chemoperception.

Time perception
Honeybees are able to register the passage of time over and within 24-hour cycles (Ribbands, 1953), and in common with many other animals they have a 'biological clock'. This is likely to be controlled by the central nervous system, and the corpora pedunculata (mushroom bodies) in the protocerebrum may play a key role. Martin et al. (1978) were successful in removing the corpora pedunculata from worker honeybees that had been trained to feed at a certain time each day, and in inserting them into other workers. A few days afterwards, more than half the implanted bees visited the feeder at the time of day learned by the donor workers.

2.35 The glandular system and glandular secretions

Glands are organs of secretion, composed of specialized cells. In insects, they are served by tracheoles that provide oxygen to them, and many of them are activated by motor nerves to secrete, when an appropriate sense organ is stimulated. Endocrine (ductless) glands secrete hormones into the blood; the most important in honeybees are the juvenile hormone (neotenin) secreted by the corpora allata, and the moulting hormone (ecdysone) of which the prothoracic glands are likely to be a major source. The part played by juvenile hormone in queen–worker differentiation is referred to in Section 2.51.

Table 2.35A Exocrine glands, and pheromones and other secretions, in the adult and larval honeybee *Apis mellifera* (worker, queen and drone)

Column 1: (2) indicates paired glands
Column 2: 9-HDA = 9-hydroxy-2-decenoic acid
10-HDA = 10-hydroxy-2-decenoic acid
9-ODA = (E)-9-oxo-2-decenoic acid
HOB = methyl *p*-hydroxybenzoate
HVA = 4-hydroxy-3-methoxyphenylethanol diacetate
W = worker, Q = queen, D = drone
Columns 3–5: w, q, d, indicates that secretion occurs at some stage in the bee's life; ( ) only slight secretion, or gland atrophied; — gland or secretion absent.

Gland	*Secretion*	*W*	*Q*	*D*
Adult, in the head				
hypopharyngeal (2)	brood food (including royal jelly), see text	w	—	—
	invertase, glucose oxidase, and other enzymes	w	—	—
salivary glands of head (post-cerebral) (2)	fatty substance	w	q	(d)
	enzymes	w		
mandibular (2)	*worker*			
	lipoid component of larval food, includes 10-HDA	w		
	alarm or alerting pheromone, alerting other W: 2-hepatanone	w		
	enzymes	w		
	queen			
	20 or more pheromones, one of more of which:		q	
	– inhibits W ovary development			
	– stimulates W to release Nasonov pheromone			
	– stimulates W to forage			
	– regulates W coherence in a swarm and attracts W to it			
	– (9-HDA) weakly inhibits queen-cup construction			
	– (9-ODA) attracts D to Q			
	– (pre-copulation) enables W recognition of Q			
	– (HOB + 9-ODA + 2 isomers of 9-HDA + HVA) induces retinue behaviour in W (see text)			
	drone			
	pheromone attracting D to D; marks congregation site			d
mandibular or postcerebral	?used in masticating wax scales and building combs	w		
postgenal (2)	?	w	(q)	?
sublingual	?	(w)	?	d
Adult, in the thorax				
salivary glands of thorax (2), derived from silk glands of larva	saliva	w	q	d
	also enzymes	w	?	?

(continued)

Table 2.35A *(continued)*

Gland	*Secretion*	*W*	*Q*	*D*
Adult, in the legs				
tarsal (?Arnhart)	*worker*			
	pheromone (at least 11 compounds), attracting other W to forage source or to the nest	w		
	queen			
	tarsal or footprint pheromone (at least 12 compounds); with 9-HDA from mandibular glands, inhibits queen-cup construction		q	
	drone			
	pheromone (at least 1 compound)			d
Adult, in the abdomen				
wax (4 × 2)	beeswax (Table 13.5A)	w	—	—
Nasonov (scent)	Nasonov pheromone attracting other W, includes terpenoids: – (Z)-citral – (E)-citral – nerol – geraniol – geranic acid – nerolic acid – (E,E)-farnesol	w	—	—
abdominal tergites (A3, A4, A5)	pheromones enabling recognition of Q by W, stabilizing 'retinue', inhibiting ovary development in W; attracts D to Q (pre-copulation)	—	q	—
rectal (6)	catalase	w	?	?
spermathecal	polysaccharide secretion containing proteins	—	q	—
mucus (2)	mucus	—	—	d
(male) accessory	peptide	—	—	d
in sting apparatus:				
venom	bee venom (Table 14.51A)	w	q	—
Dufour's	?waxy covering of egg	?	q	—
Kozhevnikov (2)	pheromones enabling recognition of Q by W and attracting W to Q	—	q	—
sting sheath (2)	several pheromones	?	q	—
'sting gland'	alarm or alerting pheromones: 6 acetates, mainly isopentyl, also (Z)-11-eicosen-1-ol and 5 other compounds	w	—	—
alkaline	?lubricant for sting	?	q	—
sting shaft	waxy esters, ?lubricant	?	q	—

Table 2.35A *(continued)*

Gland	*Secretion*	*W*	*Q*	*D*
Larvae and pupae				
silk (2), in larva	silk spun into cocoon	w	q	d
?	brood pheromone(s):			
	(*a*) incubation pheromone from pupae, enables brood recognition and clustering on it by W: glyceryl-1,2-dioleate-3-palmitate	(w)	q	d
	(*b*) inhibits queen rearing	—	q	—
	(*c*) inhibits ovary development in W	—	q	—
	(*d*) stimulates foraging, especially for pollen	?	?	?
?, in young diploid drone larvae	cannibalism pheromone	—	—	d

In addition, volatiles from empty comb are sometimes referred to as hoarding or comb pheromones, although they are not established as honeybee secretions; see Section 6.41.

Exocrine glands, which we are mostly concerned with here, have ducts through which their secretions are conveyed outside the body; these secretions include wax, venom, and a host of pheromones.

The glands are not part of one interconnected system like the alimentary, circulatory and respiratory systems. Individual glands are referred to in the different sections of Chapters 2 and 3, and Figure 2.2a indicates the locations of most of them. Table 2.35A provides some details of individual exocrine glands, which are listed according to their location, in the head, thorax and legs (there are none in the wings), and abdomen. Many substances produced by the glands are identified by name, and Melksham et al. (1988) list others.

Some glands in the worker produce substantial amounts of secretion that are used in the colony, and most of these glands are active for only part of the bee's life. In the head, the hypopharyngeal glands, shown in Figure 2.31a, *below*, produce two quite different secretions: in young bees, brood food (Section 2.41), and in older bees, enzymes essential for converting nectar and honeydew into honey (Section 13.23.1). Both the head and thorax have salivary glands (sometimes called labial glands), whose secretions play a part in the ingestion of food (Section 2.21, Figures 2.2a, 2.31a). In the abdomen, the wax glands (discussed in Sections 2.23.2 and 3.31.1) produce beeswax from which comb is built. And the venom gland in the worker and queen (Section 2.23.1) produces venom that is stored in, and used from, the venom sac.

Pheromones

The worker, queen and drone each produce many other glandular secretions that are called pheromones. A *pheromone* is a substance or group of substances secreted externally as a liquid that serves for communication between members of the same species. It may be transmitted by contact as a liquid, but many pheromones are volatile and are transmitted through the air. An animal with a sensory organ (Section 2.34) that can perceive the pheromone in question may respond to it in a behavioural or a physiological way. In honeybees, the responding bees may or may not belong to the same colony as the bee producing the pheromone, but they are usually of the same species. The sex attractant pheromone of the queen is an exception, since it contains the same compounds in all *Apis* species (Shearer et al., 1970; Koeniger, 1976*b*); see end of Section 8.23.

A number of pheromones produced by certain glands have been identified chemically, and their separate functions can therefore be studied. For instance the mandibular glands of the queen produce a number of volatile compounds that are important in ensuring both colony cohesion within the nest or hive, and the dominance of the single queen that heads the colony, and also in preventing the rearing of other queens. Slessor et al. (1988) were successful in identifying 5 compounds whose combined activity was equivalent to secretion from the queen's mandibular glands in inducing retinue behaviour in workers (Section 3.21). Forthcoming papers indicate the part these compounds play in other worker responses.

The Nasonov gland of the worker (see Figure 2.35a) produces several volatile compounds which attract bees that are outside their own colony or swarm back to it, or to a food source. The extremely small tarsal glands on the underside of the pretarus (foot) also produce pheromones (Lensky et al., 1984), which are important means of communication between bees.

Several glands in the sting apparatus produce different pheromones, with quite separate functions. One in the worker secretes the alerting pheromone that stimulates other workers to sting; see Section 3.41.

Much of the social behaviour of honeybees is regulated by the production and perception of pheromones, and this subject is discussed in Chapter 3; see also Free (1977) and Winston (1987). Many more honeybee pheromones will probably be discovered, as well as the identity of the glands that produce the pheromones and of the sensory organs that perceive and respond to them. Also the composition of some of the pheromones now known to exist, but not yet identified chemically, will be established.

Figure 2.35a Hairless front part of tergite of A7 of worker, which has pores through which pheromone issues that is produced by Nasonov gland, SEM photograph, × 56 (Erickson et al., 1986, by publisher's permission).

2.36 Reproductive systems of the queen and drone

The queen

Figure 2.36a (*right*) shows three views of the large ovaries of a queen in full lay. Each ovary consists of a bundle of 150 or more tubules, the ovarioles; these start as thin threads at the anterior (front) end of the abdomen, where they are attached to the ventral side of the heart. In the tip of each ovariole, cells are budded off from the germinal tissue, and as they move down it, some are differentiated into nurse cells and some into follicle cells; the rest are true egg cells. Nurse cells absorb food through the walls of the ovariole, and the egg obtains nutrients from adjacent nurse cells, growing until at full size it has absorbed the contents of all these cells.

The ovarioles open into the two lateral oviducts, and both of these join the median oviduct. Eggs pass from the ovarioles into the oviducts, and thence into the vagina.

The position of the valvefold in the vagina should be noted. Success in inseminating queen honeybees instrumentally was achieved only after a special hook was used to hold the valvefold out of the way of the semen injected, so that this was able to pass beyond the valvefold to the oviducts and spermatheca.

The mating process is described in Section 2.7. After mating has finished, semen from all the mated drones is in the queen's oviducts, filling them; it soon moves into the spermatheca, and is stored there. Subsequently, each time the queen lays an egg in a worker cell, a few spermatozoa pass out of the spermatheca along the spermathecal duct, and into the vagina where one of them fertilizes the egg. When she lays in a drone cell, which is larger, no spermatozoa are released, and the egg is not fertilized. Section 2.53 explains how the queen recognizes the two types of cell.

Kapil (1962*b*) published a detailed study of the reproductive system of the *Apis cerana* queen. (It is sometimes possible for workers of both species to lay eggs; see Section 3.39.)

The drone

Figure 2.36a (*left*) shows two drawings of the reproductive system of the adult drone approaching maturity. Spermatozoa are produced and become mature in the testes, which are composed of bundles of tubules. The testes are very large in a drone when it emerges from the cell, but the contents of each testis passes through the vas deferens to the seminal vesicle before the drone is about 13 days old; the testes have by then completely shrunk, and the seminal vesicles are very

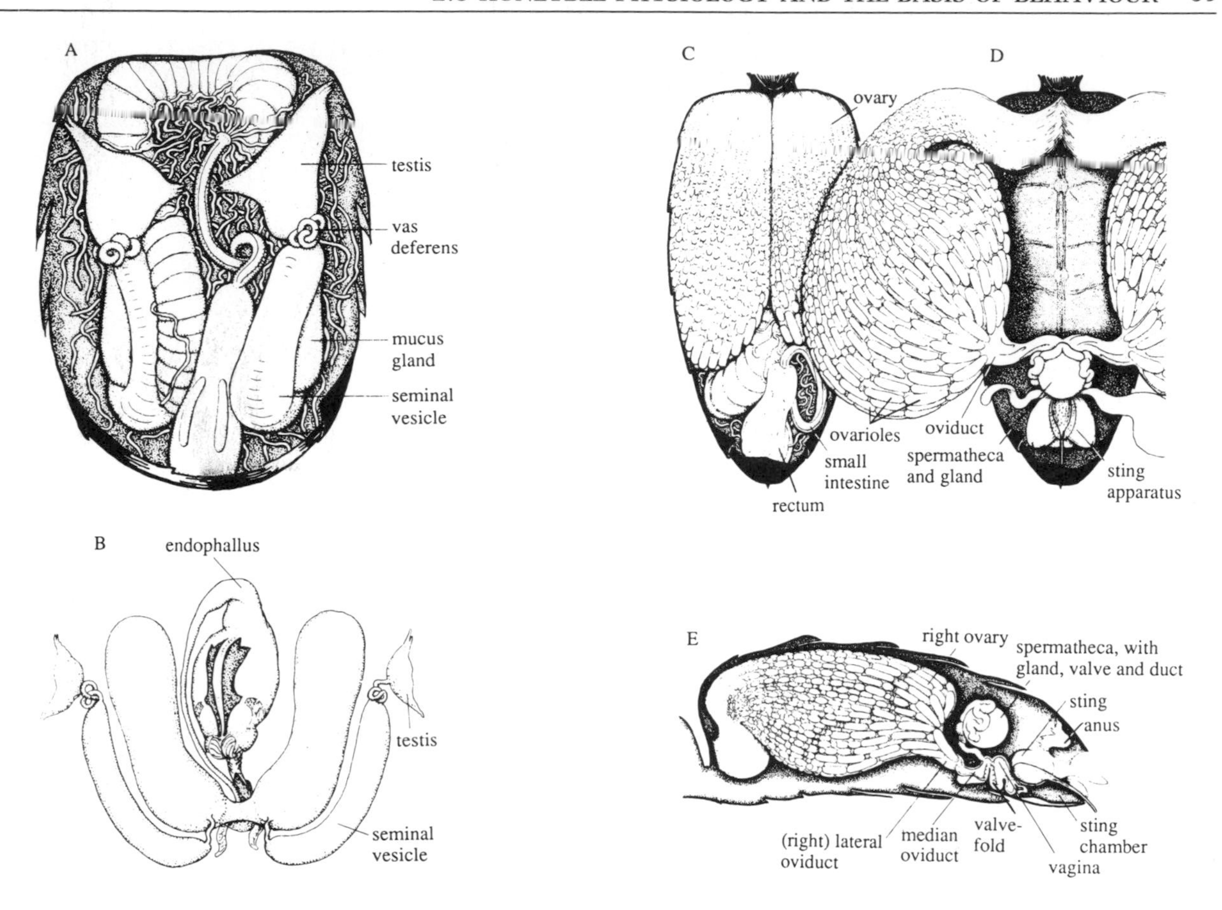

Figure 2.36a Dissection of reproductive system of queen and drone (Dade, 1962).

left Maturing drone: A, viscera undisturbed; B, reproductive apparatus removed and laid out.

right Fertile queen: C, viscera undisturbed; D, ovaries laid out; E, longitudinal 'semi-section'.

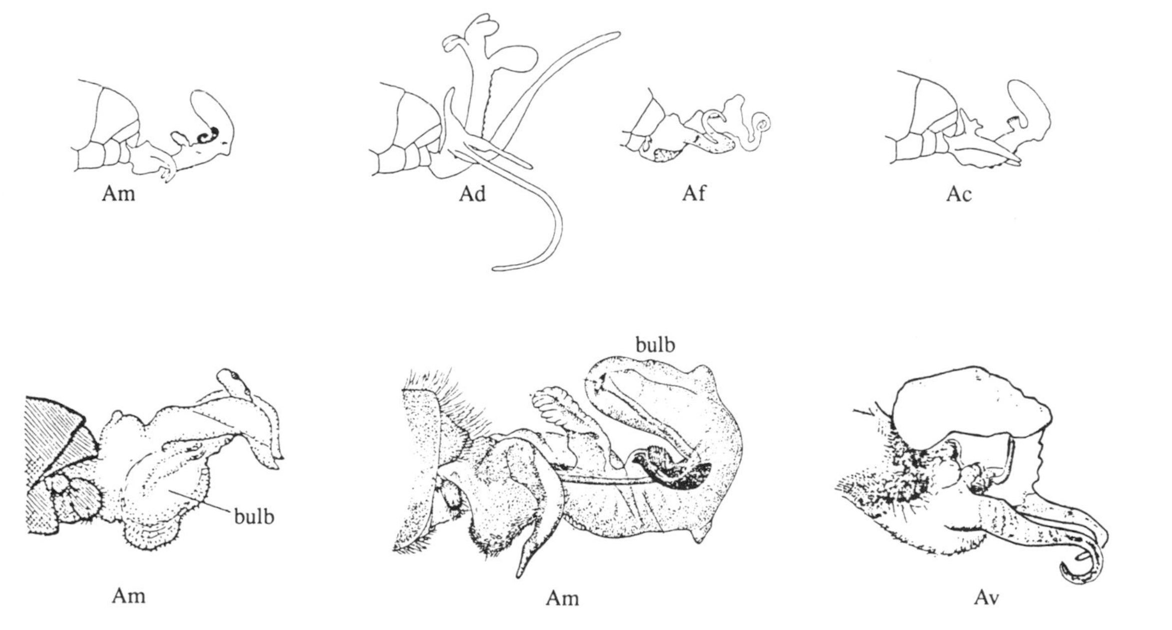

Figure 2.36b Drone endophallus.

above Fully everted, four *Apis* species (Ruttner, 1986):
Ac *A. cerana* Af *A. florea*
Ad *A. dorsata* Am *A. mellifera*

below *Apis mellifera* (Am): partly everted (Dade, 1962), fully everted (Fyg, 1952); *A. vechti* (Tingek et al., 1988).

large (as shown), since they contain all the semen, with the spermatozoa suspended in it.

In all honeybees (*Apis* species), the copulatory organ is an internal one, the endophallus. During mating, it is everted (turned inside out) by strong muscular contraction and compression of the abdomen. This mechanical action is like the blowing out of fingers of a rubber glove after they have become turned outside in. Figure 2.36a (*left*, B), in which the reproductive system is laid out, shows the uneverted endophallus, as it is stored between the two mucus glands inside the abdomen. Figure 2.36b (*below*) shows the endophallus both partially and completely everted.

The significance of the complicated structure of the endophallus in the mating of *Apis mellifera* is discussed in Section 2.7. The endophallus in other *Apis* species is shown in Figure 2.36b; their considerable differences suggest that interspecific mating is not possible (Ruttner, 1986; see also Ruttner & Maul, 1983). The reproductive system of diploid drones (end of Section 2.61) is described by Woyke (1973*a*). Kapil (1962*a*) published a detailed study of the reproductive system of the *A. cerana* drone.

Figure 2.4a Development of worker from egg to adult emerged from its cell (photo: A. Watkins, *c.* 1900, Hereford City Library).
Individual stages are indicated in Figure 2.4b.

2.4 LIFE HISTORY OF THE WORKER

Changes in the appearance of a worker honeybee as it develops from an egg to an adult bee are shown in Figures 2.4a, 2.4b and 2.41a. The larva moults four times and then becomes a prepupa without moulting. During the next moult the prepupa metamorphoses into a pupa, and a final moult changes the pupa into an adult (imago).

By the third day after the egg was laid, the embryo has increased in size although the egg as a whole has lost weight, by the embryo's metabolism of food within the egg. The larva then hatches from the egg, and its intensive feeding regime enables it to increase in weight by a factor of more than 2000 by the time it is sealed in its cell 5 days later (Table 2.41A). When the adult emerges from the cell its weight has dropped to about two-thirds of the final larval weight. Thereafter it increases again within a week to something like its earlier maximum, again as a result of feeding – especially on pollen. The bee is now fully mature, and thereafter reserves of certain nutrients in her body are used up until she dies. She needs carbohydrate to provide energy, and can carry a load of nectar up to her own body weight.

Information on the mortality of workers at different immature and adult stages is given in the discussion on the demography of the colony in Section 3.13.

2.41 Development from egg to emergence

As in other Hymenoptera, each honeybee in the course of its life passes through four stages: egg, larva, pupa, adult (imago). The duration of the three developmental or juvenile stages can be seen from the ages quoted in Table 2.41A; the Table includes data that have been located for other honeybees besides European *Apis mellifera*. Periods vary between castes; there are also differences between races of *A. mellifera*, and greater differences between species. Some of these differences have a specific importance in beekeeping; see for instance Section 11.42 on *Varroa jacobsoni*.

In the egg, the embryo develops by absorbing the egg yolk which is a rich source of proteins. The larva that hatches from the egg feeds on provisions put into the cell by nurse bees (young adult workers). It grows enormously in size and weight, as a result of the distension of body cells to many thousand times their original size, without dividing.

The workers develop in the nearly horizontal cells that form the comb (Figure 2.41a), from eggs that

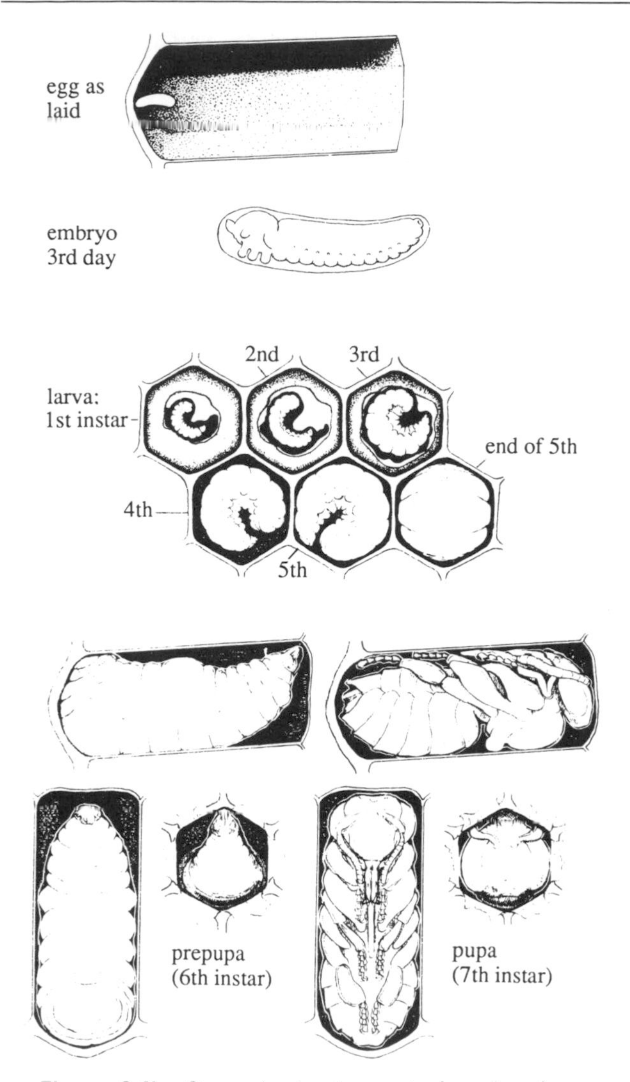

Figure 2.4b Stages in development of worker, from egg to pupa (Dade, 1962).

have been fertilized with drone spermatozoa stored in the queen's spermatheca and released in response to the queen's reaction to a worker-size cell; see Section 2.53. The developing or juvenile workers, 'worker brood', occupy most of the brood nest, in the central, most protected and thermally regulated part of the cluster of bees in the hive. The timetable of developmental stages of different honeybees is summarized in Table 2.41A, and more detail is given by Shuel and Dixon (1960).

Nurse bees repeatedly supply the cell containing the young larva with secretion from their hypopharyngeal glands. Lindauer (1953) calculated that about 143 feeds altogether were needed for one worker larva, and that one adult worker provided about 400 feeds, enough to rear 2 to 3 larvae during her lifetime.

Larvae are fed on brood food* secreted from the hypopharyngeal and mandibular glands of nurse bees

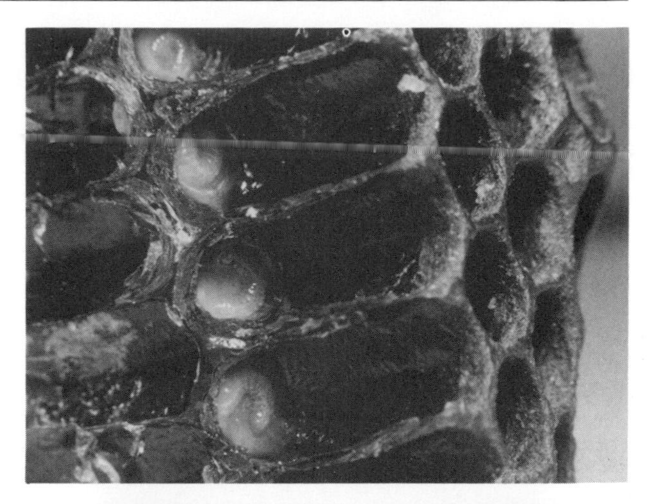

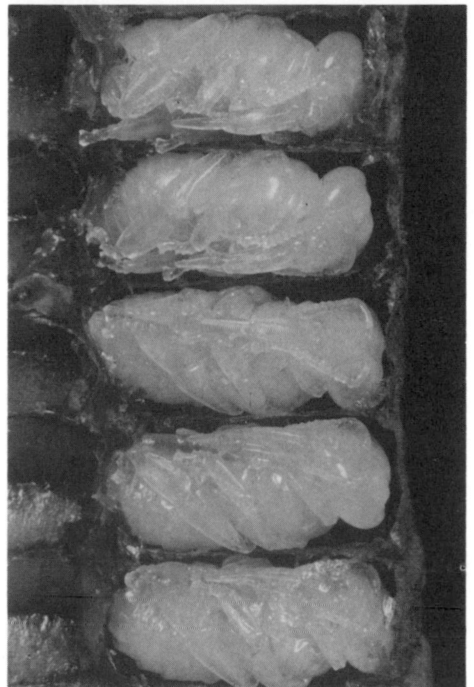

Figure 2.41a Sealed and unsealed *Apis mellifera* worker brood (photo: I. Okada).
above Young worker larvae, each on its supply of brood food.
below Pupae in their sealed cells.

(young adult workers). This brood food is high in protein, and contains vitamins and other nutrients the larvae require. Section 2.42 shows that the nutrients in brood food are derived largely from pollen eaten by the nurse bees. Worker larvae are fed at first on 'worker jelly' which is relatively low in sugars. After larvae are $2\frac{1}{2}$ to 3 days old the food is changed to

* Ribbands (1953) used 'bee milk' as a general term for larval food produced in a nurse bee's hypopharyngeal glands, reserving the term royal jelly for 'bee milk' in or from queen cells.

Table 2.41A Developmental stages recorded for *Apis mellifera* and *A. cerana* (worker, queen and drone)

Entries give the mean age in days (and/or range) after the egg is laid:
hatch = day larva hatches from egg
sealed = day larva is sealed in cell
emerge = day adult emerges from cell

	Egg	*Hatch*	*Sealed (capped)*	*Emerge*
Apis mellifera, European:				
worker	0	3	8(7–9)	19–22
queen	0	3	8(7–9)	15–17
drone	0	3	10(9–10)	24–25
(The larva pupates at 11–14, 10–12, 14–17 days for w, q, d, respectively.)				
A. m. lamarckii:[a]				
worker	0	3	7.7	19.4
queen	0	3	7.3	15.4
drone	0	3	10.3	24.3
A. mellifera, tropical African:				
worker	0	3	7–8	18–20
queen	0	3	7–8	14–15
drone	0			25
Africanized: probably similar to tropical African (above); Harbo et al. (1981) quote comparative age in hours at hatching for worker eggs: European 73.3 ± 1.14 (range 72–76); tropical African 71; Africanized 69.6 ± 1.06				
A. m. capensis:				
worker post-sealing period 9.6 days, cf. 12.0 for *A. m. carnica*	0	3	8–9	19–20
queen	0			14–15?
Apis cerana				
Kashmir[b]				
worker	0	3	8	20–21
queen	0	3	8	15–16
drone	0	3	9.5	23–24
India (hill, Nagrota)[c]				
worker	0	3	7	18–19
queen				
drone	0	3	9–10	23–24
India (plains)[d]				
worker	0	3.1	8.3	19.5
queen	0	3	(8)	(15–16)
drone	0	3		(24)
Philippines[e]				
worker	0	3		18–19
queen	0	3		15–16
drone	0	3		24

[a] El-Banby (1963)
[b] Shah (1988)
[c] Rahman & Singh (1947)
[d] Kapil (1959*b*); brackets indicate other sources
[e] Quiniones (1988)

'modified worker jelly', with some honey and pollen, although Ribbands (1953) suggests that the presence of pollen may be accidental; Haydak (1957) reported the food to be almost pollen-free. At 3–4 days the food was found to contain 47% sugar (Shuel & Dixon, 1959).

Feeding ceases when the larva is about 5 days old, and the workers build a wax capping that seals it in its cell. Unlike the queen larva, the worker larva is immured without any food. The fully grown larva secretes silk fibres for its cocoon, making a number of 'forward somersaults' during the process; workers, queens and drones observed by Jay (1964) made 27–37, 40–50 and 40–80 somersaults, respectively, each taking about an hour or less. A worker finally orients its body in the cell with the head outwards; it achieves this position not by a gravitational response as the queen does (end of Section 2.51), but by recognizing the differences in texture and shape between the capping of the sealed cell and the base which is formed by the midrib of the comb (Jay, 1963). A few days after spinning, the larva pupates, and in the pupa a second group of cells (that had remained small and inactive in the very young larva) suddenly starts to divide rapidly. These cells will form adult tissue, and eventually replace virtually all the cells of the larva. They obtain the necessary nourishment from the large larval cells, which degenerate and are digested. Development during the pupal stage is not in size but in modification of the body structures, which change to those characteristic of the adult.

While the larva was feeding, it generated heat, but during the pupal stage the control of its temperature at 35° is entirely dependent on the layer of bees covering the comb. The bees are attracted there by a pheromone produced by the pupae; see the final entries in Table 2.35A. Queen pupae have been found to contain 30 μg, drone pupae 10 μg and worker pupae only 2–5 μg of the pheromone (Koeniger & Veith, 1984). The amount produced by the many worker pupae is, however, sufficient to attract adults to cover their comb continuously.

Finally the newly formed adult worker, having cast off the pupal skin, nibbles away the cap of the cell with her mandibles, and emerges as a rather soft, greyish bee. During their whole developmental period, the bees are in their individual cells, and have no direct contact with each other. This is in marked contrast to their intensely social life as adults.

2.42 In-hive activities

By the time a worker emerges from her cell in the comb, her adult anatomical structures are defined and fixed. But the full development of the glandular and reproductive systems takes place afterwards, and depends on a high intake of protein and other non-carbohydrate food substances, obtained from pollen. The same is true of queens and drones. Young bees are 'negatively phototactic', and remain in the darkness of the hive. If they see the light at the hive entrance they turn away from it, and stay among the combs.

Most worker activities within the hive or nest are social, but what each bee can do, and does, depends on the stage of development of her various glandular systems. This is linked with her age, but in ways that depend on the condition and circumstances of her colony.

During the first few days of adult life, a worker must eat a sufficiency of pollen of good quality (Section 12.42), if her different glands are to develop their full potential for producing their specific secretions. Workers start to eat pollen within 2 hours or so of emergence from the cell, and most of them are feeding on it continuously within about 10 hours. One activity that a very young worker can carry out is cleaning vacated brood cells, and much of this cleaning is done by bees less than a few days old. After a day or so, workers may begin to feed a mixture of honey and pollen to older worker larvae (3, 4 or 5 days after hatching, according to different observers). This food is additional to the brood food provided to larvae by nurse bees.

Pollen consumption by young adult workers rises to a maximum at around 5 days, and decreases at 8–10 days. The pollen is digested by means of enzymes produced in the ventriculus (Section 2.31). At about 5 days of age the hypopharyngeal glands in the head (Figure 2.31a, *below*) secrete brood food, and this continues until the glands start to degenerate at 10 days or so. If the bee does no brood rearing and is provided with pollen, the glands may continue to develop until she is 27 days old (Ricciardelli d'Albore et al., 1987). To a great extent the worker's intake of pollen as a young adult, and the total amount of brood she rears, determine her length of life (Maurizio, 1950). Table 2.42A shows the relation between length of life and brood rearing in a region with rather warm winters and hot summers. Pollen lengthens the first phase of adult life (in the hive), and brood rearing reduces it; the second (foraging) phase is much less variable – see Section 2.43. In Ribeirão Preto, Brazil, brood rearing continued during the winter, and the life span of winter workers was not greatly extended (Terada et al., 1975). But as these authors point out, there are many complicating factors.

Beeswax is secreted mostly by workers around 10 to

Table 2.42A Life span of adult worker honeybees (*A. mellifera*) in relation to their brood rearing, Tashkent, 41°N

Data from Taranov and Azimov (1972).

Bees emerged on 20th of	*Average life span in weeks: Summer + winter + next spring*	*Average life span in weeks: Total*	*Amount of brood reared compared with April*
April	$3\frac{1}{2}$	$3\frac{1}{2}$	100%
May	4	4	96%
June	4	4	83%
July	$4\frac{1}{2}$	$4\frac{1}{2}$	56%
August	8 + 17 + 4	29	38%
September	$5\frac{1}{2}$ + 17 + $4\frac{1}{2}$	27	21%
October	2 + 17 + 5	24	5%

15 days old. It does not contain proteins, but Taranov (1959) found that young workers not allowed access to pollen lost up to 20% of their body protein after 15 days of intensive wax production. Beeswax secretion, and the wax glands that produce it, are described in Section 2.23.2, and the bee's manipulation of wax and use of wax in building comb in Section 3.31.

The sting mechanism, and venom production, are described in Section 2.23.1. The venom sac is full by the time a worker is about 2 weeks old, and from that age some bees become active as 'guards' at the hive entrance. For the rest of their lives, workers will continue to defend their colony if it is disrupted, but those older than about 6 weeks cannot produce more venom. Colony defence is described in Section 3.41.

As a worker grows older, her hypopharyngeal glands secrete increasing amounts of the enzymes invertase and glucose oxidase, which the workers use in making honey from nectar and honeydew; see Section 13.23. The secretion is greatest at about 4 weeks old, although the glands have degenerated by then. The worker is then of foraging age, and is likely both to forage and to take part in the production of honey from nectar and honeydew in the hive.

2.43 Flight, orientation and foraging

After a worker's wing muscles have developed (and some glands have already regressed), her reaction to light is reversed, and she becomes positively phototactic. This is likely to occur at an age of 5 to 15 days, depending on colony and forage conditions; Table 7.43B quotes 10–14 days for European bees and 3 days for Africanized bees. She is then attracted to the hive entrance, and makes sorties into the daylight beyond it. She cannot fly until the mitochondria of her wing muscles develop (Weaver, 1988). After this has occurred, she first takes short orientation flights, learning the position of her hive among surrounding landmarks. Some days later she starts flying to sources of nectar, pollen or water, and carries back to the hive what she collects. The social organization of the foraging bees, and the communication between them and younger bees in the hive, are dealt with in Section 3.35.

The flight is powered by muscles in the thorax (Section 2.22), and the bee is able to fly only if the temperature of the thorax is high enough for the energy-producing chemical reactions to take place. Table 3.32A suggests that the lower limit is 13° to 14° for workers of European bees, and the upper limit 43°, or 46° if the bee regurgitates liquid to produce evaporative cooling. The optimum is in the range 19° to 30°. Heinrich (1979) found that in both temperate-zone and tropical (African) workers, the temperature of the thorax, where the wing muscles are situated, was 36° to 38° on leaving the hive, 31° to 32° during foraging – and when attacking – and only 30° when returning to the hive. Bees in the hive increase their thoracic temperature before leaving it, by a mechanism explained in Section 3.32. The minimum thoracic temperature for continuous flight was 30°, and in laboratory experiments temperate-zone bees could not maintain this temperature outside the hive unless the air temperature was above 9°.

The wing-beat frequency in flight is quoted as 235 to 250/second for European workers (Dade, 1962); the frequency is higher at greater speeds. It is slightly but noticeably higher for tropical African workers, whose wings are smaller. H. T. Kerr and Buchanan (1987) measured the dominant frequency as 210/s for European bees and 270/s for Africanized bees. In Punjab, India, for workers and drones, respectively, frequencies were 235/s and 225/s for *A. mellifera*, and 306/s and 283/s for *A. cerana* (Goyal & Atwal, 1977).

The sugar concentration of the bee's blood is normally about 2%; if it is below 1% she cannot fly, and if it falls below 0.5% she can hardly move. Early studies gave relatively high results for the bee's rate of consumption of sugar during flight (e.g. 10 mg/h, Wigglesworth, quoted by Dade 1962). On this basis a honey sac full of honey would provide fuel for about 15 minutes of flight, through a distance of 6 to 8 km. More recent calculations (e.g. Seeley, 1985; Weaver, 1988) indicate that a bee can fly much further on a full load of honey, and suggest that (at any rate during long flights) sugar consumption is negligible. When the sugar supply in the honey sac of a bee is exhausted, the bee must either consume nectar or some other source of sugar, or rest for a period, during which glycogen stored in the fat-body in the abdomen is converted into sugar, providing some more fuel. Experiments in a flight room suggest that the total amount of energy a worker can expend in flight during her lifetime has an absolute upper limit; the bee's ability to synthesize glycogen from sugar slows down irreversibly as she gets older. Neukirch (1982) found that the limit of total flight distance was about 800 km. One result of this would be that during a heavy nectar flow, when foragers fly a long distance each day, they die sooner than bees foraging at other times in the active season. Nevertheless, the total duration of the second (foraging) phase of a bee's life varies much less than that of the first phase, spent within the hive.

Detailed results have been published in Switzerland of the total number of flights from a hive every day for a year, as recorded by an automatic counting device. During the week of greatest activity (the last in June) 1 141 800 flights were recorded, an average of 163 100 a day (Bühlmann et al., 1987). The colony then consisted of 22 000 bees; if half of them were foragers, each made on average 15 flights a day, which is consistent with other results obtained during a nectar flow, quoted in Section 2.43.1.

Orientation

Figure 2.34b shows that certain retinula cells of each ommatidium in the honeybee's compound eye are sensitive to ultraviolet light. These same cells perceive the plane of polarization of this light coming from the sky, even if the sky is partly overcast.* The pattern of polarized light a bee sees in the sky at any moment is unique for each different direction of view. A honeybee in flight is thus able to recognize the direction in which it is flying, and to orient itself by the pattern of polarized light it sees. Many research workers consider it likely that this is the bee's primary method of orientation (von Frisch, 1967; Dyer & Gould, 1983; Rossel & Wehner, 1986, 1987; Winston, 1987).

Important supplementary methods include visual orientation by reference to landmarks seen, including the bee's own hive or nest site. When a hive is moved to a different site, bees leaving the hive can be seen spiralling up in the air above it, as they learn the new site. Scents provide other important orientation cues, for instance scents of flowers and other sources of food, and of the bee's own colony, and possibly also scents encountered between the hive and a forage source. In addition, evidence is accumulating that honeybees can perceive the earth's magnetic field (Section 2.34), and that they are able to use it for orientation.

Two important factors in a forager's orientation to a food source – or of a swarm scout bee's orientation to a new site – are the extent of the bee's memory of it, and how much of it she can communicate to other bees. Honeybees have a good memory for scent; von Frisch (1919) found that it could last 5 days or more. Scent is easily communicated to other bees in the hive, because they smell what the forager has collected. Honeybees can also remember, and communicate to other bees, the distance and direction of a forage source (or of a new nest site) by dances performed on the comb surface (Section 3.35), unless the sun is directly overhead. As far as we know, the forager registers the distance of the *outward* journey to the food source, or more precisely her energy expenditure on this flight. But she registers the flight direction on the *homeward* journey. She can register landmarks, and her memory for them is pictorial; there is evidence that she does not simply remember landmarks along the route when foraging, but builds up some sort of 'locale map' (Gould, 1987). She cannot communicate information about landmarks. Wellington and Cmiralova (1979) found that foraging bees could register the heights of food sources, and communicate them to other bees, although the mechanism is not yet understood.

Figure 2.43a Forager (*Apis mellifera*), already carrying pollen loads, collecting nectar from a floret of Chinese milk vetch, *Astragalus sinicus* (photo: M. Matsuka).

* This is sometimes referred to as an 'e-vector compass', the direction of the e-vector being that of the electric vector of the polarized light, at right angles to the magnetic vector and to the direction of propagation of the light.

Figure 2.43a shows a forager on a flower. She recognizes flower shapes well, and is attracted by certain shapes more than by others, for instance by a flower with five separated petals (and a long outline) more than by a simple circle (with a short outline); see Ribbands (1953). A bee can remember a shape, but cannot communicate information about it to other bees.

A forager registers the colour and shape of the source of nectar as she approaches it, not when she leaves it. She learns the colour during the two seconds before and two seconds after she first dips her proboscis in the liquid food. Honeybees have good colour vision, and visit certain colours in preference to others, although colour is usually perceived from shorter distances than scent. A honeybee cannot remember a colour for long, and cannot communicate her knowledge of it to another bee. The spectrum of colours visible to honeybees (300–650 nm) extends to shorter wavelengths than the spectrum visible to man (400–700 nm); individual retinula cells in the compound eye are sensitive to blue, or green, or ultraviolet (Figure 2.34b). Blue is especially attractive to bees, and a range of ultraviolet colours invisible to man is visible, and attractive, to honeybees. Many flowers that appear plain white to us are seen in ultraviolet colour patterns by a honeybee, which uses these patterns as nectar guides (Section 12.21.1). Her learning of flower colour is reinforced many times as she takes nectar from successive flowers or florets of the same plant; her memory for it is poor, perhaps only 2 days (von Frisch, 1919). She cannot communicate colour to other bees.

Gould (1987) has summarized what is known about the honeybee's memory of landmarks, flower shapes and 'locale'.

2.43.1 Nectar and honeydew collection

Nectar and honeydew themselves are discussed in Sections 12.21 and 12.22. Nectar from flowers is the main raw material from which honey is made. Its collection by honeybees has been much studied, and many experiments have also been made on honeybees collecting at dishes containing syrup, where the sugar concentration and other factors can be controlled. Much less is known about the behaviour of honeybees collecting honeydew. But, like nectar and syrup, honeydew is primarily an energy food for bees, and the energetics of their behaviour in collecting all three substances are likely to be similar.

In most circumstances a colony's collection of nectar or honeydew takes priority over collection of other materials. (But foragers can quickly switch from nectar collection as and when necessary; see for instance Sections 2.43.2 and 3.32 on the use of water in cooling the hive.) Also, a colony commonly continues to collect nectar or honeydew, and to store honey made from it, in excess of its future requirements. It is this behaviour that makes beekeeping profitable for the beekeeper. Some honeybees have a more highly developed 'hoarding behaviour' than others (Rothenbuhler et al., 1979), and these can be of special economic value.

Section 3.36 explains how an incoming forager passes the contents of her honey sac to bees within the colony, which store and process it.

Nectar-collecting flights

Many facts about the foraging of European honeybees were established in the early part of this century. For instance Park (1922) showed that the weight of the nectar load a forager took back to the hive varied according to the plant species being worked. The average weight was 40 mg and the maximum 70 mg, nearly as much as the bee's unladen weight (80 mg). It is exceptional for a forager to collect a full load of nectar from a single flower, although it has been reported for the tulip poplar (*Liriodendron tulipifera*, Figure 12.21d). It may well occur also with some of the large tropical flowers.

Park (1928*b*) found that the average flight speed of foraging bees – in the absence of wind – was 24 km/h, whether or not they were loaded; it could reach 40 km/h for short periods. The bees used more energy when flying against the wind, and less when flying with it, so that flight speeds were not much changed. But if the wind speed reached 24 km/h, bees could make little headway against it.

How bees find flowers containing nectar

When honeybees start to forage for nectar, they find their first flowers in one of two ways. They may be recruited as a result of 'following' a recently returned nectar forager dancing on a comb (Section 3.35), so learning the distance and direction of the nectar and also its scent. They can then fly directly to the area where the flowers are, and locate them by their scent, although they may need several (up to 12) 'dance-guided search trips' to do so (Seeley, 1983). When a nectar forager reaches a flower, she is helped to find the nectar by the flower's nectar guides, which may be scented, visual, or textural (Kevan & Lane, 1985). At the flower, she also learns its shape and colour pattern (Section 2.43, under Orientation), which enable her to make fast direct flights to the flowers in future. Section 3.35 explains how the forager's behaviour when she returns to the hive differs according to the richness of the nectar supply: she may recruit other foragers, or only revisit the forage herself, or – if the supply is poor – she may not even do that.

There are, however, some bees in the colony that are not recruited by other foragers, but fly out and *search* for nectar. These are referred to as 'scout bees' or 'search bees', and their existence was understood in the early 1900s by Gaston Bonnier. The proportion of scout bees to 'established foragers' varies greatly according to circumstances. In some experiments observers have found it to be only a few per cent (see Ribbands, 1953); in others it was sometimes much higher, e.g. 5% to 35% (Seeley, 1983). There may be a genetic influence such as that referred to at the end of Section 3.38. During a heavy nectar flow it is likely that almost all nectar collectors become established foragers on the plant species producing the flow. When forage is scarce, more of the foragers become scouts, so that the colony can then take advantage of any nectar sources available. Ribbands and others cite experiments which show that established foragers can become scouts and vice versa.

Experiments have been made to compare the ability of different honeybee races and species to learn characteristics of new forage sources or to remain constant to old ones (e.g. Koltermann, 1973; Menzel et al., 1973). Tropical *Apis cerana* was quickest to learn, and to react to changes in food supply, then – in order – *A. m. lamarckii* from Egypt, *A. m. ligustica* from Italy, and *A. m. carnica* from the Alps. The 'degree of motivation' of different bees to find food was in the reverse order – highest for the most northerly bee and lowest for the tropical species. This would be expected, in view of the dominance of the flows from some of the temperate-zone plants, and the much greater variety of tropical flowering species.

Energetics of nectar collection, and choice of flowers

During the 1970s and 1980s much attention has been focussed on ways in which social animals, including honeybees, exploit their fluctuating food resources. Simpler studies in earlier years had provided relevant information on honeybees, well summarized by Ribbands (1953). For instance during Park's observations (1928*b*), he found that, on average, nectar foragers made 13.5 trips a day in a favourable year, and were out of the hive for 10 hours a day altogether; in another, unfavourable year, they made only 7 trips a day, being out of the hive for 7.5 hours. So in favourable conditions the nectar collectors made more trips, returned to the hive more quickly from a trip, and spent less time in it before the next trip. The average times per trip, including the intervals in the hive, were 45 and 64 minutes, respectively.

The more recent studies have explored further questions, for instance if, and how, foraging honeybees maximize the efficiency of their nectar collection. As a result, the constancy of foragers to one plant species (Section 2.44) does not seem to be as complete as it once did – although Ribbands (1953) cites early observations of deviations from it. New observations show many complexities in the subject.

It has been known for some time that a forager is likely to prefer a flower that gives her much nectar (high *energy gain*) per flower visit, and to prefer flowers whose nectar has a high rather than a low sugar concentration, at any rate within the limits 30% and 50%. A forager's perception of *energy expenditure* is more variable than her perception of energy gain, and individual foragers vary in their perception of both expenditure and gain. Nevertheless a forager does seem to assess her energy expenditure in obtaining a nectar load in relation to the energy gain the load provides (Waddington, 1987). The energy expenditure includes what is needed to fly from the hive to the flowers and back, and from one flower to another, and what is needed to extract nectar from each flower.

Work by Schmid-Hempel et al. (1985) suggests that a forager assesses her foraging opportunities according to her *net energy gain per unit of energy expenditure*, and that she chooses the nectar sources and foraging methods that make this net energy gain a maximum. (This may involve returning to the hive to recruit other foragers before she has filled her honey sac.) Results of experiments on the distance and direction of a forager's flight, the travel time, and the weight of the load she collects, support this hypothesis; see Schmid-Hempel, 1987.

2.43.2 Water collection

It is difficult for bees to store water (Section 3.36), and they forage for it according to their colony's immediate or imminent requirements. Water is always needed when brood is being reared, unless sufficient is available in the nectar being brought into the hive, and Figure 2.43b makes clear the relationship between the collection of water and of nectar by bees. It shows the number of water-collecting bees in an apiary in Sweden, where records have been kept for twenty years or more; 1953, the year represented, had normal weather and foraging conditions. The gains in weight of the scale hive (shown below) are due mainly to incoming nectar. There was a nectar flow during each month, and at the peak of each flow (marked ×) the number of water collectors dropped greatly – from over 100 to 20 or less. The greatest water-collecting activity was in May when brood rearing was at its height.

Water is also needed to cool the hive if there is

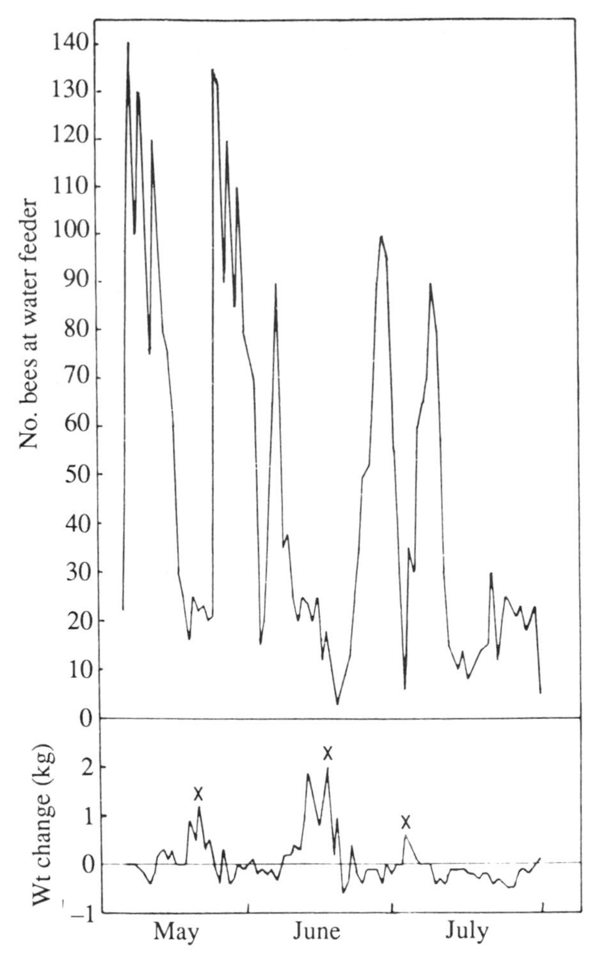

Figure 2.43b Water collection in an apiary at Ultuna, Sweden, May–July 1953 (Martinovs, 1964).
above Number of water-collecting bees counted at the water feeder during 1 minute at 12 noon each day.
below Daily changes in weight (kg) of scale hive; × marks a peak weight gain from a nectar flow.

danger of overheating, and this is the probable reason for the new increase in water collection in late June and early July, when brood rearing would be declining. Section 3.32 discusses more fully the circumstances in which a colony needs water, 12.3 the types of water sources used by bees, and 6.62.3 the beekeeper's provision of water in the apiary.

Lindauer (1955*a*) gives a vivid and detailed account of the collection and use of water by honeybees. Although most of the marked foragers he studied (1953) never collected water, some of those that did so became 'water specialists', spending their whole foraging lives collecting water and even dying at the water source. In addition to these, nectar foragers switch to water collection when much water is needed by their colony. In Park's (1928*b*) classical study in Iowa, USA, a water forager was likely to make about 50 collecting flights a day, although as many as 100 were recorded in one day. She took about 1 minute to collect her load of water (up to 50 mg, though about 25 mg was usual). The bee spent only 2–3 minutes in the hive between flights. From various published results, I would guess that in the temperate zones the water-collecting bees of a colony take around 20–25 kg of water into the hive in the course of a year, requiring up to a million flights. The water collected is carried back to the hive in the forager's honey sac, as nectar is (Section 2.43.1), and it is usually passed to another bee there, in the same way as nectar is (Section 3.36).

Gary et al. (1979) studied water collection by 2337 foragers from 399 hives in an area of California, and marked bees at 16 different water sources. The foragers tended to collect water near their hives – the mean distance flown to the source was about 90 m, and the maximum recorded 2.3 km. There are important advantages in providing a permanent water supply close to an apiary (Section 6.62.3). In early spring, bees prevented by inclement weather from collecting water for several days, will fly at quite low temperatures to fetch it. At 7° to 10°, the heat generated by their thoracic flight muscles can maintain them at the threshold temperature for flight for only a very short period, so in cold weather bees cannot collect the water they need unless it is very close to their hive. Water storage by bees is referred to in Section 3.36.

Bees collecting water from an open surface run the risk of drowning – hence the need to provide footholds for bees on artificial water sources (see Figure 6.62c). Normally the surface tension of water helps to prevent bees becoming immersed, but nowadays surfactants are widely used to reduce surface tension for domestic, industrial and agricultural purposes; detergents are a common example. Many more bees drown in water if it has been treated with a surfactant, but some of the surfactants themselves repel bees, and so reduce the number of bees at risk (Moffett & Morton, 1975).

When no water is available, honeybees have been observed to take fluid from a recently killed animal (Clauss, 1984): for instance an impala in a desert part of Botswana, and a bear in northern Canada (Crewe, 1985; Chance, 1983). Large numbers of honeybees have been recorded on the surface of seaweed mulch, in a Scottish garden during early summer (Moar, 1984). Many other unusual observations could be quoted.

2.43.3 Pollen collection

Figure 2.43c shows a forager collecting pollen. Many flowers produce both nectar and pollen (Section 12.42), and bees collecting pollen from them are able to collect nectar as well, thus providing energy for the

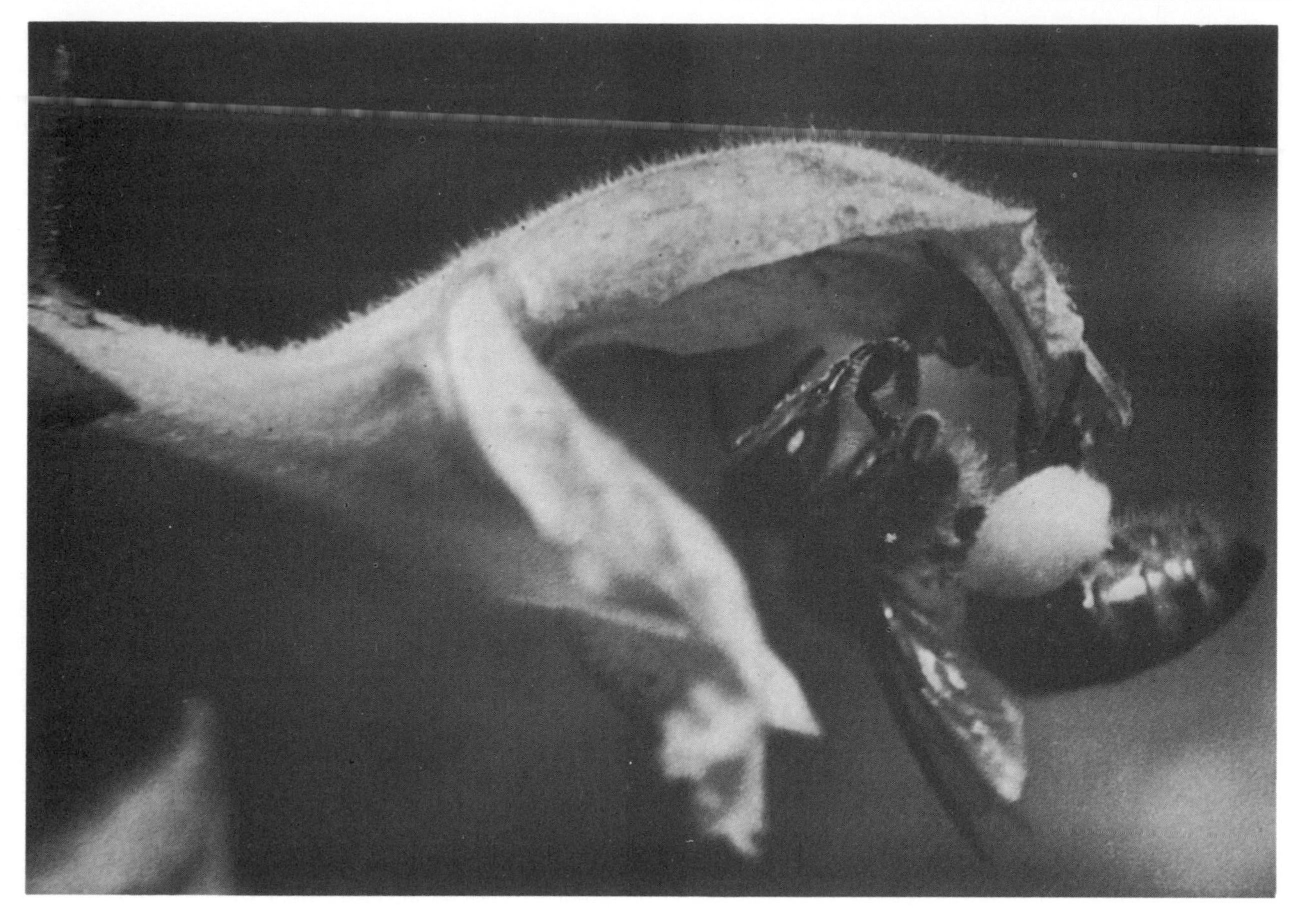

Figure 2.43c Forager collecting pollen from a flower of the sage *Salvia glutinosa* (photo: A. Fossel).
The stamen curves up and then down; the anther at the end (behind the bee's pollen load) brushes against the underside of the bee's abdomen and deposits pollen, which the bee transfers to her pollen loads.

flight home. Nectar (or honey) is also needed by a pollen forager for moistening the pollen so that she can pack it in her corbiculae; see Figure 2.43d. When bees collect pollen from plants that secrete no nectar, for example poppies and plantains (*Papaver* and *Plantago* spp.), they leave the hive well provided with honey (Hodges, 1952).

The composition of pollen varies according to plant source, and this is discussed in relation to the bees' nutritional requirements in Section 12.42. In the present context it should be noted that pollen provides honeybees with the amino acids and vitamins that are essential for their development to adult maturity; for workers, this includes the full development of their hypopharyngeal glands and venom glands. Pollen also provides the bees' requirements of lipids (fats), but these have been much less studied.

When bees start foraging, they may collect either nectar or pollen. Many that collect pollen change later to nectar, but once a bee collects nectar she is unlikely to change to pollen. Foragers seem to prefer to collect nectar, even if it takes them much longer to get a load. However, some bees forage regularly for pollen and for nectar at different times of day (Ribbands, 1953). Certain legumes and other flowers force foraging bees to come into contact with the anthers in order to reach the nectar, and most of these bees will gather both nectar and pollen. Some bees collect both on any species from which both are available.

The classical studies on the behaviour of a bee when collecting and packing pollen from some different flowers were made by Casteel (1912*a*) and Hodges (1952). The following summary relates to pollen collection from a poppy flower. The bee scrambles among the anthers and gets dusted all over with the dry pollen grains which are caught in her body hairs. She then leaves the flower and, hovering nearby, makes a complicated quick-moving action. She clears the loose pollen from her body with her legs; her proboscis is protruded, and she frequently strokes it with her forelegs, which become sticky with regurgitated honey, and as the pollen is manipulated it becomes paste-like.

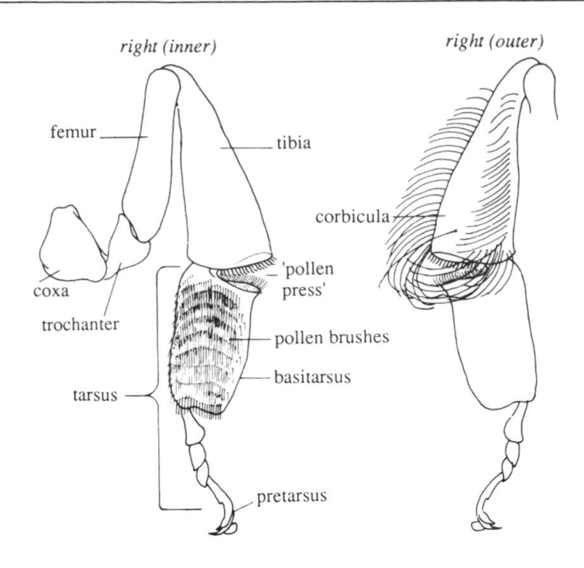

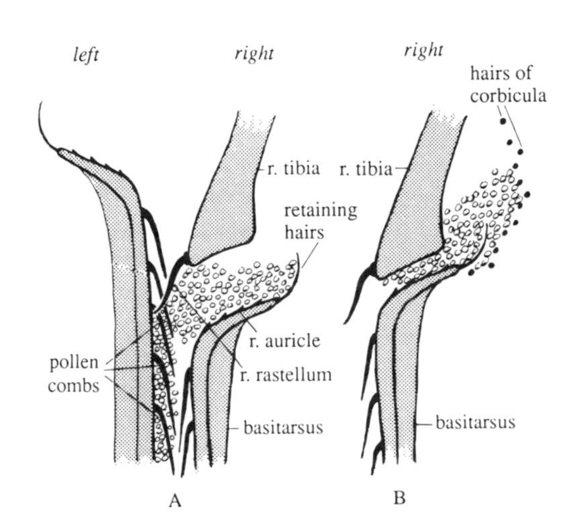

Figure 2.43d Pollen-packing structures of worker, and their operation (Dade, 1962, based on Hodges, 1952). *above* Surfaces of the right hind leg. *below* Section through hind legs, showing the action of pollen press, with press open (A) and press closed (B); see text.

The following refers to Figure 2.34d. The bee passes it backwards to the nine rows of bristles (the pollen combs or brushes) on the inner side of the basitarsus of the hind legs. She rubs these legs rapidly against each other in an up-and-down motion. As the right leg goes down, (A) its rastellum rakes some of the pollen out of the pollen combs of the left leg, (B) pressing it into the tibio-tarsal joint of the right leg, between the rastellum and auricle (the pollen press), and squeezing it onwards to the outer surface of this leg, where it collects on the corbicula, a smooth concave area (sometimes called the pollen basket) at the base of the tibia; it is surrounded by a fringe of hairs. The corbicula also has a central stiff hair (seta) which holds the pollen load in place, and probably also plays a part in enabling the bee to monitor the size of her growing load (Hepburn, 1980). As more pollen is pressed and pushed through to the corbicula (Figure 2.43d), the surrounding fringe of hairs guides it and shapes the load. The pollen load on the corbicula of the right hind leg is formed simultaneously, and in a similar way.

Different flowers require different behaviour by the collecting bees, and individual bees vary greatly in their agility and efficiency as pollen collectors.

The following results, mostly quoted by Gary (1975) from other authors, relate to pollen collection by European *Apis mellifera*.

Pollen collection occurred at temperatures 8–11° to 35°, above which it was reduced, also at wind speeds up to 34 km/h, although above 18 km/h it decreased. Pollen only was collected on 25% of flights, both nectar and pollen on 27%, and nectar only on 58%. The number of pollen-collecting flights per day varied from 6–8 to 47, and the number of flowers visited on one flight was 84 for pear (*Pyrus malus*), and up to 32 for dandelion (*Taraxacum officinale*) – although one observer recorded 100. The time spent actually collecting a load of pollen varied from 6–10 to 187 minutes. Foragers may need to visit many flowers to get a full load of pollen (Weaver, 1988), for instance about 500 florets for two clover species, which took a bee 25 minutes. Winston (1987) has published data on both nectar and pollen collection, from various sources.

On her return, a forager carrying pollen enters the hive; she may carry out a communication dance (Section 3.35), and shc is likely to feed on nectar or honey, or to be fed by other workers. She herself places her pollen loads in a cell, and the cell she chooses, after 'inspecting' a number, is usually just above or beside the brood nest. She grasps one edge of the cell with her forelegs, and arches her abdomen so that its rear end rests on the opposite side of the cell. She thrusts her hind legs into the cell, and they hang freely within it. The middle legs are raised, and each basitarsus is brought into contact with the upper end of the tibia of the hind leg on the same side. The middle leg is then pushed between the pollen load and the surface of the corbicula, so that the load is pried downwards and outwards, and falls into the cell. The process of packing the pollen tightly into the cell is continued by other, younger bees.

Section 12.63 discusses the amount of pollen collected in a year, and the seasonal cycle of its collection. The composition of different pollens in relation to the bees' nutritional requirements is dealt with in Section 12.42.

2.43.4 Propolis collection

Propolis is the name given to sticky plant material collected by honeybees and used in the construction and adaptation of their nests (Section 12.5). Table 12.51A lists reported plant sources of propolis, and Section 12.52 discusses the likely origins of material within the plants; Section 14.3 discusses propolis as a material.

Unlike food and water, propolis is not essential to most honeybees. It is commonly available during much of the active season, and in most parts of the world, and honeybees collect it *when there is no nectar flow*, provided it is of a consistency that they can manipulate (not too hard and not too sticky). Some of the details of propolis collection are outlined below.

Figure 12.51a shows a forager collecting resin on a poplar leaf as propolis. Different races of *Apis mellifera* collect propolis in varying degrees. Caucasian bees (*A. m. caucasica*) and Tellian bees (*A. m. intermissa*) in North Africa are noted for their extensive collection and use of it. Carniolan bees (*A. m. carnica*) and Ukrainian bees collect little, employing beeswax instead. Italian bees are intermediate (see Starostenko, 1968). In the tropics, both temperate-zone and tropical *A. mellifera* are known to collect and use propolis, sometimes extensively: see Figure 3.31b. In contrast, observations have shown that the other cavity-nesting honeybee, *A. cerana*, collects no propolis in a temperate climate (in Japan, Tamiji Inoue, 1985) or in the tropics (e.g. in India, Phadke, 1987; and in Sumatra in Indonesia, Tamiji Inoue, 1985). Propolis is essential to *A. florea*, at any rate in some areas, for defence of their comb against ants; see Section 3.31.4. *A. dorsata* bees have been observed to use propolis to strengthen the attachment of their comb to its support (Akratanakul, 1986*b*), but this does not seem to be very common.

In northern Italy, Marletto and Olivero (1981) made five counts of bees returning to a hive with loads of propolis from buds of poplar (*Populus* spp.), on each calm sunny day from April to November (Figure 2.43e). Propolis (resin) was abundant on the buds throughout this period, except for a short period in April/May when the old bud-scales of the leaves had fallen, but were not yet replaced by young ones. Outside this dearth period, bees were seen returning to the hive with propolis during every counting period when the air temperature was above 18°, but never when it was 18° or below; it fell to 10° during the observations. There was a great reduction in propolis collection during each nectar flow. In England Howes (1979) observed that 'propolis is never collected

Figure 2.43e Seasonal pattern of propolis collection from buds of poplar, *Populus* sp. (Marletto & Olivero, 1981).
graph Number of foragers returning to the hive with propolis between 13.00 and 13.30 h, on calm fine sunny days; broken lines indicate dates on which propolis collection dropped to a minimum.
horizontal black band Period when resin was present on poplar buds.
horizontal white bands Principal nectar flows.

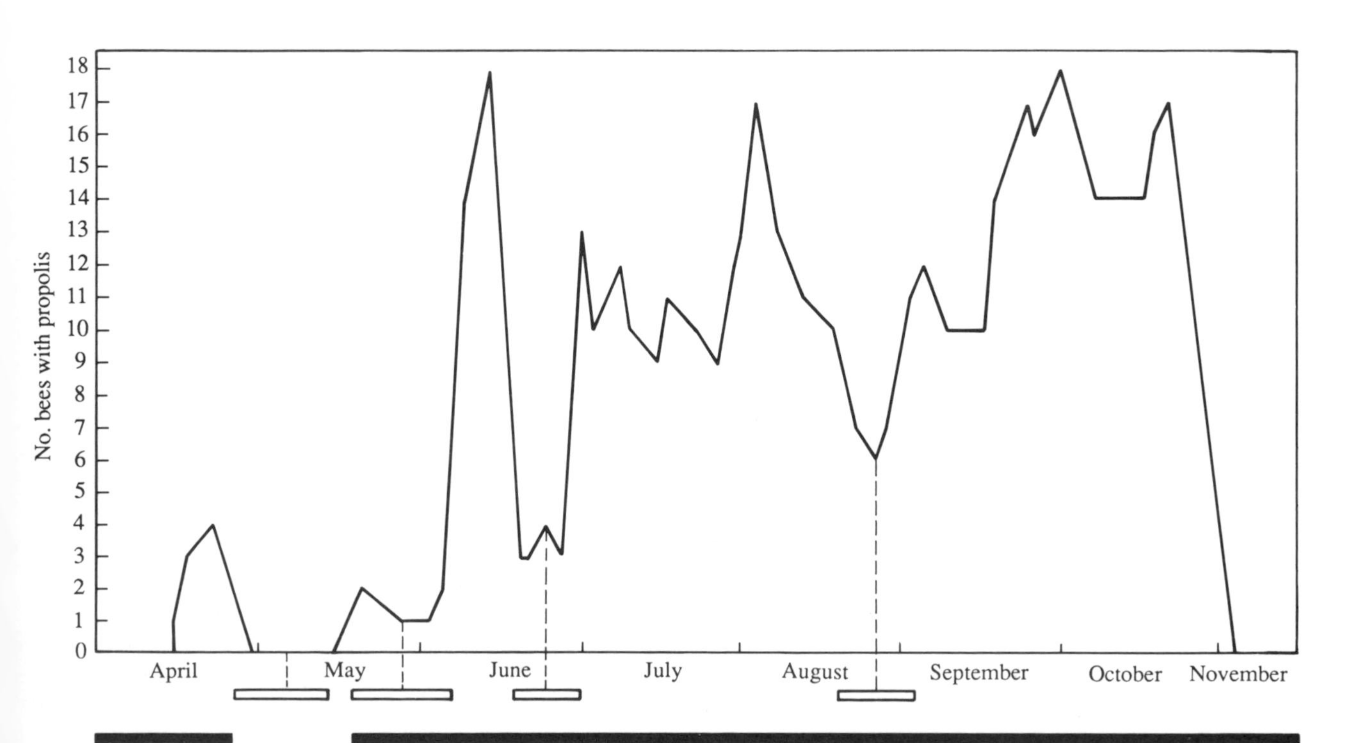

during a strong honey flow'. Meyer (1956*a*) found in Germany that propolis foragers formed a fairly small but steady group of bees in a colony; Marletto and Olivero were able to show that, during a flow, these bees change to nectar foraging. Propolis collection would thus seem to be an activity subsidiary to nectar collection. The interplay between dances for propolis and for nectar are discussed near the end of Section 3.31.4.

Meyer (1956*a*) took propolis from a hive and placed it where she could observe bees collecting it. She describes in detail the method they used in separating off a small amount of the propolis, and in manipulating and transferring it – with their mandibles and all six legs – on to the corbiculae on the hind legs that are normally used for carrying loads of pollen. The procedure was difficult for the bees because of the stickiness and thick consistency of the propolis, and it was possible only when the temperature was high enough – usually above 18°, the temperature below which the bees observed in Italy did not collect it.

During hot weather bees occasionally collect sticky, plastic non-plant materials such as tar, paint, or heavy oil, and use them in the same way as propolis.

No evidence is known that bees materially alter the propolis they collect, before delivering it in the hive (but see Section 14.3). Propolis collected is not necessarily used immediately, and Figure 3.31c shows a bee's load of propolis deposited in a cell. Bees' use of propolis is described in Section 3.31.4.

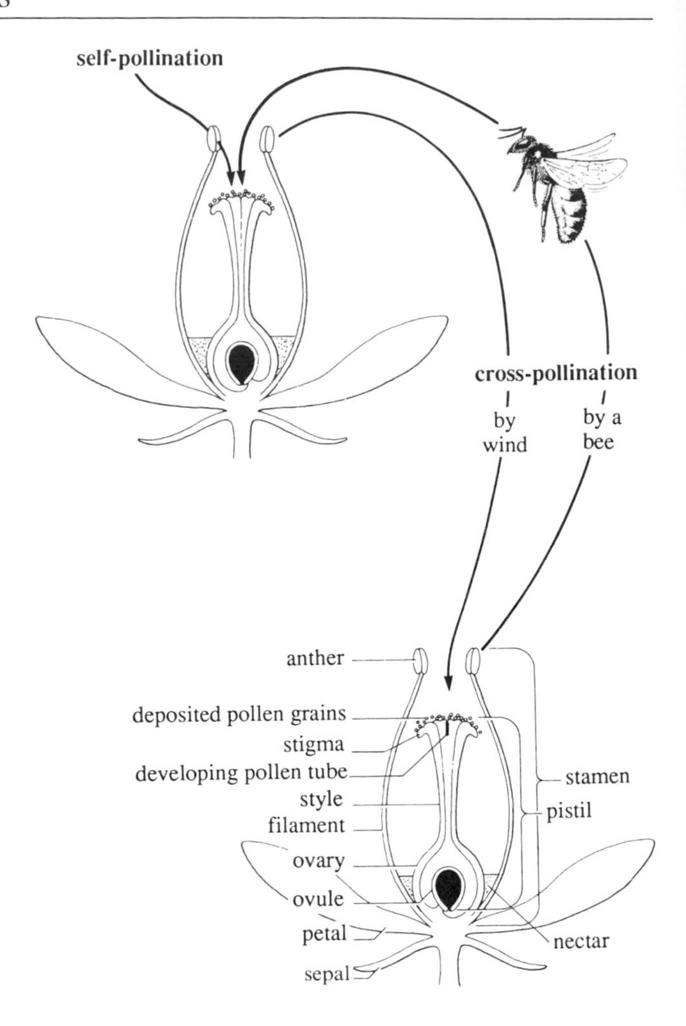

Figure 2.44a The mechanism by which a honeybee pollinates a flower.

2.44 Pollination by honeybees

In their movements among flowers, foraging bees pollinate many of them by transferring pollen grains from the anthers of one flower to the stigma of the same or another flower; see Figure 2.44a.

Pollen grains are produced by the flower's anthers, each of which is at the outer end of the stamen, the flower's male sexual organ. When the pollen grains have matured inside the anther, the anther wall opens and the ripe pollen is discharged; this is dehiscence. The flower's female sexual organ is the pistil, with an ovary at the base, and a style arising from it which terminates in a stigma. A ripe (receptive) stigma has a sticky surface to which pollen grains will adhere if they touch it, and the male nuclei subsequently travel through a 'pollen tube' (extruded from the grain), to the ovary where they fertilize the ovules.

Figure 2.44b shows the various ways in which pollination occurs. In some plants the pollen is automatically deposited on the stigma of the flower that produced it; this is known as *automatic self-pollination* or autogamy. All other plants need a pollinating agent (pollinator) to transfer pollen from the anther to the stigma of the same or another plant of the same species. Some plants, such as grasses and trees bearing catkins, produce pollen grains that are light and dry, and are borne on the wind – perhaps leaving a trail of hay-fever attacks in their wake, but some reaching other flowers. These plants are *wind-pollinated*. A few plants are pollinated by birds or small mammals, or by rain, gravity or vibration. But throughout the world, most plants not pollinated by wind are *insect-pollinated*.

Bees are the most effective pollinating insects, mainly because of their constancy, fidelity, or faithfulness – it has various anthropomorphic names – in foraging. A foraging bee, whether a honeybee or one of the 20 000 species of wild bees in the world, usually moves from one flower to another of the same species, not to a different species as butterflies and many other insects may do. A bee which collects nectar or pollen from a flower is likely to brush against the anthers;

some pollen grains adhere to her hairy body, and some of these will be captured by the sticky surface of the next stigma she touches. If the stigma is on the same plant or the same clone, or genetically identical material (e.g. in grafted apple trees), then what has happened is *self-pollination* (geitonogamy). If it is in a flower on a different plant of the same species, the bee has brought about *cross-pollination*. Self- or cross-fertilization can then follow, in which male and female gametes unite (from the same plant, or from different plants of the same species, respectively): pollination is the necessary precursor to fertilization. In many plants both *self-pollination* and *cross-pollination* occur. Some species bear male and female flowers on different plants, and for these cross-pollination is essential.

DeGrandi-Hoffman et al. (1984, 1986) studied individual pollen grains on the stigmas of apple flowers, and also those caught in the hairs of foraging honeybees. As many as 78% of the stigmas of Delicious apple flowers had pollen grains from plant species other than apple. Even workers with clipped wings, and drones, carried pollen. The authors conclude that some pollen grains must be transferred from one bee to another within the hive, and this implies: (a) that it is not necessary for *the same bee* to carry pollen between compatible cultivars; (b) that social bees, and especially honeybees, would be much more effective pollinators of self-incompatible cultivars than solitary bees, which have no possibility of in-colony pollen transfer. Karmo and Vickery (1987) describe their experiments done in 1957 and 1959, also with apple, that gave similar results.

The management of honeybee colonies for pollination is described in Section 8.4.

Outside pollinating agent	*Source of pollen*	*Description of process*
none	same flower	self-pollination (automatic)
common: insects, wind	same flower or another flower of same plant	self-pollination (by agent)
unusual: birds, mammals, rain, gravity, vibration	flower of another plant of same species	cross-pollination (always by agent)

Figure 2.44b Ways in which a flower can be pollinated.

2.5 LIFE HISTORY OF THE QUEEN

Table 2.41A shows the chronology of the queen's development up to emergence as an adult, and Table 2.5A the chronology of her adult life until she lays eggs.

2.51 Development from egg to emergence

Queens, like workers (Section 2.41), develop from fertilized eggs. They become queens as a result of the different diet secreted and placed in the cells by the nurse bees; these bees adjust the water content of the food for different larvae, and thus regulate its composition. Worker larval food was found to contain only 12% sugar up to $1\frac{1}{4}$ days (it probably remained very low up to $2\frac{1}{2}$ days), whereas the food of a queen larva contained much sugar (34%) throughout the period when it was 1 to 4 days old (Shuel & Dixon, 1959). This high sugar concentration stimulates the larva to eat more of the food, which is provided in excess; the cell is then capped (sealed) with a supply of the food inside it, and the larva continues to feed. Dietz and Lambremont (1970) found that larvae developing into queens consumed 25% more food than those developing into workers. The concentrated secretion is known as royal jelly, and its composition is given in Table 14.41A. Traces of pollen have also been reported in the food of queen larvae (Ribbands, 1953).

Sensory organs on the mouthparts of a queen larva, shown in Figure 2.51a, contain chemoreceptors that react to sugars, but have not been found to react to any other component of larval food. In a paper summarizing many of the factors involved in queen–worker differentiation, Beetsma (1979) describes these sense organs and suggests that the high rate of food intake by a queen larva regulates the activity of the corpora allata – that secrete the juvenile hormone, neotenin – and hence the level of the hormone in the larval blood. (The moulting hormone, ecdysone, which is secreted by the prothoracic gland, is discussed by Hoffmann and Hetru, 1983.) A high level of neotenin during the third day of larval development induces differentiation into a queen; a low level results in the development of a worker. The production of neotenin (juvenile hormone) is inhibited at the end of the 4th and 5th larval instar by activity of the medial neurosecretory cells of the brain.

Experiments by Weaver (e.g. 1962), Shuel et al. (e.g. 1978) and others suggest that the above summary is too simplistic. There is much evidence: for interactions among several dietary components in causing changes in hormone levels, for effects of differ-

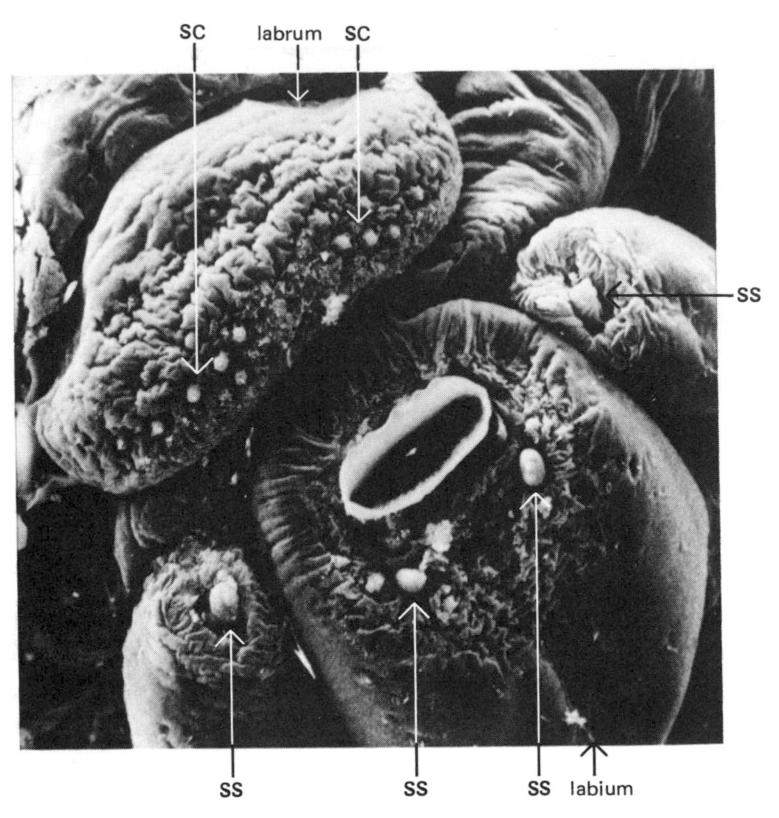

Figure 2.51a Sense organs of mouthparts of queen larva 3 days old (Beetsma, 1979).
The labrum bears two groups of five sensilla campaniformia (sc), and the labium below has two sensilla styloconica (ss) near the duct to the silk glands. Each of the two maxillae bears one sensillum styloconicum (ss).

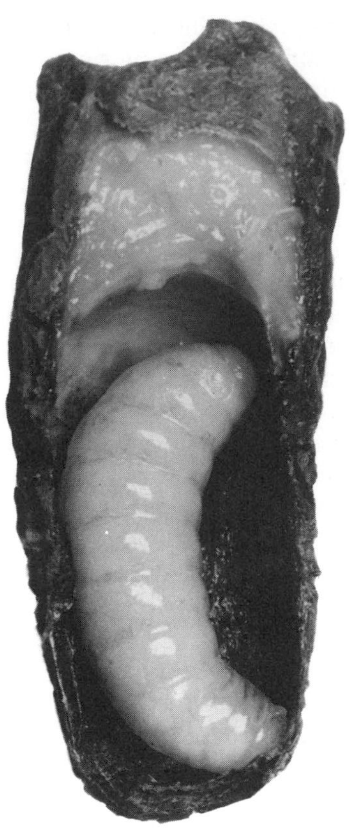

Figure 2.51b Fully grown queen larva, 9–10 days after hatching, which had been sealed in the cell with a store of brood food (photo: G. H. Hewison).

ences in diet on the physiological state of the larva when caste differentation begins, and for changes made in the larva's diet during development.

In European *Apis mellifera* 3 days is the critical age after hatching at which a larva fed as a worker *cannot* develop into a queen, however it is subsequently fed. Most beekeepers who rear queens take care to arrange for larvae to be fed as queens (i.e. in queen cells) from 1 day old at the latest, and preferably from 12 to 18 hours; see Section 8.22. The developmental stages of queen and worker (Table 2.41A) follow almost exactly the same time-scale for the 5 days of larval growth, but the pupal stage is nearly twice as long in workers as in queens. In tropical *A. mellifera* and *A. cerana* the critical age is probably less than 24 hours.

The stimulus that makes nurse bees treat a larva as a queen is the downward orientation of the cell containing it; Section 3.34.2 explains what leads the bees to build such cells. They are usually at or near the edge of a brood comb, where there is space for them to protrude slightly (Figure 3.34a). They are shaped like a thimble or acorn, and are large enough to hold the queen larva and its food reserve (Figure 2.51b). The pupa produces enough of the pheromone that attracts worker bees (end of Section 2.41), to make them cluster over the cell and thus to keep it warm.

Before the queen larva spins its cocoon inside the cell that the workers have sealed up, it orients to gravity, head down. (This results from the high level of neotenin (juvenile hormone), and external application of neotenin to worker larvae reared in experimental cells induced a similar orientation in them, Jay, 1963.) When the queen's development in the cell is complete, she is thus able to leave it – usually by cutting almost completely round its 'cap' – and to join the adult workers of the colony.

2.52 The virgin queen and her mating flight

Table 2.5A shows the ages at which a queen is likely to make her first flight, and to mate and to start to lay eggs. Throughout her life, a queen is fed by nurse bees, perhaps on glandular secretion (royal jelly) such as is given to queen larvae. The ventriculus (midgut) of the queen contains fewer and less active pollen-digesting enzymes than that of the worker, and she cannot digest pollen. However, isolated young virgin queens can feed themselves on sugar syrup or candy – and presumably therefore on nectar or honey in the hive – and live for some weeks, provided they also have access to water.

The queen's first action after emergence from her cell is to seek out any other cells in the colony that contain queens nearly ready to emerge. She opens the side wall of any such cell, and stings the queen inside so that she dies. If the emerged queen has been reared as part of a colony's swarming preparations, the workers may cluster round any other queen cell and prevent the emerged queen from attacking it. (These excess queens fly out (unmated) with separate 'afterswarms', as explained in Section 3.34.4, and in this way the workers can regulate the number of these swarms.)

For several days, a newly emerged queen remains negatively phototactic, and is repelled by the light at the hive entrance. The workers take no notice of her at first, but after she is a few hours old they touch and groom her; they may also start to harass and assault her, and repeatedly chase her round the hive until she tires; usually the chase is followed by increased feeding (Hamman, 1958). The queen also runs about spontaneously, and may walk over workers and behave aggressively towards them – even stinging (and killing) them. In these and other ways she establishes her dominance (E. Weaver & Weaver, 1980). Also, as a result of the vigorous exercise, she becomes capable of sustained flight out from the hive and back again. To a colony, the loss in flight of an individual worker does not matter, but the loss of its new queen is fatal if it has no larvae young enough to be reared as queens.

Hamman (1958) gives a graphic description of the queen's response to the workers' assaults – mostly by running away. By the time she is about 3 days old she becomes more quick and agile in her movements and starts to retaliate. She also shows signs of sexual maturity, opening her sting chamber for a few seconds, also vibrating the tip of her abdomen and dipping it into worker cells. On the day of her first flight, commonly about a week after emergence, she first rushes through the hive, with whirring wings, for 5–10 minutes. She suddenly becomes positively phototactic, seeking and finding the hive entrance. She goes out through it, and flies off, spiralling upward, until she reaches a higher level than that used by workers.

Observations in the north temperate zone suggest that queens are most likely to fly out during the afternoon between 14 and 16 h, when the temperature is above 20°, and when there is little or no wind. Many queens make short orientation flights, lasting only a few minutes, before a longer flight (up to 30 minutes) during which mating occurs. In Austrian experiments (H. Ruttner & Ruttner, 1972) the average range of a queen's mating flight was 2 km from her hive, and it was rarely more than 5 km. Section 2.63 gives the range of drone flights, and Section 8.23 discusses the combined flight range of queen and drone, which is important in determining the location of colonies from which the mating drones come; 20 km is the maximum recorded. Failure to fly is usually due to lack of suitable weather (see below), but failure to mate during a flight could result from the queen's not having flown within the visual range of any drones of the same species. Drones within visual range of a queen in flight fly towards her; when they are close enough, her sex-attractant pheromone draws them to her. The mating process is described in Section 2.7.

Table 2.5A Adult life stages recorded for queens of some races of *Apis mellifera* and *A. cerana*

Entries give the age in days after emergence; developmental stages are in Table 2.41A.
flight = day of first flight
mates = day mating occurs
lays = day egg laying starts

	Emerge	*Flight*	*Mates*	*Lays*
Apis mellifera				
temperate-zone	0	5–6	6–9	8–13
tropical	0		5–6	7–9
A. m. capensis	0		c.5?	
Apis cerana				
temperate-zone	0	4–6	5–7	9?–11
Kashmir	0	3–4	4–6	6–8
India, hill type	0	2–7	5–8	8–12
India, plain type				
Philippines	0	3–4	5–6	7–8

2.53 The mated and laying queen

When a mated queen re-enters her hive, workers are very attentive to her, and feed her. As a result, she usually starts to lay eggs 2–4 days later, although an interval as short as 24 hours has been reported. Oertel (1940) recorded the intervals for 56 queens:

no. days	1	2	3	4	5–8
% of queens	2%	20%	39%	24%	15%

Each time the queen lays an egg, she first carries out a stereotyped sequence of actions. She walks over a comb surface within the brood nest, and when she finds a suitable empty cell she inserts into it both her head and her forelegs, withdrawing them after a few seconds. She then either moves on to another cell, or curves her body and quickly (in 9–12 seconds) pushes her abdomen into the cell she has inspected, and lays an egg on the cell base; then – turning to one side – withdraws her abdomen. Koeniger (1970) showed that the queen uses her forelegs to gauge whether the cell is of worker or (larger) drone size. If the former, she releases spermatozoa from the store in her spermatheca, and one of them fertilizes the egg. If the latter, she does not release spermatozoa. The mechanism of this release is not known, but when a queen was prevented from inserting her forelegs in a drone cell before she laid in it, she released spermatozoa.

Taranov and Ivanova (1946) found that when egg laying was intensive, the queen broke off and rested for 10 to 15 minutes, and took food from 5 to 7 bees. When there were few empty cells, the queen might travel 60 cm in search of one.

The number of eggs laid in 24 hours depends on the age of the queen, the strength and the condition of the colony, the space in the brood chamber, and external circumstances – which probably include both the day length (photoperiod) and its rate of increase or decrease. It also varies between species of honeybees, and is in general lower in most *Apis cerana* than in most *A. mellifera* colonies. I do not know of any information for *A. dorsata* or *A. florea*. The highest daily rate during Nolan's (1925) detailed observations in the USA, during a 12-day period in summer, was about 1600; this was exceptional, but a rate of 2000 a day is sometimes quoted as a maximum. Gary (1975), also in the USA, says that 'a good queen in a strong colony may lay up to and over 200 000 eggs a year' (which would amount to an average of 1000 eggs a day for 200 days); 150 000 eggs a year is more usual, or half a million in a queen's natural lifetime. Moeller (1958) investigated the relationships between egg-laying capacity of the queen and honey production of the colony.

A queen's rate of laying decreases by the time she is 2 or 3 years old, or earlier if she received too little semen during her matings. As she ages, the rate of production of her various pheromones (Table 2.35A) also decreases, and she can no longer 'hold the colony together' as she did in her prime; see Section 3.21. Or an abnormally high proportion of her eggs may be unfertilized, giving rise to drones. As a consequence, the workers are likely to rear a new queen, either to supersede her or as part of the swarming process, unless the beekeeper intervenes by requeening the colony with a young queen.

2.6 LIFE HISTORY OF THE DRONE

Most of our knowledge about drones has come from research work done since 1960; 300 publications on drones were indexed in IBRA Bibliographies (1977*b,c,d*), and there are also reviews by Ruttner (1966) and Currie (1987).

2.61 Development from egg to emergence

Drones develop from unfertilized eggs laid by the queen in drone cells (Section 2.53). These cells are not capped (sealed) until 7 days old; the larvae grow larger, and receive much more brood food, than worker larvae; figures quoted are 384 mg and 159 mg, respectively (Haydak, 1957). Cocoon spinning and pupation follow, as for workers. Less is understood about the food provided for drone larvae than that for workers or queens. 'Drone jelly' fed to larvae 1 to 3 days old is known to differ in several ways from worker jelly or royal jelly. Matsuka et al. (1973) showed that 'modified drone jelly' fed at 5–6 days had the approximate composition of 2 parts drone jelly and 1 part honey, together with pollen; 15 000 pollen grains per mg were found in the food. Haydak's figures suggest a sugar content in the larval food of 7.5% and 24.9% at ages of 1–2 days and 3–5 days, respectively. Some figures for worker and queen larvae are given in Sections 2.41 and 2.51. The death rate is much higher among drone larvae than worker larvae; see Section 3.22.

Drone cells are similar to worker cells in shape and orientation, but the hexagons are wider – and their 'caps' are domed – to accommodate the larger drone; they are usually built in groups near the periphery of the brood nest. The pheromone that attracts adult workers to cluster on the cells and keep them warm is produced in greater amounts by drone pupae than by worker pupae (see end of Section 2.41). Most drones are produced during the reproductive (swarming) phase of the colony's seasonal cycle.

In many ecotypes of *Apis cerana*, there is a small central hole in the capping of sealed drone cells. Its significance is not known, but Hänel and Ruttner (1985) and Ono (1989) have studied its production.

The developmental period is longer for drones than for queens and workers (Table 2.41A), due to an extended pupation period – nearly 5 days longer than for queens. This long pupation period enables larvae of the *Varroa* mite to develop very successfully on drone larvae (Section 11.42).

Diploid drones
The drones reared in honeybee colonies are from unfertilized eggs; they are haploid, i.e. they have only one parent, the queen, and only half the normal number of chromosomes, 16 instead of 32 as in the females (females are diploid, having both a queen and a drone as parents). In a long series of experiments published in *Journal of Apicultural Research* between 1962 and 1984 (e.g. 1969, 1973*a*), Woyke showed that in colonies headed by inbred queens, *diploid* drone larvae hatch from some of the fertilized eggs laid in worker cells. While the larvae are less than 1 day old they secrete a pheromone which attracts the workers to eat them, and the bees therefore do not rear them. Woyke calls the pheromone 'cannibalism substance'. Diploid drone larvae are also produced by inbred *Apis cerana* queens, but they secrete much less cannibalism substance, and are not eaten until they are 1 to 4 days old (Woyke, 1977, 1980*b*).

Woyke (1969) succeeded in rearing diploid drones of *A. mellifera* to adults in colonies; he transferred the larvae to an incubator and fed them there during the critical period when they were producing cannibalism substance (and workers would therefore eat them), and afterwards replaced them in the colony. The adult diploid drones reared could become sexually mature (like haploid drones) at or after 20 days. Chaud-Netto (1978) showed that diploid drones of Africanized *A. mellifera* were attracted by the sex pheromone 9-ODA (see Table 2.35A), and by a virgin queen tethered 10 m above the ground; the drones tried to mount her, but experimental conditions precluded any chance of mating. These drones are in fact functionally infertile, having small testes and little semen (Woyke, 1969, 1973*a*); the cannibalism substance produced by very young larvae thus prevents the colony wasting its food resources on rearing them.

2.62 Adult drones in the colony

After emergence from their cells as adults, young drones need pollen, although they are less able than workers to digest it. Most of the time these drones stay in the brood nest, where nurse bees feed them with a mixture of pollen, honey and glandular secretions. As a result of this feeding they grow; their dry weight was found to increase by up to 28% in the first 4 days, and their nitrogen content by 38–62% (Dietz, 1975). Older drones are fed less often, and workers have been observed feeding honey to drones 7–8 days old. Still older drones spend their time outside the brood nest, and eat honey directly from the cells. Under experimental conditions, a drone lived longer, and was more active, if caged with a worker than if caged with another drone (Morimoto, 1963).

Drones become sexually mature at 12 days or older. The developing spermatozoa of a drone honeybee do not undergo a complete meiosis (cell division by which the chromosomes are reduced from the diploid to the haploid number, i.e. from 32 to 16). The spermatozoa are therefore genetically identical to the mother-queen of the drone. The drone passes on only his mother's genes to the next generation, enabling her to function as the 'father' of the offspring that arise from his mating (Laidlaw, 1986).

Only a very small minority of drones mate with a queen (Section 2.7), and they then die. The rest live on in their colony for some weeks or months. In temperate zones their average life in summer has been quoted as 21 to 32 days. A few drones have been found alive after 70 days, and drones have occasionally overwintered. But normally, when the next dearth season approaches, workers force drones to the outer combs, then off the combs altogether, and finally out of the hive, where they have no access to food, and die. This also occurs in the tropics.

2.63 Drones' flight behaviour

Kurennoi in the USSR (1954) and Drescher in Germany (1969) made detailed studies on the flight behaviour of drones. In the USSR, drones began to make orientation flights at 4 to 14 days, 82–90% of them at 6–10 days. Most drones then made between 1 and 4 flights a day (some up to 7) during the warmest hours, 11 to 17.30 h, and mostly between 14 and 16 h. In Germany (Bonn) it was established that orientation flights lasted only a few minutes, but that when the drones became sexually mature they flew for much longer – 10 to 60 minutes. At Oberursel, also in Germany, Berg (1988) recorded flights by young drones lasting 2 to 4 minutes and those by mature drones 10 to 40 (a few up to 45) minutes. The exercise of flying is necessary to strengthen the muscles of the abdomen and sexual organs, so that when mating occurs there is rapid eversion of the endophallus and ejection of semen and mucus (Kurennoi, 1953). Clear skies and absence of wind encourage drones to fly, and the temperature probably needs to be above 16°. In the USSR drones flew at 9–16 km/h, and often went no further than 3–4 km from their hive; in Austria 6 km was recorded (H. Ruttner, 1976). In Germany, when afternoon temperatures were above 20°, many mature drones spent up to 3 hours flying, making 3 to 5 flights.

Before flying out, a drone elaborately cleans his

body, especially the eyes and antennae, and is likely to eat a little honey (or much, if he is sexually mature). Drones probably orient themselves by landmarks, not by the sun's position. On their return 1% or 2% of the drones may drift to another colony, in which they continue to live.

Drones fly in a higher air space than that of the workers. Its height above the ground is fairly constant as long as weather conditions remain the same, but changes considerably when these are different. For instance the height is inversely related to the wind speed at the time. In the north temperate zone the lower boundary of the drones' air space is quoted as less than 4 m, and up to 10 m; in hotter regions as 2 to 3 m. The greatest number of drones is usually recorded at heights between 15 m and 25 m. The upper boundary of this air space has not been much studied. Drones flying within their regular air space (but not below it) are attracted by a queen; they respond to pheromones secreted by a queen's mandibular glands, whose main component is 9-oxo-2-decenoic acid; see Table 2.35A.

Within their air space, drones also attract each other, whether from the same colony or not, by a pheromone secreted by the mandibular glands; see Table 2.35A, and also Lensky et al. (1985). These glands produce a secretion which fills the central cavity by the time a drone is 3 or 4 days old. The glands then regress, but they store the secretion, releasing it at 7 days of age or more, when the drone starts to fly. A result of the pheromone's release is that many drones in flight congregate together, sometimes in quite large numbers, in specific places. Sexually immature drones are there together with the mature ones.

The term drone congregation area, or drone assembly area, has come to be used indeterminately for both the place on the ground above which flying drones congregate, and the site up in the air where the drones are – whose height varies on different occasions. For clarity, *area* is used here to indicate the specific place on the ground, and *drone congregation site* for the three-dimensional space in the air above this area, where drones are congregating at any one time.

The existence of drone congregation sites was first observed directly by Zmarlicki and Morse (1963), who used Gary's method of attracting drones to a queen tethered or caged high in the air. Currie (1987) gives later references. Drones locate a congregation site quickly, and may fly there on their first flight. Drones flying within the site are more responsive to the queen's pheromone than drones flying outside it, even at the same height.

Drone congregation areas of European *Apis mellifera* have been reported in many parts of Europe and North America, and in South America. Jean-Prost (1957) seems to have given the first clear description of them, although Funke may have done so before 1950 (see Müller, 1950). Cooper (1986) published map grid references to 60 drone congregations reported by dif ferent observers in about 30 counties of Britain and Ireland. In South Africa, Tribe (1982) found many congregation areas of drones of tropical African *A. mellifera*. The largest of them was 1.5 × 1 km, with an extraordinary maximum of 10 000 drones in the site above it at one time. In Europe and North America, sites reported have usually been less than 0.2 km across, and any estimates of the number of drones less than about a hundred.

The map in Figure 2.63a shows two drone congregation areas A and B, about 0.8 km west of an apiary. A radar beam was used to monitor the location of flying insects, including worker and drone honeybees; the display screen was photographed during each revolution of the beam. This made it possible for the first time to monitor the vertical distribution of drones. Both areas were on the slope of a hill, with no obvious topographical peculiarities; the shadow on the map at area B was caused by trees in front of the radar in that sector. Drones flew at sites above both A and B at the same time, and measurements on three separate days recorded a high density of drones at the site above A. On 6 July the site stretched from 12 to 40 m above the

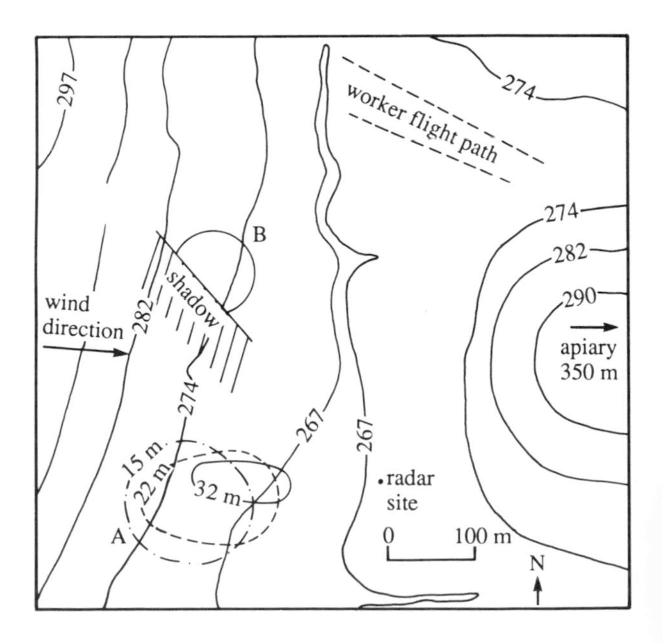

Figure 2.63a Two drone congregation areas, A and B, and worker flight path, near Lawrence, Kansas, USA, 6 July 1985 (based on Loper et al., 1987).
Contours are indicated in metres. The ellipses marked 15 m, 22 m, 32 m, indicate the extent of drone flight at three heights above area A.

ground, i.e. through nearly 30 m vertically. It was slightly more than 100 m across; its extent at three heights is shown on the map. The density of drones at any one height tended to be greatest near the centre of the site. The total number of drones in this congregation site (above 12 m) was estimated at 25 under normal conditions – or up to 60, closer together, when a cage containing two queens was flown from a balloon. Detailed observations suggested that all the drones in the site moved into the comet 'attached' to the queens (see Figure 2.63b), but comets continually and rapidly dissolved and formed again.

Workers from the apiary were foraging in an area north-west of the map. They usually flew too low to be recorded by the radar beam, but part of the worker flight path (marked) was identifiable because foragers had to fly higher there, in order to surmount a wooded hill.

There is much evidence (a) that drones within any one congregation site are likely to have come from different colonies and different apiaries, and (b) that any one drone may fly in more than one of the sites within flight range from his colony. Both (a) and (b) would reduce the chance of inbreeding.

Some drone congregation sites have been recorded above the same area over a period of many years. But in spite of much enquiry, it is still not clear what determines their location. One suggestion is thermal turbulence in the air, due to the nature or the topography of the ground surface below; another is the light intensity distribution seen by drones when they leave their hives: if this attracts them to fly towards a dip in the land-horizon, the dip would become a congregation area.

The drone pheromone released into the air at a congregation site attracts queens, as well as other drones, to the site, and when a queen enters it many drones chase her. They locate her at first by sight, and they may indeed chase anything visually resembling a queen: for instance a queen of another honeybee species (with which they could not mate), or a small inanimate object – even a piece of burned paper rising upwards from a fire.

While the queen flies within the site, up to about a hundred drones pursue her in a 'comet' (Figure 2.63b), attracted now by her sex attractant pheromone (Section 2.52). In observations with a fixed queen, the comet might be behind her, at her side, or even in front of her, but always below, and extending downwind from her. The drones do not follow the queen consistently, and they may suddenly leave her, but if so, the comet soon reforms. Finally one of the many drones succeeds in seizing the queen from behind and mating with her, and this mating is followed in rapid succession by others, as described in Section 2.7.

Figure 2.63b 'Comet' of 18 drones attracted to a caged queen suspended 10 m above ground (photo: N. Gary).

In contrast to queens, only a small proportion of drones mate, and they mate only once, dying immediately afterwards as a result of it. The rest continue to live in the colony and to fly out.

The sex pheromone that attracts drones to a mobile queen is the same for all *Apis* species, although only drones of the same species can mate with her (Koeniger, 1982); see also end of Section 8.23.

2.7 MATING BETWEEN DRONES AND A QUEEN

Much of our detailed knowledge of the mating process in honeybees has been obtained by close observations on queens kept at a suitable height in the air either by tethering them (Gary, 1963), or by fixing them at the outer end of a rotating arm, the inner end of which carried a ciné camera operated by remote control (G. Koeniger et al., 1979). The description below of the mating process follows that of G. Koeniger (1986*a*, 1986*b*); see also Woyke and Ruttner (1958).

A nubile queen in flight responds to the presence of

drones in the vicinity by holding open the entrance to her sting chamber, as in Figure 2.7a, A. One of the drones flying in the comet below the queen (Section 2.63), having succeeded in seizing her, within $\frac{1}{3}$ of a second clasps her between his legs – the forelegs and middle legs on her back, and the hind legs below her abdomen – as in Figure 2.7a, B. During the next second he bends his abdomen, everts the endophallus into the queen's opened sting chamber, and then becomes paralysed; his legs lose their hold, his wings no longer move, and he swings backwards, as in Figure 2.7a, C.

The pair remains connected; part of the drone's everted endophallus is lodged inside the queen's sting chamber, 'like a cork in a bottle', and it is impossible for the drone to drop off. So the queen carries the paralysed drone as she continues to fly, and the drone's slender cervix enters the queen's vagina (G. Koeniger, 1984). After a short pause eversion progresses further (in spite of the drone's paralysis, and through active

Figure 2.7a Three early stages in the mating of a drone and a queen honeybee (G. Koeniger et al., 1979).

participation of the queen, which is not yet understood), and the semen is ejaculated into the queen's median and lateral oviducts, eversion being completed. This final stage is shown in Figure 2.7b in which, however, the boundary between the median oviduct and the lateral oviducts is not apparent because the latter are so distended. The pressure which forces the semen into the queen's oviducts also pushes out most of the rest of the drone's endophallus. The very thin outer layers of the endophallus now become detached from the inner part (the bulb, which includes two pairs of chitinized plates). They slide away with the major part of the endophallus, with the result that the paralysed drone drops from the queen, and he dies. The rest of the bulb remains in the queen's sting chamber, together with mucus (from the drone's mucus glands) which is held between the chitinized plates and a thin membrane of the bulb; some of the mucus protrudes out of the sting chamber and coagulates on exposure to air. A plug is thus formed which prevents escape of the semen, and this is known as the 'mating sign'; see below, and Figure 2.7c.

Within a few seconds, another drone seizes the queen and mates with her in the same way as the first, except that he must remove the mating sign left by the first drone. The hairy field on the ventral side of the vestibulum of his own endophallus is well adapted to do this. The convex side of the mating sign to be removed has a thin transparent yellow layer, and this (together with mucus that is present) forms a sticky mass which probably adheres to the hairy field. The adhering hairs turn the mating sign, and its chitinized plates slide out along the tracks shown in Figure 2.23b, *above right*. The same action probably lifts the mating sign, and thus supports the whole of the sting apparatus and fixes the sting in a safe position between the queen's last tergite and the new drone's endophallus, as shown in Figure 2.7b. This reduces the danger of injury to the endophallus of the drone now mating. Moreover the queen's sting is not straight like the worker's; it is curved just sufficiently to correspond to the shape of the structures involved. (Mating between different species of *Apis* seems to be prevented by the considerable species differences in endophallus structure (Figure 2.36b); Ruttner and Maul (1983) reported that on occasions when an *Apis mellifera* drone tried to mate with a *A. cerana* queen, the chitinized plates of his endophallus appeared to hurt the queen severely.)

After the semen from the succession of drones has filled the queen's oviducts, she bends her abdomen, thus closing the vaginal orifice; the semen is now enclosed within the oviducts and vagina, by the valvefold (Figure 2.7b). Contractions of the oviduct muscles

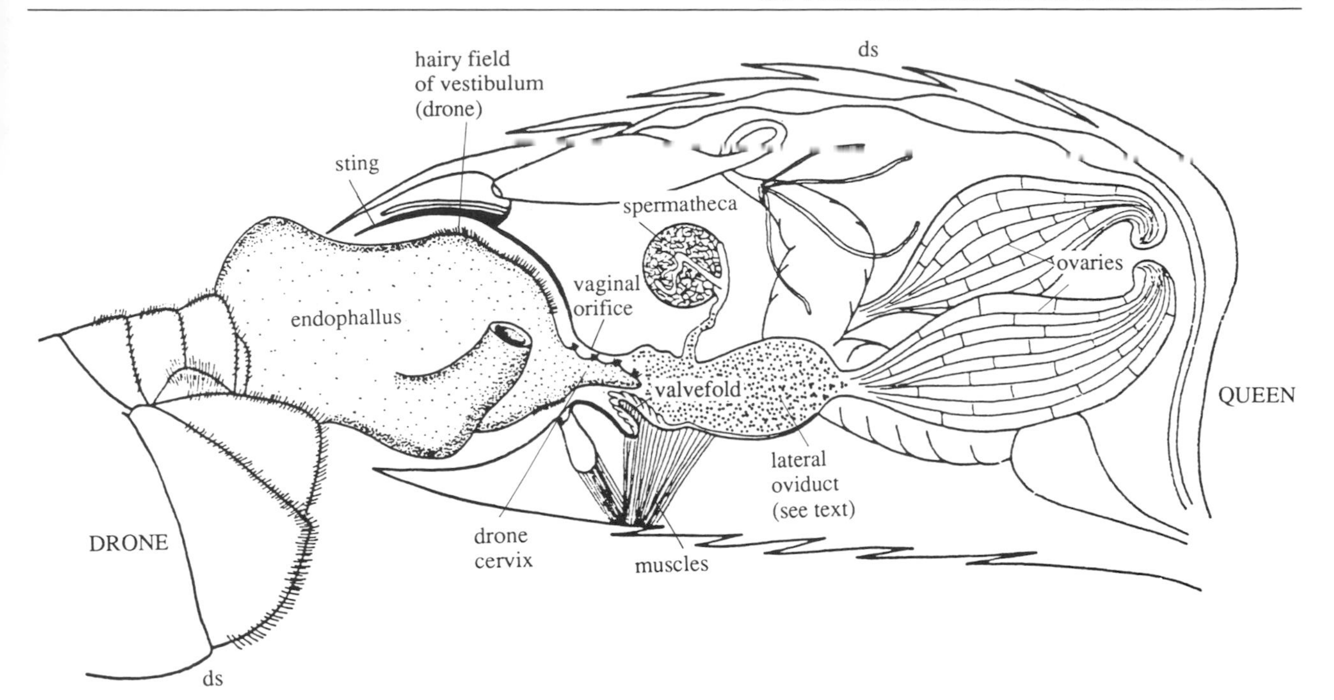

Figure 2.7b Final stage in the mating of a drone with a queen honeybee; the dorsal side (marked ds) of the drone is below, that of the queen is above (G. Koeniger & Ruttner, 1988).

Figure 2.7c Newly mated queen with mating sign still in place (G. Koeniger, 1986*a*).

push the semen into the spermatheca, any excess escaping past the valvefold and through the vaginal orifice in the form of thin threads.

The queen returns to the hive, carrying the mating sign from the final mating drone, as shown in Figure 2.7c. (Long before beekeepers knew that a queen mates many times, beekeepers looked for it in a returning queen, as a sure sign that she had mated.) She frees herself of the mating sign within 3 to 5 minutes by movements of her abdomen, especially by trailing it along a comb until the edges of the cells catch it and drag it out. The workers may pull on it, but dislodge it only with difficulty; they do, however, remove any threads of excess semen (Tryasko, 1951). From a study of 10 000 mating signs, Tryasko (1957*a*) reported that it normally comprises the two pairs of chitinized plates of the bulb of the endophallus (of the last drone to mate); these plates together form a hollow that is filled with mucus from the drone's mucus glands. Occasionally only a lump of mucus is present.

2.8 FURTHER READING AND REFERENCE

Details of publications listed will be found in the Bibliography.

Dade, H. A. (1962) *The anatomy and dissection of the honeybee*, reprinted 1985
Erickson, E. H. et al. (1986) *A scanning electron microscope atlas of the honeybee*
Faure, R. (1979) *Atlas anatomique de l'abeille*
Snodgrass, R. E. (1956) *Anatomy of the honeybee*, reprinted 1976, 1984

The following books relate to both Chapter 2 and Chapter 3.

Butler, C. G. (1954) *The world of the honeybee*, rev. ed. 1974
Chauvin, R. (1968*a*) *Traité de biologie de l'abeille*
Free, J. B. (1982) *Bees and mankind*
Gary, N. E. (1975) *Activities and behavior of honey bees*
Gould, J. L.; Gould, C. G. (1988) *The honey bee*
Hepburn, H. R. (1986) *Honeybees and wax*

Laidlaw, H. H.; Page, R. E. (1989) *Apiculture: introductory biology and husbandry*
Louveaux, J. (1985) *Les abeilles et leur élevage*
Ribbands, C. R. (1953) *The behaviour and social life of honeybees*
Winston, M. L. (1987) *The biology of the honey bee*

3

Honeybees as members of the colony

3.1 THE COLONY AS A SOCIAL UNIT

3.11 Basic colony structure

The social structure of a honeybee colony is a result of the long evolutionary history of social life among the bees. This history, and the development from primitive to highly advanced social organization, are discussed in English in Michener's book (1974) on social bees, in Wilson's (1971) on insect societies, and more briefly in many others. There are books on the social life of honeybees: Ribbands (1953) and Butler (1954), and more recent ones by Seeley (1985) and Winston (1987). Books in French include Volumes 2 and 3 of *Traité de biologie de l'abeille* (Chauvin, 1968*b*, 1968*c*), and *Les abeilles et leur élevage* (Louveaux, 1985).

As in Chapter 2, we consider here the honeybees used for beekeeping: *Apis mellifera* native to Europe, the Middle East and the whole of Africa, and *A. cerana* native to Asia. What is said applies to temperate-zone *A. mellifera*, and much of it to tropical *A. mellifera*; in principle much is true also for *A. cerana*, although often with quantitative differences. Where differences are known, they are indicated.

The terms *colony*, and *nest* and *nest site* have been variously defined. 'Colony' always comprises the adult bees and brood; Michener (1974) does not include the combs and the stores in them, whereas Butler (1954) does, and it is customary to include combs and stores when colony weight changes are discussed, as in Section 12.62. 'Nest site' is the place where the colony lives; this is normally a cavity for either *A. mellifera* or *A. cerana*, although tropical races more frequently nest in the open. (A rare example of the construction by a colony of a propolis envelope to constitute a nest cavity is shown in Figure 3.31b.) A beekeeper provides a nest site for each of his colonies in the form of a *hive*, which has been made to suit his own convenience in managing the bees; nowadays it is usually a box or a tier of boxes already containing framed combs or comb foundation. In nature, 'nest' usually refers to what is constructed by the bees, i.e. the combs (whether empty or containing brood and stores), together with any construction of wax or propolis in the cavity or at the flight entrance. For colonies in hives, combs that are provided by the beekeeper, even if made of plastic, must count as part of the nest. If the entire colony absconds or dies, the nest still remains (until it is destroyed). However, the term wild nest is often used to include a colony, its combs, and honey and pollen stored in them.

An *A. mellifera* colony normally has one queen (the reproductive), up to tens of thousands of adult workers (non-reproductive females), and in the active season some hundreds of adult drones (males), together with worker brood and a small amount of drone brood. Most, but not all, *A. cerana* colonies are considerably smaller; Figure 1.2c shows a fairly large example.

There are exceptions to the above colony composition. For instance a colony may be temporarily broodless, (a) during winter or other dearth periods, or (b) if it is a cast (afterswarm) with a young queen that is not yet mated and laying, or (c) in the course of certain colony manipulations by the beekeeper.

If the queen of a colony dies or is removed by the beekeeper, the colony can rear a new queen provided fertilized eggs are present (*Apis cerana*), or either fertilized eggs or young larvae hatched from them (*A. mellifera*), as explained in Section 2.51. Otherwise, unless the beekeeper 'requeens' the colony, i.e. gives it a new queen (Section 6.51), the colony will die out, even though some workers may undergo sufficient ovary development to become 'laying workers'. The eggs they lay give rise only to drones – except in the race

Apis mellifera capensis in the southern tip of Africa; see Section 3.39.

A colony is often without drones in those part(s) of the year when queen production would not be possible: in the temperate zones during the winter, and in most of the tropics during part of a dearth period. In some areas with year-round forage, drones are always present.

3.12 Colony characteristics in relation to beekeeping

Success for a colony of honeybees is measured by the survival and spread of its genes to future generations. The genes of a queen, or of a worker that is her full sister, are derived from their mother queen and the drone that is their father, i.e. whose spermatozoa fertilized the eggs from which the full sisters developed. On the other hand a drone develops from an unfertilized egg; his genes are derived only from his mother queen (Section 2.62).

The method of *colony* reproduction in honeybees is swarming. The first (prime) swarm is headed by the colony's own queen, and thus introduces the colony's genes to a new area, where this swarm becomes established as a colony. Any afterswarm (cast) – as well as the parent colony remaining in the old nest or hive – is headed by a daughter queen whose progeny have genes inherited from her and from a different father-drone.

Success for the beekeeper is very different, and is achieved by inducing or increasing in colonies the particular behaviour most useful to the beekeeper at any given time, and by suppressing traits that are undesirable to him. During a dearth period brood rearing will have been minimal or absent, but afterwards – as more and more forage becomes available – the colony population grows rapidly and the numbers of both immature and adult bees increase greatly (Section 3.5). A beekeeper engaged in honey production is likely to manipulate these populous colonies so that – instead of making swarm preparations – they increase their populations by continuing to rear brood. Resultant high adult populations at the time of the honey flow can then produce honey greatly in excess of the colony's requirements (Section 6.2), and the beekeeper can harvest the excess. Alternatively, the beekeeper may manipulate colonies so that they forage on the pollen of a specific agricultural crop in need of pollination, in preference to other food sources (Section 8.42).

Heritable characteristics that can increase the honey yield of a colony, and that are convenient for the beekeeper, include the following:

- a colony has a low tendency to swarm, and can achieve a large population before swarming;
- the bees have a highly developed 'hoarding behaviour', storing much honey when forage is available (see Section 6.41);
- they have a highly developed 'hygienic behaviour', which can make the colony resistant to certain diseases and parasites;
- during beekeeping manipulations they remain on the comb, and do not run about in the hive;
- they are gentle; they do not fly around the beekeeper, and are not inclined to attack and sting or to alert other bees to do so;
- they use little or no sticky propolis in the hive;
- in temperate zones, the colony has a good overwintering capability;
- with tropical bees, the colony has a low tendency to abscond from the hive;
- the queen lays eggs in a way that produces a compact pattern of brood on the combs;
- bees cap storage cells as soon as the water content of the honey is low enough.

Section 8.26 discusses the selection and breeding of honeybees with desirable characteristics.

3.13 Demography of colonies in the wild

Demography is the study of population statistics as illustrating the conditions of life. The demography of individual adult workers is considered in Section 2.43 and elsewhere, of immature workers in Section 3.33, and of drones in Section 3.22.

Here we consider briefly the demography of colonies living wild. Seeley (1985) studied the demography of temperate-zone *Apis mellifera* nesting in woods in north-eastern USA. He found that on average 76% of the swarms that established nests in summer died the next winter, whereas only 22% of colonies surviving the first winter died during the second. He postulated that the most successful swarming pattern (and therefore that favoured by evolutionary selection) would be the one in which the total number of swarms issuing from a colony yield the greatest number of daughter colonies surviving through their first winter to maturity in the next summer. Section 3.34.2 (near the end) refers to the short 'window' in the year during which swarms can issue with a good chance of survival in cool temperate zones.

If more swarms issue than can be supported by the forage available in the area, a high proportion will not obtain enough food to last them through the first winter, and they will die. In any one colony, after the prime swarm has left (with the queen that headed the colony), afterswarms with young virgin queens may also issue. The bees in the colony are, however, able to limit the number of afterswarms when conditions for swarm survival are unfavourable, by keeping some or all of the fully developed virgin queens confined in their cells, so that afterswarms cannot be initiated with them.

The production of new colonies by the issue of many swarms has been an important factor in the spread of Africanized bees in South and Central America. The colonies may swarm when they are quite small, and many more swarms issue than from colonies of temperate-zone bees (Table 7.43B). Winston (1979*a*) and Winston et al. (1981) give details.

3.2 QUEEN, DRONES AND WORKERS WITHIN THE COLONY

In honeybees, the queen is the only reproductive. She develops from a fertilized egg, and therefore has two sets of chromosomes and of the genes carried on them. Each gamete (nucleus of an unfertilized egg) she produces has one set of chromosomes, and each is a different mixture of the maternal and paternal genes she carries. Unlike the queen, the drone has no father; he develops from an unfertilized egg, and therefore has only one set of chromosomes and genes. All his gametes (nuclei of spermatozoa) are identical The result of this breeding system is that, in honeybees, full honeybee sisters share more of their genes than do full sisters of other species (including man) which develop from two parents, each having both a mother and a father. This has probably had a profound effect on bee biology and on the evolution of sociality in bees. Wilson (1971) published a useful discussion of these matters.

Some authors (e.g. Michener, 1974) refer to a queen honeybee as a queen when she heads a colony and lays eggs, and as a gyne when she is only potentially reproductive. In this book the queen is referred to as such throughout her life.

In the following text, the queen and drones are discussed first, and then the workers in Section 3.23 and the whole of Sections 3.3 and 3.4.

3.21 The queen

Section 2.35 discusses pheromones, substances that serve for communication between members of the colony. Pheromones produced by the queen (see Table 2.35A) are largely responsible for the coherence of a 'queenright' colony – the fact that the adult bees stay together as a social unit. The queen's mandibular glands secrete a number of substances, abdominal tergite glands secrete others, and glands in the pretarsi (feet) – and possibly other sites – secrete still others.

In queenright colonies, pheromone production from the mandibular glands is greatest in young mated and laying queens, and colonies headed by such queens are the least likely to supersede their queen, to swarm, or – in tropical honeybees – to abscond. The behaviour of a mated laying queen is described in Section 2.53, and Figure 3.21a shows a laying queen in the brood nest, surrounded by young workers. These bees feed her, and repeatedly lick her body and brush over it with the antennae (Allen, 1957, 1960); this is sometimes referred to as retinue behaviour. By so doing the workers obtain minute amounts of pheromones from the queen; these pheromones are translocated on the body surface of each individual bee, and the bee also spreads it by grooming her body. Subsequently she moves around among other bees in the colony. By making antennal and other physical contact, and probably by taking regurgitated food, these other workers obtain a share of the pheromone, with the result that all are aware of the queen's presence. It is likely that 9-oxo-2-decenoic acid is the main queen-presence signal within the colony; see Seeley (1985), Velthuis (1985), Winston (1987).

An immature queen still in her cell, and the recently emerged young queen, produce pheromones that induce in the workers certain queenright behaviours, but not all. Weaver (1987) found that the composition of her pheromones differs from that in a mature laying queen, and a young virgin does not elicit in workers the behaviour they show towards a mature queen. The amount of 9-ODA in the pheromones of a young queen is vanishingly small; it does not become signific-

Figure 3.21a Queen (*Apis mellifera*) with a retinue of young workers round her (photo: M. Ono).

ant until the virgin queen is about 5 days old, but increases rapidly thereafter. Changes in the composition of virgin queen pheromone probably account for the changing behaviour of workers towards her as she ages. A mated queen starts to produce large amounts of the typical adult queen pheromone mixture at about the time she begins to lay, and when she grows old the amount decreases; the colony is then likely to rear queens for supersedure or swarming.

3.22 Drones

The number of drones in a colony is controlled by the workers. The extent of drone rearing is very variable, but it is a feature of strong colonies, of colonies nearing the culmination of their population growth, and of prime swarms after 3 weeks in their new nest or hive. In both colonies in the wild and those in traditional hives, workers can increase drone rearing by building more drone cells, which are larger than worker cells (Table 4.25A). The bees seem to restrict the total number of drones in the colony, although the mechanism is not yet understood; in wild colonies the area of drone comb is reported as 13% to 17% of the total comb area (Seeley, 1985; Winston, 1987). If a managed colony is given some drone cells already built, it constructs fewer new ones than if it had been given none.

A beekeeper using movable-frame hives can greatly restrict drone rearing by providing only worker combs or foundation in hives. A normal colony may alter some of the cells for drone rearing, but it is unlikely that more than 3% to 6% of the colony population will be drones during the active season. Most drones are reared during the reproductive (swarm preparation) phase of the colony, and many of them within a few weeks. In Scotland, when Allen (1965) gave experimental colonies an unlimited area of comb with drone cells, they reared 4 times as much drone brood as colonies whose drone comb was restricted, but they did not use all the drone comb available; the greatest area used was 2580 cm² (one side of the comb in a Langstroth frame occupies 860 cm²). It used to be thought that beekeepers should prevent drone rearing, because drones consume some of the colony's honey without helping to produce it. But Allen found that unrestricted drone rearing did not decrease honey yields, or lead to more swarm queen production, or reduce the number of workers reared.

In queenright colonies, normal mortality of drone brood is always higher than that of worker brood. In Poland, Woyke (1977) found that the difference was greater under adverse conditions:

brood mortality	*worker*	*drone*
summer	11%	13%
spring	19%	25%
autumn	33%	47%

In general, brood mortality was lower in colonies with a good pollen supply or to which pollen combs were given. It was always highest in the early larval stages, and workers ate brood that was not reared; see Section 3.33.

Workers also limit the total number of adult drones in their colony, the total varying according to the season of the year. They can restrict the drone population by reducing the number reared when there are already many in the colony, and by not feeding adult drones. For instance during a 14-day experimental period, Rinderer et al. (1985*b*) found that whereas control colonies of European bees reared on average 3300 drones, colonies to which adult drones were added (e.g. 1000, 2000) reared fewer (2600, 2000); see also Page (1981). When the dearth approaches, workers finally drive the drones out of the colony, where they die (Section 2.62).

Drones take no part in social activities of the colony such as brood rearing and honey production, but they consume a certain amount of food, and their presence plays a small part in raising the temperature. Through pheromones or in some other way, they seem to make a colony more stable and 'contented'.

3.23 Workers

Workers live entirely within their colony during the early part of their life; they are negatively phototactic, shunning the light outside. Workers at this stage, sometimes referred to as 'house bees', carry out a number of activities that are needed to maintain the colony's well-being. Many of the activities involve secretions from a certain gland or glands, so the relative activity of a worker's various glands helps to determine what she can do. For instance only workers whose hypopharyngeal glands secrete brood food can feed young larvae; only workers with active wax glands can secrete beeswax with which to build comb (although others can also build with wax); and only workers that have a store of venom can defend the colony. As house bees become older, they are gradually driven outward from the brood nest by the pressure of emergence of new workers (Ohtani, 1989).

Figure 2.2a shows the location of important glands in the body, and Table 2.35A indicates what they secrete. Different glands have their greatest activity at different periods in the life of an individual worker, although conditions within the colony and outside it influence the exact age of the bee when the glands are active. Quite commonly brood food is secreted by relatively young workers, beeswax by slightly older ones; enzymes required for converting nectar into honey, and for synthesizing venom, are elaborated later. Classical studies by Rösch in Germany (1925, 1927, 1930) showed that an individual worker's activities within the colony are likely to be related to her age. He used the term *Arbeitsteilung*, which is usually translated as 'division of labour', different activities sometimes being referred to by the anthropomorphic-sounding terms 'duties' or 'tasks'. A number of biologists use the term 'polyethism' to designate the full range of activities a worker might carry out.

Many worker activities within the colony can be studied by observing marked bees in a glass hive made to accommodate a tier of single frames one above the other, so that bees on both sides of all combs are visible (Section 8.82). Except where indicated, results below refer to European *Apis mellifera*. Rösch and others (e.g. Lindauer, 1953) showed that in principle, during the part of an individual worker's life spent as a house bee, she remains in the colony and is likely to carry out a sequence of activities in more or less the same order, according to her age, most of which are exemplified in Figure 3.23a:

- first, in the centre of the nest: cell cleaning and capping (glands not yet active);
- still in the centre of the nest: brood rearing and tending and feeding the queen (hypopharyngeal and mandibular glands producing food for brood);
- later, throughout the nest: secreting wax (wax glands active) and comb building;
- still later, on the periphery of the nest: receiving nectar, packing pollen, elaborating honey (hypopharyngeal glands secreting enzymes);
- finally, becoming positively phototactic and at the flight entrance: defending the colony (the venom sac having been filled with secretion from the venom gland).

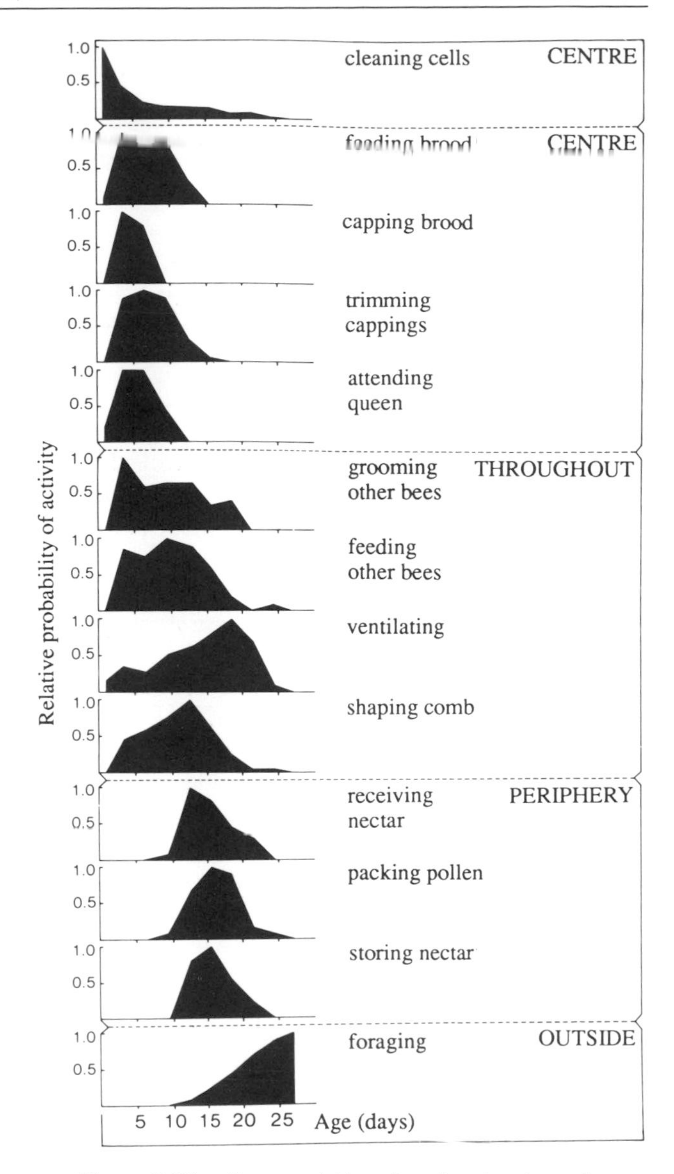

Figure 3.23a Some activities of workers in a honeybee colony (European *Apis mellifera*) in an observation hive, plotted against their age (Seeley, 1985).
All activities shown, except foraging, were carried out within the hive.

Tingek (1986/87) recorded the ages at which he observed *Apis cerana* in Malaysia carrying out different activities:

cleaning cells	11–30 days
cleaning hive	17–30
building combs	13 and on
receiving nectar	11–26 (mostly 11–16)
guarding	20 and on
foraging	23 and on

Figure 3.23b shows where in a hive some of the activities represented in Figure 3.23a occurred during one set of observations. The activities are discussed in more detail in Sections 3.31 to 3.38. During a single day any one worker usually carries out several different activities, including some within different groups in Figure 3.23a. One worker may never undertake a certain activity, and another may become a 'specialist', continuing one activity for the rest of her life; see end of Section 3.38.

Widely varying age-ranges for individual types of worker activity are reported by different observers (see Nowogrodzki (1984) for references). The sum of the worker activities in a colony changes according to the

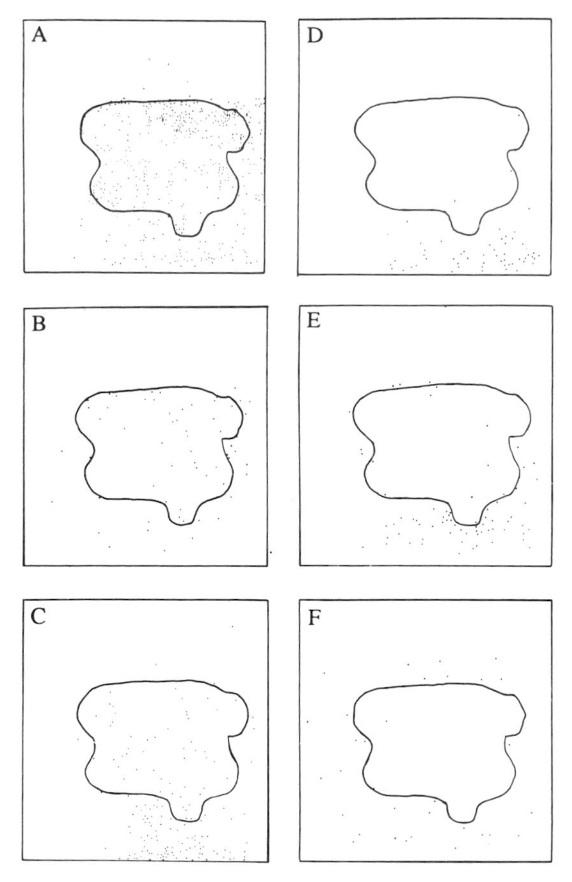

Figure 3.23b Plan of comb surface in an observation hive, showing where certain behaviours were observed (Seeley, 1985).
The flight entrance is at the bottom right, and the brood area is outlined.

A cleaning cells
B feeding other bees
C ventilating
D receiving nectar
E packing pollen
F storing nectar

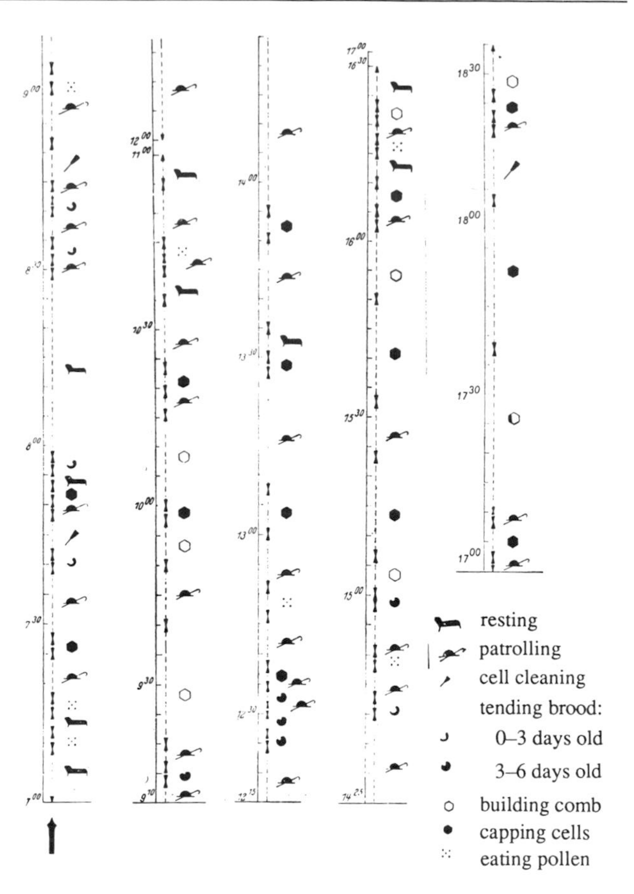

Figure 3.23c Activities of bee 107 from 07.00 h to 18.30 h, at 8 days old (Lindauer, 1953).

needs of the colony, which in turn depend on conditions currently within it and in the bee forage environment outside. It may be possible for bees of almost any age to carry out an activity if the occasion demands it (e.g. Winston & Ferguson, 1985), and regeneration of a number of the glands involved has been demonstrated under extreme colony conditions (e.g. Rösch, 1930). Kolmes (1986) suggests that the most useful comparison is not between activities of individual bees, but between the entire behaviour of a group of bees when they are at one age, compared with their activities when they are at other ages.

Workers within a colony are active throughout the 24 hours of day and night, but Lindauer (1953) found that for more than a third of the time any individual worker rests;* the word 'idle' is sometimes used but seems inappropriate, and her glands may be active in preparation for a future activity. For about a third of the time the bee 'patrols' the combs, until she comes across an opportunity to behave (sometimes under the stimulus of a pheromone) in some way that is consistent with her glandular activity. According to Lindauer's observations, a worker's activity occupies only a third of the 24 hours or somewhat less. He watched one worker, bee 107, for $176\frac{3}{4}$ hours during her life of 24 days; she was caught in a thunderstorm during her first foraging flight. Bee 107 spent all but 10 hours of her life inside the observation hive, and Lindauer was able to apportion the $176\frac{3}{4}$ hours as follows: 52 hours of activity, 56 hours of patrolling, and 69 hours of rest. Figure 3.23c shows graphically the variety of her activities during a single day, and the individual periods occupied with each. For instance she usually

* *Foraging honeybees* spend the nights in the hive, and Kaiser and Steiner-Kaiser (1987) observed their behavioural and neurophysiological responses. The bees remained in one place for long periods, during which their thoracic temperature fell to that of their environment; antennal movements declined, and a stronger stimulus was required to elicit grooming behaviour. These results 'strongly support the hypothesis that sleep occurs in bees'.

spent only short periods feeding larvae, after which she rested. The frequency with which a worker carries out certain activities may be influenced genetically through her father; see end of Section 3.38.

3.3 SOCIAL ACTIVITIES OF WORKERS WITHIN THE COLONY

This section discusses activities in the following order: nest construction and cell capping; temperature control within the nest; brood rearing; preparations for swarming; communication about forage; the handling of nectar, and the production and storage of honey.

3.31 Nest construction and comb building

3.31.1 Wax production and treatment by bees

Worker honeybees secrete beeswax 'scales' from a single layer of cells in their wax glands (Section 2.23.2). The cells are largest and most active in bees about 9 to 17 days old that are in an established colony. In a newly settled swarm – which must build comb before it can store food or rear brood – the glands remain active for a longer period. Table 3.32A quotes the commonly cited temperature of 35° to 37° as optimal for wax secretion, but the maximum rate measured by Hepburn (1986) was at 33°. Bees can work on comb building at 30° to 34°.

The full wax-producing potential of bees is achieved in a colony that has space for comb building, and an adequate continuous income of carbohydrate food; Section 6.43 refers to the benefits to the beekeeper of getting new combs built during the honey flow. Colonies with many young bees – for instance in late spring – readily produce wax, and small colonies produce more in proportion to their size than large ones. In the USSR, Taranov (1959) recorded the total wax production of colonies between 5 July and 8 October, as follows:

colony wt (kg)	0.5	1	2	3	4
wax produced (kg)	0.26	0.41	0.89	1.05	1.32
wax wt as % of colony wt	53%	41%	45%	35%	33%

Seeley (1985) quotes 1.2 kg as the typical amount of wax a wild colony must produce in order to build a nest, and discusses the energy expenditure this involves. Bees secrete 2 μg beeswax per joule of energy expended, the metabolic efficiency being about 8% (Hepburn, 1986). The energy comes from carbohydrate food (sugar), and Hepburn quotes values between 3.5 and 13 kg for the weight of sugar fed to a colony that enables it to produce 1 kg wax. The literature gives many higher figures (indicating lower efficiencies), but they are likely to refer to conditions that were unsuitable for wax production. In winter, and in any queenless colony, the wax cells are relatively inactive.

Casteel (1912*b*) was the first to describe in detail the manipulation of wax scales by bees secreting them, and he noted that there was much reworking of wax even when new combs were being built. At the brood-nest temperature of 35°, newly secreted wax scales are stronger, but less stiff and more easily distended, than wax from comb. As the bees work with wax, they add to it a salivary secretion whose composition and origin are not clearly understood. Less initial energy input is needed to work wax already built into cell walls than to work newly secreted wax scales (see below). This may account for the large amount of reworking of wax that is done by a colony, and for the fact that not all the bees building comb are secreting wax.

Other components are added to brood combs. Silk (a protein, fibroin) is secreted by larvae just before pupation. The silk fibres played out by the larvae are impacted randomly in the wax of the cell walls, so that, by the end of spinning, the walls are covered. The larvae subsequently produce from the anus a colourless material containing no pollen, and then a yellow material containing pollen, and these are applied in turn to the silk base (Jay, 1964). Hepburn and Kurstjens (1988) have shown that the silk fibres greatly enhance the strength and stiffness of brood combs, and their mechanical integrity.

Propolis is deposited on cell edges (Section 3.31.4), and also kneaded with wax for some comb building operations. Old comb may thus contain an admixture of propolis.

Hepburn's *Honeybees and wax* (1986) examines in great detail the mechanical strength of comb in relation to the bees' construction and treatment of it. When bees manipulate wax scales and build comb, they alter the mechanical properties of the wax, making it much easier to work with at the temperatures in the colony. Between 25° and 45°, for working scale wax the energy cost is more or less constant. The energy cost for working comb wax is about the same if its temperature is 30°, but very much less if its temperature is higher (and greater if its temperature is lower). Hepburn refers to the bees' 'superb compromise' in comb building, between the mechanical strength of the material used and the energy required, at a temperature of 35°.

3.31.2 Comb building and comb geometry

The combs in a natural nest of *Apis mellifera* or *A. cerana* are in general parallel, whether each comb lies in a single vertical plane (Figure 1.2c) or not (Figure 3.31a). Except for queen cells which are large, and oriented downwards (Figure 3.34a), all cells in the comb are almost horizontal, sloping up slightly to the comb surface. Drone cells (Section 2.61) are larger than worker cells; widths are given in Table 4.25A.

The pattern of hexagonal cells on each side of a comb results from the random activity of a number of 'building bees' working in close proximity. Darwin (1859) described what happened after he separated two combs in a hive and interposed a vertical sheet of beeswax between them. The bees excavated minute circular pits in the wax, which they extended into shallow round basins of about the diameter of a cell. Wherever several bees began to excavate the wax near together, they started at such a distance apart that the rims of the basins intersected by the time they had reached the cell width. A bee always stopped excavating when the remaining wax was thin, and thus avoided breaking through; Darwin considered that the flexibility of the thinned wax warned them when to stop. He concluded by saying that the work of construction seems to be a sort of balance struck between many bees, all instinctively standing at the same relative distance from each other, all trying to sweep equal spheres, and the building up, or leaving ungnawed, of the planes of intersection between these spheres. When two pieces of comb met at an angle, the bees would pull a cell down and rebuild it in different ways, sometimes reverting to a shape previously rejected. When Darwin gave the colony some wax dyed red, he could see that the bees removed particles of the red

Figure 3.31a Spacing of combs by honeybees, Zimbabwe, 1963 (photo: P. Papadopoulo).
Looking into a deep hive lid in which tropical African bees have built combs; these are more closely and more regularly spaced in the centre than near the periphery where honey is stored.

wax and worked them into the growing edges of the cells around. T. W. Woodbury, who introduced movable-frame hives to Britain in 1862, sent a sample of comb foundation to Darwin; IBRA has Darwin's letter of thanks for 'the artificial comb *which interested me very much*' (Brian & Crane, 1959).

Ulrich (1964) confirmed Darwin's observations, and showed that building bees work independently of each other. He found that comb construction begins with an irregular field of wax spots deposited by the bees on the surface above them. One of the spots is then built on, and grows: the comb is not a structure 'composed of cells'; it is a pattern, and grows as a pattern, and the holding capacity of cells in a pattern of hexagons is achieved with minimal expenditure of energy and material. (A layer of soap bubbles on the surface of water makes a similar minimal-energy hexagonal pattern.)

The modern use of movable frames is made possible by several characteristics of honeybees when building and working on comb. Unlike stingless bees (Meliponinae), honeybees repeatedly use the same combs, and their re-use of old wax has already been referred to. (Old wax tends to be darker than new wax, due to admixtures of propolis, pollen, and cocoon material from cells in which brood has been reared. Section 13.6 discusses the resulting difficulties in extracting beeswax from old combs.)

Secondly, honeybees accept and use comb that has been built by another colony, so the beekeeper can keep the same combs in use year after year. Also, the bees are adaptable enough to build comb on sheets of beeswax foundation, in which case they secrete only about a third of the wax needed for the cell walls, taking the other two-thirds from the foundation. Section 4.31 discusses these subjects further, and also the importance of cell dimensions and of spacing between combs; these differ for different honeybees, and are listed in Table 4.25A.

The use of tiered boxes in modern beekeeping is possible because of one further characteristic of honeybees. In the wild, they build each comb as a continuous unit, down from its support at the top. Nevertheless with framed combs in tiered boxes, the bees will accept and retain a gap of a bee-space between the bottom of the frames in one box and the top of those in the box below it (Section 4.2). Tiered hive boxes can be used without framed combs, but special provision must then be made; see Sections 9.22 and 10.23.

3.31.3 Cell capping or sealing

In modern beekeeping, bees are normally provided with fully built combs, and they build comb only from any foundation supplied to them. On the other hand cells must be capped (sealed) whenever brood is reared, and usually when honey is stored, so cell capping is likely to occupy more of the bees' time than comb building. As with comb building, bees use old wax for capping whenever it is available in their nest or hive. Many bees may contribute to the capping of a single cell, and one bee may even undo what another has done – as Darwin once observed with cell construction.

There are differences between the capping of brood cells and honey cells. Only young bees work on brood cells; Meyer (1952) showed that crumbs of wax were stacked round the edge of a brood cell from the time it contained an egg, so that when the cell was ready for sealing (7–10 days later, Table 2.41A), there was a rim of wax round it which could be used. The actual capping of one brood cell, watched by Meyer and Ulrich (1952), was done in 25 minutes by a single bee. For one brood cell recorded by Lindauer (1953), preparations for capping, and the capping itself, took between 6 and 7 bee-hours, shared between 657 bees, and representing about 60% of the total number of bee-hours devoted to the larva. Some of the bees worked for only 1 to 3 minutes and then moved on to other cells or to other activities. The crumbs of wax were only roughly worked, and the capping remained porous; no pollen was added. After brood cells had been capped (Figure 2.41a), bees nibbled bits from the cappings to stockpile for use elsewhere, and a capping lost 60% of its weight during the 12 days it was on the cell. This behaviour was observed with southern, but not with northern, races in the USSR (Taranov, 1968).

During a honey flow, many cells of honey are capped in a very short time. Meyer and Ulrich (1952) found that old wax was used up first, and freshly secreted wax afterwards. Honey cells were capped by bees of any age; about half of these bees secreted wax, and they did this where they were working. Newly secreted wax is usually white, and very white cappings are produced during a strong honey flow, presumably because any old wax available is soon used up. If there is space enough, honey cells are also lengthened with new wax. Walrecht (1963) suggested that the stimulus to cease lengthening cells, and to cap them, comes from the corridor between the drawn-out cells and comb (or other) surface opposite. When this becomes critically narrow (the bee-space, Section 4.2), and the bees' movements would be impeded, the cells are sealed.

Gubin (1953) reported that northern honeybee races in the USSR leave a space between the honey and the capping, which appears white and dry, whereas southern races fill the cell, and the capping appears dark and wet. He accounted for this on the basis of the larger winter temperature drop in the north, and the greater thermal contraction of wax than of honey.

After honey cells are sealed the bees polish the capping with their mandibles, and add more wax, making the seal more airtight (Meyer & Ulrich, 1952). Propolis may also be incorporated, which is likely to darken the cappings. When the flow is over, long cells are shortened again, and the new wax removed and reused.

3.31.4 Constructions apart from combs

When drawn-out combs are provided in a hive, brace comb and burr comb are built where there are gaps larger than a bee-space; Figure 4.25a (*below*) shows an example. In addition to using beeswax, workers of *A. mellifera* (but not of *A. cerana*) make certain constructions of propolis which they collect from plants (Sections 2.43.4, 12.51); they sometimes mix it with wax. Wild colonies of tropical *A. mellifera* sometimes use large amounts of propolis to provide shade from the sun, possibly where the only nest site available is a very exposed one; Figure 3.31b shows two examples. The nest below was observed from November 1974 to May 1977; the bees left the combs exposed for the first winter, and they then enlarged them and shielded them partly with propolis. In the second summer the combs were enlarged further and completely covered with propolis, except for several entrances up to 20 mm across in the lower half (Tribe & Fletcher, 1977). Air temperatures varied from 34° to −3° during the three years. Papadopoulo (1987) reports a similar propolis-enclosed nest in a tree in Zimbabwe.

In cool temperate zones many colonies use propolis to seal up cracks in their hive or nest cavity, and to restrict or baffle the flight entrance, for instance as shown in Figure 4.43a. In the wild, bees may line their nest cavity with a smooth propolis coating. Upon occasion, when a mammal that enters a hive dies there, but is too large for the bees to remove, it is similarly coated with propolis. Walrecht (1962) regarded the use of propolis by honeybees as somewhat similar to the use of salivary secretions by wasps and solitary bees, in smoothing and finishing off the walls of their nest. From this view point it indicates the 'completion' of a certain part of the nest, after which the bees leave it; the covering of an intruding mammal is seen as a continuation of this activity. Bees often deposit propolis on the edges of cell walls in combs that have been left in the hive for some time and not used. Figure 3.31c shows such cells, and also a small store of propolis just inside one cell.

Meyer (1956*a*) in Germany described in detail the behaviour of bees foraging for and using propolis. A

Figure 3.31b Extensive use of propolis by tropical African bees.
above Propolis shield built by a colony nesting in a disused monkey cage, closing most of the open front which was about 35 cm high (photo: P. Papadopoulo).
below Looking up at a wild nest 3 years old, in which the combs were completely enclosed by a propolis envelope about 85 × 60 × 45 cm, Villeria, Pretoria; some vegetation had been removed to make the envelope visible (photo: D. J. C. Fletcher).

bee returning to the hive with propolis in her corbiculae (Section 2.43.4) went to a place where propolis was being used, and remained there until her loads were taken from her by bees using it; but it was mainly the bees that collected propolis (in the mornings) that worked with it in the hive (in the afternoons). Altogether, honeybees seem to treat propolis differently from nectar and pollen, and it may be that its use by *A. mellifera* is an adventitious exploitation, stimulated by availability of the material and by lack of other foraging opportunities; see Figure 2.43e.

Marletto (1986) observed foragers dancing on the comb after returning with propolis, as they do when returning with nectar. Measurements of the direction and distance indicated by the dances corresponded with the position of a stand of poplar trees. Presumably, during a flow, the dances by nectar foragers give a greater stimulus in recruiting new foragers than dances by propolis foragers.

Hepburn (1986) discusses the relative strengths of propolis and beeswax.

Observations on both temperate-zone and tropical *A. cerana* indicate that this species does not collect or use propolis, although if *A. mellifera* is in the same area it does so (Tamiji Inoue, 1985; Phadke, 1987). (Franssen, 1931, however, refers to an observation of very

Figure 3.31c Propolis storage and use on combs (Marletto & Olivero, 1981).
left A load of propolis deposited at the entrance to a cell.
right Edges of cell walls coated with propolis.

small quantities in *A. cerana* nests in Java.) Where *A. mellifera* uses propolis in the hive if it is available – for stopping up cracks, restricting the flight entrance, and so on – *A. cerana* uses beeswax. Gontarski (1955) found that if *A. mellifera* has no access to propolis, it also uses beeswax instead. It would be interesting to know whether *A. cerana* is able to manipulate propolis, or to collect it; Michener et al. (1978) state that all Apidae can do both. *A. mellifera* experiences difficulties with both activities (Section 2.43.4) and seems able to live without propolis (McGregor, 1952).

Propolis is sometimes used by *A. dorsata* to strengthen the attachment of the comb to its supporting branch (Akratanakul, 1986*b*). Propolis seems to be essential to *A. florea* where and when ants may attack the nests. A colony then builds two rings of sticky propolis round the branch supporting its comb, one to protect each end of the comb, and the bees may 'freshen' the propolis surface each morning to maintain its stickness, so that ants cannot pass, except over bodies of ants already stuck.

3.32 Temperature control during the active season

Table 3.32A lists temperatures at which individual bees can carry out certain activities, also their preferred temperatures, and temperatures maintained in various parts of the colony and nest/hive. Most of the temperatures are likely to differ slightly under different conditions, and for bees of different genetic origin.

An individual bee can carry out certain activities only at a temperature within the range at which a necessary chemical reaction occurs at a sufficient rate. But the bees within a colony can together maintain temperatures that are above that of the air outside. This extends the bees' capabilities, and in particular enables them to secrete wax and build comb, to rear brood, and to elaborate honey. Colonies of temperate-zone races of honeybees (*Apis* spp.) are probably the only insects that can maintain a temperature above that of the air outside, throughout a temperate-zone winter.

A colony of any species of honeybee has a considerable ability to maintain a constant temperature within the brood nest, in spite of great variations in the outside temperature. Temperate-zone *A. mellifera*, which lives where outside temperatures may be very low, can do this more precisely than tropical honeybees. Heran (1952) found that honeybee workers within a colony can respond to quite small temperature changes; they settle where the temperature is at a certain preferred value, which is highest (35.1° to 37.5°) for bees up to 7 days old.

Honeybees can *generate heat* and raise the colony temperature by contracting their flight muscles. When a honeybee is flying, the thoracic flight muscles operate wing movement (Section 2.31); their action also generates heat, since the efficiency is only 10–20%. When the bee is at rest, the flight muscles are uncoupled from the flight mechanism and can be contracted isometrically so that heat is produced but no locomotion. In a colony within a cavity or hive, heat produced is not dissipated as it is from a bee in flight; Seeley (1985) gives a full and quantitative discussion of the energy conversions, and see also Heinrich (1981*c*).

Even a small group of clustered bees can achieve and maintain a higher temperature than a single bee, and this ability is used by *A. cerana* to kill hornets attacking their colony. A group of 180–200 workers formed a ball round individual hornets of one species, and generated sufficient heat to raise and maintain the hornet's temperature at 46° or more, for up to 20 minutes (Ono et al., 1987). This was lethal to the hornet, but not to the bees round it. The larger *A. mellifera* bees also ball a hornet, but cannot achieve a temperature lethal to it, and they sting it instead.

The bees can *dissipate heat* in the colony by spreading out within the nest/hive. They may cluster even outside it, as a 'beard' round the flight entrance (see below). They can also *cool* the colony if there is danger of overheating, by collecting water and increasing its surface area in the hive, and thus the rate of evaporation. Lindauer (1955*a*) showed that they do this in two ways. They spread out the water in the cells, and individual bees hold a drop of it between the two sections of the proboscis that articulate at the cardines (Section 2.21, Figure 2.21a). When the outer end of the proboscis – at its resting position tucked back under the fossa – is swung forward, the drop is stretched out and evaporation from its surface is much increased. Dilute nectar may also be evaporated to dissipate heat. (Figure 3.36a shows the same procedure being used during the conversion of nectar into honey.) For evaporative cooling to be effective, a sufficient current of air is needed, which other bees produce by fanning with their wings; this is most effective if the hive has openings to allow a through current. It is also important that the bees are able to fan outside the hive entrance as well as inside (Southwick, 1987*b*). A 'beard' of bees clustered outside the entrance may also be evaporating water from nectar (Shaparew, 1979).

The book *Insect thermoregulation* (Heinrich, 1981*c*) includes honeybees in discussions of temperature control by individual social insects, and by colonies of them.

Table 3.32A Critical and other temperatures for the honeybee (temperate-zone *Apis mellifera*)

Many of the temperatures vary according to conditions, and the genetic origin of the bees. Entries are based on Heinrich (1981*a*, etc.), Hauser (1987), and other sources.

Normal temperatures

Temperature	
Temperatures inside hive	
	preferred resting temperature:
35°–37.5°	young workers
31.5°–36.5°	older workers (7 days+)
35°	(laying) queen
	temperature maintained in brood nest:
	European bees
35°	eggs and young larvae
32°	pupae and older larvae
30°–38°	tropical African bees
35°–37°[b]	secretion of wax
30°–34°	building with wax
21°?	working with propolis
14°	hive temperature at which colony forms winter cluster
13°	lowest temperature at centre of broodless winter cluster
8°	temperature of outer bees in such a cluster
Temperatures in clustered swarm	
35°	core bees maintained at 35°
17°	mantle bees generally above 17°
Body temperatures of worker in flight	
30°	thoracic minimum (European and tropical African)
7° above air temp.	head, normally
(3° below air temp.	if air temp. = 46°)
close to air temp.	abdomen
Air temperatures outside hive	
19°–30°	optimal for foraging
	lowest for:
20°	mating flights
16°	drone flights
13°–14°	foraging
10°–12°	spring cleansing flights
8°–10°	flight in winter
9°	collecting water in spring
16°	sustained comb building in hive
11°	spring comb building in hive
7°–9°	colony forms winter cluster
5°	minimal consumption of food in winter

Damaging temperatures

Temperature	
High temperatures	
47°	combs break down
45°–60°	(according to period of exposure) workers die
40°	damage to honey-producing enzymes
38°	damage to brood
46°	upper limit for flight
43°	ditto, without regurgitation[a]
37°	bees generally cease to fly out to forage, except for water
Lower temperature limits	
31°	damage to brood
26°	damage to emerging brood
8°	workers on periphery of winter cluster become rigid/immobile
−2° to −6°	(according to period of exposure) workers die
−80°	colony survives 12 h or more

[a] regurgitation of water or nectar produces evaporative cooling
[b] see text (Section 3.31.1)

3.33 Brood rearing, brood mortality, and protein conservation

Sections 2.41, 2.51 and 2.61 deal with the growth of individual workers, queens and drones through their developmental stages, and with the food provided by individual 'nurse bees' to the larvae. The present section is concerned with the social organization of brood rearing, and interactions between brood (immature bees) and adult bees in the brood nest.

All brood rearing starts with the queen's egg laying. A small number of young bees, sometimes referred to as her 'court' or 'retinue', encircle the queen. By feeding her frequently, or not so frequently, these workers initiate and regulate the rate at which she lays eggs. Bees attending the queen do not do so for long. When they are replaced by other bees they move away among workers in the colony, thereby distributing queen pheromone (Section 3.21). The queen normally lays each egg on the bottom of a cell; if for some reason she lays one outside the cells, a worker may move it into a cell (Bee World, 1936*a*).

Honeybee brood secretes a pheromone that is attractive to workers (Table 2.35A), which therefore remain on areas of comb containing brood, and they warm it by generating heat (Section 3.32). A larva also generates heat because it is metabolizing food. An embryo in the egg feeds on the yolk of the egg, and generates only a tiny amount, but it can withstand lower temperatures than a larva or pupa. Pupae generate much less heat than larvae; however, the incubation pheromone they secrete attracts workers to cluster on the capped cells containing them, with the result that their temperature is maintained at or above 32° until they are ready to emerge as adults (Section 2.41). Workers of inbred honeybee colonies are unable to maintain the brood-nest temperature adequately, and this may be one reason why brood mortality in such colonies is high (Brückner, 1979).

Young workers (Section 2.42) secrete brood food, and to do so they need to eat large amounts of pollen which provides them with proteins, minerals, amino acids and vitamins (Tables 14.21A and 14.21B). Worker, queen and drone larvae all feed on secreted brood food (Sections 2.41, 2.51, 2.61). The food contains about 70% water, so when much brood is being reared but rather little nectar is available (Section 2.43.2) – as in early spring at high latitudes – bees secreting brood food need water as well as honey and pollen (Lindauer, 1955*a*). A high humidity must be maintained in the brood nest; workers can regulate the humidity to some extent by fanning, but not if the outside air is too dry, e.g. a relative humidity equivalent to 25% at 34° (van Praagh, 1975). Section 3.31.3 describes the workers' behaviour in capping cells containing fully developed larvae, thus sealing them in.

Some brood is likely to die even in a normal colony, especially in outer brood combs where temperature control is less precise. Amounts of sealed brood, and adult colony populations, are therefore lower than those calculated on the assumption that each egg laid gives rise to an adult bee.

Fukuda and Sakagami (1968) studied the mortality of immature workers in normal colonies during the main honey flow in Japan. Mortality was 15% in the central part of the brood nest, and markedly higher in outer combs – up to 64% in one. Nevertheless the mortality found was incomparably lower than that in most other insects, because of the protection and care the immature bees receive in the colony. In more detail, the daily mortality rate (per 1000 bees) was 14 before the egg hatched, 29 for unsealed larvae, and only 1 for larvae and pupae in the safety of sealed cells. (It then rose to 5 or so for young adult bees in the hive, which faced new dangers, but life was several times as safe for them as for eggs or unsealed larvae. After foraging started, mortality increased very rapidly indeed.)

Brood that does not survive is often eaten by workers, usually as eggs and young larvae, in which the colony has made little or no investment by feeding. This enables the workers to conserve and recycle the colony's protein reserves when these are scarce because of adverse circumstances. A few examples follow. When colonies of Africanized bees face a food shortage and prepare to abscond (Section 3.34.5), workers probably eat all eggs and larvae that remain during the last few days (Winston et al., 1979); the same may well be true for other tropical honeybees. Larvae infected with *Bacillus larvae* do not survive, and some of these are also eaten by workers. Diploid drone larvae less than a day old produce a pheromone (cannibalism substance) which makes them attractive food for workers, and they are all eaten; if these drones had been reared as adults, they would be of no benefit to the colony – see Section 2.61. In experiments by Weiss (1984*b*), groups of workers caged without pollen, but provided with brood as a protein source, were subsequently able to rear brood. In later experiments (Webster et al., 1987), workers ate larvae marked with a radioactive substance, and the royal jelly they secreted became strongly radioactive, showing that the nutrients from the larvae could be directly used for brood rearing.

3.34 Preparations for reproductive swarming (and absconding)

A seasonal period of maximal brood rearing, and consequent large adult populations, is likely to lead an unmanaged colony of any honeybee species to rear queens and initiate other preparations for reproductive swarming. The swarm, which includes a substantial proportion of the adult bees and the colony's queen, leaves the colony and establishes another colony in a new hive or other nest site; see Sections 3.34.2 and 3.42.

Tropical honeybees often succeed in surviving in adverse circumstances by flying out as an absconding swarm, which consists of all adult bees of the colony (Sections 3.34.5 and 3.43). The tendency to abscond varies between species and races of bees, and according to environmental conditions. But absconding cannot occur unless temperatures are high enough during all or most of the year for the flowering of forage plants, and for the bees' flight and comb building.

3.34.1 Definitions

The following definitions are used in this book:

- *Reproductive swarming* (*or swarming*) comprises the sudden departure of a proportion of the adult worker bees of a colony from its nest, with a queen and sometimes also some drones.
- *Absconding* comprises the departure of all adult bees of a colony from their nest, leaving behind whatever brood and stores are in it. Absconding may be due to shortage of food, to disturbance, or to other adverse circumstances.

Migration is referred to at the end of Section 3.34.5.

3.34.2 Initiation of reproductive swarming

Section 6.33.1 sets out conditions that are known from practical beekeeping experience to lead to swarm preparations in colonies. These conditions include overcrowding of bees in the hive (rather than too little space for the queen's egg laying – see Simpson, 1972); poor ventilation; and a queen older than 1 year. Also, some strains of bees are genetically more disposed to swarm than others. Many explanations have been published as to the sequence of events in a colony that determine whether or not it makes swarm preparations, and the timing is referred to below. What follows is based largely on results obtained by Lensky and Slabezki (1981), and their interpretation of these results and of those obtained by earlier researchers. The experiments were done in the subtropics (Israel), with Italian bees, but this account should apply also to temperate zones and, with some adjustments, also to the tropics.

In a normal queenright colony the workers are inhibited from initiating the building of queen cups, and the rearing of any new queens, by two pheromones produced by the queen that heads the colony (Table 2.35A). One pheromone is secreted by the mandibular glands; workers obtain it directly from the queen, while they are tending her (Section 3.21), and distribute it to other workers in the colony by moving among them. The other, footprint pheromone from the tarsal glands, is deposited on a comb surface when the queen walks over it, and is accessible only to bees in direct contact with it.

After the winter dearth, brood rearing in a colony increases as the photoperiod (day length) increases, as explained in Section 3.5. Consequently, the population, and – unless the beekeeper provides more space – the population density of bees in the hive, both increase. Population density was measured as the number of bees in the hive divided by the 'free volume', i.e. the total volume excluding the space taken up by the combs and frames. Using a single-comb observation hive, Lensky and Slabezki established that when the population density was 0.64 bees/ml, the queen moved freely over all the comb, even to the bottom, depositing footprint pheromone everywhere. But at 1.4 bees/ml the colony was becoming overcrowded, and many bees congregated round the bottom of the comb, with the result that the queen did not go there, and no footprint pheromone was deposited. Colonies then built queen cups at the bottom edge of the comb. For 5 colonies in normal hives the average number of queen cups was:

at 2 bees/ml	(threshold)
at 3 bees/ml	4
at 5.5 bees/ml	8
at 9 bees/ml	16

When the population density was higher than 2.3 bees/ml, the number of queen cups was directly related to it.

After this stage, events take one of two courses. If the overcrowding continues and intensifies, eggs are laid in some of the queen cups, and queen cells are built on them in which new queens are reared (Figure 3.34a); this leads in due course to the issue of one or more swarms, as described below. (Although most of the eggs in queen cups are laid there by the queen, workers sometimes move eggs into them; Winston, 1987; Bee World, 1936*a*.) Immediately the swarm has issued, the population density is much reduced, and no further queen cups are built. The alternative course is that the population density eases off: because of less

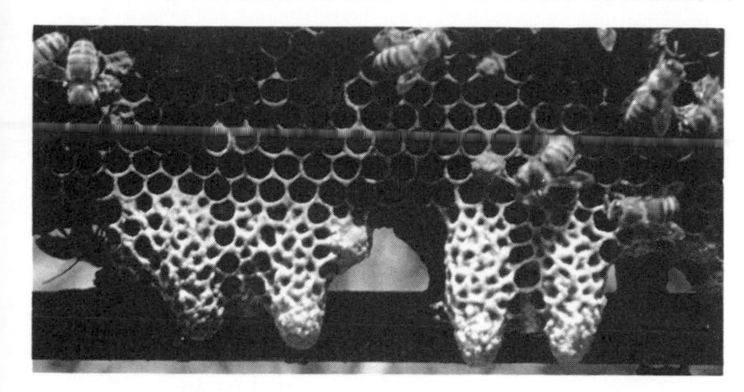

Figure 3.34a Sealed queen cells of *Apis mellifera*, built down from the bottom edge of a comb (photo: I. Okada).

brood rearing, or because the beekeeper removes bees and/or brood or provides more space in the hive. No more queen cups are built, and those present are generally destroyed. The colony does not swarm.

In regions where winters are cold and without forage for bees, there may be only one short period during the summer when swarms can issue, with a good chance of growing into large enough colonies, and storing enough food, to ensure their survival through the next winter (Section 3.13). The short 'window' for the safe issue of swarms may be earlier or later in the season, depending on the vagaries of the weather. For colonies nesting in the wild, a prolonged state of readiness to start swarm preparations (i.e. having queen cups) may be necessary, and colonies may even continue to build queen cups into late summer (Simpson, 1958*a*).

Swarm queen cells are usually built from lower edges of combs, where overcrowded bees cluster, and the queen therefore does not walk. If the queen's pheromone production is ineffective, or the queen dies or is removed, lack of the two pheromones is much more complete, and is not confined to the lower edges of combs. The supersedure (or emergency) cells then built are on the face of the comb; some other differences are referred to in Section 8.21.

3.34.3 Events up to the issue of the prime swarm

The period of swarm preparations is regarded here as that between the deposition of eggs in queen cups or cells (counted as day 0) and the issue of the prime swarm. Assuming for simplicity that all the eggs are laid on the same day, larvae hatch from them on day 3. In some honeybee races including Italian, the prime swarm may issue before day 8, when queen cells are sealed; in other races, including those from northern Europe, the prime swarm issues after day 8, i.e. after the cells have been sealed. If bad weather delays the issue of the prime swarm until after queens in the cells are ready to emerge (about day 16), workers in the colony keep the queen cells closed, except for a tiny opening through which they feed the confined queens.

The period of swarm preparation may thus last for fewer than 8 days, or for more than 16. Behavioural changes observed during this period include the following.

- The queen is fed by fewer workers, and these are very young; as a consequence she loses perhaps half her body weight, and egg laying is reduced. She may, however, continue to lay a few eggs up to the day of swarming.
- Workers increasingly keep their honey sacs filled with honey; Combs (1972*a*) found that the honey-sac contents increased by about 3 mg a day, reaching 40 to 50 mg on swarming day. In other observations (see Severson, 1984) bees flying out with a swarm were found to carry 60 to 155 mg of honey.
- Workers 'shake' the queen more and more often as swarming approaches. This was observed even before day 0, and the frequency increased to 150 times an hour just before the swarm issued (Allen, 1958).
- During the few hours preceding the issue of the swarm, a few of the workers run about excitedly among the others, vibrating their partially spread wings. This movement, which has been named *Schwirrlauf* (whirring run), is a form of communication; it initiates similar activities in workers that are contacted, and finally culminates in the emergence of the prime swarm, comprising workers, often a few drones, and the queen; workers have been aggressively manoeuvring her towards and out of the flight entrance.

Section 3.42 follows the life history of the swarm in transit until it becomes established as a new colony.

The identity of the workers that fly off in the swarm does not seem to be predetermined, although there is some indication that bees in a swarm are more likely to be full sisters than half-sisters (with different drone-fathers; see Section 3.38, under Antennation). It may be that most of the bees – except those too young to fly – leave the hive, and that some return (Simpson, 1972). Many bees engorged with honey are found in the hive after the swarm has flown off. Also, when a swarm has been put back in its hive and another swarm quickly issues from it, the new swarm is not composed of the same bees as the first.

There is a great variation in the number of bees that go with the prime swarm (see Table 3.34A), and in the proportion that these represent of the total number in the colony – which may be about half. Data on 59 swarms showed that only three swarms contained more than 3.5% drones (5%, 6%, 18%); the average

Table 3.34A Estimated number of workers in swarms issuing from 8 unmanaged colonies of European *Apis mellifera*, Kansas, USA

Based on Winston (1980).

Colony	Prime swarm	Afterswarms			Total
		1st	*2nd*	*3rd*	
1	11 676	11 076	unknown	—	>11 076
2	13 529	6 091	4086	—	35 382
3	unknown	—	—	—	unknown
4	21 818	10 608	—	—	34 426
5	14 824	13 778	3765	4296	36 663
6	17 260	13 127	unknown	—	>30 407
7	12 978	14 625	—	—	27 603
8	20 143	11 458	—	—	31 601

was only 1%. Avitabile and Kasinskas (1977) give details.

3.34.4 The parent colony afterwards

The parent colony left in the hive has perhaps half the colony's original adult worker population, and some brood. It has lost its own queen to the swarm, but is likely to have several immature queens in sealed or unsealed queen cells, which will be ready to emerge in about 6 to 10 days, according to the time interval between the eggs being laid in the queen cells and the issue of the prime swarm. If the swarm was delayed, the emergence of queens from their cells will already be overdue. The workers may allow only one new queen to emerge, in which case she will head the parent colony. If conditions are good for swarm survival, they allow several queens to emerge, one at a time, delaying others until the emerged one has flown out with an afterswarm. (If two or more virgin queens are free in the hive when an afterswarm flies out, they may all fly with it; Avitabile and Kasinskas (1977) found 2 or 3 queens in 10 out of 59 swarms.) Table 3.34A exemplifies the variation in the number of afterswarms, and Section 3.13 discusses what happens in unfavourable conditions.

The workers' behaviour is influenced by pheromones from any emerged virgin queen, and also by her 'piping'. She generates short pulses of high pitched sounds (the German verb is *tüten*, to toot) which can be heard even outside the hive. She produces the sound by vibrating her folded wings while pressing her body against the comb surface, and the vibrations are transmitted through the comb (Simpson, 1972); workers have mechanical receptors which perceive them – see Section 2.34. The immediate effect of the piping is to make the workers 'freeze' on the combs. Queens held in their cells may also respond, by piping. But the acoustics of a closed cell are not the same as those for a free queen on a comb, and the pitch of the piping is lower (the German verb is *quaken*, to croak). Simpson and Greenwood (1974) showed that piping-vibrations could induce swarming.

The queen that remains in the parent colony will fly out and mate, and start to lay eggs 1 to 2 weeks after the prime swarm issued. This colony has lost many adult bees to the swarms (Table 3.34A), and some of the remaining brood dies; Winston et al. (1981) found a mortality of 40% or more in eggs and larvae. The size and honey-getting ability of the parent colony are thus likely to remain low for several months.

Preparations for reproductive swarming are discussed in more detail by Simpson (1972 and other papers), Lensky and Slabezki (1981), Severson (1984), Seeley (1985) and Winston (1987).

3.34.5 Initiation of absconding

Absconding is a common feature of bees that evolved in the tropics, including races of *Apis mellifera* and *A. cerana*; it enables the colonies to survive even when they are disturbed by marauders, and when there is no longer any food within the flight range of their foraging bees. It is a survival strategy not available in temperate zones with a cold winter.

Absconding due to disturbance

All the adult bees of the colony leave the nest or hive, quite suddenly or within a few hours, leaving behind them any brood and such stores as they cannot carry. They usually find a new nest site not far away, generally within 10 m.

Disturbances likely to make colonies abscond include attacks by predators (frequently ants), wax moths or other pests; exposure of the nest to excessive sun or rain, or to nearby fire or heavy smoke. Excessive manipulation by the beekeeper is another possible cause, but some say only an occasional one. Most tropical honeybees may abscond if their nest becomes too cold or wet, or too hot. Verma (1984) and others mention disease as a cause of absconding. In Bangladesh, Khan and Razzaque (1981) reported the presence of *Bacillus alvei* and *B. circulans* in honey stores of *A. cerana* colonies that absconded. Colonies of *A. florea* are very likely to abscond when their nest gets too hot, and to build another in the shade. They also often abscond when their honey is taken, and build a new nest site not more than a few metres away (Section 8.62).

A colony established in a small nest or hive may well abscond when it outgrows the space (for tropical *A. mellifera*, see Fletcher, 1975/76).

Although it is quite abnormal for European *A. mellifera* to abscond, P. Martin (1963) did succeed in getting small experimental queenright (but not queenless) colonies to do so by disturbing them with dripping water or repeated application of smoke or repellents.

Absconding due to shortage of food

This is a characteristic of tropical honeybees in areas where – because of a difference in altitude, or for some other reason – plants are in flower at different places during different seasons of the year. (It does not normally occur in a region of homogeneous forage that provides food through most of the year. For instance beekeepers in the rich cultivated plains of peninsular Malaysia told me that they rarely encounter it.) In a dry season, shortage of water may induce absconding.

Absconding of this type sometimes occurs in *Apis cerana* within 24 hours of the onset of a sudden dearth (Muid, 1988), but it is usually an orderly affair prepared for in advance, as shown in Figure 3.34b. Reduction in brood rearing had started 20 to 25 days before the colony absconded, which did not occur until all the sealed brood had emerged as adult bees which could fly in the swarm. For the final 10 days no eggs had been present, any eggs laid by the queen having been eaten by the workers as described in Section 3.33. Some absconding swarms leave eggs and stores behind, and Section 7.41.4 gives details for *A. cerana*; see also Woyke (1976*b*).

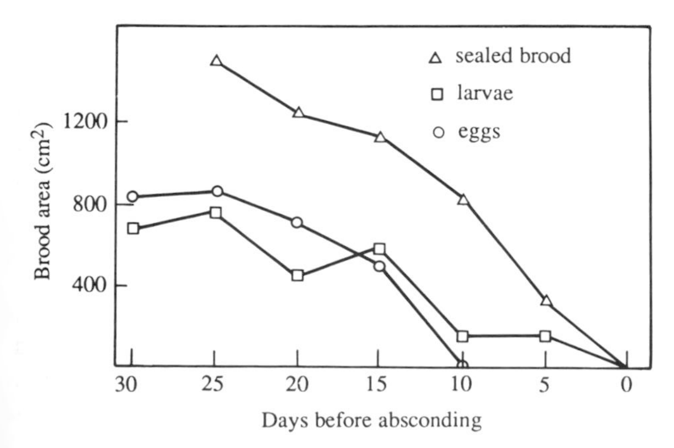

Figure 3.34b Reduction of brood rearing in a colony of Africanized bees preparing to abscond because of food shortage (after Winston et al., 1979).

It seems certain that in colonies preparing to abscond from an area where food is short, scout bees fly for long distances in search of a new area where good forage will be available. Seeley (1985) calculated that an individual scout bee of tropical *A. mellifera* could reach a point up to 55 km away, and return, on one load of honey. The absconding swarm in transit to a new nest site is discussed in Section 3.43. In a detailed study on tropical African honeybees in Kenya, Kigatiira (1984) found that the distance and direction of the route taken by the swarm appeared to be communicated within the colony by dances before the colony left. Moreover the direction of flight was the same for all absconding colonies at the same experimental site in any one season, but it was different at other seasons. Pre-absconding behaviour could be communicated to nearby colonies, with the result that they also absconded.

Temperate-zone honeybees, taken to an area where native tropical honeybees thrive, may well be at a disadvantage because they do not respond to failure of forage by absconding to a new area, as tropical honeybees do, and they are likely to starve as a result. This is true in parts of South America, where managed colonies of Africanized bees abscond, but those of European bees do not (Winston et al., 1979). In Europe P. Martin (1963) tried many times, but failed, to induce absconding in small colonies of European honeybees by starving them. On the other hand Weaver (1988) reports that in Texas, USA, when the weather gets hot, queen-rearing nuclei (very small colonies) often abscond – whether they are in very small hives or on 2 or 3 combs in a standard hive. They are less inclined to abscond if they have sealed honey, if they have larvae, and if they have fairly large populations in relation to the size of the hive. But most abscond by the time the weather becomes really hot. However, in one apiary in Yucatán, Mexico, all colonies (in standard hives) had only about 2 combs of bees, yet none absconded.

Nomadic absconding

This term is sometimes used for a periodic movement of colonies induced by food shortage. For example on the island Ko Samui in south Thailand, it is triggered in *Apis cerana* by the alternate flush of flowering on the central hills and on the coastal plains; see Section 7.41.1. Another example is the movement of tropical *A. mellifera* up and down the sides of the Rift Valley in Ethiopia and Kenya.

Migration
A periodic movement of colonies of honeybees is referred to as migration by some authors; see for example the submissions to the Third International Conference on Apiculture in Tropical Climates (Kigatiira, 1985). In the present book, after consultation with Rinderer (1987), the term is restricted to regular seasonal movements of colonies (which have probably reached a broodless state) that result from heritable responses to geophysical cues, not those that result from *direct* responses to lack of food which may also result from the geophysical changes. If true migration occurs at all in honeybees, it is confined to *A. dorsata* and *A. dorsata/laboriosa*.

3.35 Communication about forage

Honeybees find their forage outside the colony in the light of day. But the centre of communications, where a forager learns what she should collect and where to find it, is in the darkness within the colony. Information is transmitted to foragers both by house bees taking their loads of nectar or water, and by recently returned foragers performing recruiting dances on the comb.

When a foraging bee returns to her colony with nectar, honeydew or water, she passes her load to any house bee that is willing to take it from her. The entire load may be passed to one bee, although it is usually distributed among several; see Section 3.36. Lindauer (1955*a*) found that when the forager's load was taken from her quickly, within 40 seconds, she performed a communication dance that recruited more foragers to the same source (see below). If the load was not taken so speedily she did not dance, but she flew out to collect more from the same source. If she had to wait 10 minutes or longer to get rid of her load, she did not even return to the food source, and was then susceptible to the influence of recruiting dances by other foragers that had brought in loads more acceptable to the house bees. Seeley and Levien (1987) examined the co-ordinated nectar foraging of a colony, and the colony's ability to adjust the distribution of its foraging according to the ever-changing nectar supply in its surroundings. They then set out a 'systems view' of colony integration in foraging.

The recruiting dances, performed on the vertical comb surface, consist of repetitions of a stylized cycle of movements, each cycle lasting a few seconds. Bees nearby 'follow' a dance, and by performing it a forager communicates to them the direction and the distance of her food source. Provided these bees are able to fly, they can then find the forage source indicated by the dance, even if they have not themselves previously foraged or danced.

When the dancing bee was foraging, she registered the sun's position in the sky, i.e. the angle between her own flight path from her colony and a horizontal line from the colony in the direction of the sun. She remembers this, and if she found good forage on this

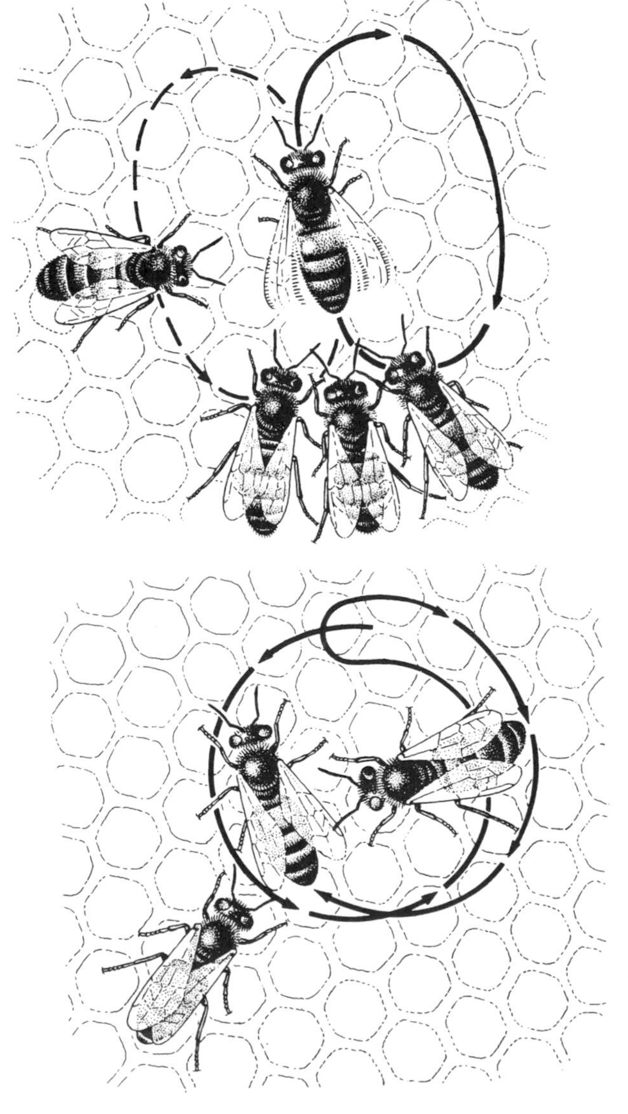

Figure 3.35a Communication dances by a honeybee returning from foraging (after von Frisch, 1967).
above Dancing bee on the 'wagtail' run of a figure-of-eight dance, with 4 'followers'.
below Round dance, with 2 'followers'.

flight path, she carries out a 'wagtail' or 'figure-of-eight' dance on the comb (Figure 3.35a, *above*). She moves her abdomen rapidly from side to side while she is on the wagtail run, and the direction of the wagtail run bears the same angle to the upward vertical on the comb as her flight path bore to the sun's azimuth. The bee 'translates' the horizontal direction of the sun to the vertical direction of (or rather against) gravity. 'Following' bees can sense and remember the vertical angle, and can fly out from the colony in the same horizontal direction as that taken by the dancing bee that found the forage. Information on the flight direction is thus communicated from bee to bee.

If the sun is overhead or within 2° to 3° of this position (which occurs within the tropics), honeybees cannot communicate the direction of forage by dancing on the comb (Lindauer, 1957*b*). Some of the resultant daytime complications for bees near the equator are discussed by New and New (1962), and night-time complications by Edrich (1981).

The dancing bee has also registered her energy expenditure on the flight. In normal circumstances, this gave a measure of the distance she has flown, but if she flew steeply up or down hill, or with or against a strong wind, the relationship would be slightly distorted. The tempo of her dance on the comb is rapid if the source of food was close by, but lethargic if it was far away, or she may not dance at all. There is a direct relationship between the time for one wagtail circuit and the distance of the food. For European *Apis mellifera* von Frisch (1967) found it to be:

duration of each wagtail circuit (seconds)	2.1	2.5	3.3	3.8	5.6	6.3
distance (m)	200	500	1000	2000	3500	4500

Information on flight distance is thus communicated from bee to bee. The efficiency of this communication is quite high. In some experiments where a rather low efficiency might have been expected, of 339 bees that 'followed' one marked dancer returning with food from 100 m, 152 were kept under continuous observation. Of the 152, 56 remained in the hive and 96 flew out, of which 58 (60%) found the food, most of them within a few minutes of leaving the hive.

A forager returning with nectar has also stored in her memory the scent, colour and shape of the flowers, and the time of day at which she found the nectar. When she dances on the comb, her followers perceive the scent of the flowers, and they produce a 'stop' signal, a sound which makes the dancing bee pause, regurgitate nectar and offer it to followers. The frequency of the sound is specific for each species: *A. mellifera* 300, *A. dorsata* 375, *A. cerana* 445, and *A. florea* 475 per second (Towne, 1985). Two characteristics of flowers that the forager cannot communicate (although she remembers them) are their shape and colour.

If a colony is in danger of becoming overheated, and the bees urgently need water to use for evaporative cooling (Section 3.32), nectar loads may be spurned by house bees, but loads of water taken very quickly (Lindauer, 1955*a*). As with nectar, if a house bee takes a forager's load of water within 40 seconds, she dances to recruit other bees to forage for water before she fetches another load. If the interval is longer, but not more than 10 minutes, she herself collects more water but does not stimulate other bees to collect it (see Section 2.43.2).

If a forager found nectar or water very close to the colony, less than perhaps 25 m, she performs a cycle of simpler movements known as a 'round dance' (Figure 3.35a, *below*). This may be regarded as communicating a general excitement, alerting 'followers' to the presence of the forage, together with the scent of the food. When the source is 25–100 m away, movements are likely to be transitional between round and wagtail dances.

Dances by returning pollen foragers also vary according to the location of the source, in the same way as those by nectar foragers (von Frisch, 1967). A bee may dance before she deposits her pollen loads in a cell, or afterwards, or before and after. Pollen has a scent which differs from one plant species to another, and the scent of pollen is not the same as that of the flower producing it. 'Followers' of pollen dances pay special attention to the pollen loads, touching them many times with their antennae and perceiving the pollen scent. They then forage on plants of the same type (von Frisch, 1967).

Foragers returning with propolis loads seem to perform dances rather similar to those of nectar, water or pollen foragers. Marletto (1986) measured their direction and tempo, and found that these corresponded with the position of a stand of poplar trees. He also showed that propolis foraging was suppressed during nectar flows, as shown in Figure 2.43e; nectar dances presumably provided a greater stimulus to recruits than propolis dances. Collecting loads of propolis takes much time, and the foragers sometimes return to the colony, where they perhaps feed, before their loads are complete, and fly out again after an interval to add more to the loads (Meyer, 1956*a*). Pollen foragers also do this, and they have been observed to dance while in the colony before they return to collect more pollen; this would recruit extra foragers without too great a delay (von Frisch, 1967).

In wagtail dances for nectar or syrup, the dance

tempo that indicates a certain distance is known to vary between different species (and even races) of honeybees. In general it is faster for bees that have a longer foraging range than for bees that cannot fly so far. For forage at 500 m (Lindauer, 1957*a*; F. G. Smith, 1958; von Frisch, 1967), results indicate the following durations of a wagtail circuit:

European	
A. m. carnica	2.3 seconds
A. m. ligustica	2.8
tropical African	
A. m. scutellata	3.6
tropical Asian	
A. cerana	5.0
A. florea	no dance; (but for 300 m, 3.8 seconds)
A. dorsata	2.7

In Sri Lanka, Punchihewa et al. (1985) studied dances and flight ranges of the last three species more extensively; *A. dorsata* was rather similar to *A. mellifera* in Europe; *A. cerana* did not forage beyond 600 m, *A. florea* not beyond 300 m.

Many books on the social behaviour of honeybees give fuller accounts of the subject of this section. The most detailed (and very readable) book in English is von Frisch's *The dance language and orientation of bees* (1967) translated from *Tanzsprache und Orientierung der Bienen* (1965). Some of the many complications in the interpretation of the honeybee dances described, and of the bees' responses to them, are discussed by Krebs (1975) and G. E. Robinson (1986). Dyer (1987) has summarized information obtained more recently on *A. florea*.

It has been known for some time that tropical African *A. mellifera* forages on moonlit nights (see e.g. Fletcher, 1978). Dyer (1987) has shown that *A. dorsata* can forage when the moon is between half-full and full, and that when she dances, her wagtail run indicates the direction not of the moon, but of the (hidden) sun.

This section presents the generally acepted hypothesis that honeybees are able to communicate information about forage by 'dances'. Among those who do not accept it, Rosin (1988) summarizes the counter-arguments.

3.36 Food and water storage, and the production of honey

Section 2.43 and its subdivisions describe how foragers bring materials to the colony. The subsequent storage of pollen, honey and water within the colony is discussed here, and propolis storage is referred to in Section 3.31.4.

Pollen

Pollen is stored in cells on the periphery of the brood nest (where it is accessible to nurse bees, Section 2.42), but not greatly in excess of requirements. The foragers themselves deposit their loads in the storage cells, as described at the end of Section 2.43.3, and house bees do the rest of the processing. They tamp the loads down tightly, which helps to exclude air; they incorporate with the pollen a little regurgitated honey, which is microbicidal (Section 13.23.1). They also add – possibly from the hypopharyngeal or mandibular glands – salivary secretions containing a substance that inhibits germination and bacterial spoilage of the pollen. The bees often place a thin layer of honey over the pollen packed in a cell, which keeps it out of contact with the air, and then cap the cell. The pollen with nectar or honey added is probably protected from spoilage by undergoing (anaerobic) lactic fermentation. Pollen processed in this way is sometimes referred to as 'bee bread', a name given to it before the significance of pollen as the bees' protein source was understood.

Honey

The colony is able to store energy-foods by converting nectar and honeydew into honey, which has a high sugar content; see Section 13.21. Park (1925) gave the first detailed description of the process, which is illustrated in Figure 3.36a, and Maurizio (1975*a*) summarized what had been discovered during the next 50 years.

When nectar or honeydew is brought into the colony, the foraging bee has already added to it secretions from the hypopharyngeal glands which contain enzymes used in elaborating honey: invertase, glucose oxidase and diastase. The vital part played by the first two in producing honey is explained in Sections 13.21 and 13.23.1. But no water has yet been evaporated. In the hive the returning forager passes the nectar or honeydew in her honey sac to one or more house bees (Figure 3.36a). The forager opens her mandibles widely, with her proboscis retracted, and a drop of nectar appears on the upper surface of the base of her proboscis (A). The receiving bee stretches out her proboscis to full length and quickly takes the nectar. During the transfer the antennae of each bee are in continual motion, stroking the antennae of the other bee. The receiving bee may also stroke the forager's head with her forefeet.

Bees usually elaborate (or 'ripen') honey from the nectar or honeydew in an uncrowded part of the hive above the brood nest, where the temperature is a few

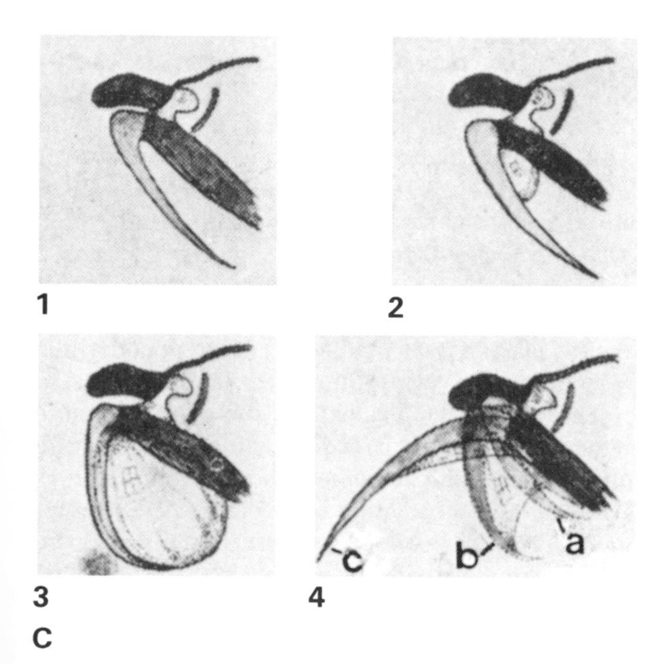

Figure 3.36a Actions of worker honeybees within the colony during the production of honey from nectar (Park, 1925).

A Forager (on left) passing nectar to a house bee (right).

B House bee evaporating water from partly made honey, by the process shown in C.

C Movements of the bee's proboscis by which a drop of liquid, held in the angle between the two parts of the proboscis (1), is expanded to form a thin film with a much larger surface (2, 3); three positions *a, b, c,* are shown in 4.

D Drops of partly made honey on right, suspended to allow further evaporation of water, in cells that may possibly contain brood; house bee on left, depositing finished honey in a storage cell, shown for convenience in the same comb.

degrees below 35°. They manipulate the liquid in their mouthparts (Figure 3.36a, B, C). Each bee rests with her head uppermost, and repeatedly unfolds and refolds her proboscis, exposing to the air an attenuated drop of liquid in the angle between the two parts of the proboscis. After 5–10 seconds, she sucks the drop back into her mouth. She repeats the whole process, with brief pauses, for perhaps 20 minutes, reducing the water content of the liquid and adding to it more salivary secretions containing enzymes. When little honey is being produced, it is much manipulated and the finished honey contains more of the enzymes than honey produced during a strong flow.

Finally the bee holds a drop of the liquid in the 'offering' position used by a disgorging forager, and deposits it in a cell, adding it directly to what is already there. If the cell is empty, she may suspend the drop from the upper surface of the cell wall – hanging it up to dry, so to speak (Figure 3.36a, D). (During a heavy flow, the house bee that receives the forager's load may deposit it in a cell immediately, and move on to receive other loads.) Many more bees are available for the conversion of nectar into honey at night than during foraging hours.

The bees evaporate the rest of the surplus water from partly made honey as it lies in the cells. During the honey flow, when much nectar or honeydew is processed, bees expedite evaporation from the cells by fanning, directing a current of air between the combs; on still summer evenings one can hear this fanning in progress. Ribbands (1953) reports that one experimenter sealed all joints and cracks in a large hive full of bees, and fitted anemometers to two small openings which formed the only entrances. More air was drawn through the hive during the day than at night, and the bees directed the flow by fanning, the direction changing at irregular intervals. On a hot July day (mean temperature 27°) air entered the hive at 200 to 400 litres/min. (A dozen strongly fanning bees, positioned across a hive entrance 25 cm wide could produce an air flow through the hive amounting to 50–60 litres/min.)

When sugar solutions of various concentrations below 80% were experimentally placed in the cells of a comb, and the comb put into a hive *but screened from the bees*, the bees were able to evaporate water until the sugar concentration was over 80%, as in honey (see Park, 1928*a*, 1946). But the speed of the process depended on the degree to which the cells were filled: it was more than twice as rapid when they were one-quarter full as when they were three-quarters full. In the former case, 60% sugar solution was concentrated to 80% in 48 h, and 20% sugar solution in 72 h. In screened combs that had been completely filled with nectar, the process took 3 or 4 days. In other experiments, the bees took 1 to 5 days to convert nectar into honey in normally ventilated hives that gave the bees adequate space for spreading out the liquid. Additional ventilation at the top of the hive could reduce a time of 5 days to 3 days; reduced ventilation increased it to more than 21 days. Park (1946) gives details.

Some nectars such as acacia (*Robinia pseudoacacia*) contain up to 60% sugar, and relatively little water needs to be evaporated, but the amount increases rapidly for very dilute nectars:

sugar concentration in nectar	*water evaporated in producing 1 kg honey*
60%	0.3 kg
40%	1 kg
20%	3 kg
13%	5.2 kg

Nectar containing as little as 13% sugar is hardly worth collecting, and indeed bees do not normally collect it unless they need the water to dilute stored honey for brood rearing.

Evaporation of water is presumably continued until the equilibrium point is reached between the water content of the honey and the relative humidity of the surrounding air in the hive; see Figure 13.22a. (Because of the cooling effect of evaporation, the bees' maintenance of constant hive temperatures is not readily compatible with maintenance of a constant humidity, Ribbands, 1953.) The bees finally 'cap' the cells with a more or less airtight beeswax seal (Section 3.31.3). Honey is hygroscopic, and the capping reduces water absorption in a damp environment, with the subsequent risk of fermentation. However, water can be evaporated from capped honey after removal from the hive if dry air is passed over it, as described in Section 13.32.

In France, Moreaux (1939) found that different colonies working the same flow capped their honey at slightly different water contents, although honey from different capped cells in the same colony had the same water content. In New Zealand the water content of honey from different parts of the same sealed comb frequently varied (Fix & Palmer-Jones, 1949). In Canada Gooderham (1938) found some differences for honey in both capped and uncapped cells in the same comb at extracting time. There was less variation in the water content of honey in capped cells (18.8% to 19.2%) than in uncapped cells (17.7% to 21.1%).

Tropical honeybees seem to cap honey at a higher water content than temperate-zone honeybees, and in hot humid areas capped honey may contain up to 28% water. Rubber (*Hevea brasiliensis*) honey is especially noted for its high water content. On the other hand in desert areas the water content of honey may be as low as 13% when capping is done.

The water content of the honey when it is capped is of great importance to beekeepers; see Section 13.22. There is much that we do not yet know about factors that interact with it – such as atmospheric temperature and humidity, hive size and ventilation, the rate at which the colony is producing honey, the water content and sugar spectrum of the nectar or honeydew, and genetic characters of the honeybees. Sections 13.32 and 13.39 discuss practical aspects of handling honeys with a high water content.

Water

Water cannot be stored in cells of the comb like other materials brought to the colony. But in 1923 Park showed that a colony can store some water, in the honey sacs of 'reservoir bees' (see Park, 1946). These bees took loads of water brought to the colony by foragers, and then remained inactive on the combs round the brood nest. If no bees took the water from them within a few hours, they often imbibed a little honey which then mixed with the water in the honey sac. During a period in early spring when some days were too cold for water collection (but water was needed constantly for brood rearing) Park found one morning that 1300 workers (about half the experimental colony) were acting as reservoir bees.

3.37 The winter cluster

Formation of a winter cluster has a function in honeybees somewhat similar to that of hibernation in warm-blooded animals. A colony is able to survive in winter conditions as a cluster in which its component bees have a lower body temperature and a much lower food consumption and metabolic rate than normal, and undertake minimal bodily activity.

Ability to overwinter was a notable achievement developed by honeybees in the north temperate zone during evolutionary times – and one not shared by honeybees in the tropics. It enabled them to survive a cold winter, and thus to extend their living range into cooler regions. Capabilities of different species are now as follows.

can overwinter clustered	*cannot overwinter clustered*
A. mellifera	
European	tropical African
A. cerana	
races of high mountains (e.g. Kashmir), and of NE Asia	races in lower hills and plains
A. dorsata	
A. dorsata/laboriosa in high Himalayas	all other *A. dorsata*
A. florea	
none	all

The description below is based on European *A. mellifera*, which are the most economically important overwintering honeybees, and about which most is known.

As long as the temperature outside the hive is higher than 18°, bees in the hive are dispersed within it. Below 18°, the bees move closer together as the external temperature decreases, and when their temperature falls to 14° (the outside temperature then being perhaps 9°) they start to form a well defined cluster, with a compact outer shell of rather cold, inactive bees. Bees in the cluster occupy the passage-ways between combs, and also empty cells in the combs. The centre of the cluster is both warmer and less crowded than the outer shell. Bees there have space to move about and feed on honey in the cells, and the temperature (13° or above) is high enough to enable them to do so. As outside temperatures drop further the cluster contracts, bees being packed more tightly. The temperature of the outside bees may be as low as 8°, but bees frequently change positions between the cold periphery and the warmer centre.

So long as the external temperature is above −5°, the bees are able to reduce heat loss from the cluster by contracting it. At external temperatures below −5°, contraction continues. (It is not enough to maintain the necessary temperature within the cluster, and more food has to be consumed so that metabolic energy can be generated.) Results of Wilson and Milum (1927) indicate the following relationship between temperature and cluster size:

external temperature	*cluster diameter*
+5°	36 cm
+2°	28 cm
−14°	26 cm
−26°	10 cm

Section 6.53.2, on preparing colonies for overwintering, shows that the proportion of bees dying in the course of the winter is much higher in small colonies than in large ones. Under given conditions, the heat production of a cluster is proportional to the number of bees in it, whereas its heat loss is proportional to the surface area of the cluster. The heat loss *per bee* is less for a cluster containing many bees than for one with only a few. (Similarly, larger mammals are more able to withstand cold than small ones.)

Many different relationships have been found between the food consumption of clustered colonies and the outside temperature (see e.g. Ribbands, 1953). There is, however, some consensus that food consumption is least when the outside temperature is about 4°, and especially when the temperature is steady and does not fluctuate. Section 6.54.2 quotes 4° ± 1° as the optimal temperature inside a building used for indoor wintering.

In Germany, Southwick (1988) measured the heat production of wintering colonies (populations about 16 000 to 20 000 bees); the hives were placed in turn in a controlled-temperature cabinet. Figure 3.37a shows that the rate of heat production per bee was least at external temperatures between −10° and +10°, and that it increased more or less linearly at lower temperatures. The lowest temperature a colony can survive depends on food availability and the number of bees; colonies of 16 to 20 thousand bees maintained the core of the cluster at 34° for at least 12 hours when the external temperature was −80°, although at great metabolic expense (Southwick, 1987*a*). It would be interesting to see results from similar measurements on colonies of tropical honeybees.

Johansson and Johansson (1979) have written a detailed summary of knowledge about the *Apis mellifera* colony in winter. *A. cerana* colonies winter in a somewhat similar way in regions with a cold winter, as in the Himalayas and north-east Asia, but they have been much less studied. Verma (1970) compared *A. cerana* and *A. mellifera* colonies wintering in northern India. At external temperatures of +2°, the temperature within the cluster was about 3° higher in *A. cerana* (although the hive temperature outside the cluster was about 3° lower). *A. cerana* also had slightly higher cluster temperatures in the spring.

Colonies of both these species live (summer and winter) in a cavity, and this has been considered necessary for overwintering. However, at altitudes in the high Himalayas above those where *A. cerana* lives, colonies of *A. dorsata/laboriosa* nest in the open; they are reported to migrate to winter sites where they cluster without building comb, but storing honey in their honey sacs (end of Section 1.23.2).

An apparent anomaly of the winter cluster is that the colony restarts brood rearing soon after the winter solstice, often before the coldest part of the winter (Section 3.51.2). Brood rearing is probably initiated by an increasing photoperiod, i.e. day length. The brood is in the central part of the cluster and must be kept at or near 35°, which necessitates increased food consumption by bees in the colony. This very early start of brood rearing, when there is no forage and when air temperatures are too low for the bees to fly, would seem to be an inefficient procedure. Seeley (1985), discussing colonies at latitudes where they winter as a cluster, suggests that this early brood rearing is necessary to enable the colony to be populous enough to swarm early in the summer season, in time for swarms that issue to store enough food for the next winer. In general, the higher the latitude, the shorter the summer season (Section 12.62), and the shorter the

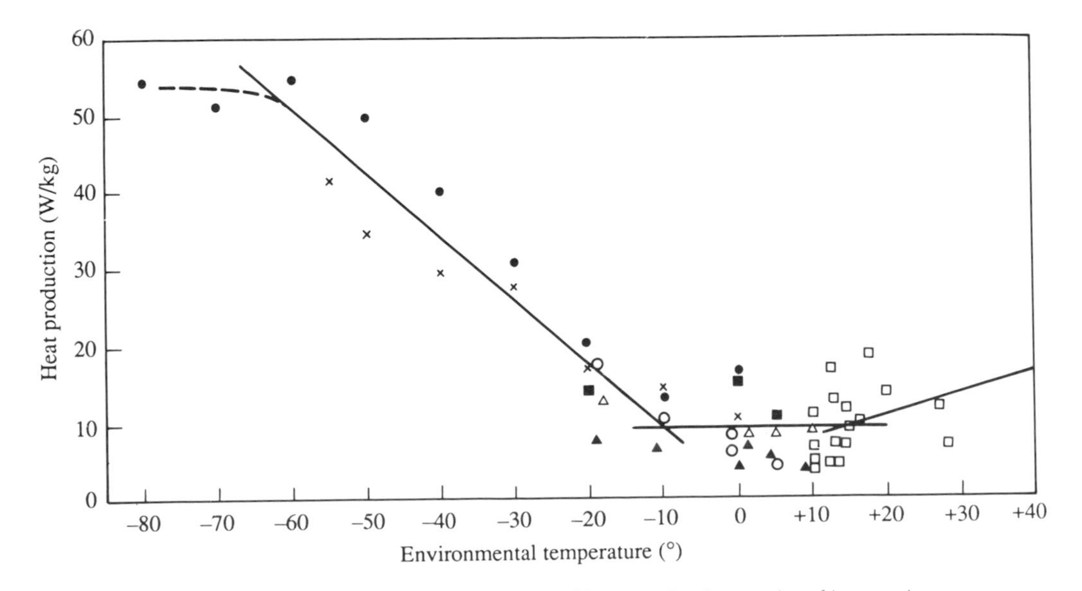

Figure 3.37a Minimum maintained rate of heat production per kg of large colonies of temperate-zone *Apis mellifera* in hives held at temperatures from −80° to +27° (Southwick, 1988).

duration of the 'window' that allows the survival of swarms (end of Section 3.34.2); also the longer the period when the colony must remain clustered. So the use of increasing photoperiod as a stimulus to initiate brood rearing would be appropriate. But it would not apply to, say, Kashmir (latitude 34°N), where *A. cerana* clusters, but the photoperiod changes little.

3.38 Other activities constantly occurring within the colony

Workers constantly communicate with each other by passing on small amounts of food, and by making antennal contact. They groom their own bodies and those of other bees. They also maintain the hygiene of the nest, cleaning cells, and removing debris from the nest or hive, including any bees that die there. Figure 3.23a exemplifies the pattern of different activities.

Food sharing (mutual feeding, reciprocal feeding)
The coherence and identify of a colony are reinforced by the transmission of small amounts of food by mouth from one worker to another. Istomina-Tzvetkova (1953) watched two workers in an observation hive for 8 hours daily, from the age of 3 to 28 days. During each 8-hour spell, these two bees on average fed another worker on 19 and 10 occasions, and received food on 16 and 9 occasions, respectively. Nixon and Ribbands (1952) were able to trace the extent of food transmission throughout the bees in a colony, by training 6 foragers to take sugar syrup containing radioactive phosphorus from an outside feeder; 20 ml was provided. Samples of bees were subsequently collected from the colony, and the following percentages of them showed radioactivity:

	after 5 h	*after 29 h*	*after 48 h*
foragers	62%	76%	
in brood box	18%	43%	
in first super	16%	53%	
in second super	21%	60%	
drones		27%	
older larvae			100%

Such food sharing is additional to the passing on of small amounts of nectar by dancing foragers to bees following their dances (Section 3.35), and it is more widespread in the colony than mutual handling of nectar in the course of its conversion into honey. It ensures that the bees in a colony have the specific odour of that colony; this is essential for defence of the colony (Section 3.41), and is important in some other aspects of colony organization (Ribbands, 1953; Free, 1959).

Antennation
Like other social insects, worker honeybees constantly touch the antennae of other workers with their antennae. This 'antennation' between workers in the colony can be observed when a hive is opened, or in an observation hive, and many close-up films show it clearly. Bees antennate others to solicit food from them, and the antennal contact helps them to position their mouthparts for food transference. They also antennate to solicit chemical information, such as the scent of other bees, and antennation enables a worker guarding the colony at the flight entrance to distinguish between a returning bee from her own colony and a bee from elsewhere (Section 3.41).

Antennation can even help a worker to distinguish between those workers of her colony that are her full sisters (with the same drone-father) and those that are only half-sisters (having a different drone-father). This ability, common in social insects, makes 'kin recognition' and 'kin selection' possible. This subject has been studied in honeybees by Getz and colleagues, who explain its importance in the transmission of an individual's genes to the next generation; they found some indication that bees in a swarm are more likely to be full sisters than half-sisters (Getz et al., 1982; Getz & Smith, 1983).

Grooming
Social insects frequently groom their own bodies, especially the eyes and antennae. The chief cleaning instruments are certain structures on the legs, and also the lower mouthparts including the proboscis; Wilson (1971) gives a detailed analysis of the movements involved. All honeybees have an antenna cleaner on each foreleg (Figure 2.22a), and the worker forelegs also have brushes on the basitarsus that remove pollen, dust, or other foreign material, from the head. The worker's middle legs clear pollen from the thorax, and specialized structures on the hind legs complete its clearance and pack it into loads on the corbiculae (Figure 2.43d).

A worker uses her proboscis to groom her own body, and also those of other workers, drones, and the queen; drones do, however, groom themselves. Some parasites can be removed by grooming, and Section 7.41.5 describes how *Apis cerana* workers use their mandibles to rid themselves of *Varroa* mites.

The workers tending the queen lick her body, and brush their antennae over it, thereby adsorbing minute amounts of pheromone which is subsequently distributed through the colony; see Section 3.21.

If a foraging bee visits a food source that proves unsatisfactory she may groom herself, but this seems to be displacement activity (Pflumm, 1968).

Cell cleaning
Workers just emerged as adults are likely to clean cells before they undertake any other activity, as exemplified in Figure 3.23a. But, as the Figure indicates, house bees of almost any age may clean cells. The cleaning makes cells ready for the next batch of brood, and is continued after eggs are laid in them (Lindauer, 1953). There has been much conflicting evidence as to whether they also apply a 'varnish' to the cell surfaces and, if so, what the material is. Ribbands (1953) discussed the question in some detail, and his conclusion is as follows. 'One could suppose that the cells are belaboured with the mandibles, and perhaps varnished with a glandular secretion. The yellow colour of the combs is probably derived from pollen pigments, but it is uncertain whether it comes via secretion or by contamination spread by the bodies of the bees.'

Removing other debris, including dead bees
Debris in a hive includes scraps of discarded wax or pollen, also mites or insects that die in the hive, and any bees that die there. All such debris is likely to fall on to the floor board, and *Apis mellifera* bees generally remove what they can; *A. cerana* bees do not, and the beekeeper must take special steps to keep the floor board clean, for instance those mentioned near the end of Section 4.5.

The removal of dead bees (necrophoric behaviour) has been studied in *A. mellifera* by Visscher (1983); no observations on *A. cerana* are known. Visscher found that workers removed dead bees more rapidly than debris in general, probably recognizing them by chemical changes immediately after death. About 1% to 2% of the workers specialized in this activity, and they carried the dead bees out of and away from the hive – sometimes as far as 100 m, and nearly always at least 10 m. Tew (1987) reported the behaviour of bees of a large swarm (7.7 kg) that he put into a hive containing a colony dead from the previous winter. Within 36 hours the new bees had removed about 4.5 kg of the dead ones; if a bee carrying a corpse became airborne she dropped it at least 25 m away, otherwise no further than 1.2 m.

Genetic influence on activity specialization by workers
A small proportion of workers may become 'specialists' in certain activities, carrying them out to the exclusion of many others. Specialist activities include removing dead bees from the hive, acting as reservoir bees for storing water, water collection, guarding, and scouting for forage. In 1988 experiments were carried out on variations between the behaviour patterns of individual workers of a colony, which showed that the likelihood of any worker carrying out certain specific activities frequently, or becoming a 'specialist' in them, is in part genetically determined. Colonies were used in which all the workers were daughters of the same queen, but the queen had been inseminated from two (or three) drones, chosen so that the worker-daughters of each drone could be distinguished. In colonies used by Frumhoff and Baker (1988), the workers had one of two drones as father: one with 'black' body colour and the other 'brown' (cordovan). Whether black or cordovan, young workers (those up to 18 days old were studied) groomed other workers, but grooming was significantly more frequent by black than by cordovan workers. On the other hand black and cordovan workers showed no difference in the frequency with which they shared food with other workers. Robinson and Page (1988) distinguished between worker-daughters of three drones by electrophoretic analysis of certain enzymes in the bee's body. They found that the frequency with which workers removed dead bees from the hive was significantly different according to which of the three drones was the father of the workers. The same was true of guarding the hive entrance (Section 3.41).

3.39 Laying workers

In a queenright colony, pheromones produced by the adult queen (and probably another from developing queens) prevent ovary development in the workers (Table 2.35A). If a colony loses its queen and has no eggs or young larvae, it has no chance of producing a future queen, and is without these queen pheromones. As a result, the ovaries in some of the workers develop, even to the stage where the workers lay eggs – which are unfertilized and give rise to drones, which are normally haploid. The process is referred to as parthenogenesis, reproduction by a virgin female. Egg laying is likely to start within 3 to 4 weeks in European *Apis mellifera*, but much sooner in tropical honeybees; figures quoted are around a week in *A. cerana* (a few days in temperate-zone races), and 5 to 10 days in tropical *A. mellifera*.

The drone eggs are usually laid in worker cells, but laying workers do not produce a regular 'brood pattern' as a queen does, and several eggs laid by different workers may be found in the same cell. A colony with laying workers does not readily accept a new queen introduced to it, probably in part because the laying workers are producing pheromones that complicate the situation. It is rather a nuisance in beekeeping practice, but Section 6.35.2 describes a method of using the adult bees of such a colony by uniting them to a queenright colony.

When all the workers of a laying-worker colony grow old, the colony will die out. Very occasionally, as a result of a genetic accident, an unfertilised egg of European *A. mellifera* contains the full diploid complement of chromosomes (32 instead of 16) and therefore develops into a female – which the bees rear as a queen. In *A. m. capensis* only, laying workers – which develop within about a week – habitually lay eggs that can develop into full females (thelytoky); Anderson (1963) and Ruttner (1988) give more details. These workers cannot mate, although some have been instrumentally inseminated with success (see Ruttner, 1986). Winston (1987) discusses possible reasons why this apparently beneficial trait is very rare except in *A. m. capensis*.

Where hives of other *A. mellifera* are kept near an *A. m. capensis* colony, laying workers from the latter are likely to enter the hives (Johannsmeier, 1983). An *A. m. capensis* laying worker secretes queen pheromones, and is treated as a queen in the hive; she lays eggs, which can develop into *A. m. capensis* queens, that mate and head a queenright colony; the original queen disappears.

3.4 SOCIAL ACTIVITIES OF WORKERS OUTSIDE THE COLONY

Activities considered here mainly take place in daylight, away from the colony or at the flight entrance, and the bees concerned are in general older than the house bees discussed in Section 3.3 and its subdivisions.

3.41 Defence

It is essentially their nest or colony that honeybees defend, exhibiting behaviour that some people interpret as aggression. Events away from the colony may also elicit an attack from an individual worker – for instance if it is handled roughly.

Stinging

The basic unit of colony defence is an individual worker whose venom sac contains venom, and is thus able to sting (Section 2.23.1). If stinging occurs, it is usually the last of the bee's responses to certain stimuli that culminate in an attack, as described below.

Venom is produced in the worker's venom gland and stored in the venom sac (Figure 2.23a), which is likely to be filled by the time the worker is 14 days old. The age distribution of the bees in a colony is thus relevant, and colonies with many bees less than 2 weeks old, whose venom sacs are not yet filled, show relatively little defensive behaviour. A worker that stings another bee can usually withdraw her sting without injury to herself, but if she stings into thick skin (as of a person or animal) she usually dies, being unable to retract her sting (Section 2.23.1). Section 5.5 discusses ways of avoiding bee stings, and of dealing with them.

Defensive behaviour

Figure 3.41a sets out the time sequence of the behaviour of a bee in defence of her colony. The bee's response to the first (alerting) stimulus strengthens her guarding stance (Figure 3.41b); for instance the abdomen is raised, possibly with the sting protruded, and the antennae are waved. In addition, the bee may *recruit* other bees to guard activity, by entering the colony with her sting chamber open and the sting protruded, thus releasing alarm pheromone (Table 2.35A). The second (activating) stimulus causes the bee to *search* for the source of disturbance. When she locates it, the third (attracting) stimulus makes her *orient* herself towards it and move there. As a result of the fourth (culminating) stimulus, she attacks the target: she *threatens* it, emitting a high-pitched buzz and making body thrusts towards it. The attack itself may consist of *biting*, *burrowing* into hair, *pulling* hairs, and *stinging*, which – if she cannot retract her sting – is her final act in defence of the colony.

At stage 1, unresponsive bees may move away from the source of disturbance, and at stage 4, many are likely to do so, running to undisturbed combs (which the beekeeper tries to make them do, by the use of smoke), or flying away from the colony.

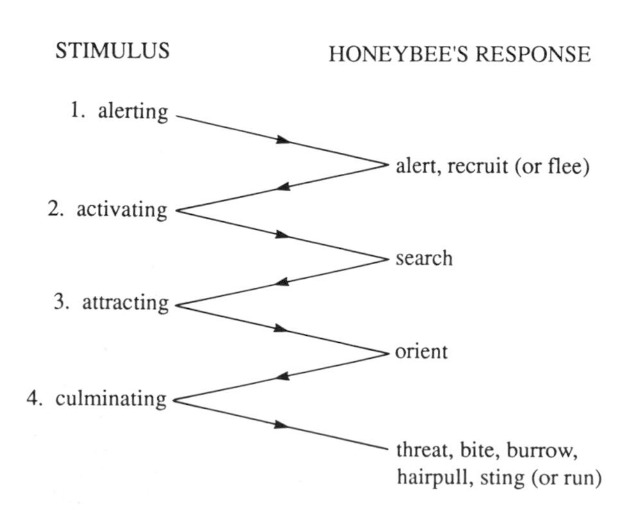

Figure 3.41a Basic sequence of honeybee defensive behaviour (adapted from Collins et al., 1980).

Some of the stimuli that are likely to elicit individual stages in the above sequence of defensive behaviour are listed in Table 3.41A. The behaviour of different colonies varies greatly, according to the colony's genetic make-up, and to certain environmental and colony conditions. Environmental factors include: temperature, and variation between day and night temperatures; light intensity and rate of change of day length; atmospheric humidity, pressure and electric potential; magnetic field. Food availability is important, and its diminution (as at the end of a flow) usually increases defensive behaviour, as does robbing by bees of other colonies, which is likely to occur when forage has become scarce. Lack of water does so, probably also contamination of bees with some insecticides, and apparently foraging on certain nectars and pollens.

Overcrowding of bees in a hive, or of hives in relation to forage available, is likely to increase defensive behaviour. On the other hand colonies weakened by disease, by a failing queen or by lack of food, cannot mount an effective defence; this situation is exploited by predators, and by honeybees from stronger colonies – especially after the last flow of the season has ended and many would-be foragers are searching for any food that they might collect.

Heritable factors are very important in determining defensive behaviour, and give rise to differences between races and also within the same race or strain. Since the 1960s the subject has been studied extensively in South and Central America, because of the impact of Africanized bees on commercial beekeeping there (Sections 1.22.4, 7.43).

Comparison of European and Africanized honeybees
Standardized tests were established for comparing the defensive behaviour of Africanized (derived from tropical African) and temperate-zone *Apis mellifera*, and Table 3.41B gives some results (Collins, 1985). The moving target was a dark suede leather square moved about in front of the hive.

The Africanized bees always showed significantly greater defensive behaviour than the Europeans. They responded 2.4 times as fast to alarm pheromone, and about 30 times as fast to a moving target; 8 times as many Africanized bees stung the moving target. Africanized bees may also 'follow' an intruder (or beekeeper) as far as 1 km when he moves away from the colony (Collins, 1988), whereas European bees do so for no more than a few metres.

Section 7.43.2 discusses problems of management caused by the defensive behaviour of Africanized bees. Many of the tropical African bees in their native habitat show similar behaviour, although this has been less well studied.

Table 3.41A Stimuli known or believed to elicit defensive behaviour in *Apis mellifera*
Data from Collins et al. (1980). — = probably *not* a stimulus for this step
? = may be a stimulus for this step. × = probably *is* a stimulus for this step

Stimulus	*Alerting*	*Activating*	*Attracting*	*Culminating*
motion	?	?	×	?
colour contrasts	×	?	×	?
shape	?	?	×	?
texture of substrate	—	—	—	×
vibration of substrate	×	?	—	?
mechanical stimulation of bee	×	×	×	×
harassing bees	?	?	?	?
scent:				
colony	—	—	?	×
mammalian:				
breath, carbon dioxide	×	?	×	—
sweat, fur	?	?	?	×
pheromones:				
2-heptanone (mandibular gland)	×	×	?	?
isopentyl acetate (sting)	×	×	?	—
other components (sting)	×	×	?	?
whole sting	×	×	?	?

Table 3.41B Comparison of defensive behaviour in European and Africanized *Apis mellifera*

Adapted from Collins (1985), who describes the methods of testing. All pairs of results are significantly different ($P < 0.01$). There were 150 commercial European colonies in the USA and 147 Africanized colonies in Venezuela, respectively.

	European (*E*)	*Africanized* (*A*)	
Period before response to:			*Speed ratio A/E*
alarm pheromone	13.1s ± 0.5s	5.4s ± 0.3s	2.4
moving leather target	9.2s ± 0.5s	0.3s ± 0.1s	31
No. bees standing or flying in front of a hive in response to moving target, hive jolted, etc.:			*Ratio A/E*
before stimulus	44.2 ± 3.5	70.6 ± 7.0	1.6
30s after	61.1 ± 4.9	121.6 ± 6.8	2.0
60s after	66.6 ± 4.0	141.7 ± 7.8	2.1
90s after	84.3 ± 4.4	172.8 ± 9.8	2.1
No. stings in moving target	10.4 ± 0.9	85.7 ± 2.6	8.2

Defence at the flight entrance

The flight entrance offers the only entry to the nest for small intruders including insects, and the easiest access for large predators (Sections 11.54–11.56). It is therefore constantly guarded, commonly by workers during their transition from behaving as house bees to behaving as field bees. A guard bee stands at the entrance in a characteristic stance (Figure 3.41b), with the forelegs slightly raised, antennae pointed forward, mandibles open, and wings sometimes spread. If the colony is disrupted – as when a beekeeper opens a hive, or when a large mammal tears open a wild nest – guards will be active wherever the combs are exposed.

Figure 3.41b Honeybees (*Apis mellifera*) in guard position at hive entrance (photo: M. Ono).

Honeybees of any colony have an odour specific to that colony, which is partly genetic in origin, and partly dependent on the colony's food (Section 3.38, see under Food sharing). Guards intercept other honeybees arriving at the flight entrance and examine them with their antennae; 1 to 3 seconds is usually long enough for a guard to identify a stranger, and she then grips one of its legs or wings with her mandibles, and curves her abdomen round so that she can sting it. The behaviour of an incoming bee provides clues as to its status; an unladen worker from another colony, intent on robbing the stores inside the hive, has a characteristic rapid jerky flight, whereas a loaded forager flies heavily, and is unlikely to be challenged. Drones are usually allowed to enter any colony, but if a queen returning from a mating flight enters a queenright colony, workers recognize her as such, and kill her.

It is relatively easy for a guard bee to distinguish an intruder that is not a honeybee, and different techniques are adopted against different predators (see e.g. Sections 11.54–11.56, and Seeley, 1985). Section 3.32 refers to *Apis cerana*'s method of killing hornets.

Defence of food sources

Honeybees do not in general defend their food sources. But where two or more species use the same forage, the smaller species are likely to attack and drive away the larger ones when both are at the same source. The small bee can reach only a limited foraging area, and its colony may starve if other bees use that particular food source. This source does not have the same importance to a larger bee, since it can fly further afield to other sources. In Sri Lanka, Koeniger and Vorwohl (1979) set up feeding dishes and recruited bees of four native species to them. The bees are listed below in increasing size, and the percentage represents the average success rate of a bee of the species in forcing a bee of another species to leave the source.

Trigona iridipennis	58%
Apis florea	33%
Apis cerana	29%
Apis dorsata	10%

3.42 The reproductive swarm in transit

Section 3.34.3 described the events leading up to the emergence of a prime swarm from its parent colony, and the emergence itself. The swarm is a viable unit which contains the colony's queen and from a few thousand to over 50 thousand workers (the range in Table 3.34A is 12 to 22 thousand), many carrying honey in their sacs; there are probably also a few drones. For *Apis mellifera* (which is considered here), and also for *A. cerana*, the swarm must normally find a cavity which can provide a safe nesting site, before it builds comb and becomes established as a colony. Ambrose (1976), Severson (1984) and others have summarized the story of the swarm in transit, and Seeley (1985) discusses site selection in detail.

The swarm flies from the hive as a swirling mass of bees, and appears to cast around for some support (in the open) on which to form a temporary cluster. The bees detect the queen's presence in the flying swarm by the pheromone 9-ODA she produces (see Table 2.35A). Butler and Simpson (1967) found that this pheromone was most attractive to workers at a height above ground of about 1.3 m; at greater heights it attracted drones (Section 2.63), but not workers.

There have been many views as to what characterizes an acceptable clustering place. Unless the swarm is flying above a featureless plain, it is usually well above ground level, although the height is of course limited by what is available. Caron (1979*a*) in the USA found most clustered swarms at heights up to 1.5 m (43%), and many at 1.5 to 3 m (28%). The swarm clusters not far away from the parent colony; Ambrose (1976) in the USA recorded 26 m as the average distance for 24 swarms. When a few workers have settled, they expose their Nasonov scent gland and fan to disperse the pheromone produced (Table 2.35A). This attracts other workers to join them, and when the queen has also done so, pheromones from both queen and workers act in concert to attract the rest of the bees in the swarm to cluster (Morse & Boch, 1971). Provided the queen is in the cluster it becomes stable, and the swarm enters a second phase.

The stabilized cluster is highly organized to conserve heat and thus food, and to afford the bees some protection from the weather and from enemies. Meyer (1956*b*) in Germany found that 2 or 3 layers of workers, clinging on to each other with their feet, formed a covering mantle which gave the cluster mechanical strength. There was a constant quiet exchange between bees on the outside and inside of the mantle; 10 minutes after the cluster had been dusted with flour, half the outside bees were unfloured ones. Inside the mantle, bees hung in loose chains. Heinrich (1981*a*, 1981*b*) studied the cluster's temperature regulation. Bees on the outside layer of the mantle had their heads uppermost. When the temperature of the air round the cluster was reduced experimentally to 3°, the bees' backs were arranged like roof-tiles, the head of each bee being tucked under the abdomen of the bee above it. At low temperatures the cluster contracted, and there were few open spaces in the centre. At an air temperature of 25° the bees in the mantle spaced themselves more widely, and their heads were visible, and at high temperatures the bees left a number of (mostly vertical) passage-ways throughout the cluster, which helped to dissipate heat. The mantle was always above 15°, even at an air temperature below 5°, and the centre of the cluster was at 35°, brood-nest temperature.

Even before the swarm had left the parent colony, a few foragers had switched from collecting food and became scouts, seeking a new nest site (Lindauer, 1955*b*). An hour or two after the swarm had clustered, Meyer (1956*b*) was able to see a flight entrance through the mantle, by which scout bees left the swarm interior to undertake search flights for a suitable permanent cavity for the swarm to live in; this is much rarer than a temporary clustering site. Often there is no foraging from a clustered swarm; the scout bees search for a nest site, not food, and all they bring back is information to communicate to other bees in the cluster, about nest sites they have found (Lindauer, 1951/1953, 1953, 1955*b*). The scout bees use much food on their flights, and other bees in the cluster serve as food reservoirs for them, remaining quiescent, and conserving the food carried in the bees' honey sacs when the swarm issued (Combs, 1972*b*). Absconding swarms of tropical honeybees sometimes forage (Section 3.43).

A scout may spend nearly an hour inspecting a possible cavity she finds. She makes forays inside the cavity that last a minute or so, alternating them with inspections outside, during which she 'scurries over the structure and performs slow, hovering flights all around the site, apparently conducting a detailed visual inspection of the structure and surrounding objects' (Seeley, 1985). Figure 3.42a shows the routes a scout took on some of her visits inside a cavity; she did not venture far in at first, but penetrated further on successive visits, occasionally flying, but mostly walking over the inside of the walls, assessing the cavity's size. By the time a scout has finished her explorations, she will have covered all its inner surface, and may have walked 50 m or more.

Seeley quotes experiments in which swarms of Italian-type bees, in various habitats in the USA, were given certain choices of nest site. Cavity preferences recorded, together with those recorded by Lindauer (1955*b*) in Germany, include the following:

- 100 to 400 m from the parent colony
- at least several metres above ground
- with a volume of 15 to 80 litres
- with a flight entrance smaller than 75 cm², near floor level, and facing south
- furnished with fully built beeswax combs
- protected from wind
- free from flooding
- free from ants.

Swarms of other temperate-zone honeybees, in other habitats, would probably show somewhat different preferences, and in tropical conditions differences would be still greater.

Lindauer (1951/1953) found that when a scout bee returns to the cluster, she performs a wagtail dance on its surface that indicates the distance and direction to the nest site, as for nectar (Section 3.35). The suitability of a cavity is indicated by what Lindauer calls the 'enthusiasm' of the dance. (He noticed that whereas foragers danced more vigorously for sources close to the colony, scouts from a swarm did this for cavities *farther away* from the parent colony, which would serve to reduce competition for forage in the future.)

Different scout bees danced to indicate the different sites they had found; the scouts from one swarm found, and danced for, 13 to 24 sites (Lindauer, 1955*b*). Those dancing unenthusiastically for a rather unsuitable site were 'persuaded' by a more vigorous dancer to visit her site, and their own dances subsequently indicated hers. A scout continually revisited her own site, perhaps reassessing its suitability at different times of day. By the time all dancers are indicating the same site, they may number about 500. The swarm is then ready to enter its third and final phase. Some of the bees make a *Schwirrlauf* (whirring run), similar to that in the parent colony before bees flew out in the swarm (Section 3.34.3). It now excites the swarm bees to fly, and all become airborne within a minute or two. Heinrich (1981*a*, 1981*b*) discusses the temperature regime of the cluster in relation to the need to minimize food consumption – which is less at lower temperatures – and yet for all the bees to be warm enough to fly off at the same moment to the new nest site.

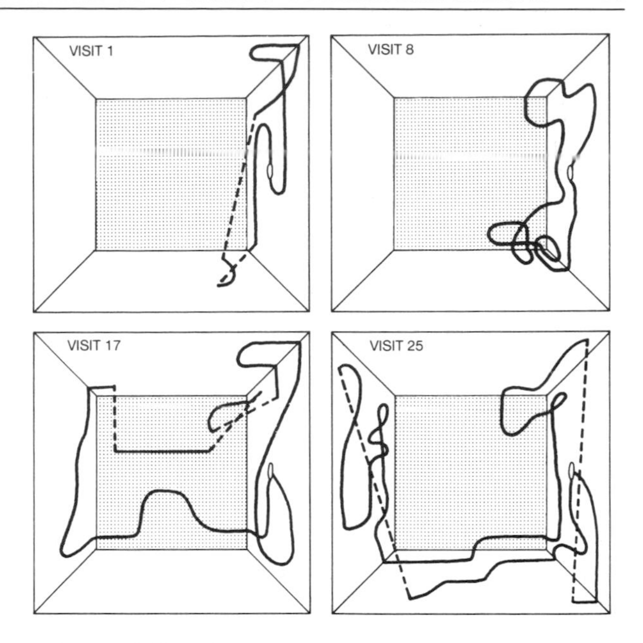

Figure 3.42a A scout bee's inspection routes on 4 out of 25 visits inside a dark box offered as a nest site (Seeley, 1985).
Walking is indicated by solid lines, and flying by broken lines. The flight entrance of the box was in the centre of the right-hand side.

The bees fly *en masse* to the cavity finally selected; they may move quite slowly as on their earlier flight, or up to 24 km/h. The 500 or so scouts that have already visited the cavity ensure that the rest fly in the right direction by their action in flight, or by communication beforehand, or in some other way. Once they reach the site, the scouts alight and fan with the Nasonov scent gland exposed, which attracts the other bees to the flight entrance, and all enter quickly.

Comb building and foraging can now proceed, and also brood rearing; if the swarm has a virgin queen, brood rearing will be delayed until after the queen has mated.

3.43 The absconding swarm in transit

Section 3.34.5 describes how a colony prepares for absconding when food is no longer available to it. When the whole colony finally flies off as an absconding swarm, its primary need is to find a nest site in an area within its flight range, where there are adequate food resources. (We do not know whether a new nest site is located before or after the swarm reaches the new forage area, but the latter would seem more likely. One would suppose that, when an absconding swarm of tropical African or Africanized *A. mellifera* clusters at a resting place near new forage, scout bees search for and assess possible nest sites, as scout bees do from a clustered swarm of temperate-zone bees (Section 3.42).)

Tropical *Apis mellifera* bees in an absconding swarm leave their old hive or nest with much more food than bees in a reproductive swarm (Seeley, 1985). In Kenya, Kigatiira (1984) also found that food is unevenly distributed among the flying bees; those in the 'lead' carried much less food than those at the back. These absconding swarms (and those of Africanized bees) may fly for many kilometres, clustering overnight at 'resting places' on the way, from which foraging flights are probably made. But Seeley (1985) calculated that the honey taken in the honey sacs of the bees provided enough fuel for the bees in an absconding *A. mellifera* swarm to fly as far as 90 km.

In Sri Lanka, Koeniger and Koeniger (1980) described 'resting places' where migrating swarms of *A. dorsata* clustered overnight – or for several nights – without building comb. The scout bees performed dances all over the surface of the cluster, indicating the direction of the migration flight, and the swarm finally flew off in that direction. In addition, where most light fell on the cluster surface, there were dances that indicated the direction in which food had been found.

Absconding swarms may amalgamate together (tropical African bees, Kigatiira, 1984; Africanized bees, Cosenza, 1970; European bees, Weaver, 1987; Section 3.34.5). In Brazil, all 55 absconding swarms examined had more than one queen, and all queens examined were mated.

3.5 SEASONAL CYCLES OF HONEYBEE COLONIES

Previous sections of this chapter have dealt with the colony and the interrelated behaviour of individual bees in it, in a wide range of circumstances. Many of these circumstances occur in an annual cycle linked with the earth's movement round the sun, and this section is concerned with the consequent seasonal cycle of the colony in different parts of the world, especially at different latitudes. Other parts of the book, especially Chapters 6 and 7, deal with the management of colonies in relation to the seasonal cycle.

3.51 The seasonal cycle with cold as the limiting factor

In land areas at latitudes above 45° (almost all of which are in the north temperate zone), cold is the factor that limits the period in the year when honeybees can be active outside their nest or hive, and forage for food. The air temperature below which temperate-zone *Apis mellifera* workers cannot make sustained flights is commonly quoted as 16° in summer. In winter they may make short flights at temperatures as low as 8°–10°. The colony itself is able to survive winter temperatures much lower than this, by clustering within the shelter of its hive or natural cavity; see Section 3.37. Almost certainly in response to the increasing photoperiod (day length) after the winter solstice, the colony starts to rear a little brood, and the metabolic energy produced by the clustered bees enables it to keep a small region inside the cluster at 35°, so that the brood can survive. Colonies overwintered in darkness indoors (McCutcheon, 1984) seem to be more tardy in rearing brood.

3.51.1 Adult population

The number of bees available for foraging (which is of the utmost importance to beekeepers) is largely determined by the population of adult bees in the colony. The annual cycle of the adult population is governed by the rate of brood rearing (Section 3.51.2), and by the length of life of the adult workers which varies from a few weeks in summer to several months in winter; see Table 2.42A. A worker that survives to adulthood emerges from its cell as an adult 3 weeks after the egg is laid, so the maximum adult population of a colony lags somewhat behind the maximum rate of brood rearing. The annual cycle of the adult population of a rather small experimental colony at 57°N, that did not swarm, is shown in Figure 3.51a. The cycle shows the following phases:

- A decrease in population in January–March, when many old bees from the previous autumn died and few young ones were reared, dropping to the minimum for the year in March.
- A steep rise during April–June, from 10 000 to 40 000; the population increased from its annual minimum to its maximum in only 3 months.
- A decrease in July; the colony started to build queen cells on 20 June and was requeened on 14 July, hence the small rise in population during August.
- A large decrease from late September to December, when many bees died but very few were reared; in the 6 winter months October–March the population dropped from about 20 000 to 10 000. The average

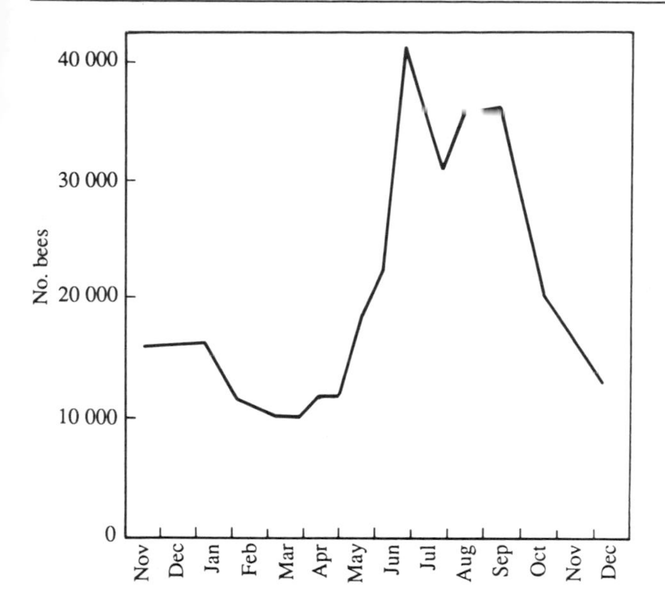

Figure 3.51a Adult population of a colony throughout the year, at Aberdeen, UK, latitude 57°N (Jeffree, 1955). Small experimental colony; see text for discussion.

adult population in Avitabile's (1978) colonies represented in Figure 3.51b dropped from 21 000 in November to 12 000 in March.

Section 6.2 discusses the adult population cycle of *managed* colonies, which the beekeeper controls so that he obtains the maximum financial return from them. Sections 6.3 to 6.5 show how he can achieve this optimal managed cycle.

3.51.2 Brood rearing

The brood-rearing cycle depends on many factors (see Table 3.51A) and on interactions between them. One is photoperiod (day length), and this follows an annual cycle determined solely by latitude. At 45° it varies by $7\frac{1}{2}$ h in the course of the year, and at 60° by nearly 14 h. Table 8.3A gives some related data. In the northern hemisphere the winter solstice, when the photoperiod is shortest, is on 21 December; the summer solstice, when it is longest, is on 21 June. In the southern hemisphere the dates are reversed.

In Figure 3.51b, curve A from Nolan (1925) provides an example of the annual brood-rearing cycle in the temperate zone (Washington, DC, USA, about 40°N). It shows a single smooth rise and fall in the amount of sealed brood, with a maximum at the end of May. Records for brood rearing in winter are very difficult to obtain, and usually a colony must be killed to get a single result. The broken line for the winter is based on Avitabile's (1978) results obtained in Connecticut, USA, at latitude 41°N; he sacrificed 4 colonies on each of several dates through three winters, so that he could count the numbers of eggs, larvae, pupae and adult bees in them.

Curve A rises abruptly after the winter solstice; brood rearing increased during the coldest months of the year, when there was no nectar and pollen and bees could not even fly out of the hive. When day temperatures became high enough for the bees to fly and early flowers were in bloom, the rate of brood rearing increased much more rapidly, and continued at a high rate until the end of July – since the colony did not swarm. (The temporary drop in June was due to lack of space in the hive and storage of nectar in the brood nest.) The highest daily rate of egg laying in any of Nolan's 22 colonies was 1587, and most maxima

Table 3.51A Factors having an influence on the annual cycle of brood rearing in honeybee colonies in temperate zones

Colonies are assumed to be healthy, and to have adequate hive space, food stores, and water.

		Annual cycle affected
1	latitude	photoperiod (day length)
2	latitude + geographical and other factors	climate
3	climate and weather	plant growth season when bees can fly and forage
4	native/wild plants	nectar/honeydew and pollen flows (series of minor flows, and one or a few major flows)
5	agricultural crops	additional major flows, which may be out of phase with cycle 4 above
6	genetic constitution of the bees	cycle is adapted to 4 above in the bees' native region
7	acclimatization of bees	cycle may show some adaptation to 4 and 5 in another region (e.g. Louveaux, 1966)

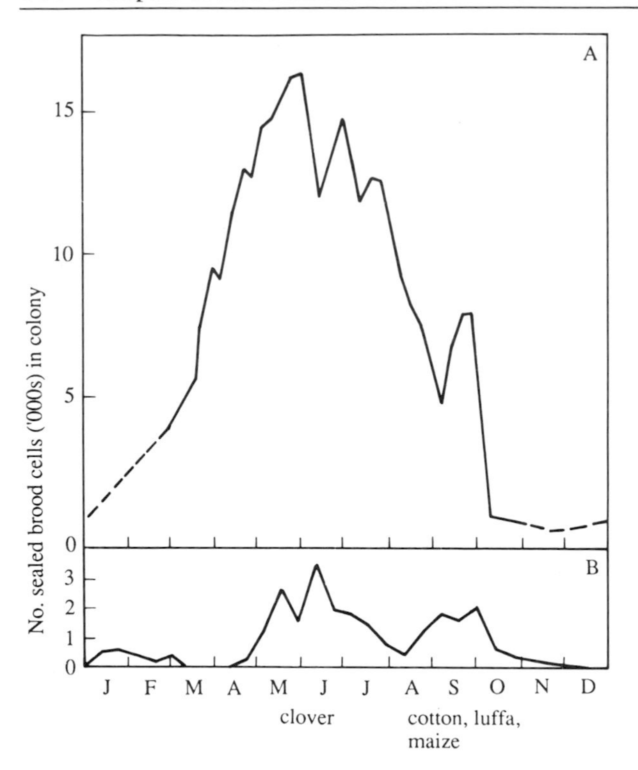

Figure 3.51b Brood-rearing cycles of (A) European honeybees (*Apis mellifera*) in the USA at 40°N, and (B) Egyptian honeybees (*A. m. lamarckii*) in Egypt at 30°N. Points plotted in curve A are the number of sealed brood cells in Nolan's (1925) colony 6, except that the broken line is based on the mean total number of brood cells in Avitabile's colonies (1978) for winter months. Points on curve B are the number of sealed brood cells in colony 3 of Hassanein and El Banby (1955), also in a Langstroth hive.

were between 1000 and 1500. At the end of the year brood rearing decreased until it ceased almost entirely in November – as Avitabile also found.

Curve B in Figure 3.51b shows the annual brood-rearing cycle of a colony of native Egyptian honeybees throughout the year. Colonies of these bees are very small, and are usually kept in the traditional mud hives (Section 1.42.1). The experimental colonies were in 5-frame Langstroth hives, except that some were transferred to a larger hive during a honey flow. The daily rate of egg laying never exceeded about 450. In the main, brood rearing occurred during the honey flows, and the pollen flow from maize. But, as in curve A, it dropped almost to zero in November to December, and showed a small rise after the winter solstice; brood rearing was not then sustained for long, presumably because of lack of pollen and nectar.

Kefuss (1978) investigated the direct relationship between brood rearing and photoperiod by using colonies in a flight room, in which he could change the photoperiod at will by altering the lighting regime. He showed that the rate of brood rearing was not affected by the photoperiod itself, but by the rate and direction of change (increasing or decreasing). Egg laying by the queen was stimulated when the photoperiod was increasing, and inhibited when it was decreasing.

Other factors besides photoperiod are important, and the interlocking effects of factors in Table 3.51A on the brood-rearing cycle were examined by Louveaux (1966, 1973) in France. He emphasizes that the annual cycle of colonies in their native habitat is adapted to the bees' successful survival there. In particular, strains in a heather region are becoming extinct, because their brood rearing cycle produced maximum adult populations in late summer, which is not wanted by beekeepers who take bees outside such regions.

3.51.3 Variation with latitude

At low latitudes, but still within the temperate and subtropical zones, the bees are active for a much longer part of their annual cycle than at high latitudes; the inactive winter part of the cycle is correspondingly shorter, or disappears entirely. This is demonstrated in Figure 12.62a, which presents scale-hive records obtained at different latitudes, including three in the temperate zones and one in the subtropics. The records are used in Chapter 12 to show the timing of honey flows in different parts of the world, and are arranged in order of latitude, starting with 60°N in Sweden and ending at 35°S in Argentina. The main factor governing the weekly changes in weight is the net amount of honey being stored in the hive – the amount added into store less the amount consumed. The peaks thus indicate honey flows from nectar or honeydew. The total weight of adult bees, brood, pollen and comb also affects the weight of the hive (see Section 12.62), and in many parts of the year, the week-to-week weight changes give an indication of the metabolic activity of the colony and of the environmental conditions, as explained below.

Colony 1, at the highest latitude (60°N, in Sweden), lost weight very slowly and steadily when the year started (from week 1 to week 15), and again from week 36 to the end of the year. On average the loss in colony weight for each of these 32 winter weeks was only about 120 g. The colony was clustered during this period, and using minimal energy to stay alive. After week 15 (mid-April), brood rearing increased rapidly and much more food was consumed. But foraging quickly started, and in time compensated for the stores used; after week 22 (mid-June) there was a net surplus; the following main flow lasted 4 weeks (25–28). For several weeks after it had finished, the colony (still large) consistently consumed more food than it collected.

Colony 2, at 40°N in Ohio, was active and consuming food during the 16 weeks up to late April. The main flow started earlier and lasted longer, and after it finished the colony remained active until the end of the year – consuming more food than it collected except for small net surpluses in 4 scattered weeks.

For colony 3 at 32°N in Arizona (just within the subtropics), and colony 6 at 35°S in Argentina (just beyond them), temperatures were high enough for flight throughout the year. Unlike colonies 1 and 2 at higher latitudes, these colonies gained weight during a few of the winter weeks, although there was a net loss in weight almost every week. Nevertheless the fluctuations indicate that nectar was collected in some weeks. Like colony 1, these colonies consumed much food for an extended period after the main flows.

These four scale-hive records show common features, and variations with latitude. All had one main flow period, and are likely to have followed cycles of adult population and brood rearing similar in principle to those shown in Figures 3.51a and 3.51b. There were additional minor flows at 32°N and 35°S, and several at 40°, but at 60° the growing season was so compressed that there was little flowering except during a single month. At 32°, temperatures were high enough for bees to fly and to rear brood throughout the year. In both hemispheres the active season for the colonies contracted as the latitude increased – and at 60° it was less than 20 weeks out of 52. At low latitudes the bees were active and consuming food even when there was no flow. At high latitudes almost the whole of the long dearth period was cold enough to keep the bees inactive, consuming very little food and maintaining themselves at a minimum cost to the beekeeper.

The records in Figure 12.62a indicate that in temperate zones the rise and fall of population in the active part of the season stretches over a much longer period at low latitudes than at high ones, and that management in spring, flow, and post-flow periods must be adjusted accordingly. Many parts of the subtropics adjacent to the warm temperate zones have great advantages in world beekeeping. First, temperate-zone *Apis mellifera*, which is amenable to management, can be used. Second, the seasonal cycle is sufficiently regular and predictable for management to be preplanned, and Figure 6.2a shows notional colony population curves for adult bees in managed colonies. Third, temperatures are high enough during most of the year for foraging and brood rearing. It is not surprising, therefore, that most of the honey exported as surplus on to the world market comes from countries with large areas in the subtropical zones; major exporters are Mexico, China, Argentina and Australia.

3.52 Seasonal cycles in the tropics

Less is known about the brood-rearing cycle in the tropics, but Hameed and Adlakha (1973) have published comparative results for *A. mellifera* and *A. cerana* in northern India.

In most parts of the tropics bees are not prevented by low temperatures from flying and foraging, or from rearing brood. Neither day length or its rate of change provide the bees with significant seasonal cues as they do in temperate zones, for nowhere does the day length change by more than 4 h during the year (compare Table 8.3A). Conditions likely to prevent flight and foraging are high temperatures (Table 3.32A) and heavy rain. Heat or excessive rain, and also drought, may prevent plants flowering, or may damage or destroy nectar, honeydew or pollen that is produced. Consequent beekeeping problems are dealt with in Section 7.3. Apart from the difficulties under which bees live in very hot dry seasons (Section 7.32.2), the variability and unpredictability of the timing of the rains present major difficulties for beekeepers. Except in permanently wet areas, the rains are needed before many of the bee forage plants can flower.

In Figure 12.62a, colony 4 was in the tropics at 5°S in the equatorial zone in Tanzania. The scale-hive record shows a double cycle within the year, characteristic of such low latitudes where the sun passes overhead twice in the year. There are two wet and two dry seasons, and honey flows (and brood rearing) follow the rains. Even after the most important flow ended in November–December, brood rearing was fairly continuous until May–June, and the only break in the year occurred in July (F. G. Smith, 1951).

Colony 5, within the tropics at 22°S in Brazil, stored some honey during 9 months out of the 12, the main flow being in the relatively dry winter. Brood rearing was greatest in autumn, after the end of the (wet) summer dearth, and the low weekly gains then – with two losses – indicate the increasing consumption of food as the colony grew.

3.6 FURTHER READING AND REFERENCE

Section 2.8 lists books that relate to both Chapter 2 and Chapter 3. In addition, the following deal especially with social biology; details about them will be found in the Bibliography.

Free, J. B. (1977) *The social organisation of honeybees*
Free, J. B. (1987) *Pheromones of social bees*
Frisch, K. von (1967) *The dance language and orientation of bees*
Michener, C. D. (1974) *The social behavior of the bees*
Seeley, T. D. (1985) *Honeybee ecology*
Wilson, E. O. (1971) *The insect societies*

PART II

Beekeeping with Movable-frame Hives

4

Movable-frame hives, their frames and accessories

This chapter deals with movable-frame hives such as are used in modern beekeeping throughout the world, based on the principles explained by Langstroth in his book on *The hive and the honey-bee* (1853). Movable-comb hives in which the combs have no frames are discussed in Chapter 10. One purpose of the present chapter is to set out requirements for movable-frame hives, and details about existing hives; a second purpose is to encourage readers to explore possible changes and developments that may be useful. Many new technologies have become available since 1853, and at the same time continued increases in labour costs have turned some cost-effective procedures into loss-making ones.

4.1 HIVES OF WOOD AND OTHER MATERIALS

4.11 Wooden hives

Ever since precision-made movable-frame hives first came into use in the 1850s, most have been made of wood. In recent years wood has become increasingly scarce in many countries, and prices have risen greatly, so hive makers sometimes have to use timber that is not the most suitable: from tree species not customarily used, sapwood instead of heartwood, and wood that has not been well seasoned.

4.11.1 Woods suitable for hives

Some of the timbers used for hives in different parts of the world are listed in Table 4.11A. In the temperate zones, most are easily worked softwoods from coniferous forests in the colder areas. One of these has special properties: arbor-vitae, now usually known as western red cedar, although it is not a *Cedrus*, but *Thuja plicata*. The wood contains oils that make it resistant to fungal decay and insect attack, and may be used without preservative treatment. It is straight-grained and can be worked to very precise dimensions. It is remarkably strong, yet its relative density is only 0.39, and it is the lightest wood of all those in Table 4.11A. Although it is expensive, this wood has been much favoured for hives in Britain, where mild damp winters promote decay, and where absorption of heat by the surface of unpainted wood can be useful. But in the region where the tree grows, in the west of North America, I have never known it to be used for hives; the price is regarded as too high, and pine or other timber is used instead. The climate there is somewhat drier, so decay is less of a problem. Also many hives are painted white to reflect radiant heat in summer (Section 4.13).

Hardwoods from the north temperate broad-leaved forests are not often used for hives, being expensive and also heavy; for instance oak (*Quercus*) has a relative density of 0.72 at 15% moisture content. Suitable softwoods are readily available from tropical forests and forests in the southern hemisphere, and a number of these are used regionally (Table 4.11A) according to availability and price. Thailand has large forests of teak (*Tectona grandis*), and this wood is used for hives and other beekeeping equipment, including frames. It has great dimensional stability after seasoning; the oils it contains make it very resistent to fungal decay and insect attack, and hence very durable. But its relative density is high (0.66).

In some countries hives are made from plywood, blockboard, or reconstituted timber. In timber-producing areas where narrow offcuts of wood can be purchased cheaply, it is possible to manufacture hives from the narrow strips bonded together laterally, and this can greatly reduce the selling price (Chapman, 1985).

Lime or basswood (from *Tilia* species) has been

much used for honey sections, because the grain is fine, straight and long, and the wood is very easily worked.

4.11.2 Hive construction

A beekeeper who makes his own hives must ensure that neither the timber, nor any adhesive used, has been treated with insecticide or fungicide toxic to bees. He must adhere exactly to dimensions of the type of hive chosen. The book *Let's build a bee hive* (Miller, 1976) gives down-to-earth details on all aspects of construction, and on materials, carpentry and working equipment, but if dimensions for the various parts of any one hive are required, other sources may be needed. Corner (1976) deals with Langstroth hives only. Difficulties with timber sizes are referred to at the end of Section 4.2.

Most small-scale beekeepers buy their hives, which are often supplied with the parts packed flat, accompanied by directions for assembly. Nails should be cement-coated or galvanized; for western red cedar the latter are essential. End-grain, whether exposed or within joints, is especially vulnerable to fungal attack, so preservative treatments (Section 4.11.3) should be carried out before the hives are assembled.

Table 4.11A Some trees whose wood is used or considered suitable for movable-frame hives in different countries

In column 1, shrinkage is from green to 12% moisture. In column 4, mean relative density is measured at 15% moisture content; the figures are from F. G. Smith (1987).

Botanical name	*Common name*	*Example of country*	*Relative density*
GYMNOSPERMAE: CONIFERAE (SOFTWOODS)			
Araucariaceae			
Araucaria spp.		S. America, S. Africa	0.55
Araucaria cunninghamii Ait. ex D. Don	hoop pine	Australia	0.56
Cupressaceae			
Cupressus spp.	cypress	Australia, S. Africa, USA	0.47
	southern cypress	USA	
Thuja plicata D. Don Ex Lamb. (shrinkage up to 2.2%)	western red cedar	UK	0.39
Chamaecyparis lawsoniana (A. Murray) Parl.	Port Orford cedar	USA	0.50
Chamaecyparis nootkatensis (Lamb.) Spach	yellow cedar	USA	0.50
Pinaceae			
Abies spp.	fir	France	0.45–0.48
Juniperus procera Hochst. ex A. Rich.	East African pencil cedar	E. Africa	0.58
Picea spp.	spruce	France, UK, USA	0.40–0.50
Picea abies (L.) Karsten	Norway spruce	Italy	0.47
Pinus spp. including:	pine	Australia, France, S. Africa, USA	0.42–0.67
Pinus lambertiana Dougl.	sugar pine	USA	0.43
Pinus patula Schlecht. & Cham.		Zimbabwe	
Pinus ponderosa Laws	ponderosa pine	USA	0.48
Pinus radiata D. Don	radiata pine	Australia, NZ, S. Africa	0.48
Pinus roxburghii Sarg.		India	
Pinus strobus L. (shrinkage up to 2.8%)	yellow pine	New Zealand, USA	0.42
Pinus sylvestris L. (shrinkage up to 4.4%)	Scots pine	France, UK	0.51
Pinus wallichiana A. B. Jacks	kail	India (sub-Himalayas), Romania	

(*continued*)

Table 4.11A *(continued)*

Botanical name	*Common name*	*Example of country*	*Relative density*
ANGIOSPERMAE (HARDWOODS)			
Bombacaceae			
Bombacopsis fendleri Pittier	pochote	Costa Rica	
Leguminosae			
Hardwickia pinnata Roxb.	piney	India	
Pterocarpus angolensis DC. (shrinkage less than 1%)	muninga	E., Central, S. Africa	0.64
Meliaceae			
Toona ciliata M. Roem	toon	India (sub-Himalayas)	
Myrtaceae			
Eucalyptus grandis W. Hill ex Maiden	saligna gum	S. Africa	0.48–0.64*
Eucalyptus saligna Smith	Sydney blue gum	Australia	0.92
Proteaceae			
Grevillea robusta A. Cunn.	grevillea or silky oak	S. Africa	0.58
Verbenaceae			
Tectona grandis L.f. (shrinkage up to 2.2%)	teak	India (C./S.), Thailand	0.66

* in Australia 0.82

4.11.3 Hive preservation

As the price of wood has risen, so it has become more worth while to do everything possible to prevent decay in hives. Decay in wood is caused by the growth of fungi, which is always initiated by a high moisture content. The life of hives can be increased by applying a fungicidal preservative to the outside of the hive parts that are exposed to the environment, and by placing them where they will be least subject to damp, and thus least liable to fungal attack; see also Section 4.48 on hive stands. Creosote or any other preservative that might harm the bees must not be applied by dipping, which coats the inside of the hive as well as the outside. Creosote is best avoided altogether, but floor boards – which are especially vulnerable to fungal attack (see below) – are sometimes soaked in creosote and very well aired afterwards.

Some wood preservatives are water repellent (e.g. waxes), and others are fungicidal (e.g. copper-8-quinolinolate which also controls both termites and powder post beetles); some mixtures are used for both purposes. The most effective preservatives are those that are well absorbed into the wood. Their application (and thorough drying) can be followed with one or more coats of paint. Painting, discussed below, gives some additional protection, but the paint layer can blister and crack, leaving the underlying wood exposed. Painting also allows a choice of colour for the hive, which can affect the temperature inside it (Section 4.13).

Some of the effective wood preservatives on sale should not be used for hives, because they can poison the bees, or contaminate honey or beeswax. They include tributyl tin oxide (TBTO), all materials containing arsenic (e.g. chromated copper arsenate, CCA), and pentachlorophenol.

Copper naphthenate is one suitable preservative for hive parts, provided the formulation does not include any insecticide. Another is copper-8-quinolinolate, which is less effective on wood in contact with the ground, but safe in contact with foodstuffs. A third is acid copper chromate; wood should be purchased that has already been treated commercially with this, and a dust mask worn when the treated wood is being sawn or machined. Many preservatives may be brushed on, or sprayed, or the wood dipped for about 3 minutes – which is a faster method. Soaking in a hot solution for several hours, or in a cold one for 2 days, gives still better penetration. Weatherhead (1987) gives a useful table of treatment schedules with copper naphthenate solutions, and stresses that airing for at least 10–14 days must follow treatment. By far the most effective impregnation is that done commercially using pressure or other equipment to apply preservatives that are soluble in an organic solvent or water. Robinson and French (1986) discuss methods appropriate to different preservatives.

An alternative is to dip the wood or hive parts in wax. Paraffin wax (Matheson, 1980) is the cheapest, and 5 minutes at 160° is sufficient; see below for painting instructions. Or microcrystalline wax may be used (Robinson & French, 1986), or a mixture of the two (Anderson, 1979). All wood must be dry before it is dipped.

Any painting follows the preservative treatment after several days of drying, except that hive parts dipped in paraffin wax should be painted with acrylic (water-based) paint immediately afterwards, because the paint is absorbed into the wood as the wax cools. A survey in Australia showed that honey supers lasted an average of 16½ years if painted only, or 22 years if a wood preservative had been applied before painting (Robinson & French, 1986). Kennedy (1986) assesses the usefulness of paints and other protective finishes, and their compatibility with wood preservatives. When painting hives, exposed parts should always be given at least two coats; some specialists recommend primer, then undercoat, and finally two coats of exterior-quality gloss paint. On hives, high quality latex paints are less inclined to blister and peel than oil-based paints.

The colony in the hive is a living organism whose respiration produces water. In cold weather this moisture can condense or freeze on the inner surfaces of the hive near the top, so these parts can be at extra risk. But floor boards are the most liable to decay: within two years if untreated hives are placed directly on moist soil, especially if vegetation is allowed to grow close around them. On the other hand hive boxes may well last twenty years – or much longer in some climates – if they are made of treated wood and placed on stands 20 cm or more from the ground, the ground nearby kept free of vegetation (see Section 6.62.2), and air flow ensured round the hives.

Termites are another important source of damage, especially to wooden floor boards on the ground. Copper-8-quinolinolate is reported to be effective against them. In Australia copper naphthenate did not seem to afford any protection (Robinson & French, 1986), but the insecticide CCA – listed above among the preservatives dangerous for bees – is effective and may be applied to wooden hive stands. The use of ferrocement hives or floor boards (Section 4.12) is one response to the termite problem. Protection of hives against attacks by mammals and birds is dealt with in Sections 11.54 and 11.55, and protection against environmental hazards in 11.8 and subdivisions.

Further information on hive preservation is given in the papers referred to above, and in others by Cross (1983), Kalnins and Detroy (1984), and Kalnins and Erickson (1986).

4.11.4 Double-walled hives

When hive boxes are kept permanently in an outer casing or in a bee house (Section 6.63), weather-resistant and heat-reflecting or heat-absorbing qualities are needed by the outer wall rather than by the boxes containing the frames. An example of 'double-walled' hives is the WBC hive which used to be common in England; it was introduced by William Broughton Carr in 1890. The inner hive boxes (made to take British Standard frames) were made of much thinner wood than normal hives, and the outer 'lifts' sloped slightly outwards to the bottom, making them telescopic; they were traditionally painted white.

Hives in German bee houses were also made of thinner wood; some had a removable cover board at the top, and others were opened from the back where there was a hinged door; see Figures 4.2c and 6.63b.

4.12 Hives of other materials

In areas where wooden hives are likely to last twenty years or more, wood will probably continue to be the generally preferred hive material. But it has serious disadvantages in some parts of the world, especially where termite attack is common, and in hot humid regions where fungal decay is rapid. Also, some tropical countries and islands are too small to support a hive manufacturing enterprise, and transport costs on imported hives of wood (or on the wood for making them) may not be the wisest outlay of money. In such places a more lasting substitute for wood is especially worth while, and the most usual are plastics, glass fibre, and ferrocement or fibrocement.

Plastic hives

Different types are on sale, for instance in European countries, Australia, Israel, and the USA. Some are designed for special types of management, but two representative hives that are manufactured to commonly accepted standards may be mentioned. Parker Engineering, 21 Shellharbour Road, Dunmore, NSW 2531, Australia, uses a very hard, high-density polypropylene with added ultraviolet-resistant stabilizers. Standard 10-frame and 8-frame Langstroth deep hive boxes (interchangeable with wooden hive boxes) are supplied, in the flat. The walls are 19 mm thick with 9 mm cores. The beekeeper assembles the boxes by using a rubber mallet, and secures them with screws inserted in pre-drilled pilot holes. Floor boards and migratory roofs are 9 mm thick. The price of a hive is the same as that of a treated and painted wooden hive, and its weight is 10% less.

I have seen many hive boxes made of rigid polyethylene in use on islands in the Caribbean area (Figure 4.12a), imported from the USA. Some of the early hive boxes and floor boards used to split at the corners, but this trouble has been overcome, and the hives were liked there, being completely resistant to termite damage and fungal rot. The white plastic hive roofs have proved very satisfactory.

Expanded plastics have a low density and a low mechanical strength, but provide good thermal insulation. The wall thickness needed for adequate mechanical strength gives too high a thermal insulation for normal colonies, but a high thermal insulation can be beneficial for very small hives, whose rate of heat loss is relatively high. Examples are queen mating hives (Section 8.23) and 'disposable pollination units' (Section 9.21). Thomas Fils SA (45450 Fay-aux-Loges, France) sell an Aluruche in both Langstroth and Dadant dimensions, made of expanded polystyrene sandwiched between aluminium sheets; the total thickness of the walls is 20 mm. The hive consists of one deep and one shallow hive box, floor board with integral stand, and a telescopic roof reinforced for use during migration. Its weight is 40% of that of a similar wooden hive, and the price 80% higher.

Hives somewhat similar to the two above are reported by various authors to be weather- and decay-resistant, and not to need maintenance (Tew, 1986). Bees chew the soft types of expanded plastic, but a facing of aluminium or wood on the interior prevents their gaining access to it. A resin coating can protect the plastic surface to some extent.

Plastic materials have been used quite widely for comb foundation and artificial comb and for special mountings for comb honey that replace wooden 'sections' (Sections 4.32, 4.33); Tew (1986) gives references. According to a survey in the USA (Ferracane, 1987), plastics are most commonly used for the following other hive parts and accessories (in order): feeders, floor boards, roofs and inner covers, bee-escapes, queen excluders.

Glass-fibre hives

Hive boxes of glass fibre were in use in the USA in the 1950s, but their manufacture was discontinued because of the high cost. A glass-fibre hive is described by Soares (1981) in Brazil, and Tew (1986) refers to others.

Ferrocement and fibrocement hives

The ferrocement used for hives such as that shown in Figure 4.12b is concrete in which wire mesh has been embedded to provide increased strength; it is really ferroconcrete, one of the types of reinforced concrete. Another type, known as fibrocement, is produced by mixing asbestos with the cement. Both materials are cheap, being made from cement, sand and water, with the additional strengthener; usually none of the constituents except the cement need to be imported. The materials are resistant to fungal decay, including dry rot, and to attack by termites and boring insects. The hives are too heavy to be moved easily by hurricane-force winds; an empty ferrocement Langstroth 10-frame hive box weighs 13 kg. In Sri Lanka cement extended with rice husks has been used.

Figure 4.12a Part of an apiary on St Kitts, Leeward Islands, 1986.
Plastic hive boxes can be identified by their struts, and wooden ones by their handholds.

Figure 4.12b Ferrocement hive box, with floor board on the left, and shallow telescopic roof on the right (Hobson, 1983).

Full instructions for making ferrocement hives are given by Hobson (1983), based on 15 years' experience on a Pacific island subject to typhoons. Fibrocement for hives has been tested in Angola and, with added cellulose, in Costa Rica (Ramírez, 1976; Ramírez & Pontigo, 1979). In a greenhouse, empty fibrocement hives tended to be somewhat cooler than wooden hives, but small colonies used in test hives could not easily control the hive temperature or relative humidity.

In New Zealand hive roofs and floor boards of ferrocement or fibrocement are being increasingly used (e.g. Johnson, 1958; Forster, 1959), the rest of the hive being made of wood as usual.

4.13 Colour of hives

Paint on a wooden hive is not an adequate preservative on its own, but it can enhance the hive's appearance, and is a useful way of changing the colour of the hive and thus of reducing or increasing absorption of heat radiation that falls on it. Reflective colours, especially white, reduce heat absorption and lower the temperature inside the hive; absorbent colours, especially black, increase this temperature. The choice of colour is most important where hives must stand out in the sun in very hot conditions.

Changing the colour of only the roof can have a significant effect on the hive temperature, as is shown by experiments made with empty hives, in places where air temperatures were high. In Israel, where they rose to 30°, a white hive with a white roof was coolest, and its internal temperature did not differ much whether it was in shade or sun (Lensky, 1958). A hive painted with silver paint and with an untreated galvanized roof was hottest. In Manitoba, Canada, Mitchener (1940) tested roofs of different colours on hives all painted blue, at shade temperatures up to 37°. The temperature inside a hive was about 5° lower if the roof was white than if it was black or unpainted galvanized metal; these latter hives were around 10° hotter than the shade temperature. White plastic hive roofs are manufactured, and they offer an alternative to painted metal roofs.

Dark-coloured wood absorbs heat, and in cold temperate areas hives are often unpainted. In northern latitudes with hot summers and very cold winters, hives may be painted white, and covered with black plastic during the winter (Section 6.54.1).

Painting hive fronts in different colours, or painting coloured patterns on them, can help the bees to locate their own hive, and thus reduces drifting (Section 6.62.1). In areas where most hives are unpainted, or painted white, a beekeeper who paints his hive boxes, roofs and floorboards in distinctive colours may well reduce the risk of theft.

4.2 SIZES AND DIMENSIONS OF HIVE BOXES AND FRAMES

The following is a much simplified account of the reasons why modern hives have the general style, shape and size they do today. Except that external fittings such as entrance porches and additional outer walls have been largely discarded for economic reasons, the hives are of a type that has hardly changed for well over a hundred years. In the interests of future progress in beekeeping, it seems important that beekeepers should bear in mind which aspects of the hive are imposed by the bees used in them (and are therefore mandatory, but may differ for different bees), and which are at the beekeeper's discretion – depending on his convenience, what he can afford, and other factors whose relative importance changes with changing prices and with the availability of new materials.

A modern movable-frame hive (Figure 4.2a) consists of precision-made rectangular hive boxes (sometimes called the hive bodies) superimposed one above the other in a tier – hence the term *super* for the upper ones. The number of boxes is varied seasonally according to requirements. In each box, a set of framed combs is suspended – like the files in a suspension filing system – from two parallel metal runners, one fixed along a rebate at the top of each end of the box. A frame is made with each end of the top-bar extending beyond the end-bar to form a lug, and the lugs rest on the two runners (Figures 4.2a, 4.24a).

For any one type of hive, the dimensions of the hive box depend on the dimensions of the frames (Section 4.24) and the number of brood frames it is required to hold. The width of any gap where bees move about in the hive between two facing surfaces must be equal to the 'bee-space' for the worker bees, which varies slightly between species and races of honeybees according to their body size. The bee-space is 8 mm for most European *Apis mellifera*.

Figure 4.2b shows where a bee-space must be left in a hive. It is needed between the outside end of each frame and the inner hive wall opposite it. It is also needed between the opposite surfaces of completed, sealed worker brood combs, and this requirement determines the distance the beekeeper must allow for the 'comb spacing', i.e. the distance between the midribs of adjacent combs (or of adjacent sheets of comb foundation); see Figure 4.2b, *right*. In addition, when one hive box is placed on top of another, a bee-space is needed between the top of the frames in the lower box

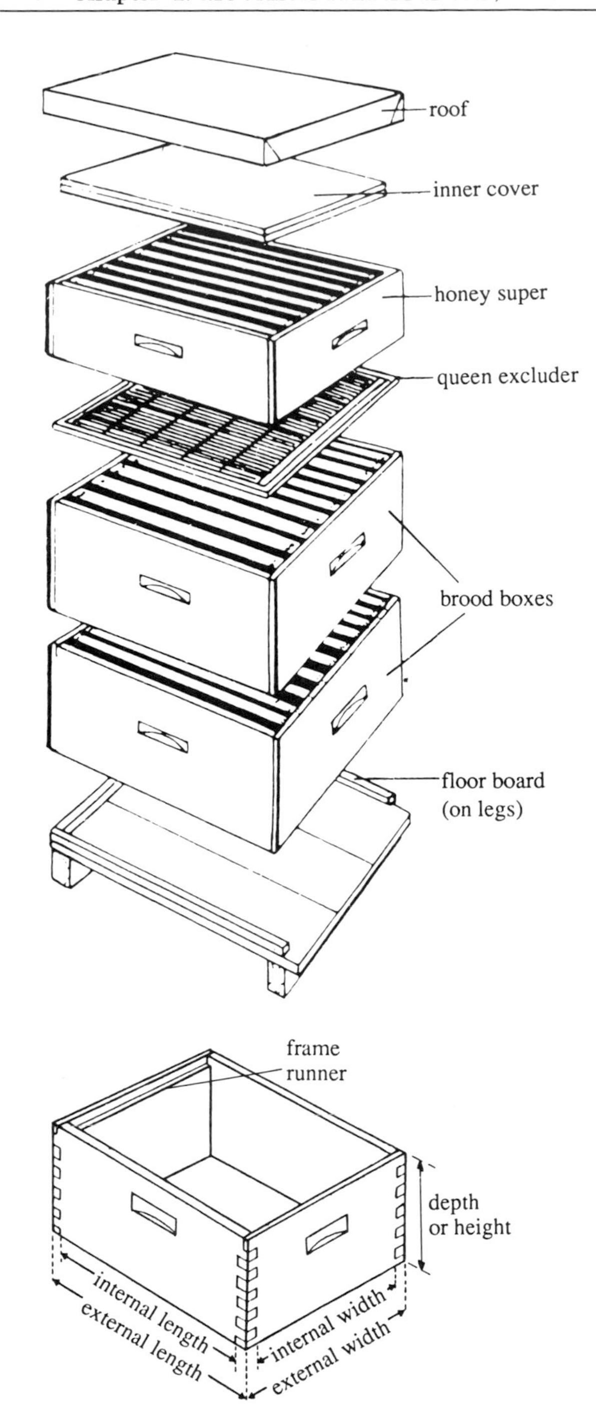

Figure 4.2a Exploded view of a movable-frame hive showing the component parts; *below*, an empty hive box to show one of the frame runners.

and the bottom of the frames in the upper box. The depth of the hive box is thus the depth/height of the frames plus the bee-space. Hive boxes are made either with a 'bottom bee-space' (the frames flush with the edges at the top) or with a 'top bee-space' (frames flush at the bottom).

It is practical for a beekeeper to choose whatever

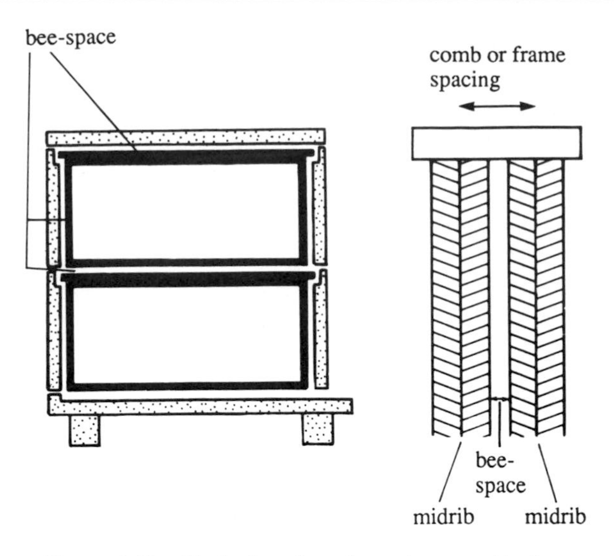

Figure 4.2b Vertical sections through a movable-frame hive, and parts of two brood combs in it, to show the bee-spaces.
This hive has a top bee-space, i.e. above the frames in each hive box. There is a bee-space between frame end-bars and the hive wall, between frames in one hive box and those in the box above, and (*right*) between adjacent sealed brood combs.

type of hive is most common in his locality, unless there is a good reason why some other hive would be better for him. He can then take advantage of help from neighbouring beekeepers, for instance purchasing equipment that is interchangeable with his own.

For several types of hive, Table 4.2A gives dimensions of the hive boxes, and whether the bee-space is at the top or bottom, in the columns numbered as follows.

1, 2 Langstroth, the type most commonly used in the world as a whole.
3, 4 Modified Dadant and USSR Dadant which, with Dadant-Blatt (5), constitute the second most common type.
6 British National (included because of the advice given in this book – which is published in the UK – that the reader is likely to be best-off if he uses the hive most common in his locality).

Table 4.2B gives dimensions of the frames for hives 1–6 above, and Table 4.2C shows where to find more complete dimensions of these hives, and of some others that are used in different countries. Dimensions of roofs and floor boards, which vary according to the individual hive style and type (Sections 4.41, 4.42), are also given in publications listed in Table 4.2C.

Whatever type of hive a beekeeper uses, all hive parts, frames and accessories must be compatible, i.e. conform to the same set of dimensions, and all hive boxes must have either a top bee-space or a bottom

Table 4.2A Dimensions of hive boxes: Langstroth, Modified Dadant (MD), Dadant-Blatt (DB), Standard (Modified) British National (BN)

Dimensions are in mm except where otherwise stated; columns 3 and 6 are converted from imperial units (1 inch = 25.4 mm). Length is used for the dimension parallel to the frames.
Comb areas quoted (cm²) include both sides of the comb; the number of worker cells in the comb is approximately 4 times this area, for a cell width of 5.35 mm (Table 4.25A).

Hive	1 *Langstroth, NZ adopted*	2 *Langstroth, range*	3 *Typical MD (USA)*	4 *Typical Dadant (USSR)*	5 *Typical DB (France)*	6 *BN (UK)*
Cross-section						
external: length	505	500 to 515	508	524	500	460
width	405	405 to 435	470 413[a]	524	500	460
internal: length	465	452 to 477	464	450	450	371
width	365	362 to 385	425 375[a]	450	450	422
Brood box						
height	238	238 to 245	298	320	307	225
total comb area (cm²)	17 200		24 442	26 892	27 216	14 157
volume (litres)	40.4		58.8	64.8	62.2	35.2
Shallow honey super						
height	185	145 to 185	168[b]	165		149
volume (litres)	31.4		33.1	31.4		23.3
Frames in the hive box						
brood box:						
no. frames	10	10 or 11	11	12	12	11
spacing	33	32 to 36	38	37.5	37.5	38
honey super:						
no. frames	8		10	12	10, 11	9, 10, 11
spacing	43		41	37.5	45, 41	46, 42, 38
Bee-space	bottom	either (see text)	top	top	top	bottom

[a] 10-frame type
[b] also 150, 137 mm and other heights
Sources: column 1, Walton (1975); 2, sources available for different countries; 3, Thorne (1987) and Dadant & Sons (1987*a*); 4, Bilash (1987); 5, Mertzig (1988); 6, BS 1300:1960 (British Standards Institution, 1960).

bee-space. Column 2 in Table 4.2A shows that dimensions of hives of a single type (Langstroth) can vary considerably from place to place. For this reason all hive parts should be purchased in the same country, and if possible from the same manufacturer.

There are some countries where almost all beekeepers use hives of the same type, especially: (a) where the beekeepers form a fairly homogeneous group (e.g. New Zealand), and (b) where modern beekeeping was introduced comparatively late, and a well tried hive that suited its conditions was adopted as standard (e.g. China).

Countries with good opportunities for full-time professional beekeeping have tended to adopt a single hive type, or two or three if the country is large and with very diverse beekeeping regions, like the USA. Beekeepers in such countries rely on migration and appropriate apiary management to get economically rewarding honey yields; they use as many hive boxes for brood as are necessary, and as many honey supers as are convenient.

At the other end of the spectrum are some countries where beekeepers value having a large variety of hives to choose from. In most of these countries modern

Table 4.2B Dimensions of frames: Langstroth, Modified Dadant (MD), Dadant-Blatt (DB), British Standard (BS)

Explanatory notes and sources for Table 4.2A apply also to this table.

Frame	1 *Langstroth, NZ adopted*	2 *Langstroth, range*	3 *MD (USA)*	4 *Dadant (USSR)*	5 *DB (France)*	6 *BS (UK)*
In brood box						
external: length	450	446 to 452	448	435		356
height	230	214 to 250	286	300		216
internal: length	430		429	415	420	337
height	200		259	270	270	191
top-bar length	482	480 to 495	483	470		432
comb area (cm^2)	1720		2222	2241	2268	1287
In shallow honey super						
external: length	450	446 to 452	446	435		356
height	177*	137 to 177	159	145		140
internal: length	430		429	415	420	337
height	147		133	115	135	114
comb area (cm^2)	1264		1141	955	1134	768

* 159 mm is more usual; New Zealand uses a deeper super than most countries.

Table 4.2C Sources giving dimensions of parts of movable-frame tiered hives (and their frames)

Many sources include a description and/or drawings, but do not always quote a complete set of dimensions.
i = in imperial units; m = in metric units; ( ) = conversion at 1 inch = 25.4 mm

Hive	*Country*	*Reference*	*Units*
Langstroth	Australia	F. G. Smith (1966*b*)	i
		Walton (1975)	i (m)
	France (Pastorale)	Merlin (1978)	m
	Greece	Walton (1975)	m
	Mexico	Walton (1975)	m
	New Zealand	Tables 4.2A, 4.2B	m
		Walton (1975)	i
	S. Africa	Voges (1982, 1983)	m
	UK	Walton (1975)	i (m)
		Walsh (1983)	m
	USA	Walton (1975)	i (m)
		Miller (1976)	i
Modified Dadant	African Dadant	Portugal Araújo (1960)	m
		F. G. Smith (1961*b*)	i (m)
	France	Alphandéry (1981)	m
	USA	Tables 4.2A, 4.2B	(m)
Dadant-Blatt	Belgium	Mátyás (1932)	m
	France	Tables 4.2A, 4.2B	m
		Merlin (1978)	m
	Italy	Mátyás (1932)	m
	USSR	Tables 4.2A, 4.2B	m
		Mátyás (1932)	m
		Avetisyan (1978)	m

Table 4.2C *(continued)*

Hive	*Country*	*Reference*	*Units*
British Standard 1300:1960 (all UK):			
National		Tables 4.2A, 4.2B	(m)
		Ministry of Agriculture (1968)	i
		Walsh (1983)	m
National Deep		Walsh (1983)	m
		Heath (1985*b*)	i
WBC		Ministry of Agriculture (1968)	i
		Heath (1985*b*)	i
Commercial/Modified Commercial		Ministry of Agriculture (1968)	i
		Walsh (1983)	m
		Heath (1985*b*)	i
Smith		Ministry of Agriculture (1968)	i
		Bielby (1977)	m
		Walsh (1983)	i
Langstroth Deep (Jumbo) UK		Walsh (1983)	m
		Heath (1985*b*)	i
Buckfast Dadant	UK	Heath (1985*b*)	i
Layens	Belgium, France	Mátyás (1932)	m
		Merlin (1978)	m
Voirnot	Belgium, France	Mátyás (1932)	m
		Merlin (1978)	m

beekeeping began early (before about 1870), and a number of beekeepers devised their own hives and wrote books about them, which perpetuated their use. Standardization of several hive types encourages their continued manufacture, often by independent suppliers of beekeeping equipment. Honey yields may not be very high in such countries, and many people who keep bees do so as an interesting sideline, not as a profession. Britain and Germany are examples. The largest British hive manufacturer offers 8 types, each with a choice of top or bottom bee-space. Germans have been even more prolific in devising hives and writing about them; Armbruster (1935) listed 95 different movable-frame hives exhibited in Leipzig in 1928. (He gave the weight of each hive per litre of space inside, which varied from 0.18 kg to 0.70 kg.) Leipzig is now in the German Democratic Republic, and centralized organization has reduced the number of hives available to two – one opened from above, and one from the back. In the German Federal Republic a single 1987 catalogue offered 13 hive types. Free-standing tiered hives did not come into wide use there until well after the 1960s, and even now represent only a few per cent of the total. Many other top-opening hives, as well as back-opening hives, are currently on sale. Englert (1986) describes beekeeping in the GFR with back-opening hives, such as those shown in Figures 4.2c and 6.63b.

Several attempts have been made to come to grips with the possibility of international standardization of hives; the Langstroth hive is discussed in Section 4.21. Townsend discussed some of the problems (1978*b*) and made certain recommendations (1982), and Kleinschmidt (1981) has written from the Australian standpoint. Wood is normally the cheapest suitable material, and one factor that affects the choice of hive dimensions is standardization of timber sizes. International timber-size standards have been established and are quoted by Townsend (1980), but national standards vary according to what is most easily available in the country, and whether metric or another system of measurement is in use. Standard hive dimensions that are set outside a country might involve inefficient use of timber, or might make its use impossible where the size required is slightly larger than the nationally adopted timber size.

Figure 4.2c German two-storey back-opening hive, with one brood frame removed (photo: E. Englert). A tray inserted level with the floor board provides a working shelf, and the smoker is hooked over its edge.

4.21 The Langstroth hive

This hive is now used in many countries, including those listed below.

more or less exclusively	*used widely*	
Argentina	Albania	Japan
Australia	Brazil	Mexico
Canada	Chile	Portugal
China	Greece	Romania
Egypt	Israel	Spain
New Zealand	Iran	Turkey
South Africa	Italy	USA
	many others in Latin America	

In Table 4.2A, columns 1 and 2 relate to the Langstroth hive, in its simplified form developed by the A. I. Root Company in USA in the 1870s. This is the most common hive, world-wide; its early spread was due partly to Langstroth's book (1853), and partly to A. I. Root's personal evangelism which led him to offer hives free to Christian missionaries anywhere in the world (Burtt, 1950).

At a meeting of hive manufacturers from different countries, held during the 27th International Apicultural Congress in 1979, it was agreed to take steps towards establishing a world standard for dimensions of the Langstroth hive. Apimondia (the International Federation of National Beekeepers' Associations) set up a Normalisation Commission to pursue the matter, but no way has yet been found to achieve this international standard (Thomas, 1987). One consideration is the difficulty with timber sizes; another is the capital investment beekeepers have in their present hives, and their need to continue to purchase (or to make) additional hive parts compatible with their existing equipment.

In the absence of any world standard, Table 4.2A (column 1) quotes dimensions adopted in New Zealand as a representative set – not necessarily better than any other, but agreed upon after much enquiry and consideration, for reasons set out by Walton (1974, 1975). Column 2 shows the variation that occurs in different countries, and Table 4.2C cites publications that give descriptions and/or dimensions of the Langstroth hive in various countries. A top bee-space is used in most countries, but a bottom bee-space in New Zealand and China.

Over many decades, and in many countries, Langstroth hive boxes have proved to be convenient for handling, but there is nothing magic about their particular dimensions; it has been suggested that Langstroth himself adapted a box that had held champagne bottles.

4.22 Other named types of movable-frame hive

In the early years of movable-frame beekeeping, a beekeeper had no prototypes to follow except hives that he saw, and descriptions that were published. When Charles Dadant emigrated from France to the USA in 1863 (see Section 1.44), he took with him the 12-frame 'Dadant-Blatt' hive; Johann Blatt was Swiss. The brood frames were 457 × 305 mm externally (Burtt, 1950), but were later adjusted to the size used in Moses Quinby's hive, a height of 286 mm, and a length variously quoted, for instance 470 and 486 mm. The frame finally settled on (448 × 286 mm) is known as the Modified Dadant, and the hive box to take 12 of these frames is the basis of the Dadant hive as now used in a number of countries (Table 4.2A). The African Dadant hive (F. G. Smith, 1961*b*) is virtually identical with the Modified Dadant except that the frame spacing, and the width of self-spacing frame end-bars, are slightly smaller to accommodate tropical African bees (Table 4.25A). The Dadant-Blatt hive

(Table 4.2A) is used in many French-speaking and some other countries; in the GFR it is more common than the Langstroth. In the USSR the majority of hives are of the Dadant type (Table 4.2A); long hives that take Dadant-type frames are also used. The Dadant brood box has a total comb area about 1½ times that of the Langstroth, and where the Dadant type is preferred, it is because of this larger size. No other movable-frame hive is used as widely as the Langstroth or Dadant types; Campbell (1951) reckoned that the Dadant type might then have been the most widely used in the world.

The only comprehensive study I know on movable-frame hives used in continental Europe was published from Hungary, in German, by Mátyás (1932). All stocks of the book, and all Mátyás's bees, were destroyed during the war, and he died in 1946. His book gives dimensions of frames for many named hives (as used in different countries in Europe) in addition to those in Table 4.2A. In the following list, back-opening hives (see Figure 4.2c) were dominant in countries marked *, and top-opening hives in the others.

Belgium: Sherlin
Czechoslovakia*: Balogh, Boczonádi, Zoller
France: Baldensperger
Germany*: Berlepsch, Dathe, Freudenstein, Gerstung, Kuntsch, Reidenbach, Schulz, Zaiss, Zander, and many regional types
Hungary*: Boczonádi, Horthy, Hunor, Mogor
Poland: Lewicki, Root
Yugoslavia: Groznadić, Zivanović

Also included are Danzenbaker, Gallup and Schenk hives in the USA, and 'national' hives of Austria*, Belgium, Denmark, Netherlands, Norway, Poland, Sweden and Switzerland*.

The many hives used in Britain are described by Heath (1985*b*), who gives some dimensions in imperial units. They are discussed in the order: WBC (named after William Broughton Carr), Smith, National, British Commercial, Langstroth, Langstroth Deep (Jumbo), Modified Dadant, National Deep, Buckfast Dadant, and a few more. Walsh (1983) provides detailed dimensioned drawings of most of the hives, as shown in Table 4.2C.

4.23 Brood boxes and honey supers

A modern hive has one or more *brood boxes* in which the queen lays eggs and brood is reared. Honey is stored above the brood nest, in *honey supers*, i.e. hive boxes superimposed on the brood box(es); some American authors, however, use the term super for any hive box. For simplicity in discussion here, it is assumed that the queen is confined to the hive box(es) to be used for brood rearing, by a queen excluder (Section 4.45).

The dimensions of the brood box determine the number and size of the frames in it, and thus the total comb area and the number of cells available for the queen to lay in. These data are given for several types of hive in Table 4.2A. Towards the end of the last century, beekeepers were trying out different sizes of frames and hive boxes. One of their aims was to have a brood box that provided the queen, at her peak rate of egg production, with enough cells to lay in for 21 days, i.e. until the cells first laid in were vacated by emerging adult bees. When a box large enough to satisfy this condition was used also for honey storage, it proved to be too heavy for most beekeepers to lift when full of honey. So 'shallow' boxes came to be used for honey storage from one-half to three-quarters of the depth of the 'deep' box used for the brood nest; see Table 4.2A.

By the late 1800s beekeepers in northern Europe and North America were seeking more prolific queens than those available to them, and they imported such queens from Italy and elsewhere. The brood box in use was too small to accommodate the peak rate of egg laying by the new, more prolific, queens. The custom therefore arose of assigning to the brood nest 'a double brood box' (i.e. two brood boxes) or 'one and a half' (a deep and a shallow). It did not seem to matter that this practice changed the shape of the space available for the brood nest, but it often gave too great an enlargement of this space.

The largest brood box listed in Table 4.2A is the Dadant used in the USSR (capacity 65 litres), and the smallest the British National (35 litres); the NZ Langstroth has a capacity of 40 litres. A worker honeybee develops from egg to adult in 21 days, and the total comb area in the different brood boxes (Table 4.2A) could accommodate the following daily rate of egg laying by the queen, assuming that this is constant:

Dadant-Blatt and USSR	1360
Modified Dadant	1220
NZ Langstroth	860
British National	700

Hive boxes of other shapes were advocated elsewhere, in line with tradition or ideology. In Germany, where frame hives were preceded by upright logs (Section 9.12), tall boxes and frames were usual. A hive had no more than two boxes, because hives were packed tightly together in a bee house, in one, two or three tiers. In some countries, individual beekeepers felt that bees needed a hive that resembled a natural cavity more closely, so octagonal, trapezoidal and other shaped hives were devised. Hives of such shapes are still promoted from time to time, but are not in wide use; they are more expensive to make in wood than a rectangular box, more expensive to furnish with frames, and more awkward to work with.

One of the greatest exponents of colony management for honey production was C. L. Farrar (1946, 1953) in Wisconsin, USA. He used rather shallow 12-frame Dadant boxes throughout his hives, because he found deep 10-frame Langstroth boxes unsuited to his two-queen management system (Section 8.29.1). He settled on a depth of about 170 mm, and used up to 8 of these shallow boxes for a single-queen hive, and 12 for two queens. These boxes were lighter to handle than the standard Dadant boxes 300 mm deep; they gave more flexibility in colony control; and the total height of the hive was usually less. Colonies were found to winter as well in several of the shallow boxes as in a smaller number of deep boxes. In 1953 I found some beekeepers in Dr Farrar's area using the same system, but none farther afield. This was due partly to the extra expense (a shallow box costs almost as much as a standard one), but I think mostly to the inability of other beekeepers to match his expertise in colony management – which produced average honey yields approaching 200 kg per hive.

Honey supers

The honey supers of a hive normally have the same length and width as the brood boxes. The size (volume) of a super – and its weight when full of honey – can thus be varied only by altering its depth (height). Table 4.2A gives a usual depth for the super of the hives listed, but other depths may also be available; most are between half and three-quarters of the brood-box depth. F. G. Smith (1966*a*) discusses the various benefits of a box for the honey super that is shallower than the brood box. Clearing a super of bees, by using a repellent or a stream of air, is more effective with shallow supers (Section 5.32), and many beekeepers find that more honey can be extracted per hour from shallow frames than from deep ones.

In Langstroth hives, a shallow super 168 mm deep is used in many countries. A. I. Root Co. in the USA sells this, and also a super 144 mm deep (for frames 137 mm deep). Full of honey, the super weighs about 20 kg, whereas a deep box can weigh up to 38 kg. New Zealand adopted a box somewhat deeper, 185 mm. Steinhobel (1972) in South Africa is a commercial migratory beekeeper who works entirely with boxes having the 10-frame Langstroth cross-section and a depth of 200 mm. The choice of super depth is a matter of convenience, and of timber sizes available; it depends also on the extent to which the supers can be handled mechanically.

Beekeepers who use honey supers of the same depth as brood boxes avoid the complication of handling and storing boxes and frames in two sizes, since every frame fits every box. Nevertheless, they should not use combs indiscriminately for brood or honey. Honey extracted from dark brood combs was found to be darker than honey from light combs in the same super (Townsend, 1974*b*); while the nectar is being converted into honey it may absorb substances from the cell walls (including some from propolis), which darken it. Also brood cell walls are slightly thicker, and their capacity is correspondingly less.

Another difference between the combs the bees use for brood rearing and for honey storage, which has nothing to do with the dimensions of the frames or hive boxes, is the cell depth, and the greater comb spacing permissible in honey supers; these are discussed in Section 4.25.

4.24 Frames

Figure 4.24a shows a frame for use in a modern hive. The length and depth (height) of the frames are determined by the length and depth of the hive box (Figure 4.2a). The number of frames in a hive box is determined by the width of the box and the spacing required between the frames. Alternatively, the frame dimensions determine the length and depth of the hive box, and their spacing and number determine the width of the box.

Table 4.2B gives dimensions of the frames for hives represented in Table 4.2A, and the comb area per frame (the total comb area in the brood box is in Table 4.2A). The comb area of a single Dadant, Dadant-Blatt or USSR Standard brood frame is about 30% more, and that of a British Standard brood frame 25% less, than that of a Langstroth brood frame.

Most frames are made of untreated wood, and are supplied for the beekeeper to assemble, together with suitable nails, and sometimes glue. Joints are often mortised; they must be strong, in order to take the strain if the frame has to be prized out of a hive box when stuck to it with propolis, and also when the frame is spun in a honey extractor.

The top-bar is designed so that foundation can be attached securely to it (see end of Section 4.31), and a lug at each end rests on the frame runner in the hive box (Figure 4.2a, *below*). The bottom-bar usually consists of two separate strips, the foundation being

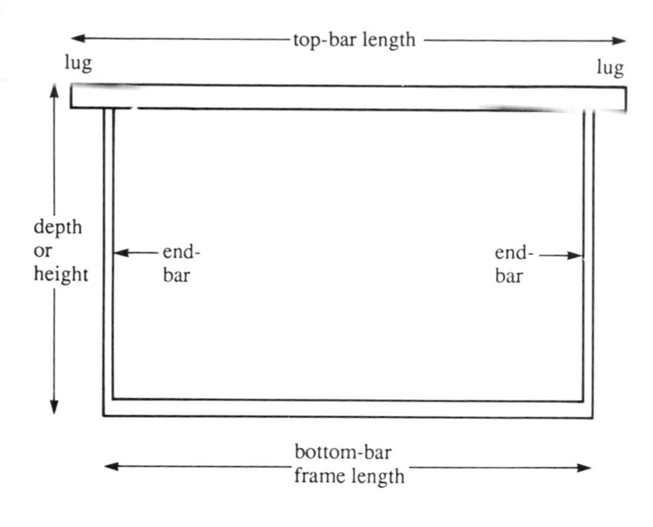

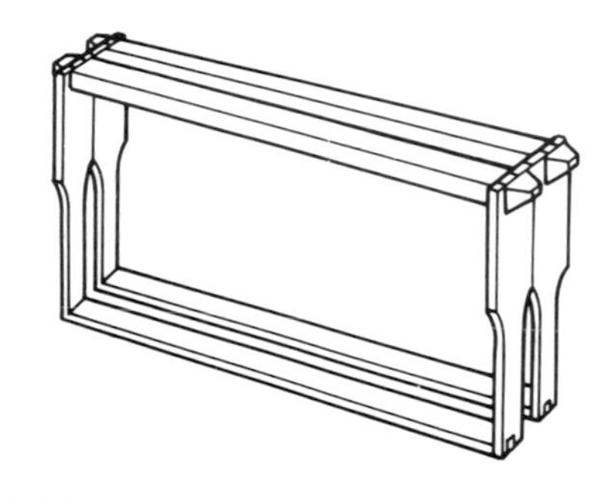

Figure 4.24a Frame showing component parts; dimensions indicated are external.
below Two frames with Hoffman self-spacing end-bars; the top-bars have short lugs, for a Langstroth hive.

inserted between them. End-bars are straight narrow strips of wood, sometimes with a central groove along the inner side to receive the edge of the sheet of foundation; they can be made to serve as frame spacers as explained in Section 4.25.

A few beekeeping suppliers sell plastic frames, or integral plastic frame-plus-foundation or frame-plus-comb with half-depth cells. These are discussed in Section 4.32.

4.25 Comb spacing and frame spacers

The 'comb spacing' or 'frame spacing' in a hive box is the distance between the adjacent comb midribs, or sheets of foundation, in their frames.

Spacing of brood combs

In the brood box the spacing should be the same as the centre-to-centre distance between adjacent worker brood combs in a wild nest built by similar bees. The distance is equal to the thickness of the comb (twice the cell depth required by the immature worker) plus the bee-space required by the adult bees to move about in the passage-way between opposite comb surfaces. The comb spacing is thus predetermined by the size of the worker bees. If too small a spacing is used, the bees cannot rear brood on both sides of the combs. If the spacing is too large, or is irregular, they are likely to build 'burr' comb in any over-large gap; see Figure 4.25a (*below*), also Figure 7.32a.

Spacing for brood combs for various bees is given in Table 4.25A; it is about 35 mm for European bees. Bees will, however, tolerate a certain latitude; for instance Charles Dadant (Section 4.22) used 38 mm because he believed that it reduced swarming. Some beekeepers with darker (smaller) bees prefer to use 11 frames, made to give 32 mm spacing. When the metricated Langstroth hive was introduced in New Zealand (Walton, 1975), the frame spacing was set at 33 mm, for use with Italian-type bees.

The beekeeper must use a frame-spacing device to space the brood combs at the appropriate distance, and most of the world's beekeeping suppliers provide spacers for European *Apis mellifera*. (It may be more difficult to obtain spacers for tropical *Apis mellifera* and *A. cerana*, which must be narrower, Table 4.25A.) Self-spacing Hoffman frames (Figure 4.24a, *below*) have the top part of each end-bar made to the full width of the comb spacing; when the frames are placed in contact, they are correctly spaced. Some spacers for frames with straight-sided narrow end-bars are shown in Figure 4.25b. Two screw-eyes may be used, each inserted through an exactly measured distance into an end-bar of the frame, but on opposite sides (A). Alternatively a single 'castellated' strip of metal (B) is used in place of each frame-runner; it has slots which house the lugs of top-bars, thus fixing their positions. Some beekeepers set the frames in position in the hive box with the aid of a hand-held spacer (C); however, this leaves the frame loose, and until the bees fasten them down with propolis, they may shift if the hive is moved from one place to another. When frames have no spacing device near the bottom, care must be taken to insert them into the hive in a vertical position.

Some types of hive are wide enough to leave a gap at one end of the brood box; the beekeeper can then more easily remove the first frame when starting manipulations, and he can prize frames apart so that a frame in any position can be removed easily. In other hives, beekeepers often use one frame fewer than the full complement, to leave a gap at both ends (Figure 4.25a, *above*). Examples are the New Zealand metricated Langstroth hive, and the British National hive. A 'follower board' (Bees & Honey, 1987) can then be used to prevent bees building comb in an end gap; it has the same cross-section as the frame, and similar lugs, and may be 9 mm thick.

Spacing of honey combs

The bees tolerate a larger comb spacing in honey-storage combs than in brood combs; during a flow they extend the depth of cells containing honey until only a bee-space is left between capped combs, as in the brood nest. If the brood box is fitted with 10 frames, the beekeeper can thus advantageously use only 9 (or even 8) in the honey supers, spaced out regularly. A test in Brazil used comb spacings of 37, 41, 46 and 53 mm, corresponding to 10, 9, 8 and 7 frames per super in standard Langstroth 10-frame hive boxes. Supers with wide-spaced frames contained up to 20% more honey, which was attributed also to the smaller number of cells the bees had to seal (Barbosa da Silva, 1967). Also, there are fewer frames for the beekeeper to uncap and handle when such supers are used. Most of the comb-spacing methods mentioned can be used to give wider spacing. However, Hoffman frames are not suitable for use in honey supers, because of difficulties in uncapping, and Manley (1946) introduced frames with both top and bottom bars 30 mm wide, so that the uncapping knife can always rest on both bars. These frames are close-ended, with the whole of the end-bar as wide as the comb spacing (40 mm).

A hive box with fixed 'castellated' metal strips that give wide spacing must not be used as a brood box.

Figure 4.25a Standard New Zealand Langstroth brood boxes (Bryant, 1984).
above 10 frames at 33-mm frame spacing, leaving a gap at each end.
below 9 frames designed for 35-mm spacing, but spaced more widely and somewhat irregularly; brace comb, burr comb, and much propolis can be seen.

Two other dimensions that vary between races and species of honeybees are included in Table 4.25A: the width of worker cells, referred to in Section 4.31, and the slot width for a queen excluder, i.e. the smallest gap that will prevent the passage of a queen, but will allow workers to pass; see Section 4.45.

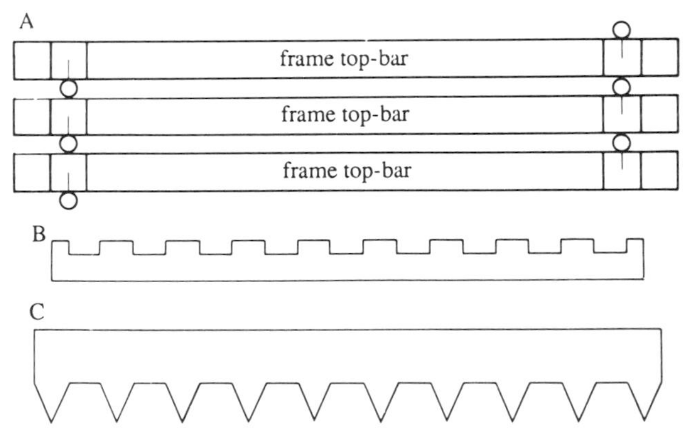

Figure 4.25b Frame spacers that are separate from the frames; A is viewed along the length of the hive from above, and B and C along the width of the hive from the side.
A Screw-eye in each frame end-bar, inserted all clockwise or all anti-clockwise.
B Frame spacer fixed along each end of a hive box, consisting of a 'castellated' metal strip; the lug of a top-bar fits in each slot.
C Hand-held spacer moved from hive to hive when manipulating.

Table 4.25A Comb spacing, worker cell width, and queen excluder slot width (all in mm), for different honeybees (*Apis* species)

Comb spacing = centre-to-centre spacing for worker brood combs, and for their top-bars and frame top-bars.
Cell width (w) is measured across parallel walls. If w is in mm, then the number of cells per cm^2 on one side of a comb = $115 \div w^2$; some values are given in Table 7.41A.
Drone cell/worker cell = ratio between widths of drone and worker cells.
Excluder slot width = width that allows passage of workers but not of queen.

	Comb spacing	*Cell width*	*Drone cell / Worker cell*	*Excluder slot width*
A. mellifera, European	35 (32–38 is used)	5.1–5.5	1.3	4.14–4.5
		5.3[a]	1.23[a]	
A. m. cypria				3.8
A. m. syriaca		4.9		
A. mellifera, African				
A. m. unicolor		5.0		
A. m. scutellata	32	4.7–4.9		4.4[b]
A. m. lamarckii	32	4.6	1.33	
unspecified 'tropical'	32	4.77–4.94	1.38	
in Zimbabwe	32	4.8		
in Angola, Tanzania	30–32	4.8		4.35
A. m. capensis	31.8	4.86		
A. m. litorea	28–30	4.62	1.33	
A. m. jemenitica		4.75	1.31	
A. m. monticola		5.0		
Africanized		4.5–5.0		
A. cerana				
Japan	30	4.7–4.8	1.13	
Nepal	30			3.5
India: Kashmir	35	4.9	1.08	4.0–4.2
India:[c] High Himalayas	30	4.9		4.0
Sub-Himalayas	31	4.7		3.75
Central	32	4.5		3.50
South	32	4.3		
Bangladesh	27–31			
Burma	31			
Java	28		1.17	
Philippines	30	3.6–4.0		3.70
A. florea				
Iran		2.9	1.59	
Java			1.55	
Oman			1.50	3.5

[a] in USA (Taber & Owens, 1970)
[b] this is the gap in square-mesh coffee wire, 0.58 mm diameter mesh, 1 mesh per 5 mm
[c] Indian Standards Institution (1976)

4.3 FOUNDATION AND COMB

4.31 Beeswax comb foundation

Most comb foundation is made from beeswax, i.e. wax secreted by honeybees. Beekeepers who have access to a supplier of comb foundation usually buy sheets of a size to fit their frames, and insert the sheets themselves. Foundation is made commercially by a continuous process, in which a long thin sheet of beeswax is passed between the two incised rollers of a foundation mill (Coggshall & Morse, 1984), making a raised pattern of hexagons on each side of the sheet. The hexagons become the beginnings of the cell walls, the patterns being offset so that the base of a cell on one side is opposite the junction of three cell walls on the other. As the embossed sheet is extruded, it is cut into pieces of the size required. Foundation is sold in several thicknesses or 'weights', according to the number of sheets per kilogram or pound. A common 'weight' gives 13 sheets per kg for Langstroth deep frames; for sections, much thinner foundation is supplied (see Section 4.33). If the bees in a hive are provided with vertical sheets of foundation, at the correct distance apart, and with embossed hexagons of the correct size for the bees concerned, they will build cell walls on the foundation. This process is referred to as 'drawing out foundation', and the resulting comb is said to be 'drawn out'.

The Indian Standards Institution (1979) published a revised standard for a mill for *Apis cerana* foundation.

Cell size

The correct size of the hexagonal brood cells varies according to the size of the bees that are to use it, as shown in Table 4.25A. Foundation is manufactured with different cell sizes, appropriate for European and for African *Apis mellifera*, and for *A. cerana*; Table 4.31A gives addresses of some suppliers. The cell size referred to is that for worker brood, although 'drone foundation' is also sold and is sometimes used in honey supers, as well as in the drone-producing colonies in mating apiaries. Worker foundation for European *Apis mellifera* could presumably be used similarly in honey supers for *A. cerana*. Rahman and Singh (1946) measured the width (and depth) of cells of four *Apis* species, at an altitude of 1350 m in the Indian Himalayas.

Many experiments were done in the 1930s–1960s in which bees were reared in cells larger than normal. Larger bees were produced, and increased honey yields were reported (e.g. Glushkov, 1956), but no sustained application of these trials seems to have been made.

Making foundation non-commercially

The beekeeper can make foundation from his own beeswax, using a mould or press, or dies. A foundation mould (or cast) has two opposite surfaces in which the hexagons are incised. Wax just molten is poured over the lower surface, and the upper surface closed on it; the two are hinged together, each being mounted on a heavy base. The wax sets into a sheet with a raised pattern of hexagons on each side. A foundation press is similar, but (instead of molten wax) a thin sheet of solid wax is placed on the lower surface; then the press is closed, and sufficient pressure is applied to emboss the hexagons.

Nowadays the heavy mould and press have both been largely superseded by a pair of inexpensive and lightweight incised plastic sheets that serve as dies, which are usually hinged together at one end. A sheet of beeswax is placed between them, and pressure exerted by passing a roller over the sandwich, to emboss the hexagons on the wax sheet. Suppliers of the dies provide instructions, and suggestions for making the necessary beeswax sheets are given below.

The sheets produced manually are up to twice as thick as those that come from commercial rollers, and contain up to twice as much beeswax. But in localities where foundation of the correct cell size cannot be purchased, they are much valued. Table 4.31A gives addresses from which rollers, moulds, presses and dies can be obtained.

Orientation of cells in foundation

Rollers for the commercial production of foundation are made so that each cell base on the foundation has two sides *across* the long beeswax sheet, and none along its length – the direction in which tension is applied as the sheet is extruded (Figure 4.31a). Conventionally, the small sheets to fit into the frames are cut *along* the roll, with the result that, when the frame is in position in the hive, each cell has two vertical sides (vertical mode). It has been widely believed that this was the 'correct' orientation for the bees, and that other orientations (e.g. the horizontal mode) were 'wrong', but in fact, it does not matter to the bees. Hepburn and Rigby (1981) found that the horizontal and vertical modes, shown in Figure 4.31a, occur at equal frequency in natural comb. They also found that comb foundation extruded from rollers currently in use is 30% stronger if the finished foundation is cut *across* the length of the roll, the cells in the comb then having two horizontal walls but no vertical ones.

Use of plain beeswax sheets

Some beekeepers find it satisfactory to insert plain sheets of beeswax in frames. Of course this will not

Table 4.31A Some addresses for obtaining beeswax foundation (and equipment for making it) for different honey-bees, also plastic foundation, comb, sections

Equipment	*Address*
***Apis mellifera*, European**	
beeswax foundation	most beekeeping suppliers
foundation dies (plastic inexpensive)	H. T. Herring & Son, 14 Severn Gardens, East Oakley, Basingstoke, Hants RG23 7AT, UK
foundation mould	Leaf Products, 24 Acton Road, Long Eaton, Nottingham NG10 1FR, UK
foundation rollers	Roger Delon, 83–83 bis, Route de Grand-Charmont, 252000 Montbéliard, France
	Bernhard Rietsche, Bienengerätefabrik, 7616 Bieberach/Baden, GFR
	Chr. Graze KG, Strümpfelbacherstrasse 21, 7056 Weinstadt 2 (Endersbach), GFR
	Tom Industries, PO Box 800, El Cajon, CA 92022, USA
***Apis mellifera*, tropical African**	
beeswax foundation	John Rau & Co. (Pvt.) Ltd, 2 Moffat St, PO Box 2893, Harare, Zimbabwe (1050 cells/dm^2)
foundation mould	Leaf Products (address above)
foundation rollers	Tom Industries (address above)
***Apis cerana*, tropical**	
beeswax foundation	Rawat Apiaries (Himalayas), Ranikhet, Dist. Almora, UP, India
	Imelda's Beekeeper Supplies, 1910 F. Tirona Benitez St, Malate, Manila, Philippines
foundation rollers	Tom Industries (address above)
Apis florea	
foundation rollers (to order)	Lotlikar & Sons, A-1/4 Pioneer Co-op Society, Panvel 410206, Kulaba, MS, India
***Apis mellifera*, European:** plastic foundation, etc.	
plastic foundation, combs and frames	Pierco Inc., PO Box 3607, City of Industry, CA 91744, USA (plastic beeswax-coated foundation in frame)
	Ets Thomas, BP 2, 45450 Fay-aux-Loges, France (plastic comb and frame combined, also frame end-bars)
	Stapla GmbH & Co, Dieselstrasse 5, 6365 Rosbach v.d.H., GFR
round and other plastic sections	Ross Rounds, PO Box 485, Massillon, OH 44646, USA (round sections) Dadant & Sons Inc., Hamilton, IL62341, USA

guarantee that the bees build only worker cells, or a regular array of cells. But it is used for *A. cerana* in some places where foundation cannot be obtained. Plain sheets can be made by using a wooden board just larger than the sheets required, fitted with a rigid handle on one long side. The board is wetted, dipped into molten beeswax, and withdrawn. Experience will show how to get a beeswax sheet of the required thickness (about 1.5 mm is usual) on each side of the board; the two sheets are peeled off when they have cooled. Or a sheet of glass can be used, dipped three times into molten beeswax at intervals of 3 seconds, and then

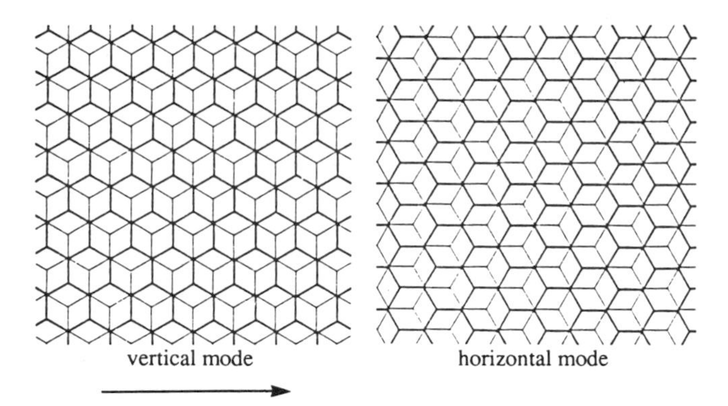

Figure 4.31a Patterns of honeybee comb construction (Hepburn, 1986).
The arrow indicates the usual direction of tension when comb foundation is extruded from the rollers during manufacture.

quickly into cold water. Mosquito netting dipped in beeswax has also been used for *Apis cerana*.

Use of other waxes
From time to time other waxes have been used for comb foundation, either mixed with beeswax (for cheapness, or for greater strength), or as a strong centre sheet coated with beeswax on both sides. Even if bees accept such foundation, there is a difficulty in that many beekeepers sell their recovered beeswax to beekeeping suppliers, or exchange it for new sheets of foundation. The supplier may well refuse to accept wax that is not pure beeswax, in which case the beekeeper loses his reimbursement.

Fixing and wiring foundation in frames
When fixing the sheet of foundation in a frame, a provision will be found for inserting the upper edge of the sheet a short distance into the top-bar, so that it can be nailed in place. The top-bar may have a slot along the bottom, into which the edge of the foundation is inserted before nailing. Or the top-bar may have a 'wedge', a strip of wood forming half the width of its lower part. This wedge is only lightly attached and can be pulled off; the foundation is then put in position, the wedge is replaced, and thin nails are driven through the wedge, the foundation, and the fixed part of the top-bar.

Sheets of beeswax comb foundation are usually strengthened by embedding tinned iron wire across them. Practices differ in different countries. Wires may be inserted in a vertical or horizontal direction (or both), or in a zig-zag pattern, and wire may be straight or crimped. In some countries foundation is sold already wired. In others the beekeeper wires the foundation himself, passing a continuous wire through eyelets inserted in the end-bars, pulling it taut and securing it, and then embedding it by passing an electric current through to heat it. Wedmore (1945) gives diagrams and details. Wired frames without any foundation are sometimes used for *Apis cerana*.

4.32 Plastic and metal foundation and comb

The search for substitutes for beeswax, to use either in making comb foundation, or as prefabricated comb itself, started in the last century. In 1870 Quinby produced a full-depth comb of galvanized tin coated with beeswax, and in 1861 Wagner had obtained a USA patent covering a wide-ranging variety of structures. In a detailed account of these developments, Johansson and Johansson (1971*b*) list among the materials used by early inventors:

- other waxes than beeswax
- cellophane, celluloid
- cloth, hard fibre
- wood veneer
- paper, cardboard
- aluminium, tin, zinc, metal foil, wire cloth.

Some of these enjoyed a temporary popularity, but none remained in permanent use. Either the bees did not work them well, or they were structurally inadequate or had some other disadvantage compared with beeswax, or they gave no advantage in cost.

The search continued, because of the drawbacks of beeswax comb: its fragility, its vulnerability to damage by wax moths (Section 11.61) and to distortion and breakage at high temperatures, and the difficulty of sterilizing it to prevent transmission of disease organisms such as *Nosema apis* on stored comb. A comb built of rubber, from which honey was extracted by squeezing full combs, was reported in April 1936, and quoted since, but it was an April Fool's Day joke (see Bee World, 1936*b*).

The development of various plastics during the Second World War gave rise to a new series of experiments, and in 1951 I saw integral comb-plus-frame of plastic in experimental use at the Bavarian Beekeeping Institute in the German Federal Republic. Weiss (1983), at this Institute, describes some of the more successful types of plastic foundation, and gives results of tests on them. Detroy and Erickson (1977) also tested various plastic combs for brood rearing and for honey storage. Ferracane (1987) surveyed the rating accorded to plastic foundation by beekeepers in the USA.

Warren and Warren (1982) operate more than 3000 hives, and are the largest beekeepers in British Columbia, Canada. They use the Pierco plastic frame-plus-foundation, moulded as one unit and coated with a layer of beeswax so thin that it is invisible. Experience with thousands of these units has shown no disadvantage compared with beeswax foundation, and many advantages. Perfect combs are easy to obtain, and any drone comb is easily removed. Each year up to 100 of the Warren hives are attacked by bears, which eat beeswax combs but learn to lick the honey off plastic foundation without doing much damage to the foundation. The Warrens' honey sometimes has a water content of only 12%, and its viscosity is consequently high; extraction of the honey is faster with plastic than with beeswax foundation, and comb breakage – up to 10% with beeswax foundation – is eliminated. The increase in costs incurred in using the plastic frame-plus-foundation is partly offset by the saving of labour costs in assembling and wiring traditional frames. Table 4.31A includes some addresses from which plastic foundation, etc., can be obtained; the different sizes of frames in different countries may limit the choice of what any one beekeeper can use.

Aluminium combs were much promoted in the 1920s but did not prove very satisfactory. Since then expanded metal has come into wide commercial use, and van Laere et al. (1985) tested sheets with 7.4 hexagonal cell structures per cm², after completely covering and sealing the metal with wax, by methods described. On the upper parts of the sheets bees normally built satisfactory comb, but on the lower parts only during a strong flow. When a 10-cm strip of the expanded metal was attached below the frame top-bar the bees used it and built normal comb down from it; this might be useful for top-bar hives (Section 10.4).

Some materials, and some methods of manufacture, give products that are not satisfactory, and only a trial of a particular product can show whether it is effective in use in a beekeeper's own circumstances. If it is, he may be able to improve his efficiency and ease of operation considerably. If not, he should not let this failure lead him to reject the important concept of a more hygienic and more manageable comb than beeswax framed in wood.

4.33 Special mountings for comb honey

In modern beekeeping, honey produced for sale in the comb has traditionally been in 'sections'. These are sold to the consumer as a double-sided comb built on very thin foundation (since comb as well as honey is eaten), in a square wooden frame; in the hive, several of the sections substitute for one frame. A common size for a section is about 11 × 11 × 5 cm, which holds one pound of honey (0.45 kg). Lime or basswood (*Tilia* spp.) wood is used, because it has such a long fine grain that it can be thinned at the four corners and bent at right angles, to make the 'section'.

Bees are not very willing to store honey in these small sections; each comb surface is opposite a flat separator of wood or metal, instead of another comb, and the passage from one section to another is through a narrow gap. There are many difficulties in getting all the sections of comb drawn out, completely filled with honey and capped, and thus saleable as a pack of standard weight. Detailed information on sections and section production is given by C. E. Killion (1975) and his son E. E. Killion (1981); Wedmore (1945) may also be consulted. The use of certain plastic materials has, however, now opened the way for the production of several attractive and more easily produced packs of comb honey, and these are discussed below.

Round packs of comb honey can be produced by using special plastic equipment in honey supers. The USA now produces more 'rounds' of honey than square wooden sections, to which they are a direct successor. Bees complete and fill the cells more readily in round than in square sections, where cells in the corners are often left unfinished. Rounds are easier to assemble, the foundation does not have to be cut to fit them, and less time is spent preparing the sections for sale after they are removed from the hive. Their use does, however, involve a greater initial capital expenditure.

As purchased, round plastic sections are moulded in two halves, one ring to contain the comb on each side of the foundation. Four half-rings are mounted in a shallow half-frame, using special fittings (Figure 4.33a, *above*), and the two half-frames are clamped back to back with a sheet of thin foundation in between. After the filled and capped rounds are removed from the hive, each is fitted with a transparent plastic lid on each side. The foundation outside the rings was not accessible to the bees (Figure 4.33a, *below*), and it is trimmed off. E. E. Killion's full description of the process (1981) is illustrated by 18 photographs.

Half-combs are also used in which a cell pattern is moulded on to the base of a square box of clear plastic. Rows of the boxes are mounted vertically in a very shallow super, all facing in the same direction and appropriately spaced. The filled boxes are ready for sale after transparent lids have been fitted to them (Berkhout, 1987; see also Hogg, 1989).

Production of *cut-comb honey* needs no special equipment. Frames or top-bars fitted with extra-thin foundation (or none) are used in the honey supers. When filled with honey and capped – although not

necessarily completely – the combs are freed from bees (Section 5.3), then cut from the frame or top-bar and cut into pieces that fit into inexpensive containers. Honey from the cut surfaces is drained off before each piece of comb is placed in its container, which has a securely fitting transparent lid. Remnants of comb and honey are strained for sale in jars.

Chunk honey is a term usually applied to a piece of cut comb (as above, but smaller) which is inserted in a transparent screw-top jar, and the jar filled up with light-coloured liquid honey. There are two special requirements. A wide-mouthed jar is needed (or only a small piece of comb can be inserted); and the surrounding honey must not start to granulate before sale, since this would spoil the visual appeal of the honey comb floating in honey. So production of chunk honey is limited to regions where non-granulating light honey can be produced; it usually has an extra high fructose content (Section 13.21.1). Sources of such honey include false acacia (*Robinia pseudoacacia*), tupelo (*Nyssa ogeche*), and sourwood (*Oxydendron arboreum*).

4.4 OTHER HIVE PARTS AND ACCESSORIES

4.41 Hive roofs

The roof or lid of a hive protects the colony from enemies, and also especially from the weather, and it must remain in place even in a strong wind. Nevertheless, when the beekeeper wants to remove it, he must be able to do so easily and quickly. The roof is usually made of the same material as the rest of thc hive, but moulded white rigid plastic is particularly suitable for tropical conditions, being resistant to decay and termite damage, and helping to keep the hive cool. In constructing the roof, adequate provision must be made for top ventilation, which is especially necessary for bees evaporating water when producing honey during a flow, and in hot weather.

Four types of wooden hive roof are described here. Starting with the simplest they are: cleated (A), migratory (B), cavity (C), telescopic (D), the letters referring to Figure 4.41a. Construction and maintenance of

Figure 4.33a Production of round sections, and equipment used (E. E. Killion, 1981).
above Four half-rings for round sections, mounted in a frame, with the fittings that secure them; the other four half-rings are faintly visible on the left, behind the foundation.
below Finished comb in three of the four round sections; the unused foundation is easily cut away.

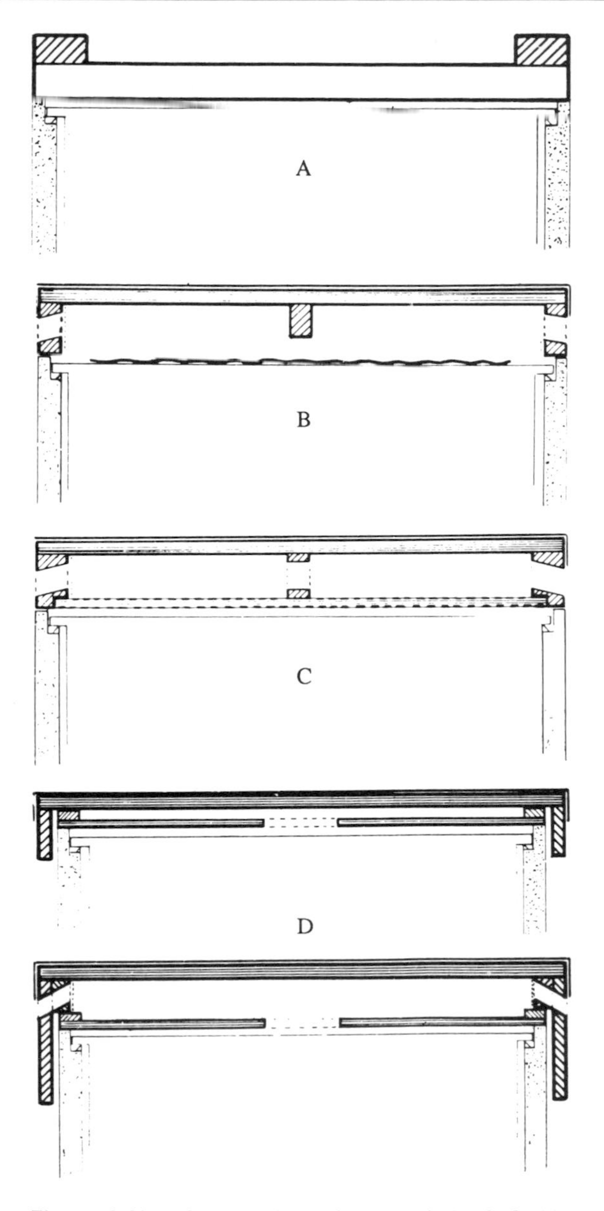

Figure 4.41a Some designs of hive roofs (F. G. Smith, 1966*b*).

A Cleated roof.
B Migratory roof.
C Cavity roof.
D Telescopic roof.
above Shallow
below Deep

hive roofs are discussed in detail by F. G. Smith (1966*b*) who gives clear diagrams, and by Morse (1976), and Australasian Beekeeper (1978).

Cleated roof (A)

This is a flat lid with the same cross-section as the hive, made of tongued and grooved boards of some durable timber, 25 to 40 mm thick. The boards are fixed together by a strip of wood (cleat) along each end, at right angles to the frames. When the roof is in position the cleats are on top, and the flat underside rests directly on the top hive box. The bees will seal the two surfaces together with propolis, but a weight may be needed on top of the roof until the joint is secure. A cleated roof is not satisfactory for hives where the bee-space is at the bottom of each box, since frames are then flush with the top and would become stuck to the roof, making its removal difficult. Unlike the roofs described below, it provides no top ventilation.

Migratory roof (B)

This is similar to the cleated roof (A), but instead of cleats on the top it has a 50-mm rim on the underside, which rests on the top hive box. A 50-mm space is thus provided inside the top of the hive, where bees can cluster when the hive becomes hot, as during transport (see Section 8.13). If the colony is short of storage space it may build comb there, but this can be prevented by placing, centrally over the top of the frames, a flexible cover (Section 4.44) about 50 mm smaller than the inside of the hive. The gap round the edges allows access to the roof space by the bees, and also by air which escapes from the hive through holes (25 mm high, 50–75 mm long) made in the rim, and covered with wire gauze. If these holes are situated in the sides that will be at the ends of the hive, the bees will not cover the gauze with propolis.

Cavity roof (C)

This is similar to the migratory roof (B) except that, inside the bottom of the rim, it has an integral inner cover which leaves a 12-mm gap along the two sides parallel to the frames. Ventilation is provided as in roof B above. Cavity roofs, much used in Australia to keep hives cool in hot weather, are described in detail by Chambers (1981), and by Purdie and Doull (1964) who include graphs to show their effect on hive temperatures.

Telescopic roof (D)

This roof has a slightly larger cross-section than that of the hive, and a rim is attached all round it, to fit loosely over the upper part of the hive sides, and this rim keeps the roof in place. An inner cover (Section 4.44) is used between the top box and the roof, which are therefore not stuck together with propolis; the roof could not be prized off if they were. The top of the roof consists of a flat sheet of wood or masonite (tempered hardboard), and a sheet of expanded polystyrene can be inserted on the underside of the roof if extra thermal insulation is required. The roof is protected on top by a sheet of galvanized iron, which is folded down (uncut) over the edges, to protect the joint between the rim and the top. The use of paint on the metal, and colours to be chosen, are discussed in Section 4.13. A 70-mm rim is usual, but a 'deep' telescopic roof has a 140-mm rim. In some types there is a strip of wood about 8 mm thick round the edge of the underside of the flat top; this keeps the roof off the inner cover, and holes are made through it and the rim (protected with wire gauze, as in roof C), to provide ventilation.

4.42 Floor boards

The hive floor board, or bottom board, is at greater risk from fungal decay and termite attack than any other part (Section 4.11.3). It is commonly made of tongued and grooved boards of a durable wood, about 20 mm thick. Alternatively, a sheet of metal, or rustproof metal mesh with holes 3 mm or less, is set in a wooden frame. Aluminium, or galvanized or stainless steel mesh may be used, but plastic mesh is not durable enough. In Norway 'netting' floor boards are quite common, and are advantageous except in spring, when thermal insulation should be inserted (Aarhus, 1981). The use of concrete hive floors is increasing (Section 4.12).

As well as different materials, different designs of floor boards are used to suit specific conditions or methods of hive management. All floor boards described here have the same cross-section as the hive boxes, but some others incorporate a 'landing board' for the bees (Section 4.43). For *Apis mellifera* the distance from the upper surface of the floor to the bottoms of the frames immediately above it should never be less than the bee-space; it can be as much as 25 mm, but if it is greater the bees may build comb down into the gap; for *A. cerana* see near end of Section 4.5.

Most of the floor boards described below are illustrated in Figures 4.42a and 4.42b. Features of the entrance are discussed in Section 4.43, and hive stands – on which the floor board is placed – in Section 4.48.

Standard American floor board (*Figure 4.42a, A*)
The upper side of the base has a rim 19–22 mm high on three sides; the fourth side is open along its full length and constitutes the flight entrance, also 19–22 mm high.

Reversible floor board (*Figure 4.42a, B*)
This is similar to A, but the underside also has a rim, 9–11 mm high, for use in winter, the more restricted entrance giving the colony more protection against wind, and against the entry of small animals such as mice.

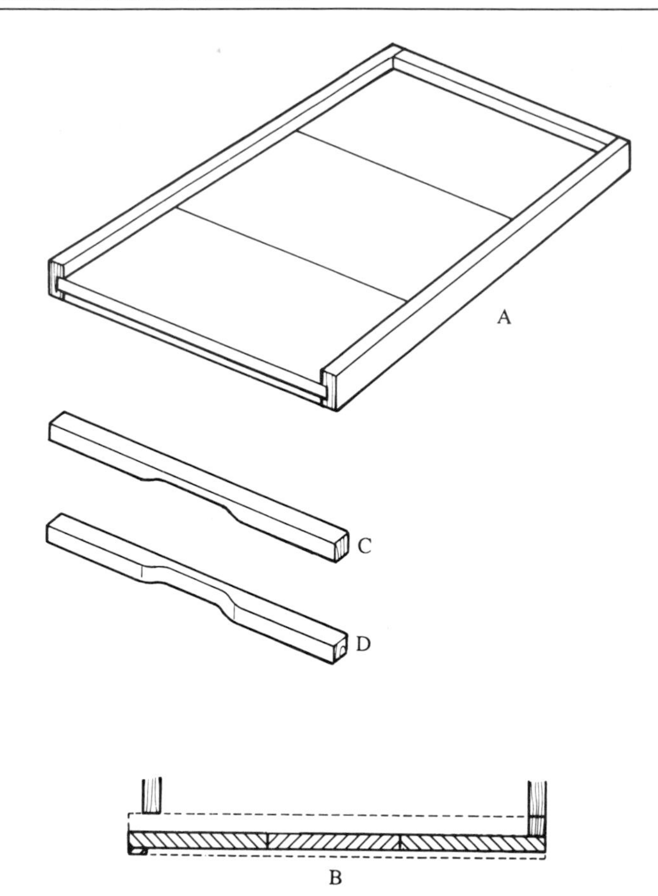

Figure 4.42a Some designs of floor boards for hives (F. G. Smith, 1966*b*).
A Standard American floor board.
B Reversible floor board.
C Entrance block as used to reduce a high wide entrance.
D Entrance block as used to close a high wide entrance.

Entrance block (*Figure 4.42a, C, D*)
The entrance block shown can be used to reduce (C) or to close (D) a high, wide flight entrance.

Deep floor board with slatted rack (*Figure 4.42b, above*)
C. C. Miller (1831–1920) and C. E. Killion (1899–1979), both successful producers of comb honey in Illinois, USA, used an extra-deep floor board which is favoured by some beekeepers in many different countries. For successful production of comb honey, hives need to be overcrowded with bees, and the higher floor-board rim (about 50 mm) gives more clustering space at the bottom of the hive, thus helping to reduce overheating and swarming. This bottom space has a somewhat similar function to the top space in the roof shown in Figure 4.41a (B). To prevent the bees extending their combs down into this space, a horizon-

Figure 4.42b Other floor boards.
above Floor board with 50-mm rim, fitted with a slatted rack, USA (photo: E. E. Killion).
below Metal floor board with 10-mm rim, and baffled entrance, New Zealand (photo: T. G. Bryant).

tal 'slatted rack' or 'slotted board' is inserted above it (at a bee-space distance below the bottom frames), either permanently or only during the swarming season. E. E. Killion (1981) gives full details. With the slatted rack, ventilation at the bottom of the hive is more evenly dispersed, and the bees consequently rear brood in lower parts of combs in the bottom hive box, which is not always so if the entrance is the usual single slot. But many beekeepers do not use this system because of expense and complication of the extra equipment.

Floor with baffled entrance (Figure 4.42b, below)
Variants of the floor board above are designated by the special feature of the hive entrance: baffled, protected, or tunnelled. Returning bees enter a slot at the front of the floor board, and access into the hive itself is below the centre or the rear of the hive, where it is protected from weather and from intruders. This type of floor board does not provide a bottom clustering space.

Pallet floor board
Some migratory beekeepers move their hives on pallets (Section 8.12), for instance in blocks of four as shown in Figure 15.5a. Instead of using a separate floor board, a section of the pallet forms the base of the hive, the rim usually being attached directly to the pallet. The hives may be fixed securely in place on the pallet with strapping. Other methods, incorporating cleats, are also used.

4.43 Entrance fittings and landing boards

The hive entrance is usually a slot running below the lower edge of the front of the bottom hive box, and is created by features in the floor-board design, as in Figure 4.42a (A, B). Such an entrance may be entirely or partly closed at will by inserting into it a strip of wood (C) the length of the entrance, with a square cross-section (the height of the entrance gap) which is cut away to a depth of 10 mm along the central 100 mm or so. (In winter a high entrance can be protected by a 'mouse guard', a strip of perforated metal sheet with holes 9 mm in diameter.)

The height of a slot entrance must be sufficient to allow easy passage for workers, drones and queens, i.e. at least 8 mm; 18 mm or more enables workers to fan within the entrance slot itself. At high temperatures it is also important that bees can fan just outside the hive entrance. But in very hot conditions, a high entrance (50 mm) is likely to increase the temperature within the hive. The less the height and width of the hive entrance, the smaller is the variety of intruders that can pass through it, and the easier it is for the bees to guard it (Section 3.41). A long restricted passage-way into the hive, such as that provided by the baffled entrance in Figure 4.42b, *below*, also helps the bees to guard their hive against intruders.

Entrance closures for use when moving hives of bees are discussed in Section 8.11, those for confining bees in their hives during pesticide application in 11.72.1, and pollen traps across the hive entrance in 14.22.

If there is no upper entrance to a hive, and no ventilation through the roof, all entry and exit of air must take place at the entrance, including what is expelled when bees evaporate water from nectar they are converting into honey – up to several kg of water each night during a heavy flow. Top ventilation is usually advantageous during a honey flow, and also during damp winter weather, when it can help to prevent the growth of moulds in the hive.

The position of the hive entrance has been the subject of discussion and enquiry. On the principle 'let the bees tell you', Chaplin (1969) in the USSR provided 10 hives with a horizontal entrance 450 × 8 mm, and 10 hives with a vertical entrance 320 × 10 mm. All entrances were covered with waxed paper containing very small holes at 5-mm intervals. The bees could chew through the paper to make flight entrance holes, where and of what size they chose. Figure 4.43a shows that the bees opened the whole length of either type of entrance as the seasonal peak of activity approached, and that afterwards they reduced it progressively with propolis until September, when only several small holes near the ends remained. With a horizontal entrance, the greatest number of bees used holes near the boundary between brood combs and honey combs. With a vertical entrance, the greatest number of bees used entrances 40–60 mm from the bottom. The bees in hives with vertical entrances were weighed: the length of the open entrance increased roughly in proportion to the colony population at the time, about 20 mm for every kg of bees. In experiments by Free and Williams (1976) in England, more of the bees using an entrance that was level with the brood nest brought in pollen than bees using an entrance (in the same hive) level with the honey supers. Also, more bees brought pollen into a hive if the brood was near the entrance.

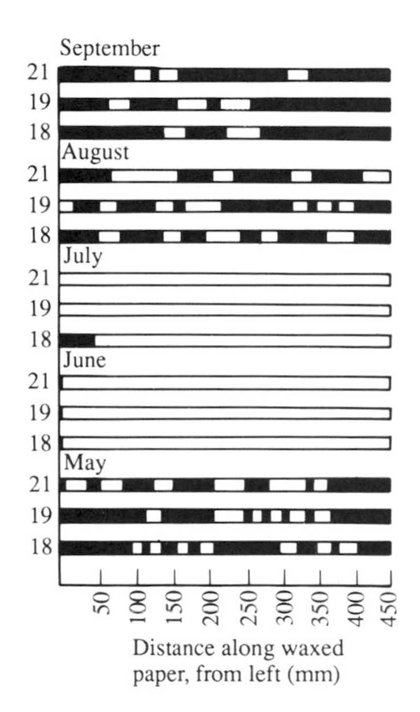

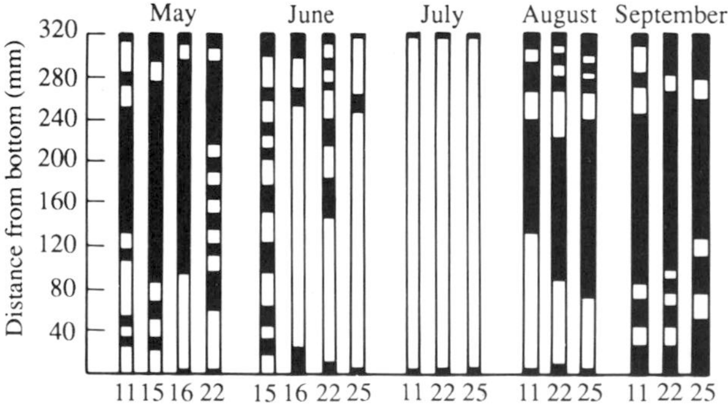

Figure 4.43a Horizontal and vertical hive entrances, showing in black the parts the bees kept sealed (or resealed with propolis) between May and September, in Kazan, USSR (Chaplin, 1969).
above Horizontal entrances (results shown each month for hives no. 18, 19, 21).
below Vertical entrances (results shown variously for hives no. 11, 15, 16, 22, 25).

In this book, where reference is made to an upper, top or additional entrance, what is usually meant is a hole – often about 20 mm in diameter – drilled (hence the term auger hole) in the side of a hive box; when not wanted, such a hole can be closed up, e.g. with a cork. Upper entrances may be provided in one or more hive boxes during a honey flow to give bees direct access to supers, or when operating two-queen systems. In winter it can provide additional ventilation and thus prevent the growth of mould, and it is also a safeguard in case the bottom entrance becomes obstructed by dead bees. Additional entrances may also be incorporated into certain types of divider boards (Section 4.44).

If a landing or alighting board is used, it is an extension to the floor board or the hive stand, adjacent to the lower edge of the entrance slot. It may be horizontal and only 3 to 5 cm wide, or it may slope up from the ground with the idea that heavily loaded foragers can climb up it to reach the entrance.

One advantage of a landing board is that it provides plenty of landing space when many foragers approach the hive together. It also gives the beekeeper a chance to watch the activity of returning bees, and check whether they are carrying pollen – and what colour the pollen is, and therefore what plant it is likely to have come from. A landing board also provides space for bees to stand outside the entrance fanning to ventilate the hive.

The main disadvantage of a landing board is that it normally protrudes laterally beyond the hive boxes, and hives cannot therefore be packed tightly together for tranpsort. This disadvantage is overcome if the board is part of the hive stand, and this need not be transported with the hive. Boards reaching the ground must not be used where ants are troublesome. And some beekeepers would say that bees needing to climb up a board from the ground to reach the entrance are old or diseased, and that the colony would be better without them.

4.44 Inner covers and divider boards

I use the term 'inner cover' as a generic name for any solid (i.e. not perforated) board or other cover having the same cross-section as the hive boxes, that is placed over the top box to enclose the bees within the hive. A cover board or crown board is normally a rigid inner cover; a cloth, quilt or mat is a flexible one, although the term 'glass quilt' is also in use. I use the term 'divider board' for a solid board placed horizontally between two hive boxes in the course of certain operations, such as the formation of two-queen colonies (Section 8.29.1). Some boards are suitable for use as either a cover board or a divider board. Both usually have a rim round the edge on one side, to give a bee-space (8 mm for European *Apis mellifera*) between the surface of the board and the top or bottom of the frames nearest to it. 'Division boards', used vertically, are described at the end of the section.

Most cover boards are made of plywood, fibreboard or masonite, and have a hole in the centre to serve as a feed hole (Section 4.47). The hole is often 8 × 4 cm to take a Porter bee-escape when required; see Section 5.31. If a sheet of glass or rigid transparent plastic is used in a wooden frame, the beekeeper can see the bees in the hive, but such a cover board needs rather careful handling. Mammo Gebreyesus in Ethiopia (1986) recommends its use with tropical African bees, so that the progress of the colony can be monitored to some extent without opening the hive. If the cover is to be used for observing the bees, the gap below the glass must not be greater than the bee-space, or bees are likely to obscure the view by depositing wax on the glass. Shaparew's ventilated inner cover (1986) is a double board that provides an upper entrance from a central hole in the lower board.

Usually a weather-resistant roof is placed over an inner cover, but a cleated roof (Figure 4.41a, A) rests directly on the top hive box, and no inner cover is used.

It is satisfactory to use a flexible cover, which is often preferred to a rigid cover board, in hives where the bee-space is at the bottom of each hive box, and the top surface of the frames is flush with the top of the walls of the box. Such a cover can be turned back little by little when the hive is opened, and the seal of propolis made by the bees between it and the top box is broken more gently than can be done if a rigid board is used. Suitable materials for a cover cloth include jute sacking or sailcloth; plastic sheet allows no ventilation. The bees are likely to nibble holes in any material that is not very strong. A feeder can be used with a cloth cover (if a hole is cut in the centre), but not a bee-escape.

A divider board, used to close off upper hive boxes from lower ones, may be a standard cover board provided any hole is covered over. Some divider boards have special fittings for a specific operation, for instance a flight entrance in a rim above and/or below the board itself. Such a board can be used when making nucleus colonies or 'divides' (Section 6.34), especially if the entrances can be opened or closed from outside. A Snelgrove board (Snelgrove, 1934) is of this type.

The term 'division board' is used for a board inserted *vertically* in a hive box to divide it into two parts, with some frames in each part, e.g. when housing two nuclei in one hive (Section 6.34, Method 3). It is suspended like a frame, but instead of having a bee-space round its edges, it fits snugly across the hive so that the bees cannot pass from one side to the other.

4.45 Queen excluders

A queen excluder is a flat grid, having the same cross-section as the hive, its holes (usually slots) being large enough to let worker bees pass, but not the queen whose body is wider. It is placed above the hive box(es) that the beekeeper wants the bees to use as the brood nest, so that the boxes above (the honey supers) are kept free from brood.

Queen excluders of a sort (e.g. a piece of wood with slots cut in it) were used by skep beekeepers, across the hole leading from the skep to a cap above, which was later harvested full of honey. 'Queen sieves' that prevented passage of queens and the still larger drones, and also 'drone sieves' that kept back drones only, were used with top-bar hives by Hannemann in Brazil in the 1870s (Whyte, 1919). The first effective queen excluder for movable-frame hives is usually attributed to Abbot Colin in France in 1865.

A wire queen excluder consists of parallel wires welded to cross supports, and is usually mounted in a wooden frame to keep it rigid and to protect it. This rim also distances the excluder by a bee-space from the nearest frame surfaces. Wire excluders are the most expensive, but well made ones are probably the best of all. A less expensive excluder (see Figure 4.45a) is a sheet of metal or plastic with slots punched all over except round the edge. Where this type is used on hive boxes with a top bee-space, it should be mounted in a frame, to prevent it sagging on to the tops of the frames. In Asian countries, I have also seen queen excluders of cut wood and of bamboo.

The correct slot width for queen excluders depends on the queen's body size, and Table 4.25A gives the slot width for a number of different *Apis mellifera* and

Figure 4.45a Corner of punched-metal queen excluder (actual size).

A. cerana, and for *A. florea*. Slots from 4.14 to 4.5 mm wide are used for European *A. mellifera*. A single slight distortion in an excluder can let the queen through, and excluders should therefore be stored in a flat pile, in a place where they will not be disturbed. Wire screen used for sorting coffee beans (5 meshes per inch) is usually satisfactory for tropical African bees in Kenya. A few beekeepers use American 5-mesh wire for European *A. mellifera*, but others find that the size of this mesh is too variable (Morse, 1982).

Many beekeepers use a queen excluder routinely during the honey flow period, whereas some others regard it as a 'honey excluder' and reject its use. I think that both views are based on the individual beekeeper's own experience, which is linked with the nature and timing of the flows in his locality, and on his management methods. Use of a queen excluder is very beneficial in that boxes of honey, being free from brood, are easier to free from bees and to harvest; also the honey is not darkened. It enables the beekeeper to reduce the brood nest to two hive boxes (or to one) in preparation for winter. On the other hand, if an excluder is inserted below a hive box that contains adult drones or drone brood, the drones are trapped and die; if there are many, they may block the excluder. This trouble can be overcome by making an extra flight entrance in one of the hive boxes above the excluder. An excluder may also reduce ventilation in the hive, and it is said to impede the movement of workers to and from the honey supers, but on the basis of his own observations Morse (1982) denies this statement – and also another, that edges of holes in punched-metal excluders are likely to damage the bees' wings.

My recommendation is that a beekeeper should use a queen excluder in the hives when honey supers are placed in position, unless or until he finds an adequate reason for not doing so. I do not know of any data that show a reduction in honey yields through the use of a queen excluder. If, $3\frac{1}{2}$ weeks before the first flow is due to start, an excluder is inserted below an upper box containing brood (which must not contain the queen), all workers and drone brood will have emerged from combs by the time the flow starts.

A queen excluder on which wax and propolis have accumulated can be scraped with a knife or hive tool provided it is resting on a completely flat surface; otherwise it may be damaged. Alternatively excluders can be cleaned by putting them in a steam bath, or in a solar wax extractor. Morse (1982) recommends that after excluders have been removed from the hives at the end of the honey flow, each should be left on the (metal) top of the hive roof; the sun's heat should melt the wax and propolis, cleaning the excluder and giving the roof a useful protective coating. A brush with radial wires, attached to the motor used to power an electric drill, may be used out of doors to clean excluders; the spinning brush should rotate so that it throws wax particles away from the operator, who should nevertheless wear goggles.

4.46 Bee-escape boards

A bee-escape board has the same cross-section as the hive, and is usually inserted immediately below one or more of the honey supers that are to be harvested, a day or two in advance. The board incorporates a bee-escape mechanism that allows worker bees to move down to the box below, but not to return to the supers. These can therefore be taken off the hive free from bees. Section 5.31 describes the use of different types of escape board, and deals with other methods for clearing bees from honey combs to be harvested.

If a queen excluder is already in place where the escape board is to go, it should be removed first.

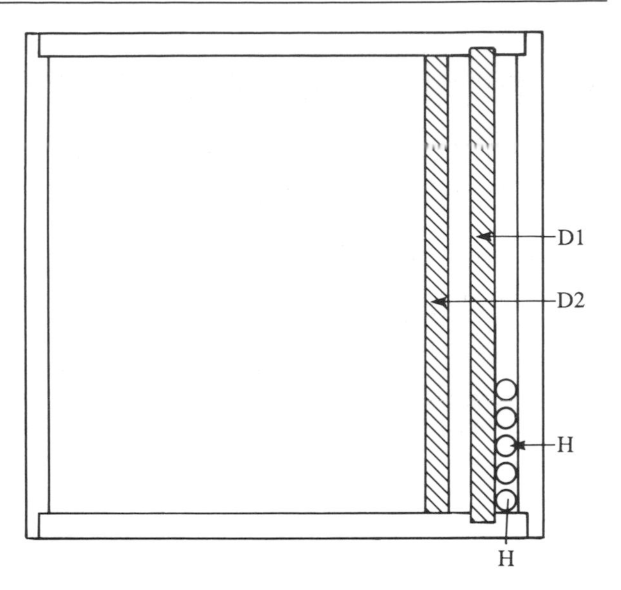

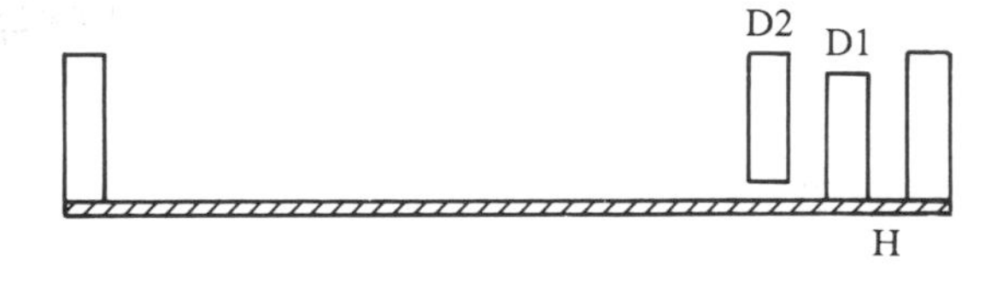

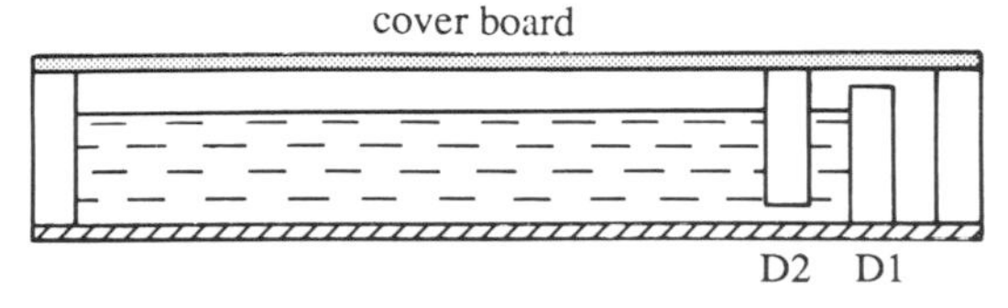

Figure 4.47a Plan of overall syrup feeder (Crundwell, 1985).
above The cross-section is that of the hive, walls and internal dividers (D1 and D2) being of 20-mm wood, and the base of plywood; about 12 holes (H) 18 mm in diameter are now (1988) used to provide access for the bees.
centre Vertical section of empty feeder.
below Feeder filled and with cover board in place; each divider leaves a gap of 6 mm, above D1 and below D2. See text for further details.

4.47 Feeders

The feeders described here are for sugar syrup, and Section 6.31 explains when this should and should not be fed to colonies. A number of different feeders are available from beekeeping suppliers, and others can be made by the beekeeper from general-purpose materials. Johansson and Johansson (1976/1977) give details of many types. What is most suitable for use on any particular occasion depends on the amount of syrup the beekeeper wants to give to a colony at one filling; whether he wants to be able to refill the feeder without putting on protective clothing; whether he wants the bees to take the syrup immediately and quickly, or slowly; and whether the price of the feeder is important to him.

Feeders 1–5 below are placed at the top of the hive under the roof, 1–4 being refilled without opening any of the hive boxes. The frame feeder (6) is placed inside the hive, and the entrance feeder (7) outside it. Where necessary, arrangements are made to prevent bees drowning in the syrup provided in the feeder.

1. Overall feeder; Miller-type feeder

Feeders of this type are based on one used by C. C. Miller in the USA. It is a tray with the same cross-section as the hive and usually 80–100 mm deep, covered tightly with a board having no rim on its underside. The feeder is placed directly above the top hive box, any cover board or cloth having been removed. For hives with a bottom bee-space, a rim is added below the base of the feeder to raise this a bee-space above the frames. The feeder should be assembled by using a good quality waterproof glue, and may be finished with a bitumastic paint, although this is not necessary. Most of the tray is a container for syrup, but near one side (in the type shown in Figure 4.47a) are two vertical dividers D1 and D2. D1, which leaves a space at the top, separates off an area where holes (or a long slot) allow the bees to come up from and return to the hive below. By going over the top of D1, the bees reach a small area of syrup between D1 and D2. Divider D2, which leaves space at the bottom, keeps them away from the main supply of syrup until nearly all of it has been taken. The bees can then enter under D2 and take the rest of the syrup, so that the beekeeper does not have to clean the feeder. The feeder is refilled while on the hive, if this is necessary, but it has a large capacity – nearly 20 litres of syrup if it is 100 mm deep. A cover board (unframed on the lower side) serves as a lid. Various versions of the feeder can be purchased, including some of plastic.

2. Rapid feeder

This name is used in some countries for a feeder on the same principle as type 1 but smaller, holding up to 5 litres, and placed on the cover board of the hive, centrally over the feed hole. This feeder is usually circular and made of aluminium or plastic, with the bees' access at the bottom of a central tube; the bees walk up into it from the feed hole, and a cap restricts their access to a small surface of the syrup. With a rapid feeder in place, an extra hive box must be added to the hive, so that the roof comes above the feeder. The name rapid feeder is in contradistinction to a 'slow feeder' of type 3.

3. *Atmospheric pressure feeder*
This is a taller container, of almost any size, with a number of small holes made in its lid; it is filled with syrup and then inverted over the feed hole on the cover board of the hive. The partial vacuum in the air space left inside the container prevents the syrup running out. The bees take the syrup from the holes; a 'slow feeder' has only a few holes, so that not many bees can feed at a time. With more holes, they take it more quickly. A friction-top tin (can, pail) will do, or a screw-top jar. A check must be made to ensure that, when the feeder is in place, the holes in it are positioned over the feed hole and accessible to the bees. An extra box is needed on the hive, as with type 2. These feeders must be removed for refilling; to prevent the bees coming out of the feed hole, the feeder is slid clear of the hole, and at the same time a piece of glass or wood is slid across it.

4. *Percolation feeder*
This feeder does away with the need to dissolve the sugar in water. As described by H. Ruttner (1962), it is an 8-litre plastic pail or can, with a friction-top lid in which a hole 6–7 cm in diameter is covered with bronze wire gauze, mesh 0.32 mm. Into the can are put 5 kg sugar and 5 litres of cold water. The feeder is closed tightly and inverted, like a type 3 feeder. The water percolates down through the sugar and forms a concentrated solution at the bottom, which the bees take through the gauze, at the rate of 1 kg or more a day. Again, an extra box is needed on the hive to accommodate the feeder.

5. *Plastic bag; zip-lock bag feeder*
This is simply a strong plastic bag with a secure fastening; a twisted wire will do, but a zip is better (Spangler, 1974). Sizes from 1 to 4 litres are recommended. The bag is filled with syrup to about two-thirds of its capacity, closed, and laid flat above the frames. To give the bees access to the syrup, a slit or a hole is made in the air pocket that appears at the top of the bag; the use of several very small holes is not advisable as they may become blocked. To keep the hive roof clear of the top of the bag, a shallow rim with the same cross-section as the hive may be inserted instead of a hive box. When removing or replacing the bag, the beekeeper is exposed to flying bees.

6. *Frame feeder; dummy feeder; division board feeder*
These names are given to a feeder that is suspended inside the hive in place of one of the frames. It has solid sides so that it holds syrup, and incorporates a float to prevent the bees drowning. The hive must be opened to insert the feeder and to refill it, and it holds only 2 or 3 litres of syrup, but it is useful for feeding small colonies as it can be placed close to the small brood nest. An alternative plastic bag feeder is on sale, slightly larger than the comb with its frame, and open at the top. The framed comb (except the top-bar) is inserted into the bag, and is secured by a strong rubber band round it and the bag. Syrup is poured into the bag, and the bees reach it by walking down the comb inside the bag, so are not in danger of drowning.

7. *Entrance feeder; Boardman-type feeder*
This is an inverted jar with holes in the lid, as described in 3, but it is supported in a recess in a block of wood on the hive entrance board, the block fitting tightly against the hive and incorporating an enclosed beeway into it. Alternatively, the block can be on an outside extension of the floor board elsewhere. The feeder can be refilled, and the amount of syrup in it monitored, without opening the hive. But the syrup is at air temperature, and unless this is high the bees are unable to take it as quickly as from a feeder inside the hive, or above it at the top. To prevent leakage of syrup (which can easily start robbing), apply to the inside of the jar lid a mixture of equal parts of beeswax and petroleum jelly, which the bees do not attempt to remove.

4.48 Hive stands

Stands that are used to support hives off the ground do not normally form a part of the hive itself, but it is convenient to consider them in this chapter. Examples are shown in Figures 7.51a and 7.51b. The ground surface in the apiary on which the stands are placed is discussed in Section 6.62.2

Hives can be raised off the ground by placing them on concrete building blocks or bricks, but a specially constructed stand of wood or metal is more satisfactory. It can be used to set the hives level and firm on uneven ground, and to raise the bottom of the hives above ground level, even if only by 100–200 mm. This allows an air flow below and thus retards decay in the hive; it also helps to ensure adequate ventilation inside the hive. Hive stands raise the hives to a level convenient for the individual beekeeper to work on them; see Section 5.41. They can be modified to keep the hive free from ants, which is necessary at certain seasons almost everywhere except in cooler regions (Section 11.51). Higher stands can be used to prevent the entry of flood water, and of certain pests and enemies – for instance hive stands 600 mm high are reported to be the most effective defence against the giant toad, *Bufo marinus* (Roff, 1975).

Busker (1973) discusses various types of stands, and some beekeepers have a local source of materials that are plentiful and inexpensive. For instance a pair of rails from discarded plantation railways that transported sugar or cotton are used in many Caribbean apiaries to support a row of hives; see Figure 4.12a, between the two nearest hives. In the western highlands of Ethiopia where safari ants are troublesome, I have seen rows of hives supported on long wooden poles suspended by wires from a gantry.

Many migratory apiaries in temperate regions have no hive stands, because of the extra work and transport that these would involve, but such apiaries are not normally in use during long spells with adverse conditions.

4.5 HIVES FOR *APIS CERANA*

Most of what has been said about hives for *Apis mellifera*, applies also to the hives for *Apis cerana*, except that all dimensions are smaller. Table 4.5A provides information similar to that given for *A. mellifera* in Tables 4.2A (hives) and 4.2B (frames). Table 4.25A gives comb spacing, worker cell width, and queen excluder slot width, for both species. If *Apis cerana* combs are too widely spaced, drone comb will probably be built. Table 4.31A provides suppliers' addresses for beeswax comb foundation and foundation rollers; I do not know of any plastic foundation or comb with a cell size for *Apis cerana*.

All Sections of Chapter 5, on beekeeping equipment other than hives and their fittings, apply to *Apis cerana* – except that dimensions for openings in bee-escapes need to be reduced slightly. Less protective clothing is usually needed when working with *Apis cerana* than with *A. mellifera*.

The body size of *Apis cerana*, like that of *A. mellifera*, varies considerably over its range, increasing from the equatorial zone northwards. *Apis cerana* is, however, unlike *A. mellifera* in that a single country, India, has ecotypes covering almost all of the whole size range. The Indian Standards Institution (1970) has published a Standard for different hives and their frames that are considered suitable for the bees in different regions. The tables in this Standard (which uses metric units) provide organized information on hive and frame dimensions.

The ISI Standard gives dimensions for two Type B hives, with 31-mm and 32-mm comb spacing, which have a similar shape to the Dadant hives but linear dimensions about 30% less; the comb area of a frame is thus about 50% less, but since cells are smaller than those of *A. mellifera* their number is not so much reduced: see Tables 4.5A and 4.2B.

The Standard also gives dimensions for three (smaller) Type A hives (comb spacing 30, 31, 32 mm) whose frames have a more elongated shape than Type B, and a smaller comb area. In Jammu and Kashmir State in the north of India, *Apis cerana* bees are as large as some temperate-zone *A. mellifera*. Here, as well as ISI Type B hives, Langstroth hives are used, and also hives with a brood frame 339 × 185 mm, rather similar to the British Standard frame.

Kapil (1971) gives a historical account of earlier movable-frame hives in India, on which the ISI Standards were based. Type A specifications were derived from Newton's hive and variants of it, which were used in southern India from 1880 onwards; the comb area in the brood box is quoted as 7897 cm². Chandran and Shah (1974) found Type A better than Newton's hive. Type B specifications were derived from Muttoo's somewhat larger Jeolikote Villagers' hive (total brood comb area 9142 cm²), which was widely used for the larger colonies in the north.

Bisht et al. (1982) made a 5-year comparison of colonies kept in Delhi in Newton's and Muttoo's hives. Brood area, and amounts of both honey and pollen stored in the brood box, were higher in Muttoo's larger hive. In Kanpur, somewhat lower down the Ganges, Pandey (1977) used Muttoo's hive, with two hive boxes for brood, for some prolific *Apis cerana*, that had 'swarmed repeatedly'. This prevented swarming, but the brood nest did not extend beyond one of the two boxes provided.

Apis cerana does not remove debris from the floor board as *A. mellifera* does, and in Sri Lanka Punchihewa (1988) finds that reduction of the gap above the floor board to 10 mm – so that the bees must walk across it – decreases wax moth infestation. In Malaka State, Malaysia, removal of the floor board altogether is reported to keep hives free from wax moths.

The ISI Standard includes hives with 10, 8 and 4 frames, the smallest being included for use as nucleus boxes. Mahindre (1983) regards all the frames in use as too large, since the queen cannot lay eggs at a rate that fills more than 5.5 frames (ISI, B), 4 (British Standard) or 3 (Langstroth), and he suggests remedies for this situation. The Indian Standards Institution (1961) has also published a specification for stands for *Apis cerana* hives.

In China, where there are at least 1 million movable-frame hives of *Apis cerana*, standard hives were

Table 4.5A Dimensions of hive boxes and frames for *Apis cerana*

Dimensions, for 10-frame hives, are taken from IS:1515–1969 (Indian Standards Institution, 1970). The Standard also includes those for Type A and Type B hives with 31-mm frame spacing (8-mm bee-space), 8-frame hives, and 4-frame hives (nucleus boxes) for both Types. Length is used for the dimension parallel to the frames, and dimensions are in mm except where otherwise stated. Comb area includes both sides.

	Type A	*Type A*	*Type B*
frame spacing	30	32	32
cross-section, length × width:			
external	286 × 356	286 × 376	356 × 376
internal	240 × 310	240 × 330	310 × 330
brood box:			
height	172	174	204
total comb area (cm^2)	6090	6090	9800
volume (litres)	12.8	13.8	20.9
honey super:			
height	92	94	114
volume (litres)	6.8	7.4	11.7
brood frame:			
length × height:			
external	230 × 165	230 × 165	300 × 195
internal	210 × 145	210 × 145	280 × 175
top-bar length	260	260	330
comb area (cm^2)	609	609	980
shallow/honey-super frame:			
length × height:			
external	230 × 85	230 × 85	300 × 105
internal	210 × 65	210 × 65	280 × 85
comb area (cm^2)	273	273	476
no. worker cells/cm^2:			
one side	6.2	5.0	5.0
both sides	12.4	10.0	10.0

established after observations on natural nests, and experiments made with different types (Yang et al., 1981). One of the Chinese Standards Institution hives (1983) has a brood box holding 10 deep frames and a super with shallow frames, whose dimensions are fairly similar to those of the Langstroth hive for *A. mellifera* (Tables 4.2A and 4.2B). Honey yields with it are reported to be 'up to 20 kg per year' (FAO, 1984).

In addition to China and India, movable-frame hives for *Apis cerana* are known to be used quite widely in Bangladesh, Malaysia, Sri Lanka (Szabo, 1988, gives details), and to some extent in Burma, Indonesia, Nepal, Pakistan, Philippines, Thailand and Vietnam. A few are used in Japan, but probably none in the Far East of the USSR; this bee is no longer common there (Bilash, 1987). In Nepal, movable-comb top-bar hives (Section 10.21) have been promoted for use with *Apis cerana*.

4.6 FURTHER READING AND REFERENCE

Details of publications listed will be found in the Bibliography.

Cross, D. J. (1983) *Preservative treatments of wood used in hives*

Heath, L. A. F. (ed.) (1985*b*) *A case of hives*

Indian Standards Institution (1970) *Specification for beehives*

Johansson, T. S. K.; Johansson, M. P. (1971*b*) *Substitutes for beeswax in comb and comb foundation*

Killion, E. E. (1981) *Honey in the comb*

Miller, W. R. (1976) *Let's build a bee hive*

Ministry of Agriculture [UK] (1968) *Beehives*

Smith, F. G. (1961*b*) *The African Dadant hive*

Smith, F. G. (1966*b*) *The hive*

Walton, G. M. (1975) *The metrication of beekeeping equipment*

5

Other beekeeping equipment

5.1 PROTECTIVE CLOTHING

Satisfactory protection of the beekeeper against bee stings is important. It can present particular problems in a hot climate, or when handling 'aggressive' bees, but various materials and techniques are now available that make protective clothing more effective and comfortable than formerly. Protection used by beekeepers in the past is mentioned in Section 1.42.3.

5.11 Protection for the face and head

This is the most important of all, because a sting in the mouth could impair breathing, and one in an eye could damage it. If the hair is left uncovered, bees may burrow into it and become entangled; even if they do not sting, they then buzz loudly, and although this is not dangerous it can be annoying to the beekeeper. In my opinion a beekeeper is wise always to wear a head covering and a veil for the face when handling bees. Even if he chooses not to do so, he should never allow learners or non-beekeepers to be present when he opens a hive, unless the head and face are protected.

All parts of the head and face covering should be of cool material, and the veil should nowhere touch the skin. There are several alternative systems.

(a) A veil with a see-through section has an elasticated top to fit round the crown of a separate hat. The hat must have a stiff brim, or one that is held taut by wire, cane or bone round the edge; it may be of plastic, with ventilation holes, or of straw – older straw hats were excellent, but many modern ones become limp when wet. Figure 5.11a shows a common type of hat and veil, worn with complete body-covering for handling African bees. Many styles of this type of bee veil can be purchased, and some can also be made at home; Sammataro and Avitabile (1986) give instructions.

(b) The hat and veil are similar to (a), but the veil is permanently attached to the edge of the hat brim. In both the USSR and Japan I have been given a combined hat and veil in which the front part of the veil had a separate elasticated hem at the top, and could be let down like a visor, to allow the beekeeper to cool off between operations.

(c) A see-through veil or mask is permanently attached to the front of a hood (Figure 5.11b).

Whatever is worn, the section in front of the eyes is made of some type of mesh: metal or plastic gauze, woven horsehair, or fine net of cotton or nylon; it should be black to give good vision. (Beekeepers working with some tropical African or Africanized bees may need to cover the outside of the black see-through section with white paint, to prevent bees flying against the mesh, and obscuring the wearer's vision. Figure 7.43c shows what conditions can be like when working with Africanized bees.) The type of black mesh chosen, and the size and shape of the see-through section, depend on the bees, the operating conditions, and the beekeeper's preference. A common choice is an inset of metal mesh round three (or four) sides of the head, arranged in flat sections for folding, with cotton cloth attached above and below it, as in Figures 5.11a and 14.22c. With gentle bees and absence of much wind, black net can be used for the whole veil; cotton is cooler than nylon. Figure 6.62c shows such veils in use in Australia.

The lower edge of the veil must be fixed in such a way that bees cannot get inside the veil. Personal preference plays an important part. Some beekeepers simply tuck the lower edge inside a jacket; others use a

Figure 5.11a Dr F. G. Smith in Tanzania, wearing hat, veil, coverall, gauntlet gloves and gumboots, *c.* 1955 (photo: F. G. Smith).

Figure 5.11b Brian Sherriff in England, wearing bee suit with attached hood and PVC veil, and gauntlet gloves and gumboots, 1985 (photo: Central Office of Information).
The hood can be unzipped at the front and thrown back.

veil with a loop of elastic under each arm. Many veils have two long strings attached at the back midcentre, which are passed through one or two loops at the front, then again round the waist, and the ends tied together tightly, so that the lower edge of the veil is taut against clothes worn underneath. A zip fastener can provide a complete bee-proof join between head and body covering; see also Section 5.12.

Many Dutch beekeepers wear a simple cotton hood, with an insert at the front of black horsehair. At the centre of the insert there may be a small patch of stronger material with a cross-cut, to take the stem of a pipe smoker (Section 5.21.1). In Spain I have used a much larger hood with an internal fitment to distance it from the head, but it was rather unstable, especially in wind. In Hungary, Greece and Turkey, I have encountered a cloth hood sewn on to a metal mask shaped like half an egg cut lengthways – similar to a fencing mask. If the metal is heavy it pulls the hood forward, especially when one bends. In the UK, B. J. Sherriff (address in Section 5.12) has designed various stiffened hoods (e.g. Figure 5.11b) which – zipped on to a coverall – are more comfortable and completely bee-proof. The hood can be unzipped to allow the wearer to cool off.

Working in any protective headgear for long periods in hot weather can be stressful. Racal Safety Ltd (1986) incorporate a fan to blow on to the face in their 'Dustmaster 2' respirator, which has been well spoken of by beekeepers. It can be worn inside a bee veil or hood, and is recharged from a unit attached to the waist.

5.12 Protection for the body

Whether or not it consists of special protective garments, the outer layer of clothing should be light in colour and smooth in texture, to reduce the likelihood of bees trying to sting; it should be close-woven or non-woven, and thick enough to prevent the sting of a bee penetrating it and the skin as well. Special care must be taken in handling the smoker if nylon is worn, as it melts readily.

In a dark enclosed space bees tend to move upwards. So skirts, shorts or loose-bottomed trousers may land the wearer in trouble, although they may be adequate for quick operations at the top of hives. In general, if the beekeeper is not covered with completely bee-tight protection, it is probably better if he wears less rather than more, as bees can become trapped in many types of clothing.

Many general-purpose white coveralls are satisfactory, provided openings are adequately dealt with, including the junctions between the veil and the coverall. Unless gloves or gauntlets are worn, the cuff must be elasticated. Three special coveralls that are useful for different purposes can be purchased.

- B. J. Sherriff, Five Pines, Mylor Downs, Falmouth, Cornwall, TR11 5UN, UK, makes a coverall that zips on to a hood. His alternative two-piece suit is shown in Figure 5.11b.
- Mrs. D. Olsen, 115 South First East, Providence, UT 84332, USA (also W. T. Kelley Co., Clarkson, KY 42726, USA) sells a coverall of Ripstop nylon for working Africanized bees, which is large enough to wear over other clothing. It is reported to be bee-secure but hot.
- North West Protective Garments Ltd, 2163B Kingsway, Vancouver, Canada V5N 2T4, makes disposable (but washable) lightweight industrial coveralls with elasticated wrists and trouser cuffs. They are useful for disease inspection, since they can be discarded after handling infected bees and hives.

I often carry one of the last type when travelling on beekeeping assignments; it weighs only 130 g. An alternative light protection is a mosquito net such as those made for Canadian anglers, which encloses the head and arms and has elasticated openings. Pchelovodstvo (1988*b*) gives patterns and directions for making a complete bee suit.

Whatever covering is worn it should be washable, and washed as often as necessary to remove dirt and any infective material, or odours to which bees might respond by stinging.

Protective clothing for children can be obtained from specialist suppliers. If necessary a net can be made to fit over the opening to a baby's pram. In one apiary of Africanized bees in a Trinidad garden, I saw a baby's outdoor playpen above which a net was fixed to be let down to the ground when hives were being opened.

5.13 Protection for the hands and legs

Many beekeepers prefer to work with bare hands, because it is cooler and allows greater dexterity, and they can 'feel' what they touch. If no coverall is worn, a pair of cotton gauntlets, elasticated at the top and bottom, can be used to seal the open end of the sleeves.

Gloves need elasticated gauntlets, long enough to put over the sleeves, and one style has a gauze insert for ventilation. Ideally gloves should fit the wearer's hand, and be sufficiently soft and pliable to 'feel' through, but stout enough to withstand stings. Stings readily penetrate pigskin; cowhide is better, and goatskin is best (Collins, 1983). The Meyer Stingless Bee Glove (POB 61, Winifred, SD 57076, USA) is made of goatskin with a gauntlet of Ripstop nylon; it is effective when working with Africanized bees, and is washable. Upon occasion industrial rubber gloves may be necessary, but they are terribly hot.

Gumboots are usually worn *over* trouser legs; sometimes it is wise to insert additional material inside a wide top opening, to prevent entry of bees. If trousers are worn over long boots, the trouser bottoms must be elasticated or closed in some other way. They can for instance be sealed with a strap – e.g. a 30-cm length of flat elastic 18 mm wide – fastened by Velcro sewn on the ends (J. Meyer, 1986). For those who need an effective seal between shoes and loose-bottomed trousers, W. T. Kelley Co. (address in Section 5.12) sell lightweight leggings of close-woven material, which are fastened with Velcro straps and have a strap to go under the shoe.

Final words of advice on protective clothing:

1. Once you are working with bees, it is not easy to make your clothing *more* protective. So when going into an apiary, it is best to overestimate the likelihood of needing protection, and the amount you will need.
2. Protect the head and face for any operation with bees, or for entering any apiary which has a history of stinging incidents.
3. If bees do penetrate your protective clothing, before you remove any of it to deal with them, *go well away from the apiary* and, if possible, inside a building or under low trees or bushes.

5.2 EQUIPMENT USED WHEN OPENING HIVES

This section deals with personal equipment that a beekeeper needs, or may find useful. A smoker and a purpose-made hive tool are essential; the other appliances mentioned are valued by some beekeepers but not necessarily by all.

5.21 Smokers, smoker fuels, and other pacifiers

Now, as in earlier times, the pacifier almost universally applied to bees is smoke; some others that may have a useful potential are mentioned in Section 5.21.3. Smoke appears to have several effects on a colony of honeybees, when properly applied:

- it makes some of them engorge honey;
- it pacifies them, for instance reducing the number that fly off the comb or behave as guards at the entrance, and making them less inclined to sting;
- it repels them, so that they move away from the smoke.

There are almost certainly interconnections between these effects, some of which may be linked with the fact that the smoke masks certain other odours, and hence reduces or stops the bees' reactions to them.

5.21.1 Smokers and fuels for them

The fuel used in smokers depends on what is available locally in different parts of the world, and on the individual beekeeper's preference. But virtually all modern beekeepers use the same type of smoker (Figure 5.21a); differences are in size and quality, and in minor points of design. Its development, which has been recounted by Morse (1954) and J. C. Dadant (1977), culminated in Bingham's design of 1877, whose important features are still used today. Inside the bellows (B) is a spiral spring, which enables them to be operated with one hand. Air is expelled from them through the hole H1, and into the firebox (F) through hole H2, which is below a perforated shelf (S) that supports the fuel, and thus does not become blocked by it. The hinged lid, which incorporates the nozzle, is opened to insert fuel. When it is closed, sufficient air enters hole H2 to keep the fuel burning slowly; when the bellows are operated, the additonal air from H1 produces a dense stream of smoke.

Many smokers are too small. However few hives a beekeeper has, his smoker should have a firebox at least 25 cm high and 10 cm in diameter – 12 cm if possible. A smaller smoker needs refuelling inconveniently often, has too small a heat capacity, and produces too little smoke at any one time. For working with Africanized bees much more smoke may be needed, and a firebox 40 × 15 cm has been recommended.

The quality of both materials and workmanship is all-important in producing an effective smoker that will last many years, and suppliers should know which of those in their catalogue will give longest service. Smokers of poor quality burn through quickly.

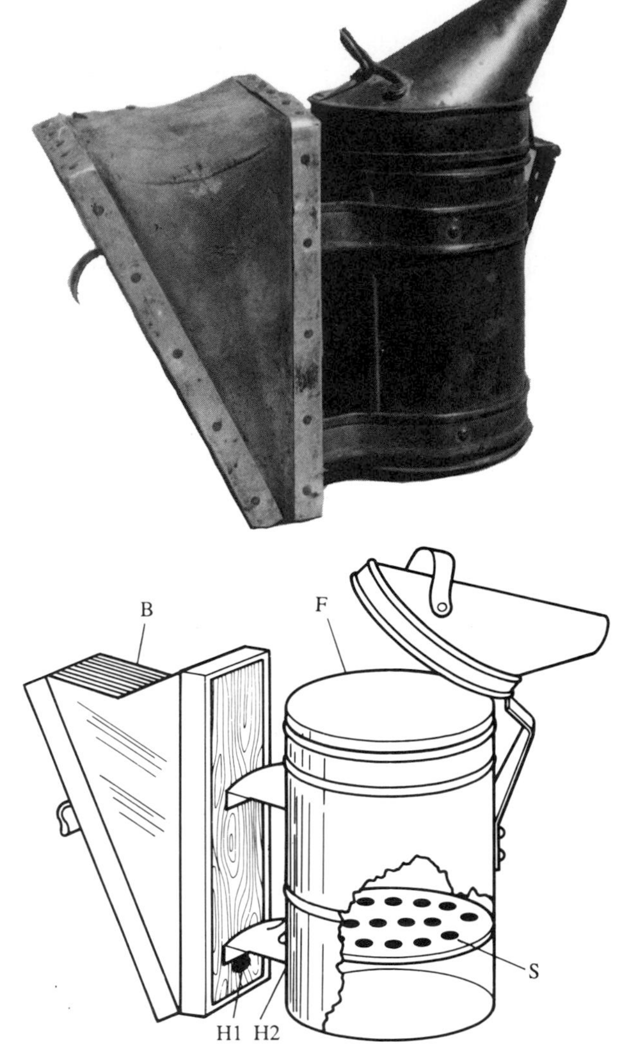

Figure 5.21a Typical smoker (photo: F. G. Smith). The sketch below shows how the smoker opens.

B	bellows	H1, H2	air holes
F	firebox	S	shelf for fuel

If a smoker has a hook (see Figure 5.21a), this can be hung over the end wall of a hive box, to avoid the mischance of standing it on a surface that could be damaged by the heat. Some models have an openwork metal casing round the firebox which stays cool, and this can prevent burning of the hands or gloves. Smokers used by paraplegic beekeepers (Section 16.83) need special insulation, as they may have to be held between the legs.

Other types of smoker are used in certain areas. Pipe smokers have been popular with some German and

Dutch beekeepers who wear a veil with a reinforced cross-cut for the stem of the pipe to pass through, level with the mouth. The pipe is shaped so that smoke is blown from the bowl on to the bees. It leaves both hands free, but continued use can damage the front teeth. In some countries smokers are sold that incorporate a clockwork or electrical mechanism to maintain a stream of smoke. A battery-operated backpack model has been designed (Laperrouzaz, 1986); it blows smoke through a flexible tube whose outlet is attached to a hive tool.

Fuel for the usual type of smoker should fulfil the following conditions:

- it is easy to ignite (unless a separate starter is used);
- it has an open texture so that air is interspersed with the fuel itself;
- when the smoker is left standing upright, it continues to burn slowly;
- when the bellows are operated, it quickly produces a steady stream of smoke;
- it has no smell objectionable to the beekeeper or the bees.

Most beekeepers start by experimenting with different fuels, and finally settle on one that is easily available to them and suits their own style of beekeeping. Some commonly available fuels are listed below.

from plants
soft rotten wood
dry bark
coconut husks
dry pine needles
tightly packed hay
dried cut lucerne
giant puffball (see text) cut into slices and dried
tobacco (in pipe smoker)

from animals
dry dung of local ruminants
dry camel dung

manufactured
old sacking/burlap made from jute or other natural fibres (not plastic)
cotton or linen rags
corrugated cardboard (which must not produce noxious fumes)

When using manufactured materials, it is essential to avoid any that have been treated with insecticide; the risk is most likely from sacking, and from glue incorporated in cardboard, which may also have been treated.

Some of the manufactured fuels listed have a high tar content, which leaves a black deposit inside the nozzle, partially blocking it. A suggested remedy is to insert into the nozzle a metal pan-scrubber made of curly copper shavings; this traps the deposit, and can be removed and washed. Various prepared fuels may be purchased from beekeeping suppliers. S. R. Taylor (1959) patented a 'hardwood smoke concentrate', which is sprayed on the bees as an aerosol; it is entered in catalogues as a 'hive bomb' – expensive, but handy for emergency use.

In general, the smoke is not intended to have a narcotic effect, i.e. to make the bees unconscious. But there are substances which do this, for instance nitrous oxide, and in the early 1950s the addition of ammonium nitrate to smoker fuel was recommended, since nitrous oxide is then produced. However, it was found that nitrous oxide shortens the bees' life (Jones et al., 1964), and that the reaction also produces hydrogen cyanide (Simpson, 1954) which is toxic, so the practice was discontinued.

The giant puffball listed above is *Langermannia gigantea* (Pers.) Rostk. (= *Calvatia gigantea*, formerly known as *Lycoperdon giganteum*), a fungus which grows as large as a football. It is known widely in Europe, and is found in many other parts of the world. Its use for smoking bees (described e.g. by Cook, 1970) was recorded as early as 1597, in Gerard's *Herball*. Millard (1987) suggests the possible use of the common puffball, *Lycoperdon perlatum* Pers. Smoke from the African puffball, *Langermannia wahlbergia* (Fr.) Dring, has a narcotic effect on honeybees, and its use by beekeepers in the Meru crater in Tanzania was recorded by John Corner (Wood, 1983). In experiments by Wood, the bees recovered after about 20 minutes, and their life was not shortened; Mollel (1987) found that about 1 to 2 g could pacify a colony, and also that larger amounts were harmful. The narcosis proved to be due partly to hydrogen sulphide, which decomposes readily but is poisonous. Wood (1983) was able to produce similar narcosis – in both European and tropical African bees – with hydrogen sulphide from burning chicken feathers or human hair.

Brenzinger (1987) reports that in NE Tanzania a 'torch' prepared from *Spirostachys africana* Sond. (Euphorbiaceae) is used to smoke bees when harvesting honey, and also to singe a newly made hive.

In Ethiopia, Mammo Gebreyesus (1986) reported that the bees' engorging behaviour was stimulated by the inclusion with normal smoker fuel of parts of certain plants – to such an extent that slight smoking once or twice a day for 2–3 weeks, during the early part of the season, led to rapid brood rearing. He used parts of the fleshy leaves of *Aloe abyssinica* Lam. (Liliaceae), together with dried stems and flowers of a

Helichrysum (*splendidum*, *transversil*, *odoratissimum* or *schimperi*) in the Compositae.

5.21.2 Use and effect of smoke on bees

Figure 5.21b shows the effect of smoking a colony on the number of guard bees at the hive entrance and on the number of bees engorging honey from cells of a comb (Newton, 1968/69, 1971). Guard bees (Section 3.41) were monitored and counted through a glass panel fixed over an extended entrance board. Curve G shows that there were around 12–14 guards before the entrance was smoked, but none immediately afterwards; guarding gradually increased again, reaching its earlier level more than 10 minutes later.

Engorging behaviour was studied by using marked bees. No bees were engorging before smoke was used. The hive cover was gently lifted and a single puff of smoke applied; then a frame was removed and observed: the number of marked bees engorging on a designated area of honey comb (curve E) rose to 30 within 2 minutes, dropping back to half within 10 minutes. (Up to 60% of the bees engorged, and they did not seem to belong to any recognizable group, such as young or old.)

The honey sac of a bee that engorged as a result of being smoked contained more honey for at least 2 h afterwards (Free, 1968). Free also found some evidence that bees stinging a test object tended to have an emptier honey sac than bees that did not sting. In these tests the bees were alerted by thumping the hive, not by smoking; in Newton's experiments such action produced engorgement that followed a similar time sequence to curve E (Figure 5.21b), although the effect was less pronounced.

In the light of both the above results, and long experience by beekeepers, the following is recommended. Before opening any hive, light the smoker, and ensure that it is burning well – some beekeepers do this 10 minutes in advance. The smoker must produce cool smoke, not flame or fragments of burnt fuel. Blow 3 or 4 puffs of smoke across the entrance of the hive to be opened, then slowly remove the cover from the frames at the top, while gently puffing smoke across the exposed frames; if the cover is a rigid board, puff smoke

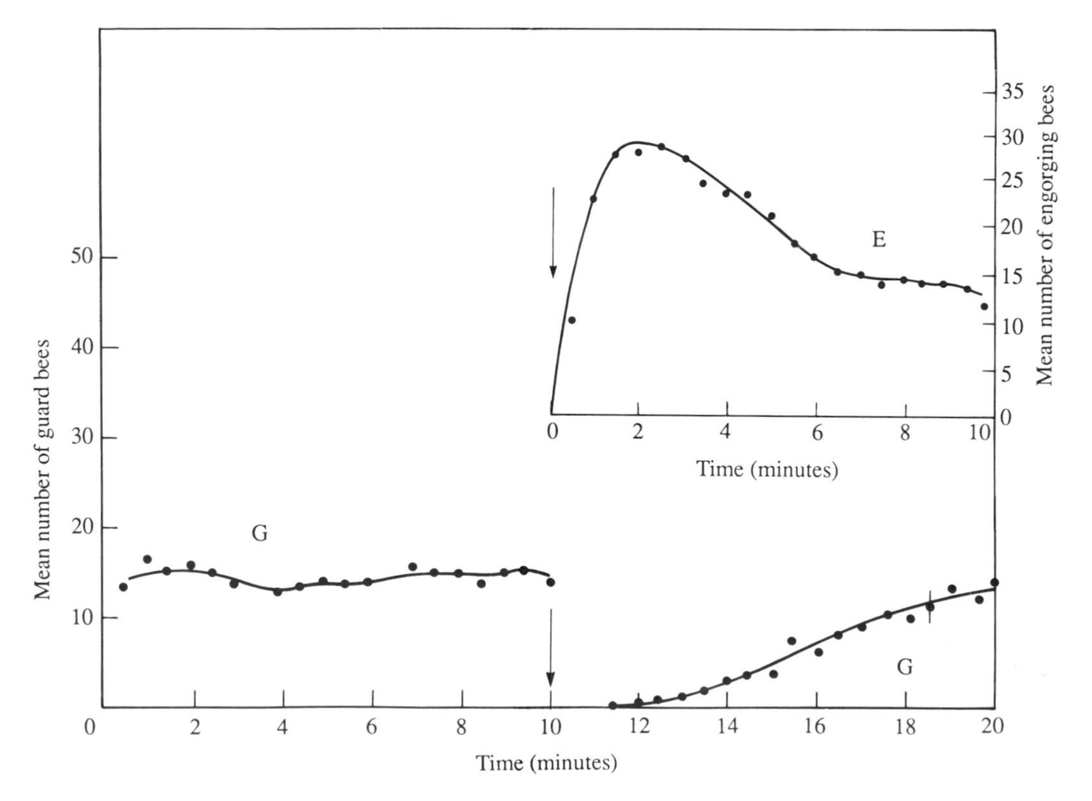

Figure 5.21b Effect of smoking a colony on the number of guard bees (G) at the hive entrance, and the number of bees engorging honey (E) from cells on a comb (Newton, 1968/69).
Arrows show when smoke was applied.

beneath it before lifting it off. Bear in mind that the number of bees on the combs engorging honey is likely to be greatest 2 minutes after applying smoke, and that half as many will be engorging 10 minutes later. So, time further operations accordingly, and apply more smoke when required. When frames are to be removed, take out an end frame first, which is likely to have few bees on it, and spread out the next ones so that smoke applied along the spaces between frames will reach more bees.

Whatever the smoker, or the operation for which it is used, certain fire precautions are necessary. When lighting the smoker, ensure that the hands are not burned, and that vegetation is not set alight. Immediately it is finished with, stand it well clear of any combustible material, and stop up the nozzle with a plug that will not burn, such as damp grass. Better still, when the smoker is purchased, make a hardwood plug to fit the nozzle and attach it permanently to the smoker, with a chain or some other tie that will not burn.

5.21.3 Other bee pacifiers and repellents

A plant extract or a synthetic substance that pacifies or repels bees when they are being handled would be valuable, especially if it were easy to apply. Some plants – mostly from the tropics – whose extracts are reported to have an effect are listed in Table 5.21A. (The first and third of them contain component(s) of Nasonov pheromone produced by worker honeybees; see Table 2.35A.) These and other such plants warrant further study, and the extracts should also be assessed for possible toxic effects on honeybees; for instance *Adenia* is used as a fish poison and contains cyanogenetic substances (Oliver-Bever, 1981). Melksham et al. (1988) list a number of compounds reported to repel bees, but not necessarily during colony manipulation.

It has been reported that N′,N-diethyl-*m*-toluamide sprayed on clothing repels bees (Sporek, 1982). From time to time proprietary products have been on sale in different countries, but they do not all seem to have the properties claimed, and none has so far ousted the smoker. A device marketed in the USA as Bee Calm is claimed to keep bees calm within a distance of 4 m. It is battery operated and produces a very high-pitched buzz (6 KHz), almost beyond the range of human hearing. Lord et al. (1985) tested its efficacy by switching it

Table 5.21A Some plant extracts reported to pacify or repel honeybees (*Apis* spp.)

Plant name and family	*Uses, effects*	*Region*
***Apis mellifera*, European**		
Cymbopogon nardus Rendle (= *Andropogon nardus* L.) and spp.[a], lemon grass (Graminae)	used for 'quietening' bees	widely reported
Lantana camara L. (Verbenaceae)	oil from leaves repelled bees	India (Attri & Singh, 1977)
Melissa officinalis L.[b], balm (Labiatae)	'rubbing the hands with the leaves is claimed to, and probably does, help in preventing stings' (Howes, 1979)	widely reported; native to Europe and Mediterranean
***Apis mellifera*, tropical**		
Warburgia longimanii[c] (Canellaceae)	used as repellent	Tanzania (Kawa, 1982)
Adenia cissampeloides (Planch. ex Hook.) Harms, bekyem (Passifloraceae)	crushed stalks are placed near hive entrance; after 3 min 'the bees appear to be dead, but 92% recover and fly normally 30 min later'	Ghana (Adjare, 1987; Yeboah-Gyan, 1988)
Lippia javanica (Burm.f) Spreng. (Verbenaceae)	used as repellent	Swaziland (Bechtel, 1989)
Aristolochia sp. (Aristolochiaceae)	paste made from juice of crushed leaves, mixed with sticky pulp of *Amomum* [or *Aframomum*?] leaves, is spread thickly over face and hands	Gabon (Coon, 1972)

(*continued*)

Table 5.21A *(continued)*

Plant name and family	*Uses, effects*	*Region*
Apis cerana		
Cymbopogon nardus (see above)	oil from leaves 6 times as repellent as benzaldehyde	India (Kumar et al., 1986)
Lantana camara (see above)	oil from leaves 1.3 times as repellent as benzaldehyde	India (Kumar et al., 1986)
Shorea floribunda G. Don. (Dipterocarpaceae)	bark placed in toddy pots repels bees from them	Thailand (Crane, 1988)
Apis dorsata		
Piper celtidiforme Opiz (Piperaceae)	leaves rubbed on body to prevent stinging	Philippines (Adey, 1985)
Orophea katschallica Kurz (Annonaceae)	chewed leaves smeared on body, and spat on comb, as repellent	Little Andaman Is. (Dutta et al., 1985)
Amomum aculeatum Roxb. (Zingiberaceae); also *A. fenzlii* Kurz, *Alpinia manii* King [added in proof, from AA 249/89]	sap from crushed stems and leaf stalks used as repellent and tranquillizer; ineffective with *Apis cerana*	S. Andaman Is. (Dutta et al., 1983, 1985); see Figure 8.61b
Zingiber squarrosum Roxb. (Zingiberaceae)	similar effect to *Amomum aculeatum*; ineffective with *Apis cerana*	Little Andaman Is. (Dutta et al., 1985)

[a] Leaves are source of citronella oil which contains citral; also used for attracting swarms to bait hives (see Section 6.35.1, under Bait hives).
[b] Oil from leaves contains 63% citral, and some geraniol and nerol; also used as above.
[c] A small family of aromatic trees (*longimanii* could not be traced at Kew). In Venda, S. Africa, bark from *W. salutaris* is placed in hives to make the bees aggressive, and thus to deter robbers (Netshiungani, 1981).

on after provoking three groups of 20 colonies by hammering their hives; they concluded from the results that the device would confer no protection on beekeepers.

A water spray as a pacifier is referred to in Section 5.23, and substances used as smoker fuel to pacify (or repel) bees are discussed at the end of Section 5.21.1. Repellents may be applied to clear bees from honey supers before these are removed for harvest (Section 5.32), and some repellents are being studied for their effect in keeping bees away from toxic pesticides on plants.

5.22 The hive tool

The English term 'hive tool' refers to an implement described in some other languages as a frame lifter or a scraper. It is a strong metal bar (usually of high quality stainless spring steel) about 25 cm long; Figure 5.22a shows two types. Type A has one end slightly narrowed for pushing between hive boxes to separate them, and for loosening top-bars before lifting them out of the hive. The other end is bent at right angles to the bar, and has a wide sharp edge for scraping wax and propolis off wooden surfaces. Type B, sometimes referred to as 'J type', has a wide flat sharp end, and the step and hook at the other end enable the tool to be used as a lever to pull up one frame by using the next top-bar as a fulcrum. The hole at that end is for hanging the tool. In another type described by Crawford (1979) there is a hook at right angles to the plane of the rest of the tool, for levering out a top-bar. Dines (1978) illustrates 8 different shapes.

In default of a specific hive tool, I have seen beekeepers use a variety of general-purpose tools for opening hives, from an old chisel to a cutlass. But a good purpose-made hive tool is so useful that one should be purchased, after trying several in the hand to see which best suits the individual concerned.

5.23 Some other tools and appliances

Among the equipment at hand when opening hives, it is important to include a tin or other *lidded receptacle*, and to collect in it all scraps of extraneous wax and propolis scraped off when examining a hive. They

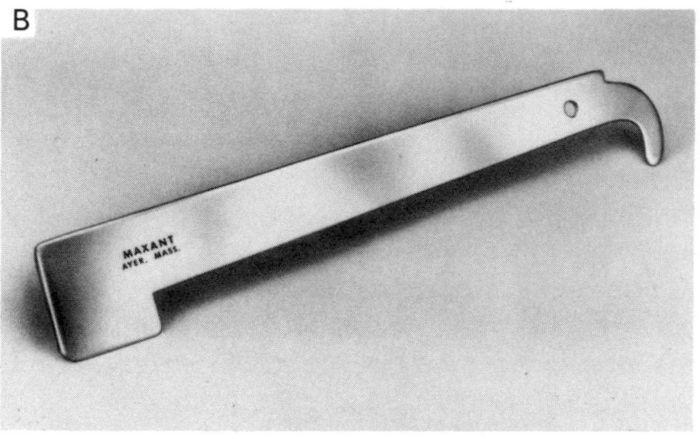

Figure 5.22a Two types of hive tool:
A common type (photo: F. G. Smith)
B Maxant J type.

should not be thrown on the ground, where they may encourage robbing, or spread infection.

A very soft *bee brush* is useful for removing bees that cling to a comb after others have been shaken off; they are usually young bees not yet able to fly. A large feather or a goose wing works well, but manufactured washable brushes are available.

Many beekeepers value a spring-loaded *frame grip*. It enables one hand to hold a frame securely by its top-bar, as long as the two pivoted handles are gripped together, and can be used for pulling out of the hive a frame that is stuck down with wax or propolis. Some beekeepers do not favour the use of this appliance, on the grounds that it might transmit disease from one hive to another, and that the bees may well be handled less gently than when frames are lifted out with both hands.

A bottle with a *pressure spray* attachment produces a fine spray of water that reduces the number of bees flying around an opened hive, and this can help to counteract robbing. It is especially useful when taking honey off a hive, and can be used to control bees during hive manipulations. Some beekeepers would not open a hive on any occasion without having it available. Such spray attachments are not listed in many beekeeping catalogues, but are readily available from gardening shops.

A pair of *manipulating cloths* is appreciated by many small-scale beekeepers working alone, and is especially useful for beginners who want to spend time handling their colonies, to observe the bees and combs. Not many suppliers list these cloths, but they are easily made at home. A piece of stout white cotton is used for each cloth, its width being the length of the hive (along the frames), and its length 20-25 cm greater than the width of the hive. A hem is sewn at each end, just wide enough to take a strip of wood or a wooden or metal roll-rod, which is inserted to weight the cloth, and kept in place by sewing up the ends of the hem. When removing frames from a hive box for examination, the two cloths are positioned on either side of the frame to be removed first, each cloth being rolled or unrolled until a few centimetres hang over the ends of the hive. All the top of the hive box is thus covered (and the bees are enclosed in the box) except the frame to be worked on. Before another frame is removed, one cloth is unrolled and the other rolled up, sufficiently to uncover the next frame. The cloths should be washed frequently; if not, they may induce robbing or spread infection.

For beekeepers wanting *to mark bees*, and especially queens, M. V. Smith (1972) gives details of various materials and methods. The most common is quick-drying paint sold for model aeroplanes or cars; type-writer correcting fluid is also used. Alternatively tiny coloured or numbered discs can be purchased from some beekeeping suppliers; they are attached with a spot of glue to the thorax or abdomen. A circular cage is also available, to press gently into the comb over a queen, so that she can be marked through the cotton threads that form the top of the cage. Many queens are sent from one country to another, and a five-year international colour code for marking them is in operation. The same code is used for the covers of many beekeeping journals, to remind readers of each year's queen-marking colour:

year ending in	1 or 6	white
	2 or 7	yellow
	3 or 8	red
	4 or 9	green
	5 or 0	blue.

Most beekeepers use some sort of *portable container* for the equipment they carry from hive to hive, and many find it worth making one to suit their purpose, which ensures that no item of the essential working kit is forgotten. In apiaries where a truck or van is driven between the hives, the container is usually kept at the back of the vehicle.

5.3 EQUIPMENT FOR REMOVING BEES FROM HONEY TO BE HARVESTED

The foremost operation in harvesting honey from hives is to get the bees away from the combs to be harvested. Thereafter, the physical removal of the frames or combs presents no special problems, except lifting them and keeping them bee-free until they are in the safety of the honey house. Further treatment is described in Section 13.31.

With any movable-comb hive, one possibility is to remove bees from each comb in turn, by lifting it up from the hive, then shaking or jolting it so that most bees drop off into the open hive below, and smoking or brushing off the remainder. This method is used by many small-scale beekeepers, and also by some commercial beekeepers in the tropics who do not have supers, but remove frames in the brood box that contain honey. There should be minimal disturbance to the colony when honey combs are removed, smoke being applied sparingly; a water spray can also be useful. Combs should be kept covered as much as possible, and those removed put in a bee-tight box. Equipment is discussed in Section 5.23.

Where whole supers of honey combs are handled, equipment is available that allows more expeditious removal of bees before the supers are taken off the hives. An enquiry in the USA (Henderson, 1987) established the percentage of commercial beekeepers that preferred different methods and the average time each required per hive. The list below includes the sections in this chapter which describe the different procedures:

brushing and shaking (above, and 5.23)	7%	12.0 min
bee-escape board (5.31)	7%	4.5 min
bee-repellent, Bee-Go (5.32)	20%	6.4 min
bee-blower to produce a stream of air that blows bees out of the supers (5.33)	52%	7.2 min
bee-blower plus repellent	15%	6.0 min

Bee World (1968) discusses and compares the methods further, and gives detailed references.

If a bee-escape board or a blower is used, the whole super has to be lifted as a unit. Beekeepers for whom this is too heavy should consider the other methods. If bees are shaken and brushed from the combs, or driven from them by a repellent, the frames in each super can be carried away in two lots, by putting half of them temporarily in an extra hive box.

Whichever method is used, every hive box from which bees have already been removed, or which is in the process of evacuation above a bee-escape board, is at risk from robbing by bees unless it is completely bee-tight, i.e. there is no gap through which bees can enter. If robbing starts, it is most likely that all the honey will be taken before the beekeeper's next visit. This risk is continuously present until the boxes are in the honey house (Section 13.31). If operations related to the removal of honey from hives can be done at dusk, the chance of robbing by bees is reduced.

5.31 Bee-escapes

The term 'bee-escape' is applied to a mechanism which allows worker bees to pass out of a hive box (in practice usually one full of honey combs to be harvested), but prevents any bees entering it. A bee-escape board, or clearer board, has the same cross-section as the hive and incorporates the bee-escape mechanism. One or two days before the supers of honey are to be removed, the board is put in position just below the lowest honey super to be harvested; the queen excluder – if present – is taken off first. When the supers are removed they should be empty of bees.

The use of bee-escapes is the only method for removing bees from combs of honey that involves two visits to the hives. There is, however, minimal disturbance to the colony on the first visit, and none on the second. The chance of people nearby being stung is less than with other methods, and bee-escapes have a special advantage in urban areas.

The original bee-escape (Figure 5.31a, *above*) was devised by Porter in 1891, and it is still in common use. It is fitted into the central feed hole of a standard hive cover board (Section 4.44). A bee can move down through the centre hole in the escape, then between a pair of very light springs and into the hive box below, but it cannot negotiate the springs in the opposite direction. The mechanism becomes ineffective if either pair of springs becomes too widely spaced through distortion, or is immobilized with propolis by the bees, or if the passage-ways become blocked, e.g. by a drone. Propolized bee-escapes can be cleaned by leaving them for a few minutes in a solution of washing soda, but if the springs are distorted it is best to buy new escapes. The Porter escape remains popular in spite of the above drawbacks, probably because of its convenient use with a cover board which is a standard piece of beekeeping equipment.

Other bee-escape mechanisms can be more certain, and more rapid, in their action. They are usually incorporated permanently in an escape board, and they work without springs or other moving parts. They provide (a) a large escape hole, A, through which the bees leave the box above; (b), distanced

Figure 5.31a Bee escapes.
above Porter escape, metal/plastic; the overall size is about 11 × 4.5 cm (Root & Root, 1940).
below Underside of an escape board with no moving parts, described in text (photo: A. Clemson).

from this escape hole, a small entrance hole, B, into the box below; (c) within the board, and close to the escape hole, an area that is covered only by wire gauze, through which the scent of the brood nest below can attract the bees above to move down. Because of the small size of the hole B and its distance from the large hole A, bees below do not re-enter the box above.

The example shown in Figure 5.31a (*below*), described by Clemson (1980), is based on a design by A. Adie in Canada in 1943 (Townsend & Burke, 1952). The escape board has the same cross-section as the hive, and on both sides has a wooden rim all round, 10 mm high and 22 mm wide. A square piece of wire mesh (8 mesh, 1 per 3 mm) is used to close off an area 90 × 90 mm at each corner of the underside – except for the entrance B cut in the mesh, which is not less than 6 mm wide; an ordinary lead pencil can be used as a gauge. The exit A, a hole 38 mm in diameter, has previously been drilled through the board at each corner, or – as in the Figure – the corner of the board has been cut off. When the board is placed underside down, below the supers, bees go down the large hole A and out of the small hole B in the corner of the mesh, and do not return to the super(s). Another type of quick-action bee-escape board (e.g. Shaparew, 1981) has two rows of conical gauze outlets, simpler and much smaller than the one shown in Figure 5.43b.

During hot weather some beekeepers experience difficulty in that the bees are less inclined to go down, and may fail to leave the supers and join the cluster in the brood nest – which could well become overcrowded and overheated if they did so. In these circumstances, bee-escapes that lead outside the hive, instead of only into the box below, seem to be more successful. Koover (1965) in California described several variants; he got all bees to leave the supers by using a board with a Porter escape fitted at two opposite corners which led bees either to the outside or to the box underneath.

Instead of inserting bee-escape boards in the hive the supers can be removed with the bees still in them, and stacked separately from the hives, with an escape board placed upside down on top of each stack, as in Figure 5.31b. The bees leaving the supers fly back to their hive; any young bees remaining can be recovered by shaking them off in front of it. This operation also enables bees to be removed from frames or top-bar combs harvested from long hives (Section 10.5), which do not have separate honey supers. The honey combs (with the bees still on them) are put into a bee-tight box, with a reversed bee-escape board on top.

If the temperature is high, honey combs (from any hive) that have no bees on them to control the temperature, may become overheated and melt. This applies both to separate stacks of supers and to supers above an escape board on the hive.

In Australia, Noel Bingley successfully uses boards fitted in each corner with an escape rather similar to that in Figure 5.31a (*below*), placing the boards upside down on the top of stacks of supers (Figure 5.31b),

Figure 5.31b Reversed bee-escape boards on stacks of honey supers on a truck in Australia, banded together for safety during transport (photo: A. Clemson).

either in the apiary or on a truck (Clemson, 1980). Clemson assesses a number of different escape boards from long experience of their use.

With most types of bee-escape, the passage of bees is speeded up if two or more escapes are used instead of one. This also allows bees to continue to leave the super if one escape becomes blocked. Any escape board that is handled as a single unit, and could be positioned upside down, should be labelled to indicate its upper side in order to prevent a serious mishap.

5.32 Bee repellents used with a fume chamber

Apart from smoke, the earliest airborne substance used for driving bees out of honey supers was probably carbolic acid (phenol), in 1868. But carbolic acid can contaminate the honey (see below), and it can burn the skin. Its use is not now recommended, and is prohibited in certain countries. Nevertheless some beekeepers still use it because it is effective except in cool weather. Addition of a few drops of methylated spirit increases its efficiency (B. White, 1987).

Phenol is applied on an absorbent board, which is placed over the open top of the super to be cleared of bees, and the vapour drives the bees down. So that their passage is not impeded, the queen excluder should be removed in advance. If too much bee repellent is used, or too high a concentration – especially at high temperatures – many bees may be driven out of the hive, as in Figure 5.32a. The phenol may also contaminate the honey, and Canadian experiments (Daharu & Sporns, 1984) showed that any of the following can increase phenol residues in honey:

Figure 5.32a Three hives (on the right) after the use of too much bee repellent.

- large area of honey combs exposed
- high surface temperature of fume board
- close proximity of combs to fume board
- overlong exposure time
- harvesting of unsealed honey (the wax cappings give some protection to sealed honey).

Section 13.27.3 shows that considerable levels of phenol have been found in honey.

In 1961 propionic anhydride, applied similarly, was suggested as an alternative to phenol; see Bee World (1962*a*). Like carbolic acid it is most effective at high temperatures (above 26°). It is not now much used.

In 1963 another proposal was made, to use benzaldehyde which is known in the food industry as artificial oil of almonds. Benzaldehyde is an effective bee repellent at lower temperatures, working best at 15°–27°. It is applied from a fume chamber – an open box 5 cm deep, having the same cross-section as the hive, with a base that is absorbent on the inner side. From $\frac{1}{2}$ to 2 tablespoons (4–15 ml) of benzaldehyde are sprinkled on the base, and when the liquid is absorbed the chamber is inverted over the open frames of the super to be harvested. It is best to smoke the bees lightly first, to start them moving. Benzaldehyde is generally regarded as unsuitable for use at high temperatures, and even in warm weather some beekeepers find that additional insulation may be needed to slow down evaporation. In Australia, B. White (1987) dilutes benzaldehyde with water as necessary, and finds it as effective as carbolic acid. In Canada, Townsend (1963) reported that benzaldehyde cleared bees from a depth within the hive box of 13–15 cm at 10° or about 24 cm at 18°. A mixture of benzaldehyde and carbolic acid has been found very effective (B. White, 1987).

Since the 1960s butyric anhydride has become popular as a repellent with some beekeepers in North America. It is sold commercially as Bee-Go, mentioned in Section 5.3, and has a very strong and objectionable rancid odour; it is not effective in cool weather (Henderson, 1987).

Reich and Reich (1969) identified a number of nontoxic organic acids and bases, and their precursors, that repel bees, but none seems to have come into general use.

5.33 Bee-blowers

No bee-escape works perfectly on every occasion, and no chemical repellent is suitable under all conditions. The use of a stream of air from a compressor to blow the bees out of each super, taken off the hive and suitably positioned nearby, is more nearly ideal than any other method. It is effective whether the weather is hot and sunny or cold and cloudy, and it does not contaminate the honey. A bee-blower does, however, involve considerable capital outlay, and most models are noisy in operation.

The blower must produce a large volume of air, at a low pressure. A petrol-driven motor is most often used, but an electric blower can be powered by a generator, or a low-power model from the mains. The following data are USDA recommendations.

	imperial units	*metric units*
power of motor	2–6 hp	1.5–4.5 kW
speed of air stream	up to 150 mph	up to 4000 m
rate of flow of air	65 cu ft/min	18 m^3/min
	1500 cu ft/min[a]	42 m^3/min[a]
air pressure	2 lb/sq inch	14 000 N/m^2
diameter of outlet tube	4 inches	10 cm
time to empty bees from 10-frame super ($6\frac{1}{4}$ inches, 16 cm, deep):		10–12 s[b]

[a] Gojmerac (1980) [b] Diehnelt (1966)

An idler is useful during intervals between blowing bees.

The blowing operation is much easier and quicker if two people work together. Each super is dealt with separately. Some beekeepers stand it on its side, on top of its own hive (Figure 5.33a, *left*) or an adjacent one. Others use a framework to support it horizontally in front of the hive (Figure 5.33a, *right*) or at the side (Figure 5.33b), with a deflector below at such an angle that bees are directed towards their hive. (But if the deflector becomes sticky with honey, falling bees may adhere to it.) The bees leave the supers more quickly if frames are shallow rather than deep, and spaced out (8 or 9 in a 10-frame super), and if combs are fully capped. They fly quickly into their hives, except young ones which walk.

Further information on different blowers will be found in Diehnelt (1966), Bee World (1968), Jones (1986), and in beekeeping supply catalogues. Most are 2-stroke, and noisy, but if a generator is used the source of noise can be distanced from the operator. Jones, in Australia, has successfully adapted an Ag-Murf Rabbit Fumigator which has a 4-stroke motor.

Figure 5.33b An experimental bee-blower operated from the side of a hive, Israel, 1969 (photo: C. Kalman).

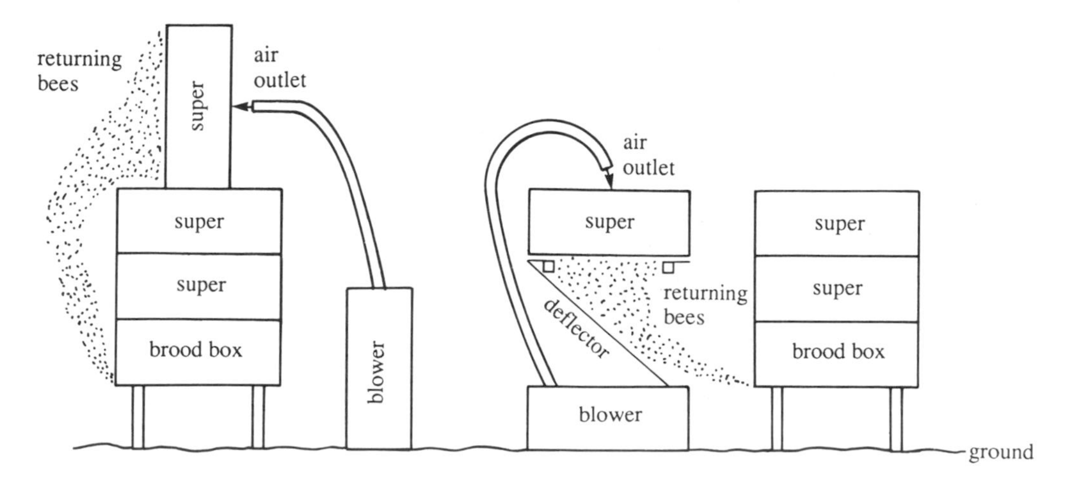

Figure 5.33a Two examples of layouts used for blowing bees out of honey supers to be harvested.
Hive entrances face left.

5.4 EQUIPMENT FOR OTHER PURPOSES

5.41 Lifting and carrying

Some appliances for moving single hives within an apiary simply hold the hive securely in a way that two people (or even one) can carry it; others are adapted hand-barrows on wheels. Most of the appliances can also be used for lifting and moving honey supers in the honey house.

A French device (Figure 5.41a, *above*) enables a beekeeper to carry a one-storey hive supported by a nylon strap round his neck; the hive is secured by two grips attached to each end of the strap, which is 4 cm wide. Most other hive-carrying devices need two people, and Figure 5.41a (*below*) shows one of the simplest.

Wheeled hive carriers are sold by some suppliers; Thomas (45450 Fay-aux-Loges, France) lists three types with a single wheel, that have a chassis to hold a Dadant hive securely. Beekeepers in Australia and New Zealand seem to be especially innovative in devising equipment for lifting and moving hives (e.g. Roberts, 1958; Brookes, 1967). The Treloar bag lifter, made in Western Australia, has been adapted for lifting hives on to trucks (Campbell, 1961). Compressed air is used, plus a little additional power from the operator, for extra heavy loads, and the air is recompressed by the operator's own weight. In New Zealand, Dawson (1959) describes 'the beekeeper's other man' designed by Stan Wilson. It has two wheels and can be used to move a hive (or by operating a winch, to lift it before moving it into position on a truck).

Power-operated equipment for moving hives, and also for lifting them on to trucks, is mentioned in Section 8.12 which deals with transport of hives, including those on pallets.

Some hive-lifting equipment has been adapted for use when opening hives, which can be especially beneficial when one person has to operate hives that have many heavy supers. For instance Dawson's (1959) can be used to raise only the top hive boxes, in order to insert a queen excluder or bee-escape board underneath them. A lifter described by E. Smith (1971) in the USA can be used similarly. A 'hive cracker and lifter' invented by Peter Pearson in New Zealand (Reid, 1976*a*) has no wheels, but rests on two curved pieces of mild steel tubing, so arranged that upper boxes of the hive (only) can be tilted back and held in place by a chain; this leaves the lower ones upright, and accessible for inspection or for the insertion of a queen excluder or bee-escape board. In the USA, Knutson (1964) uses a hive tipper without wheels, which tips the whole hive on to its side so that it rests on a horizontal platform (part of the tipper), at a convenient height for working. The hive boxes, which are side by side, can be separated, and the frames inspected, while they are in this position.

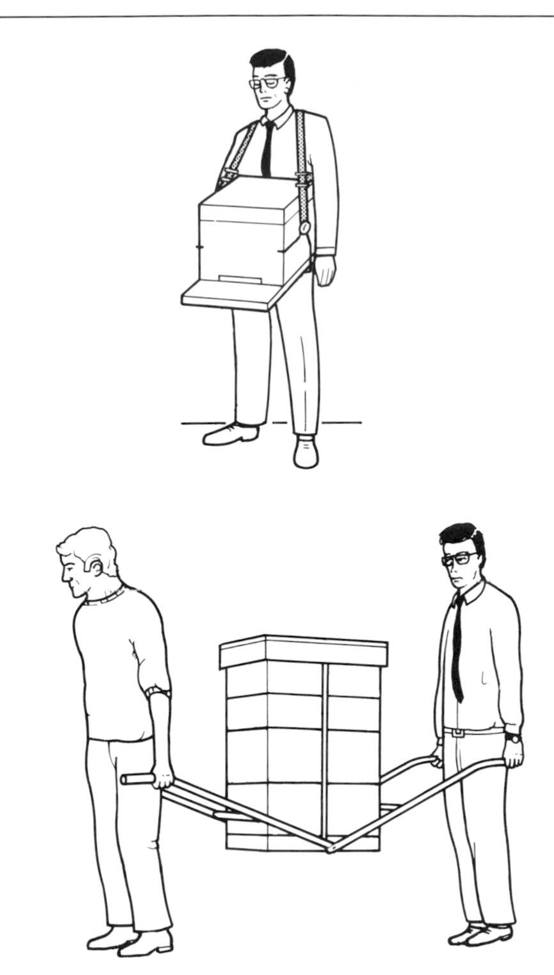

Figure 5.41a Simple carrying devices for a hive or honey super(s).
above Sangle J. B. (J. Boeuf, France).
below Hive carrier (Boone & Wright, UK).

When operating certain multiple-queen hives, 50-frame supers must be lifted, and a mini-gantry is used (Figure 8.29b). F. G. Smith (1970) describes gantry loaders for pallets supporting four hives, and other heavy loads.

According to the International Labour Office, maximum weights (kg) that should be lifted unassisted are as follows (in frequent lifting the weights should be 25% less):

age	*man*	*woman*
14–16	15	10
16–18	19	12
18–20	23	14
20–35	25	15
35–50	21	13
over 50	16	10

Mariola (1986) discusses the appropriate use of lifting devices by beekeepers from the viewpoint of preventing back strain or injury, and also assesses methods of treatment favoured after injury has occurred. He stresses the importance of routine use of an auxiliary stand when examining hives, so that boxes removed are not unnecessarily lowered and later lifted to replace them on the hive. The stand can be a light framework carried from hive to hive, or a central space left empty on a long stand for two hives, or – at its simplest – an empty deep hive box placed on the lid taken off the hive to be examined.

5.42 Weighing hives

There are many advantages in monitoring the progress of colonies in an apiary by weighing a representative hive at regular intervals. Figure 12.62a shows hive-scale records from different parts of the world, plotted as the net *increase* in weight each week, which is positive when a hive gains weight and negative when it loses weight. Any weight added by the beekeeper, such as honey supers or food, is recorded by weighing before and after the addition, and the same applies to any weight removed – usually supers of honey. If a swarm is known to leave the hive the fact is recorded; if it has left unobserved, the next weight reading is likely to indicate that this has happened.

The inexpensive equipment described here, and shown in Figure 5.42a, uses bathroom scales that record weights up to 110 kg. The scales (B) are placed upside down on top of the supporting box A, a hole being made to accommodate the lens that enlarges the reading. The hive is placed directly on the upturned scales. The light box C holds two small mirrors set at about 45° to the vertical, one just below the lens and the other at the end that protrudes outside box A at the back of the hive; the mirrors make a periscope through which the weight can be read by looking down into the outside mirror. Useful extras are a waterproof flap to keep rain out of the periscope, and drainage holes in both periscope and scales. Witters (1987) uses similar scales the right way up, with one side of the hive resting across their centre, and the scale protruding so that it can be read directly; it registers half the total hive weight.

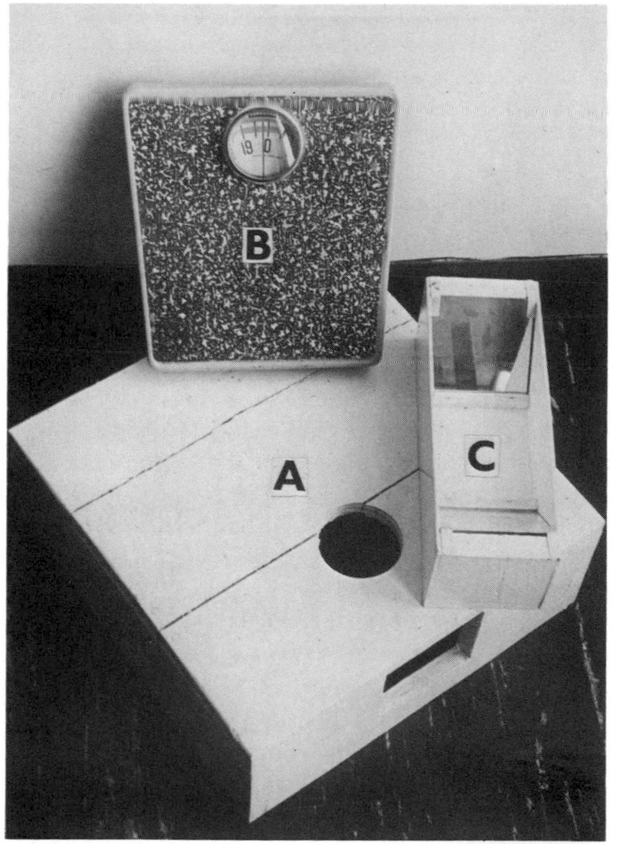

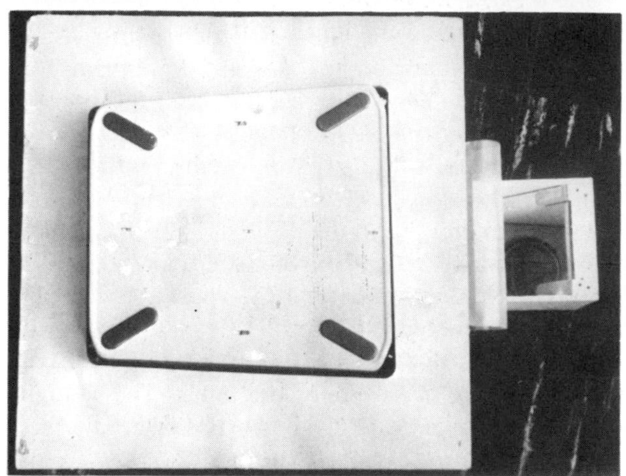

Figure 5.42a Equipment for weighing a hive throughout the year, using bathroom scales and a periscope (Hillyard, 1958).
above A Supporting box 40 × 40 × 15 cm.
B Bathroom scales.
C Light box 30 × 10 × 10 cm, housing periscope.
below Inverted bathroom scales in position, showing also view into periscope from above.

Many other types of hive scales have been devised for permanent use with one hive. Others can be moved to different hives in turn. One of the best, moved on two bicycle wheels, is built on the principle of a fork-lift and takes less than a minute to operate (Owens, 1962). Weights are recorded by either a tensioned spring or a hydraulic unit under compression. The former is cheaper and easier to use, but the latter enables the scale to be fitted about 25 cm lower down, where it is easier to read.

Scales such as those described above are sensitive enough to record weekly changes in weight. More sensitive equipment is required for measuring daily or hourly changes.

5.43 Collecting bees from inaccessible places

This section is concerned with the *removal* of bees in an unwanted place, rather than with their subsequent use.

5.43.1 Bees without combs

When a swarm must be removed that has recently settled, and has not yet built comb, only adult bees have to be dealt with, and they may be collected for hiving. If it is necessary to kill them, and the swarm has settled inside a room that can be closed off, a suitable insecticide can be sprayed on them. The maker's instructions must be followed carefully, since the insecticide is toxic. Any material contaminated with insecticide should afterwards be buried or burned, and hands must be washed thoroughly.

If a swarm has recently gone into a cavity, and it is not worth saving, the bees may be killed with an insecticidal dust or spray, provided there is an accessible entry hole. M. V. Smith (1967) lists some suitable insecticides, and up-to-date instruction leaflets – indicating what is best among insecticides currently on sale – should be available from addresses in Appendix 2. Alternatively the bees to be killed can be removed by using an industrial vacuum cleaner without any adaptation, a few crystals of paradichlorbenzene having been first put in the dust bag. The open end of the suction tube (a width of 10 cm is recommended) is placed as close as possible to the flight hole, and the bees are stimulated to fly out by rapping on the wall nearby; they are sucked through into the dust bag, where they are overcome by the fumes. The contents of the bag should be buried or burned before the bees recover. (An unwanted swarm in an accessible place out of doors can be collected in a similar way.)

If the swarm is to be collected and hived, a specially adapted container – where the bees can be assembled in safety – is interposed between the outer end of the suction tube and the dust bag of the vacuum cleaner (Figure 5.43a), and no paradichlorbenzene is used. A package bee box or a nucleus hive is sometimes used, but any box of a similar size will do. Adams (1985*b*) inserts in the box a frame fitted with netting foundation (Section 10.4), for the bees to cluster on. Two pipes are fitted to the container: an inlet pipe attached to the tube through which the bees are drawn in, and an outlet pipe to fit the tube leading to the dust bag. Inside the container, the opening of the outlet pipe is covered to prevent bees being sucked in, but this must not restrict the air flow. In order to retain full suction through the system, the box – with the two junctions between pipe and tube – is enclosed in a plastic bag which is secured tightly round the tubes.

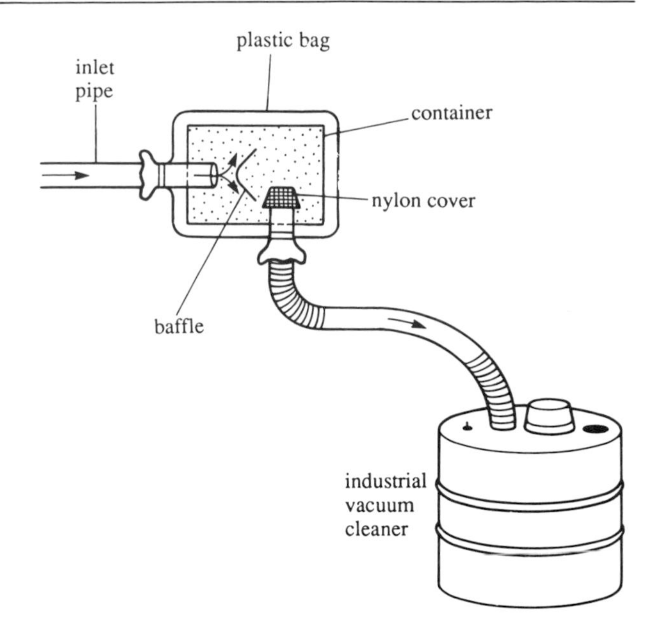

Figure 5.43a Arrangement for collecting bees, using an industrial vacuum cleaner (based on Witter, 1981). Arrows indicate the direction of air flow. The tin baffle is fitted to prevent injury to the bees as they enter the container from the inlet pipe, and the outlet is covered with a stretched piece of nylon stocking to prevent bees being sucked through.

Adapted in this way, the vacuum cleaner can be used to collect an accessible swarm in the open, or one that has fairly recently settled inside a cavity. Most models are, however, not powerful enough to be used in reverse as a bee blower for clearing honey supers (Section 5.33).

Whatever operation has been carried out to remove or exterminate bees in a cavity, it is most important that all bee entry holes are sealed before the operator leaves. Otherwise combs or traces of wax left behind will attract swarms in future, and the clearance procedure will have to be repeated.

5.43.2 When the bees have brood and honey

Beekeepers are sometimes called on to remove an established colony from an inaccessible space in a roof or wall. The person seeking help may not realize that this can be a difficult operation, and one in which the beekeeper – however well he is protected against stings – may be at risk from injury by toxic chemicals or by falling from a height. If the beekeeper undertakes the operation, he should make every effort to get out the brood and the honey as well as the bees, even if the comb is inaccessible. If the colony is killed *in situ*, there is likelihood of an objectionable smell from decaying brood, and of honey seeping or dripping through interior walls. So the adult bees should not be killed unless it will be possible to remove the brood and honey afterwards.

M. V. Smith (1967) describes the following procedure for saving the bees and brood, and for getting the honey out as well. The equipment (Figure 5.43b) includes a 'conical bee-escape', which can be made by bending and cutting wire gauze to the shape required. Close up all but one of the holes into the cavity to prevent bees entering them. Near the hole left open, put up a platform on which a hive can be placed. Fix the bee-escape over this hole, to let the bees out but not in. On the platform, place a one-storey hive containing 2 or 3 frames of bees with a queen, the rest of the combs being empty. The hive entrance should be close to the exit of the bee-escape. Bees leaving the cavity cannot enter it again, and go into the hive instead. After about a month all the brood in the cavity will have emerged as adult bees, which will have joined the colony in the hive. Now remove the bee-escape, so that foragers from the hive can collect the honey from combs in the cavity, which are now undefended. The queen in the cavity will eventually die. If there is much honey, it may be necessary to add a super to the hive. Finally, plug up the entrance hole to the cavity, to prevent future swarms getting in, and remove the hive to an apiary.

Where it does not matter that brood and bees die in their nest, poison can be used. Titěra et al. (1987) kill unwanted wild colonies as follows. Dark comb containing honey is used as a bait; when the bees are visiting it, poisoned bait is substituted: a 60% sugar solution containing 0.5% of the herbicide Gramoxone, which contains paraquat. All necessary precautions to protect the operator must be used.

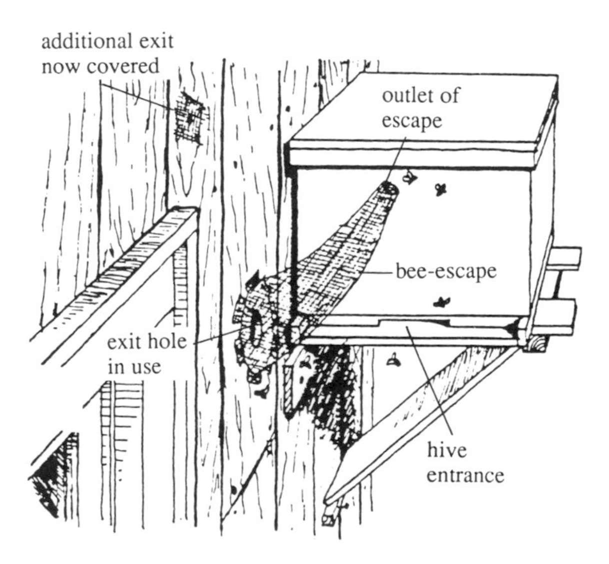

Figure 5.43b Set-up for removing all the bees and honey from a nest in an inaccessible cavity in a building (M. V. Smith, 1967).
The bees leave the cavity through the cone bee-escape, but do not re-enter its small open end.

5.5 ANNEX: DEALING WITH BEE STINGS

5.51 Avoiding stings

It is the beekeeper's responsibility to site his hives so that, even when he is working with them, bees do not fly low across neighbouring land where they might sting other people (Section 15.31). Protective clothing (Section 5.1) is designed to prevent the beekeeper himself from being stung while working at his hives, and any other people close by should be similarly protected. Anyone entering the area directly in front of a hive for any purpose should wear a bee veil; especially if he is operating a grass cutter or other machine there, he is likely to be stung if unprotected.

Bees may burrow in hair that is not covered (Section 3.41). The scent of some cosmetics and hair dressings can provoke bees to sting. If a bee appears to be following or homing in on a person, the best thing he can do is walk away, to an area of shade or protective vegetation, or better still into a building, where the bee is unlikely to follow. The worst thing he can do is make quick exaggerated movements, such as flapping the hands, which present a moving target to the bees.

5.52 Minor stinging

The honeybee's sting apparatus, and its mechanism, are described in Section 2.23.1. When a worker honeybee stings a soft-skinned animal, or the intersegmental membrane of another bee, she can often withdraw her sting. This is also true if the sting penetrates only slightly into human skin. But the two lancets of the sting are barbed (Figure 2.23b), and the bee cannot usually retract them from the relatively tough human skin. Venom from the bulb is driven through the venom canal (Figure 2.23a), and enters the puncture in the skin. When the bee has pulled herself free, part of the sting apparatus is torn away from the rest of her body and left behind; she normally dies shortly afterwards. The sting with its attachments should be scraped away from the skin with a finger nail.

Normally the sharp pain at the site of the sting lasts only a few seconds or a few minutes; it may be followed by some itching, and by local swelling that lasts one or more days. A beekeeper becomes familiar with the course of events. When working with bees, he learns to continue his operations after being stung, removing the sting as soon as possible, but meanwhile keeping his hands steady in spite of any pain. Many beekeepers do not bother to treat a sting in any way.

Most of the recommended treatments do no harm, and some do good, especially if used immediately. A cooling agent applied locally is soothing: cold water, dilute vinegar, blue bag, eau de Cologne or – better still – an ice pack. Antihistamine ointments, creams and sprays are widely available for treating insect bites and stings in general; many of them contain a local anaesthetic as well as an antihistamine, and sometimes a cooling agent as well. Antihistamines are also sold in tablet form. (Any hypersensitive person (Section 5.54) should not use antihistamine creams, etc., repeatedly, as this could cause dermatitis.) Meat tenderizer has been recommended for treating bee stings, presumably on the grounds that the proteolytic enzyme in it may break down protein in the venom. A steroid cream may be used to treat severe inflammation. The itching that usually follows a bee sting can be treated with lotions containing calamine and benadryl hydrochloride, such as are used to treat painful sunburn. Excessive swelling may be reduced by applying a pack of wet salt or Epsom salts (hydrated magnesium sulphate), which extracts water by osmosis through the skin.

Medical attention should be sought immediately if a person is stung in the eye or mouth: in the eye it may impair sight, and swelling in the mouth may obstruct breathing.

If a non-beekeeper or novice is stung, it is important that any beekeeper present should pay sympathetic attention and apply some remedy. If nothing else is available, a little honey could be applied to the site of the sting.

Table 5.52A summarizes four stages of the action of the venom on humans, and mentions some of its active components. On the left is the action on normal individuals, in which minor stinging leads to stages 1 and 2, and massive stinging also to stage 3. On the right is the action on the relatively few individuals who are hypersensitive (allergic) to bee venom; see Section 5.54. The amount of histamine in the venom is probably too small to play a direct role in the venom's toxicity to mammals, although it may make a significant contribution to its toxicity to bees and other arthropods, which have no mast cells (Owen, 1978*b*).

Amounts of the various compounds received from one sting are shown in Table 14.51A. In addition, natural venom from the venom sac contains 88% water, and also glucose, fructose and phospholipids. The composition of bee venom is discussed in Section 14.51, and also the possible pharmacological use of some of the components. Dotimas and Hider (1987) describe the important components of honeybee venom and their effects on human beings and other mammals.

5.53 Massive stinging

When a bee stings, an alarm pheromone is released, and if its concentration in the air reaches a certain threshold, other bees are likely to sting as well, especially at the site of the first sting. This alerting mechanism increases a colony's defence capability against an attacker. In bees that are described as quiet, gentle or docile, the threshold is high, and is rarely reached in practice. In bees that are referred to as bad-tempered, vicious or mean (*stechlustig* is the German term), the threshold is lower, and is more likely to be reached. In general, tropical African and Africanized bees have a far lower threshold than temperate-zone bees, and a sting by one bee is much more likely to alert others to sting, and in extreme cases a mass attack may be initiated. The subject is discussed in Section 3.41, and – in relation to the tropical bees – in 7.42, 7.43.

When handling bees that are easily alerted, a beekeeper who is stung once should try to prevent other bees being alerted, by washing or smoking the site of the sting. Upon occasion, it may be necessary to close the hive and retreat to whatever cover is available, until the pheromone has dispersed.

The effects of more than 500 stings on a single occasion may be severe and even fatal to a person, even if he is not hypersensitive. Mejia et al. (1986) describe renal failure in 5 patients in Colombia, each of whom received more than 1000 stings from Africanized bees; only one died. People have survived still more stings without permanent harm. The greatest number recorded (from Zimbabwe) is 2243, in a European man aged 30, and he did not develop an allergy to bee venom as a result (Murray, 1964). Most reports of such massive stinging relate to tropical bees – tropical *A. mellifera* in Africa (as in the case above) and their Africanized counterparts in South and Central America, and *Apis dorsata* in Asia. Mass attacks by temperate-zone *A. mellifera* are extremely rare but have been recorded, for instance when a person suffered a heart attack and became unconscious near to hives that were open at the time. Moret et al. (1983) published a detailed case history of a beekeeper 77 years old who received between 5000 and 6000

Table 5.52A General mechanism of the action of honeybee venom on humans
Adapted from O'Connor & Peck (1980).

Action in normal individuals	*Action in hypersensitive individuals*
Stage 1 involves hyaluronidase, MCD peptide, protease inhibitor, and some 'small molecules'	
Hyaluronidase breaks down hyaluronic acid polymers that serve as intercellular cement; this allows the venom to spread through the tissue. The protease inhibitor prevents enzymatic destruction of the hyaluronidase.	The hyaluronidase may participate immediately in an antigen-antibody reaction to trigger an allergic response.
Simultaneously, the mast cell degranulating (MCD) peptide penetrates the membrane of the mast cells, creating pores. This releases histamine, which – in combination with some small molecules of the venom – contributes to the weal and flare, and the local itching and burning sensation of the sting. Protective antibodies present in the serum of most beekeepers can effectively neutralize hyaluronidase, preventing the spread of the venom.	
Stage 2 involves phospholipase A and melittin	
As venom penetrates blood vessels and enters the circulatory system, phospholipase A and melittin (as a micelle) act synergistically to rupture blood cells.	Both melittin and phospholipase A can also produce allergic reactions.
When only *a few stings* are received, this action is mostly localized. After *massive stinging* or injection of venom directly into the circulatory system, it can become widespread.	
Stage 3 involves apamine, MCD peptide, melittin, phospholipase A, and probably some other components	
When only *a few stings* are received, actual toxic effects are insignificant. Toxic effects of *massive stinging* may be severe, particularly when significant amounts of venom enter the circulatory system. Apamine acts as a poison to the central nervous system, and both melittin and phospholipase A are highly toxic. Large concentrations of histamine are produced by the action of the MCD peptide on mast cells, and contribute to overall toxicity. The role of other components is still unknown.	
Stage 4 involves hyaluronidase, phospholipase A, and melittin	
	Hypersensitive individuals may encounter antigen-antibody reactions to any or all of these components. Severe reactions can result in death from anaphylactic shock.

stings, and who died three days later from kidney failure.

5.54 Allergy or hypersensitivity to stings

A very small number of people are allergic (hypersensitive) to insect venom. Three large-scale surveys in the USA showed percentages between 0.35% and 0.40% of the total population studied. Hypersensitivity to bee venom can lead to a general body reaction to a bee sting, which can be dangerous. Beekeepers therefore need to have some knowledge of it, and Riches (1982) has written a clear account, on which most of this section is based; it was updated in 1989.

Figure 5.54a illustrates ways in which an individual may react when he begins to be stung frequently, for instance when he starts to keep bees. His

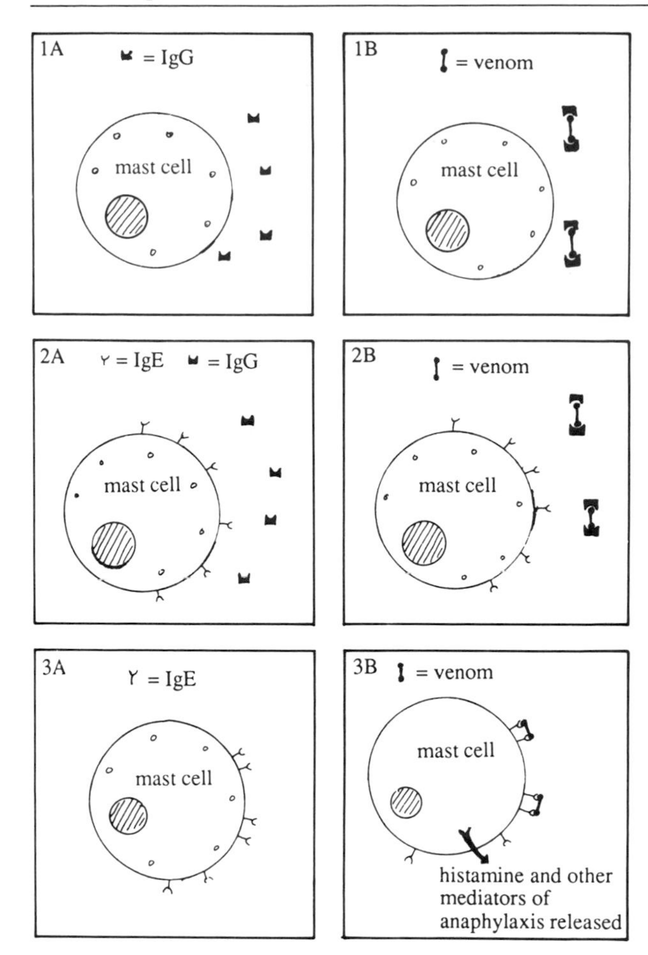

Figure 5.54a Diagram of a mast cell and the IgG and IgE antibodies, (A) before a sting and (B) immediately afterwards (adapted from Riches, 1982).
1 Immune person.
2 Previously allergic person now protected by 'blocking' antibodies.
3 Allergic person.

reaction may be one of three types, depending on the nature of the protective antibodies in the gamma globulin fraction of his blood-serum proteins.

Beekeeper 1 suffers only the expected momentary pain, and minimal local swelling where he was stung; within a few months he becomes immune to the effects of stings, through the following mechanism. Protective antibodies produced as a result of earlier stings are mainly immunoglobulin G type (IgG), with very few immunoglobulin E type (IgE). In Figure 5.54a, 1A shows a mast cell in the beekeeper's skin, with the protective IgG antibodies in the tissue fluids. When he is stung again (1B) the IgG antibodies combine with the hyaluronidase in the venom and neutralize its effects.

Beekeeper 2 is typical of most people starting beekeeping. He has only a slight reaction to the first few stings, but develops more IgE than IgG antibodies (as beekeeper 3 does, see below). Local swelling increases with subsequent stings, and may be unpleasant and unsightly. However, in the course of time, he develops more IgG, and the reactions gradually diminish, and immunity is eventually achieved. In the Figure, 2A shows IgG antibodies in the tissue fluids, and IgE attached to the mast cell; its molecules possess a region which binds specifically to mast cells and, when bound, mast cells are stimulated to release histamine. Also in the Figure, 2B shows the situation after a sting when beekeeper 2 has become immune. The IgG antibodies then take on a protective or 'blocking' action (as in 1B), and hyaluronidase in the venom combines preferentially with the circulating IgG (blocking antibody), thus preventing interaction with IgE.

Beekeeper 3 has developed serious hypersensitivity to bee venom. Local swelling becomes more and more widespread with successive stings, and general symptoms also occur after a sting. He has developed large quantities of IgE type antibodies, but never enough of the protective IgG type to produce the blocking action that prevents hyaluronidase in the venom combining with the IgE at the cell surface, with the result that too much histamine and other substances are released, possibly leading to anaphylactic shock.

There are several types of bee venom hypersensitivity mediated by IgE antibodies. In the least severe, extensive local swelling (Stage 1 in Table 5.52A) is followed 3 or 4 hours later by more extensive swelling, which may last 2–3 days. In another type, general (systemic) reactions occur within a few minutes of being stung (see Table, Stage 1, right); they may include a skin rash, wheezing, nausea and vomiting, abdominal pains, palpitations and faintness. The more quickly these symptoms appear (e.g. in only 1–2 minutes), the more serious they are likely to be.

The most severe type of hypersensitive reaction, anaphylaxis, also occurs within seconds or minutes of the sting, and constitutes a major medical emergency. The symptoms listed above are followed by falling blood pressure which leads to unconsciousness, and death may result; see Riches (1982), who also describes several less common types.

Airborne allergens from live or dead bees can also cause problems. Nine out of 595 Swiss beekeepers reported respiratory symptoms while working on their colonies (Böhny et al., 1986). In Poland, when over a million dead bees were being dissected each year for disease diagnosis (Kirkor, 1959), the women doing the preparatory work inhaled dried blood and muscle of the bees, and four became ill, showing respiratory and other symptoms which are described.

5.55 Treatment of hypersensitive reactions

Antihistamines can give some protection to a moderately hypersensitive person if taken before exposure to stings. Riches suggests 4 mg of Piriton (chlorphenir-amine), taken orally an hour or so before starting bee-keeping operations. Prime (1958) found that taking 0.5 mg diphenylpyraline half an hour before opening hives prevented 'moderate reactions, including slight fever, and local pain, itching and swelling', which had previously followed a bee sting. He was able to dispense with the treatment as the beekeeping season advanced.

Systemic reactions following a sting should be treated immediately with adrenaline, mild symptoms by inhalation and severe ones (and anaphylaxis) by injection. Extremely prompt medical treatment is essential for acute anaphylaxis; Riches (1989) gives details of drugs to be injected. In recent years, inflatable military (or medical) antishock trousers, referred to as MAST suits, have been used in treating patients suffering from shock (Wayne, 1983,) and they are recommended in the treatment of anaphylaxis following a bee sting (T. Oertel, 1984). When the trousers are in place and inflated, they exert pressure on the lower part of the body, producing mechanically some of the effects of adrenaline; they increase blood pressure and the blood supply to the brain.

At some allergy clinics, carefully regulated courses of venom injection are available which can decrease sensitivity to the venom. Riches (1989) describes some types of immunotherapy, and readers wishing to find out what might be available to them should make enquiries locally.

If a beekeeper develops serious hypersensitivity to bee stings, he should seek advice from an allergy specialist. He may be able to continue beekeeping after he has satisfactorily completed a course of desensitization. If another member of his household who may come into contact with the bees becomes hypersensitive, either the hives should be moved away, or secure arrangements made to keep bees away from the affected person, unless he is willing to undergo desensitization. Emergency treatment should be kept readily available, and all members of the household should know exactly how to use it, and what other steps to take in an emergency.

5.6 FURTHER READING AND REFERENCE

Many of the publications listed at the end of Chapters 6 and 7 include information on beekeeping equipment in general. Beekeepers can usefully enlarge their knowledge by learning about equipment used in other countries as well as their own, especially countries where beekeeping is carried out on a similar scale and the climate is also similar. Suppliers' catalogues can be a valuable resource; see Section 16.6, and addresses in Appendix 2.

The following books are expecially relevant to subjects covered in Chapter 5, and details of them will be found in the Bibliography.

Anderson, R. H. et al. (1983) *Beekeeping in South Africa*, 2nd ed.

Crane, E. (1985*d*) *Beekeeping: some tools for agriculture*

Dadant & Sons (ed.) (1975) *The hive and the honey bee*

Gojmerac, W. L. (1983) *Bees, beekeeping, honey and pollination* [USA]

Matheson, A. (1984*a*) *Practical beekeeping in New Zealand*

Morse, R. A.; Hooper, T. (eds.) (1985) *The illustrated encyclopedia of beekeeping*

Riches, H. R. C. (1989) *Bee venom hypersensitivity update*

6

Beekeeping management, especially in temperate zones

This chapter and the next are concerned mainly with colony management for honey production; management for crop pollination is discussed in Section 8.4. The present chapter deals with beekeeping management in general, and much of it applies to beekeeping with hives of any honeybees anywhere in the world. It also gives details of beekeeping management practices in temperate zones, with temperate-zone *Apis mellifera*. The seasonal cycle in temperate zones is more regular than seasonal cycles in tropical and subtropical zones, and it will also be more familiar to many readers. Chapter 7 discusses special features of beekeeping in the tropics and subtropics, with both tropical and temperate-zone honeybees, including *Apis cerana*.

The two temperate zones of the world (following Good, 1974) lie between the subtropics and the arctic zones, at latitudes from 34° to 66½°; they are indicated on the map on page 504. These zones include 39% of the earth's land area on which plants grow. In much of the warm temperate zones (latitudes below 45°), temperatures are usually high enough all the year round for honeybees to fly. In the cold temperate zones (at higher latitudes) this is usually not so, and a colony winters as a cluster inside the hive.

In the north temperate zone are: most of the USA and Canada, all Europe and the USSR, much of China and almost all of Japan; the term 'temperate' is, however, hardly appropriate for the interior and some other regions of N. America and Asia. The south temperate zone has a smaller land area: the southern half of Chile and Argentina, the Cape Peninsula in South Africa, two rather small portions of Australia, and New Zealand; there is no land south of about 55°S except Antarctica. Honeybee colonies are kept permanently at all temperate latitudes, and some even beyond the Arctic Circle.

Except for wintering procedures, the principles of management for honey production are similar in most parts of the temperate zones. But there is considerable variation in the timing of operations and their relative importance, and in the hazards that are most likely to prevent colonies achieving their full potential for the beekeeper. In any particular area, the most valuable sources of information on the local situation are likely to be local beekeeping advisors, beekeepers' meetings, and individual experienced beekeepers.

Before management itself is discussed, Section 6.1 gives general information on handling bees when performing management operations. Again, readers can learn much by watching an experienced beekeeper handle bees. The way in which individuals work is related to their general behaviour characteristics, and beekeepers must use a management system that suits them, and also that fits in with their other commitments. Those who can work their bees only at weekends need to devise very careful timetables for some operations.

Among gardeners, a few have 'green fingers' and are able to achieve exceptional results: plants flourish under their hands as nowhere else. A few beekeepers have a comparable ability to get the best from their bees. They handle bees superbly, quickly assess the qualities and needs of a colony, and also recognize which colony is headed by a queen that should be selected for use in breeding. Examples of such beekeepers are Brother Adam in England, Jim Powers in the USA, and Jim Nightingale and Penelope Papadopoulo for tropical African bees.

6.1 HANDLING HONEYBEES

This section applies to temperate-zone *Apis mellifera*, and much of it also to *A. cerana* and tropical *A. mellifera*; Sections 7.41 to 7.43 include specific information on handling these other bees. Most operations in colony management involve opening hives, and handling the open hive boxes and individual combs in them, which are usually covered with bees. Under suitable conditions, many strains of bees are remarkably tolerant of such manipulations, provided they are properly handled and controlled. A number of bees are likely to be in the air while the beekeeper is working, and some may settle on him, but they are unlikely to sting unless they are injured or alerted. The unlikely event of a colony getting out of hand is referred to at the end of Section 6.12.

6.11 Environmental and colony conditions

In cooler areas, beekeepers who can choose when they open hives should work on a warm, dry, windless, sunny day between 10 h and 16 h, when most of the older bees are out foraging. In any case, hives should not be opened (exposing the bees or frames) during cold winter months, or when the air temperature is below 15°–18°. Cold, windy or showery days, when foragers are all in the hive, should be avoided when possible. Bees are especially likely to sting when a nectar flow has suddenly ceased, and during thundery weather. Ideal conditions are not essential, however, and *necessary* manipulations should not be postponed on account of the weather.

Colony conditions which tend to make bees more difficult to handle include the following: a very large population; presence of many old bees; previous alerting of the colony by disturbance of any kind; robbing by bees from other colonies; queenlessness or a supersedure; a failing queen (end of Section 2.53); starvation. Starving colonies should be fed sugar syrup before the hives are examined.

Several studies seem to show that colony activities, including foraging, are little affected by normal hive examination. However, in one experiment Taber (1963) inspected hives during the main flow, and found that their gain in weight on that day (only) was less than that of hives not inspected, by 20% to 30% according to the extent of the disturbance. Hive manipulations in unsuitable conditions or at too frequent intervals, which put the colony under stress, can aggravate nosema infection.

6.12 Basic recommendations

Always wear protective clothing, as set out in Section 5.1; Section 5.5 explains how to deal with any stings. Use the equipment described in Section 5.2, including the smoker. Apply smoke at the start of operations, and as required throughout them.

Open hives only when necessary, and then as briefly as possible. Always work at the side or back of a hive, away from the bees' flight paths in front of hive entrances. All movements should be smooth and deliberate, not jerky or rough. Do not bump or jar hives or frames. Keep boxes of frames of bees covered, to reduce the number of flying bees and to prevent robbing; manipulating cloths (Section 5.23) can be useful for this purpose. Section 3.41 sets out the sequence of a honeybee's responses that may culminate in stinging.

When opening each hive, put the hive roof upside down beside the hive at a convenient height (Section 5.41), and place boxes taken off the hive on top of it. The queen is unlikely to drop off a comb, but if so she is then caught in the hive lid. If brood box(es) are to be examined, do this first, having removed upper boxes unexamined to be dealt with later. Otherwise, when a brood box is dealt with, it will contain many bees from boxes above, that have been smoked down into it.

With any hive box, if a gap has been left at one end (Section 4.25), it may be possible to push several adjacent frames to that end, so that there is space near the centre to allow removal of any chosen frame. If not, first take out a frame near one end, which will probably have fewer bees on it, and be easier to remove, than a central one. This frame can be stood on end in front of the hive, to provide more working space in the hive box while the rest of the frames are inspected (but bear in mind the next paragraph). When inspecting any frame, keep the comb surfaces vertical to prevent any newly collected nectar dripping out of the cells. Having examined one side, bring the second side into view by raising one end of the frame directly above the other as shown in Figure 10.23c where the comb is from a top-bar hive. Rotate the frame (keeping the comb vertical), and the end that was raised can then be lowered. Before returning the frame to the hive, remember to reverse it again. It is a good general rule to replace the frames from any box in their original position and orientation, except when there is a good reason to do otherwise.

Unless the air temperature is nearly as high as that

of the brood nest (35°), the surface of a comb containing brood not covered by bees should not be exposed for longer than a few minutes, or the brood may be chilled and thus injured. The beekeeper must keep this constantly in mind, as he will receive no visible or audible warning that brood is being injured.

Brood boxes may be left separated from each other for up to half an hour, provided each is kept covered when being inspected. Boxes containing sealed honey may be left 5–10 minutes, if covered on top and made bee-tight below. Except during a heavy flow, combs containing unsealed honey are likely to attract robbers, so do not leave such combs outside the hive unless they are in a bee-tight container. Disorganization or robbing will itself alert the beekeeper that he has had the hive open too long.

Many of the above points are included in a chapter by Johansson and Johansson (1978), who also give details of procedures when 'difficult' colonies have to be manipulated. During a complicated manipulation, a colony can occasionally get out of hand, and so many bees fly off the combs that effective operation is no longer feasible. In such a situation, reassemble the hive as quickly as possible, checking that no gaps have been left through which robber bees could enter. Leave the apiary, taking away any uncovered combs. Do not open any hives in the apiary until the next day.

6.13 Finding the queen

Many beekeepers want to find the queen in a hive, and it is sometimes necessary to do so. Queens are easier to find if they are marked with a spot of colour on the thorax. Those that are purchased may already be marked, and beekeepers can also mark their own queens; see Section 5.23.

A queen has a longer and wider abdomen than workers, and longer legs so that she often appears to stand higher off the comb; she usually moves more slowly. A queen of a 'yellow' race of honeybees is lighter in colour than the workers, and is thus more readily distinguished from them than is a dark queen from dark workers. Virgin queens are thinner than laying queens, since they contain no eggs, and are more difficult to spot; they may move quite fast, and are sometimes referred to as 'flighty'.

If the queen may be in any one of several hive boxes, these should be separated before searching, otherwise she may go from one to another. Any box off the hive should be placed on an upturned roof or other surface in case the queen falls off a comb.

It is best to look for the queen when many of the workers are out of the hive foraging. The bees in the colony should be disturbed as little as possible, and a minimum of smoke used. The queen will probably be within the brood nest on an area of comb containing eggs, young brood and vacated cells – in which she will next lay eggs. A queen often tries to move away from the light, and this may take her out of the beekeeper's sight – hiding among workers, or below the comb, or between it and the frame. As each frame in turn is removed to search for her, it is best to scan the outer part of one surface of the comb first, and then the centre, before examining the frame and the comb edges, and finally the other surface of the comb.

If you cannot find the queen and it is necessary to do so, the following procedure may be adopted. Examine each comb carefully and transfer it to a spare empty brood box placed on a floor board. Then examine the box itself. If the queen is still not found, take two or three of the combs in turn out of the spare box, shake all the bees off them back into this box, and return the combs to their original box. Now cover the original box with a queen excluder; stand on this a completely empty box, and shake the bees from each comb in the spare box on to the excluder. The workers will go down to the two or three combs below, but the queen cannot pass through the excluder, so she should be found on it, and can be caught. This process is called 'sieving' the bees, and Matheson (1984*a*) gives more details. Finally, return the rest of the combs to their box, in their original order. With large colonies, sieving is a major and disruptive operation that is best avoided, and the use of marked queens is especially valuable.

6.2 STRATEGY OF MANAGEMENT FOR HONEY PRODUCTION

Various studies have shown that large colonies produce more honey than small ones. Ratnieks (1986) found that for colony populations between 30 and 60 thousand bees, the honey production was roughly proportional to the population, i.e. a colony twice as large as another produced twice as much honey – and for relatively little additional outlay of equipment and labour. Table 6.2A exemplifies this, and Moeller (1958) gives further data. The primary aim is therefore to have large (or maximum) populations of adult bees in the colonies during the period of main honey flows. Populations must be built up at the appropriate time, and after the main flow they must be used in some other cost-effective way; see for instance C. L. Farrar's 1968 series of articles.

Seasonal cycles of honey flows at different latitudes in the temperate zones are shown by some of the scale-

Table 6.2A Effect of colony size on honey production
Data relate to New Zealand (Matheson, 1984*a*).

No. adult workers per colony	*Honey production*	
	per colony (kg)	*per bee* (g)
10 000	4	0.40
20 000	14	0.70
30 000	23	0.77
40 000	32	0.80
50 000	41	0.82
60 000	50	0.83

hive records of colonies in Figure 12.62a. At lower latitudes, honey flows are spread over much of the year, but as the latitude increases they become shorter and more pronounced; in Sweden they last hardly more than 4 weeks in the year.

Seasonal cycles of the colony show corresponding latitude differences. The seasonal cycle of the adult population in a colony is explained in Section 3.51.1, and of brood rearing in 3.51.2. Briefly, the adult population of a colony is governed by the rate of brood rearing and by the length of life of the adult bees, which varies from a few weeks in summer to several months in winter if brood rearing is then minimal. In the cold temperate zones about half the bees in a colony at the onset of winter are likely to die before spring; the mortality rate is highest in early spring, and too few new bees are then being reared to replace those dying. Figure 3.51a shows a minimum colony population (10 000 bees) in March. Brood rearing then increases rapidly, so that the adult population reaches its peak around the summer solstice; this is 40 000 in Figure 3.51a, but it may reach 60 000 or more. After midsummer the adult population decreases again, more or less continuously, until the next spring. This is a simplified picture, and assumes that the colony does not swarm.

Figure 6.2a shows three notional curves of the adult population of a colony in the course of the bees' active season in the temperate zones. Colony A does not swarm. The colony grows when forage becomes available in spring, and completes its population growth during the flow period. The beekeeper's honey harvest is lower than it might be, both because the colony population is below its maximum during the flow, and because the larger population afterwards will consume much honey. Colony B also does not swarm, but is more successfully managed by the beekeeper for honey production, so that the population is highest during the flow; the colony stores the maximum amount of honey. Another benefit is that the population starts to decrease by the end of the flow, consuming less honey than in colony A, and the beekeeper gets the greatest honey yield possible. Colony C swarms in late spring. If it swarms early enough, both the parent colony shown, and the swarm, may be sufficiently large during the main flow period to collect enough food to keep them alive through the next winter. But a beekeeper would not harvest very much honey from the colony.

In order to achieve the growth curve in B, the beekeeper needs to learn in advance the probable dates of the major flow period in the area, and then to time colony management to fit in with them. The more predictable the dates of the flow, the easier it is to time operations correctly. For this reason, temperate-zone regions with a continental climate often give more reliable honey yields than regions with an oceanic climate, where the weather is more variable from month to month and from season to season.

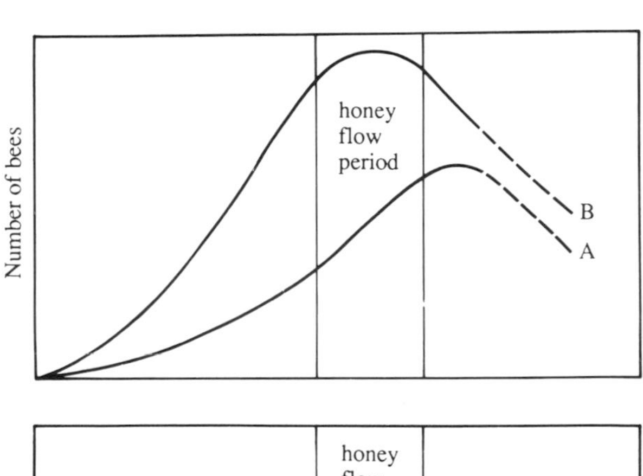

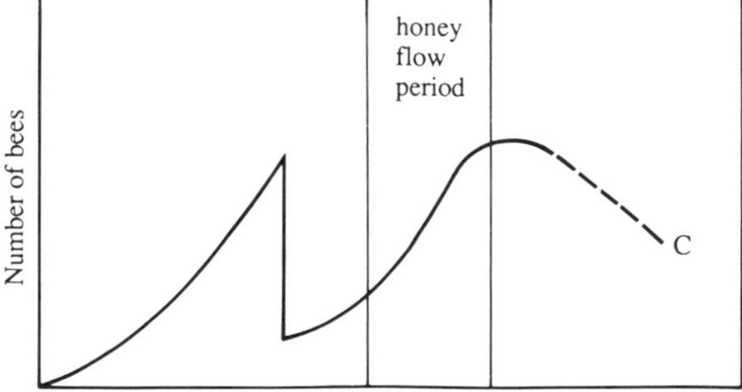

Figure 6.2a Notional diagrams indicating changes in the number of adult honeybees in colonies during the active season in temperate zones. The base line is the population at the start of the active season.

A Increase (without swarming), becoming rapid during the flow period and reaching a maximum afterwards.
B Rapid increase in spring (achieved by the beekeeper's management) to a maximum during the flow period.
C Rapid initial increase, followed by swarming and regrowth during the flow period.

Beekeepers who are observant and knowledgeable about plant growth may be able to predict the timing of the main flow from a previous phase of development of the flowers. In some areas the dates of certain earlier flowering trees are useful phenological indicators (e.g. Schröder, 1954); or accumulated temperatures can be used (e.g. Yim, 1986).

Uniform colonies in management strategy
Colonies owned by any one beekeeper often show substantial variations, because queens differ in their laying rate, or bees drift from certain hives to others, or food stores are (or have been) inadequate, or because of disease or other damage. Management, and its timing, for a particular colony therefore differ from what is needed by others. One of the main aims of a commercial beekeeper is to reach the stage where all colonies, headed by queens that are full sisters, can be managed as identical units, rather than as a group of individually different units, each of which must be examined and should then be treated according to its own specific needs (Jay, 1979). There is no reason why smaller-scale beekeepers should not also work towards this aim. Every colony will then be ready on the same date to receive the same treatment (feeding, supering, swarm prevention measures, etc.). Much time and work will be saved, colony performance will be both uniform, and higher, and the total honey yield will be increased.

6.3 MANAGEMENT DURING THE PERIOD OF COLONY GROWTH, INCLUDING SWARM PREVENTION

This Section, and 6.4 and 6.5, deal with basic principles of beekeeping management; for tropical conditions see Section 7.3 to 7.37. The details of what has to be done, and when, depend on local circumstances, and also on whether the beekeeper wants to finish the season with the same number of colonies, or more, and whether he will migrate his colonies for honey production (Section 8.1) or for crop pollination (8.4 to 8.42). Readers who stock their hives with package bees in spring should refer to Section 8.35.

For temperate-zone beekeepers, spring is the period when the timing of operations is likely to be most critical. And these operations will have a significant effect on the honey-getting ability of the colonies. Hives must not be inspected in spring until the weather is warm enough, or opened at any time if the air temperature is below 15°–18° (Section 6.11). At the first inspection each hive should be kept open for only a few minutes, to find out if the colony needs special attention to ensure its well-being.

1. If any colonies are dead, remove the hive and clean out the dead bees.
2. If any colonies, alive or dead, shows signs of disease, this must be dealt with; see Section 11.2 for brood diseases, 11.3 for diseases of adult bees, and 11.4 for parasites.
3. Check that each colony has a laying queen. It is not usually necessary to search for the queen, or for eggs. As long as uncapped brood is present, she was laying within the past week or so.
4. If any healthy colony is noticeably smaller and less active than others, unite it with another such colony, or with a normal colony, as described in Section 6.35.2.
5. If any colony is short of food (honey or sugar), feed it with syrup as explained in Section 6.31. Each colony should have at least 5 kg, and preferably 10 kg, at this time.
6. Assess the need for feeding pollen or pollen substitute (Section 6.32).
7. Check that the bees have access to water, which can be vital to them at this time of year (see Sections 2.43.2, 6.62.3).

6.31 Feeding sugar (and water)

Colonies must have adequate stores in the hive at all times. They should have been left for winter (Section 6.53.1), or other dearth period, with sufficient stores to last until they can collect substantial amounts of nectar again. Older adult bees can survive on honey alone. Larvae need food with a higher water content, and nurse bees feeding brood must have access to water (as well as to pollen, see Section 6.32). During much of the year, sufficient nectar is brought into the hive to provide the necessary water, but this is not always so in early spring, and in a colony with ample honey and pollen, brood rearing may be limited by lack of water. Provision of a permanent water supply in or near the apiary is therefore of great importance in spring; see Section 6.62.3.

In the spring, some amateur beekeepers feed dilute syrup made by dissolving sugar in its own weight of water, or in its own volume of water – which is more dilute. The assistance provided by feeding such syrup, particularly to small colonies, could be due as much to the water it contains as to the sugar. Commercial beekeepers are likely to feed a more concentrated syrup. Syrup can be fed in the hive in one of the feeders described in Section 4.47; in early spring, feeders 3, 5 and 6 are the most satisfactory, because they give the bees access to the syrup close to the brood nest.

There is no benefit in feeding syrup routinely to larger colonies in spring – it is better to feed plenty in autumn. Syrup feeding is, however, beneficial:

- for any colonies short of stores;
- for newly formed small colonies, or nuclei (Section 6.34);
- when an unseasonable cold spell occurs, after foraging and substantial brood rearing have commenced.

Provided bees have access to a water supply, combs containing honey can be inserted in the hive instead of feeding syrup; this is a usual method for feeding package bees in spring (Section 8.35). Or a soft candy prepared from invert sugar can be purchased, which is expensive but convenient to handle as it needs no extra container.

The question of honey adulteration is dealt with at length in Section 13.26, but it must also be mentioned here. Sale of honey deliberately adulterated with high fructose corn syrup (HFCS) caused concern to beekeepers in the early 1980s, and although HFCS can be detected in honey (Section 13.26.1) the process is by no means simple. Like honey, HFCS is produced by enzyme action, and its sugar composition is somewhat similar to that of honey. A sample analysed by Shuel (1982) contained 40% glucose and 25% fructose.

Beekeepers would discount, as improper practice, the feeding of *any* form of sugar to bees during or immediately before a flow from which honey will be harvested. But apart from such deliberate methods of adulteration, it may upon occasion be quite difficult for beekeepers to avoid inadvertent adulteration of honey after colonies have been fed with sucrose or HFCS syrup, or candy. An incident in which denatured sugar was fed in autumn is cited in Section 6.53.1. The subject is one in need of further discussion and exploration.

Johansson and Johansson (1978) give a full account of the provision of sugar and water to bees. It is not usually necessary to provide water to colonies individually. But if a beekeeper wishes to feed water in the hive, a syrup feeder of a type recommended above for use in spring may be used. Or a wet sponge, in a flat plastic bag with a hole at the top, can be placed above the frames.

6.32 Feeding pollen and pollen substitutes

Plant sources of pollen used by honeybees are covered in Section 12.42 and Appendix 1. Colonies need a continuous pollen income when they are rearing brood; see Section 3.33. In many parts of the temperate zones, there is a great enough variety of flowering plants to provide sufficient pollen to support brood rearing throughout the active season. But in some regions, especially where agricultural use of the land greatly limits the variety of spring flowering plants, there may be periods when insufficient pollen is available to the rapidly growing colonies. Flows later in the season, from honeydew and a few from nectar, are not accompanied by a pollen flow (Sections 12.22, 12.42). These situations can be remedied by feeding colonies with pollen previously collected by bees and harvested from the hives (Section 14.22); not all pollens are equally nutritious for honeybees (Section 12.42). Alternatively a pollen substitute may be fed, or a mixture of pollen and a substitute; the term pollen supplement is sometimes used in connection with such a mixture.

To be successful, a pollen substitute must be nutritionally adequate to support brood rearing, providing the proteins, lipids, minerals and vitamins of which pollen is normally the bees' only source. It must also contain a phagostimulant, a substance that attracts nurse bees to go to it in the hive and to eat it (Doull, 1974). The presence of sugar in a pollen substitute increases its attractiveness to bees, and its consumption by them, but it is not the only factor.

Pollen substitutes and supplements are usually fed in the hive, in the form of a soft 'patty' covered with plastic film or waxed paper, laid over the tops of the frames. The covering is needed to prevent the material drying out, in which case the bees cannot take it. Provision of a powder in an outside feeder, which bees collect in their corbiculae and carry back to the hive, is less effective.

Various formulations for pollen substitutes are recommended in different countries, according to what suitable constituents are easily and cheaply available. Many substances have been used with some success as components of a mixture, for instance dried brewer's yeast, expeller-process soya-bean flour, dried skimmed milk, egg yolk; use of raw eggs for feeding to bees is referred to in Section 8.29.1. On the Pacific coast of Canada the cheapest protein source is fish meal, which was tested with success by Chalmers (1980; also Nelson, 1982).

Many pollen substitutes are on sale under enticing trade names, but their composition is often not stated. One in the USA whose components are known is the Beltsville Bee Diet (Herbert & Shimanuki, 1980). It is based on a mixture of lactalbumin from the whey of milk (1 part) and *Torula* yeast (2 parts). This mixture contains 66% protein, and it is combined (35 parts) with sucrose (65 parts) and sufficient water to make a moist patty, which contains 13% protein. Gojmerac (1980) favours a patty made by mixing pollen and either expeller-process soya-bean flour or brewer's yeast (1 : 3) with concentrated sugar syrup. A formulation with fish meal uses powdered herring meal, brewer's yeast and sugar (1 : 1 : 2.5), mixed with water to form a patty; it does not taint the honey.

In general, if fresh pollen is available, the bees will collect and use it in preference to any substitute, in which case the provision of a substitute is useless. But Gojmerac (1980) in Wisconsin, USA, was able to stimulate brood rearing during the 6 to 8 weeks before pollen foraging started, by providing in the hive the patty described above, and renewing it as necessary; a colony might consume about 3 kg altogether, and take more even after foragers were collecting pollen from flowers.

Johansson and Johansson (1978) give many formulations, and review the subject in detail; Chalmers (1980) cites some more recent references.

6.33 Swarm prevention and control

Having managed the colonies in early spring to stimulate rapid growth, the next phase of management is the difficult one of ensuring that this rapid growth continues, so that colony populations become much larger than they would be in nature. To achieve this, the colonies must be prevented from swarming (Section 3.34.2), in spite of their over-large size, for swarming can cut the population by half within a few minutes. Simpson (1958*a*, 1958*b*) discusses some of the problems. What is referred to here is reproductive swarming; tropical *Apis mellifera* or *A. cerana* may also leave the hive in an absconding swarm (Section 3.34.5); management to prevent absconding in *A. cerana* is dealt with in Section 7.41.4, and similar considerations would apply to tropical *A. mellifera*.

With reproductive swarming, the term *swarm prevention* is often used to mean preventing preparations for swarming by a colony including the production of swarm queen cells.* The term *swarm control* is then confined to emergency measures the beekeeper carries out when a colony is found with queen cells, in order to cut short the swarm preparations and prevent emergence of the swarm. The terms are used here with these meanings.

6.33.1 Swarm prevention

Table 6.33A lists a number of colony conditions, some or all of which can stimulate a colony at any latitude to rear queens in preparation for swarming, although probably none of the conditions would do so on its own. To prevent swarming these conditions should be avoided, and they are discussed in order below. Outside the tropics, swarm prevention in general becomes more difficult as the latitude increases, at any rate between 25° and 55°N (Morse, 1972), because colony populations grow very quickly while days are lengthening rapidly (Section 3.51.3).

(*A*) *Genetic predisposition to swarming*

This involves several factors. Races and strains of bees differ in their tendency to swarm – as a result of conditions in which they evolved, and also of selection inherent in methods of bee management in past centuries, some of which favoured swarm production. Modern beekeepers who collect swarms to populate their hives may be using and perpetuating strains likely to swarm. Bee breeders, on the other hand, consistently select for non-swarming traits.

(*B*) *Colonies with an older queen*

When a queen's egg-laying ability is decreasing, colonies are more likely to swarm than are colonies with a young queen. A queen one or more years old lays sufficient eggs in the early phase of population growth, but when the population is approaching its maximum, she may not be able to lay as many as the colony could tend and rear. Also pheromones produced by the queen (Table 2.35A and Section 3.21) which inhibit the workers from rearing new queens, are produced in smaller amounts as the queen ages, so when the colony grows large there may be insufficient for all workers in the brood nest. Inhibition is then not complete, and workers build swarm queen cells. In a commercial apiary in England, swarms were at least 3 times as likely if colonies had a queen 2 years old as 1 year old (Simpson, 1960).

Professional beekeepers commonly requeen colonies at intervals varying from 6 months to 2 years (Section 6.51), and this practice alone can greatly reduce the percentage of colonies that make swarm preparations.

(*C–E*) *Overcrowding in the hive as a whole*

Such overcrowding can be prevented or relieved by adding empty honey supers above a queen excluder, although this action on its own may not be sufficient to prevent swarm preparations. Another treatment is to move an over-large colony to the site of a small one and vice versa, so that the large colony loses returning foragers and the small colony gains them (Cale et al., 1975). The average age of workers in the large colony becomes lower after the operation than before it, since the bees it lost are older ones.

* Supersedure queen cells are explained in Section 8.21.

Table 6.33A Some colony conditions contributing to the initiation of queen rearing in preparation for swarming

A to G are discussed in the text.

Entries below (from Winston et al., 1980) refer to 'average-size' colonies living wild, of European bees at 42°N and Africanized bees at 22°N. Maxima in brackets on the right refer to managed colonies.

Colony condition	*Mean limit above which queen rearing is initiated*		
	European	*Africanized*	*What was measured*
A genetic predisposition to swarming			
B old queen			
C large adult population	16 000	12 000	no. workers (max. 60 000+)
D both young and old workers present	17 days	9 days	mean age
E total (hive) space overcrowded	71%	86%	% of comb occupied
F large brood population	21 000	23 000	no. brood cells occupied (max. ?30 000+)
G overcrowded brood nest	90%	94%	% of available brood cells occupied
H poor ventilation, overheating			

(F, G) Overcrowding in the brood nest

This can be reduced by removing some frames of sealed brood; they can be used to strengthen nuclei, as suggested in Section 6.34, method 3. If this outlet is not available, it is easiest to prevent overcrowding of the brood nest in hives where two boxes are used for brood (and this in itself provides more brood space). Most bees expand their brood nest upwards, and the queen will move from a brood box to one placed above it. Three weeks afterwards, all the brood will have emerged from combs in the lower box, which is left relatively empty. The beekeeper can then interchange, or 'reverse' the two boxes. When the box with the queen (now the lower one) is filled with brood, she may again move to the box above, and the reversing can be repeated – if necessary several times in all during the period of colony growth. Where three boxes are available to the bees for brood rearing, the (full) top one and the (empty) bottom one are 'reversed'; the bees and the queen are encouraged to move up into the new top box by inserting a frame of brood in the centre of it.

A method of which many variants are used is 'demareeing', named after G. W. Demaree, who developed it in the 1880s and 1890s (see Root & Root, 1940). It is effective in that it relieves overcrowding and separates the queen from the brood, giving her a new brood nest. But it is time-consuming, and necessitates finding the queen. In one variant, the queen, on a frame with a small amount of young brood, is put into the bottom box of a hive, the rest of the box being made up with more or less empty combs. She is confined there by a queen excluder placed on top of the box. One or two hive boxes containing empty combs are placed above the excluder, and then the hive box(es) containing most of the brood. The bees in the upper box of brood, being well separated from their queen, may rear queen cells, so this box must be inspected about 7 days later and any such cells cut out. No further queen cells can be built there, since no eggs will be present.

A method described especially in French beekeeping texts is *blocage de la ponte* – restriction of the queen's egg laying for 10 days or so, for instance by caging her within her colony. Other methods involve moving one or two boxes (containing brood and bees without a queen) above a dividing board provided with a separate small entrance, to form a colony separated from the one below. The upper colony can be given a new queen or a queen cell, or allowed to rear a queen itself. In due course it is moved to a new site as an additional colony, or used as part of a two-queen colony (Section 8.29).

(H) Poor ventilation

During the height of the active season inadequate hive ventilation can result in overheating of the brood nest and other parts of the hive. It can also hinder evaporation of water from nectar during its conversion into honey. The bees increase ventilation in the hive by creating air currents, and can do so most effectively if there is a large flight entrance at the bottom of the hive (Section 4.43), and also top ventilation provided by holes in the roof and in any inner cover (4.41, 4.44), or by top entrances (4.43). Some beekeepers offset a stack of supers slightly, so that a gap is left at the front and back of the hive, or set each at a slight angle to leave gaps at the corners. Since these gaps will allow entry of bees, it is not safe to use them when there is any danger of robbing, but they should be safe during a strong flow.

6.33.2 Swarm control

This term is used to mean preventing the issue of a swarm from a colony which is found to contain swarm queen cells, i.e. it has already started swarm preparations. The necessary procedures can be rather tiresome.

Bees start treating a larva as a future queen, and building a queen-type cell round it, almost as soon as the larva hatches from the egg; the adult queen emerges from the cell 12 to 14 days afterwards (Table 2.41A). The swarm is likely to leave the hive some days before this, so colonies liable to make swarm preparations must be inspected for queen cells every 9 to 10 days during the swarming season, or swarms may issue from them.

There are several alternative treatments:

1. Having destroyed (or removed) all queen cells, interchange the hive with one containing a small colony, so that many of the flying bees transfer to the small colony.
2. Having destroyed (or removed) all queen cells, remove the queen and requeen with a young mated queen; she should be introduced initially to a nucleus containing frames of bees and brood from the colony, and set beside it. Section 6.51 discusses requeening in general.
3. Having removed the queen, leave one queen cell only, so that the colony requeens itself.
4. Split the colony to make two or more new colonies (Section 6.34), using one queen cell (or a new mated queen) in each, unless one is left with the original queen.
5. Split the colony into two by making an 'artificial swarm', which involves separating the brood from the queen (see below).
6. Separate the 'swarm' bees by using a Taranov board (see below).

Morse and Hooper (1985) give details, and explain when the different treatments are suitable; options 1 to 3 are used by many commercial beekeepers, and option 1 by Dadant & Sons in the USA (Cale et al., 1975).

To make an artificial swarm – shook swarm is an old American term – from a colony (A) which has built swarm queen cells:

- From A, select a frame with one or more good queen cells, check that the queen is not on it, and put it temporarily in a safe place, e.g. in a nucleus box.
- Move hive A off its site, and replace it with an empty hive (B) which is fitted with framed combs.
- Shake most (not all) of the adult bees from A into hive B, checking that the queen is not left behind; this is the artificial swarm. (Alternatively, shake the bees on to a board placed in front of hive B and sloping up to the entrance, and let them run in, as when hiving a natural swarm, Section 6.35.1.) Before replacing each frame in hive A, remove any queen cells from the comb.
- Place in hive A the frame with the queen cell(s), selected in operation 1.
- Move hive A to a new site. The new queen will emerge from her cell and head the colony, and the adult population will grow as the brood emerges.

The Taranov method for separating 'swarm' bees from the rest of the colony (1947) may be of interest to beekeepers with only a few hives. Two boards, 50 cm long and the same width as the hive, are joined together at one end to form a ramp with its top edge at the height of the hive entrance. Two triangular boards are cut to form the sides. The ramp is positioned so that it leads up from the ground towards the entrance, but leaving a 10-cm gap. All the bees in the colony (including the queen) are shaken in front of the hive, the first ones on the ramp and later ones on a cloth spread out beyond its bottom edge; all queen cells except one are removed before the frames are returned to the hive. Some of the bees fly directly back into the hive, and others do so after walking to the top of the ramp. But the 'swarm' bees, with the queen, do not do this; instead they move down and cluster in the (dark) space under the sloping board. The ramp, with the clustered 'swarm', is put into the shade until evening, and then hived.

A different method of attempting swarm control involves clipping one of the queen's wings after she has mated (see e.g. Dadant & Sons, 1975; Morse & Hooper, 1985). A swarm will still fly out, but it will return to the hive. However, unless the beekeeper takes appropriate action, it is likely to fly out later with one or more of the virgin queens.

No method of swarm control is a proper substitute for swarm prevention measures discussed earlier, in Section 6.33.1. By the time control measures are applied, a colony is likely to have gone through the complete sequence of swarming preparations, and this can interfere with its honey-getting ability.

Action to be taken if a swarm issues is described in Section 6.35.1.

6.34 Making new colonies

Very small colonies grow rapidly. Ratnieks (1986)

found that the population of colonies with fewer than about 10 000 bees increased roughly in proportion to their size, whereas the rate of growth of larger colonies depended rather little on their populations. New colonies made in late spring, when the natural growth of colony populations is rapid, are likely to grow into full-sized colonies by the end of the active season, ready to serve as productive units next year.

Alternatively, it may be convenient to split large colonies around the end of the honey flow, when their populations are no longer useful for honey production (Section 6.52). New colonies so formed are likely to need feeding, and those fed generously will be able to rear more brood and grow faster than those that have to forage for all their food while their populations are relatively low. In warm climates it is possible to divide colonies in most parts of the season.

Creating an artificial swarm (Section 6.33.2) is an opportunistic way of making a new colony, and more systematic methods are described below. The operation of making new colonies is referred to variously as making increase, splitting or making splits, dividing or making divides or divisions. Making nuclei (method 3 below) involves paying more attention to the composition of each new unit created.

Method 1

A colony is split between two hive boxes, each box being used to house a separate colony. The queen must be found, and the colony containing her is moved to a new site. A new queen is given to the colony without one (Section 6.51), on the old site. Alternatively this colony can be left with eggs or young larvae, to rear its own queen. But the colony may not be large enough to provide the most suitable conditions, especially in cool summers. Also, about 3 weeks will elapse before the new queen is reared, mates and lays eggs, and another 3 weeks before these develop into adult workers; see Table 2.41A. Meanwhile the colony population will decrease.

In warm climates with an extended honey flow, the date when the division is made is not critical. At higher latitudes, unless the date is at least 8 weeks before the main honey flow, the total honey harvest from the two new colonies is likely to be much less than would be obtained if the colony were not divided, or were divided during or after the honey flow.

Method 2

Adult bees are shaken gently from a colony into another hive containing only empty combs, leaving young bees and the queen on the brood combs in the original hive. This hive is placed on the site of another (strong) colony which has been moved elsewhere in the apiary, and it immediately receives a good population of adult bees because foragers from the strong colony return to it. The hive containing the shaken bees is placed on the site of the original hive, and given a new queen (Section 6.51). It has many adult bees, and will quickly create a sizeable brood nest.

Method 3 (making nuclei)

A nucleus, or nucleus colony (the abbreviation nuc is also used), is a small colony made up on 2–6 frames, with a few thousand bees. It may be housed in a nucleus box, which is like a brood box except that it holds only a few frames. Or two (or three) nuclei may be housed in a standard brood box, divided into completely separated compartments by vertical division boards (Section 4.44), each compartment having a flight entrance facing in a different direction.

It is important that the combs used when making a nucleus should include at least one full of sealed brood; the bees will not have to feed this brood, and it will soon produce adult bees. Unsealed brood is best avoided. The nucleus should have honey in at least two combs, and pollen in at least one. If all frames are taken from the same hive, the bees on the combs can be transferred with them. The queen must be left behind unless she is to be used in one of the nuclei; a nucleus is usually given a young queen (see Section 6.51). Further bees from the same hive can be shaken in, but if any combs are to be added from other hives, older bees should first be shaken off them (back into their own hive), leaving only the young bees that stay on the comb. It is possible to mix older bees from several hives together if precautions are taken, as described for instance by Johansson and Johansson (1978), who give a full account of the establishment and use of nuclei.

If newly formed nuclei are moved out of flight range of the apiary where they were made up, the older bees will remain permanently with them. If not, many are likely to return to their old hive site and enter the hive there (Section 8.15), and the nuclei will be weakened. All new nuclei should be fed with sugar syrup, which may be dilute.

Nuclei should not be made during a dearth. Colonies can usefully be divided to make nuclei towards the end of the honey flow, when they are likely to be at their strongest. Many honey producers in the USA get an added advantage from their north-south latitude range by moving colonies south in the autumn; they make nuclei from them during the earlier spring there, and return for the next season's honey flow with 3 to 6 times as many colonies as they took south.

Figure 6.35a An exceptionally large swarm, containing two queens (photo: Kopec).

6.35 Other management operations

6.35.1 Dealing with swarms

A reproductive swarm that issues from a colony usually settles in a cluster within a few minutes, on a tree, bush or other support less than 50 to 100 metres from the hive it left; Figure 6.35a shows an example. From the cluster, scout bees fly to explore cavities which the swarm might occupy as its new home, and the swarm usually flies off to the chosen cavity within a few hours or a few days (Section 3.42).

In addition to the queen of the parent colony, a prime swarm is likely to contain about half the workers of the colony, and a few drones (end of Section 3.34.3). The number of bees quoted in Table 3.34A ranges from 12 000 to 22 000, but it is sometimes more, and Figure 6.35a shows an extra large swarm. Swarm bees are loaded with honey, and 1 kg contains about 8000. If a colony swarms, the beekeeper should examine the brood nest within the next few days, and remove all but one queen cell; otherwise smaller afterswarms (casts) may issue, each headed by a virgin queen from one of the extra queen cells, and this further weakens the colony.

If the clustered swarm is accessible, it should be 'taken' as soon as possible, in a temporary receptacle referred to below as a skep – although it may be any convenient basket, carton or box – or in a bee veil, if nothing else is to hand. The swarm may be shaken directly into a hive box fitted with framed combs, or an empty hive box into which the combs are inserted afterwards. Or it may be left in the skep which is placed in the shade until evening, and then hived (see below).

A swarm clustered on a branch can sometimes be taken, after trimming the nearby vegetation, by cutting off the part of the branch supporting the cluster, and putting both the branch and most of the bees in the skep. This is then inverted and placed mouth down in the shade nearby – if possible on a piece of wood or a hive stand. A gap must be left through which the rest of the bees can enter, and – provided the queen is in the skep – all the bees of the swarm will go in. The swarm is likely to remain in the skep for the time being.

If the swarm is clustered on a support that cannot be removed, it may be possible to shake the bees into the upturned skep by giving the support a sudden jerk. If not, the bees may be ladled, or moved with the hands, into the skep; this is not difficult with well fed, newly issued swarms, because they behave in a lethargic way, and tend to cling to one another. As soon as the queen is among the bees in the skep, all others will follow.

A beekeeper with sufficient time may well enjoy the traditional taking and hiving of a swarm, and his apparent power over the bees in carrying out the operations will impress onlookers today, as it did in past centuries. He should hive the swarm in the evening of the day it is taken. An empty hive (one deep box, covered), with a full complement of frames and at least some drawn combs, is put in position. A rigid board, as wide as the hive and somewhat longer, is sloped down from the hive entrance to the ground in front; a Taranov board (Section 6.33.2) could be used. It is customary to lay a white cloth over the board before shaking the bees on to it, since bees seem more inclined to move off white than off a dark surface. With one vigorous jerk, the bees in the skep are shaken on to the board. After a few minutes of unorganized movement, the bees will start walking up the board and into the hive entrance; see Figure 6.35b. Some bees will be seen standing with the tip of the abdomen raised, exposing a clearly visible hairless area (Figure 2.35a) which has pores through which the Nasonov pheromone issues. The bees fan with their wings, which distributes the pheromone; in contains substances (citral, nerol, geraniol, nerolic acid, geranic acid, farnesol,

Figure 6.35b Penelope Papadopoulo hiving a swarm of tropical African bees, in the bee house shown in Figure 7.52c (photo: P. Papadopoulo).
The long hiving board has rimmed edges slanting inwards to the hive entrance; most of the bees have already entered the hive.

Table 2.35A) that attract other bees. A possible application of synthetic pheromones for attracting swarms is mentioned below.

Once the queen has entered the hive, the bees move up and in more rapidly, and any bees that were left in the skep will join the others. A little smoke may be used on outlying groups of bees to start them moving.

It is always best to provide food – sugar syrup or a comb of honey – in a hive that receives a new swarm. Feeding is *essential* if the weather is uncertain, or if no drawn combs are available and the bees have to build comb on foundation before they can store honey or rear brood. Feeding also helps to ensure that the swarm will remain in the hive. A more certain way is, however, to insert a comb containing some brood, which the bees are unlikely to abandon.

Bait hives

In some parts of the tropics, bait hives are used routinely to capture swarms, as discussed in Section 7.33.1. In times past, beekeepers in temperate zones encouraged swarming as part of their management system and they often set up a bait hive, having prepared it by rubbing the inside with a mixture of materials thought to make it attractive to bees, such as honey, beeswax, propolis, and certain fragrant herbs. Leaves of one such herb, balm (*Melissa officinalis*), produce an oil which contains 63% citral and small amounts of geraniol and nerol (Burgett, 1980), all substances that are also produced by the honeybee's Nasonov gland (Table 2.35A). Lemon grass is used similarly; see Table 5.21A.

It would be convenient if a swarm leaving a hive could be induced with some certainty to enter an empty hive in the same apiary. In the 1980s there was great interest in the possibility of using a synthetic substitute for the Nasonov pheromone, to attract swarms into bait hives set out to catch them. An attractant mixture available commercially is 10 mg each of geraniol, citral and nerolic acid, in 100 μl hexane (Free et al., 1984). The mixture is supplied in a sealed polyethylene vial, from which the substances escape by diffusion through the walls.

In an experiment in Arizona, 96 baited hives captured 43 swarms in a 7-week swarming season (Schmidt & Thoenes, 1987). In various other tests, swarms have been recorded more frequently in hives containing a vial of pheromone material than in hives without one, but there seems to be no very clear-cut effect. Nor has it yet proved possible to ensure that a swarm issuing from a hive clusters on a support that is baited with a source of synthetic pheromone, although bees at a cluster site are known to emit Nasonov pheromone. Free et al. (1984) and Witherell (1985) have given details of much of the research on the effectiveness of swarm attractants. Seeley and Morse (1982) and Witherell (1985) discuss bait hives in general, which should probably have a capacity of 40 litres or more. Figure 10.4b shows an example of a modern bait hive.

Miscellaneous operations with swarms

A beekeeper may well be asked to remove a clustered swarm from a neighbour's land, and he may be willing to do so provided the swarm is fairly accessible. If not, he can with some certainty assure the neighbour that the bees will fly off within a few days. Whether the beekeeper should hive the swarm in his own apiary is another matter, and will depend on his circumstances and inclination. Swarms of unknown origin may or may not carry disease, and may or may not come from a strain genetically predisposed to swarming. But some beekeepers have built up large apiaries by collecting stray swarms.

Section 7.33.2 describes procedures for transferring to a frame hive a swarm that has built a nest (a wild colony) in an accessible place, or a colony from a traditional hive. A beekeeper may be asked to remove a swarm that has settled in an inaccessible place in a neighbour's building; this can be difficult, especially if the swarm is well established, with combs containing brood and honey. Directions are given in Sections 5.43.1 and 5.43.2.

6.35.2 Uniting colonies

If a colony is found to have no queen, or to be too small to become a productive unit, it may be united with another, provided both are healthy. Such uniting may have to be done in any part of the active season, although conditions should be avoided that are unsuitable for hive manipulations in general: poor weather, or a period when robber bees are active or drones are being driven out of hives.

During a honey flow, or in a warm early spring or late autumn, it is often safe to unite the colonies by setting the boxes containing them one above the other, or to mix the frames together without any preparation (Cale et al., 1975). If both colonies have a queen, one should be removed 24 hours before uniting. Otherwise one queen will kill the other, and the beekeeper cannot choose which will survive; also the survivor may possibly be injured.

Methods described below are likely to be safe in a wider range of conditions, but if there is no nectar flow at the time it is worth feeding both colonies for a day or two beforehand.

Uniting queenright colonies
The uniting operation is best done in cool weather, and towards evening, since one colony will be temporarily enclosed in the hive afterwards and there could be a risk of overheating. To unite two queenright colonies, the weaker colony (B) is dequeened, and this hive is placed on the site of the stronger colony (A), hive A having been moved off its site. The brood box of B is uncovered, and a single sheet of newspaper is placed over the frames, enclosing the bees below. Several holes are punched in the paper with a matchstick or pencil. Then the brood box of A is placed on the newspaper, and the hive closed. Any supers of honey are added above a queen excluder, and any belonging to colony B also, above more newspaper, which is punched as above.

Foragers belonging to A that are out of the hive will return to their old site, enter the hive, and be accepted by B since they are bringing in food. Bees enclosed by newspaper will gradually chew at the holes in it, having no other means of getting out of the hive, and they will then intermingle with the bees below. (In very hot weather, extra ventilation may be necessary until the enclosed bees are released; matchsticks can be inserted between the boxes, one at each corner.) Within a few days, pieces of chewed newspaper removed by the bees will be seen in front of the hive. After about a week, the hive can be opened, the rest of the paper removed, and the frames re-arranged as desired between the boxes.

Uniting a queenless colony to a queenright colony
The above method can be used to unite a colony found queenless (treated as B) with a nucleus (treated as A). To unite a queenless colony to a full-sized colony, *treat the queenright colony as B*. Shake the adult bees from each frame of the queenless colony (A) into a container from which they can be thrown (all together) into a hive box. Place an empty hive box above the newspaper, throw the bees in, and close the hive. The bees will chew through the newspaper. Any frames containing brood can be added to other colonies.

Johansson and Johansson (1978) describe the above methods of uniting in more detail, and a number of others, including remedies for the following situations:

- a colony found to have laying workers (Section 3.39)
- a swarm that is not wanted as a separate unit
- two or more small colonies or nuclei which have served their purpose (e.g. for mating queens), and are to be made into one large colony.

6.4 MANAGEMENT DURING THE HONEY FLOW

Most honey producers in temperate zones add empty supers above the brood box(es) and remove them when full. They leave honey in the brood box(es) as a reserve of food for the colony. Where honey yields are low, and also in some parts of the tropics, beekeepers may not use supers, but extract honey from frames in the brood box, and replace them straight away (Section 7.34). This under-uses the capabilities of the tiered movable-frame hive.

In addition to the specific operations described here, Section 7.32.2 discusses ways of preventing heat damage to colonies in very hot weather, which can occur in some temperate-zone areas as well as in the tropics. Moving hives to take advantage of honey flows in different places, referred to as migration, is part of normal seasonal management for many beekeepers, and some recommendations and precautions are set out in Section 8.1. When colonies are taken to a honeydew flow, they should be given a continuous supply of pollen or pollen substitute in the hive (Section 6.32). Unless this is done, they will be unable to maintain brood rearing.

6.41 Adding empty supers to hives

During the 1970s evidence was accumulated that the presence of empty combs in a hive during a strong flow can stimulate nectar collection and honey storage by the colony. Empty supers should therefore be put on hives well in advance of their being needed for honey storage, and before the main flow starts.

In Louisiana, USA, Rinderer and Baxter (1978) had two groups of colonies (9 in group A, 10 in B) working a strong flow; hives in A had just over twice as much empty comb available for honey storage as hives in B. At the end of the flow all hives were moved to another strong flow, where the areas of empty comb were reversed between groups A and B. Amounts of honey stored (average increase in weight (kg) per colony during the flow) were as follows:

	first flow	*second flow*
larger comb area ($4.1\ m^2$)	A, 50.9 ± 4.5	B, 57.9 ± 5.0
smaller comb area ($1.9\ m^2$)	B, 36.1 ± 3.8	A, 47.5 ± 4.6

The 'hoarding behaviour' of groups of 50 worker bees in a cage showed a similar relation to the amount of empty combs made available to them. Each cage had separate feeders containing sugar syrup (to simulate a flow), water and pollen substitute, and it was provided with 1, 2 or 3 pieces of empty comb, 47 cm² in area. The more pieces of comb a group of bees was given, the more syrup they took from the feeder and hoarded in the comb:

no. pieces of comb	1	2	3
syrup hoarded/bee/day (ml)	0.11	0.14	0.19

The stimulus to collecting and storing nectar seems to have its origin in volatile substances from uncovered surface of warm comb (Rinderer, 1981), and it operates only during a strong flow (Rinderer, 1982).

Periods of strong nectar flows are characterized by high air temperatures, which raise the temperature in the honey supers, and thus the rate at which volatile substances are released from the comb surface. It is still unclear whether these substances are constituents of beeswax, or have been placed in or on the beeswax by the bees.

The presence of empty combs also speeds up the evaporation of water from incoming nectar, because bees can spread the nectar out in empty cells. (It can have yet another effect on bees, at any rate when temperatures are high: Collins and Rinderer (1985) found that bees were more defensive, reacting more rapidly to a moving target, and also stinging it more often.)

When there are already supers on a hive, it is satisfactory – and much easier – to add empty ones on top of them; this is known as top supering. If comb honey is being produced, however, empty supers for it should be inserted below others, immediately above the brood box(es). This is bottom supering, which is also used when getting new combs built (Section 6.43). Offsetting supers to increase ventilation in the hive is mentioned at the end of Section 6.33.1.

6.42 Removing full supers

Combs from the hives should not be removed until all or most of the cells are sealed (capped). This is not so important where the atmosphere is very dry, but in very humid conditions it is essential to harvest only completely sealed honey; even this may have too high a water content. Sections 7.34 and 7.35 deal with special problems encountered in hot and in humid conditions.

The beekeeper wants to harvest as many supers as possible *with full honey combs* (Figure 6.42a), not supers containing combs only partly filled. Early in the flow there was an advantage in providing more empty combs than the bees immediately needed, but after the peak of the main flow, the storage space left in a hive should be no more than the bees are likely to fill. If there is doubt as to whether a colony will need more

Figure 6.42a A fully sealed comb of honey, Mauritius.

space than that already in the hive, an empty super can be put above the inner cover as a safety measure, the central hole being left open. If the hive does become full, the bees will store any extra honey in the combs in this super; if not, it can be removed empty.

It has so far been assumed that the beekeeper has more than enough supers to accommodate all the honey stored from the main flow, and that he removes all supers to be harvested at the end of the flow. The whole honey crop can then be extracted together, which for many beekeepers gives a sufficient saving in labour costs to outweigh the capital investment in the large number of supers and framed combs required. However, the procedure is practicable only where supers of framed combs can be stored safely against damage by wax moths during the rest of the summer, as well as through the winter when low temperatures may provide a sufficient safeguard (Section 11.61). Section 7.34 discusses honey harvesting where this is not so.

When honey supers are ready for harvesting, the bees are removed from them as described in Sections 5.3 to 5.33, and the supers are taken to the extracting room free from bees. Subsequent extraction and handling of honey are dealt with in Sections 13.3 to 13.39.

6.43 Making new combs

The beekeeper can most easily get bees to make (build) perfect combs by drawing out foundation, if the foundation is provided when conditions in the colony are conducive to wax production (Section 3.31.1). These include:

- warm weather, since a temperature of 35°–37° is needed for the bees to secrete beeswax;
- a strong colony, to provide many wax-making bees,

and to maintain the required temperature throughout the comb-building area;
- an abundant nectar flow, since bees secreting wax consume an extra amount of carbohydrate.

If combs can be stored safely, it is worth getting new brood combs drawn out (in advance of requirements) during a main flow which provides the best possible conditions. Perfect brood combs can usually be obtained by placing a brood box immediately above the brood nest of a hive (and below the queen excluder). A hive box containing framed foundation will attract some bees away from the brood nest, and this helps to discourage swarm preparations. Bees will draw out framed foundation placed in the brood nest, but may not attach it firmly to all four sides of the frame, or may not build the comb to the full size of the foundation, or may chew away edges.

When package bees are hived, none of the conditions conducive to comb building are likely to apply. Syrup must be fed liberally if no drawn combs are available, and the bees are hived on foundation. A new swarm is able to produce much wax: until a swarm has built comb it has no cells in which to deposit nectar, and no brood to feed, so a high proportion of its energy can be used for wax making. But unless a honey flow is in progress, a swarm hived on foundation must be fed sugar syrup.

Morse (1974) gives recommendations for making good combs, for extracting honey from newly built combs, and for discarding old ones. Brood frames should be removed from hives and taken out of circulation when the combs are distorted or misshapen, or have more than a small area of drone comb, or when many generations of brood have been reared in them and they have become very dark from the discarded cocoons. It is usually not worth the beekeeper's time and effort to extract the wax from these dark combs; see Section 13.6.

6.44 Requeening with a queen cell

Colonies will accept a sealed queen cell more readily than an adult queen, and if a honey-producing colony is requeened with a sealed queen cell 3 weeks before the end of the main flow, the consequent interruption in egg laying will not reduce the number of foragers working the flow, or the honey yield. In Canada, Peer (1984) obtains good results by introducing sealed queen cells 9 or 10 days old at this time. (They are produced as indicated in Section 8.22.) One of the cells is lowered gently between frame top-bars of the top box below the queen excluder, or in the top box of the hive if no excluder is used. The virgin queen emerges there, and normally kills the laying queen, mates, and starts to lay a few days later. It is thus not necessary for the beekeeper to use an introduction cage, or to find and remove the old queen before inserting the queen cell.

6.5 MANAGEMENT AFTER THE FLOW, TO THE END OF THE WINTER DEARTH

6.51 Requeening and queen introduction

Although requeening and queen introduction are easiest during a honey flow, they can be done before or after the main flow according to the beekeeper's convenience, but *not during a dearth.*

Requeening is done because of certain deficiencies of older queens from the beekeeper's point of view. They lay fewer eggs, and an abnormal proportion of these may give rise to drones. Also – partly because pheromone production decreases – the presence of an older queen becomes less effective in preventing workers in the colony rearing new queens and making other preparations for swarming; see Section 3.21 and the start of Section 3.34.2.

For beekeepers aiming to get large colonies and a high honey yield, it is therefore important to replace a queen before she becomes 'old'. In the temperate zones and subtropics, I have found that beekeepers usually requeen colonies as indicated below.

– large colonies in which the queen lays rapidly through almost all the year	6 months
– large colonies in which the queen lays rapidly or through a long season	1 year
– large colonies in regions with a shorter season	2 years
– smaller colonies in similar regions	3–4 years
– colonies with breeder queens, whose laying is restricted by the beekeeper in order to prolong their life	up to 5 years
– colonies not requeened (natural life of queen)	up to 5 years

Methods of requeening and queen introduction
Towards the end of a honey flow, a colony may be requeened with a queen cell (Section 6.44); at other times a young mated queen is normally used, so that the period without egg laying is minimal. The old queen must first be found (Section 6.13) and removed; the new queen is introduced 6–12 h later, using a specific technique which protects her from being killed by the workers as being 'foreign' to them. The ease with which a mated queen can be introduced safely to a colony varies greatly according to conditions.

1. Requeening is easiest during a honey flow.
2. A new queen is most readily accepted by *young* bees (which are in the brood nest).
3. It is often safer to introduce a queen to a small colony than to a large one.

A mated queen is most easily accepted if she is in a similar physiological condition (rate of egg laying) to the queen just removed. The easiest introduction of all is of a sealed queen cell, described in Section 6.44. A newly emerged virgin queen is usually fairly easy to introduce, but an older virgin queen is the most difficult of all – it is more likely to succeed if the colony has been queenless for several days beforehand.

In view of 1, if there is no honey flow, an artificial one should be created by feeding syrup to the colony *before, during and after* the introduction. In view of 3, it can be best to put the new queen first into a nucleus (Section 6.34, method 3), and then to unite the nucleus with the dequeened colony she is destined for (6.35.2). Such precautions should also be taken if, for instance, the new queen has been caged – and therefore not laying – for some days. As an alternative to using a nucleus, some beekeepers place a divider board (Section 4.44) consisting of a double screen below the top box of the receiving hive, thus separating the bees in it from the rest of the colony. The queen is then introduced to the bees in the top box; the old queen is later removed, and the two lots of bees are united as in Section 6.35.2. Details are given by WAS (1987).

When making any queen introduction, it is best to put the queen first into a special introduction cage; many types have been developed (Johansson & Johansson, 1971*a*). Experiments in the UK by Butler and Simpson (1956) and Free and Butler (1958) showed that the cage should be of a large enough mesh for the workers of the new colony to feed and to touch the queen inside, thereby obtaining pheromones that form the basis of the colony's cohesiveness (Section 3.21). The apertures should be at least 2.5 mm wide if square, slightly more if round, and slightly less if they are long slots. A simple cage is rectangular, 90 × 20 × 13 mm, made by bending a piece of mesh 90 × 66 mm, to make the 4 sides. One end is closed permanently by a block of wood or cork, and the other is closed temporarily after the queen is inserted, by a single thickness of newspaper secured with a rubber band. The cage is inserted in the colony, suspended or wedged between two combs of brood in a position where the workers have access to as much as possible of the mesh, to touch and feed the queen. The bees will soon chew through the paper and so release the queen. An introduction cage that arranges for a more cautious release of the queen is shown in Figure 6.51a.

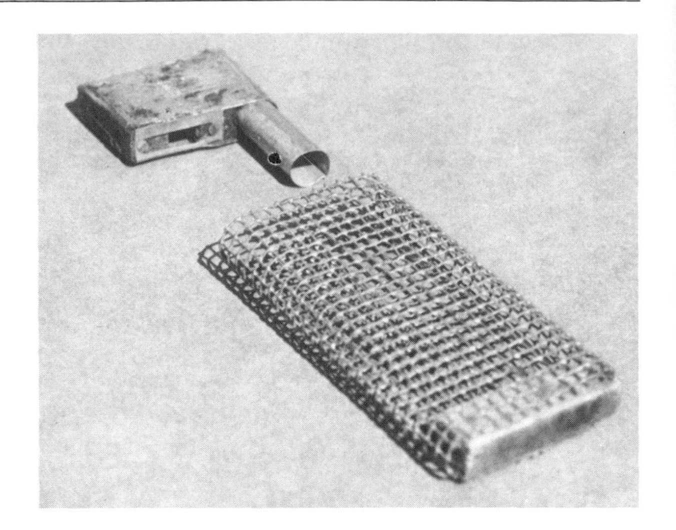

Figure 6.51a Cage in which a queen is placed (alone) for introducing to a colony (Laidlaw, 1981).
The cage is 80 × 40 × 10 mm, and is inserted between frames; see the text for mesh size. The closure for the open end has two entrance tunnels, both filled with candy. The shorter one (25 mm, left) has a piece of queen excluder across one end, and workers can enter the cage as soon as they have eaten the candy. The bees clear the longer tube (50 mm, right) more slowly, after which the queen can escape into the colony; by this time the bees have become accustomed to her.

If a queen is purchased, she will probably be received by post, in a postal (mailing or shipping) cage, which also contains up to 20 workers to feed her, and candy for food. The workers should be removed and killed before the queen is put into her new colony; they will be of no further benefit, and might possibly carry an infection such as nosema. If the queen is imported from another country, the workers should be sent for disease examination. The queen can be introduced to the colony in the postal cage, following directions supplied, but it is better to transfer her to an introduction cage. Do this indoors at a closed window. Workers that leave the postal cage will fly to the window and can be killed (or recaged to send for examination), and if the queen escapes, she can easily be recaptured.

Laidlaw (1981) gives a clear summary of the general requirements and methods for introducing queens, and Butler and Simpson (1956) present experimental evidence on the relative success of queen introduction under some of the more difficult of the conditions likely to be encountered.

6.52 Post-flow management

When the main flow is over, colonies will give the beekeeper no further honey harvest until the next season, unless there is an additional late flow. Nor is there likely to be any call for hiring out colonies for crop pollination. Management strategy for this period must therefore have other aims, and the colonies – which now have large populations – can be used to provide bees and brood for queen rearing, mating hives (Section 8.2), and making new colonies, especially nuclei (6.34).

During their first week of life young bees need to consume much pollen, so that their hypopharyngeal glands and fat-body develop (Maurizio, 1950). In summer the protein reserves are used up in brood rearing, and 'summer bees' which rear much brood live only a few weeks. But in autumn, with much less brood rearing, the reserves give the bees a greatly increased length of life, and enable 'winter' bees to survive until spring; see Table 2.42A. It is therefore important that the worker bees in the colony when the dearth starts (in temperate zones, those reared in late summer) should have had access to ample pollen. Late-summer pollen flows are useful for this reason; in some areas beekeepers may feed colonies with pollen or pollen substitutes in the period before winter.

If package bees are purchased each year for honey production, post-flow colony management consists only of killing the colonies and disposing of the bees (end of Section 8.35).

6.53 Feeding and other preparations for the winter dearth

Many beekeepers in the temperate zones would say that proper preparation of colonies for winter is the key to good management, because it largely determines the condition of colonies in spring.

6.53.1 Providing stores

In cold temperate regions the winter dearth is likely to be long, and sufficiently cold for each colony to form a cluster in the hive and to remain clustered for weeks or months (Section 3.37). The clustered bees are relatively inactive and consume little food, as exemplified by Figure 12.62a, colony 1. In warm temperate and subtropical regions, the winter dearth is shorter, and not cold enough for colonies to cluster. (In the tropics the dearths are shorter, but not associated with cold, and the rate of food consumption is much higher.) Before any *cold* dearth starts, colonies must have sufficient carbohydrate stores in the hive to last them until they can obtain substantial amounts of nectar by foraging next spring. Winter stores should consist of honey (a few unsuitable honeys are referred to below), with or without additional sugar fed by the beekeeper. In countries where the cost of sugar to beekeepers is appreciably lower than the price they receive for honey, it is economical to feed sugar in autumn. Recommendations as to the total amount of stores to be left in hives over winter vary from region to region; in general they have increased steadily over the years. In many parts of Europe 20 kg – sometimes as little as 12 kg – is considered adequate; in North America recommendations range from 30 kg to 40 kg.

The sugar is normally fed as concentrated syrup, which the bees concentrate further by evaporating water from it, and store in combs. A usual concentration for feeding is 1.6 kg to each litre of water (62%), or in British Imperial units 2 pounds to 1 pint. Ribbands (1950) recommended 1.8 kg per litre as the maximum concentration (64%) at which there is no risk of granulation in the feeder. It is wasteful to feed dilute syrup because bees then consume much food to supply energy for evaporating the excess water. Ribbands found that his concentrated syrup produced about one-third more stores than the same weight of sugar fed as dilute syrup (35% sugar).

Feeders for syrup are described in Section 4.47; types 1–4 have a large capacity and are most suitable for autumn feeding. Alternatively, if the bees have access to water, the sugar can be fed dry, which does not need a special feeder; Johansson and Johansson (1976/1977) give details. But bees can take syrup more quickly than dry sugar, and this can be important: they must take all syrup from the feeder, process it and seal it in the cells while the temperature is still high enough for them to do so. On the other hand if syrup is fed too early, colonies are likely to use some of the sugar for brood rearing instead of storing it. In the south of England Ribbands (1950) found that concentrated syrup fed between 13 and 20 September produced 10% more stores than similar syrup fed between 24 and 31 August.

With regard to the types of honey and sugar suitable for winter stores, much depends on whether the winters are cold (with long spells when bees cannot fly out of the hive to defaecate), or mild and with no long spells when flight is impossible. Where winters are mild, various forms of unrefined sugar may be fed, but for cold winters refined sugar, such as white table sugar, must be used. For a similar reason, honeys containing material that cannot be absorbed by the honeybee gut are unsuitable as the sole food for colonies in cold winters – for instance most honeydew honeys, especially if crystallized (Imdorf et al., 1985), and heather honey (*Calluna vulgaris*). Combs of such honeys can, however, be left in the hive for winter, provided at least 5–10 kg refined sugar is also fed. There is quite a lot of evidence that colonies wintered entirely on sugar syrup, with no honey, perform less satisfactorily in the next active season than colonies wintered with some honey; this is one reason for not harvesting honey combs from the brood box(es); see also Section 7.34.

In some countries, high fructose corn syrup (HFCS) is on sale for feeding to bees (Section 6.31). Rinderer and Baxter (1980) found that bees took HFCS more readily than sucrose syrup at the same (high) concentration, and they suggest that unintentional adulteration of honey might result from feeding with HFCS. Denatured sugar is sometimes available at a concessionary rate for feeding to bees. The denaturing is designed to make the sugar unsuitable for human consumption; a dye may be added, or a substance that has a bitter taste. In autumn 1969 many beekeepers in England fed colonies with concessionary syrup that had been dyed green. The colonies were in brood boxes only, and next spring empty honey supers were added as usual. The honey stored in some of them was green (Crane, 1980*a*), showing that the bees had moved stores from brood box to honey supers many months after feeding. This also raises the question of inadvertent honey adulteration resulting from feeding bees with sugar, even in the autumn.

6.53.2 Other preparations for winter

It is essential to winter only healthy colonies, and before preparations for wintering are started, colonies should be examined for diseases – especially American and European foul brood (Sections 11.21, 11.22). One of the first preparations, before colonies are fed, is to reduce the size of the hive to the number of boxes that will hold the bees and the food to be stored for their winter provisions. Any queen excluders in the hives are removed at the same time.

It is important to winter large colonies. Counts made in Scotland (Figure 6.53a) show that colony populations after winter were on average about half those before winter, and that a much higher proportion of bees died in small colonies than in larger ones. The largest colony shown, with 18 000 bees in November, lost 6300 (35%), and the smallest, starting winter with 4500 bees, lost 3800 (85%), leaving only 700 in spring. Many beekeepers would regard all these colonies as too small to winter; in the USA Johansson and Johansson (1978) recommended 25 000 to 35 000 as a minimum population (about 4.5 kg of bees, covering 6 or more frames in October). Colonies smaller than this should be united into larger units, as explained in Section 6.35.2. The subject is discussed further in Section 3.37.

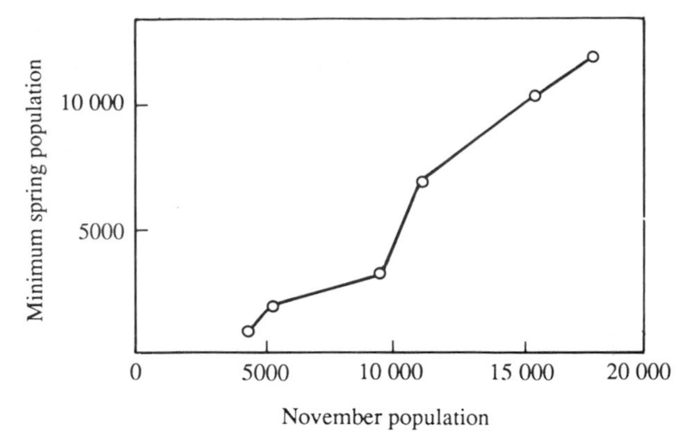

Figure 6.53a Populations of 6 colonies at the start and finish of winter, Aberdeen, UK (Jeffree, 1955).

Especially where winters are damp, provide hives with much ventilation; use a wide flight entrance, and possibly a floor of metal mesh instead of solid board (Section 4.42). At the top, use a roof and inner cover that allow ventilation (4.41, 4.44), or remove the inner cover so that there is a free space between the frames and the (ventilated) roof. An additional top entrance should be provided (4.43) in areas where winter mortality of bees is so high that dead bees may block the bottom flight entrance; a stiff wire can be used in a very cold period to check for and clear any blocking there.

Before winter, put in place any necessary protection against animals or birds in the locality that might endanger bees in winter. Place hives on stands, out of danger of any flood water. An ideal winter site is a slope facing the sun, not damp, provided with a windbreak against prevailing winds, and close to early spring pollen sources and a suitable water supply for the bees. Many sites that are good for honey production are not good wintering sites.

6.54 Management for long, very cold winters

Methods for keeping colonies through long and very cold winters have been worked out in some western parts of Canada, and *The prairie beekeeping manual* (1986) gives details of both outdoor and indoor systems. Such areas are often characterized by snowfalls followed by clear sunny weather. The combination of bright sunshine and snow-covered ground can present a visual hazard to bees. Colonies that are buried in snow (see Figure 6.54a) seem to suffer remarkably little damage, but if snow covers the ground without blocking the flight entrances, and the sun shines brightly, bees are stimulated to fly out; absorption of sunlight by the hive walls increases the temper-

Figure 6.54a Hives partly buried in snow, Hungary, 1966 (photo: Z. Örösi-Pál).

ature inside the hive, and reflected light from the snow at the flight entrance has a smaller effect. A flying bee normally receives most light from the sky above, but because the snow is much brighter than the sky, the angular light distribution (ALD) above snow-covered ground is reversed (Velthuis & Verheijen, 1963). The bee therefore flies upside down, in a disoriented way, and crashes on to the snow, where she dies of cold. While snow is on the ground, a board placed so that it darkens the full width of the entrance can prevent most of these deaths. In certain circumstances, not well understood, honeybees and other insects can adjust to the reversed ALD, and fly without difficulty over sunlit snow. But flying bees have been observed to crash land on white coral sand beaches in Wake Island during bright sunshine (Hitchcock, 1986), probably also because the ALD is reversed.

6.54.1 Outdoor hive protection

In very cold northern regions, across Canada, Scandinavia and Russia, there are long traditions of (a) wintering hives in an insulating outer cover, singly or several together, and (b) wintering large numbers in a half-underground 'cellar'. Modern versions of (a) are discussed here, and those of (b) in Section 6.54.2. A conference in Finland, on beekeeping in cold climate zones (Apimondia, 1975), discussed some other aspects of beekeeping in long cold winters.

Nowadays the outer layer of hive packing is usually a black plastic sheet, which is waterproof and absorbs heat radiation from the sun during the day (Figure 6.54b). Inside this, thermal insulation is provided by glass fibre, expanded polystyrene or foam plastic. Heat conservation and economy of material are both increased by packing a block of four hives together; with a greater number the bees in the central hives can be at risk from overheating. However, in Nova Scotia, Canada, Karmo (1984) has been successful with blocks of 4 hives stacked 6 high, i.e. 24 hives in each unit. In October 1980 I watched the packing process near the Peace River in British Columbia, Canada. Two pairs of hives, with entrances wide open, were back to back on pallets as in Figure 15.5a. Each had two hive boxes, with a Miller-type feeder (Section 4.47) above them. A plastic tube was led out through the packing from the feeder, so that this could be refilled when necessary without unpacking the hives,

Figure 6.54b Two pairs of hives side by side, packed together for winter, with insulation inside a black waterproof cover, Canada (photo: IBRA Collection).

by using a compressor pump. A plastic sheet and several thin sheets of glass fibre insulation were placed on top of each hive; then a thick sheet of insulation was tied on to each side of the block of four hives. The thick insulation stopped well short of the entrances, so that the air there would remain cold (and the bees would not leave the hive) when sunshine warmed the interior but the air temperature was so low that bees would die if they left the hive. Finally, a strong black plastic sheet was wrapped round the whole, and secured by a batten along each side, just above the entrance.

The packing is removed in spring, when temperatures are high enough for the bees to fly, and forage becomes available – but not while the ground is still covered in snow.

In March 1984 I saw Dave Tozier's colonies that had been successfully overwintered at a latitude of 65°N, near Fairbanks in Alaska. There was still deep snow on the ground. Each hive had one deep box of brood, and above it two of honey, with pollen in the centre. Two hives were packed together with thick insulation and an outer wooden casing; at the top was a super (without frames) packed with glass fibre, and Shaparew's ventilated cover board (1986).

The idea of providing hives with background heating during the winter or spring has been promoted from time to time. Heating a hive can affect cluster behaviour in winter, and egg laying in spring, as well as levels of general bee activity. In small colonies it can probably reduce winter mortality, and increase early spring brood rearing (Lunder, 1950). But there is no firm evidence that applying purchased heat to normal colonies is a cost-effective procedure. Seven years of experiments in Sweden showed no useful effect of spring heating on colony development or honey yields (Schwan, 1955).

A strong colony can be used to warm a small colony placed above it in a hive box with its own flight entrance, the two colonies being separated by a small-mesh screen the size of an inner cover. The small colony may even form a single cluster with the large one (Gojmerac, 1980).

6.54.2 Indoor wintering

In those northern regions of the world where beekeeping in summer can be profitable but winters are long and very cold, some beekeepers use special buildings for wintering hives at an equable temperature and in the dark. Half-underground 'bee-cellars' were used by the early 1900s in Canada and northern USA, and probably before this in Russia. In the Swedish valley Klarälven, a number of granite structures were built into the hillside many years ago for storing potatoes, and in 1981 I found some of them in use for wintering hives. In Britain and Ireland, windowless buildings erected one or two hundred years ago still exist, with recesses in the interior walls to accommodate overwintering skeps (Crane, 1983*a*); some ice houses were probably also used.

Modern buildings constructed for wintering hives of bees have equipment for automatic control of temperature and ventilation. They are used routinely by some of the commercial beekeepers in parts of Canada and northern USA. In Canada they are mainly in the Prairie provinces, and in Quebec province where 90% of hives are overwintered inside. The practice has thus become one of the standard operations in commercial beekeeping. McCutcheon (1984) discusses the construction and use of the buildings and gives much practical information, some of which is quoted below.

Single-box or double-box hives are prepared in late August or early September. Sugar syrup is provided for winter stores, and feeding must be finished by early October. Hives are placed in the building in early November. On average, colonies in single-box hives consume 12 kg, and those in larger hives 23 kg, during the winter. Colonies overwintered indoors are reported to produce 40% more honey next season than package colonies, and the overwintering cost per colony is almost equal to the price of a 0.9-kg package. Losses of colonies overwintered indoors vary from 5% to 10%.

When calculating how large the winter bee house should be, allowance must be made for spacing the hives sufficiently to ensure an adequate air flow between them, and ventilation within them. If no cooling unit is incorporated in the building, 0.42 to 0.9 m^3 has been suggested for each single-box hive, and 0.75 to 0.85 m^3 for each double-box hive. If the building has a cooling unit, packing can be tighter – 0.25 m^3 per single-box hive. But high-density packing should not be attempted by anyone who has not yet gained experience with the whole procedure. As with hives packed in a container for transport, colonies that become active produce much metabolic heat, and adequate temperature control is essential if colonies are not to be lost.

Heat insulation must be built into the structure: expanded polystyrene 15 cm thick for the ceiling and 13 cm thick for the walls, and 5 cm thick for the floor, which should be of concrete. A polyethylene 'vapour barrier' is necessary on the inside of the walls. In order to maintain complete darkness in the building, any doors must be tight-fitting and insulated, and each opening for air intake and discharge must be protected with a light-trap (McCutcheon gives a diagram).

During the winter a recirculated air flow of 0.10 litres per second per 1 kg of bees is generally sufficient. In autumn and spring, outside temperatures are higher, and up to 1.6 litres per second per 1 kg of bees may be required, or even more upon occasion.

For the winter period, $4^\circ \pm 1^\circ$ is considered to be the optimal temperature inside the building: food consumption by the colonies is then minimal, and bee mortality is also comparatively low; at 8° it was found to be twice as high. Control of humidity in the building is usually found to be unnecessary in Canada, where a relative humidity of 50% to 75% is considered appropriate; in Saskatchewan 30% to 44% proved satisfactory. Very high humidities can be dangerous, because parts of the control system may become iced up.

Pirker (1978*b*) has described the system he uses in northern Alberta.

6.6 APIARY SITES AND THE PLACEMENT OF HIVES

6.61 Selection of apiary sites

From necessity or choice, a beekeeper may have his apiary on his home plot, i.e. in the immediate vicinity of his dwelling house. This may allow no possibility of avoiding deficiencies due to the locality, such as limited bee forage, or the prevalence of strong winds where the bees must fly. But within the area at his disposal the beekeeper can probably choose the exact site that is most to his liking, and best suited to the bees. He can plant bushes or erect fences to provide shelter for the hives, and also to ensure that the flight lines of the bees are above head height wherever people will walk, work or sit. He can provide shade for the hives if this is necessary (although they should not be placed where rain will drip on them), and can erect a barrier to keep animals away from the hives, or use high hive stands if the ground is liable to flooding. Additional requirements for apiaries in the subtropics and tropics – especially where tropical honeybees are used – are discussed in Section 7.5 and subsections.

An apiary away from the home plot (referred to here as an out-apiary) offers a wider choice of site, but the site chosen may have fewer features that are at the beekeeper's disposal to alter. On a migratory site that is used only for a few weeks in the year, it may not be practicable to attend to such details as preparing a smooth and level surface for the hives to stand on, or to provide hive stands.

6.61.1 Desirable features for an apiary site

The following features are highly desirable for all apiaries:

1. easy accessibility to all hives, in all weathers, by the beekeeper, and by his vehicle if he uses one
2. a reasonably flat surface, without rocks or other impediments to easy and safe movement by the beekeeper within the apiary, especially when carrying apiary equipment or bees
3. shade from the sun and shelter from excessive wind
4. water permanently available to the bees (see Section 6.62.3)
5. minimal danger from fire (see 11.82)
6. minimal danger from flood water high enough to enter the hives
7. protection from attacks or incidental damage by animals (see below)
8. safety from vandalism and theft
9. minimal danger of nuisance from the bees, to members of the public and to their animals (see 15.3 to 15.32).

Figure 6.61a exemplifies some of these features.

In mountainous terrain where no flat sites are available, apiaries may have to be situated on steep slopes, with access up or down hill, and the beekeeper learns to adapt his working methods accordingly. With regard to point 7 above, the wild animals that present the greatest danger to apiaries in widespread areas of the world are bears. Where they occur in any numbers, it is essential to protect every apiary against them, which is most effectively provided by an electric

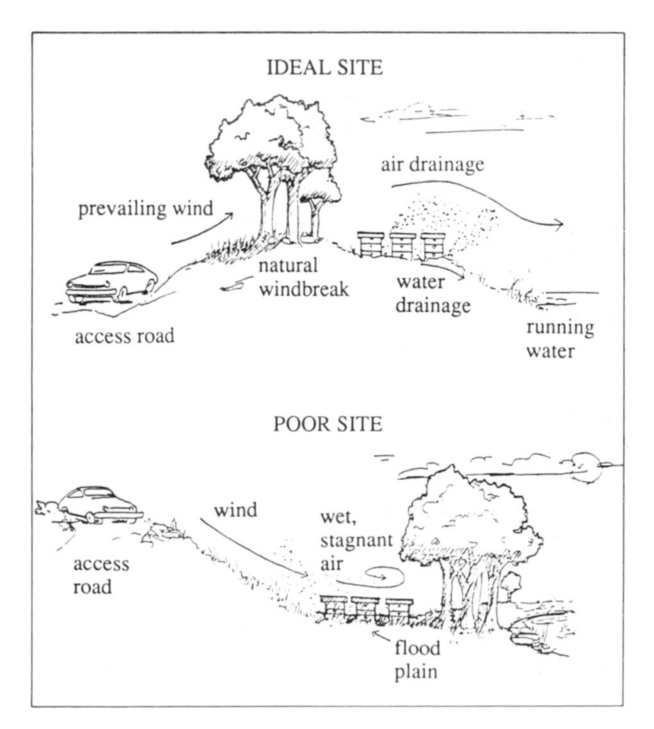

Figure 6.61a Diagram showing features of a desirable and a poor apiary site, for temperate climates (from a USA newsletter in 1978).

fence (Figure 8.23c). Caron (1978*c*) describes and compares different types, and their erection. They are usually powered by a 12-volt battery, housed in a hive – which should contain bees as well, to prevent theft. In Saskatchewan, Canada, solar-powered batteries produced by Margo Supplies are in use.

6.61.2 The apiary site in relation to its surroundings

The most important feature of the surroundings is the bee forage during the period when the apiary is in use, for on this the honey harvest depends; Figure 6.61b shows a good site in this respect. If the apiary is on the home plot, there is little a beekeeper can do to improve the forage unless he has the use of an extensive area of land; Section 12.83 suggests how this could be planted. Where bee plants are grown in even a small garden or on the edges of fields (12.84), it will help the beekeeper to monitor the activity of his bees through the seasons, and also give him much pleasure.

Wild colonies of honeybees (except *Apis dorsata*) nest singly, not in groups, but it is usually not practicable to site hives singly, and this is not necessary. The total number of colonies per hectare that a tract of land can support is determined by the amounts of nectar and pollen per hectare that the surrounding land produces. On the other hand the number of hives the beekeeper groups together in one apiary is usually governed by his convenience, and practical considerations such as the size of his truck. Current practice is to place no more than 50, or even 30, hives in one apiary, although some photographs taken early in the twentieth century show apiaries of a thousand hives or more.

With migratory out-apiaries, the number of hives is certainly related to the capacity of the vehicle used to move the hives. It may well be that one apiary of 50 hives exploits the food resources of the land less efficiently than 5 apiaries each of 10 hives, but the beekeeper's extra labour costs in running 5 small apiaries would greatly outweigh his possible extra honey yield. In addition, it would be much more difficult to acquire 5 suitable sites than one.

Bees from different apiaries separated by less than 3 to 5 km are likely to compete for forage at some season of the year. If honey sources are prolific, apiaries can be much closer together without reduction of the honey yield per colony. In jungle areas of Yucatán, Mexico, I have seen apiaries only 1 km apart, each

Figure 6.61b Part of an apiary in New South Wales, Australia, to which hives had been moved to work a eucalyptus flow, 1967.

with 50 hives, whose annual honey yield per hive was 140 kg. In a few countries there is a legal requirement that an apiary shall not be established within a specified distance of any existing apiary. Much more common is legislation concerning the placement of apiaries in relation to buildings or roads, to prevent nuisance caused by the bees to people or their animals; see Section 15.31, and also end of 7.51 for tropical bees.

Vehicles using roads can kill bees by colliding with them, but honeybees do not often fly low across roads. In Germany, Dreher (1978) showed that when vehicles were travelling at 50–60 km/h along a busy road, honeybees were killed only where the road ran between an apiary and a heavy nectar flow or a source of honey such as combs left out in the open.

6.62 Arrangement of the apiary

6.62.1 Placement of hives to minimize drifting

If bees returning from a flight enter a hive other than the one from which they originated (referred to as drifting), this has many disadvantages from the beekeeper's point of view, which are set out below. Most of the measures taken to prevent drifting are related to the placement of hives in the apiary.

Bees are likely to drift from any queenless hive to a neighbouring one with a queen. Apart from this, most drifting occurs because the returning bee (worker, drone or queen) does not recognize the hive it flew from, and young bees need a period of learning before they can locate their own hive with certainty. Drifting is most likely to occur when hives in the apiary look alike (in shape and colour) and are placed in straight rows, especially if they are close together with their entrances facing in the same direction. When returning to the apiary, bees from inner hives in a row tend to enter the end hives they first encounter, and to join those colonies. If hives are in groups which look alike, returning bees may enter a hive in another group that is in the same position as their own. The presence of landmarks such as trees or buildings near the hives makes drifting less likely. Section 2.43 discusses honeybee orientation in general.

One disadvantage of *any* bees drifting is that they may transmit diseases or parasites to the colony they join. If *workers* drift consistently into certain hives from others, the depleted colonies may become too weak to prosper, and the receiving colonies become excessively large, so that they need extra space for bees and for honey storage. A proper assessment of colony performance, and the evaluation of queens, become impossible because colony populations are altered to an unknown extent. Also, since colonies are no longer uniform, systematic colony management becomes im-

Figure 6.62a Jim Powers in one of his apiaries on Molokai, Hawaii. Hives are well scattered under light shade from kiawe (*Prosopis pallida*) that gives high honey yields.

possible. Significantly more honey in total is produced in groups of hives where drifting is minimized (Jay, 1987). If *drones* drift to another hive, drone rearing by the depleted and the receiving colony may be put out of balance (see Section 3.22). A *queen* that returns to a hive other than her own after flying will almost certainly be killed by the bees in it.

Drifting can be minimized by taking several fairly simple measures.

1. Use or provide landmarks within the apiary – windbreaks, trees, bushes, fences, etc. Figure 6.62a shows a honey-producing apiary with natural landmarks; in Figure 8.23b mating hives are in an apiary with trees and bushes to give both shelter and orientation aids to returning queens.
2. Give individual hives some distinguishing appearance, by using paint of different colours (black, white, yellow or blue) to make stripes or patterns near the flight entrances; see for example Figure 7.52c. Free and Spencer-Booth (1961) found that bees learned the height of their hive and its entrance, and that they learned to distinguish between different patterns painted near the entrance, although not colour combinations. They registered the colour below the entrance more than that above it, and colour higher than the hive box just above it was not registered.
3. Do not place hives in straight rows unless this is unavoidable, in which case face entrances in different directions. (Unfortunately the most practical layout in many apiaries is a series of straight rows with hive entrances facing the same way.)
4. Avoid repetitive patterns of both hive positions and entrance directions. Figure 6.62b shows five satisfactory layouts for 8–12 hives; ideally, adjacent layouts should be different ones.

Cook (1962) found that drifting (5% when hives were in two rows) was reduced (to 0.8%) when the hives were arranged in a circle with entrances facing inwards. Jay gives further practical information (1971), and details of the experiments on which it is based (1965–69). Section 8.35 discusses drifting after package bees are hived.

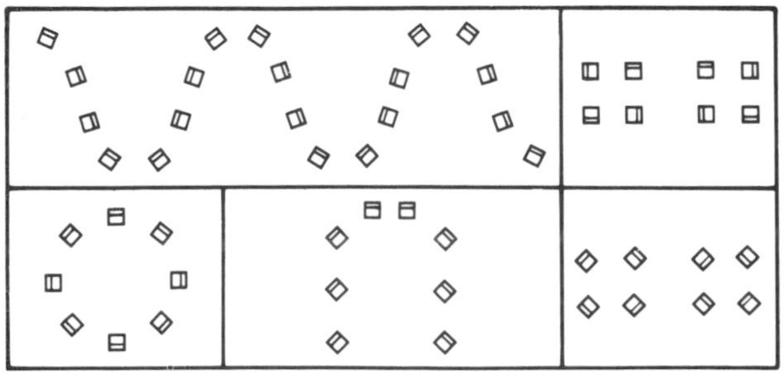

Figure 6.62b Five useful hive layouts to prevent bees drifting (Jay, 1971).
The double line indicates the hive entrance.

6.62.2 Placement of hives in relation to other factors

The particular layout used can be chosen according to the apiary site, to make colony management easy for the beekeeper, with enough space for him to move between hives, and to put down the parts removed from a hive, in a place from which they can be conveniently replaced on the hive without too much lifting (Section 5.41). The layout must allow access by truck, loader, or whatever is necessary. It should also take into consideration the main lines of bee flight.

It is best to have a clean hard surface under the hives, or at least in front of them, so that the beekeeper can see if dead bees accumulate there, which may be a sign of disease or pesticide poisoning. On a smooth surface, hive parts can be put down without picking up debris, and tools will not be hidden as in long grass. It is especially important that there is a clear air flow round the hives, so the ground where the hives are placed should be free from high-growing vegetation. Such vegetation can speed up fungal attack on the hives, especially floor boards, and also provide access for ants and other nuisances to the bees. If the vegetation cannot be cut regularly, it must be killed each year with herbicide that is safe for the bees (Section 11.73). Ferguson (1987) suggests glyphosphate. In hot conditions in Israel, Lensky (1958) found that the best ground cover (for keeping hives cool) was green grass, which transpires – producing moisture – and also absorbs solar radiation. The temperature of bare ground can be several degrees above shade air temperature.

6.62.3 Provision of water in the apiary

Colonies cannot store water in the hive except in the honey sacs of some of the bees (end of Section 3.36), so they must collect it when they need it: (a) to dilute honey when this is used for brood rearing; (b) to cool the hive by evaporation if there is danger of overheating; (c) to dilute nectar which has a very high sugar concentration (see Section 2.43.2). Colonies obtain water in nectar, so long as it has a low sugar content (or in dilute sugar syrup fed in the hive), and they then collect less water outside the hive, as shown in Figure 2.43b.

In general, provision of a safe water supply in or close to the apiary is cost-effective – sometimes highly so – because it minimizes the bees' energy expenditure in collecting the water. Figure 7.41a shows an apiary with its own natural water supply.

In hot dry conditions a water supply near the hives (within 0.5 km) is essential for colony survival. When I visited Miel Carlota south of Mexico City they had 20 000 hives, in 450 apiaries, one of which is shown in Figure 7.51c. It was not practicable to truck water regularly to the apiaries, so each had to be sited near a permanent water source; if at some season one dried up, the apiary had to be relocated. In cooler climates, early spring is the time when water is needed close to the hives.

A water feeder in the apiary should be of a size commensurate with the rate of water usage and opportunities for refilling. A colony can use as much as 4 litres of water a day in very hot weather – i.e. 200 litres for an apiary of 50 hives. Water containers should be clean, and certainly not contaminated with oil or pesticides. They should be maintained in a hygienic condition, and there must be facilities for washing as necessary, and for easy refilling; the feeder must never be allowed to become dry, and evaporation should therefore be minimized. A feeder should not be located in front of the hives under the main line of flight, where the water or container might become contaminated with faeces; if there is danger of such contamination, a suitable cover must be provided. Floats or other access should be provided for the bees, to prevent them drowning.

In cool weather, bees more readily take warm water than cold; a very shallow source (or water trickling over a surface) is likely to be warmer than a deep one. In early spring in temperate climates, the feeder must be in place and filled well before it is needed, so that the bees learn to use it and not some less desirable source.

For a small home apiary in an area with moderate or high rainfall, it may be adequate to have a shallow trough containing sand or a layer of pebbles which is refilled by the beekeeper whenever the rain does not do this. One recommendation is a cellulose sponge 30–50 mm thick, in a trough 100 mm deep. Or a type of poultry waterer can be used, which has a central container standing in a shallow tray. Owens and McGregor (1964) describe and illustrate two long shallow troughs containing coarse gravel; each is continuously provided with water through a perforated tube running along the bottom, supplied from a (raised) covered tank. A common installation – which is not protected from evaporation – is half a 220-litre drum, cut parallel to the ends, with a layer of cork or other floats on top of the water, and a mesh cover above to prevent farm or wild animals using the water. Figure 6.62c shows such a water feeder in Australia, just before it was netted to protect it against kangaroos, which can quickly empty it.

Figure 6.62c Water tank for bees, with floats, in the dry Australian bush.
It is often necessary to fit netting over the top of the tank as a 'roo guard', otherwise kangaroos will soon use the water.

Natural water sources used by bees are discussed in Section 12.3, and the behaviour of foragers collecting water in 2.43.2. A survey of the provision of water for bees, both inside and outside the hive (Johansson & Johansson, 1978) also discusses bees' preferences for water containing different substances, from salt to malodorous contaminants. It may be that *any* scent can increase recruitment of other foragers to a water source. Free and Williams (1970) found that some foragers collecting pure water exposed their Nasonov scent gland; the water itself provided the foragers with no scent signal by which they could recruit other bees to the source.

When hives are to be closed to protect the bees against pesticide application, or in other circumstances where there is a temporary risk of overheating, water should be provided inside the hive, e.g. using a syrup feeder, or a wet sponge in a perforated plastic bag. Alternatively, wet sacking may be placed over the hive and the entrance so that it is accessible to the bees. Section 11.72.1 discusses these situations.

The benefit to crop growers of providing water close to hives moved in for pollination was tested in California, USA (Sheesley & Atkins, 1986). Seed yields of lucerne were higher in fields with hives in which water was provided for the bees, than in control fields without a water supply.

6.63 Bee houses

6.63.1 Fixed bee houses

A bee house is a building inside which hives are placed against the walls, a flight entrance for each hive being provided through the wall. Arrangements must be

Figure 6.63a Bee house of the firm Mack at Illertissen, GFR, about 1950 (photo: F. Löw).
Section 14.52 describes the firm's production of bee venom.

made for ventilation, and if necessary for cooling. The beekeeper can move about, and operate his hives, inside the bee house. When a hive is opened some bees fly off the combs, but the building has windows with a louvre device or baffle board at the top, which enables these bees to fly out but prevents their re-entry. A bee house is usually a permanent structure, which cannot be shifted as an outdoor apiary can (mobile bee houses are dealt with in Section 6.63.2). The capital cost of a bee house is much higher than that of an outdoor apiary, although this is offset to a slight extent in that the hives need not be as substantial as those out in all weathers, and they last longer.

Bee houses can have special advantages in certain circumstances. Hives in them are more secure against damage by animals, and against vandalism and theft. The temperature in the hives is more equable than in hives out of doors and – perhaps mainly because of this – bees in a bee house are in general better tempered. Also in all weathers the beekeeper works under shelter and in the shade; there are relatively few bees flying around him, so he needs less heavy protective clothing and is cooler and more comfortable. These features make a bee house especially advantageous for housing tropical African bees; see Section 7.52.2. Some features of the bee house are sufficient to make it an attractive proposition to a few individuals in many countries, and some beekeepers may have a building that they can adapt to the purpose. Nevertheless to many beekeepers, one of the attractions of beekeeping is that it is an outdoor occupation.

In the interests of economy, hives in a bee house are placed quite close together, and this can encourage drifting of bees, with the likelihood of transmitting diseases or parasites. To reduce drifting, entrances of adjacent hives are usually distinguished by painting the external wall round them in different colours and patterns; see Figure 7.52c. A common practice is to place hives together in pairs, pairs being separated by a hive-width working space.

A bee house can provide convenient facilities for many queen-rearing operations. But natural light is restricted, so close examination – e.g. for the presence of eggs – may have to be done outside, or under good artificial light which may not easily be provided in a bee house. Crane (1981*b*) has discussed the pros and cons further, and Spiller's book *The house apiary* (1952) gave details and illustrations of bee houses and their benefits in the uncertain English climate.

In one area of central Europe, a bee house was the

Figure 6.63b Interior of a traditional German bee house (drawing: H. Preusse, in Zeiler, 1984).

normal type of apiary until very recently, and the change to siting hives in an open apiary did not begin until about 1960. This area can be identified as that in which the German language has been widely known and beekeeping texts in German were influential: Germany and Austria, with adjacent parts of Denmark, Poland, Czechoslovakia, Hungary, Yugoslavia and Switzerland; Crane (1983*a*) gives a map. German beekeeping in bee houses is described by Graze (1979), and Figures 6.63a and 6.63b show examples.

Where a bee house was the norm, it sometimes took on an additional function: it became a retreat – for the beekeeper alone, or for the family during weekends and holidays – in a way impossible in an apiary of hives set out in the open. If bee houses in their traditional area are ever completely replaced by apiaries such as are used in the rest of the world, beekeeping there will lose a special characteristic that has added much to the beekeeper's way of life.

6.63.2 Mobile bee houses

In the European area where bee houses are used, hives that are migrated for honey production or pollination may be stood out in the open, packed together in a tight row with weather protection secured above and around them. This is not always satisfactory, and an alternative, still widely used in some parts of Eastern Europe, is a 'mobile bee house' or 'bee wagon' such as that shown in Figure 6.63c, which can be towed behind a motor vehicle. The hives remain permanently in this bee house, so there is no loading or unloading. Space is provided inside for the beekeeper to operate the hives, and sometimes for honey-extracting equipment as well. Laffers (1981) in the German Democratic Republic describes these vehicles; he regards a unit for 30 to 60 hives, arranged in two or three tiers, as the most economic. But the capital cost per hive of such systems must be quite high. In North America, large open trailers holding 40–50 hives are used similarly.

Figure 6.63c Mobile bee house in Hungary, for 48 back-opening hives, 1983.

6.7 ECONOMICS OF BEEKEEPING

6.71 Commercial and non-commercial beekeeping

At first sight it may seem to be a financially attractive proposition to set bees to work, collecting nectar which does not have to be paid for – so that they both feed themselves and produce honey, a non-perishable product that can be harvested and sold. A small-scale beekeeper may regard the honey he produces as an ample reward – provided he does not keep a detailed account of his costs, including his labour. However, to succeed commercially as a beekeeper is a very different affair, possible only in certain parts of the world.

6.71.1 Non-commercial beekeepers

Small-scale beekeepers with a few hives of bees use the honey they produce within the family, selling extra produce to friends and neighbours, and regarding the money received as a contribution towards their outgoings. Included in the costs, they would reckon purchases of queens, and sugar for feeding, any expenditure on transport of hives in a vehicle used mainly for other purposes, and interest lost on the capital invested in equipment. They may not expect to be financially recompensed for their labour.

Sideline beekeepers who keep 50–100 hives must spend most of their spare time on the bees, during much of the bees' active season. But their economic income is derived from other sources. If they are fortunate, they can call upon a certain amount of free labour from members of the family, especially with extracting and bottling honey. They are likely to supply jars of honey on a systematic basis to local shops for resale, regarding receipts from sales as a bonus, or as a reward for their labour – and for their close attention to the bees at critical times, which may preclude social and other spare-time activities. Some sideline beekeepers have their honey extracted under contract by a commercial beekeeper, and sell it in bulk, thus avoiding the honey-handling part of beekeeping altogether.

Sideline and small-scale beekeepers operate in all countries; together they constitute 95% or more of the total number in some. In the USSR they produce over half the honey, and the same may now be true in China; see Section 6.71.2. Although such beekeeping is not run as an economic commercial activity, it is an integral and significant part of the world's beekeeping. Table 13.8B shows the world's honey production in 1984 as 992 000 tonnes, of which 269 000 tonnes (27% of the total) was exported. From data for different countries, I estimate that around half of the world total (i.e. about 500 000 tonnes) was produced by sideline and small-scale beekeepers. Without them, the reduction in the supply of honey could well lead to increased retail prices, and honey production would be limited to the high-yield regions of the world, leaving many crop-growing areas without an effective population of honeybees to pollinate crops dependent on insects for setting seed or fruit (Sections 2.44, 8.4).

6.71.2 Professional or commercial beekeepers

Beekeeping is the full-time occupation of these people, and their beekeeping must make a sufficient income to support the family. In the USSR and other socialist countries of Eastern Europe, many beekeepers belong to a collective or state farm, receiving regular wages. In China they may be members of a state enterprise, or beekeeping members of communes, but current policy encourages individual enterprise. In keeping with this policy, beekeepers can sell direct to the consumer on local markets, but hive products for export are marketed by government agencies. In countries with a capitalist economy, a commercial beekeeper probably has to provide his own capital for purchasing hives and other equipment, perhaps with the aid of a commercial bank loan. Financial support available in some developing countries is discussed in Section 7.6.

An independent commercial beekeeper working single-handed is likely to run 200–300 hives, or 500 to 1500 if the operation is highly mechanized. Income, and return on any capital invested, are derived from revenue after all costs of the operation have been met, including the beekeeper's own labour and that of other family members. If permanent outside labour is employed, this would probably be paid for not only at busy times, but also during dearths when work may be restricted to dealing with honey and other products, together with the making and maintenance of equipment and buildings. Alternatively, where the winter dearth is long, both beekeeper and employees may need to find other work for 4 to 8 months each year. There are a few who commute annually between the northern and southern hemispheres, e.g. between Canada and New Zealand, and thus work through two summer seasons in the year.

In general, it is possible to set up a successful commercial beekeeping enterprise (i.e. one that will give the beekeeper an adequate financial return) only where the honey yield per hive is substantially higher than in other areas – in either the same or another country. The key to profitability is the yield per hive, together with the scale of activity within the constraints of the labour market. The beekeeper must also find an area where the nectar and pollen resources are not already fully exploited by the bees of other beekeepers. Beekeeping is likely to be most successful commercially where large areas of land, accessible by road, carry plants that are good honey sources – and that flower at specific and predictable times of year, so that colony management can be organized to take full advantage of their honey flows (Section 6.2). Beekeeping is labour-intensive work, and the rate of pay for labour can be a critical factor in determining the financial success or failure of a beekeeping enterprise. So the law of diminishing returns often operates when the amount of labour required exceeds the capacity of the family. Because of this factor, the number of owner-beekeepers is large relative to the number employing outside labour.

A commercial beekeeper in a country where labour costs are high may be attracted to invest in an enterprise in another country – usually in the subtropics or tropics – where labour costs are much lower; see the final paragraph of Section 7.6. But all large-scale commercial beekeepers need to have markets among affluent societies, since this is where the main purchasers of the world's honey now live.

6.72 Making an analysis for an individual beekeeping enterprise

An analysis of the profit or loss of a commercial beekeeping enterprise must take into account the items below. Results of several such analyses are quoted in Section 7.6 under heading 1.

Annual revenue Receipts from the sale of honey, beeswax and any other bee products, and queens and bees sold; also fees for hiring colonies for crop pollination.

Annual expenditure (a) Fixed costs, which do not vary, and (b) variable costs, which vary according to the amount of honey, etc., produced.

(a) *Fixed annual costs* include:

- (i) Rent paid for apiary sites, insurance, taxes (e.g. on property).
- (ii) Interest forgone on capital invested, and interest paid on capital borrowed, i.e. total capital × current lending rate of interest.

(iii) Depreciation of capital assets (items 1–5 below), i.e. the cost of the assets less the salvage value, divided by the estimated number of years of use, which is entered on the right. The salvage value may be taken as 10% of the cost, except that the total salvage value of 1 is only the value of the beeswax in the hives.

	no. years of use
1. hives of bees with their furnishings	20
2. other beekeeping equipment	25
3. extracting and processing equipment for honey and wax, also any other hive products sold	25
4. honey house and other buildings used for beekeeping	25
5. vehicles used for beekeeping	10

Cost of labour for assembling, painting, installing, etc., is included here.

(b) *Variable annual costs* include especially labour (except the labour cost just cited). Sanford (1983) shows how to keep the records, and Stephen (1971) provides estimates for use in calculating the time a beekeeper is likely to spend in the course of a year on different types of work, such as apiary visits and hive manipulations, and extracting, bottling and selling honey. Costs in addition to labour include:

- sugar and other foods or drugs fed to bees
- queens (averaged over e.g. 2 years, if each colony is requeened every 2 years)
- containers for honey
- electricity, fuel, petrol, oil
- other running costs of vehicles
- any other transport charges, e.g. freight on honey sold
- administration, selling and marketing.

In Figure 6.72a, both costs and revenue are plotted against total honey production. Fixed costs are represented by the horizontal line, and total costs (fixed + variable) by the sloping line above it. Revenue, assumed to be proportional to the amount of honey produced, is represented by a straight line through the origin of the graph, its slope depending on the price at which (all) the honey is sold. The total honey production needed to break even (revenue = expenditure) can be read from the graph. The annual honey yield per hive needed to break even is the total annual honey production at the break-even point divided by the number of hives operated. (The diagram is oversimplified in that both variable costs and revenue are represented as directly proportional to honey production.)

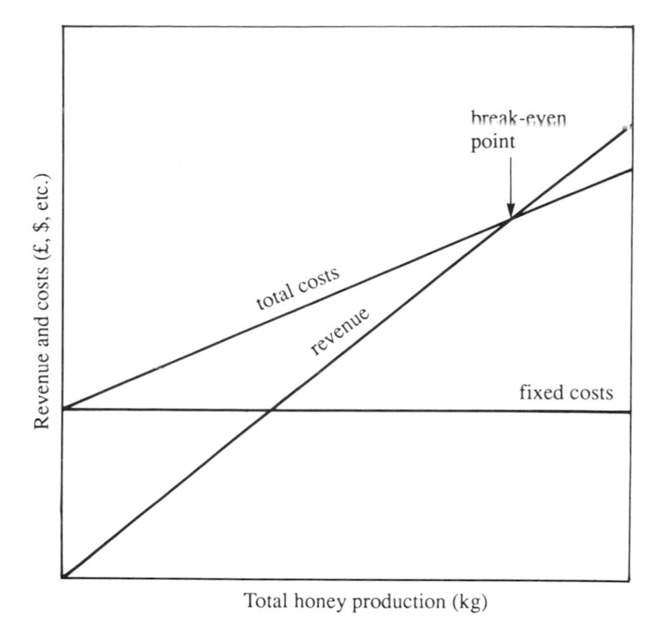

Figure 6.72a Simplified graph relating the revenue from a beekeeping enterprise, and its fixed and total costs, to its honey production (based on Stephen, 1971).

One item not taken into account in the above analysis is governmental support to beekeepers, in advisory, technical and research services, which can be considerable in temperate zones; these are among the resources for beekeepers that are the subject of Chapter 16. In some countries the basis for this support is the minor but real value of the honey industry to the national economy. In others, where honey yields are lower, the rationale is likely to be the value of honeybees in crop pollination. For other types of support available in developing countries, see Section 7.6.

Sanford has published an economic analysis of a beekeeping operation with 500 hives in Florida, USA, using traditional methods of accounting (1986*a*). Elsewhere (1986*b*) computer methods are used, and a model analysis on disc is provided with the publication. Based on the analyses, recommendations are made for reducing costs and increasing financial returns.

6.73 Economic studies of honey production

The economics of beekeeping, and especially of honey production, has been studied mainly in areas where honey production is an industry of some importance. The following reports are among those published since 1970.

Australia: Income variation in the Western Australia beekeeping industry (Paust, 1975)
Investing in commercial honey production (Gulliford, 1981)
Inquiry by the Industries Assistance Commission into the honey industry (New South Wales Department of Agriculture, 1983)
An economic survey of the honey industry in Victoria, 1980–81 (Evans, 1984)
Business opportunities in agriculture and horticulture: bees (Goodman & Oldroyd, 1988)

Canada: Western Canada's beekeeping industry (Nelson, 1976); wintering v. package bees
The economics of beekeeping in Alberta (Andruchow, 1983)

New Zealand: Beekeeping: honey and other hive products (Martin, 1982); includes pollination

USA: An economic analysis of beekeeping operations (Stephen, 1971)
Bee industry economic analysis for California (Reed & Horel, 1976)
Economic trends in the US honey industry (Garoyan & Taylor, 1980)

USSR: Beekeeping organization and economy, pp. 247–267 from *Apiculture* (Avetisyan, 1978); in English
[Factors affecting production costs and how to diminish them] (Prokof'eva & Savel'eva, 1984)

Other studies have been made in Europe, for instance:

France: [Technical economic study . . .] (Merle, 1985)
[Improvement loans from the Crédit Agricole] (Abeille de France et l'Apiculteur, 1986)

GFR: [Economic analysis of beekeeping management] (Scherhag, 1983)

Norway: [The economics of bee management: hobby or professional] (Haavi, 1973)
[The economics of honey production] (Søbstad, 1988)

Poland: [Profitability of beekeeping production in some types of apiaries] (Pidek, 1977)

World production, trade and consumption of honey and beeswax are discussed in Sections 13.8 to 13.82, and at greater length by International Trade Centre (honey, 1986; beeswax, 1978). The economics of hiring colonies of bees to growers for crop pollination are incorporated in a study in British Columbia, Canada (British Columbia Ministry of Agriculture, 1984).

6.8 FURTHER READING AND REFERENCE

Details of publications listed will be found in the Bibliography.

Böttcher, F. K. (1985) *Bienenzucht als Erwerb*, 5th ed.
Dadant & Sons (ed.) (1975) *The hive and the honey bee*
Gojmerac, W. L. (1980) *Bees, beekeeping, honey and pollination*
Jay, S. C. (1965–69) *Drifting of honeybees in commercial apiaries*
Jean-Prost, J. (1987) *Apiculture*, 6th ed.
Johansson, T. S. K.; Johansson, M. P. (1978) *Some important operations in bee management*
Manley, R. O. B. (1946) *Honey farming*
Matheson, A. (1984*a*) *Practical beekeeping in New Zealand*
McCutcheon, D. M. (1984) *Indoor wintering of hives*
Moeller, F. E. (1976) *Two-queen system of honey bee colony management*
Morse, R. A. (1972) *The complete guide to beekeeping*
Morse, R. A.; Hooper, T. (eds) (1985) *The illustrated encyclopedia of beekeeping*
Prairie beekeeping manual (1986)
Sammataro, D.; Avitabile, A. (1986) *The beekeeper's handbook* [written for beginners]
Spiller, J. (1952) *The house apiary*
Townsend, G. F. (1981) *Introductory apiculture*, and (1985) *Advanced apiculture* [text, filmstrips, taped commentary]
USDA (1980) *Beekeeping in the United States*
Wedmore, E. B. (1945) *A manual of beekeeping for English-speaking beekeepers*, 2nd ed.

Books on beekeeping in individual countries are cited in Appendix 2. Section 16.82 lists International Apicultural Congresses, and their Proceedings are published.

7

Special features of beekeeping in the subtropics and tropics

7.1 THE SUBTROPICS AND TROPICS IN RELATION TO THE TEMPERATE ZONES

The map on page 504 identifies the following zones (Good, 1974):

	latitudes (N and S)
subtropics	$23\frac{1}{2}°$ to $34°$
tropics	$0°$ to $23\frac{1}{2}°$
equatorial (within the tropics)	$0°$ to $15°$

In any latitude zone, as the height above sea level increases, conditions become more suitable for plants characteristic of a higher latitude zone. Thus at certain altitudes within the tropics, plants can grow that are characteristic of subtropical or even temperate zones.

The subtropics are already widely used for honey production (Section 7.2). Tropical regions have been much less studied in relation to beekeeping than those in the temperate zones and subtropics; one early comparative study – confined to the Americas – was by Sechrist (1922). The tropics include nearly 40% of the earth's land surface on which plants grow, with large areas of nectar and pollen resources that have not yet been exploited, and beekeeping could probably be a profitable undertaking in some of them.

Table 7.1A summarizes differences between conditions in the temperate zones and tropics from the beekeeping point of view, and Table 7.1B shows which honeybees are used for beekeeping in different countries.

Most of the management described in Chapter 6 is applicable in the subtropics, and much of it also in the tropics; reference is made as appropriate in this chapter.

7.2 BEEKEEPING IN THE SUBTROPICS

As defined above, the subtropics include about 22% of the earth's land surface on which plants grow. The northern subtropics include the north of Mexico and southern USA, most of Africa north of the Sahara, parts of the eastern Mediterranean area and of Iraq, Iran, Afghanistan, Pakistan and India, and the southern part of China. The southern subtropics include about half of Chile, Argentina and Paraguay, all Uruguay and the southern tip of Brazil; South Africa and parts of Namibia, Zimbabwe and Mozambique; and most of the southern part of Australia.

Beekeeping manuals are available for regions all or partly in the subtropics:

- Mediterranean area (Bailo, 1980)
- south-eastern USA (Sanford, 1983)
- Argentina (Persano, 1980)
- Australia (New South Wales Department of Agriculture, 1983)
- South Africa, with African honeybees (Anderson et al., 1983)
- India and China, with *Apis cerana* (Shah, 1983; Singh, 1962).

Publications on beekeeping in individual countries are included in Appendix 2.

Table 7.1A Some differences between temperate and tropical zones from the beekeeping point of view

+ indicates more; – indicates less; subtropical zones are usually intermediate between temperate and tropical zones.

		Temperate	*Tropical*	*Section in book*
1	sun overhead	never	twice a year	7.31, 12.62
2	droughts, excessive rain, high winds	no	yes	7.36, 11.81
3	regional variations in annual rainfall	–	+	12.72
4	high temperatures	–	+	7.32.2
5	cold winters	+	–	6.53, 6.54
6	relevant to honeybee orientation:			2.43, 3.35
	inclination of earth's magnetic field	+	–	
	summer sun sweeps through large angle daily, low in the sky	+	–	
7	annual variation in photoperiod (day length)	+	–	3.5 and subsections
8	long summer days during main flow	yes	no	12.21.3
9	no. species of plants yielding nectar (and pollen)	low	high	12.71, 12.72
10	native honeybees store much honey	yes	no	1.2, 7.31
11	native honeybees abscond, and swarm readily	no	yes	3.34.5
12	bee diseases and parasites more significant than 'enemies'	+	–	11.2–11.4
13	bee 'enemies' more significant than diseases	–	+	11.5, 11.6
14	honeybees originated	no	yes	1.23
15	no. species of indigenous honeybees now	–	+	1.1

7.21 Features that affect beekeeping in the subtropics

Some parts of the subtropics are among the world's most valuable areas for honey production, and also for queen rearing. The flowering season usually lasts for many months instead of weeks, with a great variety of flowering plants giving major or minor honey flows, and there is often a year-round active beekeeping season.

During the year the day length in the subtropics changes much less than at higher latitudes, and it is never as long as during high-latitude summers. The subtropics are similar to the temperate zones in having a distinct seasonal rhythm with a well marked summer and winter, but the climate is usually warmer, and the winters mild, temperatures being high enough all the year round for honeybees to fly. *Dry subtropical regions* have a long dry summer and a relatively rainy winter. *Wet subtropical regions* have a higher rainfall, most rain falling in the summer. Dry and wet regions are often found on eastern and western sides of continents, respectively.

Bee diseases are generally less prevalent than in temperate zones, owing partly to the absence of many of the stresses that bees are subject to in colder regions. Such stresses as exist are likely to be due to heat and drought, and to some of the predators and enemies described in Chapter 11 – including the ubiquitous wax moths, which can be especially troublesome to *Apis cerana* (Section 7.41.5). But the subtropics are free from many of the difficulties that beset beekeeping in parts of the tropics.

7.22 Colony management in the subtropics

Unlike some important beekeeping areas in colder parts of the north temperate zones, the subtropics do not have a cold winter, and except at high altitudes, bees do not form a cluster. However, the higher temperatures make the bees more active, and colonies therefore consume more food than those quiescent in a winter cluster. In Figure 12.62a, the scale-hive records for colony 1 at 60°N, and colony 3 at 32°N, illustrate the difference between the beekeeping season in the cool temperate zone and the subtropics. The higher level of food consumption in the subtropics makes it essential for the beekeeper to check that colonies always have a reserve food store in the hive – at least 10 kg for *Apis mellifera*, and somewhat less for *Apis cerana*. Sections 4.47, 6.31 and 6.53.1 deal with syrup feeders and feeding.

In the subtropics summer temperatures are high, and the provision of water is important, as are the shading of hives and other methods of keeping colonies from becoming too hot, as indicated in Sections 7.32.2 and 7.5 to 7.52.2.

Productive strains of European *Apis mellifera*, about which much is known, are used almost everywhere in the subtropics. Honey yields are often very good indeed, as witness the large exports of honey from some countries in the subtropics. But the beekeepers as well as the bees must work hard for almost 12 months of the year. They must tend the colonies throughout the year, and may also migrate them to a number of honey flows. Colonies may need to be requeened more frequently than in temperate zones (Section 6.51), because the queens lay eggs through almost the whole year. Although honey yields are high, the yield per month of beekeeping work may not compare so favourably with that in the cold temperate zones.

What are broadly referred to as tropical *Apis mellifera*, and Africanized bees derived from them, are used for beekeeping in some subtropical areas, and *Apis cerana* in others (see Table 7.1B). Sections 7.41 to 7.43 should be consulted for such areas.

7.3 BEEKEEPING IN DIFFERENT TROPICAL CONDITIONS

7.31 Features that affect beekeeping in the tropics

Both *Apis mellifera* and *Apis cerana* are kept in hives in the tropics (Table 7.1B), and the colonies pass through a similar development cycle to that in temperate zones: growth and reproductive swarming (if this is not prevented) and honey flow, and dearth quiescence; Section 3.52 gives details. Notable differences from the

Table 7.1B Honeybees used for beekeeping in countries with subtropical and/or tropical regions

all = all countries where beekeeping is done.
Years (1957 and on) are those in which Africanized bees were first reported.

		A. mellifera			
Tropics in:	*A. cerana*	*tropical*	*Africanized*		*temperate-zone*
Africa	—	all	—		—
Americas	—	—	Argentina	1969	Caribbean islands (but not Trinidad)
			Belize	1987	Chile
			Bolivia	1967	also some of the countries (on left) with Africanized bees
			Brazil	1957	
			Colombia	1978	
			Costa Rica	1983	
			Ecuador	1981	
			El Salvador	1985?	
			French Guiana	1974	
			Guatemala	1985?	
			Guyana	1975?	
			Honduras	1984/5	
			Mexico	1986	
			Nicaragua	1984	
			Panama	1981	
			Paraguay	1965	
			Peru	1977	
			Surinam	1975	
			Trinidad	1979	
			Uruguay	1971	
			Venezuela	1977	

(continued)

Table 7.1B *(continued)*

		A. mellifera		
Tropics in:	*A. cerana*	*tropical*	*Africanized*	*temperate-zone*
Asia	Afghanistan Bangladesh Bhutan Burma China India Indonesia Iran Japan Kampuchea Korea Laos Malaysia Nepal Pakistan Philippines Sri Lanka Taiwan Thailand Vietnam	—	—	Afghanistan Bhutan* China India* Indonesia Iran Iraq Israel Japan Jordan Korca Lebanon Malaysia* Oman (also *A. florea*) Pakistan Philippines Saudi Arabia Singapore Syria Taiwan Thailand UAE Vietnam Yemen AR Yemen PDR
Oceania	1985/86: Irian Jaya (Indonesia) Papua New Guinea	—	—	all islands

* relatively few colonies

temperature zones are listed below. Points 1 to 3 apply to all bees, and points 4 to 7 to tropical bees.

1. It is hot during most or all of the year, and the sun is always high in the sky at midday. Therefore hives need to be kept in as cool a position as possible, and under permanent shade; see Section 7.51.
2. The sun is directly overhead at midday during two seasons of the year, so there are likely to be two colony development cycles every year. The scale-hive record in Tanzania (Figure 12.62a, colony 4) demonstrates this.
3. The seasonal pattern of brood rearing is likely to be less simple than in temperate zones, and more closely related to forage available. Where brood is reared all through the year, colonies should be requeened annually; see Section 6.51.
4. If food sources are depleted, and there is no forage within reach, the adult bees of a colony of tropical bees may escape starvation by moving to another area where forage is available (see Sections 3.34.5 and 3.43).
5. In many parts, temperatures are high enough all the year round for bees to fly, and for colonies to

secrete wax and build comb (Table 3.32A), so absconding colonies can build a new nest in a different area, as well as forage there.

6. Colonies do not generally grow as large as they do at higher latitudes, or store as much honey, and reproductive swarms are produced from relatively small colonies (see Table 7.43B).
7. The bees' daily foraging cycle is attuned to cycles of nectar and pollen production by tropical flowers. Bees forage in the early morning and in the evening – occasionally even by moonlight – rather than in the heat of the day.

7.32 Colony management in the tropics

7.32.1 General considerations

Sections on beekeeping management which are relevant to the tropics include especially:

6.1 Handling honeybees
6.2 Strategy of management for honey production
6.3–6.5 Seasonal management, except 6.54 on cold winters
7.41–7.43 Beekeeping with different honeybees
10.2–10.5 Use of top-bar hives and long hives; see also 7.37

After the end of a dearth period when no plants flower, the queen starts, or increases, her egg laying as soon as flowers appear again and the bees begin to forage. Colonics then need to consume an ever-increasing amount of food: nectar or honey for all adults, and pollen for nurse bees secreting brood food. If the amount of flowering increases and the bees can always fly out to forage, colonies may need little attention from the beekeeper. But a check must be kept on food reserves in the hives throughout the period of population growth, as well as during dearths. Each colony of *Apis mellifera* should always have at least 10 kg of honey in the hive, and each colony of *Apis cerana* 7 kg or more. Any colony with less must be fed without delay; it is possible to feed bees during a tropical dearth, although not in a cold dearth. Sections 4.47, 6.31 and 6.53.1 deal with syrup feeders and feeding.

I have met beekeepers in the tropics to whom feeding colonies is not a normal part of beekeeping practice. Or, if they do provide food, a single feed of a litre of syrup is considered adequate. The wide variety and almost constant succession of plants in flower may make it seem that food is always available to bees, but this is not necessarily so; in addition, there can be periods when plants in flower produce pollen but not nectar.

In parts of the tropics refined sugar may not be available, or other suitable food for bees may be much cheaper; see e.g. Adjare (1984). Provided the bees can fly out to defaecate, colonies can be saved from starvation by feeding unrefined sugar, or syrup made from it, or 'residue syrup', or dates, or juice from sweet fruits such as orange and mango; bees will feed on damaged ripe mango fruits on the ground.

A constant water supply is essential for colonies rearing brood, and also for those at risk from overheating. During the rains, water is usually available to bees, and during a nectar flow the nectar taken into the hive may well provide sufficient water. Colonies can store a little water (Section 3.36), but an apiary – especially an out-apiary – must be set up near a permanent water source, or a water supply provided in the apiary, as explained in Section 6.62.3. Evaporation, and the likelihood of the water being used by other animals, can present considerable problems. Water can be fed in the hive, but this is time-consuming work.

In the tropics, an adequate supply of pollen for colonies during their growth period is usually assured by the great variety of flowering plants, but pollen substitutes can be fed if necessary; see Section 6.32. In Ghana the yam bean (*Sphenostylis stenocarpa*), which has a fairly high protein content, is used for feeding bees (Adjare, 1984). It is cooked, then mashed smoothly and cassava powder added; the mixture is made into 'yam balls' which are placed inside the hive.

Section 6.41 explains that bees are stimulated to collect nectar if their hive contains combs with empty cells. For this reason, honey supers should be added to hives that contain strong colonies before the main flow is expected to start. The extra space also helps to keep the hives cooler, and to prevent the bees making swarm preparations. But in weak colonies, combs not covered with bees are vulnerable to damage by wax moths.

Honeybees originated in the tropics (Section 1.23), and they themselves are very successful in many tropical regions. But certain tropical conditions, including some of those that make for the *bees'* success, make it difficult for a beekeeper to get large honey harvests from them. For instance colonies of honeybees that evolved in the tropics are not inclined to store large quantities of honey, having adopted a different strategy for survival through dearth periods: absconding to a neighbouring area where there is forage. Absconding leaves the beekeeper with empty hives. Also, efficient honey-getting for the beekeeper depends on colony management to build large populations, and tropical honeybees are generally less amenable to handling and management, and they swarm readily.

Figure 7.32a Results of careless comb spacing in the hive (photo: IBRA Collection)
above Brood box with brace and burr comb built where frames have been spread out because two are missing.
below Comb misshapen where spacing in the hive was too wide.

Whatever bees are used, it can be difficult to work profitably at year-round high temperatures, where colonies are constantly active and constantly consuming food. Nevertheless, in many areas the loss of colonies could be much reduced, and honey yields per hive much increased, if more knowledge were acquired and applied.

In my experience, beekeepers in the tropics who lack adequate training or knowledge often suffer poor results because they pay insufficient attention to two details. One is the feeding of colonies whenever their food stores are dangerously low (see above). The other is the proper spacing of frames, especially those containing brood combs, which is explained in Section 4.25; Figure 7.32a shows what happens when too few frames are used, spaced out too widely.

7.32.2 Management for very hot conditions

Section 3.32 explains how the bees of a colony regulate the temperature in the hive, and Table 3.32A lists a number of temperatures that are critical to individual honeybees, and to their colonies. At very high ambient temperatures, a colony must cool the brood nest so that its temperature does not rise above 35°. This uses up energy, and therefore food, even if the excess temperature is not so great that the bees are stressed. Beekeepers should therefore do all they can to reduce the hive temperature by providing shade, by treating the outside of the hives so that heat is reflected from them (Section 4.13), and by placing the hives on suitable ground (Section 6.62.2). The risk of overheating is greatest for large colonies.

Hives should not be opened during the heat of the day, but in the early morning or the evening. It is not practicable to inspect colonies after dark, but some beekeepers remove honey from their hives at night.

The importance of providing food for colonies during a hot dearth, and of providing water in apiaries, is referred to in Section 7.32.1. If the humidity is high the bees have difficulty in evaporating water, and adequate ventilation is important. Beekeepers' problems connected with honey produced at high humidities are dealt with in Sections 7.35.1, 13.32, 13.39. Overheating of honey in the hive presents other problems to the beekeeper, discussed in Section 7.35.2. Both bees and honey confined in hives are exposed to greater dangers of overheating during transport, than in the more stable environment of the apiary. In many different parts of the world there may be a hot period in the course of the year, but in the tropics the heat is most likely to be extreme and the period to last for many months. Lensky made a long series of studies and experiments on colonies of Italian *Apis mellifera* in a subtropical climate in Israel; most are published in a thesis (1963) and in journals (1964*a*, *b*, *c*). At high external temperatures (up to 48°) in an insulated 'hot house', the bees controlled only the temperature of the brood nest in the hive, and this did not exceed 37.6° in any experiment. Just below the inner cover the temperature was near that outside the hive, but honey proved to be a good thermal insulator. The temperature of the brood nest – not that outside the hives – affected ventilation activity by fanning bees, and also water collection. But the outside temperature determined activity in foraging for syrup. This decreased during the hottest hours, and the bees then preferentially collected water instead; they foraged for water at temperatures up to 47°. There are many other reports of brood-nest temperature control in hot conditions, for instance from Jamaica (Jay & Frankson, 1972). Lensky found that colonies in the hotter hives consumed more food than others, because of the greater activity needed to regulate the temperature of the brood nest. Their honey yields were correspondingly lower: the average August yield was 13–15 kg from shaded hives and from white-painted hives in the open, and only 7.6 kg from aluminium-painted hives in the open.

Lensky (1963) found the following experimental method effective for cooling a hive. The hive roof was supported at each corner, so that a 50-mm gap was left between it and the upper surface of the top hive box. The wooden inner cover on this box was replaced by mosquito netting over half the area and by woollen cloth over the other half. Water from a closed container on the roof was led through a 2-mm polyvinyl tube to the woollen cloth. At a supply rate of 1 litre per 24 h, water dripped continually from the saturated cloth on to the brood nest below. Brood rearing was consistently greater in the 'watered' hives than in others, the difference increasing as the summer heat intensified; in August the watered hives had on average 5700 cm² of brood, and control hives 3900 cm². At a temperature up to 35°, Nye (1962) found that the provision of empty supers on its own reduced hive temperatures. It also increased honey production, but this may have been influenced by the empty combs in the added supers; see Section 6.41.

Studies were also made during two drought years (Lensky & Golan, 1966). Under drought conditions, feeding syrup with different sugar concentrations, at different times during the winter (when colonies had no brood), did not affect either colony populations or honey yields during the next main flow, which was from citrus in spring. Nor did feeding pollen and pollen substitutes. Honey yields were, however, increased by greatly restricting the area of comb in which the queen could lay during the flow, so that little brood was reared. If the queen had access to a large area of comb at that time, the colony reared much brood – and consumed much food in doing so – but the adult bees produced were too late to forage on the flow. (Section 6.44 shows how to induce a useful break in brood rearing by requeening during the main flow.)

In extreme conditions of heat and drought in Iraq (maximum temperatures 45°, RH 10%), a 4-litre tray made to fit under a perforated floor board, and kept filled with water, allowed 'normal' egg laying and brood rearing (Abdellatif & Abou-El-Naga, 1983). But foraging was minimal under these conditions.

One separate effect of constant high temperatures is the near impossibility of storing combs away from the hive during dearths between honey flows, because greater wax moths (*Galleria mellonella*) would quickly destroy them. Use of only a single hive box, for brood and honey (Section 7.34), obviates the need for comb storage, as does the use of relatively inexpensive top-bar hives (Section 10.5), from which the combs are harvested with the honey.

7.33 The period of colony growth and swarming

Wherever bees are kept during the period of colony growth it may sometimes be necessary to feed colonies while they are small; later, when they grow larger, swarming must be prevented if high honey yields are to be obtained. Both these subjects are dealt with in Section 6.3 and its subsections, together with other operations during this period. Two other operations described below are important for stocking hives with tropical bees: the use of bait hives to collect swarms, and the transfer into frame hives of existing colonies in wild nests or in traditional hives.

7.33.1 Using bait hives to catch swarms

Because of the high rate of reproductive swarming and absconding by colonies of tropical honeybees in many regions, beekeepers can often populate their hives by capturing swarms. In some areas swarms of bees fly along the same route during the same season each year. Beekeepers learn these routes from observation and experience, and at the approach of the season they set out large numbers of bait hives or catcher-boxes where flying swarms are most likely to encounter them.

Many of these bait hives are at some distance from dwellings, and may be stolen or damaged. The less the cost of the hives, the lower the amount of capital at risk, and from this point of view simple boxes or baskets are best. But a new swarm quickly builds comb, and unless bait hives can be inspected every few days, and any captured swarms removed and hived, combs and brood will have to be transferred to a frame or top-bar hive, by a procedure such as that explained in the next section. Such transference is much easier if bait hives have frames or top-bars, but this represents more capital investment. Guy's top-bar hives shown in Figure 7.33a are an example of a cost-effective compromise; they are described in Section 10.23. When a swarm builds combs in the hive, these can be moved, with their top-bars, directly into a Langstroth brood box containing some empty drawn-out combs; later on, frames can be inserted, and the top-bar combs moved to the ends of the brood box; they are removed when they no longer contain brood.

Figure 7.33a Using a truck as a platform when suspending bait hives in trees to attract swarms, South Africa, 1976 (photo: R. Guy).
Figure 10.23a shows Guy's top-bar brood boxes.

Section 6.35.1 refers to bait hives in temperate zones, and to the use of synthetic pheromones to attract swarms into the hives, which could be of special value with tropical honeybees. In an experiment in Kenya (Kigatiira et al., 1986) a row of 18 new empty top-bar hives (volume 80 litres) was set up at each of three locations. Some hives were baited with the synthetic pheromones, and some were not. During the 19 months of the experiment, the 54 hives captured altogether 27 swarms; 67% of the hives containing pheromones were occupied by a swarm, and 17% of those without. Witherell and Lewis (1986) discuss the use of bait hives and attractants in eradicating unwanted swarms of Africanized bees.

7.33.2 Transferring a colony into a hive

In the tropics, many swarms establish nests in trees or rock crevices, and these nests may be a main source of colonies used by beekeepers to populate their hives. Beekeepers with traditional hives may also want to transfer colonies from them into frame hives.

It is comparatively easy to get a captured swarm of adult bees into a hive; the bees can be shaken in at the top – or if deposited on a board leading up to the hive entrance, they will walk in of their own accord; see Section 6.35.1. There is more difficulty with an established colony, whether it is in a natural cavity or a traditional hive, because it has brood which the adult bees will be unwilling to leave; also, some brood comb must be obtained from the source colony and fixed in the new hive. This section describes the operation of transferring such colonies. The receiving hive is assumed to be an empty brood box of a frame hive; empty frames (containing no foundation or comb, but preferably wired) are needed, as well as frames with foundation or drawn-out comb, and if possible some frames of honey.

Transferring a colony from a wild nest

The colony may be easily accessible, like that of *Apis cerana* shown in Figure 1.2c. Or it may be inside a tree, as in Figure 7.33b (A), and access must first be made so that the combs can be taken out without breaking them, smoke being applied as necessary.

Honey comb, drone brood comb and any pieces of empty comb, with any bees on them, are put gently into a bee-proof container, and brought near the new hive. Larger areas of comb with worker brood are carefully removed, trimmed if necessary to fit into the empty frames, and tied in (B); the wires will help to support them. Fibres from a local plant, such as strips of dried banana stalk, are often used for tying in.

When all combs have been removed, the bees still in the nest hole are smoked gently so that they walk (not fly) out of it, and since the queen should be among them, great care must be taken. The bees will start to form a cluster just outside the hole, and one or more frames containing tied-in pieces of brood comb are then fastened just above the cluster (C), so that the bees can walk up on to the comb. When most have done so, the frames and the bees on them are carefully placed in the brood box, together with several pieces of honey comb tied into other empty frames. Figure 7.33c shows the process of tying in pieces of comb. If there is only a little honey, a frame or two of honey from another hive, brought for the purpose, should be inserted near to the combs of brood. (It is not worth tying in empty comb, or any small pieces, and it is useless putting them loose in the hive; they should be taken away.) The brood box is filled up with frames fitted with combs or, failing that, with foundation. The hive, complete with roof and floor board, is placed as close as possible to the nest, so that foraging bees return to it; it should be on a stand and protected from ants. Some beekeepers place a queen excluder between the brood box and floor board, so that the queen can-

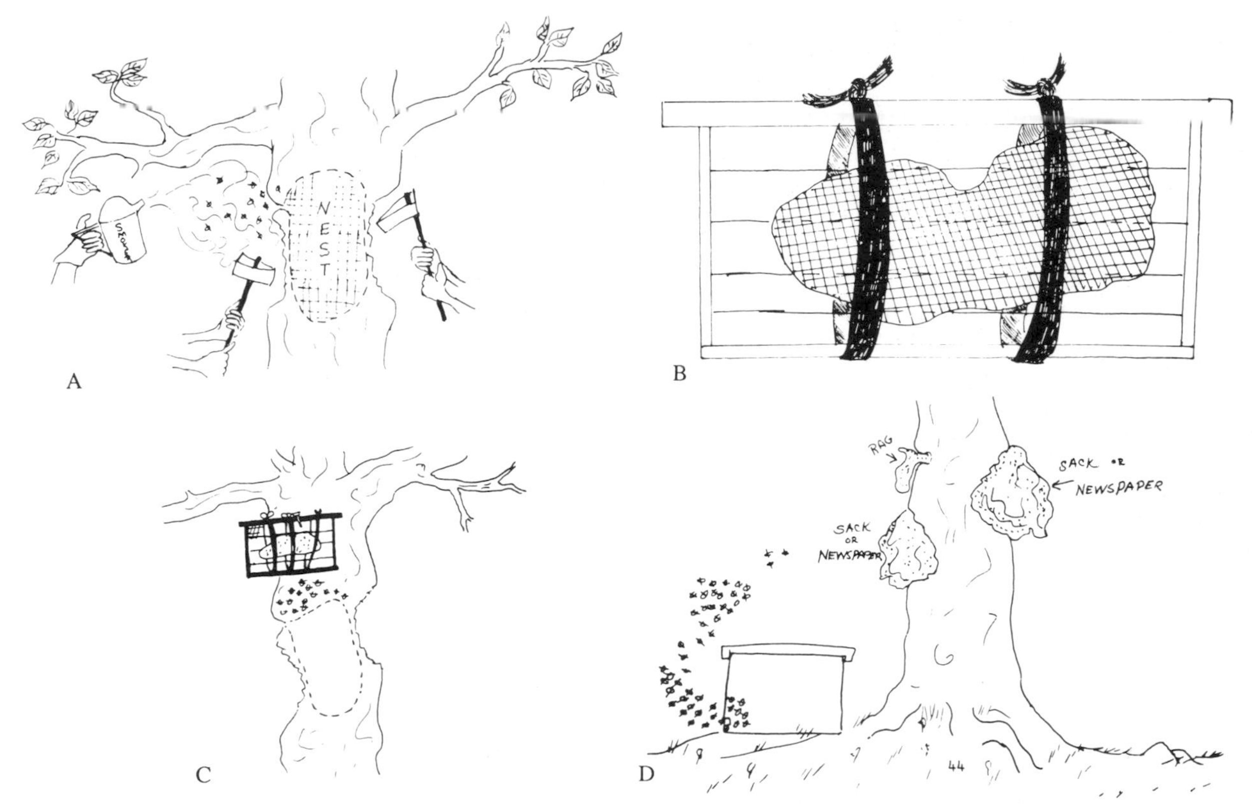

Figure 7.33b Transferring a wild colony to a hive (examples of the many drawings in a book for beginner beekeepers in the tropics, Bianco, 1985).
A Colony nesting in a tree, being smoked while the wood on either side is cut away with machetes, to allow access to the nest.
B Wired empty frame with a piece of brood comb tied in.
C Frame and brood comb fixed above the cluster of bees.
D The colony is now in the hive, and the nest hole sealed up.

not leave the hive. The empty nest hole should be sealed up (Figure 7.33b, D) so that bees cannot return to it.

In the late evening a day or two later, when the bees will have fixed the combs securely, the hive entrance is closed and the frame hive moved to its new site, where the entrance is opened again; see Section 8.15. As soon as the brood has emerged from the patchwork combs, the beekeeper should replace these by regularly built combs.

Transferring a colony from a traditional hive

The traditional fixed-comb hive is assumed to be a horizontal cylinder which can be opened by removing the closure from one end. (If it is supported in a tree, as is usual in many parts of Africa, it must first be lowered gently to the ground; some smoke is blown into the hive entrance(s) before the hive is touched, to pacify the bees, and again just before and after it reaches the ground.) The traditional hive is supported on the ground, with the closure end slightly higher than the other. The receiving hive, containing a few empty drawn-out combs, and with its roof in place, but no floor board, is supported close to the higher end of the traditional hive, and in contact with it. The closure of the latter hive is removed, and its sides are 'drummed' with the hands; the drumming must not be so vigorous that combs are broken. This should induce most of

Figure 7.33c Tying pieces of natural comb into a frame, Rodrigues, Indian Ocean, 1982.

the bees, including the queen, to move from their hive up into the dark interior of the other, where they will cluster on the frames provided. Smoke may be applied to the flight entrance(s) of the traditional hive to encourage the bees to leave it. The combs in the traditional hive are finally cut out, and dealt with as explained in the second paragraph about wild nests.

Any small clusters of bees remaining outside the new hive are shaken or brushed in, and the hive, with its roof and floor board, is placed on a stand as close as possible to the original hive, which is then removed. The final steps to be taken are the same as when transferring from a wild nest.

If the bees to be transferred are on combs in a box (without frames), the process is easier. The lid is prized off, and the combs cut out in convenient pieces and dealt with as above. Most of the bees can be scooped out of the box and put into the hive. Throughout the operation, great care must be taken not to make any move that might injure the queen; the sooner she is found and transferred (with the bees around her) the better. The box is finally removed, and the frame hive stood in its place, the entrance being at the same height and facing in the same direction.

Whatever the source of the bees, a queen excluder may be placed between the brood box and the floor board, so that the queen cannot escape.

Fuller details are given e.g. Root and Root (1940) and F. G. Smith (1960). The more complicated procedure needed to get a colony into a hive from a nest in an *inaccessible* place is dealt with in Section 5.43.2.

7.34 Harvesting honey from hives

Some beekeepers in the tropics and subtropics remove full supers after the end of a main flow, as is customary in temperate zones. There are, however, two reasons why this may be inappropriate or unwise. If temperatures are very high, honey left in the hive for several weeks may deteriorate (Section 7.35.2). Also, in permanently hot areas wax moths are active all the year round and storage of empty combs until the next flow can be very risky. Some beekeepers may not even attempt it, but return supers to the hives as soon as the honey is extracted from them. This is also done where a beekeeper travels with his bees – as some in Pakistan and Australia do – and has spare time to extract honey on the site, in a tent or a mobile extracting plant such as that in Figure 7.34a.

Sections 6.4 to 6.44, on colony management during the honey flow in temperate zones, stress the benefit of providing plenty of ventilation through the hive, and plenty of comb space in which bees can spread out the

Figure 7.34a Caravan containing both honey extracting equipment and living accommodation built by A. Brookes, Guildford, Western Australia, 1967.
Full supers are pushed into the caravan on rollers, through the rear door, and empty supers are offloaded through the door at the side.

nectar they are converting into honey, and thus increase evaporation of water from it. The section also stresses the importance of leaving honey in the hive until the bees seal it in the combs. These points are also important in the tropics.

Honey is harvested from a few combs at a time in parts of south-east Asia where beekeepers use a single-box frame hive and no supers. They remove a few combs, as soon as they are filled or even before, and – having shaken the bees off – spin them in an extractor, and then replace them in the hive. In Vietnam, during a flow *all* combs are removed, centrifuged, and replaced every few days; eggs and brood are said not to be much damaged. The harvest from this system is not honey, but nectar in the process of conversion into honey. It contains much water and, if fermentation is to be prevented, water must be removed from the bulk liquid harvested (Section 13.39), or this liquid must be heated to 'pasteurizing' temperature to kill yeasts (Section 13.35). Beekeepers trained in most parts of the world would disapprove of this harvesting practice. Reasons stated for using it include:

- colonies do not thrive as well in supered hives
- in smaller colonies, mite control is more effective if brood and honey are in the same box
- single-box hives are easier to handle, with the transport available
- capital is not available to buy supers
- out-of-hive comb storage is very difficult because of wax moths
- for the honey flow, many small colonies are pre-

ferred to fewer larger ones, on a variety of grounds: they thrive better; they are easier to manage; the total honey yield is higher; the system increases the sugar allocation to the beekeeper (e.g. in Vietnam).

In general, unless a beekeeper has much experience in the district in which he operates, removal of honey from the brood box can put the colony at risk: if flow conditions change for the worse, the food he leaves behind may be insufficient to maintain the colony's brood rearing, or even to ensure its survival, until he attends to it again.

Whether honey is harvested in supers or in single combs, the bees must be removed from them before the honey is taken out of the apiary. Sections 5.3 to 5.33 describe various methods of removing bees from supers. Some beekeepers in the African tropics remove honey at night, when it is cooler, and the bees are less inclined to fly. Use of a bee-escape involves enclosing many bees in a super until they leave through the escape; meanwhile there may be a risk of overheating the confined bees. Special care must be taken, because the bees may be killed if the escape mechanism becomes blocked, or does not function properly – or if it is put in place wrongly, with the result that bees cannot leave the super. Bee-escapes usually lead the bees down to the hive box below; but in hot conditions, escapes that lead the bees directly outside the hive are likely to be more effective. Section 5.32, which discusses bee repellents commonly used for clearing bees from supers, indicates their effectiveness and safety at different temperatures. Bee blowers (Section 5.33) are effective at high or low temperatures; Figure 5.33b shows one in operation in the subtropics, and their use in the tropics is likely to increase.

When combs are harvested individually from hives, the usual way of removing the bees is to brush them off with a feather or very soft brush, or to shake them off (Section 5.3) or, if the bees are too much inclined to sting, to smoke them (5.21.2). The following plants (among those listed in Table 5.21A as pacifying honeybees) have also been reported to repel one or more species:

A. mellifera (European)	*Lantana camara*
A. mellifera (tropical African)	*Warburgia longimanii*
A. cerana	*Shorea floribunda*
A. dorsata	*Amomum aculeatum*
	Orophea katschallica
	Zingiber squarrosum

Section 5.21.3 gives further details, and Figure 8.61b illustrates the effectiveness of *Amomum aculeatum*. I think that further searches for tropical plant materials having a repellent effect on honeybees might be rewarding.

7.35 Problems with the honey harvested

It is often more difficult for beekeepers to produce good quality honey in the tropics than in the temperate zones. Equipment for extracting, handling and storing honey is not always available, and there can be difficulty in obtaining containers suitable for retail sale. It may not always be easy to maintain high standards of cleanliness in the honey offered for sale. Lack of facilities for extracting, handling and bottling honey has been overcome in some countries by the establishment of co-operatives with a central plant (see e.g. Townsend, 1976*a*) that serves beekeepers in the surrounding area. If beekeepers are paid a premium price for honey that meets required standards of cleanliness and composition, honey quality can be upgraded and maintained.

If and when beekeepers want to export honey on to the world market, they are faced with additional difficulties. They must produce a large quantity of honey each year, preferably of the same type (flavour, appearance and composition), and the honey must conform to standards laid down by the importing country. In the present century the long-standing major importers of honey have been countries in the north temperate zone, especially in Western Europe. Many components of current honey standards were initiated there (Section 15.71), on the basis of data for honeys produced from temperate-zone plants, in areas not subject to excessive temperatures and humidities. (Standards are now being revised to have world-wide application.) Honeys produced in some tropical regions may not conform to these standards. Water content is dealt with in Section 7.35.1, and other constituents in 7.35.2.

Morse (1984) discusses some of the problems in honey harvesting. Separate problems arise when honey from unframed combs is extracted by pressing and squeezing (Section 10.61; Vorwohl, 1976), instead of by centrifuging.

7.35.1 Obtaining honey with a low enough water content

Some components of a honey standard have absolute validity – for instance a water content low enough to prevent the honey fermenting. In humid parts of the tropics, where the relative humidity of the air is very high, even sealed honey may contain too much water, and it is then likely to ferment, especially at tropical storage temperatures.* Section 13.22.2 discusses factors that can lead to fermentation, and also refers to

* In Malaysia, some consumers even *prefer* fermented honey, so beekeepers have little incentive to produce honey with a lower water content.

the water content of honeys produced at high humidities.

In humid areas, beekeepers should always pay great attention to management practices that will enable them to market honey with a low enough water content. It is most important, although it may be tiresome, not to remove any comb for harvesting until all the cells of honey in it are completely sealed, as in Figure 6.42a. Where whole supers are harvested, they should not be removed until the end of the flow. Any incompletely sealed combs should be kept separate, and the honey spun out of the unsealed cells first, before the sealed cells are uncapped. This honey, which must not be mixed with honey to be marketed as table honey, can be used for feeding to colonies. When combs are removed individually while the flow is in progress, in order to extract the honey and return them to the hive to be refilled by the bees, there is a greatly increased risk of harvesting honey with too high a water content.

An example of honey that often has a high water content is that from rubber (*Hevea brasiliensis*); the nectar is produced in extrafloral nectaries. This honey has increased in commercial importance in south-east Asia since the mid-1970s; in India it accounts for nearly half the total amount of honey produced from hives. The water content of honey from sealed combs was found to be 25.0% in Sri Lanka (Fernando, 1978) and 21.5% to 23.0% in southern India (Nair, 1983). Methods have been developed in temperate zones for 'drying' honey (reducing its water content) while the sealed combs are still in the supers (Section 13.32). Where supers are not used, water can be removed only from bulk honey after extraction and this is much more difficult; see Section 13.39.

7.35.2 Obtaining honey with acceptable levels of enzymes and HMF

Honey contains enzymes, including amylase and invertase (Section 13.23.1), in amounts that decrease if the honey is exposed to high temperatures. Honey also contains HMF (hydroxymethylfurfural), whose level increases at high temperatures; see Section 13.26.1. The three substances are regarded as indicators that can show whether honey has been overheated, limits laid down as acceptable in the ECC Directive (European Communities Council, 1974) being:

amylase	not below 8 ppm
invertase	not below 4 ppm
HMF	not above 40 ppm

Section 13.26.1 discusses the suitability of these and alternative values as limits of acceptability.

Overheating of honey may occur in the course of processing and bottling (Sections 13.3 to 13.39), or in hot climates during transport, or if the honey is left in the sun as in Figure 7.35a. Also, and immediately relevant to beekeeping management, honey may become hot in an unshaded hive, and undergo alterations if it is left there for a period before removal. Alterations in honey caused by heat are increased according to the period for which it is exposed to high temperatures.

Honeybees do not regulate temperatures in honey supers as closely as those in the brood nest, which are about 35°; see Table 3.32A. In temperate zones the air temperature is generally below that of the brood nest, between 30° and 35° according to the outside temperature. In the tropics, air temperatures are often above 35°, and temperatures in the supers may then be above 35°, by an amount depending on the efficiency of steps taken to keep the hive cool. The question then arises as to whether this honey should be regarded as overheated, even if no overheating has occurred after it was removed from the hive: honey importers in countries of the cool temperate regions would argue that overheating has taken place. A further point to be considered is the bees' *production* of honey in supers that are at a temperature above 35°. The bees' ability to produce the extraordinarily high sugar concentration in honey (Section 13.21) depends on a combination of factors which include the temperature at which they work, and the solubilities of the sugars involved at that temperature. Many analyses of tropical honeys show them to have a higher water content and lower sugar content than honeys from temperate zones, but the conditions under which these samples were pro-

Figure 7.35a Drums holding 300 kg of honey, stored out of doors in full sun.
Where temperatures are high, honey deteriorates if stored in this way, especially if the metal is damaged.

duced are not known. It might be worth investigating the water content and the sugar composition of honeys produced by tropical and temperate-zone bees at different temperatures.

7.36 Post-flow management and the dearth period

Sections giving information on management after the honey flow, and through to the end of the dearth, include:

6.51	Requeening and queen introduction
6.52	Post-flow management
8.2	Queen rearing
6.53.1	Providing food for a dearth period
11.5, 11.6	Protection against enemies
11.81	Preventing damage by floods (e.g. due to monsoon rains)
4.13, 7.32.2	Protecting bees against excessive heat
6.62.3	Providing water for bees

In most tropical regions there are likely to be two main dearth periods a year, one taking the place of the temperate-zone 'winter dearth', and one due to drought or to excessive rain – during a very hot part of the year. Where the latter dearth is due to heat and drought, high temperatures and wind dry up nectar sources. Where it is due to the rainy season, rain may fall continuously for several days at a time, and even during short intermissions the relative humidity is very high, and the bees do not fly. Any nectar in flowers becomes much diluted, and pollen is soaked or washed away.

Whether this dearth is due to excessive drought or rain:

(a) The bees are active because temperatures are very high, and although they rear little brood and may not fly much, they consume much food. Feeding may therefore be necessary throughout the dearth. Adjare (1984) reckons that in Ghana an average colony consumes 1.4 kg of honey a day during the dearth periods, which are due to excessive rain.

(b) Colonies of *tropical* honeybees may abscond, abandoning whatever brood and stores are in their hive, and moving to a second area, which provides forage at this time. The beekeeper may be able to prevent absconding by transporting his colonies to the second area in advance, and back again to the first area when the time approaches for the colonies to return from the second area. Mammo Gebreyesus does this successfully up and down the Rift Valley in Ethiopia.

(c) Colonies of temperate-zone *A. mellifera*, and of Kashmir *A. cerana*, do not have this habit of absconding. Beekeepers may, however, be able to migrate colonies of these bees to other areas where bee forage is available, and so keep them alive and possibly obtain more honey.

Management of *A. cerana* during the dearth, in so far as it differs from management of temperate-zone *A. mellifera*, is described in Section 7.41.4.

7.37 Circumstances favouring the use of frame or top-bar hives

Most of this book is concerned with keeping bees in movable-frame hives that consist of tiered boxes, but Chapter 10 deals with movable-comb hives which are fitted with top-bars instead of four-sided frames. A comparison is set out below (for use in the tropics) of the Langstroth-type tiered hive with movable frames (MF), and the long single-storey hive with movable top-bars (MTB) described in Section 10.21.

1. MF hives allow colony management at a higher level of technology, with larger colonies, and can give higher honey yields.
2. MF hives are likely to be the only economic alternative if capital is already invested in them, and in equipment for extracting honey by centrifuging.
3. If no capital has been invested, MTB hives require less financial outlay; their construction is cheap (especially if local materials can be used), and involves much less precision work.
4. If the local bees readily fly off the combs and sting (like many tropical African bees), it is easier to handle them in a single-storey MTB hive with full-width top-bars than in a tiered hive with narrow top-bars; see Section 10.21. Full-width top-bars can also be used in MF hives (see Section 10.23, under (b)), but this is not usual.
5. It is easy to harvest a few combs of honey at a time from MTB hives.
6. Where wax moths are very active, as in most of the tropics, it can be virtually impossible to store empty combs through the inactive season away from the hives. Two possibilities are:
   (a) with MF hives, to keep the colonies strong enough, even throughout dearth periods, to protect empty combs stored in supers above the brood box in the hives;
   (b) to store no combs from season to season, but to harvest the combs with honey, as is usual with MTB hives, either selling honey in the comb, or processing honey and wax separately.
7. MTB hives can be suspended out of reach of enemies that attack from the ground (hanging

wires are shown in Figure 10.4c), whereas MF hives must be supported from the ground, on stands.

8. Empty MTB hives can be suspended high up during the swarming season, as catcher-boxes; see Section 7.33.1.

MTB hives have been developed especially in tropical Africa, where intermediate technology and low cost are appropriate, and suspending the hives on wires gives them essential protection.

MTB hives are less common in tropical America, where movable-frame hives were often in use before the bees became Africanized. There are a few MTB hives in Asia, where *Apis cerana* and European *Apis mellifera* are used, but neither of these bees presents management problems in movable-frame tiered hives as tropical *Apis mellifera* does. In Oceania, with no native honeybees, European *A. mellifera* is used in Langstroth-type hives.

7.4 BEEKEEPING IN THE TROPICS USING DIFFERENT HONEYBEES

Tropical honeybees have evolved with the ability to overcome certain difficulties that are inherent in their life in the tropics (see e.g. Table 7.1A):

- at the approach of a food dearth, the whole colony may abscond to an area nearby where flowering is commencing, temperatures being high enough for flight and for comb building;
- the bees can exploit the great variety of forage sources, because they do not become 'fixed' on a single food source to the extent that temperate-zone bees do;
- they fly early and late in the day (when temperatures are lower), and when flowers are producing nectar and pollen;
- they are better able to overcome attacks by wasps and other enemies that flourish in the tropics;
- probably, also, appropriate changes in their rate of brood rearing do not depend much on changes in the photoperiod (day length), which undergoes little seasonal change in the tropics.

Temperate-zone *Apis mellifera* evolved in different conditions, and these bees are not so well adapted to the climate, environment and seasonal cycle of flowering in most regions of the tropics.

7.41 Beekeeping with *Apis cerana*

Apis cerana is native to large areas of Asia; it extends from Iran in the west to Japan in the east, and from the Far East Province on the Pacific Coast of the USSR in the north almost to the southern extremity of the continent in the south – but Asia does not extend beyond the south equatorial zone. Figure 1.2b, Table 2.5A and Sections 1.23.1 and 4.5 provide further details. *A. cerana* is of less economic importance in the north where introduced European *A. mellifera* can be kept successfully, than in the tropical south where European bees do not do well in many regions. (As far as is known, tropical *A. mellifera* has never been introduced to Asia, and in view of experience in the Americas, it is most important that this should not be attempted.) Beekeeping with *A. cerana* was not extended beyond Asia until colonies in hives were taken to Irian Jaya (in the island of New Guinea) in 1985 or 1986; see Section 1.23.1.

Movable-frame hives for *A. cerana* (shown in Figure 7.41a) are discussed in Section 4.5, the end of which lists countries where *A. cerana* is known to be kept in them; see also Table 7.1B. Much of Chapter 4, and most of Chapters 5 to 7, apply to *A. cerana* as well as to *A. mellifera*, provided account is taken of a number of differences between the two species that are relevant to their management. These differences, and those between various types of *A. cerana* in India, are summarized in Table 7.41A. 'Plains type' are characteristic of lower altitudes and the south of the country, and 'hill type' of higher land, especially in the north.

Colonies of most *A. cerana* produce smaller populations and smaller honey yields than those of European *A. mellifera*. They often respond to lack of food or to disturbance by absconding (defined in Section 3.34.1) – a habit which they share with tropical *A. mellifera*. Colonies of most *A. cerana* are liable to invasion by wax

Figure 7.41a An apiary of *Apis cerana* in a coconut plantation, with abundant water, near the mouth of the Mae Khlong river, Thailand, 1986.

Table 7.41A Comparison of some colony characteristics of *Apis cerana*, and of introduced temperate-zone *A. mellifera*, in India

Based on Shah and Shah (1982*a*), Phadke (1987) and other data.
Table 4.25A gives cell widths for many types of honeybee.

	A. cerana			European *A. mellifera*
	plains	*hill*	*Kashmir*	
no. worker cells/cm² (both sides of comb)	12.4	10.0	9.5	8.6
peak adult population (thousands)		18–22	60–70	60–70
relative size of colony at which swarming occurs	small	small	medium	large
occurrence of absconding	yes	yes	no	no
robbing tendency	high	high	moderate	low
response to examination of combs	strong	strong	slight	slight
response to smoke		'irritation'	quiet	quiet
honey yield	v. low	low	high	medium
queenless period before laying workers develop	?	1 week	20 days	1 month
effective defence against wasps	yes	yes	yes	no
attacks by wax moths	frequent	frequent	none (see text)	rare
Queen				
age at mating (days)	?	4–8	4–6	6–13
introduction easy	?	yes	yes	no
daily egg-laying capacity	300–500	500–800	1400–2000	900–1400
drone laying if unmated	?	rare	rare	common

moths if they are not large enough to cover all the combs, and their response to this may also be absconding. They do not respond to smoke or beekeeping manipulations as temperate-zone *A. mellifera* do, but tend to run on the comb, and excessive interference may lead the colony to abscond.

Brood mortality can be much higher in colonies of *A. cerana* than in those of temperate-zone *A. mellifera*. In India, Woyke (1976*b*) found that during a major flow up to 95% of the larvae were reared and sealed (as in *A. mellifera* colonies in Poland), but during minor flows only 50%, and during a dearth none, because the workers ate some or all of the larvae. Such behaviour recycles the colony's protein reserves in times of food shortage. (The eating of young larvae of diploid drones that are produced in inbred colonies has a different basis; see end of Section 2.6.1.)

Data from Kashmir in the north of India are included in Table 7.41A to draw attention to the unusual *A. cerana* of that region. Kashmir has a large sheltered valley at an altitude about 2000 m, almost surrounded by high mountains of the Himalayas and ranges to the west. In many ways the bees native to the valley are more similar to temperate-zone *A. mellifera*, whereas *A. cerana* farther south have more similarities to tropical *A. mellifera*. The Kashmir bees are relatively large, and build large colonies (as populous as those of temperate-zone *A. mellifera*), which can be kept in Langstroth hives and produce 15 to 25 kg of honey a year. Like European *A. mellifera*, the colonies do not swarm until their populations are quite large, and they do not abscond; their responses to smoke and to colony examination are also similar to those of *A. mellifera* (Shah & Shah, 1982*a*; Verma, 1987).

Proceedings of a conference on beekeeping with *A. cerana* (Kevan, 1989) provide much information relating to south-east Asia. Verma (1987) discussed in some detail the biology of *A. cerana* in relation to beekeeping development programmes. General information on beekeeping management with *A. cerana* in certain countries is available:

Burma	Maung Maung Nyein (1984)
China	Xiao (1982)
India	Thakar (1976)
Malaysia	Phoon (1983)
Nepal	Saubolle and Bachmann (1979)
Sri Lanka	Wickramasinghe (1983)

Accounts that include descriptions of the annual beekeeping cycle are referred to below.

7.41.1 The annual beekeeping cycle

For effective beekeeping with *A. cerana* in any particular area of the tropics, it is necessary to understand the seasonal cycle of the colonies there. Examples are quoted below from India, based on information provided by Phadke (1982), and also from Ko Samui, an island I visited in 1986, when Dr Siriwat Wongsiri translated during our discussion with the beekeepers.

India stretches from about 8°N to 37°N, and the greater part of the beekeeping area lies within the tropics. The following annual cycle is representative, although the timing varies from place to place. There is a short 'winter' dearth around the end of the calendar year; at higher latitudes and altitudes this may be cold, in which case colonies form a winter cluster. The dearth ends in early February; it is followed by nectar and pollen flows which stimulate increasing colony growth, leading to reproductive swarming if this is not prevented, and to honey storage and a harvest for the beekeeper. The monsoon dearth ('summer' dearth) starts towards the end of May, and continues through June, July and part of August. Colonies are likely to abscond from their hives in June–July if the beekeeper has not taken effective action to prevent it. Towards the end of August pollen may become available again, but there is usually no nectar until some time in October, after which flows continue, and the colony is again large enough to store honey, which is harvested. By the end of December the winter dearth starts, and it lasts through January.

In the hot dry plains of the Punjab in northern India, the summer dearth (June to October) is not due to rain, but to drought and heat with temperatures up to 45°. *A. cerana* does not survive in the wild, and for this reason European *A. mellifera* colonies were introduced (Atwal & Goyal, 1973). Provided the beekeeper fed them as necessary during the dearth season, they were successful – because they could exploit the prolific flows from agricultural crops grown there; see Section 7.44. Salvi (1975) introduced *A. cerana* from nearby foothills; he found that colonies followed a seasonal cycle similar to that described above, but it proved difficult to keep the colonies alive through the hot dry period except by migrating the hives to the cooler hills. Left in the plains, colonies in hives tended to abscond, as wild colonies do, but unless they could reach long-lasting forage, they would not survive.

In Guangdong Province in southern China, about 90% of the colonies are *A. cerana*. Wongsiri et al. (1986) reported that *A. cerana* bees foraged for 2 to 3 hours longer each day than introduced *A. mellifera*, that they made better use of forage where this was scattered, and were better able to survive hot summer dearths.

Ko Samui lies off the east coast of peninsular Thailand, at 9°N. It is only about 11 km across, but is noted for its (traditional) beekeeping with *A. cerana*. The apiaries are on the cultivated coastal plain around the edge of the island, where the people live. Coconut palms and other plants, both wild and cultivated, pro-

Figure 7.41b Harvesting honey from a box hive of *Apis cerana* on Ko Samui, an island off the east coast of Thailand, 1986.
In his right hand the beekeeper holds the long knife used for cutting combs out, and in his left a smouldering piece of coconut husk, on which he blows to regulate the production of smoke and to direct it where he wants it.

Figure 7.41c The beekeeper in Figure 7.41b offering his honey combs for eating.

vide flows, and colony populations grow from February right through to April and May, when honey is harvested (Figures 7.41b and 7.41c). As in India, the colony growth period culminates in reproductive swarming. Minor flows and honey storage may continue until the start of the rains in October. This heralds the dearth period on the coast, which lasts until January. In October colonies respond to the dearth by absconding to the jungle-covered hills in the centre of the island, where forage is becoming available. They build nests and live in the jungle until February, when the forage fails there, but pollen and nectar become available again on the coast, and beekeepers put out bait hives to catch the swarms returning from the jungle.

In Samui Island there is thus forage at different periods in the year, in areas close enough for the colonies to move from one to the other, and so take advantage of both of them. In each area (plains and jungle) the forage period is sufficiently long for the colonies to build new nests there, and to rear brood and maintain colony strength; when the forage fails, it is restarting in the other area and the colonies abscond to it.

Other descriptions of seasonal beekeeping cycles with *A. cerana* include:

India (general overview)	Singh (1962)
India (Kashmir)	Shah and Shah (1976)
Thailand (coastal plain west of Bangkok)	Wongsiri and Tangkanasing (1986).

In the foothills of northern India, S. Verma (1983) and Mattu and Verma (1985) studied the annual foraging cycle of *A. cerana*, and Hameed and Adlakha (1973) the annual brood-rearing cycle of both *A. cerana* and *A. mellifera*.

7.41.2 Management for colony growth, including swarm prevention

The broad principles of management to encourage rapid colony growth after the winter dearth are similar to those for temperate-zone *A. mellifera* (Section 6.3 and subsections), except that temperatures are almost everywhere high enough for colony inspections and manipulations. As colonies grow larger, management by the beekeeper is needed to prevent or control reproductive swarming. Colony conditions that contribute to the initiation of queen rearing in preparation for swarming are set out on the left of Table 6.33A.

The beekeeper's actions to prevent swarm preparations by *A. cerana* colonies are similar to those described in Section 6.33.1. They include using young queens; preventing the overheating of hives by shading them, ensuring adequate hive ventilation, and supplying water near the apiary; also providing ample space in brood box(es), and in the hive as a whole, well before the bees are able to fill it, in so far as this will not endanger colonies through wax moth damage. Removal of combs of brood, honey and pollen is part of the normal procedure of making new colonies (Section 6.34), which should be done during this period – as early as possible according to Salvi (1975), so that they have time to build up populations and stores before the next dearth. This is also a good time for queen rearing (Section 8.28).

There often seems to be great difficulty in obtaining *A. cerana* colonies large enough for efficient honey production. Colonies tend to make swarm preparations while they are still quite small. Also, the peak population of some *A. cerana* quoted in Table 7.41A is only one third that of European *A. mellifera*. In Pakistan, Hameed and Adlakha (1973) found that at the end of April, well on in the brood-rearing season, *A. cerana* colonies contained only 38% as much brood as those of *A. mellifera*. Latif et al. (1960*b*) found that the production and use of two-queen colonies helped to check swarming, and increased honey production – the more so, the earlier in the season the two-queen colonies were established. But there were difficulties in obtaining the extra queens, and in establishing the two-queen colonies.

Another method is recommended by Fang (1981) in China, but it involves much work. In spring, half the colonies are set aside as laying colonies (L), the rest being either for rearing (R) or caretaking (C) colonies. A succession of empty combs is put in L which has a prolific queen, and each is moved to R when full of eggs. When the resulting brood is sealed, the comb is moved to C where the quality of the queen is not important. It is reported that none of the L, R or C colonies swarmed.

Among the swarm prevention measures detailed in Section 6.33.1, the following are probably often not sufficiently attended to in *A. cerana* colonies:

- providing ample space for both brood rearing and honey storage;
- requeening colonies regularly, so that all colonies have queens less than 1 year old, which produce more brood than older queens; see Section 8.28 for rearing *A. cerana* queens, and 6.51 for requeening procedures.

7.41.3 Management during the honey flow

The major honey flow season usually follows after the main colony growth period; it may occur in April/May in the northern tropics, or may last much longer. Section 6.41 explains the best timing for adding empty honey supers, and for removing full ones, and 7.35 to 7.35.2 describe the management necessary to obtain good quality honey in areas where the temperature or the humidity is high.

A. cerana bees are much inclined to rob weak colonies (Table 7.41A), and for this reason Shah (1983) recommends harvesting combs of honey while some nectar is still coming in – but removing only frames in which the comb is more than three-quarters sealed – after smoking and brushing bees off the combs. Sections 5.3 to 5.33 describe this method, as well as those normally used for temperate-zone bees where complete supers are cleared, and then taken off the hive. In India, Naim and K. G. Phadke (1972) compared the results of removing litchee honey (a) all at the end of the flow, and (b) in two supers as soon as they were full, replacing them in the hive after extracting honey from them. In all three years, (b) gave higher colony yields than (a): means were 12.8 kg and 8.9 kg, respectively.

With *A. mellifera* colonies, the presence of empty combs stimulates nectar collection and storage during a prolific flow (Section 6.41). With *A. cerana* many authors stress the need to remove full combs of honey frequently during a flow, in order to make colonies produce more, but it is not clear whether the stimulus to further storage is the colonies' consequent lack of honey, or the presence of empty combs used to replace the full ones.

The honey flow period is the best time to get the bees to draw out new brood combs as well as honey combs; see Section 6.43.

Migration of hives is often a profitable way of extending the honey flow period, provided there are roads and vehicles that make it possible. Recommendations and precautions for moving the hives are set out in Sections 8.1 to 8.15. If temperatures are high, combs full of honey in the brood box should be removed beforehand, or they may break during transport (Salvi, 1975). Ahmad et al. (1984*a*) compared results of four migration schedules within Pakistan, to take advantage of different flows:

mean honey yield per surviving colony (kg)	16.1	4.6	3.3	2.8
no. of colonies (of 5) that died owing to dearths	0	2	3	3

Less is known about differences between the honey-getting ability of strains of *A. cerana* than of *A. mellifera*. Ahmad et al. (1983) compared the performance, in various parts of Pakistan, of colonies that had different geographical origins. In general, colonies from Swat (hills) performed better than those from Margalla (foothills), producing 40% to 70% more brood and slightly more honey. Section 8.28 refers to procedures for selecting strains of *A. cerana* that are more profitable for beekeeping than others.

7.41.4 Management after the flow, to the end of the next dearth

Section 6.5 and subsections discuss post-flow management, including profitable ways of using the peak colony populations the beekeeper has obtained for the flow, which can be an economic liability afterwards because of their high rate of food consumption. Some of the bees may be useful for making new colonies and for queen rearing, if such operations were not done in the period of colony growth. Migration of hives to further honey flows may be economic in some areas.

In the summer dearth in India, when air temperatures varied from 36° to 46°, Kapil (1960) found that colonies maintained the central part of the space they occupied at temperatures between 35.0° and 36.0°. In the winter dearth, brood temperatures were between 34.5° and 35.5°. He concluded that the decrease in brood rearing during the hot summer dearth is not due to the bees' inability to control the brood-nest temperature, but to effects of wind and high temperature in drying up nectar sources and preventing foraging.

Absconding

Colonies of *A. cerana* are likely to respond to hazards that endanger their survival by absconding (Section 3.34.5). Absconding is especially likely to occur during the summer dearth period, after most of the stores left in the hive at the end of the main flow have been used up, and most of the brood has emerged from the comb as adults – which can fly. In northern India, Rana (1985) found that absconding colonies left behind them small amounts of brood and stores, occupying areas of comb less than:

49.5 cm² eggs	704 cm² honey
1.0 cm² larvae	17.5 cm² pollen
8.0 cm² sealed brood	

The eggs left are viable, but they represent only a small investment by the colony compared with brood that has been fed to an advanced stage.

In India, up to 75% of colonies may abscond each year, according to Woyke (1984*b*), 30% to 40% (Verma, 1987), 50–75% in southern India (Olsson, 1989). Contributory factors include: 6 to 10 days without foraging; infestation by wax moths or beetles, or other pests; brood comb more than 1 year old, or the provision of too much foundation; disturbances or manipulation; unsuitable hive location; other factors causing stress in the colony (Olsson, 1989).

The first main step to prevent absconding is to ensure that colonies always have a sufficient reserve of honey in the hive; Singh (1962) suggests a minimum of 3 to 7 kg, and Olsson 5 kg. A reserve of pollen is also important, so that brood rearing can continue; colonies are less likely to abscond if they would have to desert brood. Perhaps the provision of pollen substitutes in the hives might help; see Section 6.32. Some of the conditions likely to prevent absconding are similar to those likely to prevent preparations for reproductive swarming (Section 7.41.2), including especially the presence of a young queen. A Thai beekeeper told me that he requeens colonies every six months to prevent absconding. Olsson (1989) suggests that the flight entrance of hives should be reduced by using vertical bars leaving only 6-mm gaps.

Winter dearth

In preparing for the winter dearth it is necessary first to unite small colonies as explained in Section 6.35.2; colonies for wintering need to be strong before the dearth starts, with many young bees. Adequate feeding is essential; Singh (1962) recommends that each colony should start the dearth with at least 7 to 9 kg of stores in the hive, but this may not always be enough. Special care must be taken that syrup feeding does not initiate a robbing attack: feed late in the day, and not just after hives have been opened and many bees are flying; do not spill syrup outside the feeders; if possible, reduce the size of hive entrances while feeders are on the hives. Hives should be well ventilated during the dearth, and colonies should not be disturbed or examined. Packing with thermal insulation is not usually recommended.

7.41.5 Diseases, parasites and enemies

Diseases, parasites and enemies of honeybees, including *Apis cerana*, are the subject of Chapter 11; a summary is given in Table 11.1A. Akratanakul (1987) has written a practical guide to them that refers specifically to Asia. Some ways in which *A. cerana* differs from *A. mellifera* are indicated here.

Certain viruses infect *A. cerana* but not *A. mellifera*: *Apis* iridescent virus, the Indian strain of Kashmir virus, and the Thai strain of sac brood virus – which was first reported from Thailand in 1976, and reputedly killed more than 90% of the colonies there and in Burma, Nepal, Pakistan and parts of northern India (Verma, 1987); it started to decline again after 1 to 2 years. *A. cerana* is, however, said to be less susceptible than *A. mellifera* to sac brood, as well as to European foul brood, in China (Wongsiri et al., 1986). Nosema disease (Section 11.31) is common in *A. cerana* in some areas, including northern India, and may be a response to stress.

Table 11.4A lists mites parasitic on *A. cerana*. *Euvarroa haryanensis* and the pollen-feeding mites *Neocypholaelaps* spp. have been found in large numbers on individual bees, and *Pyemotes herfsi* can attack the bees' mouthparts (Section 11.43). *Pseudoacarapis indoapis* has also been reported. In general, the beekeeper faces a special difficulty in dealing with heavily parasitized colonies of *A. cerana*, because they tend to abscond, whereas those of *A. mellifera* do not. However, colonies of *A. cerana* are much less injured by infestation with *Varroa jacobsoni* than colonies of European *A. mellifera*. Peng et al. (1987) found a reason for this, in addition to the shorter development period of *A. cerana* (see Section 11.42). Infested *A. cerana* workers exhibit a specific cleaning behaviour to rid their bodies of the mites: they groom themselves and other workers, biting and killing the mites and removing them from the hive, all within a few minutes (or even a few seconds). *A. mellifera* does not show this active cleaning behaviour, and Peng and her colleagues have been experimenting with the insertion of (a) adult *A. cerana* workers into infested colonies of *A. mellifera*, to remove the mites from them, and (b) infested *A. mellifera* brood into colonies of *A. cerana*. (It has been known for many years that mixed colonies can be formed under certain conditions; see e.g. Muttoo, 1957.)

A. cerana colonies must be protected against predators. Insects such as ants and wasps (Sections 11.51–11.53) may develop very high populations, especially during the honey flow season. Certain mammals and birds in an area may present minor or major hazards (11.54–11.56).

Infestation by wax moths

Infestation of combs by the greater wax moth, *Galleria mellonella* (Section 11.61), can be very troublesome with *A. cerana* beekeeping, especially when colony populations decrease during the summer dearth, or as a result of the beekeeper's operations. It is one of the main causes of absconding. An important protection in the summer dearth is the maintenance of strong enough colonies throughout the period, housed on only as many combs as the bees can cover. Wild colonies remove comb they cannot cover, but it is more difficult for hived colonies to do this (Muid, 1988). In Punjab, India, Singh (1940) reported that during a dearth, especially in summer, hived colonies commonly tried to do so. They gnawed holes in them, starting with the outside combs and proceeding from the bottom upward; old dark combs were the first to be attacked. *A. cerana* leaves such gnawed-off pieces on the hive floor, and methods for getting rid of them are mentioned near the end of Section 4.5. During the winter dearth, colony populations are usually smaller, but, the weather is not so hot, so wax moths are less active.

Protection of stored combs against wax moths is discussed in Section 11.61, and see also 7.37, item 6.

7.42 Special features of beekeeping with tropical African *Apis mellifera*

General features of beekeeping in the tropics are set out in Section 7.3 and its subsections, and most of what is said there applies to tropical *Apis mellifera*. Seasonal cycles in the tropics are discussed in Sections 3.52 and 7.41.1. Beekeeping with tropical African *A. mellifera* has specific features discussed here; some of them apply also to beekeeping with Africanized *A. mellifera* in the Americas, the subject of Sections 7.43 to 7.43.2.

7.42.1 A beekeeping tradition from early times

Egypt in North Africa may well have been the first place where honey was harvested from bees kept in hives (Section 1.42.1). It is linked by the Nile valley to tropical Africa, where there has been a rich and varied tradition of beekeeping, many of the hives having a similar shape to those of Ancient Egypt (Seyffert, 1930). Large numbers of hives were kept, and much honey and wax produced. Beekeepers had a special status in their tribe, and their ownership of hives was respected; some of them used quite clever methods of management (Nightingale, 1983). Honey was also collected from wild nests, especially by some tribes, and it had an important place in community life – for instance for the production of fermented drinks (*tej*, *pombe*, etc.).

In tropical Africa the widely scattered human population was accustomed to the presence of bees, and to the fact that colonies were readily alerted to sting (Section 1.22). In much of the lower land, at altitudes below 400 m – especially where there was tree cover – there were tsetse flies, and cattle rearing was severely constrained where infestation was high. However, many of the trees provided forage for bees, and sites for hanging hives, and a harvest could be got from beekeeping even in areas where cattle – and horses – could not be kept. So, compared with tropical America, there were few valuable domestic animals at risk from bee stings; see Section 7.43.1.

Empty hives were set out in trees by traditional beekeepers, in appropriate places to attract swarms. The beekeepers knew when to expect the swarms, which included both reproductive swarms and absconding swarms that moved from one forage area to another (Sections 3.42, 3.43). In an area of poor bee forage, many of the hives would remain unoccupied. In Kenya, ten-year records of traditional hives by Mwangi (1985) showed that, at any one time, 98% were occupied by bees in a cultivated area but only 36% in the arid north-east.

7.42.2 Modern beekeeping experience in tropical Africa

Beekeeping with traditional hives – described for instance by F. G. Smith (1960), Silberrad (1976) and Nightingale (1983), in Tanzania, Zambia and Kenya, respectively – has now decreased. Much effort has been expended in encouraging more modern types of beekeeping, and the choice of hives has varied according to the region, the type of bees used, and also the background, training (for instance in development programmes), and inclination of individual beekeepers.

Movable-frame hives

F. G. Smith's *Beekeeping in the tropics* (1960) deals extensively with the use of Dadant and Langstroth hives in Tanzania, and with bee houses as well as outdoor apiaries; Silberrad (1976) describes the use of Dadant hives in Zambia. In the present book, general management of tropical African bees in tiered frame hives is included in Sections 7.3 to 7.37, and Sections 7.52 to 7.52.2 deal with the placement of the hives in outdoor apiaries and in bee houses, which are greatly valued by beekeepers who use them. In much of South Africa, tropical African bees have been kept commercially in Langstroth hives since the 1920s or earlier. The manual *Beekeeping in South Africa* (Anderson et al., 1983) is based entirely on beekeeping with these hives, and the world's record honey yield from a single hive (960 kg in 8 months, Section 8.29.1) was obtained with a multiple-queen colony in a Langstroth hive.

Large colonies in Langstroth hives in a hot climate need much ventilation, and therefore large entrances. One South African beekeeper uses an inverted floor board at the top of the hive (instead of the normal insulated roof plus inner cover); as well as allowing ventilation this provides an upper entrance, and the bees do not close it with propolis although they may cluster to form a curtain across it (Fletcher, 1979). Section 4.41 discusses a number of other roof arrangements for hot conditions.

Even with modern hives and management methods, the annual loss of colonies in South Africa averages 20% to 30% and may reach 50% due mostly to absconding, queenlessness and theft (Fletcher, 1979). The hives are restocked by setting out bait hives (Section 7.33.1) to catch swarms. (In South Africa both reproductive and absconding swarms occur mainly during the heavy flow from saligna gum, *Eucalyptus grandis* introduced from Australia, and one beekeeper catches 2000 swarms each year.) Such a beekeeping system, which follows the traditional style, allows no strain selection or maintenance. A better system would be to kill the queens in the swarms, and replace them by reared queens, of strains that show low rates of swarming and absconding. Section 8.27 refers to Coleman's successful use of this system in Zimbabwe.

Top-bar hives

In tropical Africa there have been many failures with Langstroth-type tiered hives, and Chapter 10 describes an alternative: movable-comb hives with top-bars. These hives have been used especially in beekeeping development programmes, and have proved successful in some areas where tiered frame hives were not. In tropical Africa a number of small holes are nowadays often used for the hive entrance instead of a long slot (Section 10.4), to prevent the entry of the large hive beetle.

Section 7.37 discusses relative merits of top-bar and frame hives. One important benefit of top-bar hives for tropical African bees is that colonies can be kept in a single long box, with full-width top-bars, and when any comb is being inspected, the bees are enclosed except for the slot from which the comb has been withdrawn. Paterson (1988) has shown that such full-width top-bars, without frames, can be used in a tiered Langstroth hive, as explained in Section 10.23.

Handling tropical African bees

Some beekeepers seem to find it best to smoke colonies heavily when inspecting them, and others smoke very lightly. Smoking the flight entrance after reassembling the hive, as well as before opening it, helps to prevent bees 'following' the beekeeper when he leaves the site (Fletcher, 1979). One guiding principle is to move and operate quietly and gently, and in a planned way so that what must be done in one apiary or bee house is finished as quickly as possible, which is usually easier if two or more beekeepers work together. The chance that bees in the apiary will become aroused is then less. Clauss (1982, 1983), in Botswana, has been a strong advocate of extremely gentle, patient and 'respectful' working, keeping the bees comfortable, and opening hives only when the weather and other conditions are just right for it. He finds that little protective clothing is needed. Not all his advice is practicable for commercial beekeepers with many hives, and full protective clothing is commonly necessary (Figure 5.11a).

Peterson (1985) has stressed the great variability of the bees' behaviour: between different regions, different seasons of the year, and as a result of other factors. For instance in Sierra Leone, and generally in Ghana, she found that bees were more likely to attack in the dry season, but in Mali at the start of the rains. In Uganda the bees were relatively quiet and only a veil was needed for protection; with Africanized bees in Venezuela, full protective clothing was needed, and even then there were difficulties. Two common features of all the above bees were their greater 'aggressiveness' around midday (11.30 to 14.30 h, or longer in hotter climates), and a difficulty in bringing them back under control once they were aroused. Such arousal may well reduce the honey production of the colonies (Fletcher, 1979), in addition to making things unpleasant for the beekeepers.

Some of the beekeeping experience with Africanized bees included in Section 7.43.2 may also be useful to readers working with tropical bees in Africa. In particular, the comparative data on Africanized bees and European bees in Table 7.43B are likely to be relevant.

7.43 Special features of beekeeping with Africanized *Apis mellifera*

7.43.1 A beekeeping history dating from 1957

In 1956 W. E. Kerr had made a search in Southern Africa for honeybees that might do better in Brazil than those descended from temperate-zone European bees introduced in earlier years (Kerr, 1957). He succeeded in bringing 63 queens alive to Piracicaba, S. Paulo, Brazil; all were tropical African bees now commonly referred to as *Apis mellifera scutellata* (Section 1.22.2). Of queens that successfully headed colonies, one was from Tanzania and 48 were from South Africa. Colonies headed by 26 of these queens proved to be 'the most prolific, productive and vigorous bees' that Kerr had ever seen.

In 1957 some colonies swarmed (Gonçalves, 1974/1975), and as a result there was hybridization between *A. mellifera* from tropical Africa, and those already present. The resultant bees, being different

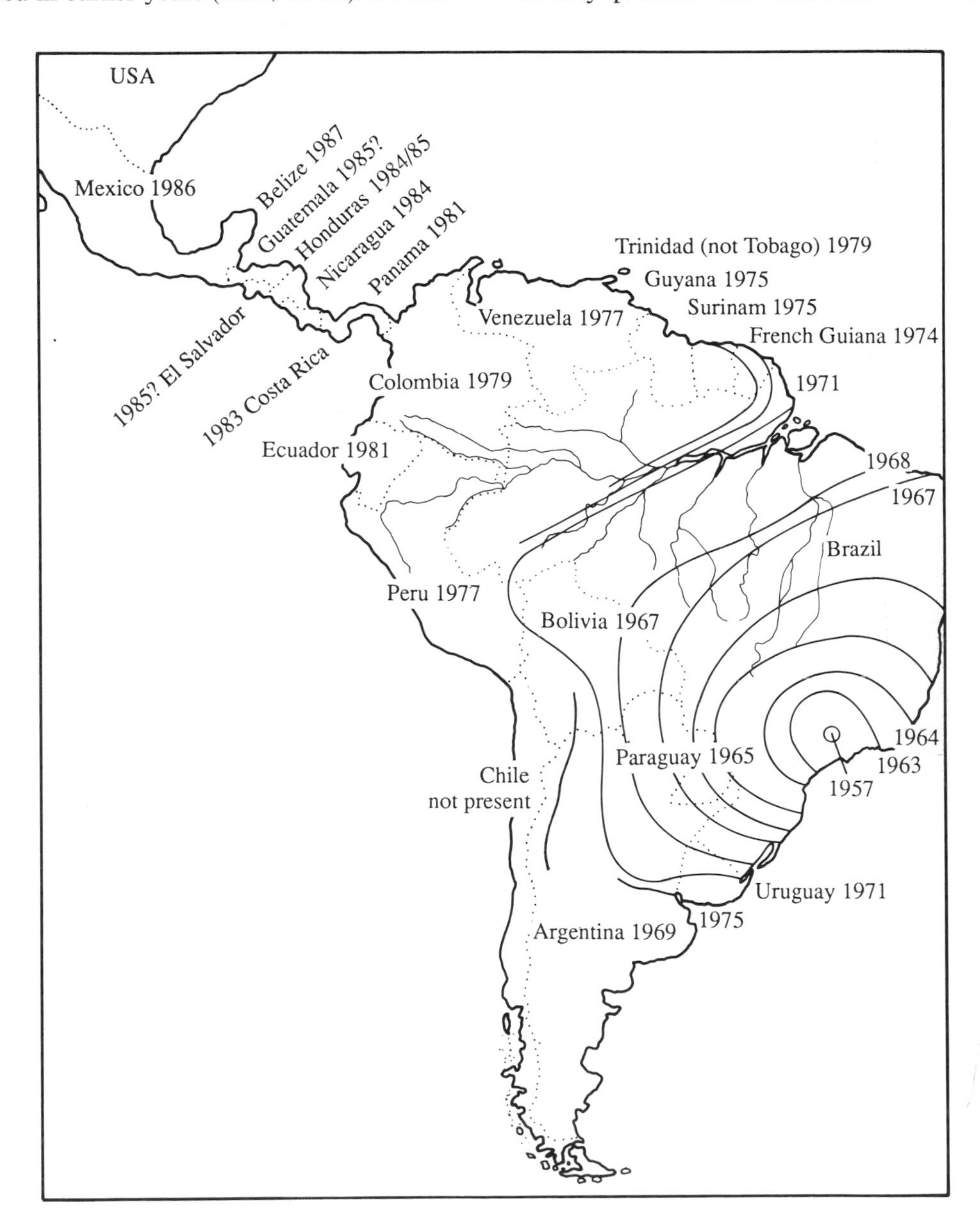

Figure 7.43a Chronology of the spread of Africanized bees in South and Central America.
Lines of advance until 1975 (Taylor, 1977), and reported year of subsequent entry into further mainland countries (Table 7.1B). Of the Caribbean islands, the bees are in Trinidad only (not Tobago).

from the local ones, were referred to by beekeepers as 'Africanized', and the term has been generally adopted. In South America many of the enemies in Africa that attacked the bees were absent. So was the tsetse fly, and cattle and horses were kept by many people, who were unaccustomed to bees that were readily alerted to sting. Moreover a comparatively high proportion of beekeepers were of European descent, and kept their bees in movable-frame tiered hives. These are not ideal for Africanized bees, and colonies in top-bar hives of the type used in tropical Africa (Section 10.21) do not sting as much. But colonies have smaller populations, and produce less honey, in the top-bar hives (Rinderer, 1987). Beekeepers are therefore discouraged from changing, and many have capital invested in tiered frame hives; see Section 7.37.

For reasons explained below (Section 7.43.2), the progeny from the hybridization formed a population that achieved dominance over the European bees, in the immediate neighbourhood and in a continually increasing area from which the bees spread by swarming, the 'front' advancing by 300 to 500 km a year. In the next thirty years these Africanized bees spread throughout almost all South and Central America, as shown in Figure 7.43a. By 1988 they had reached 21 countries, from Argentina in the south to Mexico in the north; Table 7.1B lists the year in which they were first reported in each. The dramatic story of the spread of Africanized bees through South and Central America has been told many times (e.g. Gonçalves, 1974/1975; Taylor, 1977). Information for individual countries will be found in publications cited in Appendix 2.

The southward spread of Africanized bees is limited by the less tropical climatic conditions, and the bees are not found in Chile. In Argentina they seem to be confined roughly to areas: where there are no more than 6 to 8 weeks with mean temperatures below 10°, and only short periods with temperatures as low as −10°; and not more than 60 days a year with freezing temperatures, or 150 days between the first and last frosts (Taylor & Spivak, 1984). However, at the edge of the bees' suggested southern distribution, Krell et al. (1985) found that Africanized colonies survived the winter as well as European colonies at 1400 m, so they may be able to live farther south than had been predicted, provided other conditions are suitable. Africanized bees have also prospered in the mountains of Colombia at 2000 to 4250 m, where temperatures were below freezing every night and below 10° in the day (Villa, 1986).

The spread of Africanized bees northward, into and through Central America, has not been impeded by cold. Winston (1979*b*) discusses some of the bee-

Table 7.43A Effects of 'Africanization' of honeybees in Trinidad

The first entry refers to feral swarms destroyed by the Apiaries Unit. $TT = Trinidad and Tobago dollar.

	1978	*1979*	*1980*	*1981*	*1982*	*1983*	*1984*	*1985*	*1986*[a]	*Total*
no. swarms destroyed		12	50	133	613	737	3777	2993	2186	10 441
no. people stung				10	200	150	473	528	386	1850
no. people died					1		2		1	4
no. animals died				2	17	27	362[b]	40	29	477[b]
										Change
no. beekeepers	419						373		295	−30%
no. colonies	7209						4920		3985	−45%
honey:										
production (tonnes)	20.2						12.1		9.1	−55%
yield (kg/colony)	18						16		15	−18%
retail value (1000 $TT)	1453						1679		1379	−5%
price per kg ($TT)	72						140		150	

[a] up to 20 November only; [b] includes 300 head of poultry

keepers' problems with these bees: for instance much swarming and absconding from hives, with a consequent reduction in hive populations and of honey crops, and the beekeepers' difficulty in obtaining apiary sites because the bees are more 'aggressive'. Also, beekeepers in many of the countries lacked experience in queen rearing and requeening; Wiese's book (1985) has a chapter on these operations for Africanized bees.

Table 7.43A gives an indication of the course of events in Trinidad, where detailed records have been kept. It is an island with no land borders, and is a closed system except that two promontories reach within 15 and 19 km of Venezuela in mainland South America, and there are small islands in between. Swarms of Africanized bees first arrived from the mainland in 1979. It was official policy of the Ministry of Agriculture to destroy feral swarms, and the Apiaries Unit carried out the work, with the results shown. Honey production has decreased since 1979, due to the reduction of beekeeping rather than to a lower colony productivity; however, as a result of price increases, the commercial value of the island's total honey production has not altered much. Tobago is 50 km NE of Trinidad, and Africanized bees have not yet been reported there, or in any other island in the Caribbean.

In the USA there has been much consternation at the probable arrival of Africanized bees across the border from Mexico, which is expected to occur by 1990. The impact of the bees has been assessed (e.g. Taylor, 1977, 1985; McDowell, 1984), and it is considered that the bees might well spread east along the warm coastal area to Florida and beyond, and north along the Pacific coast at least to San Francisco. In September 1987 the Africanized Honey Bee Cooperative Program in Mexico was established, funded by USDA and Mexican authorities (Rinderer et al., 1987*b*; Tew et al., 1988). The Sierra Madre mountains channel the bees' movement along the coastal strips, and a designated Bee Regulated Zone has been established on both the Pacific and the Gulf coasts (Figure 7.43b), each with an Operational Unit. Bait hives are used to monitor bee movement west of the Isthmus of Tehuantepec, and between the two coastal zones. The aims of the Program are to slow down the northward movement of the bees through the Bee Regulated Zones, and to modify the genetics of the bees as they cross them. Each zone covers some 5500 km², and together (in 1988) they contained about 38 000 colonies of European bees. Various techniques are being employed to reduce the population of Africanized bees, e.g. catching swarms of Africanized bees in bait hives, and destroying them. Populations of European bees are being increased, and Africanized colonies being replaced by European ones: by good management of European colonies, by rearing large numbers of European drones, and in other ways. It is still too early to assess the success of this scheme.

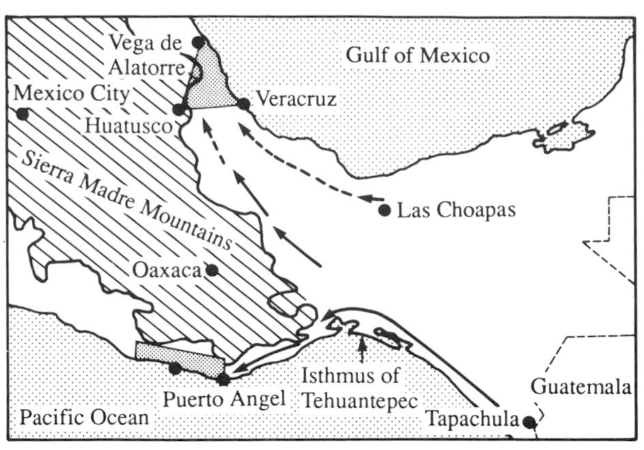

Figure 7.43b Bee Regulated Zones in Mexico, each with an Operational Unit (adapted from Tew et al., 1988). Arrows show the movement of Africanized bees in May 1987 (full lines, verified; broken lines, projected).

7.43.2 Differences between beekeeping with Africanized and European bees

Table 7.43B shows ways in which the Africanized bees that spread out from Piracicaba in Brazil differ from European bees. Characteristics of the Africanized bees marked S enabled these bees to supplant the European bees in the warm parts of the Americas, and those marked B make beekeeping with Africanized bees different from beekeeping with European bees. Winston et al. (1983) provide further information, especially on social behaviour, from 69 references cited.

How Africanized bees can supplant European bees

Some of the entries in Table 7.43B relate to swarms and to feral colonies (colonies living wild) established from swarms, which are the chief vehicles for the spread of Africanized bees.

Swarms of Africanized bees establish nests in smaller cavities than European bees (A), and even in sites that are partly open, e.g. between branches of a large tree. Huge areas of South America, including for instance the Amazon basin, offer few cavities large enough for European bees, but many that are acceptable to Africanized bees.

Africanized queens lay more eggs (B), so colonies established from these swarms grow more quickly. Also, the colonies swarm when they are much smaller, even within 5 to 7 weeks after their establishment, and produce at least ten times as many swarms

Table 7.43B Some behavioural differences between temperate-zone (European) and tropical (Africanized) honeybees

Many of the data are extracted from *Africanized Bee News* 1(2): 5–7 (1985)
S = enables Africanized bees to supplant European bees
B = affects beekeeping with Africanized bees

		European bees	*Africanized bees*	*S*	*B*
A	Size of typical wild (feral) nest:				
	volume (litres)	45	22	S	
	total comb area (1000 cm²)	23.4	8 to 11		B
	honey storage area (1000 cm²)	2.8	0.9		B
B	Queen's egg laying:				
	max. no. per day	2 500	4 000	S	B
	no. per year	58 000	105 000	S	B
C	Swarming:				
	no./colony/year	1 to 4	5 to 10	S	B
	increase/year in no. colonies	× 4	× 16	S	
	unassisted spread/year (km)	14	200–500	S	
	max. distance travelled to new nest site (km)	5	perhaps 75*	S	
	Table 6.33A gives data on the initiation of swarming.				
D	Absconding	no	yes	S	B
E	Age (days):				
	worker – first flight	10–14	3	S	B
	queen – mating flight	7–10	5–6	S	B
	– egg laying starts	3 days later	3 days later		
	drone – mating	13	7.5	S	B
	For development periods see Table 2.41A.				
F	Drone rearing (see text)	suppressed	stimulated	S	B
G	Stinging test:				
	time before first sting (seconds)	229	14		B
	no. stings/min	1.4	35		B
	pursuit distance (m)	22	160		B
	colony recovery time after arousal (min)	3	28		B
	See also Table 3.41B.				

* with overnight stops; Africa 100

as European bees, according to Ruttner (1986). As well as sending out reproductive swarms (Section 3.42), Africanized colonies may abscond (Section 3.43), and bees in absconding swarms (D) carry more food with them, and can fly much further than reproductive swarms; by resting overnight in a cluster (Section 3.42) they may be able to establish a new colony 75 to 100 km from the parent colony (C). Virgin queens, reared as part of the preparations by the parent colony for reproductive swarming, mate and start to lay eggs at an earlier age in Africanized bees (E).

Where colonies of both Africanized and European bees are present in the same area, drones of the former drift into colonies of the latter, whereas European drones rarely drift into Africanized colonies (Rinderer et al., 1985*b*). The amount of drone rearing in a colony depends on the number of adult drones already there (Section 3.22), so this one-way drifting stimulates drone rearing in Africanized colonies and depresses it in European colonies (F). In a 'mixed' area, many more Africanized than European drones are therefore present (in 'mixed' apiaries, 91% were Africanized, Rinderer et al, 1987*c*); a virgin queen of either Africanized or European bees is much more likely to mate with Africanized than with European drones, so almost all the workers of the next generation are likely to have an Africanized drone father. For the beekeeper's use where mating of European queens occurs in a mixed area, Helmich (1986) published a table to show how many European drone colonies are required to ensure that the queens mate with European drones, according to the density of Africanized drone colonies in the area. Section 8.27, on queen rearing with tropical African bees, is relevant here.

How 'Africanization' affects beekeeping

Referring again to Table 7.43B, colonies of Africanized bees tend to be smaller, and to store less honey, than those of European bees (A). They build up their populations quickly (B), and may then swarm while still fairly small, sending out many swarms in a year (C). Also they may abscond (D) unless they are given adequate food when a dearth starts. These are characteristic features of honeybees native to the tropics, which evolved survival strategies successful in regions where temperatures were high enough for flight and foraging all the year round, but not in temperate regions with a cold winter. Tropical dearths due to heat and drought, or rain, were rather short, and regional variations in altitude or rainfall might well provide another forage area within 100 km of an area entering a dearth period.

In experiments carried out in Venezuela, European bees collected more and larger nectar loads than Africanized bees; colonies grew larger, produced more honey per day (Rinderer et al., 1984, 1985*a*), and consistently stored more honey in total; a higher proportion of European bees foraged for nectar only (Pesante et al., 1987). When little nectar was available Africanized bees were marginally superior in certain ways, and there are a few reports that Africanized bees are more productive than European bees (e.g. Gonçalves, 1974/1975; De Jong, 1984).

Africanized workers and queens develop from egg to adult in a shorter time than European bees (Table 2.41A). This gives colonies of the tropical bees an advantage in resisting infestation by the mite *Varroa jacobsoni*; see Section 11.42. The bees also progress more rapidly in their development when adult (E).

Africanized bees receive much adverse publicity on account of their arousal to mass stinging (G), and Table 7.43A provides some actual data. Section 3.41 describes the behaviour of honeybees in defending their colony, and Figure 3.41a sets out the time sequence of a honeybee's responses when her colony is threatened. In particular Table 3.41B shows that Africanized bees respond more rapidly than European bees to alarm pheromone (which recruits other bees to attack), and very much more rapidly to a moving target. Because of these characteristics, beekeepers need to wear much protective clothing – which adds to the discomfort of their work and can take a good deal of pleasure out of it. A very large smoker is recommended by Wiese (1985), with a fire box 40 × 15 cm.

Climatic factors, especially high humidity, have a considerable effect on 'aggressiveness'; see e.g. Brandeburgo et al. (1982). Colonies are also more easily handled during the cooler hours than in the heat of a day.

Africanized bees as crop pollinators

Danka (1986) and Danka and Rinderer (1986) carried out experiments in Venezuela in which 15 colonies of Africanized and of European bees were moved 6 times from crop to crop for pollination, on a similar system as that used with European bees in North America. The work was done during two months of hot, dry (stressful) conditions. Africanized bees tended to collect more pollen, and individual behaviour suggested that they should be efficient pollinators. However, in Venezuela the Africanized colonies had a smaller proportion of foragers than European colonies, and they swarmed frequently, thereby reducing the total colony population, and some colonies were much weakened; colonies stored less honey. Moreover, Africanized colonies were not as amenable to management as European colonies are, and moving and inspecting them was hard, unpleasant work. The photograph in Figure 7.43c was taken during these experiments. Running, festooning or stinging bees hampered inspections three times as frequently with Africanized as with European colonies, and locating Africanized queens was often difficult and time-consuming.

Identification of Africanized bees

Where colonies of both Africanized and European bees are present, it is very important to be able to distinguish between them – for instance in order to requeen Africanized colonies with specially reared European queens. (In Africa, where the tropical bees are the native ones, the need does not arise.) Behavioural characteristics, such as tendency to sting, are certainly not reliable for differentiation. Some advanced laboratory methods are referred to in Section 1.22.4. The following more widely applicable methods are based on measurements of length and weight; Africanized bees are intermediate in size between European and tropical African bees. Body measurements, which must be subjected to statistical analyses,

Figure 7.43c An example of difficult working conditions with Africanized *Apis mellifera* in hives close together (photo: J. E. Tew).

include lengths of the forewing and a specified part of the hind wing, and of the femur of the hind leg (Sylvester & Rinderer, 1986; Rinderer et al., 1986, 1987*a*). A simple and rapid method, which cannot, however, always differentiate successfully, is to make three measurements across the parallel sides of 10 cells of natural worker comb; results of Rinderer et al. (1986) predict that an average total of 49 mm or less indicates comb built by Africanized bees, and of 52 mm or more, by European bees. Identification is not possible if the distance is 50 to 51 mm, but Africanization might be suspected.

7.44 Experience with introduced temperate-zone *Apis mellifera*

Chapter 6 covers beekeeping management with temperate-zone (European) *Apis mellifera* bees in temperate zones, and Sections 7.2 and 7.3 refer to their management in the subtropics and tropics, respectively. The present section discusses the usefulness of these bees (for brevity called European bees) in different regions of the tropics. Unlike tropical honeybees, colonies cannot respond to a dearth of forage by absconding (Section 3.34.5), but they are more amenable to management, an essential part of which is feeding during dearth periods. Beekeeping with *A. mellifera* and *A. cerana* are sometimes referred to as high-input and low-input, respectively.

European bees were introduced into all countries of the Americas in the present century or earlier, and in some, for instance in Argentina and Mexico, they have given large honey yields. Woyke (1981*b*) found their seasonal management much easier in El Salvador (14°N) than in Poland (52°N), largely because of the continuously high temperatures and the absence of the constraints of a cold Polish winter. European bees have now been ousted from much of the tropical and subtropical regions of the mainland by the Africanized bees discussed in the previous section. But they continue to produce 100 kg per colony or more in some of the Caribbean islands, without very sophisticated management.

European bees are used exclusively in Australia, New Zealand, New Guinea and Pacific islands (Figure

7.44a), and they do well there; these land areas were never part of an Old World continent, and have no native honeybees.

In Africa, European bees are used for beekeeping in parts of the region north of the Sahara, but south of the desert where there are native tropical races of *A. mellifera* they have not thrived (in spite of many attempts to introduce them), and as far as I know they are not used for commercial beekeeping anywhere in Africa south of the Sahara.

European bees can be very successful in the subtropics of Asia. Tropical Asia has other *Apis* species, but no native *A. mellifera*, and European bees have been introduced and are used with success within certain regions. Table 7.1B lists the countries concerned, and reports on many introductions are cited by Crane (1978*a*, No. 8). In tropical areas, European bees are likely to be most productive where much land is planted with agricultural crops producing extensive honey flows, which colonies of *A. cerana* do not fully exploit. In upland areas where the tropical climate is mitigated, European bees are likely to do better than in the lowlands.

One area already referred to, where these bees are successful at getting honey, is the Chiangmai region of northern Thailand, where longan (*Euphoria longan*) is the main honey source. Another is Burma, where some of the dry hilly slopes are planted with niger (*Guizotia abyssinica*) and some flatter parts with sunflower (*Helianthus annuus*); see Zmarlicki (1984*a*). In tropical and subtropical parts of China, Fang Yuezhen's observations (1984) echo experience in many other parts of the tropics: 'On the plains with convenient transporting facilities and large-scale rich forage, the imported European bees prevailed and almost replaced the local Chinese bees [*A. cerana*]. In hilly

Figure 7.44a Part of an apiary of European honeybees on Niue Island, at 19°S in the Pacific, 1975 (photo: F. Ikimau).

regions, however, on account of climatic conditions, nature of nectar resources and natural enemies, the local bees survived better and remained dominant.' More than 4 million colonies of bees are kept in the 13 provinces he refers to – 60% of all colonies kept in China – and more than 2.5 million are of European bees.

In the Philippines, with 2000 colonies of *A. mellifera* (Cadapan, 1984), the average honey yield per colony from the different honey-producing bees is estimated as:

European *A. mellifera*	28 kg
A. cerana	2 kg
A. dorsata (wild nests)	10 kg
A. florea	not harvested
Trigona spp.	0.3 kg

Introduction of European bees to India has been restricted to parts of the north: the plains of Punjab and hill country below 1500 m in Himachal Pradesh, where there are 1500 and 1000 colonies, respectively (Verma, 1984); in these areas *A. mellifera* has been reported to produce 3 to 4 times as much honey as *A. cerana*, and some other comparisons with *A. cerana* are quoted in Sections 7.41 to 7.41.5. In Pakistan 'a beekeeper with *A. cerana* rides a bicycle, but if he keeps *A. mellifera* he drives a car'.

None of the above discussion implies the safety or wisdom of introducing European *Apis mellifera* into a tropical (or other) area where another *Apis* species already exists; Sections 8.7 to 8.75 discuss the factors involved. Section 8.73 explains that tropical African or Africanized bees should *never* be introduced to a new area.

There are many tropical regions in Asia where European bees cannot be kept profitably, and may not even survive. In spite of what the beekeeper can do, they may succumb to disease, enemies or parasites that do damage locally; see Chapter 11. Akratanakul's (1987) book on diseases and enemies of honeybees in Asia discusses their implications for beekeeping with European *A. mellifera*, including the transmission of diseases and parasites to it from other honeybee species.

Information on many aspects of beekeeping with European bees in the tropics is published in the proceedings of meetings that were held in the 1970s and 1980s, as a result of the rapidly increasing interest in the subjects. They include:

1972 in Queensland, Australia (1st Australian Bee Congress, 1973)
1977 in Adelaide, Australia (26th International Apicultural Congress)
1984 in Thailand, on tropical and subtropical Asia (FAO, 1984)
1985 in Japan (30th International Apicultural Congress)
1988 in Queensland, Australia (2nd Australian and International Bee Congress)

7.5 APIARY REQUIREMENTS IN THE SUBTROPICS AND TROPICS

Sections 6.6 to 6.62.3 deal with apiary requirements in the temperate zones, and almost all of what is said there applies to the subtropics and tropics. The present section emphasizes some of the factors that are especially important in hot and dry areas, and introduces others that do not arise in temperate zones.

7.51 Apiary sites

Section 6.61.1 lists 9 features as being desirable in temperate zones. Of these, 1 and 2, dealing with access to the site and freedom of movement within it, are similarly desirable in the subtropics and tropics. Features 3 to 9, repeated below, are sometimes likely to be very much more important:

3. shade from the sun, and shelter from damaging winds
4. water permanently available to the bees (see Section 6.62.3)
5. minimal danger from fire (see 11.82)
6. minimal danger from flood water high enough to enter the hives
7. protection from attacks or incidental damage by animals
8. safety from vandalism and theft
9. minimal danger of nuisance by the bees to members of the public and their animals (see 15.3)

Shade from the sun is essential in many parts of the subtropics and tropics. It may be provided by trees, as in Figures 6.62a and 7.41a, although some trees lose their leaves during drought, when the shade is most needed. In Egypt a wooden framework is erected over many apiaries as shown in Figure 7.51a, and leafy branches are laid over it as required. The high roof over an apiary in Thailand (Figure 7.51b) is effective and airy, but more expensive. Even a large cactus can provide a little shade, as in Figure 7.51c.

Provision of a constant water supply for the bees is essential in the subtropics and tropics. In dry tropical areas there is also a great hazard from grass fires. In addition, vegetation can grow much more quickly

Figure 7.51a Apiary near Assyut, Middle Egypt, 1978. In the hot season, shade is provided by laying vegetation over the wooden structure above the hives.

Figure 7.51b Hives of European *Apis mellifera* shaded by a high roof insulated with thatch, Chantaburi, eastern Thailand, 1986.
Legs of hive stands (e.g. of hive on left) stand in a container that is filled with oil in the season when ants are active.

Figure 7.51c Miel Carlota apiary, with an organpipe cactus providing the only shade, Morelos Province, Mexico, 1957.

where temperatures are high, so it is doubly important to keep apiary sites clear – and also their immediate surroundings, to provide a firebreak. Heavy tropical rain may put hives at risk from floods unless proper precautions are taken.

With any tropical bees, small apiaries help to limit interactions between colonies that make manipulations difficult. In Asia, *A. cerana* foragers have a smaller flight range than *A. mellifera* foragers, so fewer hives should be sited in each apiary.

Protection from attacks by mammals is vitally important in the African tropics, but often less so in Asia and South and Central America; Section 11.54 gives details of species likely to cause damage. In contrast to parts of the north temperate zone, bears do not usually constitute the greatest danger, and they are absent from the African continent. Protection may be needed against thieves and vandals; Figure 10.3b shows a well fenced apiary, and Figure 7.52b shows one with a trip wire round it; I do not remember having seen electric fencing used in the tropics.

In South and Central America, the danger of nuisance to members of the public – and to the beekeeper and his family – becomes much higher when Africanized bees replace European bees in the area (Section 7.43.2). An established apiary site near a house may no longer be usable, yet not all beekeepers can find another site for their hives; this is certainly the case in Trinidad.

7.52 Placement of hives of tropical African and Africanized bees

7.52.1 Outdoor apiaries

Where the bees kept are temperate-zone *Apis mellifera* or *A. cerana*, recommendations in Sections 6.62 to

6.62.3 should suffice. Where the apiary contains hives of tropical African bees or Africanized bees, it is usually necessary to follow more stringent recommendations.

1. The apiary should not be near a built-up area, a road or track used by people, or a place where farm animals are kept, especially if they are tethered.
2. The apiary should be well screened from its surroundings, by trees or bushes high enough to ensure that bees fly out of the apiary at a height of 3 m or more.
3. Hives should be farther apart than is necessary for temperate-zone bees. The minimum separation recommended varies from about 3 m, as in Figure 7.52a, to 9 m in Zimbabwe (Papadopoulo, 1967). Individual hives should be screened from each other by bushes, as in Figure 7.52b, or by some other form of screen.

Recommendations 1 and 2 aim to minimize nuisance to the public and to animals. Recommendation 3 aims to prevent bees from one hive (opened for examination or manipulation) alerting bees from adjacent hives, with the result that they are also likely to sting people or animals.

Figure 7.52a Apiary of tropical African bees near the west African coast at Dakar, Senegal, 1973.

Figure 7.52b One of J. Nightingale's apiaries in the Rift Valley, Kenya, 1967.
Hives are well separated and screened from each other, and only two hives are visible. The apiary is surrounded by a trip wire (arrowed).

By no means all tropical *Apis mellifera* need to be treated with such caution, and I have seen hives close together in rows, where colonies were manipulated without ill effect. But what may be safe on many occasions is not always so.

7.52.2 Bee houses

Bee houses are of special value in the tropics, for housing bees that are difficult to manage, and for security. The first part of Section 6.63.1, on bee houses in temperate zones, applies also to those in the tropics.

My first-hand knowledge is mostly confined to bee houses in Africa (Ethiopia, Kenya, Uganda, Tanzania, Zimbabwe, South Africa), where they accommodate from 20 to 80 hives. Figures 7.52c and 6.35b show a bee house in Zimbabwe, painted white to reflect heat radiation, and with a light-coloured roof. F. G. Smith built and used one in Tanzania, and published full details and dimensioned drawings (1960). It has space for 24 hives, 12 along each of the two long walls, each hive having its own flight entrance through the wall. For ventilation and coolness, the upper half of these walls consists of windows of wire gauze, and the height from floor to ceiling is 2 m or more. The windows incorporate a gap or louvre at the top through which bees fly out, and re-entry is prevented by a baffle board inside, distanced a bee-space from the gauze. The 24 flight entrances are about 0.5 m above the floor; a short length of bamboo or piping leads through the wall from the hive entrance itself, the rest of which is blocked off, so that bees returning to their hive do not enter the building instead. All the rest of the building is made bee-tight, and the door fits tightly. When hives are opened, bees flying off the comb go to the windows through which they can see daylight, then walk up the gauze and escape at the top. Papadopoulo's bee house (1964, and Figure 7.52c) has a corrugated asbestos roof, and the bees escape through gaps formed where the corrugations meet at the top of the walls. Guard bees outside the flight entrance will fly around outside the house, but no bees should be able to enter. Drifting of bees is minimized by coloured designs on the outside wall. If bees from a hive examined become too active, the hive can be covered and the bee house left closed and dark; bees flying within the building will leave it and re-enter their hive through the flight entrance.

The hives are relatively cool in the heat of the day, and are protected to a great extent from fluctuations

Figure 7.52c Bee house at Henderson Research Station, near Mazoe, Zimbabwe, 1964 (photo: P. Papadopoulo).
The part to the left of the water tanks contains the hives; the multicoloured designs help bees to locate their hive entrance and thus to reduce drifting.

between day and night temperatures. They are also less subject to disturbance than hives outside. All the beekeepers I know who have used a bee house for tropical African bees maintain that it has considerable benefits over an outdoor apiary. They comment on the greater comfort and lower level of stress in which the bees live and in which the beekeeper works, and on the greater security against theft, vandalism and attacks by predatory animals. But the capital cost is considerable.

7.6 ECONOMIC SUPPORT FOR BEEKEEPING IN THE SUBTROPICS AND TROPICS

Until 1950, the beekeeping resources of the subtropics and tropics were relatively unused. Since then, and especially since 1970, there has been a flow of aid and expertise from technologically advanced countries to those less developed. Section 16.7 refers to beekeeping development programmes, and the 'Directory of beekeeping aid programmes' (in Drescher & Crane, 1982) has entries for 85 countries. The standard of beekeeping, as well as its extent, has increased enormously since 1950. Nevertheless there is still much land that is under-used for honey production, because many inhabitants who could obtain an income from beekeeping lack the capital for equipment, as well as the knowledge, that would enable them to do so.

Beekeeping operates under a wide variety of economic systems in the subtropics and tropics, many of which involve some form of aid from within or outside the country. In the following outline, the bees used are *Apis mellifera* unless *Apis cerana* is named.

1. System based on private capital

Most of Sections 6.7 to 6.73 on the economics of beekeeping apply here. Virtually all the beekeeping in countries such as Australia and South Africa, and some in Argentina, Brazil and Mexico, is done by private enterprise which also provides the necessary capital investment. Argentina and Mexico have half their lands in the subtropics, and they are among the three largest honey-exporting countries in the world. South of Mexico City, A. Wulfrath and J. J. Speck (both Germans) built up their business, Miel Carlota, into one of the largest beekeeping enterprises in the world (Willson, 1955). They had up to 50 000 hives at one time, and their 1955 book includes some account of the operations. Yucatán, farther to the east in Mexico, is referred to under Co-operatives (3, below).

Private-enterprise beekeepers also developed the industry in countries of Central America (Kent, 1979). In South America, Iparraguirre (1966, 1967) published detailed reports on the economics of beekeeping in Argentina, and Persano's book (1980) gives more recent information. In Brazil, Juliano (1975) assessed the costs, profitability and capital turnover of a unit of 250 hives in Rio Grande do Sul. Figures compared well with those for more traditional enterprises in the region such as growing soya beans and wheat, or raising cattle. In Colombia a detailed study has been made of the profitability of different types of apiary (Alvarado Velasquez, 1975).

An economic analysis on similar lines to that in Section 6.72 has been made in various tropical countries. In Africa, F. G. Smith (1960) set out the revenue and expenditure account of bee farms for 200 and 500 hives, kept in bee houses in Tanzania. Analysis showed that the farms must produce an average of 55 kg honey per hive in order to break even. In South Africa du Toit and du Toit (1987) made a study of operations with 100 and with 400 hives, some migratory and some not.

In Asia, O. P. Sharma and Thakur (1982) made estimates of the income and expenditure for a 50-colony apiary of *Apis cerana* or *A. mellifera* in northern India; Delima and Roberts (1987) published estimates for *A. mellifera* in the Philippines, and Ismail and Shamsudin (1987) for *A. cerana* in Malaysia, on similar lines. Shyu (1983) made an economic study of the beekeeping industry in Taiwan. China is mentioned in Section 6.71.2. In Bangladesh, a very poor country with few resources and rather little state support for beekeeping, Mohammad (1984) made an analysis of a peasant farmer's operation with 10 hives of *Apis cerana*; he costed the beekeeper's labour at Taka 1.5 (6 US cents) per hour. The annual expenditure and revenue are estimated at Taka 253 and 1180 per hive, respectively. Most farms are smaller than 0.2 ha, and the average family's income from its farm is about Taka 5000 a year, so the net return from bees (Taka 927 per hive) would be 15% to 20% of the total farm income. Svensson (1989) has calculated a very much lower return from the same operation, with an initial deficit, but shows that the beekeepers could obtain a net income from the first year onwards if they use home-made top-bar hives, with (purchased) precision-made top-bars; see Section 10.21.

2. Sponsored credit

Credit may be provided at a reduced rate of interest by a government-sponsored agricultural bank, either directly or through an intermediate agency, to start or to expand private beekeeping enterprises. The system operates to some extent in many countries of Latin America, including Colombia, Costa Rica, Cuba, El Salvador, Guatemala, Honduras, Nicaragua and Panama. Espina and Ordetx (1984) give some details. The bank may also enable beekeepers to obtain equipment at much reduced prices. An FAO proposal for Jamaica (Matheson, 1985*b*) includes a 50% grant and 50% interest-free loan to be repaid within 5 years, to new beekeepers at the end of their training period.

Sometimes a revolving fund is provided by an internal or external agency, to allow a succession of loans to be made, one after another, as previous ones are paid off.

3. Co-operatives

Beekeeping co-operatives are likely to receive some government support, although a few are independent. Support may include technical instruction, provision of honey-processing premises and equipment, honey-marketing facilities, and arrangements for the purchase of equipment at reduced prices. Co-operatives are or have been active in many countries, for instance: in Africa, Kenya, Tanzania (see papers in Crane, 1976*a*); in the Americas, Belize, Mexico, St Lucia; and in Asia, Thailand. In Yucatán, which has some of the best beekeeping country in Mexico, 9000 Maya beekeepers produce on average 10 000 tonnes of honey a year for export. Many belong to co-operatives, for which the Government does most of the marketing and makes advance payments that act as a partial loan to the beekeepers (Patty, 1978). A comprehensive anthropological study of the commercial beekeeping of these beekeepers has been published (Sands, 1984).

4. Support from an autonomous statutory body

India is an outstanding example of a country where beekeeping receives its support from an autonomous organization, established by an Act of Parliament. The Khadi and Village Industries Commission was set up to plan, organize and implement programmes for the development of village industries such as weaving and pottery. It was the culmination of the rural reconstruction programme envisaged by Mahatma Gandhi. The programmes are executed through the various State K & VI Boards, Registered Institutions (including those for research) and Co-operatives, funds being released to these agencies. Village Industries programmes are implemented by local institutions and co-operatives, under the State Boards through which they are accountable to the Commission. Shri C. V. Thakar was largely responsible for building up the beekeeping programme, which is concerned almost entirely with the keeping of *Apis cerana* in hives. It was started with a few villages and 232 beekeepers in 1953. By 1974 the Beekeeping Directorate had an annual budget of Rs. 5 million (£250 000 or $700 000), and nearly 40 000 villages were involved (Thakar, 1976).

5. Bilateral aid from another government

Economic and technical aid has been provided under a bilateral agreement by the governments of a number of individual countries in a temperate zone to a country in the tropics or subtropics with which there have sometimes been special links in the past. Examples are given below.

aid from:	*to:*
Belgium	Rwanda
Canada	Kenya*, Sri Lanka, Tanzania
France	Gabon, Haiti, Togo
German Federal Republic	Costa Rica, Panama, Tanzania, Turkey
Israel	Senegal
Japan	Paraguay
Netherlands	Surinam, Vietnam
New Zealand	Niue Island, Papua New Guinea
Sweden	Guinea-Bissau
UK	Ghana, Malawi, Nepal, Oman
USA	Egypt, Pakistan, Thailand, Tunisia

Drescher and Crane (1982) give details of these and other beekeeping projects funded by bilateral aid, and by aid from international and other agencies (6 below).

6. Support from an international or other aid agency
The Food and Agriculture Organization of the United Nations (FAO) has supported beekeeping projects in a number of countries; in 1986/87 (Paltrinieri, 1987) they were:

Africa: Algeria, Burkina Faso, Cape Verde, Djibouti, Egypt, Ethiopia, Gabon, Ghana, Guinea, Guinea Bissau, Libya, Madagascar, Mali, Mauritius, Morocco, Mozambique, São Tomé
Americas: Brazil, El Salvador, Guyana, Honduras, Mexico
Asia: Afghanistan, India, Indonesia, Iraq, Republic of Korea, Lebanon, Maldives, Pakistan, Saudi Arabia, Turkey, Vietnam
Pacific: Vanuatu

Other United Nations agencies, including the International Trade Centre, UN Development Programme and UNICEF, also undertake beekeeping projects, as do non-governmental agencies such as CARE, Lutheran World Relief, Oxfam, Salvation Army, and World Council of Churches, and many agencies based in individual countries; a list has been published by IBRA (1985*a*, No. 6).

*Kenya has received greater, and more effective, bilateral aid for beekeeping than most countries. It came from the Canadian International Development Agency, after the initiation of a project by Oxfam; see Section 10.21.

7. Private capital invested from another country
The outside investment usually comes from an individual (or a company) in a technologically advanced country in a temperate zone (Section 6.71.2), who provides both capital to finance the undertaking and expertise and technological equipment to which he has access. In general the foreign undertaking is set up in a country not too far from the investor's homeland, but where labour costs are lower and honey yields higher than at home. For instance a beekeeper in the USA might establish a beekeeping undertaking in Mexico. The honey produced must be readily saleable at an acceptable price, in the investor's own country or elsewhere.

7.7 FURTHER READING AND REFERENCE

Details of publications listed will be found in the Bibliography.

General

Commonwealth Secretariat; IBRA (eds) (1979) *Beekeeping in rural development*
Crane, E. (ed.) (1976*a*) *Apiculture in tropical climates*
Crane, E. (1978*a*) *Bibliography of tropical apiculture* Nos 1–11, 16
Drescher, W.; Crane, E. (1982) *Technical cooperation activities: beekeeping. A directory and guide*
FAO (1986) *Tropical and sub-tropical apiculture*
Gentry et al. (1985) *A manual for trainers of small scale beekeeping development workers*, 2nd ed.
Segeren, P. (1988) *Beekeeping in the tropics*
Townsend, G. F. (1984) *Tropical apiculture* [text, filmstrips, taped commentary]

European *Apis mellifera*

Akratanakul, P. (1986) *Beekeeping in Asia*
Espina, D.; Ordetx, G. S. (1984) *Apicultura tropical*, 4th ed.
FAO (1984) *Proceedings of the expert consultation on beekeeping with* Apis mellifera *in tropical and subtropical Asia*
Ormel, G. J. (1987) *Guide concis d'apiculture avec référence spéciale à l'Afrique du Nord*
Redma Consultants Ltd (1981) *The development of beekeeping and hive products in the Caribbean community*

Many books cited in Section 6.8 gives details of operations that are necessary for management in the subtropics and tropics as well as in the temperate zones.

Tropical African *Apis mellifera*

Adjare, S. (1984) *The golden insect: a handbook on beekeeping for beginners*, 2nd ed.

Anderson, R. H.; Buys, B.; Johannsmeier, M. F. (1983) *Beekeeping in South Africa*, 2nd ed.

Fletcher, D. J. C. (1977) *African bees: taxonomy, biology and economic use*

Smith, F. G. (1960) *Beekeeping in the tropics* [also tropics in general]

Africanized *Apis mellifera* in the Americas

Espina, D. (1986) *Beekeeping of the assassin bees*

Gomez Rodriguez, R. (1986) *Manejo de la abeja africanizada*

Wiese, H. (1985) *Nova apicultura*, 6th ed.

***Apis cerana* in Asia**

Akratanakul, P. (1986*a*) *Beekeeping in Asia*

Kevan, P. G. (ed.) (1989) *Beekeeping with* Apis cerana *in tropical and subtropical Asia*

Rawat, B. S. (1982) *Bee farming in India*

Shah, F. A. (1983) *Fundamentals of beekeeping*

Singh, S. (1962) *Beekeeping in India*

Series of Conference Proceedings

International Conference on Apiculture in Tropical Climates, 1–4 (1976, London; 1980, Delhi; 1984, Nairobi; 1988, Cairo)

Congresso Brasiliero de Apicultura 1–7 (1970, 1972, 1974, 1976, 1980, 1984, 1986)

8

Special types of bee management

8.1 MOVING HIVES OF BEES

In countries with a good road system, or railways following suitable routes, many beekeepers make their living by moving hives from one honey flow to another, or to a succession of crops that need bee pollination. Migratory beekeeping can be especially worth while where there is a succession of flowering seasons because a country extends over a range of latitudes, such as Japan and Sweden, or of altitudes, such as Pakistan and Turkey. Restrictions are imposed on the movement of hives from one country to another, in order to prevent the spread of bee diseases and parasites, and this limits most migratory beekeeping to a single country of operation.

Transport is mainly by road in Europe, the USA, Canada, Australia, New Zealand and the USSR. In China rail is also extensively used, special trains being run to transport the hives. Figure 8.1a shows preparations for long-distance transport in Australia. Hives have been taken 3000 km across this country, to the honey flow from karri (*Eucalyptus diversicolor*) in Western Australia; this flow occurs only once every 4–12 years, and is possibly the longest and heaviest in the world, giving yields of 150–300 kg/hive. In some people's minds there is a great romance in the idea of a nomadic life spent migrating with one's bees from one honey flow to the next, and several books have been written on the theme. *The honey flow* (Tennant, 1956) is a story set in Australia. *Bees are my business* (Whitcombe & Douglas, 1955) is a popularly written account of a large beekeeping enterprise in California, in which migration played an important part. Ortega Sada (1987) in Spain deals with both wild plants and crops needing pollination, in relation to migratory beekeeping.

A good many beekeepers, whether they have only a few hives or a large number, move them once a year to an outstanding honey flow, and this is often the occasion for memorable days – and sometimes nights – spent in unusual country surroundings. In Britain, heather (*Calluna vulgaris*) is the flow to which bees have traditionally been migrated and *Bees to the heather* (Whitehead, 1954) gives a straightforward account of what needs to be done before, during and after the migration.

In many languages the word for migratory beekeeping is *transhumance*, used for the old practice of moving livestock to new pasture at certain seasons of the year, and an international symposium on *apicultura transhumante* was held in Spain in 1980 (Apimondia, 1981). Nowadays the amount of migratory beekeeping any one operator undertakes is closely governed by economic considerations: income receivable from the sale of honey harvested and from pollination contracts, in relation to outlay on fuel, labour and other expenses. Thompson (1976) made a detailed economic analysis applicable to Australian conditions.

Sections below deal with ways of moving bees in hives from place to place, without harming them; Morse (1970), Detroy et al. (1975) and Thurber (1983) give information on many aspects of the operations. Problems associated with moving hives through short distances – from a few kilometres down to a few metres – are dealt with in Section 8.15.

Figure 8.1a Hives being loaded by hand on a 'semi-trailer' in Australia, about 1960 (photo: Australian News and Information Bureau).
The 285 hives loaded were transported 1600 km by rail.

8.11 Preparation of hives and bees

However long or short the journey will be, the frames in the hives must be sufficiently secure to prevent shifting, and all parts of each hive must be securely fastened together. If hives are transported inside a vehicle with the driver, their entrances must be securely closed, and it is essential that no bees will be able to escape.

The main aim of many of the operations described below is to prevent the hives becoming overheated. If this occurs bees will die; also, combs may melt and the spilt honey kill even more bees, so it is best to leave rather little honey in the hives. In hot conditions, much trouble can be prevented by travelling in the coolest part of the 24 hours – usually during the night.

The following applies to the preparation of hives for transport with the bees enclosed in them. (Transport with the entrances left open is discussed in Section 8.13.) Arrangements must be made in advance for adequate hive ventilation while the bees are confined. Ventilation can be increased by replacing the hive roof by a top-screen mounted on a deep rim (Yakovlev, 1975); even if the depth is only 5 cm (Morse, 1970) it gives the bees extra clustering space. The bees will have had no chance to attach the rim to the top hive box with wax or propolis, so measures must be taken to prevent slippage, for instance by using four nails that fit loosely into the holes in the screen frame (Figure 8.11a); the screen shown is of 8-mesh (3-mm) hardware cloth, which is stronger than metal or plastic. Experiments on more than 1000 colonies (Yakov-

Figure 8.11a Top ventilating screen of hardware cloth (3-mm mesh) for use when moving hives; see text (Morse, 1970).

Figure 8.11b Tuck-in screen of stiff wire mesh partially in place; it is pushed into the hive entrance to close it (Morse, 1970).

lev, 1975) showed that the volume of the empty box above the frames was an important factor in ensuring the bees' well-being during transport. For strong colonies (5 kg or more), the best results were obtained when the empty box was 25 cm deep.

The hive entrance should be closed after dark, so that all or most of the bees are transported with the hive. Wire screen (14-mesh, 1.8 mm) can be tucked into the full width of the entrance (Figure 8.11b), to keep bees inside but to allow a flow of air. Experience shows that the screen remains in place during transport without any special fixing, and it is easy to remove (Morse, 1970). Some beekeeping equipment suppliers sell special migratory floorboards which allow sure and easy entrance closure, but types with moving parts that might be propolized together by the bees should be avoided.

Many beekeeping equipment suppliers sell steel or plastic strapping suitable for securing the parts of each hive together, and also a tool or device for tensioning and securing it. Yerly (1980) gives details of types made by Signode Corporation in the USA, and their use. Often only one strap is put on, but a second strap at right angles to the first gives greater security. Metal staples are sometimes used instead of strapping, but their insertion and removal take time, and damage the wood of the hive. Various devices are marketed for permanent attachment to the hive parts, and some of them can be operated very quickly. Any attachment that protrudes beyond the side walls of a hive will prevent the hives being loaded tightly together, and this is also true of telescopic roofs and projecting floor boards.

8.12 Loading and unloading hives

The cost effectiveness of mechanical and powered aids for moving hives on and off vehicles depends on individual circumstances – how much they would be used during the year, their cost in relation to the cost of labour, the number and weight of the hives to be shifted, and so on. Manual loading is sometimes the most practical (see Figure 8.12a).

Most beekeepers with only a few hives do not get sufficient use from powered machinery to warrant its purchase. But I have seen many ingenious home-made hive lifters around the world, that are operated by hand. Some, designed also for use when opening hives, are referred to in Section 5.41. Others are adapted from general purpose equipment, such as a bag lifter (Campbell, 1961) designed to give the optimal mechanical advantage for the loads to be moved. Some beekeeping equipment suppliers sell loaders that are suitable for hives or supers, without any alteration.

Powered equipment is not much used in tropical and subtropical Asia, where labour is cheap and efficient, and many hives consist of only one box; Section 7.34 explains the beekeeping system used.

At the other end of the scale, I remember a beekeepers' meeting in Phoenix, Arizona, where the only types of equipment on display were powered fork-lifts, booms, and articulated loaders (Figure 8.12b). No beekeeper present had fewer than a thousand colonies, and most looked after their hives unaided. Beekeeping involved moving hives from one site to another, rather than handling bees, and it was made possible only by investment in the most effective powered mechanical aids. One beekeeper told me 'If I didn't have a boom on my truck, I don't believe I could keep bees any more'. Haynie (1973) gives working drawings for the assembly of a hive loader with a boom that can be fitted

Figure 8.12a Miel Carlota hives in Mexico being carried over rough ground to a truck, 1957.

Figure 8.12b Hive-moving equipment, Phoenix, Arizona, 1957.
Three trucks with boom loaders; on the extreme left a fork-lift; on the centre truck a ground-plane for levelling apiary sites.

to a truck. Detroy et al. (1975) describe various American loaders, trucks and trailers, and Murphy (1980) covers Australian hive handling equipment.

In Western Australia a gantry loader was being used in preference to a boom by 1967. It is supported at both ends, and more weight can be lifted than by a boom supported at only one end. Rails along the top of the gantry, between the supports, carry the hoisting trolley (F. G. Smith, 1970).

The use of pallets has greatly simplified the handling of loads in beekeeping, as in many other industries. (A pallet is a platform about the same size as the load standing on it, its upper surface being raised sufficiently for the flat arms of a fork-lift truck to be inserted underneath, in order to lift it and the load.) Hives have been moved on pallets since the early 1950s, and many beekeepers in Florida and California were using them by the 1970s (Detroy et al., 1975). Each hive must be accessible for inspection but need not be on a separate pallet. A common arrangement is to mount two pairs of hives back-to-back, with entrances facing outwards (Figure 15.5a). The hives are often kept permanently on the pallets, which are sometimes used also as floor boards, cleats being attached to keep each hive in place. Carpenter (1978) and Reid (1976*b*) describe the use of pallets in the USA and New Zealand, respectively.

Hives are usually transported upright, if necessary stacked up to four high, with a gap of 5–8 cm between hives in the same tier, as shown in Figure 8.1a. Smaller colonies can be placed in the centre of the load, as they are less likely to overheat. The frames should be parallel to the length of the vehicle, so that any swing is from side to side, and if entrances are open, these should face forward for maximum ventilation. Alternatively a ventilating screen is fitted at the top and/or bottom of each hive, and the hives laid on one side, with the frames in them vertical. This arrangement can give better ventilation during transport, but extra fittings inside the hive are necessary to hold the frames firmly in place (Morse, 1970).

A device by Wichman (1976) may be useful if hives often have to be removed from apiaries during daylight when bees are flying. A block that closes the hive entrance is cut to accommodate a Porter bee-escape (Section 4.46) from which the springs at one end have been removed. The escape is mounted so that the end with springs intact is directed into the hive; returning bees can enter (and most will have returned within 30 minutes), but no bees can leave.

8.13 During the move

By the 1960s beekeepers in Australia and North America came to realize the bees would remain inside hives on a truck by day as well as by night, even if the entrances were left open, so long as the engine of the truck was running. One beekeeper successfully moved 228 two-queen hives with open entrances, through a distance of 2500 km, a journey that took 4 nights and 3 days.

The Australasian Beekeeper (1979) gives detailed recommendations for the operation. Each hive is well smoked just before loading, which is accomplished in daylight and finished half an hour before dark; the many bees flying around will be seeking their hive entrances, and will not be inclined to sting. The truck is driven in the dark, and the engine must not be stopped (unless a bee net is used, see below). Hives that are inside a vehicle with the driver must not be moved with their entrances open. On arrival, hives transported with open entrances are left until sunrise, and smoked before unloading – especially those to be unloaded first.

At first it was assumed that the bees remained within their hives because vibration from the engine 'froze' them into immobility. But Spangler and Owens (1975) watched bees in a glass observation hive mounted on a truck; they saw that when the engine was started during the day, most of the bees moved upwards to the top of the hive. The effect was similar to that induced by traditional beekeepers when they 'drummed' bees upwards out of a hive by thumping its sides rhythmically with both hands (Section 7.33.2). The fact that bees subject to vibration tend to move upwards may explain the benefit of an empty clustering space at the top of the hives being transported (Section 8.11). During a 200-km journey in Hungary, Örösi-Pál (1976) found that vibration from the engine caused many bees to leave the brood area, and to move up into an empty box that had been added at the top of the hive for ventilation. In three experimental hives, 30%, 56% and 71% of all the bees were found in the top box, and brood rearing was much neglected. Yakovlev (1975) in the USSR showed that bees in hives being transported became excited initially, and the temperature in the hive rose, but that as the journey continued most of them were to be found rigid, hanging in a cluster from the hive cover.

When hives are transported with open entrances, a 'bee net' of strong nylon reinforced with webbing is often secured over the load on the vehicle, as a safety measure (Donnelly, 1976). If the engine must be stopped for some reason, and bees do emerge, they are contained. The net is also useful for covering a truckload of honey supers, to keep robber bees out during an enforced stop.

During a stopover, hives can be cooled temporarily by playing water over them from a hose. Or Kramer's (1985) method for watering confined colonies (Section 11.72.1) can also be used, provided hives – with entrances closed – have been stacked on the truck with a gap between top screen and roof, parallel to the length of the truck. The nozzle of the hose, moved along each side of the truck, directs water into the hives through these gaps.

A more systematic way of keeping bees cool during transit is to use a refrigerated trailer; a large one for package bees is shown in Figure 8.34b. Carpenter (1978) describes one in which the hives are transported on pallets, with entrances open; she regards 10° as the optimal temperature inside the trailer.

8.14 After arrival

For hives of bees being transported, the moment of arrival can be the most dangerous. Yakovlev (1975) found that, once the engine was stopped, the bees changed from rigidity to excitement within 2–3 minutes. The temperature could then rise to 45° or more, and if hive entrances were not opened quickly enough the colonies could die. A rise to 40° caused no damage, but even at 41° some adult bees and brood died. With a refrigerated trailer, Carpenter stresses that the refrigeration *must not* be switched off until the hives are removed from the trailer, or colonies will quickly overheat and die. Kramer (1985) recommends giving hives an extra heavy watering before the bees are released, to reduce their overactivity and their tendency to sting when they fly out.

Once the bees are free to fly, they will be seen foraging within a few minutes. But they cannot forage at full efficiency from a new site until they have learned which forage sources are the best, and where these are. Moeller (1975) compared colony weight gains during the 7 days following a move to a new site. He found that colonies moved there gained less weight than colonies left there undisturbed, and less than colonies at the site which had been moved away from it and back again during the same night.

At their new, unfamiliar site, bees may drift from one hive to another, and this is a nuisance to the beekeeper in that it can lead to unequal hive populations. There is less drifting if hives are opened in late evening, but if they are delivered early in the day it is not possible to leave the bees confined so long. Morse (1970) recommends painting parts of hives in several different colours. Placing hives in irregular patterns can greatly reduce drifting (Section 6.62.1), but this may not be practicable at a short-stay migratory stand.

8.15 Moving hives short distances

If it is necessary to move one or more hives within an

apiary or not far beyond it, lifters such as those referred to in Section 8.12 can be used, or simple devices such as are shown in Figure 5.41a.

If a hive is moved well beyond the bees' flight range, the bees do not encounter remembered territory on subsequent flights, and do not return to the old site. If the new site is within or near the bees' old flight range, some of the bees may visit the old site and enter a hive there, but if there is no hive the bees are not necessarily lost. In southern England, Free (1958*a*, 1958*b*) obtained the following results when he used individually marked bees for experiments in which hives were moved 5, 15, 400 or 1600 m, all probably within the bees' earlier flight range. When an isolated hive was moved, most of its bees returned to it – 94% after a move of 5 m, down to 83% after a move of 1600 m. If an empty hive was placed at the old site, it was visited by a number of the bees (from 71% down to 19% for the longest move), but many of these bees subsequently found their hive in its new site, in addition to those that returned directly to it. When all hives in a row of 5 were moved, and no other hives were put in their place, most bees found the new site (from 100%, down to 75% for the longest move), although a number visited the old site first (76% down to 19%). When only the centre hive of the row was moved, many of its bees returned to the old site and joined the colonies that had been adjacent to their own. An empty hive placed at the old site seemed to attract bees from the moved hive – although uselessly – and all the marked bees then ended up in adjacent hives.

In Poland, Bobrzecki (1973) moved groups of up to 35 hives through distances of 0.9–3.4 km, to honey flows. The number of bees returning to the old site depended less on the distance than on the season, and the weather in the next few days. It was lowest in May and highest in July, varying from 0 to a few thousand bees. A honey flow near the new site tended to 'hold' the bees there. In contrast, bees working a flow near the old site might return to it after their hive has been moved away, for instance to prevent bees being killed by an insecticide application.

8.2 QUEEN REARING AND BEE BREEDING

Beekeepers rear queens to requeen existing colonies, to make new colonies, and to improve the genetic quality of their bees. They may alternatively purchase young mated queens from specialist 'queen breeders'. There are several recent books on queen rearing (Laidlaw, 1979; Morse, 1979; Ruttner, 1983; Cook, 1986) to which this section serves as a brief introduction. In addition, I hope that it may stimulate some beekeepers to rear queens for the first time.

Instrumental insemination of queens (Section 8.24) and queen storage (8.25) use special techniques, likely to be of interest to readers who do not use them in their own beekeeping.

Several subjects in other Sections are directly related to queen rearing: development of the queen from egg to emergence (Section 2.51); marking queens (5.23); requeening and queen introduction in general (6.51) and requeening with a queen cell during a honey flow (6.44); and making new colonies (6.34).

8.21 Small-scale queen rearing

Johansson and Johansson (1973) give clear instructions for various methods. The simplest is to use sealed queen cells found in a colony preparing to swarm (Section 3.34.2) or to supersede a queen which is deficient in some way. These two conditions can be differentiated as follows:

swarming	*supersedure*
many queen cells (3–30)	few queen cells (1–5)
queen cells hang down from a comb, often from the lower edge, sometimes several in a group	isolated queen cells, protruding slightly from the face of the comb
the cells may be at different stages of development	the cells are of a similar age
healthy queen	queen in poor condition or missing
brood in compact pattern on combs	irregular brood pattern on combs
probably no eggs in worker cells	eggs probably present in worker cells

A frame containing one or more queen cells can be transferred to a colony to be requeened (whose own queen has been removed), or a hive in which frames of bees, worker brood and stores have been placed, in order to start a new colony. The queen will emerge from her cell in her new colony, and will start to lay eggs a few days after mating (Section 2.53).

If a queen dies, or is killed (as when a colony is dequeened), the supersedure cell is sometimes referred to as an emergency cell. (In tropical African bees the differentiation between development into a queen or into a worker probably occurs within 24 h after the larva hatches from the eggs; Section 8.27). Swarm cells are much more often found than supersedure cells, but many experienced beekeepers prefer not to use them, on the grounds that it may lead to selection for swarming tendency. They would use methods discussed below.

It is possible to induce queen rearing in a colony that is not preparing to swarm or to supersede its queen, and the following simple method does not involve having to find the existing queen. Assuming that the colony is in a single brood box, half the frames of brood are put into the centre of a second box, and in each box the extra space on either side of the brood frames is filled by inserting frames containing drawn-out comb – or, failing that, foundation. The second box is placed above the first, but separated from it by a board having a feed-hole which has been covered with a queen-excluder screen. After 8 days, eggs will be found in one box, indicating the presence of the queen, and it is placed on a new stand. In the other box the bees will have started building queen cells if eggs or newly hatched larvae were present; if not, a frame containing them is inserted to enable these bees to rear a queen.

More systematic methods of rearing queens for requeening colonies are discussed by Lensky (1971), Forster (1972), and the authors mentioned earlier, who also cover large-scale queen rearing summarized below. Mangum (1987) describes queen rearing using a Kenya top-bar hive.

8.22 Large-scale queen rearing

Professional operations, in which hundreds or thousands of queens are reared at the same time, take account of the colony conditions required at each stage of development, since – apart from genetic considerations – these are the factors crucial to the quality of the queens produced. Woyke (1980*a*) has discussed some international aspects of queen rearing.

Each queen may be passed through some or all of the following five environments during rearing:

1. the 'genetic' colony, in which the (fertilized) eggs are laid
2. the cell-starter colony
3. the finishing colony
4. an incubator
5. the mating colony.

The reasons for this arrangement are set out here because they are not always made clear in queen-rearing instructions. Detailed instructions for the procedures are provided in the text books mentioned above in Section 8.2.

1. The 'genetic' colony

The larvae to be used for rearing queens are taken from worker cells of a colony headed by a breeder queen *selected for chosen genetic characters*; this is the first step in selective breeding.

2. The cell-starter colony

Very young worker larvae are transferred (with a grafting tool) into cell cups mounted mouth down on wooden bars; the cups are sometimes primed beforehand with a little royal jelly. The bars are placed in a cell-starter colony (see Section 14.42 for details, and Figure 14.42a) that has been queenless for 2 to 4 hours. This colony is made up with many (young) nurse bees, and little or no other brood to rear. Its bees build the cell cups further and feed the young larvae, and the colony can 'start' the rearing (as queens) of 45–90 larvae a day.

3. The finishing colony

As the larvae grow (and develop into queens, Section 2.51), they receive more food and are better cared for if the number of nurse bees per larva is high. So it is usual to put only about 15 cells in each 'finishing' colony, which is queenright, the larvae being separated from the colony's queen by a queen excluder. The presence of a queen inhibits the initial phase of queen rearing (2 above), but Vuillaume (1957) showed that once the bees have started to work on queen cells, the larvae in them fare better in a queenright colony.

4. The incubator

When the bees have finished feeding the larvae and these are sealed in their cells; the only requirements for the next 7 days are appropriate conditions of temperature and humidity. These can be provided as well in an incubator as in the finishing colony. (In Canada a portable incubator has also been developed for transporting 40–70 queen cells (Davis et al., 1985), for instance to an out-apiary.) Whether the queens will emerge from their cells in an incubator or in a colony, each must be in a separate cage to give protection from attacks by other queens already emerged.

5. The mating colony

The next environment for each queen is a separate small colony containing a few hundred or more workers, but no other queen, in a 'mating hive' from which the virgin queen flies out and mates (Section 2.7). Arrangements for mating are discussed in Section 8.23.

Each queen is left in her mating hive until she has started to lay eggs, and thus shows that she is ready to head a new colony. This usually takes about 12 days, and the queens from a single starter colony will occupy 500–1000 mating hives for this period.

8.23 Arrangements for mating

The process of mating between a queen and a number of drones, in quick succession, is described in Section 2.7. Before a queen flies out and mates she must be in a colony where the workers make her take exercise that prepares her for sustained flight (Section 2.52). The colony can be quite small – as few as 500 bees – and mating hives are often boxes of styrofoam (which provides good thermal insulation), containing a few top-bars fitted with wax starter-strips for comb building, and a feeder (Figure 8.23a). The type shown has the entrance below, which may protect the queen from predators while she is re-entering the hive. Each of the small mating hives must be identifiable by the returning queen; if she tries to enter the wrong hive, she will be killed. It also needs protection from marauders such as ants. Mating hives are therefore commonly mounted on separate poles or stands, and distributed among trees or bushes (Figure 8.23b); they are colour-marked, with entrances facing in various directions. Alternatively 4 compartments are made in a standard hive box, each entrance facing in a different direction.

If queens are being reared without any attempt at selective breeding, and well into the bees' active season, hives in the same or nearby apiaries will probably provide sufficient drones for mating. If drones from specific colonies are to father the next generation of workers, the drone-producing colonies need to be carefully prepared: queenright, populous, with an abundance of young nurse bees, with plenty of honey and pollen stores, and containing frames fitted with

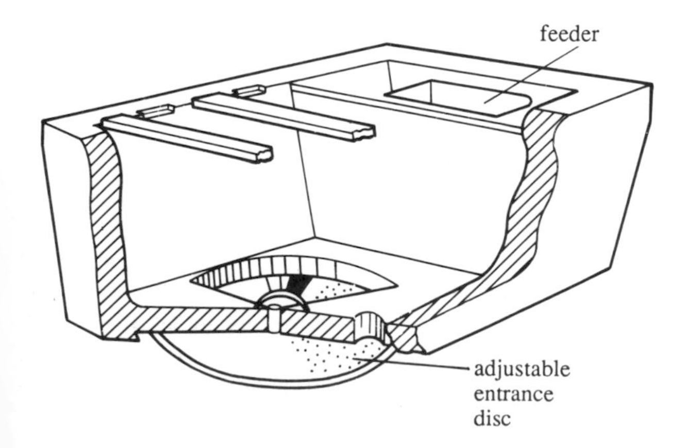

Figure 8.23a Miniature styrofoam mating hive with top-bars to support combs, a syrup feeder, and an adjustable bottom entrance.

Figure 8.23b Brother Adam's mating apiary in an isolated Dartmoor valley, SW England, 1967.
The randomly placed hives appear white in the photograph.

drone comb or foundation. The mating apiary (with the drone colonies) must be isolated from all other hives, and from any colonies nesting wild, by a distance greater than the flight range of the queen plus that of drones from outside. In Austria the maximum found under normal conditions was 12 km, and about 6 km was common (H. & F. Ruttner, 1972). To achieve isolation, a distance of 15 km has been quoted as safe (Bee World, 1971), but in Canada Szabo (1986) found that even 20 km was not. In his 1984 experiments 22 mating hives were within 14 km of the only drone-producing colony, and all the queens mated; in addition one queen from a hive 18 km away, and one from 20 km, mated. Separate flight ranges of queens and drones are quoted in Sections 2.52 and 2.63.

It is possible to achieve effective isolation only in some exceptional places. One is a desert oasis, which has been cleared of all colonies of bees not wanted. (The Dakhla oasis in Egypt was used when Carniolan queens were reared for use in modern hives instead of the native Egyptian bee (*Apis mellifera lamarckii*), whose colonies do not grow large enough to cover more than 5 Langstroth combs (El-Badawy, 1969). Carniolans are dark bees, and any accidental cross-mating with local drones is recognizable by yellow colouration.) Islands in a lake or in the sea, sufficiently far off shore, are more widely available, but many islands are too windy for the purpose. The isolation distance can be reduced by using a place where the topography of the land precludes flight in one or more directions. In a deep mountain valley, the isolation would be needed only along the valley. A clearing in a dense forest can be satisfactory (Figure 8.23c); in northern Canada, forest beyond the edge of cultivation is used, where winters are too cold for the survival of wild colonies.

A commonly used but less sure precaution is to flood

Figure 8.23c Everett Hastings's mating apiary in an isolated forest clearing with drone colonies on the left, in N. Saskatchewan, Canada, 1963.
Posts beyond the hives support an electric fence to keep out bears.

the whole area over which the queens might fly with drones of the required strain. An example is cited in Section 8.27.

Attempts have been made in several countries to get queens to mate in a large cage (Bee World, 1972); the one built in Zimbabwe was 14 m high. Only very occasional successes have been reported, and the method is not at present a practicable one.

A special mating problem may arise when queens of one *Apis* species are to be mated in an area with a preponderance of another *Apis* species. This occurs, for instance, after the introduction of a few colonies of *Apis mellifera* into tropical Asia. (Sections 8.7 to 8.75 comment on the wisdom or otherwise of such an action.) Drones are attracted to a nubile queen, within their joint flight space, by the sex pheromone which is the same for all *Apis* species, 9-oxo-2-decenoic acid (Shearer et al., 1970; Koeniger, 1976*b*). So drones of any species that are in the air when the queen flies may locate her and try to mate – possibly crowding out drones of her own species, which alone can be successful in the mating. (In Sri Lanka, where three *Apis* species co-exist naturally, drones of different species fly at separate times of the day, thereby avoiding inter-species competition in mating, Koeniger & Wyjajagunasekera, 1976.)

Whatever arrangements the beekeeper makes for a queen's mating, she is left in the mating hive until she has started to lay eggs.

8.24 Instrumental insemination

The difficulties of ensuring pure mating were appreciated long before details of the mating itself were understood. An inviting approach to the problem was instrumental insemination of the queen with drone semen. A technique was first devised in the USA by Watson (1927), and has been improved by many others since, in the USA and elsewhere; Laidlaw (1987) has related the history. The procedure is now a routine one which can be carried out by a technician, and many thousands of queens are inseminated annually in bee-breeding programmes.

Figure 8.24a shows how the queen, anaesthetized with carbon dioxide, is immobilized in a tube from which the tip of her abdomen protrudes. A specially designed syringe is mounted above; it is charged with semen from several of the drones selected, which is injected into the queen's vagina, beyond the valvefold which is held out of the way by a hook; Figure 2.36a (E) shows the valvefold.

For readers who want to learn the technique, a manual with 125 photographs is available (Laidlaw, 1977), and a set of 111 colour slides (Laidlaw, 1976) for use with the manual. Other books in English are by Ruttner (1975*a*) and Holm (1986).

Instrumental insemination guarantees the identity of the drones that fertilize the queen, but in normal circumstances it does not produce *better fertilization* of the queen, or better colony performance, than natural mating. It should not be regarded as an essential component of a queen-rearing operation, or as one which gives it a high status.

8.25 Queen storage

In normal circumstances a colony will tolerate only one queen. Since the 1960s methods have been explored whereby a large number of queens could be stored for some weeks or months – and overwintered – more economically than by using a separate colony to house each queen. In early experiments, especially by Foti in Romania, each queen was caged with 40 to 60 workers, and the cages held in an incubator; a paper in German (Foti et al., 1962) describes the work and refers to the Romanian reports. This method achieved some success, but subsequent efforts have been directed towards keeping a number of queens (alone in individual cages) in a queenright or a queenless colony, containing plenty of workers, which feed and tend all the queens. These efforts led to methods for transporting, storing and overwintering queens in large numbers, that are successful on a commercial scale.

Queen banks

These were developed for the transport of young

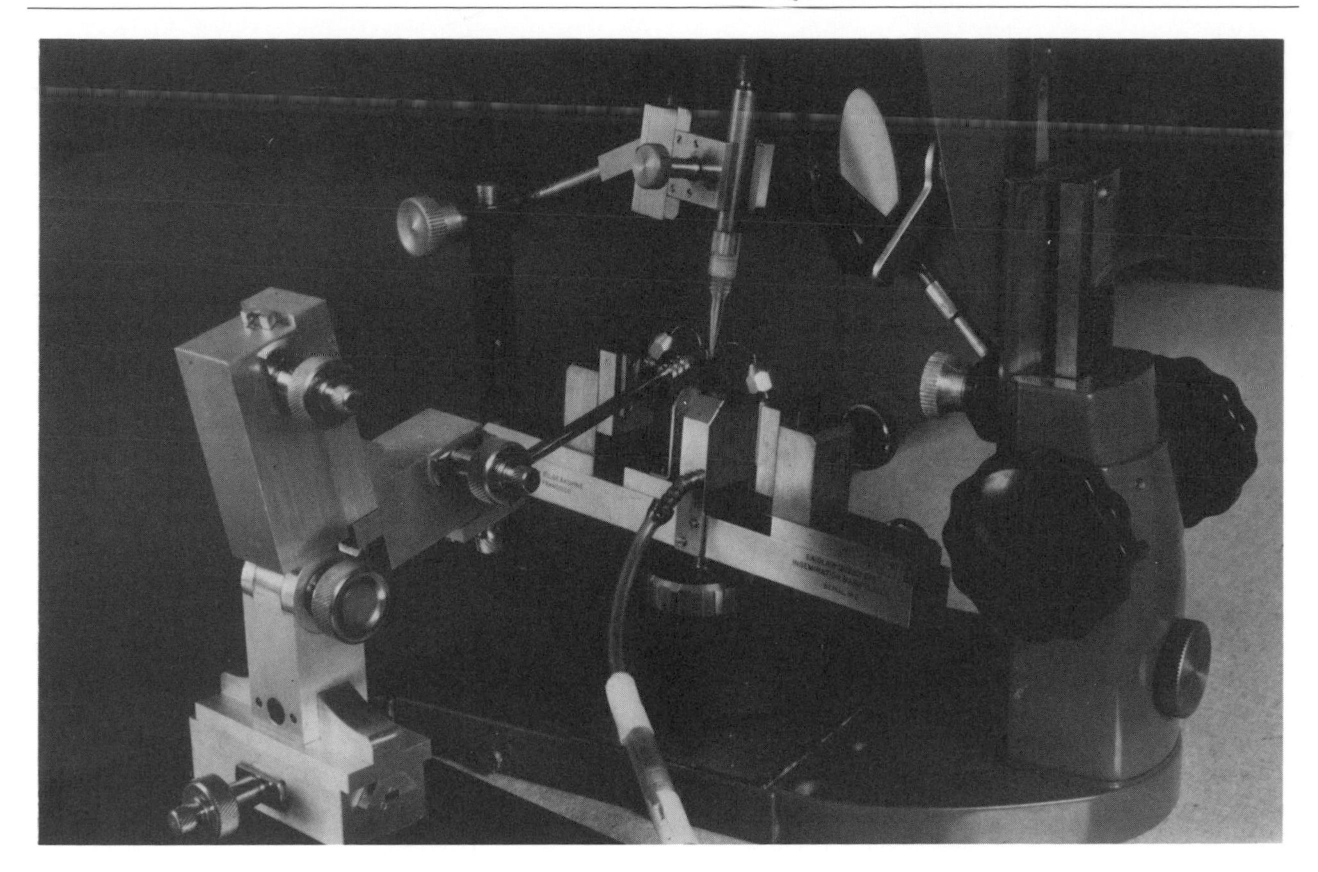

Figure 8.24a Laidlaw insemination apparatus, for use with a binocular microscope, 1968 (photo: J. M. F. de Camargo).
On the left is the valvefold manipulator; in the centre the queen is held in position below the (vertical) Mackensen syringe containing semen; the plastic tube (centre front) is for carbon dioxide (anaesthetic).

mated queens from the queen rearer to the purchaser, who could also store the queens received in the 'bank' for a short time during the active season, until he could conveniently introduce them to colonies. I first saw queen banks in 1967 in David Gear's apiary in New South Wales, Australia; see Figure 8.25a. An extra-thick brood frame holds three rows of 13 queen cages on each side, 78 in all. The outer (open) face of each cage is covered by wire gauze with holes large enough to allow workers to feed and lick the queens through them. The holes may be 2 mm across or less, or up to 3 mm. The frame is suspended in a ventilated expendable cardboard box, and on either side of it is a framed comb containing honey and pollen, and covered with bees. The box has a flight entrance at one end that can be fixed either open or closed.

No workers are put inside the cages with the queens, but an ample number must be present in the box to keep all the queens warm and well fed.

Reservoir (queen storage) colonies
Queens may be stored for several months during the active season in individual cages, mounted in extended brood frames (as above), one or more frames per colony. In Israel, Levinsohn and Lensky (1981) had quite good success in storing up to 56 young queens per colony, at any time of year. Both queenright and queenless colonies were used, and both vir-

Figure 8.25a Queen bank showing 'frame' for 78 queens (each in a cage), New South Wales, 1967.

gin and mated queens were stored. An ample supply of young bees was essential, and after the end of the main flow the colonies had to be fed with syrup. Protection of the colonies against heat and cold was also necessary.

Overwintering stored queens
This is a much more difficult operation, but in temperate zones the most important commercially, because it can enable queen rearers to sell queens before a purchaser can rear his own queens in spring, i.e. when queens are in great demand but short supply. Walsh's early experiments (1967) showed that in New Zealand conditions no more than two rows of cages should be stored over winter; with more, workers neglected queens low down in the hive. Nevertheless 65 queens out of 100 survived. Szabo (1977) found it much more difficult to keep queens alive in (queenless) storage colonies through the long winters of Alberta in Canada. Most queens survived one month, but only 11% were still alive after 6 months.

In 1985 I was taken to see a small building in the Wilbanks apiary in Georgia, USA, designed by Dr A. Dietz for commercial overwintering of queens. The building was thermally insulated, with a controlled heating unit; it is most important that the colonies are kept warm enough to prevent winter cluster formation. The building held 24 queenless storage colonies, each in two hive boxes; they accommodated altogether 8640 individually caged queens. In each box (which contained no brood) there were 3 storage frames, each with 3 rows of 10 compartments on each side (180 per box). The outer end of each compartment was covered with 3-mm mesh, and the inner end with masonite (pressed wood fibre board) which allowed insertion or removal of queens. A small amount of sugar syrup was fed to the colonies every 4 or 5 days. Queens were withdrawn for sale as required; 3680 queens were put into storage in autumn 1982, and 2189 sold between 13 November and 5 February (Dietz et al., 1983). Of the others, 10% had died, and of those still in storage 54% died by 15 March. There are inherent difficulties, which the authors discuss. All virgin queens survived storage for 6 months under the same conditions.

Information on queen storage is also given by Reid (1975), and by Woyke (1988) who draws attention to a possible danger to queens when many are caged in the same colony. If the mesh of the queen cages is as large as 3 mm, a worker can push through part of her head, and use her mandibles to bite the claws, tarsi, antennae, and even the wings of the queens. Woyke therefore recommends a mesh size of 1.2 to 1.5 mm, rather than 2.5 mm or larger.

8.26 Bee breeding

Breeding honeybees is much more difficult than breeding mammals or birds, because mating cannot be controlled in the same way, and a queen mates with a number of drones. Use of an isolated mating apiary (Section 8.23) ensures that the male parents are chosen by the breeder, but only by instrumental insemination (8.24) can the male parent be restricted to a specific individual.

Many beekeepers practise a limited form of bee breeding, to the extent that they rear daughter queens from those heading their best-performing colonies, but this gives little or no control over the drones that mate with the new queens. In some countries in continental Europe there have been schemes whereby beekeepers send selected queens they have reared to an isolated mating apiary where selected drone-producing colonies have been installed.

Professional bee breeders select for characters that are most important to them and their clients; of the characters listed below, all beekeepers would consider 1–4 important, and 7–9 are specialized aims for certain circumstances.

1. Increased honey production (but this is not a simple genetic character).
2. High potential rate of egg laying by the queen, and long-lived workers, giving populous colonies with a low swarming tendency.
3. Gentleness, bees remaining on the combs when colonies are inspected or manipulated – as they must be with modern management – and not flying round the operator or stinging him.
4. Colony characteristics such as good overwintering, early spring development, and a compact brood nest.
5. Low variability in colony performance, so that all colonies in an apiary can receive the same management at the same time (Section 6.2).
6. Resistance to brood diseases (Section 11.2), adult bee diseases (11.3), and parasites (11.4).
7. For package bee production, capability for much early brood rearing.
8. Effectiveness in pollinating a difficult crop, such as lucerne (alfalfa).
9. Light (golden) colour of queens; the only merit of this character is that a queen is easy to spot on the comb, but it has found inordinate favour among beekeepers.

Selected characteristics cannot be maintained in subsequent generations through natural mating, unless these matings also occur in isolated conditions. Moreover, bee breeding can impose a heavy burden on the operator in the maintenance of colonies with the required genetic characteristics. Advances in the storage of frozen spermatozoa (Harbo, 1979) may be helpful here.

The idea of crossing different strains of honeybees to exploit hybrid vigour is attractive, and heterotic effects have been well studied. Experiments have been done with inter-racial crosses, for instance in France and in the Soviet Union. In the USA, hybrids have been produced as a commercial operation since 1949. In this system four inbred lines A, B, C, D are produced from colonies selected after several years of evaluation and breeding. The lines are crossed in pairs whose characteristics complement each other (e.g. AB, CD), and the resultant hybrids are themselves crossed to give a 'double hybrid' (AB × CD). Two such double hybrids have been developed by Dadant and Sons, as Starline (from all Italian stock) and Midnite (from Carniolan and Caucasian stocks). References to further information are given by Crane (1984*d*).

The beekeeper who buys these hybrid queens in spring, and uses them to replace ordinary Italian and Caucasian queens, can get up to twice as much honey from the colonies concerned while they are headed by the purchased queens. But only F_1 hybrid queens benefit from heterosis, and in subsequent generations genetic reshuffling (recombination) produces variable results. Some beekeepers keep F_2 queens but replace F_3 queens.

Further information on breeding *Apis mellifera* can be obtained from Bee World (1964), Laidlaw (1977), Laidlaw and Page (1986), and Rinderer (1986). Page and Laidlaw's (1982) closed population breeding programme has been used in commercial practice with considerable success (Cobey & Lawrence, 1988).

8.27 Rearing queens of tropical *Apis mellifera*

Rearing queens of tropical *A. mellifera* in South Africa is based on methods for temperate-zone bees such as are described in Sections 8.21 to 8.23; see Anderson et al. (1983). F. G. Smith (1960) gives a shorter account of methods he used in Tanzania. C. J. Coleman, a commercial honey producer with about 1000 hives in Zimbabwe, is unusual in that he operates a planned queen-rearing programme with local African bees. The account here is based on what I learned from him in Zimbabwe in 1984, on information he has provided since then, and on an article by Parker (1986). In the 1970s only a third of his colonies were productive in any one year; a third just maintained themselves, and a third died out. Having learned of methods of selective breeding used with temperate-zone bees, he adapted them to improve on the strain of bees available to him locally.

Coleman has based his selection on the honey-producing ability of the colonies, and on the laying performance of the queen and her size and colour – because conspicuous queens can be spotted quickly in the colony. A ready source of local bees is available in swarms, which are abundant; see Section 7.42.2. Five hundred 5-frame bait hives, set up over a wide area to collect swarms, are inspected 7 times a year. The colonies so obtained are transferred to 10-frame hives placed in a 'nursery area', and their queens are killed; they are requeened with mated queens, or with queen cells from which queens are due to emerge, of the selected strain. This nursery area is within the 'gene zone' containing the selected large colonies in which queens are reared, small nuclei from which they mate, and the drone-producing colonies.

The numbers below refer to the operations described in Section 8.22.

1. The hive containing the breeder queen consists of two brood boxes, the queen being confined to a central 4-frame section which is separated from the other frames by wire queen excluders. Twice a week the frames in this queen section are replaced with empty drawn-out worker combs. The outer parts of the brood boxes contain 4 frames of pollen; all the rest of the frames contain sealed brood, and more frames of sealed brood (and of pollen) are introduced as necessary throughout the season, so that the colony population is very high and always consists predominantly of young bees.
2. The hive (1) containing the breeder queen is also used for rearing larvae. As soon as the colony has become large enough, a frame holding 30 queen cells with grafted larvae 3 days old is placed in each brood box. These cells are inspected after 3 days, and another similar frame is inserted in each brood box. Sealed queen cells are removed every 7–8 days. A cycle is established on this basis, and continues as long as queens are required.

3 to 5. The sealed queen cells are placed separately in 3- or 4-frame nuclei which have been queenless for at least 24 hours. Here the queens emerge from their cells, and in due course fly out to mate; during this time the hives are not inspected or otherwise disturbed. (These mating hives are relatively large, and Helmich et al. (1986) found that those for Africanized queens need to be twice as large, and with twice as many bees, as those used for European queens.)

The new queens are usually ready to fly about 15 days after grafting. To prevent the queens drifting to another hive, the hives are at least 1 m apart, with entrances facing in different directions. At 21 days, the nuclei are inspected to monitor the laying performance of the mated queens. Where this is inadequate, the hive is marked for reinspection after a further 5 days; if performance is still below standard, the queen is culled.

The mating apiary is on the windward side of a drone congregation area and within 500 m of it. The whole district is flooded with bees of the selected strain, and all wild colonies in it are killed, so virtually all drones there are of this strain, and drone-producing colonies are fed constantly to ensure a plentiful supply. The colony density is high, resulting in a dearth of nectar, and any wild colonies that remain do not produce drones. A strict check is maintained that bait hives brought into the nursery area contain no drones.

Coleman reared 6500 selected queens in 1985, and by using the above programme he has more than doubled his honey yield per colony.

Near Pretoria, South Africa, Fletcher and Tribe (1977) experimented with queenless colonies in observation hives, and showed that the larval period before queen-worker differentiation is probably less than 24 hours in the bees there – much shorter than in European *A. mellifera*; see Section 2.51. This reduces the ability of tropical African bees to rear emergency queens, and relatively few dequeened colonies were able to rear new queens. Fletcher and Tribe also give many data on the number and timing of queen flights.

Queen rearing with *A. m. capensis* presents some peculiar problems (Anderson et al., 1983). Cooke (1986/1988) describes in detail the system developed and practised by Ilse Harmann.

Methods used in Brazil with Africanized bees are described in Wiese's book (1985) and by Camazine (1986*a*).

8.28 Rearing queens of *Apis cerana*

Queen rearing with *Apis cerana* is well established in India and China (Phadke, 1987; Wongsiri, 1989), and follows similar principles to those used with *A. mellifera* (see Figure 8.28a). Queen cups should have a smaller internal diameter (7.5 mm) than for *A. mellifera* (9 mm) (Koeniger, 1986). Fewer queen cells should be reared per colony (16–20, Phadke, 1986). The best results seem to be obtained by 'double grafting', i.e. inserting temporary larvae (of any origin) in the queen cups, leaving them in the rearing colony for 24 hours, and then replacing them with larvae selected for rearing; these are put on the royal jelly the bees have supplied to the temporary larvae. Some beekeepers choose colonies in a swarming condition, and without eggs, to rear the larvae. For rearing drones, *A. mellifera* worker foundation is sometimes used, as the cell width is suitable. Some developmental stages do not, however, last as long (Table 2.41A), so the work is done to a slightly shorter time scale. The time intervals between the queen's emergence, mating and laying also show some differences from *A. mellifera*, as shown in Table 2.5A. If a colony is dequeened, an emergency queen is rarely reared, and laying workers are produced quite rapidly (Section 3.39).

Mating can present problems. It may be difficult to get mating nuclei established, on account of absconding, drifting and robbing, but inserting a frame of sealed brood can help to reduce drifting (Koeniger, 1986). Woyke (1973*b*, 1975) developed techniques for instrumental insemination of *A. cerana* queens.

Shah and Shah (1980) give information on Kashmir *A. cerana* queens, and F. A. Shah (1980) was able to keep a number of mated queens caged in a colony, as in a queen bank (Section 8.25). The system was used for transporting queens one year old to outapiaries. It seems unlikely that transport and introduction to colonies present much greater difficulties than with *A. mellifera* (Sharma, 1944), and queen introduction may be easier (Adlakha & Sharma, 1975).

Work has been started towards selection of strains more economically profitable for beekeeping than some in use; see e.g. Rana (1985), Verma (1987).

8.29 Multiple-queen colonies

The terms two-queen and multiple-queen colony are used here for colonies with one or more extra queens, whether throughout the year or only during the honey-producing season. Even before the present century, the idea of 'breaking' the natural unit of a honeybee colony – tens of thousands of workers headed by a single queen – was attracting attention. It is possible to induce worker bees to maintain a colony with more than one laying queen (a polygynous colony), the queens occupying different parts of the same brood nest (Darchen, 1960). In the 1870s F. A. Hannemann in Brazil operated enormous colonies by uniting many swarms, each queen being confined by queen excluders, and his largest colony gave 730 kg of honey in a season (Whyte, 1919).

When more than one queen per colony is used, each queen must have her own brood nest, separated off by queen excluders from that of any other queen. Requeening and swarm control can be integrated into the system. The benefit to the beekeeper of a multiple-queen colony is that its large population enables it to

Figure 8.28a A frame of *Apis cerana* queen cells lifted out from a queen-rearing colony (photo: Central Bee Research Institute, Pune, India).

store more honey than could be done by its component parts separately, each with one queen. But working even two-queen hives is a skilled operation not suited to all areas – or to all beekeepers, or to very large enterprises – and the six-queen colony referred to below is an interesting exception, not the rule.

Each hive box must be accessible for the beekeeper's inspection and the manipulation of frames. One option (Section 8.29.1) is to use movable boxes stacked in a single tier, the queen in each brood box being confined to it by a queen excluder above and below. This does not in fact need a special hive, only an extra high one, and a ladder or other device to reach high boxes. The beekeeper must be able to gain access to frames in any hive box as and when necessary, and Section 5.41 includes useful devices for tall hives. Another method (8.29.2) is to have an extra-long hive with brood nests in separate compartments side by side, each with a queen excluder between it and the honey super above – which is accessible to all workers of the colony. Special lifting gear is needed for hives of this type.

The bees discussed below are temperate-zone and tropical African *A. mellifera*, but multiple-queen colonies of *A. cerana* are also used (Shah & Shah, 1982*b*).

8.29.1 The multiple-queen colony in tiered boxes

The most important scientific exponent of the use of multiple-queen colonies was C. L. Farrar in the USA, who used two queens per colony during the active part of the season. In many papers from 1936 onwards (e.g. 1946, 1953) he demonstrated the substantial benefits of using such colonies. Two-queen colonies give more honey, and more honey per unit number of bees, than single-queen colonies; they also require less equipment per unit of honey produced. Colonies enter winter in very good condition, and with twice as much pollen as colonies that have had a single queen. The second queen can be reared at a time when colonies are strong and a heavy flow is in progress. A colony with two queens grows very rapidly, giving an optimal population during the honey flow. And if one queen fails partly or completely, the colony is still an effective honey-producing unit. Farrar used shallow Dadant boxes, 170 mm deep, throughout his hives, but this is in no way necessary.

Moeller continued Farrar's work, and in 1976 published the results of trials in the years 1967–1974. Average annual honey yields per hive (from packages) were 127 kg from two-queen colonies and 53 kg from single-queen colonies. Banker, who built up a successful commercial enterprise of 1500 two-queen colonies in Minnesota, USA, set out the system's disadvantages, as well as its advantages (1975). Proper timing of each stage in the colony manipulations is essential if the high potential honey yields are to be achieved; this imposes a severe discipline on the beekeeper. It limited Banker's scale of operations, and it would create difficulties for anyone not working full time at beekeeping. From an operational point of view, tall hives are inherently unstable, even when they are on level stands placed on a firm foundation, and manipulation of heavy top boxes is not easy.

The procedure can be understood from Figure 8.29a, which is based on 10-frame deep Langstroth boxes (Gojmerac, 1980). At stage A, in spring, the colony occupied three boxes, and had one queen. At a date between A and B, an additional flight entrance was provided in the top box, by drilling a hole in the hive wall. This box was placed above a solid divider board, and a young mated queen was introduced to the bees in it, which then behaved as a separate colony; in B it contains brood. By C, at the beginning of the main honey flow, each queen had brood in two boxes. The divider board separating the two brood nests was then replaced by a queen excluder, and in due course supers were put in place at the top of the hive (5 are shown in Figure 8.29a), with extra flight entrances. By D, 4 weeks before the end of the flow, it was already too late for any eggs not yet laid to give rise to foragers in time to increase the honey harvest. The queen excluder was therefore removed, so that one queen would kill the other and the survivor would continue to lay, making a single brood nest. (Usually the younger queen survives, so the colony is usefully requeened.) Finally at E, after the honey supers have been removed, and the queen has stopped laying, the colony is left with ample food for winter. At this time the extra flight entrances are closed.

Farrar's two-queen system is in use in a number of countries. Among amateur beekeepers in Europe, popular interest in multiple-queen colonies was aroused by Dugat's book *La ruche gratte-ciel à plusieurs reines* (3rd ed. 1947), translated as *The skyscraper hive* in 1948. A book was also published in Norway (Jensen, 1959).

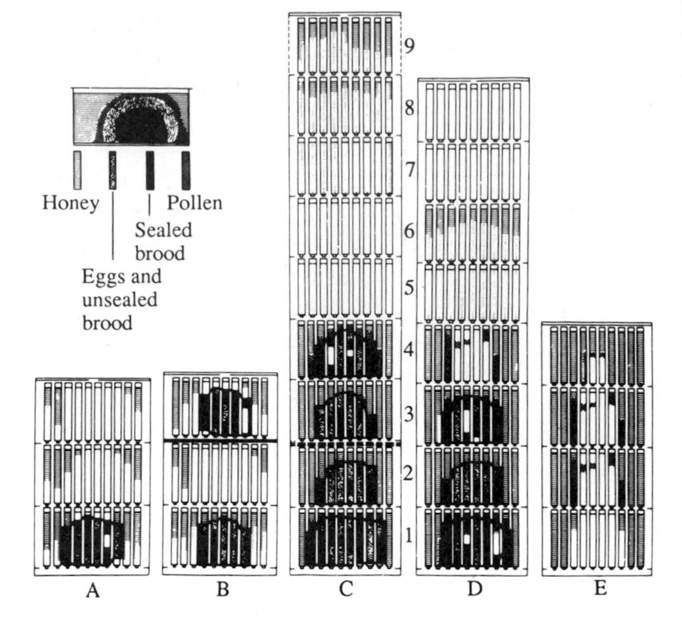

Figure 8.29a Steps in the management of a two-queen colony (Gojmerac, 1980).
See text for further details; dates relate to S. Wisconsin and N. Illinois, USA.
A 1–15 April
B 20 April to 5 May
C Start of main honey flow
D 4 weeks before end of main flow
E Return to single-queen colony

A world record honey crop

The ultimate in skyscraper hives (Kotze, 1949; Papadopoulo, 1987) was one used to create a world record for the honey yield of a single hive in one year. The hive, owned and operated by E. A. Schnetler, was kept on the same site near Pretoria in South Africa (in an area with honey flows from different eucalypts), from 20 March 1948 to 19 March 1949. A platform and a ladder were built to give the beekeeper access to the upper boxes for manipulations, and a system of stays prevented the hive being blown over by strong winds. On the starting day, all honey was removed from the hive, which was then placed on scales, where it remained throughout the test period. The hive was weighed daily after sunset, when the flying bees had returned, and Officers of the Transvaal Beekeepers' Association acted as scrutineers. The bees were selected local tropical African bees. A queen was in the lowest hive box with a queen excluder above it. As the year progressed, extra queens were introduced, each in a separate box with an excluder above and below it. Six queens were used to build up the colony population for the honey flow, and the hive was extended to 16 deep boxes, a height of 4 m. Additional entrances were provided, and also drone escapes which allowed exit but prevented re-entry of drones. Only 3 queens were present during the main flow.

During the period before the flow (only) the colony was fed with pollen substitute, in powder form provided in the apiary, and as raw egg mixed with sugar syrup fed at the hive in a Boardman feeder (Section 4.47, type 7); Papadopoulo (1987) had tested raw egg on 500 colonies in Greece, during three springs in the 1930s, with good results.

When supers of honey were removed, they were replaced by similar supers containing only empty combs, so that the total reduction in the weight of the hive was the weight of honey harvested. Weights of honey removed were:

1948	March 20	starting date
	August 15	42 kg
	September 4	30 kg
	October 3	131 kg
	October 24	159 kg
	November 21	236 kg
1949	February 20	247 kg
	March 19	115 kg
Total in 8 months		960 kg

8.29.2 The multiple-queen colony with lateral brood nests

As early as the 1890s, G. Wells in England used two queens in separate brood compartments side by side, with a common honey super above. It enabled him to harvest honey from fruit trees which flowered early, when normal colonies were too small to get enough honey to store (Kidd, 1921). Hives of a similar type have been used by a few beekeepers in various countries, and in some impressive commercial operations in Western Australia, initiated by Sid Murdock. The hives I saw in 1967 were 2.5 m long, the lowest storey being a unit comprising six brood compartments, each with 8 Langstroth deep frames and a queen excluder above it. Entrances to alternate brood boxes are usually faced in opposite directions, to minimize drifting. In the simplest form of the hive, the second storey (directly above the queen excluders) was a 50-frame honey super, accessible to all worker bees in the hive, but this had two disadvantages. If one of the brood nests became queenless for any reason, the bees tended not to rear a new queen in it. Also, when a repellent was used to clear bees from the honey super prior to harvesting (Section 5.32), the bees tended to flood out of the entrances, and queens might also leave – with the risk that they might return to a different entrance, and be killed.

Ken Gray's migratory apiary contained twenty of these hives, with all entrances facing the same way, so that he could work the hive from the back, free from flying bees. The hives had a second storey of six separate hive boxes, and the common super as a third storey, as in Figure 8.29b. This gave more flexibility in management; for instance a nucleus containing a new queen could be inserted in the second storey by using an excluder above, and newspaper both above and below. Although Figure 8.29b shows two beekeepers, all operations could be done by one using a small purpose-built rubber-wheeled gantry for lifting. To open the hive, the gantry is wheeled into position so that it straddles the hive lengthways. The top storey is winched up, and wheeled just in front of the hive, so that bees flying out from the underside return through one of the hive entrances. When harvesting honey, the gantry supporting the 50-frame super is wheeled to the truck and offloaded there.

Whether extended upwards or sideways, these two-queen and multiple-queen colonies are not for everyone. The beekeeper who operates them must be experienced, and also sufficiently dedicated to the system to accept the discipline involved in carrying it out. He must have access to prolific and predictable honey flows. The tiered system is more flexible, and of more general application, than the lateral system with long hives. And these long hives are very different from the simple single-storey long hives described in Section 10.3.

8.3 PRODUCING AND USING PACKAGE BEES

At high latitudes, large honey crops can be obtained in the long summer days (Section 1.52), but winters are so long that it may not be easy to keep bees alive through them, and honeybees cannot survive in the wild. One way of harvesting the honey is to transport bees each year from a lower latitude, where it is warm enough for bees to rear brood through most of the year, and queens can be produced early. There are difficulties and hazards in such transport of colonies, but most can be circumvented by taking 'package bees' with no brood or combs. Each unit consists of young adult workers, usually about 1 kg, with a cage containing a young mated queen, in a well ventilated disposable box which incorporates a supply of food. The packages are used to populate empty hives, and the colonies are tended until the end of the final honey

Figure 8.29b Multiple-queen hive with mini-gantry, W. Australia, 1967. The hive has 6 lateral brood compartments in the lowest (dark) box, then 6 separate honey supers, topped by a common super with 50 frames, which is being lifted up; the roof is in 6 sections.

flow, when they are killed, or overwintered to become permanent colonies.

A package bee industry is most likely to be viable where a single country stretches over a sufficient north-south distance (up to 2000 or even 2500 km, and at least 1000 km), to give a worthwhile seasonal advantage, i.e. the season where the packages are produced is earlier by two months or more than where the bees are used. The midsummer day length is much longer at high than at low latitudes. So there are special advantages where the country has honey-producing areas at high latitudes – above say 50° to 55°. Table 8.3A illustrates this effect. The extra-long day length (photoperiod) has two benefits. It allows the bees more flying hours each day, and the high insolation (radiation received from the sun) produces a greater intensity of plant growth, including nectar secretion; see Figure 12.21e.

At any one latitude, the interior of a temperate-zone land mass is likely to have a continental climate with hot dry summers, giving good honey yields, and long cold winters that make overwintering difficult. Package bees produced in milder coastal areas may make it possible for bees to be kept profitably in the interior.

On the above basis, areas suitable for the development of a package bee industry would be the USSR, and North America. In Norway and Sweden the latitude range for beekeeping is shorter. A package bee industry might be developed in China, but the country does not extend to latitudes much beyond 50°, nor does Japan. A tempting range of operation would be from North Africa to northern Europe, especially Scandinavia. But such a development is unlikely in the foreseeable future, because of the risk of transmitting diseases and parasites, and undesirable genetic traits, in bees sent north, and the few possibilities for organizing package bee production in North Africa.

In view of the spread of mites that parasitize honeybees, and of Africanized bees, the world's beekeepers may well look to the larger islands of the southern

Table 8.3A Length of midsummer daylight (photoperiod) at different latitudes in the northern hemisphere, with examples of beekeeping locations

Daylight lasts while the sun is above the horizon, or not more than 6° below it (limit of Civil twilight). 14.05 = 14 h 05 min.

Latitude	*Sunrise-sunset interval*	*Duration of daylight*	*Examples of locations*
30°	14.05	15.01	Houston, Texas; Shanghai; N. Africa, to 35°
40°	15.01	16.07	N. California; S. Caucasus; Beijing (= Peking)
50°	16.22	17.52	Winnipeg; Vladivostok
52°	16.44	18.22	Saskatoon; Saratov
54°	17.09	18.58	Edmonton; Kuybyshev
56°	17.37	19.43	Peace River; Moscow
58°	18.11	20.43	Sitka (Alaska); S. Sweden
60°	18.52	22.26	Helsinki; Leningrad
62°	19.45	24.00	Palmer (Alaska)
64°	21.02	24.00	Dawson City (Yukon); Fairbanks (Alaska), 65°

hemisphere for package bees. Export of packages from the southern to the northern hemisphere has an additional advantage in that the northern spring, when the packages are needed, is the southern autumn, when colony populations are high and bees can well be spared. In the southern hemisphere, New Zealand's package bee industry is already developing rapidly (see Section 8.32), and Australia has sent packages to Canada (Guilfoyle, 1988) and to Israel (New Zealand Beekeeper, 1987).

8.31 The USA package bee industry

The package bee industry was created by beekeepers in the USA, and until the 1980s served both the USA and Canada. The first boxes of bees without combs were produced by A. I. Root Co. in 1879, and a large-scale commercial shipment was first sent north in 1911, from Alabama (Cutts, 1961). Early packages contained only 1 or 0.5 lb of bees [450 or 230 g], in a box with several sides made of metal mesh to allow ventilation, and containing slats of wood for the bees to cluster on, although these were soon found to be unnecessary. Packages were sent by parcel post, but – understandably – boxes were required to be substantial, and double screening had to be used. At first a section of comb honey was provided for food. This was soon replaced by a box of candy, but the bees then suffered from dehydration, and finally an inverted can of sugar syrup (with small holes in the lid, Section 4.44) was fixed in each box. It is of course vital that the can does not come loose, and that the syrup does not spill. (For air transport a gel is used, such as that described in Section 8.34.)

As time went on, packages were increasingly transported by truck (Section 8.34), and boxes were standardized so that they could be packed without wasting space. They had to be spaced out to allow ventilation, and were often fixed together in sets of four (Figure 8.33b), each holding 2 lb (about 1 kg) of bees. Boxes are usually expendable, and most are made of wood, with metal mesh.

8.32 The package bee industry elsewhere

Canadian beekeeping has developed strongly since 1950, especially in the Prairie Provinces – Alberta, Saskatchewan and Manitoba – where honey yields are very high (annual average about 70 kg per hive); 80% of Canada's honey is now produced there (Nelson & Jay, 1982). Because of difficulties in overwintering, beekeeping depended heavily on packages from the USA, and 350 000 were imported in 1984. But by the 1960s there was already dissatisfaction with the quality of bees and queens, and some carried disease, especially nosema, which reduced their productivity; see Section 8.35. Also, Africanized bees are expected to arrive in southern USA in the late 1980s, and *Varroa* was found there in 1987. Canada closed her border with the USA to honeybees in 1988 (end of Section 8.72).

During the 1980s methods had been devised in western Canada for overwintering colonies, both out of doors and in special buildings (Section 6.54.2).

Efforts were also made to produce package bees, especially in southern British Columbia near the Pacific coast, which has relatively mild winters although the latitude is 50°N. The Trans-Canada Highway, opened in 1962, provided a hard-surface road across the Rocky Mountains between this area and the honey-producing Prairie Provinces. Queens were imported from New Zealand for use in locally produced packages; they presented relatively few disease problems, and were found satisfactory (Pankiw, 1974). In 1988 Canada imported 5000 two-queen packages from Western Australia (Guilfoyle, 1988). In a trial with Canadian reared queens in 1984, Winston (1986) produced 943 packages in April/May, 1 to 4 from a colony, and the net profit from these colonies was higher than from colonies producing honey alone. In Alberta, Pirker (1978*a*) was able to make packages in spring, at a profit, from colonies overwintered in a specially built winter bee house.

Southern Republics of the USSR, such as the Ukraine, Georgia and Azerbaidzhan, have a geographical position *vis-à-vis* northern beekeeping regions similar to that of the southern package-producing states of the USA. I saw package bee production in Georgia (south of the Caucasus mountains near the Black Sea) in 1962, when it was in its early stages, and the packages were heavy and cumbersome, but there has been much progress since then. Many packages are sent by post to farms and individuals who have ordered them, and air transport is also used (Mertsin, 1980). But package bees have not played as important a part in beekeeping in the USSR as in North America.

Air transport makes it possible to send bees as packages to distant islands, although this may not always be a wise or safe operation (Section 8.7). In New Zealand most package bees are produced in the North Island, north of latitude 38°S. They have been sent to Fiji and Tuvalu in the Pacific, and to Papua New Guinea, Thailand and Pakistan. Those successfully introduced to Vaitupu, Tuvalu, had been 48 h in transit on a seaplane (Thin, 1984). Air transport also makes possible New Zealand's largest and most distant operation, which spans both hemispheres: the regular export of packages to Canada, over 10 000 km away. In 1986 a total of 16.5 tonnes of bees were exported, mainly in April, which is New Zealand's autumn and Canada's spring. Within New Zealand, packages are distributed throughout both Islands, about 1200 a year being used north of latitude 38°S.

Package bees have many other uses besides supporting a honey-producing industry based on them, as in Canada (Section 8.35). For instance in New Zealand, where colonies can be overwintered in the temperate island climate, packages are used in spring to provide extra colonies or to make up winter losses. Later in the season they can strengthen populations of colonies that continue to work the honeydew flow from *Nothofagus* (Section 12.22.2) in the autumn and winter (Bryant, 1987*b*).

8.33 Colony management for producing package bees

Colonies for package bee production must have large numbers of young workers during the whole of the period when the packages are needed – in North America this lasts from late March to early May. Much brood must therefore be reared from January/February onwards, and package bee apiaries are usually situated where much nectar and pollen is available throughout the relevant period. If not, much pollen and syrup must be fed. In many ways colony management is not very different from that for honey production, and the removal of so many bees in spring effectively prevents swarming. However, timing is critical throughout the operation.

Large-scale queen rearing is carried out in the normal way (Section 8.22), and here also timing is critical. Many hundreds of caged queens, one for each package ordered, *must* be ready and on site on the day the packages are made up. The following procedure to obtain the bees is used in a Weaver package-bee apiary near Navasota, Texas. Every box of every hive in the apiary is shaken in turn, over a large square funnel leading down through a queen excluder to a bee-storage box below, fitted with a chute (Figure 8.33a). Most of the bees fall from the frames into the funnel, and the queen, held back by the excluder, is returned to the hive. (Alternatively the comb with the queen on is set aside before shaking is begun.) Many bees escape, and they return in sufficient numbers to keep the colony going. The storage box holds the bees from 3–4 hives; when it is full, it is pulled on its sledge-runners to the 'filling station' (Figure 8.33b) where the empty package boxes are filled. Those shown were part of an order for 850 which took about 1½ days to collect, and were loaded into a truck to be driven 2500 km north to Minnesota.

Colonies are shaken two, or occasionally three, times in the season, removing altogether 3–4 kg of bees from each. Many bees get crushed, and the air can become pungent with alarm odour (Table 2.35A); apiaries used must be well away from livestock and residential areas. The shaking operation is hard on the operators, who must work for a long period in the heat – repeatedly moving heavy weights – in full protective

Figure 8.33a (above) Equipment used in Figure 8.33b.
upper left Funnel viewed from above, showing queen excluder, through which bees are shaken from hive boxes.
right Funnel viewed from below.
lower left Sledge used to take bees from hives to weighing machine (Figure 8.33b), showing the hole to receive the funnel, and the chute (with closure) for pouring bees out.

clothing and with many bees flying around them and ready to sting.

York (1975) gives some general details of package bee production in the USA.

8.34 Transport of package bees

In the USA, packages of bees were carried by rail (Railway Express, or post) until the early 1970s, when road and air replaced rail for freight transport. Many beekeepers then found that bees suffered because ventilation was poorer in the commercial or postal trucks, and some lots suffocated; also packages were sometimes handled roughly, and bees escaped. A number of beekeepers in the north therefore drive south to collect their packages from the producer, so that the safety of the bees during the journey is in their own hands. Sometimes one beekeeper collects for a local group. A truckload of 1000–2000 packages represents a considerable capital investment in itself; also, if the bees

Figure 8.33b (below) Filling package boxes with bees, Texas, USA, 1957.
Sets of four packages are stacked on the left. On the right, one set is on a weighing machine, monitored by the seated beekeeper. The standing beekeeper is pouring bees into a round funnel (Figure 8.33a) that is supported by the feeder-hole of the package being filled.

die, the year's honey production is lost. The truckload can generate a great deal of heat, and one of the greatest risks is death of the bees through overheating. Specially adapted vans are used, with ventilation holes at front and back, covered with hinged or sliding doors (Hamilton, 1960). Slight overheating may be reduced by sprinkling water over and between the rows of packages. The doors must be continually adjusted as the air temperature changes, and closed completely if it is abnormally low. There are benefits in using a refrigerated container such as that in Figure 8.34a, but the bees are then in a closed environment, and accidental overheating can still kill them. In New Zealand the whole of an early shipment of 1200 packages was lost because stray bees were sucked into the controls that monitored the temperature, and put it out of action (Bryant, 1987*b*).

Figure 8.34a Container load of 1200 packages (on the left) being attached to a truck with a refrigeration unit, for transport to railhead, N Island, New Zealand, 1985 (photo: T. Bryant).

Air transport has the advantage of speed, and has been used within the USA (e.g. Ashby et al., 1982) as well as for intercontinental transport (Section 8.32). Air transport is, however, subject to a number of hazards that are unpredictable in their occurrence: extreme temperatures in the aircraft or during delays on the ground, insecticide sprays, and mishandling through lack of knowledge by staff involved. Successful transport of package bees is even more difficult if bees are sent without queens (Bryant, 1987*b*).

Ashby et al. (1986) set out to develop a better type of container, and solved many of the problems by using a cylindrical tube of fibreboard, shown in Figure 8.34b. The ventilated polyethylene end-closures (which are moulded) have external fins to prevent air blockage by other cargo, and are also designed to prevent bees stinging a handler through the ventilation holes. Inside the tube is a piece of fairly rigid wire mesh (1.3-cm^2 mesh) which is made Z-shaped by folding it twice so that it fits into the tube. Bees cluster on it, or spread out to avoid overheating, and near its centre the queen cage and feeder are attached. The feeder is a soft-drinks can with a circular opening, filled with a gel instead of liquid syrup (granulated sugar 1.4 litres, agar 10 g, cold water 1.7 litres). A can holding 330 ml is sufficient for 2–3 days, and for a longer period a 500-ml can is used.

Another type of tube (Arataki) has been developed by Berry (1987) and used for transporting packages by air from New Zealand to Canada; the tubes travel upright, 704 per pallet.

8.35 Colony management with package bees

Once colonies from package bees are well established in their hives, their management differs little from that of permanent colonies – over which, however, they have several advantages. The package has a young queen, reared early in the season. Packages are certified free from brood diseases, and should be free from adult bee diseases. If they are not, nosema is the most likely infection; in different years, it was present in 26–53% of packages received in Canada from USA (Jay & Dixon, 1982). New infection may be prevented by using hives and combs that have been sterilized while free from bees.

Troubles with package bees, mentioned only briefly here, are likely to be associated with the first few days

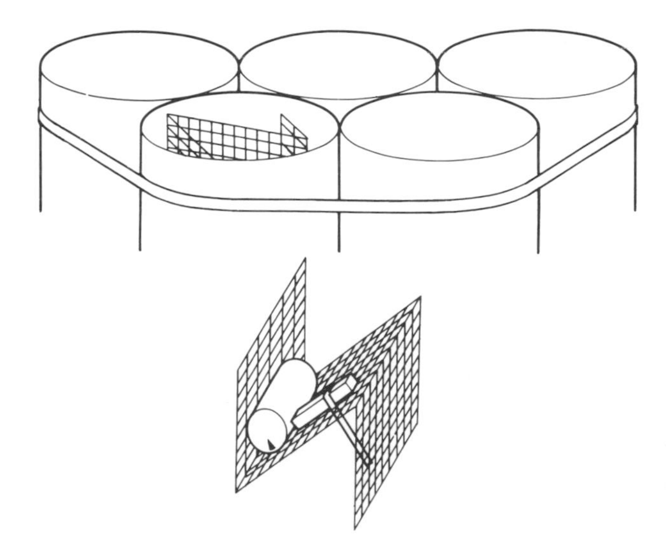

Figure 8.34b Fibreboard tubes for transporting package bees, as described in the text (Ashby et al., 1986). Internal diameter 20 cm; length 36 cm for 0.9 kg bees, 41 cm for 1.4 kg of bees. The five tubes shown are strapped together ready for transport (*above*). A feeder and queen cage are attached to the wire mesh, shown *below*.

Figure 8.35a Hiving package bees, Colorado, USA, April 1957.
The beekeeper will invert the package over the space left in the hive between the frames (here, fitted with foundation). The bees are jerked and shaken out through the feeder hole of the package and, when the remaining frames are inserted into the space, the queen cage will be attached to the centre one.

after the bees arrive, and with their end-of-season disposal (see below). Ideally, packages should be installed in hives 8–12 weeks before the first main honey flow, and when some nectar and pollen are already available. If the weather is bad, it may be best to feed the bees in the packages and to store these in a cool dark place until it improves.

As soon as the bees are in the hives, sugar syrup should be fed to them. This is essential if they are hived on foundation. In Figure 8.35a the beekeeper is shaking syrup from a feeder on to the bees still in the package, so that they take it quickly. If packages were used in the previous year, and killed at the end of the season, combs of food from their hives will have been stored through the cold winter, and each new colony is then usually given the equivalent of 4 frames of honey and 2 of pollen.

Many bees may be lost when they start flying, up to 25–40% of the colony population on the first day. Jay (1969) identified some factors that lead to high losses of flying bees at this time; they include snow-covered ground, a hive position central in a long or short row, and high winds. Losses were reduced by placing hives in a clearing within windbreaks, but were not affected by air temperature, landmarks near the hives, hive spacing, age of bees, and whether or not the colonies were fed, or placed inside the hive or on top of the frames, smoking after hiving (Jay, 1970), and shutting the bees in the hive until evening (Jay, 1980). In addition, some bees drifted to other hives, but this was rather similar to the drifting of bees between hives in general (see Section 6.62.1), and continued after the losses of bees had dropped off. Both losses and drifting could be reduced by putting the package bees in hives in late evening instead of during the day (Jay, 1980).

Nelson and Jay (1982) have described honey production with package bee colonies in Canada, and Avetisyan and Gaidar (1974) published results from Siberia, where package colonies gave somewhat more honey than overwintered ones. Once the last honey flow has finished, and the supers of honey have been removed, colonies that will not be overwintered are killed, heartless though this seems to those not accustomed to it.

Hydrogen cyanide has been commonly used for killing the bees, produced by exposing calcium cyanide to the moisture of the air. The gas is effective, but is extremely poisonous, and strict precautions must be taken if the operator is not to risk death. In many countries the use of calcium cyanide is not permitted. Following a 1979 study on residue levels in honey and wax, it has been given full registration by Agriculture Canada for killing bees, after the honey for human consumption has been removed. Instructions issued by the British Columbia Department of Agriculture and Food (Manitoba Beekeeper, 1986) are summarized here because the information is not readily available. But full instructions must be obtained and read in detail before cyanide is used, or handled in any way.

Calcium cyanide may be transported in case lots by truck, but may not be sent by post, and should not be carried in a car. It must always be kept locked in a secure place with external ventilation, and never in a place occupied by people. When taken to the hives, it should have packing round it, inside a larger container with a tight-fitting lid which is locked. Goggles and impervious gloves that can be washed must always be worn, and a special type of respirator of which details are given. The maximum exposure allowable for a person is very low, 10 ppm for an 8-hour period. No one should ever work alone with calcium cyanide.

It is best to kill colonies in late evening. For each hive, spread 1–2 tablespoons (12–25 g) of calcium cyanide on a large sheet of cardboard which can easily be slipped inside the hive entrance and on to the floor board. Then block up the entrance, and jar the hive to make the bees active. Do not open the hive until next morning. A pump or dust gun can be used if there are many hives.

A much safer agent for the operator would be an insecticide that is relatively nontoxic to man. In the USA use of the insecticide resmethrin is permitted for killing diseased colonies (Moeller & Corley, 1977). However, resmethrin contaminates beeswax combs, and the residue is not removed by normal wax rendering. Carbaryl is also registered by the USA Environmental Protection Agency for killing bees. Neither is regarded by everyone as suitable for killing bees on combs in hives, on account of the residual toxicity (Killion, 1986). The University of California (1983*a*) suggests using the insecticide Phostoxin (aluminium phosphide), whose residue in wax combs disappears after they are aired for a few days. For killing bees in hive equipment before this is irradiated to sterilize it, Bruce White (1987) in Australia is using a mixture of 93% ethyl acetate, 3% carbon disulphide, 2% dimethyl sulphoxide and 2% acetic acid; the mixture is blown on to the combs, and it leaves no residue (Bryant, 1987*b*).

By whatever method the bees are killed, the brood boxes will contain combs containing honey, pollen and dead brood, and several kg of dead adult bees. The honey and pollen are stored for use by bees hived from new packages next spring, which will also clean up the dead brood. The adult bees are scattered over the ground, or buried.

8.4 MANAGING HONEYBEES FOR CROP POLLINATION

Section 2.44 explains how bees pollinate flowers in the course of their foraging activities. In an area of mixed bee forage, such as occurs in small-scale agriculture or suburban gardens, there are likely to be scattered distributions of both wild bees and honeybees. These bees may well pollinate any crops, and the mixed bee forage may provide them with nectar and pollen throughout the active season. It may also enable honeybee colonies to store sufficient for the next dearth period, whether or not any surplus honey is left for the beekeepers. Table 8.4A gives information on bee-pollinated crop plants, many of which are included in Appendix 1 as important honey sources. So it may be mutually advantageous for a beekeeper within easy reach of such a crop to arrange with the grower that hives should be placed on the crop during its flowering period. However, colonies prepared specifically for pollinating work – for instance with extra unsealed brood (Section 8.42) – consume nectar or honey as well as pollen, and the beekeeper is unlikely to harvest much honey from them. He must therefore be recompensed by the grower for his work in preparing, moving and tending the hives, and for his capital investment. The moving of hives is covered in Section 8.1 and its subsections.

Beekeepers who hire out hives of bees for crop pollination need to have a sound legal contract with the crop grower; a recommended text is given in Figure 15.63a. They should also be aware of the risks of insecticide poisoning (Section 11.7). Section 15.5 gives details of laws in relation to pesticide damage to bees, and the impact of pest management on bees in relation to pollination is discussed by Crane and Walker (1983).

Most of the present section deals with European *Apis mellifera*, which has been much studied as a crop pollinator, but Africanized bees as crop pollinators are discussed in Section 7.43.2. Management of *A. cerana* is very similar (Figure 8.42a); but its flight range, and often its colony population, are smaller, and few data are available on the most effective number of hives to be used per hectare for different crops. *A. dorsata* and *A. florea* are both very efficient pollinators; for instance in one study *A. florea* constituted 75% of foraging insects on a *Brassica* crop in India. But neither of these bees can easily be managed for crop pollination.

Other bees are reared specifically for pollination including bumble bees (*Bombus*), leafcutter bees (*Megachile*), and alkali bees (*Nomia*); see Section 8.5 and its subsections.

Table 8.4A Some important crop plants pollinated by honeybees and/or other insects

Data extracted from Crane and Walker (1984*a*). Plants marked * are important honey sources and are included in Appendix 1. For plants in List 1, production of the *crop harvested* depends on pollination; for those in List 2, pollination is needed only *if seed is required* for propagation or in plant breeding. All plants are entirely or partly insect-pollinated, and the right-hand column indicates:

(a) the value of honeybees
(b) if known, the number of hives (of *Apis mellifera*) required per hectare
(c) other remarks.

Common and botanical names of crop	*Value of honeybees; other remarks*
List 1 Plants whose crop production depends on pollination	
acerola – *Malpighia glabra*	more information needed
allspice – *Pimenta dioica*	increase yield
almond – *Prunus dulcis*	increase yield, 5–8 hives/ha
* angelica – *Angelica archangelica*	known to pollinate
anise – *Pimpinella anisum*	more information needed
* apple – *Malus domestica*	increase yield, 2+ hives/ha
apricot – *Prunus armeniaca*	increase yield, 2.5 hives/ha
aubergine = eggplant	
* avocado – *Persea americana*	increase yield, 2–4 hives/ha
balsam pear – *Momordica charantia*	increase yield
bean, adzuki – *Vigna angularis*	more information needed
bean, broad = bean, field	
bean, butter – *Phaseolus lunatus*	known to pollinate
* bean, field or faba – *Vicia faba*	increase yield, hives recommended
bean, haricot or French or kidney – *Phaseolus vulgaris*	known to pollinate
bean, horse or jack – *Canavalia ensiformis*	known to pollinate
bean, Lima = bean, butter	
bean, mung – *Vigna radiata*	more information needed
bean, rice – *Vigna umbellata*	more information needed
* bean, runner – *Phaseolus coccineus*	increase yield, hives recommended
* bean, soya – *Glycine max*	increase yield
bean, sword – *Canavalia gladiata*	known to pollinate
* ber – *Ziziphus mauritiana*	known to pollinate
* bergamot – *Citrus bergamia*	*see* citrus, general
* blackberry – *Rubus fruticosus*	increase yield, 7–10 hives/ha
blueberry, highbush – *Vaccinium* spp	increase yield, 10+ hives/ha
blueberry, lowbush – *Vaccinium* spp	increase yield, 2.5 hives/ha
* buckwheat – *Fagopyrum esculentum*	increase yield, 2.5–8 hives/ha
cacao = cocoa	
cajan = pea, pigeon	
carambola – *Averrhoa carambola*	known to pollinate
caraway – *Carum carvi*	known to pollinate
cardamom – *Elettaria cardamomum*	increase yield
* cashew – *Anacardium occidentale*	known to pollinate
castor – *Ricinus communis*	known to pollinate
chayote – *Sechium edule*	more information needed
cherimoya – *Annona cherimola*	more information needed
cherry, sweet – *Prunus avium*	increase yield, 2.5–3 hives/ha
cherry, sour – *Prunus cerasus*	increase yield
* chestnut, Japanese – *Castanea pubinervis*	more information needed
* chestnut, sweet or Spanish – *C. sativa*	increase yield, 1.5–100 hives/ha

(*continued*)

Table 8.4A *(continued)*

Common and botanical names of crop	*Value of honeybees; other remarks*
Chinese goosberry = kiwi fruit	
* citron – *Citrus medica*	may be major pollinator; *see also* next entry
* citrus, general – *Citrus*	great variation; 1–2 hives/ha for some varieties
cocoa – *Theobroma cacao*	midges are important pollinators
* coconut palm – *Cocos nucifera*	increase yield
* coffee – *Coffea arabica*	increase yield, hives recommended
colza, Indian = sarson	
* coriander – *Coriandrum sativum*	increase yield
* cotton – *Gossypium* spp	increase yield, 0.5–12 hives/ha
cranberry, American – *Vaccinium macrocarpon*	increase yield, 0.5+ hives/ha
cress, rocket – *Eruca sativa*	probably pollinate
cucumber and gherkin – *Cucumis sativus*	increase yield, (up to) 10 hives/ha
cucumber, bitter = balsam pear	
currant, black – *Ribes nigrum*	increase yield, 6 hives/ha
currant, red – *Ribes rubrum*	increase yield, 8 hives/ha
date palm – *Phoenix dactylifera*	insects minor pollinators; hand pollination used
dill – *Anethum graveolens*	increase yield, hives recommended
* durian – *Durio zibethinus*	more information needed
eggplant – *Solanum melongena*	increase yield
feijoa – *Feijoa sellowiana*	increase yield
fennel – *Foeniculum vulgare*	known to pollinate
fig – *Ficus*	each sp. pollinated by different wasp sp.
gherkin, *see* cucumber	
gooseberry, Chinese = kiwi fruit	
gooseberry, European – *Ribes grossularia*	increase yield, 0.5–3.5 hives/ha
* gourds; pumpkin, squash, marrow – *Cucurbita* spp	increase yield, 2–4 hives/ha
gourd, ridge (and smooth or sponge) = loofah (and loofah, angled)	
gram, horse – *Macrotyloma uniflorum* v. *uniflorum*	increase yield
gram, red = pea, pigeon	
granadilla = passion fruit	
granadilla, giant – *Passiflora quadrangularis*	hand pollination recommended
grape – *Vitis* spp	increase yield, 1 hive/ha
* grapefruit – *Citrus paradisi*	more information needed
* guava – *Psidium guajava*	known to pollinate
* jujube, Chinese – *Ziziphus jujuba*	known to pollinate
jujube, Indian = ber	
kapok = silk-cotton tree	
kiwi fruit – *Actinidia chinensis*	increase yield, 8 hives/ha
kola nut – *Cola acuminata* and *C. nitida*	more information needed
* lemon – *Citrus limon*	known to pollinate
* lime – *Citrus aurantifolia*	increase yield
* linseed – *Linum usitatissum*	known to pollinate
litchi = lychee	
* longan – *Euphoria longan*	increase yield
loofah – *Luffa cylindrica*	may be important
loofah, angled – *Luffa acutangula*	hawkmoths are one pollintator
* lychee – *Litchi chinensis*	known to pollinate
macadamia – *Macadamia integrifolia*	increase yield, 5–8 hives/ha
* mandarin orange – *Citrus reticulata*	increase yield, 4 hives/ha

(continued)

Table 8.4A (*continued*)

Common and botanical names of crop	*Value of honeybees; other remarks*
* mango – *Mangifera indica*	increase yield, 8–15 hives/ha
* melon – *Cucumis melo*	increase yield, 0.5–7.5 hives/ha
melon, water – *Citrullus lanatus*	increase yield, 0.5–3 hives/ha
* milkweed – *Asclepias syriaca*	known to pollinate
mu tree = tung [*Aleurites montana*]	
* mustard, black or brown – *Brassica nigra*	increase yield
* mustard, field – *Brassica campestris*	increase yield
* mustard, Indian – *Brassica juncea*	increase yield
* mustard, white – *Sinapis alba*	increase yield, 2.5 hives/ha
nectarine, *see* peach	
* niger – *Guizotia abyssinica*	increase yield
nutmeg – *Myristica fragrans*	more information needed
* oil palm – *Elaeis guineensis*	pollinators include thrips and weevils
okra – *Abelmoschus esculentus*	known to pollinate
* orange, sweet – *Citrus sinensis*	*see* citrus, general
* orange, Seville – *C. aurantium*	*see* citrus, general
* papaya – *Carica papaya*	increase yield
passion fruit – *Passiflora edulis*	known to pollinate
pawpaw or papaw – *Asimina triloba*	known to pollinate
pawpaw [*Carica papaya*] = papaya	
pea, garden or field – *Pisum sativum*	known to pollinate
* pea, pigeon – *Cajanus cajan*	known to pollinate
peach and nectarine – *Prunus persica*	increase yield,1–2.5 hives/ha
pear – *Pyrus communis*	increase yield, 1–5 hives/ha
persimmon – *Diospyros kaki*	known to pollinate
pili nut – *Canarium ovatum*	more information needed
plum – *Prunus domestica*	increase yield, 2.5 hives/ha
* pomelo – *Citrus grandis*	*see* citrus, general
pyrethrum – *Chrysanthemum cinerariifolium*	more information needed
quince – *Cydonia oblonga*	known to pollinate
* rambutan – *Nephelium lappaceum*	increase yield
rape, Indian = toria	
* rape, oilseed – *Brassica napus* v. *oleifera*	increase yield, 2–6 hives/ha
* rape, turnip – *Brassica rapa* subsp. *oleifera*	increase yield, 2.5–5 hives/ha
* raspberry – *Rubus idaeus*	increase yield, 0.5–2 hives/ha
rocket-salad = cress, rocket	
safflower – *Carthamus tinctorius*	increase yield, 2 hives/ha
* sarson – *Brassica campestris* v. *sarson*	increase yield
sarson, brown = toria	
* satsuma – *Citrus unshiu*	*see* citrus, general
saunf = fennel	
scarlet runner = bean, runner	
* sesame – *Sesamum indicum*	known to pollinate
* silk-cotton tree – *Ceiba pentandra*	bats pollinate, also insects probably including honeybees
soya bean = bean, soya	
strawberry – *Fragaria* × *ananassa*	increase yield, 25+ hives/ha
* sunflower – *Helianthus annuus*	increase yield, 1–4 hives/ha
tamarugo – *Prosopis tamarugo*	known to pollinate
tangerine – *Citrus deliciosa*	*see* citrus, general

(*continued*)

Table 8.4A *(continued)*

Common and botanical names of crop	*Value of honeybees; other remarks*
tangerine [*C. reticulata*] = mandarin orange	
tomato – *Lycopersicon esculentum*	known to pollinate
* toria – *Brassica campestris* v. *dichotoma*	known to pollinate
tung – *Aleurites fordii*	known to pollinate
tung – *A. montana*	known to pollinate
yicib or ye-eb – *Cordeauxia edulis*	more information needed

List 2 Plants whose propagation by seed depends on pollination

Plants in single brackets are usually propagated by seed only for breeding purposes; plants in double brackets are not usually propagated by seed at all.

alfalfa = lucerne	
asparagus – *Asparagus officinalis*	increase yield, 5 hives/ha
beet, sugar; beetroot; mangold	most pollination by wind, some by honeybees
* berseem – *Trifolium alexandrinum*	known to pollinate
birdsfoot trefoil = trefoil, birdsfoot	
* borage – *Borago officinalis*	known to pollinate
box elder – *Acer negundo*	more information needed
cabbage, Chinese (pak-choi) – *Brassica chinensis*	known to pollinate
cabbage, Chinese (pe-tsai) – *B. pekinensis*	increase yield
* carrot – *Daucus carota*	increase yield, hives recommended
cassava – *Manihot esculentus*	more information needed
cauliflower – *Brassica oleracea* v. *botrytis*	increase yield
celery and celeriac – *Apium graveolens*	known to pollinate
chicory – *Cichorium intybus*	known to pollinate
chives – *Allium schoenoprasum*	known to pollinate
cinchone = quinine	
* clove – *Syzygium aromaticum*	known to pollinate
* clover, alsike – *Trifolium hybridum*	increase yield, 2–8 hives/ha
clover, arrowleaf – *T. vesiculosum*	increase yield, 2.5 hives/ha
clover, ball – *T. nigrescens*	increase yield
clover, bush = lespedeza	
* clover, crimson – *T. incarnatum*	increase yield, (up to) 12 hives/ha
clover, Egyptian = berseem	
clover, hop = medick, black	
* clover, Persian – *T. resupinatum*	increase yield, 4–5 hives/ha
* clover, red – *T. pratense*	increase yield, 3–15 hives/ha
* clover, strawberry – *T. fragiferum*	known to pollinate
* clover, sweet – *Melilotus* spp	increase yield, 2–5 hives/ha
* clover, white – *T. repens*	increase yield, 2–3 hives/ha
cole crops – *Brassica oleracea*	increase yield, hives recommended
endive – *Cichorium endivia*	more information needed
eucalypts – *Eucalyptus* spp	importance varies between species
* flax – *Linum usitatissimum*	known to pollinate
hemp, sunn – *Crotalaria juncea*	more information needed
henequen – *Agave fourcroydes*	known to pollinate
kale, sea – *Crambe maritima*	known to pollinate
kenaf – *Hibiscus cannabinus*	increase yield, hives recommended
* lavender – *Lavandula angustifolia*	known to pollinate
* lavender, spike – *L. latifolia*	known to pollinate

(continued)

Table 8.4A *(continued)*

Common and botanical names of crop	*Value of honeybees; other remarks*
leek – *Allium porrum*	known to pollinate
* lespedeza – *Lespedeza* spp	increase yields, 2 + hives/ha
lettuce – *Lactuca sativa*	more information needed
* lucerne – *Medicago sativa*	increase yield, 4–8 hives/ha
lupin – *Lupinus* spp	known to pollinate
manioc = cassava	
marjoram – *Origanum vulgare*	known to pollinate
* medick, black – *Medicago lupulina*	more information needed
melilot = clover, sweet	
mint, garden – *Mentha spicata*	more information needed
onion – *Allium cepa*	increase yield, 12–36 hives/ha
pak-choi = cabbage, Chinese [*Brassica chinensis*]	
parsley – *Petroselinum crispum*	more information needed
parsnip – *Pastinaca sativa*	known to pollinate
peppermint – *Mentha* × *piperita*	more information needed
pe-tsai = cabbage, Chinese [*Brassica pekinensis*]	
* potato, sweet – *Ipomoea batatas*	known to pollinate
quinine – *Cinchona* spp	more information needed
radish – *Raphanus sativus*	increase yield, 5 hives/ha
rhubarb – *Rheum rhaponticum*	more information needed
* rosemary – *Rosmarinus officinalis*	known to pollinate
rutabaga = swede	
saffron – *Crocus sativus*	known to pollinate
* sage – *Salvia officinalis*	known to pollinate
* sainfoin = *Onobrychis viciifolia*	increase yield, (up to) 10 hives/ha
salsify – *Tragopogon porrifolius*	known to pollinate
sann hemp = hemp, sunn	
sisal – *Agave sisalana*	known to pollinate
spearmint = mint, garden	
sulla = vetch, sweet	
sunn hemp = hemp, sunn	
swede – *Brassica napus* v. *napobrassica*	increase yield, 2.5 + hives/ha
sweet clover = clover, sweet	
tapioca = cassava	
tea – *Camellia sinensis*	bees important in one area and flies in another
* tobacco – *Nicotiana tabacum*	known to pollinate
* trefoil, birdsfoot – *Lotus corniculatus*	increase yield, 2 hives/ha
trefoil, yellow = medick, black	
turnip – *Brassica rapa* subsp. *rapa*	increase yield, 2.5 hives/ha
* vetch, hairy – *Vicia villosa*	increase yield, 1 hive/ha
vetch, kidney – *Anthyllis vulneraria*	known to pollinate
* vetch, sweet – *Hedysarum coronarium*	increase yield, 5–8 hives/ha
wattle, black – *Acacia mearnsii*	known to pollinate
yucca – *Yucca filamentosa*	*Pronuba* moths pollinate wild plants

Substantial books on crop pollination by bees include those by Free (1970), McGregor (1976) and Pesson and Louveaux (1984); Crane and Walker (1984*a*) give brief details indexed under crop.

8.41 When special honeybee management for pollination is needed

A certain amount of crop pollination occurs wherever bees are kept. In some circumstances, however, it is necessary to prepare colonies of honeybees, and to move them to the crop specifically for the purpose (Figure 8.41a).

(a) Where the area covered by the crop to be pollinated is so large that many colonies are needed during the flowering period, although there may be little or no forage for these colonies during the rest of the year. One example is the cultivation of lucerne (alfalfa, *Medicago sativa*) for seed production in 13 000 sq km of the San Joaquin Valley in California, USA (Motter, 1981). As many as 69 000 hives of bees (14% of all those in California) are moved on to this crop for pollination each year. Another is the production of kiwi fruit (*Actinidia chinensis*) in New Zealand; 80 000 colonies were used for its pollination in 1986 (Bryant, 1987*b*).

(b) Where pesticide applications have killed the wild bees, or where the land is newly reclaimed from desert and there are no wild bees. In Egypt, the construction of the High Dam across the Nile at Aswan in 1971 has enabled many such 'new lands' to be irrigated and brought into production. It is essential to include a beekeeping component in each such enterprise.

(c) Where the crop flowers for only a short period, which comes early in the bees' active season. Forager populations in normal colonies are then small, and are able to fly only during occasional short warm periods in the day. Early-flowering crops in the north temperate zone include almond (*Prunus dulcis*), pear (*Pyrus communis*), apple (*Malus domestica*) and cherries (*Prunus avium*, *P. cerasus*); also some cultivars of oilseed rape (*Brassica napus* v. *oleifera*) known as winter rape, which are sown in autumn. In Europe the area used for winter rape increased rapidly in the 1970s–80s (Williams, 1985). Large numbers of flowers are produced, and give a prolific honey flow, but early in the season. Where winters are colder, spring-sown cultivars are used which do not flower until summer.

Figure 8.41a Groups of hives on site for pollination of lucerne (alfalfa), California, USA, *c.*1960. (photo: IBRA Collection)

(d) Where the crop is grown in a greenhouse. Strawberries are one such crop, which is very important in Japan (Figure 8.41b). Usually one small colony is sufficient for each enclosure, and it is kept there for the whole flowering period, which lasts from November to March.

Aircraft have occasionally been used where the ground is unsuitable for vehicles; see e.g. Cantwell et al. (1971), and Herron (1988) who describe the benefits of a helicopter.

8.42 Factors in special bee management for crop pollination

Colonies managed for pollination should have a large population of foraging bees during the flowering season of the crop, whenever this occurs; Section 6.3 describes management methods. In general, pollen-collecting bees are better pollinators than nectar collectors (Free, 1970), and colonies can be stimulated to collect pollen by feeding them with sugar syrup. Also, pollen collection by the bees quickly increases if very young brood is added to a colony (Free, 1967), because pollen is used in feeding the extra brood (Section 2.42). Frames of brood are usually inserted from other colonies that are not being used for pollination. The foragers themselves deposit their pollen loads into store in cells immediately around the brood nest, and direct contact between foragers and brood seems to stimulate pollen collection. The beekeeper may also create a pollen dearth in colonies by removing combs containing stored pollen, or by trapping the pollen before it enters the hive (Section 14.22). Detailed instructions for evaluating honeybee colonies for pollination have been published by Illinois Cooperative Extension Service (1983) and by Burgett et al. (1984).

Hives should be placed *on* the crop (Figure 8.42a) – not just near it. The number of hives required varies from crop to crop, and Table 8.4A gives the recommended hive density (number per hectare) for 53 crop plants. On a large field, hives should be spaced throughout it; for convenience a few are often grouped together (Figure 8.41a), and it is best if their entrances face in different directions, to minimize drifting (Section 6.62.1). Recommended numbers are 5 to 7 per group.

Crop irrigation can affect hive management. Flood irrigation may prevent access to the hives for more than a week at a stretch, and it also necessitates standing the hives on pallets to keep the entrances dry (Motter, 1981). On the other hand if irrigation is carried out frequently, tracks giving access to the hives may be so muddy for much of the time that they are impassible to vehicles.

Also, except perhaps for a large area of monoculture, it is important *when* hives are taken to a crop. If they arrive too early – before sufficient flowers are open to attract bees – the bees will visit other flowering plants in the neighbourhood, and may continue to forage on these when the crop flowers, especially if it is not attractive to the bees.

Even when the crop is in full flower, a good nectar or pollen flow from other plants during the same time of day can attract the bees away from it. Flowering weeds may produce nectar with a higher sugar concentration than that of the crop. An example is dandelion (*Taraxacum officinale*) in fruit orchards; its

Figure 8.41b Strawberry pollination.
left Honeybee pollinating a strawberry flower in the open, USA (photo: L. J. Connor).
right Design for a 1985 60-yen postage stamp, to mark the importance of honeybee pollination for strawberry growing in Japan (design: Takeo Amano).

Figure 8.42a Inspecting a colony of *Apis cerana* pollinating sunflower, Maharashtra State, India (photo: Central Bee Research Institute, India).

nectar sugar concentration may reach 73%, whereas nectar from pear (*Pyrus communis*) is likely to contain only 8–10% sugar. A similar situation can arise with other bees reared for pollination (Section 8.54).

Bees newly on a site, searching for forage but without fixed patterns of flight and flower visitation, can often be more effective pollinators than bees whose foraging patterns are already fixed. A procedure of 'switching' colonies from one site to another, even interchanging them if necessary every few days, can sometimes prove beneficial. There is evidence that foragers can recruit others to a specific area of an apparently uniform crop, on a day when for some reason the flowers there are yielding better than those in other areas (Weaver, 1979).

In *Spatial management of honey bees on crops*, Jay (1986) evaluates various techniques for increasing the numbers of pollen and nectar foragers on crops to be pollinated, by actively attracting them to the target crop and encouraging them to continue visiting it. He discusses the use of pheromones to attract honeybees to crops, also substances that make pollen attractive to bees, and repellents for use on nontarget plants.

Sometimes the fruit blossom in an orchard must be pollinated with pollen from a specific (different) cultivar, and one way of achieving this is to use a 'hive insert', also called a 'pollen dispenser'. This device is inserted in hive entrances, with a supply of the required pollen arranged so that a little of it is deposited on each bee as she leaves the hive. Pollen dispensers, which are used especially for fruits, have been assessed by Legge (1976).

In hives used for pollination in greenhouses, 4–8 frames of bees per colony are often enough. The beekeeper must check regularly that the colonies have enough food. It is often better if each hive has two flight entrances, one into the greenhouse and one outside, so that the bees can get access to other plants in flower. Bees from a hive just taken into a greenhouse are likely to become disoriented, flying against the glass or plastic, but young bees that fly only inside the greenhouse become accustomed to it. Free (1970) gives a useful discussion of the use of bees in greenhouses and other enclosures.

If bees placed on crops sting or harass agricultural workers to an unacceptable level, growers will not want to hire colonies. Difficulties of this kind with Africanized bees taken to crops for pollination are described in Section 7.43.2. Tropical African bees were found to be good pollinators of lucerne (*Medicago sativa*) in Kenya, tripping on average 54% of flowers visited (Chandler & Mdemu, 1975).

8.43 Crops that benefit from honeybee pollination

In Table 8.4A, Lists 1 and 2, identify 177 crops that are known to benefit from honeybee pollination. The 112 plants in List 1 are fruit and seed crops, and pollination is needed for producing the crop to be harvested. The 65 plants in List 2 require pollination, not for crop production but when grown routinely to produce seeds. Seeds may be needed for sowing in the next season, as with forage legumes (Figure 8.43a), or for plant breeding, as with tea. Both Lists 1 and 2 include recommendations as to the number of hives of honeybees (*Apis mellifera*) to be used per hectare, where figures are available.

Figure 8.43a Flower heads of white clover, *Trifolium repens* (photo: G. H. Hewison).
above No pollination has occurred.
below The lower florets droop, showing that they have been pollinated.

There is no simple way of calculating the economic value of crop pollination by honeybees. In the USA, the total annual value of crops grown with the aid of honeybees' pollinating activity has been estimated at 19 000 million dollars, which is 143 times the value of the honey and wax produced by the honeybees that pollinate the crops (Levin, 1983). The estimate takes into consideration fruit, cucurbits, etc., from bee-pollinated flowers, also vegetables and fodder crops grown from seeds of bee-pollinated flowers, and meat and milk from the use of such fodder crops. The total yield (and value) would be lower if there had been *no* benefit from honeybee pollination, but the amount of the deficit (i.e. the value of the honeybees' contribution) is not easily estimated.

8.5 MANAGING OTHER BEES FOR CROP POLLINATION

Certain bees besides honeybees are nowadays reared for pollinating crop plants. Table 8.5A includes the few that are already reared commercially, together with other candidate species that have been reared on an experimental basis.

Bees that are commercially useful for crop pollination must fulfil certain requirements.

- They must be social bees that nest as a colony, or solitary bees that nest gregariously and can be mass-reared, so that large numbers can be made available at flowering time.
- They must be amenable to rearing treatments which produce a maximum adult bee population during the flowering of the crop to be pollinated.
- They must be sufficiently resistant to diseases and parasites, under the conditions of mass-rearing.
- They must rear brood successfully in artificial nests.
- They must forage on, and pollinate, the commercial crop, preferably even when other forage plants are in flower nearby.

New types of beekeeping have been developed since about 1950, with species of bees that are effective pollinators of specific crops (see e.g. McGregor, 1976).

In general, social species that form large colonies, and solitary bees suitable for large-scale mass-rearing, are the most worth consideration. As with the rearing of any animals in artificial and often crowded conditions, an ability to control diseases, parasites and predators of a species may be a determining factor in its selection for commercial pollination, and this question is dealt with by most of the authors quoted in this section.

Table 8.5A Bees other than honeybees which have been 'managed' for crop pollination, a few commercially and the others experimentally

Management of many other bees is still at an experimental stage. Where no crop is entered, the reference describes management in general. Entries for which no reference is given are from Table 2/3 in Crane and Walker (1983), which cites the reference.

Bee	*Example of crop*	*Country*	*Reference*
Andrenidae			
Andrena ilerda	sarson	India	
A. ovatula	berseem	Egypt	Rashad (1985)
Anthophoridae			
Anthophora spp.	forage legumes	Poland	
A. parietina	forage legumes	Poland	
Ceratina binghami	lucerne	India	
Peponapis pruinosa	cucurbits	USA	
Pithitis smaragdula	lucerne	India	Batra (1979)
Xenoglossa spp.	curcubits	USA	
Xylocopa brasilianorum		Puerto Rico	
X. caffra	lucerne	S. Africa	Watmough (1987)
X. fenestrata	gourds	India	
X. mordax	yellow passion fruit	St Vincent	
X. pubescens	luffa	India	
X. sonorina	passion fruit	Hawaii	
Apidae			
Bombus borealis		USA	Medler (1962)
B. hortorum		New Zealand	Donovan & Wier (1978)
B. lapidarius	red clover	Finland	Valle & Aaltonen (1969)
B. mixtus	cranberry	USA	Johansen (1967)
B. pascuorum		Poland	Bornus (1975)
B. ruderarius		Poland	Bornus (1975)
B. ruderatus		New Zealand	Donovan & Wier (1978)
B. rufocinctus		USA	Medler (1962)
B. terristris	tomato	Poland	Pinchinat et al. (1979)
Melipona spp.		Angola	
	litchee	India	
Trigona spp.	spondias	India	
T. ruficrus	gourds	Brazil	
Halictidae			
Nomia melanderi	lucerne	USA	
N. ruficornis		Egypt	
Rophitoides canus	lucerne	Ukraine (USSR)	
Megachilidae			
Chalicodoma spp.	lucerne	Egypt	
C. mucorea	onion	Egypt	Rashad (1985)
Eumegachile pugnata	sunflower	USA	Parker & Frohlich (1985)
Heriades (Eriades) spp.		Poland	Wójtowski (1971)
Megachile spp.	forage legumes	Poland	
M. flavipes	lucerne	India	

(continued)

Table 8.5A *(continued)*

Bee	*Example of crop*	*Country*	*Reference*
M. gratiosa	lucerne	S. Africa	Watmough (1987)
M. rotundata	lucerne	USA, Chile	
M. submucida	berseem	Egypt	
Osmia spp.	apple	Poland	
O. coerulescens	lucerne	France	Tasei (1972)
O. cornifrons	apple	Japan	
O. cornuta	tree fruit	Spain	Asensio de la Sierra (1984)
O. fulviventris	cotton	USA	Waller et al. (1979)
O. latreillei	berseem	Egypt	Rashad (1985)
O. lignaria	almond	USA	
O. rufa	brassicas (greenhouse)	Denmark	Holm (1973)
O. sanrafaelae	lucerne	USA	Parker (1985)
O. submicans		Egypt	Wafa & El-Berry (1971)

Artificial rearing has important advantages over reliance on wild bee populations. The bees can be reared near the crop, or moved to it for the flowering. The number of adult bees, and the date when the population reaches its peak, are under control. Since the bees are on the crop during flowering, it is easier to protect them from insecticide applications. Where possible, it is also economically advantageous to maintain and increase wild populations of the bees: by fostering habitats suitable for nesting sites, and plants that provide forage for the whole of the bees' life cycle, and by using only pest control methods that safeguard the bees. On reclaimed desert land which has no wild bees, it is essential to introduce honeybees or to rear other bees.

8.51 Managing bumble bees

Bumble bees (*Bombus*), social bees native to the north temperate zones, were the first to be reared as alternative crop pollinators to honeybees. Many bumble bees pollinate a wide variety of crops, and Table 8.5A includes 9 species that have an actual or potential use for rearing as crop pollinators. Long-tongued species – such as *B. terrestris* and *B. hortorum* – are especially useful for pollinating flowers whose corolla is so long that honeybees cannot reach the nectar in it. Red clover (*Trifolium pratense*), grown in northern Europe, is an example. There, red clover florets have a longer corolla, and the native honeybees have a shorter tongue than in the south of Europe. So bumble bees have a double advantage over honeybees. In addition, species of bumble bees with a large body size can fly at a lower temperature than honeybees.

The bumble bee colony lasts only through the summer; when queens and males have been produced in late summer, the colony dies out. After mating, each queen overwinters separately, below ground; in spring she emerges and starts a new colony. Alford (1975) and Prŷs-Jones and Corbett (1987) provide much information on bumble bees in Britain.

To rear bumble bees, wild nests may be searched for in spring, and transferred to boxes (Alford, 1975; Donovan & Wier, 1978); see Figure 8.51a, *above*. Alternatively queens are captured in spring, after they emerge from hibernation, and installed in prepared nest boxes indoors (Figure 8.51a, *below*) which are placed outside later on (Valle & Aaltonen, 1969). Or the boxes may be placed outside to attract the queens that are searching for a nest site. In one experiment (Donovan & Wier, 1978) 50 queens reared in a nest in an upper-floor laboratory were allowed to fly out, and next spring at least 25 flew into the same room.

Procedures and equipment for rearing bumble bees have been worked out for native species in European countries and North America, and in New Zealand where species have been introduced. Papers at the Second International Symposium on Pollination (1966) deal with bumble bee rearing for pollination. In addition to species listed in Table 8.5A, Morgan and Percival (1967) and Alford (1975) describe the rearing of British species in general. In Canada, Plowright and Jay (1966) give techniques for some bees listed in Table 8.5A and some others, and Hobbs (1967) deals with still further Canadian species.

Donovan and Wier (1978) describe nest boxes for species used in New Zealand, in which colonies of

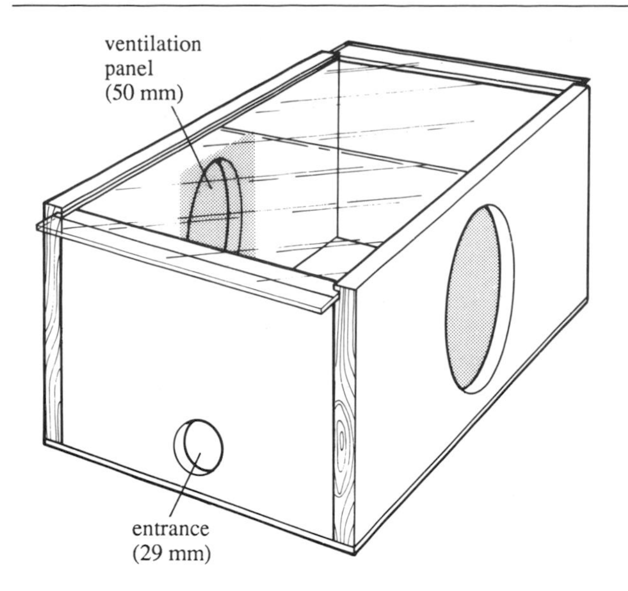

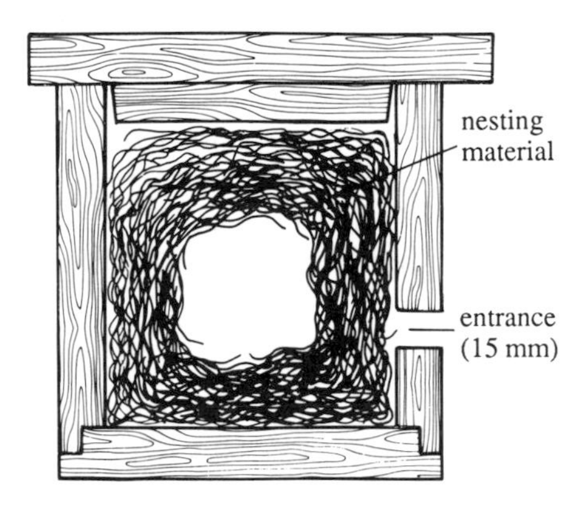

Figure 8.51a Two types of nest box for bumble bees (Alford, 1975).
above Nest box 17.5 cm wide and 23 cm long, with glass top, into which an established colony will be transferred.
below Section through nest box, 16 cm wide, with hinged lid, where a queen will establish a new colony.

introduced *B. hortorum* and *B. terrestris* produced 2–5 times as many individuals as they do in Europe, probably because of relative freedom from enemies and from competitors for food; New Zealand has no native bumble bees.

8.52 Managing *Megachile rotundata*

To pollinate a lucerne flower (alfalfa, *Medicago sativa*), an insect must 'trip' the flower. This is brought about when a bee inserts her tongue inside the flower and in so doing exerts pressure on the keel petal, releasing the sexual column, which may hit the underside of the bee's head with some force. It is sometimes difficult to get *Apis mellifera* to forage on and pollinate lucerne, but two other species of bees, which can be mass-reared, are highly effective pollinators of this crop: *Megachile rotundata*, and *Nomia melanderi* (Section 8.53). The biology and management of both are described by Johansen et al. (1982). *M. rotundata* is a highly gregarious solitary bee from the Old World, known in North America as the alfalfa leafcutter or leafcutting bee. It nests in hollow plant stems, and these can be simulated by such man-made structures as bundles of 4-mm straws or of blocks of grooved wooden or plastic boards, as in Figure 8.52a. The pupae are wintered in dry cool storage (1°–4°); then in spring, about three weeks before flowering is due, they are incubated at about 29° and 50–70% relative humidity. Conditions are regulated so that the bees start to emerge when flowering starts.

At this stage, trays of late-stage pupae are taken to the lucerne fields and housed in shelters that also contain many thousands of empty horizontal artificial

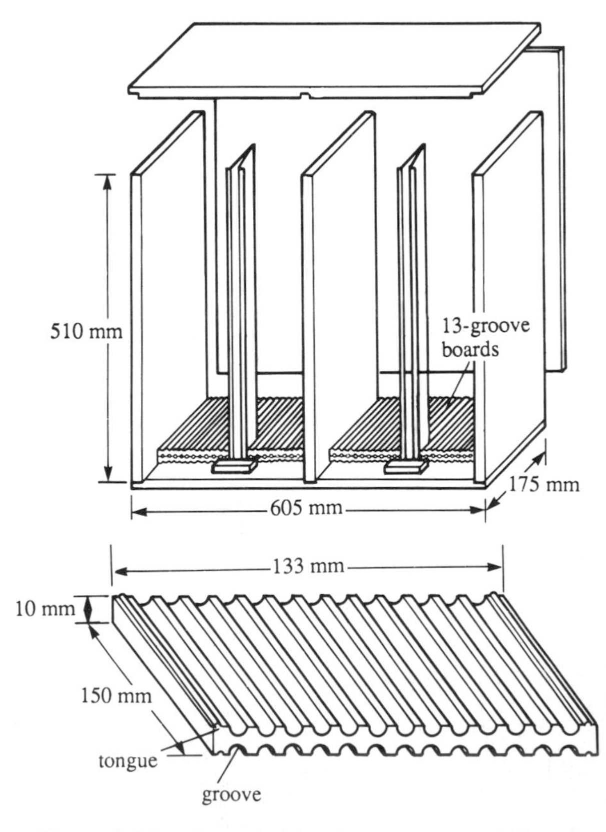

Figure 8.52a Expanded drawing of a module for housing nests of *Megachile rotundata* in simulated hollow stems formed from grooved wooden boards (Richards, 1984).
One of the grooved boards is shown in detail below; 200 of them (2600 'stems') are accommodated in the module.

'stems' (Figure 8.52b). After emergence, the adult females mate, and then make their nests in the stems, lining each cell with pieces of lucerne leaf. The bees forage for pollen and nectar on the lucerne flowers, which are pollinated in the process. A shelter with about 60 000 nesting females can pollinate 1.2 ha of lucerne (50 000 females per ha).

Richards (1984, 1987) gives information on rearing and maintaining the bees, and the necessary control of diseases and parasites. The First International Symposium on Alfalfa Leafcutting Bee Management was held in Canada, and a collection of 33 papers published (Rank, 1982).

8.53 Managing *Nomia melanderi*

The alkali bee (*Nomia melanderi*) is a highly gregarious solitary bee native to some parts of North America, where it nests in large numbers in certain fine saline soils. Large-scale methods have been developed for preparing and stocking suitable new nesting sites or 'bee beds' near lucerne fields. Blocks of undisturbed soil containing pupae are then introduced from established beds. Soil conditions, including temperature and drainage, are maintained at optimal levels, and salts are added as necessary. The bees may colonize other prepared bee beds up to 1.5 km away, by migration; at further distances pupae must be introduced. Alkali bees pollinate lucerne most efficiently and, provided diseases and enemies can be kept under control, very high seed yields can be produced. A bee bed 250 m² in area is sufficient for 30 ha of lucerne. Bohart (1972) give details, and Johansen et al. (1978) discuss problems encountered, and their solutions.

Figure 8.52b Shelter containing artificial stem nests for *Megachile rotundata* in a field of lucerne (alfalfa), British Columbia, Canada, 1965.

8.54 Managing other bees

Many of the megachilid bees are suitable for mass rearing, because their nesting sites (in soil, wood, stems, etc.) may be simulated by materials that can conveniently be moved from an incubator out to the crop to be pollinated, as is done with *Megachile rotundata* (Section 8.52). Williams (1972) gives an account of the biology of *Megachile* and *Osmia*, and of important factors in achieving success with artificial nests and management techniques. Torchio (1985) 'managed' 100 000 nests of *Osmia lignaria propinqua* for apple pollination each year, and increased the population 3–4 times annually in commercial orchards. Only 600 female bees per hectare were needed for full pollination of the apple flowers. *Osmia rufa* is an example of species that have proved especially suitable for rearing and using in greenhouses; its use in Denmark has been described by Holm (1973). Like honeybees, megachilid bees should not be moved too early to crops; in Haryana, India, *Megachile* bees taken to pollinate lucerne foraged instead on Jerusalem thorn (*Parkinsonia aculeata*) already in flower along the roadsides (Sihag, 1982).

Other species in Table 8.5A have also been reared on an experimental basis. In the Anthophoridae, most *Ceratina* are small, brightly coloured 'metallic' bees, of which some species have been used to pollinate lucerne in India. *Peponapis* and *Xenoglossa* are New World bees that forage on flowers of cucurbits, and are especially well adapted to pollinate them. Michelbacher et al. (1968) recommend them for introduction for pollination in new areas (with the precautions that are always necessary in such operations). Many of the large carpenter bees (*Xylocopa*) are useful for the crops indicated, and for others; Kapil and Dhaliwal (1968/69) made successful trials with them in India. The search is continuing for new bees that are both well adapted to pollinate specific crop plants and highly efficient at doing so, and for which methods of mass-rearing can be devised that allow the beekeeper adequate control of diseases, parasites and predators.

8.6 MANAGING OTHER BEES FOR HONEY PRODUCTION

The most important honey producers after the hive bees *Apis mellifera and A. cerana* are *A. dorsata* (giant or rock bee) and *A. florea* (little bee)* in the tropics of Asia, and stingless bees (Meliponinae) which occur in the tropical regions of all continents. Distributions of the bees are shown in Figures 1.2b and 1.23e, and Figure 1.23a indicates the relative sizes of *Apis* species.

The management of colonies of these bees has nowhere advanced to the level of beekeeping with *A. mellifera* and *A. cerana* (Section 7.41). Management of *A. dorsata* and *A. florea* is unusual, although much honey is collected from wild nests of *A. dorsata*. On the other hand stingless bees can be managed without much difficulty; some of the larger species have been kept in hives for many centuries, and archaeological evidence is dated to periods before AD 300 (Crane & Graham, 1985). Honey yields are mostly lower than those from *Apis* species, except *A. florea*. Stingless bees and their distribution are discussed in Section 1.24; all are tropical, and their importation to places outside their native areas has not very often been attempted. They are all good pollinators, but probably none are yet routinely managed for crop pollination.

Publications on the above bees are listed by Crane (1978*a*): No. 12 (*Apis dorsata*), No. 13 (*Apis florea*), No. 14 (Meliponinae).

8.61 Managing *Apis dorsata*

It is very difficult to 'manage' this most productive bee in something equivalent to apiaries. Progress that has been made is described here, and I hope it will stimulate some readers to follow up what has been started.

Nests of Apis dorsata

An *A. dorsata* comb is up to 2 m from side to side and 1.2 m from top to bottom. The part of the comb where brood is reared is about 4 cm thick, with cells 2 cm deep; at the top where honey is stored the comb is much thicker – even up to 46 cm, i.e. with honey cells up to 23 cm deep (Mahindre, 1987). Many combs are much smaller. A colony may contain 30 000 to 50 000 bees or sometimes more; most are not foragers, but workers that remain motionless on the comb with their bodies oriented upwards, maintaining an insulating and protective curtain 3 bees deep over the occupied part of it.

Apis dorsata is the only honeybee that builds its nests gregariously. The primary requirement of a nest site is a very secure support, and large nests are built only where the support is strong enough to bear the great weight of the comb when full of honey. An uninterrupted flight area round and below the nests is also necessary. The outermost bees in the curtain are able to reduce their body temperature by flying off the nest and regurgitating their honey-sac contents (Mardan, 1987). In relatively inaccessible sites within reach of much forage there may be a great many nests. Schneider (1908) recorded 65 nests in one tree in virgin forest in Malaysia, and aggregations of up to 100 have since been referred to. Figure 8.61a shows a more modest example. Large aggregations do not seem to occur in the Philippines, or in the Sundarbans, a tidal swampy forest at the mouth of the Ganges. The heights at which nests are built also vary greatly in different areas. It seems likely that there are genetic differences in the bees; see Section 1.23.2.

Colonies of *A. dorsata* are easily alerted to attack and sting, especially when someone first approaches a nest; subsequent approaches may be easier (Mahindre, 1987). People tend to be very wary of them, yet in some circumstances the bees appear to become accustomed to human activity nearby. For instance groups of nests are commonly found in trees in city streets, and in temples – which provide man-made equivalents of the nesting sites on rock faces. Mahindre et al. (1977) refer to a group attached to a railway bridge subject to frequent vibration.

Honey collection from wild nests

At the end of a honey flow, the nest may contain much honey, sometimes up to 50 kg. In India, 13 500 tonnes of *A. dorsata* honey are produced in a year, three times the amount obtained from hives of *A. cerana* (Table 13.8A, Part 6). The rock painting in Figure 1.41b shows how the honey was collected in prehistoric times, and similar methods are still used. Many descriptions of the procedure have been published, e.g. by Singh (1957), Trehen and Longia (1962), Strickland (1982) and Valli and Summers (1988).

Most of the men who collect honey from *Apis dorsata* are extremely poor, and have no power to control the price at which they sell it. There is almost no extension

* See Sections 1.23.2 and 1.23.3; *A. vechti*, *A. laboriosa* and *A. andreniformis* are here considered with *A. cerana*, *A. dorsata* and *A. florea*, respectively.

Figure 8.61a Twelve or more nests of *Apis dorsata* supported by thick branches of a tree in India (photo: Central Bee Research Institute, India).

teaching on handling honey collected from *A. dorsata*, as there is on handling honey from hives. For these and other reasons, *A. dorsata* honey is often badly prepared and presented, and usually fetches a low price. Honey stored by *A. dorsata* is clear and transparent, and it remains so if it is extracted by straining combs without squeezing them; squeezing forces pollen through with the honey, and thus impairs the flavour. A centrifugal extractor might be designed for the pieces of comb; see Section 10.62.

When working with *A. dorsata*, full protective clothing might be considered highly desirable, although many experienced honey collectors do not use it and are indeed unable to afford it. To pacify the bees, some honey collectors use parts of specific plants or smoke from burning them; see Section 5.21.3 and 5.21.1, and Figure 8.61b.

Attempts at managing Apis dorsata

'Management' is difficult, because a colony cannot be kept in an enclosed space such as an ordinary hive, and because colonies usually move in the course of the year, following a seasonal pattern governed by honey flows in different areas; see end of Section 3.34.5. So a managed colony may fly off and leave its nest after a

Figure 8.61b *Apis dorsata* on South Andaman Island (Dutta et al., 1983). The comb is supported from a very thick branch. Much of it has recently been cut away to harvest honey, and most of the remainder is covered by a curtain of bees. Dr R. Ahmed has smeared his hand with sap from *Amomum aculeatum*, which is repelling the bees from a small piece of newly built comb, and tranquillizing them (see Section 5.21.3). He carries a squeezed stem of the plant on his shoulders. (See Figure 8.62a for the relative size of an *A. florea* comb.)

few months. Also, the bees may be readily alerted to attack and sting.

Following earlier work by Muttoo (1939, 1952/53), a method for managing *Apis dorsata* was published in 1956 by Ghotge from the Rock-Bee Research Station at Ujjain, Madhya Pradesh, in central India. It described a long 'clip' to be fitted across a comb (a wild nest), so that the part above the clip – containing most of the honey – could be harvested, leaving the lower brood area safely supported by the clip. In the same region Mahindre followed up this work, and he has had twenty years' experience in handling *A. dorsata* colonies. The account below is based on Mahindre's publications (1968, 1983*a*) and on further detailed information from him (1987).

Management using a clip

Letters in brackets below refer to the sequence of diagrams in Figure 8.61c. A comb that contains honey as well as brood (A) is found attached to a branch. At night, a clip (see B) is fixed horizontally across the comb, below the honey cells and above the brood nest; this is shown also in Figure 8.61d (*above*). The clip is formed from a piece of bamboo, 5 cm in diameter and slightly longer than the comb; it is cut lengthways into two halves, which are hinged together at one end with a piece of flexible leather. In C, and in Figure 8.61d (*below*), the upper part of the comb (containing honey) has been cut off just above the clip, for harvesting; the lower part of the comb, attached to the clip, is subsequently transferred to a new 'managed' site in shade, immediately under an 'attraction plank'.

Management using a top-bar (attraction plank)

An attraction plank is a board 25 cm wide and long enough to shelter a comb, which has been smeared on the underside with beeswax. It serves as a wide top-bar as shown in Figure 8.61c (D); the bees will soon attach their comb to the underside of it, as *A. mellifera* or *A. cerana* does when wild comb is tied into a frame. The lower part of the comb from C is supported with its clip just below the attraction plank, and this clip is removed when the comb is firmly attached to the plank. Later on, when the bees have stored honey at

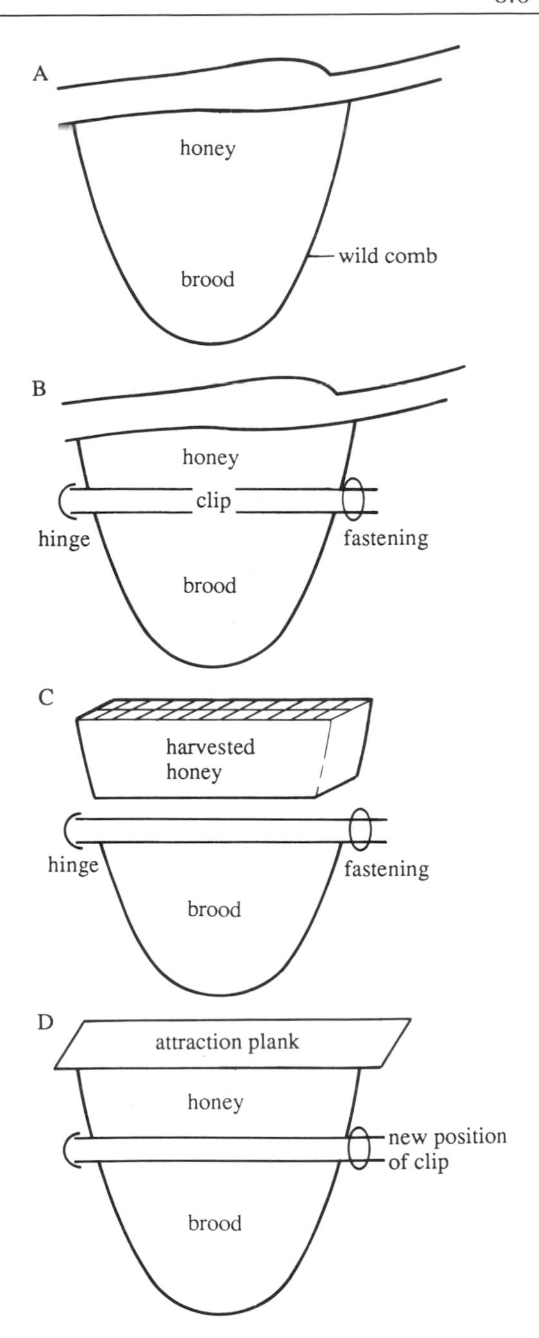

Figure 8.61c Management of *Apis dorsata* using a clip and an attraction plank (Mahindre, 1977). Diagrams A to D are explained in the text.

Figure 8.61d *Apis dorsata* combs with their clips (photo: D. B. Mahindre).
above Comb without bees, showing brood cells below the clip and honey storage cells above it.
below Mrs Mahindre holding one of her colonies by its clip, after removal of the honey comb, 1964.

the top of the comb, the clip is refixed farther down the comb (as in D), and the process repeated.

Another way of getting combs built on to attraction planks is to take action before the bees are due to move in from another area, by tying planks underneath a tree branch where *A. dorsata* colonies customarily settle, so that combs are attached to them instead of to the branch. When a comb is fully built, the plank complete with comb and bees can be removed to a more convenient location. Mahindre has maintained up to 150 such colonies, mostly 6 to 10 m above ground, but sometimes only 1 to 3 m.

Mahindre found that a wild colony taken to a new site could be prevented from deserting its comb as follows. Only adult bees, with the queen, were collected, in a screen box similar to those used for package bees. They were confined for 2 or 3 days, and then released through a small hole, close to an attraction plank fixed in a convenient site. The bees then clustered on the underside of the plank, built their comb, and did not desert it. Incoming swarms could be made to settle beneath an attraction plank by a similar procedure.

A feeder is attached to the upper side of the plank, and is kept filled with syrup during the dearth period. The colony population decreases to perhaps a quarter of its previous size, and the cluster of bees contracts, leaving much of the most recently built comb uncovered and unprotected. This comb must be cut away; if left, it is likely to become infested with wax moths (Section 11.61) – especially in dry areas – in which case the colony may well abandon the nest after rearing the brood present, and consuming any stores.

A comb built from an attraction plank can be migrated to honey flows, by fixing the attraction plank in a screen box; the wide attachment of the comb prevents the comb swaying during transport.

Management in partly enclosed hives

If colonies are sited near ground level, there are advantages in keeping them as fully enclosed as the bees will tolerate. Mahindre found it possible to keep *A. dorsata* colonies in hives with the back and sides enclosed, and even part of the front. Figure 8.61e shows a somewhat similar hive. The brood box below (95 × 75 × 30 cm) has a hinged front shutter of wire gauze, normally left open (as shown) but closed for transport; here, the brood box is fitted with a frame. The shallow honey super above has its own wire-gauze shutter. In Pakistan, Ahmad et al. (1984*b*) used a hive 100 × 100 × 30 cm, made of 16-mesh wire gauze (6 per cm) fixed on a wooden frame, with two doors 30 × 30 cm near the top, also covered with gauze; they were kept open except during the absconding/migrating season. Colonies were always fed throughout dearth periods, and absconding could be prevented by doing this; they were also moved in the hives from crop to crop.

Flight cages

Koeniger et al. (1975) were able to keep colonies enclosed in flight cages 4 × 5 × 2 m, for 5 months. The colonies reared brood all the time and fed on sugar syrup and pollen provided in the cages.

8.62 Managing *Apis florea*

Apis florea is a small, gentle tropical bee, which builds its single-comb nest in a fairly open position, although in summer it needs shade from the sun. The comb may be up to 50 cm high and 35 cm wide, and the colony may contain up to 30 000 bees. Nests are often within a few metres of the ground, and are usually in rocks, or in vegetation; I have seen many attached to a branch in an open network formed by a dead spiny shrub, which itself was covered by the foliage of a living climbing plant. Free (1981) gives a summary of the biology of *A. florea*, and theses by Akratanakul (1977) and Whitcombe (1984) report detailed observations on this bee in Thailand and Oman, respectively.

Figure 8.61e An open-fronted hive containing a small *Apis dorsata* colony (Thakar, 1973). See text for details.

Combs of *A. florea* are easily taken for their honey, and the bees shaken off are likely to resettle nearby and build a new comb, in which case the colony is not lost. (In north-east Thailand combs are collected during the honey season and displayed for sale on conical racks.) In many regions a person may cut a nest off its branch, remove the upper part containing honey to eat, and tie the rest to a suitable branch close to his dwelling house, where he can check its well-being from time to time. Figure 8.62a shows a comb which has recently been attached to a twig. Later on, the comb will have deep honey cells at the top which

Figure 8.62a 'Managed' nest of *Apis florea* (photo: Central Bee Research Institute, India).
There are comparatively few bees on the comb, and a patch of larger drone cells is visible at the bottom. The size of this comb (in relation to the hand) may be compared with that of *Apis dorsata* in Figure 8.61b.

entirely surround the supporting twig; any drone or queen cells (which are larger than worker cells) are at the bottom.

Like *A. dorsata*, *A. florea* cannot be kept in an ordinary hive or other enclosure. Wild colonies may well abscond because of shortage of food or disturbance (see Section 3.34.1), also from an unshaded site when this becomes too hot in spring or summer, and to an open sunny site in autumn (Iran; Tirgari et al., 1969). From the 1940s onwards, occasional attempts were made to keep *A. florea* in cages or flight rooms (see Crane 1978*a*, No. 13). Then an existing form of beekeeping with *A. florea* was found in the mountains of northern Oman (Dutton & Simpson, 1977; Free, 1981; Whitcombe, 1984). It is somewhat similar to that described above for *A. dorsata*, but considerably easier. The branch or other support for the *A. florea* comb is replaced by the two halves of a split date palm frond somewhat longer than the width (length) of the comb; they form a clamp across the comb above the brood, and the part above the clamp, containing honey, is harvested; see Figure 8.62b. The comb supported by the clamp is placed in a sheltered position, open on one side but shaded from direct sunlight. This may be a shallow recess made in a wall, with a rebate near the top of each side to take the ends of the palm stalk (Figure 8.62c, *above*). Or a pit is hollowed out in the ground to take the comb, and subsequently covered sufficiently to provide shade. The aims are to protect the bees, to prevent the comb being overheated by day or chilled by night, and to be able to harvest the honey conveniently by moving the split palm stalk down the comb.

Whitcombe (1982) has developed an experimental hive (Figure 8.62c, *below*) and described its use. In many ways it is similar to Thakar's for *A. dorsata*

Figure 8.62b Beekeeper in Oman fixing the lower part of an *Apis florea* comb between two halves of a length of split date palm frond, which are tied together to prevent the comb slipping out (Dutton & Simpson, 1977).

(Figure 8.61e). The upper frame is designed to contain honey; when the comb has been fully built and filled, the honey can be harvested and a new frame inserted. Each frame is in a separate box, the upper one covered at the front and back with a piece of wire mesh (31 × 3 cm) which allows workers through but not the queen. Above the mesh on each side is a full-width entrance block 2 cm wide, which is removed when absconding is unlikely, to allow the bees easier passage in and out of the hive. A horizontal roof extends several cm on either side, to provide some shade. The lower box has a solid (removable) front and back. If the bees are provided with starter comb at the top of the lower frame, they may sometimes build down from it, i.e. accept the enclosed lower part of the hive. The top-bar of each frame is suspended from rebates in the side walls of the hive.

8.63 Managing stingless bees (Meliponinae)

Where both honeybees and stingless bees are present, there may be little incentive for beekeeping with stingless bees because of their smaller honey yield per colony, and no Old World civilization is known to have developed beekeeping of this type. But in the New World, which had no honeybees in antiquity, some of the stingless bees were exploited extensively. They were the only source of honey and wax in the Americas until honeybees were introduced in the 1700s and 1800s. Stingless bees are widely distributed in the tropics of all continents (Figure 1.2b) and are described briefly in Section 1.24. It is often fairly easy to take the honey from their nests in the wild, and this is done in Asia, Africa, Australia and – especially – in the Americas, where there are more native species than in any other continent. Roubik (1986) estimates that, in Panama alone, honey is harvested each year from 100 000 to 200 000 nests, of species which include *Melipona* (*fuliginosa, fasciata, compressipes*) and *Trigona* (*pectoralis, capitata, frontalis, angustula*).

Quite large amounts of honey can be obtained from colonies of some species, and beekeeping with stingless bees – sometimes referred to as meliponiculture – has been widely practised with them. *Melipona beecheii* is kept by Maya beekeepers, in Yucatán (a low peninsula on the east coast of Mexico) and adjacent regions. The Maya use horizontal logs with end closures of clay or stone, the flight entrance being midway along the log (see e.g. Weaver & E. C. Weaver, 1981). This beekeep-

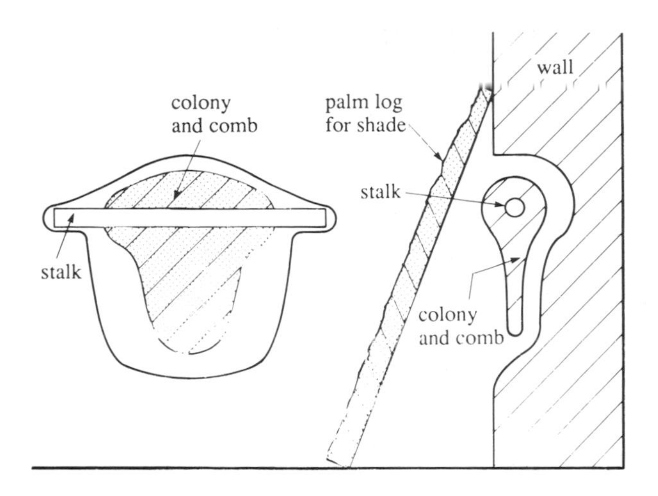

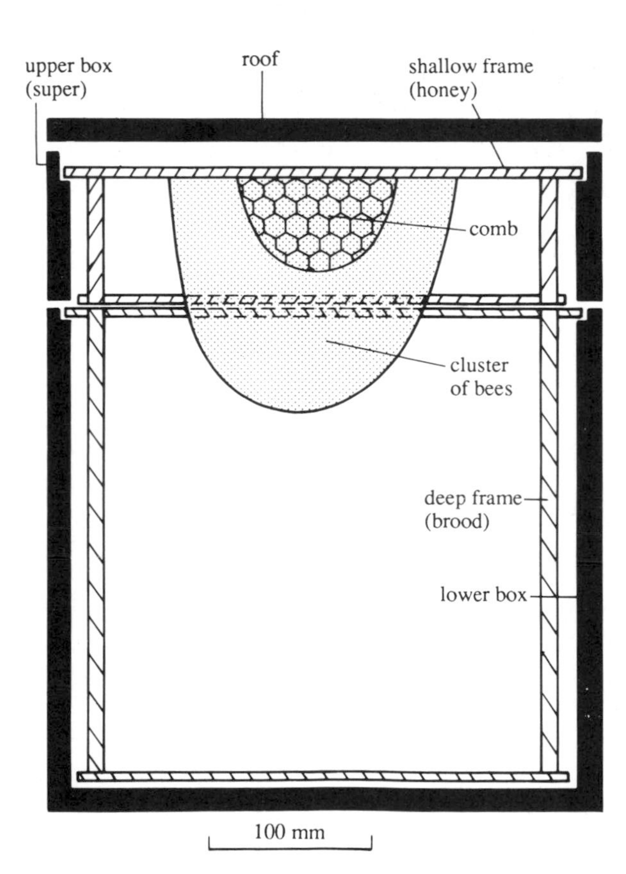

Figure 8.62c Housing *Apis florea* (Whitcombe, 1984, 1982).
above Vertical cross-sections (along the wall on left; through the wall on right), showing the shape of a typical recess and a completed *A. florea* comb, built round a split palm frond. The height and width of the recess are about 57 cm, and its maximum depth 19 cm.
below Experimental hive for *Apis florea*. This vertical section shows the size of the cluster of bees, and of the separate honey and brood combs, 4 weeks after a comb with brood had been attached to the lower frame; a starter comb had previously been attached to the top-bar of the upper frame.

ing is very old; hive end closures dated to between 300 BC and AD 600 have been excavated (Crane & Graham, 1985). Elsewhere, pots or gourds are also used, and sometimes wooden boxes. Schwarz (1948) gives information for many areas.

There is currently a resurgence of interest in keeping stingless bees (Nogueira-Neto, 1987), and characteristics of nests are described briefly here, together with a modern hive that facilitates the harvesting of honey.

Stingless bees build most parts of their nests from cerumen (a mixture of wax and propolis), and other parts from wax alone. The wax is secreted by the bees, and the propolis collected – often in large quantities – from plants; see Section 12.5. Figure 8.63a shows one nest, together with a 'rational' hive appropriate for the same bees. The whole of the nest (*below*) is enclosed by a layer of batumen, B (cerumen plus propolis, and sometimes also plant materials or mud). This seals the nest cavity from the outside world, except for its flight entrance (E), some small ventilation holes, and possibly also a drainage tube. The entrance is made or lined with propolis or cerumen; some of the species that build a long entrance tube close it each night.

Honey and pollen are stored in 'pots' of soft cerumen (P); the honey and pollen pots may be separate, or intermixed. Brood cells are mass-provisioned, i.e. all the food for the larva is put into the cell, then the egg is laid in it, and the cell closed. Many features are common to nests of all species, but the enclosure of brood cells in a separate cavity, surrounded by an involucrum of soft cerumen as in Figure 8.63a, *below*, is not usual.

Most species build the brood cells on the upper side (only) of combs that are horizontal or in the form of a spiral, and the cells are open at the top. Other species build brood cells in a rather amorphous cluster, reminiscent of those of bumble bees, and still others use arrangements intermediate between clusters and regular combs. One species, *Dactylurina staudingeri*, builds vertical double-sided combs (F. G. Smith, 1954) something like those of *Apis mellifera* and *A. cerana*. In all species the various structures of the nest are supported firmly by pillars or other interconnections of cerumen. Some species keep stores of their building materials in the nest, wax and propolis being in separate places; propolis may be used also to immobilize enemies. The 'beeswax' harvested from nests of stingless bees contains a large admixture of propolis (e.g. F. G. Smith, 1954; references in Crane, 1978*a*, No. 20).

In most species worker and drone cells are of the same size, queen cells being larger and on the edge of the brood nest. Some species also produce queen cells interspersed with those of workers and drones. In most

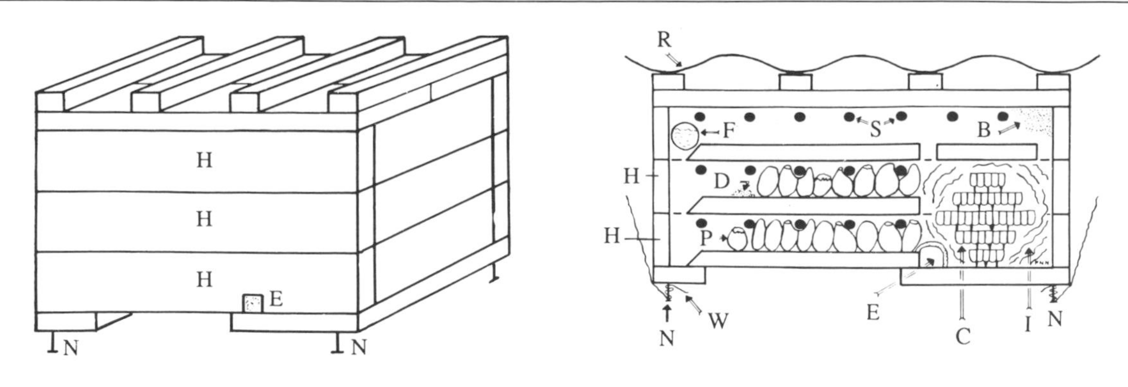

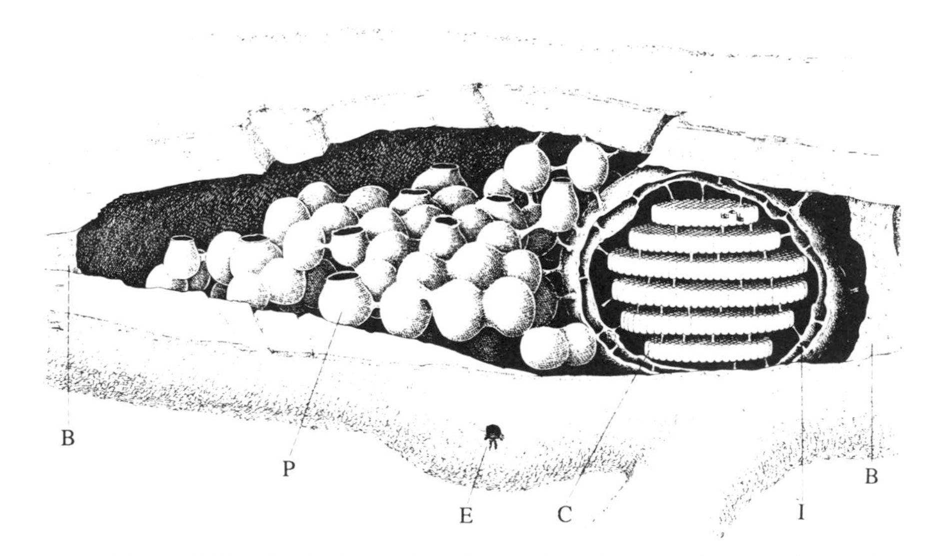

Figure 8.63a Rational hive for stingless bees, incorporating superimposed 'drawers' for honey storage (Nogueira-Neto, 1970).
left Exterior, viewed from the side.
right Interior, vertical section

B batumen
C brood combs
D debris
E entrance tunnel
F feeder
H honey-storage drawers
I involucrum
N nail
P honey storage pots
R roof
S bamboo sticks
W wire

below Nest of *Melipona interrupta grandis* (Camargo, 1970).

species mating takes place outside the nest.

Both editions of Nogueira-Neto's book (1953, 1970) on the rearing of stingless bees should be consulted, because of their different contents and the large amount of information in each, especially on Brazilian species and conditions. The 1970 edition gives full details of the hive shown in Figure 8.63a, in which three 'drawers' (H) are provided for honey storage; extraction of honey from them is described below. The drawers allow only sufficient space vertically for the bees to build a single layer of honey pots. The thin bamboo sticks (S in the drawing on the right) are necessary to keep the honey pots upright and in position; their ends are fixed in holes in the front and back walls of each drawer.

Table 8.63A lists some species of stingless bees that have been kept in hives in different countries. Examples of dimensions to be used are indicated by the following external widths (cm):

	total	*brood chamber*
Trigona (Tetratrigona) clavipes	50	14.5
T. (Scaptotrigona) postica	50	12.5
Melipona quadrifasciata	40	10
Trigona (Plebeia)	34	7
Melipona marginata	34	6

Table 8.63A Some species of stingless bees (Meliponinae) that are, or have been, kept in hives for honey and/or wax production

Names as used by Michener (1974); *Trigona* divided into a number of subgenera, given in brackets.

Species	*Example of region where kept*	*Reference*
Lestrimelitta (Cleptotrigona) cubiceps	Angola	PA 55
Melipona		
beecheii	Central America	S 48
compressipes	Brazil	K 67, NN 70
fasciata guerreroënsis	SW Mexico	S 48
fasciata rufiventris[1]	Brazil	NN 53,70
interrupta[1]	Colombia	S 48
marginata[3]	Brazil	NN 53,70
nigra	Brazil	NN 70
pseudocentris pseudocentris	Brazil	K 67
quadrifasciata[1]	Brazil	NN 53,70
schencki picadensis[1]	Brazil	NN 53
schencki schencki[1]	Brazil	NN 53
scutellaris[1]	Brazil	NN 53,70
seminigra merrillae[1]	Brazil	K 67
Meliponula bocandei	Angola	PA 55
Trigona		
(Axestotrigona) erythra togoensis	Angola	PA 55
(Cephalotrigona) capitata[1]	Brazil	NN 53,70
(Friesella) schrottkyi[4]	Brazil	NN 53,70
(Hypotrigona) gribodoi	Angola	PA 55
(Nannotrigona) testaceicornis perilampoides[2]	W. Mexico	Bennett (1964)
(Nogueirapis) mirandula	Costa Rica	Wille (1964)
(Oxytrigona) tataira[1]	Brazil	NN 53
(Partamona) sp.	Yucatán (Mexico)	S 48
(Plebeia) emerina[3]	Brazil	NN 53
(Plebeia) mosquito[3]	Brazil	NN 53
(Plebeia) remota[3]	Brazil	NN 53
(Scaptotrigona) depilis	Bolivia	KM 66
(Scaptotrigona) pectoralis	Yucatán (Mexico)	S 48
(Scaptotrigona) postica[1]	Brazil	NN 53
(Scaptotrigona) tubiba[1]	Brazil	NN 53
(Tetragona) clavipes[1]	Brazil	NN 53,70
(Tetragona) jaty[2]	Bolivia	KM 66
(Tetragona) mombuca[1]	Brazil	NN 53,70
(Tetragona) nigra	Yucatán (Mexico)	S 48
(Tetragona) silvestrii[4]	Brazil	NN 53
(Trigona) fulviventris	Honduras	S 48
various spp.	Australia	DD 85

[1, 2, 3, 4] Size of nest (NN 53): large (1), medium size (2), small (3), very small (4)

DD 85 Dollin & Dollin (1985)
K 67 Kerr et al. (1967)
KM 66 Kempff Mercado (1966)
NN 53,70 Nogueira-Neto (1953, 1970)
PA 55 Portugal Araújo (1955*b*)
S 48 Schwarz (1984)

Other dimensions are in proportion; the entrance tunnel is 1 or 2 cm across according to the size of the bees.

For species that build queen cells on the edge of the brood nest (all except *Melipona*), removable transparent inspection windows are fitted across the full width of the brood chamber, so that queen cells can be seen and removed. When a queen cell is available, a colony can be divided into two by placing a second hive above the first and transferring to it some brood combs, including queen cell(s) and storage pots. The parent colony remains in the lower hive, and for a few days the two hives are left with internal access but with only one entrance; they are then separated.

Other 'rational' hives for stingless bees have been described, for instance by Kempff Mercado in Bolivia (1966). Portugal Araújo's hive (1955*a*) used in Angola has the storage chamber above the brood chamber, and 'drawers' can be used in it. There is a passage-way accessible to bees between the outer walls and the inner compartments, and the entrance can be sited either at the top or bottom of the brood chamber. The paper includes detailed dimensioned drawings of the component parts.

Nogueira-Neto (1970) gives the fullest information available on the different operations in meliponiculture. There are instructions for obtaining wild colonies for use in rational hives, and for transferring into a hive the brood combs, storage pots, stores of wax, cerumen and propolis, and the bees and queen. Apart from this – and dividing colonies as already mentioned – management consists of feeding colonies when necessary, protecting them against attacks by robbers and other enemies, and extracting the honey. In English, Schwarz's (1948) monograph on stingless bees discusses robbing and enemies, and also wax and honey of stingless bees; honey is also referred to by Crane (1975*d*).

To collect the honey from rational hives such as that in Figure 8.63a, the drawers to be harvested are removed, and the honey storage pots are opened, a check being made that no brood cells are among them. Then the pots are placed upside down on a strainer above a receptacle. Adhering bees that are sticky with honey are carefully washed; they, together with any young (pre-flight) bees, are returned to the hive. In traditional Maya beekeeping, an end closure of the log hive is removed, that end of the hive tilted down, and accessible honey pots are broken so that the honey runs out into a straining basket (Weaver & E. C. Weaver, 1981). Honey is sometimes taken from wild nests by piercing a hole through the nest wall into some of the honey pots inside it; the honey is collected as it comes out of the hole. A rock painting showing this has been found in Queensland, Australia.

8.7 INTRODUCING HONEYBEES TO NEW AREAS

8.71 General considerations

The world's present honey industry would not exist if honeybees (*Apis mellifera*) had not been introduced from Europe into continents that had no native *Apis*, notably the Americas and Australasia. It is therefore understandable that countries in the tropics just starting to develop their own honey industry look to the introduction of similar bees as a first step towards getting large honey crops. However, the question of introducing exotic bees is not a straightforward one. If a country has native *Apis mellifera* or other *Apis* species that are economically valuable, importing foreign bees can have a negative result; it may even prevent the successful development of apiculture based on native bees, or destroy what already exists, through disease or other factors. Legislation has been passed in at least sixty countries to control or prohibit the import of honeybees (Table 15.4A), but it is difficult to make it completely effective.

Imports should not be made until the characteristics of the native honeybees have been adequately assessed. If it is decided to introduce new bees, expert advice is needed as to sources of bees that should, and that should not, be used. Before any major introduction is undertaken, a pilot scheme should be carried out in an isolated area, so that the behaviour and performance of the introduced bees under local conditions can be studied.

The most common transport of bees between countries is of young mated queens (produced as in Section 8.2), each in a cage with a few accompanying workers. The cage must be destroyed on receipt, and the workers (labelled with the reference number of the hive into which the queen is introduced) sent to a competent authority for confirmation that they carry no disease. Only the queen should be allowed contact with bees or hive materials in the new area.

8.72 Transmission of diseases and parasites

One of the greatest dangers in importing new bees into an area is the introduction of diseases and parasites, and bees should not be imported unless they are known to be healthy and free from parasites. European and American foul brood diseases, and tracheal and varroa mites, have been gratuitously introduced in this way into many countries, to the lasting detriment of their beekeeping. No combs or used beekeeping equipment should be imported, because some diseases can be transmitted by them.

Bees should be imported only from areas – or if this is impossible, from apiaries – that are free from major bee diseases, and where bees have no undesirable genetic characteristics. They should be obtained from sources that maintain effective quarantine and disease control programmes, and that can certify the health status of export shipments. Pacific islands (including Hawaii), and New Zealand, may well be the first choice. But the situation must be constantly reassessed, as bee diseases and parasites, and Africanized bees, extend their range.

As far as is known, adult bee diseases are not transmitted by immature stages of honeybees, and for this reason methods have been developed for transporting honeybee brood in an incubator, without comb, from one continent to another (M. V. Smith, 1962). Also, drone semen can be deep-frozen and transported for use in instrumental insemination of queens (Taber, 1961), without risk of disease transmission.

If introductions are made, they should be under rigid quarantine conditions, and permission for importations should be granted only to institutions and individuals that can maintain strict adherence to the quarantine requirements. Experimental quarantine procedures for imported honeybees, as used in New South Wales, Australia, have been described by the Director of Animal Quarantine (Skillman, 1979). Unfortunately in some countries where quarantine is mandatory, no bee expert is available at the airport, and bees arriving may be let through.

The spread of the mite *Varroa jacobsoni* through much of the world is due almost entirely to introductions of bees from one country to another (Section 11.42). Where bees in the receiving country are kept in traditional hives, it is often several years before colonies that become infested show outward signs of it. By this time the mite is widely distributed and many colonies are irretrievably lost. For instance the mite was first recorded in Turkey in 1977, which then had over a million fixed-comb hives, and nearly as many modern ones. By 1985, between 80% and 95% of the hives in many traditional apiaries – which had been well organized and cared for by experienced beekeepers – contained no live bees.

Varroa was first reported in the USA in the summer of 1987, and within a few months it was found in many states. One result was the closing of the Canada–USA border to honeybees on 1 January 1988 (Canadian Beekeeping, 1988).

8.73 Introduction of new genetic characters

There are genetic risks and complications in introducing into an area honeybees of the same species as those already present.

Where the bees already in the area are *native* to it, they constitute a gene pool that cannot be recovered if hybridization with the imported bees has taken place. In the last hundred years northern European strains of *Apis mellifera* have been largely lost, through indiscriminate importations of Italian, Carniolan, Caucasian, Cyprian and other races. However desirable the bees to be introduced may seem to be, arrangements should be made beforehand for preserving the gene pool of the local bees. (And it is important that mistakes made in the past with *A. mellifera* are not now repeated with *A. cerana*.) In no circumstances should new bees be taken to a 'sanctuary' area already established for native bees, or to such an area for bees previously introduced, if these have been isolated and now form a genetically valuable resource. For example Kangaroo Island, off South Australia, has purer 'Italian' bees (*A. mellifera ligustica*) than now exist in Italy or anywhere else (Woyke, 1976*a*).

Initial introductions should probably be confined to one small area sufficiently isolated to prevent inadvertent hybridization with bees outside it. The introduction must be planned and executed in co-operation with all authorities concerned. After obtaining sufficient experience, and after competent consultation, it may be decided to replace the (unsatisfactory) native bees of a country by others that are more productive and more amenable to modern management. Where this has been successful the introductions were systematic, thoroughly planned, and on a large scale. The effort needed is considerable, as the following examples show.

In Israel, Italian bees (*A. m. ligustica*) were used to replace the native *A. m. syriaca* bees which are 'aggressive' and in other ways not easy to handle, and also not very productive. Dr Y. Lensky reported as follows (see Crane, 1982*a*). Prior to the large-scale replacement of the native *A. m. syriaca*, observations on the introduction of Italian queens into *syriaca* colonies were made for several years. Introductions were successful when either mated queens or queen cells were placed into small *syriaca* colonies (2–3 frames covered with emerging workers). But the easiest way proved to be to split Italian colonies and to strengthen them with sealed brood from *syriaca* colonies. Within ten years 80% of colonies were 'Italianized'. Hybrid queens were constantly being replaced by Italian queens, because of the influx of *syriaca* drones across the frontiers with neighbouring countries. To preserve the Italian genes, every year several hundred Italian queens were imported from California, USA, and Australia, for distribution among queen rearers and beekeepers.

The replacement of *A. m. syriaca* was achieved here because of the persistent efforts of the beekeepers and of the Extension Service of the Ministry of Agriculture. Final success was achieved when the mite *Varroa jacobsoni* (Section 11.42) killed wild-nesting native bees which received no treatment.

Germany provides a second example. By the 1930s, the bees there were mixed progeny of native north European bees and bees indiscriminately imported in the past. This country, with many small-scale beekeepers, has traditionally had strong Beekeepers' Associations, and after the war the Deutscher Imkerbund in the Federal Republic set out to complete the replacement of bees throughout the country by the Carniolan bee from the eastern Alps (*A. m. carnica*). This bee was selected because colonies develop early in spring, which suited the seasonal pattern of honey sources consequent upon changed agricultural practices. The task of making the change-over involved much planning, much attention to detail, and much co-operation from the beekeepers, in order to ensure that no pockets of the old bees remained. The story has been set out by Dreher (1962), and brought up to date by Bienefeld (1988). Any such operation leaves a continuing problem at the national boundaries, where queens may mate with drones from across the border.

If the area into which introductions are to be made has only previously introduced honeybees of the same species, loss from the gene pool is unlikely to be important. Nevertheless the most cataclysmic introduction in the history of beekeeping occurred as a result of such an introduction – into the Americas which have no native honeybees. European honeybees had done well when introduced into temperate and subtropical America, but less well near the equator. In an attempt to improve productivity, in 1956 tropical African queens were introduced to Brazil, and the resultant 'Africanized' bees subsequently spread over nearly all South and Central America; the story is told briefly in Section 7.43.1. They had been introduced into Europe on various occasions in the past, but do not seem to have survived (Morse et al., 1973). Oldroyd (1987) has discussed results of their possible future introduction to Australia. *In no circumstances should tropical African bees, or Africanized bees from the Americas, be introduced elsewhere.*

8.74 Relations between different honeybee species

If the honeybees already present are of other species, the problems associated with introductions are not genetic. They arise through the transmission of new diseases and parasites, and also through competition: in mating, and in foraging – including robbing from hives (see e.g. Atwal & Dhaliwal, 1969). According to circumstances, either the native or the introduced species may win.

Where *A. cerana* is native (in Asia), introducing *A. mellifera* can have a variety of results. It may lead to the extinction of *A. cerana*, if the introduced bees carry a new disease or parasite, or through competition in foraging, and robbing between colonies. Or *A. mellifera* colonies may die out because they do not exploit the native food resources as *A. cerana* can. *A. mellifera* queens may fail to mate, because *A. mellifera* drones are outnumbered by *A. cerana* drones; see end of Section 2.63 and end of 8.23. In a few areas, introduced *A. mellifera* and native *A. cerana* both do well, and complement each other (Goyal, 1974).

In 1985/86 two Asiatic honeybee species were introduced into regions which previously had only *A. mellifera*. *A. florea* was found in Sudan in Africa, and *A. cerana* was taken to the island that comprises Irian Jaya and Papua New Guinea; see Sections 1.23.3, 1.23.1.

8.75 Environmental considerations

There is no apicultural objection – apart from those referred to above – to the introduction of honeybees to an isolated area entirely without honeybees, such as some Pacific islands. There may, however, be environmental objections, for instance the likelihood of reducing populations of native bees or of other nectar and pollen foragers, through competition for food. One example concerns the island of Yap in the South Pacific, where several species of native nectar-feeding birds, and a fruit bat *Pteropus yapensis*, forage on (and pollinate) flowers of certain native plants. These animals could well be disadvantaged if honeybees were imported, and competed for their forage. *Apis mellifera* has already been introduced in Guam, and similar species of birds and the fruit bat have become very rare (Falanruw, 1983). The subject is discussed by Sugden (1986).

In Australia, concern has recently been expressed that in some areas introduced *A. mellifera* may be reducing populations of native nectar and pollen feeders, and also of plants of which this bee is an inefficient pollinator.

8.8 MANAGING HIVES FOR EDUCATION AND STUDY

8.81 General considerations

This subject is an important one to those concerned

with teaching bee biology and beekeeping, and with interesting the public in beekeeping, especially children – from whom the next generation of beekeepers will be drawn. A few general comments are given here, and references are provided to detailed publications on various types of hive for education and study; flight rooms are also described briefly.

A high standard of hygiene and cleanliness should always be maintained with hives used for teaching purposes. Where normal hives are concerned (i.e. not observation hives), it is best to use whatever type is most common locally. Teaching beekeeping to beginners is best started in spring when the colony is small, and no colony should ever be allowed to grow so large that it is frightening to the learners. If the hive is moved to a new site (Section 8.15), so that the older flying bees join other colonies, a smaller colony will be left whose bees are young and less inclined to sting. This operation should certainly be done with any hive to be taken to a show or fair for demonstrations.

In the course of learning operations in bee management, students will be shown normal colony activities (Chapter 3): they will be able to see workers building comb, brood rearing, dancing on the comb, and processing honey and pollen. They will also see the queen laying eggs, young bees emerging from their cells, and interactions of workers with others, and with drones and the queen.

If it is possible, circumstances should also be arranged to allow learners to see events that beekeepers try to avoid in modern bee management: the bees' swarm preparations inside a hive, a swarm emerging, and being 'taken', and hived. Bees in a swarm secrete wax readily, and if a swarm is hived temporarily in a box with transparent sides, the bees can be seen building their combs. The box should first be covered for a day or two, as darkness is needed for the initiation of comb building (Section 3.31.1). If the base of the box is also transparent, and a mirror is mounted beneath it at 45° to the horizontal, the combs can be viewed from below.

An *observation hive* (i.e. one with transparent walls) should be mounted at a height where the people looking at it can conveniently watch the bees in it. It can be a normal-size hive with transparent walls, but the most common type of observation hive, discussed below, has the thickness of a single comb plus a bee-space on either side. Several framed combs are mounted one above another between the transparent side-walls of the hive, as in Figure 8.81a. Such a hive enables people to see many activities of bees for the first time, including those mentioned above that can otherwise be seen only when combs are taken out of a hive. It delights both learners and the general public – especially the chance to see the queen. A light-coloured queen should be used, as it is more difficult to see a dark one among the workers. The queen should be colour-marked (Section 5.23).

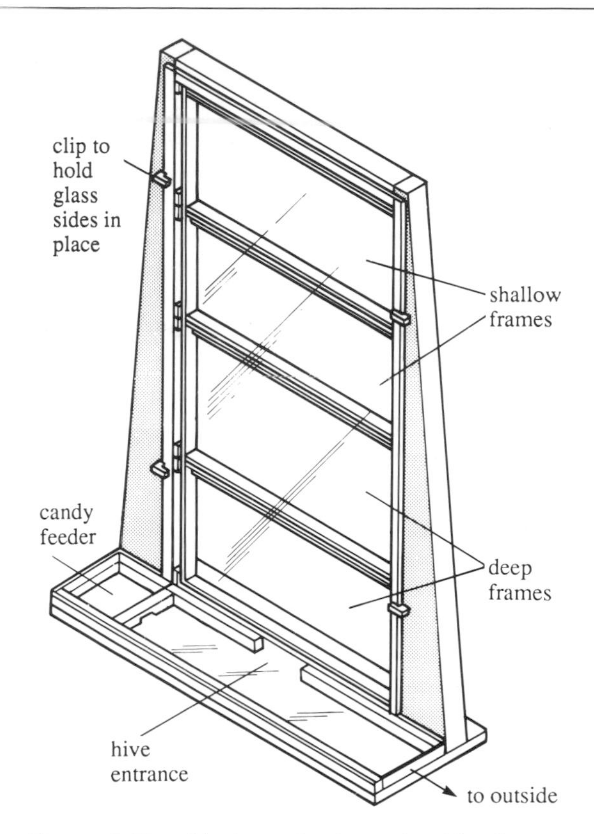

Figure 8.81a Single-comb observation hive for two deep and two shallow frames (drawing: University of California).
Bees can also be watched as they leave and enter the hive through the wide glass-topped passage at the bottom; one end of this gives access to the outside.

These observation hives present special management problems because the combs are housed in a way that is abnormal for a colony of bees. A hive of this type should never be left in sunlight; even so, the bees are easily overheated through difficulties in ventilation, especially if the hive is indoors and the only access to fresh air is through a long tube leading to a flight exit outside. Brood may get chilled at night, so an insulating cover should be available. If the spacing between the two walls is slightly too large, the bees may deposit wax on the glass which obscures the view; if it is slightly too small anywhere, the bees cannot get access to the cells there.

A colony put into an observation hive for an event lasting only a few days should be in peak condition for observation: with brood at all stages, pollen, honey, some drones, and a laying queen, but rather few adult bees so that the cells are not obscured. The colony can, however, become much less interesting within a few days, for instance if the queen has laid in all the cells. An observation hive installed in a school or museum needs frequent attention by a beekeeper if its full interest is to be maintained; this work cannot be left in the charge of non-beekeeping staff. School summer holidays present problems unless regular attention can be ensured, and it may be best to install the hive early in the summer term and dismantle it at the end.

In the cold temperate zones it is rarely possible to keep a colony in an observation hive in satisfactory condition through the winter. It is best to unite the colony with another at the end of summer, and start afresh next spring.

In the Museum at Mutare, Zimbabwe, I saw a 3-frame single-comb observation hive, containing tropical African bees in good condition. It was connected to the back of a nucleus hive, which was hidden from sight and had a flight entrance outside the building.

Some readers may want to make a single-comb observation hive, but do not have previous experience of managing one. They are strongly recommended first to study such a hive in operation, learning from the owner what he finds satisfactory and unsatisfactory about it; what situation he finds best for it; how long he can keep the colony in it in good condition; and so on. Many beekeeping supply firms sell observation hives but, again, try to see an example in use, and learn its capabilities and shortcomings, before making a purchase.

Observation hives for stingless bees are described by Sakagami (1966), Nogueira-Neto (1970) and Sommeijer (1983).

8.82 Instructions for specific observation hives

Whether an observation hive is purchased or made, it is important to ensure that the hive is for frames of the same size and style as those in your normal outdoor hives, so that frames can be interchanged between the observation hive and others, whenever this is desirable.

There are many leaflets and bulletins, and several books, on observation hives. Some of the most recent are referred to below with an indication of topics dealt with. All are listed in the Bibliography.

General

Gary and Lorenzen (1980) – construction and maintenance of a hive containing single Langstroth frames, one above another.

van Laere (1971) – as above, but containing 3 Dadant frames side by side; also a closed hive, 12 × 6 × 10 cm in which 50–200 bees can be kept for up to 2 weeks.

Gojmerac (1973) – a modular hive in which each module holds a Langstroth shallow frame, and can be removed from the hive temporarily (with bees inside) by shutting it off with metal slides.

von Frisch (1967) – a 20-page chapter in *The dance language and orientation of bees*, describing his observation hives (for German frames) and their use.

Caron (1979*b*) – installation of hives, their care, and instructions for making observations on comb building, egg laying, brood rearing, cell cleaning, removal of intruders, communication dances, and foraging.

Rothenbuhler et al. (1968) – includes descriptions of observation hives and of a portable building providing a controlled environment for them.

Special types of observation hive

Robb (1951) – heated hive which includes half-combs with the cell bases against the glass, and some combs set at right angles to the glass so that the interiors of cells can be seen, sideways on.

Kolb (1973) – a hive without frames; bees build natural comb on starter-strips of foundation. With the empty hive, the supplier dispatches bees and a queen in a canister, and they can be transferred to the hive without any bees escaping to the outside. (This applies only within the USA, where the hive is produced.)

Mangum (1987) – two types of top-bar observation hive.

Österlund (1986) – construction and use of a transparent queen cell in which queen development can be observed.

Books

Showler (1985); Khalifman (1960, in Russian) – both entitled *The observation hive* and including information on the history of observation hives in the nineteenth and twentieth centuries.

8.83 Flight rooms

A flight room is an enclosure within a building, where research workers can keep hives of bees in an environment that is programmed, with complete control of temperature, humidity, photoperiod and bee forage. In temperate zones it extends the bees' active season, and free-flying colonies can be maintained throughout the year.

The first flight rooms were built in the 1950s, but they were far from satisfactory. They were brightly lit

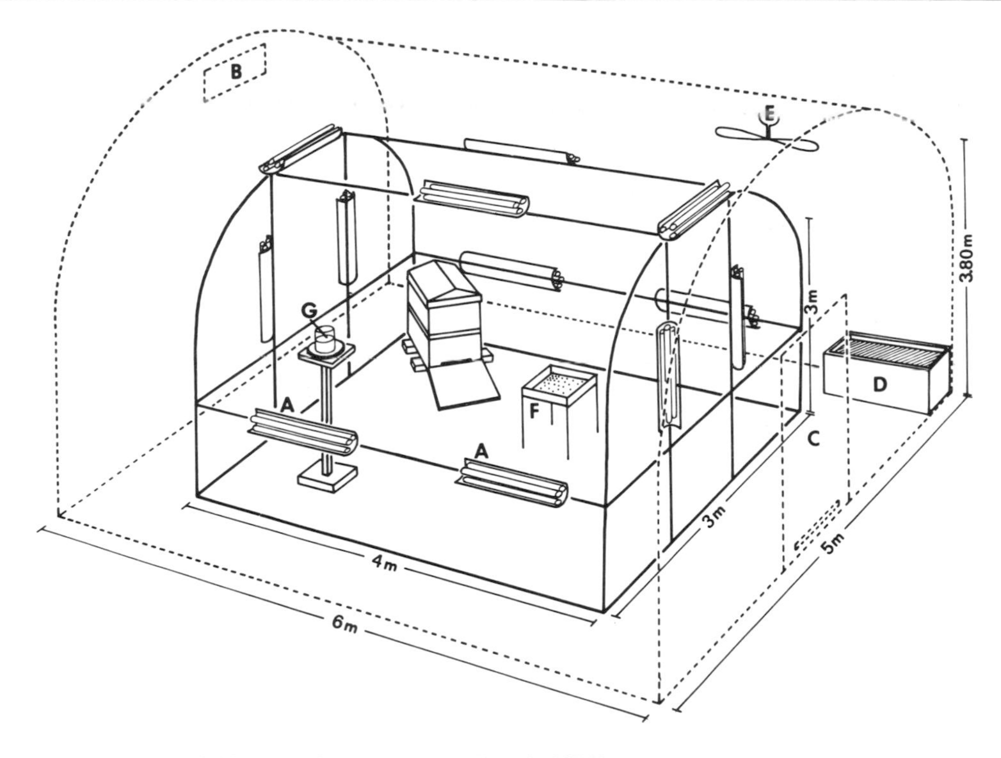

Figure 8.83a Bee flight room (van Praagh, 1972).
A set of 3 fluorescent tubes with reflector
B air exit
C door with air entrance
D water-bath for humidity control
E fan
F pollen feeder
G syrup feeder

to simulate daylight, and bees became disoriented and collided with the walls. There was also difficulty in getting the bees to collect the pollen provided, so brood rearing was not maintained. Many bees died, and the faeces and dead bees made these early flight rooms rather unhygienic and unsavoury places.

In the Netherlands, van Praagh (1972) discovered how to overcome the above difficulties, and the resulting flight room, shown in Figure 8.83a, has been the model on which many others have been constructed. An iron framework, with a fine gauze fitted over it, constituted the enclosed flight space. Immediately outside the gauze was a coarse net with a 15-mm mesh; the threads were 4 mm thick, and the bees could see them, and so did not collide with the gauze walls. Illumination was arranged to simulate natural conditions out of doors *as perceived by the bees*. To do this, special attention had to be paid to the ultraviolet component of the light, and also to its ALD, 'angular light distribution'. The walls and ceiling of the building containing the flight room were lined with wrinkled aluminium foil, making a huge diffusing reflector for ultraviolet light. (The original flight room had a curved ceiling because it was built in the vault of an old fortress; in fact this proved to be beneficial, because with a flat ceiling the corners reflect too much light, van Praagh, 1987.) The light sources were 12 sets of 3 fluorescent tubes distributed round the room. Each set was fitted with its own aluminium reflector, which reflected the light to the aluminium-covered walls and ceiling, so direct lighting was eliminated. Colonies can be maintained in such a flight room for one or more years.

In any flight room, humidity must be kept very high, and nosema disease must be continuously controlled, e.g. by feeding fumagillin. The period of lighting in each 24 hours (photoperiod) can be kept constant throughout the year, or varied to simulate the annual cycle of day-length at whatever latitude is desired. Many aspects of flight room techniques were discussed at a symposium on the subject (Ruttner & Koeniger, 1977).

8.9 FURTHER READING AND REFERENCE

Useful books to supplement references in the text are available on only some of the subjects covered in this chapter. Details of all publications cited will be found in the Bibliography.

Queen rearing and bee breeding

Laidlaw, H. H. (1979) *Contemporary queen rearing*

Ruttner, F. (ed.) (1983) *Queen rearing: biological basis and technical instruction*

Cook, V. A. (1986) *Queen rearing simplified*

Laidlaw, H. H. (1977) *Instrumental insemination of honey bee queens*

Rinderer, T. E. (ed.) (1986) *Bee genetics and breeding*

Taber, S. (1987) *Breeding super bees*

Vaillant, J. (1986) *Initiation à la génétique et à la séléction de l'abeille domestique*

Weiss, K. (1986) *Zuchtpraxis des Imkers in Frage und Antwort*

Managing honeybees and other bees for crop pollination

Entries in the Bibliography under International Symposium on Pollination give details of Proceedings of Symposia held as follows:

1st, Copenhagen, 1960
2nd, London, 1964
3rd, Prague, 1974
4th, Maryland, 1978
5th, Versailles, 1983

McGregor, S. E. (1976) *Insect pollination of cultivated crop plants* – detailed practical discussions of individual crops.

Free, J. B. (1970) *Insect pollination of crops* – especially botanical aspects, and bee behaviour.

Pesson, P.; Louveaux, J. (eds) (1984) *Pollinisation et productions végétales* – includes much information on botany and on pollination of crops by wind as well as by insects.

Crane, E.; Walker, P. (1984*a*) *Pollination directory for world crops* – brief details for many crops, temperate, subtropical and tropical

See text references for the use of individual species of bees.

Managing stingless bees

Nogueira-Neto, P. (1970) *A criação de abelhas indigenas sem ferrão*

Crane, E. (1978*a*, No. 14) *Beekeeping in the tropics with stingless bees* – bibliography.

PART III

Beekeeping with Simpler and Cheaper Hives

9

Traditional and modern fixed-comb hives and their management

Chapters 9 and 10 explore the potential of well tried hives that are simpler, and involve fewer precision measurements, than movable-frame hives. They may also be much cheaper, since they can be made from a wide variety of local materials. Although in most countries none of these hives is likely to replace movable-frame hives in large-scale beekeeping, some – especially among those described in Chapter 10 – have advantages in certain circumstances. Their use is certainly worth further exploration.

9.1 FIXED-COMB HIVES: THE HERITAGE

Bees have been kept in hives for at least 4500 years (Section 1.42), and traditional hives were simple purpose-built containers for the bees and their combs. They had no fittings such as frames, and the bees secured the tops of their combs to the interior of the hive. These hives are nowadays referred to as *fixed-comb* hives, in order to differentiate them from the modern *movable-frame* hives (Chapter 4), and the intermediate *movable-comb* hives (Chapter 10). With fixed-comb hives, the beekeeper could remove the combs – for honey harvesting or for any other purpose – only by breaking or cutting them away. With movable-frame and movable-comb hives, any comb can be removed and replaced, or exchanged for another.

This section gives a very brief summary of fixed-comb hives with which the craft of beekeeping was carried forward through several millennia, supplying honey to the world's people when this was the only sweetener available to them. Much fuller information has been published in *The archaeology of beekeeping* (Crane, 1983*a*) and other publications (e.g. Crane & Graham, 1985), from which most of the following is taken.

9.11 Horizontal traditional hives

The most universal type of traditional hive, known to have been in use as early as 2500 BC (see Section 1.42.1), is a cylinder, sometimes tapered or ovoid, or a box with a square cross-section, laid horizontally. In its most primitive form, only one end of the hive could be opened, but in more advanced forms, each end of the cylinder was fitted with a removable closure. The beekeepers who were most practised and skilful in using these hives could do many operations now customary with modern movable-frame hives, although with less facility. Because this is not widely known, a few examples are given below.

The flight entrance for the bees was in (or at the edge of) the front end-closure. To harvest honey the beekeeper removed the back end-closure, and blew smoke through the opening into the hive to drive the bees towards the front, and some probably out through the flight entrance. He then cut away the combs of honey one by one from the back (by now relatively free from bees). The beekeeper also took out the back closure if he needed to place food inside the hive.

In 1978 I watched traditional beekeepers in Egypt

remove brood combs from their hives, and replace them, as follows (Figure 9.11a). The brood nest is commonly near the flight entrance, so the front end-closure was taken out, and – after smoking – a specially shaped long-handled metal knife was used to cut the first brood comb away from the hive where it was attached at the top. Subsequently combs were carefully removed in the same way. They could be inspected and, if wished, used with their (young) brood and some adult bees to start a new colony, by putting them into an empty hive. Or they could be freed from bees, and used to prime an empty hive to receive a swarm. In either case, the beekeeper fixed each comb in the hive at the correct spacing, with a small forked twig selected from nearby vegetation, and the bees would soon attach it to the hive. Alternatively, he could replace the combs in the original hive in the same way. I was told that such replaced combs are more easily removed next time, because the bees' second-time attachments are less strong. By fixing these combs *across* (not along) an empty hive, the beekeeper ensures that the colony will build the rest of the combs across the hive. From ancient times, beekeepers seem to have preferred this.

We have no direct evidence that these operations were carried out in Ancient Egypt, only the indirect evidence that, in several tomb paintings, harvested honey combs are clearly shown to be round (e.g. Figure 1.42a). A similar practice was used in recent times in Crete (Ruttner, 1979), but no depiction or descriptions of honey harvesting in antiquity are known from the island.

Figure 9.11a A beekeeper in Egypt (1978) has used a knife to remove a brood comb from the front of a mud pipe hive (in a stack of several hundred), and is lifting it out with a long-handled spatula.

Recent excavations in mainland Greece, and on some Aegean islands, have yielded hives dating from 400 BC to AD 600. They are tapered cylindrical pottery vessels closed at one end; pottery rings were also found, that could be used as extensions for honey storage (Figure 9.11b). Along the Mediterranean coast from Morocco to Turkey, on some Mediterranean islands, and as far east as Kashmir, extensions are still used with traditional hives, to serve as a separate honey chamber (Figure 9.11c).

Many beekeepers in East Africa use a horizontal hollow log for a hive, suspended from a tree to protect it from enemies. In a few tribes, beekeepers cut the log lengthways, at or below the centre line (Figure 9.11d), and tie the two halves together by vines, rope or wire. When harvesting honey, the two parts of the log are prized apart, and the upper half is lifted up with the combs attached to it. Combs of honey can easily be cut from the wood, and the hive is then closed together again with the brood nest intact (Nightingale, 1983). Similar hives are used for *Apis cerana* in the Borobudur region of Java (Franssen, 1931; also Bijenteelt, 1986).

Ko Samui, a small island in the Gulf of Thailand, has several thousand traditional hives of *Apis cerana*. Samart Panchawang, one beekeeper I visited in 1986, has for fifty years used his own method for making new colonies, and also for preventing swarming. As a young man he had noticed that bees were attracted to a queen and would cluster round her. So when colonies start swarm preparations he carries out the following procedure. He removes the closure from the end of the hive (a horizontal cylinder) and cuts away the combs one by one, until he finds the queen. He catches her, and tethers her on a fine thread, placing one loop of a double bow round her abdomen and then tightening it gently between the abdomen and thorax

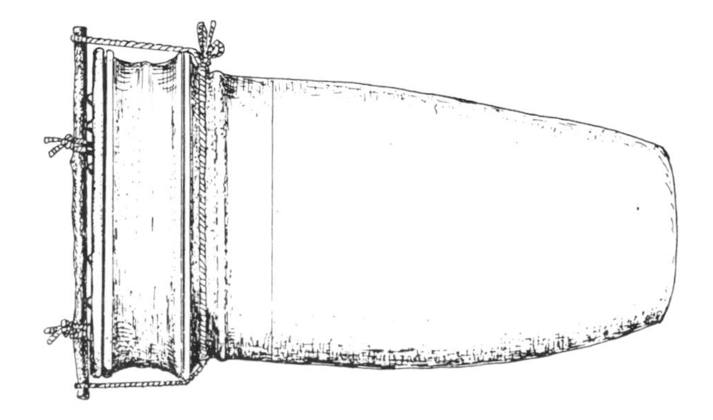

Figure 9.11b Longitudinal section through a reconstructed pottery hive from Trachones, Athens, Greece (*c.* 400 BC), showing the hive extension on left (drawing: J. E. Jones).

(round the petiole). He fastens the queen by her tether inside an empty hive placed with its open end very close to the other hive; most of the adult bees then move into the hive with the queen. The queen is later released by pulling undone the other loop of the bow (I was given a demonstration), and the hive is placed at a new site. Maung (1984) reports a similar procedure by a beekeeper in the north of Burma; in Vietnam a hair was used to tether the queen to the branch where the swarm would cluster (Toumanoff & Nanta, 1933). By following the procedure whenever colonies start to rear queens for swarming, the Ko Samui beekeeper said he found that his colonies do not swarm or abscond, although other colonies on the island do so. (Swarming and absconding are relatively unusual in *Apis cerana* colonies with young queens; see Section 7.41.4, under Absconding.)

Figure 9.11c Pottery hive with extension, and closure with flight entrance, Gharb, Morocco, 1963.

Figure 9.11d Log hive, split just below the centre line (at arrows), Arusha, Tanzania, *c.* 1970 (photo: F. G. Smith).
The combs are attached by the bees only to the upper part of the hive and so are easily harvested by separating the two parts.

9.12 Upright traditional hives

In Northern Europe, the traditional hives in forests were hollow logs placed upright; elsewhere they were skeps – baskets open end down, made of woven wicker or cane or, later, of coiled straw, reeds or grass. Many such hives are illustrated and discussed by Crane (1983*a*), and those known from the Ancient World by Crane and Graham (1985). In general it was much more awkward to harvest honey from hives of this type than from horizontal hives, in which smoke could be used to drive the bees from the honey combs and through to the other end of the hive.

By the 1600s or earlier, an extension was used with some straw skeps to serve as a honey chamber during the main honey flow. A cap, like a small skep, was placed on top of the skep, and a hole allowed access by the bees, although hopefully not by the queen. Alternatively an extension ring (eke) was placed below the skep, the diameters of skep and eke being the same.

In Poland and some neighbouring regions, a narrow vertical door was cut in upright log hives to give access to the combs (Figure 1.42d). This practice was inherited from 'forest beekeeping' in which bees in tree cavities in living trees were regularly tended, as explained at the end of Section 1.41. In these forested areas, early movable-frame hives – such as that devised by J. Dzierzon (1848) – were also provided with a door at the back, and were not worked from the top as Langstroth hives are. The back-opening hive (*Hinterlader*) is still common in Germany, mostly kept in bee houses; see Figures 4.2c and 6.63b.

9.13 Contradistinctions between horizontal and upright hives

In antiquity, whereas horizontal hives were very widespread, upright hives were common in only a relatively small region in northern Europe; later they were used also in Asia, especially in the north. Almost every type of tradional hive used upright in the north was also used horizontally elsewhere – and vice versa. The type and construction of a hive were determined by local materials, but whether it was used upright or horizontally depended on the geographical locality. I have put forward the following tentative explanation (Crane, 1983*a*), on the basis of direct and indirect effects of latitude on honeybees.

At high latitudes there is a relatively long winter period when bees cannot collect food. Colonies of bees can survive beyond it only if they start to rear brood while the weather is still cold, store a large amount of food in summer, and in winter form a cluster within the hive so that their metabolic heat is conserved: their energy requirement at low temperatures is minimal. The bees can best achieve this by storing honey *above* the brood before winter begins, which they can do in an upright hive. The honey then provides a thermal buffer that reduces loss of heat from the colony. In early spring the cluster almost always moves upward and the first brood is reared in cells adjacent to stored honey (Corner, 1988, in British Columbia). In summer, the rising heat helps the bees to evaporate water in the storage cells when they are making honey.

The only high-latitude temperate area where beekeeping was practised early was the relatively small European region where traditional hives were upright. (There were no native hive honeybees at latitudes much beyond 50°N in Asia, or beyond 55°S in any continent.)

At low latitudes, it is usually warm enough for bees to forage during most of the year, and heat presents a greater hazard than cold. Bees in a horizontal hive may be able to avoid overheating by spreading out to the ends of it. Also, because of the high temperatures in the tropics, bees can secrete wax and build combs in much of the hive, not only where it is warmed from the brood nest. They can continue comb building at one or both sides of the nest. Being small, such combs are less likely to break in hot weather under the weight of honey in them, whereas this can happen with upright hives, even in temperate climates. In the tropics and subtropics honey may be produced over a long period in the year. Early beekeepers could remove a few honey combs at a time, from one or other end of a horizontal hive, and this practice continues. The dichotomy into upright and horizontal traditional hives seems to have persisted throughout recorded times. With upright hives, the awkwardness in getting access to the brood combs, and in removing honey, led to much experimentation to find a better type of hive. Removable top-bars made operations much easier, and Chapter 10 discusses the efforts to use them to advantage, which led to modern movable-frame hives. One might say that these efforts were the outcome of the *difficulties* of using upright hives.

When seeking to develop beekeeping with tropical honeybees, the introduction of modern upright hives is not the only alternative. Improvements and innovations in horizontal hives should also be sought, whether these are traditional or with top-bars or frames.

9.2 FIXED-COMB HIVES IN BEEKEEPING TODAY

Modern beekeeping with movable-frame hives involves many manipulations in which hive boxes with their complement of framed combs of brood and adult bees, and also individual framed combs, are transferred between hives. In many management systems, complete hives are moved frequently from one site to another. Any of these operations can transmit pathogenic micro-organisms between colonies. It is possible to transport bees between countries, and in spite of legislation regulating imports, this is done (Section 15.45). From time to time diseased or infested bees are imported into a country, and there is always a risk of introducing a disease or parasite not previously present. The consequences of this are especially serious where there are both movable-frame hives and fixed-comb hives. Colonies in movable-frame hives are then likely to be moved about within the country, and the pathogens spread widely; when they reach fixed-comb hives they are usually not detected, so cannot be treated.

In areas with *only* fixed-comb hives, the causes of ill health in bees may not be understood, but there is little opportunity for the transmission of micro-organisms from one colony to another. Combs are not often transferred, and if migratory beekeeping exists, it is likely to be limited to one honey flow a year. So beekeepers may lose comparatively little by their lack of easy access to inspect brood combs.

Even today, beekeeping is practised with many millions of fixed-comb hives in the remoter areas of less developed countries. For the period until modern frame hives are introduced, with the attendant dangers of disease transmission to bees in the traditional hives, these fixed-comb hives can yield a modest amount of honey, and also about 10% of its weight in beeswax. This harvest is achieved with minimal cost and labour, and it is valuable to peoples living a marginal existence, whose diet may be dull and uninteresting.

Portugal is one country with a rich tradition of beekeeping with upright fixed-comb hives, and various modest improvements are used (Moreira, 1968). An intermediate hive was designed for tropical Africa (Rosário-Nuñes, 1973); this hive can lead on to hives with separate top-bars (Section 10.22). It is a long rectangular box that holds 28 combs, and the roof is in short sections each with several top-bars fixed on the underside. Internally the hive is 900 mm long and 290 mm wide and high, and the centre-to-centre top-bar spacing is 32 mm. Table 4.25A lists comb spacings for different honeybees.

I do not think any knowledgeable person would countenance the use of fixed-comb hives in technologically advanced countries. Some people who are interested in the back-to-nature 'simple' life may be tempted to favour beekeeping with logs, skeps or boxes, as being more 'natural' than the use of precision-made modern hives. But there will be no local expertise in the management of bees in these hives, and a beginner beekeeper trying to use them is likely to end up with many dead bees, and with no honey. In an area of movable-frame hives, beekeeping with fixed-comb hives is anti-social, and indeed is prohibited by law in some countries (Section 15.42).

9.21 Transitory fixed-comb hives

Several specialist operations in modern beekeeping involve short-term housing of bees, in containers to which the bees fix their combs as in fixed-comb hives. If the bees are in the hive for long enough, they build whole combs – but their orientation to the hive walls is unimportant to the beekeeper. Such hives are used only in situations where hazards of disease transmission are minimal.

Bait hives or catcher boxes, sited in strategic positions for flying swarms to occupy, do not need frames or top-bars if they are inspected every few days and any swarms in them removed. They are discussed in Sections 6.35.1 and 7.33.1. Some bait hives are designed for use either with or without frames (Egea Soares, 1985; Adams, 1986).

In the USA, expendable boxes – usually of waxed cardboard or styrofoam – have been used as 'disposable pollination units' for almonds and other crops. Erickson et al. (1977) give details and further references. Each box contains bees (up to 6 lb, 2.7 kg), without a queen, or with a caged queen, or with a virgin treated with carbon dioxide so that she lays only drone eggs. The purpose is to *prevent* development of a normal colony, so that the bees die out after the crop has finished flowering.

9.22 Fixed-comb honey supers for modern centrifuges

By the 1870s, the use of Langstroth's movable-frame hive had been followed by the invention of comb foundation and the development of a centrifuge that could spin honey out of the combs after they had been uncapped, i.e. their wax lids removed (Section 1.44). Ever since then, the norm for modern beekeeping has been the removal of honey supers from the hive, followed by individual handling of each frame: to uncap the comb, place it in a centrifugal extractor, and after spinning to replace it in the super for subsequent use or storage.

With the more recent development of large and powerful centrifuges, the unit for handling need not be the individual frame; it can be a super complete with fixed combs full of honey. The use of fixed-comb

Figure 9.22a Fixed-comb honey supers in Roger Blackwell's apiary, Australia, 1967.
A hive is opened to show the (identical) top and bottom of a super, with the wooden strips that hold the top and bottom bars in place.

supers, above a brood chamber with movable combs in frames, is therefore worth exploring. Provided the queen does not have access to the supers, there is no risk of undetected brood disease. With appropriate equipment, honey extraction can be both easier and quicker than with movable frames. I record two examples to encourage readers to think about possibilities for extending their repertoire of hive and honey management.

The first example is an operation I encountered in NSW, Australia, in 1967. Roger Blackwell was using hives with an 8-frame Langstroth brood chamber, a queen excluder, and supers (about $\frac{2}{3}$ the depth of the brood box) with 7 frames permanently fixed in each (Figure 9.22a). The frames had identical top and bottom bars, clamped in place by strips of wood at right angles to them, across their ends (like frame runners, but outside, not inside, the frame bars); the strips were used at both top and bottom of the super. Frames had no side bars, but each was fitted with a sheet of Plasticore (drone) foundation, which has a metal strip along each side to strengthen it. The frames were not handled individually at any time. A Fox-Harrison uncapper was used (B. N. Anderson, 1966), which jiggled the supers to and fro mechanically as they moved down by gravity, so that each frame passed between its own pair of stationary knives. The extractor was simple but very large; see Figure 13.33a (D).

The second example involves a large high-speed centrifuge of the type used for separating cream from (higher-density) milk. Such centrifuges have been used for several decades to separate wax cappings from (higher-density) honey; the first one I saw was Carl Meilicke's in Saskatchewan, Canada, in 1965. In Sweden, Lagerman (1985, 1986) has now developed a honey-production system using fixed-comb shallow honey supers without frames or top-bars, above a standard 12-frame brood chamber and a queen excluder; it is described in English by Österlund and Lagerman (1986). Above and below each super is fixed a board with 11 slots, each slot leading to a gap between strips of wax foundation suspended from two pieces of angle-aluminium, at right angles to them (Figure 9.22b, *above*). The bees start building combs on the strips of foundation and extend them to the full width and depth of the super (Figure 9.22b, *below*). When harvesting honey, each super is freed from bees by using a bee escape board (Figure 5.31a, *below*), and taken to the honey house still fixed between its two slotted boards. There, both boards are ripped away, the aluminium-angle supports are pulled out, and a jerk is sufficient to knock the unsupported combs out of the super into the large funnel of an industrial meat

Figure 9.22b Fixed-comb honey super for producing extracted honey (Österlund & Lagerman, 1986*b*).
above Honey super, fitted with sheets of foundation, standing on its lower slotted board; its upper slotted board has been lifted off. The boards are made of 3-mm masonite, and have a rim 3.2 mm deep round the edges on both sides, giving a bee-space 6.4 mm. Slots are 10 mm wide.
below Filled super, freed from bees, and with the upper slotted board removed.

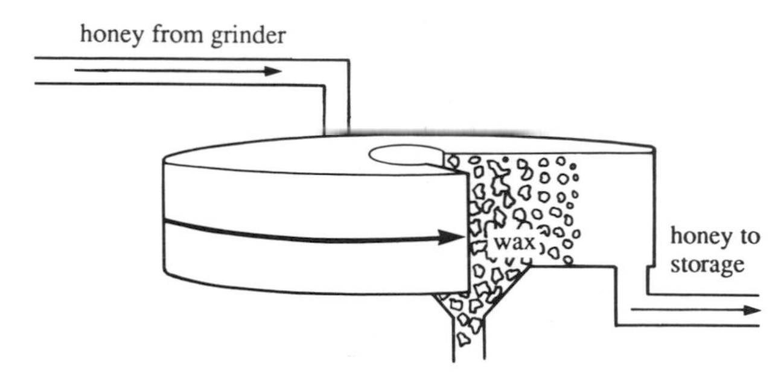

Figure 9.22c Diagram of separator to extract honey from fixed-comb supers (based on drawing by T. Rydén). Honey and wax from the grinder enter through the pipe on the left, honey leaves through the pipe leading from the outer wall on the right, and dry wax drops down the central hole.

grinder. From the grinder, which breaks up the combs, the honey is pumped via a buffer vessel into a centrifugal separator (Figure 9.22c), diameter 60 cm, whose rotor spins at 1000 rpm; the honey separates from the wax continuously, at 500 kg/h. Heather honey (*Calluna vulgaris*) can be separated at 300 kg/h, provided the temperature is 30°–35°. Honey is pumped to storage tanks, and needs no further straining before bottling. Dry wax collects in a removable container below the centre of the separator. With this system, no combs are stored, and no frames are required.

A beekeeper in New Zealand uses a somewhat similar system with normal frames in supers (Jaycox, 1980). There seems to be no reason why something similar should not also be used with top-bar hives. The separator would be operated at the larger processing stations, individual combs being cut or broken off their top-bars above a grinder which would not need a very large funnel opening.

9.3 FURTHER READING AND REFERENCE

Details of publications listed will be found in the Bibliography.

Crane, E. (1983*a*) *The archaeology of beekeeping*

Crane, E.; Graham, A. J. (1985) *Bee hives of the Ancient World*

Seeley, T. D.; Morse, R. A. (1982) *Bait hives for honey bees*

Witherell, P. C. (1985) *A review of the scientific literature relating to honey bee bait hives and swarm attractants*

10

Other movable-comb hives and their management, especially top-bar hives

The hives in Chapter 4 (such as Langstroth's) consist of tiered boxes fitted with movable frames in which the bees build their combs. The hives discussed in Sections 10.1 and 10.2 of this chapter have movable *combs*, i.e. any individual comb can be removed from the hive and replaced, but instead of four-sided frames, the hives have only top-bars, similarly spaced, and they are often referred to as top-bar hives. The bees build down from each top-bar, and the beekeeper can remove any comb from the hive by lifting up its top-bar (Figure 10.23c). A long single box is commonly used for such hives, and in many of them the top-bars fit tightly against each other, so that bees cannot pass between them, and the bars together form a solid cover over the space below that contains the combs and the bees.

A single long box can alternatively be used with frames, and a minority of beekeepers in many different countries do this; some of them have access to a source of boxes that can be adapted at almost no cost. Section 10.3 deals with this type of long hive, which usually holds 30–35 frames.

Whether a long single-box hive is fitted with top-bars or with frames, every comb in it is accessible without lifting off any super, and all combs can be inspected for disease. This was not possible with fixed-comb hives (Chapter 9), or with early frame hives such as Prokopovich's (*c.* 1806) and Dzierzon's (1848); see Section 1.44.

10.1 THE HERITAGE OF MOVABLE-COMB HIVES

At some stage – we do not know when – beekeeping developed from what was possible with fixed-comb hives (Chapter 9) to the more advanced management achievable by using hives in which individual combs were freely removable. Our earliest knowledge of this development is in Sir George Wheler's book *A journey into Greece* (1682). The author described the hives he had seen at St Cyriacus's monastery on Mount Hymettus in Attica (Figure 10.1a).

> The Hives they keep their Bees in, are made of Willows, or Osiers, fashioned like our common Dust-Baskets, wide at the Top, and narrow at the Bottom, and plaister'd with Clay, or Loam, within and without. They are set the wide end upwards, as you see here. The Tops being covered with broad flat Sticks, (as at C.C.C.) are also plaistered with Clay on the Top; and to secure them from the Weather, they cover them with a Tuft of Straw, as we do. Along each of those Sticks, the Bees fasten (a) their Combs; so that a Combe may be taken out whole, without the least bruising, and with the greatest ease imaginable. To increase them in Spring-time, that is, in *March* or *April*, until the beginning of *May*, they divide them; first separating (b) the Sticks, on which the Combs and Bees, are fastened, from one another with a Knife: so taking out the first Combs and Bees together, on each side, they put them into another Basket, in the same Order as they were taken out, until they have equally divided them. After this, when they are both again accommodated with Sticks and Plaister, they set the new Basket in the Place of the old one, and the old one in some new Place. And all this they do in the middle of the day, at such a

time as the greatest part of the Bees are abroad; who, at their coming home, without much difficulty, by this means divide themselves equally.

(c) This Device hinders them from swarming, and flying away. In *August* they take out their Honey; which they do in the day-time also, while they are abroad, the Bees being thereby, they say, disturbed least. At which time they take out the

(d) Combs laden with Honey, as before; that is, beginning at each out-side, and so taking away, until they have left only such a quantity of Combs in the middle, as they judge will be sufficient to maintain the Bees in Winter; sweeping those Bees, that are on the Combs they take out, into the basket again, and again covering it with new Sticks and Plaister.

With these hives the monks could (a) inspect and manipulate colonies while foraging bees were out of the hive, (b) divide colonies, (c) carry out swarm control, and (d) harvest combs of honey without brood or adult bees, and ensure that adequate winter stores were left for the bees. In 1979 Penelope Papadopoulo took me to an apiary containing several hundred hives of this type, at Phyli near Athens. Fewer than a hundred then contained bees. In 1985 in Crete, I saw similar hives, some woven and others of pottery (like giant flower pots), but none were then in use.

It has been suggested that the woven hives, or their pottery counterparts, were known in Ancient Greece, and Infantidis (1983) has suggested how they may have arisen. But evidence is lacking, and certain Ancient Greek pottery vessels once thought to have been hives of the type described by Wheler were not;

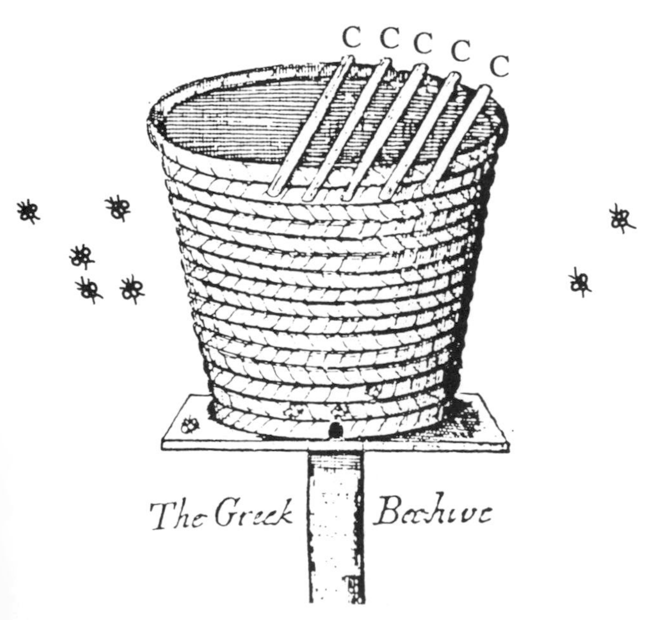

Figure 10.1a Sir George Wheler's drawing of a Greek top-bar hive (1682).

they were more probably *clepsydra* (water clocks) or sand clocks, the presumed flight entrance being the hole through which the water or sand escaped (Crane & Graham, 1985). Roman authors described nine types of hive, but not this one.

In the two centuries after Wheler's book was published, a few beekeepers with enquiring minds experimented with top-bar hives such as he described. Wheler's drawing shows thin sticks with spaces between, but he referred to the top-bars as 'broad flat Sticks, (as at C. C. C.)', along which 'the Bees fasten their Combs; so that a Combe may be taken out whole . . . with the greatest ease imaginable'.

Langstroth owned or knew the key publications that paved the way for the advance from various movable-comb hives to movable-frame hives (Johansson & Johansson, 1972*b*). An essential feature in both is that the centre-to-centre distance between adjacent top-bars is the natural spacing at which the bees built their combs.

10.2 THE MODERN RESURGENCE OF TOP-BAR HIVES

10.21 Long top-bar hives with sloping sides

In the 1960s, interest in top-bar hives was reawakened by the need for cheap and simple hives that would enable beekeepers to manage the bees in tropical Africa (Section 7.42.2), in a more efficient way than was possible with the fixed-comb traditional hives (log or bark cylinders) which were lodged or suspended in trees. Important developments (Crane, 1976*b*) are mentioned below, and dimensions of some of the hives are given in Table 10.21A.

(1) *1965* In Rhodesia (now Zimbabwe), Penelope Papadopoulo from Greece used and demonstrated traditional Greek top-bar hives like those described by Sir George Wheler (1682), but woven from local materials (Papadopoulo, 1965).

(2) *1965* At the Hampshire College of Agriculture in England C. J. Tredwell and Peter Paterson – a student from Kenya – built and used a hive on the same principle as the Greek one, but made as a long rectangular wooden box instead of a round basket. Its sides were sloping, and the top-bars across the top had the same width as the British National hive. All top-bars were of the same length, and so were interchangeable (Figure 10.21a); they were full-width, i.e. they fitted close against each other, leaving no space between them. A triangle of foundation wax, and a wire loop, are shown.

Table 10.21A Dimensions of horizontal top-bar hives (movable-comb frameless hives)

The number on the left refers to the entry in Section 10.21.
Dimensions are in mm. 'Angle of sides' is from the vertical, and where possible is calculated from internal dimensions. Top-bar spacing = centre-to-centre distance = width of full-width top-bar.

		Length	*Depth*	*Top width*	*Bottom width*	*Angle of sides*	*Total length of top-bars*	*Top-bar spacing*	*No. top bars*
Apis mellifera, European:									
(1) traditional Greek, 1958[a]	*int.*	400[a]	490	400[a]	250[a]	9°	230–460	36	11
(2) Tredwell-Paterson, UK, 1965[b]	*ext.*	1067	254	356	203	17°	360	35[c]	29
W. Virginia Dep. Agric. (1975)[b]	*int.*	1143	283	457	229	22°	495	32	35
Bielby, UK (1977)	*ext.*	508	304	381 (*int.*)	—	—	432	38[d]	10
Apis mellifera, tropical African:									
(3) Linder, Senegal (1966)	*ext.*	700	400	480	400	6°[e]	480	31	24
(4) Paterson, Kenya (Kenya Min. Agric., 1973)[b]	*int.*	914	305	457	229	21°	480	32	26
Kenya top-bar hive (Townsend, 1976*a*)	*int.*	889	305	443	189	23°	483	32	28
Tanzania transitional hive (Ntenga, 1972)[b]	*int.*	762	229	445	445	0°	533	32	12 (24 combs)
Apis cerana:									
Temple (1986)	*int.*	600	200	360	170?	19°?	400	30	20

[a] hive B58/1 in IBRA Collection; the hive is round, and entries under length and width are diameters.
[b] dimensions are quoted to the nearest mm, converted from inches; 1 inch = 25.4 mm.
[c] 32 mm proposed for tropical African bees.
[d] increasing to 50 mm for outer top-bars, fitted with starter-strips of drone foundation (Bielby, 1986).
[e] author says about 10°.

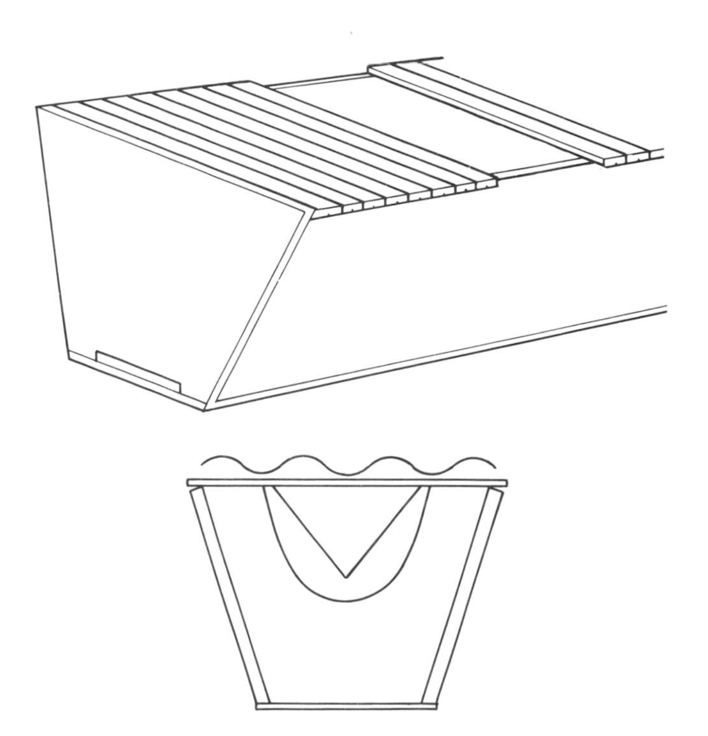

Figure 10.21a Sketch of Tredwell-Paterson top-bar hive, England, 1965 (Tredwell, 1976).
Transcription of notes on original sketch (1" = 25.4 mm): Top width 14", bottom width 8", depth 10", length 3' 6". These are the dimensions used here, but it is not suggested that they are ideal or critical. The 'roof' is a piece of corrugated asbestos sheeting cut to size, with heavy stones to prevent it blowing off in rough weather; something more positive could easily be devised. End and side entrance have both been tried, but no preference was determined.
Width of bars 1⅜" (1¼" would probably be appropriate for *adansonii*). Each bar is grooved centrally on the under side to ensure central fitting of the foundation [Section 10.4]. The wire loop must hang vertically, so that it becomes incorporated in the comb. [The wire proved to be unnecessary, and its use was discontinued.]
The tops of the sloping walls are left square, so that the inner edge forms a 'hive runner', with minimal contact with the bars. Complete freedom from attachment to the side walls was not achieved, but there was little, and after a few manipulations during which a knife was used to free the combs, it ceased.

(3) *1966* In Senegal, Joseph Linder introduced a 'David' hive named after his grandfather and his son (Linder, 1966). This was based on the Greek hive (1 above) but was made of local materials. Its horizontal cross-section was rectangular, and the top width was that of a Langstroth hive. Top-bars were 31 mm wide. (In 1965 Linder had introduced, as a first stage for beginners, a 'Rivka' hive of wooden boards, having the dimensions of a Langstroth hive, but with no top-bars or frames. The bees built their combs down from the flat wooden roof, with little or no attachment to the side walls (Linder, 1965). His David hive was to be used as a second stage; the third stage incorporated the use of Langstroth frames.)

(4) *1967* The Kenya Beekeeping Pilot Project was started, funded by Oxfam, UK. It was run by Paterson who used, among other hives, the one shown in Figure 10.21b. It was similar to hive 2 above, but the top width was that of a Langstroth hive (Kenya Ministry of Agriculture, 1973); dimensions are given in Table 10.21A. Nightingale (1976, 1983) has described its use, and his detailed instructions for inspecting a colony of tropical African bees in it are included in Section 10.5. The top-bars fit tightly together, and this has a special merit when used for tropical African bees – which readily fly off the combs when their hive is opened, and may try to sting the beekeeper. The only opening through which bees can come out is the gap where a single top-bar plus comb has been removed, and this can be constantly smoked.

(5) *1971* The above Kenya Project was succeeded by the Kenya-Canada Beekeeping Project, which lasted until 1982 and greatly expanded the use of a slightly modified form of hive 4; Kigatiira (1974) tested a number of variants. This hive became known as the Kenya Top-Bar or KTB hive (Figures 10.21c, 10.4c), and it is now widely known and used, especially in Kenya and some other African countries, including Ghana (Adjare, 1984). (The name KTB hive is sometimes used incorrectly for top-bar hives with different dimensions, or for top-bar hives in general.)

Section 7.37 discusses circumstances favouring the use of top-bar or frame hives in the tropics.

In Asia, Saubolle and Bachmann in Nepal (1979) published drawings of an 'African hive' with somewhat different dimensions, and of a smaller version for

Figure 10.21b Paterson's top-bar hive in use in Kenya, 1967 (photo: P. D. Paterson)

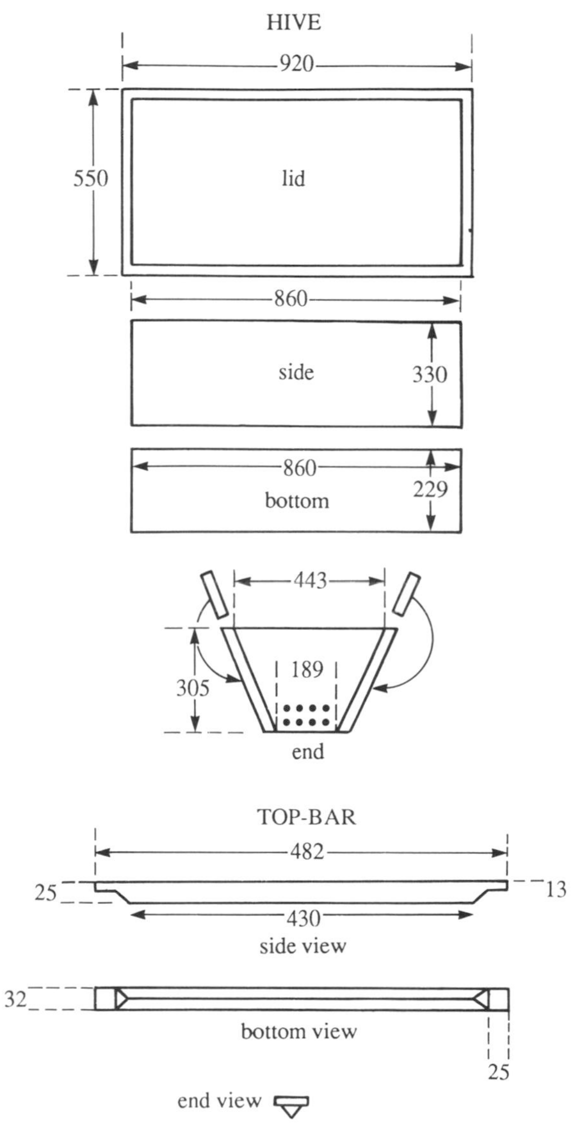

Figure 10.21c Diagram showing parts of the Kenya Top-Bar (KTB) hive for 25 top-bars (Townsend, 1976*a*). Dimensions are in mm. The other end has no entrance holes.

use in Nepal with *Apis cerana*. The full-width top-bars are retained, although the special reason for them, explained in (4) above, was irrelevant since *A. cerana* bees are very gentle. Gordon Temple has been active in distributing hives of this type in Nepal, under various aid projects. I saw them in use in 1984 – beautifully made, with vertical queen excluders framed in metal and shaped to fit the sloping sides. Currently used dimensions of the hives are given in Table 10.21A. Nakamura (1987) introduced members of the Chepang tribe to beekeeping with these hives, but they encountered difficulties, and seem to have given up using the hive. An apparently independent development of top-bar hives, for *A. cerana* in Vietnam, is referred to in Section 1.44.

Argyle (1984) had used a long hive for over thirty years in South Africa, fitted with top-bars

> suitably equipped to cause the bees to build the combs along the under-side thereof, fitted in the desired transverse positions. These bars are loose, to facilitate movability of the combs, and are specially spaced ... The sides [of the hive] are [very slightly] sloping inwards from the top, to facilitate easy removal of the combs, and also minimize crushing bees in the manipulations. This also discourages the bees (though not completely) from fastening the combs to the sides of the hive body, which is frequent in vertical-sided hives; while also retracting the sides from the noon-day sun.

Argyle's primary objective seems to have been to devise a hive in which the bees used separate combs for honey, pollen and brood.

Mangum (1987) rears queens in Kenya top-bar hives in the USA, and has also constructed glass-sided observation hives with one, and with ten, top-bars.

Preliminary experimental evidence on comb attachment in relation to the angle of slope of hive sides was obtained in England by Free and Williams (1981). A colony of bees (without combs) was put into a British National brood box (Table 4.2A), and measurements were made 'when the bees had finished building comb beneath each top-bar'. Bees were less inclined to attach combs to the hive walls at any of the slopes tested (10° to 45° to the vertical) than to vertical walls. Also, as the angle to the vertical was increased, any combs that *were* attached, had their highest point of attachment lower down:

angle	0°	10°	20°	30°	40°
mean ± se (mm)	25 ± 2	25 ± 2	28 ± 3	39 ± 6	48 ± 3

Up to 33% of the ends of the combs were completely unattached, whatever the angle of the hive sides. Perhaps such combs contained brood; it had been commonly observed since the 1960s that bees rarely attach brood combs to sloping hive walls, whereas this is more common with honey combs, especially if the hive walls are nearly vertical.

Budathoki and Free (1986) carried the 1981 experiments further, showing for instance that with hive sides vertical or sloping at no more than 10%, comb attachments were only half as wide for combs containing brood as for honey combs; with angles of 20° to 40° there was little difference between brood and honey combs, but the depth of free comb above the attachment was more than doubled. In general, there was rather little difference between hives with vertical sides and those with sides sloping at only 10°; the most marked differences occurred between 10° and 20°. In the USA, Brock (1980) used 15°, which 'worked very well'. Budathoki and Madge (1987) have studied the distribution of brood, and stored honey and pollen, in hives with sides sloping at different angles.

10.22 Long top-bar hives with vertical sides

In some circumstances bees do not attach unframed combs to the hive sides even if these are vertical. This may be true only for hives wider than a certain (undetermined) distance, and for some types of honeybees. A systematic study of such matters in relation to tropical African bees could be of considerable practical value.

In the Kalahari region of southern Botswana, which lies between South Africa and Namibia, Clauss (1982) consistently taught beekeeping with top-bar hives made of the cheapest possible materials. A cardboard carton, or an open framework of sticks, was plastered with cow dung or clay. The hive sides were vertical, but apparently the bees did not attach combs to them. Phiri (1985) says that the honeybees in this region differ from those farther north in being very gentle, and in other ways too, showing some similarities to the Cape bee (*Apis mellifera capensis*).

The next section, describing the use in Kenya of Langstroth hive boxes containing combs built from top-bars (without frames), demonstrates the practicability of using this system for tropical African bees in vertical-sided hives.

10.23 Hives with top-bars in tiered boxes

To beekeepers familiar only with tiered movable-frame hives, or with the long top-bar hives described above, it may seem surprising that top-bars can be used instead of frames in hives built up of tiered boxes, but

Figure 10.23a A stack of Guy's top-bar brood boxes, showing the top-bars and spacers, 1976 (photo: R. Guy). Figure 7.33a shows wooden hives of a similar type.

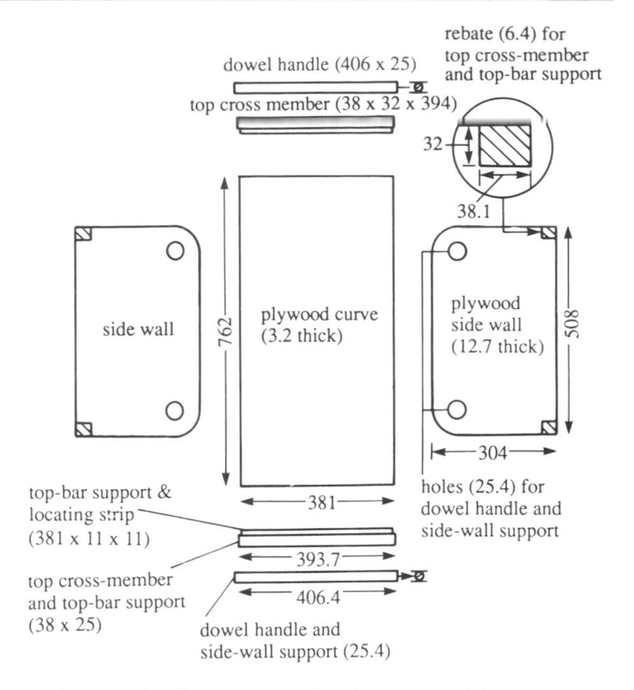

Figure 10.23b Diagram showing parts of Bielby's top-bar brood box, (Bielby, 1977).
The plywood curve fits inside the dowel supports. Dimensions are in mm; British Patent 1,182,987.

this can be done (a) in the brood box only, or (b) in both brood box and honey supers.

(a) Top-bar brood boxes

Robin Guy, a professional beekeeper in South Africa, showed me top-bar brood boxes in his out-apiaries in 1973 (Figure 10.23a). The bottom and two sides of the brood box were formed from a single curved piece of 3-mm hardboard, i.e. the floorboard was an integral part of the brood box. The boxes were cheap to make and were used as a means of reducing capital investment in equipment left for long periods out in the bush, where it was vulnerable to theft and to damage from animals – wild pigs when I was there. They were fitted with round wooden top-bars, as cut from suitable trees, uniformly spaced by a metal strip on each side of the brood box which had rounded depressions to take the bars. The top of the box had the same horizontal cross-section as a Langstroth hive, and normal Langstroth honey supers fitted with frames were used above it. All four sides of the brood box sloped in slightly to the bottom, so that empty boxes could be stacked one inside another for storage and transport. Similar top-bar units were hung in trees as bait hives to attract swarms (Figure 7.33a). Guy has published information on these hives (1975), and on a minimal-cost version of woven wicker (1971).

In the mid-1960s in England, Bielby (1977) independently devised a top-bar brood box that he refers to as a catenary hive. Its cross-section at the top is that of a British National hive, and the base and two sides are made from a single curved sheet of plywood. It is extra deep, and has a capacity about 50% greater than that of the British National brood box. Dimensions are given in Figure 10.23b and Table 10.21A. Attridge, also in England, uses hives with a top-bar brood box

Figure 10.23c Attridge's top-bar brood box, with a comb removed, 1986.

(based on Bielby's), which he makes almost entirely from materials that he, as a builder, is able to obtain as scrap material (Figure 10.23c). He modified Bielby's design by incorporating glass wool insulation, sandwiched between two pieces of plywood for the curved part of the hive, and by painting the inside of the hive with black paint to retain heat.

(b) Complete supered hives with top-bars

In Kenya (Paterson, 1986), a top-bar system with Langstroth hive boxes has been used regularly since 1975 (Figure 10.23d); each box is fitted with full-width top-bars 480 × 32 × 11 mm deep. The top-bar has a rectangular cross-section, with a 3-mm saw cut along the underside to receive a starter strip of beeswax. The bars fit tightly up against each other, as in the long top-bar hives used in Kenya (Section 10.21).

The honey supers are shallow, 166 mm deep (F. G. Smith, 1960). In the brood box, and in supers above it, there are only 10 top-bars, centrally placed to leave an equal gap at each end, beyond the outermost bars. The bees can move between boxes only through these end gaps, but the colony behaves as a single unit in spite of the solid barrier between adjacent boxes except at the ends. The barrier of top-bars at the top of the brood box eliminates the need for a metal queen excluder (and indeed its use may cause problems), and the queen very rarely moves into the super above. (The uppermost super can have the full complement of 11 top-bars, which together form a bee-tight cover so that only a simple roof is needed above it.)

Figure 10.23d Top-bar system used with Langstroth hive boxes in Kenya (photo: P. D. Paterson).
The photograph shows (below) the brood box with top-bars and end-gaps, and (above) the full honey super which has been tipped up to show the bottoms of the combs, almost fully sealed.

Without the solid barrier provided by the full-width top-bars, the bees would join the bottom of the comb in one box to the top-bar immediately below unless some barrier (e.g. horticultural screen plastic) is used over the central area. Brock (1980) in the USA describes several other variants of division boards between supers that prevent this. In Lagerman's fixed-comb supers (Section 9.22) the division boards are solid above and below each comb, with slots in the inter-comb spaces. In France, Kemp (1987) also dispenses with frames, using only top-bars (24 mm wide), by making rather rigid comb foundation from sheets of aluminium, paper board, polyethylene foam (2.5 mm thick), or other plastic material. The sheets are dipped in liquid beeswax, which is then embossed with a mould or die as described in Section 4.31. Combs built on such sheets of foundation are occasionally attached to hive walls, but can easily be separated. A narrow bottom bar seems to be necessary for honey supers.

Paterson establishes when a super is full by prizing it up at one end and looking along the combs from below (Figure 10.23d), or by the method used with the long top-bar hives – tapping top-bars in turn with a hive tool and listening to the resonance. He uses an escape-board to get the bees out of the full super. In his experience, the bees do not attach brood combs to the floor board or to the side walls. Honey combs are occasionally attached during a heavy flow, but not across their full thickness, and they can be freed easily with a thin knife. Paterson (1988) has described his method.

As a third alternative, it would be possible to use top-bars in honey supers and frames in the brood box, providing the frames had full-width top-bars, and a gap was left at each end as described above.

10.3 LONG MOVABLE-FRAME HIVES

Towards the end of the last century, Langstroth's movable-frame hive was being increasingly used in North America (Section 4.21). But a few beekeepers preferred to have the frames in a single horizontal box instead of in a tiered stack of squarish boxes. O. O. Poppleton, one of the chief exponents of this system (e.g. 1898), was at one period the largest successful migratory beekeeper in the world. He moved his hives 'on a gasoline launch, hiring cheap coloured help to move them on and off the boat' (Gleanings in Bee Culture, 1917). Poppleton's writings are quoted extensively by Adams (1984) who tells the story in full; it is summarized by Root and Root (1940). Similar developments took place in other countries, in many of which a minority of beekeepers still prefer to use these long hives, and some bees seem to do better in them. Certain types are referred to as long-idea hives, coffin hives, trough hives, *ruches horizontales*, *Lagerbeuten*, *Trogbeuten*, and so on. In the USSR a long hive is used especially in southern regions: Ukraine, N. Caucasus, and Asian republics. It suits the southern bees – especially Caucasian – which are unwilling to expand the brood nest into an upper hive box, or to store food there (Avetisyan, 1978). Kasianov (1978) gives dimensions of the standard 16-frame long hive.

Movable-frame long hives are also in common use in some tropical countries. In parts of Africa they have been introduced in beekeeping development programmes as a transitional hive between traditional fixed-comb hives of log or bark (Section 9.11) and a tiered frame hive. The 'Modified African Long hive' (Figure 10.3a) is fitted with Langstroth-size frames that have several non-standard features. The frame top-bars are wide, as in the hives described in Section 10.2, so that no space is left between them. Also, instead of a bottom-bar, a thin bar is fixed between the end-bars (half-way down), which provides support for the comb and makes it possible to operate without wiring or foundation. (Tropical African bees build comb readily enough in frames not fitted with foundation.) Top-bars with their combs from a KTB hive can be transferred to this hive, and the modified 'frames' from this hive can be transferred to a Langstroth (tiered) hive. So beekeepers can graduate from one type of hive to the next more advanced type, transferring their own combs of bees. Beekeeping with the different hives in Ghana is described in a book by Adjare (1984), which includes dimensioned drawings.

In Tanzania a long hive was based on the Modified Dadant hive instead of the Langstroth, because its larger frame size was preferred. To reduce costs a 'half-frame' was used: a top-bar and the top half of the two end-bars; this proved sufficient to support the comb, without additional wires or struts. There seem to have been no problems with attachment of comb to the hive sides below the end-bars. From this hive, two long hives were developed – a 'Tanzania Dadant hive' with 22 (Modified Dadant) frames or half-frames, and a 'Tanzania Transitional hive', made as cheaply as possible and fitted with 12 double-width top-bars each having starter-strips for *two* combs (Ntenga, 1972). Table 10.21A gives dimensions of the latter hive. Top-bars and combs were interchangeable between all three of the Tanzanian hives, as in the sequence of hives referred to in the preceding paragraph.

The Zambia Forest Department (1979) favours a

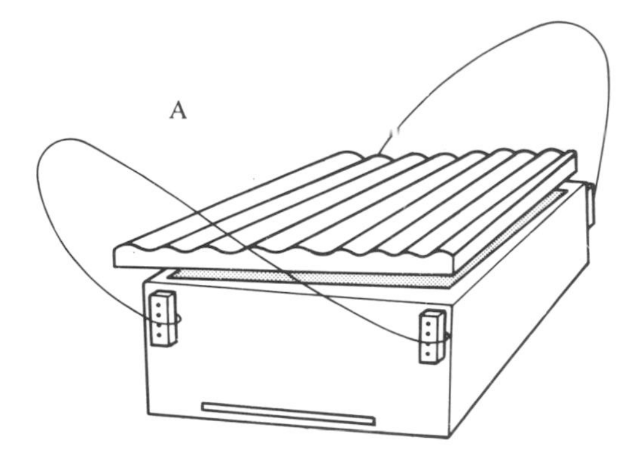

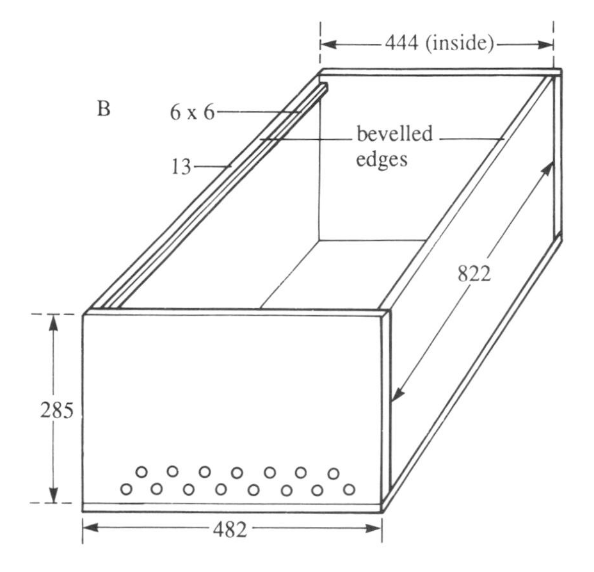

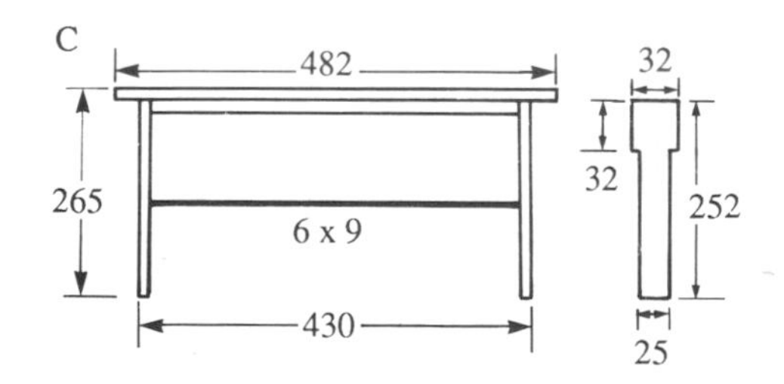

Figure 10.3a Diagram showing parts of the Modified African Long hive, Townsend's measurements (mm), as FAO (1986).
A Complete hive with roof and hanging wires.
B Hive box, made from 19-mm timber, for 25 frames.
C Frame.

'Dadant Transitional hive' on similar lines; see also Silberrad (1976). The Dadant frame is used, the top-bar being fitted with a starter strip of beeswax sheet or foundation (about 25 mm wide) along the central two-thirds of their length. Eight frames in the middle of the hive constitute the brood chamber, being separated by vertical queen excluders (Section 10.4) from 10 frames for honey on either side, making 28 frames in all. The brood frames are wired horizontally, to prevent breakage during handling, comb inspection, or moving.

A long hive that was used in Uganda (described in 1962, see Johnson, 1976) had normal frames, with top-bars 24 mm wide; bees would thus be able to fly out as soon as the frames were uncovered. Kellogg (1943) used a long hive with frames in a development programme in Mexico. In England, Tredwell (1964) used and described a 'Westley hive', named after the location of the college where he worked. It was this hive that gave rise to his interest in developing the long horizontal hive with top-bars (Section 10.21).

Long hives have advantages, though nowadays they are not very widely used for commercial beekeeping except in southern Republics of the USSR, where they suit the Caucasian bees kept there. The hives hold 16 or 20 USSR Dadant frames (Table 4.2B), and they have the same dimensions as the USSR Dadant brood box (Table 4.2A) except that the depth is 10 mm greater, and the width (at right angles to the frames) is 675 mm for 16 frames and 870 mm for 20 frames; internal widths are 615 and 810 mm. I also saw long hives used commercially in Turkey in 1985. In France the best known is the de Layens, with 20–30 frames (Layens & Bonnier, 1897). Weber (1982) describes a German type, and Dartington (1985) one in the UK. Some beekeepers using a long hive superimpose a tier of standard-size honey supers above it (e.g. Weber, 1982; Adams, 1984; Dartington, 1985).

The apiary shown in Figure 10.3b, in the Andes near Quito in Ecuador, had thirty long hives, not as wide as a Langstroth but holding 30 frames, 10 for brood in the centre, and 10 for honey on either side. Vertical queen excluders (Section 10.4) separated the brood frames from the others, and each compartment had a flight entrance. *Chicas*, small hives fitted with similar frames, were used to produce brood, which was added periodically to the long (honey-producing) hive, and frames of honey were transferred to the *chicas* as necessary. The Zarria brothers who ran the apiary harvested an average of 270 kg of honey annually from a long hive with its two *chicas* – 30 kg every ten days during the three-month flow period, in the dry season. Using tiered hives, they produced 40 kg per hive in a year.

For movable-frame beekeeping, advantages of long hives include any or all of the following.

1. If home made, the hives are cheaper than tiered hives, and most dimensions are less critical.
2. The hive length (and width) can be chosen to suit individual conditions, although in practice it is convenient to use frames with dimensions that are standard in the beekeeper's own locality.

Figure 10.3b Apiary containing long hives at 3000 m in the Andes, Ecuador, 1983. The apiary is enclosed by a secure fence, and a stream runs below it.

3. Every frame is directly accessible without removing others; if the frame top-bars are full-width, relatively few bees fly around when the hive is opened sufficiently to remove the frames one at a time.
4. No lifting is required (except of frames), the hives thus make beekeeping possible for a person who cannot move heavy supers.
5. All frame top-bars are at the same level, which can be chosen to suit individual beekeepers, including any that are disabled (paraplegic); see Section 16.83.
6. The size of the brood compartment can be adjusted a frame at a time, and according to many authors (e.g. Bukharev, 1964) swarming is reduced by the lateral expansion of the brood nest that is possible with a long hive.
7. Where necessary for protection against enemies, the hive can be suspended by wires above the ground, at a convenient height for operation.

Disadvantages of the long hive include the following:

1. The hive cannot be reduced in total size, even when it contains little honey, and it is heavy and unwieldy for one person to lift.

2. For the above reason it is generally unsuitable for migratory beekeeping.
3. Although satisfactory for cut-comb honey, the hive cannot be used for producing good 'section' honey, because the section combs need to be directly above the brood nest.
4. It may well be more expensive to purchase a few long hives than the same number of tiered hives, if the latter are mass-produced.

Benefits of using hive equipment that is the most common local standard pattern often outweigh advantages of a particular type of equipment *per se*, and for the majority of beekeepers the standard is a tiered hive (Chapter 4). But for a minority, the advantages of the long hive will outweigh the disadvantages.

10.4 ANCILLARY FEATURES AND EQUIPMENT FOR LONG HIVES

Only a protective roof is needed above full-width top-bars. With narrower top-bars it is best to use an inner cover as well; this may be of cloth, or of hardboard or thin wood, in which case it is convenient to divide it into three sections.

Top-bars must be exactly spaced in the hive, at the correct distance for the type of bees used. Sperling and Caron (1980) seemed to regard the use of sloping hive walls as 'all that matters', and the bee space as irrelevant. This is untrue: the centre-to-centre distance between adjacent combs (i.e. between the centre lines of the top-bars) is always important, whatever the type of hive. The slope of the side-walls may affect the attachment of combs to them, but it has nothing to do with the bees' orderly building of parallel combs, whether from the hive roof, from top-bars, or in frames. Table 4.25A gives the natural centre-to-centre distance for different honeybees.

It is necessary that the bees build each comb down from the centre line of its top-bar. The top-bar can be made wedge-shaped (V-shaped) underneath, and beeswax smeared on the edge of the V. Or (e.g. Bielby, 1986) it can be grooved or split along its length, and a starter strip (or triangle) of beeswax sheet or comb foundation inserted along the whole length, or along only its middle portion; full sheets of foundation are unnecessary, but Kemp's system (near end of Section 10.23) might be practicable. It is beneficial to provide a support for the comb at its junction with the top-bar, so that the comb does not break off if the whole is tilted into a horizontal position. Nightingale (Figure 10.4a) used three bamboo sticks to support the comb. Vertical wooden sticks will do, but vertical wires were found to

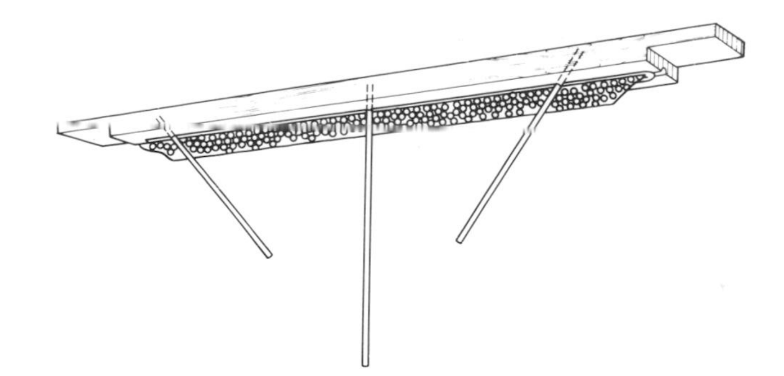

Figure 10.4a Nightingale's use of bamboo sticks to support the comb built from a top-bar, fitted with a starter-strip of foundation.

be rather unsatisfactory by Budathoki and Free (1986). A strip of expanded metal foundation, referred to in Section 4.32, might be a solution to this problem.

In many countries with *Apis cerana*, comb foundation for frames is not available with the correct cell size (see Table 4.25A). Nylon net is a possible alternative that is cheap. In the USA (with *A. mellifera*), Adams (1985*a*) used $\frac{3}{4}$-inch (19-mm) plastic fishing or bird net, and says that it stands up well during centrifugal extraction. I saw it in use in 1985, attached to the frames with staples (Figure 10.4b). In England, Bielby (1977) used fine net of the type manufactured for window curtains, but some people found that the bees tried to chew it away, and could get entangled in it. In Thailand, frames for *Apis cerana* (without foundation, but with a narrow starter strip at the top) are strengthened by two horizontal wires. In Mexico Kevan and Bye (1982) coated such wires with beeswax before use. I found that in many Thailand apiaries honey was harvested by cutting out the part of the comb above

Figure 10.4b Langstroth frame with partially built comb on nylon fishing net, 1986 (photo: J. Adams) Behind is a corrugated cardboard bait hive for swarms (Section 6.35.1).

the upper wire, leaving the brood which was below it; the bees rebuild the upper part of the comb satisfactorily.

The size and position of the flight entrance are important in some circumstances, although I suspect not in all. A top-bar hive for 35 combs used in North America (West Virginia Dep. Agriculture, 1975) has a slot entrance 150 × 22 mm, and in the modified African long hive it is 250 × 25 mm (Townsend, 1976*a*). In some Kenya Top-Bar hives the entrance now consists of a group of round holes, to prevent entry of hive beetles. Diameters of 8 mm to 10 mm have been recommended for keeping out large hive beetles, but with such holes a sufficient number must be provided to allow bees to fly freely in and out (Corner, 1988).

The entrance is often at one end. On the other hand some hives, including those used in Nepal (Temple, 1986), have a slot entrance in a long side near one end. Brock (1980) found that if the hive had a side opening 'the bees cross-combed, with the combs running towards the entrance'.

Queen excluders are in general much less necessary in a hive where the brood nest expands laterally, and many long hives are used successfully without them. Any queen excluders must be vertical unless a long hive incorporates a superimposed box for honey, and they must fit the hive exactly round their edges, to prevent passage of the queen. In Italy they are on sale with draught-excluder strip round the edges to ensure a good fit. Especially if the hive sides are sloping, excluders must be framed, as in Figure 10.4c, and they are thus less easy and cheap to make than the flat sheets placed horizontally in tiered hives. In East Africa, queen excluders have been made from coffee wire (mesh used for screening coffee beans), with wires at 5-mm intervals; this leaves a gap which is

Figure 10.4c Kenya Top-Bar hive, showing vertical queen excluder.
The alighting board is on the right, and wires for hanging the hive are in place.

satisfactory for the bees there, but many batches of coffee wire have proved defective (Corner, 1988). The screen is mounted in a wooden frame, or is stapled directly on to the hive walls. Jim Nightingale, referred to in Section 10.5, used some top-bar hives with a slot entrance half way along the long side, and a vertical queen excluder positioned to divide the entrance in half. When a swarm occupied the hive, the half the queen entered became the brood chamber, and honey was harvested from the other half, beyond the queen excluder (Corner, 1988).

For long frame hives, frame (or division-board) feeders (Section 4.47, type 6) are usual, but other types are also used. Adams (1985*b*) makes a 20-mm hole in the hive roof to fit the (perforated) cap of a large plastic bottle. The hole is so small that it does not lead to any trouble if the wind blows an empty bottle off a hive. Feeders for long top-bar hives are discussed in Section 10.5.

10.5 MANAGEMENT OF BEES IN LONG TOP-BAR HIVES

Much of the general bee management covered in Chapters 6 and 7 is common to all hives with movable combs, whether the combs are supported by frames or only top-bars. It is governed by the geographical location and by the bees in the hive, rather than by the pattern of the hive. The present section refers briefly to operations with bees in long hives of the Tredwell-Paterson type, including those used in Kenya and elsewhere in Africa, and also locally in the Americas and Asia. Operations are dealt with in the order: stocking the hive with bees, opening and inspecting a hive, harvesting honey, feeding. Adjare (1984) gives detailed instructions on these and other operations, based on his experience in Ghana.

It is best to stock a top-bar hive with a colony on combs, by one of the methods for making new colonies that are described in Section 6.34. In tropical regions where colonies swarm readily, and a swarm must be used, it will be more likely to remain in the hive if this contains a comb of brood, or failing that a comb of honey, near the hive entrance. If no comb is available, sugar syrup should be provided in a feeder, as explained at the end of this section.

By instinct, the bees will build their combs parallel to each other, and spaced apart at a distance determined by their own body size. The beekeeper wants them to build a comb underneath each top-bar, with the midrib along its centre-line. So (a) the bars must be at the correct spacing for the bees; if they are full-width they must touch each other; if narrower, a proper spacing device (Section 4.25) must be used to ensure that they are equally and exactly placed. Also, (b) the centre line of each bar must attract the bees to start comb building along it, i.e. to use it for the midrib of the comb. One way is to have each bar wedge-shaped (V-shaped) underneath and to smear beeswax on to the edge of the V; other ways are explained in Section 10.4. Unless these details are attended to, combs are likely to be built slanting across several bars, and the value of using top-bars is lost.

Nightingale (1983) gives clear instructions for opening a long top-bar hive to inspect a colony of tropical African bees, which will also be helpful with Africanized bees.

> Give a little smoke at the entrance one or two times, until no more attacking bees seem to come out. Shortly the bees will start to cluster near the entrance, and this is the time to open the hive, because returning bees will join the cluster, and other bees will also do so and not try to attack.
>
> Remove the lid, tap each bar lightly with a hive tool, and from the sound you can tell how far along the hive the bees have built comb, and that is where to remove the first top-bar. Blow a little smoke over the top of all of the bars, and then slowly remove one comb containing honey and introduce a small amount of smoke. All the time combs are being removed,* pass a continuous flow of light smoke over the openings. This is very much easier if two people work together.
>
> Do not work bees in one apiary for more than 45 minutes; after this period, the first colony operated will start preparing to attack. Do not work them before 4 o'clock in the afternoon, and a time nearer 6 o'clock is preferable. If robbing does start, darkness will soon put an end to it.
>
> When examining the colony, smoke should be necessary to control only the flight entrance and the single opening from which the comb being examined has been removed.

For a hive containing quieter bees, there is no need for a second person, or the continuous smoking, or for the limitation of the working period. If top-bars are narrower than full-width, the bees can be seen through the gaps between the top-bars, and it will be clear which top-bar is to be removed first. Manipulating cloths (Section 5.23) can be useful with these narrower top-bars.

When taking honey from the hive, only light-coloured combs containing no brood should be harvested; as with framed combs, only those with all or most of the honey cells capped should be taken. Instead of removing the bees by jolting and brushing, the combs being harvested (along with any bees not easily shaken off them) can be placed in a secure box with a bee-tight cover incorporating a bee-escape (Section 5.31) that will prevent bees entering, but will allow those inside to leave and return to their hive. Bee-escapes, repellents or blowers (Sections 5.31 to 5.33) cannot easily be used to remove bees from honey combs in single-storey hives.

Adjare (1984) recommends that, when honey is harvested, each comb is cut 5 cm from the top, and the upper part returned with the top-bar to the hive. In any case at least 7–10 combs with brood, honey and pollen should be left behind, as this is the best way to ensure that the bees have enough food to last them through the forthcoming dearth season. In hives with no separate brood chamber, the beekeeper may be tempted to take too much honey at harvest time.

Honey is usually extracted by straining or pressing (Section 10.61). The production of cut-comb honey is, however, well suited to beekeeping with top-bar hives. The most completely capped combs should be used for cut-comb honey, and care must be taken not to damage them in any way. As each top-bar + comb is removed from the hive, and jolted and brushed free of bees, it should be placed carefully in a collecting box with a bee-tight cover, in which the comb can be suspended as in the hive. The extraction of honey from unframed combs, and the preparation of cut-comb honey, are described in Sections 10.6 to 10.62.

Feeding sugar syrup can save new or small colonies from starvation, and there are several options for doing this in top-bar hives. Unlike frame hives, which are fitted throughout with framed combs, top-bar hives have an empty space below any top-bars from which comb has not yet been built, and a feeder can be placed here. For instance a glass jar can be filled with sugar syrup and inverted in a saucer, with a match-stick or twig to tilt the jar very slightly. Or one of the top-bars can be removed to accommodate a frame feeder of a shape modified to fit into the hive; feeders in general are described in Section 4.47. Feeders placed inside the hive are usually safe from robbing bees from other hives. The use of external feeders can lead to robbing if syrup leaks out, but they are convenient in that they can be refilled without opening the hive. In Nepal, top-bar hives have a fitment for a Boardman (inverted jar) feeder, on a platform outside one end. Feeders for long hives with frames are mentioned at the end of Section 10.4.

If it is necessary to provide water in the hive, an external feeder can safely be used. Or a feeder – or even a

* If combs of honey are to be harvested, as each is removed, most of the bees are jolted off it by a quick blow to the arm, and the rest brushed off with a very soft grass brush.

clean wet sponge – may be put into the back of the hive.

10.6 LOW-COST EXTRACTION OF HONEY AND WAX FROM COMBS

Production of cut-comb honey offers an inexpensive method of preparation for marketing, provided suitable packaging is available. It is discussed briefly near the end of Section 4.33. The present section is written for the benefit of beekeepers who want to extract their honey from the combs, but do not produce enough honey to warrant investing in a centrifugal extractor (Section 13.33), or who are too poor to do so. It is also addressed to readers who wish to centrifuge unframed combs, for instance because they are in charge of beekeeping in a country where honey is harvested from top-bar or traditional hives. In addition, Section 10.63 offers some ideas for beekeepers with frame hives who want to make a centrifuge at lower cost than the normal purchase price.

10.61 Strained honey and squeezed honey

The equipment need cost little or nothing. Strained honey is obtained by putting broken-up honey combs in a straining cloth or basket, or on a mesh screen, so that the honey drips through without any pressure; this 'run honey' used to be considered the finest honey of all.

Nightingale's method (1983) for obtaining first quality strained honey from top-bar hives is as follows:

> Only light-coloured combs containing no brood are harvested, and only combs that are at least half-capped. Efforts must be made to keep the harvested combs intact, and they should not be pressed into a sack or barrel. Where the honey will be extracted, a bucket or similar container is draped with fine-mesh cloth such as cheesecloth or nylon, and a light-coloured board is placed across the top. Each comb is held up to the light, and all cells containing pollen are cut out with a sharp knife. Any impurities in the comb, or in the honey as it begins to ooze out of the comb on to the board, can be detected. The clean comb is then broken up slightly with the knife, to allow the honey to flow, and the combs are pushed from the board on to the cloth over the bucket. They must not be chopped up finely, or squeezed, or pressed through the cloth.
>
> When all the combs have been lightly crushed, the edges of the straining cloth are gathered and tied together, and the bundle hung above the bucket to drain overnight. If particles of wax have passed through the strainer, they can be skimmed off the surface after the honey has settled for about a week. When the honey is finally poured into jars, it should be exposed to the air as little as possible.
>
> The clean honey remaining with the crushed wax can be heated over water in a double-boiler arrangement to separate them. After the cake of wax on top has cooled and hardened, it is removed from the honey below; since this has been heated, it is slightly inferior in quality to the strained honey, and should be kept separate for marketing.

A larger gravity strainer used experimentally in Sri Lanka (Mulder, 1988*a*) is a 200-litre metal barrel-shaped container, painted black on the outside and with a black lid, to take advantage of solar heating. A protective coating is necessary on the inside to prevent contact between honey and metal. Two horizontal grids are fitted: one, a third down from the top, has about 1 mesh per cm, and the other – half way down – is a finer mesh screen. Broken honey comb is placed on the upper grid; the lower one then retains small wax particles. The temperature inside the barrel must be controlled so that it does not exceed 60°, or wax will melt, and grids must be of plastic, not metal. About 50 kg of honey can be extracted in 6 hours, and it should be marketed separately from unheated honey. The wax left on the grids is almost dry.

Another method of straining is to lay uncapped combs flat on stretched mosquito netting, and then to turn them upside down to drain the other side. Adams (1985*b*) puts pieces of comb, broken up lightly, into a large strong plastic bag, which is pegged on to a line in a warm but not a hot place. A number of small holes are pierced in the bag, and the honey drips into a container below. Progress is visible, and the combs can be shifted within the bag so that all cells are finally drained out. Hooker (1983) describes some other variants.

Squeezed honey is obtained by pressing honey through a strainer. It is of a lower quality, and should not be mixed with other honey. It may contain pollen (unless this has been removed as in Nightingale's method), as well as other materials that may impair its flavour and transparency. The Indian Standards Institution (1977) published a Standard for squeezed honey, including directions for determining the content of pollen and other plant elements in it. The Standard was designed for use with honey from *Apis dorsata*, which is commonly obtained by squeezing.

Thixotropic honeys (Section 13.27.1) have peculiar flow properties and cannot be extracted by normal centrifuging. Good quality honey can be obtained by pressing honey combs (containing no pollen) in a special honey press (described e.g. by Whitehead, 1954). Heather honey (*Calluna vulgaris*) is an example.

10.62 Centrifuges for unframed combs

When honey is extracted from framed combs by centrifuging them, the combs are undamaged and can be re-used. It is possible, although more difficult, to centrifuge and re-use combs attached only to top-bars. With unframed pieces of comb, the sole purpose of centrifuging is to separate the honey from the wax. At its simplest, a centrifuge for honey can be a pail or bucket containing a metal grid that will support a piece of honey comb. The whole is swung round by hand in either a vertical or a horizontal plane. Scheuermeier (1988) describes such a centrifuge used in Nepal, with two shallow woven baskets containing grids, suspended at the ends of a horizontal wooden pole. The pole is spun round by means of a cranked handle.

The type of tangential extractor described in Section 13.33 can be used for unframed comb, if the normal support for each frame is replaced by a lidded shallow box of wide metal mesh; pieces of comb are laid in each box, and are held in position when the lid is closed. Such an extractor is illustrated by FAO (1986). Thomas in France manufactures a sturdy centrifuge in which 6 pairs of shallow horizontal metal baskets are mounted, one above another, and are spun about the vertical axis which passes through their centres, as shown in Figure 13.33a (E, F). Pieces of comb are placed in each basket, which can alternatively be used to hold framed combs. Both sides of the combs are extracted at once.

The combs must be uncapped on both sides before centrifuging (Section 13.33), and this is not always easy; it is best to lay them one by one on a flat board. Combs from the Swedish fixed-comb hive described in Section 9.22 are an exception, and uncapping is not necessary for them.

10.63 Inexpensive centrifuges for framed combs

Where there is no electricity supply, a hand-operated centrifuge (Section 13.33) can be adapted for operation by a bicycle mechanism. A simple hand-operated extractor for two frames can be made by constructing a cradle and spinning mechanism for a tangential extractor (as in Section 13.33), to fit a 115-litre plastic dustbin (garbage can). The cradle is attached to the two dustbin handles by screws with wing nuts, and is removed before the bin is tilted for pouring out the extracted honey. I have seen such an extractor in use in Malaysia, and Miller (1976) gives details of it.

Two mini-extractors operated by electric motors are marketed in the UK that extract both sides of the combs at once. That from Brinsea Products spins two frames supported horizontally in a plastic bowl fitted with a honey gate; it is like one tier of the Thomas extractor mentioned above. The Honeyspinner spins a single (vertical) frame inside the user's own receptacle. A device like an electric drill is fixed to the midpoint of a clamp over the length of the uppermost end-bar of the frame.

These examples may suggest still other ways of applying centrifugal force to framed combs, as alternatives to using the purchased extractors described in Section 13.33.

10.64 Beeswax rendering

With framed combs, only the cappings wax is harvested, because the empty combs are put back in the hives. Where the combs are unframed the whole of the comb is harvested, and about 5 times as much beeswax per kg of honey is likely to be obtained. The pieces of comb are first freed from residues of honey by washing in water. If any dark brood comb is present it should be removed and dealt with separately, otherwise it will darken the wax from the light-coloured honey comb, and lower its quality. The wax is rendered (melted and converted into clean wax blocks) by processes described in Section 13.6; those described first use the least expensive equipment.

10.7 FURTHER READING AND REFERENCE

Details of publications listed will be found in the Bibliography.

Adams, J. (1984) *The long box hive and its operation in America*

Adjare, S. (1984) *The golden insect: a handbook on beekeeping for beginners*, 2nd ed.

Bielby, W. (1977) *Home honey production*

Crane, E. (ed.) (1976*a*) *Apiculture in tropical climates*, especially papers by:
- pp. 9–13 K. I. Kigatiira
- pp. 15–22 J. Nightingale
- pp. 31–37 R. Guy
- pp. 181–189 G. F. Townsend

Kenya Ministry of Agriculture (1974) *Bee keeping in Kenya*

Root, A. I.; Root, E. R. (1940) *The ABC and XYZ of bee culture*, pp. 374–376

Saubolle, B. R.; Bachmann, A. (1979) *Beekeeping: an introduction to modern beekeeping in Nepal*

Zambia, Forest Department (1979) *An introduction to frame hive beekeeping in Zambia*

PART IV

Maintaining Honeybee Health

11

Health, sickness, and injury

11.1 INTRODUCTION

At least as early as Roman times, beekeepers recognized that their bees could suffer from *disease*, although the causes were not known. The Roman authors also cited many *enemies* of bees, which were large enough to be visible; they included sheep 'which entangle bees in their wool', frogs and toads, wasps, wax moths and spiders, and birds – tits, bee-eaters and swallows. In general, mammals, birds, amphibia and insects that attack bees or their nests are of more significance in the tropics; in the temperate zones, where most beekeeping development took place in the past few centuries, diseases caused by pathogenic microorganisms are more important. Their pathogens could not be seen – let alone identified – until microscopes had been developed. As time went on, smaller and smaller organisms were made visible by new techniques, and their significance could be assessed. The latest group to be studied, during the last thirty years, are viruses (Bailey, 1975, 1982).

Many beekeepers are interested in a 'new' bee disease, and from time to time great importance is attached to a pathogen that is relatively easy to see, especially if there is some action the beekeeper can take against it. Bee specialists may differ as to the significance of the pathogen, or the need for its control. An example is the tracheal mite *Acarapis woodi* in the early part of the present century (Bailey, 1963; Adam, 1968*b*).

Until well into the 1800s any outbreak of disease – induced, perhaps, by climatic or other stress to which the bees in the area were subjected – tended to be confined to the one district. But in the course of the last century two new methods of transmitting disease from one colony to another became common. After 1851 the use of movable-frame hives made it possible to move combs of bees and brood from one hive to another, and if the bees or comb were infected the disease was transmitted too. Later, with the possibility of transport by rail and then by road, hives of bees (and with them any diseases) could be moved over quite long distances. Since the development of air travel in the mid-1900s, bee diseases have been transmitted directly with bees sent from one continent to another. Man is now the greatest agent in the spread of bee diseases affecting either brood or adult bees, and also of parasites of bees.

We shall see again and again that a good climatic environment with rich food resources, and absence of stress, are important factors in preventing honeybee colonies from succumbing to disease. Stress may be caused by keeping bees where food resources are inadequate, for instance by overcrowding colonies, or where there is an insufficient nectar flow to stimulate activity within the colony; by siting the hives in damp, windy, or otherwise poor locations; by effects of pesticides; and by many other factors.

Nowadays colonies are kept by beekeepers in many areas where conditions would not allow them to survive in the wild. As in other animals, poor health in honeybees can arise from genetic factors, and beekeepers' colonies may be nurtured with the result that they live and pass on defects to their progeny. By inbreeding, the beekeeper may introduce defects that cause brood to die. This subject, and those mentioned in the next paragraph, are discussed in some detail by Tucker (1978).

Another cause of injury and death, not necessarily related to disease, is exposure to temperatures too high or too low. Adult bees are more likely to die from overheating than from chilling, and ways of preventing this are described in Sections 7.32.2 and 11.72.1. Brood is liable to chilling, especially in small colonies in unseasonably cold spring weather; the 'chilled brood' can be found dead in the cells. Starvation also leads to death, but adult bees in a colony may survive even if they cannot feed all the brood; nurse bees eat the larvae that starve.

Diseases and parasites of honeybees are listed in Table 11.1A. They can cause economic loss in many countries, and leaflets on their diagnosis and control are widely available. Several modern text books have been published: *Honey bee pests, predators, and diseases* (Morse, 1978), *Honey bee pathology* (Bailey, 1981), *Nozione pratiche sulle malattie della api* (Giordani et al., 1982), *Bienen-Pathologie* (Weiss, 1984*a*), *Krankheiten der Bienen* (Zander & Böttcher, 1984). Toumanoff's *Les ennemis des abeilles* (1939) includes many observations from south-east Asia, which is otherwise rather little documented. Akratanakul's smaller book (1987) provides an introduction to both diseases and enemies in Asia as a whole. These books may be consulted for a far greater wealth of detail than can be given here, including information on control methods. The first two break with tradition by discussing in turn the disorders caused by different classes of organisms. Here we revert to the sequence commonly used in books on beekeeping, which is more relevant to the beekeeper's experience: first, transmissible disorders affecting brood, then those affecting adult bees; parasitization; various 'enemies' of bees; pesticide poisoning; finally natural environmental hazards of a cataclysmic nature which can be disastrous for beekeeping, but are little discussed in the literature. Since many publications include photographs of pathogens, of brood and adult bees affected by them, and of bee 'enemies', none are included here except those of mites of special current concern (Section 11.4).

Fifty years ago, it was enough if a beekeeper knew about bee diseases existing in his own country. Now that the long-distance transport of queens and bees is relatively easy, he needs to know the status of other countries as well. The world distribution of diseases and parasites is thus of great importance to beekeepers, and is shown in ten maps in Sections 11.21 to 11.44; the note to Figure 11.21a explains the basis on which the maps were compiled. They have been updated by Bradbear (1988) from those published by Nixon (1982, 1983). All three papers include tables indicating which diseases and parasites have been recorded for each country, and cite the references.

At least 48 countries have laws or regulations for the control of bee diseases as shown in Table 15.4A, and Bradbear (1983) summarizes the position in individual countries. Disorders most commonly covered are American and European foul brood, and the presence of the tracheal (acarine) mite and the varroa mite. But regulations may relate also to at least one of the following: sac brood, chalk brood, stone brood, nosema, bee lice, wax moth.

Some control methods can be effective against a number of diseases. Since 1970, methods for large-scale fumigation of hive parts – especially combs – have been developed for preventing the transmission of infection between one active season and the next. Fumigation with ethylene oxide has been widely used in the control of American foul brood, chalk brood and nosema disease, also wax moth infestation (Section 11.61); see Cantwell et al. (1975). A fumigation chamber and ancillary equipment represent a large capital outlay, and construction and operation are usually on a regional basis, so that a great many combs can be processed. Recommended dosages (Morse, 1978) are 450–700 mg of ethylene oxide per litre of space, at 38° ± 10° and RH at least 80%, for 8–24 hours. The vapour is toxic, so fumigation chambers must be used only by specially trained personnel; also all honey must be removed from combs before they are fumigated. Ethylene dibromide fumigation has also been used, but there may be problems with residues in honey (Section 13.27.3).

Irradiation offers a safer treatment against AFB, nosema and wax moths. In a large-scale test in Australia, each load of hives received 1 megarad of gamma-radiation, which took 2.5 minutes. Although the equipment was transported 1900 km to a cobalt-60 source, the operation was cost-effective, because it saved 70% of the replacement cost of contaminated equipment (Bee World, 1984).

11.2 DISEASES OF HONEYBEE BROOD

A honeybee spends its development period – as egg, larva, prepupa and pupa – in the very sheltered environment of a cell in the comb, where the temperature and humidity are kept constant by the nurse bees of the colony. Even so, there is some mortality: in healthy colonies in Japan, during June and July, 4% of embryos died before the eggs hatched, and 14% of the larvae died in the 5 days before their cells were sealed – the stage when they may succumb to various infections. During the 12 days between cell sealing and the emergence of the adult insect only 1% died (Sakagami & Fukuda, 1968); this is the safest period in the whole of a bee's life.

Hansen (1980) has published diagnostic colour photographs of brood infected with different diseases. Two diseases that can be very damaging are caused by bacteria affecting young larvae. They are known as American foul brood and European foul brood, from the fact that when they were first being investigated in the early 1900s, scientists in the USA paid most attention to the first and those in Europe to the second. Both diseases are widespread in the temperate zones and subtropics (Figures 11.21a and 11.22a) but – in common with most bee diseases – less so in the tropics. AFB is absent from much of South America and Africa.

11.21 American foul brood (AFB)

This disease is caused by *Bacillus larvae*, whose characteristics are described by Bailey (1981) and others. The spores are very resistant to heat, to disinfectants, and to drying out. They can remain viable, and able to cause re-infection, for a very long period – certainly decades. AFB may occur at any time of year when colonies contain brood, and the infected colony usually dies. Spores are often transmitted to larvae by adult bees that have cleaned out cells containing infected larvae, when these bees start to feed larvae; as few as 10 spores can cause infection in a larva under 24 hours old. The nurse bees can recognize infected larvae, and remove them from the cells before or after these are sealed. If the larva dies after the cell is sealed, the capping over it becomes moist, dark coloured, and sunken. At this stage, if a matchstick is pushed into the remains of the larva and withdrawn, it brings with it a brown, semi-fluid ropy thread.

AFB is highly contagious, and modern beekeeping practices in which combs are moved from one hive to another can actively transmit the disease from an original source of infection for many years.

Legislation is in force in many countries (Table 15.4A) to provide apiary inspection and to enforce destruction (usually by burning) of hives, frames, combs and bees, where infection is found. Notification of the disease is compulsory. Any beekeeper who suspects the presence of AFB in his colonies should quickly get in touch with the appropriate local authorities, and carry out their instructions or recommendations. It is essential that control methods within each country should be co-ordinated.

Efforts to control AFB are based on one or other of the following concepts.

- The view that compulsory destruction of infected colonies is the only proper course, and that no remedial treatment should be given to infected colonies, for reasons (a) to (c) given below.

- The view that appropriate therapeutic chemicals (usually sulphathiazole or oxytetracycline) should be fed to diseased colonies in powdered sugar or sugar syrup, or in 'extender patties' made up with sugar and dehydrogenated cooking oil. The sugar makes the food attractive to the bees, and the oil 'extends' the period over which the antibiotic is administered. The antibiotics may also be administered as a preventive measure. This is the policy in the USA and Canada and some other countries. There are objections to this procedure:
  - (a) it may lead to the development of strains of *B. larvae* that are resistant to the drugs;
  - (b) it may mask infection which could become active when treatment is stopped or, alternatively, treatment would have to be applied permanently;
  - (c) the agents used for treatment may contaminate honey harvested from treated colonies (Hansen, 1984).

- The view that, since infected combs are the source of contagion, control of AFB should be carried out by sterilization of infected combs and other equipment, or even by sterilization as a routine procedure. Ethylene oxide is the currently preferred fumigant.

- The view that the use of bacteriophage specific to *B. larvae* should be further explored as the treatment of choice.

- The view that bee breeding can develop strains in which infection with *B. larvae* is minimized because of the workers' hygienic behaviour in cleaning out cells (Rothenbuhler, 1964).

There is an environmental indication that, apart from differences in strains of bees, the bees in some areas are prone to AFB while those in others are not. In the USA, where the average incidence is high (20% of all colonies), it is much lower in Florida (above 1% in only two of the last 43 years) and in California (1% for 27 years). In general, diseased colonies that recover seem to do best during a good nectar flow, a suggested reason being that the incoming nectar dilutes the spores so that larvae are less likely to be infected.

The world distribution of AFB is indicated in Figure 11.21a.

Table 11.1A Honeybee diseases and other pathological conditions caused by micro-organisms and parasites, and insect predators

Entries refer to *Apis mellifera*; notes in italics give information on *A. cerana*, which is less complete.
The order followed is that used in the text, which discusses also organisms not in the table, and other hazards to honeybees.

Disease	*Causative organism*	*Text section*
Diseases of honeybee brood		
American foul brood (AFB) *reported also in A. cerana*	bacterium: *Bacillus larvae* White	11.21
European foul brood (EFB) *reported also in A. cerana*	bacterium: *Melissococcus pluton* (White)	11.22
sac brood *Thai strain in A. cerana only*	virus	11.23
black queen cell virus	virus	11.23, 11.34
chalk brood	fungus: *Ascosphaera apis* Maassen ex Claussen (Olive et Spiltoir)	11.24
stone brood	fungus: *Aspergillus flavus* Link and other species	11.25
Diseases of adult honeybees		
nosema disease	protozoan: *Nosema apis* Zander	11.31
amoeba disease	protozoan: *Malpighamoeba mellificae*	11.32
gregarine disease	protozoans: Gregarinidae	11.33
septicaemia	bacterium: *Pseudomonas apiseptica* Burnside; other organisms possibly involved	11.33
melanosis	fungus (yeast): *Torulopsis* spp.; other organisms possibly involved	11.33
viral diseases		11.34
	chronic bee paralysis virus	
	chronic bee paralysis virus associate	
	black queen cell virus	
	filamentous virus	
	bee virus X	
	bee virus Y	
	cloudy wing virus	
	acute bee paralysis virus	
	slow paralysis virus	
	Arkansas bee virus	
	Egypt bee virus	
	Kashmir bee virus; *mainly in A. cerana*	
	Apis iridescent virus; *only in A. cerana*	

(continued)

Table 11.1A *(continued)*

Harmful attack by pests	*Text section*
Parasitization by mites (Acari)	
internal (tracheal) mite in adult bees: *Acarapis woodi* (Rennie); *also in A. cerana*	11.41
external mites of both brood and adult bees:	
Varroa jacobsoni Oudemans; *also in A. cerana*	11.42
V. underwoodi n.sp.; *reported only in A. cerana*	11.43
Tropilaelaps clareae Delfinado & Baker; *also in A. cerana*	11.43
Table 11.4A gives more information on parasitic mites, and refers to all *Apis* spp.	
Parasitization by insects	
Diptera: (external) bee louse, on adult bees: *Braula coeca* Nitzsch and spp.	11.44
Diptera: (internal) fly larva, in adult bees: *Senotainia tricuspis* (Meigen)	11.45
Strepsiptera: (internal) larva, in adult bees and brood: *Stylops* spp.	11.45
Coleoptera: (internal) triungulin larva, in adult bees or brood: *Meloe* spp. (oil beetles or blister beetles)	11.45
Predation by ants (Hymenoptera) on adult bees and brood in the hive	
Formicidae (*also attack A. cerana*)	11.51
Dorylinae; army/driver/safari/legionary ants	
Dolichoderinae: Argentine ant, *Iridomyrmex humilis* (Mayr)	
Predation by wasps (Hymenoptera) on adult bees	
Vespidae:	11.52
giant hornet, *Vespa mandarinia* Smith; *also on A. cerana*	
oriental hornet, *Vespa orientalis* L.; *also on A. cerana*	
hornet, *Vespa crabo* L.; *also on A. cerana*	
yellowjackets or wasps, *Vespula* spp.	
Sphecidae:	11.53
bee wolf, *Philanthus triangulum* (Fabricius) and spp.	
yellow bee pirate, *Philanthus triangulum diadema* Fabricius	
banded bee pirate, *Palarus latifrons* Kohl	
Infestation of honeybee comb and contents	
Lepidoptera:	11.61
greater wax moth, *Galleria mellonella* (L.); *also in A. cerana nests*	
lesser wax moth, *Achroia grisella* (Fabricius); *also in A. cerana nests*	
adult death's head hawk moth, *Acherontia atropos* (L.)	
Coleoptera:	11.62
large hive beetle, *Oplostomus fuligineus* (Olivier)	
small hive beetle, *Aethina tumida* (Murray)	
fungus:	11.62
pollen mould, *Bettsia alvei* (Betts) Skou	

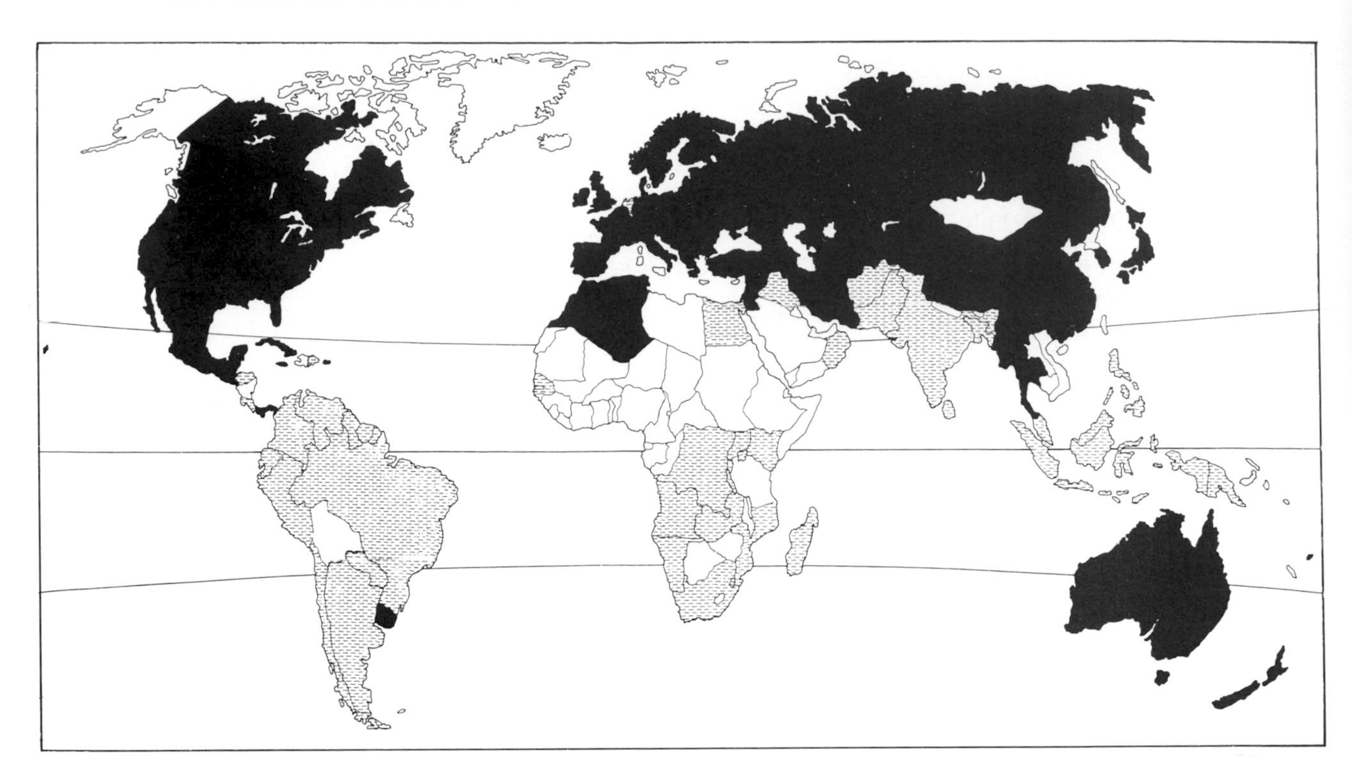

Figure 11.21a Countries in which American foul brood occurs (Bradbear, 1988).
By December 1988, AFB was also reported in St Vincent, Windward Islands; it has also been present in Costa Rica since 1985.

The following details apply to all maps in this chapter:
black present
stippled not found in samples from hives
white no information
vertical hatching suspected present (on some maps only)
All distribution maps in this chapter show *countries* in which the pathogen has been found, or is reported not to have been found in hive bees, or for which there is no information. Some maps also show countries where the presence of the pathogen is suspected. The distribution within the country is not indicated, so maps do not show whether the whole country is affected, or only part.

11.22 European foul brood (EFB)

This disease is caused by another bacterium, *Melissococcus pluton*, known earlier as *Streptococcus pluton*. It affects larvae up to 48 hours old, and usually kills them when they are 4 or 5 days old. Unlike AFB, EFB follows a pronounced seasonal cycle, with a peak early in the active season, when colonies are developing rapidly. Before this period, an infected colony can often maintain a balance between the increase and spread of *M. pluton* and its elimination from the hive when nurse bees remove infected larvae. With the subsequent increase of brood rearing, there are more infected larvae than can be removed, so they die in the cells, and this increases the spread of infection. The outbreak subsides when the number of larvae being reared decreases again. The disease especially affects small colonies, which are developing faster than larger ones. Secondary bacteria are associated with EFB, and the fact that *M. pluton* is the causative organism of EFB was not established until 1957. Bailey (1981) gives details of this and many other aspects of the disease.

Unlike AFB, EFB is not a serious threat to beekeeping generally, and it may occur and disappear without the beekeeper knowing about it. It may be regarded as a stress disease, and it occurs repeatedly in certain locations, where it presents a major problem. Colonies moved about for pollination are especially liable to suffer from it. Inadequate colony nutrition may be a predisposing cause, and a good nectar flow will sometimes lead to the elimination of the disease. Requeening has been advocated for infected colonies; the consequent break in the brood-rearing cycle gives the nurse bees an opportunity to remove infected larvae. Treatment in North America includes feeding with the antibiotic oxytetracycline, but nowhere should this be done except in consultation with the appropriate authorities who know the local legislation and regulations.

EFB (Figure 11.22a) is more widely distributed in the southern hemisphere than is AFB, although it is absent in New Zealand.

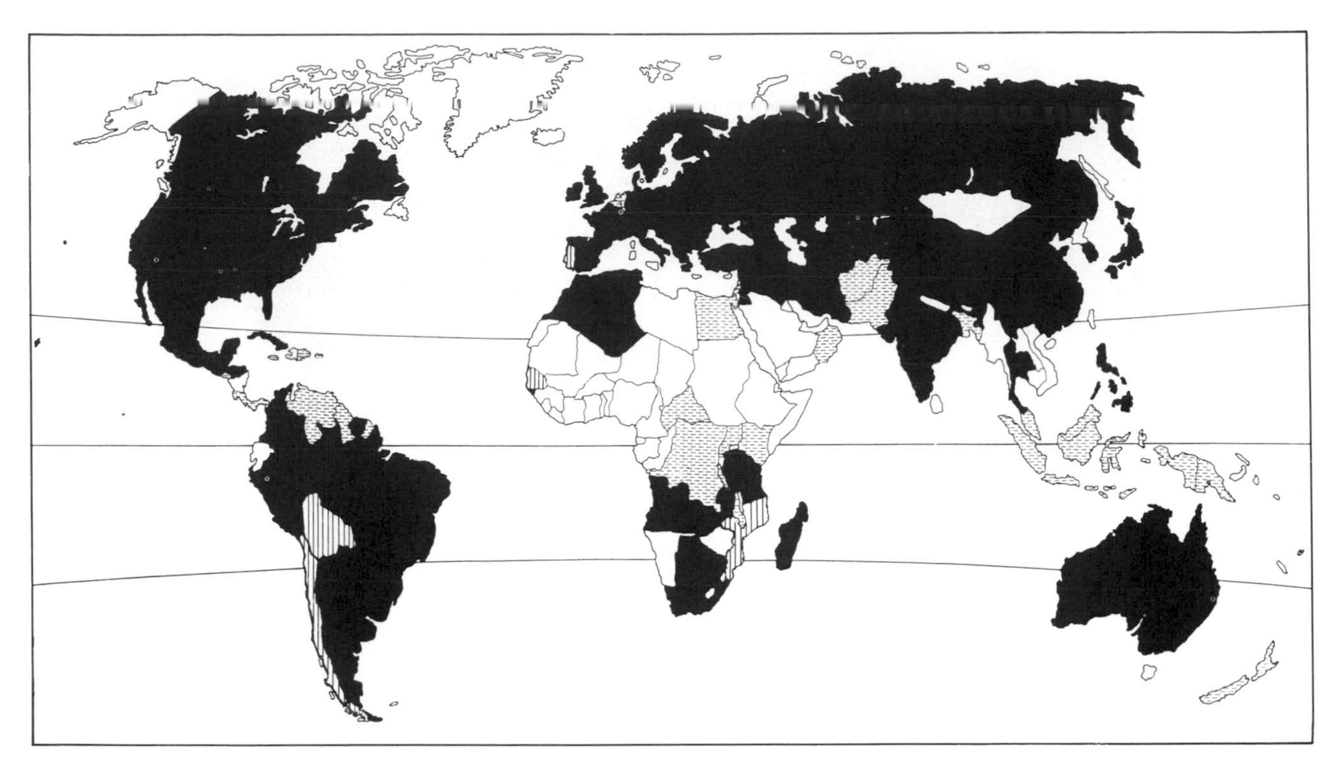

Figure 11.22a Countries in which European foul brood occurs (Breadbear, 1988).
Details as in Figure 11.21a. By August 1988, EFB was also reported in Tasmania, Australia.

11.23 Viral brood diseases

One of the few viral honeybee diseases that produces clear-cut symptoms is sac brood. Larvae 2 days old are most susceptible. An infected larva does not pupate as a healthy larva does, 4 days after being sealed in the cell. It remains stretched out, fluid accumulates between the body and its unshed skin, its colour darkens to dark brown, and it dies a few days later, drying to a 'flattened gondola-shaped scale'.

Sac brood infection tends to occur in the period before colony populations are highest, and to decrease markedly later in the active season. The virus is carried over from year to year in adult bees, in which it causes no obvious disease; it is probably transmitted to new generations of larvae in their food. Adult workers recognize and remove larvae in the early stages of infection, and the virus quickly loses its infectivity in the dried remains of larvae left in the cells. There is no accepted treatment for sac brood, but in many countries it is not regarded as a serious disease.

Sac brood is widespread in the north temperate zone (Figure 11.23a). In the southern hemisphere it is present in Brazil, South Africa, Papua New Guinea, Australia and New Zealand. A strain 'Thai sac brood virus' was identified from *Apis cerana* in Thailand (Bailey et al., 1982), and since then in north-east India and in Nepal, where many colonies died (Kshirsagar, 1982). Our knowledge of tropical countries is very incomplete.

Black queen-cell virus (Bailey, 1982) can kill queen larvae or prepupae after they have been sealed in their cells, especially during the first half of the active season. The pupae darken as they decompose, and the cell walls are often almost black in patches. The disease becomes particularly evident in queenless colonies that have been used for rearing many queens. Unlike sac brood, this virus rarely attacks worker larvae, nor does it usually multiply in healthy adult bees (see Section 11.34).

11.24 Chalk brood

Many different fungi can be found in hives and on bees, and *Ascosphaera apis* is one of the few that causes a disease. This is known as chalk brood because diseased larvae often turn into 'mummies' to which the mycelium of the fungus gives a white appearance. Young larvae are infected, and death usually occurs within 2 days of the cells being sealed. The disease is not usually considered serious, but it has a number of

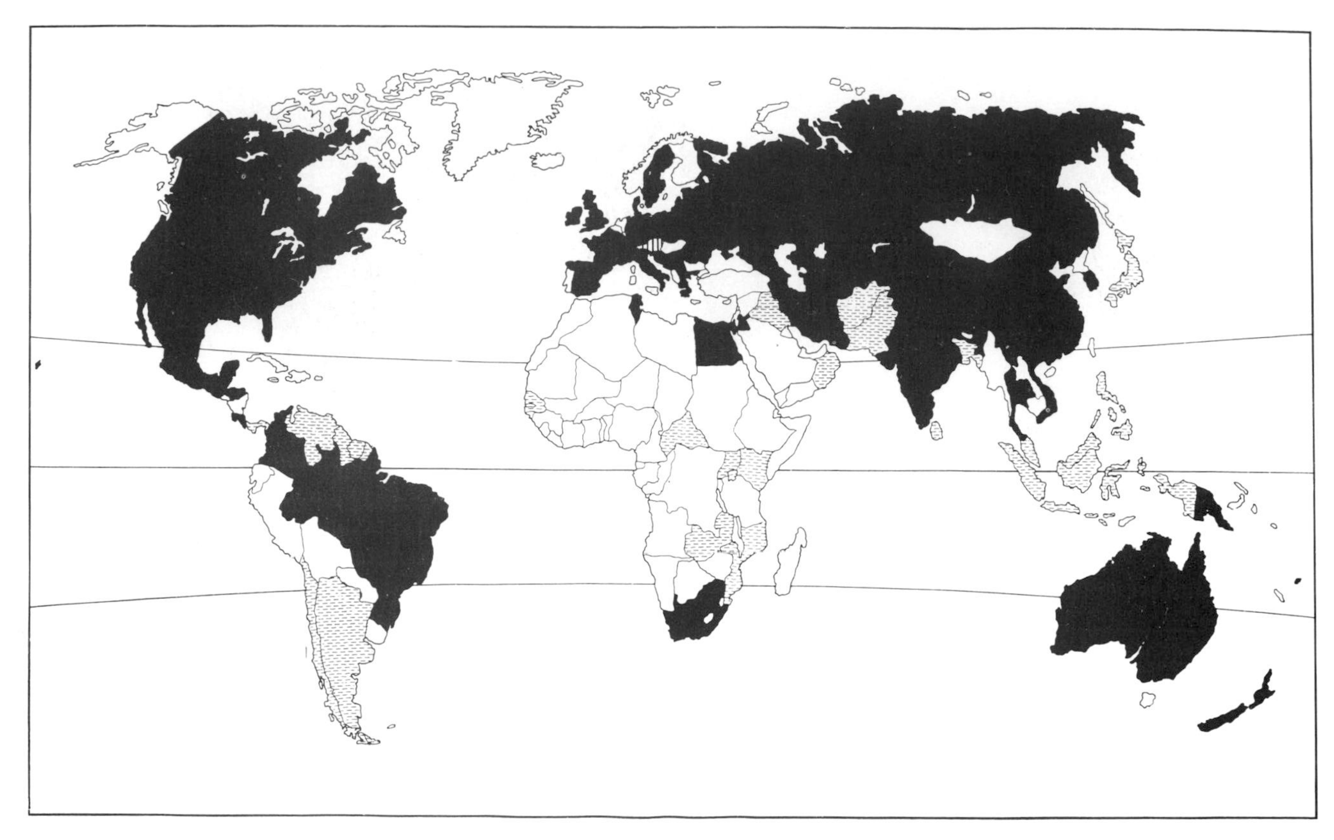

Figure 11.23a Countries in which sac brood occurs (Bradbear, 1988). Details as in Figure 11.21a. By August 1988, sac brood was also reported in Tasmania, Australia.

puzzling aspects. Outbreaks occurred in a number of different countries in the 1970s, including some in Europe, and the USA and Canada (Gochnauer & Hughes, 1976). It seems more likely that viable spores of the fungus were present, and germinated because conditions were suitable, than that they were transmitted from elsewhere. It is possible that a mutation of the fungus may have appeared. Chalk brood appears to be a stress-related disease. There is much evidence that damp conditions and poor hive ventilation favour the disease, which also occurs in colonies that are weakened by some other disease. Strong, balanced, well fed colonies, in which the brood is warm enough, and which are kept in a healthy situation, are less likely to succumb than weak colonies, with too few adult bees to care for the brood and to cover the combs in the hive. The disease, and the pathogen and related fungi that may be implicated, have been dealt with fully by Heath (1982*a*, 1982*b*).

Chalk brood is a disease characteristic of the cooler parts of the world (Figure 11.24a). It occurs in many countries of the north temperate zone, but in only a few of those in the subtropics or tropics; *Apis cerana* seems relatively unaffected. Heath (1985*a*) discusses the distribution in detail.

Ascosphaera apis can infect other species of bees, and there is a possibility that some of them play a part in infecting honeybees.

11.25 Stone brood

This disease is caused by the fungus *Aspergillus flavus* or other species of the same genus. They are common, occurring for instance in the soil, and they also infect other insects and animals. Bailey (1981) suggests that the fungus may grow in honeybees only if they are weakened by other factors. No method of treatment is known, but colonies are rarely affected severely. Larvae killed become hard mummies – hence the name stone brood.

The disease is encouraged by damp conditions and poor hive ventilation. It occurs in Europe and North America, and has been recorded in Venezuela.

11.26 Other disorders

A condition of brood known as 'half-moon disorder' is widespread in New Zealand, although very few col-

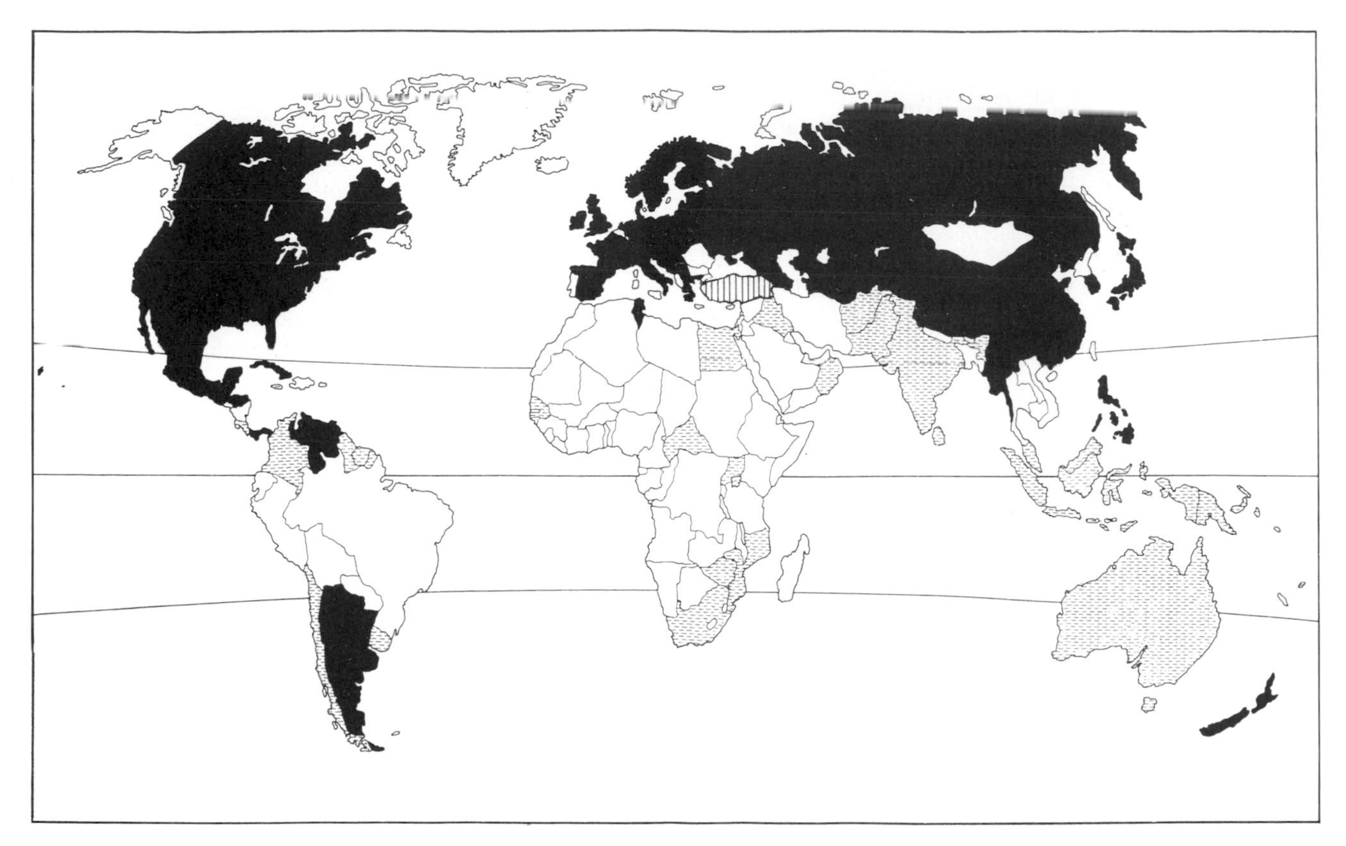

Figure 11.24a Countries in which chalk brood occurs (Bradbear, 1988). Details as in Figure 11.21a. Chalk brood is also in Thailand.

onies are affected. Larvae usually die when 4 days old, lying in the cell in the shape of the letter C. An extensive investigation by Shimanuki failed to find any causative organism, although *Bacillus coagulans* appeared to be present in some dead bees; see Matheson (1985*a*), who discusses information so far available. The cause is still unknown.

Bald brood is a name that sometimes refers to damage done to cell cappings by the lesser – or occasionally the greater – wax moth (Section 11.61), but similar damage may have other causes.

11.3 DISEASES OF ADULT HONEYBEES

Microscopic examination is essential for the diagnosis of diseases and parasites in adult bees, since general symptoms may be similar for different pathogens (Bailey, 1969; see also Hansen, 1983).

11.31 Nosema disease

Nosema disease is the most widespread of adult bee diseases. The pathogen, *Nosema apis*, belongs to the group of microscopic unicellular or acellular organisms known as Protozoa. It develops only within the epithelial cells of the midgut of adult honeybees (workers, and to a lesser extent queens and drones). The proventricular valve provides some protection by filtering out some of the spores ingested; see Section 2.31. The disease is transmitted by spores in faeces from infected bees, usually when young bees clean contaminated combs. Bees normally fly out of the hive to defaecate, and contamination is most likely to occur late in a dearth season when they are unable to fly and they therefore defaecate on the combs. There is no further contamination of the combs after flight becomes possible, and young bees are thus less likely to become infected. The disease apparently clears up, but recurs during the next season when flight is restricted, since the spores remaining on combs are a permanent focus of reinfection. It may appear in a colony that is confined to the hive for a spell, for instance by unseasonal bad weather. But if the combs were replaced by clean ones after the first infection, there is less chance that the disease will recur.

In common with other Protozoa, *Nosema apis* has a much longer life cycle than other micro-organisms affecting honeybees. Infected bees therefore do not die suddenly, and the presence of *Nosema* is often not apparent to the beekeeper. Nevertheless the bees are debilitated, and their life is shortened. A worker bee goes through her age-linked sequence of activities more quickly; she is younger than normal when changing from brood rearing to foraging, and when she dies.

Nosema is one of the few bee diseases that can be successfully treated with a drug: the antibiotic fumagillin, derived from the fungus *Aspergillus fumigatus* and marketed under the trade name Fumidil B.

Nosema disease occurs in all parts of the temperate zones. Figure 11.31a suggests that it may be absent from most parts of the Old World tropics, but not many systematic searches have yet been made. In the New World, Wilson and Nunamaker (1983) showed that it was comparatively rare in Mexico; only 3.8% of 840 bees examined – from 17 locations – contained spores. Experiments have failed to infect *Apis cerana* with *Nosema apis*.

11.32 Amoeba disease

This disease is caused by *Malpighamoeba mellificae*, another protozoan, which affects the Malpighian tubules of adult bees (the tubules perform something of the same function as the kidneys of mammals). It is not nearly as harmful as nosema, but infection follows a similar seasonal cycle, and the two infections are rather likely to occur together. The disease is considered to be detrimental to bees, although there are no well established symptoms. No chemical treatment is known, and control must depend on good management practice and hygiene.

Amoeba disease is less widespread than nosema, and may be largely confined to higher latitudes in temperate zones, but it is known to occur in the subtropics and tropics (Figure 11.32a).

11.33 Other non-viral diseases

A number of micro-organisms live in adult honeybees. Some are known to cause disorders, and others may do so in certain circumstances; still others are probably harmless.

Gregarines, another group of Protozoa (Gregarinidae), have been found in France, Italy and Switzerland, and in Canada, USA and Venezuela (Morse 1978; Bailey, 1981). They attach themselves to the epithelium of the honeybee midgut, and the presence of 1000–3000 in one bee is likely to be lethal. If a colony has many heavily infected bees, its population dwindles.

Bacteria infecting adult honeybees include a soil organism *Pseudomonas apiseptica*, which is found in the blood. It leads to the condition known as septicaemia, although Shimanuki (Morse, 1978) rejects the term since it describes a symptom only; see also Bailey (1981). It kills the infected bees rapidly. *P. apiseptica* has been reported from Europe, North America and Australia.

Morse (1978) and Bailey (1981) have given information about yet other disorders, and Fyg (1959) provided details of many abnormalities in queens, drones and workers.

11.34 Viral diseases

Finally we come to the important subject of viruses that infect adult bees. A number have been identified and found to cause diseases in them, in addition to the sac brood virus and black queen-cell virus that infect immature bees (Section 11.23).

Almost all the investigations and identifications of these viruses were carried out by Dr L. Bailey and his colleagues at Rothamsted Experimental Station in the UK. Several summaries of the work are available (Bailey, 1975, 1981, 1982), and details of the world distribution of the viruses (Bailey, 1967; see also 1981). The viruses infecting adult bees that have been identified are listed in Table 11.1A. Arkansas bee virus and Egypt bee virus were found in bees in the USA, *Apis* iridescent virus and Kashmir bee virus in *Apis cerana* in Kashmir, and the latter also in *Apis mellifera* in Australia, and in New Zealand (and also Canada and Fiji to which it exports bees). Kashmir bee virus could well be transmitted by the varroa mite, as other viruses are (Bailey, 1988).

There is evidence that a significant proportion of colonies may be infected with a virus, and that viruses can cause damage and death which had previously been attributed to other pathogens. For instance, about 15% of samples of bees from a number of colonies in Britain were infected with cloudy wing virus, and this virus was significantly associated with mortality of colonies in winter (Bailey, 1982). It has also been detected in Egypt and Australia. Black queen-cell virus, mentioned in Section 11.23, was detected in about 15% of over 300 samples of adult worker honeybees from all parts of England and Wales in 1979. It multiplies abundantly in bees that are also infected with *Nosema apis*, and shortens their life further. The virus causes no outward signs of disease, but in combination with *Nosema*, it is much more strongly associated with winter deaths of colonies than is *Nosema* infection alone.

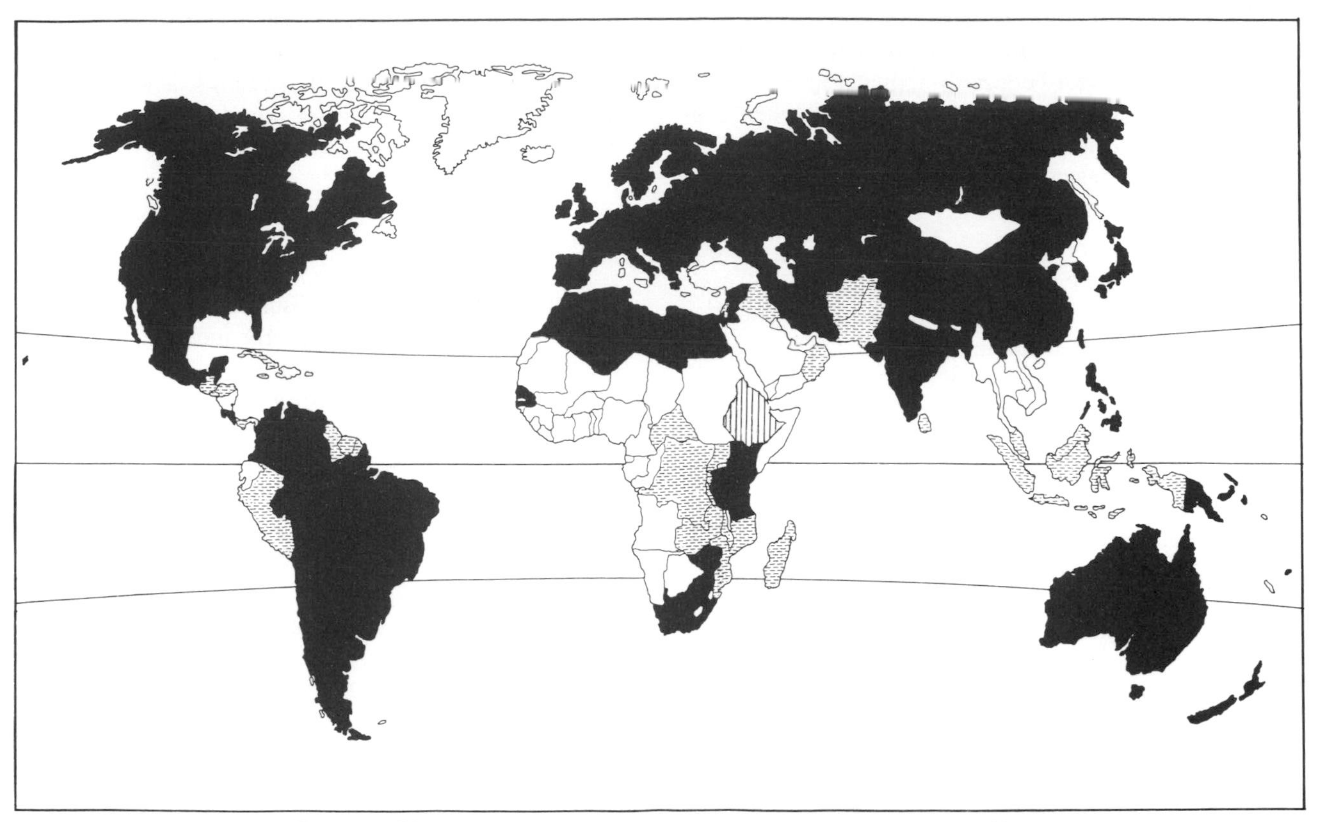

Figure 11.31a Countries in which nosema disease occurs (Bradbear, 1988). Details as in Figure 11.21a.

Figure 11.32a Countries in which amoeba disease occurs (Bradbear, 1988). Details as in Figure 11.21a. By August 1988, the disease was also reported in Australia.

Viral infections are widespread, and cannot be 'cured' by drugs or other direct treatment. Damage caused by them can be greatly reduced by good bee-keeping practice using strong colonies, under conditions that do not induce stress, and kept free from such treatable infections as *Nosema apis*. Bailey says (1982):

> Many viruses commonly infect honeybees, several of them often in the same colony, and they may cause no sign of infection for long periods. This fact underlines the inadequacies of attempts frequently made to explain disorders of bees solely in terms of the comparatively few but common, long-known and relatively easily identified non-viral parasites. The outstanding example of these is *Acarapis woodi* which has been blamed on two separate occasions, in two entirely different parts of the world [Britain and India], for spectacular symptoms in adult bees, although the available evidence indicates that they were caused by two unrelated viruses.

Bee World (1983) quotes Bailey further:

> The knowledge that there are at least ten viruses in bees in Britain, and some 18 throughout the world, all of which have been identified and characterized at Rothamsted and many of which are very common, has, with other findings in recent years, considerably modified concepts of honeybee pathology. Until about the mid-century, the comparatively few pathogens that were known were believed to lead inexorably to severe losses or the death of whole populations once they became established. This attitude is still encountered and has not been without influence on views about the properties of insect pathogens in general. In fact, most pathogens of bees, including their viruses, are endemic; and, although they sometimes cause severe disease and are always damaging, their incidence and the disease they cause usually pass unnoticed or are accepted as 'normal'. Consequently their eradication, once believed to be the only solution, although desirable, is unlikely. Nevertheless, their control in many instances is possible, if only by avoiding husbandry practices that aggravate them. Devising and improving control measures require information about the very diverse properties and natural histories of the pathogens. Much of this is now known, but past experiences should indicate that far more probably remains to be learned.

11.4 PARASITES

Table 11.4A lists mites parasitic on different species of honeybee, including two that were first reported in the 1980s, *Varroa underwoodi* and *Euvarroa haryanensis*; they are referred to in Section 11.43. The table also mentions some of the mites found in honeybee colonies which are not parasitic on the bees.

11.41 The tracheal mite *Acarapis woodi*

'Acarine disease' in adult bees is caused by infestation of the prothoracic tracheae by the mite *Acarapis woodi* (Tarsonemidae). Only a young adult bee can become infested, the chance decreasing day by day after emergence. The mite (Figure 11.42b) is too small to be seen without a microscope. It feeds on blood by piercing the tracheal wall with its stylet-like chelicerae. The behaviour of the female mites in transporting themselves from an old or dead bee to a young one passing by in the hive has been engagingly described (Hirschfelder & Sachs, 1952). It has been assumed that the mites can partially block the tracheae near one or both of the first spiracles (Section 2.33), and the inflow of air is then so much impeded that the oxygen supply is insufficient to allow the bee to fly. Bailey (1981), however, could find no evidence for this.

Acarapis woodi is well described by Delfinado-Baker and Baker (1982). It was discovered by Rennie et al. (1921) during searches for the cause of the 'Isle of Wight disease' that was believed to account for many colony deaths, especially in Britain, in the early part of this century (Bee World, 1974). On the other hand Bailey (1963) has proposed that viral infections, together with over-stocking of colonies, were more likely determining factors. He estimated (1981) that in England and Wales, something less than 2% of colonies could then be expected to suffer measurably from infestation. He linked the decrease in *A. woodi* infestation, from about 17% in the early 1950s, with a decline in the number of colonies of bees kept and a concomitant increase in forage available per colony (1985). It has, however, been shown that infestation with *A. woodi* shortens the life of overwintered bees (Maki et al., 1988); severely infested colonies are smaller in spring, and less productive (Eischen, 1987).

Early methods of controlling *Acarapis woodi* infestations (Crane, 1953) were superseded by the fumigation of colonies with chemicals developed for use as acaricides. Of these, Bailey and Carlisle (1956) found Ovotran and chlorobenzilate to be the most effective in cool English summers. Bromopropylate (see Section 11.42), dichlorvos and menthol are reported to be effective; one fumigant strip is used per hive weekly, 6 times in all. Dinabandhoo and Dogra (1979) used dichlorvos on *Apis cerana* with success. IBRA (1986*a*)

Table 11.4A Mites parasitic (and some non-parasitic) on honeybees (*Apis* spp.)

Mites listed above the centre line feed on honeybee fluids.

Am = *A. mellifera* *Ad* = *A. dorsata*
Ac = *A. cerana* *Af* = *A. florea*

Year 1 = year the mite was described and named.
Year 2 = year of first known record of the mite in a different continent.

	Am	*Ac*	*Ad*	*Af*	*Year 1*	*Year 2*
Laelapidae						
Tropilaelaps clareae	×	×	×	×	1961	—
Tropilaelaps koenigerum			×		1982	—
Tarsonemidae						
Acarapis woodi	×	×	×		1921	1951
Acarapis dorsalis	×				1934	1961?
Acarapis externus	×				1934	1959
Varroidae						
Varroa jacobsoni	×	×			1904	1970
Varroa underwoodi		×			1987	—
Euvarroa sinhai				×	1974	—
Euvarroa haryanensis		×		×	1985	—

Entries below include some of the phoretic mites which are found on honeybees, but which do not feed on them (see text for *P. herfsi*); the final column gives the entry number to a reference in *Apicultural Abstracts*.

	Am	*Ac*	*Ad*	*Af*	
Ameroseiidae					
Neocypholaelaps indica	×	×	×	×	247/85, 439/85
Neocypholaelaps apicola		×			247/85
other species					1338/82
Laelapidae					
Melittiphis alvearius	×				688/79, 281/84
Macrochelidae					
Macrocheles glaber	×				260/80
M. muscaedomesticae	×				260/80
Pyemotidae					
Pyemotes tritici	×				768/74
Pyemotes herfsi		×			810/81
Pyemotes sp.	×				587/82
Tarsonemidae					
Pseudoacarapis indoapis		×			693/68
Tarsonemus apis	×				Rennie & Harvey (1921)

summarizes many chemical and other methods, some of which are useful in the control of both the tracheal mite and the *Varroa* mite.

Figure 11.41a shows countries in which *A. woodi* is known to infest honeybees. Infestation is a feature of poor beekeeping areas and of poor beekeeping seasons where food resources are insufficient. It is noticeably absent – or reduced in intensity – when or where there is a succession of copious honey flows. It has also been absent where winters are long and severe, and fewer infested bees consequently survive until brood rearing is resumed in the spring. The mite is absent in Australia, New Zealand and much of Africa, and (within Europe) in Scandinavia. It was recorded as entering the USA from Mexico in July 1984, and 13 months later had been reported from 17 states – as a result of the great 'mobility' of American beekeeping (Delfinado-Baker, 1985). It reached Canada by 1988. *A. woodi* has probably been searched for more than any other honeybee pathogen, hence the relatively large proportion of countries where its absence has been reasonably attested.

11.42 Varroa jacobsoni

This mite, in the family Varroidae, is an endemic parasite of *Apis cerana*, and was first described from a specimen found on this bee in Java in 1904. By 1968 it had been found on *Apis cerana* in a number of countries in Asia. More significantly for world beekeeping, it was successful in parasitizing *Apis mellifera* in places where both species of honeybee were present, near the Pacific coast of the USSR (Crane, 1968). It reached European USSR from there, on bees transported across the country, and was spread quite quickly to other European countries. It was found in North Africa by 1980, and in the USA in 1987. Also in 1987, an Australian Quarantine Station discovered *Varroa* mites on bees imported (by the approved procedure) and was thus able to prevent their entry into the country.

V. jacobsoni is now in at least 56 countries, shown in Figure 11.42a, and virtually all its spread (Griffiths & Bowman, 1981) has been brought about by the transport of bees infested with the mite – from one region, country or continent to another – and viruses can be

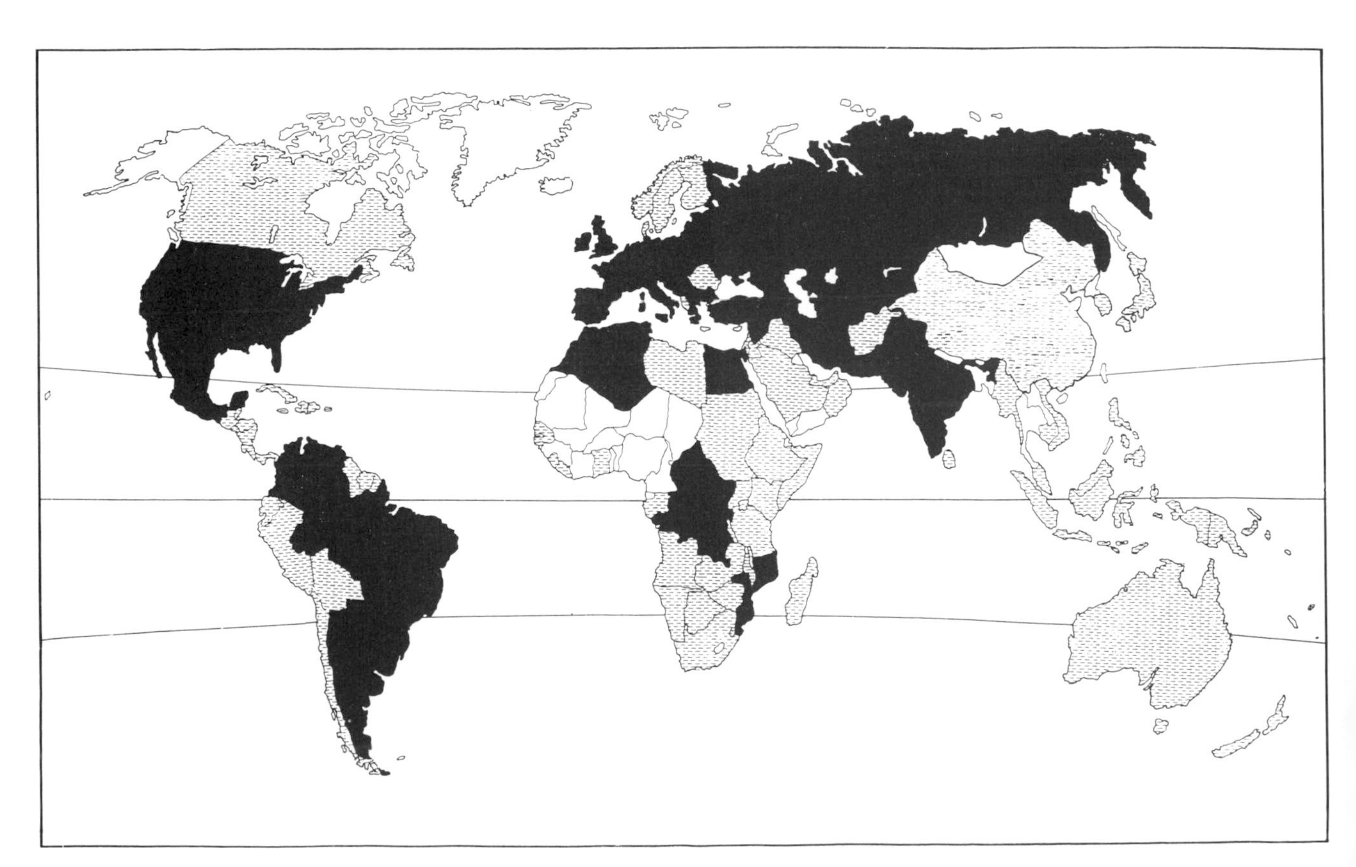

Figure 11.41a Countries in which the tracheal mite *Acarapis woodi* occurs (Bradbear, 1988).
Details as in Figure 11.21a. By August 1988, the mite was also reported in Canada.

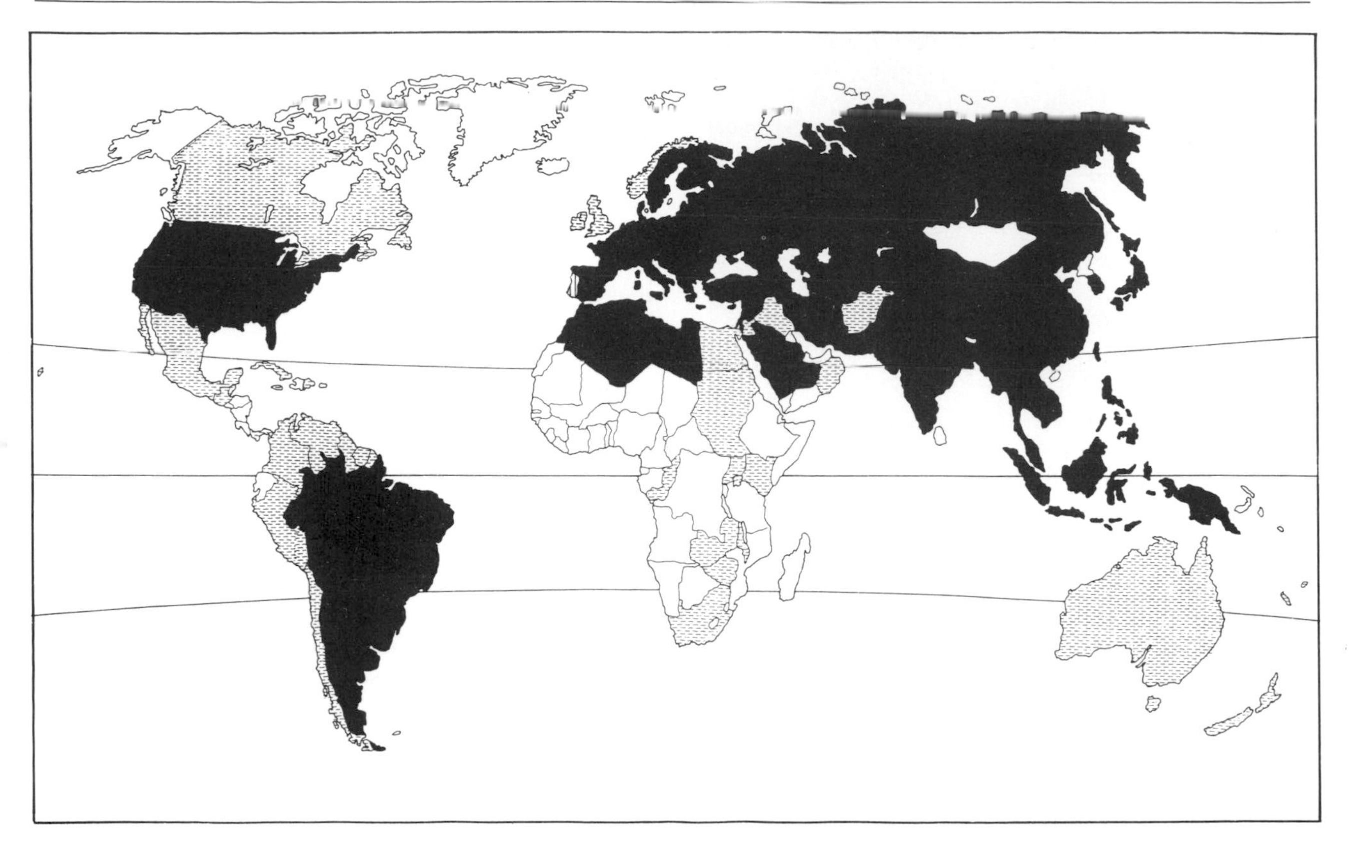

Figure 11.42a Countries in which *Varroa jacobsoni* occurs (Bradbear, 1988). Details as in Figure 11.21a. By August 1988, the mite was also reported in Iraq, Portugal, Sri Lanka and United Arab Emirates.

transmitted with the mite. Experience has often shown that, by the time mites were identified in a new area, they had already been present for several years, and in due course all or most colonies become infested. Regions where *Varroa* has been sought and not found include the islands of Britain, Ireland, Australia, New Zealand, and some in the Pacific including Hawaii.

The efforts in many countries to monitor the incidence of the mite, and to limit the damage when it is present in honeybee colonies, have resulted in a large number of publications, especially in German, Russian and other eastern European languages. (Few English-speaking countries were affected by *Varroa* before 1985.) Bibliographies on *Varroa* cite publications in many languages (Crane, 1985*c*; IBRA, 1986*a*; Wienands & Madel, 1987). A number of countries issue their own official information on diagnosis and control of the mite, and the proceedings of a meeting of European Community experts on *Varroa* have been published (Cavalloro, 1983).

Ritter (1981), Ramirez and Otis (1986) and Dietz and Hermann (1988) have given general accounts of the mite's life history. The female (Figure 11.42b) enters a brood cell shortly before the bees seal it; she feeds on the larval blood, which enables her to produce eggs (usually 2–5), laid after the bee larva has finished spinning its cocoon. The larval mites feed on the blood of the developing bee, and the longer the period before the adult bee emerges, the more successfully the mites develop. So within the same colony, drone brood is parasitized in preference to worker brood. And the fact that the larval period of each caste is longer in *Apis mellifera* than in *A. cerana* seems to account, at least in part, for the greater success of the mites – and the greater damage they do – in *A. mellifera* than in *A. cerana* colonies. Within *A. mellifera*, differences in duration of developmental stages in various geographical races may be significant in overcoming the problem of *Varroa* infestation. For example the development period is shorter for tropical African and Africanized workers than for European workers; see Table 2.41A, and Harbo et al. (1981) also give figures. Camazine (1986*b*) found only 49% of brood cells in Africanized colonies were infested with the mites, compared with 75% in European colonies. Also, the post-sealing period is recorded as only 9.6 days for *A. m. capensis*, and 12.0 days for European *A. m. carnica*; *Varroa* was able to reproduce on the latter, but

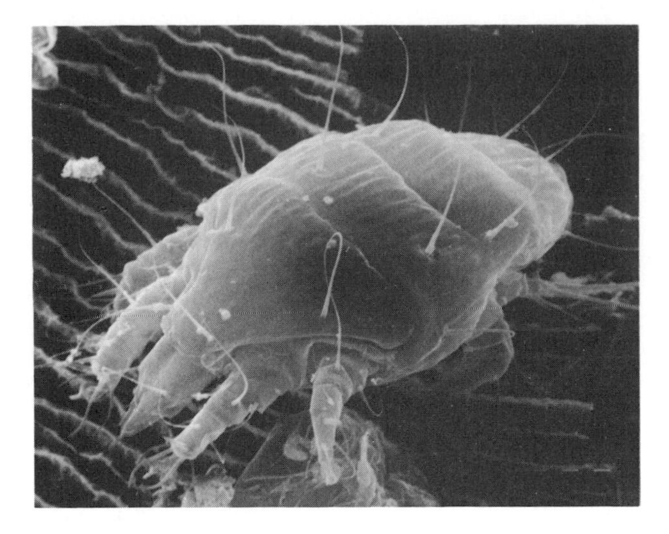

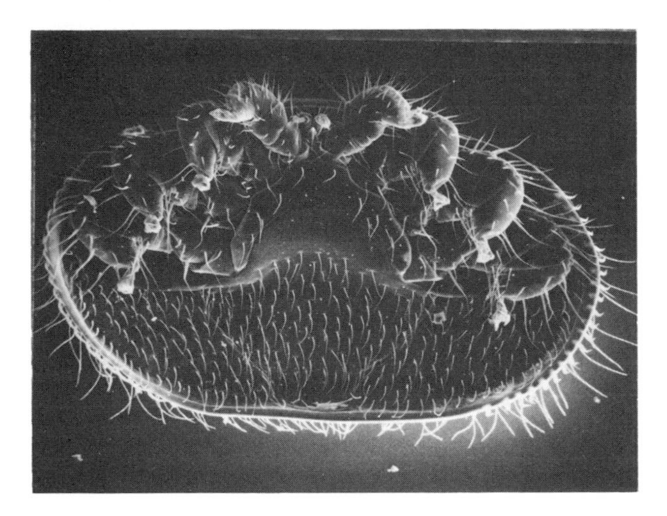

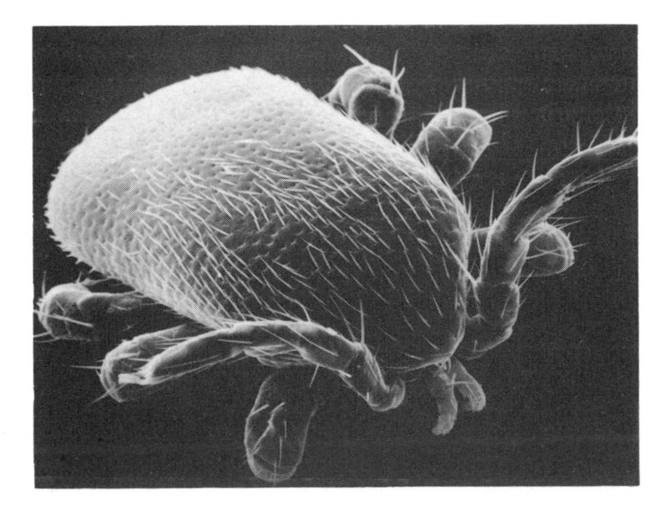

Figure 11.42b Three parasitic mites of the honeybee (SEM photos: M. Delfinado-Baker & W. E. Styer).
above Female *Acarapis woodi* in feeding position, dorsal view, × 500.
centre Female *Varroa jacobsoni*, ventral view, × 50.
below Female *Tropilaelaps clareae*, dorsal view, × 50.

possibly not on the former (Moritz & Hänel, 1984). In 7 colonies of *Apis cerana* studied by Koeniger et al. (1981), the mites reproduced only on drone brood, so they were less harmful to these colonies than to colonies of *A. mellifera*. (*Varroa* has, however, been reported to reproduce on worker *A. cerana* brood in Korea, De Jong, 1988.) *A. cerana* workers have another important advantage over *A. mellifera*, in that they are able to kill adult *Varroa* mites, as described in Section 7.41.5.

Since the mid-1970s, much research has been done on methods of controlling *Varroa* infestation in honeybee colonies. One type of treatment is based on the use of combs of drone brood to attract female mites away from worker brood; the 'bait' drone combs are removed after the cells are sealed. An alternative is a regime of colony management that causes a complete break in brood rearing by the colony, so that female mites have no access to bee larvae; they cannot then lay eggs, and the mite population dies out. To achieve this, the queen may be caged, or else confined to different frames in turn and the capped brood produced in them removed and destroyed. Such methods are very laborious, and interfere with the normal population growth of the colony. Another treatment, used experimentally, disrupts brood development by feeding colonies with a hormone that inhibits the synthesis of chitin (BAY SIR 8514); the bees removed or ate the eggs laid, and no brood was reared (Herbert et al., 1986).

Stimuli in the honeybee colony currently being investigated for their possible effect on the behaviour of female *Varroa* mites include air currents, temperature gradients, and also the size of the cell containing a larva.

Another type of treatment is the fumigation of colonies with a chemical acaricide. One of the most successful fumigants has been bromopropylate (isopropyl-4,4′-dibromobenzilate), sold under the trade names that include Folbex-VA, Folbex-VA-Neu, Folbex-Forte, Neoron and Varroatex. In 1982 it was approved by the West German Health Authority for treating colonies infested with *Varroa* or *Acarapis woodi* (Bee World, 1982); there were, however, a few subsequent reports of colony deaths after its use. To prevent contamination of honey, it may be used only when there is no honey flow. The fumigant strips purchased can be burned in the hive, or in a smoker with an extra-long nozzle that can be pushed through a slit in a sheet of plastic with which the hive entrance is closed. Ritter et al. (1984) recommend treating 8 colonies together, using 8 strips in the smoker, and giving each colony 8 puffs of smoke at 5-second intervals. Treatment is repeated weekly, 4–6 times. Other fumigants in use include amitraz (sold as Fumilat A, Taktic, TCL, Varrescens) and phenothiazine or thiophenylamine (sold as Krka, Varitan, etc., and in preparations mixed with other agents).

It is also possible to disperse an acaricidal vapour through the hive at a controlled concentration, by evaporating a liquid in a specially designed apparatus. The most widely tested evaporant is formic acid (e.g. Ritter & Ruttner, 1980). Care must be taken to prevent contamination of honey.

Fluvalinate (sold as Klartan and Apistan) and amitraz are contact acaricides that show considerable promise for *Varroa* control. Plastic strips impregnated with the active ingredient are placed in the brood nest, and bees walking over them spread it as they carry out their various activities. As a result, mites are killed within the brood cells, and as adult bees emerge from them.

An alternative chemical approach is to use a systemic acaricide; see e.g. Moritz (1982). The acaricide is fed to honeybee colonies in winter when there is no brood, at the dosage specified for the acaride used; it must be too low to harm the bees. The acaricide is absorbed into the adult bee's blood, from which the mites ingest it and are killed by it. (The colonies must be without larvae, because larval blood would not contain enough of the acaricide to kill mites.) Systemic acaricides in use that are reported to be successful include: chlorphenamidine (sold as Galecron); coumaphos (Perizin) which is reported to be better tolerated by the bees (Ritter, 1986); Apitol, sold as a replacement for Folbex-VA.

Some of the chemicals are useful also for diagnosis, and tests should be made constantly in areas at present free from the mites, so that beekeepers can be alerted immediately if they are found. Tobacco smoke can be used (Ruijter, 1982), by methods described in many advisory leaflets (e.g. Cook & Griffiths, 1985, for the UK). After the bees have stopped flying in the evening, a sheet of clean stiff white paper (the size of the floorboard) is slid into the hive entrance. Smoke from pipe tobacco is applied after the entrance has been blocked up with newspaper except where the smoker nozzle is pushed in. The hive is left closed until early next morning, when the paper sheet is carefully removed. If mites are present, dead ones will have dropped on to the sheet, which is sent with its contents for analysis. In some countries a similar white sheet is kept in the hive, and debris on it examined at regular intervals.

Many diagnostic methods are described which involve collecting samples of adult bees in a glass jar, and adding to it an anaesthetic, detergent or other liquid, which kills and dislodges the mites.

11.43 Other external mites

Table 11.4A lists the mites discussed here; Delfinado-Baker and Styer (1983) give a fuller list, with scanning electron micrographs of each mite; they also identify locations on the body where the mites are found. Since then, further mites have been identified. *Varroa underwoodi* Delfinado-Baker & Aggarwal 1987 was found with *V. jacobsoni* in material from a nest of *Apis cerana* in Nepal. In describing the mite, Delfinado-Baker and Aggarwal (1987*b*) comment that in view of its morphology and feeding behaviour, it may be of importance to beekeeping. *Euvarroa sinhai*, another member of the Varroidae, was identified in 1974, but its only known host is *Apis florea*, where the mites live on drone and worker brood and adults (Mossadegh & Komeili Birjandi, 1986). In Haryana, India, another species, *Euvarroa haryanensis* Kapil, Putatunda & Aggarwal 1985 was found associated with *A. florea* and subsequently with *A. mellifera* (Delfinado-Baker, 1987).

A much more important mite is *Tropilaelaps clareae* (Laelapidae), whose natural host is *Apis dorsata*. It was first described from specimens taken from a rodent in the Philippines in 1961 (De Jong et al., 1982). It is distributed throughout mainland China, where it lives on both worker and drone brood of *A. mellifera* (Griffiths, 1988*a*). It has been reported on *Apis cerana* worker brood (see also Burgett et al., 1983). The known world distribution of the mite is shown in Figure 11.43a.

Tropilaelaps clareae (Figure 11.42b) is about half the size of *Varroa*. The mites cannot live more than 2 days on adult bees, but they may parasitize up to 90% of the brood, and several of the authors quoted suggest that they may be more destructive than *Varroa* to *Apis mellifera* colonies. This is certainly so in mainland China, where *Tropilaelaps* can kill an untreated colony within a few months (Griffiths, 1988*a*). However, Woyke (1984*a*) reported that colonies wintered in warmer parts of Afghanistan were infested, whereas those wintered in colder areas were not. He ascribed the difference to a broodless period of 2 months in the colder area, and proposed a control method in which colonies are made broodless for a period. Chlorobenzilate (Folbex) and powdered sulphur have been indicated as chemical control agents (De Jong et al., 1982).

Griffiths (1988*b*) suggests that *T. clareae* is an opportunistic predator which can exploit honeybee colonies, possibly surviving broodless periods by living in the nests of small rodents, feeding on the blood of the young. A second species, *T. koenigerum*, has been found in Sri Lanka and Nepal, on *Apis dorsata* only.

Mellitiphis alvearius Berlese, another species in the Laelapidae, was found in 1982 on *A. mellifera* workers

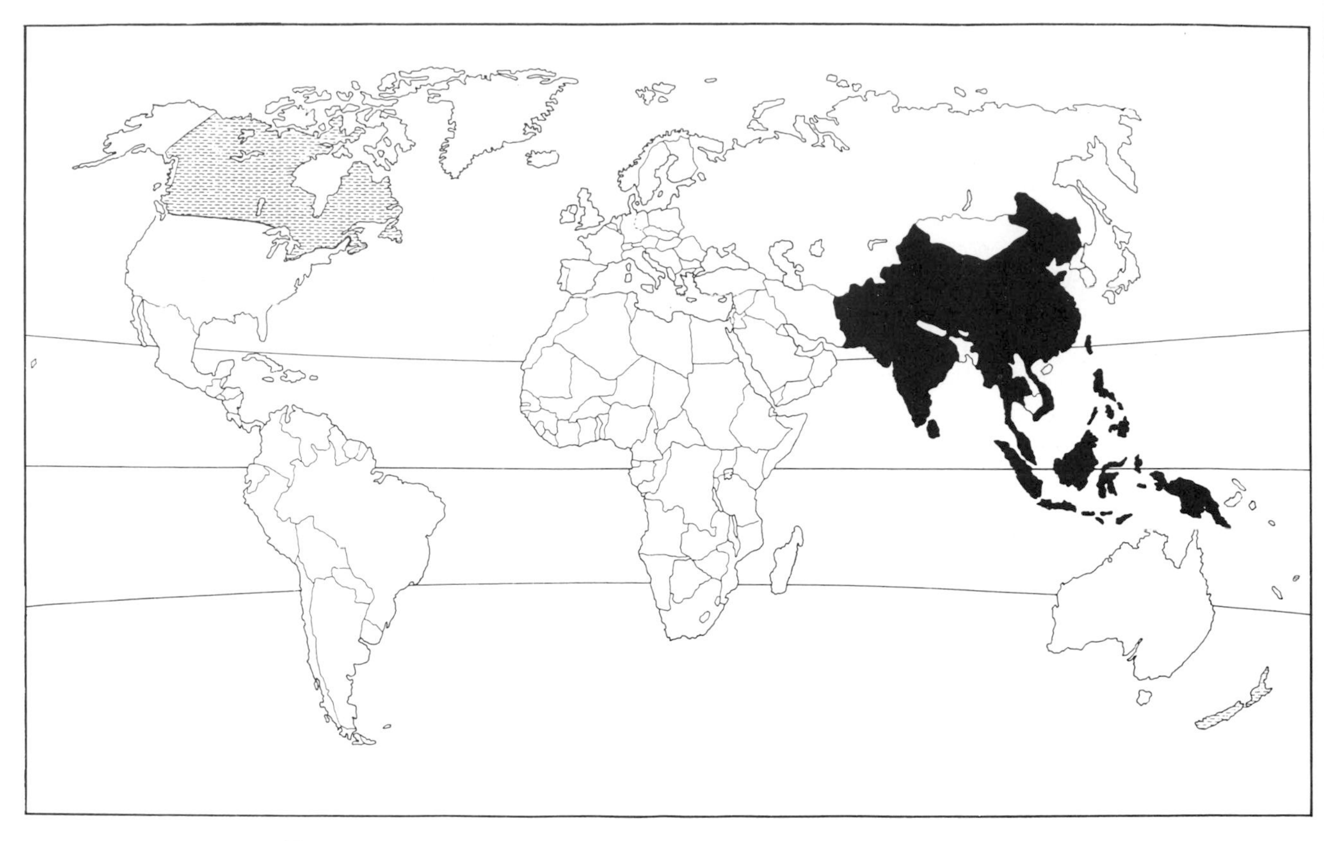

Figure 11.43a Countries in which *Tropilaelaps clareae* occurs (Breadbear, 1988).
Details as in Figure 11.21a. The mite may well be absent from many of the countries left white.

accompanying queens imported into the UK from New Zealand (Cook & Bowman, 1983), and in 1985 in package colonies sent to Canada from New Zealand. The mite had been present in as many as 5.5% of debris samples from hives in Czechoslovakia (Samsinak et al., 1978). There seems to be no evidence that it is detrimental.

Two species of *Acarapis* (Tarsonemidae) live externally on *A. mellifera*, appearing to do little or no harm to them. *A. dorsalis* is found in a groove on the thorax, and at the base of the wings (and on the forewings), and on the fore part of the abdomen. *A. externus* lives on the 'neck' at the junction between the head and thorax. Some mites described in 1941 as a third external species, *A. vagans*, were probably *A. dorsalis* or *A. externus* (Delfinado-Baker & Baker, 1982).

Pyemotes herfsi (Pyemotidae) is a mite found on *Apis cerana* in India. It prevents normal functioning of the mouthparts, and also occurs on other parts of the body. Dichlorvos fumigation has been used as a control method (Dinabandhoo & Dogra, 1979). Species of *Pyemotes* are reported on *Apis mellifera*, from Chile and Poland (IBRA, 1986*a*).

Neocypholaelaps (Ameroseiidae) is an Old World genus of mites that feed on pollen. When the mites are feeding, they use foraging insects to transport them between flowers and may be found on honeybees. They are not parasitic and seem to do no damage to the bees, although in Japan up to 3000 individuals (of *N. favus*) have been found in one hive of *A. mellifera* (De Jong et al., 1982).

For information on still other mites that live on or with different honeybees, see De Jong et al. (1982) and IBRA (1986*a*). There are a number of new scientific findings on mites that parasitize honeybees in the Proceedings of an International Conference in Ohio in 1987 (Needham et al., 1988).

11.44 Bee lice (*Braula*)

Species of *Braula*, commonly known as bee lice, are wingless flies. Figure 11.44a shows that the genus is very widely distributed, although records are lacking for much of the north tropical region. *Braula* is apparently absent in New Zealand, the Australian mainland, and Papua New Guinea to which bees were introduced from Australia, but it occurs in Tasmania. It probably does little harm to *Apis cerana*.

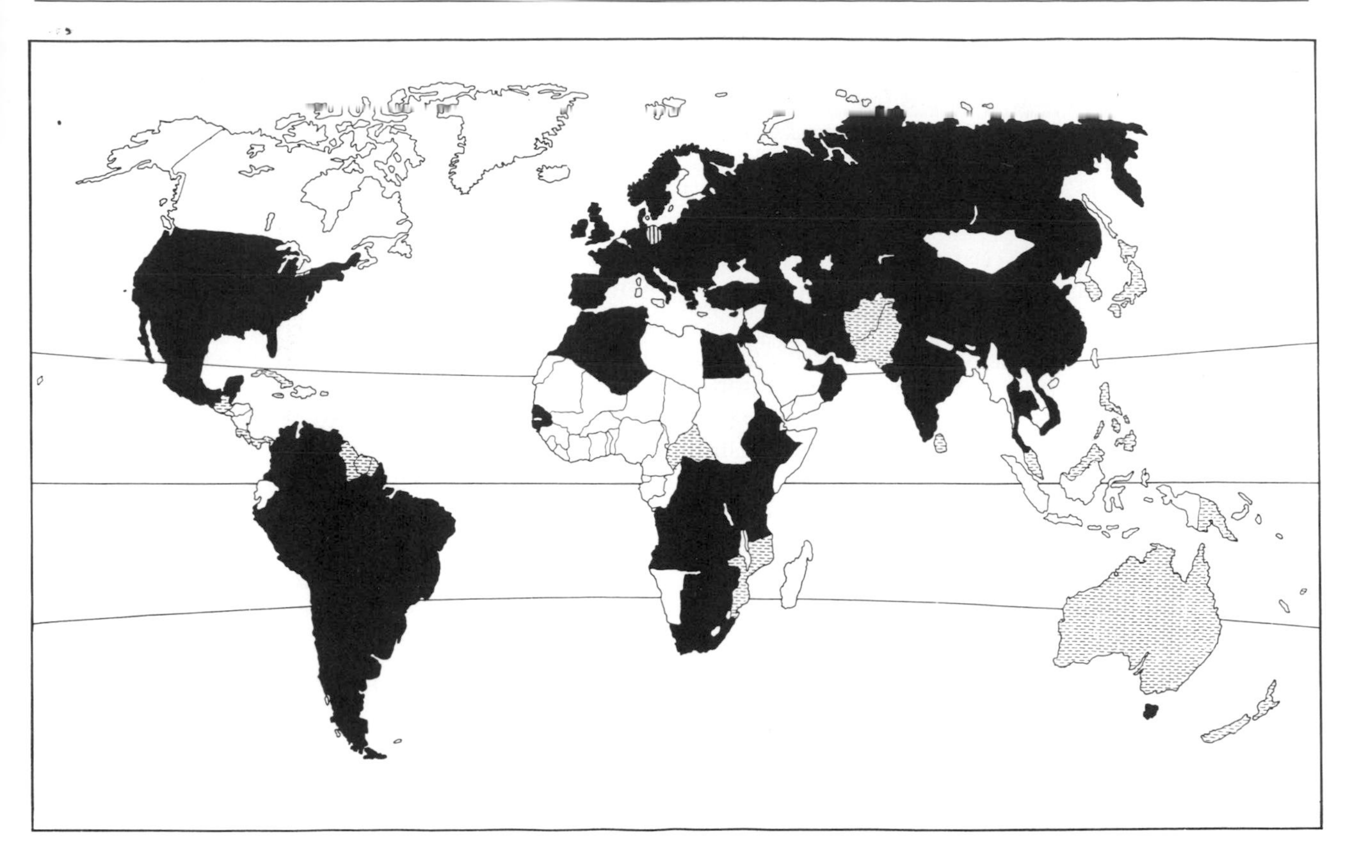

Figure 11.44a Countries in which bee lice (*Braula* spp.) occur (Bradbear, 1988).
Details as in Figure 11.21a.

Much of our knowledge of the taxonomy and differentiation of *Braula* species is due to the work of Örösi-Pál in Hungary. His latest paper (1980) names the following species, which are present in the regions below (Knutson, see Morse, 1978):

Braula angulata Örösi – Southern Africa, Lipari (Italy)
Braula coeca Nitzsch – Europe, Africa, Tasmania, USA, Caribbean, S. America
Braula kohli Schmitz – Congo
Braula orientalis Örösi – Pacific, USSR, Turkey, Israel, Egypt
Braula pretoriensis Örösi – Africa south of the equator
Braula schmitzi Örösi – widely in Asia, southern Europe, S. America
two (unnamed) species – Africa.

Örösi-Pál found that the species in different parts of the New World corresponded to those in the part of the Old World from which honeybees had been introduced. Several species may live together in the same colony.

The adult insects attach themselves to the bee's hairs, usually on or near the head and thorax, and apparently feed on nectar and pollen at the bee's mouth. Bee lice are often referred to as harmless, but if there are a large number they certainly incommode their host. A queen is likely to collect more bee lice than a worker (perhaps simply because she lives much longer), and her egg laying may be reduced. As many as thirty have been found on one queen; they can be made to leave her by blowing tobacco smoke briefly on them.

The female bee louse lays eggs on the inner capping-surface (or on the wall) of full honey cells. The developing insects remain within the honey cells, and may disfigure the comb by making tunnels through the cell walls.

11.45 Some parasitic insects of local importance

Parasitization of drone brood of *Apis mellifera* by *Melittobia acasta* (Chalcidoidea) is reported from Poland (Jeliński & Wójtowski, 1984).

Adult *A. mellifera* may be parasitized by the larval stage of a viviparous fly, *Senotainia tricuspis* (Sarcophagidae). Usually during late summer, the female deposits a newly hatched larva on the back of a bee entering the hive after a flight. The larva's mandibles pierce the integument and enable it to enter the bee's blood where it remains; the bee is then said to suffer from apimyiasis. When the bee dies, the larva eats the soft parts of the body and in due course pupates. It may emerge as an adult within about two weeks, or it may not do so until the next summer. The presence of the fly larvae is easily detected, but bees in affected colonies may die from other, less visible causes, so damage caused by *Senotainia* is not necessarily as great as observations might suggest. Control methods are discussed by Smirnov and Luganskiĭ (1987).

Adult *A. mellifera* is parasitized by larvae of some other viviparous flies, in the families Conopidae and Tachinidae; examples are *Physocephala fascipennis* and *Rondaniooestrus apivorus*, respectively, both found in South Africa (Anderson et al., 1983). Such an infestation is also referred to as apimyiasis; when other insects are the hosts it is myiasis. Knutson (see Morse, 1978) and Bailey (1981) give further details, but I have no information on the effects of these flies on species of *Apis* other than *mellifera*.

Other viviparous parasitic insects include species of *Stylops* (Strepsiptera). *Stylops* females are wingless, and live within the host's abdominal cavity. The males have wings, and during their life of only a few hours they find and mate with the almost enclosed females. A newly born larva emerges and clings to the body of the host honeybee until it reaches a honeybee larva and can complete its life cycle in it, feeding on blood inside its body; the bee develops to maturity but may be deformed. There have been reports of honeybees as hosts of *Stylops mellittae*, and also of *S. aterrimus* (Caron, 1978*a*). *Stylops* has been found parasitizing *Apis cerana* in both hills and plain regions of N. India (Adlakha & Sharma, 1976), where heavily infested colonies are likely to die.

At least 8 species of oil or blister beetles parasitize the honeybee during their first larval stage. After hatching from eggs laid in the soil, the larvae – known as triungulins – climb on to open flowers. When opportunity occurs, a larva attaches itself to the body of a foraging bee and is thus carried into a hive, where one of two behaviour patterns is followed. The triungulin of *Meloë variegatus* and some other species perforates the bee's abdominal integument and feeds on the blood, attacking another live bee when its first host dies. In *M. cicatricosus* and some further species, the triungulin does not attack its carrier-bee, but finds and eats an egg in the hive and develops inside the brood cell, possibly even being fed by nurse bees. Beljavsky (1933) stated that *M. proscarabaeus* was the most widespread species in the USSR, with *M. variegatus* in central and southern regions, as well as in Europe and North America. He also listed *M. cavensis* and *M. hungarus* in Armenian SSR, and Caron (1978*a*) cites *M. cicatricosus*, *M. faveolatus*, *M. tuccius* and *M. violaceus* from various European authors.

11.5 PREDATION BY INSECTS, MAMMALS AND BIRDS

Predators normally use their prey as food, for themselves or their offspring. The strategy of the potential prey must be to evade contact with the predator, or to make it withdraw from attack by inflicting pain or by disabling it (as with bee stings). In a natural environment a balance is maintained between populations of bee predators and of the bees that are their prey. If a beekeeper moves hives into a predator's territory, they provide a new food supply which it will exploit.

Readers whose bees have not been subject to attacks by any of the predators considered here may find it difficult to comprehend their effect. In the most severe circumstances, the enemy in question is likely to be the major constraint to profitable beekeeping, or even to beekeeping at all. Attacks can be more immediately devastating than almost any disease or parasitization of bees. Damage to an apiary may be widespread within a few minutes, for instance by bears, honey badgers, or woodpeckers – or there may be relentless attacks lasting days or weeks, as by certain wasps (e.g. *Philanthus*) or birds (e.g. bee-eaters, *Merops*).

With infectious diseases, and parasites that live entirely within the honeybee colony (Sections 11.2–11.4), a dominant factor in increasing incidence of attack is transmission between colonies as a result of modern beekeeping operations. With predators, this factor does not arise, attack being initiated by the presence of hives of bees and of the predator in the same locality. In many cases killing the predatory animals (including birds) is irrelevant, because so long as the bee prey is present, other predator individuals are likely to move in to exploit it. If the predator's area is localized, as during nesting, the most useful action is to move the hives away while the danger lasts.

11.51 Ants

In my travels in the tropics during the past thirty years I have found ants to be one of the most widespread sources of trouble for beekeepers.

Only a few ant species kill a colony immediately, and these species kill even larger animals that cannot get away. They are carnivorous poneroid ants in the subfamily Dorylinae 'which represent the savage or hunting stage' in evolution (Richards & Davies, 1977). Names given to them exemplify their behaviour; species recorded as killing honeybee colonies include army, driver, safari, or legionary ants – especially *Eciton burchelli* in South America and African driver ants *Anomma* and *Dorylus*. An 'army' of army ants may contain up to 700 000 individuals, and the migrating column raids and kills along its path. Among myrmecioid ants (subfamily Dolichodermae) is *Iridomyrmex humilis*, the Argentine ant. *Hypotrigona* bees are better than honeybees at defending themselves against these ants (Darchen, 1986).

A great many species of ants harass honeybees, and may eat or carry off any comb contents – honey, pollen and brood (F. G. Smith, 1953). This harassment is a contributory cause of 'aggressiveness' in tropical *Apis mellifera*, and also of absconding by colonies, of tropical *A. mellifera* and *A. cerana* (Section 7.41.4), which is one of the major constraints to beekeeping with these bees. Much less is known about ants in relation to *A. dorsata* and *A. florea*.

In south-east Asia, Toumanoff (1939) mentions especially *Monomorium pharaonis*, *Oecophylla smaragdina*, *Cremastogaster dohrui*, *Pheidole megacephala*. He lists altogether 84 ant species in the Camponotini, Dorylini, Myrmicini and Ponerini, identified in the part of Asia that now includes Laos and Vietnam. I learned in Thailand that *O. smaragdina* harasses *Apis cerana*, and that it also attacks *A. florea* nests: the first ants to reach the protective bands of propolis (end of Section 3.31.4) remain stuck to it, and later arrivals can then climb over them into the nest. In South Africa, Anderson et al. (1983) include *Anoplolepis custodiens* among ants that can cause honeybee colonies to abscond.

Preventive measures against ants start with clearing vegetation from the apiary site; a persistent insecticide such as dieldrin can be sprayed with a watering can to a strip of ground 1 m wide round the perimeter of the site, but not near the hives. An ant-killing insecticide dust may be applied carefully round individual hives, at least 20–30 cm from them. Chlordane (or, if that is not available, diazinon) can be used for controlling Argentine ants in the apiary, after the ground has been cleared of vegetation. It should be applied as a dust or wettable powder, when the bees are not flying (De Jong, 1978). To destroy ant nests, carbon disulphide is poured down the entrance holes.

An ant-proof hive stand is essential during the season when local ants are most active. The stand for each hive, or a shelf for several hives, is mounted on legs, each standing in a shallow container of some sort, filled with discarded diesel or other oil; see Figures 7.51a, 7.51b. In many Thai apiaries in 1986, I found precast concrete pedestal bases being used for individual hives; the base incorporated a circular channel filled with oil. Painting legs of hive stands with a 4 : 1 mixture of engine oil and rubber latex is said to give protection for 3 weeks. Any of the above precautions becomes useless if vegetation is allowed to grow so that it provides the ants with a bridge across the oil.

Alternatively, hives are suspended individually from trees or between posts, or are mounted together on a suspended shelf; in Africa this can afford protection against many other enemies too.

In temperate zones, a few ants may frequently be found beneath hive roofs, especially when sugar syrup is provided there. They can generally, although not always, be ignored. In the USA, Tack Trap is recommended (Bees and Honey, 1980); it is smeared inside small plastic cups which are mounted (inverted) on nails driven into hive stands.

In parts of Europe which have honeydew flows from conifers (Section 12.22.2), certain species of ants found on the trees are beneficial to beekeepers: by helping to protect the honeydew-producing insects from attack by predators, and by stimulating the insects to produce honeydew. Examples are *Camponotus herculeanus* and *C. ligniperda*, *Lasius niger*, and wood ants such as *Formica rufa* and *F. polyctena*; these and other species are discussed by Gleim (1985) and Kloft and Kunkel (1985).

11.52 Vespid wasps

These social wasps, especially *Vespa*, are among the most widespread of honeybee predators. They attack bees at the hive entrance and also well away from it and – when the colony is sufficiently weakened – inside it. They carry the prey (the thorax of the adult bee) to their colony to feed to their own larvae. De Jong (1978) gives many details about individual species, with records from countries in all continents. On a world basis, species causing most damage are probably:

Vespa mandarinia (giant hornet) – Middle East and Asia
Vespa orientalis (oriental hornet) – most of Asia
Vespa crabro (hornet) in temperate zones
Vespa mongolica in Japan.

In tropical Asia other troublesome species of *Vespa* include *affinis*, *analis*, *auraria*, *basalis*, *cincta*, *ducalis*, *magnifica*, *tropica*, *veluntina*.

Vespa mandarinia is a very large and powerful hornet, with a body length of 3–4 cm. It attacks en masse, especially colonies of honeybees – but also those of other hornets – hunting and killing the adults and finally taking over and guarding the hive; both immature and adult bees are then transported to feed larvae in the hornet's own nest. *Apis cerana* bees counterattack with considerable effect, clustering on individual hornets and heating them to a lethal temperature (Section 3.32). On the other hand European *Apis mellifera*, which has been introduced into this hornet's territory, cannot do this, and is likely to succumb (Matsuura, 1985); a colony may be killed within 2 hours. In Israel *Vespa orientalis* killed 3000 *Apis mellifera* colonies in one year.

Vespula is a temperate-zone genus of wasps, known as yellow jackets in the USA. Although smaller than *Vespa*, *Vespula germanica*, *Vespula vulgaris* and other species can nevertheless do severe damage to bees in countries to which they are introduced. In one year *V. germanica* killed 3900 colonies in New Zealand. Rye (1986), who assessed effects of its introduction to Australia, recommends control measures; these include a glass baffle across the hive front, the entrance being central and reduced to a height of 8 mm. For mild attacks, traps at hive entrances are often useful, and can be as simple as a jam jar containing dregs of jam, beer, etc., with a paper funnel leading into it. Wasps captured in the jar seem to increase its attractiveness to others.

Control is essential where there is severe predation, which is common in many warm or hot regions. In many tropical countries the most effective method is considered to be continuous daily hand-killing of individuals during their active season. In December 1978 in Egypt, I met a little boy in an apiary near Assyut who was employed full-time to swish a long besom of palm stalks to kill all hornets (*Vespa orientalis*). A more sophisticated method, also in common use, is to set out a bait of cheap protein food impregnated with an insecticide, so that foraging hornets carry poisoned food back to their nest where it kills the individuals in it. Traps and screens of various types are sometimes placed across hive entrances. Wasps' nests are destroyed if they can be found, and Rye (1986) lists suitable insecticides. Biological methods are also being tried, for instance by Ahmad et al. (1985) in Pakistan. The mite *Sitotroga cerealella* – found in a hornets' nest – was reared on larvae of the greater wax moth. Hornets were captured with a net and released to return to their nest after mites had been transferred to them with a brush. Between 40% and 100% of hornet brood was infested with mites, and no adults were found in any nest after 3 months. The mite *Pyemotes ventricosus* also attacks hornets, but was not an effective control agent.

11.53 Sphecid wasps

The Sphecidae do not live in colonies, but individual females build their nests gregariously (close together), and nesting areas are likely to develop close to the wasp's specific prey insect. Of the 135 species of *Philanthus* wasps (bee wolves) that prey on different bee species, *P. triangulum* preys only on *Apis mellifera*. As far as is known, other species are much less important to beekeepers; six are known to attack *A. mellifera* in Africa and North America (De Jong, 1978), and *P. remakrishnae* preys on hill *A. cerana* in India (Singh, 1962). Evans and O'Neill (1988) have written a monograph on North American bee wolves.

P. triangulum wasps nest gregariously in sandy soil, and in certain circumstances they can make beekeeping impossible. Simonthomas and Simonthomas (1980) gave a vivid account of the sequence of events in the Dakhla oasis in Egypt. This oasis was used for a bee breeding programme (Section 8.23), to replace the native Egyptian bees with Carniolans, in order to increase honey production. But these are less aggressive, and consequently less successful at defending themselves against the wasps. After the change, populations of the wasps near to apiaries increased greatly, and the wasps not only caught bees on flowers – which is normal – but also attacked them close to their hives when there was little foraging, which is less usual. Numbers of wasps counted near hives during 5 or 6 winter months were:

1975–76	14 000	
1976–77	24 000	(premium paid for wasps captured)
1977–78	17 000	(premium doubled, but winter temperatures often too low for wasps to fly)

Capturing wasps, and attempting to destroy their nesting areas, were both ineffective in controlling them. After a detailed investigation, the method recommended was based on changing apiary sites annually. All hives on the oasis are first to be concentrated on three sites. The soil is sandy, and suitable for nesting, throughout the oasis, so the wasps (which prey exclusively on honeybees) will nest near the three sites. Next year, all hives are moved to three new sites, as far away as possible from the first three. If necessary yet other sites are used in the third year.

In parts of Southern Africa *Philanthus triangulum diadema*, known as the yellow bee pirate, mainly catches bees on flowers. The banded bee pirate, *Palarus latifrons* is regarded as the most destructive of all honeybee predators there (Anderson et al., 1983). *Palarus orientalis* Kohl is a predator in India (F. G. Smith, 1960), and other species of *Palarus* also prey on honeybees, doing much damage. *Palarus latifrons* nests gregariously in sandy soil. A female usually catches a returning forager near the hive entrance; she stings and thus paralyses it, then either takes it to her nest or squeezes the nectar out of the honey sac, imbibes this and probably body fluid, and leaves the empty bee on the ground. Anderson et al. (1983) describe various traps for use in front of hive entrances, but conclude with the advice quoted above for *Philanthus* control: in extreme cases all hives must be moved out of the area, to be re-introduced when the *Palarus* population has died out. This method is effective only for predators that prey exclusively on honeybees.

Predation by insects on honeybees is not confined to Hymenoptera. For instance some of the 4000 species of robber flies (Asilidae) prey on honeybees, although none exclusively; Toumanoff (1939) gives a fairly comprehensive account of them. Most reports are from temperate zones, especially from Yugoslavia and Argentina. There seem to be no methods of control that are both effective and economic, but in general the flies do not seem to be important pests of honeybees. They take many pest insects, and their usefulness may well outweigh any damage done to bees.

Although not predation, it may be mentioned here that other species of bees may rob honey from honeybee colonies, and one honeybee species may rob from another. *Apis mellifera* and *A. cerana* may rob colonies of the other species; *A. dorsata* may rob colonies of *A. cerana*, but apparently not those of *A. mellifera* (De Jong, 1978).

11.54 Mammals

Whereas many insect predators of honeybees catch adult bees outside the hive, mammals that prey on bees use their strong claws and forelegs to break into a hive or nest, so that they can eat its contents. The best known and most widely distributed are the bears (Ursidae), which range from the tropics to beyond the northern extremity of *Apis mellifera*. At least 5 of the 10 species can be a source of costly trouble to beekeepers in some parts of Asia, Europe, or North America. Caron (1978*c*) cites reports of damage in countries listed below:

- black bear (*Ursus americanus*) in North America: Canada, USA (Florida, California, Pennsylvania)
- brown bear (*Ursus arctos*) in Europe and Asia: various parts including USSR, India, Japan; not N. America
- sloth, honey, or Indian bear (*Melursus ursinus*) in Asia: India, Sri Lanka
- Asiatic black bear (*Selenarctos thibetanus*) in Asia: Afghanistan to USSR (Siberia)
- Malayan sun bear (*Helarctos malayanus*) in Asia: Burma to peninsular Malaysia, Indonesia.

Bears hunted bees' nests in the wild long before hives were used, climbing high trees to do so. A single-comb nest of *A. dorsata*, which is built down from a strong branch, is knocked to the ground. Nests in cavities (of *A. cerana* and *A. mellifera*) present more difficulty, and bears will suffer many stings in order to get combs of honey from them (Crane, 1975*e*). Their liking for sweet food is often quoted, and they are one of the few wild animals susceptible to dental caries; they also eat adult bees and brood along with the rest of the comb contents. In Spain and Turkey, one can still find traditional apiaries with elaborate defences against bears.

Bears tear modern hives to pieces, and may carry supers of honey off into the woods, in their efforts to escape mass stinging. Many estimates of the cost of bear damage are available from North America; Caron (1978*c*) quotes up to US$200 000 in 1973 in Alberta, Canada, 'which would have been about $500 000 without a trapping/shooting programme'. Such a control measure may lower the average age of bears, but does not seem to reduce their number. Trapping individual rogue or pest bears (to be shot or permanently relocated) may, however, be effective. The only reliable protection for colonies is a strong electric fence (Figure 8.23c), and Caron (1978*c*) describes several types and variations. Alternatively, hives may be sited in bee houses, or on high platforms, or on pallets, to which each individual hive is attached with steel strapping (McCutcheon, 1988). A site in the open near a stream, or in any other place likely to be frequented by bears, should be especially avoided.

In Africa there are no bears, but beekeepers much fear the smaller honey badger or ratel (*Mellivora capensis*, Mustelidae), which is fairly widespread south of the Sahara; it is now mostly nocturnal (F. G. Smith, 1960). The animal has very powerful front legs and claws, and does great damage to hives, usually breaking them up and carrying combs away before eating them. It is very difficult to catch or to control. Ratels feed on honeybee nests in the seasons when they contain most bees and honey (Figure 11.54a). Kingdon (1977) gives a detailed account of their life, including the methods they use for locating honeybee nests and hives, stupefying the bees with a secretion from the anal glands, and then getting at the contents and consuming or carrying them off. A cattle herder is quoted as saying that he saw a ratel 'put its anus to the nest cavity, swirl its tail and then rub its anus all around. The bees were all chased away, and then by biting and clawing the tree the ratel was able to scoop out the comb and eat the honey. We saw later that many of the bees had died.'

The nocturnal Indian ratel or honey badger (*Mellivora indica*) is sometimes referred to as being *Martes* in

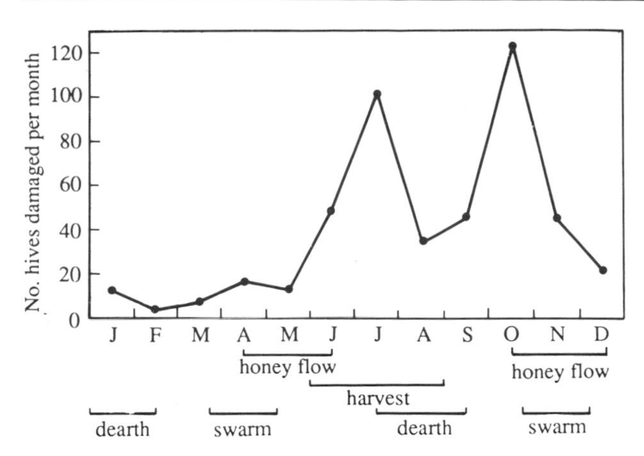

Figure 11.54a Number of hives reported damaged by ratels in central and western Tanzania (1967–1968), in relation to honey-flow and colony conditions (Kingdon, 1977).

India. Species of marten attack hives in different parts of Asia. The Himalayan yellow-throated marten (*Martes flavigula*) is in Nepal, and its range extends from south China to Malaysia; in Manchuria it is known as the honey dog. In Java the subspecies *robinsoni* is well documented as attacking hives and eating the contents (Wegner, 1949). The stone marten *Martes foina* is widespread in Europe and USSR, and certainly attacks hives in northern Spain. Toumanoff (1939) cites other Mustelidae.

Skunks (*Mephitis* and *Spilogale*) are another group of Mustelidae, common in North America, frequently described as bee enemies. Trapping, together with fencing apiaries or placing hives on platforms, seems to be effective (Caron, 1978*b*). In the same family, Wiese (1985) lists the *tayra* (*Eira barbarta*) among several mammals that attack honeybee colonies in Brazil; he also includes a marsupial, *gamba* (*Didelphis* sp.).

The kinkajou (*Potos caudivolvolus*, Procyonidae), a nocturnal racoon-like animal, raids bees' nests in the Central American rain forest. In Brazil an armadillo, possibly *Dasypus novemcinctus* (Dasypodidae), will push a hive over, after scuffling earth into the entrance to block it (Nixon, 1985).

Many primates eat honey and brood from bees' nests. Rhesus monkeys (*Macaca mulatta*) have been seen taking brood from an *Apis florea* nest. Gautier (1976) discusses baboons, gorillas and chimpanzees. Chimpanzees have been observed getting honey combs from *Apis mellifera* nests with their hands, and also using a twig to extract honey (Van Lawick-Goodall, 1968). Many species handle hives in a destructive way in order to get at the combs in them. Baboons may break up hives or tip them over repeatedly until they come apart, and then carry the combs away for consumption, out of reach of most of the bees. A beekeeper in India told me that black-faced monkeys – probably common langurs, *Presbytis entellus* (Bertram, 1986) – watched him working his hives and then copied him, opening hives and removing the frames, which they carried up a tree, and threw to the ground to knock the bees off; they then ran down and ate the honey. This behaviour was initiated only if a beekeeper left pieces of comb about, so that the monkeys learned that they were good to eat.

Other mammals such as mice seek the sheltered environment of a hive for the winter, and damage combs in making their nest; they eat pollen, honey and bees. Their entry can be prevented by placing across the entrance a strip of perforated metal with holes 9 mm in diameter. Toumanoff (1939) and Caron (1978*b*) give details of these and some other mammalian enemies of honeybees and of control methods used for them. But I do not know of any systematic study of tropical species.

11.55 Birds

Some birds interfere with beekeeping because, like most flying insects that prey on honeybees, they catch bees on the wing or near the hive entrance. Other birds attack the hive (or, in the wild, the nest) to eat its contents, but since birds lack the strong legs and claws of successful mammalian bee predators, they are dependent on some other specialization that helps them.

In addition to specialist predators on bees, there are many insectivorous bird species that sometimes eat honeybees, and do not seem to be greatly affected by their stings. Bee-eaters (Meropidae) in the Old World, jacamars (Galbulidae) in tropical America, and some shrikes (Laniidae) in many regions, take action to prevent a bee or wasp they have caught from stinging them, by rubbing its abdomen against a perch to make the bee eject its venom (Fry, 1984). One bird reported to have morphological adaptation against stings is the honey-buzzard *Pernis apivorus* (Accipitridae). Its face is covered with small scale-like feathers, forming an armour that is apparently impregnable to bee or wasp stings. Although the bird is known as the honey buzzard, it is little recorded as attacking honeybee nests, and feeds mainly on wasp and bumble bee nests; perhaps it is unable to broach nests in a tree or hive. Libert and Rotthier (1987) have described *P. apivorus* in Belgium; the oriental paraspecies *Pernis ptilorhynchus* almost certainly has a similar adaptation.

The best known bird predators of honeybees are the bee-eaters (*Merops*), and the 21 species are the subject of a monograph (Fry, 1984). These birds are sentinel feeders, using an exposed perch to keep watch for insects flying past. In spite of their name, they prey on a wide variety of insects, ranging in length from under 1 mm to 80 mm. Fry (1983) has published a list of over 300 insect prey species of the European bee-eater, *Merops apiaster*, which is native in Europe, Asia and Africa, and his Table 2 summarizes prey records for 28 500 insects examined. In different countries, from 4% up to 52% (in Spain) were honeybees. Other bee-eaters especially regarded as pests of honeybees are the blue-cheeked (*M. persicus*) and the little green (*M. orientalis*) in Africa and Asia, and the blue-tailed (*M. superciliosus*) in Asia, together with the rainbowbird (*M. ornatus*) in Australia, whose predation on honeybees is discussed by Goebel (1984). Fry's Table 1 gives information on insects found in most species, 47 000 insects in all. The number of honeybees in one bird varies from 0 to very high percentages.

Bee-eaters make large-scale attacks on honeybees in two different situations. In one, an apiary happens to be in the feeding territory of a nesting colony, which is usually situated in an earthen cliff, often with water nearby. There is no evidence to suggest that bee-eaters *choose* a nesting site that is near a supply of bees, but if hives are present, the birds will prey on the bees. As with the bee wolf *Philanthus*, killing individual bee-eaters is unlikely to reduce predation effectively. The best course is to avoid siting apiaries near a bee-eater nesting site, at any rate for the duration of the nesting season, although where water is scarce it may not be easy to do this. In Thailand, Sawang Piyapidrat told me that he had cleared an area of bee-eaters by letting red-thighed falconets (*Microhierax caerulescens*) fly there.

The other situation occurs when a flock of bee-eaters on migration happens to find an apiary. Migrating flocks, which contain more birds in autumn – after breeding – than in spring, are likely to concentrate their feeding activity for some days on any apiary they find. *M. apiaster* and *M. ornatus* reach the highest latitudes, and are the most migratory species; they can be especially troublesome because of this, and their attacks are difficult to prevent. The bees' response, as to attacks by *Philanthus*, is to remain in the hive. Fry (1983) cites one observation in which the number of bee flights dropped to 4% when an attack occurred. Presumably, losses of bees in the wild are minimized by the fact that (with the exception of *Apis dorsata*) any one honeybee nest is usually at some distance from the next; there is never a concentration of fifty or more together, as in an apiary, to attract large numbers of bee-eaters. *A. cerana* is sometimes able to escape attack because it can turn quickly, and fly into vegetation.

Bee-eaters prey on many insect species that themselves prey on honeybees. These include *Philanthus*, *Vespa crabro* and *V. orientalis*, sphecoid wasps, and robber flies. Fry (1983) adduces evidence that bee-eaters as a whole eat at least one honeybee predator insect for every 16 honeybees, so that in areas where such predators are active, the presence of bee-eaters may actually reduce the total predation on honeybees.

Swifts (Apodidae) also take insects in flight; most forage in the open, not in the shade of trees, and thus may take drones and queens. Losses of workers should be fewer where hives can be kept in wooded areas. Larger species of swifts tend to fly nearer the ground, some, whose weight is more than 80 g, only a metre above it (Brooke, 1986). In the Philippines, the Philippine spine-tailed swift (*Hirundapus celebensis* = *Chaetura gigantea dubia*) is reported as a major constraint to beekeeping; see Morse and Laigo (1969*b*). Attempts are made to capture the birds as they feed around an apiary, by people stationed there who wield fish nets on long poles. About 450 birds were captured in this way in Batangas in 9 weeks in 1968. The numbers of honeybees found in 14 birds, caught on different dates, were identified as follows:

Apis mellifera	0–162	mean 54
Apis cerana	0–234	mean 33
Apis dorsata	0 in 12 birds, 5 in 2	

The same or a related bird occurs in Malaysia and some islands in the region, but I do not know to what extent it preys on honeybees.

Still other birds are noted for catching bees in flight, and some can be troublesome to beekeepers in certain areas. They include some kingbirds (Tyrannidae) in the Americas, and flycatchers (Muscicapidae) and drongos (Dicruridae) in the Old World. Some of the shrikes (Laniidae) prey on honeybees upon occasion.

Birds in other families attack honeybee nests in hives or natural cavities. Woodpeckers (Picidae) have a strong, sharp-pointed bill for excavating insect holes in trees, and a very long sticky tongue for extracting their prey. In parts of Europe, especially, the green woodpecker (*Picus viridus*) can cause considerable economic damage by drilling holes through hive walls, to reach the bees inside. The damage is done during a very cold spell in winter when the bird's other food sources are not available, and in southern England I have seen piles of hive boxes that a beekeeper had discarded because they were so badly damaged. Hives can be protected to a certain extent by a 'skirt' of thick polythene sheeting, or by wire netting, but action must be taken before the cold spell starts. Without such protection, a green woodpecker can pierce through a hive wall in less than half a minute.

The most intriguing birds that eat the contents of

honeybee nests are the honeyguides, small brood-parasitic birds related to the woodpeckers. They include 11 species of *Indicator* and 6 in three other genera, all native to Africa (Short & Horne, 1985), and most information is about the greater and the lesser honeyguides (*I. indicator*, *I. minor*). Honeyguides have the most unusual capability of digesting beeswax – by means of a micrococcus and a yeast in the gut (Friedmann, 1955; Friedmann & Kern, 1956). Culturing the microccus and transferring it to chicks enabled them to digest beeswax.

Some of the honeyguides take advantage of an animal's actions that make a bees' nest accessible to them; in the past, the animal was likely to be either a human being or a ratel, but ratels are now mainly nocturnal (Short, 1986). Only the greater honeyguide, *Indicator indicator*, 'guides' the animal to the nest. The guiding behaviour is described by Friedmann (1955), Short (1986) and – in detail – by Isack and Reyer (1989). Having found a honeybee nest, the bird seeks a (larger) mammal, then accompanies it, fluttering and keeping close to it, perching and calling, 'guiding' it. This behaviour continues until the bird is stopped by a second stimulus: the sight or sound of bees, which usually occurs near a nest. The mammal is likely to find the nest and to open it up for its own consumption. The bird can then feed, taking adult bees and comb containing brood, pollen and honey.

Because the honeyguide needs a stronger partner to enable it to get access to honeybee nests, and the partner gains thereby, the bird tends to be regarded as a helpmate to the honey hunter rather than as an enemy of bees. Many tales and legends were – and still are – told about the partnership. For instance, unless the honey hunter gave the guiding bird its proper share of food, on the next occasion he would be guided to a fierce wild animal instead of to a source of honey.

In south-east Asia is *I. archipelagicus*, and in the Himalayas and other mountains *I. xanthonotus*. Neither is known to show 'guiding' behaviour, but there is a most unusual behavioural link between the latter and one *Apis* species, which was recorded by Chang Hua who lived between AD 200 and 300 (Friedmann, 1955). The male bird establishes and defends a territory that includes one or more occupied nests of *Apis dorsata*, which provide it with a continual source of beeswax. In a forested canyon in east Nepal, where the bees nested on a sheer rock cliff, Cronin and Sherman (1976) found that the birds ate the comb where it was attached to the cliff face, apparently with little or no interaction between birds and bees. During the birds' nonbreeding season, a male would allow his female mate(s), and their progeny, to eat wax from his *A. dorsata* nests. Hussain and Ali (1983) give other details.

11.56 Other animals

Most animals that attack bees or their nests are in the families already mentioned. Of the others, some toads (Bufonidae) prey on bees, and frogs (Ranidae) are troublesome to a lesser extent in a few areas.

Spiders (Arachnidae) prey almost exclusively on insects, but are usually no more than a minor nuisance to honeybees. Bees can be found in webs of social spiders. In parts of the Kalahari desert, webs (probably of *Stegodyphus* sp.) are up to 5 m across, and catch many bees at dusk (Clauss, 1986). The orb weavers (Argiopidae) and crab spiders (Thomisidae) have also been mentioned. I was told in Thailand that the golden spider (*Nephila cravipes*) and *N. maculata* build a web across a hive entrance, which can catch a queen going out to mate.

11.6 INFESTATION OF HONEYBEE COMB AND ITS CONTENTS

The sheltered environment of a hive, or a cavity occupied by a wild-nesting colony of honeybees, may be utilized by a variety of insects and other organisms, especially in winter. Banaszak (1980) lists 150 species found in hives in different parts of Poland, of which only 3 were parasites and 17 others were harmful. There are sometimes very large numbers of individual harmless mites; Banaszak found over 10 000 in one hive.

Wax moths (Section 11.61) are potentially very troublesome to beekeepers all over the world. No other hive inhabitants have a similar importance except in certain restricted localities, and only a few need be mentioned (11.62).

11.61 Wax moths

One or both of the wax moths (greater, *Galleria mellonella*; lesser, *Achroia grisella*) occur naturally or have been introduced by man in almost all regions of the world where bees are kept, and the larvae feed on wax comb of all honeybee species. In 1985 a lesser wax moth *Achroia innotata obscurevittella* was found in wild nests of *Apis cerana* in Japan (Okada, 1986, 1988).

In a natural environment with only wild-nesting colonies, or where only traditional fixed-comb hives are present, wax moths may serve a useful purpose in destroying combs of dead or dying colonies that could otherwise become foci of infection. But throughout much of the world today they are an ever-present cause of economic loss to beekeepers who have capital invested in framed combs built on comb foundation. The combs are vulnerable to wax moths when in store, as well as in parts of hives not occupied by bees. Williams (1978) estimated the annual loss in the USA as $500 000, and the damage is likely to be relatively more in hotter parts of the world. Nevertheless in temperate zones storage of framed combs from one active season to the next, away from the hive, is usually cost-effective. In the tropics such stored combs are likely to be destroyed by wax moths. Supers of empty combs are sometimes left above the brood box for storage, but they are safe from wax moths only where the colony can be kept strong enough *throughout the year*.

Greater wax moths emerge as adults in the hive, and the female may live a few days or a few weeks. After dark she flies out to a tree and mates. She then re-enters a hive, and in due course lays about 500 eggs, inserting each into a crevice in the hive from which the bees cannot remove it. Unusually for moths, the newly hatched larvae can run very fast, and thus distribute themselves around the hive; they then burrow into a comb, damaging it in a characteristic way by constructing a feeding tunnel of silk through it. The tunnel is enlarged as the larva grows, and lengthened to 15 cm or more (Burges, 1978); a mature larva is up to 28 mm long. About 250 can develop on one Langstroth brood comb, and 10 000 cocoons have been found in a two-storey hive. Additionally, eggs may be laid around apiary buildings and wherever there is comb debris. Damage may be done to combs containing honey, to freshly extracted combs, and to stored combs.

Only the larvae feed. In the hive most of them eat beeswax, but they cannot complete their development without other food, which normally includes pollen and honey stored in the hive. Adults require neither food nor water.

The life history of the lesser wax moth is fairly similar. Adult females may live for about 7 days and lay 250–300 eggs. Larvae are up to 20 mm long, but adults weigh only 10–17% as much as those of the greater wax moth.

In both species, different stages of the life cycle are shortened if the temperature is higher. The adult life span of female *Galleria* has been recorded as 19.6 days at 20°, up to 7 days at 30°–32°, and 3.8 days at 40°. The total life cycle may be completed in 4 weeks, or it may last 6 months. With a long development period, only one generation can be reared during the bees' active season, and dormancy occurs in the prepupal stage.

The two species may co-exist; *Galleria* is then likely to do the greater damage, and if it becomes too numerous, *Achroia* will probably disappear. *Galleria* is present throughout the tropics and subtropics and well into the temperate zones, but only *Achroia* can survive in hives through the cold winters at higher latitudes. In South America greater wax moths are attacked more vigorously by colonies of Africanized bees than by those of European bees. Both bees attack female moths in preference to males (Eischen et al., 1986).

No method for killing wax moths can take the place of management that prevents infestation – in the hive, the apiary, and comb storage. The most important actions in the apiary are to ensure that all colonies are strong and healthy, to replace floor boards each year and sterilize those removed, and to pay special attention to preventing infestation of any hives containing colonies that are weak or have just died.

Several wax moth control measures are available for combs and other equipment not occupied by bees. Most physical methods involve exposure to heat or cold. The shortest times found for killing *Galleria* larvae by cold, quoted by Burges (1978), are:

−17°	1.5 hours	0°	4 hours
−18°/−15°	2 hours	+2°	6 days
−12°	3 hours	+5°	10 days
−7°	4.5 hours	+10°	15 days

At 15°, all large larvae survived 8 weeks, but they did not grow. In Finland where winters are very cold, I found all boxes of empty combs in an apiary closely stacked together in the open for the winter, covered tightly with a plastic sheet; the temperatures there are low enough to kill all moths.

With heat treatment, all stages of *Galleria* were killed by 24 h at 50% RH and 60° (Burges, 1978), but this is likely to damage honey, and beeswax combs. A small-scale heat method for *Achroia* infestation of hives and combs containing no bees or brood has been described in India (Naim & Bisht, 1972).

Another physical control method, effective in some conditions, is to store combs in full light, for instance spaced 2 to 3 cm apart on racks in an outbuilding, unenclosed so that there is ample ventilation. Popolizio and Pailhe (1973) in Argentina described and illustrated a store built for this purpose. It was 3.2 × 3.2 × 2.5 m high, and consisted of a roof mounted on posts, fitted with four tiers of racks. In 1986 I saw a somewhat similar arrangement at 45°N in France, where it was effective at altitudes up to 800 m, above which wax moths present no problem. There was no wax moth damage in ten years, except in two with a mild winter and spring (Ratia, 1986). Combs may be new or old, but must not contain pollen. This method does not kill wax moths, but females do not lay eggs (or else the eggs do not hatch) unless the light intensity is low enough and the humidity high enough – perhaps corresponding to requirements of immature stages of the moths (Borchert, 1974). Ratia also reports that neither wax moth attacks combs in hives with transparent walls. In Melaka State, Malaysia, I was told that hives without a floor board were free from wax moths.

A more advanced physical method is irradiation with gamma rays. This can be used to kill the wax moths, or possibly to sterilize males so that mated females produce no offspring, or to induce a genetic defect which would – for instance – greatly reduce egg viability. Operational costs are high, but the end of Section 11.1 refers to one satisfactory trial.

The following chemical methods can be used on empty combs only. Contact insecticides are not effective since they do not penetrate into crevices, or throughout the comb. Any chemical treatment must therefore be by fumigation (see Section 11.1). The use of ethylene oxide, methyl bromide and carbon dioxide are large-scale operations discussed by Burges (1978) and Williams (1978). Paradichlorbenzene is one of the least hazardous fumigants to handle, crystals being placed in a stack of boxes containing combs, which is then enclosed against entry by further wax moths; the vapour is produced by slow evaporation. But the chemical does not kill wax moth eggs, and it may be absorbed by honey. In Australia phostoxin has been registered for use (Australasian Beekeeper, 1986), but it is highly toxic and must never be used inside inhabited buildings. The operator must wear a safety mask.

A biological material used for *Galleria* control is the microbial insecticide *Bacillus thuringiensis* (Burges, 1978; Cantwell & Shieh, 1981), and various commercial preparations of the useful strains are available. A suspension in water is sprayed on to empty combs according to the directions, and repeated each season. *B. thuringiensis* has also been used to impregnate comb foundation, but in the hive environment its effectiveness deteriorates. Residues of the material in honey and beeswax have been declared exempt from any requirement of tolerance in the USA Laws and Statutes (1980).

Other biological methods being investigated include use of a virus, and of juvenile hormone. The virus is multiple embedded nuclear polyhedrosis virus of *Galleria mellonella*, which is sprayed in suspension on empty combs; at present its use would be very expensive. Muslin dipped in juvenile hormone, inserted in test hives, reduced the number of moths developing into adults, but it also affected the bees. Williams (1978) and Matheson (1984*b*) discuss these and other methods under consideration.

Whatever measures are adopted to kill wax moths, continuous vigilance is necessary to prevent reinfestation. In each species, adults communicate by ultrasonic pulses, and males of *Achroia* uses these pulses to guide females directly to them for mating. So the use of ultrasound to manipulate the behaviour of this moth may have potential for controlling it. Spangler (1984) developed a simulator of male sound which attracted females. Such a device could be used to monitor and detect reinfestation of comb stores after they had been treated to kill the wax moths originally present.

11.62 Other organisms

In Southern Africa especially, hive beetles can do much damage to combs and comb contents. The large hive beetle, *Oplostomus fuligineus*, does not breed in hives, but it takes honey. It is slightly more than 20 mm long, and the bees cannot pierce its strong integument to sting it. Reducing the height of the hive entrance to 9 mm, or placing a metal screen over it, can keep the beetle out of hives.

The small hive beetle, *Aethina tumida*, is only 7 mm long; larvae pupate in the soil, and adults cannot be prevented from entering hives. The bees do not seem able to sting them or to remove them. Females lay eggs in any combs not defended by bees, and thus do most damage in hives where colonies are small or weak. Pollen combs are reduced to dust, and if honey is contaminated by the beetles it ferments. No way seems to be known for preventing or curing infestation (F. G. Smith, 1960; Anderson et al., 1983). The best advice is to keep strong colonies, in hives without any cracks or crevices. This beetle can also infest honey combs before extraction, or in store, which must therefore be secured against them. Smith suggests fumigating stacked boxes of empty combs with paradichlorbenzene, then sealing them up with scotch tape.

Other hive beetles listed by Smith are *Diplognatha gagates*, *Coenochilus bicolor* and *Rhizoplatys trituberculatus* in Africa, and *Platybolium alvaerium* and *Dermestes vulpinus* in tropical Asia.

Pollen mites are often referred to in the literature, but none of them is harmful to bees. *Carpoglyphus lactis* is widely distributed in the USSR; it lives in wax and pollen debris on the hive floor – Grobov (1977) found 3500 per gram of debris – also in stored pollen and honey. *Neocypholaelaps indica* in India reaches a hive by attaching itself to a bee foraging on a flower.

Pseudoscorpions (Chelonethi) are sometimes reported to do damage in hives they inhabit, especially the bee scorpion or chelifer (*Ellingsenius fulleri*) in Southern Africa. But they may well do more good than harm, for instance by eating wax moth larvae

(Anderson et al., 1983). They also prey on mites, and their possible use in *Varroa* control has been suggested (Prehn, 1986).

The pollen mould *Bettsia alvei* received much attention in the 1920s and 1930s. Found in the north temperate zone, this fungus lives on pollen stored in comb cells; the pollen dries into hard plugs that are difficult for the bees to remove (Gilliam; see Morse, 1978). It does not attack brood.

11.7 INJURY TO BEES BY INSECTICIDES AND OTHER PESTICIDES

11.71 The background

Ever since pesticides have been applied to crops, bees have been poisoned by some; almost all pesticides that kill bees are insecticides. The early poisons were few, arsenic being one of the most harmful. During the creation of large-scale mechanized operations in agriculture after the Second World War, chemicals were increasingly used for pest control, in ways that killed large numbers of bees. Beekeepers in many countries were faced with a crisis, and some felt that beekeeping was doomed. Education and public relations have achieved much in converting the crisis into a hazard amenable to improvement by co-operation. As a result of persuasion and legislation, producers of pesticides in many countries now label their materials according to the degree of toxicity to bees; in some countries legislators have banned the use of the most toxic materials at critical times for the bees. Growers accept, and often act upon, the concept that since live bees are needed to pollinate their crops, it is counterproductive to kill these bees. On their side, beekeepers have accepted the fact that crop yields are being increased through the operation of pesticide programmes.

Effective co-operation has not been quick or easy to achieve, but it is the key to the long-term well-being of beekeeping in agricultural areas. It can also lead to legislation that usefully protects bees in recognition of their value as crop pollinators, and to indemnity or compensation to beekeepers for damage to their bees by pest control measures.

Hardest hit areas have included the Pacific states of the USA. The number of colonies killed in California was 82 000 in 1962, but was reduced to 36 000 by 1973, i.e. from about 17% to 7% of the total. Many other colonies lost their foraging bees several times during the season. In August 1979, on cane-fruit farms in the Columbia River valley in Oregon, I found that beekeepers expected their field force of bees to be killed at each insecticide application. One application was made on the day before harvesting – although many flowers were still open – because the presence of any insect parts with the harvested fruit disqualified the crop.

Honeybees are still at very high risk in some developing countries, where agricultural programmes often include insecticide applications carried out in ignorance of, or indifference to, the indispensability of bees. In 1978 Egyptian beekeepers told me that they no longer dared to take their bees to the cotton, which used to give their main honey crop. The cotton was sprayed from the air 5 times in all, at 2-week intervals, which is too short a period to allow the forager populations to build up again, because an adult worker is not produced until 3 weeks after an egg is laid.

Crane and Walker (1983) have assessed the impact of pesticides on bees and pollination with special reference to tropical and subtropical regions, and Adey et al. (1986) provide some detailed information on pest control that is safe for bees. The situation in north temperate regions has been described in many publications, including Stevenson et al. (1978) in the UK; Canada National Research Council (1981); Johansen (1979), Atkins et al. (1981) and Johansen and Mayer (1985) in the USA.

Pesticides include insecticides, acaricides, fungicides, and toxicants applied to control various other pests. In what follows, the term 'insecticides' is used to cover also the many acaricides (applied to control mites) that are toxic to bees. The great majority of bee deaths are due to poisoning by insecticides applied to crops *while the crop is in flower*.

The following are some of the other hazards to bees from pest management practices, which can be of local or minor significance. Publications already referred to give information on ways of reducing their effects on bees.

- Killing by insecticides on *non-crop plants in flower*; for instance beekeepers in Crete told me that in 1984 their bees were killed because aircraft, on returning from flights applying insecticide to olive groves, flew over a hillside covered with thyme (*Thymus*) where the bees were foraging, and the applicator failed to switch off the spray.
- Killing by insecticides applied to control *non-crop pests*, for instance mosquitoes or tsetse flies.
- Killing by insecticides applied to *crop plants not in flower*, for instance to cereal crops infested with aphids that produce honeydew on which bees forage; a 1975 French decree prohibits the application of insecticides in these circumstances, except those labelled 'not dangerous to bees' (Louveaux, 1985).

- Killing by agents introduced for *biological control*; this is unusual, but occurs where the giant toad (*Bufo marinus*) is used to control pests on sugar cane, as in Queensland, Australia.
- Killing by *systemic insecticides*, which are absorbed by the plant and make its sap toxic to (pest) insects feeding on it, and also to bees that collect nectar or pollen; see e.g. Ferguson (1987).
- Killing by certain *herbicides applied to kill weeds* – especially any of the systemic herbicides that make nectar toxic to bees.

Far more serious than this last direct damage to bees is the destruction, by weed control programmes, of plant communities that are nectar and pollen sources. This destruction decreases honey yields obtainable by the beekeeper (Oertel, 1966). More significantly for agriculture, it can lead to the disappearance of existing populations of wild pollinating insects – often irreversibly – which has detrimental effects on the yields of many crops, and on the environment generally.

No antidotes to insecticidal poisoning of bees are currently available. So the main aim must be to prevent poisoning, and the next section describes ways in which beekeepers may be able to protect their bees.

11.72 Beekeepers' actions to protect bees from pesticide poisoning

In order to ensure that a beekeeper can protect his bees from insecticide poisoning, he must receive, from the grower or insecticide applicator, advance warning of the time of a toxic pesticide application close to his hives. Circumstances will then usually determine whether he need not take any action, or whether he should confine the bees in their hives, or move the hives away for a few days or longer.

11.72.1 Confining colonies to prevent bees flying

In some circumstances it is possible to confine bees while insecticide is being applied to a crop in flower, and until its toxicity has disappeared. This action may well be worth the effort involved.

The bees can be shut inside the hive by closing the entrance, after certain essential preparations. Extra space is provided by adding an empty super (hive box) to the hive. Extra ventilation is needed, so the inner hive cover is replaced by a securely attached wire screen, and the roof above it is positioned to allow good ventilation. Provision can also be made for external evaporative cooling, by draping sacking (burlap) over the hive and keeping it wet. Use of a wire screen instead of a wooden block to close the hive entrance allows more ventilation, but it may get blocked by dead bees falling to the hive floor, or (unless it is darkened) by bees congregating there.

In hot conditions, or with strong colonies, or if the bees will be confined for more than 12 hours (Jaycox, 1963), the bees must have access to water as well. A wet sponge or absorbent pad can be placed on the top screen, or water can be supplied in a frame syrup feeder, or in an inverted lever-lid tin with holes punched in the lid. The bees collect the water, spread it out in the hive, and by fanning bring about evaporative cooling within the hive, so long as this is adequately ventilated (Section 3.32). Kramer (1985) advocates providing much more water, as follows. He uses aluminium screen to close the hive entrance, and as a top cover. The flat hive roof has a strip of wood attached under opposite edges, so that there is an 18-mm gap along the other two edges, which must run *along* the row of hives. A vehicle is driven along the row, and a jet of water from an adapted garden hose nozzle is directed at each hive in turn, through the gap; it is showered off the underside of the roof into the hive. The flow of water is powered by a petrol pump. The operation delivers $\frac{1}{2}$ to 1 litre of water per hive, and is repeated every 2 hours while the hives are closed; it takes 15 minutes per 100 hives. A heavy watering is recommended just before releasing the bees; it reduces their agitation and their tendency to sting.

In experiments on confining colonies for 12 hours, Jaycox (1963) found that a container of chipped ice, over (but not closing) the entrance, produced a 'cold curtain' that satisfactorily confined the bees without blocking the entrance. In Israel, Eisikowitch and Lupo (1984) prevented bees flying by using a spray of water played on the front of the hives, so that it dripped down across the entrance, and the hive temperature did not rise above normal. It was essential to start the spray each day before any of the bees flew out, and it was left on all day.

The difficulty in keeping colonies alive in closed hives for more than one day led to searches for a way of confining them that allows more space and more air, as well as easy access to water. In one method, the hive entrance is left open, and sacking – not plastic sheeting – is draped over the hive, reaching the ground all round, where it is secured by stones (Figure 11.72a) to prevent the escape of bees. The sacking forms a tent in which the bees can cluster outside the hive, in darkness. It is kept soaked with water – for instance from a sprinkler – and the bees can collect water from it. Such a tent may be left in place for up to two days.

In Arizona, Owens and Benson (1962) experi-

Figure 11.72a A method for confining bees during pesticide application (Sanford, 1980). See text for details.

mented with three-storey hives when the outside temperature was up to 41°. In a control hive the temperature was everywhere above 35°; in about half the hive it was above 36°, and in one part above 40°. But in a hive under wet sacking (without the clustering space described above), it nowhere exceeded 36°. The confinement with wet sacking was reckoned to be safe for up to 2 days under these conditions.

11.72.2 Moving colonies away from areas to be treated

Bees in hives moved well away from the crop to be treated with insecticide are safe, but steps must be taken to prevent damage during transport. If hives are to be closed, this must be done at night when all or most of the bees arc in. The risk of overheating is much reduced if hives are transported with entrances left open. This can be done, provided: hives are removed from the site at night, the complete journey is made during hours of darkness, and the engine of the vehicle is kept running for the whole of the time, since this keeps the bees clustered. A 'bee net' may be used to enclose the hives (and bees) on the vehicle. Section 8.1 gives details.

Section 8.14 explains the behaviour of bees after their hives have been moved to a new site within their old flight range. Provided no hives are left at the old site, any bees that return to it are likely to find their hive at the new site eventually, and they are not lost. However, if hives are moved away from a crop in flower before insecticide is applied, to a site *with no forage nearby*, some bees might fly back to the flowering crop made toxic by insecticide. If the hives are moved well beyond the old flight range, none will return. So a distance of 8 km would be safer than one of 4 km.

11.73 Ways of reducing pesticide poisoning of bees

In any country, pesticide poisoning of bees is reduced only as a result of advances upon several fronts, all of which need much patience and dedication. The advances, which are discussed at length in books mentioned in Section 11.71, may be summarized as follows:

- knowledge of the toxicity to bees of different pesticides
- knowledge of the relative hazard to bees of different formulations, and of different methods and times of application
- co-operation by manufacturers in labelling pesticides according to whether they are toxic to bees or safe for them
- co-operation by crop growers in safeguarding their crop pollinators, by using non-chemical methods of pest control where possible, or by using the safest pesticides, in the safest ways, at the safest times
- action by extension workers in the education of crop growers and others
- action by governments in passing legislation to protect bees and other pollinators.

Most of the publications cited in Section 11.71 include lists of pesticides grouped into use-classes according to their toxicity to European *Apis mellifera* in temperate zones:

Use-class 1, pesticides most toxic to bees. These are insecticides (including acaricides) which are so toxic that they should *never* be applied to flowering crops or weeds visited by bees; their residual toxicity is usually still high after 10 hours.
Use-class 2, pesticides very toxic to bees. These are insecticides which should be applied to flowering plants only during late evening.
Use-class 3, pesticides less toxic to bees. These are insecticides which should be applied only during late evening, night, or early morning; their residual toxicity is usually low after 3 hours.
Use-class 4, pesticides least toxic to bees. These include certain insecticides, and most herbicides and fungicides, which can be applied at any time with reasonable safety to honeybees. (Some herbicides can be toxic to bees under certain conditions.)

Much attention has been paid to comparing standardized methods used in different countries for testing the toxicity of different pesticides to European *Apis mellifera*, and internationally agreed recommendations are available (see International Commission for Bee Botany, 1982, 1985). Similar methods would be applicable to tropical honeybees, but pesticide toxicities are not necessarily the same to these bees, or to any bees in hot climates; Crane and Walker (1983) summarize what is known for tropical honeybees of all four species.

Formulations of insecticides are listed below, with their common abbreviations, in decreasing order of their toxicity to bees:

most toxic:	dust (D), microcapsules
intermediate:	emulsifiable concentrate (EC), flowable (F), liquid suspension (LS), soluble powder (SP), ultra-low-volume (ULV) spray or aerosol, wettable powder (WP)
least toxic:	granular (G)

An application of insecticide is much less likely to kill bees if it is made close to the crop, and in still weather. If it is made from the air, or in a wind, the insecticide can fall or drift on to plants it is not intended for, or on to hives; no pesticide should ever be applied directly on hives. The safest time of day is late evening or early night, to allow as long as possible for any residual toxicity to die away before the bees next fly. Melksham et al. (1988) discuss the use of pesticides that are repellent to foraging honeybees, and the incorporation of repellents in pesticide formulations; they also list repellent compounds, and two pyrethroids were studied by Rieth and Levin (1988).

Pesticides on sale in many countries are now labelled according to their toxicity to bees, but much improvement in this respect is needed in developing countries – where, also, labels are not always in a language that is widely understood, and they may sometimes be missing from containers holding pesticides.

In many countries the situation has improved markedly after suitable legislation has been introduced, and can be enforced. Table 15.4A lists 38 countries with legislation designed to protect bees from pesticide poisoning, and a further 7 countries with a code of practice (or similar) for the purpose; see Section 15.51. A report by the Ministry of Agriculture in the UK (1985) shows some of the complexities involved in drafting legislation.

11.74 Action if pesticide poisoning is suspected

It is usually foraging bees that are poisoned. Some die at a distance from the hive, and the beekeeper is unlikely to see them; but many return, and a common sign of poisoning is a pile of dead bees outside the hive entrance (Figure 15.5a). Another sign is a great reduction in flight activity in front of all or most hives in an apiary. Swarming, which might lead to a similar reduction, would not affect all hives on the same day. If some colonies are found dead, the beekeeper must try to ascertain whether starvation or disease has been the cause.

If pesticide poisoning is suspected, it may be important to the beekeeper to get confirmation of this, and to identify the pesticide responsible. In many countries, samples may be sent to a government laboratory for analysis. For any one incident, samples should include at least 200 bees, plants treated with pesticide, and if possible the pesticide suspected. The analysis has to be done very soon after the bees died, because in the open air the amount of pesticide residue in or on bees decreases rapidly. The body weight of bees, which also decreases rapidly, provides an indicator as to the usefulness of sending a sample. When alive, 200 bees weigh about 20 g; if a sample of 200 weighs more than 15 g, there is still a good chance of residual pesticide being found in them, but if it weighs less than 10 g, the sample will be useless (Lewis, 1985). Evidence of a positive diagnosis is essential if application for compensation is considered; Section 15.52 deals with the legal aspects.

In 1986 EnzyTec Inc. in the USA marketed a $5 pesticide detector that has a sensitivity of 0.1 to 10 ppm (Gleanings in Bee Culture, 1986). The enzyme cholinesterase is used to produce a change in the colour of a test paper, in the presence of organophosphates or carbamates; these compounds account for some 85% of the insecticides sold in the USA.

11.75 Injury to bees by environmental pollution and other factors

Bees may be killed directly by pollutants such as toxic metals, and there has been high mortality near factories emitting effluents containing arsenic (Svoboda, 1962). Most countries have now curbed this form of high local pollution, but pollution from toxic metals has nevertheless become more widespread. Bees forage over a relatively large area round their hive (7 km² if they fly up to 1.5 km from the hive), and pollution of their nectar and pollen sources can kill them. Since 1980 the possibility has been explored that the bees themselves, or the materials they collect or produce, might be used as indicators of environmental pollution – especially with heavy metals; see end of Section 14.23. Contamination of honey with heavy metals, and with radioactive and other elements, is reported in Section 13.27.3.

Effects of pollution as a result of lead emission from vehicles was studied in eastern USA (Pratt & Sikorski,

1982). Along a 'moderately busy' road, both bee forage and bees were found to be contaminated. Results there were significantly different from those at a control site about 850 m from the road, the lead contents being:

	road site	*control site*
bee forage (flowers)	13.6 ppm	0.2 ppm
foraging honeybees	28.1 ppm	1.4 ppm

Fluorine is another atmospheric pollutant that is toxic to bees. In the USA Bromenshenk (1979) found significantly increased fluoride levels in honeybee tissue near coal-fired power plants. He refers also to a 1978 report suggesting that sulphur dioxide pollution may interfere with the olfactory system of honeybees.

In an industrial area of the GDR, Börtitz and Reuter (1977) found fluorine in willow catkins and in flowers of apple, dandelion (1.1–41.8 µg/blossom), red clover and tulip. In Romania, Lopez and Nicotra (1975) found a wide range of phosphates and phenol compounds, and light saturated and unsaturated hydrocarbons, in the air of an industrial zone and in pollen from hives nearby. There are many other examples.

Bees are also indirectly at risk from all forms of pollution that damage plants producing nectar, honeydew or pollen. There have been many reports of widespread deaths of trees as a result of atmospheric pollution. For instance Eisenmann (1982) refers to the death of 50% of the pine trees in parts of Bavaria, GFR. In so far as the trees provide bee forage, this must affect bees adversely. Rousseau (1972) discussed general interactions between environmental pollution and honeybees.

A number of 'problem products' that may injure bees are dealt with by Ellis (1987); they include wood preservatives and carbohydrate foods that honeybees cannot digest.

11.8 NATURAL ENVIRONMENTAL HAZARDS

Environmental catastrophes that can cause loss of human life, and economic loss, also cause damage to bees and hives in the region affected, and the whole of a beekeeper's livelihood may be endangered.

Bees and their hives are at risk from high winds, floods, fire, volcanic ash or lava flow, and earthquakes. The concentration of many hives in one apiary is likely to increase a beekeeper's total loss on any one occasion. And even if damage to hives and bees is minimal, flowers providing bee forage are likely to be destroyed, and the plants so much defoliated or otherwise injured that there is no forage for months to come.

11.81 Tempest and flood

Seasonal storms with sustained winds of 120 km/h or more are generated over oceans in the earth's tropical belt. One common system for naming them is: north of the equator, the storm is a hurricane or a typhoon, according to whether it is east or west of the Date Line; south of the equator (and anywhere in the Indian Ocean) it is a cyclone.

Bees and hives are liable to suffer from these violent storms in several ways. An ocean surge can send waves 10 m or higher inland – sometimes covering low-lying islands – with the result that hives are flooded or washed away. An extended period of high winds may lead to excessive heat loss and dehydration of adult and larval bees. The bees may also be affected by the rapid changes in barometric pressure that accompany tropical storms (Hitchcock, 1986).

The storms are so unpredictable in their course and effects that some of the calamities they cause cannot be prevented. But the following precautions form a necessary part of beekeeping management in areas at risk:

- hive sites are chosen to provide shelter, and to continue to do so after bushes and trees are defoliated by the wind;
- hives are firmly placed on concrete blocks, and raised off the ground above likely flood water;
- hives are provided with a flat roof weighted down with a large concrete or other block;
- hives slope slightly forward, so that any water entering runs out;
- frame-spacers are adequate to prevent movement of frames by the force of the wind.

These steps must of course be taken before the storm season commences. It is said that a hurricane/typhoon/cyclone sorts out good beekeepers from bad ones.

Detailed records made on St Lucia in the Lesser Antilles, just after hurricane Allen in 1980 (Duggan, 1984), suggest that 419 of the 985 hives in the 37 reporting apiaries (43%) were destroyed. Many of the colonies not killed were starving soon afterwards. A hurricane-force wind rips off all leaves, flowers and flower buds from trees in its path. In 1982 I experienced the immediate after-effects of a cyclone on the island of Rodrigues in the Indian Ocean, and saw the defoliation of the eucalypts that provide much of the honey. Flower buds, which are formed long before blooming, were torn off, with the loss of any honey flow for many months to come.

Wake Island, west of Hawaii in the Pacific, has been well documented by Hitchcock (1986). In 1981, when typhoon Freda blew sea water inland to a depth of 60 cm, well prepared hives and colonies survived, but bee forage suffered great damage, and much of the ground flora that had survived storm Nadine in 1978 was killed. The wind blew the flowers and buds off tree heliotrope (*Tournefortia argentea*) which is the main honey source. Many limbs were also broken off these trees because they became dehydrated by the high winds, and the wood was not flexible. Ironwood (*Casuarina equisetifolia*) fared better. Flowering vines were much damaged by dehydration, in spite of the heavy rain.

When hurricane Juan flooded commercial apiaries in Louisiana, USA (American Bee Journal, 1986), brood and bees submerged in the salt water were killed. About half the colonies were completely destroyed, and the hive boxes scattered. Parts of California are also liable to flooding. In 1983 floods due to very heavy rain destroyed 8000 hives in the Sacramento valley (Soiffer, 1983). In 1986 mud slides accompanying floods did more damage to colonies than the water itself (Taber, 1986). Whatever their cause, floods can kill colonies and ruin hives, and make it very difficult for the beekeeper to reach his hives and start rescue operations.

11.82 Fire and lightning

During the dry season in tropical forest and bush, fire is always a hazard, and in parts of Africa it was perhaps a factor in determining behaviour characteristics of native honeybees. When honey hunting with the aid of smoke became customary, nests were increasingly damaged by fire.

In February 1967 an extensive fire devastated 4500 km² in Tasmania, Australia. The fire travelled at about 2 km/minute, and one beekeeper with three apiaries 15–30 km apart, lost all his equipment and colonies (by smoke if not by fire). The trees in the region were mainly eucalypts, which are not necessarily killed by a fire that burns off all their leaves; they were already putting out fresh growth in the following October when I was on the island. During five weeks in different states of Australia, I saw a bush fire almost every day. Hives that had been in the path of a fire were sometimes no more than scorched, and the colonies had survived. In Western Australia 181 beekeeping sites were affected by fires in two years (1984–1986), and 1200 km² of beekeeping country were damaged. Bee Briefs NSW (1986) and Burking (1986) make recommendations for fire control management and for limiting fire damage; Burking also sets out a system by which beekeepers can assess the amount of damage to their bees, equipment and future honey crop.

The smoke from some trees, e.g. black walnut (*Juglans nigra*) and poison oaks (certain *Rhus* spp.), is toxic to man, but its toxicity to honeybees is not known (Hitchcock, 1986).

Lightning seems an unlikely hazard to hives, but there are quite a few records. In a typical one, from eastern USA in July 1985 (Rodrigues, 1986), the hive in question was 1.2 m from a tree which the lightning struck, and the four hive boxes were scattered 'as though by bears', over an area within 2.5 m of the hive site, boxes and frames being split apart at the joints. The bees clustered over the scattered frames of brood and honey, and the beekeeper successfully re-assembled the colony.

11.83 Volcanic eruption and earthquake

Mount St Helens, near the Pacific seaboard of the USA, erupted several times in 1980, and the after-effects on bees were documented in some detail (Johansen, 1980). There was an immediate killing of foraging bees from the fall of ash; the particles scoured the outer surface of the integument so that the bees were dehydrated, and smaller particles blocked their breathing apparatus. Eventually about 12 000 out of 15 000 colonies died where the ash fell, in a belt to the east of the eruptions. Many of the surviving colonies reduced brood rearing, and some were subsequently weakened by repeated swarming or supersedure. Bees in some colonies became very aggressive. The sheer weight of ash damaged much vegetation, and blown ash sealed the soil surface over a wider area; water could not penetrate the soil, and much bee forage was lost.

The following is a verbatim extract from a beekeeper's account of effects of an earthquake in New Zealand in 1931 (Ashcroft, 1986).

> A hive supered up for honey is not the most stable structure. One moment there were orderly rows of hives, and the next there was a jumbled mass of boxes and combs. The reek of honey from the smashed combs started the bees robbing, and soon the combs lying about were black with bees fighting. There was the prospect of traffic being held up on the nearby road. Some hives had gone over in a block, separating when they hit the ground. These were easy to re-erect, but some appeared to have been tossed in all directions by a vertical lift. Combs were often thrown clear of the boxes and lay in ruins on the grass. The damage was not as great as in other apiaries in Hawke's Bay, where hives were knocked down several times by succeeding shakes.

There is the grave danger of disease picked up from some unregistered hive, upset by the 'quake, and left open to be robbed.

According to Tributsch (1982) honeybees have been known to leave their hive completely prior to a major earthquake.

11.9 FURTHER READING AND REFERENCE

There is a great wealth of reading matter, and entries here have been limited to certain books, and a few shorter publications, in English, that provide much up-to-date information and are likely to be obtainable. Details of them will be found in the Bibliography. Most books on beekeeping include a chapter on diseases.

Akratanakul, P. (1987) *Honeybee diseases and enemies in Asia*

Bailey, L. (1981) *Honey bee pathology*

Bailey, L. (1982) *Viruses of honeybees*

Bradbear, N. (1988) *The world distribution of major honeybee diseases and pests*

Burges, H. D. (1978) *Control of wax moths*

Crane, E. (1978*a*, No. 18) *Bee diseases, enemies and poisoning in the tropics*

Crane, E. (1978*b*, S35) *Bee diseases and pests in specific regions of the tropics*

Crane, E.; Walker, P. (1983) *The impact of pest management on bees and pollination*

De Jong, D., et al. (1982) *Mite pests of honey bees*

Fry, C. H. (1983) *Honeybee predation by bee-eaters, with economic considerations*

Griffiths, D. A.; Bowman, C. E. (1981) *World distribution of the mite* Varroa jacobsoni . . .

Hansen, H. (1980) *Honey bee brood diseases*

Johansen, C. A. (1979) *Honeybee poisoning by chemicals*

Morse, R. A. (ed.) (1978) *Honey bee pests, predators, and diseases*

PART V

Honeybees' Plant Resources, and Products from the Hive

12

Honeybees' plant resources

12.1 INTRODUCTION

Honeybees collect everything they need (except water) from plants. Nectar and pollen from flowers can provide all their nutritional requirements. Other sweet materials are also collected from plants when available: extrafloral nectar (from parts other than flowers), honeydew produced by insects from plant sap (Section 12.22), and exposed sap as from cut sugar cane.

Adult worker bees whose glandular system has matured require only carbohydrate food – the sugars in sweet plant materials – which provide them with energy for flight (Section 12.2). If no carbohydrate food is available to a colony, it will starve to death. But when it is plentiful, a colony may collect much more than it needs at the time, and it then converts the excess into honey, which does not spoil on storage: it is this behaviour of bees which enables the beekeeper to get a honey harvest from them.

A colony needs water to counteract overheating. If the bees cannot collect water, and it is not provided by the beekeeper, they will suffer heat stress; under extreme conditions, all adult bees and brood will die. Section 12.3 discusses sources of water used by bees.

Pollen (Section 12.4) consists of minute pollen grains which are the male fertilizing elements in a flower. Pollen provides the proteins, vitamins and other nutrients needed by larvae – the feeding stage of developing bees (Section 2.41) – and therefore by immature adult bees that feed the larvae (2.42). Without pollen, or an adequate substitute supplied by the beekeeper, a colony will not starve immediately, but will slowly die, because it cannot rear new bees to replace those that die.

Honeybees also collect sticky plant materials known as propolis (Section 12.5). They use this in building work in the nest or hive, some bees much more than others, and *Apis cerana* probably not at all.

In the world today there are around 250 000 species of flowering plants. Many produce nectar and pollen accessible to honeybees, and honeybees pollinate a large number of them, but some tropical plants have nectar lying too deep in the flower for honeybees to reach it, and they are pollinated by long-tongued insects, birds and mammals. About 40 000 plant species have some importance to honeybees as food resources, and 4000 are the source of most of the world's honey – far more than can be encompassed in a book on beekeeping. Appendix 1 lists 464 important world honey sources, and summarizes their geographical distribution; 449 of them yield nectar (Section 12.21), and the majority pollen as well (12.4); 14 produce honeydew only (12.22); from cut sugar cane, the plant sap is collected (12.23).

Section 12.6 explains the seasonal pattern of the plant food resources for bees in different parts of the world, and 12.7 their geographical pattern. Section 12.8 discusses possibilities of planting specifically to increase the food resources for honeybees.

12.2 ENERGY FOODS: NECTAR AND HONEYDEW

The carbohydrate (energy) foods collected by bees from plants are derived from the sap of the phloem, the living food-conductive cell system of plants known as vascular plants. The energy the plant needs to make and transport the nutrients is obtained by photosynthesis: solar energy is absorbed by chlorophyll in the leaves and other green parts and converted into chemical energy by the combination of carbon dioxide in the atmosphere with water already in the plant, to form carbohydrates. The carbohydrates most commonly formed are sugars, for instance:

$$6CO_2 + 12H_2O \longrightarrow C_6H_{12}O_6 + 6O_2 + 6H_2O$$

In plantations of sugar cane, the canes are harvested and sugar (sucrose) extracted directly from the phloem sap in them – hence the name cane sugar for sucrose.

12.21 Floral and extrafloral nectar

As the sap moves through the phloem of the plant, the sugar concentration tends to become equal throughout; sugar is transported away from the sources where it is being photosynthesized, towards 'sinks' where it is involved in a specific process such as the ripening of fruit or secretion of nectar. Nectar is secreted by glands known as nectaries; they are usually in the flower, and attract pollinators to it, although some plants have extrafloral nectaries. Nectaries are more specialized glands in some plants than in others.

12.21.1 Production of nectar in the plant

Figure 12.21a exemplifies the position of the nectary, well protected inside the flower. A visiting bee is 'led' in the direction of the nectar by the shape of the flower, and especially by nectar guides on the petals; these may be patterns of colour (visible or ultraviolet), of texture, or of scent. Floral nectaries may be situated on any part of the flower; Shuel (1975) gives information, with examples, and nectaries are discussed in detail by Bentley and Elias (1983).

Most of the sugar in nectar is likely to have come from leaves close to the flower. In herbs (herbaceous plants) the photosynthesis that produced the nectar was probably quite recent, and sunshine in the previous few days is important for nectar production. In trees and shrubs, and in plants that flower before leaves appear, the sugar secreted in nectar may have been stored in the plant since the previous growing season, and it was the sunshine then that provided the energy for its production.

High magnification electron micrography makes it possible to see the ultrastructure of nectaries in fine detail, and thus to envisage nectar secretion as a cellular process (Figures 12.21b, 12.21c). Most cells shown in Figure 12.21c are interconnected by fine strands of protoplasm known as plasmodesmata (singular = plasmodesma), and nectar migrates in a number of planes, probably in response to gradients of concentration or pressure (Erickson, 1986).

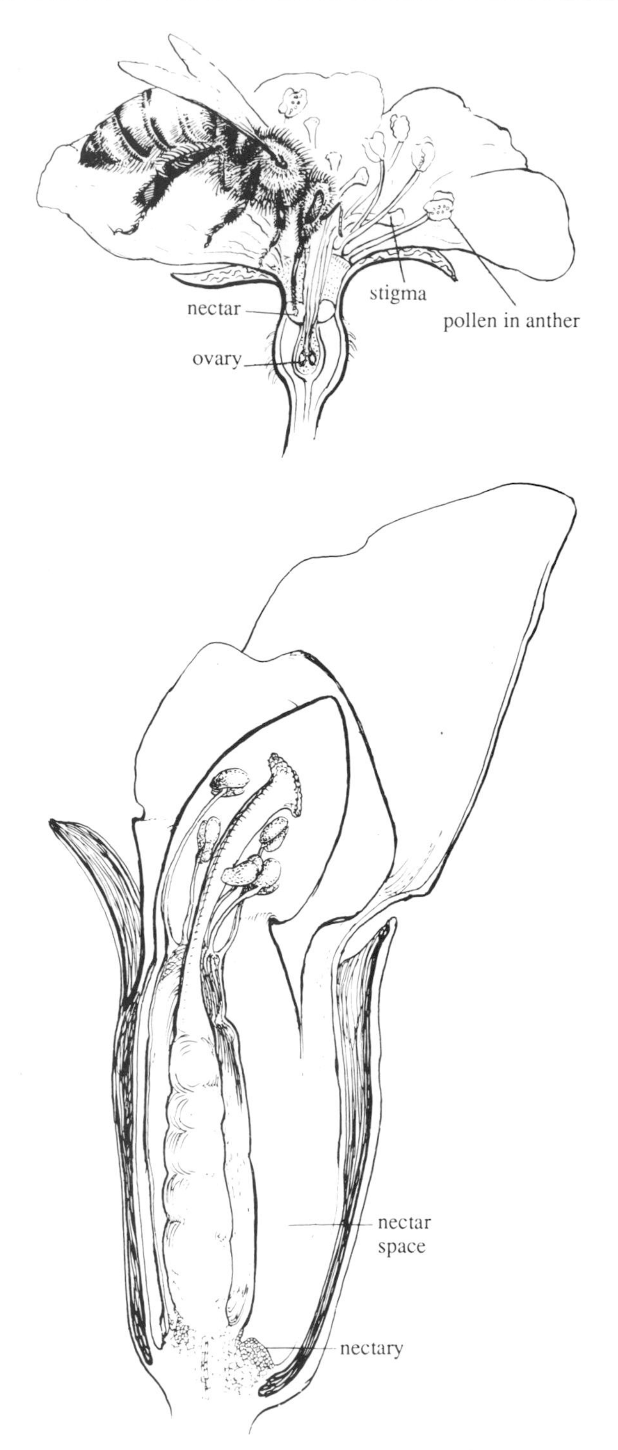

Figure 12.21a The location of nectaries in flowers.

above Section through a pear flower, which is about 25 mm wide (drawing: Dorothy Hodges). The nectary is easily reached by a bee standing on the flower petals.

below Section through a floret of white clover, *Trifolium repens*; the floret is about 7 mm long (Crane, 1980*a*). The nectary is at the base of the deep corolla, between the petal and the staminal column; secreted nectar rises up the narrow 'nectar space' if not collected, and can then be taken by an insect whose tongue is too short to reach the nectary.

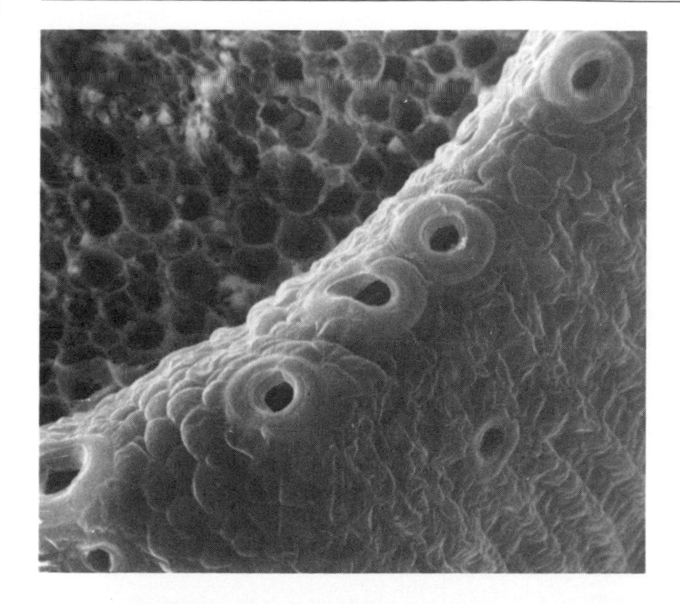

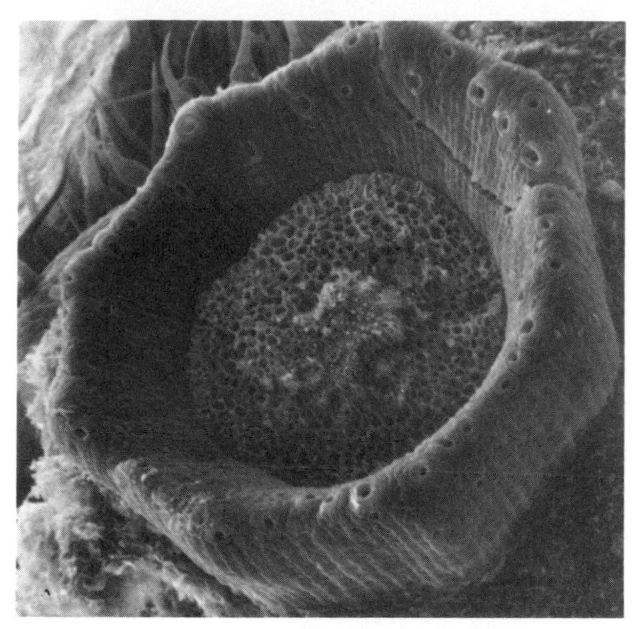

Figure 12.21b Scanning electron micrographs of a nectary of sunflower, *Helianthus annuus*, cultivar CMS 209 (Sammataro et al., 1985).
above 8-sided nectary, showing the openings (stomata) in the rim, through which nectar is secreted, × 85.
below Part of the rim showing four stomata, × 420.

The rate of nectar secretion may be assessed in terms of 'sugar value', the weight of sugar in mg secreted per flower or floret per 24 hours. Demianowicz (e.g. 1960/1963) has made many measurements in Poland, and Maurizio (1975*a*) lists sugar values for 66 plants foraged by honeybees; they range from 0.0005 to 8 mg/flower/24 h. The amount of nectar contained in different flowers varies by a factor approaching half a million: from 5 μg in a floret of white clover, *Trifolium repens* (Figure 12.21a, *below*), to nearly 2 g in a flower of the tulip tree, *Liriodendron tulipifera* (Figure 12.21d). *Banksia ornata* in Australia has been reported to drip 300 g in 24 hours from a single flower spike (Berkin, 1987).

12.21.2 Nectar composition

The composition of nectars from different plants is similar in that the main constituents are sugars and water. The total concentration of sugars varies widely from species to species, and within a species, according to many factors (Section 12.21.3); it may reach a value as high as 92%, for instance in dwarf mistletoe, *Arceuthobium abietinum* (Brewer et al., 1974). Nectar characteristics of some of the honey plants listed in Appendix 1 are discussed by Crane and Walker (1986*a*).

There seem to be three main patterns of nectar sugar composition (Percival, 1961, 1965): with sucrose dominant, with fructose + glucose dominant, and 'balanced', i.e. with the three sugars present in similar amounts. At one time it was believed that honeybees showed a preference for the third type of nectar, but it now seems more likely that the total sugar concentration is usually the main factor (Waller, 1972). In addition to these three and other sugars, nectar contains small amounts of chemical compounds that include aromatic substances, minerals, organic acids and amino acids. Any or all may be important in determining individual characteristics of the honey produced from the nectar, and some of the compounds are useful indicators of the botanical origin of the honey. Certain amino acids present in honey, especially proline, are introduced by the bees during their elaboration of the honey, and are thus not plant-specific.

12.21.3 Factors affecting nectar secretion and composition

Secretion of nectar may be determined, in a complex way, by heritable factors within the plant. For instance in plants with separate male and female flowers, only one type may secrete nectar, or one type may secrete more than the other: male in willow and banana (*Salix* spp., *Musa paradisica*), and female in cucumber and melon (*Cucumis* spp.). Some other plants whose flowers pass through separate male and female stages (e.g. *Tilia*, *Phacelia* spp.), usually show differences in nectar secretion between the stages.

Nectar secretion follows a daily cycle, usually with a peak at some time during daylight for plants pollinated by day-fliers, or during the night for plants pollinated by night-flying moths or bats. In the USSR, Dolgova (1928) assessed the daily cycle of richness of nectar supply in local herbaceous plants by recording the

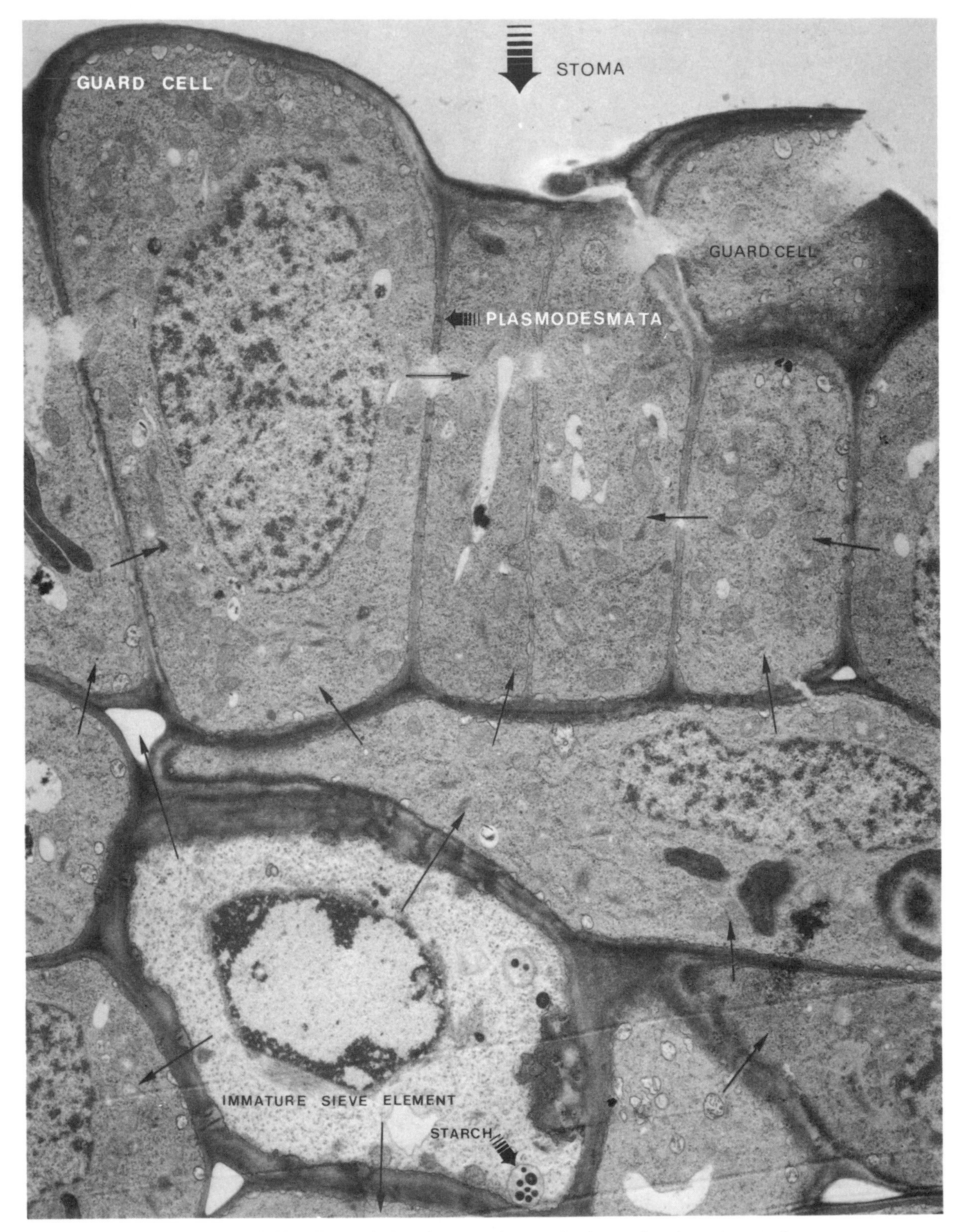

Figure 12.21c Transmission electron micrograph of a section through part of the rim of a sunflower nectary, cultivar CMS 89, × 6080 (Sammataro et al., 1985).
One opening (stoma) in the surface of the nectary tissue is shown, with a guard cell on either side. The cell system linking a sieve tube to the stoma is included, and the arrows indicate possible routes taken by nectar from the sieve tube to the surface of the stoma where it is secreted.

Figure 12.21d Flower of the tulip poplar, *Liriodendron tulipifera* (photo: R. L. Parker).
The large flowers secrete so much nectar that it may drip out of them; the two bees here are collecting pollen.

number of honeybee visits at different times of day, under standard conditions. She grouped the plants according to whether they received most visits at 9, 11, 13 or 15 h. Table 12.21A gives Dolgova's results for those of her plants that are among the honey sources included in Appendix 1.

In Hungary, Pesti (1976) measured the rate of nectar secretion directly in grams per hour throughout the day, for 22 temperate-zone Compositae. The time of day when it was most rapid varied from one species to another:

time of day	7 h	9 h	11 h	13 h	15 h	17 h
no. species	1	8	7	3	2	1

Pesti gives much information, for instance on plants with more than one peak time of nectar secretion, and on the daily cycles of the total amount of nectar and of dry matter (mostly sugar) in the nectar. The results obtained by Dolgova and Pesti exemplify the range of timing of nectar secretion by temperate-zone herbaceous honey sources. In the tropics, many plants secrete nectar in early morning and/or late afternoon, but not in the greatest heat of the day. Flowers that are pollinated by bats or moths secrete nectar during the night, and if they still contain some nectar in the morning, honeybees can collect it.

Some of the external factors affecting nectar secretion are of great potential importance to beekeepers. Shuel (1955) was able to demonstrate a striking cor-

Table 12.21A Time of day when the nectar supply is richest in certain plants

Data from Dolgova (1928). Fifteen of the temperate-zone honey sources listed in Appendix 1, grouped according to the time of day when the highest number of honeybee visits was recorded, i.e. when the nectar supply was richest.

9 h	*11 h*	*13 h*	*15 h*
Brassica juncea, Indian mustard			
Brassica nigra, black mustard			
Fagopyrum esculentum, buckwheat			
Sinapis alba, white mustard			
Taraxacum officinale, dandelion			
	Centaurea cyanus, cornflower		
	Hyssopus officinalis, hyssop		
		Borago officinalis, borage	
		Melilotus alba, white sweet clover	
		Onobrychis viciifolia, sainfoin	
		Phacelia tanacetifolia, phacelia	
		Trifolium incarnatum, crimson clover	
		Trifolium pratense, red clover	
		Trifolium repens, white clover	
			Echium vulgare, viper's bugloss
9 h	*11 h*	*13 h*	*15 h*

relation between nectar secretion and the amount of radiation or solar energy received by the plant (the insolation) during the preceding 24 hours (Figure 12.21e). In Canada the most northerly apiaries are at a latitude of 57°, where daylight lasts for 20 hours a day during the honey flow from legumes. The consequent high insolation leads to rapid nectar secretion, and large honey yields, from legumes. It was calculated that nectar secreted by alsike clover (*Trifolium hybridum*) contained 883 kg sugar per ha (Szabo & Najda, 1985). This would be equivalent to nearly 1100 kg honey per ha – more than for any of 200 high-yielding plants listed by Crane (1975*b*, pp. 47–55). Oertel (1973) gives further figures.

Shuel (1975) suggests that the association between honey yield and temperature, often reported in temperate climates, is more correctly based on an association between nectar production and sunshine. High day and low night temperatures are reported to lead to high yields, e.g. from white clover (*Trifolium repens*), and they are likely to occur in periods of fine sunny weather with clear skies – i.e. high insolation.

Threshold temperatures below which little nectar is secreted are cited by Shuel (1975) for four plants. He suggests that they may be temperatures at which the enzymes involved in nectar secretion are activated:

Prunus avium, sweet cherry	8°
Cucumis sativus, cucumber	17°–21°
Tilia americana, basswood	18°
Prunus laurocerasus, cherry laurel	18°–20°

If temperatures are high during a period of insufficient rainfall, water stress can occur in plants. The transport of sugar is then reduced and, more importantly, its photosynthesis, so that nectar secretion is reduced. On a world-wide basis, water supply is the factor that most frequently limits nectar secretion – as it does plant growth (Shuel, 1975). This is commonly true in the tropics; in temperate regions insolation during flowering may be of comparable importance to water supply. Nectar secretion is an active process, and may be stimulated if nectar is removed from the flower.

Minerals in the soil, or added in fertilizers, can have a considerable effect on nectar secretion. In England, Gloucestershire beekeepers traditionally took their hives up on to the Cotswolds for the heavier flow from white clover there, said to be due to the higher calcium content of the soil.

The sugar concentration of nectar may change after secretion has occurred. It can be increased by evaporation of water from nectar at low atmospheric humidity (especially at a high temperature), and decreased by absorption of water at high humidity, or by dilution if rain falls into the flower.

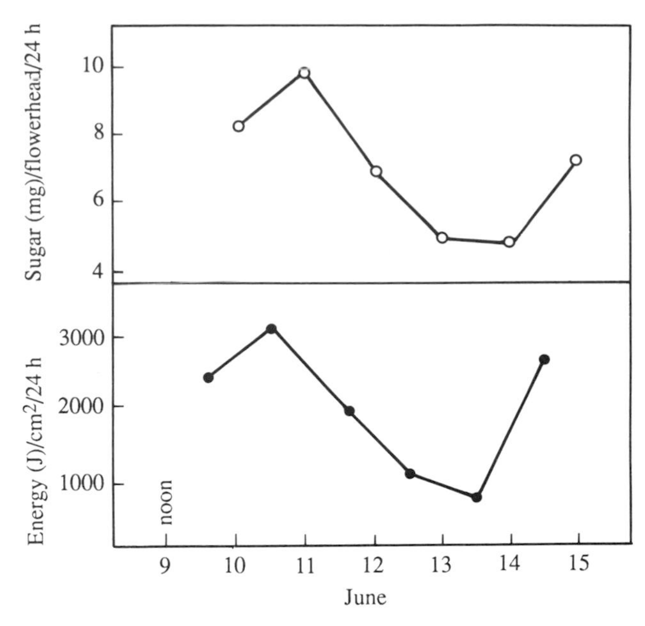

Figure 12.21e Relationship between nectar secretion and insolation in alsike clover, *Trifolium hybridum* (Shuel, 1955).
upper curve Mean weight of sugar secreted in nectar per flowerhead per 24 h.
lower curve Energy in joules received by the plant from solar radiation per cm^2 per 24 h.

Shuel (1975) and Maurizio (1975*a*) give further information on effects of internal and external factors on nectar secretion, and earlier papers by Boetius (1948) and Beutler (1930, 1953) describe detailed classical studies.

12.21.4 Extrafloral nectar

Flowers have been used for most of the research on nectar and its secretion, but (except for aspects related to pollination) most of the findings are likely to be relevant to extrafloral nectar. This is commonly secreted at the same time as the nectar in the flower, i.e. at the same stage of development of the plant. Whatever the timing, beekeepers need to be alert to a possible insufficiency of pollen for brood rearing when bees are working an extrafloral nectar flow. This situation arises with pará rubber, *Hevea brasiliensis*, although bees do collect some pollen from the flowers. Neither male nor female flowers are recorded as producing nectar, but the extrafloral flow is intense; half the honey produced in India is said to be from rubber trees, and the honey is a major crop in wet equatorial regions in Malaysia and elsewhere; see Figure 12.7a, map 2. The nectaries are between the young trifoliate leaves, on bud scales, and possibly elsewhere. They are active for 2–4 weeks, secretion ceasing when the leaves become mature. Bakar et al. (1986) give details of the production and sugar concentration of the nectar (see also Crane, Walker & Day, 1984).

Both long-staple and short-staple cotton (*Gossypium barbadense*, *G. hirsutum*) have extrafloral nectaries on leaf veins, within and outside the bracts, and minute ones on flower peduncles and young leaf petioles; they are described in detail by McGregor (1976). Extrafloral cotton nectar is more copious, and it has a higher sugar concentration (up to 40–50%) than that of floral nectar, so bees prefer it. These two species give important honey flows (Crane, Walker & Day, 1984), and honey yields of 300 and 75–90 kg/ha, respectively, are reported from them in Uzbekhistan, USSR (McGregor, 1976).

Other plants of some importance to beekeepers that secrete extrafloral nectar include field bean (*Vicia faba*) and castor (*Ricinus communis*).

12.21.5 Nectar-producing plants as honey sources

For the 449 nectar-secreting plants in Appendix 1, the *Directory of important world honey sources* (Crane, Walker & Day, 1984) quotes figures available for the sugar value and sugar concentration of the nectars where these are available. It notes factors shown to affect secretion, and cites references to sugar analyses of the nectars. Several authors have used measurements of the amount of sugar produced in nectar to calculate the 'honey potential', expressing the honey-yielding capability of a plant species. It is the quantity of honey (in kg) that could theoretically be obtained in the course of a season per hectare of land (or in pounds obtainable per acre, which is nearly equivalent). Optimal – or at any rate good – growing conditions are assumed, and an adequate force of foraging bees to collect all the nectar secreted. The honey potential is listed for 200 plants in Crane (1975*b*), using six classes:

Class 1 = 0–25 kg/ha	Class 4 = 101–200
Class 2 = 26–50	Class 5 = 201–500
Class 3 = 51–100	Class 6 = over 500

Only 9% of the plants are in Class 6, and 45% are in Class 3 or 4. Actual honey yields per hectare are often lower than the calculated potential, because the area is not fully occupied with the plant in question, or for other reasons.

12.21.6 Toxic nectars

Toxicants in poisonous honey (Table 13.27B) are derived from the nectar, with the exception of a honeydew source whose toxicant is introduced from the insect concerned (Section 13.27.2).

Very few nectars are toxic to bees. In abnormally dry years in the north temperate zone, bees may be seen dead under certain lime trees (*Tilia*); the main cause is the presence of mannose in the nectar, which disturbs the carbohydrate metabolism of the bees so that they cannot digest fructose and glucose (Crane, 1977, 1978*d*).

The nectar of any plant may become toxic to bees if a systemic insecticide is applied to the plant. Ferguson (1987) gives details of the toxicities of some of them, and of the rate of decay after application. Among those very toxic to bees are acephate, dimethoate, formathion, heptenophos, methamidophos, mevinphos, monocrotophos, omethoate and vamidothion.

12.22 Honeydew

Honeydew flows are the main source of honey in some parts of the world; for instance in Greece where they provide some 65% of the honey produced (Santas, 1983). Honeydew flows are unlikely to have an accompanying pollen flow, and pollen or a substitute should be fed to the colonies so that they can continue to rear brood (Section 6.32). These flows are a feature of temperate zones rather than of the tropics, and comparatively localized: many beekeepers have no experience of them.

12.22.1 Honeydew production and composition

Like nectar, honeydew has its origin in the sap of plants, but whereas nectar is actively secreted by the plant, honeydew is produced by the mediation of plant-sucking insects (Hemiptera), which include coccids (scale insects), lachnids and aphids. These insects have mouthparts capable of piercing the foliage or some other part of the host plant, and they feed on the sap inside the plant; the sap is released and forced out through the puncture by internal pressure, which is reinforced by the insect's own pumping. The ingested sap passes into the insect's gut; excess liquid by-passes the midgut and, after passing directly from foregut to hindgut through special filter chambers, is excreted as honeydew (Figure 12.22a).

Droplets of honeydew fall on to the plant surface, and are collected by other insects, especially bees, wasps and ants. Ants can play an important part in honeydew production, and their presence helps to protect the plant-sucking insects against predators.

The magnitude and timing of a honeydew flow are much less predictable than those of a nectar flow, because they depend on the build-up of populations of the appropriate plant-sucking insects, which is affected by past weather conditions. They often depend also on conditions governing the populations of ant species.

Figure 12.22a Production of honeydew by *Physokermes hemicryphus* on a stem of spruce, *Picea abies* (photo: A. Fossel).
The insect attaches itself to the stem between bud scales, and is covered with wax threads so that it hardly looks like an insect. A young female here has excreted a drop of honeydew, which a bee will collect directly from her.

The complex process of honeydew production, by different insect species on different parts of different host plants, has been largely unravelled in the past few decades. It is the subject of books by Gleim (1984) and by Kloft and Kunkel (1985), and the latter gives many further references. Earlier shorter accounts in English are by Kloft (1963) and Maurizio (1975*a*).

The characteristic composition of honeydew is somewhat different from that of nectar, because honeydew contains enzymes derived from the gut and saliva of the plant-sucking insect. Some honeydews contain much melezitose (see Section 13.28), a sugar which is so insoluble in water that the honeydew may crystallize on the plant.

12.22.2 Plant sources of honeydew

The most important sources of honeydew are trees, and of these, conifers (which produce no nectar) give the highest yields. A few deciduous trees give important flows of honeydew only, and others of both nectar

Table 12.22A Some important sources of honeydew honey

From Crane and Walker (1985*a*), who give details of the plant-sucking insects involved. For further information see Crane, Walker and Day (1984).

	Deciduous trees	
Conifers, honeydew only	*honeydew only*	*nectar + honeydew*
Abies alba, silver fir	*Fagus sylvatica*, beech	*Acer platanoides*, Norway maple
Abies borisii-regis; A. cephalonica, Greek fir	*Nothofagus solandri* v. *cliffortioides*, mountain beech	*Acer pseudoplatanus*, sycamore
Calocedrus decurrens, incense cedar	*Populus* spp., poplar	*Castanea sativa*, sweet chestnut
Larix decidua, larch	*Quercus robur*, English oak	*Robinia pseudoacacia*, false acacia
Picea abies, Norway spruce	*Quercus suber*, cork oak	*Salix alba*,white willow
Pinus halepensis, Aleppo pine	*Quercus virginiana*, live oak	*Salix caprea*, goat willow
Pinus sylvestris, Scots pine		*Tilia cordata*, small-leaved lime
		Tilia platyphyllos, large-leaved lime

and honeydew (Table 12.22A). Many species of plants may produce a small amount of honeydew; they are often disliked by beekeepers, especially where there is atmospheric pollution, because sticky honeydew on broad leaves traps dust, soot and other materials; if these are present in honey stored in the hive, they reduce its quality.

In addition, crops – and some wild plants – are hosts to aphid pests that may produce honeydew. Cotton, lucerne, sugar cane and sunflower are examples. If pest control measures kill all the aphids, honeydew production will cease. (Whether or not the measures do this, they may kill bees foraging on the honeydew.)

One example of a valuable honeydew honey source must suffice here: the insect *Marchalina hellenica*, living on the Aleppo pine *Pinus halepensis*, including the former variety *brutia*, referred to by Mirnov (1967) as *P. brutia*. It provides the most important honeydew flow in many parts of the Aegean region of the Mediterranean, including mainland Greece (Santas, 1983) and Turkey and in Crete and many other islands. Figure 12.7a, map 5 shows the distribution of *P. halepensis*. In late summer, Turkish beekeepers take 500 000 colonies of bees to the south-west coastal region, especially around Muğla, for this flow. There is a record from 259 BC of honey being exported to Egypt from Theangela in the same region (Crane & Graham, 1985), and this would probably have been from *P. halepensis*.

In parts of Turkey, and of Greece (including Crete), beekeepers extend the range of the honeydew flow by introducing *M. hellenica* into areas of *P. halepensis* without it. In late summer, the mature females lay

Figure 12.22b Twig of *Pinus halepensis* infested with *Marchalina hellenica*, showing the white fluffy egg cases (photo: L. Santas).

their eggs under the bark of the trees, and the fluffy white egg cases (Figure 12.22b) are easily visible to anyone passing by. Beekeepers remove a few branches carrying egg cases, and take them to an area where the pine grows but the insect is absent. The branches are tied into a few trees, and if conditions are suitable the insects will spread and multiply (Crane & Walker, 1985*a*). I travelled through a good deal of *Pinus halepensis* country in Turkey and Crete in 1985, and nowhere did I find much objection to this practice, which is known as grafting. It is, however, unsuccessful in some areas.

An important broad-leaved honeydew source, the New Zealand beech *Nothofagus*, has been exploited for honey production only during the past few decades (Walton, 1979; Cook, 1981). Its distribution is shown in Figure 12.7a, map 5.

12.23 Other sugar-containing materials

Bees are likely to collect any sweet liquid that does not have a flavour or aroma obnoxious to them. The only such plant source of economic importance in honey production is cut sugar cane (*Saccharum officinarum*), a crop widely grown in the tropics. Olbrich (1982) lists the harvesting months in different regions. Where no nectar is available, bees forage on the stems after they have been cut, or – as on Réunion in the Indian Ocean – when they are burned to facilitate mechanical harvesting, and consequently burst; bees may also forage on material in the sugar mills. The sugar content of the sap is reported to be around 10% in Mauritius, and 18–24% in Bolivia (Crane, Walker & Day, 1984). What the bees collect lacks aromatic and other constituents that characterize nectar. In Mauritius, a peasant told me: 'Most of the year the honey is good, but when they cut the sugar cane [between July and December] it is not good'. In South Africa samples of honey from sugar cane contained no amylase, but twice as much phosphate as flower honey (Guy, 1976).

If the bees collect only 0.1 ml of sap from a cut cane, using the Mauritius figure of about 100 000 canes per hectare, the honey potential would be 1.25 kg/ha; with 100 000 ha under sugar in Mauritius, as in 1982, the yield on the island could be 125 tonnes a year.

Sap of various palms is tapped for making a drink known as toddy, and bees may forage on the liquid that accumulates in the collecting pots, although many become intoxicated and are drowned (Kannagara, 1940).

Honeybees collect the juice inside fruit where the skin has been damaged. Their mandibles cannot reach the contents of smooth-skinned fruit, but apparently can pierce the crinkly porous skin of Manaresi figs (Giordani, 1953). There have been assertions from time to time that bees damage grapes, but bees in cages died of starvation when grapes were the only source of food.

Bees will forage on *honey* in any accessible place that is not adequately defended: 'empty' barrels or combs left outside; hives or nests of other bees that cannot repel the robbers; honey houses, or carelessly aligned hive boxes containing honey above an escape board, to which robber bees can gain entry. Upon occasion bees collect sweet materials from food-manufacturing plants. I have seen well-filled sections of bright red honey from a hive near a factory making raspberry jam in Scotland, and 'chocolate' honey has appeared in hives near a confectionery factory. In a detective story by Sheila Pym (1952) which involved poisonous honey, samples providing essential evidence were left inadequately secured in a police station over a weekend; bees found them, and by Monday the jars were empty. Beekeepers will find no difficulty in giving credence to this fictional incident.

Finally, it is necessary to consider the sugar (sucrose) fed to colonies of bees in the course of beekeeping practice. According to the legal definition, honey must be produced by bees from materials they collect *from plants*, so it would be fraudulent to feed sugar to bees in order to harvest it from the hive to sell as honey. This may, however, be done inadvertently, as is shown at the end of Section 6.53.1.

12.3 WATER NEEDS IN RELATION TO FOOD SUPPLY

Nectar and pollen are the bees' foods, and a colony can store honey and pollen in cells of the combs. Water is also essential to the colony, for several reasons, and a little can be stored in the honey sacs of reservoir bees; see Section 3.36. (There is a record from the Kalahari desert (Clauss, 1984) of colonies recycling water between nests: bees from a wild colony collected water condensed in the tree cavity from which the colony had been removed.)

Many natural sources of water are edges of streams, lakes and other waterways which afford the bees a foothold, also damp patches of ground, vegetation wet with rain or dew, and open drains. In early spring, some sources of water may be too cold for the bees to use. Requirements for a natural water supply are:

- permanence
- nearness to the hives (the closer the better, and preferably not more than 0.5 km)
- clean water acceptable to the bees
- safe access to the water
- in cool spring weather, a shallow source that warms quickly, in a sheltered sunny position
- no contamination from surfactants; see near end of Section 2.43.2.

Figure 2.43b shows the interrelation between the bees' collection of water and of nectar.

12.4 FOOD FOR GROWTH AND DEVELOPMENT: POLLEN

Pollen-producing plants have existed for over 200 million years. Fossil pollen from water lilies (*Nympha*) has been dated to 140 million years ago, and these flowers are today a main source of pollen and nectar in part of the Okavango Swamp in Botswana (Cooks, 1984).

Unlike nectar, pollen contains a wide variety of nutrients that support growth and development in living organisms. Many of the nutrients, and especially certain amino acids, are essential for the development of honeybees from the egg to the mature adult, and pollen is the bees' only source of them. The larvae obtain them in brood food produced by young worker 'nurse' bees, which themselves eat pollen (Section 2.42). Foraging bees collect the pollen, and in the process transfer some grains to the stigma of the same or other flowers, thus pollinating them (Section 2.44). A pollen forager 'packs' the pollen into a compact load on each hind leg (Section 2.43.3, Figure 2.43d), and flies back to the hive with it. Colonies do not normally store pollen greatly in excess of their requirements, and pollen-producing plants are important throughout the bees' active season. A beekeeper who wants to harvest pollen must use a 'pollen trap' to remove pollen carried by the bees *before* they enter the hive (Section 14.22); the colony continues to need pollen, and foragers therefore continue to collect it.

12.41 Pollen production by plants

The complex processes within the plant that lead to the formation of pollen grains are described by Stanley and Linskens (1974). The pollen grains are produced in the flower's anthers (Figure 2.44a), each of which is at the outer end of the stamen, the flower's male sexual organ. Flowers that are wholly female produce no pollen, and those of some hybrids, e.g. lavandin, *Lavandula angustifolia* × *latifolia*, produce sterile pollen.

When pollen grains have matured inside the anther, the anther wall opens and the ripe pollen is released or 'presented'; this is dehiscence. Percival described the process in many plants (1950), and gives a shorter account in *Floral biology* (1965). Table 12.41A here quotes some of her results, showing that, for instance, a flower of oriental poppy produces over 1000 times as

Table 12.41A Amount and daily cycle of pollen production in 8 temperate-zone plants

Data from Percival (1950, 1965). Plants, all in South Wales, UK, are listed in order of the time of day at which pollen collection by bees was greatest. The figure in column 2 is the number of flowers a honeybee must visit to collect her load of pollen.

		Daily period of pollen production (GMT)	
	No. visits per load	*peak*	*range*
Epilobium angustifolium, fireweed	1	before 9	all 24 hours
Papaver orientalis, oriental poppy	10 loads/flower	8 to 9	6–9 to 13–17
Heracleum sphondylium, cow parsnip	2	9 to 10	before 8 to ?12
Hypochaeris radicata, catsear	20–149	before 10	10 to 13
Taraxacum officinale, dandelion		10	9 to 15
Centaurea nigra, knapweed	45–50	10 to 13	a.m. to 13.30–17
Vicia faba, field bean		13	10 to 17
Trifolium repens, white clover	106–166	13 to 17.30	7–12 to 18

(flowers open from 7 to 12 h GMT and close after 15 h; anthers dehisce in the bud, and pollen is presented on stimulation) much pollen as the much smaller individual floret of white clover. The table also shows that pollen, like nectar, is rarely produced throughout the daylight hours; bees are thus likely to visit a particular plant for pollen during only certain parts of the day.

12.42 Plant sources of pollen

Experiments have been done in which pollen loads brought to the hive during a whole season have been identified as to plant source (Louveaux, 1954, 1958/59). They show that each colony collects pollen from a great variety of plants, and that different selections are made by individual colonies, even within the same apiary. The reasons for this are not entirely understood. It may well be that collection from many different plants helps to ensure that the colony obtains all the constituents required in the bees' diet; if so, this is achieved by a behavioural mechanism different from that which makes honeybees species-constant in nectar collection.

The majority of angiosperms (flowering plants) are insect-pollinated, and their pollen is the most nutritious for bees; European examples include fruit (*Pyrus, Prunus*, etc.), willows (*Salix*) and white clover (*Trifolium repens*). Dandelion (*Taraxacum*) is an exception, apparently because many of the grains remain intact in the bee's midgut, or are only partly digested (Peng et al., 1985). Pollens of wind-pollinated angiosperms are less good for bees; examples in Europe are poplar (*Populus*) and elm (*Ulmus*), these being better than alder (*Alnus*), birch (*Betula*) and hazel (*Corylus*). Gymnosperms (conifers and related plants) are wind-pollinated, and their pollens, e.g. *Pinus, Picea* and *Abies*, are the least nutritious to bees. Solberg and Remedios (1980) found that pollens from 3 conifers contained much more crude fibre (mean 29.2%) than pollens from 13 angiosperms (mean 3.3%).

Mixtures of pollens are more effective than single-species pollens. In Arizona, USA, a mixture of bee-collected pollens from *Populus, Prosopis* and *Rubus* greatly increased the length of life of caged bees (Schmidt et al., 1987).

Some *Eucalyptus* species are pollinated in their native regions by birds which eat the flesh of the flowers. The pollens of a number of them are not nutritious to bees; some are not even collected. This situation presents difficulties to the beekeeper: colonies might be able to produce much honey from a eucalyptus nectar flow, but without a supply of adequate pollen they would be unable to rear a new generation of bees to keep the colony alive. He can overcome this difficulty by providing pollen (or a substitute) in the hive; see Section 6.32. In some regions, pollen from lucerne (alfalfa) contains insufficient protein to maintain brood rearing. Where colonies of bees are used to pollinate large areas of alfalfa in California, safflower (*Carthamus tinctorius*) is grown as a pollen source in neighbouring fields (Motter, 1981).

Most plants that secrete nectar also produce pollen, and honey sources listed in Appendix 1 are marked accordingly. Pollen production may well not accompany nectar secretion from extrafloral nectaries and honeydew production by plant-sucking insects. Colonies working honey flows from such plants should therefore be fed with pollen or pollen substitute to enable them to maintain brood rearing.

On the other hand honeybees collect pollen from many plants that produce no nectar – especially agricultural crops such as sorghum (Figure 12.42a), maize and rice. Maize produces much pollen (Section 14.23), and in some parts of the tropics beekeepers plant a maize plot for their bees.

A few pollens contain substances that are toxic to bees, for instance some species of *Rhododendron, Ranunculus* (due to anemonine content), *Aesculus* and *Tilia* (saponin), *Hyoscyamus* (an alkaloid) and *Asclepias*

Figure 12.42a Tropical African honeybee collecting pollen from sorghum in Botswana (photo: B. Clauss).

(galitoxins). Pollen on crop plants treated with a systemic insecticide can become toxic to bees, and can then damage or kill brood in the hive, as well as adults. Section 12.21.6 lists some of the very toxic substances, and Atkins and Kellum (1986) give further information.

Section 2.43.3 describes the honeybee's behaviour in collecting pollen from flowers, and 12.63 the seasonal cycle of pollen collection; Section 14.2 deals with pollen as a hive product, and Tables 14.21A and 14.21B give details of its composition.

12.5 MATERIAL FOR NEST CONSTRUCTION: PROPOLIS

The most important materials collected by honeybees – nectar and pollen – are referred to by these botanical terms. 'Propolis' is a bee-oriented term that does not have a botanical derivation; it was used by authors in Ancient Greece: *pro* (in front of, i.e. at the entrance to) and *polis* (city or community). Bees were seen using the material to restrict, adjust or protect their hive entrances; see Section 3.31.4.

An appropriate definition of propolis might be 'material that honeybees and some other bees can collect from living plants, and use alone or with beeswax in the construction and adaptation of their nests'.* This includes substances actively secreted by plants, and exuded from wounds in plants. But it excludes materials not of plant origin, such as bitumen and paint, that are occasionally collected by bees and used in the same way as propolis. In Europe propolis is soft enough to be collected and worked by bees only at temperatures of 18° or above (Section 2.43.4).

Section 14.3 discusses the composition and properties of propolis and its use, as a hive product.

12.51 Plant sources of propolis

The propolis available to bees is produced by a variety of botanical processes, in different parts of plants. Much of it is produced high up in trees, where it is very difficult to make observations on the bees' activities. Many authors refer to bees collecting sticky material on leaf buds of certain trees; the tree species vary from country to country, e.g. poplars (*Populus*) in France and birches (*Betula*) in USSR.

Using the definition above, Table 12.51A lists 67 plants from which honeybees have been reported to collect material referred to as propolis, with an indication of the probable type of plant substance collected. Only a few studies (e.g. Popravko, 1976; Marletto, 1983; Greenaway et al., 1987) have succeeded in comparing the composition of propolis brought to the hive by foraging bees with the secretion or exudation of the plant from which the propolis was collected. Figure 12.51a shows the material on the plant, and also a honeybee collecting it.

The botanical origins of the plant substances are discussed below. Only sources used by *Apis mellifera* are considered; other *Apis* species (except *cerana*) also collect propolis.

12.52 Production by plants of materials collected as propolis

The production of these materials within the plant, and the routes by which they may become accessible to bees, have been rather little studied. In the following summary, based on Fahn (1979, 1986), reference is made to plants in the families listed in Table 12.51A.

Tissues secreting sticky lipophilic substances (including flavonoids)

From secretory epidermal cells of leaf buds, for instance in poplars (*Populus*), the secreted material is first eliminated into a space between the outer walls of the palisade-like epidermal cells (known as prismatic cells) and the cuticle covering them, forming a blister. Later the cuticle bursts (Figures 12.52a, 12.52b), and the secreted material collects between the leaves and stipules of the bud.

Glandular trichomes (hairs) on buds of species of *Alnus* and other trees secrete a substance containing flavonoid aglycones, terpenes and mucilage (see under Tissues secreting mucilages and gums); Fahn (1979) describes the secretion of flavonoid aglycones in *Aesculus hippocastanum*.

Secretory cavities and ducts

Several types of secretory tissue eliminate essential oils or resins, lipophilic substances (containing terpenes) or gum-resins (containing terpenes and polysaccharides) into intercellular spaces, which are often just below the epidermis, and the material may become accessible to bees as a result of injury, etc.

Spherical spaces of this type occur, mostly in leaves or fruits, in many families. Those in Table 12.51A

* Some authors define propolis as the material collected by beekeepers from hives (which may contain 50% or more of wax); the propolis fraction is then referred to as balsam (e.g. Ghisalberti, 1979). Stingless bees (Section 8.63) do much of their nest building with a mixture of wax and propolis, called cerumen. The term cerumen might well be used also for mixtures of wax and propolis used by honeybees in their building operations (Section 3.31.4), as Michener does (1974).

Table 12.51A Plants reported to be sources of propolis collected by the honeybee *Apis mellifera*

All plants are trees or shrubs unless marked *herb*.
Column 3 indicates the reference listed at the end of the table, and a country or region in which the collection of propolis from the plant by honeybees was reported.
An entry in the final column indicates the probable type of substance according to the author quoted, or Fahn (1979, 1986) or Howes (1974). Lipophilic substances (lipo.), mucilage and gum are plant secretions; resin and latex are wound exudates.

Botanical name	*Common name*	*Region*		*Substance*
Anacardiaceae				
Mangifera indica L.	mango	C4	Thailand	resin
		C4	Montserrat	
Schinus terebinthifolius Raddi	Brazilian pepper	K1	Hawaii	resin?
Apocynaceae				
Landolphia capensis Oliver	wild apricot	A1	S. Africa	latex
Plumeria rubra L.	frangipani	K1	Hawaii	latex
Araucariaceae				
Araucaria heterophylla (Salisb.) Franco	Norfolk island pine	C4	Barbados	resin
Betulaceae				
Alnus sieboldiana Matsumura		M2	Japan	lipo.?
Alnus viridis (Chaix) DC.	green alder	G1	USA	lipo.
Betula pendula Roth	silver birch			
as *B. alba*		G2	USSR	lipo.
as *B. verrucosa*		P2	USSR	lipo.
Burseraceae				
Bursera excelsa (Knuth) Engel.	copal	O1	Mexico	lipo.
Bursera simaruba (L.) Sarg.	gumbo limbo	O1	Mexico	lipo.
	turpentine tree	C4	Montserrat, St Kitts-Nevis	resin
Protium copal Engl.	copal	O1	Mexico	lipo.
Caryophyllaceae				
Lychnis viscaria L.	German catchfly (*herb*)	F1	USSR	
Casuarinaceae				
Casuarina equisetifolia J. R. Forst. & G. Forst.	casuarina	C4	Barbados	resin
		E2	Egypt	
		S1	Seychelles	
Compositae				
Cynara cardunculus L.	cardoon (*herb*)	C1	Chile	
Helianthus annuus L.	sunflower (*herb*)	H1	Britain	mucilage
Cornaceae				
Cornus mas L.	dogwood	C2	Romania	
Euphorbiaceae				
Hevea brasiliensis Muell. Arg. (from leaves)	pará rubber	C3	Botswana	wax

(*continued*)

Table 12.51A *(continued)*

Botanical name	*Common name*	*Region*		*Substance*
Fagaceae				
Castanea sativa Mill.	sweet chestnut	L1	France	
Quercus robur L.	English oak	L1	France	
Guttiferae				
Calophyllum inophyllum L.	takamaka	S2	Seychelles	
Hippocastanaceae				
Aesculus hippocastanum L.	horse-chestnut	G2	USSR	lipo.
		L1	France	
Labiatae				
Salvia officinalis L.	garden sage (*herb*)	G2	USSR	
Lauraceae				
Laurus nobilis L.	bay laurel	R2	Italy	
Leguminosae				
Acacia (some spp. are sources of gum arabic)		A1	S. Africa	mucilage
Acacia karroo Hayne	karroo thorn	C3	Botswana	mucilage
Brachystegia utilis Burtt Davy & Hutch.	false mfuti	D1	Zimbabwe	
Hymenaea courbaril L.	locust	E1	Mexico	resin
Julbernardia baumii (Harms) Troupin		D2	Zaire	resin?
Schizolobium parahybum (Vell.) Blake		D2	Zaire	resin?
Sophora japonica L.	pagoda tree	C2	Romania	resin?
Liliaceae				
Xanthorrhoea spp. including *johnsonii*	grass trees	B1	Australia	resin
Xanthorrhoea australis R. Br.	grass tree	G3	Australia	resin
Xanthorrhoea preissii Endl.	grass tree	G3	Australia	resin
Malvaceae				
Althaea rosea (L.) Cav.	hollyhock (*herb*)	H1	Britain	mucilage
Hibiscus rosa-sinensis L.	Chinese hibiscus	D2	Zaire	mucilage
Moraceae				
Artocarpus altilis (Parkinson) Fosberg	bread fruit	C4	Barbados, Montserrat	latex
Artocarpus heterophyllus Lam.	jack fruit	R1	Seychelles	latex
Ficus carica L.	edible fig (from surface of fruit)	C3	Botswana	resin
Ficus elastica Roxb. ex Hornem.	Indian rubber	C4	Thailand	latex
Ficus thonningii Blume		P3	Angola	latex
Myoporaceae				
Myoporum laetum Forst. f.	ngaio	W1	NZ	
Myrtaceae				
Eucalyptus sideroxylon A. Cunn. ex Woolls	black ironbark	K1	Hawaii	
Psidium guajava L.	guava	K1	Hawaii	

(continued)

Table 12.51A *(continued)*

Botanical name	*Common name*	*Region*		*Substance*
Oleaceae				
Fraxinus L.	ash	P1	USA	
Fraxinus excelsior L. ash		V1	unknown	
Onagraceae				
Epilobium angustifolium L.	rosebay willow-herb (*herb*)	G2	USSR	
Palmae				
Cocos nucifera L.	coconut	S2	Seychelles	
Pinaceae				
Abies alba Miller	silver fir	C2	Romania	resin
Pinus wallichiana A. B. Jacks	Bhutan pine	C2	Romania	resin
Rosaceae				
Prunus armeniaca L.	apricot	C2	Romania	gum
		E2	Egypt	
Prunus avium (L.) L.	sweet cherry	F1	USSR	gum
Prunus cerasifera Ehrh.	cherry-plum	C2	Romania	gum
or *Prunus cerasus* L.	sour cherry			
Salicaceae				
Populus alba L.	white poplar	M1	Italy	lipo.
Populus angustifolia E. James	narrow-leaved cottonwood	L2	USA	
Populus balsamifera L.	balsam poplar	L2	USA	
Populus canadensis Moench.	Carolina poplar	M1	Italy	lipo.
Populus deltoides Bartr. ex Marsh.	cottonwood	A1	S. Africa	lipo.
Populus × *euroamericana* (Dode) Guinier		G4	UK	
		W2	USA	
Populus nigra L.	black poplar	M1	Italy	lipo.
Populus tremula L.	aspen	M1	Italy	lipo.
Populus trichocarpa Torr. & A. Gray	black cottonwood	H1	Canada	lipo.
Salix alba L.	white willow	C3	Romania	lipo.
Salix babylonica L.	weeping willow	E2	Egypt	
Salix caprea L.	goat willow	G2	USSR	lipo.
Salix cinerea L.	grey willow, common sallow	C2	Romania	lipo.
Sterculiaceae				
Sterculia tragacantha Lindl.	gum tragacanth	D2	Zaire	gum
Zygophyllaceae				
Larrea tridentata (DC.) Cov.	creosote bush	L3	USA	

A1 Anderson et al. (1983)
B1 Blake & Roff (1972)
C1 Cárdenus (1939)
C2 Cîrnu (1980)
C3 Clauss (1983)
C4 Crane (1988)
D1 Dawson (1986)
D2 Dubois & Collart (1950)
E1 Espina & Ordetx (1983)
E2 Elsharawi (1988)
F1 Fedosov (1955)
G1 Ghisalberti (1979)
G2 Glukhov (1955)
G3 Goodman (1973)
G4 Greenaway et al. (1988)
H1 Howes (1979)
K1 König (1985)
L1 Lavie (1960)
L2 Lovell (1926)
L3 Lovell (1957)
M1 Marletto & Olivero (1981)
M2 Mochida et al. (1985)
O1 Ordetx et al. (1972)
P1 Pellett (1947)
P2 Popravko (1976)
P3 Portugal-Araújo (1974)
R1 Ratia (1984)
R2 Ricciardelli d'Albore & Tonini d'Ambrosio (1981)
S1 Silberrad (1969)
S2 Silberrad (1970)
V1 Vanhaelen & Vanhaelen-Fastré (1979)
W1 Walsh (1978)
W2 Wollenweber (1987)

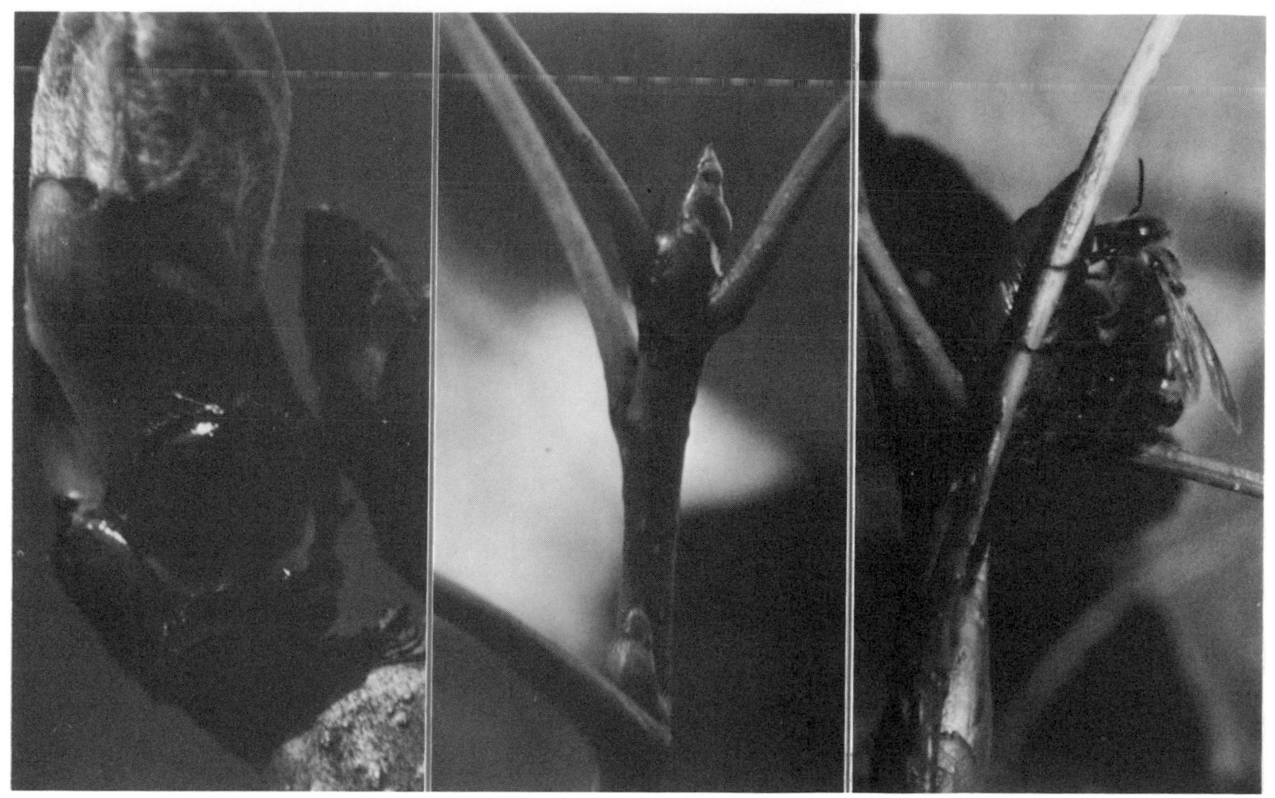

Figure 12.51a Propolis and its collection by honeybees (Marletto & Olivero, 1981).
left Surface of a horse chestnut leaf bud shining with colourless resin.
centre Poplar leaf bud covered with a heavy deposit of yellow resin.
right Honeybee collecting resin from a poplar leaf bud.

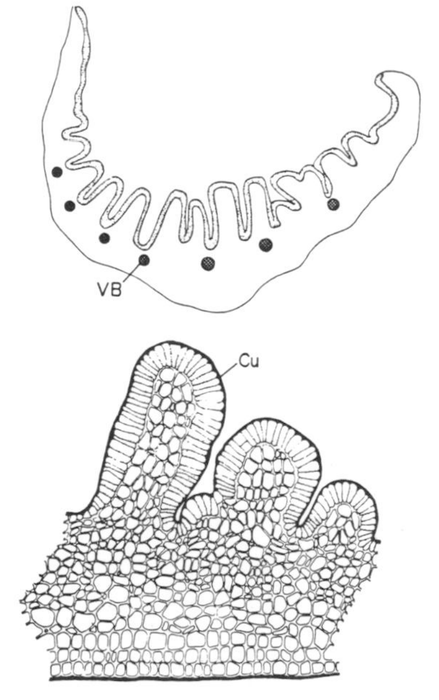

Figure 12.52a Secretory adaxial epidermal cells (stippled) of *Populus pyramidalis* stipules, before elimination of the secreted substance (Fahn, 1979).
above Stipule cross-section (× 120); VB = vascular bundle.
below Portion of stipule cross-section (× 525); Cu = cuticle.

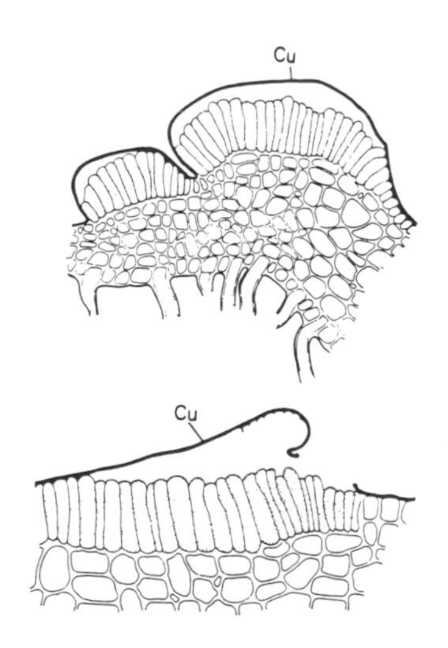

Figure 12.52b Portion of a cross-section of a stipule of *Populus alba* showing detached cuticle (Fahn, 1979).

above Secretory material was eliminated from the epidermal cells to the space below the cuticle, Cu (× 525).

below Portion of a cross-section of a leaf tooth of *Populus pyramidalis* showing ruptured cuticle (Cu) through which the secreted substance passes to the leaf surface (× 525).

include Anacardiaceae, Leguminosae, Myoporaceae and Myrtaceae. In stems, the spaces (ducts) are more elongated; they occur in members of the Anacardiaceae, Compositae, Leguminosae, Myrtaceae and Pinaceae.

Such resin ducts are common in conifers, the secreted resin containing a variety of terpenes. In some trees, e.g. *Picea*, *Pinus* and *Larix*, resin ducts are normal, but in others, e.g. *Abies* and *Cedrus*, they are produced only as a result of an injury.

Tissues secreting mucilages and gums

Plant mucilages and gums contain complex polysaccharide polymers of high molecular weight, and they have a wide variety of origins within plants. Among reported sources of propolis are gums or mucilages from the heads of *Helianthus annuus* and the buds of *Althaea rosea* (Howes, 1979).

Traumatic gum ducts

In some plants, organized cell-wall materials may be converted into unorganized amorphous substances such as gums (polysaccharides) as a result of injury by micro-organisms, insects, mechanical injury, or physiological disturbances. The process occurs mainly in the Prunoideae, a subfamily of Rosaceae. Gum arabic is produced in the bark of *Acacia senegal* and other species including *A. nilotica*. Kino veins, which occur in the cambial region of *Eucalyptus* wood, consist of a special type of traumatic duct, and – unlike gum – kino contains polyphenols.

Laticifers

These are secretory tissues that mostly produce and store latex, of which rubber from *Hevea brasiliensis* (Euphorbiaceae) is the best known example. The families Apocynaceae, Compositae, Liliaceae, and Moraceae include genera (not in Table 12.51A) that have laticifers.

Other possible sources

In some plants a wax coating is produced on leaves, stems or other parts. In Botswana Clauss (1983) saw honeybees scraping wax off the surface of leaves of rubber, *Hevea brasiliensis*, which were thereby severely damaged.

Propolis thus has a much more complex origin than any other material collected by honeybees, and samples are likely to show great differences in composition unless they are from the same plant source. This fact should be taken into account when making pharmaceutical formulations based on propolis (Section 14.33). Propolis from certain *Populus* and *Betula* species, in their respective areas, seems to be more collected by honeybees than any other. Popravko (1976) found that the composition of propolis produced by the birch *Betula pendula* differs at different seasons, i.e. when the stage of plant growth – and the temperature – are different. Figure 2.43e shows the seasonal pattern of propolis collection by honeybees in northern Italy; the propolis (from poplars) is soft in the summer months, but hard in March/April and in the autumn.

Section 2.43.4 describes the bees' collection of propolis, and 3.31.4 their use of it in the hive.

12.6 SEASONAL PATTERNS OF FOOD RESOURCES

Earlier chapters deal with beekeeping management in regions where limiting factors to bee activity are cold (Chapter 6), and heat, drought or rain (Chapter 7). The same constraints affect plant growth and flowering, and are reflected in seasonal patterns of the bees' food resources.

12.61 General seasonal patterns

Many studies – most of them in the north temperate zone – have been made in attempting to unravel the patterns of the honeybees' food and other resources through the year. In the main, honeybees collect food (nectar, pollen and honeydew) whenever it is available, and research workers often estimate availability from amounts that the bees bring to the hive. Nectar and pollen production are associated with flowering, and honeydew production depends on the presence of a large population of the insect that produces it, feeding on the appropriate host plant (Section 12.22.1); it usually occurs in late summer. Propolis and water are in a different category. Propolis may be available throughout most of the bees' active season, but its collection by honeybees drops off during a nectar flow; Figure 2.43e demonstrates this, and the subject is discussed in Section 2.43.4. Bees also collect much less water during nectar flows, because nectars contain water; see Figure 2.43b.

Sections 12.62 and 12.63 below discuss the availability of nectar and pollen, respectively, at different latitudes where studies have been made. Availability and collection of nectar (plus any honeydew) are usually estimated from weekly changes in the total weight of a colony of honeybees, kept in a hive on scales; for one type of scales, see Section 5.42. Availability and collection of pollen are estimated by weighing the pollen collected in a pollen trap fitted to a hive entrance, at regular intervals. Neither method measures exactly what is wanted, and the deficiencies are pointed out below, but each uses inexpensive equipment to provide useful comparative data: in different locations, in different seasons, and for different colonies.

12.62 Seasonal patterns of honey sources

An indication of some of the world's different seasonal patterns of food resources for honeybees – and of the potential honey production of a region – may be obtained by studying 'scale-hive records'. These are usually presented in the form of weekly changes in the weight of a colony of bees, including the adult bees and brood, stored honey and pollen, and the combs; honey is usually the largest and the most rapidly changing component of the total weight.*

Figure 12.62a shows 6 scale-hive records for colonies at different latitudes, in a sequence from north to south. Weights shown are corrected for the issue of any swarms, and for the addition or removal of other weights such as hive supers or syrup. A point below the horizontal axis indicates a net consumption of food by the colony. At the start of a cold dearth, if the beekeeper does not leave an amount of food in the hive that exceeds the total net consumption until the next flow period, the colony will starve.

Colonies 1, 2 and 6 were in temperate zones; 3 just in the subtropics; and 4 and 5 in the tropics, 4 being within the equatorial zone.

Colony 1, at the highest latitude (60°N in Sweden), lost weight – or did not gain more than 1 kg – during every week in the year except the 4 weeks 25–28, mid-June to mid-July. But the greatest weekly gain of 31 kg (in week 27) was the highest achieved by any of the colonies represented. Fries (1980) gives details of the flows; the most important were from varieties of rape (*Brassica campestris*, *B. napus*).

At 40°N in Ohio, colony 2 gained weight for 10 consecutive weeks (17–26) between late April and late June. The low early peaks were due to flows from dandelion (*Taraxacum officinale*) and fruit, and black locust (*Robinia pseudoacacia*); the high peak (25 kg) was from legumes. After the end of the flows in June, temperatures were still high and colony 2 was large; it thus consumed much honey – always more than it collected, except during late minor flows; the last of these (from golden rod, *Solidago* spp.) continued until weeks 38 and 39 in late September. The colony was active throughout the winter months and was consuming up to 3 kg honey in a week.

Still farther south and just in the subtropics at 32°N in Arizona, colony 3 had an 18-week period with a weight gain almost every week (weeks 7–24, mid-February to mid-June); mesquite (*Prosopis juliflora*) was the main source, and there were other less important ones. August gave a smaller flow from tamarisk (*Tamarix* spp.). The frequent weight changes throughout most of the year show that minor sources were available, and that the colony was almost continuously active, with bees flying.

The record for colony 1 shows one short prolific main honey flow. At lower temperate latitudes, records 2 and 3 show in addition a late summer flow, as does record 6 in the warm temperate south. At lower latitudes the flows, and the whole active season of the colony, become more spread out as temperatures allow bee activity for a longer period.

Colony 4 was in miombo woodland in the equatorial zone, and quite near the equator at 5°S in Tanzania; the sun was overhead twice a year, and there were two honey-flow cycles during the year, with peaks at the end of May and the end of November. This is characteristic of the equatorial zone; see Section 7.31. Starting in January, there was occasional flowering of species of minor importance until April. *Isoberlinia globiflora* bloomed from late April (the end of the rains) until mid-June; other species also flowered intermittently. For the rest of the dry season there was a dearth, and the other main flow came when the rains started at the end of October; it was from species of *Brachystegia*, *Combretum*, *Terminalia* and other trees. Nectar secretion ceased with the December dry spell, and the colony weight decreased until the next main flow started, during April. The colony was clearly active throughout the year. Full details of the seasonal changes, and some complications that cannot be entered into here, are given by F. G. Smith (1951).

Record 5 is the mean for three colonies in the tropics, at 22°S in Brazil. A wide variety of plants in flower gave a continuous nectar flow for the whole of the dry winter, which lasted from June to October. With minor flows during a further 4 months in autumn and spring, the colonies made a net weight gain in almost every week. The dearth period, with net weekly losses, lasted through the rainy season, January to March. This record illustrates very well the variety and the continuity of tropical honey sources.

Colony 6 was in the south temperate zone, at 35°S in Argentina, where there was a six months' shift in the seasons from the north temperate zone (colonies 1

* In Scotland, McLellan (1977) measured the separate components of the weight for the 140 days of the active season during which nectar flows occurred. Honey always represented more than 50% of the total weight of the colony, and pollen never more than 10%. At D days after the start of the nectar flow period, when the total colony weight in grams was C, he found that the weight of honey in the colony in grams could be expressed as:

$$-1416.0+0.7604C-57.142D+0.487D^2+0.00142CD.$$

and 2). The main flow period was in summer (mid-November to the end of February), starting with sources that included viper's bugloss (*Echium lycopsis*) and white clover (*Trifolium repens*). The highest peak (14 kg) was from cardoon (*Cynara cardunculus*) and other thistles. Golden rod (*Solidago* spp.) gave a late minor flow, as for colony 2 in Ohio; Crane (1975*b*) gives some further details.

The maximum weekly gain in weight during the main flow shows a fairly consistent decrease from high to low latitudes in the temperate zones, and less change within the tropics. However, individual colony performance varies widely according to circumstances, and it would be wrong to place too much emphasis on the figures here.

The scale-hive records are discussed in relation to the seasonal cycle of the honeybee colony in Sections 3.51.3 and 3.52. The seasonal cycle of beekeeping management is covered for temperate zones in Sections 6.2 to 6.54.2, for tropical zones in general in Sections 7.3 to 7.37, and with different tropical honeybees in Sections 7.4 to 7.44.

12.63 Seasonal patterns of pollen sources

The seasonal pattern of pollen collection by honeybees is studied by using a hive fitted with a pollen trap, different types of which are described in Section 14.22. The pollen is removed from the trap daily or weekly, and weighed and examined; the weight is customarily expressed in terms of g/day. This pollen is the income of pollen *into the pollen trap* during the previous interval, which O'Neal and Waller (1984) estimated to be about 60% of the amount the bees brought to their hives. The percentage varies according to a number of factors, investigated by Levin and Loper (1984); they include the design of the trap, the size of the worker bees, and of the pollen loads they carry in relation to the mesh of the grids in the trap, and the bees' agility and learning ability in manoeuvring their loads through the grids. Louveaux's traps (1954, 1958/59) were designed to collect only 10%.

It is much easier to establish the identity of pollen sources than of nectar sources, because individual pollen grains can be identified as to plant genus or species; see Section 13.26.2. A record of the variety of the sources used by the bees can thus be built up as the season progresses; see Figure 12.63b, *below*.

Figure 12.63a shows records of the weight of pollen trapped each week from hives at 7 latitudes, all north of the equator; 4 of them in the temperate zone and 3 in the subtropics. The highest latitude represented is 50°, and the lowest 30°. In Figure 12.62a only colonies 2, 3 and 6 were within this range; nevertheless, some useful comparisons can be made. Like the weight records in Figure 12.62a, the pollen records show an extended foraging season at lower latitudes: at 50° pollen was collected only during 6 months of the year, and at 30° during 10 to 11 months. The total annual period of pollen collection during the year was consistently longer than the period showing net gains in colony weight at a comparable latitude. (In many weeks without a net weight gain, much nectar may have been *collected*, but consumption outweighed collection because of the amount used in the colony.)

The maximum weekly gain in weight (honey stored) was highest in the scale-hive records at high latitudes. The maximum pollen collection (g/day) was also higher in the temperate zone than in the tropics and subtropics. During at least one week traps on 3 out of 4 hives in the temperate zone showed an average gain of 220 g/day or more, whereas none in the subtropics or tropics gained more than 160 g/day.

In Figure 12.63a some of the records for lower latitudes show signs of a double cycle in the year with two well separated peaks, as for colony 4 in Figure 12.62a. The seasonal pattern of pollen collection at different

Figure 12.62a (*opposite*) Scale-hive records (weekly weight change in kg) obtained at different latitudes. Max. gain = maximum weekly gain recorded.

Colony no.	*Latitude*	*Zone*	*Location*	*Max. gain*	*Year*	*Author*
1	60°N	N temperate	Uppsala, Sweden	31 kg	1975	I. Friese (p)
2	40°N	N temperate	Columbus, Ohio, USA	25 kg	1965–66	K. Fondrk (p)
3	32°N	N subtropics	Tucson, Arizona, USA	18 kg	1978	J. Moffett (p)
4*	5°S	S equatorial	Tabora, Tanzania	7 kg	1950–51	F. G. Smith (1951 + p)
5	22°S	S tropics	Piracicaba, Brazil	10 kg	1957	E. Amaral, 3 hives (1957)
6*	35°S	S temperate	La Plata, Argentina	14 kg	1968–69	L. G. Cornejo et al. (1971 + p)

* 48 weight records in the year not 52; p = personal communication

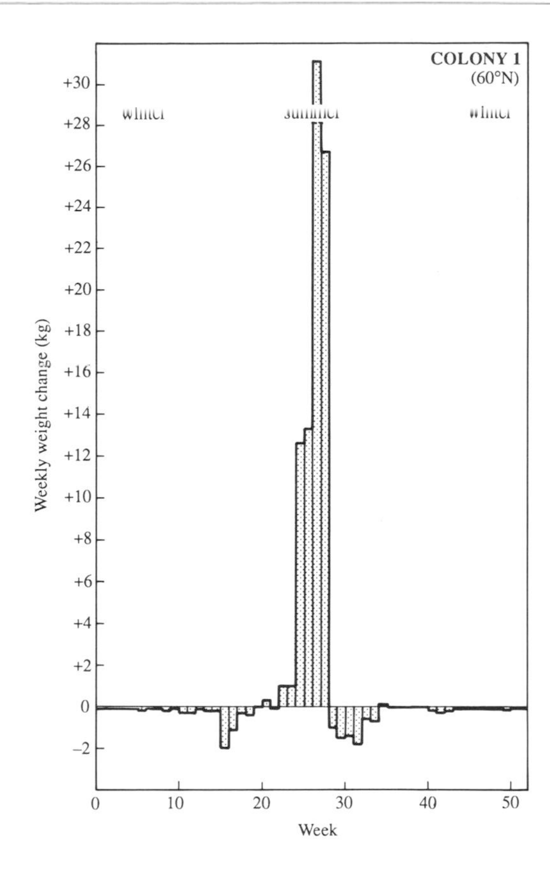
COLONY 1
(60°N)
Weekly weight change (kg)
Week

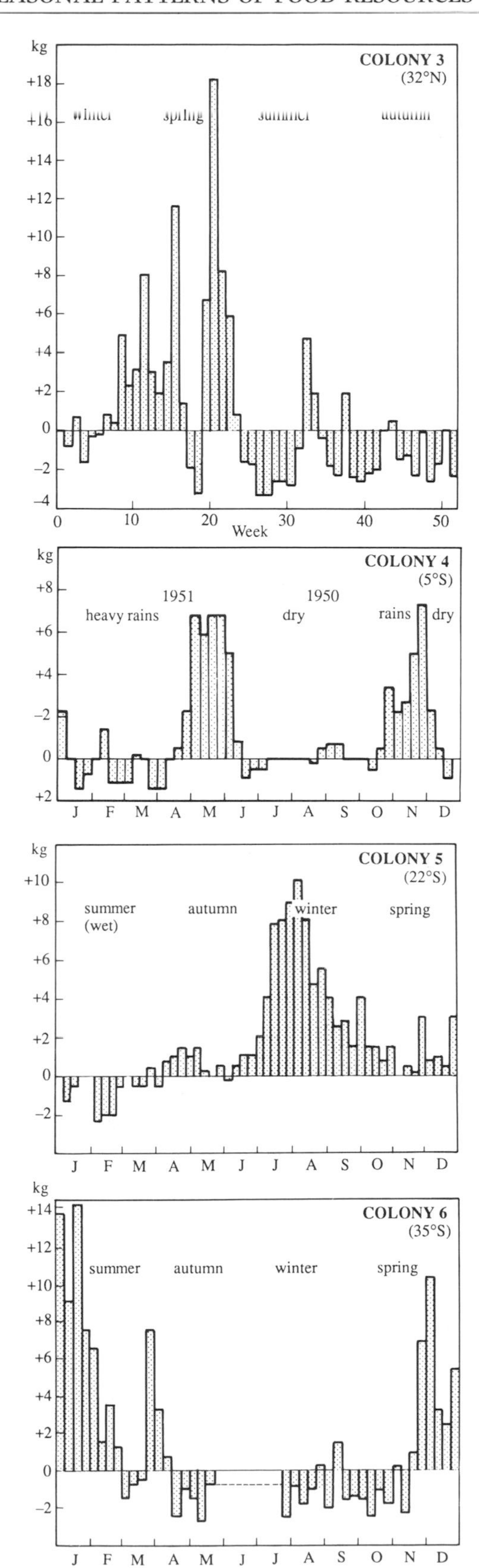
kg
COLONY 3
(32°N)
Week
kg
COLONY 4
(5°S)
1951
1950
heavy rains
dry
rains
dry
J F M A M J J A S O N D
kg
COLONY 5
(22°S)
summer
(wet)
autumn
winter
spring
J F M A M J J A S O N D
kg
COLONY 6
(35°S)
summer
autumn
winter
spring
J F M A M J J A S O N D

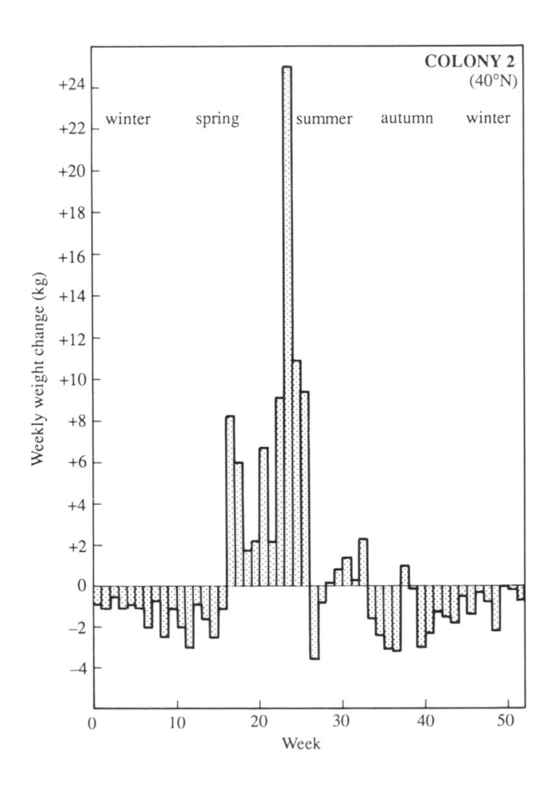
COLONY 2
(40°N)
winter
spring
summer
autumn
winter
Weekly weight change (kg)
Week

Figure 12.63a Seasonal variation in pollen collection by colonies of honeybees at different latitudes (data from O'Neal and Waller, 1984).
Diagrams show weights of pollen (expressed as g/day) taken from pollen traps on hives throughout the year, mostly in the USA. (Two further records are omitted because they do not seem sufficiently comparable.) Diagrams 2, 5 and 6 relate to more than one year, and vertical lines indicate standard deviations.

Colony no.	*Latitude*	*Location*	*Max. gain (g/day)*	*Year*	*Author**
NORTH TEMPERATE ZONE					
1	50°	Vernon, BC, Canada	300	1943	W. H. McMullen
2	49°	Munich, Germany	130	2 yrs	Hirschfelder (1951)
3	44°	Corvallis, OR	250	1949	Vansell & Todd (1949)
4	43°	Guelph, Ont., Canada	240	1977	M. V. Smith
NORTH SUBTROPICS					
5	33°	SW Arkansas	160	1941–51	Thompson (1960)
6	32°	Tanque Verde, AZ	100	1976–80	O'Neal & Waller (1984)
7	30°	Baton Rouge, LA	140	1978	N. M. Kauffeld

* Quoted in the 1984 paper; those giving a date were published previously.

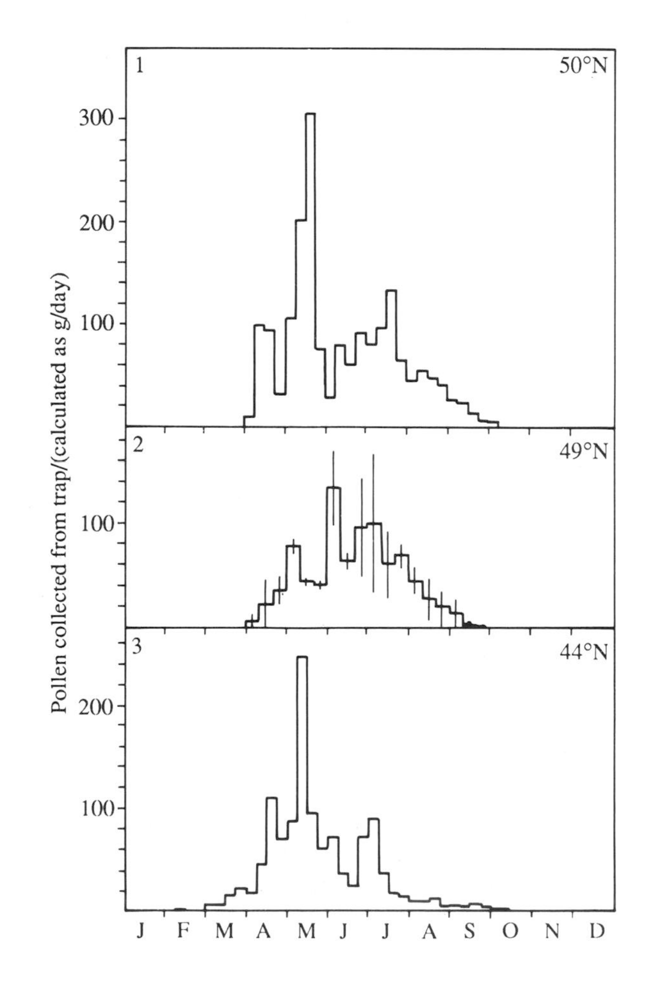

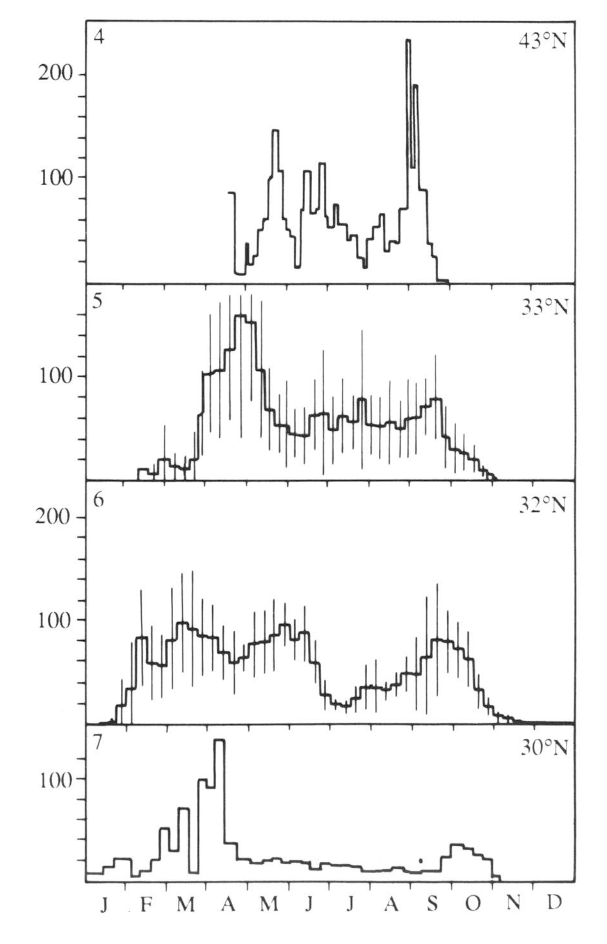

latitudes can be extended to the American tropics by using results from Mexico and Costa Rica. In Mexico at 19°N, Villanueva (1984) used traps fitted with a front grid (see Figure 14.22a); they had three rows of round holes. The amount of pollen collected (Figure 12.63b, *above*) shows a double cycle in the year, typical of the tropics; the rainy season lasted from May to October, and one dearth occurred in June/July, during the hottest part of it. The number of plant species whose pollen was represented in the trap contents each month also showed two maxima (Figure 12.63b, *below*). The first, in February, came in the latter part of the dry season and the first part of the rains, when a great variety of tree pollen was collected. The second came in November when herbaceous plants came into flower after the rains.

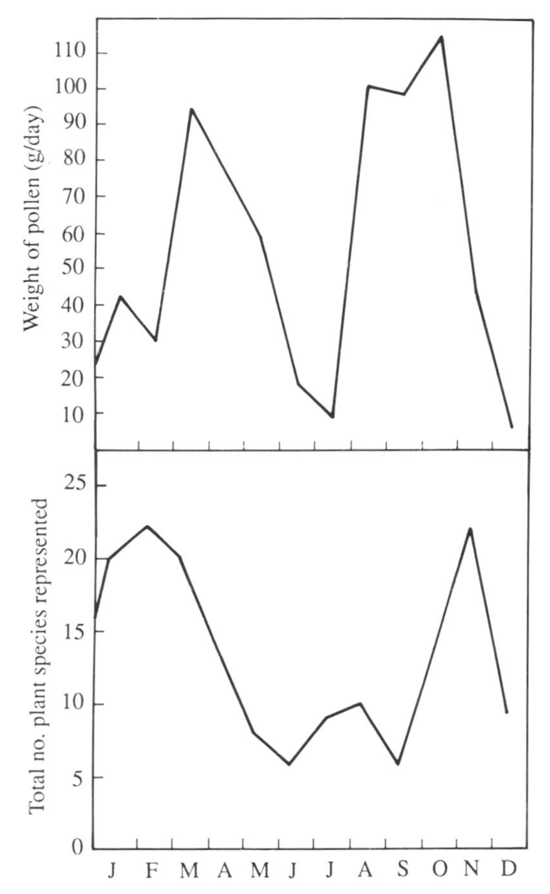

Figure 12.63b Seasonal variations in pollen collection by colonies of honeybees in Vera Cruz, Mexico, based on monthly records (Villanueva, 1984).
above Weight of pollen (g/day) collected in a pollen trap.
below Number of plant species whose pollen was identified in the trap.

Ramírez (1980) has recorded pollen yields at 10°N in Costa Rica.

12.7 GEOGRAPHICAL PATTERNS OF FOOD RESOURCES

The geographical patterns of the honeybees' food resources, and of types of honey, are closely linked with the distribution of plants producing nectar, honeydew and pollen. Apart from extrafloral nectar and honeydew, virtually all the natural foods of honeybees are produced by flowers, and most of the flowers that produce nectar also produce pollen. Section 12.42 mentions some plants that produce only nectar, or only pollen.

Figure 12.7a presents distribution maps for 14 plants that exemplify the wide variety of the world's important honey sources. Each plant, or its honey, has some special interest that is referred to in this book, and in the present section the map number is indicated where appropriate.

All the honey sources mentioned in Sections 12.71 and 12.72 are included under their botanical families in Appendix 1, three families – Leguminosae, Labiatae, and Myrtaceae which includes the eucalypts – being especially well represented. Appendix 1 also indicates broadly where the plants grow: in temperate (M), subtropical (S), tropical (T) zones of the 'continental' regions:

Europe	C. America and Caribbean
Asia	S. America
Africa	Oceania, including Australia and New Zealand
N. America, with Mexico	

12.71 Some important honey sources in temperate and subtropical regions

In the map in Appendix 1 (page 504), temperate zones are designated as those between latitudes 34° and $66\frac{1}{2}$°, and subtropical zones as those between $23\frac{1}{2}$° and 34°. There are far fewer species of flowering plants at the higher latitudes than in the tropics (Section 12.72), and the honeybees that evolved in the temperate zones developed a foraging behaviour referred to as constancy, or fidelity: they continue to forage on flowers of the same plant species as long as these last; see Section 2.43.1 (Energetics of nectar collection), and Bee World (1976*b*, 1979).

In temperate and subtropical zones, more of the land is used for large-scale agriculture than in the tropics, and many of the important honey sources are therefore cultivated plants, which are considered here first.

Cultivated crops

A number of the cultivated Leguminosae are good honey sources, provided they are not cut before flowering, especially:

Trifolium (clover):
- *T. repens*, white
- *T. hybridum*, alsike
- *T. alexandrinum*, Egyptian
- *T. incarnatum*, crimson
- *T. pratense*, red

Melilotus (sweet clover, melilot):
- *M. alba*, white
- *M. officinalis*, yellow

Lotus corniculatus, birdsfoot trefoil
Medicago sativa, lucerne or alfalfa (map 2)
Glycine max, soya bean (subtropics)
Vicia (vetches):
- *V. sativa*, common
- *V. villosa*, hairy

Among the Cruciferae many *Brassica* crops are prolific nectar sources, but those grown for use as vegetables (like cabbage) are not grown to the flowering stage except for the production of next year's seed. *Brassica* crops cultivated for their seed include *B. nigra* (black mustard) and, especially, oil-seed crops. These include: in the temperate zones *B. napus* v. *oleifera* (rape, including Canola varieties developed in Canada, which have a low content of erucic acid), and in the subtropics varieties of *B. campestris* such as *dichotoma* (toria) and *sarson* (sarson), also *B. juncea* (Indian mustard).

Other crops that give good honey yields include sunflower (*Helianthus annuus*, Compositae), and in the subtropics *Citrus* species (map 1) in the Rutaceae, especially:

aurantifolia	lime
aurantium	Seville orange
deliciosa	tangerine
limon	lemon
paradisi	grapefruit
sinensis	sweet orange

A wet subtropical climate is sometimes referred to as 'cotton belt climate'. Egyptian cotton (*Gossypium barbadense*) and upland cotton (*G. hirsutum*), in the Malvaceae, produce nectar in the flowers and also at five extrafloral locations on the plant. Cotton used to be a reliable honey source, but nowadays pesticides are applied to the crop so frequently that colonies of bees usually have to be moved away for the duration of the flowering period, and no honey is obtained.

Forest, woodland and other trees

Trees can provide a three-dimensional array of flowers, and a number of species besides fruit trees are important honey sources, especially in the cooler regions:

Tiliaceae
Tilia (limes, map 5):
- *americana*, basswood
- *cordata*, small-leaved lime
- *japonica*, Japanese lime
- *platyphyllos*, large-leaved lime
- *tomentosa*, silver-leaved or white lime

Aceraceae (maples)
- *Acer platanoides*, Norway maple
- *A. pseudoplatanus*, sycamore

In the Fagaceae, sweet chestnut can yield much honey; *Castanea sativa* is widespread and *C. pubinervis* grows in Asia. The honey has a strong flavour; upon occasion it is described as bitter or pungent, but it is much appreciated by some people. The tulip poplar (*Liriodendron tulipifera*, Magnoliaceae) – see Figure 12.21d – is less widely distributed, except in the USA; I have seen a forest of it in Turkey, and there may be others not shown in map 3.

There are a number of important tree honey sources in the Leguminosae. In warm temperate regions and the subtropics, false acacia or black locust (*Robinia pseudoacacia*) is widely planted. It tolerates a fairly low rainfall, as do some other good honey sources in the dry subtropics, which are mentioned again in Section 12.81 on planting bee forage: certain *Acacia* species including guajilla and catsclaw (*A. berlandieri*, *A. greggii*) and mesquites (*Prosopis*, map 4).

Trees so far mentioned are deciduous. In some temperate-zone forests of coniferous trees, honeydew flows (Section 12.22) give good honey yields. Many of the important trees are in the Pinaceae, and include especially:

- *Abies alba*, silver fir
- *Larix decidua*, larch
- *Picea abies*, Norway spruce
- *Pinus halepensis*, Aleppo pine (map 5)
- *Pinus sylvestris*, Scots pine

In New Zealand, forests of a deciduous mountain beech (*Nothofagus solandri* v. *cliffortioides*, Fagaceae, map 5) are important.

Lightly wooded areas of the cool temperate zones support species of *Rubus* (Rosaceae); *R. fruticosus* is blackberry and *R. idaeus* raspberry. Both are also cultivated for their berries, and are good honey sources. A herbaceous perennial, fireweed or rosebay willowherb (*Epilobium angustifolium*, Onagraceae) grows where woodland has been felled or burned, and it can give prodigious honey yields for the next few years.

Eucalypts (Myrtaceae) are native to Australia, but widely planted elsewhere in the subtropics. Some are

Figure 12.7a World maps indicating the geographical distribution of a few of the important honey sources listed in Appendix 1. Maps 2–5 are on pages 380–383. The map on page 504 shows continental regions, and boundaries between latitude zones.

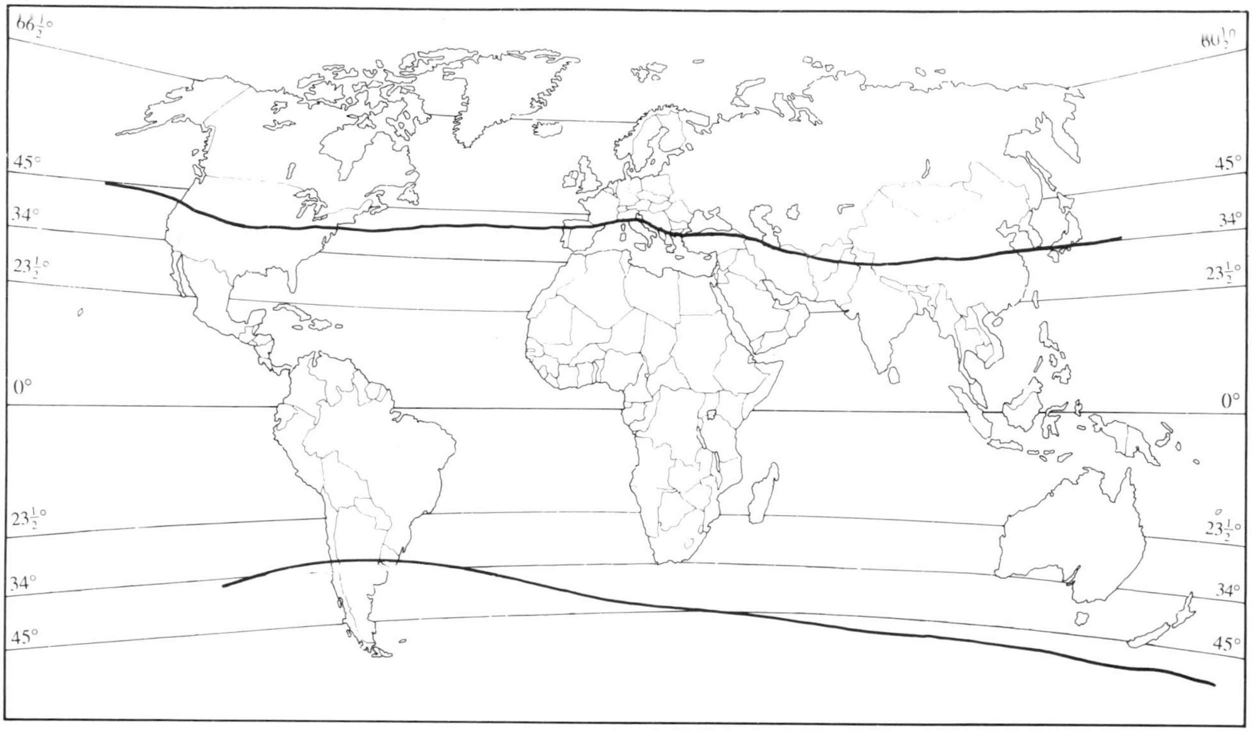

Map 1

Rutaceae (Heywood, 1978). Appendix 1 includes 12 *Citrus* species – orange, etc. The tropical and warm temperate 'citrus belt' round the world includes especially southern N. America, the Mediterranean region and, south of the equator, Southern Africa and Australia.

important honey sources in many countries, and Appendix 1 includes 42 species. Many do well in rather dry conditions.

Heathland and scrub

The renowned heather or ling honey, mentioned in Section 13.27.1, is obtained from *Calluna vulgaris* (map 4), which grows wild on acid soils in parts of north-western Europe. It belongs to the Ericaceae, as do several species of *Erica* in the same region, and grows as far south as the Mediterranean area, especially tree heath (*E. arborea*) and bell heather (*E. cinerea*).

In the Mediterranean and other warm dry areas a number of the Labiatae are important sources of aromatic honeys, for instance:

Lavandula angustifolia, lavender
L. angustifolia × *latifolia*, lavandin
Rosmarinus officinalis, rosemary
Salvia mellifera, black sage
S. officinalis, garden sage
Satureia montana, winter savory
Stachys annua, annual yellow woundwort
Thymus capitatus, Mediterranean wild thyme
T. serpyllum, wild thyme
T. vulgaris, garden thyme

Other subtropical and warm-temperate regions

Important herbaceous sources here include Chinese milk vetch (*Astragalus sinicus*, Leguminosae), Spanish needle (*Bidens pilosa*) and cardoon (*Cynara cardunculus*), both in the Compositae, the last especially in Argentina.

Swampy ground in the American subtropics produces good honey crops from several members of the Nyssaceae: *Nyssa aquatica*, *N. ogeche*, *N. sylvatica* (black and white tupelo, black gum). Another source is *Avicennia germinans* (black mangrove) mentioned in Section 12.72 under Permanent swamps.

Isolated island plant communities sometimes include a single species that surpasses all others as a honey source, for instance in the Pacific: Wake Island, tree heliotrope, *Tournefortia argentea*, Boraginaceae (Hitchcock, 1986); Molokai in the Hawaiian Islands, kiawe, *Prosopis pallida*, Leguminosae (Section 12.82); Norfolk Island, the Chinese tallow tree, *Sapium sapiferum*, Euphorbiaceae (Stratford, 1987); and Saba in the Caribbean has coral creeper *Antigonon leptopus*, Polygonaceae.

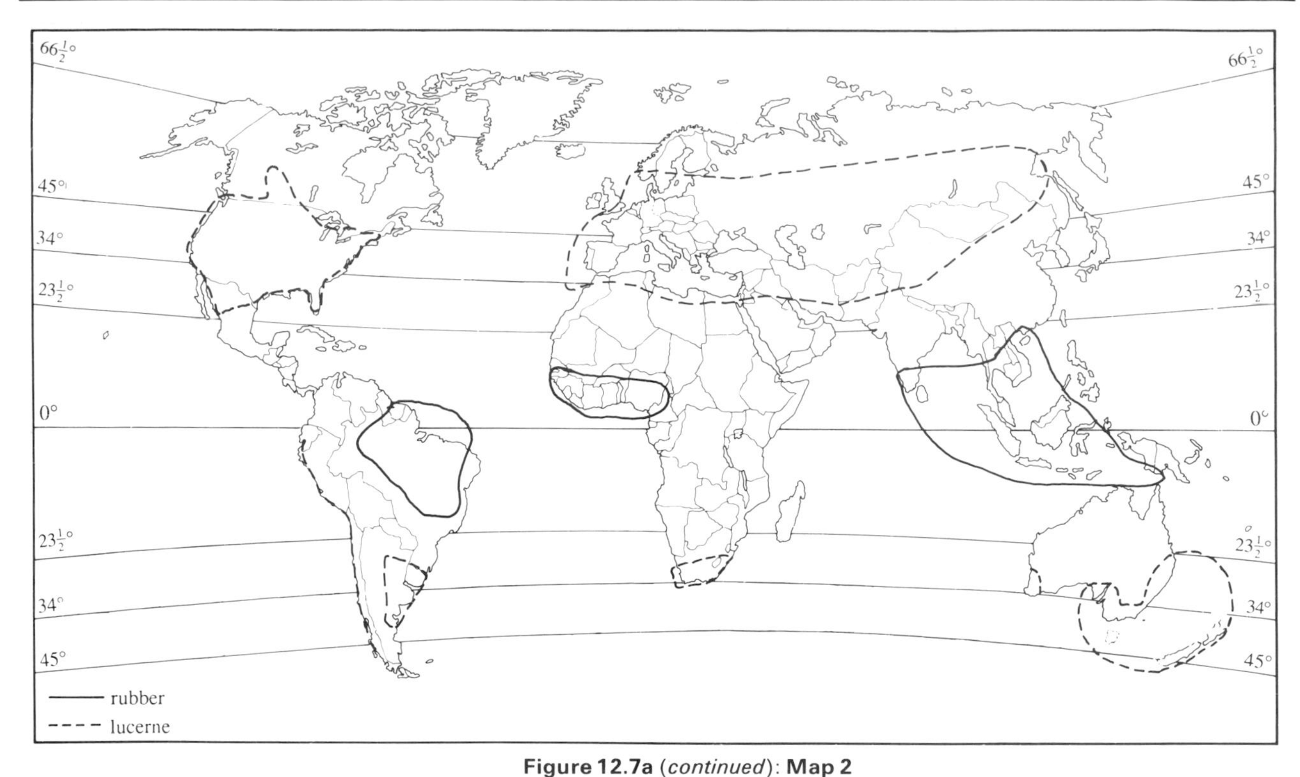

Figure 12.7a (*continued*): **Map 2**

Rubber, *Hevea brasiliensis* (Euphorbiaceae), is cultivated in the humid tropics between 20°N and 20°S, Malaysia contributing 35% of the world supply (Bakar, 1988); the trees yield much extrafloral nectar, but little pollen.

Lucerne or alfalfa, *Medicago sativa* (Leguminosae) (Bolton, 1962). This fodder legume is an important honey source in many temperate regions, provided it is not cut before flowering. Honeybees and certain wild bees are important for pollination of crops grown for seed.

12.72 Some important honey sources in tropical regions

The tropics include all the land between the two subtropical zones, i.e. between the Tropic of Cancer (23½°N) and the Tropic of Capricorn (23½°S), shown on the map on page 504. The area is about the same as that of the two temperate zones together, but the tropics have many more species of flowering plants: 90 000 in Central and South America, 30 000 in Africa – and another 10 000 in Madagascar. In contrast, Greece has 6000 and Britain only 1800 (Huxley, 1984). Understandably, therefore, honeybees that evolved in the tropics, with their rich variety of food, respond quickly to changes in the value of individual food sources when foraging, and show less 'constancy' than temperate-zone honeybees.

Some types of tropical region important for beekeeping are indicated below, with examples of important honey sources in them; they are described in more detail by F. G. Smith (1960). Certain sources occur in only one continent, and others (especially cultivated plants) in several. In general, native wild plants provide more of the honey in the tropics than in temperate zones, and a greater proportion of the honey sources are trees; in the lists, shrubs or herbaceous plants are marked S or H.

Lowland equatorial rain forest

This type of forest occurs between 15°N and 15°S, from sea level to altitudes between 750 and 1050 m. The temperature is high (mean 27°, minimum 18°–21°) throughout the year, and so is the annual rainfall (at least 1750 mm, optimum 3000 mm). In the Old World the forests grow especially in SE Asia and near the coast of W. Africa, and in the New World, in and around the Amazon basin. They contain a large variety of plant species, especially trees, and nectar and pollen may be available almost all the year round. Where cultivated plants are grown, some are important honey sources, for instance:

Anacardium (Anacardiaceae)
Eugenia (Myrtaceae)
Hevea (Euphorbiaceae), map 2
Nephelium (Sapindaceae)

Honeybees indigenous to these forests tend to store rather little honey; when forage is no longer available,

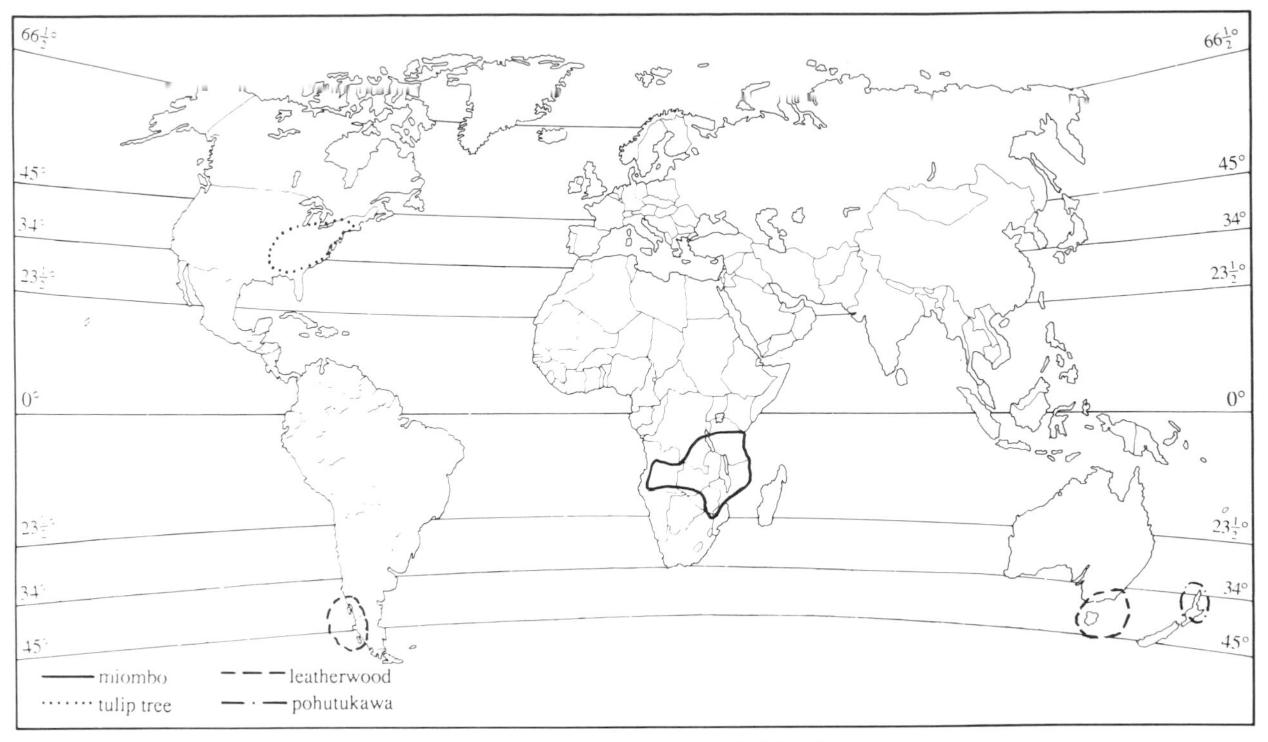

Figure 12.7a (*continued*): **Map 3**

Miombo, woodland characterized by *Brachystegia* and *Julbernardia* species (Leguminosae), stretches in a belt across tropical Africa, practically all of which is good honey country (Chandler, 1980).

Leatherwood, *Eucryphia lucida* and *E. milliganii* (Eucryphiaceae) in Tasmania, Australia, has a counterpart, ulmo, *E. cordifolia*, in S. Chile (Heywood, 1978).

Tulip tree, *Liriodendron tulipifera* (Magnoliaceae), occurs in part of eastern USA (Phillips, 1922).

Pohutukawa, *Metrosideros excelsa* (Myrtaceae), is a tree native to Auckland Province, New Zealand (Allan, 1961).

colonies abscond and move elsewhere (Sections 3.34.5, 3.43).

Upland equatorial (cloud) forest

This forest grows at the same latitudes, but higher (750 to 2700 m), and where the annual rainfall is over 1250 mm. There are two wet dearth seasons a year, when the sun is overhead, and two short dry honey flow seasons, when the sun is to the north or south. In the lower parts of the zone, with a rainfall 1250 to 1750 mm, important honey sources include many trees:

Coffea (Rubiaceae)
Croton (Euphorbiaceae)
Dombeya (Sterculiaceae)
Eucalyptus (Myrtaceae)
Grevillea (Proteaceae)
H *Musa* (Musaceae), map 5
Vernonia (Compositae)

Dearth seasons are relatively long, and native honeybees have a more developed storing instinct.

Lowland tropical rain forest

Occurring between latitudes 15° and 23½°, this forest has fairly marked wet and dry seasons; the wet season occurs when the sun is most nearly overhead (summer), and nectar and pollen are mainly available in the cooler dry season (winter). This type of forest is present in all continents, and can give very high honey yields. Important honey sources include:

Lonchocarpus (Leguminosae)
Pithecellobium (Leguminosae)
Gymnopodium (Polygonaceae)
SH *Ipomoea* (Convolvulaceae)

Upland forest

In the same latitude range, upland forest includes trees whose main flowering – and honey flows – come after the rains, although there may be some plants in flower almost all through the year. Coffee-growing areas of Brazil, hill areas of Cuba, and Morelos Province in Mexico (where Miel Carlota operates) are examples of good honey country of this type. Important trees and herbaceous sources of honey include:

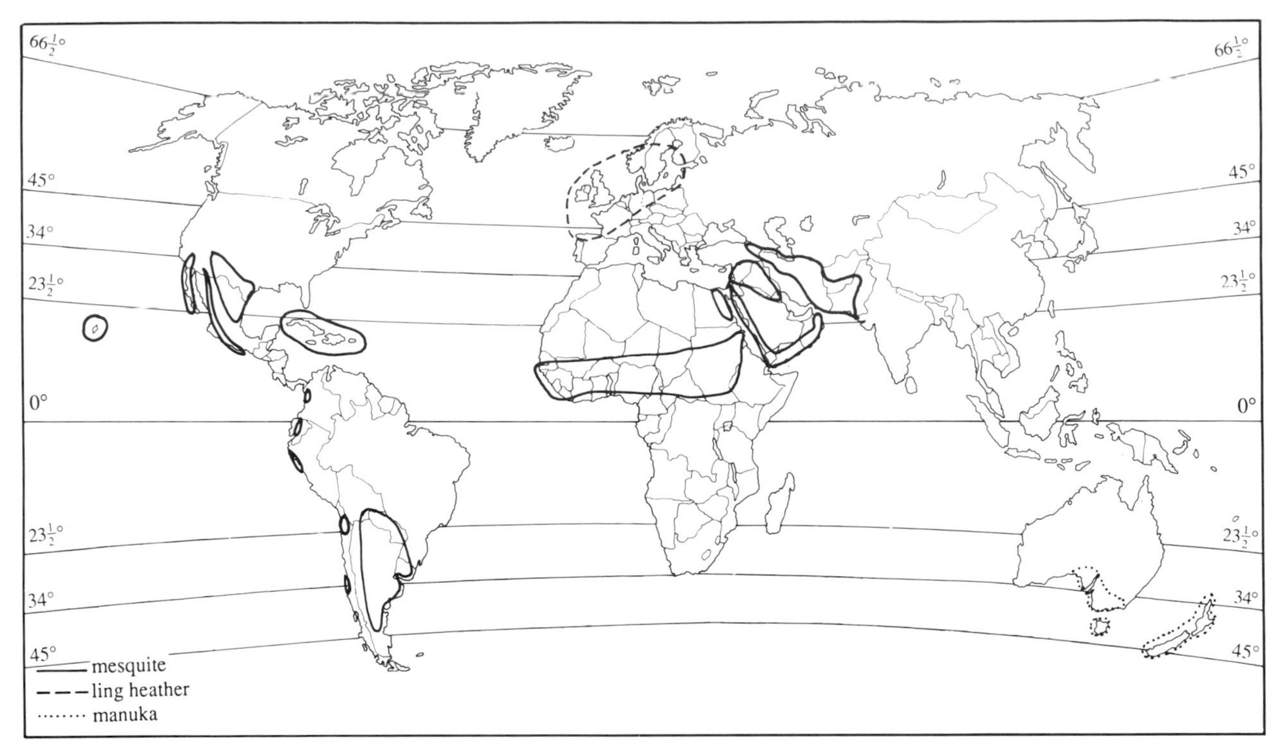

Figure 12.7a (*continued*): **Map 4**

Mesquite is a name used for a number of *Prosopis* species (Leguminosae), 6 of which are listed in Appendix 1. They are quick-growing drought-resistant trees or shrubs of the very dry tropics and subtropics (Simpson, 1977). Recent introductions, not shown, have been made in Southern Africa and Australia.

Honeys from the next two sources are thixotropic (Section 13.27.1).

Ling heather, *Calluna vulgaris* (Ericaceae), grows on moorlands and in some open forest lands in NW Europe (Beijerinck, 1940). Small amounts elsewhere, e.g. in N. America, N. Africa, New Zealand, are not marked.

Manuka, *Leptospermum scoparium* (Myrtaceae), with many varieties and hybrids, occurs in North Island and other parts of New Zealand (Allan, 1961) and in part of S. Australia, Tasmania and Victoria.

Acacia (Leguminosae)
Ceiba (Bombacaceae)
Citrus (Rutaceae)
Coffea (Rubiaceae)
Cordia (Boraginaceae)
Eucalyptus (Myrtaceae)
Gliricidia (Leguminosae)
Trichilia (Meliaceae)

H *Inga* (Leguminosae)
H *Lippia* (Verbenaceae)
H *Musa* (Musaceae)
H *Sesamum* (Pedaliaceae)
H *Tithonia* (Compositae)
H *Vernonia* (Compositae)

Tree or bush savannah

This type of dry forest is mainly deciduous woodland with a closed canopy and no grass; the rainfall is between 650 and 1250 mm in most areas. In the prolonged dry season the trees lose their leaves. Ground cover includes herbaceous plants, with scattered shrubs. Important species are:

S *Antigonon* (Polygonaceae)
Brachystegia (Leguminosae)
Calycophyllum (Rubiaceae)
Ceiba (Bombacaceae)
Cochlospermum (Cochlospermaceae)
Combretum (Combretaceae)
Cordia (Boraginaceae)
Gilibertia (Araliaceae)
SH *Ipomoea* (Convolvulaceae)
Julbernardia (Leguminosae)
S *Leucas* (Labiatae)
Lonchocarpus (Leguminosae)
H *Sesamum* (Pedaliaceae)

Miombo (map 3) is an African woodland of this type dominated by *Brachystegia* and *Julbernardia*, which flower at the end of the rains, and a dearth follows afterwards. The scale-hive record for colony 4 in Figure 12.62a was made in this region.

Wooded grassland

Where the annual rainfall is less than 650 mm, the ground is covered with grasses and other herbs. Although trees and shrubs occupy less than half the land, many species of the native trees are good honey sources, for instance in Africa:

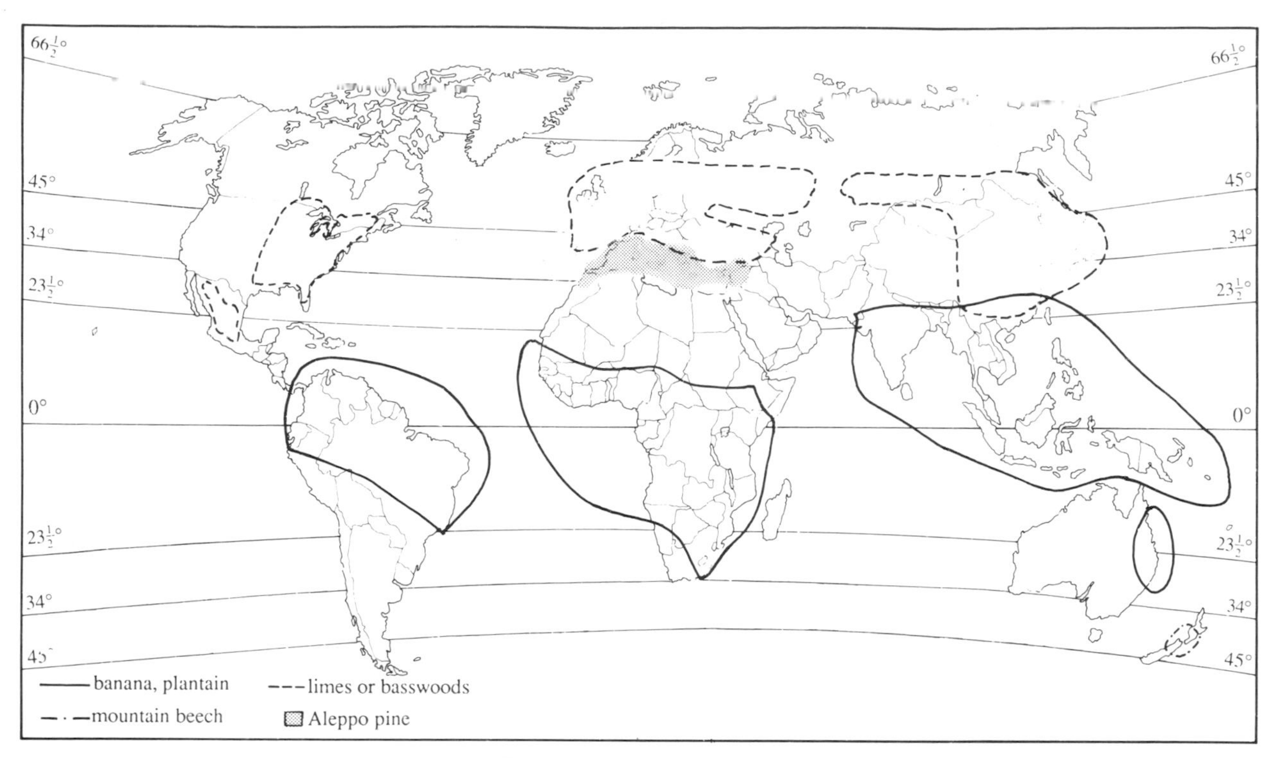

Figure 12.7a *(continued)*: **Map 5**

Banana, plantain, etc., *Musa sapientum* (Musaceae), is indigenous to SE Asia, and is also grown in many other parts of the tropics (Edlin, 1973).

Limes or basswoods, *Tilia* spp. (Tiliaceae), are deciduous trees of the north temperate zone, in all three continents (Kim & Chung, 1986).

The next two sources yield honeydew honey (Section 12.22).

Aleppo pine, *Pinus halepensis*, *P. brutia* (Pinaceae), grows in many regions near the E. Mediterranean coast (Mirnov, 1967); the chief honeydew-producing insect, *Marchalina hellenica*, is mainly in the east.

Mountain beech, *Nothofagus solandri* v. *cliffortioides* (Fagaceae), grows in New Zealand in subalpine and montane forest and scrub between 38° and 44°S (Allan, 1961); the chief honeydew-producing insect, *Ultracoelostoma assimile*, is in some northern parts of South Island.

Acacia (Leguminosae)
Combretum (Combretaceae)
Dombeya (Sterculiaceae)
Syzygium (Myrtaceae)
Terminalia (Combretaceae)

Dry scrub and bushland

Areas with a low rainfall and a prolonged dry season, and an extreme range of temperature, support vegetation that includes low thorny trees and spiny shrubs, many of which are honey sources, including:

Acacia (Leguminosae)	cacti
Combretum (Combretaceae)	Compositae
Dombeya (Sterculiaceae)	other Leguminosae

Permanent swamps

Different plants occur according to the salinity of the water. Important honey sources include the salt-tolerant *Avicennia* (Avicenniaceae) and *Rhizophora* (Rhizophoraceae); black mangrove (*Avicennia germinans*, syn. *nitida*) is the most important honey source in French Guiana, Guyana and Surinam (1° to 8°N in South America). Otis and Taylor (1979) have described the background to beekeeping there.

Coastal plains

In many parts of the tropics, coastal plains are used for agriculture, and cultivated plants form a much higher proportion of honey sources than in tropical regions of other types. Many of the cultivated plants are insect-pollinated; Appendix 1, Table 8.4A, and Crane and Walker (1983) give some details. Most of the following cultivated plants occur in several continents:

Anacardium occidentale (Anacardiaceae)
Citrus spp. (Rutaceae)
Cocos nucifera (Palmae)
Durio zibenthinus (Bombacaceae)
H *Ipomoea batatus* (Convolvulaceae)
H *Musa* spp. (Musaceae)
Nephelium lappaceum (Sapindaceae)
Roystonea regia (Palmae)

Other important honey sources include:

Bucida (Combretaceae)	H *Lippia* (Verbenaceae)
Coccoloba (Polygonaceae)	H *Viguiera* (Compositae)

12.73 Guide to published information on food resources

Most publications deal with a particular country rather than with what Good (1974) refers to as a 'floristic unit', a characteristic type of vegetation. In *Honey: a comprehensive survey*, Crane (1975*b*) cites over 250 of these publications by country, together with the few that cover a larger area. Most of them are still among the best sources of information, but the following later publications can now be added.

Temperate zones and subtropics
Europe: general, Maurizio & Grafl (1982); Bulgaria, Luchanska & Kolev (1984); Denmark, Christensen (1984); German Federal Republic, Gleim (1985); Italy, Ricciardelli d'Albore & Persano Oddo (1978); Malta, Farrugia (1979); Romania, Cîrnu (1980); Spain, Ortega Sada (1987)
Asia: Israel, Eisikowitch & Masad (1980); Korean peninsula, Kim & Lee (1986), Yim (1986)
Africa: South Africa, Anderson et al. (1983)
North America: general, Robinson & Oertel (1975), Lovell (1977); Canada, Ramsay (1987)
Oceania: New Zealand, Walsh (1978).

Tropics
Asia: Bangladesh, Dewan (1984); Burma, Zmarlicki (1984*b*); Malaysia, Mardan & Kiew (1985); Pakistan, Ahmed (1984), Wali-ur-Rahman (1985); Sri Lanka, Fernando (1979); Thailand, Pyramarn & Wongsiri (1986)
Africa: Botswana, Clauss (1983); Chad, Gadbin (1980)
Central America: general, Espina & Ordetx (1983) and Ordetx (1952); El Salvador, Woyke (1981*a*); Mexico, Villanueva (1984)
South America: general, Espina & Ordetx (1983) and Ordetx (1952); Belize, Mulzac (1979); Colombia, Echeverry (1984); Uruguay, Castro (1974); Venezuela, Thimann & Aymard (1982), Lopez-Palacios (1986)
Oceania: New Caledonia, Dutant (1973)

Additionally, publications since 1975 cited in the Gazetteer (Appendix 2) for several of the less well documented countries include information on their nectar and pollen sources. Some of the contributions in the following published collections include a list of plants foraged by bees in individual tropical countries:

Commonwealth Secretariat & IBRA (eds) (1979) *Beekeeping in rural development: unexploited beekeeping potential in the tropics with special reference to the Commonwealth*
International Conference on Apiculture in Tropical Climates 1–4, Proceedings (1976, London; 1983, Delhi; 1985, Nairobi; 1989, Cairo).
Kevan, P. G. (ed.) (1989) *Beekeeping with* Apis cerana *in tropical and subtropical Asia*

Collected world data
In *Honey: a comprehensive survey*, Crane (1975*b*) included a list of 200 plants considered to be important world honey sources, with brief details – as far as available – of type of plant, world distribution and importance; also characteristics of the honey, including colour, density, viscosity, flavour and aroma, natural granulation and any special features. The compilation of this material showed the great need for fuller information on the world's important honey plants, and thanks to financial support – especially from the International Development Research Centre, Ottawa – IBRA was able to construct a data-base with 51 search fields, from which a *Directory of important world honey sources* (Crane, Walker & Day, 1984) was published with entries for 467 plant species. The methods, and preliminary work funded by the International Union of Biological Sciences, have been described (Crane, 1983*c*).

Appendix 1 of the present book lists plants in the *Directory* under their botanical families, with an indication of their distribution. Following publication of the *Directory*, interest was expressed in honey sources with certain specific characteristics, and searches of the data led to preliminary assessments of:

- multipurpose trees for the tropics and subtropics (Crane, 1985*b*); see Section 12.81
- drought-tolerant and salt-tolerant plants (Crane & Walker, 1986*e*)
- plants for exploiting arid land resources (Crane, 1985*a*; Crane & Walker, 1986*e*); see Section 12.82.

12.8 PLANTING FOR BEES

It is not generally an economic proposition to cultivate plants solely as forage for bees. Nevertheless some species can yield unusually high amounts of honey per hectare of land, and this makes them good candidates for planting if they have other characteristics the planter wants for a particular purpose. The subject is considered here under the headings:

12.81 Multipurpose and other trees that give high honey yields
12.82 The special need and potential of arid lands

12.83 Turning unused temperate-zone land into bee pasture
12.84 Amenity and garden planting for bees.

Plants selected for any purpose should be suitable for the local conditions of temperature, rainfall, etc., and quarantine regulations must be observed for plants to be imported. Special caution needs to be exercised with plants that may prove to be invasive; see end of Section 12.82, and also 12.83. Within a group of candidate plants chosen for their primary purpose, any that are outstanding as honey sources, or as pollen producers, can make available an additional resource from the land through beekeeping. Where such honey sources are plants grown for seed or fruit, the bees may also increase the crop yields by their pollinating activities; see Section 8.4.

About 80 plant entries in the *Directory* by Crane, Walker and Day (1984) cite a recommendation from a specific country that the species is suitable for planting on account of its high honey-yielding capability.

Some beekeeping journals include a short article in each issue that describes an important bee plant and gives cultural instructions; those from Eastern Europe, especially, also provide useful quantitative data on nectar, honey and pollen yields.

12.81 Multipurpose and other trees that give high honey yields

Different species of *Eucalyptus* (Myrtaceae) are planted widely in the world except in cold temperate regions, for timber, cellulose production and other purposes, and a number are high honey yielders. Albisetti (1981) discusses the value of different species to beekeeping, and makes recommendations for planting. General books on the genus by Penfold & Willis (1961) and Jacobs (1981) include some information on honey production. Multipurpose trees that are important honey sources in the tropics and subtropics are listed by Crane (1985*b*); over a quarter are Leguminosae. Eucalypts form the largest group of the trees that have been recommended for planting.

False acacia or black locust (*Robinia pseudoacacia*, Leguminosae) is a multipurpose tree native to North America, which is widely valued for planting in warm temperate and subtropical zones; it was introduced into central Europe to fix nitrogen and to prevent soil erosion. The honey yield increases to a maximum when the tree is about 16 years old, for instance:

6 years	371 kg/ha
15, 16 years	418 kg/ha
26 years	358 kg/ha
36 years	192 kg/ha

In Hungary, where this tree is the main honey source, plant breeding has led to the production of a great many varieties with different economic characteristics (Halmágyi & Keresztesi, 1975; Keresztesi, 1977), and with different flowering periods. Whereas the control species *R. pseudoacacia* flowers for only 12 days, 33 other varieties have flowering periods from 13 to 20 days; the total period of the honey flow can be doubled by planting several selected varieties, to the economic advantage of beekeepers.

In cool temperate regions, some species of *Tilia* (Tiliaceae, limes in Europe, basswoods in North America) can give high honey yields. Muir (1984) has published a survey of the genus, and its world distribution is shown in Figure 12.7a, map 5. Certain species present drawbacks such as aphid infestation, and it is very important to choose the most appropriate species for planting. In a useful discussion of their honey production, Stroempl (1977) gives a diagram showing the flowering periods of *Tilia* species in England, GFR, USSR and Yugoslavia.

Many willows (*Salix*, Salicaceae) also grow in cool temperate regions, and are notable for their pollen production in early spring. Warren-Wren (1972) describes the different species, and Resch (1974) discusses their relative usefulness to beekeeping. Mottl et al. (1980) list the flowering periods of 13 species and clones that beekeepers can plant to give a succession of early pollen flows, and describes a further 50 promising ones; cultural details are included.

12.82 The special need and potential of arid lands

Droughts present serious problems in large areas of the world, and land made agriculturally unproductive by deforestation or improper farming usually becomes drier. In the resultant arid conditions, many of the species that thrive provide bee forage. This is no new observation. Around 400 BC, Plato bemoaned the deterioration of land by excessive farming, in a passage in *Critias*. He referred to 'mountains in Attica which can now support nothing but bees, but which were clothed, not so very long ago, with fine trees . . . [and] produced boundless pasturage for cattle.'

Crane and Walker (1986*e*) give details of important world honey sources that are drought-tolerant, and Crane (1985*a*) lists multipurpose plants among them that are recorded in the tropics and subtropics; they mostly provide food, fodder, shelter or shade; many are trees that can also be used as firewood (Crane, 1985*b*).

Some of the extremely drought-tolerant plants that are valuable honey sources in arid land can become invasive and impossible to eradicate in areas where the rainfall is higher. It is dangerous to introduce these plants into such areas, or where they might spread – or be spread by animals – into such areas. Eisikowitch and Dafni (1988) emphasize this danger in relation to certain species of *Prosopis* (Leguminosae), many of which are known as mesquite. Other plants that yield much honey in the tropics, but can be invasive, include *Melaleuca leucadendron* (Myrtaceae), *Schinus terebinthifolius* (Brazilian pepper, Anacardiaceae) and *Chromolaena odorata* (snake root).

The general usefulness of *Prosopis* species is described, for example, by B. B. Simpson (1977), and their honey-yielding capability by Espina and Ordetx (1983) and Crane, Walker and Day (1984). Good honey yielders include *P. chilensis, cineraria, farcta, glandulosa, julifora, pallida, pubescens, tamarugo*. The roots of some species, including *P. glandulosa*, grow to lengths of 20 or even 30 m to reach underground water. Their value in arid lands is great.

As part of a development programme in the Piura coastal desert in Chile, *P. juliflora* was planted over 1000 ha, drought-resistant grasses being grown under the trees (Townsend, 1983). Families were living on the project area within three years of its establishment, keeping livestock (sheep were best), and producing honey on a commercial scale. There were two harvests in the year of both honey and *Prosopis* seeds. Another project based on *P. juliflora* (Kigatiira et al., 1988) is in the Garissa district of Kenya. In Hawaii, the introduction of *P. pallida* into Molokai made this island an important honey-producing area (Figure 6.62a), giving yields of 120 to 150 kg a year per hive (Townsend, 1983).

Among other drought-tolerant multipurpose trees recommended for planting that give good honey crops are *Azadirachta indica* (neem, Meliaceae) and *Dalbergia sissoo* (sissoo, Leguminosae) especially in India; *Eucalyptus camaldulensis, E. cladocalyx* and *E. melliodora* (Myrtaceae) in Australia and other countries; and *Pithecellobium dulce* (chiminango, Leguminosae) in tropical America. The *Dalbergia* and *Pithecellobium* species, and *E. camaldulensis*, are also salt-tolerant. Many others, from which a selection can be made according to specific needs, are listed in the papers cited, and in one by the Indian Standards Institution (1973). For instance some other eucalypts are drought-tolerant, as are many *Acacia* species, including especially *A. caffra* and *A. senegal*, and also *A. berlandieri, A. greggii, A. mellifera, A. seyal* and *A. tortilis*; these are in the Leguminosae, and able to fix atmospheric nitrogen. They are not invasive.

12.83 Turning unused temperate-zone land into bee pasture

The term 'fixed-land honey production' (Ayers & Hoopingarner, 1987; Ayers et al., 1987) is sometimes used for a beekeeping system that incorporates the growing of plants specifically for honey production on a sizeable area of land round the apiary. It can make use of land whose owners or occupiers do not have the time – or the money or strength – to cultivate it intensively. The following (not listed in Appendix 1) are some of the shrubs that have been strongly recommended from time to time for such planting in the north temperate zone:

Amelanchier spp., service berry (Rosaceae)
Caragana arborescens Lam., pea-tree (Leguminosae)
Lespedeza cryobotrya Mig. (Leguminosae)
Vitex negundo incisa Clarke (Verbenaceae)

A few good temperate-zone honey sources can be invasive where the growing conditions suit them too well (Crane, 1981*a*). For instance in arid South Australia the herbaceous plant *Echium lycopsis* (Boraginaceae) acquired the name Salvation Jane because of its value as fodder in times of drought, but in regions with a higher rainfall it is known as Paterson's curse. There has been a long drawn-out battle in Australia, between farmers who want the plant to be compulsorily eradicated as a noxious weed, and beekeepers to whom it is a valuable resource for honey production. Other potentially invasive plants include:

Centaurea solstitialis, star-thistle (Compositae)
Polygonum persicaria, smartweed (Polygonaceae)
Taraxacum officinale, dandelion (Compositae)

In many temperate-zone areas, profits from beekeeping have decreased latterly as a result of the 'cleaner' and more mechanized agricultural methods, which include effective weed control, the uprooting of hedges, and early cutting for silage. Such areas provide little for the bees during most of the year, and beekeepers want action to increase bee forage during dearths in the bees' active season. In countries of Eastern Europe especially, it is frequently recommended that edges and corners of fields and other cultivated land should be used for producing bee forage. The plant most often grown is *Phacelia tanacetifolia* (Hydrophyllaceae), an annual which has been much studied in different growing conditions. In Poland, calculations based on its nectar secretion gave an estimated honey yield up to 1130 kg/ha (Zimna, 1959). The same species is also recommended for intersowing with crops. In Bulgaria (Petkov, 1966), sowing phacelia with fodder crops gave honey yields of 76 kg/ha when the main crop was spring peas, and 92 kg/ha when it was vetchling (*Lathrys sativus*). The yields of green fodder were only marginally reduced.

In Sweden, Hansson (1988) has published profiles of alternative crop plants (legumes and others) that can reduce the use of fertilizers and pesticides. A three-year study in the GFR (Griesohn, 1982) led to several recommendations: undersowing cereal crops with suitable clovers; where fields are large, inserting strips of plants such as phacelia or rape; sowing catch crops useful to bees after early crops are harvested; neither cutting nor treating with herbicide the flowering plants on field edges and under fruit trees in early summer. A German study by Bauer (1985) recommended that pollen-producing plants such as plantain (*Plantago*) should also be allowed to grow round field edges.

12.84 Amenity and garden planting for bees

As agricultural land becomes less hospitable for bees, so gardens and amenity planting become of greater importance to beekeeping. In Europe, the area involved is not negligible; in the UK gardens cover about 300 000 ha, and some of the best honey yields per hive are now obtained by urban beekeepers. Plants used for gardens, parks and roadsides are often chosen to give a long flowering season, and they can provide a continuous, if modest, supply of nectar and pollen – such as is no longer available in hedgerows, or in pastures and verges that are cut before they come into flower. 'Wild gardens' designed for ease of management are valuable in providing day-to-day food supplies for bees and other insects. However, most gardens I have seen that are devoted *entirely* to bee plants have looked rather unattractive. It is a minor misfortune for beekeepers that selection for showy and double flowers results in a reduction of the reproductive parts and a diminution in fertility – and less food for the bees.

When planning gardens and amenity planting, many publications may be consulted as to species and varieties that provide forage for bees, especially during dearth periods; for instance in Europe, books by Hensels (1981), IBRA (1981), Hemptinne et al. (1985), and Hooper and Taylor (1988), and in North America one by Ramsay (1987). In Israel, after a two-year study, Eisikowitch and Masad (1980) selected 19 trees and shrubs as suitable for planting to span the dry and warm winter dearth which lasted 6 to 7 months. Most books on bee plants (see Sections 12.73, 12.9) have a short section on planting for bees, often with gardens especially in mind. Books in French (Rabiet, 1984) and German (Ministerium für Ernährung . . . , 1985) extend the discussion and selection of plants to grow for bees in the north temperate zone. Out of 170 multi-purpose plants selected as being suitable for making hedges in France (Guinaudeau, 1984), 96 are designated as providing bee forage. To these may be added many plants useful in reclaiming, landscaping and screening land after mining operations. Some tolerate acid or alkaline soil material, or high metal toxicity; others are legumes that fix nitrogen. Richardson (1977) details trees and shrubs suitable for such specific needs.

12.9 FURTHER READING AND REFERENCE

Details of publications listed will be found in the Bibliography.

General

Crane, E. (1978*a*, No. 15) *Bee forage in the tropics*
Crane, E. (1978*b*, S34) *Bee forage in specific regions of the tropics*
Crane, E.; Walker, P.; Day, R. (1984) *Directory of important world honey sources*
Hemptinne, J. et al. (eds) (1985) *Une gestion de l'environnemènt pour une apiculture florisante*
International Symposium on Melliferous Flora (1977) *Honey plants – basis of apiculture*
Percival, M. S. (1965) *Floral biology*

Nectar

Bentley B. L.; Elias T. S. (eds) (1983) *The biology of nectaries*
Fahn, A. (1979) *Secretory tissues in plants*
Gleim, K.-H. (1985) *Nahrungsquellen des Bienenvolkes 1. Die Blütentracht*
Maurizio, A.; Grafl, I. (1982) *Das Trachtpflanzenbuch*
Shuel, R. W. (1975) *The production of nectar*

Honeydew

Kloft, W. J.; Kunkel, H. (eds) (1985) *Waldtracht und Waldhonig in der Imkerei*
Gleim, K.-H. (1984) *Nahrungsquellen des Bienenvolkes 2. Die Honigtautracht*
Crane, E.; Walker, P. (1985*a*) *Important honeydew sources and their honeys*

Pollen

Stanley, R. G.; Linskens, H. F. (1974) *Pollen: biology, biochemistry, management*

Propolis

Fahn, A. (1979) *Secretory tissues in plants*

13

The traditional hive products: honey and beeswax

13.1 INTRODUCTION

Getting honey is the primary goal of beekeeping today, as it has been in the past. Published statistics give an annual world production around 1 million tonnes, worth about 1000 million US dollars on the world market. Statistics are not collected systematically for any other hive product, and none of them is likely to reach 5% of this value.

In the present chapter, Section 13.2 deals with the composition of honey and its characteristics and properties, subjects which were covered in detail in *Honey: a comprehensive survey* (Crane, 1975*a*). Section 13.3 gives an account of the treatment of honey after the combs have been freed from bees and removed from the hives to the honey house, including its extraction from the combs. In Section 13.4 some examples of uses of honey are given, linked with the constituents or characteristics that make it especially suitable for the use in question. Sections 13.5, 13.6 and 13.7 deal with beeswax in a similar way, although less fully in view of its lesser importance, and Section 13.8 gives figures for honey and beeswax production in the various regions of the world, and for world trade, both exports and imports.

13.2 HONEY: DEFINITIONS, COMPOSITION AND PROPERTIES

In the course of evolution, honeybees developed the ability to elaborate honey from nectar and honeydew they had collected from plants (Chapter 12). The honeys of the world show great variety, especially in flavour and aroma, because the different plants contribute their own characteristic constituents; also, honeys are produced under many different climatic conditions. But the main constituents in all honeys are the same. There is a very high concentration of certain sugars dissolved in water, and as a result honeys are usually safe from spoilage by micro-organisms when properly stored, even for a long period. (An enzyme secreted by the bees keeps the raw materials safe while they are being converted into honey in the nest or hive.) Honey is not easy to define, and definitions are usually formulated only for standards or legislation relating to honey as an article of trade. The proposed World-wide Standard (Codex Alimentarius Commission, 1983/84) has the following definition:

> Honey is the unfermented, sweet substance produced by honeybees from the nectar of blossoms or from secretions of or on living parts of plants, which they collect, transform and combine with specific substances, and store [and ripen (or mature)] in honey combs. Honey shall not have any objectionable flavour, aroma or taint absorbed from foreign matter during its processing and storage and shall not contain natural plant toxins in an amount which may constitute a hazard to health.

An attempt at a biological definition might be as follows:

> Honey is a substance produced by bees and some other social insects* from nectar or honeydew that they collect from living plants, which they trans-

Table 13.2A Mean amounts (% of total honey) of important constituents of honey produced in four countries in the temperate zones and subtropics

The upper part of the table lists major constituents. Column 2 gives the mean of the four country means, each calculated from the number of samples in brackets. The country means are quoted from Chudakov (1963), White et al. (1962), Pelimon and Baculinschi (1955), and Chandler et al. (1974), respectively. The final column gives the lowest and highest individual values for the 1063 samples, mostly (but not necessarily entirely) from nectar. The lower part of the table lists minor (but also important) constituents which are detailed in further tables and sections, as indicated; amounts in USA honey samples are quoted where available.

	Mean	*USSR* (217)	*USA* (490)	*Romania* (257)	*Australia* (99)	*Range*
Major constituents (about 99% of the honey)						
water	17.0	18.6	17.2	16.5	15.6	13.4–26.6
fructose	39.3	37.4	38.2	38.4	43.3	21.7–53.9
glucose	32.9	35.9	31.3	34.0	30.2	20.4–44.4
sucrose	2.3	2.1	1.3	3.1	2.5	0.0–7.6
other sugars:	*Table 13.21A*					
disaccharides (as maltose)			7.3			2.7–16.0
higher sugars			1.5			0.1–8.5
Minor constituents (about 1% of the honey)						
total acids (as gluconic)	*Table 13.24B*		0.57			0.17–1.17
minerals	*Table 13.25A*		0.17			0.02–1.03
nitrogen (in amino acids and proteins)	*Table 13.26A*		0.04*			0.00–0.13
enzymes	*Table 13.23A*					
aroma constituents	*Table 13.24A*					
other substances	*Section 13.27*					

* Mean protein content of 740 USA samples = 0.17% (White & Rudyi, 1978*b*)

form by evaporating water and by the action of enzymes they themselves secrete. As a rule, honeybees seal the finished honey in cells of their comb.†

The composition and properties of honey are dealt with quite fully in the following chapters of *Honey: a comprehensive survey* (Crane, 1975*a*):

2 How bees make honey (Maurizio, 1975*a*)
5 Composition of honey (White, 1975*a*)
6 Physical characteristics of honey (White, 1975*b*)
7 Microscopy of honey (Maurizio, 1975*b*)
13 Honey standards legislation (Fasler, 1975).

Der Honig by Maurizio et al. (1975) covers similar ground to the above chapters. White (1978) summarizes chemical and physical properties of honey, and its production and processing.

Table 13.2A summarizes data on the major constituents of honey, based on analyses of 1063 honey samples from four countries, the great majority of which were from nectar. The upper part of the table relates to sugars and water, which together comprise 99% of most honeys. The lower part is concerned with the remaining 1%, and relates to various groups of substances that are present in honey in small or minute amounts, but are nevertheless responsible for a number of its important characteristics. These groups (enzymes, aroma constituents, organic acids, minerals, amino acids, and other substances) are dealt with in the subsections indicated, which also explain the properties of honey related to the constituents in question.

Some constituents of honey are the subject of a Directive of the European Communities Council (1974) and of a proposed World-wide Standard quoted above which was part of the FAO/WHO Food Standards Programme. The proposed Standard, designed to be more widely applicable than the Directive, sets the following limits for all except certain named honeys:

* See Section 13.23.3 for the inclusion of some other social insects in addition to honeybees.

† Honeybees do no more active work on the honey after they cap (seal) it in the cells.

water	not above 21%
reducing sugars	not below 65%
sucrose	not above 5%
ash	not above 1.0%
free acid	not above 40 meq/kg
amylase (diastase)	not below 3 on the Gothe scale
HMF	not above 80 mg/kg

Honeys from temperate-zone *Apis mellifera* form the basis of the world's honey industry. Honeys from tropical *Apis* species show wider variations in composition, which also reflect the great range of tropical plants, climates and seasons (Crane, 1975*d*). The most extensive studies have been made in India (Phadke, 1962, 1967*a*, 1967*b*, 1968; Phadke et al., 1970, 1973). Vries (1988) gives analyses of some *A. cerana* honeys in Java, Indonesia.

An index to information on the composition and properties of some of the world's honeys is provided by the *Directory of important world honey sources* (Crane, Walker & Day, 1984), and in its *Satellites* 3, 4, 5 (Crane & Walker, 1986*b*, 1986*c*, 1986*d*). The *Directory* has detailed entries for 467 plants, with available information on the composition and properties of their honeys. A number of honeys produced in temperate zones are well documented, whereas many important honeys from the tropics and subtropics have not yet been analysed. Information cited in the *Directory* suggests conformity with the proposed Standard (and the Directive) for content of total reducing sugars, ash, amylase and HMF (Crane & Walker, 1984*b*). But three out of 56 honeys contained more free acid than the limit proposed, and 7 out of 65 contained more sucrose than the proposed 5% limit, most of the 7 being produced in the tropics. Out of 75 honeys, 13 contained more water than the proposed maximum of 21%, and even this limit is too high to ensure security against fermentation (Section 13.22.2).

13.21 Sugars in honey: granulation, sweetness to man

Honey contains about 4.5 times as much sugar as water, and its sugar composition is given in Table 13.2A. The very high sugar concentration gives honey its keeping quality, its energy value (which is about 3040 kcal/kg, White, 1975*b*), and its high viscosity. In making honey from nectar or honeydew (Section 3.36), bees add the enzyme invertase (13.23.1) secreted by their hypopharyngeal glands, which 'inverts' sucrose in the nectar or honeydew, i.e. converts it into fructose and glucose.

The consequence of this inversion of sugars is as follows. At the temperature of the honey combs in the hive (30°) the solubility of glucose in a solution of fructose increases abruptly if the fructose concentration is raised above 1.5 g per g of water (Lothrop; see White, 1975*b*). The gram of water can then hold in solution, as well as the fructose, 1.25 g of glucose, which is 50% more than a dilute fructose solution can carry. This high glucose solubility does not operate at higher or lower temperatures. Moreover, the sugar sucrose does not have this property of high solubility. By inverting sucrose into glucose and fructose, at hive temperatures, the bees are thus able to produce a more concentrated solution of sugars than could otherwise be obtained – on average, honeys contain over 80% sugars, and some over 85%.

Glucose is a relatively insoluble sugar, and its amount in honey largely determines the tendency of the honey to granulate (crystallize) or to remain liquid (Section 13.21.1). Fructose is a very sweet sugar (13.21.2), and is also hygroscopic, absorbing moisture from the air (13.21.3).

Table 13.23B gives the sugar composition of honeys from *Apis cerana*, *A. dorsata* and *A. florea*.

13.21.1 Granulation

If honey is stored at temperatures much below those in the hive, some of the sugar in it may crystallize; the honey is then said to have – or to be – granulated. According to the nature of the honey, crystallization may occur within days, months or years. The length of this period is important to beekeepers and honey buyers and sellers. It depends largely on the relative amounts of the main sugars in the honey, which themselves depend on the sugar composition of the nectars/honeydews involved.

Most of the sugar in honey is fructose and glucose, the mean percentages in Table 13.2A being 39.3% fructose and 32.9% glucose. Most honeys contain slightly more fructose than glucose, and some have $1\frac{1}{2}$ times as much; the latter remain liquid for long periods, and may never granulate. Examples are false acacia (*Robinia pseudoacacia*) which is grown in many regions, and tupelo (*Nyssa ogeche*) and sourwood (*Oxydendron arboreum* in North America. A few honeys contain more glucose than fructose, and they granulate very quickly indeed. Examples are rape (*Brassica napus*); dandelion (*Taraxacum officinale*); also ivy, *Hedera helix* – in which 70% of the sugar was found to be glucose, 22% fructose and 1% sucrose (Greenway et al., 1975, 1978). In Tasmania, honey from leatherwood (*Eucryphia* spp.) granulates so firmly that it is poured into moulds, the solid blocks produced being then wrapped in paper.

Honey also tends to granulate if the glucose content

itself is high, and the ratio between the glucose and water contents of a particular honey has been used by White (1975*b*) to predict its granulating behaviour. The extent of granulation observed in honeys (which had previously been heated to clarity, see below), after 6 months' undisturbed storage in glass jars at 23°–28°, was as follows:

glucose/water ratio	*extent of granulation*
1.58	none
1.76	a few scattered crystals
1.79	1.5–3 mm layer of crystals
1.83	6–12 mm layer of crystals
1.86	a few clumps of crystals
1.99	$\frac{1}{4}$ of depth granulated
2.06	$\frac{3}{4}$ of depth granulated
2.16	complete soft granulation
2.24	complete hard granulation

The glucose/water ratio has not been found satisfactory by all authors, or for all honeys (e.g. Hadorn & Zürcher, 1974; Tabouret, 1979). In particular, honeys with a very low water content (about 13%) are much less likely to granulate than predicted by White's ratio. Tabouret (1979) developed a more complex empirical formula which gives a better prediction of granulation for these honeys, and is also applicable to honey dehydrated to contain as little as 8% water.

According to Dyce (1975), honeys 'of average consistency' granulate most rapidly at 14°, and those containing more water at slightly lower temperatures; see also White (1975*b*).

The *onset of granulation* in liquid honey can be delayed by removing from the honey all fine particles held in suspension, which could act as nuclei on which crystals could form. These may include pollen, wax particles, dust, minute crystals already in suspension, and air bubbles. Such particles cannot be seen under normal illumination, but when honey is viewed between two polaroid sheets in a polariscope, they become visible (White & Maher, 1951). Filtering the honey under pressure through diatomaceous earth removes these particles, which is one reason why this is done in large honey-processing plants (Section 13.35).

Once granulation has started, its extent and spread depend on the sugar composition and the water content of the honey, as already described. 'Frosting' – irregular surface granulation – is discussed at the end of Section 13.36.

13.21.2 Sweetness to man

In Finland, Hyvönen (1980) studied variations in the relative sweetness of different sugars under various conditions, especially in relation to fructose (Crane, 1982*b*). Fructose is an exceptionally sweet sugar, so high-fructose honeys are likely to be extra sweet. However, sweetness depends on a complex of factors, including the concentration of other sugars and of acids present, and the temperature at which the food is tasted.

Results quoted by Hyvönen and Koivistoinen (1981) show that a low (5%) fructose concentration, at low temperatures, gave the greatest enhancement of sweetness compared with that of sucrose. In the following figures, 100 represents the sweetness of a 5% sucrose solution at the same temperature:

temperature	5°	18°	22°	25°	37°	40°	50°	60°
relative sweetness of fructose	143	129	125	125	100	103	88	80

Honey is thus likely to be relatively sweeter than sucrose when taken cold or at room temperature than in hot food or drinks, and when taken at a low rather than a high concentration in the food. In cold solutions, a slight acidity enhances the sweetness of fructose, although the presence of acids in general depresses it.

The sweetness of fructose-glucose mixtures showed a synergistic effect of about 20–30%; effects with other mixtures of sugars were less marked.

13.21.3 Hygroscopicity, and baking properties

Fructose is very hygroscopic, and at room temperature its rate of absorption of water from the air rises rapidly when the relative humidity reaches about 60%. Neither glucose nor sucrose is hygroscopic to the same extent, and it is likely that honey owes its hygroscopicity entirely to its fructose content. Experiments on the use of fructose in baking show that its hygroscopic nature preserves the moistness of foods such as cakes and breads. The same property of honey may thus be ascribed to the fructose in it.

When breads and cakes are being baked, browning occurs as a result of reactions at high temperatures between sugars and amino acids. Fructose has a stronger reaction than glucose with amino acids (Kilpi & Hyvönen, 1982). Breads and cakes containing fructose brown more intensely than those containing glucose or sucrose; also browning occurs at lower concentrations of amino acids, and at an earlier stage of baking (Hyvönen & Espo, 1981; Hyvönen & Koivistoinen, 1981). A properly baked fructose cake was always darker than a similar sucrose cake, and at the same baking temperatures it had to be removed from the oven earlier, to prevent burning. There was also more crust browning. Readers who use honey in baking will recognize these characteristics (see L. B. Smith & Johnson, 1952).

13.21.4 Sucrose and other minor sugars

Honey contains small amounts of several other sugars (Table 13.21A), but none determine its characteristics as much as fructose and glucose do. In making honey, bees do not invert all the sucrose in nectar or honeydew, and honey is likely to contain about 2% or 3% sucrose; the average in Table 13.2A is 2.3%. Honeys from many plants have a sucrose content somewhat above this mean value. If honey contains substantially more than this, it may have been removed from the hive before the bees finished making it. But during a heavy nectar flow from any plant, bees convert the nectar into honey very quickly, and thus with less of the enzyme action that inverts sucrose (Section 13.21); the honey then produced is likely to contain more sucrose than normal. Direct adulteration with sucrose (Section 13.26.1) may give a sucrose content of 30% or more.

Gas chromatography has been used to make detailed studies of the sugar composition of honeys from different sources (e.g. Battaglini & Bosi, 1973; Patetta & Manino, 1978). Another method for studying the sugar composition of honeys is to measure their specific optical rotation (White, 1975*b*; Battaglini & Bosi, 1973).

Enzymes are still present in honey after extraction from the comb, and the sugar composition of honey therefore continues to change. In particular, reducing disaccharides are produced, and also trisaccharides and higher sugars. White (1975*a*) provides information on the rather complex reactions involved.

Section 13.28 discusses minor sugars in honeydew honey.

13.22 Water in honey: viscosity, fermentation

Table 13.22A compares limits for the water content of honey, on physicochemical and biological grounds (left) and according to regulations in different countries (right). A lower limit to the water content exists because of the solubilities of the constituent sugars of honey. An upper limit to the water content of what is allowed to be sold as honey is commonly set in honey regulations, and should be low enough to ensure that the honey is safe from fermentation during future storage; 18.6% is a figure commonly used.

The water content of honey may be changed deliberately by exposure to a dry or humid atmosphere, usually before extraction from the comb (Section 13.32), and – with more difficulty – it can be reduced after extraction (13.39). Apart from this, absorption of water must be guarded against, because it increases the chance of fermentation. Honey absorbs water more rapidly from a damp environment than it loses water to a dry one; equilibrium values for a legume honey are shown in Figure 13.22a.

In a region with high atmospheric humidity, it seems likely that the bees may sometimes be unable to reduce the water content of the honey to a safe level; honey from rubber (*Hevea brasiliensis*) is an example, and in Sri Lanka the water content of sealed honey from rubber may be 25% (Fernando, 1978). Killion (1975) in the USA also reports 25% in sealed honey produced in high humidity. Even if the humidity is not very high, honeys from certain plant sources characteristically have a high water content, for specific

Table 13.21A Minor sugars in honey (% of total oligosaccharide fraction)

Amounts quoted by White (1978) from 1967 and 1968 results of Siddiqui and Furgala. The oligosaccharide fraction itself constituted 3.65% of honey.

Disaccharide	%	*Trisaccharide*	%	*Higher oligosaccharide*	%
maltose	29.4	erlose	4.5	isomaltotetraose	0.33
kojibiose	8.2	theanderose	2.7	isomaltopentaose	0.16
turanose	4.7	panose	2.5		0.49
isomaltose	4.4	maltotriose	1.9		
sucrose	3.9	1-kestose	0.9	*Acidic fraction*	6.51
maltulose (and 2		isomaltotriose	0.6	(not investigated)	
unidentified ketoses)	3.1	melizitose	0.3		
nigerose	1.7	isopanose	0.24		
α, β-trehalose	1.1	centose	0.05		
gentiobiose	0.4	3-α-isomaltosylglucose	trace		
laminaribiose	0.09				
TOTALS	56.99		13.69		

Table 13.22A Limits for the water content of honey

Physicochemical and biological limits		*Limits set by regulations*
lowest commonly recorded	13%	
maximum for complete safety from fermentation (Lochhead, 1933)	17%	
	18%	maximum in Argentina
working maximum for safety from fermentation (Killion, 1975)	18.6%	maximum for USDA grading classifications
maximum for safety from fermentation if yeast count < 1/g (Lochhead, 1933)	19%	
	20%	maximum in Australia, Brazil, Canada, Latin American Codex, New Zealand, Switzerland
	21%	maximum in Central American Standard, and in ECC[b] and proposed World-wide Standard[a]
	22%	maximum in Austria
	23%	proposed maximum in ECC[b] and in World-wide Standard[a] for heather honey (*Calluna vulgaris*), whose gel-like structure hinders fermentation; (in ECC[b] also for clover (*Trifolium*) honeys)
	25%	maximum in ECC[b] for *Calluna* and *Trifolium* honeys that are 'the result of natural conditions of production'
reported values for honey in capped cells, at high atmospheric humidities in the tropics	up to 28%	

[a] see Codex Alimentarius Commission (1983/84)
[b] see European Communities Council Directive (1974)

reasons. An example is heather, *Calluna vulgaris* (Section 13.27.1), for which Crane, Walker and Day (1984) quote values ranging from 19.2% to 26%.

The determination of the water content of honey is important. In addition to the use of laboratory methods (White, 1975*b*), the water content can be obtained from the relative density measured with a hydrometer, or from the refractive index measured with a refractometer. Both instruments must be models especially designed for use with honey, because this contains much less water than most aqueous solutions. A portable refractometer is easy to use. Instruments may be calibrated to a give a direct reading of water content.

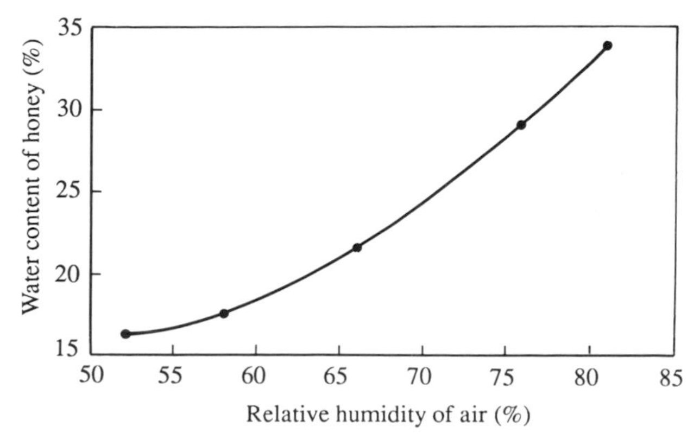

Figure 13.22a Approximate equilibrium between the relative humidity of the air and the water content of legume honey (data from Martin, 1958).

Entries in Table 13.22B give the refractive index of honeys containing from 13% to 40% water, and also the correction factor which must be applied to the reading unless the temperature is 20°. (The very high water contents are included for use when dealing with partly made honey in the humid tropics; see Section 13.39.)

13.22.1 Viscosity and fluidity

Viscosity is the internal friction which slows down the flow of honey, for instance through pipes in honey-processing equipment. All honey has a high viscosity, due to its high sugar content, as is shown by the following values (at 20°) for coefficients of viscosity

Table 13.22B Refractive index (RI) and water content of honeys containing from 13% to 40% water

If the refractive index is measured at a temperature higher than 20°, add 0.00023 per degree C above 20° to the reading, before using the table.

Figures between 13% and 22% are from White (1975*b*); between 22% and 40% from White et al. (1988).

RI	*Water* (%)	*RI*	*Water* (%)	*RI*	*Water* (%)
1.5044	13.0	1.4915	18.0	1.4815	22.0
1.5038	13.2	1.4910	18.2	1.4789	23.0
1.5033	13.4	1.4905	18.4	1.4763	24.0
1.5028	13.6	1.4900	18.6	1.4740	25.0
1.5023	13.8	1.4895	18.8	1.4717	26.0
1.5018	14.0	1.4890	19.0	1.4693	27.0
1.5012	14.2	1.4885	19.2	1.4670	28.0
1.5007	14.4	1.4880	19.4	1.4647	29.0
1.5002	14.6	1.4875	19.6	1.4624	30.0
1.4997	14.8	1.4870	19.8	1.4600	31.0
1.4992	15.0	1.4865	20.0	1.4577	32.0
1.4987	15.2	1.4860	20.2	1.4554	33.0
1.4982	15.4	1.4855	20.4	1.4531	34.0
1.4976	15.6	1.4850	20.6	1.4509	35.0
1.4971	15.8	1.4845	20.8	1.4487	36.0
1.4966	16.0	1.4840	21.0	1.4465	37.0
1.4961	16.2	1.4835	21.2	1.4443	38.0
1.4956	16.4	1.4830	21.4	1.4421	39.0
1.4951	16.6	1.4825	21.6	1.4299	40.0
1.4946	16.8	1.4820	21.8		
1.4940	17.0				
1.4935	17.2				
1.4930	17.4				
1.4925	17.6				
1.4920	17.8				

and fluidity; the latter is the reciprocal of the coefficient of viscosity.

	viscosity	*fluidity*
honey	180–190 poise	0.005 rhe
castor oil	9.9	0.1
glycerine (26.5°)	4.9	0.2
water	0.01	100

The rate of flow is directly proportional to the coefficient of fluidity.

Water content is the most important characteristic of different honeys in determining their viscosity (and thus fluidity); other factors are discussed by Pryce-Jones (1953). Honey with a low water content flows relatively slowly (Figure 13.22b, A); but its fluidity, and rate of flow, are increased by a factor of about 3 if the water content is raised from 15.0% to 18.6%.

The other factor that greatly affects the viscosity and fluidity of a liquid is its temperature (Figure 13.22b, B): fluidity, and rate of flow, are increased by a factor of 4 if the temperature is raised from 20° to 32°.

Too high a temperature can damage honey, and 35° is the preferred maximum. But honey is often heated to 40° for easy flow through pipes or strainers. In Figure 13.22b (B), temperatures above 35° are indicated by a dotted line and those above 48° by a broken line, as a reminder that a very rapid rate of flow is achieved only by excessive heating.

A few honeys show exceptional flow properties (thixotropy) because their protein content is unusually high; see Section 13.27.1.

13.22.2 Fermentation

Alcoholic drinks are produced by the controlled fer-

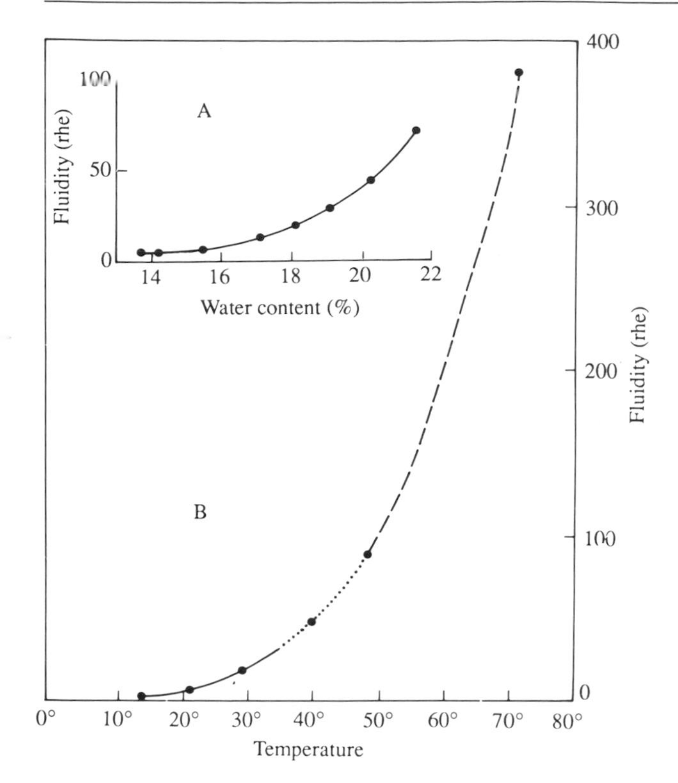

Figure 13.22b Increase in honey's fluidity (in rhe) with water content (upper curve) and with temperature (lower curve); 1 rhe = 1000/viscosity in poise (curves based on figures in White, 1975*b*).
A Honey from white clover (*Trifolium repens*) at 25°.
B Honey from sweet clover (*Melilotus*), water content 16.1%. See text for significance of dotted and broken lines.

mentation of honey (Section 13.42), but in all other circumstances fermentation spoils honey, and steps should be taken as necessary to prevent it, by storing honey only if its water content is low enough, and by using a storage temperature below 11°. All honeys contain certain osmophilic (sugar tolerant) yeasts which will multiply, with concomitant fermentation, if the water content is sufficiently high and the temperature favours growth of the yeasts.

The following 1933 figures obtained by Lochhead are quoted by White (1975*b*); see also entries in Table 13.22A:

moisture content	*liability to fermentation*
below 17.1%	safe regardless of yeast count
17.1–18.0%	safe if yeast count < 1000/g
18.1–19.0%	safe if yeast count < 10/g
19.1–20.0%	safe if yeast count < 1/g
above 20%	always in danger

The chance of fermentation is much increased if the honey granulates during storage, because crystals of the relatively insoluble glucose hydrate separate out, and they contain only 9.1% water, so the water content of the liquid portion is increased. White (1975*b*) gives further information. In temperate zones, honey that has granulated in cold winter storage is especially liable to fermentation when the temperature rises in spring. In commercial processing, honey is heated to a temperature between 62.8° and 65.5° to destroy yeasts in it, but this can damage the honey (Section 13.35).

13.23 Enzymes in honey

Enzymes are among the most interesting and important components of honey, not because they have nutritional significance in human diet, but because they play a vital part in the production of honey from its ultimate raw plant material, phloem sap. Enzymes are heat-sensitive, and an extra-low level may indicate that the honey has been overheated. Table 13.23A lists enzymes known to be in honey.

13.23.1 Enzymes from honeybees: *Apis mellifera*

The conversion of nectar or honeydew into honey is achieved by the action of certain enzymes from the hypopharyngeal glands of the bee (Table 13.23A).

Invertase (α-glucosidase) 'inverts' sucrose present in the nectar and honeydew obtained by foraging bees from plants, into fructose + glucose (Section 13.21). Some invertase is still present in the finished honey, with the result that the inversion may be carried further after the honey has been extracted from the combs, and during storage. In 1986 β-glucosidase was reported in honey by Low et al., but evidence is not yet available as to whether it is of bee or plant origin.

Glucose oxidase is active only in dilute or unripe honey, and is most active when the sugar concentration is 25–30%. The enzyme oxidizes glucose, producing gluconic acid, the principal acid found in honey, and also hydrogen peroxide which is microbicidal. Hydrogen peroxide quickly breaks down into water and oxygen, and its production and decomposition are continuous while the nectar is being converted into honey. The hydrogen peroxide concentration remains stable under a given set of conditions of temperature, sugar concentration, etc., and is sufficiently high to give good protection against some harmful micro-organisms, by a biochemical mechanism which disrupts their metabolism. The same system operates when honey is diluted with water, and for this reason honey can be effective as a microbicidal wound dressing.

The activity of the glucose oxidase system is reduced as the sugar concentration increases during the conversion of nectar into honey, disappearing altogether

Table 13.23A Enzymes in honey

Based mainly on White (1975*a*); White (1978) cites activities of the enzymes present.

Enzyme	*Characteristics*
From worker honeybees (hypopharyngeal glands)	
invertase[a]	'inverts' sucrose to glucose and fructose; more heat-sensitive than amylase
glucose oxidase	oxidizes glucose to gluconic acid and hydrogen peroxide in the presence of water, e.g. in unripened honey in the hive; more heat-sensitive than invertase
amylase (diastase)	breaks down starch; heat-sensitive; function in honey production not established – may take part in bees' digestion of pollen
From plants (nectar/honeydew)	
catalase[b]	presence demonstrated by Schepartz (1966); regulates activity of glucose oxidase (above), by controlling hydrogen peroxide equilibrium
acid phosphatase	occurs in pollen, and in nectar and honey, but very little in 'honey' stored by bees fed sucrose (Zalewski, 1965)
amylase (diastase)	a small proportion in some honeys is from plants
Shown to be absent in honey	
lactase	
protease	
lipase	
possibly inulase	

[a] The invertase in honeybees, and in honey, is an α-glucosidase (a glucoinvertase); honey contains very little fructoinvertase, which is of plant origin.

[b] Catalase is also produced in the honeybee gut, which is however not the source of this enzyme in honey (Gontarski, 1954; Zherebkin, 1964).

when the conversion is complete. In honey, a physical system comes into operation which does not involve enzymes: the high sugar concentration exerts a physical effect on many micro-organisms, causing osmosis which dehydrates and ultimately kills most of them. Exceptions include osmophilic micro-organisms that tolerate very high sugar concentrations, and several species of bacteria in spore form. The boundary between the two systems is not precise, but is likely to be around 18–20% water content; it is within this range of water content that honey becomes safe from fermentation (Table 13.22A).

Amylase (diastase) breaks down starch. It does not appear to be involved in chemical reactions that occur in the elaboration of honey. Its main relevance – to the honey trade – is that it is very heat-sensitive, and a low level of amylase in honey can be used as an indication that the honey has been overheated (Gothe, 1914). Some exporting countries have objected to this, since fresh honeys from different plant sources, and gathered under different conditions, differ much in their enzyme contents. For instance honeys from orange and other citrus, and from eucalypts, can contain less amylase than the permitted minimum, even when taken directly from the hive.

In general, honeys from very heavy flows, when the colonies have much incoming nectar to process, have lower enzyme levels than those from less profuse flows, which the bees have time to process intensively. Similarly, nectars with a high sugar content need less manipulation to convert them into honey than more dilute nectars, and so honeys from more concentrated nectars tend to have low invertase and amylase levels. (On the other hand 'honey' made by bees from dry sugar fed to them has a high enzyme content, because the bees must moisten and dilute the sugar before they can store it; in so doing, the bees add enzymes to it.) In the nature of things, heavy flows (which may be associated with low enzyme contents) are common in honey-exporting countries, but honey standards were set in honey-importing countries, early in the history of the world honey trade.

White (1978) has published a detailed examination of the actions and kinetics of the various honey enzymes, including their electrophoretically distinct forms (isozymes), and their heat inactivation.

13.23.2 Enzymes from honeybees: other *Apis* species

Table 13.23B quotes figures indicating enzyme activity of honeys from *Apis cerana*, *A. dorsata* and *A. florea*. *A. cerana* honeys are characteristically low in amylase activity compared with *A. mellifera* honeys; Wakhle et al. (1983) give values for both amylase and invertase, and their experiments confirmed that the enzymes are derived from the bees. Crane (1975*d*) gives further information about these honeys.

13.23.3 Enzymes from social insects other than honeybees

Both invertase and glucose oxidase are required in the production of honey from sugar-containing plant fluids, and both the enzymes are produced in hypopharyngeal glands of honeybees. In the food laws of many countries, honey is limited to what is produced by honeybees (or even by the honeybee *Apis mellifera*). Nevertheless, many other insects forage on similar plant fluids, and have capabilities for 'honey' production which, from a biological viewpoint, is linked with the actions of invertase and glucose oxidase on sugars from plants.

Table 13.23B summarizes positive evidence so far available, and includes examples from social bees, wasps and ants that store 'honey' (see Section 1.25). Colonies of all these insects build cells or combs for the purpose, except those of the honey ants, e.g. *Myrmecocystus*, which store honey in the abdomen of individuals known as repletes. Columns 2–4 of the table give ranges recorded for the concentrations of fructose, glucose and sucrose in the honey. They show that all insects studied converted as much, or nearly as much, plant sucrose into fructose and glucose as *A. mellifera* did, and must therefore produce sufficient invertase to do so. The wasp *Polistes gallicus* was an exception, and it would be interesting to have results for other Polistinae, and for Vespinae.

The columns on the right relate to glucose oxidase. The rate of hydrogen peroxide production depends on the concentration of the honey in water, and the rate quoted here is that for the 'optimal' concentration, which is cited in the final column. In Burgett's results (1973), hydrogen peroxide production (and also water content) was higher in honey from stingless bees and ants than in honey from *Apis*, and the highest rates of production in stingless bee honey occurred at lower concentrations of honey solids than in *Apis* honey.

Burgett (1974) comments that it is within biological reason to believe that most, if not all, species of honey-storing bees, wasps and ants possess a glucose oxidase enzyme system that protects the colony's food supply against contamination with harmful microorganisms, while the food is still dilute. It appears likely that both invertase and glucose oxidase systems were developed – under evolutionary pressure – by all 'honey'-storing Hymenoptera in the Apidae, Vespidae and Formicidae (see Section 1.25). It would seem reasonable, biologically, to use the term honey for the carbohydrate food stored by any social insect that has invertase and glucose oxidase systems; see the biological definition in Section 13.2.

13.23.4 Enzymes from plants

Honey may also contain enzymes of plant origin from nectar or honeydew, and possibly from pollen (Table 13.23A). Their amounts vary according to the plant source, and they are not essential to the honeybee colony as the bee-produced enzymes are, and they are not known to be important in honey production. *Catalase*, which breaks down hydrogen peroxide, occurs in both nectar and pollen. Some honeys contain no catalase, and they have high microbicidal activity (high hydrogen peroxide levels); examples are white clover (*Trifolium repens*) from nectar and Scots pine (*Pinus sylvestris*) from honeydew. Honeys with much catalase, e.g. bilberry and heaths (*Vaccinium myrtilis*, *Erica* spp.), showed the least microbicidal activity (Dustmann, 1971). Nectar contains *acid phosphatase* (Zalewski, 1965), but this enzyme is not known to have a function in honey production, and it is perhaps an adventitious constituent of honey. Honeydew, especially, contributes various invertases from the gut and saliva of the insect producing it (Maurizio, 1975*a*), and nectar may contribute a small amount of *amylase*. These and other minor honey enzymes are discussed further by White (1975*a*).

13.24 The aromas and flavours of honey

13.24.1 Aroma constituents

Some of the substances that give honey its aroma (fragrance, odour or scent) are common to all honeys, whereas others are derived from specific plants and occur only in the honeys from them. All the substances are volatile, and evaporate more rapidly at higher temperatures. Some of the most appreciated and delicate aromas in honey are the most evanescent, with low boiling points. *Any* heating above 30°–35°, the temperature of honey in the hive, may therefore degrade its aroma.

Identification of many aroma constituents became possible in the early 1960s by the use of gas chromatography. Substances with comparatively high boiling points were shown also to contribute to the general 'honey' aroma (Merz, 1963). Hydroxymethylfurfural (HMF) was the main volatile component of honeys

Table 13.23B Figures indicating invertase and glucose oxidase activities of honeys from honeybees, stingless bees, bumble bees, and some wasps and ants

In columns 2–4, figures marked M are summarized from Maurizio (1964) and those marked C from a table compiled by Crane (1975*d*). The high values for fructose and glucose, and the low values for sucrose, indicate invertase activity.

In the 3 final columns, figures are from Burgett (1973, 1974, 1986). The final column gives the concentration of honey in water at which hydrogen peroxide (H_2O_2) was produced most rapidly, i.e. at which glucose oxidase activity was highest, and the rate of production refers to this optimal concentration; '?' indicates that the concentration for the maximum rate of production is unknown. Rate A is in weight of $H_2O_2 \times 10^{-6}$ produced per hour per unit weight of honey solids, and comparisons are with *Apis mellifera* honey from nectar. Rate B is in weight of $H_2O_2 \times 10^{-3}$ produced per hour per unit weight of honey protein, when tested against a 0.4-M glucose substrate.

		Fructose conc. (%)	*Glucose conc. (%)*	*Sucrose conc. (%)*	*Comparative rate of H_2O_2 production*		*Opt. conc. (% honey solids)*
					A	*B*	
Honeybees (*Apis*, Apinae)							
A. mellifera	M	41–54	30–48	1–3	mean 284; highest (326) in honeydew honey, lowest (0–64) in 'honey' from bees fed sucrose	56	*c.* 33
A. cerana	C	33–54	27–39	0–8	somewhat lower (29–269)	22	25–36
A. dorsata	C	32–40	23–38	0–2	much lower (23)		48
A. florea	C	34–41	28–36	0–2	higher (574)	26	28
Stingless bees (Meliponinae)							
Melipona sp.	M	49	44–46	0	lower (1–23)		7
Trigona (Scaptotrigona) sp.						400	
Trigona fuscipennis						210	
Trigona sp.	M	50–55	28–44	0–3	much higher		13–25
	C	28–35	14–33	0–2	(406–2784)		
Bumble bees (*Bombus*, Bombinae)							
B. fervidus						125	?
B. griseocollis					(190)		17.5
B. perplexus						943	?
Bombus sp.	M	37–79	5–54	0.6–3			
Wasps (Polistinae)[a]							
Protonectarina sylveirae (honey wasp)					much higher (1075)	460	14.5
Honey ants (Formicinae), replete insects							
Myrmecocystus mexicanus[b]		11.9 (0.01)	18.7 (0.04)	trace (0.58)	much higher (1011)	272	?

[a] In 'honey' from sucrose-fed *Polistes gallicus*, which also foraged on honeydew, Maurizio found 9–18% fructose, 1–5% glucose and 31–53% sucrose.

[b] 'Dark amber crop contents'; in brackets: 'clear crop contents' (Conway, 1977).

Table 13.24A Aroma constituents of honeys

Most of the constituents below are listed by Cremer and Riedmann (Maurizio et al., 1975). Only some of them are likely to be present in any one honey. Those marked * are mentioned in the text.

Alcohols	*Ketones and aldehydes*	*Esters*
methanol	*formaldehyde	methyl formate
		ethyl formate
ethanol	acetaldehyde	methyl acetate
		ethyl acetate
		propyl acetate
		*iso*propyl acetate
propan-1-ol	*propionaldehyde	ethyl propionate
propan-2-ol	*dimethyl ketone (acetone)	
butan-1-ol	butyraldehyde	methyl butyrate
		ethyl butyrate
		*iso*amyl butyrate
*iso*butanol (2-methylpropan-1-ol)	*iso*butyraldehyde	
butan-2-ol	methylethyl ketone (butan-2-one)	
pentan-1-ol	valeraldehyde	methyl valerate
		ethyl valerate
pentan-2-ol		
2-methylbutan-1-ol		
*iso*pentanol (3-methylbutan-1-ol)	*iso*valeraldehyde	methyl *iso*valerate
3-methylbutan-2-ol		
	caproaldehyde	
		methyl pyruvate
	methacrolein	
	*diacetyl	
	acetoin	
*benzyl alcohol	benzaldehyde	methyl benzoate
		ethyl benzoate
*2-phenylethanol		methyl phenylacetate
		ethyl phenylacetate
3-phenylpropan-1-ol		
4-phenylbutan-1-ol		
furfuryl alcohol	furfural	
	*5-hydroxymethylfurfural (HMF)	
		*methylanthranilate

imported into the UK that had a 'satisfactory' flavour; in others, the amount of other higher boiling point components exceeded that of HMF.

Table 13.24A lists aroma constituents found in honey, some of which may have been present in only a few samples. Volatile components of honeys from Piedmont in Italy have also been listed (Bicchi et al., 1983). Probably all honeys contain formaldehyde, propionaldehyde and acetone, and most also contain benzyl alcohol – and phenylethanol with phenyl acetic acid which it gives on oxidation. Chemical changes during storage bring about changes in the amounts of certain constituents.

One of the aroma components associated with honeys from specific plant sources has been identified: methyl anthranilate which contributes to the distinctive aroma of citrus honey. A comparative test showed 0.84–4.37 ppm in 14 citrus honeys and 0.0–0.28 ppm in 12 others. Honey from lavender (*Lavandula angustifolia*) also contains a trace of the substance (White, 1975*a*). Diacetyl (or other diketoalkane) has been isolated from heather honey (*Calluna vulgaris*).

13.24.2 Organic acids and other substances affecting flavour

The flavour of any particular honey is produced by a number of components acting in concert, and our assessment of the flavour of honeys, as of other food, is linked with their aroma.

Organic acids are one group of constituents that contribute to the flavour of any one honey. All honeys are somewhat acid, the pH ranging from 3.1 to about 4.5, with a mean of 3.9 (pH 7 represents a neutral solution, and higher values alkaline ones). Table 13.24B lists 19 organic acids certainly or probably present in honey. By far the most important is gluconic acid, produced by the action of glucose oxidase on glucose (Section 13.23.1).

Flavour preferences of consumers are affected by upbringing and conditioning, and are not the same everywhere. Many people find the less acid honeys rather bland and uninteresting, or excessively sweet, but in North America and some other regions such honeys are customary and preferred. Honeys that are extra sweet, probably because of a high fructose content, include sainfoin, false acacia and Himalayan balsam (*Onobrychis sativa, Robinia pseudoacacia, Impatiens glandulifera*).

One of the few quantitative studies on the flavour of honey in relation to its composition was made by Mayama et al. (1982). Honeys from six plant sources in Japan were evaluated for flavour preference by 21 students (men and women), in comparison with a standard solution consisting of 40 g fructose + 35 g glucose + 50 mg gluconic acid + 50 mg proline in 100 g water. The honey taster used a scale of 7 points, −3 to +3, in order of increasing personal preference. No relation was found between the mean preference rating for each of the six honeys and any of the following: total sugar content, sugar composition, total content of organic acids, organic acid composition, total content of amino acids, amino acid composition. There was, however, a relation with the ratio between the total amount of organic acids and of sugars in each honey: the higher this ratio, the more the honey was preferred (Figure 13.24a).

Japanese sweet chestnut honey (*Castanea pubinervis*) was rated lowest; it is unusual in having a bitter flavour – attributed to its high content of certain amino acids, of which Mayama et al. (1982) give details.

Proteins (Section 13.27.1) are partly responsible for the marked flavour of honeys from heather, manuka and buckwheat (*Calluna vulgaris, Leptospermum scoparium, Fagopyrum esculentum*); see Pryce-Jones (1950), and Crane, Walker and Day (1984). Any of these honeys is likely to be a prime favourite of many people brought up in an area where it is produced, but to be regarded as objectionable by many others. Tannins, which are bitter, are among the many other constituents of some honeys that contribute to their flavours (White, 1975*a*). Bitterness of honey from the strawberry tree (*Arbutus unedo*) is due to the glucoside arbutin (Pryce-Jones, 1944); chestnut is mentioned above.

Table 13.24B Organic acids in honeys

Based on White (1975*a*), with more recent data. Only some of these acids are likely to be present in any one honey.

Known to be present	*Probably present*
acetic	α- or β-glycerophosphate
butyric	glycollic
citric	glucose-6-phosphate
formic	2- or 3-phosphoglyceric
fumaric	pyruvic
gluconic	
α-ketoglutaric	
lactic	
maleic	
malic	
oxalic	
pyroglutamic	
succinic	
tartaric	

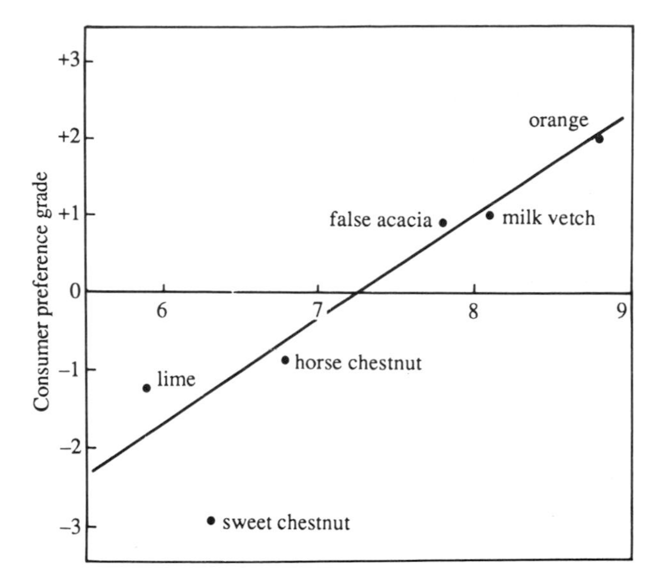

Figure 13.24a Preferences for six Japanese honeys (rated −3 to +3) in relation to the ratio between their total content of organic acids and that of sugars (based on Mayama et al., 1982; Echigo, 1986).
The ratio is plotted ($\times 10^{-4}$) along the horizontal axis.
Japanese sweet chestnut (*Castanea pubinervis*)
lime (*Tilia japonica*)
horse chestnut (*Aesculus turbinata*)
false acacia (*Robinia pseudoacacia*)
milk vetch (*Astragalus sinicus*)
orange (*Citrus*)

At a honey-tasting session held in South Africa for members of the International Food and Wine Society, 'the diversity of likes and dislikes was as individual as the honeys themselves' (E. J. Hepburn, 1982). All 13 honeys offered were considered above average by certain participants, and some of the honeys with strong flavours were no less liked than mild honeys. The honey that no-one ranked as below average for flavour was from *Eucalyptus cladocalyx*; it was the only one described as familiar. The honey least regarded as above average was from kidney vetch (*Anthyllis vulneraria*) which was described as bland, nondescript, coarsely sweet.

Flavour descriptions of *sweet* substances seem peculiarly difficult. Many of the terms used assess the acceptibility of a flavour (e.g. pleasant) rather than attempting to describe it. In France, Gonnet and Vache (1984) have published a technique for honey *dégustation*, and offer a vocabulary for describing aromas and flavours of honey. There is also a short section on honey in the French technical vocabulary of organoleptic characters of foods (Louveaux, 1962).

Little has been published on the aromas and flavours of tropical honeys from any *Apis* species.

13.25 Minerals in honey: colour

The nonvolatile inorganic residue after ignition of honey is referred to as ash, and its separate components as minerals. Most of these are metals, some of which are present in only minute amounts and are called trace elements. Table 13.25A lists 11 minerals and 17 trace elements that have been identified in normal honeys; the list includes elements as heavy as bismuth and lead, but one-half to three-quarters of the total amount is potassium. The minerals originate in the soil, and get into honey via the plants and the materials bees collect from them. Table 13.27A compares the amounts of mineral elements in honey with amounts required for human nutrition.

Minerals are among the many components that affect honey colour. Very light-coloured honeys often contain little mineral matter, and dark honeys may well contain much – although not necessarily, since the colour also depends on other factors which are largely unknown. The total weight of the mineral elements (total ash) varies from 0.02% to 1% of the honey in Table 13.2A; it is commonly 0.1–0.3%. The following honeys often contain unusually little (less than 0.1%):

water-white and very light: lucerne, sainfoin, rape, false acacia (*Medicago sativa, Onobrychis viciifolia, Brassica napus* v. *oleifera, Robinia pseudoacacia*)
light and light amber: some tree fruit, white clover, rosemary (*Prunus*/*Pyrus, Trifolium repens, Rosmarinus officinalis*)
dark: buckwheat (*Fagopyrum esculentum*).

Dark honeys that have high mineral contents include heather (*Calluna vulgaris*) and, especially, some from honeydew (Section 13.28, under Mineral content, and colour).

The colour of liquid honey can range from water-white to nearly black, with variants towards tints of green or red, or even blue. Pearly white honey is produced from species of *Ipomoea*, known as campanilla or morning glory; light golden honey from viper's bugloss, *Echium vulgare*; deep golden yellow from golden rod, *Solidago* spp.; and intense golden yellow from dandelion, *Taraxacum officinale*. Of more unusual honey colours, there is greyish-yellow from borage (*Borago officinalis*), greenish from lime (*Tilia* spp., basswood or linden) and greenish brown from tree of heaven (*Ailanthus altissima*) and maple (*Acer* spp.).

Many factors that damage honey also darken it, and are within the beekeeper's control: contact of cappings with metal at a high temperature, e.g. in a cappings melter; contamination of extracted honey with iron from damaged metal containers; storage at high tem-

Table 13.25A Amounts of mineral elements in honeys

Elements are listed in order of importance, as identified in an extensive study by Schuette and co-workers in the USA (1932, 1937, 1938, 1939). On the right are trace elements shown in one or more qualitative spectrographic assays.

Mineral element	*Average in light honey (ppm)*	*Average in dark honey (ppm)*	*Trace elements*
potassium	205	1676	chromium
			lithium
chlorine	52	113	nickel
			lead
sulphur	58	100	tin
			zinc
sodium	18	76	osmium
			beryllium
calcium	49	51	vanadium
			zirconium
phosphorus	35	47	silver
			barium
magnesium	19	35	gallium
			bismuth
silicon (as SiO_2)	9	14	gold
			germanium
iron	2.4	9.4	strontium
manganese	0.3	4.1	
copper	0.3	0.6	

peratures (F. G. Smith, 1967; White, 1975*b*); and use in honey supers of combs darkened by brood rearing (Townsend, 1974*b*). Other factors that make honey dark are related to the plant source of the honey; they include a high content of amino acids (perhaps especially tyrosine and tryptophan) and the presence of polyphenolic compounds. For further discussion see Pryce-Jones (1950), White (1975*a*) and Crane (1980*a*).

Honey colours, like honey flavours, lack a satisfactory vocabulary that is accepted in all countries. Honey 'colour' is commonly measured on a Pfund colorimeter (White, 1975*b*; Crane 1980*a*), but the readings are more closely related to optical density (degree of opaqueness, non-transparency) than to colour. Aubert and Gonnet (1983) have applied a three-dimensional system incorporating absorption of light of all spectral colours (Figure 13.25a). Of the 20 honeys, 15 were most opaque to light of about 575 nm (yellow/light-brown). At that wavelength, false acacia (Figure 13.25a) transmitted most light and dandelion least. Five honeys are represented by points for longer wavelengths, i.e. they have colours more towards the red end of the spectrum, buckwheat being the most extreme. Rodriguez López (1985) has used a spectrophotometric method to establish a classification for colours of Spanish honeys.

13.26 Diagnostic characters of honeys

The honey trade has encountered a succession of problems related to adulteration or misrepresentation. As each problem arose, diagnostic characters or indicators were sought that could detect the new type of adulteration or mistreatment of honey, or that could differentiate between honeys of different origins.

It should be stressed that, according to legislation in force, what is sold as 'honey' *may not contain any additive*. The term 'pure honey' is therefore meaningless, because by definition any honey sold is pure.

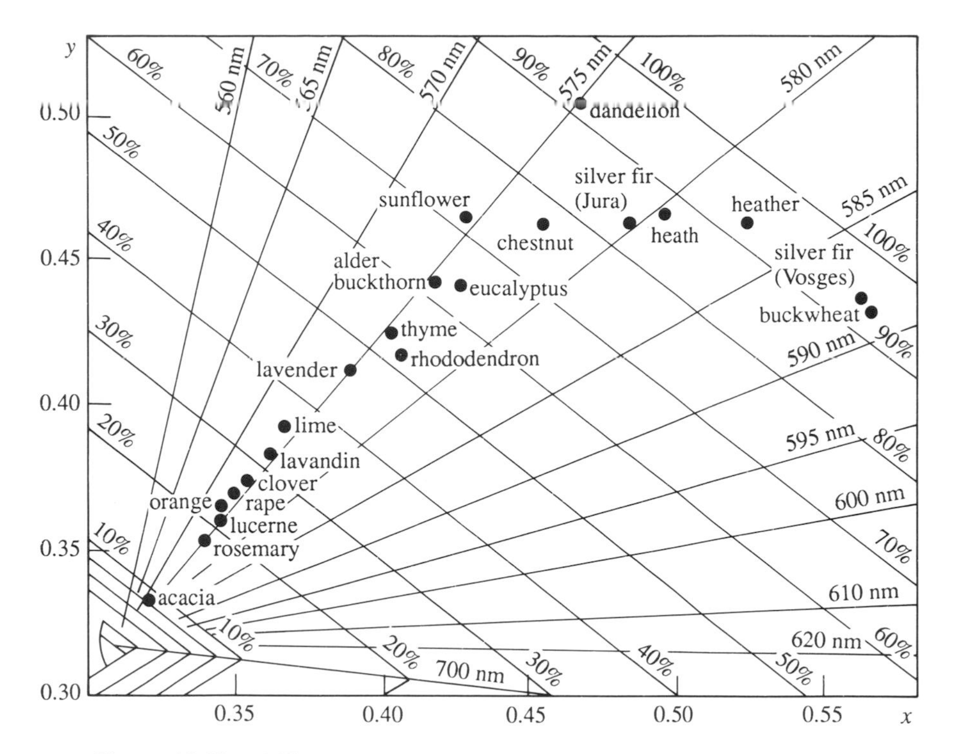

Figure 13.25a Different liquid honeys represented on an x, y chromaticity diagram (Aubert & Gonnet, 1983).
As x or y increases, the colour changes towards red or green, respectively (Clydesdale, 1969). Lines radiating from the bottom left relate to light of different colours, whose relative absorptions are indicated by the percentages marked; wavelengths are in nm, i.e. 10^{-9} m.

acacia (*Robinia pseudoacacia*)
alder buckthorn (*Rhamnus frangula*)
buckwheat (*Fagopyrum esculentum*)
chestnut (*Castanea sativa*)
clover (*Trifolium* sp.)
dandelion (*Taraxacum officinale*)
eucalyptus (*Eucalyptus* sp.)
heath (*Erica* sp.)
heather (*Calluna vulgaris*)
lavender (*Lavandula angustifolia*)
lavandin (*L. angustifolia* × *latifolia*)
lime (*Tilia* sp.)
lucerne (*Medicago sativa*)
orange (*Citrus aurantia*)
rape (*Brassica napus* v. *oleifera*)
rhododendron (*Rhododendron* sp.)
rosemary (*Rosmarinus* sp.)
silver fir (*Abies alba*)
sunflower (*Helianthus annuus*)
thyme (*Thymus* sp.)

13.26.1 Detection of honey adulteration with sugars, and overheating; HMF content

The centrifugal extractor came into use around 1870, and liquid extracted honey was put on the market. Adulteration with cheaper products soon occurred in many countries (Crane, 1975*e*), principally with starch syrup but also with sucrose and invert syrup. All were easily detected by chemical analysis. 'Honey' produced by feeding colonies of bees with large amounts of sucrose, which does not come within the legal definition of honey, contains more sucrose than honey does (above 8%).

A later adulterant was invert sugar (fructose + glucose) which, if added to honey, does not much change its gross sugar composition. This type of adulteration can be detected by measuring the level of hydroxymethylfurfural (HMF). HMF is formed during the normal industrial method of producing invert sugar from sucrose by acid hydrolysis, especially at high temperatures, and a high level of HMF in honey can thus indicate adulteration with invert sugar.

A little HMF is present in honey freshly produced by the bees, and if the honey is heated, the amount increases according to the temperatures and periods involved. Tabouret (1980) made an extensive study on the ways in which the honey is affected, and Tabouret and Mathlouthi (1972) tabulated the HMF content of the honey before and after treatment in a plate-type heat-exchanger processing up to 600 kg an hour, and also before and after 30 and 60 days' storage.

The European Communities Council Directive (1974) allows a maximum HMF content of 40 ppm. In the proposed World-wide Standard (Codex Alimentarius Commission, 1983/84) the maximum is raised to

80 ppm. On the other hand, White (1980) suggests that 200 ppm should be regarded as an action level for selecting honeys to be examined further for possible adulteration with invert sugar, and that intermediate values may imply overheating of the honey; see also White and Siciliano (1980). In Italy, the mean HMF contents for 400 honey samples obtained direct from beekeepers and for 400 commercial samples were 13 and 60 ppm, respectively (Fini & Sabatini, 1972). Both high-performance liquid chromatography (HPLC) and the bisulphite-UV method (White, 1979) give more accurate values for HMF content than the photometric method in the European Directive (1974); however, the latter method is a more sensitive indicator of heat damage to the honey (Wootton & Ryall, 1985). Levels of the enzymes amylase and invertase below 8 and 4 ppm, respectively, have also been used as indicators of overheating (Rodgers, 1975). White (1975*a*) provides further information on the above indicators, methods for estimating their amounts, and their applications.

By the early 1970s, enzymatic conversion of cheap carbohydrates, especially of starch from corn (*Zea mays*), was used to manufacture 'high fructose corn syrup' (HFCS). In spite of its name, amounts of glucose and fructose in the syrup are rather similar, and HFCS can have a sugar composition in many ways similar to that of honey – which is also produced by enzymatic action. Unlike the old invert sugar manufactured by acid hydrolysis, HFCS does not contain HMF, and a new method had to be sought for detecting adulteration of honey with it. An indicator was finally found in the isotopes of carbon present in the honey. The common carbon isotope is carbon-12, but a much less common carbon-13 also exists, and the two can be differentiated by mass spectrometry. When plants use carbon dioxide in photosynthesis, one of two metabolic pathways is followed: the Calvin cycle (C_3) which occurs in nectar and honeydew plants, and the Hatch-Slack (C_4) cycle which occurs in many other plants, two of which are corn (*Zea mays*) and canes (*Saccharum* spp.). An index used to express the ratio between the amounts of ^{13}C and ^{12}C in a plant has a value (0/00, i.e. per 1000) between -10 and -20 for C_4 plants and between -22 and -33 for C_3 plants. White and Doner (1978) investigated 84 representative honeys from the USA and 35 from 13 other countries, and found remarkably uniform values for this index, between -22 and -27, all within the C_3 range; the mean was -25.4. The mean value for samples of HFCS was -9.7, in the C_4 range. This method is thus able to detect adulteration of honey with HFCS and cane sugars; its reproducibility is well within ± 0.15 or ± 0.2 per thousand. It has also been used successfully on confectionery which was alleged to be made with honey, but in fact was not (Doner et al., 1979), and on many other food products.

Cane sugar was the earliest adulterant, and is still used for this purpose in some countries. White et al. (1988) report very high sucrose contents of some 'honeys' on sale in Sumatra, Indonesia, in 1981. The 8 highest sucrose values out of 22 were between 28% and 65%, and these 8 honeys had a water content between 27% and 37%.

13.26.2 Identification of pollen grains

In the early 1930s another problem arose for honey producers in the countries of Europe where there was a strong demand for honey. Honeys were imported from warmer parts of the world where the cost of production was lower, such as Australia and Latin America. Sometimes these honeys were improperly sold as home-produced honey, at the higher price this commanded. This was not the sale of adulterated honey, but the sale of honey at an extra-high price based on a false description.

At that time honeys from different plants could not be differentiated by any chemical methods. But honey normally contains a few grains of pollen from flowers the bees have visited in collecting nectar. If pollen grains in the honey could be compared with those collected by hand from a range of known plants that are honey sources, it should be possible to prove, for instance, that the contents of a jar labelled English clover honey – but containing many grains of *Eucalyptus* pollen and none from *Trifolium* – was falsely described.

Much painstaking work led to a considerable fulfilment of this hope, and 'pollen analysis' or 'melissopalynology' has expanded to produce a large data bank, valuable in research as well as in the detection of misdescribed honeys. At first, identification of pollen grains from different plants depended mainly on drawings, which needed much skill and time if competently done. Then photographs using light microscopy improved greatly in quality, and they superseded drawings. Scanning electron microscopy (SEM), which was developed as a routine method in the 1950s, can now give photographs showing pollen grains as three-dimensional objects at a very high magnification (Figure 13.26a).

Opposite
Figure 13.26a Photographs of pollen grains found in honey, taken with a scanning electron microscope (van Laere et al., 1969).
A *Cirsium* sp., thistle, × 1600.
B *Fragaria* sp., strawberry, × 2350.
C *Plantago lanceolata*, plantain, × 1870.
D, E, F *Taraxacum* sp., dandelion, × 1020, × 2350, × 4930, respectively.

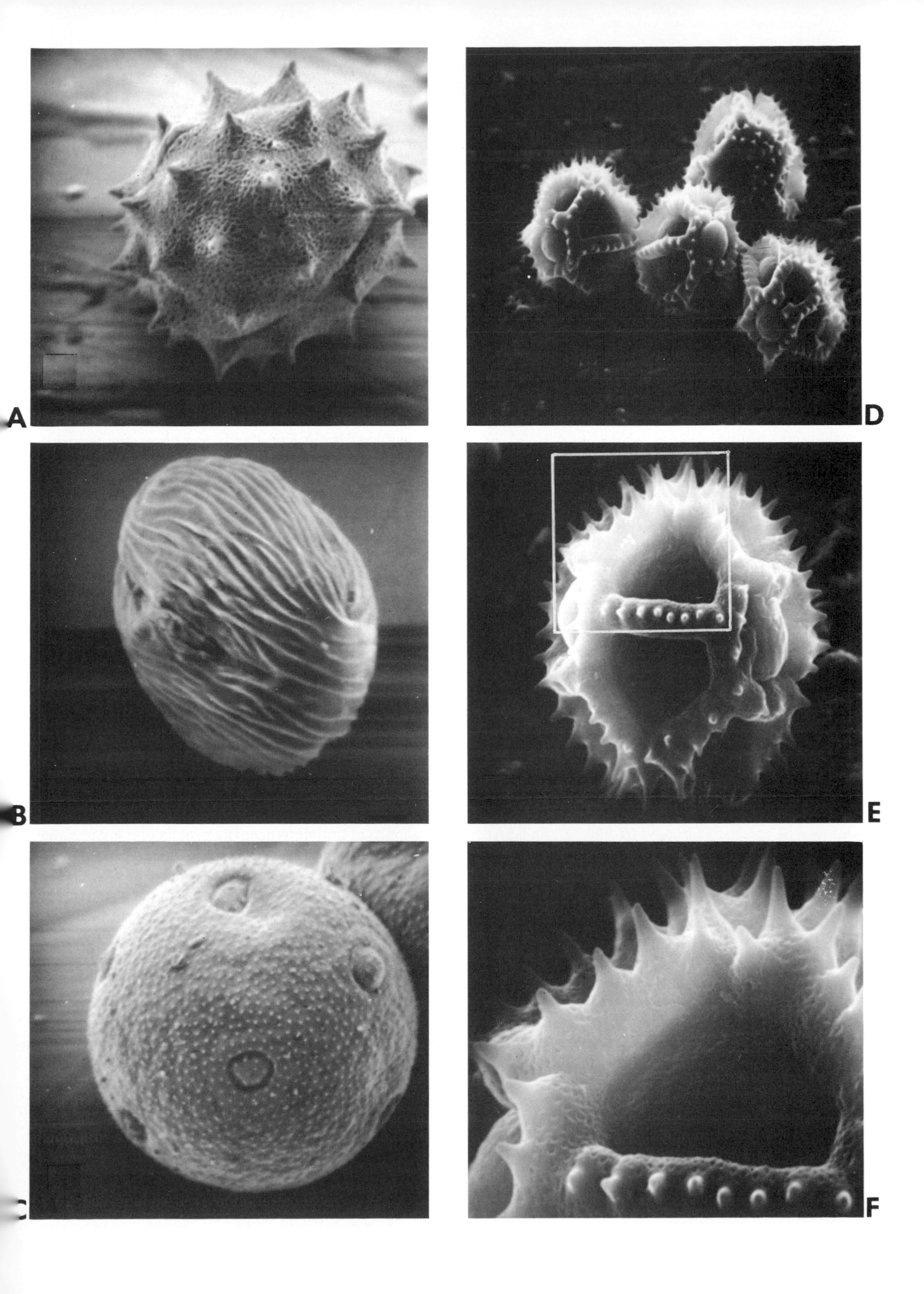
A
D
B
E
C
F

Much has been published on the identification of pollen grains in honey, from which interested readers can learn more. By far the best drawings are by Dorothy Hodges (1952). A 'photographic atlas' of pollen grains found in honeys sold in France has been published (Louveaux, 1970), and standard methods for melissopalynology (Louveaux et al., 1978). Sawyer's *Pollen identification for beekeepers* (1981) and *Honey identification* (1988) include photographs and a key to morphological features of British pollen grains. *Pollen: illustrations and scanning electronmicrographs* by Iwanami et al. (1988) has excellent photographs and other illustrations, as has the *Atlas of airborne grains and spores in Northern Europe* by Nilsson et al. (1977). Crane, Walker and Day (1984, 1986) provide indexes to pollen grain information for a number of important world honey sources.

13.26.3 Determination of amino acid composition

Since the 1960s, chromatographic techniques for chemical analysis have been applied to the detection and identification of individual amino acids in bees, their food sources and the substances they produce. Amino acids are breakdown products of proteins, and pollen is by far the most important source of them for bees, although some amino acids are also present in nectar and honeydew. Of the hive products, royal jelly and bee brood, as well as pollen, contain relatively high amounts of amino acids; honey and bee venom smaller quantities, and beeswax none. Table 13.26A gives some information on amino acids in honey, and Davies (1975) provides a fuller table. Certain amino acids are derived from the bees and are common to all honeys (White, 1975*a*). Others originate in plants, and the possibility has been explored of using the amino acid 'spectra' of honeys to differentiate between those from different plant species, or from geographical regions with different floras. Like the identification of pollen grains in honey, such a technique could be of economic value in honey-importing countries.

German studies have been discussed by Bergner (1977), and Italian ones by Bosi and Battaglini (1978), who concluded that it was not possible to use amino acid spectra to characterize honeys from different plants, since only 2 of the 28 honeys they studied gave results typical of plant source. In the UK, research concentrated on a possible differentiation between honeys of dissimilar geographical origins, and this has been more successful. Gilbert et al. (1981), following work by Davies (1975, 1976; see also Davies & Harris, 1982), used computer-operated canonical variates analysis to translate full data on the amino acid composition of each honey to a single point on a two-dimensional plot. Points for honeys with little variation between their amino acid spectra were found to be grouped close together on the plot. Two plots of these points, used in conjunction, appeared to give a good separation between honeys from the four countries studied.

Proline is the major free amino acid; honey almost invariably contains more than 200 ppm, and the mean for 740 samples was 503 ppm (White & Rudyj, 1978*a*). Proline is contributed by the bees, and must originate in the pollen they consume early in life. Davies (1978) has suggested that it is secreted by the hypopharyngeal glands and serves to regulate the secretion of invertase into nectar during its conversion to honey. Proline has a relatively low molecular weight, and the regulatory mechanism would equalize the osmotic pressure – allowing the enzyme to be transferred from the glands to the nectar or unripe honey when its sugar concentration is low, but preventing further (unnecessary) transfer when its sugar concentration increases, i.e. as the honey ripens.

13.27 Some minor constituents of honey

13.27.1 Normal minor constituents

Table 13.27A shows the amounts of vitamins and minerals in 100 g honey – which would be quite a high daily consumption – in relation to the recommended daily intake. Constituents mentioned here occur in honey only in minute quantities, but some of them may be significant under certain conditions.

Proteins and colloids in honey are of plant origin; they are discussed in some detail by Pryce-Jones (1950) and White (1975*a*). Honeys from a few plants are thixotropic, because they have an unusual amount of protein whose high molecular weight gives the honeys a viscosity much above normal (Pryce-Jones, 1944, 1950). Honey from heather (*Calluna vulgaris*) is the best known example; it contains about 1.5% (occasionally up to 1.85%) protein, whereas the norm is around 0.2%. Honey from manuka (*Leptospermum scoparium*) in New Zealand has a protein content of 1.0–1.2%. Neither of these honeys is fluid enough to be extracted by centrifuging the combs in the usual way. Pryce-Jones (1944) was able to remove the protein from heather honey and add it to clover honey; the former then flowed normally whereas the latter did not.

The following vitamins have been identified in some honeys: vitamin B_1 (thiamin), vitamin B_2 (riboflavin), niacin (nicotinic acid), B_6 (pyridoxal), pantothenic acid; and vitamin C (ascorbic acid). Their presence in honey is interesting, though the amounts are too small to be of nutritional significance (Table 13.27A).

Table 13.26A Amounts of important amino acids in honeys

Mean values for 24 unifloral nectar honeys and 4 honeydew honeys from Bosi and Battaglini (1978), who give individual values for the 28 named plant sources.
E = essential*, N = non-essential, to man.

		Nectar honeys (ppm)	*Honeydew honeys (ppm)*
Free amino acids			
proline	N	850	1057
phenylalanine	E	559	110
aspartic acid + asparagine	N	55	113
glutamic acid + glycine	N	49	195
TOTALS		1746 (0.17%)	1784 (0.18%)
Protein amino acids			
aspartic acid	N	252	177
glutamic acid	N	139	101
leucine	E	115	79
phenylalanine	E	86	69
valine	E	84	67
isoleucine	E	80	62
TOTALS		1204 (0.12%)	858 (0.09%)

* The other essential amino acids are arginine, histidine, lysine, methionine, threonine and tryptophan, all of which are present, in smaller amounts, in some honeys.

Some quantities reported for individual honeys (in ppm) and quoted by Crane, Walker and Day (1984) are:

total	260	false acacia (*Robinia pseudo-acacia*) in Japan
vitamin B_1	0.08	gela (*Catunaregam spinosa*) in India
pantothenic acid	0.7–11.5	Japanese horse-chestnut (*Aesculus turbinata*) in Japan
vitamin C	40–52	heather (*Calluna vulgaris*) in Poland
	41–82	buckwheat (*Fagopyrum esculentum*) in Poland
	113.5	karvi (*Carvia callosa*) in India

Hydrogen peroxide, produced by enzyme activity (Section 13.23.1), gives honey a microbicidal effect when this is diluted with water. Lavie (1960) found in addition a group of antibacterial factors introduced by the bees; they are light-sensitive, but are much more heat-stable than the enzyme-produced material. They inhibited the growth of certain bacteria potentially harmful to a honeybee colony, including:

Bacillus subtilis
Bacillus alvei
Escherichia coli
Proteus vulgaris
Pseudomonas pyocyanea
Salmonella
Staphylococcus aureus

In an Australian study of the inhibitory action of solutions of honeys against *Staphylococcus aureus*, honeys from different plant species showed different levels of activity, the best *Eucalyptus* honey giving complete inhibition at a concentration of 4% (Wootton et al., 1978). Lavie (1960) could find no antifungal factor in honey.

Some of the gluconic acid in honey may have a bacterial origin; ripening honey can contain the bacteria *Gluconobacter* and *Lactobacillus viridescens*, and their action is discussed by Ruiz-Argueso and Rodriguez-Navarro (1975).

Micro-organisms that survive well in honey are the osmophilic (sugar-tolerant) yeasts. White (1975*b*) tabulates a number, most belonging to the genera *Saccharomyces* and *Zygosaccharomyces*. They can be troublesome to honey handlers because when conditions are suitable they produce fermentation, especially in honey with a high water content (Section 13.22.2).

Table 13.27A Amounts of vitamins and minerals in honey in relation to human requirements

From Crane (1975*a*). Figures in Columns 4 and 5 relate to an adult man; those for an adult women, or for a person under 20, are in general the same or lower.

	Unit	*Average amount in 100 g honey*	*Recommended daily intake UK**	*Recommended daily intake USA*
Vitamins				
A (retinol)	i.u.		2500	5000
B complex:				
B_1 (thiamin)	mg	0.004–0.006	1.1–1.4	1.5
B_2 (riboflavin)	mg	0.02–0.06	1.7	1.7
niacin (nicotinic acid)	mg equiv.	0.11–0.36	18	20
B_6 (pyridoxal)	mg	0.008–0.32	(1–2)	2.0
pantothenic acid	mg	0.02–0.11	(10–20)	10
biotin	mg			0.3
folic acid	mg		0.05–0.1	0.4
B_{12} (cobalamin)	μg		3–4	6
C (ascorbic acid)	mg	2.2–2.4	30	60
D (calciferol)	i.u.		100	400
E (α-tocopherol)	i.u.		(10 mg)	30
Minerals				
calcium	g	0.004–0.03	0.5	1.0
chlorine	g	0.002–0.02	(5–9)	
copper	mg	0.01–0.1	(2.0–2.5)	2.0
iodine	mg		0.15	0.15
iron	mg	0.1–3.4	10	18
magnesium	mg	0.7–13	(150–450)	400
manganese	mg	0.02–10	(5–10)	
phosphorus	g	0.002–0.06	(1.2–2.0)	1.0
potassium	g	0.01–0.47	(2–4)	
sodium	g	0.0006–0.04	(3–6)	
zinc	mg	0.2–0.5	(10–15)	15

* Figures in brackets indicate actual daily intakes, *not* recommended daily intakes.

Pollen grains are important microscopic constituents in that they can help to indicate the plant origin of the honey (13.26.2). Algae and sooty moulds may indicate a honeydew origin of honey (13.28), and spores of other fungi can also occur.

13.27.2 Undesirable natural constituents

A few plants yield nectar containing substances toxic to man, and their honeys may therefore be toxic to some degree; see Table 13.27B. More are in the Ericaceae than in any other family. One plant listed, the New Zealand tree *Coriaria arborea*, is unusual in that honey from honeydew on its leaves is toxic, whereas honey from its nectar is not.

In South Africa 'Noors' honey from several species of *Euphorbia* produces a strong burning sensation in the throat; several honeys are notably bitter, and some others have disagreeable flavours, for instance *Melaleuca* (Myrtaceae), and *Agave* (Agavaceae).

There have been discussions of infant botulism in relation to honey, especially in California, from 1976 onwards. The causative bacterium *Clostridium botulinum* is very common in soils, and Californian soil is one of the most highly contaminated in the world. A 1982 survey of infant foods identified the bacterium also in corn syrup, so it is unlikely to have any specific relationship with bees or bee products (Crane, 1979*b*, 1983*b*; Huhlanen, 1987). Surveys of honeys from some other countries have shown no trace of the toxin, e.g. France (Colin et al., 1986) and Norway (Hetland, 1986).

13.27.3 Undesirable introduced constituents

Apart from the above naturally occurring microscopic

Table 13.27B Plant sources of honey reported toxic to man or another vertebrate, and the toxic agents

Family	*Genus/species*	*Toxic agent*	*Ref.*
Compositae	*Senecio jacobaea* (ragwort)	pyrrolizidine alkaloids (senecionine, seneciphylline, jacoline, jacobine, jacozine)	8
Coriariaceae	*Coriaria arborea* (tutu) honeydew honey, not nectar honey	picrotoxin, tutin, hyenanchin (mellitoxin)	1
	C. japonica		2
Ericaceae	*Andromeda*		1
	Arbutus unedo (strawberry tree)	arbutin (a glucoside)	1
	Kalmia latifolia (mountain laurel)	acetylandromedol (grayanotoxin I, andromedotoxin, rhodotoxin, asebotoxin)	8
	Ledum palustre		1
	Rhododrendron spp. including:		1
	R. anthopogon		3, 5
	R. ponticum	andromedotoxin	1
	NOT *R. arboreum*		3, 5
	NOT *R. campanulatum*		3, 5
	NOT *R. ferrugineum*		1
	NOT *R. hirsutum* (Alpine rose)		1
	NOT *R. thomsonii*		3, 5
	Tripetaleia paniculata (an azalea)	grayanotoxin analogues and grayanotoxin V	6
Euphorbiaceae	*Euphorbia segueirana*	tri-acylates of the diterpene ingenol, which are co-carcinogens	7
Loganaceae	*Gelsemium sempervirens* (yellow jessamine)	gelsamine	8
Ranunculaceae	*Aconitum*	aconitine	4
Solanaceae	*Datura metel* (Egyptian henbane)	scopolamine	8
	D. stramonium	atropine	8
	Hyoscyamus niger	atropine	1

References leading to further information:
1 Crane (1975*a*)
2 Ikuse et al. (1981)
3 Kafle (1984)
4 Saito et al. (1980)
5 Sharma (1984)
6 Tsuchiya et al. (1977)
7 Upadhyay et al. (1980)
8 White (1981)

constituents of honey, others can be introduced by improper apiary or honey management. Tysset and Rousseau (1981) found spores of the following pathogenic micro-organisms, whose significance is explained in Chapter 11:

Bacillus larvae
Bacillus alvei
Aspergillus
Ascosphaera
Nosema apis

Beekeepers in honey-importing countries sometimes express fears that their bees could become infected with American foul brood (AFB) as a result of foraging on imported honey containing *Bacillus larvae*, left in discarded barrels at honey-processing centres. Studies in Denmark (Hansen, 1984) showed that *Bacillus larvae* was present in 81% of samples of honeys imported from a number of countries, and in a somewhat lower proportion of Danish samples from packaging centres; only 5% of honey samples produced and bottled by Danish beekeepers were infected. These results do not necessarily imply large-scale infection of colonies.

Several other types of possible honey contaminants must be mentioned. In many countries antibiotics or other 'therapeutic' substances are routinely fed to colonies as a disease control measure. Residues of such substances must not be allowed to enter honey in the hives. It is therefore generally recommended that such treatment should be stopped 4 weeks before a flow from which honey will be harvested, and that any honey in supers then on the hive should be removed (e.g. Lehnert & Shimanuki, 1981). Antibiotics and other drugs should be used only when needed, and at the prescribed doses. Similar recommendations would apply to chemicals applied for mite control.

The microbial insecticide *Bacillus thuringiensis* is used to control wax moth (Section 11.61), and residues of this material in honey (and in beeswax) have been declared exempt from any requirement of a tolerance in the USA Laws and Statutes (1980). On the other hand ethylene dibromide, used as a comb fumigant for the same purpose, has caused many problems in the honey industry. In the mid-1980s the permitted limit in the USA was 0.03 ppm, and some of the many thousands of honey samples analysed contained up to 0.8 ppm. The situation was, however, under control by 1986 (White, 1986).

Secondly, there is a chance of more direct contamination of honey from a volatile chemical substance applied on a fume board at the top of the hive, to drive bees away from honey combs to be harvested (Section 5.32). Carbolic acid (phenol) is the substance most likely to cause trouble in honey, and its use is prohibited in some countries for this reason. In Alberta, Canada, tests by Daharu and Sporns (1984) showed that the amount of phenol in honey samples ranged from 1.5 to 14.2 ppm according to experimental conditions; see Section 5.32. Of the samples from Alberta, 60% contained more than 1 ppm. One contained 11 ppm, one from Florida in USA 12 ppm, and one from Mexico 4 ppm. The threshold for detecting phenol by taste was 7 ppm. I have not found a figure for an acceptable level, only statements that phenol should 'not be present'.

Thirdly, is there any danger of contamination of honey with chemicals used for pest control? In Table 13.27C, column 2 summarizes results of tests on honey made in a number of countries for the presence of commonly used pesticides. The amounts found were no more than 1–2% of those permitted in fruit and vegetables.

The possibility has been explored of using the mineral content of honey as an indicator of environmental pollution of the air and soil by heavy metals. Tong et al. (1975) identified altogether 47 elements in samples of honey from apiaries near an industrial area or a major highway. Meyer (1977) found 19 radioactive isotopes in honey from the vicinity of a nuclear power station. Concern about the substantial leakage of radioactive materials at the Chernobyl nuclear power station on 26 April 1986 led to the determination of radioactivity in many foods, including honey and pollen. With regard to normal levels, the Austrian Ministry of Health and Environmental Protection (Österreichischer Imkerbund, 1986) quotes values of 10–18 nanocuries/kg in 1985 (370–666 disintegrations/sec/kg) for both honey and pollen, and values up to 16 times as high after nuclear weapon tests. Safe limits are given as 50 and 100 nanocuries/kg (1850 and 3700 disintegrations/sec/kg) for honey and pollen, respectively.

A wind from the east carried the radioactive fallout from Chernobyl over western Europe, and samples of honey and pollen were taken for analysis at Tübingen, GFR, about 1500 km west of Chernobyl (Kaatz, 1986). Results are given in Table 13.27D; although much above normal, the total radioactivities on 14 May were well below the Austrian limits, except for rape pollen. Iodine was the greatest contributor, but its half-life is only 8 days, so even rape pollen would soon fall within the allowed limit. Bienenvater (1986) quoted some results obtained at about the same time. They are somewhat higher (Austria is nearer than Tübingen to Chernobyl), with iodine-131 as the greatest contributor, making pollen about 10 times as radioactive as honey. Beekeepers were recommended to store honey harvested in May for a month before selling it. The long-lived caesium-137 was found in honey harvested in May in two localities in Sweden.

13.28 Honeydew honeys

The production of honeydew by certain plant-sucking insects is described in Section 12.22, and some important sources of honeydew honey (a number of them conifers) are listed in Table 12.22A. A fair number of honeys harvested in various parts of the world contain a small proportion of honeydew honey, including some from deciduous trees and other plants that are primarily important nectar sources. Such honeys exhibit the characteristics of honeydew honey to a degree dependent on the proportion of the honeydew contribution, and if this is small the beekeeper may be unaware of it.

Honeydew honey differs from nectar honey principally because honeydew is derived from the internal phloem sap of the plant instead of from actively secreted nectar, and – especially – because enzymes from the honeydew-producing insects are involved in its production; see below. But the main constituents are mostly similar, and many characteristics of honeydew honey cover as wide a range as those of nectar honey: according to source, the honey may be lightish or dark, mild or strong in flavour, and it may granulate

Table 13.27C Residues of pesticides found in honey in various countries, and proposed Maximum Residue Limits in fruit and vegetables

Data for residues in honey are from Bech (1983*a*, 1983*b*) and for MRLs from FAO (1978)

	Max. found in honey (ppm)	*Max. permitted in fruit and vegetables (ppm)*
Herbicides		
2,4,5-T	not found (i.e. <0.005)	not laid down
2,4-D	0.3	2–5
Insecticides		
DDT (including DDE, DDD)	0.011	3.5–7
α- + β-HCH (BHC)	0.006	not laid down
lindane	0.010	0.5–3
dieldrin/aldrin	0.001	0.05–0.1
heptachlor	0.001	0.01–0.2
parathion	not found[a]	0.5–1
toxaphene	not found (i.e. <0.1)	2 (guideline only)
trichlorfon	0.005	0.1–1
dichlorvos	0.006	0.1–1
malathion	0.04–0.05[b]	
coumaphos	0.006[c]	
ethylene dibromide	0.8[c]	
Fungicide		
hexachlorbenzene	0.001	not laid down

[a] detection limit not known
[b] Thrasyvoulou et al. (1985)
[c] White (1986)

very quickly or not at all. To me, honeydew honey in SW Turkey from the Aleppo pine, *Pinus halepensis*, has a rather characterless flavour somewhat reminiscent of 'Golden syrup', and with a similar colour. In contrast, the very dark German honey of the Black Forest, from the silver fir, *Abies alba*, seems so strong that I find it unpalatable. Nevertheless both these honeys are the most favoured ones where they are produced.

Constituents and properties of honeydew honey that differ significantly from those of nectar honey are mentioned below, mainly in the order followed in 13.21–13.27. No consistent differences are reported between the water contents of honeys from nectar and honeydew.

Sugars, and optical rotation

Honeydew honey typically contains appreciable amounts of trisaccharide sugars such as melezitose, erlose and raffinose, some higher sugars, and also dextrans (polyglucoses). Figures for the composition of both honeydew and nectar honey have been collected together by Maurizio (1985), and Table 13.28A gives mean values calculated from them. The total amounts of sugar are similar in both, but on average melezitose, raffinose and maltose represent higher percentages of the total in honeydew honey, and the reducing sugars fructose + glucose a smaller percentage. The Codex Alimentarius Commission (1983/84) minimum for reducing sugars in honeydew honeys is 60%, compared with 65% for nectar honeys.

Melezitose, an important trisaccharide, was named by Berthelot in 1859 from *mélèze*, the French word for larch, when he found this new sugar in Briançon manna, the crystallized honeydew on young larch shoots. Melezitose is less soluble in water than most honey sugars, so honeys containing much of it granulate very rapidly, particularly that from the insect *Cinara laricis* feeding on the larch *Larix decidua* (Crane, Walker & Day, 1984), which has been found to contain about 42% melezitose, dry weight. Manino et al.

Table 13.27D Radioactivity of honey and pollen at Tübingen, GFR, in May 1986, after the accidental leakage at Chernobyl, USSR

Data from Kaatz (1986). Radioactivity is quoted as the number of atomic disintegrations/sec/kg, i.e. in bequerel/sec/kg; 1 nanocurie = 37 bequerel/sec.

	iodine-131	*caesium-137*	*caesium-134*	*ruthenium-103*
Honey				
control				
1984	<0.5	<0.2	<0.5	<1
1985	<0.5	<0.2	<0.5	<1
14.5.86				
rape	884	149	73	103
tree fruit	303	66	27	34
15–28.5.86				
rape	22	15	7	4
tree fruit	27	24	8	2
Pollen				
control				
1985	<0.5	<0.1	<0.5	<0.5
14.5.86				
dandelion/rape	3580	606	572	<20
tree fruit	314	52	<10	<10
Half-life	8 days	30 years	2 years	39 days

Table 13.28A Some differences between amounts of individual constituents in honeydew honeys and nectar honeys

Calculated from Tables 4 and 5 in Maurizio (1985).

Constituent	*Mean content*	
	Honeydew honey	*Nectar honey*
free acid	33.5 meq/kg	22.4 meq/kg
pH	4.5	3.9
minerals (ash)	0.58%	0.26%
fructose + glucose	61.6%	74.0%
	(415 honeys)	(1156 honeys)
Sugars as % of total sugars		
melezitose	8.6	0.2
raffinose	0.84	0.03
maltose + isomaltose	9.6	7.8
fructose + glucose	73.6	81.7
	(216 honeys)	(726 honeys)

(1985) were able to measure the sequential changes in sugar composition when phloem sap of this larch was converted into honeydew by aphid enzymes, and then into honey by honeybee enzymes. Several sugars were formed during the first process, including turanose, trehalose, and astonishingly large quantities of melezitose, and these were carried over into the honey.

The optical rotation of honey depends on its component sugars. Honeydew honeys are usually dextrorotatory, largely due to the presence of melezitose or erlose as well as glucose, whereas nectar honeys are laevorotatory because of their relatively high fructose content (White, 1975*b*).

Enzymes

Honeydew honey contains the enzymes listed in Table 13.23A, and others derived from the gut and saliva of honeydew-producing insects, including a peptidase and a proteinase (Maurizio, 1975*a*).

Organic acids, and electrical conductivity

The electrical conductivity of honey depends on both organic acids and mineral salts in it, and also on proteins and some other substances. Honeydew honeys contain more free acid than nectar honeys; Table 13.28A gives means of 33.5 and 22.4 meq/kg, respectively, both being less than the upper limit of 40 meq/kg set by the Codex Alimentarius Commission (1983/84). The electrical conductivity of honeydew honey (6.3–16.41 $\times 10^{-2}$ S/m) is higher than that of honey from most nectars (0.6–1.46 $\times 10^{-2}$ S/m), although nectar honey from heather (*Calluna vulgaris*) is an exception (7.7 $\times 10^{-2}$ S/m); see White, 1975*b*.

Mineral content, and colour

Honeydew honeys have a higher total content of minerals (ash) than nectar honeys; Table 13.28A gives means of 0.58% and 0.26%, respectively. Among the 1063 honeys in Table 13.2A, the mean mineral content ranged from 0.02% to 1.03%, so some of these were probably derived partly from honeydew. The maximum allowed by the Codex Alimentarius Commission (1983/84) is 1.0%. It is often stated that honeydew honeys are darker than nectar honeys, and this is true in general. Many of the honeydew honeys that have been most studied are dark, although that from *Marchalina hellenica* on *Pinus halepensis* is not. The presence of algae may impart a greenish tinge to some honeys; for instance samples of honey from the honeydew of *Cinara pectinatae* on *Abies alba* have been described as 'black-brown with a greenish tinge', and 'dark green'.

Antibacterial activity

This is higher in honeys from honeydew than in those from nectar (Buchner, 1967).

Microscopic constituents

Just as pollens in a sample of nectar honey can provide clues as to its plant source, so other microscopic constituents can be used to distinguish honeydew as a source of the sample, and sometimes also to indicate the plant source of the honeydew honey. The chief honeydew indicators in honey are algae and sooty moulds (Maurizio, 1975*b*). Also light dry pollen grains of wind-pollinated plants – and sometimes soot and dust – may get trapped in exposed honeydew, especially on flat leaves.

Amino acids

Certain amino acids that originate in honeydew-producing insects are also characteristic of honeydew honey.

13.3 EXTRACTING, HANDLING AND PROCESSING HONEY

Sections 13.3 to 13.39 here are concerned with the extraction of honey from combs, and the subsequent treatment of the bulk honey. Production methods for cut-comb honey, chunk honey, and comb honey in sections or other types of container, are the subject of Section 4.33.

Honey processing is taken to include all the handling of the honey during which its physical or chemical properties are purposely altered, temporarily or permanently, in order to facilitate handling or to improve certain qualities of the honey. Once the bees have completed making honey (Section 3.36), the most usual system for handling it comprises the following stages, all of which except 1, 7 and 9 involve processing. Stages 3, 4, 5 and 7 are represented in Figure 13.3a.

1. Clearing bees from the honey supers or framed combs to be harvested (equipment in Section 5.3), and taking the combs to the honey house (6.42); this is the last *beekeeping* operation.
2. Warming the combs to 32°–35°.
3. Uncapping the combs, and dealing with the cappings.
4. Extracting the honey from the combs in a centrifuge.
5. Clarifying the honey by passing it through a strainer and/or baffle tank.

(6.) In large processing plants in some countries, flash-heating and pressure-filtering.

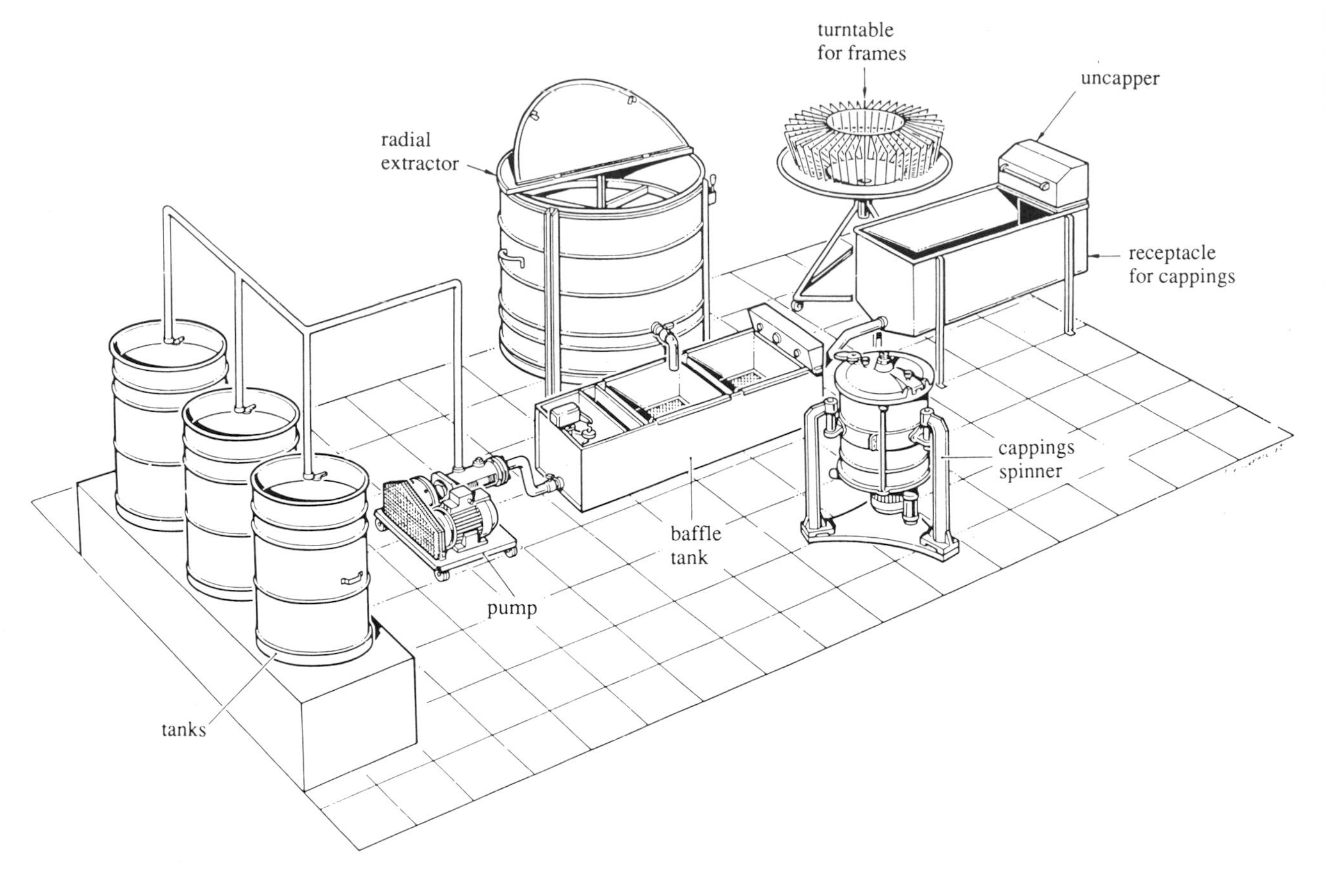

Figure 13.3a Layout of honey house equipment for stages 3 to 7 in the text, suitable for extracting up to 200 kg honey per hour (Ratia, 1984).

7. Storing the honey in bulk containers.
(8.) If desired, initiating controlled granulation, on a large or small scale, then storing.
9. Bottling in retail containers.

Since honey is hygroscopic, it should not be exposed to humid air – with a relative air humidity above 60% – at any temperature (Martin, 1958). If the water content of honey is to be reduced, this is done during stage 2 or between stages 5 and 7 (Sections 13.32, 13.39). Figure 13.22a shows the water content of honey in equilibrium with air at different relative humidities.

In all processing of honey, including its flow through pipes, pumps and strainers, the honey should be held at temperatures below 40°, or ideally below 35°, as in the hive. At lower temperatures its rate of flow would be unnecessarily slow. At higher temperatures the honey may be damaged in various ways, which are mentioned throughout Section 13.2. Higher temperatures are often used in large processing plants, reaching 71° or even up to 77° at stage 6, to kill yeasts and thus control fermentation. In any such excessive heating, both heating and cooling must be as rapid as possible, to minimize the period at high temperatures. If honey is subjected to direct heat, fructose in it is likely to 'burn', and the flavour of the honey is impaired for this reason.

During handling and processing, honey should not be in contact with the human body, or with metals that can react with it, or with surfaces such as unglazed earthenware that can absorb dirt, or with any contaminating substance.

In *Honey: a comprehensive survey*, Townsend (1975) contributed a chapter on processing and storing liquid honey (2–7 above). He gives plant layouts for:

- batch processing, up to 200 kg/h
- continuous-flow processing, up to 200 kg/h, between 200 and 750 kg/h, and above 750 kg/h.

A chapter by Dyce described the production of finely granulated or creamed honey (8 above). Both chapters have been used world-wide, by commercial enterprises setting up large honey-processing plants, and by beekeepers who handle their own honey. (Townsend (1978*a*) and Morse (1984) summarize North American practice.) This book gives a much shorter general account, but Section 13.39 refers especially to the removal of water from honey in bulk after extraction. In addition, Section 10.6 discusses the handling of honey from hives without frames, and also home-made and inexpensive devices for handling honey in general.

Organizations that buy home-produced or imported honey in bulk usually carry out stages 5 to 9 above. In some honey-producing countries, a beekeeper with a large-capacity honey plant may extract honey under contract for others, either returning the honey in the producer's own bulk containers (2–5 and 7 above), or selling it on behalf of the producer. The system is used in New Zealand (Cook, 1987), and I have seen it in successful operation in the USA.

13.31 Honey houses

All operations after the combs have been taken off the hives, and bees removed from them, involve handling food, so appropriate standards of hygiene must be applied (e.g. University of California, 1983*b*). For this reason, and to exclude bees, a separate enclosed space should normally be used, away from any storage of outdoor beekeeping equipment. It is referred to here as a honey house, whether or not it is a separate building.

Flying bees will be attracted to the honey house, and in order to prevent them entering, doorways and vehicle entrances must be protected, for instance with close-fitting hanging strips of fairly heavy plastic sheet. Any bees that do get inside must be speedily evicted. Windows may be provided with a louvre device as a bee-escape, as described in Section 6.63.1 on bee houses, except that the exits can be smaller. Reid (1981) describes a simple type in which bees are directed to the exit by a dark-light boundary on the window. Stray bees may also be removed by an industrial vacuum cleaner with a long nozzle. Insecticidal sprays should not be applied in the honey house.

In designing the layout of a honey house, advantage is often taken of sloping ground: entry of honey supers takes place at a higher level than dispatch or storage of the extracted honey, and the flow of honey through the system is then aided by gravity. In Australia, Stace and Bond (1985) published a book of honey house layouts to suit different circumstances in temperate and subtropical zones, and Hall (1977) described his mobile extracting plant. A honey house layout for tropical Africa, with ground plan and elevation, has been published by Townsend (1976*b*), and both honey house and processing equipment for use in India by Deodikar (1965), Deodikar et al. (1966) and Indian Standards Institution (1975). Paysen (1977) deals with lifting and moving devices, self-loading extractors, and other equipment.

Several references are made below to the safe warming of honey, to speed up its flow. Figure 13.22b shows that, for instance, honey flows about 3 times as fast at 30° as at 20°. Many beekeepers find electric soil-heating cable or electric heating 'tape' useful for jacketing honey containers, etc. (see e.g. Thurber, 1972; Schmechel, 1981).

13.32 Pre-extraction treatment of honey

In temperate zones, the sealed combs of honey (usually still in their supers) are warmed to 32° to 35° before extraction; this is the highest safe temperature, and at 38° combs full of honey may start to break down. A large extracting unit incorporates a thermostatically controlled 'warm room' where supers are held until they are warm enough. An area 2.3 × 1.8 m can accommodate 72 Langstroth supers, in 12 stacks each 6 supers high. Warm air is blown upwards through the supers; individual supers in each stack may be offset (staggered) to allow effective circulation of air.

If the water content of honey removed from the hive is above 18.6%, the honey may well be at risk from fermentation (Section 13.22.2; Table 13.22A). Also, if the water content is above the minimum that qualifies the honey for the highest grade in the country concerned, it will not fetch the maximum price. In either of these circumstances, it is common practice to reduce the water content of the honey to the required level. The term dehydration is preferred here, although dehumidification and drying are also used, but vacuum-drying of honey to a powder is a different process (Section 13.42, Industrial uses).

Reducing the water content of honey before extraction
Beekeepers who use honey supers on their hives normally dehydrate the honey (if necessary) while the sealed combs are still in the supers in the warm room. Water can evaporate through the cell cappings, as these are not completely impermeable. Alternatively the water content can be reduced after extraction (Section 13.39), although with very much greater difficulty and risk of damage to the honey.

Removal of water from honey in the combs has been practised at least since 1940. Air, dried with a dehumidifier and then heated, is blown through the stacks of supers in the warm room; the colder this air is before it is heated, the smaller the amount of moisture it holds, so – after it has been heated – the more it can take up from the honey. The warm moist air leaving the supers, and carrying off the excess water, is pumped out by exhaust fans. Stephen (1941) made a detailed study of the process and of ways of increasing its efficiency. Townsend (1975) found that passing air at 38° over uncapped combs reduced the water content 4 or 5 times as rapidly as from capped combs, but combs are much easier to handle while still capped. Marletto and Piton (1976) developed a rather sophisticated enclosed system for dehydrating honey in 90 supers, arranged in 9 stacks. Paysen (1987) built a warm room in which half the heat required for warming and drying the honey is collected in solar panels. E. E. Killion (1981), who produces section honey, reduces the water content if it is over 18.6%, as soon as the combs are taken off the hives.

Murrell and Henley (1988) give much quantitative information on the removal of water from combs of honey taken from hives incompletely capped, as is now customary in Saskatchewan, Canada. They emphasize the need to maintain a relative humidity in the warm room below 58%, by adjusting the air flow using exhaust and circulation fans, a dehumidifier with a humidistat, and a heat supply which is connected to a thermostat set at the temperature desired. A table sets out programmes for removing, for example, 45 kg water from 160 supers, or 180 kg from 640 supers, in 24 or 36 hours; it is necessary to use different regimes by day and by night, because outside air temperatures and humidities differ.

Townsend (1975) discusses *increasing* the water content of honey where this is desired. For instance it is difficult to process honey with a very low water content because it flows so slowly (see Figure 13.22b). Also, a producer may have honey containing only 13% or 14% water, but commanding only the same price as honey with 17% or 18%.

Uncapping the combs

In many modern beekeeping systems, combs in which the bees have stored honey are not removed from the hives until they are mostly or entirely capped (sealed). It is not a legal requirement for honey that the combs must be sealed before the honey is harvested, but capping is the outward and visible sign that the bees will do no more to the honey, and in particular that they will not reduce its water content further.

The first stage in extracting the honey from the combs is to uncap them. Many beekeepers uncap each side of each comb individually, using a hand knife that has been heated – simply by dipping it in boiling water, or continuously by steam or by an electric heating element; Detroy et al. (1979) describe one controlled at 79.4°. In parts of continental Europe the capped comb surface is sometimes scarified with an unheated fork about 7 cm wide, with fine tines.

A more rapid method, if many combs are handled, is to let each comb surface pass (usually by gravity) against a vibrating knife or a chain flail, or to use two cutters, to uncap both sides of the comb at once. Beekeeping supply firms in different areas stock types of uncapper that are the most in demand, and the most suitable for the scale on which beekeeping is done locally. Large automatic uncappers include the Bogenschutz (Sioux Bee) developed in the USA and the Penrose developed in Australia. The Dakota Gunness uncapper (Speedy Bee, 1988) moves the combs horizontally, very fast. Uncappers that tear up the wax incorporate much air into the honey, and after extraction the honey must be left to 'settle' for 5–7 days (Morse, 1984).

If the wax is torn off the combs in very small particles, an additional process must be used later to remove them from the honey. Whatever method is used, an 'uncapping tray' or other receptacle is placed below the position where the uncapping is done to catch the cappings which are separated from adhering honey by methods described below.

A different process has been used in Finland by Ahonen (1983) for uncovering the honey in the cells, provided they are capped 'dry', i.e. with a space between the honey surface and the capping (Section 3.31.3). The side of a flame from a cylindrical propane torch (e.g. Sievert 2491) is passed rapidly across the capped comb surface, so that the cappings form into beads round the cell edges (Figure 13.32a). It is a very quick process, and the honey needs little straining, since no separate 'cappings' are produced; all the wax is returned to the bees.

Time and money could be saved if the uncapping of capped combs could be eliminated. On an experimental scale Detroy and Erickson (1977) made plastic combs with cell walls less than the normal height; in the hive the bees extended the walls, filled the full-depth cells with honey, and capped them. The combs were held at 40° for 24 hours before extraction; when the combs were spun in the extractor, the wax extensions of the cell walls – with the cappings – sheared off. These combs do not seem to have come into common use.

The very large Fox-Harrison uncapper is referred to in Section 13.33 below, and in Section 9.22.

Dealing with the cappings

In traditional beekeeping, the cappings with the honey adhering to them were put in water to make mead or honey wine. This is still done in modern beekeeping in some countries, for instance Vietnam, but in the great majority of enterprises world-wide the honey and wax are separated. Many small-scale beekeepers strain most of the honey from cappings by gravity, and then put the wet cappings on hives (in an extra box, above the inner cover) for the bees to 'clean up'. A system de-

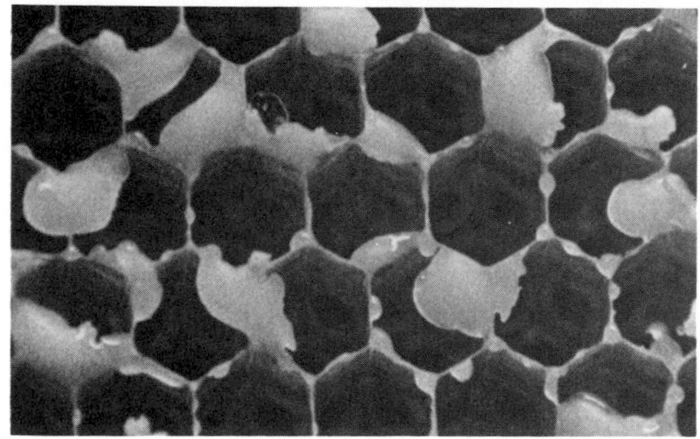

Figure 13.32a Use of a propane torch for uncapping honey (Ahonen, 1983).
above The flame being passed over the comb.
below Enlarged portion of uncapped comb.

veloped in New Zealand is described by Williams (1981). Straining is the simplest method, but the force of gravity is not large enough to remove all the honey from the wax.

Centrifugal force can quickly separate cappings from adhering honey. Beekeeping equipment suppliers sell various types of centrifuge that are suitable. Wet cappings are put in nylon netting in the centre of the centrifuge, and at high speed (1400–2000 rpm) the honey is forced away from them, leaving the (almost dry) cappings in the netting. Some types of centrifuge, with variable speed regulators, can be used for straining honey in a continuous-flow system; see Section 13.34. In New Zealand some beekeepers make their own cappings spinners (see e.g. Matheson & Reid, 1981). Jaycox (1986) has described a combined extractor and cappings spinner.

'Cappings melters' or 'wax melters' that produce a quick separation of honey and wax can be purchased, and there are models designed for continuous operation. But they are likely to overheat the honey and the wax. The cappings drop directly on to a surface or grid hot enough to melt the wax, and in a tank below, the liquid wax settles on top of the honey; honey and wax leave the tank through outlets at different levels. Since the honey from such a melter is damaged, it must be kept separate from the extracted honey, or it will lower its quality. A less drastic separator, which uses a current of heated air, is also marketed (JHB Farm Machinery, POB 100, Biniamina, Israel 30550).

13.33 Extracting honey from the comb

In modern beekeeping, after the framed combs are uncapped they are normally spun in a fairly low-speed centrifuge designed for the purpose, which is called simply an extractor. It is a cylindrical container inside which is a framework (reel, cradle) that spins about the vertical axis. The speed of rotation required for efficient honey extraction depends on the viscosity of the honey at the temperature used, and the mode of extraction (tangential or radial). For radial extractors Townsend (1975) gives a speed of 300 rpm and Girotti et al. (1977) up to 350 rpm. For tangential extractors Girotti et al. quote 380 to 400 rpm. The framework supports the frames in a vertical position, and leaves a clear depth (usually 35 to 50 cm) at the bottom of the extractor, in which honey accumulates as it leaves the combs. Honey is drawn off from a honey-gate in the side, as close to the bottom as is practicable.

Figure 13.33a shows cross-sections of different

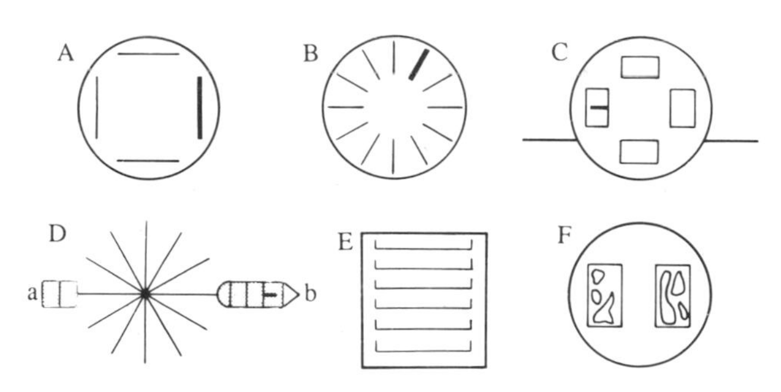

Figure 13.33a Types of centrifugal extractor; the scales are shown by a thickened line indicating a frame.

A Plan of tangential extractor for 4 frames (diameter 45 cm).
B Plan of radial extractor holding 10 to 90 frames (diameter 1 to 3 m).
C Elevation of radial extractor holding 4 honey supers complete with frames (diameter 1 to 2 m); floor level is indicated.
D Plan of Fox-Harrison extractor; the 6 supers at (a) are stationary, and the 6 at (b) are in motion, with a honey receptacle in view.
E Elevation of extractor holding 6 pairs of horizontal baskets that are filled with pieces of comb (without frames); see Section 10.62.
F Plan of extractor in E.

types of extractors. The smallest are tangential (A), holding 2, 3 or 4 frames, each at right angles to the radius. During spinning, honey from the outer side of the combs is thrown on to the wall of the extractor by centrifugal force. The combs are then reversed by hand, and honey from the other side is extracted. If combs are not built on wired foundation, they should be spun three times in all, the first at a lower speed, then reversed, then to finish extracting the first side. Some extractors incorporate a device which simplifies the reversing process. Distortion or breakage of the combs is prevented by a vertical grid forming part of the framework, that supports each comb during spinning.

Larger extractors are radial (B). The axis and the frames are also vertical but the frames are supported on radial lines, with the top-bars outwards. The extractor must be wider than twice the frame height (depth), but it can accommodate many more frames than a tangential extractor of the same diameter, and supporting grids are not necessary. A higher speed is used than with a tangential extractor, and the centrifugal force empties cells on both sides of the comb together. Radial extractors may hold from 10 to 90 frames, or even more.

Small extractors of either type are turned by hand; the gearing allows the operator to turn the handle in a vertical plane, and to rotate the combs at various speeds; Wedmore (1945) quotes safety limits. Larger models are powered by an electric or other motor, and have a continuous gearing device so that the highest speed is approached slowly; otherwise the combs would be broken.

Extractors are stocked by all beekeeping suppliers; they are usually made of stainless steel or plastic. Section 10.62 refers to extractors for unframed combs (see also Figure 13.33a, E, F), and also to some inexpensive homemade extractors, powered by hand or by a bicycle mechanism. There are also extractors for removing extraneous honey from cut comb (C. E. Killion, 1975).

Still larger extractors accommodate the uncapped frames in their supers. Those I have seen have a horizontal axis somewhat above floor level, so that supers can be placed in position from a single loading point at a convenient height. The frames in them are in a radial plane, with top-bars facing outwards; see Figure 13.33a (C).

The largest extractor I have come across (D) – in Australia in 1967 – was designed by Arthur Harrison and built by Norman Fox of California. It was used for special fixed-comb supers, described in Section 9.22; combs were uncapped while in the supers. B. N. Anderson (1966) and Harrison (1966) give descriptions. There was no large enclosing container; supers were mounted in groups of six, 2 × 3 high, each 3 being 'locked' to their own honey receptacle. Twelve groups (72 supers) were spun as a unit, at one horizontal level round a vertical axis. When the load started to spin, the receptacle rose to a horizontal position level with its supers, where it received the honey from them. When the spinning stopped, the receptacles fell back and the honey could be drained from them. Up to 8 units could be used, one above the other, and 2000 supers extracted in 8 hours. Harrison (1966) stressed that the extractor was suitable only for establishments with 10 000 colonies or more.

13.34 Straining and clarifying honey

Evaporative cooling in the extractor will have reduced the temperature of the honey, and further warming may be necessary to speed up the next process: removing small pieces of cappings wax suspended in the honey. On a small scale, straining can be done by straining the honey first through a coarse sieve and then through a fine cloth (or a nylon stocking fastened across the underside of the sieve); if only a fine strainer is used, it quickly gets clogged up. On a larger scale, the wax particles are removed by letting them rise to the surface of the honey, that has a higher density. In a continuous-flow process the honey is clarified by passing it through a baffle tank, as in Figure 13.3a; the jacket of the tank may be warmed, by a method that is safe for the honey. At right angles to the flow of honey is a series of 5 or 6 baffle plates; alternate plates leave a 2.5-cm gap at the bottom and the top. The honey is thus forced to flow under and then over a baffle, and wax particles settle at the surface between adjacent upper baffles, and can be skimmed off. This baffle tank can be used before straining in a rapid continuous flow process, in which case the honey is pumped up into an OAC strainer, that contains only the two finest strainers.

The OAC strainer (le Maistre, 1936; Townsend, 1978*a*) is effective, and is suited to continuous-flow operation. It is shown in Figure 13.34a: an upright cylindrical tank containing four graded coaxial screens, with the finest in the outside position; meshes are 5, 12, 20, 30 per cm, and diameters 25, 30, 35, 40 cm. The honey enters at the centre, and passes through the increasingly fine screens to the outer wall, whence it flows out. Each screen, as well as the tank, has a gate at the bottom to facilitate drainage; they can be opened from the top.

In large processing plants centrifugal strainers are sometimes used; coarse pre-straining is required to

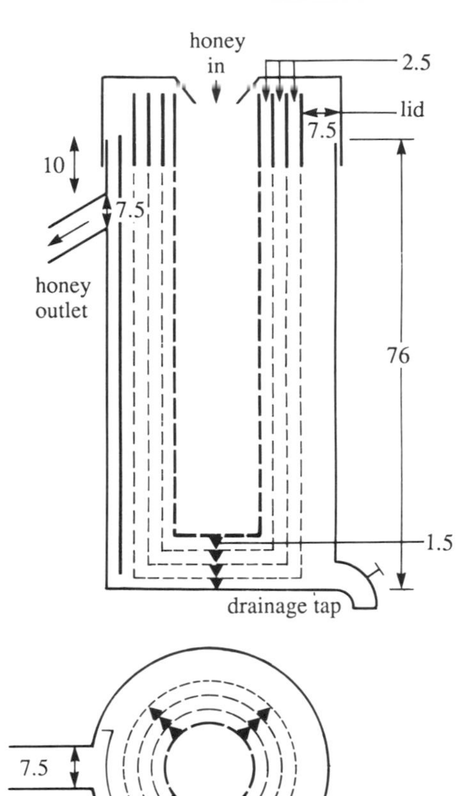

Figure 13.34a OAC honey strainer: *above*, elevation, and *below*, plan (Townsend, 1978*a*).
Measurements are in cm; the text gives mesh sizes and screen diameters, and describes the strainer.

avoid blockage of the flow process. In a continuous-flow system honey is piped into the centre of the centrifuge (capable of speeds from 2000 to 4000 rpm), where it is forced through fine nylon cloth and immediately piped out of the centrifuge.

Tubular pressure strainers are sometimes used, to give an extra 'safety' straining to honey that is already clarified, for instance after long storage or possible contamination by dirty equipment. (These are not pressure filters of the type described in Section 13.35.) A honey pump forces the honey through two tubular pressure strainers connected in parallel in a honey pipe-line. Regular cleaning of the nylon strainers is required, and this can be done by opening one of the strainers at the inlet.

Subsequent settling in the storage tanks clears honey further. In Zimbabwe, where the climate is hot, C. J. Coleman finds that honey in the storage tanks clears faster if its only previous treatment has been (coarse) straining through mosquito netting.

13.35 Heating honey to kill yeasts, and filtering

In commercial processing plants, honey is usually heated to 60° or above *to kill yeasts* in it. This is commonly referred to as pasteurization, but the purpose of the process with honey is quite different from that with milk, which is pasteurized to reduce the number of micro-organisms that are harmful to human health, such as *Bacillus tuberculosis*. Milk must be held at a temperature between 62.8° and 65.5° for 30 minutes, but the same temperature applied for a much shorter period to honey kills yeasts in it, and thus inhibits fermentation during storage. White (1975*c*) quotes the following times necessary to kill yeasts in honey at different temperatures:

temperature	60°	63°	66°	68°
time (minutes)	11	7.5	2.0	1.0

Commercial practice is flash-heating for a few seconds (or up to 1 minute), commonly to 70° or 71° but even up to 77°. The heating also dissolves any small crystals that might promote granulation. The risk of honey fermentation depends on the water content and yeast count of the honey, its natural granulation behaviour, and the processing, if any, it undergoes before storage.

Honey is damaged in various ways if heated above 35° or 40° (Townsend, 1975; White, 1975*a*). In order to minimize heat damage to honey, heating and cooling must take place as rapidly as possible. In continuous-flow systems a tubular heat-exchanger may be used, in which the heating water and the cooling water pass through a revolving coil. Larger systems use a plate-type heat-exchanger, in which the honey flows in a very thin layer through the plates. Townsend (1975, 1978*a*) gives details of the commercial processes, and Bryant (1987*a*) describes a smaller plant built by a New Zealand beekeeper that processes $2\frac{1}{2}$ tonnes an hour. Nair and Chitre (1980) published detailed results of heating *Apis cerana* honeys to 63°–68°.

Honey that has been 'flash-heated' to a pasteurizing temperature is usually pressure-filtered while it is still at the high temperature, or even after heating it further, to 77°. The honey is forced through a filter-press of diatomaceous earth, which removes material that would pass through a strainer such as pollen and colloids, and small air bubbles. Townsend (1978*a*) recommends using a temperature of 79°, for 4 minutes only, then filtering or straining at this temperature.

Pasteurizing and pressure-filtering constitute the most cost-effective bulk treatment of honey, but flavour and aroma are impaired, and HMF content is increased. The resultant honey is very clear (transparent), and is likely to remain liquid because particles that might provide nuclei for crystal formation are removed. In some countries such honey competes well with clear syrups and jellies on supermarket shelves, and it has a long shelf life without granulating.

Small-scale trials have been made in Malaysia, using a microwave oven to heat 250-ml lots of *Apis cerana* honey containing 21.3% water, to 71° for 100 seconds, to destroy yeasts and inhibit fermentation during storage (Ghazali et al., 1987).

Liebl (1977) reported the inhibition of fermentation and granulation by subjecting honey, held at 10° to 38°, to ultrasound (18 to 20 Khertz) for less than 5 minutes. Kaloyereas and Oertel (1958) used ultrasound (9 Khertz) for 15–30 minutes to delay granulation in honey. They and Townsend (1975) refer to some further methods.

13.36 Controlled granulation, and production of 'creamed' honey

The classical study on this subject was made by Dyce (1931), and the treatment he devised for honey is often referred to as the Dyce process. In his description of the process (1975), he explains the circumstances that led to his study. In Canada in the 1920s, honey was being held in bulk from year to year because of low prices, and much was lost through fermentation. Liquid honey is a supersaturated solution of sugars, of which glucose is the least soluble, and crystallizes out first (Section 13.21.1). The remaining liquid part then has a higher water content, and is likely to ferment. Dyce found that if the honey could be very finely granulated throughout, it was much safer from fermentation. It was also more attractive to many consumers, and the term creamed (or spun) is applied especially to honey with a very fine and soft texture. The process can be carried out on any scale, from a single jar to a continuous-flow operation. In large-scale processing the honey is heated at 49° (at which temperature wax particles in it are still solid), strained, then heated to 66° in a heat-exchanger and strained again. Pressure-filtering is unnecessary.

When honey crystallizes (granulates), the crystals form on nuclei suspended in the honey – unless these have all been removed by pressure-filtering. If all the crystallization occurs rapidly, only small crystals are formed, and the honey feels smooth to the tongue. Rapid crystallization can be achieved by 'seeding' cool liquid honey with 5% to 15% of a 'starter' of finely granulated honey obtained in a previous controlled process. The seed honey must be very thoroughly mixed in immediately, to disperse the small crystals throughout the liquid honey. Such mixing is not easy, since the honey is viscous at a temperature low enough to ensure that the tiny crystals added do not dissolve (14° is optimal, and 24° the maximum permissible). Mechanical stirrers of various sizes can be purchased or contrived, many operated by electricity, but they must not incorporate air. Stirrers fitted to the motor of an electric drill are not very satisfactory. A mixer with steel rotor blades, used in Finland for mixing 400 kg of honey in an upright tank, is described by Niiranen (1985). Continuous-flow operations cool the honey with a Votator, which is especially suitable for viscous liquids. They also grind seed honey to reduce the crystal size even further, and pump it at a metered rate into the cooled honey. Smeltzer (1988) gives a concise summary of his operations for creaming 1.2 tonnes of honey in a batch.

The completely mixed honey is usually packed into retail containers straight away; if this is done later, the heating needed to make the honey flow again will alter its crystalline structure irreversibly. The containers should then be maintained at a temperature between 10° and 15° until granulation is complete, which takes a few days. The optimum is 14°, and granulation is slower at 13°.

Frosting of granulated honey

Granulated honey has much less visual impact than liquid honey, and there are few disadvantages, but some benefits, in packing it in opaque retail containers. Plastic ones are inexpensive, and when empty can be stacked one inside another. Moreover granulated honey may be subject to frosting (*givrage*, *Blütenbildung*) at the interface with the container. This is likely to be caused by the incorporation of air bubbles unevenly in a part or parts of the honey. If the honey is subjected to severe cold after packing (e.g. during transport), honey shrinks and pulls away from the side walls; air enters the space formed and can become trapped when the honey is warmed again. Surface frosting (*Schaumbildung*) can also occur at the interface with the air. Associated problems, and possible solutions, were discussed in *Bee World* in the 1930s (Morland, 1931; Bee World, 1933; Manley, 1933).

13.37 Liquefying and softening granulated honey

These operations are often done immediately before the honey is put into retail packs. Bulk liquid honey that has inadvertently started to crystallize must be liquefied; hard granulated honey must be made soft enough to flow into the small containers, and if packs of chunk honey show any granulation, it may be possible to rehabilitate them.

Liquefying granulated honey
Small bulk containers can be placed in a warm room (see Section 13.32) or if not too heavy to lift, in a purpose-built insulated box or a water-bath. A box of the shape and size to accommodate one bulk container is useful for small-scale operations, and it can be insulated with expanded plastic. Or an old refrigerator can be adapted. A heat source is housed in a separate compartment, under a shelf on which the honey stands. It can be a 60-W light bulb for 12.5 kg of honey, or a 1-kW heating element for 200 kg – preferably with a thermostat incorporated.

Schley and Büskes-Schulz (1987) discuss the use of low-frequency electromagnetic radiation, and of ultrasound which is mentioned at the end of Section 13.35. Microwaves have frequencies (1 to 300 × 10^9 Hz) that are low, but higher than those of radio waves, and they transfer heat efficiently to water molecules by bringing them into resonance. Many beekeepers who have a microwave oven use this to liquefy single jars of honey (having first removed any metal lid); it takes only a minute or two, but there is a danger of overheating, and such honey is reported to regranulate quickly. In Australia, Vismatic Pty Ltd (1988) have applied 'microwave' frequency heating to soften or liquefy bulk honey. The honey can be transported – and exported – in standard tanks holding 24 000 litres (34 tonnes); on arrival at the processing or packing plant, the heating system is plugged in and the surface honey flows out continuously as it becomes warm enough. The need for return or disposal of the usual 200-litre drums is thus eliminated.

Softening granulated honey for bottling
Section 13.36 explains that honey treated by the Dyce process should be bottled straight away. There are, however, occasions when hard granulated honey must be heated sufficiently to pour it into other containers. Heat transmission in the solid honey is very slow, and heating must be done with care, because of the danger of liquefying some of the honey. On a large scale, drums of granulated honey are placed on their side on a sloping rack in the 'melting room', which is heated by hot air to 60°–70°. As soon as any honey in a drum is warm enough, it gradually flows out into a receptacle below, and a screen over the opening of each drum prevents the remaining solid core falling out in a block.

Dealing with chunk honey
A jar of chunk honey has a piece of honeycomb floating in liquid honey of a type that does not granulate. If granulation occurs, the pack is not saleable, and C. E. Killion (1975) suggests warming the jars at a temperature well below the melting point of beeswax (about 63°) until the surrounding honey is liquid again.

13.38 Storing and bottling honey

Honey should be stored at as low a temperature as possible; it is as much damaged by long storage at 30° as by heating to 70° for a few hours. At temperatures below 11° yeasts do not grow, and fermentation is thus prevented. Above 10°, darkening of honey increases rapidly. Also, HMF production increases exponentially with temperature, 3-fold for a rise of 5° to 9°. Honey granulated by the Dyce process may be stored at about 20° for perhaps a year or more; the period depends on the plant source and the previous treatment of the honey.

Hase et al. (1973) studied changes in the quality of honey during storage, and also Wootton et al. (1976, 1978), who measured effects of prolonged storage at 50° on the chemical composition and properties of different honeys: most volatile components, particularly those with honey-like aromas, were reduced in amount, and HMF and similar compounds were increased.

Containers must be airtight. In the past, much damage to honey was caused by ferrous containers with a lacquer coating that had become scratched. Tanks, drums, pails, etc., used for storing honey must be made of material that will not react with the honey, and stainless steel and plastic containers are now available. Some beekeepers find it satisfactory to use other containers with a food-grade plastic liner – which can be made airtight – and honey is now exported in such containers holding 200 litres. Containers for honey can be purchased from beekeeping supply firms, and it is worth exploring any local source of airtight containers that might be suitable. Containers must not be contaminated with any material from previous use, and they must have a securely fastened lid. If there is any risk that the honey will ferment during storage, for instance of extracted but unprocessed honey with a high water content, the airtight containers should be fitted with an air-escape valve in the top.

Bottling honey into retail containers on a large scale is carried out with automatic equipment. Townsend (1978*a*) recommends bottling at 57° to 60° to dissolve any small crystals – which can be picked up from the air. Full bottles must be cooled before they are stacked.

On a small scale the honey is also warmed to increase its rate of flow, but not above about 30°. Where honey is stored in tins (pails) without an inbuilt honey gate, there is a difficulty in transferring the honey to jars without spillage. If bottling is done from tins holding 10 to 20 kg fitted with press-in metal lids, a spring-loaded honey gate and an air escape valve can be soldered into one of the lids (Hillyard, 1960). With the tin supported on its side, sufficiently above the working surface to allow a standard jar to be slid underneath, bottling is quick and simple.

If honey is given away, there are no restrictions on its packaging, but if it is sold, all relevant legislation in force where the sale is made must be complied with. Section 15.71 gives information on laws and standards governing honey offered for sale, and 15.72 on laws relating to containers in which it is sold. Legislation varies from country to country, but its aim is to protect the purchaser against buying honey that is of an unacceptable quality, described in any misleading way, or underweight.

Honey may usually be sold only in certain quantities specified by weight. If, when bottling honey, the specified amount reaches a level in the container that is easy to identify, it is not necessary to check the weight of every container of each batch. The level will, however, change if honey of a different density is bottled.

Each retail container must be labelled, usually with the following information:

- name and address of supplier
- weight of contents
- description of contents
- statement about origin if imported.

Honey labelled according to the processing applied to it

In the past few decades there have been increasing demands from some beekeepers and some consumers in honey-importing countries for labelling legislation that recognizes the need to identify honey which has not been heat-processed and pressure-filled. (In some of the countries, all home-produced honey would come into this category.) Two examples of results of these demands are cited below.

In 1985 the US Standards for Grades of Extracted Honey included 'strained honey' (Sanford, 1985), which refers to centrifuged honey that has not been filter-pressed. (Honey *extracted* by straining is referred to in Section 10.61.) In the Netherlands, honey – either home-produced or imported – may be marked as 'biological grade' honey if it is accompanied by an appropriate certificate. This guarantees that the honey has not been heated above 40°, and that pesticides (or other substances such as antibiotics) have not been applied in the hives, or for the past 5 years to crops foraged by the bees, or to the soil on which they grow. For 'biological dynamic' honey, in addition to the above there is a guarantee that no sugar has been fed to the bees except as honey of the same grade. Rather similar labelling is used in some other countries. While it seems proper that consumers should be able to buy honey with the characteristics stated – and setting aside the difficulties of testing the validity of the guarantee – the honey should be labelled with an explicit rather than an esoteric description.

13.39 Annex: reducing the water content of honey after extraction

Section 13.2 defines honey, and Section 13.21 explains the enzymatic processes by which honeybees and some other social insects are able to convert nectar and honeydew into honey. The sugar composition of the resulting honey is such that the total sugar content can become very high, and the water content low enough to *prevent fermentation of honey* during storage in the nest or hive; see Section 13.23.1. Table 13.22A, on the water content of honey, includes the following entries:

working maximum for safety from fermentation (Killion, 1975)	18.6%
maximum in European Community Directive, and in proposed World-wide Standard (Codex Alimentarius Commission, 1983/84)	21%
reported values at high atmospheric humidities in the tropics (for honey in capped cells)	up to 28%

Most honeys with a water content too high to ensure safety from fermentation (and even too high for the proposed World-wide Standard) are produced in hot, humid regions of south-east Asia. In some of the areas concerned, what is sold as honey is outside both definitions in Section 13.2, and this high water content of the honey is a major constraint to the development and improvement of beekeeping. Coupled with it is a wide acceptance of fermented honey, whereas in most of the world honey that is properly produced and treated – by its very nature – *does not ferment.*

Factors leading to the harvesting of honey with a high water content

Constraints to the production of honey safe from fermentation include the following, of which those in the first group are in special need of detailed study.

1. Causes likely to be outside the beekeeper's control:
   - high atmospheric humidity and temperature
   - honey flows from certain plants (e.g. rubber)
   - flows at certain seasons of the year, or in certain years
   - species of bee used.
2. Beekeeper's failure to provide proper conditions for the bees:
   - inappropriate hives, e.g. too small, necessitating removal of honey before it is sealed (see below); or not allowing adequate ventilation
   - inappropriate siting of hives (e.g. without a clear air flow round them).
3. Inappropriate beekeeping management:
   - removing honey before it is sealed, for whatever reason.
4. Inappropriate treatment of honey taken out of the hive:
   - exposure to humid atmosphere.

With regard to species of bee, figures quoted for the water content of individual *Apis cerana* honeys (Crane, 1975*d*) vary from 14.3% to 28.4%; those in Table 13.2A for *A. mellifera* honeys range from 13.4% to 26.6%. (In Yucatán, Mexico, Weaver and Weaver (1981) recorded the water content of honeys from *Melipona beecheii* and from European *Apis mellifera* as 26–27% and 19–21%, respectively; they say that the former 'does not spoil'.)

Factors 2 to 4 are considered in Section 7.34 and elsewhere. For south-east Asia in general, investigations are needed to establish: (a) the exact conditions that give honey containing too much water, in relation to constraints 1 to 4 above; (b) the best ways of changing conditions that can be changed. For countries using a single-box hive in the way referred to in Section 7.34, an enquiry should be made into the validity of the stated reasons for its use. Some alternative system is needed, which is practicable in the circumstances in which the beekeepers have to work, and by which *sealed* honey can be harvested.

Methods for removing water from extracted honey

Operations that remove water from bulk honey are described elsewhere (Crane, 1989*c*). They involve evaporating water from the honey surface, and are either cumbersome, or relatively energy-intensive, according to the level of technology used. The low-technology method of boiling honey in an open pan is rejected on account of the amount of damage to the honey. Fix and Palmer-Jones (1949) experimented with circulating hot air through a 'tank drying room' holding 2.3 tonnes of honey, but this method was inefficient and has not been followed up. Other methods are also liable to damage the honey, especially if improperly carried out.

Intermediate technology offers several systems for increasing the surface area of the honey and warming it. The honey may be allowed to trickle under gravity through a large number of holes in a series of trays, heat being applied by a fan heater (Mulder, 1988*b*). More commonly, a thin film of honey flows by gravity across a series of large metal trays. The honey may be pre-heated (Maxwell, 1987*a*), or heated while in flow (Kuehl, 1988). At a higher technological level, Paysen (1987) incorporates solar heat panels, and a total tray area of 360 m²; the hourly water loss from 300 kg of honey is 3.6%, e.g. from 22.6% to 19%. Platt and Ellis (1985) use a drum along whose horizontal axis is a rotating shaft to which are attached discs or other surfaces that spread out the surface area of honey in the drum; a fan heater is incorporated. Platt (1988) and White et al. (1988) refer to similar systems which have been tried in Malaysia and Sumatra, Indonesia.

High technology systems for removing water from heat-sensitive liquid food products are discussed by Mannheim and Passy (1974), and some of them have been applied to honey. The Centri-Therm made in Sweden (Alfa-Laval, 1988*a*) can operate at atmospheric pressure or at a reduced pressure, referred to as 'under vacuum'. Low-pressure operation is to be preferred, because the temperature need not then be so high. There is still likely to be a detrimental loss of volatile substances that contribute to the flavour and aroma of honey (Girotti et al., 1977), although Tabouret (1977) reported no such loss. Experimental low-pressure systems for honey were developed in New Zealand (C. R. Paterson & Palmer-Jones, 1954, 1955; Roberts, 1957), and later in France (Tabouret, 1977) and the USSR (Stirenko, 1983). In the USA, Dadant & Sons (1987*b*) advertise a batch system; another system is used in Korea (Akratanakul, 1988), and commerical systems are available in Sweden – Centri-Therm (Alfa-Laval, 1988*a*), Convap + Contherm (Alfa-Laval, 1988*b*, 1988*c*) – and in the UK (APV, 1988; APV Crepaco, 1988).

Reverse osmosis, or ultrafiltration, is used commercially to reduce the water content of certain solutions; water under high pressure is forced through a membrane within the liquid to be concentrated (Madsen, 1974). However, the method does not seem to be practicable for honey in view of its very high sugar content.

13.4 USES OF HONEY

In almost all the Ancient World honey was the only sweetener for food, and the amount of sweetening used must have depended on the supply of honey. In England the annual consumption was probably not above 2 kg per person, and until after the Middle Ages cane sugar was a rare and expensive garnish. But as the world sugar industry grew, the price of sugar in relation to that of honey decreased, as shown in Table 13.4A. The two achieved parity at different times in different countries, in England probably between 1760 and 1860. In the end, sugar changed from a rare luxury to a cheap food; in the UK the annual consumption per person increased greatly after 1873 when the excise duty on sugar was removed. It rose to just over 50 kg in the 1960s, but has since declined to below 40 kg.

The present high consumption of sweet foods and drinks in many countries is a concomitant of the growth of the sugar industry and the wide availability of sugar. It has little to do with honey, which is now usually several times as expensive as sugar. In most parts of the world the consumption of honey depends on an adequate income as well as on a liking for honey. Section 13.81 discusses the geographical pattern of honey consumption today. Honey is a widely acceptable and easily digested food, which can be obtained in a form as near to 'natural' as almost any other, and many consumers greatly value this.

The history of honey has been outlined elsewhere (Crane, 1975*a*), and the same source gives information on present-day uses of honey.

13.41 Honey served at the table

Nowadays a major part of the world's honey production is used on its own as a sweet spread, or is added to food in other ways or to drinks, as a sweetener. There are not many other foods that are eaten so largely in their natural, unaltered state, and honey gives us a unique opportunity to sample the fragrance of the flowers round about us – or of some distant and exotic part of the earth. Where bread is a staple food, having honey to eat with it has been an indication of good living for many centuries. And at least as early as 1600 in England, a nursery rhyme had the lines 'The queen was in the parlour, Eating bread and honey'. Bread that is hot or toasted will release the honey's aroma.

Table 13.4A Chronology of the displacement of honey by cane/beet sugar
From Crane (1975*d*) and more recent data.

	Price in England (old) pence per lb				
Year	*honey*	*sugar*	*honey/sugar*	*UK sugar consumption per person (kg)*	*World sugar production (million tonnes)*
1410	1.17	24	5%		
1460	1.13	14.25	8%		
1480	1.23	8.67	14%	(estimate: 2 or less)	below 1.5
1530	1.64	6.75	24%		
1719	3.4	5.0	68%		
1750				2	
			100% probably at some time between 1760 and 1860		
1850					1.5
1890				32	5+
1895	8.5	1.8	470%		
1900				38	11+
1925					23
1930	26	3.5	740%		
1950					35
1965	70	10	*c.* 700%	50	68
1984					99
1987	150 new (360 old)	22 new (53 old)	680%	39	

Honey can be bought in different forms. It may be liquid (clear) or granulated (set, crystallized) in jars, and granulated honey can be reliquefied by standing the container in hot water, or by warming it for a very short time in a microwave oven. Honey may alternatively be prepared in 'sections', or as cut comb in a shallow container, or the cut comb may be floated in liquid honey in a glass jar; this is known as chunk honey (end of Section 4.33). Honey not in the comb should always be kept in a pot with a tightly fitting lid. A honey pot for the table should not be more than about 9 cm deep, so that the honey can be served with an ordinary jam spoon or teaspoon. In many countries honey is now sold in standard squat glass jars of this depth, holding 500 g or 1 lb. They are rather inelegant, and tall thin jars show liquid honey to better advantage. Plastic containers are taking the place of glass jars in some countries: they are lighter in weight, do not break so easily, and can be made in shapes that stack together when empty, for transport and storage. Opaque plastic does not show up the attractiveness of clear liquid honey, but it can hide any 'frosting' in granulated honey. Plastic packaging is very versatile, and lends itself to individual packs of honey for a single meal. There are sachets and small pots with peel-off covers, small tubes (of the type that holds toothpaste) for school lunches, and 'honey sticks': biodegradable transparent drinking straws filled with liquid honey and closed at each end with a 'pop-top' seal (Dunham, 1988).

13.42 Honey in foods and drinks

Domestic use

The flavour of honey is retained in uncooked dishes; some that are improved by the use of honey include: hard sauces, made for instance with butter, honey and lemon juice; liquid sauces, made with honey and orange juice; honey and cream cheese icing for cakes; mayonnaise; and hot or cold lemon drinks. Honey-fruit spreads can be made at home, and are also on sale in some countries.

I have dealt elsewhere with uses of honey in cooked dishes (Crane, 1980*a*). There are special benefits in using honey when baking certain breads and cakes, due largely to the fructose in the honey (Section 13.21.3): the hygroscopicity of honey improves the moisture-retention and keeping qualities of cakes such as gingerbreads, and bread or cake made with honey has an attractive brown crust.

When honey is used instead of sugar for baking, the temperature should be 15° lower (about 1 gas mark), or the baking time shorter. Also, if a recipe using sugar is being adapted, it is best to retain half the amount of sugar at the first trial. Honey contains about 80% sugar and 20% water, but is sweeter than sugar; Section 13.21.2 discusses the effect of temperature on sweetness. For baking, 10% to 25% more honey than sugar by weight should be used according to taste, and the amount of added liquid reduced slightly. Honey has a higher relative density than almost any other food; 1 US 'cup' holds 225 g of water, about 200 g of sugar, but about 325 g of honey; it is often unsatisfactory to use honey in a recipe which mixes units based on volume (cups) with units based on weight.

Among the more interesting honey cookery books are the following, which use the units indicated: *A taste of honey* (Wittich, 1981) and *Gale's honey book* (Jones, 1983), both metric and imperial; *Plaisir du miel* (Blaunac, 1987), metric; and *The honey cookbook* (Elkon, 1955) and *The honey kitchen* (Dadant & Sons, 1980), USA cups. Many national Beekeepers' Associations, and some local ones, publish cookery booklets, as well as leaflets available free of charge.

Industrial uses

Ingleton (1976) discusses physical and chemical requirements of honey for use in the confectionery industry. Honey is incorporated in at least one commercial brand of chocolate, and in some caramels and boiled sweets, especially those sold to soothe sore throats. It is a substantial ingredient in certain traditional sweetmeats, possibly all of Ancient Persian or Arabic origin, although sugar is used commercially for them nowadays. Directions for making the true versions at home, with honey, are available (Crane, 1980*a*): *turrón*, 8000 tonnes of which are manufactured for Christmas each year, mostly in Spain; Italian *torrone*, French *nougat*; and Greek *halvah* and *pasteli*. *Baklava* from Greece and Turkey is a cross between a sweetmeat and a cake; but *pain d'épice* from France, and *Lebkuchen* from Switzerland are baked cakes, and have a more northern origin.

Honey is used commerically in the manufacture of various food products, and their labels usually indicate this in large letters, as being likely to increase sales. It is used for coating nuts, and breakfast cereals – especially buckwheat honey or other honey with a strong flavour, so that rather little is required – and in making cakes, cookies and biscuits; in fruit preserves and other spreads (Berthold, 1988), including honey butter; also in flavoured yoghourt and in ice cream. A study of ice creams in Norway (Steinsholt, 1983) showed that samples containing Norwegian honey (up to 10%) scored higher for flavour than those without honey, and that summer honey scored higher than honeydew or heather honey. Ice creams with

7.5% or more honey were significantly softer than others, because the freezing point of fructose is much lower than that of sucrose. At least 75 to 100 tonnes of honey a year have been used by meat packers in the USA for tinned (canned) 'honey-baked' ham (Willson & Crane, 1975).

Vacuum-dried honey is produced by passing honey through an evaporator under vacuum. It is too hygroscopic to be used on its own, but a stable powder is produced if the dried honey is blended with about 55% of a starch or a non-hygroscopic sugar, or both. Yener et al. (1987) in Turkey studied different techniques. The powder is used in dry mixes for cakes and breads, and it gives an improved flavour without any need to handle honey as such. Lüpke (1980) discusses the use of dried honey in baked goods in Germany. A coating of granular dried honey is said to reduce the shrinkage of meat products by 19%. In the USA, Speedy Bee (1983) reported commercial production of dried honey in powder form 'with no additives'.

Alcoholic drinks

In many parts of tropical Africa, and especially in Ethiopia, much honey has traditionally been fermented to make beer. In Europe, honey was fermented to give mead, with an alcoholic content up to 12%. These drinks are still made, but much less than formerly. The traditional fermentation of honey to make mead or honey wine, metheglin which includes herbs, and higher alcohol drinks, is the subject of a book by Gayre (1948). This has been reprinted (Gayre & Papazian, 1986), as has the part of Sir Kenelme Digbie's 1669 book that describes *Several ways for making of metheglin ...*; it contains about a hundred recipes for honey-based alcoholic drinks, including 'hydromel as I make it weak for the Queen Mother'. Morse and Steinkraus (1975) provide a scientific exposition on the modern production of such drinks, with instructions.

In the production of apple juice, the addition of 4% to 5% honey has been found to speed up the clarification process, over a wide range of temperatures (Lee & Kime, 1984).

13.43 Honey in medicine and pharmacy

Honey is used regularly today in some specific circumstances in hospitals, and as a component of various proprietary and dispensed medicines.

The honeybee enzyme systems that protect honey from spoilage while it is being made by bees, and during storage afterwards, are explained in Section 13.23.1. The same systems form the basis of the use of honey as an antiseptic dressing for wounds, burns and some other conditions. The high osmotic pressure of honey dehydrates most micro-organisms, which inhibits their growth and may lead to their death. It can also be helpful in removing some of the fluid from soggy oedematous tissue. Diluted honey (like honey in the making) produces hydrogen peroxide, which is toxic to most pathogens. Hydrogen peroxide is unstable and decomposes, but as it is being produced all the time, a continuing concentration is maintained. Effects of the hydrogen peroxide were earlier attributed to unidentified substances in honey named 'inhibines'; like hydrogen peroxide, they were destroyed by exposure to light. Other substances are reported from time to time as sources of antibiotic or other pharmacological properties attributed to honey, for instance heat-resistant 'inhibines' (Bogdanov, 1984), and volatile components in honeys from certain plants (Tóth, 1986). Molan and Russell (1988) reported that in New Zealand honeys with *very high* antibacterial activity, a large part of the activity (on *Staphylococcus aureus*) was due to a factor other than hydrogen peroxide, which was considered to be of floral origin.

External (topical) uses

In some ancient civilizations, especially in Asia, honey was used more in medicine than as a food. In Ancient India a honey-butter paste was a favourite of Ayurvedic surgery, and it was used for dressing pierced or stretched ear lobes that were painful after surgery. In parts of Asia today, including China, honey is still much used in pharmacy.

In Ancient Egypt a usual salve for wounds contained one-third honey and two-thirds fat or grease, and its effectiveness as a wound dressing has been well proved (Manjo, 1975). This use of honey was carried over into Ancient Greece – although it was more or less absent from Hippocratic medicine – and into Ancient Rome. In mediaeval leech books, honey was cited quite frequently for external use. An early eleventh-century Anglo-Saxon manuscript in the Wellcome Historical Medical Library in London has one recipe that starts: 'To make yourself an ointment for tumours, one shall take pure honey such as it is used to lighten porridge ...'; a list is then given of juices of various herbs which may be incorporated. Among mentions of honey in the eleventh-century *Leechbook of Bald* are: as a component of 'the best eye salve'; for treating styes, dirty wounds and internal wounds, and for use after amputations and to help the removal of scabs.

In modern times honey has been used successfully in some hospitals as a surgical dressing for open wounds, proving to be more comfortable than most dressings because – surprisingly – it does not stick to the skin or the wound, and the solubility of honey in water allows it to be removed easily. There are reports of successful hospital treatment with honey of chronic and infected wounds (Armon, 1980; Dumronglert, 1983), and also of amputations and badly infected wounds, bed sores, and skin and varicose ulcers and, especially, ragged wounds difficult to dress (e.g. Bulman, 1955; Blomfield, 1973), and dental affections (Attar, 1981).

It is not easy to apply dressings to a damaged area of mucous membrane after hospital surgery or minor injuries, and there have been reports of dramatic success after undiluted honey was poured into extensive and intransigent wounds made in operations for carcinoma of the vulva and of the uterine body (Cavanagh et al., 1970). The wounds became bacteriologically sterile within 3–6 days, and remained so. Honey was found to be non-irritant, and to promote rapid growth of healthy granulation tissue. Liberal application of undiluted honey has been found to bring relief from the tiresome symptoms of pruritus vulvae, after other treatments had failed. Honey has also been used to treat diseased mucous membranes of the mouth, and in veterinary medicine for teat lesions in cows.

Honey has often been used as a first-aid dressing for wounds and burns. There are many empirical reports that burns dressed immediately with honey heal well, without infection and with less scarring than after most treatments. Burlando (1978) has published evidence from experiments on skin burns in rats. However, a casualty department would probably say that burns should receive no treatment at all until they have been seen by a doctor.

Internal (systemic) uses

At least 200 tonnes of honey are purchased annually world-wide for use in the manufacture of cough mixtures and sweets; menthol and eucalyptus are among other ingredients added. Honey has a soothing effect and is taken, often with lemon juice, in household treatments to ease coughs and sore throats. Some sufferers from hay fever have reported that their symptoms are alleviated if they eat the cell cappings sliced off honey combs before the honey is extracted from them, but I know of no controlled trials on this use of honey and beeswax together in cappings. Riches (1987) does not consider that claims for honey in hay fever treatment have been properly substantiated.

Honey is widely used as a general pick-me-up; its main sugars, fructose and glucose, are absorbed directly into the blood and provide a rapid source of energy. Honey has been used with success in treating gastrointestinal disorders (Salem, 1981; Haffejee & Moosa, 1985). It is beyond the scope of this book to assess the validity of the many reports of beneficial effects of taking honey by mouth for specific disorders. Honey is much used in general medicine in some countries, especially in Eastern Europe and Asia, and in Islam. Books on honey therapy have been published in Romania (Mladenov, 1972; Apimondia, 1976) and France (Donadieu, 1983). The use of honey is sometimes regarded as part of 'apitherapy', referred to in Section 14.1, and further information will be found in publications cited there.

13.44 Some other uses of honey

Honey is used, although not in large quantities, for a variety of other purposes. These include the preservation of bull semen, and of human corneas for transplanting. Low-grade honey is sometimes used in feed for pigs and cattle. Honey has also been used for treating various veterinary ailments.

Honey has sundry traditional uses in cosmetics. Krochmal (1985) gives a number of home recipes (in USA cups) for skin softeners and moisturizers, face packs and hair conditioners, and Crane (1980*a*) quotes some from Russia, where the composition of cosmetics is not a closely guarded trade secret. Elsewhere the inclusion of honey is usually stated, but not the amount, or what the main ingredients are.

13.5 BEESWAX: COMPOSITION AND PROPERTIES

The term beeswax is often confined to wax produced by honeybees (*Apis* species), and many commercial users would specify *A. mellifera* as the source; see Section 13.82. But wax from stingless bees (Meliponinae) is the beeswax of regions where these bees are the main source, and was the wax used by the Incas in pre-Columbian times, for casting their magnificent gold ornaments.

Among insect waxes, beeswaxes have comparatively low melting points; in pharmacopoeias the melting point is listed as 61° up to 66° (Table 13.5B), and is usually quoted as 63° to 65°. Beeswax is unusual in being plastic and supple at temperatures as low as 35° – the temperature of the brood nest – or even 32°, and this property enables honeybees to build the large combs that constitute their nest. At temperatures between 25° and 35° beeswax is so strong that a comb can support as much as several kilograms of honey (Hepburn, 1986). A colony can thus store enough food to sustain its highly organized social life throughout the year.

Beeswax is secreted by worker honeybees from four pairs of wax glands on the underside of the abdomen, which are functional when the bees are about 9–17 days old; see Section 2.23.2. The composition of beeswax from a single species of honeybee, though very complex, is relatively constant wherever it is harvested. And indeed, the characteristics needed for the bees' building material – great plasticity and great strength at temperatures in the nest or hive – are similar everywhere. The most extensive studies on the mechanical properties of beeswax were made by Hepburn and his colleagues in South Africa in the early

Table 13.5A Fractions of *Apis mellifera* beeswax, with information on their components

'Major' components constitute more than 1% *of the fraction*. Percentages are given for substances that constitute more than 1% *of the total*. C_{40}, C_{42}, etc., indicate the number of carbon atoms in long-chain compounds. Data from Tulloch (1980).

Fraction	*Percent of total*
Monoesters	
10 major components including:	
5 saturated:	
C_{40}	6
C_{42}	3
C_{44}	3
C_{46}	8
C_{48}	6
2 unsaturated:	
$C_{46:1}$	2
$C_{48:1}$	2
10 minor components	
Diesters	
6 major components including:	
C_{56}	2
C_{58}	2
C_{60}	2
C_{62}	3
C_{64}	1
24 minor components	
Hydrocarbons	
10 major components including:	
3 saturated:	
C_{27}	4
C_{29}	2
C_{31}	1
2 unsaturated:	
$C_{31:1}$	1
$C_{33:1}$	2.5
66 minor components	
Free acids	
8 major components including:	
C_{24}	6
C_{26}	1
C_{28}	1
10 minor components	
Hydroxy polyesters	8
5 major components	
20 minor components	
Hydroxy monoesters	4
5 major components including:	
C_{46}	1
20 minor components	
Triesters	3
5 major components	
20 minor components	
Acid polyesters	2
5 major components	
20 minor components	
Acid esters	1
7 major components	
20 minor components	
Free alcohols	1
5 major components	
? minor components	
Unidentified	6
7 major components	
? minor components	
	100%

1980s (Hepburn & Rigby, 1981; Hepburn, 1986); see end of Section 3.31.1.

Like many fats and waxes in living organisms, beeswax is a mixture of different fractions (groups of compounds), each consisting of a series of long-carbon-chain compounds. Analytical techniques for determining the detailed composition of beeswax were developed around 1960. Table 13.5A, which lists 10 fractions of beeswax, indicates the presence of at least 284 components in all; see also Hepburn (1986). The 21 named components in the table account for 59.5% of beeswax. The remainder includes a large number of minor components, whose properties are partly responsible for the relatively low melting point of the wax, and are the principal reason for its plasticity.

Official specifications for beeswax in seven countries (Table 13.5B) are fairly similar in most respects. Callow (1963) pointed out that they are considerably out of date, as did Tulloch (1980), who suggested 70–80, or perhaps 70–77, as a better ester value. British specifications (1 and 8) are quoted in full by the International Trade Centre (1978), and the USA specification (10) by Coggshall and Morse (1984). Because the composition of beeswax does not vary very much, adulteration can be detected fairly easily. Tulloch (1973) describes the tests to be used and discusses complicating factors.

Beeswax is an inert material which can retain its shape and quality for many centuries, as exemplified by the model of a bird in Figure 13.71a. Table 13.5C gives values for physical properties of beeswax which make it useful for purposes described in Section 13.7. The values are quoted from various sources, and the previous history of most of the beeswax samples examined is not known. The relative density becomes higher as the temperature decreases, i.e. the beeswax shrinks; the wax also shrinks as it changes from a liquid to a solid – by about 10% in volume as it cools from 74° to 25° (Warth, 1956).

Beeswax does not mix with or absorb water, and can form a protective layer impermeable to water. (It can also act as a boundary lubricant where monomolecular layers of the wax are adsorbed on opposing surfaces.) However, a solution of beeswax in a solvent can form an emulsion with water, and beeswax can be used as a stiffener in creams and soft polishes. The electrical resistance of beeswax makes the wax useful as an insulator. Beeswax also has dielectric properties, which have led to its use in the production of electrets.

Waxes from bees other than European *Apis mellifera* have been studied to some extent. Figure 13.5a shows chromatograms of waxes of four honeybee species. Waxes from European and tropical African *Apis mellifera* are very similar. Waxes from tropical Asiatic *Apis* species (*cerana, dorsata, florea*) differ from them, but these three resemble each other fairly closely, and are known collectively as Ghedda wax in the trade. Detailed studies of the composition and properties of

Table 13.5B Official specifications for 'yellow beeswax'

From Tulloch (1980), who cites publications that give full specifications; the wax is generally specified as from *Apis mellifera*, occasionally as 'possibly from other species of *Apis*'.

	Pharmacopoeia	*Melting point*	*Acid value*	*Ester value*	*Ratio number*[c]
1	British (1948)[a]	62–64°	17.0–23.0	70–80	3.3–4.2
2	Deutsches Arzneibuch	61–66°	17.0–22.0	66–82	3.0–4.3
3	Farmacopea Argentina	62–65°	17.0–23.0	70–80	
4	United States XV[a]	62–65°	18.0–24.0	72–79	
5	Pharmacopée Française VII[a]	62–66°	16.8–22.4	72–80	
6	Farmacopéia Brasil	62–66°	17.0–24.0	72–79	
7	State Pharmacopeia USSR[a]	63–65°	17.0–20.5	66–76	3.42–3.9
	Other specifications				
8	British Pharmaceutical Codex 1973[b]	62–65°	17.0–23.0	70–80	3.3–4.2
9	National Formulary [USA] XIV[b]	62–65°	18.0–24.0	72–77	3.3–4.2
10	American Wax Importers and Refiners Association[b]	62–65°	17.0–24.0	72–79	3.3–4.2

[a] more recent editions give specifications for bleached beeswax only
[b] also specifies maximum cloud point 65°
[c] ester value/acid value

Table 13.5C Physical properties of beeswax

Data here are presumed to be for *Apis mellifera* beeswax. Most are from early references, and do not record the sources of beeswax within the hive (except for 4 of the melting point entries), or whether the beeswax studied was as collected, or refined, or bleached. New measurements are needed that also record such information.

Property	*Value*
relative density at 15°	0.958–0.970
at 25°	0.950–0.960
temperature at which beeswax becomes plastic	32°
melting point:	
quoted in various pharmacopoeias	61°–66°
of wax from cappings	63.7°
of wax from old combs	63.4°
of wax from caged bees (not comb):	
fed honey, 2 samples	62.7°
fed syrup, 2 samples	63.1°
solidifying point (for melting point 61°–64°)	60°–63°
flash point	242°–250°
refractive index at 75°	1.4398–1.4451
electrical resistivity:	
yellow, unrefined	8×10^{12} ohm m
ditto	20×10^{12} ohm m
white, at 22°	5×10^{12} ohm m
dielectric constant (specific inductive capacity)	2.4
ditto	3.1–3.3

Table 13.5D Melting points of beeswaxes in relation to those of other insect waxes

Insect	*Melting point*	*Reference*
bumble bees (*Bombus*)	34°–35°	Tulloch (1970)
honeybees:		
A. dorsata	60°	Phadke et al. (1969)
A. florea	63°	Phadke et al. (1969)
A. cerana	65°	Phadke et al. (1969)
A. mellifera (European), *A. mellifera* (tropical African)	63°–65°	various
Meliponinae:		
Trigona spp. (India)	66.5°	Phadke et al. (1969)
T. beccarii (Africa)	64.6°	F. G. Smith (1954)
T. denoiti (Africa)	66.4°	F. G. Smith (1954)
Coccoidea:		
Indian lac insect (*Laccifer lacca*, Lacciferidae)	72°–82°	quoted by Tulloch
Icerya purchasi (Margarodidae)	78°	quoted by Tulloch
Coccus ceriferus, *Brahmaea japonica* (Brahmaeidae) which produce Chinese insect wax	82°–84°	quoted by Tulloch
cochineal insect (*Dactylopius coccus*)	99°–101°	quoted by Tulloch

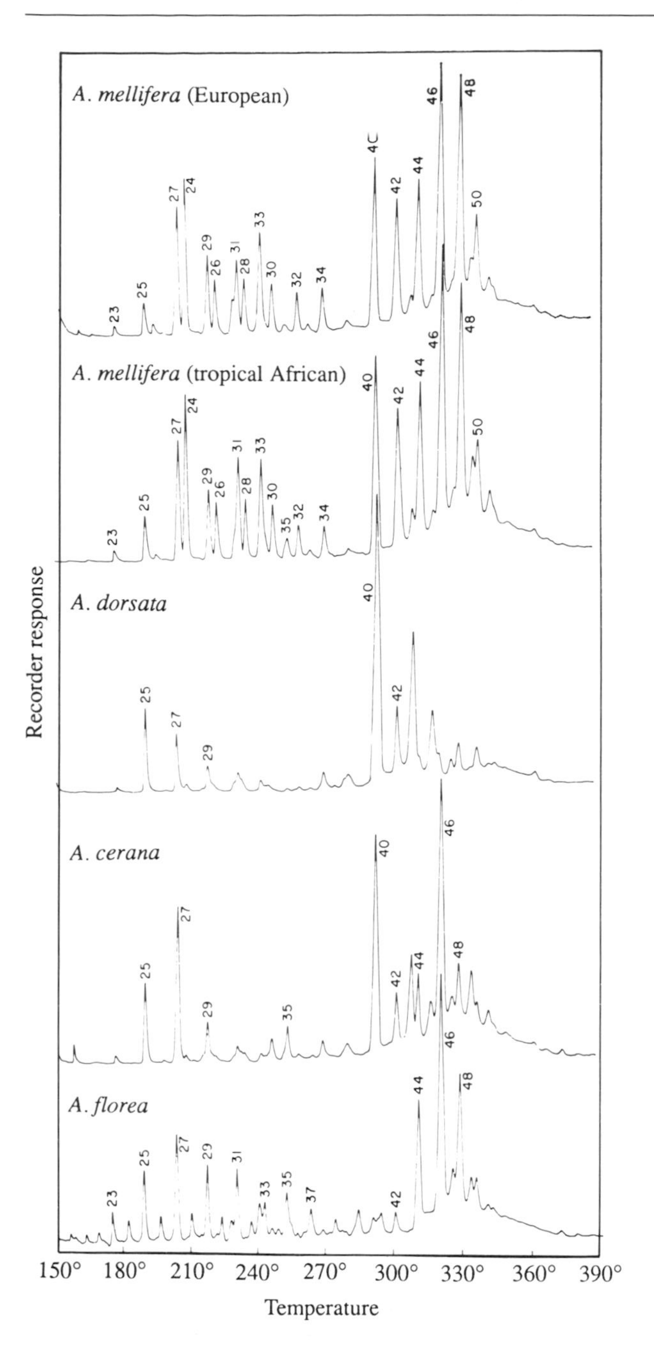

Figure 13.5a Gas liquid chromatographic analysis of beeswaxes from four *Apis* species (after Tulloch, 1980, who gives further details).
Peaks marked with odd numbers (23–35) are hydrocarbons, and those with even numbers (24–34) are free acids – as methyl esters – and (40–50) monoesters.

these waxes have been made in Pakistan (Latif et al., 1960*a*) and in India (Phadke, 1961; Phadke et al., 1969, 1971; Phadke & Nair, 1970; Phadke & Phadke, 1975). Some of the studies also include waxes from *Trigona* species, which belong to the Meliponinae (stingless bees). Rather varied results have been obtained for the composition and properties of waxes of stingless bees obtained in different areas, perhaps partly because of the large number of species in different parts of the tropical world. However, waxes from nests of four *Trigona* species in Tanzania were found to be similar (F. G. Smith, 1954).

Waxes produced by different species of bumble bees (*Bombus*) are fairly similar to each other, but as a group they show considerable differences from honeybee wax, for instance in having a much lower melting point. Nearly all *Bombus* species live in a temperate zone; their nest is smaller than that of honeybees, and combs are not suspended. These bees do not store a large weight of honey, because the colony does not last beyond the summer – the queen overwinters alone.

Table 13.5D shows the melting points of beeswax from different honey-producing bees, and of some other insect waxes. Wax secreted by certain insects serves as a protective body coating, and this may have to withstand high temperatures in the tropics.

13.6 HANDLING AND PROCESSING BEESWAX

The process by which wax from combs is converted into blocks of clean beeswax by melting is commonly known as rendering. In modern beekeeping, beeswax for processing may come from various sources. Cappings (Section 13.32) give the best quality beeswax – and the highest yield. Any new 'wild' comb that is removed from hives, or recently built comb that has been broken or crushed, is also satisfactory. Wax scraped off hive walls or frame bars is less good, since it may be contaminated with propolis. Dark combs, culled from hives because of their age, are of least use, and they may yield relatively little beeswax. When lower grade wax is processed, the dross that has settled at the bottom of the solidified blocks must be scraped away, and the wax remelted and strained through fine mesh or a thick close-woven cloth.

In many countries it is illegal to save beeswax from colonies infected with American foul brood disease; the wax may be contaminated with spores of *Bacillus larvae*, and these could subsequently infect any colonies hived on comb foundation made from it; see Section 11.21.

Methods

The beekeeper may choose from the following methods for rendering beeswax from cappings or combs; their effectiveness is compared later:

– simple straining of melted wax (Figure 13.6a)

- solar wax extractor (Figure 13.6b)
- several types of hot-water (e.g. Figure 13.6c, *above*), and steam processors
- hand steam press (Figure 13.6c, *below*).

Specialist firms that reclaim wax from old combs use much more powerful steam presses and centrifuges. Except in very large beekeeping operations (for which see Coggshall & Morse, 1984), wax rendering is done on only a few occasions a year, and expensive equipment is likely to be inappropriate for the beekeeper. Equipment described here is on sale from suppliers, but some can be made fairly easily, and many beekeepers devise their own variants of it.

All methods of rendering beeswax involve melting the wax. Since beeswax is inflammable, care must be taken not to start a fire, and methods in which the wax is heated in water are safer than those in which it is heated alone, as for direct straining (below); see e.g. FAO (1986). It is unsafe to heat beeswax over an open flame indoors, and a few beekeepers have suffered severe injuries through doing this. Nor should superheated steam be allowed to play directly on to beeswax, because this may partly saponify the wax. Vessels used in processing beeswax may be of stainless steel, aluminium, unchipped enamel, or galvanized or tinned iron, but beeswax is discoloured if heated in contact with zinc, brass, copper or iron.

When beeswax is melted alone (not in water or steam), a 3-mm mesh is commonly used for straining it. When it solidifies again the dross is scraped off the bottom of the solid block, and further melting and straining through a finer mesh may be needed. In many African countries, the strainer is a long bag of strong cotton cloth or closely woven vegetable fibres; when the wax ceases to run through, the bag is twisted, and squeezed by being pulled slowly between two horizontal sticks (Figure 13.6a).

The solar wax extractor (Figure 13.6b), which is completely safe, is useful for rendering small amounts of beeswax. It needs no power supply or source of heat, has no operational costs, and can produce very clean wax. Radiant heat from the sun melts pieces of wax placed on the metal base of a shallow box; its lid consists of two sheets of glass 5 mm apart. The whole is tilted so that the lid faces the sun. Below the base that reflects radiant heat is an insulating layer, and the heat is trapped inside the box. Melted beeswax runs down the base – leaving most of the dross behind – through a screen, and into a container where it solidifies, further dross sinking to the bottom. In Figure 13.6b a second container receives even cleaner wax from the first one.

In methods where the wax is put in water, rain water should be used; if only hard water is available 0.1% of commercial vinegar should be added. It is best to soak the combs in water for several hours before rendering, to wash out any honey present; if the combs are dark or contain pollen, they should be broken up and soaked for 1 or 2 days.

Figure 13.6a Primitive open-air beeswax extraction, Tharaka, Kenya, 1967.
Comb pieces are heated in the open air; melted wax has been strained, and what is left in the strainer is being squeezed out into a receptacle in a pit dug in the ground.

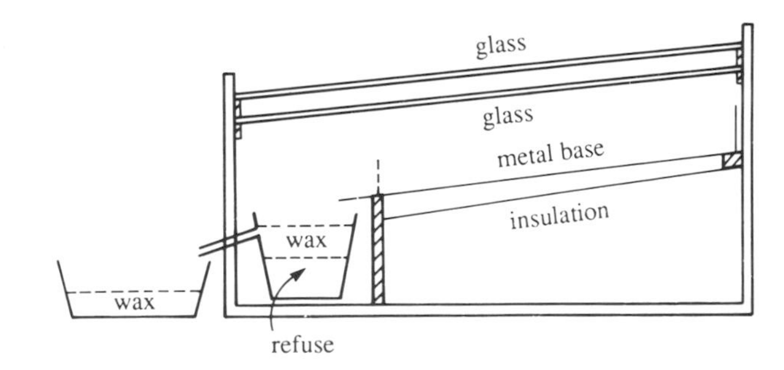

Figure 13.6b Solar wax extractor.

In the simplest type of hot-water processor (Figure 13.6c, *above*), cappings or combs, etc., are heated in a vessel containing water. The melted beeswax rises to the top and is drained off through the outlet tube shown on the left, leaving behind the dross and dirty water. Hot water is added to the bottom through the filling tube shown on the right, to keep the beeswax at the level of the outlet tube.

The Mountain Grey (MG) beeswax extractor is used especially in England. Cappings or combs are put with water in a 10-litre steel vessel, and a coarse straining cloth secured in position over the top of the container. The water is heated until all the wax is melted. More water is then added to the bottom of the vessel through a long filling tube, and this exerts enough pressure to force the wax up through the straining cloth; it flows into a collecting channel surrounding the container and thence into a receptacle below. Relatively clean wax is obtained, but the yield is lower than that obtained by some methods; see Table 13.6A. Berthold (1987) describes a home-made processor that produces refined wax of a quality suitable for candles.

Some beekeepers use a microwave oven to render small quantities of beeswax. This can be done only with wet cappings, or if water is added to dry wax; the honey is then heated because water molecules in it are energized by the microwaves. When the honey or water becomes hot enough, it will melt the wax in the cappings, which rises to form a layer at the top. A microwave oven cannot melt beeswax alone in the same way.

Figure 13.6c (*below*) shows a German steam press. Steam generated from water at the bottom of the container reaches the combs (wrapped in canvas bags) through the perforated basket into which they are placed; the bags are separated by wooden boards. The cross-arm of the screw plunger is locked in position so that the combs are under pressure, and the melted wax runs through the basket and out of the tube. When the wax flow ceases, the screw is turned back and the bags shaken and replaced, then further pressure is applied. After this has been done 2 or 3 times, only the dross is left behind.

If the container (mould) in which the wax is to be collected has sloping sides, the block of wax is easier to remove. If it is rinsed with soapy water before use, the wax will not stick to it. Cooling of the melted wax should be slowed down as much as possible, or cracks may develop.

Most comprehensive books on beekeeping management describe in some detail several methods for rendering beeswax, e.g. Wedmore (1945), Dadant & Sons (1975), Root and Root (1940); see also FAO (1986).

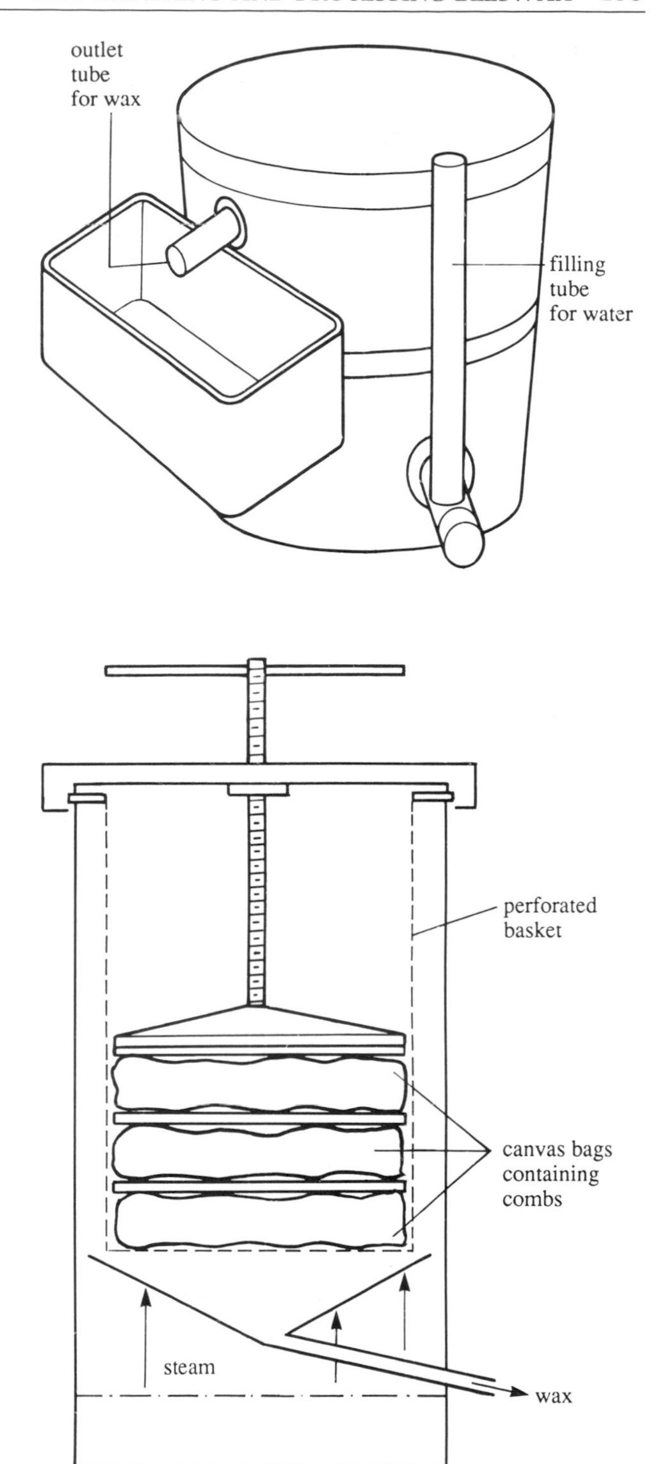

Figure 13.6c Wax extractors.
above Simple extractor using hot water and gravity.
below More efficient extractor: a steam press.

Yields by different methods and from different sources of beeswax

In traditional beekeeping, whole combs are melted down after the honey has been drained, squeezed or soaked from them, and the beeswax is part of the beekeeper's harvest. It may amount to 8% or 10% of the honey harvested. Until a few decades ago, traditional beekeepers in tropical Africa provided large amounts of beeswax on the world market; current trade figures are given in Table 13.8A. For some of these beekeepers, beeswax is an easier hive product to handle than honey, because it can be transported to a trading centre without any need for a container. However, Section 13.82 refers to much wastage of beeswax.

Modern beekeeping is geared to honey production, and combs from which honey has been extracted are used again in the hives. Beeswax is harvested from cell cappings (end of Section 13.32) after the honey has been drained from them, and the yield is likely to be 1½% to 2% of the honey extracted from the combs. Cappings give high quality beeswax, light in colour, and this wax (together with any from newly built combs) should be rendered separately from any other. ('Cappings melters' referred to in Section 13.32 heat both cappings and honey in order to separate them.) A second possible source of beeswax is clean wax (not mixed with propolis) that is scraped off frames and hive parts in the course of the year; Matheson (1984*a*) estimates that the amount could be about 0.5 kg per hive. Combs culled from hives constitute another possible source, and if they are replaced every year or two, the value of the wax obtained from them exceeds the cost of fitting new foundation to the frames. But very old combs, darkened by cocoons of many generations of brood reared in them, are usually not worth the beekeeper's time and effort in rendering them. In fact they contain more wax than new combs, but much of it is left in the residue after rendering (Simpson & Fairey, 1964) unless commercial methods are used.

Table 13.6A shows that between 90% and 99% of the wax in cappings can be extracted, according to the method used. The percentages from culled combs are lower, and from old combs very much lower, unless some force stronger than that due to gravity is applied.

Comparing the methods, simple straining of molten wax seems to be as effective as methods involving hot water that do not use extra force. But the other methods are probably safer, and some may be more convenient to carry out; see Table 13.6A, and Simpson and Fairey's discussion (1964).

The residue left when the beeswax has been rendered is sometimes referred to as slumgum. It is usually thrown away, but if a large amount is available it can be sold to a commercia. wax-reclaiming firm. Residue from combs that contained pollen and brood has some nutritive value. In Poland it has been used experimentally for feeding to poultry, and it improved egg colour (Faruga et al., 1975). The residue from the operation shown in Figure 13.6a was eaten by dogs as soon as it was cool enough.

In theory it might be possible to develop a system of modern beekeeping for wax production, by collecting 'virgin' wax secreted by the bees before they expend energy in building comb with it. Morse (1965) explored the idea of harvesting the wax scales into which secreted beeswax hardens as it solidifies. He used artificial swarms clustered on wooden supports, each with a caged queen but without comb. Initiation of comb building was prevented by keeping the swarms in full daylight. Beeswax scales were produced and fell to the ground.

13.7 USES OF BEESWAX

Beeswax has had an extraordinarily rich history, with a far wider ranger of uses than other bee products. Books that may be consulted by readers interested in this history include *Wax craft* (Cowan, 1908), *The Wax Chandlers of London* (Dummelow, 1973), *Beeswax* (Brown, 1981), *Beeswax* (Coggshall & Morse, 1984), and *Viel köstlich Wachsgebild* (Hansmann & Hansmann, 1959). The most comprehensive source is the superbly illustrated *Vom Wachs: Hoechster Beiträge zur Kenntnis der Wachse* (Büll, 1959–1970); Walker's bibliography (1983*e*) gives details of the 12 parts of this book, and of many publications not mentioned here. There is a historical account of the operations of Koster Keunen (1977), a firm founded in 1852 and probably the world's largest supplier of bleached beeswax.

The present section is concerned with modern uses of beeswax. Many are continuations or adaptations of earlier uses, but a few are connected with new industries. Beeswax is obtained from the combs that have contained the honey. It has not usually been the primary hive product, except in parts of tropical Africa (Section 1.57), and in Europe during the centuries when the Christian church used large quantities of beeswax for candles (Section 1.46). Table 13.7A gives an indication of what happens nowadays to the beeswax produced in the world (48 000 tonnes in 1984). The exports and imports, as well as the wastage, are referred to in Section 13.82. The amounts of *imported* beeswax used in different ways have been estimated by the International Trade Centre (1978) as follows; the uses are discussed here in the sections indicated:

		Section
for cosmetics	35–40%	13.72
for pharmaceutical preparations	25–30%	13.72
for candles	20%	13.71
for other minor uses	10–20%	13.73

Table 13.6A Percentages of beeswax extracted by different methods

From Simpson and Fairey (1964).

Method	*Cappings (combs had been used for brood)*	*Culled combs (ages not known)*	*Brood combs (20+ years old)*
Without water:			
solar wax extractor	probably similar to the next two methods		
simple straining	94%	59%	15%
With water or steam:			
submerging combs, or steam (mean)	94%	64%	24%
MG extractor	90%	37%	0
With extra force:			
hand press	97%	83%	64%
commercial reclaiming by centrifuge	99%	95%	90%

13.71 Uses of beeswax itself as a material

Most comb foundation is made from beeswax, and also some high quality candles. Beeswax is also used for modelling, in the lost-wax process of metal casting, and as a 'resist' to protect surfaces being treated in various ways.

Comb foundation

In countries where movable frames fitted with comb foundation are used in hives, the production of foundation is likely to be a major use of beeswax. Section 4.31 describes various methods for making beeswax foundation, and Section 4.32 refers to the use of beeswax for coating plastic foundation.

Candles

A candle is a source of light consisting of a cylinder of wax or other fuel – which must burn away completely, leaving no deposit – enclosing a wick along its axis. The function of the wick is to transport the melted wax to the flame.

Beeswax has many attributes that make it suitable for candles, and in past centuries it was widely available. Beeswax burns with a bright, steady flame, and has a higher melting point (63° to 65°) than cheaper waxes such as paraffin wax (48° to 68°); beeswax candles therefore remain upright in hot weather when paraffin wax candles bend. Some microcrystalline waxes have still higher melting points, up to 93°, but melting points above 88° are too high for good burning.

Candles are made by various methods. The first cited below is probably always a hand operation, the next two can be mechanized to some extent, and machines are now used for all the others.

Pouring Molten wax is poured down and over the wick, and the process is repeated as soon as the previous layer has solidified, many times in all. Candles up to 8 cm in diameter can be produced.

Dipping The wick is dipped into molten wax, and the procedure repeated as above. Altar candles for the Christian church, and high quality candles for formal and ceremonial occasions are made in this way.

Table 13.7A What happens to the annual beeswax production

Regions with traditional hives and harvesting from wild colonies

beeswax production is about 10% of honey production:
- much is wasted
- some of the rest is used in traditional ways
- most of the rest is exported (Tables 13.8A, 13.8E)

Regions with modern hives

beeswax production is about 2% of honey production:
- much is wasted
- much of the rest is used in comb foundation
- some is used by beekeeping families
- a little is used in industry

beeswax is also imported (Tables 13.8A, 13.8E):
- nearly all is used in industry
- some is used in comb foundation

Rolling A sheet of wax is rolled round the wick until the required diameter is achieved.

Moulding Molten wax is poured into candle-shaped moulds, in which the wicks are already fixed. Beeswax shrinks by about 10% when it solidifies, and may crack if the wax adheres to the mould as it sets, but hard plastic moulds are now available that do not stick to the wax.

Extruding A partially hollow continuous tube of wax, and the wick, are extruded from holes in a metal sheet, and then cut into the lengths required.

Pressing The wax is 'atomized' in a cooling chamber, and the resultant powder pressed into a candle shape round a wick.

Drawing A continuous wick coated with wax is drawn through a series of holes of ever-increasing size, in metal plates; the coated wick is dipped into molten wax between successive passages through the holes, each time acquiring another wax layer.

Coggshall and Morse (1984) provide details of the different methods, and of other aspects of modern candle manufacture. Instructions for making candles at home are published, e.g. by McLaughlin (1973), and Furness (1984) who deals specifically with beeswax candles and also includes a section on judging them at shows.

Models and casting

These uses of beeswax are nowadays little more than vestiges of a glorious past, products of which can be seen in national museum collections all over the world; examples are cited by Crane (1983*a*). Beeswax was the only material available that could be both modelled and cast. Wax from all honeybee species has been used, and in the Americas and Australia wax from stingless bees.

Life-like models of famous people can be seen in Madame Tussaud's Exhibition in London; the mixture used is 3 parts of beeswax and 1 part of a harder wax that raises the melting point. For the head – the most intricate part – a plaster mould is first made directly round the person's own head. (It is said that originally, in Paris in the 1790s, Mme Tussaud was forced at sword point to use heads fresh from the guillotine.) The hollow plaster mould is soaked in water to reduce its porosity, and filled with the molten wax. After 15 to 30 minutes a layer of wax about 12 mm thick has solidified inside the mould, and the rest of the wax, still liquid, is poured out. After the layer in the mould has become quite solid, the mould is chipped away and the hair and eyes are inserted (Coggshall & Morse, 1984).

The First International Congress on Ceroplasty in Science and the Arts was held in 1975, and its 728-page *Proceedings* (Congresso Internazionale . . ., 1977), give details of uses of beeswax for anatomical and other scientific models, and for many other purposes.

Beeswax was also one of the few materials available in past centuries that could easily be sculpted to make a permanent small model with detailed features. The statue of a falcon shown in Figure 13.71a is an example from Ancient Egypt. This model may have been intended for use in making a metal casting by the lost-wax (*cire-perdue*) method, and never used. In 1984 I visited Pancha Raj Sakya, a lost-wax caster in Patan, an ancient city in Nepal, whose premises were behind a temple with a thousand images of Buddha. I watched his men carving the models, many only a few centimetres long. When completed, each model was encased in mud – a cheaper and more primitive material than plaster – and left in the sun for the mud to dry out. Then the whole would be heated so that the wax melts and runs out, leaving a mud 'negative' of the model. Molten metal – usually brass or bronze – is poured in, replacing the beeswax, and when this has solidified the mud casing is broken off, leaving a metal positive. For very small objects, the wax and the metal 'positives' are solid, otherwise they are usually hollow.

In West Africa, Ashanti 'gold weights' – used for weighing gold dust – are cast from brass by the lost-wax method, and in the 1970s beekeeping in Ghana was expanded in order to provide the wax required. A modern manual on lost-wax casting is available (Feinberg, 1983).

Figure 13.71a Small model made of solid beeswax in Ancient Egypt; it portrays the falcon-headed god Horus (photo: William Meyers Museum, Eton College).

Beeswax as a 'resist'

A 'resist' is applied to protect certain areas of a surface from reaction during a subsequent process. One of the traditional uses of beeswax as a resist, dating from ancient civilizations in Asia, is in the batik method of dying cloth, which was brought to high levels of design and execution in Java. Before the cloth is immersed in dye of a specific colour, all areas not intended to receive that colour are painted with molten wax, which may be beeswax or another wax or a mixture; these areas resist the dye and are not coloured. Martin's book on batik (1977) gives both historical information and practical instructions.

Beeswax can also be used as a resist when etching a surface of glass or metal, or of china in the acid-gold process. The whole surface is dipped in wax, or warmed and coated with it. When the wax has solidified, the design to be etched is inscribed through the wax with a stylus, and the surface below is thus exposed. An appropriate acid is applied which etches the design into the surface, and the wax is subsequently removed by heat.

13.72 Beeswax as a component of cosmetics and pharmaceuticals

One of the most important uses of beeswax nowadays is in ointments, emollient skin creams and lotions, used for both cosmetics and pharmaceutical purposes. This continues a tradition dating at least as far back as the time of the Greek physician Galen (AD 130–201). Beeswax is valued as a stiffening agent, and also because it considerably increases the water-holding ability of an ointment; according to the American Pharmaceutical Association an ideal ointment base should be able to hold at least 50% of water. Beeswax also mixes readily with a large number of organic compounds; it is saponifiable, and forms stable emulsions, both ionic and nonionic. Because of these properties, beeswax is used in many emulsified and dispersed pharmaceutical products for applying to the skin.

Originally, skin creams were made with beeswax and vegetable oil, but mineral oil is now used instead of vegetable oil, because the cream then does not become rancid. The type and quantity of the wax and oil determine the viscosity, smoothness, consistency and other properties of the cream. Beeswax is sometimes replaced partly or wholly by spermaceti (a white wax from the sperm whale), or paraffin wax – especially a type known as ceresin (ceresin wax), a bleached form of ozocerite which has rather similar properties to beeswax.

In creams, lotions, and other products containing water, the wax is usually emulsified by reaction with an alkali such as borax to achieve water dispersibility; the borax saponifies acids in beeswax to form a product which is technically a soap, and in cleansing creams is used as such. It also increases the whiteness and the stability of the cream. The solubility of borax in water is low (14 g/100 ml at 55°), and sufficient water must be used to dissolve the borax. Only 6.8 parts of borax are needed to neutralize the fatty acids in 100 parts of beeswax; if a higher proportion is used, excess borax may crystallize out and give the resultant cream a rough texture.

An old and widely used recipe for skin cream, which can be made in the home, contains the following parts by weight:

beeswax (see text below)	18%
liquid paraffin	61%
distilled water	20%
borax	1%

The beeswax and liquid paraffin are heated to about 70° until the beeswax is just melted, and then blended; the borax is dissolved in the water at the same temperature, and the two mixtures blended together with rapid stirring. Jars are filled when the cream has cooled to 42°. In England, Marjorie Townley has initiated the production of this cream by three pharmaceutical chemists since 1980, supplying them with fresh beeswax from cappings, etc., untreated except for heating it sufficently to allow straining. She stresses the importance of using beeswax that has not been bleached. Over the years, the cream has been found effective in alleviating cracked and chapped skin, dermatitis, skin rash and irritation from various causes, and some forms of eczema. It has also been successful in treating certain similar conditions in domestic animals.

A specialized modern product that contains beeswax is black camouflage face cream for commandos and SAS troops (Brown, 1981).

Beeswax is the wax most frequently used in lipstick, which contains 20% to 25%. Its relatively high melting point is suitable; it gives the lipstick a good consistency and high sheen; together with lanolin, it aids in binding the castor oil which is another ingredient. Beeswax contracts on cooling, making the stick easy to expel from its holder, and free melissic and cerotic acids in the beeswax help to stabilize the final colour. Shiny, brilliant lipsticks contain less beeswax – even under 10%. Ceresin wax is often mixed with beeswax for lipsticks, and may even replace it.

Eye shadow contains 6% of beeswax, mascara 12%, hair creams 8%, and epilators 20%. Epilatory waxes, which uproot the hair, contain beeswax or microcrystalline wax, and resin; softer systems used at room temperatures contain oil, resin and wax, and are cloth backed. They form a eutectic mixture, whose melting point is lower than that of its constituents. For example a blend of 75% resin (melting at 80°–90°) and 20% beeswax (melting at 62°) has a melting point of only 45°–50°, which can be reduced even further by the addition of 1–2% of Japan wax. Human skin can therefore be coated with the liquid wax mixture at its melting point, without burning or pain.

The use of beeswax in the cosmetics industry is as much affected by its cost as by any of its properties. Paraffin wax is often substituted for beeswax because it is cheaper and readily available; also – like beeswax – it does not become rancid. But the desirable qualities of beeswax are well known in the industry, and it is difficult for those who manufacture cosmetics to avoid its use. Sometimes only 1% to 3% will impart the required final quality to a product (Coggshall & Morse, 1984).

Beeswax is soft, flexible and inert, and for these reasons is used for coating pills to delay the ingestion of their contents. It is not digested or absorbed.

13.73 Beeswax as a component in other products

Polishes and other protective coatings

Waxes blended in manufacturing polishes may include beeswax, carnauba, candelilla, paraffin (melting point not lower than 58°–60°), and microcrystalline waxes. Polishes can be made by dissolving the wax in white spirit (a petroleum fraction with a boiling point about 195°), or in turpentine, or in a mixture of the two, to make a paste of a consistency that is easily spread over a surface. It should leave a very thin film when the solvent evaporates, and this can be rubbed to give a smooth, shining finish. Floor polishes contain less of the lustre-giving waxes, but certain resins are added to reduce the risk of floors becoming slippery when wet. Furniture and shoe polishes contain more natural waxes and less solvent.

The following is an old and well tried home recipe for beeswax furniture polish with the consistency of a cream; in common with others, it is best used sparingly. At the percentage of wax used (10%), the cream spreads easily and cleans well. Ingredients are:

beeswax	225 g
turpentine (not substitute)	1 litre
shredded dry pure soap	25 g
distilled or rain water	1 litre

The method is as follows. In a bowl of china, glass or stainless steel, holding at least 3 litres, warm the beeswax (over hot water, because it is inflammable) until just melted, then remove from the heat, add the turpentine and stir. Meanwhile warm the distilled water with the soap until the soap has dissolved. Add the warm soap solution very slowly to the wax-turpentine mixture, stirring continuously. The mixture should emulsify; if not, standing the bowl in cold water may help. This polish is most easily used from a wide-mouthed jar.

In the manufacture of spray-on aerosol polishes, hard waxes are emulsified with an agent, using water as a solvent, and mixed with an emulsion of shellac or synthetic resin, and sometimes silicones. A few of these aerosol polishes on the market contain some beeswax, but their total wax content is only 2–4%.

Many other preparations used as protective coatings contain beeswax, usually between 10% and 60%. Types described in modern patents and publications (see e.g. Walker, 1983*d*) include the following:

furniture varnishes
water-resistant preparations for walls
air-permeable waterproofing compositions
finishes for leather, textiles, wood, paper
anticorrosion preparation for terminals of lead accumulators.

Grafting wax, used in horticulture to protect new plant grafts, must be pliable, have no effect on cell and tissue growth, and give protection over the whole of the cut surfaces for about two months. Cheaper waxes are now used commercially, but grafting wax was traditionally made with beeswax, and can be produced at home. One formula contains 1 part each of beeswax and resin, enough lard or tallow to make the mixture pliable, and some finely ground charcoal to shield the growing plant cells from sunlight. The mixture is applied warm, or it may be formed into strips and wrapped round the graft so that it covers all wound surfaces.

Lubricants

Beeswax is a component of a number of lubricants patented or otherwise published; examples cited by Walker (1983*d*) are:

- as a drawing lubricant for tantalum and uranium
- for slips used in hot casting of glass under pressure
- in a lubricating and fuel-additive composition with anti-knock properties
- in a solid paste for wear-resistance enhancement of cutting tools.

Uses of beeswax as a lubricant (and in other ways) in armament industries may be considerable, but details are not available.

Electrets

The use of beeswax in electrical engineering depends mainly on its dielectric properties. When beeswax is solidified in a strong electric field, it becomes polarized; after the electric field is removed the polarization remains, although it does decay slowly, and more rapidly at increasing temperatures. Such a quasi-permanently polarized dielectric material is an electret, the analogue of a magnet. Beeswax is only one of a number of widely different materials from which electrets can be made; others include certain other organic waxes and also some metallic salts. Electrets are used in transducers, devices that produce electrical energy from, or transform it into, another form of energy; this may be acoustical as in microphones and earphones, or mechanical as in impact or touch-of-key transducers.

13.8 HONEY AND BEESWAX: WORLD PRODUCTION, TRADE AND CONSUMPTION

Tables 13.8A to 13.8F, together with Tables 1.5A and 1.5B in Chapter 1, give comprehensive data on the extent and results of beekeeping in different countries of the world. They include data in unpublished print-outs from the Statistics Division of FAO in Rome and from the Trade Analysis System of the International Trade Centre in Geneva. This co-operation is very much valued, as is that from many individuals in different countries who have provided additional information. The tables give the following data:

Number of hives, hive density (per km²), mean honey yield per hive:

Table 1.5A individual countries
Table 1.5B continental regions, with world totals; includes total honey production

Honey production and trade:

Table 13.8A individual countries
Table 13.8B continental regions, with world totals

Honey consumption per person:

Table 13.8C selected countries
Table 13.8D continental regions, with world totals

Beeswax production, imports and exports:

Table 13.8A individual countries
Table 13.8E continental regions, with world totals; includes beeswax/honey ratio

Beeswax imports and consumption:

Table 13.8F

Entries in all tables are incomplete in that data from many family apiaries do not enter official records, and that in the statistics available some data are not recorded at all for certain (mostly small) countries. But more countries have entries than in previously published tables, so totals more nearly represent the true state of affairs; the greatest under-representation is for beeswax production and exports, as explained in Section 13.82. Figures for honey production and yield/hive vary from year to year according to whether the season is good or bad, and may also be reduced sporadically as a result of damage to bees by disease, pests, or injury.

13.81 Honey

Figure 13.81a shows annual recorded figures for the world production of honey and the amount exported on to the world market, from 1971 to 1987; production in the early 1970s is detailed by Crane (1975*c*). Over these 17 years production has increased by 30%, and world exports have doubled.

The total recorded world honey production (Table 13.8B) is now nearly a million tonnes a year, and the total amount harvested by beekeepers is almost certainly more than this. The largest honey-producing

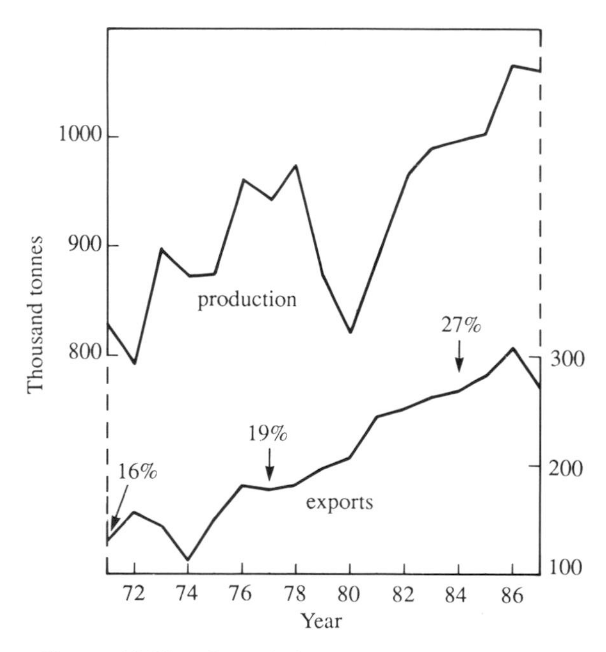

Figure 13.81a Recorded annual world production of honey in thousand tonnes (upper curve), and exports on to the world market (lower curve), 1971–1987.
Data are from International Trade Centre (1977, 1986), *FAO Yearbooks* (1987). The entries 16%, 19% and 27% indicate the proportion of the world honey production that was exported in 1971, 1977, 1984, respectively.

countries, with their production recorded for 1984 (Table 13.8A), are:

USSR	193 thousand tonnes
China	161
USA	75
Mexico	67

These four countries together produced 496 000 tonnes, just one half the world's honey crop. Next in order were Canada, Argentina, Turkey and Australia.

An appreciable amount of honey is also harvested from wild colonies of honeybees and from stingless bees, and honey production data do not usually specify whether or not this honey is included. Some countries in tropical Asia produce much more honey from wild nests of *Apis dorsata* than from hives, and where possible separate figures are quoted (Table 13.8A, Part 6). In countries where *Apis mellifera* has not been introduced, all the honey is from other *Apis* species.

About three-quarters of the recorded world production is consumed within the country of origin, and one quarter is now exported (Table 13.8B). In the world at large, honey is still mainly a home-produced – often locally produced – food. Of the honey that is traded around the world, most of the 'flow' is from countries with subtropical regions to countries in the north temperate zone. Until the early 1970s, Argentina was the world's leading honey exporter, but it was then overtaken by Mexico, which exported 54 000 tonnes in 1984, almost twice as much as Argentina. China started to become important in the late 1960s, and in 1980 to 1982 it outstripped every other country. China, Mexico and Argentina are at present the largest exporters. The average honey consumption per person is low in all three countries (Table 13.8C), but whereas exports from Mexico and Argentina represent over 80% of their honey crops, China exports only 28%. Australia and Canada export 44% of their honey production, but both also have a high per capita consumption; see Table 13.8C and its discussion later in this section.

The largest honey importers are at present the German Federal Republic, the USA and Japan, and then the UK. Figures 13.81b and 13.81c show how honey imports into these four countries have risen over the years. Up to 10 000 tonnes a year had been imported into Germany between 1925 and the outbreak of war in 1939. As soon as it became possible after the war, the GFR started to import honey again (Figure 13.81b), and it has consistently imported more than any other country. The amount was 74 000 tonnes in 1984, of which some 10 000 tonnes were re-exported; production within the GFR was 16 000 tonnes.

In Japan honey was not traditionally eaten, but American influence after the war created a new market there. Imports started in 1962, and by 1972 Japan had become the second largest importer in the world; in 1984 it imported 33 000 tonnes. Its own beekeeping industry has also flourished. In 1984 Japanese imports had been overtaken by those into the USA (Figure 13.81c), for reasons explained below. Figure 13.81c also shows continuing steady imports of honey into the UK, the level being about twice as high in the 1970s as in the 1950s.

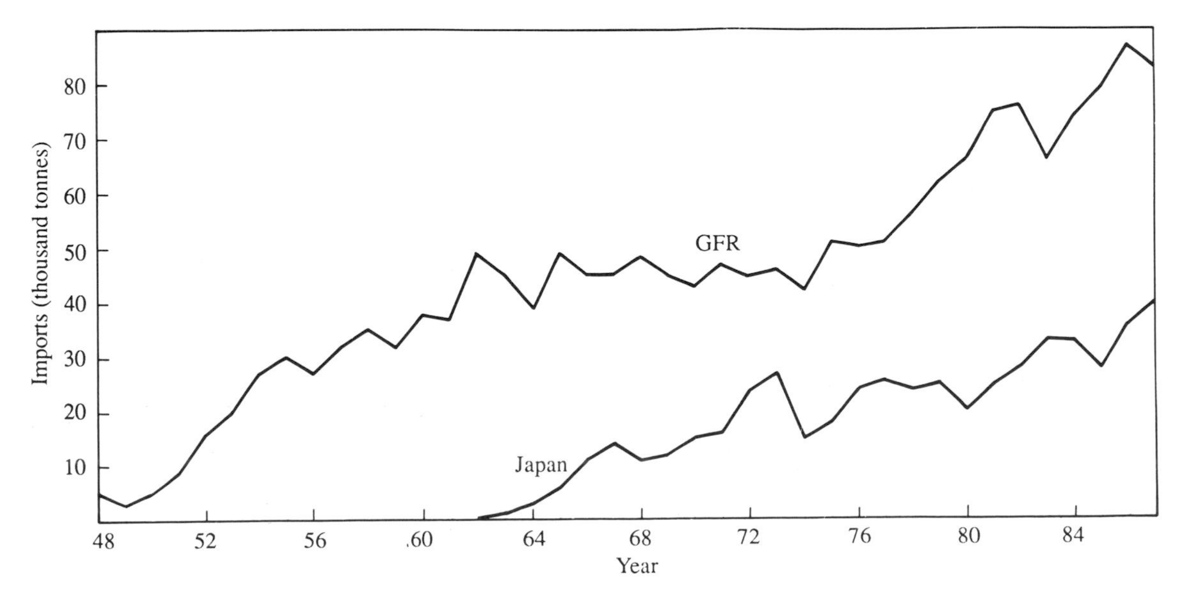

Figure 13.81b Gross annual imports of honey into the German Federal Republic and Japan, 1948–1987.
Data are from International Trade Centre (1977, 1986), Willson (1975), *FAO Trade Yearbook* (1987).

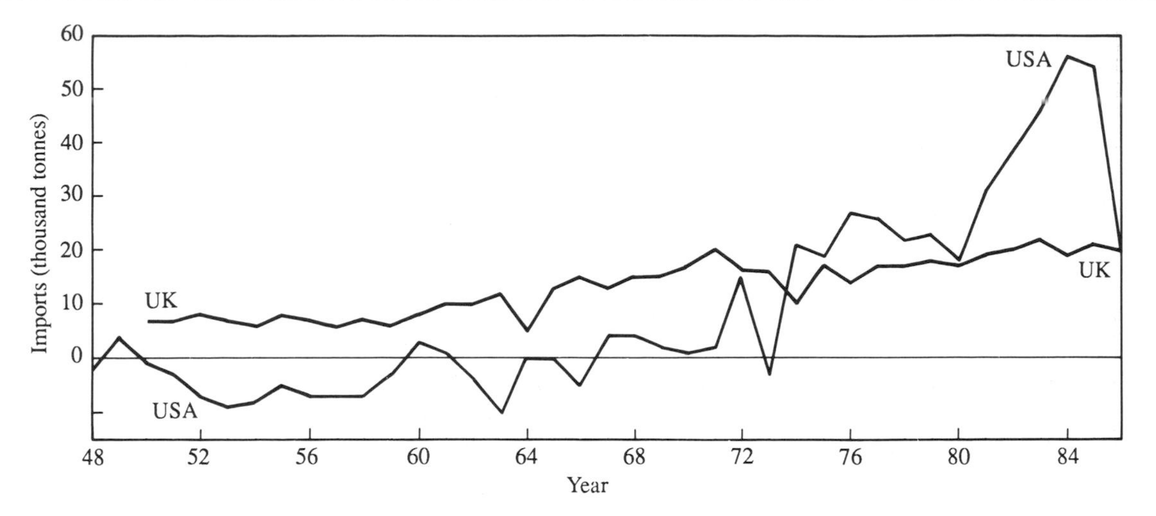

Figure 13.81c Annual imports of honey into USA and UK, 1948–1986. Data are from International Trade Centre (1977, 1986), Willson (1975), *FAO Trade Yearbook* (1987), Willett (1988). For the USA *net* imports are plotted, because of the substantial exports; see text.

For many years the USA both exported its own honey and imported honey from elsewhere; changes in net imports are shown in Figure 13.81c. An unprecedented excess of imports over exports started in 1967, and by 1984 gross imports has risen to 59 000 tonnes. The reason for this situation is as follows. A price-support programme was established in the USA in 1949/50, which enabled a beekeeper to obtain a loan from the Commodity Credit Corporation (CCC) of the Department of Agriculture, and to use his unsold honey crop as collateral security. The loan was calculated on a 'support price' which was (per tonne) $198 in 1950, $560 in 1975 and $1450 in 1984. Willson (1975) pointed out that although the scheme safeguarded US producers against ruinous honey prices, it would attract heavy imports into the USA if the world market price became low, and these would compete with domestic production. In recent years the relatively strong dollar, and other factors, have induced just this situation; in 1980–84 the CCC acquired 160 000 tonnes of honey, paying more than the world market price for it. Many of the US consumers then ate honey from other countries (whose price was lower) instead of their usual home-produced honey, and honey consumption also decreased somewhat. From 1986, the price-support programme was modified in ways that helped beekeepers in the USA (see International Trade Centre, 1986), and honey imports were halved within a year (Figure 13.81c). Willett (1988) gives more details.

Table 13.8A includes one new country as a honey importer: Saudi Arabia, which produces very little honey of its own, but is affluent. A market for honey

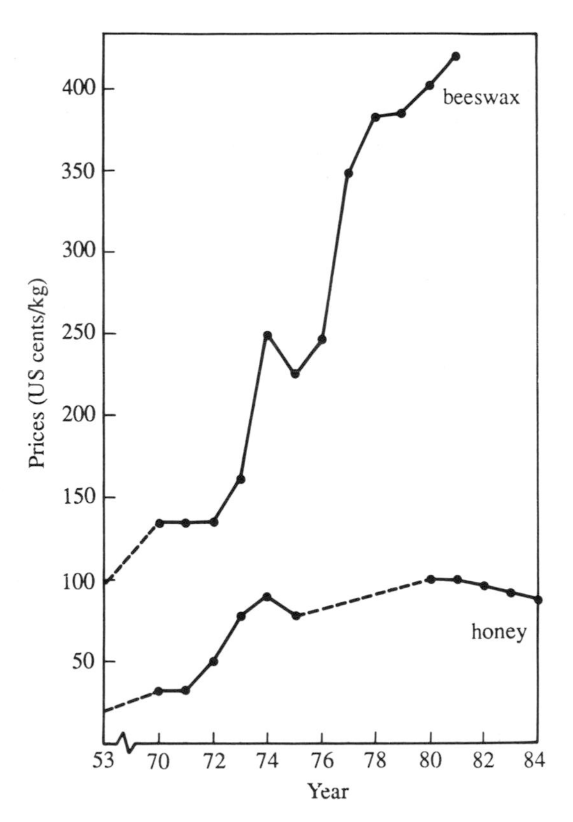

Figure 13.81d Estimated world prices of honey and beeswax, as available, 1953–1984. Prices are average bulk import prices on entry into the USA (US cents per kg), and exclude cost of transport from the exporting country. Honey data are based on International Trade Centre (1975, 1983), and beeswax data on Morse (1983), both taken from official USA figures.

has been created there, as in Japan; imports started in 1975, and grew to 33 000 tonnes by 1984.

The *price of honey* has risen over the years (and centuries) as a result of inflation. It varies from country to country, so the world price is not easy to define. An approximation to the world price of honey is the average price of honey – from ten or more countries – that is imported into the USA, and this can be calculated for a number of recent years. Figure 13.81d gives this price between 1953 and 1984. There has been a general rise, the lowest price ($0.20/kg) recorded being in 1953 and the highest ($1.00/kg) – five times as high – in 1980. The large increase shown in the Figure during the early 1970s was due to a greater world demand for honey, coupled with a shortage of it. Between 1975 and 1984 the availability of export-quality honey rose considerably, and the world supply easily met the growth in demand. By 1984 world market prices for honey were lower in real terms than in 1975 (Kortbech-Olesen, 1986).

The world price is influenced by international political and economic situations, by government support programmes and tariffs, and by fluctuating exchange rates, as well as by yearly production and demand, plant source and quality, and the sale of competing products (sugar is referred to in Section 1.46, and in Table 13.4A and its discussion in the text). International Trade Centre (1986) discusses the interlocking effects of these and other factors. Willson (1975) has told the story of the export-import honey trade of world regions in some detail up to 1974, and International Trade Centre (1977, 1986) continued it up to 1984.

We turn now to the *consumption of honey*. The total world population (1984) was about 4760 million, so – if all the honey were equally distributed – each man, woman and child would have got a little over 200 g of honey a year. This amount is decreasing year by year, as the world's population is growing slightly faster than its honey production.

Although honey is produced in almost every country of the world, its consumption is largely concentrated in affluent societies. Many of the people in such countries are descended from ancestors who lived in northern Europe, where there has been a very long tradition of beekeeping and honey consumption. These affluent countries are now able to import honey, mostly from less rich countries to make good any deficit between what they themselves can now produce and what they want to eat.

The average consumption of honey per person in a country is calculated as the country's total production *plus* imports *minus* exports, divided by the population. It includes what is used in manufactured foods, and also the small amount in non-food products. Table 13.8C shows this average consumption in major importing and exporting countries in 1975 and 1984.* It increased during the 9 years in almost all countries listed, although it has been decreasing in some other countries, and the world average has dropped slightly. Countries with the highest honey consumption (all above 600 g per person per year in 1984) are importing countries in Europe, together with Canada, Australia and the USSR which are net exporters of honey (see below). Countries with the lowest consumption are in Asia and Central and South America – continents where the average honey consumption is less than a fifth of that in Europe. (India, which imports and exports very little honey (Joshi, 1982), is included to exemplify the low honey consumption in Asia.) China, Mexico and Argentina are the three highest exporters in the world. Canada, Australia and New Zealand are also substantial exporters, but they retain a higher proportion of their total production. They have many inhabitants of northern European extraction, and their consumption of honey per person is about twice as high as in Europe. This is shown in Table 13.8D, which gives the average consumption in continental regions; regions with the lowest consumptions are now Central and South America and Asia, where consumption has decreased somewhat. Regions with the highest consumption are Oceania, USSR, Europe and North America; a survey of beekeepers' families in the USA (Jeremiah, 1932) showed some high consumptions by individuals: 11% ate 46 kg or more a year, and one family of four ate an average of 83 kg each.

The following broad conclusions may be drawn from the data presented here on the world production of honey, trade between countries, and honey consumption.

Countries with a high honey production per colony are likely to include regions in or around the subtropics, or at high latitudes. Among these, only very large countries achieve a high total production or net export of honey. Most countries exporting large amounts of honey consume relatively little themselves. Some others with populations that traditionally eat honey also have large honey-producing regions, which enable them to satisfy the home demand for honey, and also to export on a substantial scale.

Countries with a high honey consumption are all affluent. With the exception of the post-war honey-importing countries in Asia, all have populations that

* Pre-1975 figures for many further countries, together with sugar consumptions, given by Crane, 1975*c*, are from diverse sources and are not used for comparison here.

inherit a tradition of beekeeping and honey consumption from past centuries, through their ancestors in northern Europe.

13.82 Beeswax

Table 13.8A gives the beeswax production, exports and imports of individual countries, and Table 13.8E summarizes these for world regions. Tropical and subtropical regions produce, and export, more beeswax than temperate regions, for several reasons.

1. Bees secrete wax more readily in hot climates; also, tropical honeybees produce many swarms, and colonies migrate to new nest sites, so bees generated by one colony may build many new combs in a year.
2. Traditional hives are still widely used in developing countries – which are mostly in the tropics and subtropics – and the beeswax yield from these hives is 8–10% of the honey yield, compared with 1.5–2% or less from modern hives. Annual production figures for Egypt in the years 1973 to 1979 are quoted by Page and Laidlaw (1980); the average wax production was 9.3% to 11.4% of the honey production for traditional hives, and only 0.4% to 0.6% for frame hives. (It seems likely that the modern beekeepers, at any rate, did not harvest all the beeswax available.)
3. Containers are not needed for marketing beeswax, whereas lack of them can be a real constraint to the sale of honey in developing countries. In Uganda in 1984 I found that villagers were very ready to sell their beeswax, but they would not part with the containers that held their honey.

In many parts of the world, much of the beeswax produced by bees which could be harvested by beekeepers is wasted, and I have seen much evidence of this in many countries, in the tropics, subtropics and temperate zones. The wax is left or thrown away because beekeepers do not bother to collect it and render it into marketable blocks. In less developed countries, beekeepers in areas distant from a large town may have no possibility of selling the wax, and no incentive to collect and prepare it. In 1960 F. G. Smith said 'it is my estimation that only one-third, or at the most one-half, of the world's production of beeswax comes on to the market, the rest being thrown away or lost.' I cannot offer a quantitative estimate of the total amount now wasted, or of the amounts that are used in different ways, but Table 13.7A indicates the probable position.

The final column of Table 13.8E shows, for world regions, the average beeswax production as a percentage of the honey production. Referring to point 2 above, in general the higher percentages (characteristic of traditional beekeeping) are associated with exporting regions, and lower percentages (characteristic of modern beekeeping) with net importing regions. Of the exceptions, for many years Australia and New Zealand (Oceania) have exported beeswax produced in the course of modern beekeeping – showing what can be done in conserving beeswax. With regard to the apparent net imports into Asia shown by the table, Japan certainly imports considerable quantities, but an export figure is lacking for China, by far the largest producer. A net import of beeswax is recorded into a number of tropical Asian countries that should themselves have a good capability for producing and exporting it, e.g. Indonesia, Malaysia, Pakistan and Thailand. One problem for these countries is that many commercial formulations specify *Apis mellifera* wax, and waxes from other honeybee species differ somewhat from this (Section 13.5). So wax wholly or partly from other bees may not be accepted in importing countries unless suitability has been tested. However, Joshi (1982) reports that India has exported beeswax to 33 countries, and gives details since 1957.

Figures for beeswax are less complete, and almost certainly less reliable, than those for honey. Total recorded world imports are 7421 tonnes, but recorded world exports only 4360 tonnes. Import figures are thus fuller than those for exports, presumably because the documentation in exporting countries is less complete or is not published. Of the high beeswax producers, China (12 800 tonnes) shows no figure for exports, Mexico (9150 tonnes) shows only 210 tonnes, with 260 tonnes imported. Of countries with many traditional hives, Turkey, Angola, Ethiopia and Kenya all produce over 1000 tonnes, but none of these countries is recorded as exporting more than a sixth of its production. Morocco and Tanzania export over a third of the beeswax they produce.

Europe receives 65% of the total recorded imports; the USA and Japan import about 10% each, and the only other countries outside Europe that import more than 1% of the world total are Mexico and, in Asia, Hong Kong, UAE and the Korean Republic. Recorded beeswax imports into the 7 major importing countries (Table 13.8F) increased by nearly a third between 1972 and 1984, and in 1984 these 7 countries together accounted for 75% of the total recorded world imports. Most imported beeswax is used in industry, as discussed in Sections 13.7 to 13.73.

(*Text continues on page 450*)

Table 13.8A Honey and beeswax: production and trade in individual countries

Figures are in tonnes; those undated relate to 1984.

FAO honey production figures and trade figures are quoted from *FAO Production Yearbook* and *Trade Yearbook*, respectively (both 1984, published 1985), with some updating from FAO Statistics Division printouts.

Beeswax production figures are for 1982, from Table 6/4 in *Tropical and sub-tropical apiculture* (FAO, 1986). Beeswax trade figures are extracted from Trade Analysis System (TAS) printouts kindly prepared by International Trade Centre, Geneva; they relate to 'insect waxes', but it is believed that these are comprised almost entirely of beeswax.

In honey columns:

[a] FAO 1984 figure from government or national institution
[b] FAO 1984 figure from some other source
[c] FAO most recent figure, 1980–1983
FAO figure estimated (no symbol)
[n] information from another source, usually within the country (not FAO), latest date available

(*Notes continue opposite*)

Part 1 Europe and USSR

Country	Honey			Beeswax		
	Production	*Imports*	*Exports*	*Production*	*Imports*	*Exports*
Albania	714					
Austria	4 500	+6 259[a]	−397[a]	+42		−1
Belgium-Luxembourg	1 000	+4 172[a]	−346[a]		+41	−7
Bulgaria	10 030[a]	+375[a]	−5 854[a]	290[f]		
Czechoslovakia	9 633[a]	+2 000	−5 223[a]			
Denmark	2 400[n]	+2 188[a]	−683[a]		+29	−21
Faroe Islands		+6[a]			+1[c]	
Finland	1 349	+20[a]			+33	
France	20 126	+5 693[a]	−1 753[a]		+1563	−207
German DR	5 980[a]	+120[b]	−1 920[b]			
German FR	16 300[a]	+73 951[a]	−10 378[a]		+1275	−413
Greece	11 958[a]	+520[a]	−807[a]	264[a]	+127	
Hungary	14 048[a]	+2 035[a]	−18 407[a]			
Iceland		+37[a]			+1[c]	
Irish Republic	180	+865[a]	−81[a]		+25	−12[c]
Italy	6 000[a]	+9 028[a]	−274[a]	66	+195	−1
Malta		+5[b]			+1[c]	
Netherlands	500[a]	+8 199[a]	−1 336[a]		+553	−621
Norway	1 212		−72[a]		+3	−41
Poland	18 018[a]		−10[bc]			
Portugal	3 106[a]	+802[a]	−7[a]	375[a]	+3	−14
Romania	15 382[a]		−4 700[b]			
Spain	10 100	+8 000	−1 100[b]	730[a]	+96[c]	−89[c]
Sweden	3 272	+1 645[a]	−38[a]	52[f]	+39	−9
Switzerland	2 340[a]	+5 500[a]	−61[a]		+170	−3
UK	1 500	+19 324[a]	−1 376[a]		+431	−111
Yugoslavia	5 612[a]	+4 000	−1 650[b]		+169[c]	
TOTALS FOR EUROPE	165 260	+154 744	−56 473	1777	+4797	−1550
USSR (in both Europe and Asia)	193 000[a]		−24 226[a]	(3860*)		

* estimated arbitrarily as 2% of honey production

In the beeswax columns:

TAS 1984 figure (no symbol)

[c] TAS most recent figure, 1980–1983

[d] figure estimated from TAS entry giving value in US dollars

[f] 1984 figure from 1986 FAO printout

Part 2 North America

Country	Honey			Beeswax		
	Production	Imports	Exports	Production	Imports	Exports
Canada	43 298[a]	+196[a]	−18 871[a]	1327	+34	
USA	75 000[a]	+58 608[a]	−2 942[a]	1700	+813*	
Greenland		+7[a]				
TOTALS	118 298	+58 811	−21 813	3027	+847	

* includes Puerto Rico

Part 3 Central America

Country	Honey			Beeswax		
	Production	Imports	Exports	Production	Imports	Exports
Antigua/Barbuda	6[n]					
Bahamas		+5[b]				
Barbados	1[n]	+10[b]			+29[c]	
Belize	250[a]		−230[b]		+6[c]	
Bermuda		+39[a]				
Costa Rica	900		−560[b]	116	+4[c]	
Cuba	8 840[n]		−12 062			
Dominica	2[n]		yes			
Dominican Republic	1 300		−955[a]	350	+2[c]	−208[c]
El Salvador	2 600	+9[c]	−2 576[a]	130		−9[c]
Guadeloupe	4	+36		2[f]	+3[c]	
Guatemala	2 650		−3 500[b]	371	+7[c]	−4[c]
Haiti	320		−110	46[f]		
Honduras	1 200	+4[c]	−800[b]	120	+1[c]	−1[c]
Jamaica	1 000		−10	230	+1[c]	
Martinique	106	+5[c]		11[f]		
Mexico	67 095[a]		−54 030[a]	9 150	+260[f]	−210[f]
Montserrat	7[n]					
Netherlands Antilles		+20				
Nicaragua	150[n]				+7[c]	
Panama		+15[c]			+8[c]	
Puerto Rico	173					
St Kitts/Nevis	11[n]		yes	0.2[n]		
St Lucia	69[n]		yes	1[n]		
Trinidad/Tobago	57[n]			1[n]	+4[c]	
TOTALS	86 741	+143	−74 833	10 528	+332	−432

(*continued*)

Table 13.8A *(continued)*

Part 4 South America

Country	Honey			Beeswax		
	Production	*Imports*	*Exports*	*Production*	*Imports*	*Exports*
Argentina	35 000[b]		−29 000[b]	1500	+1[c]	−122[c]
Bolivia	1 400			140		
Brazil	7 500	+500	−35[a]	750	+1[c]	−78[c]
Chile	5 000		−1 427[a]	500[a]	+1[cd]	−265[c]
Colombia	2 200	+52[c]	−24[c]		+17[c]	−1[c]
Ecuador	1 120			128	+6[c]	
French Guiana		+9				
Guyana	70					
Paraguay	1 050			104		
Peru	80	+3[c]	−43[c]		+4[c]	−1[c]
Surinam	72					
Uruguay	3 500		−1 722[a]		+1[cd]	−15[c]
Venezuela	320	+100		146	+16[c]	−1[c]
TOTALS	57 312	+664	−32 251	3268	+4	−483

Part 5 Oceania

Country	Honey			Beeswax		
	Production	*Imports*	*Exports*	*Production*	*Imports*	*Exports*
Australia	24 963[a]	+67[a]	−10 791[a]	482[a]	+10[d]	−327
Fiji	5	+7[c]			+1[c]	
French Polynesia	13	+5[c]				
Guam	8	+12[bc]				
Hawaii	280[n]					
New Caledonia		+13[c]				
New Zealand	5 818[a]	+8[a]	−825[a]	110	+1	−116
Niue	20[a]		−6[b]			
Papua New Guinea		+24			+1[c]	
Samoa (Western)			−5[c]		+3[c]	
Tonga	18					
Tuvalu	2					
Wallis/Futuna Is.	10					
TOTALS	31 137	+136	−11 627	592	+16	−443

Table 13.8A *(continued)*

Part 6 Asia

Country	*Honey*			*Beeswax*		
	Production	*Imports*	*Exports*	*Production*	*Imports*	*Exports*
Afghanistan	3 000					
Bahrain		+15[a]				
Bangladesh	2[n†]				+1[d]	−15[d]
Brunei		+5[b]			+6[c]	
Burma (now Myanmar)	5					
China	160 605	+213	−45 059	12 800		
Cyprus	407[a]	+714[a]	−922[a]		+5	
Gaza Strip			−40[b]			
Hong Kong		+1 325[a]	−254[a]		+97	
India	4 500[n] 13 500[n*]	+3[n]	−3[n]	12[n‡]		−19[n‡]
Indonesia	100[n] 160[n*]	+100[b]	−6[c]		+14[c]	−2[c]
Iran	6 000	+10[b]			+45[f]	
Iraq	52	+2 000				
Israel	2025[a]		−100		+10[d]	−1[c]
Japan	6 798[a]	+33 178[a]	−5[b]		+766	−170
Jordan	300	+200	−8[c]			
Korea, Republic	6 300	+21[a]	−1[a]	600	+81	−2
Kuwait		+450[b]	−6[c]			
Lebanon	300	+500[b]				
Macau		+50	−9[c]			
Malaysia		+500			+26[c]	−4[c]
Mongolia	360					
Oman		+169[a]				
Pakistan	688*	+311[n]		80[f]	+12[c]	−2[c]
Philippines	70[n] 20[n*]	+67[c]			+17[c]	
Qatar		+40[b]				
Saudi Arabia		+2 600[b]	−95[c]		+22[c]	
Singapore		+1 004[a]	−164[a]		+8	−6
Sri Lanka	39[n]				+1[c]	
Syria	683[a]	+6		66[f]		
Taiwan	3 247[a]	+213[a]	−59[a]			
Thailand	750?[n]	+111	−70[c]		+29[c]	−2[c]
Turkey	35 620[a]		−1 500[b]	2 302	+40	−39[c]
United Arab Emirates		+500[b]	−35[b]		+96[c]	
Vietnam	316			10[n]		
Yemen Arab Rep.	300	+260[b]				
Yemen, PDR	70	+300	−35			
TOTALS	246 217	44 865	−48 371	15 870	+1276	−262

* from wild *Apis dorsata* nests, but for Pakistan includes honey from hives
† from *A. cerana* only
‡ *Apis* species unknown

(continued)

Table 13.8A *(continued)*

Part 7 Africa

Country	*Honey*			*Beeswax*		
	Production	*Imports*	*Exports*	*Production*	*Imports*	*Exports*
Algeria	1 584	+2000			+17[c]	
Angola	15 000			1500		
Botswana		+3[c]				
Burkina			−3[c]		+1[c]	−2[c]
Burundi	910			93[f]		
Cameroon	2 300	+8[b]		225	+2[c]	
Cape Verde		+2[c]				
Central African Rep.	6 800			500	+1[cd]	−83[c]
Chad	960					
Egypt	7 500		−81[c]	140	+37[c]	−5[c]
Ethiopia	21 000		−15	2100		−210[c]
Gabon		+10[b]			+2[c]	
Guinea-Bissau	300			100		
Ivory Coast		+10[b]			+6[c]	−32[c]
Kenya	12 000		−34	1050	+1[c]	−165[c]
Libya	540[a]	+500				
Madagascar	3 570			351		−76[c]
Malawi		2[c]				
Mali	4 212[n]			30[f]		
Mauritius		+45				
Morocco	3 100			350	+2	−149
Mozambique	260			52[f]		
Nigeria		+150[bc]				
Réunion	53	+1[c]			+1	
Rwanda	8			11[f]		
Senegal	202	+2[c]	−9[c]	60[f]	+1[c]	−6[c]
Seychelles		+2[a]				
Sierra Leone	600			100		−2[cd]
South Africa	900		−6[b]		+27[c*]	
Sudan	600			100	+8[c]	−23[c]
Tanzania	11 500		−80	1050		−437[c]
Tunisia	900		−11[b]	90[f]		
Uganda	170[a]			780		
Zaire	170[n]					
TOTALS	95 139	+2735	−239	8682	+106	−1190

* includes SA Customs Union

Table 13.8B Honey: world production and trade by continental region (1000 tonnes)

The 1984 figures are totals from Table 13.8A, Parts 1–7, derived from FAO data. Those for 1971 are from International Trade Centre (1977), also based on FAO statistics.

Continent	Production		Imports		Exports	
	1971	1984	1971	1984	1971	1984
Europe	*122*	165	*103*	155	*29*	56
USSR	*220*	193			*5*	24
(N. America)		(118)		(59)		(22)
(C. America)		(87)		(0.1)		(75)
N. + C. America	*158*	205	*6*	59	*40*	97
S. America	*37*	57	*0.4*	0.6	*15*	32
Oceania	*25*	31	*0.1*	0.1	*12*	12
Asia	*241*	246	*18*	45	*31*	48
Africa	*25*	95	*3*	3	*0.5*	0.2
WORLD TOTALS	*828*	992	*131*	263	*133*	269

Table 13.8C Honey consumption per person in some importing and exporting countries (in units of 100 g)

Total consumption is calculated as production plus imports less exports. Entries for importing countries are from International Trade Centre (1986), and other data from FAO sources, *Britannia World Data* (1985), Table 13.8A, and Thakar (1976).

Importing countries	*1975*	*1984*	*Exporting countries*	*1975*	*1984*
		Europe			
Austria	11.2	18.3	USSR	6.6	6.4
German FR	9.3	13.3	Hungary	1.8	2.6
Switzerland	8.6	12.0			
Denmark	5.6	6.5			
Netherlands	2.6	4.8			
Belgium/Luxembourg	3.9	4.7			
France	2.8	4.4			
UK	3.5	3.4			
Spain	1.2	3.0			
Italy	0.8	2.9			
		North America			
USA	5.1	5.5	Canada	7.2	10.1
	South/Central America:		Mexico	1.3*	1.7
			Argentina	*	1.3
		Oceania:	Australia	9.1	9.2
		Asia			
Japan	2.2	3.4	China	2.2	1.2
Kuwait	1.1	2.5			
Saudi Arabia	0.2	2.4	Virtually no imports or exports:		
Hong Kong	1.2	2.0	India	0.004	0.02

* For Mexico the mean 1975–77 consumption is entered, as unusually little honey was exported in 1975. For Argentina exports exceeded production in 1975, and existing stocks of honey must have been used.

Table 13.8D Honey consumption per person in continental regions (in units of 100 g)

Honey data are from International Trade Centre (1977, 1986) and other official sources, and populations from *FAO Production Yearbook* (1984).

	1971	*1984*
Europe	4.3	5.4
USSR	8.9	6.1
N. America	4.6	5.9
C. America	2.2	0.9
S. America	1.2	1.0
Oceania	8.4	10.1
Asia	1.1	0.9
Africa	0.8	1.8
WORLD TOTALS	2.3	2.1

Table 13.8E Beeswax: world production and trade by continental region (tonnes)

Figures are totals from Table 13.8A Parts 1–7, where their source is stated. The final column gives beeswax production as a percentage of honey production.

Continent	*Production*	*Imports*	*Exports*	*Net imports*	*Net exports*	*Beeswax / Honey*
Europe	1777	4797	1550	3247		1.1%
USSR	3860*					
N. America	3027	847		847		2.6%
C. America	10 528	332	432		100	12.1%
S. America	3268	47	483		436	5.7%
Oceania	592	16	443		427	1.9%
Asia	15 870	1276	262	1014		6.4%
Africa	8682	106	1190		1084	9.1%
WORLD TOTALS	47 604	+7421	−4360			4.8%

* estimated arbitrarily as 2% of honey production

The world price of beeswax is always higher than that of honey. It varied between 2.1 and 4.2 times the price of honey between 1973 and 1981 (Figure 13.81d). The economics of the beeswax market have been discussed by Morse (1983) in relation to the USA, and by International Trade Centre (1978) in *The world market for beeswax*, a publication that gave much useful information and discussion on the world beeswax market not obtainable elsewhere. It included addresses of importers in individual countries. An updated list has also been published (IBRA, 1985*a*, No. 2).

Table 13.8F Beeswax imports and consumption (tonnes) in the 7 major importing countries

Consumption is calculated as production plus imports less re-exports. Figures for 1972 and 1976 are from International Trade Centre (1978), and others from Table 13.8A.

	Imports				*Consumption*			
	1972	*1976*	*1980*	*1984*	*1972*	*1976*	*1980*	*1984*
France	491	411	668	1563	347	328		1356
German FR	728	1004	1229	1275	620	747		862
Netherlands	376	364	416	553	2	−8		−68
Switzerland	105	183	220	170	103	158		167
UK	488	741	558	431	369	547		320
TOTALS, 5 EUROPEAN COUNTRIES	2188	2703	3091	3992	1441	1772		2637
USA (N. America)	1298	1412	1269	813	3105	2560*	3040	2513
Japan (Asia)	780	689	670	766	712	547		996
TOTALS	4266	4804	5030	5571	5258	4879		6146

* 1975

13.9 FURTHER READING AND REFERENCE

Details of publications listed will be found in the Bibliography.

Honey

Couston, R. (1985) *Co-operative honey handling and extracting techniques*

Crane, E. (ed.) (1975*a*) *Honey: a comprehensive survey*

Crane, E. (1980*a*) *A book of honey*

Gonnet, M.; Vache, G. (1984) *Le goût du miel*

Kloft, W. J.; Kunkel, H. (eds) (1985) *Waldtracht und Waldhonig in der Imkerei*

Maurizio, A. et al. (eds) (1975) *Der Honig*

New Zealand Ministry of Agriculture and Fisheries (1981) *Honey export certification manual*

Townsend, G. F. (1978*a*) *Preparation of honey for market*

White, J. W. et al. (1962) *Composition of American honeys*

White, J. W. (1978) *Honey*

Beeswax

Coggshall, W. L.; Morse, R. A. (1984) *Beeswax: production, harvesting, processing and products*

Hepburn, H. R. (1986) *Honeybees and wax*

Tulloch, A. P. (1980) *Beeswax – composition and analysis*

Production and trade

International Trade Centre (1978) *The world market for beeswax*

International Trade Centre (1986) *Honey: a study of major markets*

Periodical publications include *FAO Production Yearbook* (honey production), *FAO Trade Yearbook* (honey exports and imports), and *USDA Agriculture Circular*, from 1957 or earlier (honey production and trade).

Bibliographies

Crane, E. (1978*a*, Nos 19, 20, 21) *Honey in the tropics; Beeswax and other hive products in the tropics; Descriptions of pollen grains in tropical honeys.*

14

The newer hive products: pollen, propolis, royal jelly, bee venom, bee brood

14.1 INTRODUCTION

Honey and beeswax have constituted the harvest from beekeeping since ancient times. Other hive products were hardly considered until after the Second World War, when various factors led beekeepers to attempt further diversification. Factors included low prices for honey, replacement of beeswax by other substances for many industrial purposes, and a widespread new interest in what are called health foods and dietary supplements among people able to afford them. New technologies made it possible to harvest, handle and store perishable substances from hives on a commercial basis. So attempts were made in many countries to find markets for the newer hive products: pollen, propolis, royal jelly, bee venom and bee brood.

All hive products have a biological origin, and their compositions are complex. Products secreted by a single species of honeybee (royal jelly and bee venom in this chapter, and beeswax in the previous one) are rather constant in composition, each fulfilling a specific function in the life of the bees. Adulteration is therefore relatively easy to detect. On the other hand products collected by bees from plants (pollen and propolis) have a very variable composition, characterized by the plant of origin. Honey is intermediate in variability; nectars and honeydews that the bees collect vary according to the plant source, but the bees then elaborate the honey, and contribute their own enzymes, giving the honeys certain common characteristics that have a special function in the colony.

The compositions of the hive products treated in this chapter are outlined in Tables 14.21A to 14.61A. The tables are constructed in different ways, according to what is known about each product and what is likely to be most useful to interested readers. Tables and text relate to products harvested from hives of European *Apis mellifera*, but where information is available on the product from other *Apis* races and species, it is included.

The application of hive products in general to treating disorders or to promoting health is sometimes referred to as 'apitherapy' or 'apiotherapy'. Its materia medica range from plant products such as pollen to insect secretions, and include honey and beeswax. Its practitioners have a common bond in believing that substances collected or produced by honeybees are essentially beneficial to human health. Many contributions to this theme can be found in publications under the following entries in the Bibliography:

Chauvin, R. (1968*d*)
Yoirish, N. (1959)
North American Apiotherapy Society Proceedings. USA, Vols 1–8 (1978–1985)
Apimondia (International Beekeeping Technology and Economy Institute) (1976)
International Symposium on Apitherapy
1st, Madrid, in 1974, Proceedings (1975)
2nd, Bucharest, in 1976, Proceedings (1977)

The following International Symposia on Apitherapy have also been held: 3rd, Portorož, Yugoslavia, in 1978; 4th, Herzliya, Israel, in 1983; 5th, Kraków, Poland, in 1985; 6th, Portorož, Yugoslavia, in 1988.

In most countries the sale of pharmaceutical products is governed by a Food and Drug Act, which strictly limits claims that can be made for them. This does not necessarily apply to natural health foods, which in Denmark for example are exempt by law from the need of proof that a preparation has a specific property claimed for it.

14.2 POLLEN

Flowers of wind-pollinated plants produce light, dry pollen grains that can be blown by the wind, and commercial harvesting of such pollen is referred to in Section 14.23. Flowers of most insect-pollinated plants produce both nectar and pollen, the pollen grains having a sticky coating (Hesse, 1981) which helps to attach them to hairy insect bodies. Pollen from some wind-pollinated plants, as well as from most insect-pollinated plants, is collected by bees. The former contributed 24%, 9% and 13%, respectively, of the pollen harvest in an Arizona apiary in 1976, 1978 and 1979 (O'Neal & Waller, 1984).

Beekeepers need no specific skills to harvest pollen as a hive product, but this can be done only where and when pollen is in plentiful supply. Commercial pollen production is especially successful in dry areas such as Arizona in the USA, Western Australia and southern Spain. In humid areas the collected pollen is at risk of deterioration through contamination with moulds, but hive design can help to overcome this problem (Villeneuve et al., 1988).

In the diversification of hive products that was sought by beekeepers in the 1950s, the exploitation of pollen as a commercial product was stimulated partly by the fact that royal jelly production had proved to be very labour-intensive (Section 14.42). Harvesting pollen seemed an easier alternative, yielding large quantities of brightly coloured pellets which could be promoted and packaged attractively for sale.

14.21 Composition of pollen

The composition of pollen varies according to the plant source, and shows a greater diversity than that of any other hive product. Figures for the gross composition of pollen (Table 14.21A) are quoted from the early results of Todd and Bretherick (1942), who named the plants they studied, and the individual results they quote enable both ranges and means to be determined. Other species could have given somewhat different results.

'Bee-collected' pollen comes from plant species selected by the bees, and contains additional sugar derived from the honey or nectar with which bees moisten the pollen when packing it into 'pollen loads' on their hind legs (Section 2.43.3).

Bee-collected pollen has antibiotic properties (Lavie, 1960), and the germinating ability of pollen grains is likely to become impaired unless the pollen is suitably treated after harvesting; a method is mentioned at the end of Section 14.22.

Table 14.21B lists substances that have been identified in pollens from various (recorded or unrecorded) plant species; many are present in only minute quanti-

Table 14.21A Gross composition of pollens from different plant species

The 6 and the 27 samples (air-dried) were analysed by Todd and Bretherick (1942). Results for the 7 samples are quoted by Stanley and Linskens (1974), as % dry weight (3 of them were air-dried).

	Hand-collected (6)		*Bee-collected (27)*		*Hand-collected (7)*
	mean (%)	*range (%)*	*mean (%)*	*range (%)*	*range (%)*
water	9.7	3.9–17.1	11.2	7.0–16.2	
crude protein	20.4	11.4–35.5	21.6	7.0–29.9	5.9–2.83
ether extract	5.0	1.3–17.1	5.0	0.9–14.4	1.2–3.7*
carbohydrates:					
reducing sugar	*3.1*	*0.0–7.5*	*25.7*	*18.8–41.2*	
non-reducing sugar	*8.2*	*0.1–18.9*	*2.7*	*0.0–9.0*	
starch	*8.0*	*0.0–22.4*	*2.6*	*0.0–10.6*	
total	19.4	1.2–36.6	31.0	20.5–48.4	13.2–36.6
ash	3.5	2.4–6.4	2.7	0.9–5.5	1.8–3.7
undetermined	42.9	31.3–57.2	28.6	21.7–35.9	

* (lipid)

Table 14.21B Substances identified in pollens

Adapted from Bee World (1975), based on Stanley & Linskens (1974).

lipids	polar lipids, monoglycerides, diglycerides, triglycerides, free fatty acids (palmitic, stearic, oleic, linoleic, linolenic); hydrocarbons; alcohols; sterols (β-sitosterol, cholesterol, fucosterol, 24-methylene-cholesterol, campesterols, sigmasterol, C_{29}-di-unsaturated sterols)
carbohydrates (sugars)	sucrose, fructose, glucose
compounds related to carbohydrates	callose, pectin and other polysaccharides, cellulose, sporopollenin, lignin
ash	major minerals K, Na, Ca, Mg, P, S; trace elements Al, B, Cl, Cu, I, Fe, Mn, Ni, Si, Ti, Zn
organic acids, including phenolic acids	*p*-hydroxybenzoic, *p*-coumaric, vanillic, protocatechuic, gallic, ferulic
free amino acids	alanine, arginine, aspartic acid, glutamic acid, glycine, histine, leucine/iso-leucine, lysine, methionine, phenylalanine, proline/hydroxyproline, tyrosine, valine
nucleic acids	desoxynucleic acid, riboxynucleic acid
terpenes	
enzymes	24 oxidoreductases, 21 transferases, 33 hydrolases, 11 lyases, 5 isomerases, 3 ligases, and others
vitamins	B complex: B_2 (riboflavin), niacin, B_6 (pyridoxal), pantothenic acid, biotin; C (ascorbic acid); E (α-tocopherol)
nucleosides	
carotenoids	at least 11
flavonoids	at least 8
growth regulators	auxins, brassins, gibberellins, kinins, also growth inhibitors

ties, and some probably occur in only a few pollens. Stanley and Linskens (1974) quote comparative amounts of some of the constituents, from a number of plant sources, and they cite many studies on pollen composition.

14.22 Harvesting and handling pollen

Unlike honey, pollen is not usually stored in the hive greatly in excess of the colony's requirements. Pollen can be harvested from bees by fitting a 'pollen trap' across the flight entrance. In the earliest type of trap (Böttcher, 1941), returning foragers were made to scramble through holes of the shape shown in Figure 14.22a in order to enter the hive, and the pollen loads on their hind legs were knocked or scraped off as they did so.

Subsequent traps have used a wire grid as a simpler alternative, and one which allows more bees to enter the hive at once. Sometimes two parallel grids close together are used. Grid(s) can be made from 5-mm wire mesh (wire diameter 0.58 mm); if there are two grids they should be 6 or 7 mm apart. Underneath is the pollen tray, which is covered by 4.2-mm or finer mesh; the pollen loads drop through, but it is too fine to let bees pass. The floor of the pollen tray is of cloth, plastic fly wire, or other material that allows any water to escape, and air to circulate round the pollen, to prevent deterioration. The beekeeper must remove the pollen from the tray regularly.

In commercial pollen harvesting, grids are usually horizontal, and placed underneath the hive, with the collecting tray below. The whole is conveniently incorported in an adapted floor board. Horizontal screens can be made much larger than vertical ones, since the bees enter the hive from below. The trapped pollen is well protected from weather damage, although not from condensation within the hive. A type widely used for large-scale pollen collection is the OAC pollen trap, designed at Ontario Agricultural College, Canada, in the early 1960s (M. V. Smith, 1963).

Figure 14.22a Early pollen trap (Böttcher, 1941; photo: W. P. Stephen).
The trap is fixed across the hive entrance, and the bees' pollen loads fall into the removable tray below, which is screened to prevent access by the bees.
Inset close-up showing shape of holes.

Two parallel grids are mounted together on wood, the unit fitting over a standard floor board, on which the pollen-collecting tray slides. In this type, debris from the hive can fall directly into the pollen trap. Figure 14.22b shows a variant widely used in Australia (Chambers, 1975), designed to overcome this defect by fixing a 'dirt tray' which, however, reduces the area of the pollen-removing screens. The tray is usually removed from the back of the hive, which does not disturb the bees (Figure 14.22c). Some other pollen traps are referred to in Bee World (1976*a*).

A colony cannot rear brood without pollen, and if efficient traps are fitted for commercial harvesting, the beekeeper must pay special attention to the well-being of his colonies. While a hive has a trap fitted across the flight entrance, a proportion of the pollen brought to the hive is still carried in. This proportion varies according to the design, and Section 12.63 quotes 10% and 60% for different traps. Pollen foragers may adapt to constant trappings by collecting small loads that they can carry through the grid(s) into the hive (Chambers, 1985).

It is usually recommended that traps should be removed every 10 days or so, to enable the bees to collect pollen for brood rearing. In Western Australia, Chambers (1985) finds it best to leave traps on for 3 weeks and remove them for every fourth week. During 3-year experiments in the south of France, hives fitted with traps for 40 days gave 24% less honey than others (Lavie, 1967). Traps should not be left on in bad weather, or when pollen is in short supply. Even if the value of the pollen produced compensates the beekeeper for loss of honey, the colonies will not prosper without sufficient pollen intake. Experience is the best guide in any given situation.

The average amount of pollen trapped per year per colony in Arizona is 5–15 kg (Robson, 1986). Section 12.63 shows the weekly yields throughout the year at 7 different latitudes. Using 17 hives in Quebec, Canada, Villeneuve et al. (1988) obtained from 1.5 to 2.7 kg per hive, with a honey yield (from 55.5 to 125 kg per hive) that was only marginally reduced; there was no correlation between pollen and honey yields from individual hives.

Care must be taken that the pollen does not deteriorate during or after transport to the processing plant, and the pollen should be air-dried immediately on arrival. Shallow boxes, seen in Figure 14.22d (Chambers, 1980), are used for both transport and drying in Western Australia, where the moisture content of pollen may upon occasion be over 18%. The boxes are made of wood (not cardboard), with a ventilated base of fibreglass screen such as is used against flies. Even so, unless the pollen is very dry, the layer in each box should not be more than 13 cm deep, or it may coagulate and ferment. Drying can reduce the moisture content to 6%.

Air-dried pollen can be kept at a cool temperature for a short period, but it deteriorates after some months. Deterioration can be prevented by mixing the pollen with a substance that will absorb moisture from it; if it is to be fed to bees, soya-bean flour (Whitefoot & Detroy, 1968) or sugar (Townsend & Smith, 1969) can be used.

It is easiest for the beekeeper who produces pollen to sell it directly to a local buyer. In 1985 when I arrived at a pollen-processing plant in Phoenix, Arizona, a local beekeeper was delivering a truck load of freshly collected pollen in covered plastic buckets; the company had lent a hundred or so pollen traps to him and to a number of other beekeepers.

Healtheries of New Zealand Ltd (1981) list the following requirements for pollen they purchase:

- clean, free from moth eggs, larvae and insect fragments;
- moisture content 8–10%, and temperature during drying not to exceed 49°;
- dried pollen stored in poly-lined 20-litre lever-lid tins, or doubly lined corrugated cartons, and sealed to exclude air, light, and the possibility of moisture uptake;
- contract price per kg for lowest quality NZ$11, for best quality NZ$12.

If intended for human consumption, the bees' pollen loads are normally cleaned by passing them over a screen to remove small particles, and also through a

Figure 14.22b Pollen trap: design used in the Department of Agriculture, W. Australia (Chambers, 1975).

- B dirt baffle
- D drone escape (tube 6.5 mm internal diameter)
- E1 entrance to pollen trap
- E2 free flight entrance
- H handle
- M metal plate with holes (5 mm diameter) to remove pollen loads from bees' legs
- P pollen collection tray
- S 6-mesh screen (2.4 per cm) allowing pollen to fall into collecting tray
- T dirt tray
- ↕ bee-ways

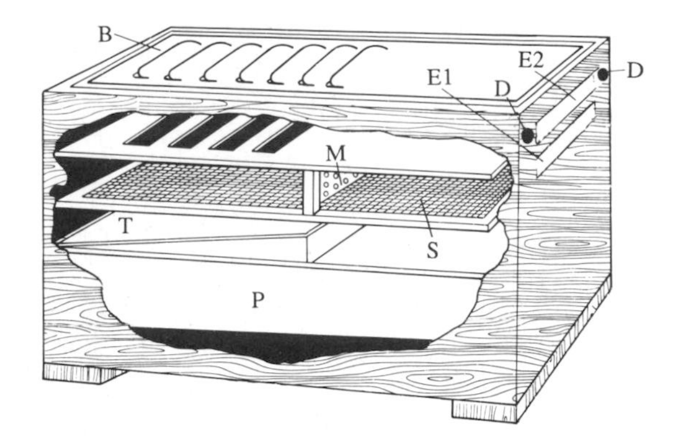

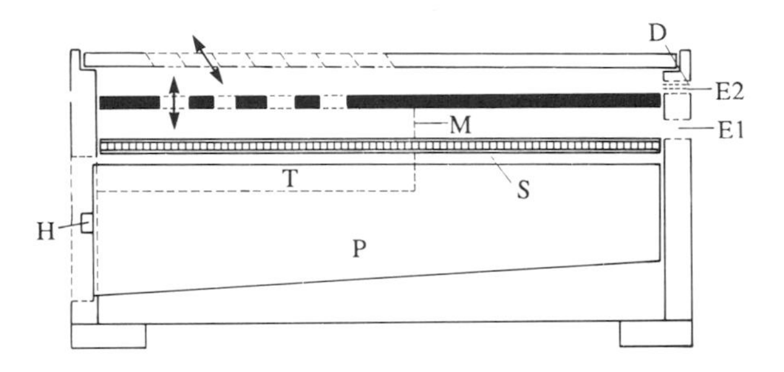

Figure 14.22c The beekeeper has pulled out the tray in Figure 14.22b, to monitor the pollen collected (photo: S. R. Chambers).
In this photograph the dirt tray is without a bottom board, so material falls on to the ground.

Figure 14.22d Air drying pollen (photo: S. R. Chambers).
Five collection boxes (made interlocking for stability when stacked) are placed above an air-inlet baffle box connected to an industrial vacuum cleaner used as a blower. A vent fitted to a lid at the top allows the moist air to escape.

winnowing machine to blow off pollen dust and lighter debris. If the pollen is to be sold in tablet form, a lubricant must be incorporated, or the tablet-forming machine will become clogged.

Pollen to be used for pollination must be stored in such a way that its viability (germinating ability) is retained. It is normally lyophilized (i.e. freeze-dried) before storage: frozen slowly at $-60°$ to $-80°$, and the air pressure then reduced to 50–250 mm of mercury so that water is removed by sublimation. Pollen treated in this way can be stored at cool room temperature under nitrogen or *in vacuo*. Stanley and Linskens (1974) give full details, and discuss problems of maintaining and testing germinating ability. Sakai et al. (1983) have developed a method for treating pollen for storage without freeze-drying, but retaining its viability, by repeatedly washing it in sucrose solutions and then pulverizing it and storing it frozen.

14.23 Pollen: world production, trade and uses

No figures are available for the world's commercial production of bee-collected pollen, but the amount must be larger than that of any other hive product discussed in this chapter. The following countries are known producers:

- Europe: France, German Federal Republic, Hungary, Romania, Spain, USSR, Yugoslavia
- Americas: Argentina, Chile, Mexico, Uruguay, USA
- others: Australia, China, Israel, Taiwan, Tunisia, Vietnam.

Western Australia produces 60 to 130 tonnes a year; the price was US$5 to 6.50 per kg according to quality in 1985 (James, 1986). The UK price was £4.20 per kg in 1986, but fell to £3.50 in 1987. Taiwan produces 150 tonnes; tea bushes are allowed to flower there, and 50 tonnes are from tea (Chang, 1988). Vietnam produces 15 to 20 tonnes.

Bee-collected pollen is bought and sold by honey traders, who are said to prefer pollen that is yellow or orange, as having a sweeter and better flavour than dark pollen. Much of the commercial pollen is, however, not bee-collected pollen, but the dry pollen produced in large quantities by certain wind-pollinated plants and easily dislodged from them. Pollen from corn (*Zea mays*) can be collected quite cheaply by machine when the plants are detasselled; in Romania, for example, a yield up to 50 kg/ha has been reported. It has been calculated that many sweet corn varieties in the USA produce over 170 kg/ha (Nowakowski & Morse, 1982). The dry pollen is released in the morning and bees collect it then; they do so most easily if it has been wetted by rain. In the USSR the calculated maximum pollen yield per hectare from plants in the Compositae was 8 kg from *Inula britannica* (Rudnyanskaya, 1981); this is much easier for bees to collect.

Lists of pollen buyers in various countries have been published (Stanley & Linskens, 1974; IBRA, 1985*a*, No. 2; CCI, 1987). Not all buyers would be willing to accept pollen from an unknown source in another country. Importing countries include Benelux, France, German Federal Republic, Italy, UK and USA; the UK price paid in 1985 was about £4 per kg of raw material (Kjaersgaard, 1985). Exporting countries include Spain, Hungary, Argentina and Taiwan. No figures are known for the amounts of bee-collected pollen traded within or between countries. Pollen from wind-pollinated plants has been used by some cultures as a folk remedy (McCormick, 1960), and both this and bee-collected pollen are likely to have a continuing place on the world market.

Commercially produced pollen is used for a variety of specialized purposes, which include the following:

1. plant breeding programmes
2. fruit pollination

3. for feeding to bees, alone or with supplementary material
4. as a source of certain compounds it contains
5. as a dietary supplement for human beings and for domestic animals
6. in cosmetics
7. in the study and treatment of allergic conditions such as hay fever
8. for monitoring environmental pollution
9. in 'prospecting' for minerals.

Pollen of specific (or even of varietal) plant origin is required for the uses 1, 2 and 7 and sometimes for 3; for 1 and 2 the pollen must be viable. In 2, a pollen dispenser or 'insert' may be used; it is a hive entrance device rather like a pollen trap in reverse; pollen adheres to a bee's hairy body as she leaves the hive. Townsend et al. (1958) describes its use in Canada, and Bee World (1976*a*) gives further references. The value of different pollens for feeding to bees (3 above) is referred to in Section 12.42.

Much of the pollen sold commercially is used for human consumption (5), as an additive to a diet which may already be adequate. The protein content of different pollens is relevant – and also their digestibility, which has been less fully studied (Bell et al., 1983). Stanley and Linskens (1974) conclude their section on nutritive supplements with the comments: 'Pollen probably is truly beneficial, although the benefits, we suspect, can be equalled by many other less expensive, more readily available foods.' The possibility of its benefit to underfed peoples in the tropics – where pollen could also be harvested – has so far received much less attention.

Experiments have shown that pollen harvested from hives may have potential uses in animal rearing (5). With piglets, calves and broiler chickens, for instance, the incorporation of a small percentage of pollen in the rations was found to lead to increased weight gains and other beneficial effects. Whether bee-collected pollen is the most cost-effective additive of its kind depends on the comparative cost of alternatives.

Since 1980 the possibility (8) has been explored that the materials honeybees collect, or the bees themselves, might be used as indicators of environmental pollution – especially by heavy metals (Crane, 1984*b*). In England, pollen collected by honeybees reflected the environmental soil content of manganese, zinc, copper and lead, although not of magnesium (Free et al., 1983). Experiments in the Puget Sound area of western USA also gave some promising results (Bromenshenk et al., 1985). In British Columbia to the north, two large mining companies have participated in experiments with hives of bees (9) to sample the pollen in areas of interest (Lilley, 1983). It is too early yet to tell whether this use of pollen will develop into a commercial enterprise.

14.3 PROPOLIS

Propolis is defined in Section 12.5 as 'material that honeybees and some other bees can collect from living plants, and use alone or with beeswax in the construction and adaptation of their nests'. Honeybees collect propolis (Section 2.43.4) from a variety of plants, mostly trees and bushes (Table 12.51A). It has generally been considered that bees probably do not alter the composition of propolis after they collect it (McGregor, 1952), although Greenaway et al. (1987) now suggest that some of the sugar in propolis may be derived from a glucoside in bud exudate, which has been hydrolysed enzymatically by the bees.

For various building operations in the hive, bees use mixtures of beeswax and propolis (Section 3.31.4). Unless proper precautions are taken (Section 14.32), propolis collected from hives may therefore contain admixtures of beeswax. Beekeepers harvesting propolis should also be alert to the possibility of other contamination, since honeybees occasionally collect some other substances – such as bitumen and paint – and treat them as propolis. Lowe (1980) mentions a number of such contaminants.

The use of propolis by different species of honeybees is explained in Section 3.31.4.

14.31 Composition and properties of propolis

In view of the many disparate plant substances that are grouped together under the name propolis, and the difficulty of observing bees high up in trees and bushes when they collect the material, it is perhaps not surprising that most of the published analyses do not identify the plant source or the type of plant substance. In general we know comparatively little about species variation in the composition of propolis.

Analyses in the 1970s categorized about half the propolis as resin and balsam, and a quarter as wax (suggesting that samples contained both propolis and beeswax). The remaining quarter included fatty acids and polyphenols (e.g. Cizmarik & Matel, 1970; Tikhonov et al., 1977; also Ghisalberti, 1979). In the USSR Popravko et al. (1969) pioneered the use of modern separation and identification techniques and, since then, increasingly advanced analytical methods have identified more and more substances in propolis. Walker and Crane (1987) list 149 compounds and 22 minerals from different samples, and Table 14.31A gives some details. The group of 38 flavonoids probably constitutes a greater part by weight than any other single group; in Bulgarian samples it accounted for 30% to 40% of the total (Marekov et al., 1984), in British samples 10% to 29% (Greenaway et al., 1987). Popravko et al. (1969) had found 6 flavones and 2 flavanones, and isovanillin, each in amounts of 1% to 4%.

Some constituents, including certain flavonoids, are probably present in all samples of propolis, and contribute to its characteristic properties. Others may occur only in propolis from certain plant sources. Popravko (1976) was able to identify 20 of the compounds found in propolis also in extracts from *Betula verrucosa* (= *pendula*). In a quantitative analysis in the UK, Greenaway et al. (1987) identified 56 compounds that were present in a sample of propolis from hives and also in bud exudate from *Populus* × *euramericana* trees nearby. A total of 104 compounds were identified in one or more of 3 samples from hives and the bud exudate. However, the hive samples contained from 20% to 50% of beeswax, and a small amount of honey.

Table 14.31A Groups of substances identified in samples of propolis (based on Walker & Crane, 1987)

No sample contained all the substances listed. The plant origin of most samples was unknown.

Group	*No. substances identified*
flavonoids	38
hydroxyflavones	27
hydroxyflavanones	11
chalcones	2
benzoic acid and derivatives	12
acids	8
esters	4
benzaldehyde derivatives	2
cinnamyl alcohol, cinnamic acid and its derivatives	14
other acids and derivatives	8
alcohols, ketones, phenols and heteroaromatic compounds	12
terpene and sesquiterpene alcohols and their derivatives	7
sesquiterpene and triterpene hydrocarbons	11
aliphatic hydrocarbons	6
minerals	22
sterols and steroid hydrocarbons	6
sugars	7
amino acids	24

The plant origin of any particular commercial batch of propolis is usually not known, and it would seem wise to give high priority to setting standards for materials to be accepted as propolis. Standardization has been discussed in Bulgaria (Bankova & Markov, 1984); in Hungary there is a standard, MSZ-080202-79, and in the USSR there is a regional Standard (RSFSR, 1977). This describes the appearance, colour, odour, flavour and consistency, and sets the following limits, which seem to allow considerable admixture with beeswax:

upper	
wax	30.0%
oxidizability value	22.0%
mechanical impurities	20.0%
lower	
phenolic compounds	20.0%
iodine number	35.0
qualitative reaction for flavonoid compounds:	positive

The analytical basis for standardization of propolis has been discussed by Popravko (1977) and Tikhonov et al. (1977), and a book by Popescu et al (1985) deals with a commercial standardized extract of propolis.

Some solvents for propolis are acetone, benzene, and a 2% solution of sodium hydroxide. They should not be allowed to come into contact with the skin; ethyl alcohol is safer, but somewhat less effective (Witherell, 1975). Waterglass (sodium silicate) mixed with its own volume of water has also been mentioned (Bee World, 1947).

Pigments in propolis from various plants may give it a reddish, brownish or yellowish colour, but water-clear propolis has been reported (Coggshall & Morse, 1984).

Propolis is sticky and soft when warm, but hard and brittle when cold, a fact which is used in freeing it from hive equipment on which it has been collected (Section 14.32). Like beeswax, it is an entirely plastic material at temperatures between 25° and 45°. Hepburn and Kurstjens (1984; see also Hepburn, 1986) studied the tensile and ductile properties of propolis in the above temperature range. They showed that it is highly ductile, tending to retain its strength, and to remain free from cracking, when its shape is altered.

Propolis has various pharmacological properties, and it seems important that the constituents responsible for both beneficial effects and side effects should be identified. Some of the flavonoids help to account for these properties. The constituents which have antibiotic activity are soluble in water and in alcohol; they are thermostable and stable on ageing. In extensive studies on antibiotic properties of different hive products, Lavie (1960, 1968) found propolis to be the most active, and the only one that was strongly fungicidal. Propolis can also inhibit germination of pollen grains and seeds of higher plants. It is very active against *Proteus vulgaris* and *Bacillus alvei*, but has no inhibitory action against *Escherichia coli* or *Pseudomonas pyocyanea*. Ghisalberti (1979) lists over 30 bacteria and other micro-organisms on which propolis extracts have an inhibitory effect, and cites the original studies on them. Lavie (1960) was able to show that the resin on buds of poplar (*Populus*) in France gave an extract with an activity similar to that of propolis. The most active extracts were from leaf and flower buds of *Populus nigra*; the wood also showed some activity.

There are reports of activity of propolis against a herpes virus (König & Dustmann, 1985), and of skin surface or local anaesthetic properties of extracts of propolis. Effects of propolis on wound healing and tissue regeneration, and on certain dermatological disorders, seem fairly well established, although again the active ingredients are not known (Ghisalberti, 1979).

14.32 Harvesting and handling propolis

Bees often mix wax and propolis together in minor comb-building operations in the hive (Section 3.31.4). Marletto (1983) found that purer propolis was deposited above the top hive box than on the hive floor or near the entrance, where the propolis could become mixed with beeswax and with other materials. Beekeepers usually harvest propolis in the course of normal colony management, arranging that the bees deposit propolis without admixture of wax on a device which they can easily remove from the hive. Figure 14.32a shows slotted sheets that may be used at the top of the hive; slots (or holes) to be filled with propolis must be less than a bee-space wide. Such a sheet is sometimes placed instead at the side of the hive where the bees will also regard it as part of the surface forming the boundary to their nest, and close the slots with propolis.

It is not necessary to use a specially made board. With European *Apis mellifera*, ordinary plastic net with holes 3 mm internally is used in Hungary. Ochi (1981) in Japan recommends nylon net or wire gauze (10–15 mesh, i.e. with holes 2.4–1.6 mm), laid across the tops of the frames; he reported a harvest of 50 g per hive between June and November. In Italy Andrich et al. (1987) quote yields per hive in summer from 10–31 g to 62–158 g.

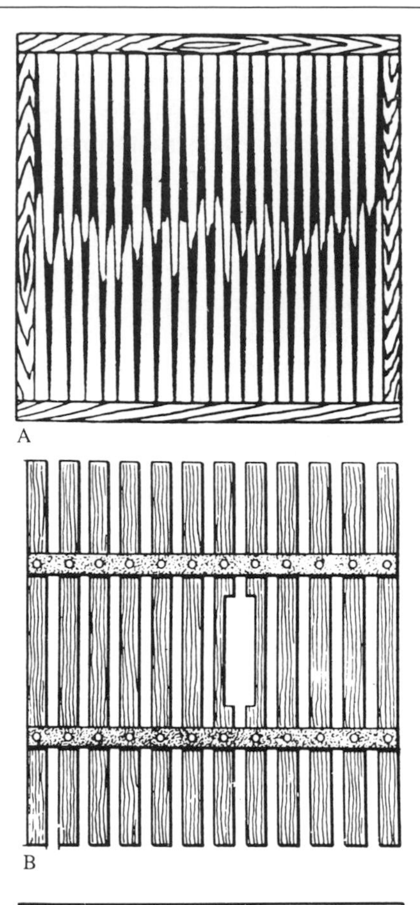

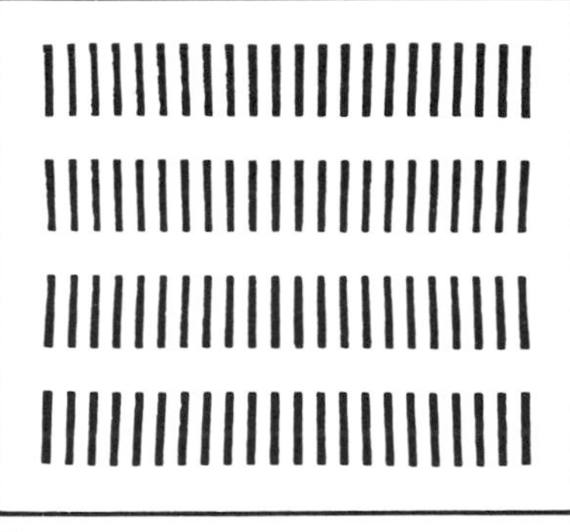

Figure 14.32a Propolis collectors.
A From Cuba; wood, covers the whole of the top hive box (Asis, 1979).
B From Hungary; wood, size as A, with gaps 2 to 3 mm wide; incorporates feed-hole.
C From Hungary; plastic, actual size 135 × 115 mm, gaps 2 mm wide.

When the sheets, net or gauze are removed from the hive, they are cooled in a deep-freeze; the propolis then becomes brittle, and a sharp blow will fracture it off the substrate. In the USSR, the usual canvas inner covers are sometimes kept after removal from the hives until freezing winter weather, when they are passed between rollers to fracture off the propolis (Sadovnikov, 1981). An alternative method uses compressed air (Pechhacker & Hüttinger, 1986).

Flakes and pieces of propolis harvested can be stored for up to a year in plastic bags, in a reasonably dry place. A simple purity test is to crunch up a sample and shake it with warm water in a glass jar; any material that floats is waste. Beekeepers usually sell propolis to an agent or to a business enterprise (Section 14.33) which will itself use it, probably in pharmaceutical preparations.

In a survey in the UK (Bunney, 1968), about 1 beekeeper in 2000 was found to be hypersensitive to propolis, contact with it leading to dermatitis. In patch tests in Italy, 6 out of 612 patients showed a strong positive reaction to 10% propolis solution (Tosti et al., 1985). There are various other reports, but the identity of the allergens is still not clear. Beekeepers affected could try using a silicone barrier cream; one sufferer found that a beeswax skin cream, whose formulation is given in Section 13.73, alleviated the effects. Anyone who develops a serious rash should wear rubber or plastic gloves which are impermeable.

14.33 Propolis: world production, trade and uses

In modern beekeeping up to the 1950s, propolis was generally regarded by beekeepers as a nuisance; it spoilt the appearance of section honey, and made frames and hive parts difficult to prize apart. On the other hand propolis makes possible the use of non-telescopic hive roofs (see Section 4.41, A), and it helps to prevent movement between frames and hive parts while hives are being transported.

Propolis was one of the potential hive products explored in the 1950s and 1960s. Its antibiotic properties were established, and many clinical trials were carried out with preparations made from it, especially in Eastern Europe (Walker, 1976). At present, commercial production of propolis seems to be confined mostly to temperate regions, although propolis is also collected by *Apis mellifera* in the tropics (see Figure 3.31b), and Cuba has published one of the few specialist books on it (Asis, 1979).

Amounts of propolis exported in 1984 (Kjaersgaard, 1985) were:

China	55 tonnes
Argentina	7 to 8 tonnes
Chile + Uruguay	7 to 8 tonnes
Canada	3 to 4 tonnes

Other countries producing propolis commercially include:

- Europe: Austria, Bulgaria, Denmark, France, German Federal Republic, Spain, USSR
- elsewhere: Australia, Brazil, New Zealand, USA.

In the 1970s advertisements were appearing in the beekeeping press of Western Europe and North America, inviting beekeepers to sell propolis to the advertisers at prices that seemed attractively high, for example US$160 per kg. Expectations were not always fulfilled: some dealers received propolis contaminated with wax or debris, and some beekeepers found that the dealer to whom they sent propolis had gone out of business. The flurry died down, and few of the same advertisers still advertise in the beekeeping journals.

Propolis is nevertheless regularly bought and sold on the world market, mostly by specialist traders, between 100 and 200 tonnes a year probably being exported and imported. The price of propolis on the world market fluctuates quite widely, but it never regained the 1974 level; in 1986 it was US$40–50 per kg in North America and US$30–35 in Europe (Kjaersgaard, 1988). Importing countries include Benelux, Denmark, France, German Federal Republic, Hungary, Sweden, UK and USA, and lists of importers, agents and users in different countries are available (IBRA, 1985*a*, No. 2; CCI, 1987).

Propolis is sold in capsules and tablets, as granules to be chewed or swallowed, in throat pastilles and in chewing gum. Particularly in Eastern Europe, it is used in cosmetics and healing creams, and as an ingredient of medicaments for treating digestive disorders.

Some other applications of propolis are mentioned by Ghisalberti (1979). It is no longer used in varnishes, as in earlier times. Propolis from the Cremona area of northern Italy was an ingredient in the varnish for violins made by Stradivari and others, and was supposed to have given the instruments special qualities, but this claim has not been substantiated (Jolly, 1978).

14.4 ROYAL JELLY

Royal jelly is a secretion of the hypopharyngeal or brood food glands in the head of young worker honeybees (Figures 2.2a, 2.31a). These bees provision queen cells with an excess of it, up to the time the cell is sealed. The larvae that develop into queens consume 25% more food than larvae developing into workers, and the food of older worker larvae differs in various ways from it; see Section 2.51.

The first person to use the term *gelée royale* (royal jelly) may well have been François Huber (1792). Suppliers of this hive product refer to it by one of these names, which are useful for promotion purposes. In some socialist countries, the egg-laying female bee is not royal, and a more biologically appropriate term is used, derived from mother. In Russian, mother is *matka* and royal jelly is *matochnoe molochko*, mother 'milk'.

As a result of the queen larva's high consumption of royal jelly, its body weight increases by 1300 times in 6 days. This dramatic effect created an interest well before the twentieth century, and in the 1950s royal jelly became the first of the 'newer' hive products to be exploited. Johansson and Johansson published detailed accounts of its composition, properties and applications, in 1955 and 1958. Royal jelly won, and has maintained, a place in the world market, as a specialized dietary supplement for human consumption and in cosmetics; between 500 and 600 tonnes are probably produced annually (Section 14.43).

14.41 Composition of royal jelly

Table 14.41A lists the main constituents of royal jelly, with amounts of some of them, and indicates sources of more detailed results. As a material that bees secrete – in contrast to those that they collect from plants – royal jelly has a relatively constant composition, and

Table 14.41A Composition of royal jelly (*Apis mellifera*)

Most data quoted by Takenaka, 1982; see also Takenaka, 1984.

Constituents		*Further details*		
water	66.9%			
protein[1]	11.4%			
		of total protein:		
		water-soluble protein comprises		75–85%
		protein nitrogen	65–70%	
		non-protein nitrogen	9.4–13.5%	
		free amino acid*	0.6–1.5%	
		* as % of total free amino acid:[2]		
		proline	50.6%	
		lysine	21.3%	
		glutamic acid	10.0%	
		others	18.1%	
			100%	
		water-insoluble protein comprises		15–25%
sugars[1]	9.1%	most important: glucose, fructose, sucrose		
ether extract[1]	6.2%	major acid:[3] 10-hydroxy-2-decenoic acid		
ash	0.94%	minerals (in order of importance):[4] K, Na, Mg, Ca, Zn, Fe, Cu, Mn		
enzymes		glucose oxidase		
		phosphatase		
vitamins, etc.		vitamin B complex:		
		B_1 (thiamine)	pantothenic acid	
		B_2 (riboflavin)	biotin	
		niacin	folic acid	
		also inositol, acetylcholine		

[1] Lercker et al. (1982) also give values for total protein, sugar and lipid, and of 31 free organic acids and 6 sterols; see also Pain et al. (1962). Simo and Christensen (1962) give amounts of individual sugars.

[2] As % of total amino acid, contents of important amino acids in royal jelly are: aspartic acid 17.1, lysine 8.9, serine 8.9, glutamic acid 8.6, others 56.5. Howe et al. (1985) give details of 17 amino acids, and Johansson and Johansson (1955) indicate the functions of each, in insects and in invertebrates.

[3] Robinson and Nation (1970) give values for 15 carboxylic fatty acids, and Echigo et al. (1982) for 13.

[4] Rembold and Lackner (1978) give values for other samples of royal jelly.

adulteration of it can therefore be detected. A likely form of adulteration is dilution with honey or some other source of sugars. This is a harmless procedure, but a fraudulent one: weight for weight, royal jelly costs over a hundred times as much as honey to produce. In the USA, when some samples sold commercially were compared analytically with royal jelly harvested directly from hives, 6 of 11 were found to be adulterated (Howe et al., 1985).

The major acid of royal jelly is 10-hydroxy-2-decenoic acid. The acid fraction shows wide-spectrum activity against bacteria, although none against fungi (Lavie, 1960, 1968). The antibacterial activity may fulfil an essential role in the colony, by inhibiting microbial growth in royal jelly lying uneaten in the cells, and by protecting larvae from bacterial infection, but the fraction becomes inactive as soon as the larvae eat it. Vitamins of the B complex are present in royal jelly; vitamins A, C, D, E and K have usually been reported as absent; if any are present, the amounts must be very small.

Standards are proposed for royal jelly to be used as food. Standards for medical purposes should be more stringent, and the efficacy of the royal jelly must be demonstrated (Nakamura, 1985). One presumed property of royal jelly has been that it slows down the ageing process in human beings, perhaps on the grounds that queen honeybees live several years, whereas the life span of a worker is only several weeks (or over a cold winter, several months). It is an idea very difficult to substantiate or to refute. There would seem to be no *a priori* reason why a dietary component important in the development of reproductive individuals in social insects should increase life-span – or sexual vigour, another claim made for royal jelly – in human beings or mammals. Increased 'well-being' is quite often reported, but is also difficult to evaluate. Tamura (1985), reviewing Japanese work since 1974 from the standpoint of clinical pharmacology, includes reports of positive effects of royal jelly on anorexia, headache and fatigue, and – like bee venom – on untoward effects of irradiation, and also therapeutic (but not preventive) effects on sarcoma-180 and Ehrlich ascites tumours in mice. He also points out that royal jelly can have a mutagenic effect, and for this reason should not be taken in large amounts for a long period.

14.42 Harvesting and handling royal jelly

One factor underlying the growth of the royal jelly industry in the 1950s was the possibility of producing commercial quantities of it in areas where plant sources of nectar and pollen are insufficient for profitable honey production. Provided colonies are fed sugar syrup and pollen, they can be managed so that they produce royal jelly for the beekeeper.

On the other hand, royal jelly production involves labour-intensive procedures, carried out by skilled operators to a rigorous timetable. Procedures are those for the early stages of queen rearing (Section 8.22). Colonies are organized so that a large number of young female larvae are reared as queens instead of workers, in cells which nurse bees supply lavishly with royal jelly (Figure 14.42a). Then, the royal jelly is harvested (Figure 14.42b), and the larvae die instead of being reared to maturity as queens. Figure 14.42c shows an early apiary for royal jelly production in Cuba.

The following method is based on a Canadian one (M. V. Smith, 1959), with a 3-month production period. There is continuous daily operation, 1 man-hour of labour being required per day for every 2 to 3 hives. On average it should be possible to produce 100 g of jelly per day from 13–17 hives, or about 500 g per hive per season. In order to obtain the maximum production of royal jelly per hive, with minimal expenditure of time and labour, there is a regular schedule of grafting larvae which allows royal jelly to be harvested from large numbers of cells of the same known age.

A colony in use for royal jelly production over a prolonged period must be queenright; the queen is confined to the brood box by a queen excluder. A similar box containing honey combs (i.e. without brood) is placed immediately above the excluder. A third similar box filled with combs of honey, pollen and brood is placed above. As the season advances, boxes of empty honey combs are inserted immediately above the queen excluder, to provide more space.

Cells containing grafted larvae (see Section 8.22) are placed directly into the top brood box for 'starting' (acceptance and feeding by the bees). One outside comb of the top box is replaced by a division-board feeder, and syrup is fed when there is no nectar flow. Once every 7–10 days the colonies are checked, and the brood combs rearranged to ensure that a maximum number of young nurse bees are present to provision the cells. Frames of young worker brood from beneath the excluder are raised to the top brood box and placed adjacent to the frames carrying the cells. Colonies used for three months appeared to suffer no ill effects from this management, and their honey production was not reduced.

The greatest amount of royal jelly was obtained when the frames of queen cells were removed 3 days after grafting. On day 2 there was still not sufficient

Figure 14.42a Royal jelly production: nurse bees secreting royal jelly into artificial queen cells in the hive (photo: M. Matsuka).
A frame containing about 50 of the queen cells is shown in Figure 14.42b.

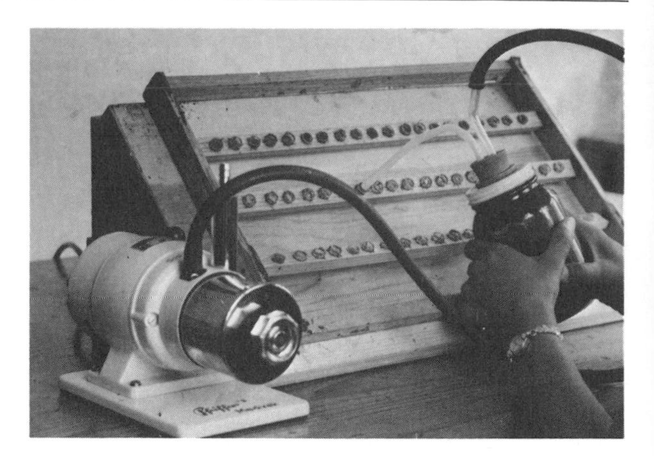

Figure 14.42b Use of a vacuum pump to transfer royal jelly from queen cells into a container, Miel Carlota, Mexico.
(The bars supporting the queen cups/cells have been swivelled through 90°.)

Figure 14.42c Dr Espina's apiary for producing royal jelly, Cuba, 1957.
Dr D. Espina (on right) and Dr Ordetx (left) were the authors of *Flora apícola tropical* (1983) and *Apicultura tropical* (1984).

royal jelly in the cells, and by day 4 the larvae were so large that they were consuming the royal jelly too rapidly:

	2 days	*3 days*	*4 days*
mean weight of royal jelly per cell (mg)	147	235	182
mean no. cells to produce 1 g royal jelly	6.79	4.25	5.49

A 3-day schedule is therefore followed:

day 1 – a frame of grafted cells is placed in the top brood box
day 2 – an additional frame of cells is inserted
day 3 – as day 2
day 4 – frame 1, now containing larvae 3 days old, is removed, the royal jelly collected, new larvae grafted in the cells, and the frame replaced in the colony.

Thus each colony is feeding three frames (40–45 cells per frame) at any one time, and one lot of cells is collected and replaced each day.

Each cell from which royal jelly is to be harvested is pared down with a sharp razor blade to just above the level of the jelly; the larva is then easily removed with a pair of forceps, and after the jelly has been collected new larvae can be rapidly grafted into the shallow cells.

Royal jelly is withdrawn from the cells by suction (Figure 14.42b) into a glass collection tube, whose lower end is closed by a cork cut to fit snugly inside it. The tube is emptied by using this cork as a piston, pushing it up from the bottom with a rod or plunger. Freshly collected jelly is strained through 100-mesh (4 per mm) nylon cloth into airtight vials, and tightly stoppered. It must be placed under refrigeration as soon as possible after collection. Ordinary refrigeration at 2° will preserve it for up to a year, although some of the organic acids tend to crystallize. If kept in a deep-freeze at −18° royal jelly shows little deterioration for periods up to several years. Fresh royal jelly which has been lyophilized (freeze-dried under vacuum) may be stored as a dry powder for years, with very little apparent loss of biological activity.

14.43 Royal jelly: world production and trade

Royal jelly production was already proving profitable in France in 1953, and by 1958 the country was producing 1.5 tonnes a year. Production was established in Cuba before 1957; see Figure 14.42c. In a survey of the industry in the early 1960s, Inoue and Inoue (1964) reported production in 12 countries: France and Japan each 1.5 tonnes; Canada 200–250 kg; Israel, Taiwan and Korea 50 kg or less; unspecified amounts in Czechoslovakia, German Federal Republic, Italy, USSR, USA and Yugoslavia. The price was from US$220 to 500 per kg (Kjaersgaard, 1985).

By 1984 the world price had dropped to US$70 per kg. China and Taiwan had become established as the main exporters. In 1984, 400 tonnes were produced by China (Fang Yue-zhen, 1984), 234 tonnes by Taiwan, and 12 by Thailand. Other countries known to produce 1 tonne or more annually are France, Italy, Japan (see below), Uruguay and Vietnam; little information has been found for Latin America. Much of the royal jelly produced is exported to Japan, and small amounts to Hong Kong, the German Federal Republic, Benelux, Italy and France. Importers and users of royal jelly in some European countries are listed by CCI (1987).

Eastern Asia is still the main centre of the world's royal jelly production, and Japan has shown spectacular growth in production, imports, and consumption; the following figures are in tonnes.

	production	*imports (mainly from China and Taiwan)*	
1972	9	24	(Sakai & Matsuka, 1982)
1978	23	99	(Sakai & Matsuka, 1982)
1982	18	100	(CCI, 1987)
1984	33	182	(CCI, 1987)
1985	46	210	(CCI, 1987)

14.5 BEE VENOM

When a bee stings, venom enters the puncture in the skin (Sections 2.23.1 and 5.52). The use of stinging bees has been a traditional folk remedy for rheumatic complaints, but venom cannot be administered in a very controlled way by this method. The 1930s–1950s saw the beginnings of an attempt to overcome this problem by making special preparations for medical use. Beck (1935) described the production of 'injectable bee venom' by manipulating individual bees, and he used the venom extensively in his medical practice. Commercial harvesting and handling of bee venom (Section 14.52) seems to have stemmed from a Czech paper by Marcovič and Molnár (1955) describing an electrical harvesting method. Broadman (1958) reviewed medical uses of the venom.

14.51 Composition and uses of bee venom

Some components of bee venom were identified and

studied in the 1950s (see Hodgson, 1955). Others have been isolated and examined since then, by more and more advanced analytical techniques (see e.g. Habermann, 1972). Some of the individual components are of considerable interest to pharmacologists, and more scientific research is probably now being done on bee venom and its components than on any other hive product. By the mid-1980s the journal *Apicultural Abstracts* was reporting over 50 scientific publications a year on the composition and components of worker honeybee venom, in addition to others on allergic responses to the venom and immunization against them.

Table 14.51A indicates the composition of venom from *Apis mellifera* workers, and Section 2.23.1 discusses physiological aspects of its production by bees. The composition of bee venom – or even of 'whole bee venom' – harvested from bees by different methods (Section 14.52) is not necessarily the same. By incorporating a cooling system into the standard collecting apparatus, the liquid fraction of the venom is prevented from evaporating, and is present in the material harvested (Gunnison, 1966). Pence (1981) makes further distinctions, between:

1. freshly discharged liquid from the sting lancet tip
2. venom from the sac
3. as 1, but collected under water so that the volatile fraction is not lost
4. as 2, but collected under water.

He established that the volatile fraction is produced in the alkaline gland, and that 3 and 4 are more potent

Table 14.51A Composition of venom from *Apis mellifera* workers

Data are for commercial freeze-dried venom, from Shipolini (1984) who cites the references in the final column. Constituents marked * are mentioned in the text here and/or in Section 5.5. Glucose, fructose and phospholipids are also present, at similar concentrations to those in the blood (Shipolini, 1986). Natural venom from the venom sac contains 88% water.

	Percentage in dry venom	*Amount (nmol) in venom from one sting*	*Reference*
Enzymes	15–17		
α-glucosidase	0.6		Shkenderov et al., 1979
acid phosphomonoesterase	1.0		
*hyaluronidase	1.5–2.0	0.03	Habermann & Neumann, 1957
lysophospholipase	1.0	0.03	Ivanova & Shkenderov, 1982
*phospholipase A_2	10–12	0.23	Habermann & Neumann, 1957
Small proteins and peptides	48–58		
adolapin	1.0	0.06	Shkenderov, 1982
*protease inhibitor	0.8	0.07	Shkenderov, 1973
*melittin	40–50	10–12	Neumann et al., 1952
secapin	0.5	0.13	Gauldie et al., 1976
*MCD or 401 peptide	2	0.6	Fredholm, 1966
tertiapin	0.1	0.03	Vernon, 1976
'melittin F'	0.01	0.003	
*apamine	3	0.75	Habermann & Reiz, 1965
*procamine A, B	1.4	2	Nelson & O'Connor, 1968†
Physiologically active amines	3		
*histamine	0.6–1.6	5–10	Reinert, 1936
dopamine	0.13–1.0	2.7–5.5	Owen, 1971
norepinephrine	0.1–0.7	0.9–4.5	Owen, 1982
Amino acids	0.8–1.0		Nelson & O'Connor, 1968†
γ-aminobutyric acid	0.4		
β-aminoisobutyric acid	0.02		

† Venom was obtained by electrical stimulation of single bees under laboratory conditions.

Table 14.51B Comparisons between worker venoms from *Apis* species

Characteristic	*Sources of venoms and differences between them*
toxicity	Similar in *A. mellifera* and *A. dorsata*; less in *A. florea*; twice as high in *A. cerana* as in *A. mellifera* (Benton & Morse, 1968).
protein pattern	All species show differences; within one species, *A. mellifera* from Europe and tropical Africa are similar (Mello, 1970); *A. cerana* from India and Thailand are similar to each other, but that from the Philippines is somewhat different (Benton & Morse, 1968).
melittin	*A. mellifera* and *A. cerana* appear identical; *A. dorsata* and *A. florea* both show differences, those between *A. florea* and *A. dorsata* being greater than those between *A. dorsata* and the first two species (Kreil, 1975).

than 1 and 2 in producing neuromuscular responses.

Most of the results quoted here refer to 1, as obtained by electrical stimulation (or to 1 after evaporation to dryness); 2 has been studied by O'Connor and Peck (1978). It differs in several respects – for instance in protein composition – from venom obtained by electrical stimulation (Hsiang & Elliott, 1975). Papers by Shipolini (1984), Banks and Shipolini (1986), and Dotimas and Hider (1987) give much further information, and further references.

Queen venom differs in composition, and also in its age-related pattern of change, from that of workers (Marz et al., 1981; Owen and Bridges, 1976; Owen et al., 1977; Owen, 1978*b*; H. Inoue, 1984). Venoms of different honeybee species also show differences, and Table 14.51B summarizes results obtained so far. The most detailed analysis of *Apis cerana* venom is by H. Inoue and Nakajima (1985).

Most studies have been made on venom of *Apis mellifera* workers. Some of its components have notable pharmacological properties, and administration of carefully controlled amounts can give beneficial results; see Kreil (1978). Experiments with animals have indicated anti-arthritic properties of bee venom, in which several components of the venom have been implicated (Dotimas & Hider, 1987). Vick and Brooks (1972) found that both melittins and apamine produced an increased plasma cortisol in dogs, and believe that they are probably responsible for many of the reported beneficial effects of bee venom in alleviating arthritic conditions. Bodnaryk (1978) has examined melittin, MCD peptide, apamine and other peptides. Dotimas and Hider (1987) refer to uses of components of bee venom in molecular biology.

Effects of a bee sting, and the part played by individual components of the venom in causing them, are discussed in Sections 5.52 and 5.54. There is a long-held popular belief that bee stings can relieve certain rheumatic complaints. Broadman published *Bee venom: the natural curative for arthritis and rheumatism* in 1962, and groups of beekeepers and general practitioners in different countries have made special studies of the applications of bee stings and of the injection of bee venom. Section 14.1 lists proceedings of conferences on apitherapy held by North American and international groups.

One belief that seems to have been disproved is that stings give beekeepers a certain immunity against cancer. McDonald et al. (1979) analysed the cause of death (as stated on death certificates) of 580 'occupationally exposed' beekeepers who died in the USA between 1949 and 1978. Results showed neither adverse nor beneficial effects of intense exposure to bee stings. The number of deaths from lung cancer was in fact lower (at a moderate significance, $P < 0.05$) among male beekeepers than among the general male population, but the authors suggest that this may have resulted from less cigarette smoking by (rural) beekeepers. There were only 60 female beekeepers in the survey, too few to yield statistical data.

14.52 Harvesting and handling bee venom

The venom sac of a worker honeybee (*A. mellifera*) becomes full when she is about 2 weeks old. Commercial production of bee venom, and of pharmaceuticals from it, were pioneered by the firm Mack in 1930, at Illertissen in south Germany; their bee house is shown in Figure 6.63a. They kept large apiaries, and also purchased bees from skep beekeepers in the Lüneburg region of north Germany after the flow from heather (*Calluna vulgaris*). This was the end of the beekeeping season, and the bees would otherwise have been destroyed.

I am indebted to Dipl.-Ing. Hermann Forster (1985) for the following information. In the first method, used between 1930 and 1937, girls were stationed in front of hives; each picked up worker bees one at a time, and pressed the bee so that she stung into a fabric tissue. This absorbed the venom, which was extracted with a solvent such as distilled water. After removing the solvent by freeze-drying, the venom was left as a crystalline powder. Mack's second method (1936–1974), being used when I visited the firm in 1951, applied a mild electric shock to previously moistened bees. These were in a shallow box (50 × 40 cm) which had a wooden lid covered on the underside with absorbent filter paper and a thin protective rubber sheet. The bottom of the box was similarly covered, and alternately charged electric wires were laid 2.5 cm apart on it. The lid was closed gently so that the bees were in a single layer; this produced a slight pressure which, together with the electric shock, made them sting through the rubber. They were unable to retract their stings, and subsequently died; the venom was absorbed by the filter paper and extracted as above.

Mack's third method is somewhat similar to one I saw in operation in 1960, at the Czechoslovak State Bee Venom Apiary near Radošina. A bare wire was stretched to and fro across a membrane mounted on a horizontal frame which could be placed directly in front of a hive entrance. Benton et al. (1963) describe the set-up, which is shown in Figure 14.52a. When a low voltage is applied to the ends of the wire, a few guard bees are shocked; they sting, and also quickly alert others to sting by releasing the alarm pheromone isopentyl acetate. Gałuszka (1970) in Poland made a detailed behavioural study of individual bees in a colony after they were 'milked' for venom. In autumn almost all bees were alerted, but at other seasons only guard bees.

The membrane through which the bees sting has to be very thin (e.g. of nylon taffeta), so that they can withdraw their stings. The drops of venom released remain on the underside of the membrane and are removed by the operator; in hot weather they dry and can be scraped off. Since 1974 Mack have used a similar method, but one which dispensed with a membrane altogether: bees are forced to walk over a glass plate to which alternatively charged wires are attached. The electric current stimulates a bee to protract her sting (which she can retract) and to release about 0.02 mg venom; this dries rapidly and is scraped off the plate with a razor blade.

In Gałuszka's experiments (1972) a bee stung the membrane 10 times on average during a 15-minute venom collection period. About 0.011 mg of dried venom was produced altogether per bee, corresponding to about 0.1 mg of whole venom, a figure also quoted by Morse and Benton (1964*a*). On the basis of 0.3 mg in the venom sac when it is full (Section 3.42),

Figure 14.52a Collection of bee venom (Morse & Benton, 1964*a*).
above The board is the full width of the hive, wired ready for placing in front of the entrance.
below The board covered by bees after current has been passed through the wires and stimulated bees to sting.

the bee would have used one-third of her venom in the 10 stingings, and she would thus be capable of 30 stingings in all provided she could withdraw her sting on each occasion. Gałuszka (1972) found that the most efficient programme for collecting venom from a colony was a series of impulses lasting 15 minutes, administered 3 times at 3-day intervals, and repeated after 2–3 weeks. By that time there would be a new generation of bees with developed venom glands. Venom yields were highest between May and September in Poland.

In Bulgarian experiments on 8 colonies (Mitev, 1971), milking was done every 3 days by two treatments lasting 45 minutes each, with an hour between. The colonies used produced 14% less honey than control colonies. Morse and Benton (1964*b*) reported that honey production was not affected, but give no figures; they say that colonies with at least 50 000 bees should be used. They also (1967) describe similar venom collection from *Apis cerana*.

Milking colonies of bees for their venom by electrical stimulation is not to be undertaken lightly. The operation is not a pleasant one for the operator, or for people in the vicinity of the apiary, since large numbers of bees are alerted by alarm pheromone released at the site of stinging, and may then sting without further provocation.

A more refined harvesting method for small amounts of venom was developed in the UK (Palmer, 1961) and patented in the USA (no. 3,163,871, 1965). If light is let into the top of the hive, the young house bees retreat from it, but older sting-ready bees surge upwards in defence. In Palmer's method these latter bees were collected in a magazine at the top of a hive by using a shutter. While still in the magazine, an electric current stimulated them to sting through a fine mesh and a sheet of silicone rubber 0.005 inches (0.13 mm) thick. The bees consumed much honey afterwards if it was provided, and were then safely returned to the hive. No bees died as a result of the treatment.

Bee venom produced commercially may be delivered direct to the purchaser without further treatment.

14.53 Bee venom: world production and trade

Bee venom is a specialized product that does not enter the world market, and the amount produced is not known. It is used in a number of countries, and probably also produced in them. From reports on the use of bee venom since 1973, and other information, they include:

- Europe: Austria, Bulgaria, Czechoslovakia, France, German Federal Republic, Italy, Poland, Romania, Spain, UK, USSR, Yugoslavia
- Americas: Brazil, Canada, USA
- Asia and Africa: China, Israel, Japan, Taiwan, Egypt.

Most bee venom is used for pharmaceutical preparations in the treatment of arthritis, etc., and for desensitization of patients allergic to bee stings.

14.6 BEE BROOD

Both immature and adult insects are a valued form of 'meat' in many non-European cultures (Bodenheimer, 1951), and a wild nest of honeybees can yield larvae and pupae, which are less painful to harvest than adults since they do not sting.

14.61 Composition of bee brood

Table 14.61A compares the composition of mature larvae and pupae of the honeybee *Apis mellifera* with that of silkworm pupae, and of several common foods. Bodenheimer (1951) quoted analyses of other commonly eaten insects including locusts and termites, mostly adult specimens. Analysis of adult honeybees (Ryan et al., 1983) gave crude protein 49.8%, total lipids 7.54% and reducing sugars 27.1% (dry weight). These were end-of-season Canadian bees; see below.

14.62 Harvesting and handling bee brood

As part of the drive made between 1950 and 1970 to diversify hive products – and also to utilize a wasted resource – Hocking and Matsumura (1960) developed a method of harvesting mature honeybee larvae at the end of the active beekeeping season in western Canada, from colonies that were in any case to be killed because it would not be practicable to overwinter them (end of Section 8.35).

Brood cells freed from adult bees are uncapped with a thin, cold serrated knife, and the brood is then removed from each frame by impact, so that it falls into a tray below. Figure 14.62a shows the uncapping process, and the brood harvested, in Japan. Another Japanese extraction method flushes the brood out of the cells with a jet of water (Gary, 1961). Where brood is reared specifically for use as a food, drone-size foundation is used for the combs, since the larvae and pupae in them are larger.

Table 14.61A Composition of honeybee brood (*Apis mellifera*) in comparison with some other foods

The number of determinations for honeybee brood is in brackets (from Hocking & Matsumura, 1960, who also quote the vitamin contents, and the values for other foods, from national Canadian figures). Values for silkworm pupae (*Bombyx mori*, 8 days old) are calculated from figures quoted by Bodenheimer (1951).

	Honeybee		*Silkworm*				
	mature larvae	*pupae*	*pupae*	*Beef*	*Milk*	*Egg yolk*	*Cod liver oil*
			Percentage of fresh weight				
water	77.0 (2)	70.2 (2)	76.7	74.1	87.0	49.4	
ash	3.02 (3)	2.17 (3)	1.21	1.1			
protein	15.4 (4)	18.2 (4)	13.4	22.6	3.5	16.3	
fat	3.71 (4)	2.39 (3)	5.83	2.8	3.9	31.9	
glycogen	0.41 (2)	0.75 (2)		0.1–0.7	<4.9		
			International Units per g fresh wt				
vitamin A	119	49.3					
	89			0	1.6	32.1	1000–6000
	112	53.3					
vitamin D	7030						
	6130	5070			0.41*	2.6	100–600
	7430	5260					

* fortified

14.63 Bee brood: world production, trade and uses

From a nutritional standpoint, bee brood should not be too despised as a food. A reporter who described eating fried bees in a Toyko restaurant in 1956 (*Saturday Evening Post* for 10 March) said that country children in Japan made a game of finding wild bees' nests; families picked the young bees [brood] from the combs, fried them and stored them in crocks. He thought that this 'strange preserve' helped to satisfy the protein hunger then common among the poorer Japanese. Japan is now the world's centre for commercial production of bee brood as a hive product, although eating it is no longer of necessity, but by choice. As a delicacy for human consumption, the brood may be boiled and canned – for instance in soy sauce. The Canadian experiments referred to above used a tasting panel which ranked fried or baked pupae highest, then smoked, brandied, and pickled, in that order. Brood can be preserved by drying, freezing or canning.

Japanese research workers have pioneered the use of powdered drone brood for rearing insects used in biological pest control, such as ladybirds (Coccinellidae). It is not always adequate on its own, but one of the few papers in English (Niijima, 1979) reports rearing 11 successive generations of the coccinellid *Eocaria muiri* on drone brood, without any loss of viability and fecundity.

Brood or adult bees have been used with some success as a replacement for some of the normal feed for pigs (Dietz et al., 1976), small mammals and birds (Witherell, 1975). It is more difficult to find outlets for adult bees than for brood, although large quantities of them are available at the end of the beekeeping season in western Canada (Section 8.35). Such adult bees,

Figure 14.62a Production of brood as a hive product in Japan (photo: M. Matsuka).
above Uncapping a comb of drone brood; harvested brood is in the tray on the ground.
below Harvested drone brood, showing mature larvae (white) and pupae of various ages, the darker being the oldest.

killed by sulphur dioxide, gave less satisfactory growth in young turkeys (when fed dried and ground in their food) than the turkeys' normal diet (Salmon & Szabo, 1981).

International trade in bee brood is quite small. China and Taiwan may be producers, exporting for instance to Japan. Japan itself has exported brood packed in cans for human consumption, and also powdered drone brood for testing as a protein source in France.

14.7 WHOLE-FOOD HARVESTING FROM HONEYBEE COLONIES

In affluent societies, hive products discussed in the present chapter are sold separately at relatively high prices. Their high cost is due to both the labour and equipment used in their production and harvesting, and to elaborate packaging and promotion. Honey consumers accustomed to a well strained product are inclined to regard any other hive product present as a contaminant, rather than as a nutritional enrichment, of the honey.

In developing countries of the tropics many colonies of bees nest in the wild, and the nest contents (honey, pollen, royal jelly, bee brood), with beeswax comb and some propolis, are harvested together – at no cost except the bee stings suffered. The dietary value of the mixture is, however, not less than that of its constituent parts, and in my view peoples living at the margin of survival in the tropics should be encouraged to harvest and eat it. I found this whole-food harvesting in operation in the mid-1980s in both the Nepal Himalayas and Southern Africa, and it doubtless occurs in many other areas.

The bees are driven out of the nest with smoke. If the nest is in a man-made cavity (e.g. *Apis cerana* kept by Gurungs in Nepal), a few combs with some brood are left so that the colony can regenerate. Some brood comb is often left in nests of *Apis dorsata* (e.g. in Nepal and Malaysia). (If the whole colony is destroyed, another swarm will probably occupy the nest site later.) In Nepal the honeycomb taken is set aside to strain, as honey can be stored for future use. The rest of the comb contents will not keep, so must be eaten fresh. Gurung families and neighbours gather together to enjoy it as a special treat: beeswax comb with pollen, propolis, royal jelly and brood. In the New World, similar whole-nest utilization of nests of stingless bees is made for example by the Yukpa-Yuko Indians of Venezuela and Colombia (Ruddle, 1973). But they kill the brood before eating it, by passing the combs momentarily through a flame.

14.8 FURTHER READING AND REFERENCE

Details of publications listed will be found in the Bibliography.

Hive products, general

CCI (1987) *Note sur les marchés pour les produits sélectionnés de la ruche au Royaume-Uni, en France et en Italie*

Chauvin, R. (1968*d*) *Traité de biologie de l'abeille*, Vol. 3

Lavie, P. (1960) *Les substances antibactériennes dans la colonie d'abeilles* (also unpublished English translation)

Pollen

Caillas, A. (1976) *Le pollen: sa récolte, ses propriétés et ses usages*

Dany, B. (1975) *Pollensammeln heute*

McCormick, M. (1960) *The golden pollen*

Stanley, R. G.; Linskens, H. F. (1974) *Pollen: biology, biochemistry, management*

Propolis

Apimondia (1978) *A remarkable hive product: propolis*

Asis, M. (1979) *El propóleo, un valioso producto apícola*

Ghisalberti, E. L. (1979) *Propolis: a review*

Walker, P. (1976) *Annotated bibliography of propolis*

Walker, P.; Crane, E. (1987) *Constituents of propolis*

Royal jelly

Caillas, A. (1960) *Manuel pratique du producteur de gelée royale*

Johansson, T. S. K.; Johansson, M. P. (1955, 1958) *Royal jelly*

Smith, M. V. (1959) *The production of royal jelly*

Bee venom

Banks, B. E. C.; Shipolini, R. A. (1986) *Chemistry and pharmacology of honey-bee venom*

O'Connor, R.; Peck, M. L. (1978) *Venoms of Apidae*

Shipolini, R. A. (1984) *Biochemistry of bee venom*

Bee brood

Hocking, B.; Matsumura, F. (1960) *Bee brood as food*

PART VI

Beekeepers

15

Law in relation to bees and beekeeping

15.1 INTRODUCTION

This chapter is concerned with the many ways in which the law impinges on a person because he keeps bees, or because he sells hive products. The system of law applicable depends on the country, and sometimes the immediate locality, where he operates. If he wishes to export almost anything used or produced in beekeeping, and especially bees themselves, he will need to conform also with the law of the country of import. Such law is less for economic protection than to prevent the introduction of new bee diseases or pests into the importing country.

This book is concerned with the present and future rather than the past, but some of the laws that relate to the beekeeper's current rights and liabilities are rooted in the distant past, when legal systems were first developed. Many readers are likely to live in countries where the legal system belongs to one of the following families.

(a) Common law, in which judgements in individual cases are used as precedents for later judgements. Much of the civil law of England, and of countries colonized from Britain, is of this type.

(b) The Romano-Germanic family of codified law, which was formed in universities of Latin and Germanic countries in continental Europe, on the basis of Roman civil law. This system is used in much of continental Europe, and in countries colonized from there, especially ex-colonies of Belgium, France, the Netherlands, Portugal and Spain – in Latin America, Africa and Asia. It also influenced Scots law, which developed in close contact with the scholarship in continental Europe, especially through the alliance between Scotland and France.

Other important families of law are Islamic, Hindu, and those of China, Japan and other countries of eastern Asia and of Africa south of the Sahara. Books by Zweigert and Kötz (1977) and David and Brierley (1978) give details.

Legal beekeeping texts surviving from ancient and mediaeval times indicate that considerable importance was attached to beekeeping. The law was invoked to protect various rights of the beekeeper (Section 15.2) and to protect the public in the case of injury from his bees (15.3).

Sections 15.3 to 15.7 are concerned largely with beekeeping legislation passed in recent times (much of it since 1950), which represents a minute part of the vast amount of modern legislation on agriculture and commerce. Much of the legislation is concerned with disease control, and Table 15.4A lists 96 countries in which there are known to be specific laws and regulations that relate to beekeeping; for example on the control of bee diseases and pests (Section 15.4), and on the protection of bees – because of their value as crop pollinators – from damage by pesticides (15.5). Section 15.6 discusses other aspects of law that are important in modern beekeeping, from road accidents to contracts when bees are hired for pollination. Finally, Section 15.7 considers legal protection of the consumer purchasing honey or other hive products, together with national standards and regional ones such as those of the European Community.

There are not many comprehensive books on law in relation to beekeeping. The USA, where many laws vary from state to state, is covered by a book published by the American Honey Producers' League (1924), and *Bees and the law* (Loring, 1981; see also Michael, 1980). A useful article by Doutt (1959) includes a summary of cases. There are books for Norway (1956), Switzerland (Rüedi, 1955), the German Federal Republic (Gercke, 1985), and the German Democratic Republic (Pech, 1958; see also Borchert, 1974). In Spain, the First National Congress on Apiculture (Marquina Olmedo & Sáenz Barrio, 1984) included detailed papers on Spanish laws. The history of law in relation to beekeeping is discussed by Moulis (1908) for France, and by Bessler (1886) and Glock (1891) for German-speaking countries. Ancient laws in Ireland and Wales, which dealt very fully with certain aspects of beekeeping, have been explored by Charles-Edwards and Kelly (1983) and those of Wales also by Crane and Walker (1985*b*).

In the 1980s, more and more things that a beekeeper does, or wishes to do, have become subject to the law in one way or another. Beekeepers must therefore be alert to changes in the law that might affect them. By being alert to *proposed* changes, they may succeed in preventing amendments that seem unreasonable to them. A single 1985 issue of the monthly *American Bee Journal* included five articles concerned with actual or proposed changes in the law affecting beekeepers somewhere in the USA. Three were on honey labelling and one was on disease regulations; the fifth (Walsh, 1985) concerned a proposed prohibition on town beekeeping, a subject discussed in Section 15.31.

15.2 LAW IN RELATION TO BEEKEEPERS' TRADITIONAL RIGHTS

Among beekeepers, as among any section of the community, a few individuals are especially prone to enter into litigation. A painting used to ornament a hive in 1864 (Figure 15.2a) provides a cynical comment on the possible outcome.

15.21 Ownership of swarms

In the past, more has probably been written about the law in relation to swarm ownership than about any other legal aspect of beekeeping. This is a reflection of the very great importance of swarms in traditional bee management. Catching swarms was the sole way of obtaining new colonies to populate empty hives, and swarms had a relatively high monetary value. (Only since the use of movable-comb hives have beekeepers been able to produce new colonies at will – and also to control swarming.) Arguments could easily arise as to ownership of the bees in a swarm, which depended on the identity of the hive they came from – sometimes not easy to establish – or where they settled.

Law on swarm ownership had many common

Figure 15.2a Traditional painted wooden hive-front from northern Yugoslavia. This 1864 design symbolizes the unsatisfactory outcome of litigation: two farmers are quarrelling over ownership of a cow; the person who benefits is neither of the farmers, but the lawyer – who laughs as he milks the cow.

factors, whatever the origin of the law, probably because it had to be related to both the biology of colony reproduction in honeybees and the capabilities and limitations of beekeepers in dealing with swarms. In Roman civil law, bees were wild animals, *animalia libera* or *ferae naturae*; they were not *mansuetae naturae*, i.e. tamed or domesticated wild animals. Bees *in a hive* were nevertheless the property of the owner of the hive, and their unauthorized removal was theft. Bees that issued from a hive as a swarm were deemed to have reverted to their original wild state. Gaius (AD 130–180) included an exposition of the subject in the Second Commentary of his *Institutiones* (Sections 67 and 68); it is quoted by Swan (1956).

Justinian, who was Roman Emperor from AD 527 to 565, codified the earlier laws, and in many countries whose law belongs to the Romano-Germanic family, Gaius's exposition has continued to be the basis of law on swarm ownership. A beekeeper is still allowed to go on to a neighbour's land to collect a swarm that flew from one of his hives – the swarm still belonging to him – although he must make good any damage done to the neighbour's property. Frimston (1966) cites as examples of such countries Austria, Bulgaria, Czechoslovakia, Finland, France, Germany, Italy, Portugal, Sweden and Switzerland, 'now or in the past. ... In Spain, Argentina and Chile the right has been limited to uncultivated and unhedged land.' Brazil has a detailed law (Apiário, 1977). The 1734 Swedish law, still valid, is shown in Figure 15.21a; the first sentence reads: 'Should bees fly to another man's forest and the owner follows them to tree and hole [which they occupy], marks the tree and notifies the village council, no one can deny him his rights.'

Sweriges Rikes
LAG

Gillad och antagen på Riksdagen
Åhr 1734.

XXI. Cap.
Om Bi.
1. §.

Flyga bi bort i annars skog, och följer ägaren them til stock och hol, märker samma trä, och gifwer thet byamän tilkänna; hafwe ingen wäld honom them förtaga. Hafwer biswerm satt sig i bärande och fridlyst trä; tå skal then stockas, och ej trä huggas, eller spillas, wid bot, som i 13. Cap. är sagdt. Är then i annat trä; hugge neder, och tage bi sin saklöst.

2. §. Hittar man bi å egen bolstad, eller then han äger lott i; ware hans, som hitte. Är thet landbo; niute han tridiung, och jordägaren twå lotter. Hittar man them inom annars hägnad; niute ingen lott theraf: hittar man them utom hägnad, i annars skog och mark; äge tridiung, och jordägaren twå lotter. Säga twänne sig samma bi hittat; niute han thenna hittelön, som först lyste. Om then, som å annars äger hittar, och ej lyser, utan borttager, och om then, som med mat och bete til sig lockar annars mans bi, urskils i Mißgiernings Balken.

Figure 15.21a Swedish law of 1734 on the ownership of swarms (see text), which is still effective.

Figure 15.21b Tanging a swarm. Part of an engraving by Hans Bol, 1534–93 (G. Hirth, *Kulturgeschichtliches Bilderbuch aus drei Jahrhunderten*, 1882).

The practice of 'tanging' a swarm, i.e. making a noise by striking two metal objects together while following the swarm in flight, has sometimes been believed to make the bees settle (Figure 15.21b). But it may well have originated in the need for the beekeeper to make his claim to the swarm apparent to all around.

In England, judgment similar to that in Justinian law was given as late as 1935, in a County Court case, Hallett v. Jones; the defendant conceded that his beekeeper neighbour was entitled in law to follow his swarm on to neighbouring land (Frimston, 1966). However, in 1938 a judgment in Hull County Court (Kearry v. Pattinson) made it clear that there was no such entitlement, and the Court of Appeal upheld the Judge's decision that the beekeeper had no special right to go on to his neighbour's land (*Law Reports, King's Bench* 1: 471–482, 1939; Bee World, 1939). In 1949 a Paris Court heard the appeal of a beekeeper against a sentence of two days' imprisonment and a fine of 5000 francs for the theft of a swarm; civil damages of 4000 francs had also been awarded to the swarm's owner. The case was reported in great and vivid detail (Apiculteur, 1950); in the end the Court upheld the verdict and the damages, but quashed the imprisonment and reduced the fine.

It is astonishing that these cases occurred nearly a century after the development of the movable-frame hive, but they may well mark the end of an era when a high value was placed upon swarms. Beekeepers' property considered worthy of litigation nowadays consists of other things than swarms – for instance colonies of bees killed by insecticides. Nevertheless many beekeepers cannot resist collecting a swarm available to them (Figure 15.21c).

Figure 15.21c David Smith, QC, Secretary of IBRA, collecting a stray swarm in Middle Temple, London; he is using a legal brief to move outlying bees towards a waste-paper basket where most have clustered, with the queen (photo: Brian Clarke).

15.22 Ownership of hives and bees

Ownership of a hive, and thus of the bees that are inside it, is easier to establish than ownership of swarms, and law relating to hives and bees is relatively straightforward.

Theft of hives and bees is the subject of the earliest known law concerned with beekeeping (Figure 15.22a). This is part of the Hittite Law Code which dates from 1500 BC or earlier (Crane & Graham, 1985). It set the fine for stealing an empty hive or a 'swarm'; from the context this probably meant a hive with bees in it. The Hittites lived in the steppes of Central Anatolia, and the hives are most likely to have been cylinders made of mud.

Complexities arose, and still arise, in countries where tree beekeeping is practised, i.e. where beekeepers own and tend colonics of bees nesting in trees, which may grow on someone else's land. For the past situation in continental Europe, some details are given by Galton (1971) and Crane (1983*a*), and for Ireland and Wales by references cited in Section 15.1. The situation in an African country (Tanzania) is described by Crosse-Upcott (1956). Ownership marks on trees used to be notched with an axe; modern hives are usually marked by branding.

In many countries hives are still stolen from time to time, and combs or bees – including queens – are also taken from them. This latter type of theft is greatly disliked by beekeepers; they know that the thief must himself be a beekeeper, and quite often are fairly sure of his identity. Legal action is usually difficult, because it is impracticable to establish the original ownership of bees, combs or frames.

15.23 Other legal rights of beekeepers

In the past, ordinances and other documents have set out the status of beekeepers in the community, rather than their legal rights. One right that most beekeepers would claim without question is that his bees should forage on neighbours' land, and he would also assume that honey produced by the bees became his own property. This right was, however, questioned in theory in Ancient Rome, on the grounds that the bees were injuring the neighbours' flowers by foraging on them (Swan, 1956). This concept is no longer current,

Figure 15.22a Copy of lines 32–36 of the Hittite Law Code, 1500 BC or earlier, containing the earliest known beekeeping laws (from Museum of Anatolian Civilizations, Ankara); see text.

but it gave rise to a number of court actions in earlier years; in *The case of the trespassing bees* Doutt (1959) examined cases in the USA.

15.3 LAW IN RELATION TO BEEKEEPERS' LIABILITY

15.31 Codified law: restrictions on placing hives

Stinging, or fear of stinging, is the most usual cause of legal action brought against a beekeeper. The likelihood of bees stinging a member of the public depends greatly on where the hives are kept, how well screened they are, and how and when the hives are manipulated. In some countries the places where hives may be situated are restricted under Codified law; for instance beekeeping in towns, or within a certain distance of a public highway, may be prohibited. The law in force in Denmark, which dates from 1793, prohibits the use of hives within 6.5 m of a public road (Bang & Jørgensen, 1983).

There are other reasons why restrictions are placed on certain sites for hives: prevention of the spread of disease, for instance, or of unwarranted competition for bee forage with colonies already established. In one area of New Zealand there are restrictions where the shrub tutu (*Coriaria* spp.) may sometimes yield toxic honeydew.

Laws governing the siting of hives are nothing new. In Ancient Greece one of the laws codified by Solon (*c.* 600 BC) decreed that a man who was establishing colonies of bees must not do so within 300 feet [100 m] of hives already established by another beekeeper, on the grounds that the new bees would compete with the others for forage. Such laws still exist; in Spain the distance is currently 500 m, and in one state of the USA it is 2 miles [3.2 km]. A 'Beekeepers' code of ethics' on territorial rights has been drafted in New Zealand (B. White, 1987).

Where there is no legislation, a modern beekeeper could perhaps bring an action against another beekeeper who placed hives close to his own apiary, on the grounds of a resultant increase in competition for bee forage. If the introduced bees are diseased, their being brought into the area may be a breach of criminal law, whether or not damage is caused by it.

The placing of hives is subject to legislation in 24 countries marked S in column 2 of Table 15.4A, and possibly in others. In the USA it comes under zoning regulations (Loring, 1981) and varies in different areas. The City Council of Glen Falls, NY, introduced a ban on beekeeping in the city in 1983, but this was amended as a result of well organized representations by beekeepers (Walsh, 1985).

South Africa is one of the countries where the laws are explicit (Anderson et al., 1983); in 'peri-urban' areas:

> Hives must be kept at least 5 m from the boundary of the stand [apiary area] and at least 10 m from any public place or building used by people. The hives must be hidden from view by a wall, fence or hedge of at least 2 m, permitting the bees to forage over this barrier only. The barrier must leave a clearance of 5 m around the hives and must be at least 1.5 m from the boundary of the stand. Hives must be kept in the shade and bees in some areas must be provided with water within 5 m of the hives ... The regulations are not applicable to bees kept further than 150 m from the boundary of the property and from any public place.

Municipal South African by-laws usually prohibit beekeeping in urban areas, but a temporary permit can often be obtained to keep swarms caught in the town, until they are transferred to an apiary.

National or local laws prohibit the keeping of bees in towns in some other countries, for instance Dutch law (from which South African law is largely derived). Earlier this century Dutch beekeepers responded to such law by establishing 'bee parks', which now exist near twenty towns (Beetsma, 1977). Urban beekeeping families own or rent a plot of land in their local bee park, where they have their hives of bees, a summerhouse, and an allotment or garden – which must conform with the general landscaping plan of the park. The bee park may also have a communal landscaped area, club house and exhibition space. An Amsterdam Bee Park was established in 1923, and the present site is at Osdorp. The park had to be moved to a new site several times, as the city expanded (Kroes, 1951); Figure 15.31a shows the site purchased in 1936.

One fairly common practice among city beekeepers is to keep a few hives on the flat roof of a building. The chance of their causing a nuisance is minimal, and indeed few people know that the hives are there.

15.32 Common law: bees as a source of nuisance

In English law, a beekeeper may be liable for the actions of his bees if they injure people or property; in this context injury includes actual damage, loss of benefits, and loss of normal enjoyment of one's property. Most of the relevant cases have been brought under the legal wrong of 'nuisance', which is the use or occupation of land so as to cause damage to, or

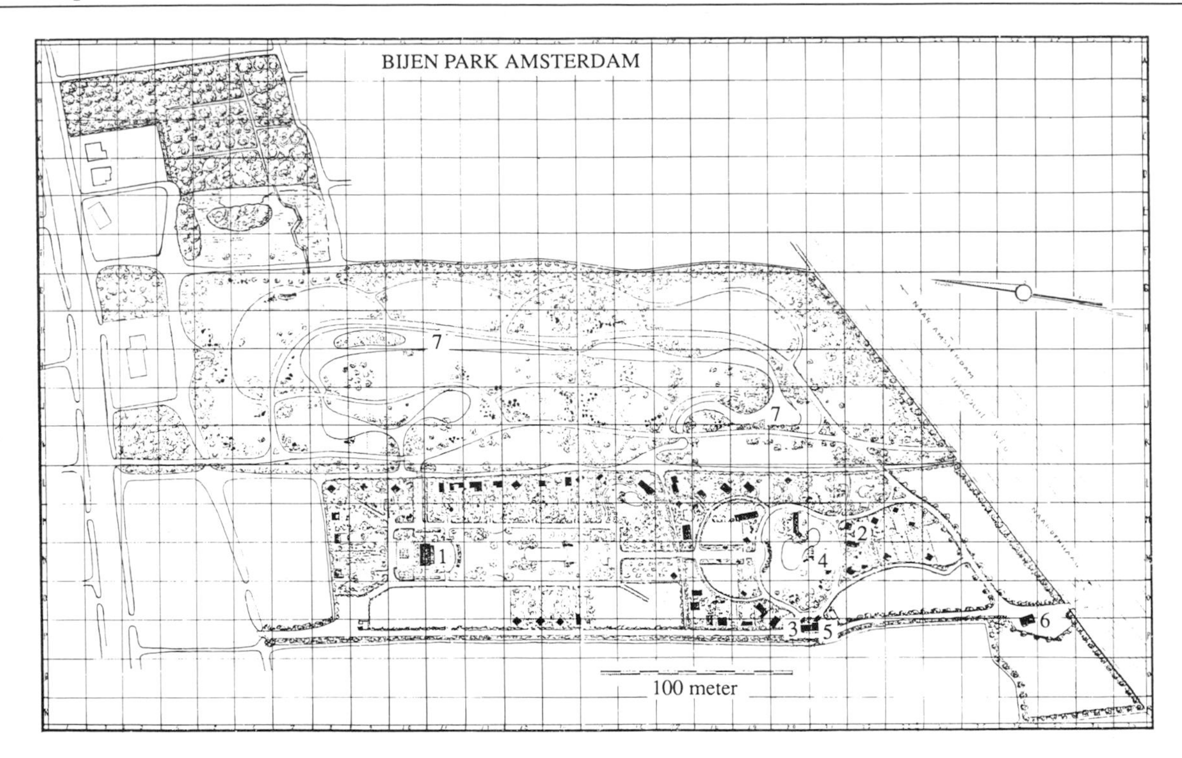

Figure 15.31a Ten-hectare Amsterdam Bee Park, 1938 (design: J. J. Bosma). The individual plots are in the lower part of the plan, and general facilities include:
1 beekeeping museum
2 general apiary
3 sugar store
4 observation hive
5, 6 worksheds
7 lake

interfere with, another person's use or occupation of his own land. An action for nuisance may be brought when bees are kept in such a way that a neighbour is badly stung, or is unable to lead a normal life because of the bees. The beekeeper may be liable in nuisance whether or not he has been negligent.

A case much referred to by those making judgments under English law is O'Gorman v. O'Gorman, which relates to events that took place in September 1900 at Gurtnalougha, Co. Clare, now in the Irish Republic; Swan (1956) gives details. A case in England, reported in *The Times* for 17 December 1906, involved two neighbours, Parker and Reynolds, who were both beekeepers. The Court decided that Mr Reynolds's bees substantially interfered with the reasonable comfort of Mr Parker, his wife, children and servants, thus constituting a nuisance.

The difference in England between a claim for nuisance and one for the legal wrong of negligence was illustrated in 1950 in a case heard at Cambridge County Court. Mr Martin kept 45 hives of bees in a copse beyond his garden, and goats belonging to Mrs Johnson – in a field adjacent to the copse, beyond her garden – were stung. The court found no evidence that the bees were managed in other than a proper manner, and that there was no substance in Mrs Johnson's allegation of *negligence*. In order to establish liability for legal *nuisance* it was necessary to show that the bees were kept in an unreasonable place, or that an unreasonable number of them were kept. 'The experts were of the opinion that . . . a comparatively small number of bees, a number which could not be said to be unreasonable, could have caused identical damage'; Mr Martin was thus not liable for nuisance (Johnson v. Martin, *Law J.* 100: 541 (1950)).

In a more recent case in the English West Midlands, the beekeeper agreed to remove all his bees from a garden whose neighbours were 'sorely troubled by the activities of the bees' (Möbus, 1985).

Less commonly, an action may be brought against a beekeeper when droppings from his bees on defaecating flights fall on washing hung out to dry, or on cars – especially new cars awaiting sale. The subject is fully discussed in an article concerned with United States law (Loring, 1984).

15.4 LAW IN RELATION TO CONTROL OF BEE DISEASES AND PESTS

Laws and regulations for the control of bee diseases and pests have been introduced in many countries. Sites of hives may be regulated; beekeepers or their apiaries may have to be registered; for certain diseases and pests, notification and specific control methods may be mandatory; the import of bees, beekeeping equipment (including beeswax) or honey may be prohibited or restricted. Table 15.4A indicates countries where such laws and regulations apply, and also countries with laws or regulations protecting bees from pesticide injury, or governing the sale of honey or beeswax.

15.41 Registration of beekeepers and apiaries

Registration is often regarded as a first necessary step towards disease control, for unless the sites of apiaries are known, the hives in them cannot be inspected. Requirement to register is sometimes linked with benefits such as cheap sugar for feeding bees. Some beekeepers are opposed to registration for various reasons, for instance a belief that it would result in increased liability for tax on income from beekeeping. The 21 countries known to have compulsory registration are marked R in column 3 of Table 15.4A.

15.42 Restrictions on type of hive used

Movable-frame and movable-comb hives were devised so that a single comb could be removed and replaced. Where such hives are used, an apiary inspector can inspect each comb for signs likely to indicate disease. He can take samples of bees for further examination as necessary, and advise the beekeeper of steps he should take to treat any disease found. The same procedure cannot normally be followed with traditional hives in which the bees 'fixed' their combs to the walls and/or roof; for this reason their use is disapproved of in many countries. It is prohibited in some, including Argentina, New Zealand, Trinidad and Tobago, Uruguay, and most states of the USA.

In many parts of Africa and Asia, some beekeepers still customarily use fixed-comb hives. The fact that they *do not* move combs or frames from one hive to another provides quite an effective form of control for many diseases and parasites which are largely spread from colony to colony by beekeepers' operations (Sections 11.2 to 11.45).

15.43 Actions required in finding and dealing with disease

Disease may be found and recognized by a beekeeper, or by a Bee Disease Officer who is normally empowered by law to visit any apiary and to examine any colonies.

The 48 countries marked D in column 4 of Table 15.4A have enacted bee disease legislation, and Bradbear (1983) lists many of the laws. These normally name diseases that are notifiable to the appropriate Government authority, state the procedure that the beekeeper must take if a notifiable disease is confirmed in his colonies, and indicate any prohibitions. Chapter 11 describes characteristics of individual diseases. Especially with American foul brood (AFB), destruction of the colony – and even of the hive – may be compulsory. In some countries compensation may be payable for hives and colonies so destroyed. In others the national BKA may organize an insurance scheme to compensate members for all or part of their financial loss (Section 16.2). AFB and EFB (European foul brood), and infestation with certain mites, are disorders commonly specified in bee disease legislation. The UK and New Zealand may be cited as examples of countries with minimal and comprehensive bee disease legislation, respectively, based on long experience. In the United Kingdom Laws and Statutes (1982), the Bee Diseases Control Order is concerned only with AFB and EFB, and the varroa mite which has not yet been found in the country. Bees infected with AFB, and those heavily infected with EFB, must be destroyed by fire. Colonies deemed by the Ministry of Agriculture to be lightly infected with EFB may be treated with oxytetracycline, but the drug may be administered only by a Ministry Bees Officer.

The 1969 New Zealand Act, which has had several amendments, lists three schedules of diseases and pests to which the Act applies (Table 11.1A gives information on their causative organisms).

- Schedule 1. Serious diseases for which compensation may be payable: acarine, varroa.
- Schedule 2. Serious diseases which are notifiable; colonies and appliances must be destroyed or treated according to instructions: AFB, EFB.
- Schedule 3. Bee diseases; colonies and appliances must be destroyed or treated according to instructions: nosema, amoeba, chalk brood, and four virus diseases (Kashmir, sac brood, acute and chronic bee paralysis).
- 'Pests and predators': bee louse, greater and lesser wax moths, and the mite *Tropilaelaps clareae*.

Table 15.4A Laws and regulations that relate to beekeeping in 95 countries

Column 1: Name of country; (EC) if member of the European Community.
* indicates updated entries; see final paragraph below.
Columns 2, 3: Laws or regulations, S, governing the placing (siting) of hives; R, requiring the registration of beekeepers or apiaries. For discussion see Sections 15.31 and 15.41.
Column 4: Laws or regulations, D, governing the notification and/or control of bee diseases and pests. For discussion see 15.43, 15.44.
Columns 5–7: Laws or regulations governing import, B, of bees; E, of beekeeping or honey-processing equipment; H, of honey. For discussion see 15.45.
Column 8: Laws or regulations, P, whose purpose is the protection of bees from pesticide injury; (P) a code of practice or similar. For discussion see 15.51.
Columns 9, 10: Laws or regulations, H, governing honey offered for retail sale and/or containers in which it is packed (but excluding general laws on adulteration, fraudulent description, etc.); (H), a code of conduct; h, one or more other standards for honey. For discussion see 15.71 and 15.72.
Column 11: w, a standard for beeswax; (w) no standard, but specifications in pharmacopoeia. For discussion see 15.73.
Entries in Columns 2–7 are based on Bradbear (1983), 8 on IBRA (1983*b*), 9 and 10 on IBRA (1977*a*), 11 on IBRA (1979*b*). Certain updatings have been added from various sources, and entries for countries marked * are based on information obtained from the country in 1985.

1	*2*	*3*	*4*	*5*	*6*	*7*	*8*	*9*	*10*	*11*
	S	*R*	*D*	*B*	*E*	*H*	*P*	*H*	*h*	*w*
Afghanistan						H				
Algeria	S		D	B		H				
Angola				B						
Argentina	S	R	D				P	H		(w)
Australia	S	R	D	B	E	H	P	H		
*Austria	S	R	D	B			P	H	h	
Bahrain				B						
Belgium (EC)	S		D				P	H	h	
Bermuda			D	B			P			
Bolivia								H		
Brazil								H		(w)
*Bulgaria	S	R	D	B		H	P	H	h	w
Canada	S	R	D	B	E		P	H	h	
*Chile	S		D	B		H	P	H	h	w
*China			D						h	w
*Colombia		R	D			H		(H)	h	
Costa Rica									h	
Cuba				B					h	
Cyprus				B		H		H		
*Czechoslovakia	S	R	D	B		H	P	H	h	w
*Denmark (EC)	S		D	B		H	P	H	h	(w)
Dominican Republic				B	E					
Ecuador								H		
*Egypt			D	B			P	H	h	w
El Salvador									h	
Fiji			D	B						
*Finland			D	B		H	P	H		†
France (EC)	S	R	D	B	E	H	P	H	h	(w)
*German DR		R	D	B	E	H	P	(H)	h	w
German FR (EC)			D	B		H	P	H	h	w

(continued)

Table 15.4A *(continued)*

1	*2* *S*	*3* *R*	*4* *D*	*5* *B*	*6* *E*	*7* *H*	*8* *P*	*9* *H*	*10* *h*	*11* *w*
Greece (EC)								H	h	
Guatemala								H		
*Honduras			D	B	E	H	P	H	h	
Hong Kong				B			(P)			
Hungary			D				P		h	
Iceland								(H)		
*India				B			(P)	H	h	w
*Iran				B	E	H			h	
Iraq				B		H		(H)		
Irish Republic (EC)			D	B			P	H	h	
Israel	S	R	D	B			P		h	
Italy (EC)	S		D	B		H	P	H	h	
*Japan	S	R	D	B			P	(H)	h	
Jordan						H				
*Kenya			D	B	E	H			h	w
Korea, Republic			D	B		H	(P)	H	h	
Lebanon				B						
Liberia								(H)		
Luxembourg (EC)							P	H	h	
Madagascar			D	B				H		
*Malawi	S	R	D	B	E	H	P	H		
Malta				B						
Mexico	S	R	D	B				H	h	
*Montserrat			D	B	E					
Morocco								H		
Mozambique									h	
Nepal				B	E	H		(H)		
*Netherlands (EC)	S		D	B	E		P	H	h	
*New Zealand	S	R	D	B	E	H	P	H		w
Nicaragua									h	
Nigeria						H				
Norway	S	R	D	B			P	(H)		
*Oman									h	
*Pakistan				B	E	H				
Panama						H		H		
Papua New Guinea				B?						
*Peru						H				
Poland							P		h	
Portugal (EC)							P	(H)	h	
*Puerto Rico				B						
Qatar				B						
Romania	S		D				P		h	
*Saudi Arabia			D						h	
Sierra Leone				B	E	H				
Singapore								H		
South Africa	S		D	B	E	H	(P)	H		
Spain (EC)	S		D				(P)	H	h	
Sri Lanka									h	
*Sweden		R	D	B	E		P	H	h	

(*continued*)

Table 15.4A *(continued)*

1	*2*	*3*	*4*	*5*	*6*	*7*	*8*	*9*	*10*	*11*
	S	*R*	*D*	*B*	*E*	*H*	*P*	*H*	*h*	*w*
Switzerland		R	D	B			P	H	h	(w)
Syria				B	E	H				
*Taiwan									h	
*Tanzania				B	E		(P)	H		w†
Tonga			D	B						
Trinidad & Tobago		R	D	B	E	H				
*Turkey				B	E	H	(P)	(H)	h	w
*Tuvalu				B	E	H				
Uganda				B	E					
UK (EC)			D	B			P	H	h	(w)
Uruguay			D	B	E		P	H		
USA	S	R	D	B	E	H	P	H	h	(w)
USSR		R	D				P		h	w
Yugoslavia				B			P			
*Zambia	S	R	D		E	H	P	H	h	w
Zimbabwe			D				P		h	

* see note at head of Table
† for law, see text

In countries of Eastern Europe, the monitoring and treatment of bee diseases is closely integrated with general beekeeping practices and regulations (e.g., Borchert, 1974; Galton, 1971; USSR Laws and Statutes, 1981). Certificates are issued annually to beekeepers who have cleaned and disinfected their beekeeping equipment by recommended methods. *Any* disease must be notified, and if confirmed, the fact is entered on the certificate, which must be shown when beeswax is exchanged for foundation, and when honey is sold.

In parts of Canada and the USA, beekeepers customarily apply drug treatment for certain diseases, especially AFB. In some other countries most such drug treatment is proscribed, for instance in New Zealand where the only drug permitted is fumagillin for nosema disease. There are several objections to the use of certain drugs; if improperly used, residues may be present in honey harvested, and such use of a drug may also lead to the development of resistant strains of the pathogenic organism: it may suppress manifestations of the disease, and thus hinder its diagnosis. In the UK 'it is an offence for beekeepers to administer any substance which may have the affect of disguising the symptoms of AFB or EFB.'

15.44 Restrictions on moving bees and hives

Whereas restrictions on *placing* hives (Section 15.31) are usually made to prevent nuisance to people and their animals, restrictions on *moving* hives may be imposed to contain an outbreak of disease. The modern practice of moving hives of bees from one place to another – for honey production or for pollination – is one of the most common causes of the spread of bee diseases and parasites. In many countries legislation therefore prohibits absolutely the movement of bees, hive products, hives and hive fittings, within a certain distance (e.g. 200 m) of an apiary known to be infected with a specified disease. Within a large country, the movement of hives of bees from one state or province to another is strictly controlled.

15.45 Import restrictions

Table 15.4A lists 60 countries (marked B in column 5) which control imports of bees, and 26 which control imports of used equipment; 34 control imports of honey (H), and a few also control beeswax imports. Some details of the legislation are given by Bradbear (1983), and Section 8.7 discusses various aspects of the importation of bees into new areas.

Legislation restricting imports does not always prohibit imports of bees altogether. A common form in which bees are sent from one country to another is a mailing cage containing a queen with attendant workers. Several diseases and parasites of adult bees can be transmitted in this way, even if the workers are removed before the queen is introduced to a colony of local bees, and in many countries the workers must be examined for disease by the appropriate authority.

Used combs and hives can transmit nosema and amoeba diseases, pests of comb and of stored food, and some of the brood diseases – including AFB, since spores of *Bacillus larvae* can survive in both honey and beeswax. Emptied drums of imported honey, left out of doors where bees find and forage on the honey residue, have been blamed for various outbreaks of AFB. Trinidad and Tobago prohibit entry even to single jars of honey.

In Australia, New South Wales has pioneered the use of experimental quarantine procedures for imported bees (Skillman, 1979), and in Fiji certain areas have been officially designated as quarantine areas for bees.

Legislation on the import of bees is also important for another reason besides exclusion of disease: exclusion of undesirable genetic material; see Section 8.73. The far-reaching results of introducing tropical African honeybees to South America in 1956 are referred to in Sections 1.22.4 and 7.43.

The Honeybee Act, [USA Federal] Public Law 94–319, passed in 1922, restricted the import of live adult honeybees. It was amended in 1947 and 1962, and again in 1976 to include the passages:

(a) In order to prevent the introduction and spread of diseases and parasites harmful to honeybees, and the introduction of genetically undesirable germ plasm of honeybees, the importation into the United States is prohibited, except . . .
(b) Honeybee semen may be imported into the United States only from countries determined by the Secretary of Agriculture to be free of undesirable species or subspecies of honeybees, and which have in operation precautions adequate to prevent the importation of such undesirable honeybees and their semen.

The full text is given by Michael (1980) and by Loring (1981).

15.5 LAW IN RELATION TO PEST MANAGEMENT DAMAGE TO BEES

By far the greatest hazard to bees from pest control measures is the use of *insecticides* for crop protection (Figure 15.5a). Very few of the insecticides are specific to a single insect species, and many that are effective against pest insects also kill bees.

Apart from crop protection, the most common cause of insecticide damage to bees is mosquito or tsetse fly control (Crane & Walker, 1983). In Manitoba, Canada, the Provincial Mosquito Abatement Program has incorporated a Beekeepers' Compensation Program. The following payments were made in 1984 as compensation for 1983 losses (Manitoba Beekeeper, 1986):

	Canadian $
to beekeepers (with honeybees)	746 250
to beekeepers (with *Megachile* bees)	73 013
to seed producers affected	39 403
	858 666

There is unlikely to be any legal protection in the tropics, where mosquito and tsetse fly control measures are on a much larger scale.

Herbicides can kill bee forage, but most herbicides – and fungicides – are not dangerous to bees themselves, nor are the biological methods now used for pest control.

Figure 15.5a Bees killed by insecticide, Washington State, USA (photo: Washington State University).
The four hives are mounted back to back on a single pallet, as described in Section 8.12.

15.51 Damage from insecticides used for crop protection

Insect pests destroy a considerable proportion of the world's crops; estimates vary from 5% to 21% for different continents (Cramer, 1967). Many effective insecticides were developed after the Second World War, and they are widely manufactured. When they were first used, many beneficial insects – notably foraging, pollinating bees – were killed along with the pest insects. Education on the need for bee pollination of crops proved insufficient on its own to prevent this killing of bees, and Table 15.4A includes 38 countries (P in column 8) in which legislation has been passed to protect the bees; Crane and Walker (1983) give further information, and their Bibliography 'Laws and regulations (world-wide) to protect bees from pesticide poisoning', is published separately (IBRA, 1983*b*). Legislation may regulate the method or the time of applying insecticides on certain crops; it may identify substances whose use is not permitted, or it may regulate the types of labelling to be used by manufacturers for different pesticides according to their toxicity to bees.

France was one of the first countries to take action, in 1945, and the wording of French legislation is very explicit. According to the 1975 decree, updated by one in 1985 (Louveaux, 1985), all insecticides are presumed to be dangerous to bees (honeybees) except those which by authorization carry a label 'not dangerous to bees'. Use of products that are presumed to be dangerous to bees is prohibited, by any type of applicator: on fruit trees and all other crops visited by bees, during flowering; on trees in forests and plantations, during the period of honeydew production resulting from aphid attack; and on cereal crops, during the period of honeydew production resulting from aphid attack, between ear formation and harvest. If plants visited by bees grow beneath the trees or in the surroundings of crops to be treated with the products, these plants must be cut down or pulled up before the treatment.

Some other European countries are similarly protective towards bees. For instance in Sweden, insecticide treatment of crops in flower requires prior government approval which is seldom given. In the Netherlands, some areas of rape are reserved for the siting of hives, and if treatment is necessary it must be with substances of low toxicity to honeybees.

In New Zealand, Section 35 of the Apiaries Act 1969 prohibits the application of pesticides toxic to honeybees: to cruciferous and leguminous crops when they are in flower, or to flowering plants within the crops, between 1 September and 31 March, unless a permit is obtained from the Ministry of Agriculture and Fisheries. As a rule permission is not given to treat flowering crops with persistent, highly toxic substances. But there is not a blanket ban on growers spraying flowering crops, and they may spray if they have a permit.

15.52 Compensation and indemnity for damage to bees

If bees in a hive (i.e. not feral, nesting in the wild) are killed as a result of the use of an insecticide in a way prohibitied by law, the person responsible may be prosecuted by the State and, if found guilty, fined or otherwise punished. In some countries there is a special compensation fund against which the owner of the bees can make a claim. He does not always win his case, but recent successful claims have been reported from various countries, including Denmark, France and the USA, where the situation differs between states. A Federal Bee Indemnity Program in USA operated with effect from 1967 to 1979. Beekeepers who through no fault of their own suffered losses after 1 January 1967, stemming from the use of economic poisons registered and approved by the Federal Government, could apply for and receive indemnity payments. Payments to beekeepers during the period totalled $34 million.

In countries which have no public legislation establishing compensation schemes for damage to bees by pest control measures, a beekeeper's only remedy is to sue the person responsible for damages in the courts, for the value of both his bees and the honey harvest he loses because they are dead. It is difficult for a claim for damages to succeed. In England and Wales no case reached the High Court until 1984, in which year 161 suspected cases of poisoning were reported to the Ministry of Agriculture, and 81 of them (representing a total of 61 incidents) were found to be connected with pesticides. Five Chichester beekeepers – who did not have legal aid – won a High Court case at Lewes against a farmer A. D. Walter Ltd, who had sprayed a crop of oilseed rape with Hostaphion in 1984, using a tractor and boom (Law Reports, 1986). The basis of the claim was the allegation that the crop was still in flower at the time, whereas Ministry of Agriculture leaflets and a code of practice published by Hoechst (the manufacturer) stated that spraying should be done only when the field had an overall green appearance. The judge awarded damages later assessed at £7000, and costs of £12 000, against the farmer. This case set a legal precedent, but legislation was finally enacted in the UK, in the Control of Pesticides Regulations 1986. According to these, users of pesticides must comply with conditions of approval listed, which include protection of bees.

In a number of countries beekeepers can get insurance cover for pesticide damage, as well as for other hazards to bees (Section 16.2). Beekeepers hiring out bees for pollination should ensure that their contract covers the possibility of pesticide damage to their bees, whether through negligence or not; Figure 15.63a shows a specimen contract.

15.6 LAW IN RELATION TO CERTAIN OTHER ASPECTS OF MODERN BEEKEEPING

15.61 General

Sundry other aspects of beekeeping are subject to legislation or regulations, from the control of bears in Canada (where they can do much damage to hives of bees but are protected by law) to the denaturing of sugar sold at a discount price for feeding to bees, as in the German Federal Republic and Italy. There are regulations concerning honey processing – relating especially to hygiene – for instance in France, New Zealand and USA, and in Australia where mobile extracting plants also have to be registered.

A New Zealand study (Martin, 1982) concludes that from the legal point of view it would appear quite feasible to accept hives as security under the Chattels Transfer Act. It is, however, suggested that – in view of current variable resale value of hives – they should be used as security only as a last resort, and in circumstances that are favourable in other respects.

A beekeeper sometimes receives a request for help in removing a colony of bees from an awkward site (Section 5.43.2) – inside a house wall cavity, or in a chimney which may or may not be in use. Experience suggests that he would be wise to ask the person requesting his help to sign a document accepting liability if anything goes wrong. But even that may not cover his actions if an accident occurs and legal action is taken against him, particularly if it is shown that the beekeeper has been negligent.

On the positive side, there are special clauses *exempting bees* (along with leeches and silkworms) from the general prohibitions and restrictions that bar living creatures from being sent through the post. A queen with attendant workers is the most common unit. For the UK inland post, the exemption is conditional on the creatures being 'enclosed in boxes so constructed as to avoid all risk of injury to officials of the Post Office, or damage to other packets'. For overseas post, the boxes 'must be so constructed as to avoid all danger and to allow the contents to be ascertained'. The latter ruling is likely to apply to other international posts. There is a standard giving recommendations for transport of bees by air (British Standards Institution, 1964), but this does not have the force of law. Section 8.34 discusses air transport of package bees.

As far is known, law on cruelty to animals does not apply to bees in any country. However, in 1959 an Austrian who stopped up the entrances of 15 hives belonging to a neighbour was imprisoned and fined for 'malicious and cruel killing of useful animals' (Bindley, 1959). In Buddhist Tibet the religious prohibition against killing animals extends to bees. In 1984 I was told by Tibetans in Nepal that Tibetans did not keep bees, or eat honey, except that children would take honey from wild nests. Honey hunters in Tibet are of Nepali (Hindu) extraction. On the other hand in the Muslim religion, and in Islamic law, killing bees is specifically permitted, although eating them is forbidden (Fahd, 1968), as it is in Jewish law.

15.62 Road accidents

If a beekeeper's vehicle transporting hives of bees is involved in a road accident, many million bees may be released, and Figure 15.62a shows the aftermath of one such incident in 1984. The load was being moved by a commercial haulier, who was insured for only $35 000 per truckload, a sum much less than the beekeeper's total loss (Moffett, 1985). In addition to any losses or injuries sustained by the beekeeper or driver, either person may be sued for negligence by other drivers and bystanders at the time of the accident. Referring to the USA, N. E. Bailey (1970) explains it thus:

> There may be no existing statute or regulation or ordinance which requires you to cover your load of beehives [with a bee-proof net] when trucking them. Nevertheless, if you drive on the highway with your load uncovered you may be violating a rule of negligence and, as a result, you may be required by a court to pay a large sum of money to another person who claims injury because of your failure to cover your truckload of beehives.
>
> The law of negligence provides that every person has the right to be free from injuries to himself or his property caused by others who do not exercise the proper degree of care. If a beekeeper transports beehives or honey in a way that a reasonable person (*as defined by a jury of non-beekeepers*) ought, in the exercise of ordinary care, to have anticipated was likely to result in injury to others, then he is liable for all

Figure 15.62a An overturned truck that carried 720 hives of bees, near Perry, Oklahoma, USA, 24 May 1984 (photo: J. O. Moffett).
This sort of accident can give rise to a costly legal action.

injuries proximately resulting from his doing the act. It is not what you and other beekeepers consider to be safe beehive-moving procedures that counts; it is what that jury thinks the reasonable beekeeper in the exercise of ordinary care would do that is determinative.

R. Bailey (1969) warned that carelessness in areas where many loads of bees are moved on the roads could lead to the imposition of rigid regulations on beekeepers. In fact a bill *Transporting of bees* (HB 1675) was introduced to the Florida House of Representatives in 1969:

1. It shall be unlawful to transport or ship honey bees without providing adequate prior safeguards against their escape.
2. The Commissioner of Agriculture shall adopt regulations as required to carry out the intent of this section.

Beekeepers succeeded in killing the bill at the committee level.

15.63 Hiring bees for pollination

When a grower hires colonies of bees to pollinate his crop, his harvest may depend on the strength and suitability of the colonies. The beekeeper, on his side, has to entrust the safety of his bees to the grower, who could kill them by improper application of insecticide to the crop. It is therefore highly desirable that the grower and beekeeper sign a contract (an agreement enforceable by law) in advance. McGregor (1976) quotes an unsatisfactory contract, and sets out its deficiencies:

'I, (*beekeeper's name*), agree to supply __ colonies of bees to (*grower's name*) to pollinate __ acres of (*crop*) for the year ____. I (*grower's name*), agree to pay (*beekeeper's name*) $__ per colony for __ colonies of honey bees to pollinate my (crop) for the year ____.'
(*Date*), (*Beekeeper's signature*), (*Grower's signature*).

Much more detail is required – for example Figure 15.63a shows a contract recommended in the USA for use where most of the pollination work is in tree fruit orchards (Mayer, 1981). A French contract is published by Borneck (1981), and a somewhat fuller version of it by Ouest Apiculture (1983). In the Soviet Union, Chetaikin (1982) describes a contract between a collective farm (*kolkhoz*) and a State bee farm (*sovkhoz*) for the hire of 260 hives of bees to pollinate 160 ha of clover. The *kolkhoz* obtained a seed yield of 2600 kg/ha, while the *sovkhoz* harvested 3 tonnes of honey and received a fee of about 300 roubles.

Powers (1979) gives details of the few laws in states of the USA, relating to bee pollination of crops. Questions that may arise concerning liability, and of indemnity or compensation to the beekeeper in case of damage to his bees, are discussed by Loring (1978, 1985) and Crane and Walker (1983).

PROFUSE POLLINATORS, INC.
P.O. Box Nothing
Various City, WA 00000, USA

Pollinator Rental Contract

Profuse Pollinators, Inc. (PPI), agrees to rent ... colonies of bees to .... in 1978 at a cost of ..... per colony for .... crop.

Colony Strength
All colonies shall meet a minimum strength of 8 frames, 2/3 covered with bees at 65° F [18%]. Renter is not obligated to pay for colonies not meeting this minimum. If colony strength is in doubt, the renter or PPI may request an inspection by the chief apiary inspector of the WSDA or a private consulting entomologist. Should some colonies be found to be under minimum, PPI agrees to pay the cost of the inspection. Should the colonies meet or exceed the minimum standard, the inspection cost shall be paid by the renter. All questions regarding colony strength shall be brought to the attention of PPI within 3 days following delivery.

Notice
At least 72 hours advance notice shall be given to PPI prior to delivery or pick up date. Site location and access shall be free from mud, irrigation lines, locked gates, ditches and farm or orchard equipment.

Colony Distribution
Colonies will be delivered on pallets - ... colonies per pallet - and be placed on the site by PPI. Site location may be made by conferring with PPI regarding warmest area where bees will commence flying as early as possible. Colonies shall not be moved by renter without PPI's permission and shall not be removed from pallets.

Pesticides
Renter agrees to refrain from applying pesticides 5 days prior to delivery of colonies and during their stay on the site. Any necessary pesticide application program shall be brought to the attention of PPI for evaluation prior to application.
Any removal of colonies made necessary during the term of this contract shall be paid for by the renter at $ ... per colony.

Rental Fee
The rental fee per colony shall be $ ... for each placement of colonies. Additional colonies must be rented if used for different crops.

Deposit
A $4 per colony deposit for ... colonies, making a total deposit of $ ..., shall accompany a signed copy of this contract and be returned to Profuse Pollinators, Inc. to reserve colonies. The balance is due and payable upon receipt of invoice from PPI approximately 2 weeks after delivery of colonies.
The above terms and conditions have been read and agreed upon by renter and PPI.

Renter .... PPI .... Date ....

Figure 15.63a Sample pollination contract prepared in Washington State, USA, by the State Department of Agriculture and the State University (Mayer, 1981).
Most of the pollination work is in tree fruit orchards.

15.7 LAW IN RELATION TO CONSUMERS OF HIVE PRODUCTS

15.71 Laws and standards relating to honey

Honey is the only hive product of sufficient commercial importance to be subject to legislation, and there are laws relating specifically to honey in 42 countries (marked H in column 9 of Table 15.4A). Honey is also subject to general laws on foods: it must not be harmful, and its presentation and labelling must not be deceptive.

In 48 countries (h in column 10 of Table 15.4A) official standards for honey have been established over the years, and these often form the basis for the country's legislation. In some others, codes of practice have been agreed. Fasler (1975) gave a clear description of the world situation, and of the varying powers of the different types of official documents concerned with honey (laws, regulations, standards, codes, etc.); such documents have been listed in a bibliography (IBRA, 1977*a*).

Some groups of countries conform to mutually agreed legislation. The European Communities Council (ECC) issued a Directive on honey in 1974, based largely on the Recommended European Regional Standard for Honey prepared in 1969 by the Codex Alimentarius Commission of the Joint FAO/WHO Food Standards Programme. The Directive required member states (indicated in Table 15.4A) to incorporate its provisions into their own national honey legislation, and all did so. However, the UK honey legislation applies to honey *offered for retail sale*, whereas the ECC Directive applies to honey *arriving in the EC*.

Most member states of the EC also have national standards – enforceable by law – that are more rigorous than the Codex used in the Directive.

In 1983 the Codex Alimentarius Commission issued a *Proposed Codex Standard for Honey* (*World-wide Standard*), which is still at the draft stage and under discussion. In principle it is similar to the 1964 Regional Standard, but a number of requirements have been relaxed to accommodate honeys from a wider range of plants and climates than those occurring in western Europe.

The *Código latinamericano de alimentos* (see Latin America, 1964) is a set of internationally agreed standards for the whole range of foods. Its requirements for honey have been set out by Fasler (1975), who also describes honey standards legislation in some other countries: Austria and Switzerland; Canada and USA; Argentina, Brazil and Mexico; Australia and New Zealand. Fasler also compares the various standards used, and makes concise suggestions for improved legislation.

15.72 Laws relating to honey containers

Honey packers, and beekeepers who sell honey retail, need to be conversant with their national laws and regulations. Most countries have their own rules for the packaging of food, and there is as much legislation about the containers in which honey is sold retail as about the honey inside it. UK legislation on both honey and containers was summarized in BBKA News (1981). Along with many other foods, honey might be sold retail in containers holding 6 or 12 ounces (oz), or 1 pound (lb) i.e. 16 ounces, or 1 kg, or in certain larger ones. There is no restriction for containers holding small amounts, although the weight must be stated.

The 1974 European Directive referred to in Section 15.71 includes additional Community regulations, including weights that may be sold. In most EC countries an allowed weight unit is 500 g, but in the Netherlands and Denmark it is 450 g, which is close to their old pound.

Ways in which honey may or may not be described on the label are laid down in detail. The type-face used for printing various parts of the wording on the label is also specified, so that information considered necessary is in print large enough to be read. There are also restrictions on the kinds of illustration on the jar or label; if flowers are shown, they must not mislead the consumer as to the floral source of the contents. Any more direct misrepresentation is of course also prohibited. In honey-importing countries, an unscrupulous honey packer might make an illegal extra profit by labelling imported honey as home-produced, which sells at a higher price; an expert's identification of pollens in the honey can, however, sometimes provide sufficient evidence for a prosecution.

15.73 Laws and standards relating to other hive products

It seems that beeswax as a commodity is not commercially important enough to be the subject of specific legislation except in major exporting countries. *The Produce Export Rules* of Tanzania (1957, amended to 1965) include detailed rules for beeswax. There are legal restrictions on the *import* of beeswax into some countries, on the grounds that it might carry spores of *Bacillus larvae* which, if the wax is used for comb foundation, could transmit American foul brood (Section 11.21). A number of countries regulate the import of all used beekeeping equipment on the same grounds, and this would presumably include beeswax foundation and comb.

In column 11 of Table 15.4A, 14 countries with a

standard for beeswax are marked w, and 7 with an official specification for beeswax in a pharmacopoeia are marked (w). Ten such specifications are summarized in Table 13.5B, which are considerably out of date in the light of beeswax analysis using modern techniques.

Very few countries have official standards or specifications for any hive products other than honey and beeswax, and no legislation is known that relates to them. Actions may, however, be brought for misrepresentation, such as in Civil Action No. 1042–58, USA v. Jenasol (Bee World, 1962*b*); claims made for royal jelly in Jenasol capsules were not allowed. Japan has a code of conduct for royal jelly, comprising a description (including standards), methods for determining the components, and the description of the contents on labels. In the Soviet Union, there is a regional Standard for propolis (RSFSR, 1977), and a national (GOST) Standard was discussed in 1979; see Chudakov et al. (1976) and Popravko (1977).

Products of honey fermentation such as mead are, like other drinks containing alcohol, subject to legal control in many countries.

15.8 FURTHER READING AND REFERENCE

Details of publications listed will be found in the Bibliography

Law in relation to bees and beekeeping

Bradbear, N. (1983) [Bibliography, disease laws, etc.]
Doutt, R. L. (1959) *The case of the trespassing bees*
Frimston, J. D. (1966) *Bee-keeping and the law* [UK]
Gercke, A. (1985) *Das Bienenrecht* [GFR]
Loring, M. (1981) *Bees and the law* [USA]
New Zealand Laws and Statutes (1969–1983) *An Act to consolidate and amend the Apiaries Act 1927*
Pech, E. (1958) *Das Bienenrecht . . .* [DDR]
Rüedi, J. P. (1955) *Rechtsfragen des Imkers* [Switzerland]

Law in relation to hive products

BBKA News (1981) *Summary of the laws . . . UK*
Codex Alimentarius Commission (1983/84) *Proposed draft Codex Standard for Honey (World-wide Standard)*
European Communities Council (1974) *Directive . . . on the harmonisation of the laws . . . relating to honey*
IBRA (1977*a*) [Bibliography on honey]
IBRA (1979*b*) [Bibliography on other hive products]

16

Resources for beekeepers

16.1 INTRODUCTION

Many resources are available for the beekeeper, although more in some countries than in others, and some are not as widely known as they deserve.

This chapter provides an introductory guide to them, and Appendix 2 gives separate entries for 177 countries and islands. The most important sources of help in individual countries are likely to be the Government Beekeeping Extension Service and the Beekeepers' Associations (BKAs). Both may provide educational services (Section 16.3), and may publish bulletins and periodicals (16.41). Beekeeping insurance schemes may be organized by BKAs (16.2), and certain government services are provided in connection with bee health and disease (16.5). Government laws, regulations and standards are discussed in Chapter 15. Firms manufacturing and selling equipment (16.6) constitute a useful resource in a beekeeper's own country, and also in countries he may visit, where beekeeping museums and various other resources (16.8) may also be of interest.

The Gazetteer in Appendix 2 gives one or more contact addresses for almost all countries listed in it: a Government department responsible for beekeeping – most commonly within the Department of Agriculture – and/or a national BKA, which exists in about 70 countries. The reader may find it useful to look at the introduction to the Gazetteer (page 517) before continuing with this chapter. Apimondia is the International Federation of Beekeepers' Associations (101 Corso Vittorio Emanuele, 00186 Rome, Italy). The International Bee Research Association (16–18 North Road, Cardiff CF1 3DY, UK) has as its members individuals and organizations in over a hundred countries; for its activities see e.g. Sections 16.4, 16.7, 16.82.

16.2 BEEKEEPING EXTENSION OFFICERS AND BEEKEEPERS' ASSOCIATIONS

Advice on a beekeeper's special problems can often be obtained from a national or regional Extension Officer, or from the appropriate Government agency. Whether or not the beekeeper has to pay for it depends on the condition of the Officer's appointment.

Many of the national Beekeepers' Associations (BKAs) listed in Appendix 2 have regional BKAs as members, which themselves may have local branches, and BKAs at all three levels may arrange regular meetings. These are usually lectures or visits to apiaries, where a demonstration is given of a topical beekeeping operation. Whether regular or occasional, these meetings are pleasant social occasions (Figure 16.2a). In some countries residential meetings are organized from time to time, by Government agencies, universities or BKAs. They are often held in a college during the vacation, and are likely to offer programmes catering for different levels of beekeeping expertise, together with appropriate entertainments.

Beekeepers are most strongly urged to join their local BKA, to support it actively, and to pay their

Figure 16.2a Beekeepers on the Tharaka Plain, Kenya, assembled for a meeting, 1967.

membership dues promptly. In many countries the officials who maintain an Association's programme do so without payment, and they help members in other ways as well.

Many BKAs were formed between 1850 and 1900, to spread knowledge of the new 'rational' beekeeping with movable-frame hives and the centrifugal honey extractor (Section 1.44). They undertook much teaching and instruction on the use of this new equipment, and on the production of clean, good quality honey. At honey shows, jars of honey prepared by different beekeepers were judged by those with more experience who specialized in this work. The same system continues, and new beekeepers can find out how their honey is rated in comparison with that from other entrants. Local, regional and national shows are held, and many include classes for comb honey as well as for liquid and granulated honey in jars, and also for beeswax, mead, foods and drinks prepared with honey, and so on.

Honey shows are valuable as a place for meeting other beekeepers, and for seeing and handling equipment at trade stands. Further, they afford an opportunity for publicity by beekeepers about their bees and honey. This applies especially to the honey or beekeeping section of a more general agricultural or other show.

A number of BKAs and Government agencies issue promotion material about honey (particularly home-produced), and about the need to conserve bees and forage, and especially to protect bees from pesticide poisoning (see also Sections 11.7–11.73). Films and videos (Section 16.33) are a valuable tool for promoting public interest in bees.

Some national BKAs, for instance the Deutscher Imkerbund in the German Federal Republic, offer their members comprehensive insurance facilities. Large beekeeping concerns will usually make independent arrangements. Insurance cover is most commonly sought for the following eventualities; points of law involved are dealt with in Chapter 15.

1. Public liability for injury to persons by stinging as a result of accident, or by food poisoning from products sold; damage to property arising from beekeeping activities.
2. Loss of or damage to the owner's hives and other beekeeping equipment or stores, and to stocks of honey.
3. Loss of bees through pesticide poisoning.
4. Personal accident during beekeeping activities.
5. Compulsory destruction of colonies for disease control.

Where appropriate, cover should include the value of honey harvests lost, as well as that of bees and/or equipment that suffer damage. In the case of liability, many points must be considered, as set out by Ely (1985).

16.3 EDUCATION AND TRAINING

16.31 Training courses

Beekeeping is a small and specialized branch of agriculture, and rather few courses provide systematic training to a specific level of beekeeping expertise, or in a particular part of it. A leaflet *Opportunities for training in apiculture world-wide* (IBRA, 1985*a*, No. 4) gives details. There are brief entries in Appendix 2 for courses that are open to students from countries other than the one where they are held. Courses lasting a year or more, leading to a professional diploma or degree, include the following:

Diploma or Certificate in Apiculture
Australia: Queensland Agricultural College
Canada: Fairview College
New Zealand: Bay of Plenty Community College
Tanzania: Beekeeping Training Institute
UK: University College, Cardiff
USA: Agricultural Technical Institute, Ohio

MSc or PhD on an apicultural subject
Brazil: Universidade de São Paulo
Canada: Simon Fraser University; University of Guelph
Egypt: Ain-Shams University; Cairo University
German Federal Republic: Institut für Bienenkunde, Universität Frankfurt
Israel: Hebrew University of Jerusalem
UK: University College, Cardiff
USA: Cornell University (NY); Michigan and Oregon State Universities; Universities of California (at Davis), Florida and Georgia

Evening classes with weekly lectures are held fairly widely where there is a demand for them. In some countries beekeeping is taught to children's groups (Figure 16.31a).

Figure 16.31a Children learning beekeeping in the German Democratic Republic, 1985 (photo: Garten und Kleintierzucht).

16.32 Examinations and correspondence courses

Apart from the courses above that lead to a diploma or degree, some BKAs (and sometimes other bodies) conduct beekeeping examinations at levels, from 'elementary' upwards, that are not linked to any specific course of instruction. For instance in England the British BKA has for many years conducted a graded series of examinations: Junior, Preliminary, Practical, Intermediate, Senior. There are also BBKA examinations leading to the qualification: BBKA Show Judge, BBKA Lecturer, Fellow of the BBKA. After passing the Senior BBKA Examination or an approved equivalent, a candidate can sit an examination for the (UK) National Diploma in Beekeeping, NDB, awarded by a Board comprised of representatives from appropriate Government departments, and from the British, Welsh and Scottish BKAs, and other bodies such as IBRA. The NDB is the appropriate qualification for many Beekeeping Extension Officers, and the examination is open to candidates from other countries who already hold qualifications recognized by the Board as making them eligible.

A Master Beekeeper examination is conducted annually by the Eastern Apicultural Society in the USA.

To satisfy the need of examination candidates for guided instruction, correspondence or home study courses are organized in some countries. Details have been published for Canada, France, South Africa, Sweden, UK, USA and USSR (Bee World, 1981); see also the New Zealand entry in Appendix 2. Three complete teaching-course packages, including filmstrips, viewer and commentary are: *Introductory apiculture*, *Tropical apiculture* and *Advanced apiculture* (Townsend, 1981, 1984, 1985).

16.33 Films and other audiovisual material

Bees in close-up, and beekeeping in the countryside of different parts of the world, are splendid subjects for colour photography, and excellent films have been made in a number of countries. There are also instructional films on special subjects. (The provisional *World list of films on bees and beekeeping*, BRA, 1973, is out of print.)

In any one country, the Extension Service or the BKA is the most likely source of films or videos for hire. Systematic international lending of films does not

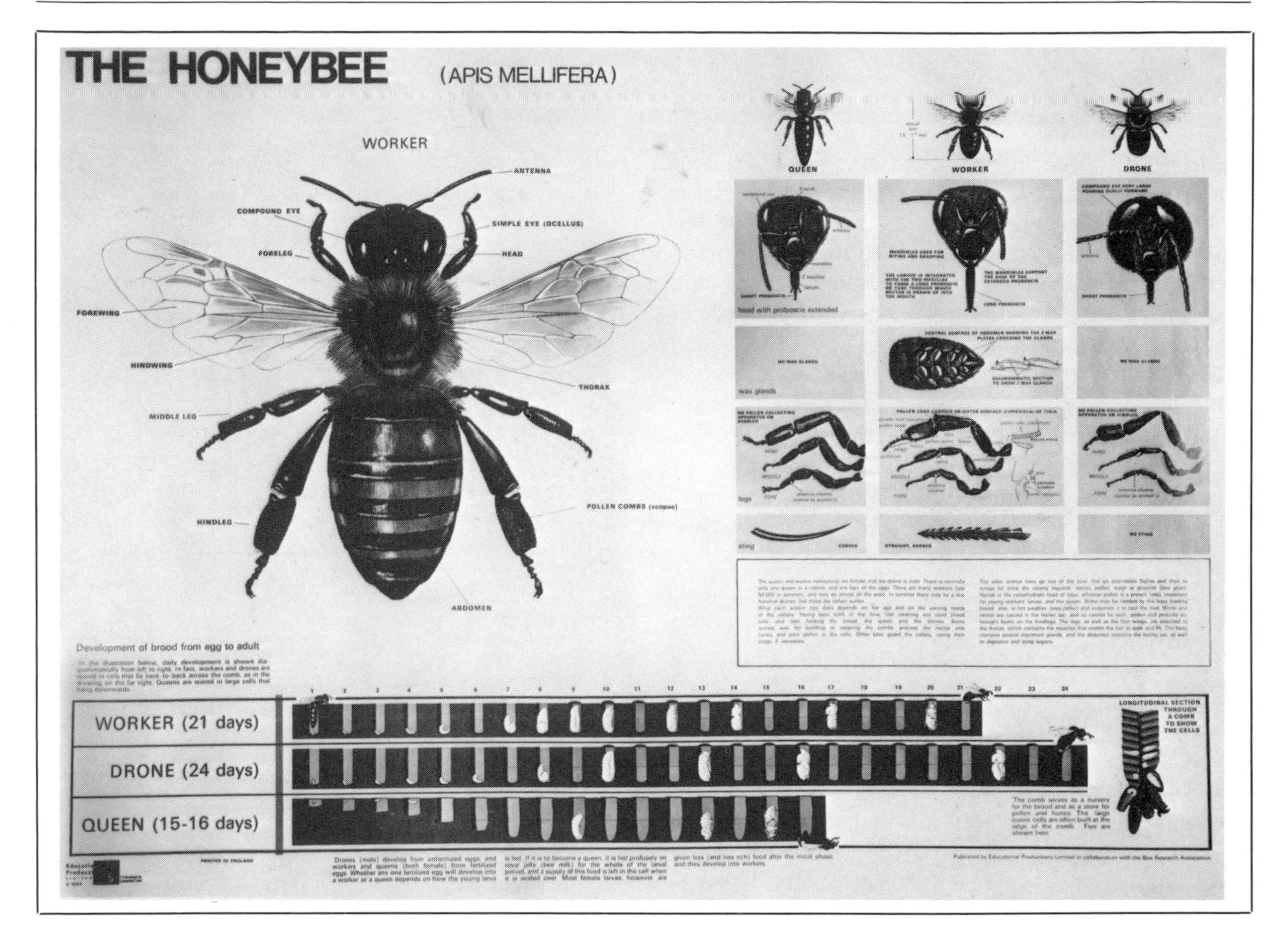

Figure 16.33a A widely used educational wall chart on the honeybee (Educational Productions Ltd in collaboration with IBRA, 1971).

seem practicable, and films or videos from one country may not be compatible with equipment available in another. If agencies within one country can co-operate, a copy of a film or video can be purchased for multiple use within the country.

Sets of slides are much easier to distribute, and notes or a taped commentary are often included. They can be purchased, for instance from IBRA in the UK and from Beekeeping Education Service (PO Box 817, Cheshire, CT 06410, USA). Also many national BKAs lend sets for local meetings. Posters and wall charts are also available (IBRA, 1985/86); Figure 16.33a shows an example.

16.4 PUBLICATIONS AND LIBRARIES

16.41 Books, journals and other publications

There are many publications on bees and beekeeping, and some are of a high standard. The beekeeping literature is well documented and indexed. The entry for each country in Appendix 2 cites a recent book on beekeeping, or other publications, and a beekeeping journal if one exists. Bibliographies have been issued of books (or of all publications) on beekeeping in certain countries: Canada and USA (Johanssn & Johansson, 1972*a*), France (Casteljau, 1983), Germany (Droege, 1962), New Zealand (Reid et al., 1988), UK and Irish Republic (IBRA, 1979*a*). In the last decade or so, increasing attention has been paid to beekeeping in the tropics and subtropics, and guides of various sorts have been published; see Section 16.7.

All beekeepers are advised to subscribe to one or more of the many beekeeping journals published, and IBRA issues a world list of current journals, with addresses (latest edition 1983*a*); publishers of most journals will send a specimen copy free of charge on request. There are several international journals. *Apicultural Abstracts*, *Bee World* and *Journal of Apicultural Research* are published by IBRA; the last, and *Apidologie*, are for scientific research papers. Apimondia, the

International Federation of BKAs, publishes *Apiacta*. *Apicultural Abstracts*, started in 1950, is the journal for those who need to consult the world's scientific and technical literature on beekeeping and bee science. There were 36 000 entries up to the end of 1988, and computer-produced cumulative indexes have been produced (Crane & Townsend, 1976; IBRA, 1985*b*). Many publications referred to in the present book have been summarized in *Apicultural Abstracts*, and the appropriate reference number is added to the citation in the Bibliography.

IBRA provides a world-wide book supply service, and will send a current catalogue on request (IBRA, 1988). Beekeeping Education Service (see Section 16.33) sells books, as do some BKAs and some equipment suppliers; Appendix 2 gives addresses for different countries. Multilingual beekeeping dictionaries are referred to in Section 16.82.

Figure 16.42a Headquarters and Library of the International Bee Research Association from 1966 to 1986, at Hill House, Gerrards Cross, Buckinghamshire, near London (photo: E. Greenwood).

16.42 Libraries

Libraries can provide access to a wider range of books than an individual beekeeper would be likely to purchase. Since its foundation in 1949, IBRA built up one of the most important existing libraries on bees and beekeeping (see Figure 16.42a). In 1986 IBRA headquarters and Library were moved to 16–18 North Road, Cardiff, Wales, UK, where it holds the publications marked B in the Bibliography of this book. There are also four Branches of the IBRA Library:

for Latin America: Universidad Nacional, Medellín, Colombia
for Tropical Asia: Central Bee Research Institute, Pune, India
for Oriental Asia: Institute of Honeybee Science, Tamagawa University, Japan
for Africa: InterAfrican Bureau of Animal Resources, Nairobi, Kenya.

The following are some other libraries with important holdings of scientific and of older works on bees and beekeeping; addresses are given here or in Appendix 2.

Canada: University of Guelph
France: Laboratoire de Neurobiologie Comparée des Invertébrés, 91440 Bures-sur-Yvette; private library of the Alphandéry family, 84140 Montfavet
German Democratic Republic: Drory Bibliothek, Berlin
UK: Moir Library, Edinburgh (Scottish BKA); Ministry of Agriculture Library, London, which includes the Cotton, Cowan and early BBKA collections
USA: Cornell University (NY); Universities of Wisconsin, Minnesota; private library of Dadant & Sons, Hamilton, IL 62341
USSR: Central Bee Research Institute, 391110 Rybnoe; Ministry of Agriculture (All-Union, and those of individual Republics).

These libraries are not freely open to the public, but a number of them would be willing to receive someone with a genuine reason for visiting, who asks in advance for an appointment. Inter-library loans are available in some countries. Subject to copyright regulations, many libraries provide (against payment) photocopies of journal articles. In the future, reference material may be available on compact discs.

Some BKAs run their own lending libraries, books being brought to meetings by a volunteer librarian, and some old-established BKAs have valuable collections.

16.5 ASSISTANCE IN MAINTAINING BEE HEALTH

Sections 15.4–15.45 deal with law in relation to the control of bee diseases and pests. Legislation may require the destruction of colonies of bees – and even of their hives – when an appointed 'bee disease inspector' finds bees that are infected with American foul brood or, in some countries, with certain other organisms.

This can prevent the infection spreading to apiaries of neighbouring beekeepers. Compensation may or may not be paid to the beekeeper whose bees are destroyed, according to the terms of the legislation. It is likely to depend on the extent to which the economic importance of bees is officially recognized in the country, and also on pressure from beekeepers' organizations. Some BKAs arrange group insurance for their members to cover bees and hives compulsorily destroyed.

Any disease that is the subject of effective law must be identifiable, and in most countries the authorities concerned with enforcing the law make provision for free examination of samples of adult bees or brood submitted by beekeepers; the beekeeper is sent a report detailing what disease is found, if any. Additionally, in some BKAs there is a member who examines samples of bees for other members as a voluntary service.

Many countries have enacted legislation to protect bees from damage due to pesticide applications (Table 15.4A). Section 15.52 sets out details, and quotes some cases where beekeepers have been awarded compensation for losses through improper use of pesticide on a crop near their hives. Beekeepers may also be able to take out insurance cover for damage.

16.6 EQUIPMENT SUPPLIERS AS A RESOURCE

Manufacturers and other suppliers of beekeeping equipment constitute a valuable and valued resource to beekeepers. Some are family concerns now in the charge of the fourth or fifth generation since their founder. Others have been established recently by specialists in a modern technology such as applications of plastic materials, who have a special interest in beekeeping. Many suppliers have a wide experience of beekeepers' problems and their solutions.

Suppliers advertise in beekeeping journals, and most issue catalogues or leaflets describing their products. Many welcome visitors, and some serve regularly as hosts to BKA meetings. There is world list of suppliers (IBRA, 1982), giving 265 names and addresses in 40 countries, and a shorter updated list *Suppliers of equipment for tropical and subtropical beekeeping* (IBRA, 1985*a*, No. 1). Both include addresses from which specialized equipment of various types can be obtained, as does *Beekeeping: some tools for agriculture* (Crane, 1985*d*), which deals with equipment especially suited for developing countries. Table 4.31A gives addresses for obtaining comb foundation with different cell sizes, and the equipment for making it.

Appendix 2 of the present book lists a beekeeping supplier in every country where one is known.

16.7 RESOURCES FOR BEEKEEPING DEVELOPMENT PROGRAMMES

This section provides a brief summary of resources for readers:

- wanting to initiate a beekeeping development programme in their own country;
- seeking funds to support such a programme or project;
- working for an aid agency that is or might be involved with a beekeeping project;
- making a feasibility study for a possible future project;
- wanting to take part in a project;
- seeking training that will qualify them to work on a project.

It is important to find out what programmes and feasibility studies have already been carried out in the country concerned, and to consult whatever reports are available. On many occasions in the past, work has been done in ignorance of recent activities of a similar nature in the same or a nearby country. Contact should be made with the appropriate Government department (see Appendix 2 of this book). If there is no relevant feasibility study, one should be made, and *Technical cooperation activities: beekeeping. A directory and guide* (Drescher & Crane, 1982) gives fairly detailed instructions. Both this book and *Tropical and sup-tropical apiculture* (FAO, 1986) make recommendations for carrying out projects.

The primary aim of many development programmes is to produce and sell good quality honey and other hive products in the country of origin. If the programme becomes sufficiently advanced to consider exporting, *Honey: a study of major markets* and *The world market for beeswax* (International Trade Centre, 1986, 1978) should be consulted. These include names and addresses of importers, as does the leaflet *Marketing bee products . . .* (IBRA, 1985*a*, No. 2).

Section 7.6 gives details of economic support for beekeeping in the subtropics and tropics. Some countries – India is one – have planned, funded and executed their own development programme (item 4). Many countries obtain financial aid and expertise through bilateral aid (item 5). The United Nations funds beekeeping projects (item 6) through FAO and TCP and through the United Nations Development Programme (UNDP). The Government department responsible for beekeeping in a specific country may request assistance through the local FAO/UNDP Representative.

Information on obtaining new beekeeping equipment useful in development programmes is given in

Section 16.6. *Used* equipment should never be imported, because of the risk of importing bee diseases. Importation of bees themselves can carry diverse risks, discussed in Sections 8.7–8.75, which should be consulted by any readers considering such action.

For beekeeping specialists who want to gain experience in developing countries, *Sources of voluntary workers for apicultural development* (IBRA, 1985*a*, No. 5) provides details of agencies which organize work by volunteers. Those who already have experience in developing countries, as well as appropriate qualifications, may apply to FAO (Via delle Terme di Caracalla, 00100 Rome, Italy) for consultancy work. Apiservices (Coulaures, 24420 Savignac-les-Eglises, France) may be able to help with information about vacant posts, and IBRA (address below) also maintains a register of consultants.

Opportunities for training in apiculture world-wide (IBRA, 1985*a*, No. 4) gives details of teaching institutions that accept foreign students. Many give courses oriented to tropical conditions, and some also provide grant-aid.

Other publications useful to interested readers include:

- *Bibliography of tropical apiculture* and *Satellite Bibliographies* (Crane, 1978*a*, 1978*b*), with 6445 entries and relating to all developing countries
- series of leaflets *Source materials for apiculture* (IBRA, 1985*a*; also in French and Spanish, IBRA, 1984) including No. 6, *Sources of grant-aid for apiculture development*, which lists agencies that have been active in beekeeping
- *Proceedings of the expert consultation on beekeeping with* Apis mellifera *in tropical and sub-tropical Asia* (FAO, 1984), with papers on development programmes, some under FAO and some under other auspices
- *Beekeeping training in developing countries* (Mulder, 1986)
- *Information for beekeepers in tropical and subtropical countries* (IBRA, 1985*c*)
- *Beekeeping in the tropics* (Segeren, 1988; French ed. 1983)
- *L'apiculture en Afrique tropicale* (Villières, 1987)
- *Newsletter for beekeepers in tropical and subtropical countries*, published twice a year by IBRA, and supplied free to developing countries, as are certain other publications and posters
- *Cornucopia*, quarterly newsletter of IAAD, 3201 Huffman Blvd, Rockford, IL 61103, USA.

Figure 16.7a Opening session of the Second International Conference on Apiculture in Tropical Climates at New Delhi, India, 1980 (photo: Bee World).
Dr M. S. Swaminathan, then Secretary of State for Agriculture, India, and President of IBRA, is third from the left; the author is third from the right.

Enquiries relating to beekeeping in developing countries may be sent to the Information Officer for Tropical Apiculture (IBRA, 16–18 North Road, Cardiff CF1 3DY, UK), a post funded by the UK Overseas Development Administration.

IBRA arranges a series of International Conferences on Apiculture in Tropical Climates (Figure 16.7a), and full proceedings are published; they are cited in the Bibliography under the references in brackets:

1st, London, 1976 (Crane, 1976*a*)
2nd, New Delhi, 1980 (International Conference on Apiculture in Tropical Climates, 1983)
3rd, Nairobi, 1984 (International Conference on Apiculture in Tropical Climates, 1985)
4th, Cairo, 1988 (International Conference on Apiculture in Tropical Climates, 1989)

Apimondia, the International Federation of (National) Beekeepers' Associations, has a Standing Commission on Apiculture in Developing Countries. It also organizes biennial International Apicultural (formerly Beekeeping) Congresses; see Section 16.82. All these international occasions afford good opportunities for locating specialists from other countries with similar interests, and also provide a meeting ground between potential employers and those seeking employment in beekeeping projects.

16.8 RESOURCES FOR BEEKEEPERS WITH SPECIAL REQUIREMENTS

16.81 For beekeepers interested in history

Section 1.4 gives a very brief summary of the long and rich history of beekeeping, and some leads are given here for interested readers to follow up.

Historical books on beekeeping have been written for the Ancient World (Fraser, 1951), and for individual countries, e.g. Britain (Fraser, 1958), France (Marchenay, 1979), Germany and Austria (Bessler, 1886), Ireland (Watson, 1981), USA (Pellett, 1938) and USSR (Galton, 1971). More general studies on man's relationship with bees through the ages are by Beck (1938), Ransome (1937), Free (1982) and others. These books give details of many other publications; the most prolific writer was Armbruster, many of whose papers are listed in a bibliography (Armbruster, 1956).

Of early English books on beekeeping, published between 1500 and 1750, some are available from IBRA, either reprinted in facsimile or as typed transcripts. Extensive bibliographies that include early material from certain countries are mentioned at the start of Section 16.41, and to these should be added Droege's large bibliography (1962) of publications in German.

The archaeology of beekeeping (Crane, 1983*a*) studies material evidence of past beekeeping. Traditional hives and other equipment can be seen in ninety beekeeping museums in different countries; world lists of the

Figure 16.81a Cav. Angelo Cappelletti's Beekeeping Museum at Bregnano, Como, Italy, 1981 (photo: Economia Lariana).

museums have been published (Crane, 1979*a*, 1983*a*), and some information is included in Appendix 2 entries; see also Apimondia (1979). Traditional beekeeping equipment and methods, still used in some of the less developed parts of the world, illustrate many aspects of past beekeeping in a living and fascinating way. Figure 16.81a shows part of one museum.

16.82 For travelling beekeepers

Beekeepers today are able to move about the world with a facility undreamed of by previous generations. Travel in other countries can greatly enrich a beekeeper's life, especially if he takes the trouble to obtain, in advance, details of beekeeping contacts, and of visits that might be made. Once personal contacts are established, doors of hospitality are often opened that lead to interesting and memorable visits, such as an ordinary tourist would never achieve.

Information useful for travellers is given in the country entries in Appendix 2. These include details of a beekeeping journal, BKA, equipment supplier, and so on. If the visit can be timed to take in a beekeepers' meeting, this affords the best chance for making contacts with different beekeepers, some of whom will enjoy showing a visitor their own apiaries and honey houses, and much else of interest. Write to Education Officer, c/o Government contact, or (as a member of your BKA) to the BKA in the country concerned, asking if there are special meetings during the periods when you could be available, and try to book at one held early in your stay in the country. Courses, often in attractive suroundings, are by no means dry-as-dust affairs and, as with other bee meetings, Government support may help to keep the charges modest. You could learn about beekeeping in general, or about special aspects of it.

A series of short articles on *Places to visit* was started in *Bee World*; those so far published cover France (Marchenay, 1981), the Netherlands (Beetsma, 1977) and Japan (Crane, 1984*c*). IBRA organizes a 'Meet the Beekeeper Scheme' for its members, providing them with addresses of beekeepers in different countries who have offered to act as hosts to them.

International Apicultural Congresses are now held in alternate years, each in a different country. The Congresses are organized by the national BKA of the host country in conjunction with Apimondia, the International Federation of BKAs (Secretary, 101 Corso Vittorio Emanuele, 00186 Rome, Italy); they are announced in bee journals. No qualification is required for attendance except payment of the Congress fee. A number of national BKAs arrange group travel to each Congress, usually with a beekeeping tour of the host country or a neighbouring one. Venues have been:

1st	1897	Belgium	18th	1961	Spain
2nd	1900	France	19th	1962	Czechoslovakia
3rd	1902	Netherlands	20th	1965	Romania
4th	1910	Belgium	21st	1967	USA
5th	1911	Italy	22nd	1969	German Federal Republic
6th	1922	France	23rd	1971	USSR
7th	1924	Canada	24th	1973	Argentina
8th	1928	Italy	25th	1975	France
9th	1932	France	26th	1977	Australia
10th	1935	Belgium	27th	1979	Greece
11th	1937	France	28th	1981	Mexico
12th	1939	Switzerland	29th	1983	Hungary
13th	1949	Netherlands	30th	1985	Japan
14th	1951	UK	31st	1987	Poland
15th	1954	Denmark	32nd	1989	Brazil
16th	1956	Austria			
17th	1958	Italy			

Yugoslavia is proposed for 1991. Simultaneous translation is provided at Congresses, and the Proceedings are published by Apimondia Publishing House, in several languages. Some other international meetings are mentioned in Section 16.7.

Language can present difficulties when travelling abroad, but beekeepers can use the *IBRA Dictionary of beekeeping terms* (Crane, 1951–88). The following volumes are in print, and between them they include all 16 languages covered:

for Western Europe
Volume 5, English–French–German–Spanish (1977)
Volume 6, English–Finnish–Hungarian (1978)
Volume 7, English–German–Dutch–Danish–Norwegian–Swedish (1978)
Volume 8, English–French–Italian–Spanish–Portuguese–Romanian (1979)

for Eastern Europe
Volume 3, English–French–German–Czech–Polish–Russian (1964), out of print
Volume 5, (above) includes Russian
Volume 6, (above) includes Hungarian
Volume 8, (above) includes Romanian

for the Americas
Volume 5, (above) includes French and Spanish
Volume 8, (above) includes Portuguese

for Asia
Volume 9, English–French–Japanese (1985)

for the Middle East and North Africa
Volume 10, English–French–Arabic (1988)

16.83 For disabled beekeepers

This section is included less because resources are available than because they are much needed. The only instructions I know were addressed 'to the disabled soldiers, sailors and marines [of the USA] to aid them in choosing a vocation' (Quick, 1919). Accompanying courses of vocational training were held, to help the disabled to become efficient and financially successful beekeepers. But the booklet contains almost no direct advice on beekeeping operations for men with disabilities of various sorts.

The greatest asset to any disabled person is without doubt an able and willing partner to share jobs that cannot be managed alone. Apart from this, a person who has full use of his hands and arms should be able to carry out many beekeeping operations, even if he is paraplegic and therefore confined to a wheelchair. If he is able to hold the body of the smoker between his legs, it should be very well insulated. But this must not be attempted if the legs are incapable of sensation. Special care is needed to prevent stings in the legs if the circulation is poor.

Two paraplegic beekeepers in Southern Africa keep bees on quite a large scale, using Langstroth-type tiered hives. Terry Dawson in Zimbabwe finds that he can work up to four shallow supers by siting the hives low enough. In South Africa, Arthur Wise (1985) manages up to 500 hives, with some help. The hives have shallow supers, full ones being removed and replaced by empty ones, as required, in one operation.

Paraplegic beekeepers who do not already have a capital investment in standard tiered hives would be well advised to consider using long single-storey frame hives such as are described in Section 10.3. They should be in an appropriately designed apiary, set at the most suitable height, and at a spacing that makes each hive easily accessible. The long hives are made to standard dimensions except that the length is about three times the standard, accommodating three times as many standard frames. The use of two vertical queen excluders enables the stored honey to be kept separate from the central brood chamber, and the three parts of the hive should be made accessible separately, by using an individual cover board or cloth for each part.

Use of long single-storey hives should be helpful to the less severely handicapped, to whom lifting presents problems. Those who can use only one arm usually devise hooks or supports to take the place of the missing hand in certain operations. Frames can be picked up with one hand provided they have long lugs. (Also, an individual comb from the hive shown in Figure 10.23c can be pivoted round one end of the top-bar to lift it out of the hive.)

If tiered hives are used, shallow honey supers are best because they can often be lifted with one hand, whereas deep ones cannot. Nelson Murrell in England suggests smearing vaseline on the top and bottom edges of the hive boxes, so that they can easily be slid apart to separate them. When manipulating, cover cloths are especially helpful (Section 5.23). If both hands are functional, a cover cloth is hung over the far edge of the hive (which is parallel to the frames), and is pulled across the top of the (open) box below, as the upper box is being slid off towards the operator.

A beekeeper who can walk only with crutches may be able to operate certain types of vehicle and lifting machinery. One such beekeeper, a polio victim in Connecticut, migrated 80 hives for many years, using a truck and a boom loader.

In La Plata, Argentina, an experiment has been conducted in teaching blind people to make and assemble hives, frames, etc., and to carry out beekeeping operations (Schelotto et al., 1973). The beekeepers learned 'how to check and handle hives, including: to estimate the hive condition and the honey ripening potential; to determine the moment of harvesting; to recognize the capped and uncapped brood, drone and queen cells; to estimate the general condition of a colony; to estimate the presence of wax moth'. The authors do not explain how this was done. In Mauritius the International Labour Organization launched a Honey Production Project in the early 1980s, as part of the Vocational Rehabilitation of the Disabled, emphasis being on the need to generate income (Tugwell, 1985).

A specialization that might be attractive to a wheelchair beekeeper is queen rearing, and adaptations to

Figure 16.83a Obverse (*left*) and reverse (*right*) of a medal issued by the Food and Agriculture Organization of the United Nations in 1981 to mark the International Year of Disabled Persons (design: Giorgio Massini and Lusiano Zanelli).

overcome his limitations should be possible. However, queen rearing involves much patience and delicate work, and it should not be undertaken until experience has been gained with ordinary beekeeping.

A disability of a different kind is allergy (hypersensitivity) to bees or bee venom, and Section 5.54 deals with this.

The United Nations designated 1981 as the International Year of Disabled Persons, and marked it by issuing a medal (Figure 16.83a) for which the theme of apiculture was chosen, 'symbolizing the collaboration of human society by the collaboration of bees ... The dignity and ability of disabled persons who work is portrayed on the obverse of the medal by the beekeeper, whose protective clothing creates a symbol independent of race or sex.' The beekeepers' disability is indicated by the crutches leaning against the hive.

16.9 FURTHER READING AND REFERENCE

Details of publications listed will be found in the Bibliography.

Beekeeping Extension Officers and Beekeepers' Associations

Apiacta, each issue has a list of Member Associations

Gleanings in Bee Culture, each April issue includes *Who's who in apiculture* for Canada and USA, which also lists disease inspectors

Beekeeper's annual (many organizational details for the UK)

Publications and libraries

Crane, E. (1978*a*, 1978*b*) *Bibliography of tropical apiculture* and *Satellite Bibliographies*

Crane, E. (1984*a*) *Bibliographical tools for apiculture*

IBRA (1985*a*, No. 8) *Apicultural reference books for developing countries*

Equipment suppliers as a resource

IBRA (1982) *Directory of suppliers of beekeeping equipment world-wide*

IBRA (1985*a*, No. 1) *Suppliers of equipment for tropical and subtropical beekeeping*

Development programmes

See Section 7.6, and publications cited in Section 16.7

Resources for those interested in history

Apimondia (1979) *Bienenmuseum und Geschichte der Bienenzucht*

Crane, E. (1979*a*) *Directory of the world's beekeeping museums*

Crane, E. (1983*a*) *The archaeology of beekeeping*

Appendixes

Appendix 1

Important world honey sources and their geographical distribution

The 464 plants in this table are in *Directory of important world honey sources* (Crane, Walker & Day, 1984). Here they are listed in alphabetical order of their botanical families; a common name is added where one is known. Many are crop plants (Table 8.4A), and crops recorded as pollinated by honeybees in the *Pollination directory for world crops* (Crane & Walker, 1984*a*) are here marked * after the common name.

The plant source of the honey is nectar, except for 14 entries marked 'honeydew, no nectar'. Eight of the nectar sources may also be important sources of honeydew honey; they are marked 'honeydew as well as nectar'.

A letter on the left of each plant name refers to its pollen production:

P = plant produces pollen reported to be valuable to honeybees
p = plant produces pollen collected by honeybees but less valuable to them, or its value is not known
× = pollen not produced, or in some way inadequate or toxic for honeybees
not marked = no information

Letters on the right of the table refer to the geographical distribution of the plant – in 7 continental regions (see map) – based on Crane, Walker and Day (1984) and some additional information from Johannsmeier (1984). For any of the continental regions referred to at the head of the 7 columns, M, S or T indicates that the plant grows in its temperate, subtropical, or tropical zone. There is no space to consider arid areas, or high altitudes. Many entries are based on a record for the plant as a honey source in a single region, and it may not be a honey source in all regions entered.

Both Europe and Asia include areas in the USSR. For convenience, the whole of Mexico is included with Central America (CA; Ca = Caribbean area). Oceania is taken to include all Pacific islands, Papua New Guinea and adjacent islands, and Australia and New Zealand. The islands of the Philippines and Indonesia are included with Asia.

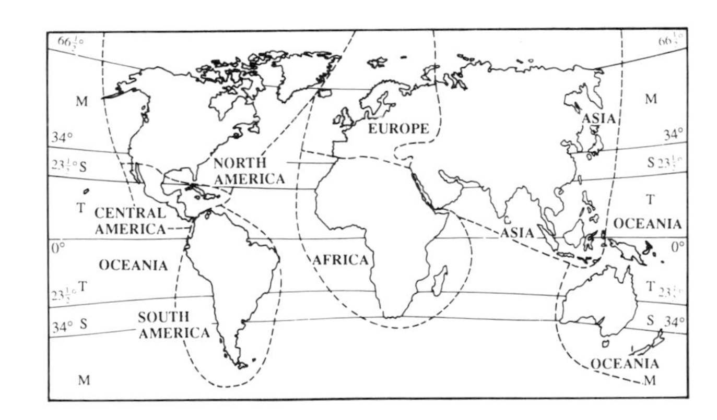

Map of the world showing continental regions.
The map (based on Good, 1974) also shows the boundaries between latitude zones used to indicate the distribution of important honey sources, and in Chapter 6 and 7.

	latitudes (N and S)	*land area on which plants grow* 1000 km²	%
tropics (T)	0° to 23½°	47	39
equatorial	0° to 15°		
subtropics (S)	23½° to 34°	27	22
temperate zones (M)	34° to 66½°	47	39
warm temperate	34° to 45°		
cold temperate	45° to 66½°		
		121	100

Botanical name	Common name	Europe M	Asia MST	Africa MST	N Am MS	CA+Ca ST	S Am MST	Oceania MST
Acanthaceae								
p *Adhatoda vasica* Nees	bhaikar		ST					
P *Carvia callosa* (Nees) Brem.	karvi		T					
p *Isoglossa deliculata* C. B. Clarke	hlalwane			S				
Mackenziea integrifolia (Dalz.) Brem.	wahiti		T					
Monechma australe P. G. Meyer	perdebos			ST				
Petalidium linifolium T. Anders	lusernbos			ST				
P *Phlebophyllum kunthianum* Nees	kurinji		T					
P *Thelepaepale ixiocephala* (Benth.) Bremk.	darmori		T					
Thunbergia grandiflora (Roxb. ex Rottl.) Roxb.	conejitos		T					
Aceraceae								
p *Acer circinatum* Pursh	vine maple				M			
p *Acer macrophyllum* Pursh	Oregon maple				M			
p *Acer platanoides* L. (honeydew as well as nectar)	Norway maple	M	M		M			
p *Acer pseudoplatanus* L. (honeydew as well as nectar)	sycamore	M	M				M	
p *Acer tataricum* L.	Tatarian maple	M	M					
Agavaceae								
p *Agave americana* L.	century plant	M		ST	S	ST		
Anacardiaceae								
Anacardium excelsum (Bert. & Balbis ex Kunth) Skeels	espavé					T	T	
p *Anacardium occidentale* L.	cashew nut		ST	ST	S	ST	ST	T
p *Holigarna grahamii* Hook. f.	mothi ranbibata		T					
P *Mangifera indica* L.	mango*		ST	ST	S	T	T	
Rhus glabra L.	red sumac				MS			
Rhus taitensis Guill.	tavahi		T					T
Rhus typhina L.	staghorn sumac				M			
Schinus terebinthifolius Raddi	Brazilian pepper			T	S	T	T	
p *Sclerocarya caffra* Sond.	marula			ST				
p *Spondias mombin* L.	hog plum			T		T	T	
Aquifoliaceae								
Ilex glabra (L.) A. Gray	lowbush gallberry				S			
Ilex integra Thunb.	yerboso					T		
p *Ilex theezans* Mart.	congonha						S	
Araliaceae								
Gilibertia arborea (L.) March.	—					T	T	
p *Schefflera wallichiana* Harms.	doddabettu		T					
Asclepiadaceae								
× *Asclepias syriaca* L.	milkweed*	M			M			
Avicenniaceae								
p *Avicennia germinans* (L.) L.	black mangrove			T	S	T	T	
p *A. marina* (Forssk.) Vierh. v. *resinifera* (Forst.) Bakh.	white mangrove			S				ST
Berberidaceae								
P *Mahonia trifoliata* (Moric.) Fedde	agritos				S	S		

| Botanical name | Common name | Europe M | Asia MST | Africa MST | N Am MS | CA+Ca ST | S Am MST | Oceania MST |

Botanical name	*Common name*	*Europe M*	*Asia MST*	*Africa MST*	*N Am MS*	*CA + Ca ST*	*S Am MST*	*Oceania MST*
Bignoniaceae								
Rhigozum trichotomum Burch.	driedoring		ST	S				
Bombacaceae								
p *Bombax ceiba* L.	silk-cotton tree		ST					ST
P *Ceiba pentandra* (L.) Gaertn.	silk-cotton tree*		T	T		T	T	
p *Durio zibethinus* Murr.	durian*		T			T		
Boraginaceae								
p *Anchusa officinalis* L.	bugloss	M						
p *Borago officinalis* L.	borage*	M						M
Cordia alba (Jacq.) Roemer & Schultes	tiguilote					T	T	
p *Cordia alliodora* (Ruiz & Pavon) Cham.	cypre					T	T	
Cordia gerascanthus L.	Spanish elm					T	T	
× *Echium lycopsis* L.	viper's bugloss	M					S	MS
p *Echium vulgare* L.	viper's bugloss	M	M	M	M		M	M
p *Ehretia acuminata* R. Br.	puna		ST					
p *Tournefortia argentea* L. f.	tree heliotrope		T					T
Cactaceae								
P *Carnegiea gigantea* (Engelm.) Britton & Rose	saguaro				S	S		
P *Opuntia engelmanii* Salm-Dyck	prickly pear				S	T		
Caprifoliaceae								
Lonicera caerulea L.	blue honeysuckle		M					
p *Symphoricarpos albus* (L.) S. F. Blake	snowberry	M			MS			
Caricaceae								
p *Carica papaya* L.	papaya*		T	T			T	T
Clethraceae								
p *Clethra alnifolia* L.	pepper bush				MS			
Cochlospermaceae								
Cochlospermum insigne St. Hil.	algodonillo						T	
p *Cochlospermum vitifolium* (Willd.) Spreng.	bototo					T	T	
Combretaceae								
p *Bucida buceras* L.	guaraguao				S	T		
Combretum celastroides Laws.	savanna bushwillow			T				
Combretum fruticosum (Loefl.) Stuntz	chapamiel					T	T	
Combretum imberbe Wawra	leadwood			ST				
Combretum zeyheri Sond.	large-fruited bushwillow			T				
Guiera senegalensis J. Gmelin	n'guer			T				
p *Terminalia arjuna* (Roxb.) Wight & Arn.	mathi		ST	T	S	T		
p *Terminalia bellerica* (Gaertn.) Roxb.	behda		T					
Terminalia chebula Retz.	inknut		T					
p *Terminalia tomentosa* Wight & Arn.	ain		T					
Compositae								
p *Baccharis dracunculifolia* DC.	alecrim						ST	
P *Bidens pilosa* L.	Spanish needle				S	T		
P *Calea urticifolia* (Miller) DC.	jalacate					ST		
P *Centaurea cyanus* L.	cornflower	M		M	M			
P *Centaurea solstitialis* L.	star-thistle	M			S			M

Botanical name	*Common name*	*Europe M*	*Asia MST*	*Africa MST*	*N Am MS*	*CA + Ca ST*	*S Am MST*	*Oceania MST*

Botanical name	Common name	Europe M	Asia MST	Africa MST	N Am MS	CA + Ca ST	S Am MST	Oceania MST
× *Chromolaena odorata* (L.) R. King & H. Robinson	white snake root		T		S	ST	ST	
Coreopsis borianiana Schultz-B.p.	adey-abeba			T				
p *Cynara cardunculus* L.	cardoon	M					MS	MS
(*Eupatorium odoratum* L. now *Chromolaena odorata*)								
p *Guizotia abyssinica* Cass.	niger*		T	T				
P *Helianthus annuus* L.	sunflower*	M	MST	ST	MS		MST	MS
p *Mikania scandens* Willd.	snowvine				MS		T	
P *Taraxacum officinale* Weber	dandelion	M	M	M	M	T	M	MS
p *Tithonia tubaeformis* Cass.	acahual					T		
P *Tridax procumbens* L.	dagad-phul		T	T		T		
Vernonia polyanthes Less.	suquinay						ST	
× *Vernonia poskeana* Vatke & Hildebrandt	—			S				
P *Viguiera helianthoides* Kunth	tah					T		
Convolvulaceae								
p *Ipomoea acuminata* (Vahl) Roem. & Schultes	blue dawn flower		T					
Ipomoea batatas (L.) Lam.	sweet potato*	M	T	T	S	T	T	
Ipomoea nil (L.) Roth	campanita					T	T	
P *Ipomoea sidifolia* Choisy	campanilla					T		
Ipomoea triloba L.	campanilla morada					T	T	
p *Jacquemontia nodiflora* G. Don	campanitas					T		
p *Turbina corymbosa* (L.) Raf.	aguinaldo blanco				S	T	T	
Cruciferae								
P *Barbarea vulgaris* R. Br.	yellow rocket	M			M			
P *Brassica campestris* L.	field mustard*	M	MST		MS		MS	M
P *Brassica campestris* L. v. *dichotoma* Prain	toria*		S					
P *Brassica campestris* L. v. *sarson* Prain	sarson*		ST					
P *Brassica juncea* (L.) Cosson	Indian mustard*	M	S				S	
p *Brassica napus* L. v. *napobrassica* (L.) Reichenb.	shaljam	M	S					
P *Brassica napus* L. v. *oleifera* DC.	oilseed rape*	M	M		M			M
p *Brassica nigra* (L.) Koch	black mustard*	M	MS		S			
P *Brassica rapa* L. subsp. *oleifera* DC.	turnip rape*	M						
P *Diplotaxis erucoides* (L.) DC.	white wall-rocket	M						
p *Sinapis alba* L.	white mustard*	M	M					
P *Sinapis arvensis* L.	charlock	M	M		M			M
Cucurbitaceae								
P *Cucumis melo* L.	melon*	M	T	ST	M			
p *Cucurbita pepo* L.	pumpkin*	M				MS		
Sicyos deppei G. Don	bur cucumber				MS	S		
Sicyos laciniatus L.	chayotillo					S		
Cunoniaceae								
Caldcluvia paniculata D. Don	tiaca						M	
p *Weinmannia racemosa* L. f.	kamahi							M
Weinmannia trichosperma Cav.	tineo						M	
Cupressaceae								
Calocedrus decurrens (Torr.) Florin (honeydew, no nectar)	white cedar	M			MS			

| Botanical name | Common name | Europe M | Asia MST | Africa MST | N Am MS | CA + Ca ST | S Am MST | Oceania MST |

Botanical name	Common name	*Europe* *M*	*Asia* *MST*	*Africa* *MST*	*N Am* *MS*	*CA + Ca* *ST*	*S Am* *MST*	*Oceania* *MST*
Dilleniaceae								
p *Curatella alata* Vent.	bejuco chachaco						T	
p *Dillenia pentagyna* Roxb.	toothed dillenia		T					
Dipsacaceae								
p *Dipsacus fullonum* L.	teasel	M	M	M	M			
Dipsacus pilosus L.	small teasel	M	M					
Dipterocarpaceae								
Marquesia macroura Gilg	muvúca			T				
Ebenaceae								
Diospyros batocana Hiern	sand jackal-berry			T				
Diospyros virginiana L.	American persimmon	M			MS			
Elaeagnaceae								
Elaeagnus angustifolia L.	wild olive	M	M					
Ericaceae								
P *Calluna vulgaris* (L.) Hull	heather, ling	M						
P *Erica arborea* L.	tree heath	M		M				
P *Erica cinerea* L.	bell heather	M						
p *Erica herbacea* L.	spring heath	M						
Erica manipuliflora Salisb.	heather	M		M				
× *Oxydendron arboreum* (L.) DC.	sourwood				M			
p *Vaccinium uliginosum* L.	bog whortleberry	M	M					
Eucryphiaceae								
p *Eucryphia cordifolia* Cav.	ulmo						M	
Eucryphia lucida (Labill.) Baill.	leatherwood							M
Eucryphia milligannii Hook. f.	leatherwood							M
Euphorbiaceae								
p *Croton floribundus* Spreng.	capixingui						ST	
× *Hevea brasiliensis* Muell. Arg.	Pará rubber		T	T		T	T	
Julocroton triqueter Baill.						T	ST	
p *Manihot glaziovii* Muell. Arg.	Ceará rubber		T	T			ST	
Fagaceae								
P *Castanea pubinervis* (Hassk.) C.K. Schn.	Japanese chestnut*		M					
P *Castanea sativa* Mill. (honeydew as well as nectar)	sweet or Spanish chestnut*	M	M	M				M
Fagus sylvatica L. (honeydew, no nectar)	beech	M						
Nothofagus solandri v. *cliffortioides* (Hook. f.) Poole (honeydew, no nectar)	mountain beech							M
P *Quercus robur* L. (honeydew, no nectar)	English oak	M	M	M				
p *Quercus suber* L. (honeydew, no nectar)	cork oak	M	M					M
Quercus virginiana Mill. (honeydew, no nectar)	live oak				MS	T		
Geraniaceae								
p *Geranium pratense* L.	meadow cranesbill	M						
Goodeniaceae								
Scaevola frutescens (Mill.) Krause	veloutier		T					

Botanical name	Common name	*Europe* *M*	*Asia* *MST*	*Africa* *MST*	*N Am* *MS*	*CA + Ca* *ST*	*S Am* *MST*	*Oceania* *MST*

	Botanical name	Common name	Europe M	Asia MST	Africa MST	N Am MS	CA + Ca ST	S Am MST	Oceania MST
	Gramineae								
p	*Saccharum officinarum* L. (sap from cut cane)	sugar cane	M	T	ST	S	T	T	ST
	Heliconiaceae								
	Heliconia aurantiaca Ghiesb.	platanillo		T				T	
	Hippocastanaceae								
×	*Aesculus hippocastanum* L.	horse-chestnut	M	M					M
p	*Aesculus turbinata* Bl.	Japanese horse-chestnut		M					
	Hydrophyllaceae								
p	*Phacelia tanacetifolia* Benth.	phacelia	M			M			M
	Labiatae								
	Dracocephalum moldavica L.	Moldavian balm	M	M					
p	*Dypsophylla stellata* Benth.	gomani		T					
	Hyptis suaveolens (L.) Poit.	oregano					T	T	
P	*Hyssopus officinalis* L.	hyssop	M		M				
P	*Lavandula angustifolia* Miller	lavender*	M	M	T	M			
	Lavandula dentata L.	toothed lavender	M		M				
	Lavandula latifolia Medicus	spike lavender*	M		M				
	Lavandula stoechas L.	French lavender	M		M				
	L. angustifolia Miller × *latifolia* Medicus	lavandin	M		M				
p	*Leonurus cardiaca* L.	motherwort	M	M		M			
	L. cardiaca L. subsp. *villosus* (Desf. ex Sprengel) Hyl.	motherwort	M	M		MS			
	Leucas aspera Link.	wild ocimum		T					
p	*Marrubium vulgare* L.	white horehound	M	M		M			MS
p	*Plectranthus gerardianus* Wall. ex Benth.	—		S					
p	*Plectranthus striatus* Wall. ex Benth.	—		S					
p	*Rabdosia coetsa* (Buch. Ham. ex D. Don) Hara	shain		MST					
p	*Rabdosia rugosa* (Wall. ex Benth.) Hara	shain		MS					
P	*Rosmarinus officinalis* L.	rosemary*	M	M	MST				
p	*Salvia apiana* Jepson	white sage					S		
p	*Salvia leucophylla* Greene	purple sage					S		
p	*Salvia mellifera* Greene	black sage					S		
p	*Salvia nemorosa* L.	—	M	M					
p	*Salvia officinalis* L.	garden sage*	M						
p	*Satureia montana* L.	winter savory	M		M				
	Stachys annua (L.) L.	annual yellow woundwort	M						
	Thymus capitatus (L.) Hoffm. & Link	mediterranean wild thyme	M	M					
p	*Thymus serpyllum* L.	wild thyme	M			M			
P	*Thymus vulgaris* L.	garden thyme	M						M
p	*Trichostema lanceolatum* Benth.	blue curls				MS			
	Lauraceae								
	Actinodaphne angustifolia Nees	pisa		T					
P	*Actinodaphne hookerii* Meissn.	pisa		T					
P	*Litsea stocksii* Hook. f.	pisa		T					
P	*Machilus macrantha* Nees	kardel		T					

Botanical name *Common name* *Europe M* *Asia MST* *Africa MST* *N Am MS* *CA + Ca ST* *S Am MST* *Oceania MST*

	Botanical name	Common name	Europe M	Asia MST	Africa MST	N Am MS	CA + Ca ST	S Am MST	Oceania MST
	Lauraceae (*continued*)								
p	*Persea americana* Mill.	avocado*			ST	S	T	ST	
	Persea caerulea (Ruiz & Pavon) Mez.	aguacatillo						T	
	Leguminosae								
	Acacia berlandieri Benth.	guajillo				S	S		
p	*Acacia caffra* (Thunb.) Willd.	common hook-thorn			ST				
	Acacia decurrens (Wendl.) Willd.	black wattle			ST			ST	ST
P	*Acacia greggii* A. Grey	catsclaw				S	S		
p	*Acacia mellifera* (Vahl) Benth.	blackthorn			ST				
p	*Acacia modesta* Wall.	phulai		S					
	Acacia polyphylla DC.	guarucaia						T	
	Acacia senegal (L.) Willd.	gum acacia		ST	ST				
	Acacia seyal Del.	spiny mimosa		T	ST				
	Acacia tortilis (Forssk.) Hayne	umbrella thorn		T	S				
	Andira inermis (Wright) DC.	angelin			T		T	T	
P	*Astragalus sinicus* L.	Chinese milk vetch		MS					
	Baikiaea plurijuga Harms.	Rhodesian teak			ST				
p	*Brachystegia floribunda* Benth.	—			ST				
	Brachystegia laurentii (De Wild.) Louis ex Hoyle	eko			T				
p	*Brachystegia longifolia* Benth.	—			ST				
p	*Brachystegia speciformis* Benth.	panda			ST				
	Brachystegia tamarindoides Welw. ex Benth.	mussamba			T				
p	*Butea monosperma* (Lam.) Taub.	bastard teak		S					
p	*Caesalpinia coriaria* (Jacq.) Willd.	dividivi					ST	T	T
	Cajanus cajan (L.) Millsp.	pigeon pea*		ST	ST		T	T	
P	*Calliandra calothyrsus* Meissn.	red calliandra		T			T		
	Cassia siamea Lam.	yellow cassia		T	T	S	T	T	
p	*Cercidium floridum* Benth.	palo verde				S	S		
p	*Cicer arietinum* L.	chick pea		T	ST		T	T	
	Cryptosepalum pseudotaxus Bak. f.	mucube			T				
	Cynometra alexandri C. H. Wright	tembu			T				
p	*Dalbergia sissoo* DC.	sissoo		ST					
	Dalea revoluta S. Watson	popote chiquito					S		
	Daniellia oliveri (Rolfe) Hutch & Dalz.	santan			T	M			
	Dialium engleranum Henriques	Kalahari podberry			ST				
	Gilibertiodendron dewevreii (De Wild.) Leonard	mbau			T				
p	*Gleditsia triacanthos* L.	honey locust	M	S	ST	MS			S
×	*Gliricidia sepium* (Jacq.) Walp.	madre de cacao		T	T	S	T	T	
p	*Glycine max* (L.) Merr.	soya bean*		MST	ST	MS	T	ST	
×	*Haematoxylum campechianum* L.	logwood			T		ST	T	
P	*Hedysarum coronarium* L.	sweetvetch*	M		M				M
	Hymenaea courbaril L.	locust					T	T	
	Hymenaea stilbocarpa Hayne	paquio						T	
P	*Inga laurine* (Sw.) Willd.	guamá					T	T	
	Inga micheliana Harms.	cushin					T		
	Inga vera Willd.	guava					T		
	Isoberlinia angolensis (Welw. ex Benth.) Hoyle & Brenan	kapane			T				
	Julbernardia globiflora (Benth.) Troupin	munondo			ST				

| Botanical name | Common name | Europe M | Asia MST | Africa MST | N Am MS | CA + Ca ST | S Am MST | Oceania MST |

	Botanical name	Common name	Europe M	Asia MST	Africa MST	N Am MS	CA + Ca ST	S Am MST	Oceania MST
	Julbernardia paniculata (Benth.) Troupin	munsa			T				
	Julbernardia unijugata J. Leon.	—			T				
	Lespedeza bicolor Turcz.	lespedeza*	M	M					
	Lespedeza cyrtobotrya Miq.	lespedeza*		M					
	Lonchocarpus pictus Pittier	majomo						T	
P	*Lotus corniculatus* L.	birdsfoot trefoil*	M	M	M	M		MS	M
	Machaerium eriocarpum Benth.	tusequi						T	
p	*Medicago falcata* L.	lucerne	M	MS					
p	*Medicago laciniata* (L.) Mill.	—	M	MS	M				
p	*Medicago lupulina* L.	black medick*	M	MS	M	M		M	M
×	*Medicago sativa* L.	lucerne or alfalfa*	M	MST	ST	MS	S	MS	MS
p	*Melilotus alba* Desr.	white sweet clover*	M	MS	MS	M		MST	MS
P	*Melilotus officinalis* (L.) Pall.	yellow sweet clover*	M	MS	M	M			
p	*Mimosa scabrella* Benth.	bracatinga						ST	
	Myrospermum frutescens Jacq.	tarara						T	
P	*Onobrychis viciifolia* Scop.	sainfoin*	M	MST	S	M		M	
	Parkia biglobosa (Jacq.) Benth.	locust bean tree			T				
p	*Parkinsonia aculeata* L.	horsebean		S	ST	MS	ST	MST	
p	*Phaseolus coccineus* L. (was *P. multiflorus* Willd.)	runner bean*	M		S			T	
	Piscidia piscipula (L.) Sarg.	dogwood				S	T	T	
	Pithecellobium arboreum (L.) Urb.	—					T		
p	*Pithecellobium dulce* (Roxb.) Benth.	Madras thorn		T	T	S	ST	T	T
	Pithecellobium unguis-cati (L.) Mart.	cat's claw				S	T	T	
P	*Pongamia pinnata* (L.) Pierre	sour fruit		T		S			T
p	*Prosopis cineraria* (L.) Druce	mesquite		ST					
p	*Prosopis farcta* (Sol. ex Russell) J. F. Macbride	mesquite		MS	MS				
P	*Prosopis glandulosa* Torrey	honey mesquite		ST	S	S	ST		S
P	*Prosopis juliflora* (Sw.) DC.	algaroba, mesquite		ST	ST		T	T	S
	Prosopis pallida (Humboldt & Bonpl. ex Willd.) Kunth	kiawe		S			T	T	T
	Prosopis pubescens Benth.	screwbean			S	S	ST		
P	*Psoralea pinnata* L.	blue pine weed			S				MS
p	*Pterocarpus rotundifolius* (Sond.) Druce	round-leaved bloodwood			ST				
P	*Robinia pseudoacacia* L. (honeydew as well as nectar)	false acacia	M	MS	S	MS		M	M
P	*Tamarindus indica* L.	tamarind		T	T	S	T		T
	Tipuana tipu (Benth.) O. Kuntze	tipa			S			T	
P	*Trifolium alexandrinum* L.	berseem*		MS	S				
p	*Trifolium fragiferum* L.	strawberry clover*	M						M
P	*Trifolium hybridum* L.	alsike clover*	M	M	M	M		M T	M
p	*Trifolium incarnatum* L.	crimson clover*	M		M	MS		M	M
P	*Trifolium pratense* L.	red clover*	M	MS	M	MS		MST	M
P	*Trifolium repens* L.	white clover*	M	MS	M	MS		MST	MS
P	*Trifolium resupinatum* L.	Persian clover*	M	MS		S			MS
p	*Vicia faba* L.	field bean*	M	ST	MST		T	ST	
P	*Vicia sativa* L.	vetch	M	MS		MS	T	M	
	Vicia villosa Roth	hairy vetch*	M	MS	S	M			
	Liliaceae								
P	*Aloe davyana* Schonl.	davyana aloe			S				

| Botanical name | Common name | Europe M | Asia MST | Africa MST | N Am MS | CA + Ca ST | S Am MST | Oceania MST |

Botanical name	Common name	*Europe M*	*Asia MST*	*Africa MST*	*N Am MS*	*CA + Ca ST*	*S Am MST*	*Oceania MST*
Liliaceae (*continued*)								
p *Aloe dichotoma* Masson	quiver tree			ST				
P *Aloe mutans* Reynolds	mutans aloe			ST				
Linaceae								
p *Linum usitatissimum* L.	flax or linseed*	M	M	T	M	T	M	
Lythraceae								
p *Lagerstroemia parviflora* Roxb.	nandi		T					
P *Lythrum salicaria* L.	purple loosestrife	M	M		M			M
Magnoliaceae								
p *Liriodendron tulipifera* L.	tulip poplar	M			MS			MS
Magnolia grandiflora L.	magnolia				M			MS
Malpighiaceae								
Byrsonima crassifolia (L.) DC.	chaparro manteco					T	T	
Malvaceae								
p *Gossypium barbadense* L.	long-staple cotton*	M	MS	T	S	T	T	
p *Gossypium hirsutum* L.	upland cotton*	M	M	T	S	T		
p *Hibiscus rosa-sinensis* L.	shoe-flower					T		
Meliaceae								
p *Azadirachta indica* A. Juss.	nim		ST	ST				
Khaya senegalensis (Desv.) A. Juss.	caïlcédrat			T				
p *Toona ciliata* M. Roem.	toon		S					
Trichilia havanensis Jacq.	bastard lime					T	T	
Moringaceae								
P *Moringa oleifera* Lam.	horseradish tree		T	T		T		
Musaceae								
P *Musa* spp	banana, plantain		T	T		T	T	
Myrsinaceae								
P *Aegiceras corniculatum* (L.) Blanco	river mangrove		T					T
Myrtaceae								
p *Callistemon citrinus* (Curt) Skeels	crimson bottle brush		ST					M
P *Eucalyptus accedens* W. Fitzg.	paper-barked wandoo							S
p *Eucalyptus alba* Reinw. ex Blume	poplar gum		T	M		T	ST	ST
× *Eucalyptus albens* Benth.	white box			S				S
p *Eucalyptus anceps* (Maiden) Blakely	white mallee							S
× *Eucalyptus caleyi* Maiden	Caley's ironbark							MS
P *Eucalyptus calophylla* R. Br.	red gum			ST				S
P *Eucalyptus camaldulensis* Dehnh.	river red gum	M	ST	MST			ST	MST
p *Eucalyptus citriodora* Hook.	lemon-scented gum	M	ST	ST		T	ST	S
p *Eucalyptus cladocalyx* F. Muell.	sugar gum			MS				MS
p *Eucalyptus cornuta* Labill.	yate			S				S
P *Eucalyptus crebra* F. Muell.	narrowed-leaved red ironbark			S				S
× *Eucalyptus diversicolor* F. Muell.	karri			S				S
× *Eucalyptus drepanophylla* F. Muell. ex Benth.	grey ironbark							S
p *Eucalyptus falcata* Turcz.	white mallet							S
× *Eucalyptus fasciculosa* F. Muell.	pink gum							MS
p *Eucalyptus ficifolia* F. Muell.	scarlet gum	M		S	S		S	MS

Botanical name | Common name | Europe M | Asia MST | Africa MST | N Am MS | CA + Ca ST | S Am MST | Oceania MST

Botanical name	Common name	Europe M	Asia MST	Africa MST	N Am MS	CA + Ca ST	S Am MST	Oceania MST
P *Eucalyptus globulus* Labill.	blue gum	M	ST	ST	S	T	ST	M
p *Eucalyptus gomphocephala* DC.	tuart	M	S	S				S
× *Eucalyptus gracilis* F. Muell.	white mallee							MS
P *Eucalyptus grandis* W. Hill ex Maiden	rose gum		ST	MST		T	S	ST
P *Eucalyptus incrassata* Labill.	yellow mallee							MS
p *Eucalyptus jacksonii* Maiden	red tingle							S
× *Eucalyptus leucoxylon* F. Muell.	white ironbark	M	M	MST	S			MS
× *Eucalyptus loxophleba* Benth.	York gum							S
p *Eucalyptus macrorhyncha* F. Muell. ex Benth.	red stringybark							MS
P *Eucalyptus maculata* Hook.	spotted gum	M		MST			M	MS
× *Eucalyptus melliodora* A. Cunn. ex Schauer	yellow box	M	MS	ST				MS
× *Eucalyptus moluccana* Roxb.	grey box							S
p *Eucalyptus oleosa* F. Muell. ex Miq.	giant mallee							MS
p *Eucalyptus panda* S. T. Blake	corky ironbark							S
× *Eucalyptus paniculata* Smith	grey ironbark			ST			S	S
P *Eucalyptus platypus* Hook.	moort							S
× *Eucalyptus polyanthemos* Schauer	red box		S	MS				MS
P *Eucalyptus propinqua* Deane & Maiden	grey gum							MS
p *Eucalyptus robusta* Smith	swamp messmate	M	ST	ST			ST	S
p *Eucalyptus rubida* Deane & Maiden	candle bark gum			S				MS
p *Eucalyptus saligna* Smith	Sydney blue gum			M		T	ST	S
× *Eucalyptus sideroxylon* A. Cunn. ex Woolls	black ironbark		S	MST				MS
p *Eucalyptus socialis* F. Muell. ex Miq.	Christmas mallee							MS
P *Eucalyptus tereticornis* Smith	blue gum	M	ST	ST			ST	MST
P *Eucalyptus viminalis* Labill.	blue gum			MS	S		S	MS
p *Eucalyptus wandoo* Blakely	white gum			M				S
Eugenia spicata Lamk.	bhedas		T					
Eugenia zeylanica Wight	gudda panneralu		T					
p *Leptospermum scoparium* J. & G. Forst.	manuka							MS
p *Melaleuca leucadendron* (L.) L.	broad-leaved tea-tree			T		T	T	S
p *Melaleuca preissiana* Schau	flat-leaved paperbark							S
P *Melaleuca quinquenervia* (Cav.) S. T. Blake	cajeput			T	S			S
p *Metrosideros excelsa* Sol. ex Gaertn.	pohutukawa							MS
p *Metrisideros robusta* A. Cunn.	northern rata							M
Metrosideros umbellata Cav.	southern rata							M
P *Psidium guajava* L.	guava*		ST	T	S	T	S	T
p *Syzygium aromaticum* (L.) Merrill & Perry	clove tree*		T	T				
p *Syzygium cordatum* Hochst. ex Krauss	water berry		T					
P *Syzygium cuminii* (L.) Skeels	Java plum		T	T	S	T		ST
p *Syzygium jambos* (L.) Alston	rose apple		T		S	T		
Nyssaceae								
Nyssa aquatica L.	black tupelo				S			
× *Nyssa ogechea* Bartram	white tupelo				S			
Nyssa sylvatica Marshall	black gum				MS			
Oleaceae								
P *Ligustrum walkeri* Decne.	pungalam		T					
Onagraceae								
P *Epilobium angustifolium* L.	rosebay willow-herb	M	M		M			
P *Fuchsia excorticata* (J. & G. Forst.) L. f.	konini							M
Ludwigia nervosa (Poir.) Hara	cariaquito						T	

Botanical name | Common name | Europe M | Asia MST | Africa MST | N Am MS | CA + Ca ST | S Am MST | Oceania MST

	Botanical name	Common name	*Europe* *M*	*Asia* *MST*	*Africa* *MST*	*N Am* *MS*	*CA+Ca* *ST*	*S Am* *MST*	*Oceania* *MST*
	Palmae								
P	*Cocos nucifera* L.	coconut*		T	T	S	T		
P	*Elaeis guineensis* Jacq.	oil palm*		T	T		T		
P	*Roystonea regia* (Kunth) O. F. Cook	royal palm		T		S	T	T	
	Sabal florida Becc.	palma cana					T		
	Sabal palmetto (Walt.) Lodd. ex Schultes	palmetto				S	T		
p	*Serenoa repens* (Bartr.) Small	saw palmetto				S			
	Pedaliaceae								
P	*Sesamum indicum* L.	sesame*		ST	ST	MS	T	T	
	Pinaceae (all honeydew, no nectar)								
	Abies alba Miller	silver fir	M						
	Abies borisii-regis Mattf.	—	M						
	Abies cephalonica Loudon	Greek fir	M						
	Larix decidua Miller	larch	M	M					
	Picea abies (L.) Karsten	Norway spruce	M						
	Pinus halepensis Miller	Aleppo pine	M	M	M				
	Pinus sylvestris L.	Scots pine	M	M					
	Polygonaceae								
P	*Antigonon leptopus* Hook. & Arn.	coral creeper		T	T	S	ST	ST	
	Coccoloba belizensis Standl.	—					T		
	Coccoloba uvifera L.	sea grape				S	T	T	
p	*Fagopyrum esculentum* Moench.	buckwheat*	M	MS	ST	M			
	Gymnopodium antigonoides (Robinson) Blake	dzidzilché					T		
	Polygonum persicaria L.	smartweed	M			M			
	Triplaris surinamensis Cham.	long jack						T	
	Proteaceae								
P	*Banksia serrata* L.f.	red honeysuckle							MS
p	*Dryandra sessilis* (Knight) Domin	parrot bush							S
p	*Faurea saligna* Harv.	beechwood			ST				
p	*Grevillea robusta* A. Cunn. ex R. Br.	silky oak		T	T	S	T	ST	S
p	*Knightia excelsa* R. Br.	rewarewa							M
	Rhamnaceae								
	Berchemia scandens (Hill) K. Koch	rattan vine				S			
p	*Gouania lupuloides* (L.) Urban	bejuco de indio					T	T	
p	*Gouania polygama* (Jacq.) Urban	rattan					T	T	
p	*Paliurus spina-christi* Mill.	Christ's thorn	M	M	M				
	Rhamnidium glabrum Reiss.	turere						T	
p	*Ziziphus jujuba* Mill.	Chinese jujube*	M	MS					
p	*Ziziphus mauritania* Lam.	ber or Indian jujube*		ST	T		T	T	T
p	*Ziziphus mucronata* Willd.	buffalo thorn			S				
p	*Ziziphus nummularia* (Burm. f.) Wight & Arn.	kokan ber		S					
p	*Ziziphus oxyphylla* Edgew.	amlai		ST					
p	*Ziziphus spina-christi* (L.) Desf.	Christ's thorn	M	ST	ST				
	Rhizophoraceae								
	Rhizophora mangle L.	red mangrove					T		
	Rosaceae								
P	*Eriobotrya japonica* (Thunb.) Lindl.	loquat*	M	MST	M				
p	*Malus baccata* (L.) Borkh.	Siberian crab-apple		MS					

	Botanical name	Common name	*Europe* *M*	*Asia* *MST*	*Africa* *MST*	*N Am* *MS*	*CA+Ca* *ST*	*S Am* *MST*	*Oceania* *MST*

	Botanical name	Common name	Europe M	Asia MST	Africa MST	N Am MS	CA + Ca ST	S Am MST	Oceania MST
P	*Malus domestica* Borkh.	apple*	M	MS	S	M		MS	MS
p	*Prunus* × *yedoensis* Matsum.	Japanese cherry	M	M					
p	*Rubus* spp [*R. fruticosus* L.]	blackberry*	M	M	M				M
P	*Rubus idaeus* L.	raspberry*	M	M	M	M			M
p	*Rubus ulmifolius* Schott.	bramble	M	MS	M				
	Rubiaceae								
p	*Borreria verticilata* (L.) G. Mayer	vassoura branca					T	S	
	Calycophyllum candidissimum (Vahl) DC.	lemon wood					T	T	
p	*Canthium coromandelicum* (Burm. f.) Alston	karegida		T					
p	*Catunaregam spinosa* (Thunb.) Tirvengadum	karegida		T					
p	*Coffea arabica* L.	(Arabian) coffee*		T	T		T	T	T
P	*Wendlandia notoniana* Wall.	renda		T					
	Rutaceae								
p	*Citrus aurantifolia* (Christm.) Swingle	lime*		ST	T	S	T	ST	
p	*Citrus aurantium* L.	Seville orange*	M		T	S	T	T	S
p	*Citrus bergamia* Risso & Poiteau	bergamot*	M	S				S	
p	*Citrus deliciosa* Ten.	tangerine*	M	MST	T				
p	*Citrus grandis* (L.) Osbeck	pomelo*	M	ST		S	T		
p	*Citrus limetta* Risso	sweet lemon		ST					
p	*Citrus limon* (L.) Burm. f.	lemon*	M	S	ST	S			M
p	*Citrus medica* L.	citron*	M	S		S	T	S	
p	*Citrus paradisi* Macfad.	grapefruit*	M	ST	ST	S	T		
p	*Citrus reticulata* Blanco	mandarin orange*	M	MST	ST	S		S	
P	*Citrus sinensis* (L.) Osb.	sweet orange*	M	MST	ST	S	T	ST	S
×	*Citrus unshiu* (Mak.) Marc.	satsuma*		M					
	Phellodendron amurense Rupr.	Amur cork tree	M	M					
	Salicaceae								
p	*Salix alba* L. (honeydew as well as nectar)	white willow	M	M	M	M		T	M
P	*Salix caprea* L. (honeydew as well as nectar)	goat willow	M	M				M	M
p	*Salix nigra* Marshall	black willow	M			S	S		
p	*Populus* spp. (honeydew, no nectar)	poplar	M	MS	M	MS	S		
	Sapindaceae								
	Dimocarpus longan Lour.	longan		ST					
	Euphoria longan (Lour.) Steud.	longan*		T					
p	*Litchi chinensis* Sonner.	lychee*		S	S	S			S
	Melicoccus bijuga L.	Spanish lime					T	T	
	Melicoccus lepidopetala Radlk.	motoyoé						T	
p	*Nephelium lappaceum* L.	rambutan*		T					
p	*Sapindus emarginatus* Vahl	soapnut		T					
p	*Sapindus laurifolius* Vahl	soapnut		T					
	Sapindus mukorossi Gaertn.	soapberry		S					S
	Sapindus saponaria L.	soapberry				S	T	T	
	Sapindus trifoliatus L.	antuwal		T					
	Serjania triqueta Radlk.	bejuco cuadrado					T		
	Sapotaceae								
p	*Madhuca longifolia* (Koenig.) J. F. Macbr. v. *latifolia* (Roxb.) A. Chev.	mahua		T					
p	*Mimusops elengi* L.	moulsari		S		S	T		

Botanical name	Common name	Europe M	Asia MST	Africa MST	N Am MS	CA + Ca ST	S Am MST	Oceania MST

	Botanical name	Common name	Europe M	Asia MST	Africa MST	N Am MS	CA + Ca ST	S Am MST	Oceania MST
	Saxifragaceae								
	Ixerbe brexioides A. Cunn.	tawari							M
	Scrophulariaceae								
p	*Scrophularia nodosa* L.	knotted figwort	M		M				
	Solanaceae								
P	*Nicotiana tabacum* L.	tobacco*		T	ST	MS	T	ST	M
	Sterculiaceae								
p	*Dombeya rotundifolia* (Hochst.) Planch.	wild pear			ST				
	Symplocaceae								
	Symplocos spicata Roxb.	porinelli		T					
	Tiliaceae								
	Tilia americana L.	basswood				MS			
p	*Tilia amurensis* Rupr.	Amur lime	M	M					
P	*Tilia cordata* Mill (honeydew as well as nectar)	small-leaved lime	M			M			
p	*Tilia japonica* (Miq.) Simonk.	Japanese lime		M					
	Tilia koreana Nakai	Korean lime		M					
p	*Tilia mandschurica* Rupr. & Maxim.	Manchurian lime		M					
p	*Tilia maximowicziana* Shiras	(Japanese) lime		M					
P	*Tilia platyphyllos* Scop. (honeydew as well as nectar)	large-leaved lime	M			M			
p	*Tilia taqueti* C. Schn.	take lime		M					
p	*Tilia tomentosa* Moench	silver-leaved lime	M			M			
p	*Tilia* × *europaea* L.	common lime	M			M			
	Umbelliferae								
P	*Angelica archangelica* L.	angelica*	M	M					
p	*Coriandrum sativum* L.	coriander*	M	ST	T				
p	*Daucus carota* L.	carrot*	M	MST	M	M			
	Verbenaceae								
	Aloysia gratissima (Gill. & Hook.) Troncoso	whitebrush				S	S		
	Aloysia virgata (Ruiz & Pav.) A. L. Juss.	acerillo				S	S	ST	
P	*Gmelina arborea* Roxb.	kumil		T	T				
p	*Lippia nodiflora* (L.) Michx.	cape vine		S		S	T	T	
	Lippia triphylla (L'Her.) Kuntze	—						T	
	Vitex agnus-castus L.	chaste-tree	M		M	S			
	Vitex cymosa Bert.	tarumá						T	
	Vitaceae								
	Ampelopsis arborea (L.) Koehne	peppervine				S			
p	*Parthenocissus quinquefolia* (L.) Planch.	Virginia creeper	M			MS	T		
	Zingiberaceae								
	Hedychium coronarium Koen.	camia		T			T		
	Zygophyllaceae								
	Guaiacum officinale L.	lignum vitae					T	T	

| Botanical name | Common name | Europe M | Asia MST | Africa MST | N Am MS | CA + Ca ST | S Am MST | Oceania MST |

Appendix 2

Beekeeping gazetteer of individual countries

Notes on entries

Contact(s) Government or other department dealing with beekeeping, and any Beekeepers' Association (BKA).

Supplier(s) supplier of beekeeping equipment

Honey importer/packer

Publication(s) a useful recent publication – a book (in the language of the country, or in English, unless otherwise indicated) and/or a useful shorter account, usually in English. Entries are in the same style as in the Bibliography (page 539). Where no useful recent publication is known, an entry 'BOTA 2 (6 refs)' indicates that the *Bibliography of tropical apiculture*, Part 2 (Crane, 1978*a*) cites 6 publications; in addition '+ Satellite 25 (143 refs)' refers to BOTA Satellite 25 (Crane, 1978*b*).

Journal(s) name [transliterated if necessary] and address of a beekeeping journal (published in the language of the country unless otherwise stated)

Education brief information on beekeeping courses and training

Library name and address of library with a beekeeping collection

Museum(s) brief details

Tables elsewhere give information on individual countries, as follows:

Table 1.5A Number of hives and whether movable-frame
Hive density (per km²)
Average honey yield per hive
Bees used for beekeeping; see also Table 7.1B

Table 13.8A Production, imports and exports of honey and beeswax

Table 15.4A Aspects of beekeeping on which legislation exists.

Chapter 11 includes 10 maps showing countries in which different diseases and parasites of honeybees are known to be present.

Afghanistan

Contact (Government) Apiculture Dep., Ministry of Agriculture, Kabul

Publications C. Hoffman (1972) *Practical ideas for Afghan beekeepers* (Kabul: USAID) 92pp. [B]; J. Woyke (1984) Beekeeping in Afghanistan (Pp. 124–130 in book by FAO (1984), *see* Bibliography)

Albania

Contact (Government) Ministria së Bukqesisë, Tiranë

Publication M. Selenica and S. Alikaj (1959) *Bleta* (Tiranë: Ministria së Bukqesisë) 475 pp. [B]

Algeria

Contact Centre Universitaire de Recherches, d'Études et de Réalisations, PB 385, Constantine

Publications K. Skender (1972) *Situation actuelle de l'apiculture algérienne et ses possibilités de développement* (Algiers: Institut National Agronomique) 102 pp. [B]; BOTA 1 (17 refs)

Angola

Contact Instituto de Investigação Agronomica de Angola, CP 406, Huambo
Publications BOTA 2 (24 refs)

Antigua and Barbuda

Contact Organization for Agricultural Development, St John's
Publication Organization (above) *Beekeeping in Antigua and Barbuda, West Indies* (St John's: author) 12 pp. [1988]

Argentina

Contacts (Government) Secretaria de Agricultura y Ganadería, Avda Paseo Colon 922, 1305 Buenos Aires
BKA Sociedad Argentina de Apicultores, Rivadavia 717, Piso 8, 1392 Buenos Aires
Supplier El Panal, Humahuaca 4229, 1192 Buenos Aires
Publications A. L. Persano (1980) *Apicultura práctica* (Buenos Aires: Editorial Hemisfero Sur) 300 pp. [B]; Argentina, Ministry of Economy (1976/77) General picture of apiculture in the Argentine Republic. *Econ. Inf. Argentina* (67/68): 40–43 [B]; BOTA 6 (14 refs) + Satellite 31 (95 refs)
Journal *Gaceta del Colmenar*, pub. BKA above
Education Courses 4–10 days (honey quality analysis) at Universidad Nacional de Santiago del Estero (CEDIA), Avda Belgrano (S) 1912, 4200 Santiago del Estero
Museum Museo Apícola Nacional e Internacional, BKA above (no details known)

Australia

Contacts (Government) Apiculture Officer for individual States, c/o Dep. Agriculture
BKA Federal Council of Australian Apiarists' Associations, Glenrowan, Vic. 3675
Supplier John Guilfoyle (Sales) Pty Ltd, 772 Boundary Rd, Darra, Brisbane, POB 18, Qd 4076
Publications Government Bulletins for individual States, recent eds: D. F. Langridge and C. D. Ilton (1981) *Beekeeping in Victoria* (Melbourne: Victoria Dep. Agric.) 139 pp. [B]; H. Ayton (1981) *Bee keeping in Tasmania* (Bull. Dep. Agric. Tasmania No. 58) 40 pp. [B]
Journals *Australasian Beekeeper*, Pender Beekeeping Supplies Pty Ltd, 19 Gardiner St, Rutherford, NSW 2320; *Australian Bee Journal*, POB 426, Benalla, Vic. 3672
Education Courses 1 week and 2 yrs at Queensland Agricultural College, Lawes, Gatton, Qd 4343; others at Hawkesbury Agricultural College, Richmond, NSW 2753

Austria

Contacts (Government) Bundes-Lehr- und Versuchsanstalt für Bienenkunde, Grinzinger Allee 74, 1996 Vienna
BKA Österreichischer Imkerbund, Georg-Coch-Platz 3/11a, 1010 Vienna 1
Supplier Stefan Puff GmbH, Neuholdaugasse, 8011 Graz
Honey importer/packer Dr F. Kernstock Waren GmbH, Boersegasse 11, 1010 Vienna
Publications W. Wallner (1986) *Imker-Praxis* (Vienna: Oesterreichischer Agrarverlag) 242 pp. [B]; bibliography by Droege (1962), *see* Bibliography of this book
Journal *Bienenvater*, pub. BKA above
Education Vacation courses at 1st Contact above
Museum Österreichisches Bienenzuchtmuseum, 2304 Orth an der Donau, open daily except Mon.

Azores (Portugal)

Contact (Government) Estação Agraria de Ponta Delgada, Quinta de S. Gonçalo, Ponta Delgada, S. Miguel
Publication V. C. Paixão (1966) A apicultura na ilha de S. Miguel. *Abelhas* 9: 34, 61, 90, 110–111, 122–123 [B]

Bahamas

Contact (Government) Ministry of Agriculture, San Andros PO, N. Andros
Publications BOTA 5 (2 refs)

Balearic Islands (Spain)

Contact (BKA) Asociación Apícola Baleares, Blanquerma 53, 2–1, 07003 Palma de Mallorca

Bangladesh

Contact Bangladesh Institute of Apiculture, 23/12 Khilji Rd, Shyamoli, Dhaka 1207
Publication B. Svensson (1988) *Beekeeping technology in Bangladesh* (Sala, Sweden: author) 40 pp. 2nd ed. [B]
Education Courses 15 days to 3 months at Contact above

Barbados

Contact Ministry of Agriculture, Bridgetown
Publication J. K. Rindfleisch (1979) Beekeeping in Barbados, West Indies. *Am. Bee J.* 119(2): 131, 135 [B]

Belgium

Contacts (Government) Informatiecentrum voor Bijenteelt, Rijksstation voor Nematologie en Entomologie, Burg van Gansberghelaan 96, 9220 Merelbeke
BKA Chambre Syndicale d'Apiculture Belge, 9510 Oosterzele
Supplier Raymond de Bie, Mechelsbroekstraat 21, 2800 Mechelen
Honey importer/packer Meli B. V., De Pannelaan 68, 6478 Adinkerke
Publication E. Leysen (1961) *Handboek van de imker* (Turnhout: J. van Mierlo-Proost & Co.) 358 pp. [B]
Journals *Belgique apicole*, Chausée de Thuin 283, 6500 Anderlues; *Maandblad van de Koninklijke Vlaamse Imkersbond v.z.w.*, Steenweg naar Gierle 351, 2300 Turnhout
Museums Bijenteeltmuseum (International Museum of Apiculture of Apimondia), Antwerpsesteenweg 92, 2800 Mechelen, open daily except Mon.; Musée de l'Abeille, rue du Bihet 9, 4040 Tilff, open daily in summer

Belize

Contacts (Government) Chief Apiary Officer, Agriculture Department, Orange Walk Town
BKA Belize Honey Producers Federation of Co-operatives, Orange Walk Town
Publication Belize Ministry of Agriculture, Beekeeping Section (1987) *A guide to beekeeping* (Belmopan: author) 82 pp. [B]

Benin

Publication R. Potiron (1972) Apiculture au Dahomey. *Gaz. apic.* 73 (779): 151–152, 148 [B]

Bermuda

Contact (Government) Dep. Agriculture & Fisheries, POB 834, Hamilton 5
Publication R. G. Hargrove (1975) How many bees in Bermuda? *Glean. Bee Cult.* 103(1): 7–8, 27 [B]

Bolivia

Contacts (Government) N. Kempff Mercado, CC 211, Santa Cruz
BKA Associación Apícola Boliviana, CC 266, Calle 24 de Septiembre 469, Santa Cruz
Publications A. M. Stearman (1981) Working the 'Africans' in eastern Bolivia. *Am. Bee. J.* 121(1): 28, 30–35, 43–44 [B]; BOTA 6 (11 refs)

Botswana

Contact (Government) Beekeeping Officer, Ministry of Agriculture, Private Mail Box 003, Gaborone
Publication B. Clauss (1982) *Bees and beekeeping in Botswana* (Gaborone, Botswana; Ministry of Agriculture) 122 pp. [B]

Brazil

Contact (*BKA*) Confederação Brasileira de Apicultura, CP 428, 88000 Florianópolis, SC
Supplier CAPEL, Exposição de Animais-DPA, Avda Caxangá 2200, CP 50000 Recife (PE)
Publications H. Wiese (ed.) (1985) *Nova apicultura* (Porto Alegre: Livraria e Editora Agropecuária) 493 pp. 6th ed. [B]; BOTA 5 (24 refs) + Satellite 30 (159 refs)
Journal *Apicultura no Brasil*, pub. BKA above
Education Courses at Universidade de S. Paulo, Dep. Genetica, Faculdade de Medicina de Ribeirão Preto, SP; Programa de Zootecnia da Coordenadoria de Assistencia Tecnica Integral, Avda Brasil 2340, CP 960, 13100 Campinas, SP

Brunei

Contact (Government) Dep. Agriculture, Bandar Seri Begawan
Publications BOTA 4 (2 refs)

Bulgaria

Contact (*BKA*) Central Bulgarian Council of Apiculture, Ministry of Agriculture and Food Industry, Blvd Botev 55, Sofia
Publications B. Bizhev (1985) [*Manual for amateur beekeepers*] (Sofia: Zemizdat) 199 pp. [B]
Journal [*Pchelarstvo*], POB 80, 1113 Sofia

Burkina

Contact (Government) Ministère du Développement Rural, BP 7010, Ouagadougou
Publication R. A. Swanson (1976) The case for beekeeping development programmes in West Africa (Pp. 191–197 in book by Crane (1976*a*), *see* Bibliography)

Burma (now Myanmar)

Contact (Government) Ministry of Defence, Rangoon
Publication R. A. Morse (1982) The beekeeping program in Burma. *Glean. Bee Cult.* 110(8): 454–455, 467 [B]

Burundi

Contact Christian Rural Service Development Programme, EPEB Buya, POB 58, Ngozi
Publications R. Bauduin (1956) Apiculture au Ruanda. *Bull. agric. Congo belge* 47: 141–162; *see also* Zaire

Cameroon

Contact (Government) Ministry of Agriculture, Yaounde
Publication J. Douet (1982) Apiculture en Afrique tropicale. *Ouest apic.* (3): 6–7 [B]

Canada

Contacts (Government) Research Branch Canada Agriculture, Entomology Section, Ottawa Research Station, Ottawa; each Province has an Apiary Officer; University of Guelph, Dep. Environmental Biology, Ont., N1G 2W1, is active in apiculture, also Simon Fraser University, Burnaby, BC
BKA Canadian Honey Council, Box 1566, Nipawin, Sask., SOE 1EO
Suppliers Bee Maid, 625 Roseberry St, Winnipeg, Man., RH3 OT4; F. W. Jones & Son Ltd, 44 Dutch St, Bedford, PQ, JOJ 1AO
Publications Government Bulletins for separate regions; recent eds: D. C. Murrell and D. N. MacDonald (1982) *The Alberta beekeeping manual* (Edmonton: Alberta Agriculture) 98 pp. [B]; G. F. Townsend and P. W. Burke (1976) *Beekeeping in Ontario* (Publ. Min. Agriculture and Food, Ontario, No. 490) 38 pp. rev. ed. [B]; bibliography by T. S. K. and M. P. Johansson (1972*a*), *see* Bibliography of this book
Journal *Canadian Beekeeping*, POB 128, Orono, Ont., LOB 1MO
Education Courses 9 months to 3 yrs at University of Guelph (2nd Contact above); 11 months at Fairview College, POB 3000 Fairview, Alta, TOH 1LO; 6 weeks at Malaspina College, 900 Fifth St, Nanaimo, BC, V9R 5S5
Library University of Guelph has many bee books
Museum Ontario Agricultural Museum, POB 38, Milton, Ont., L9T 2YE, has small collection

Canary Islands (Spain)

Contacts (Government) Ministerio de Agricultura, Las Palmas
BKA Asociación Apícola Canarias, Agustin Espinosa 6, 3°, pta 11, 380007 Santa Cruz de Tenerife
Publications M. Espinar Martínez and S. Hernández Sánchez (1987) *Apicultura para Canarias* (Las Palmas: Asociación Apícola Canaria) 81 pp.; BOTA 1 (3 refs)
Journal *Canarias apícola*, 6–3°, 38007 Santa Cruz de Tenerife

Caroline Islands *see* Mariana Islands

Cayman Islands

Publication J. Chalmers (1984) Bee keeping in the Cayman Islands. *Scot. Beekpr* 61(1): 11 [B]

Central African Republic

Contact (Government) Dep. Agriculture, Bangui
Publication K. J. Debold (1983) *Manuel d'apiculture* (République Centrafricaine: Service de l'Apiculture) 104 pp. [B]

Chad

Contact (Government) Département d'Agriculture, Ndjamena
Publication C. Gadbin (1976) Aperçu sur l'apiculture traditionnelle dans le sud de Tchad, *J. Agric. trop. Bot. appl.* 23(4/6): 101–115 [B]

Chatham Islands (New Zealand)

Contact (Government) Dep. Agriculture
Publication Apiarist (1987) Chatham Island honey for sale. *Apiarist* (52): 4 [B]

Chile

Contact Instituto Chileno de Apicultura, Casilla 3686, Santiago
Supplier CRATE, Casilla 6122, CC 22, Santiago
Publication J. Peldoza Vergara (1975) *La industría apícola* (Santiago: Sociedad National de Agricultura) 27 pp. [B]; BOTA 6 (18 refs)

China (People's Republic)

Contacts (Government) Institute of Apicultural Research, Chinese Academy of Agricultural Sciences, Xiangshan, Beijing
BKA Chinese Apicultural Association, c/o address above
Publications Peking Agricultural Research Institute (1980) [*Beekeeping*] (Peking: Peking Agricultural Research Institute) 94 pp. [B]; Ma Deh-Feng and Huang Wen-Cheng (1981) Apiculture in the new China. *Bee Wld* 62(4): 163–166 [B]; BOTA 4 (22 refs)
Journal [*Chinese Apiculture*], pub. Contact above

Colombia

Contacts (Government) Ministerio de Agricultura, Bogotá
BKA Asociación Colombiana de Apicultores, Calle 13, 8–23 y 8–27, Oficina 818, Bogotá, DE
Supplier Cooperativa de Especies Menores Ltda (COOMENORES), Calle 27, Carrera 20, Edificio Mercafé Piso 3, AA148, Tuluá-Valle
Publications R. B. Kent (1976) Beekeeping regions and the beekeeping industry in Colombia. *Bee Wld* 57(4): 151–158 [B]; BOTA 5 (16 refs)
Education Library at Universidad Nacional de Colombia, Medellín, has collection of bee books, and Branch of the IBRA Library for Latin America

Congo (People's Republic)

Contact (Government) Comité National pour le Développement de l'Apiculture, Brazzaville
Publication G. Onore (1980) Apicoltura in Congo. *Apic. mod.* 71(1): 1–4 [B]

Cook Islands

Contact (Government) Dep. Agriculture, Avarua
Publications A. H. Simpson (1969) A queen rearing venture on Rarotonga in 1933. *N. Z. Beekpr* 31(4): 36–37 [B]; *see also* Tonga

Costa Rica

Contact (Government) Centro Nacional de Apicultura, AP 25, Turrialba
Publications D. Espina and G. S. Ordetx (1984) *Apicultura tropical* (Cartago: Editorial Tecnológica de Costa Rica) 506 pp., 4th ed. [B]; BOTA 5 (27 refs)

Crete (Greece)

Contact (Government) Dep. Agriculture, Candia
Publication C. Zymbragoudakis (1979) The bee and beekeeping of Crete. *Apiacta* 14(3): 134–138 [B]
Museum C. Zymbragoudakis, Palama 2, Khania, has a museum collection

Cuba

Contacts (Government) Ministerio de Agricultura, Havana
BKAs Grupo Apícola Nacional, Lamparilla 110 e San Ignacio y Cuba, Havana 1; Asociación Apícola de Cuba, CC 259, Santos Suárez, Havana
Publications *see* 1st ref. Costa Rica; BOTA 5 (16 refs)

Cyprus

Contacts (Government) Plant Protection Section, Dep. Agriculture, Nicosia (*continued*)

Cyprus (*continued*)

BKA Co-operation of Beekeepers' Associations of Cyprus, POB 106, Larnaca

Publications N. Papaionou (1973) [*Beekeeping in Cyprus*] (Nicosia) 44 pp. *In Greek* [B]; BOTA 1 (4 refs)

Czechoslovakia

Contacts (Government) Výzkumný Ústav Vceláršky Dol, Libčice nad Vltavou

BKAs Czechoslovensky Svaz Včelařů, Křemencová 8, Nove Mesto, Prague 1; Slovensky Zvaz Včelarov, Svrckova ul. c. 24, Bratislava

Supplier 1st BKA above

Publications V. Veselý et al. (1985) *Včelařství* (Prague; Státní Zemědělské Nakladatelství) 365 pp. [B]

Journals *Včelařství*, Václavské náměsti 47, 11311 Prague 1; *Včelár*, Križkova 9, 81534 Bratislava

Museum Zemědělske Muzeum (Agricultural Museum), Kačina, nr Kutná Hora, has small collection

Denmark

Contacts (Government) Landbrugsministeriet, Slotsholmsgade 10, 1216 Copenhagen K

BKA Danmarks Biavlerforening, Nymarksvej 24, 7000 Fredericia

Supplier Swienty's Biavl, Hørtoftvej, Ragebøl, 6400 Sønderborg

Publications A. Christensen (ed.) (1986) *Vejledning i praktisk biavl* (Arnøje: author) 200 pp. [B]

Journal *Tidsskrift for Biavl*, pub. BKA above

Museum Herning Museum, Museumsgade 1, 7400 Herning, has small collection

Dominica

Contact (Government) Ministerio de Agricultura, Roseau

Publication *see* 1st ref. Costa Rica

Dominican Republic

Contact (Government) Instituto Dominicano de Tecnologiá Industrial, AP 329–2, Santo Domingo

Supplier Apicultura Industrial, Apdo 1844, km 9, Antigua, Carr. Duarte

Publications H. A. Santana Pion (1978) Beekeeping in the Dominican Republic. *Apiacta* 13(3): 143–144 [B]; BOTA 5 (8 refs)

Easter Island (Chile)

Publication T. I. Szabo (1987) Queen rearing in Chile. *Am. Bee J.* 127(8): 568–571 [B]

Ecuador

Contact (Government) INCCA, Ministerio de Agricultura y Ganaderia, 7 Piso, Av. Eloy Alfaro y Amazonas, Quito

Publication A. Isola and J. Ortiz (1985) Ecuador. *Città delle Api* 3(14): 4–7 [B]

Egypt

Contacts (Government) Beekeeping Dep., Plant Protection Research Institute, Ministry of Agriculture, Dokki, Cairo

BKA Bee Kingdom, Residence of Agricultural Syndicate, Galaa St, Cairo

Supplier House of Bees and Agricultural Activities, 6 Sekket El Manah St, Opera Square, Cairo

Publications R. E. Page and H. H. Laidlaw (1980) Egyptian beekeeping. *Am. Bee J.* 120(11): 776–779 [B]; BOTA 1 (22 refs) + Satellite 25 (143 refs)

El Salvador

Contact (Government) Ministerio de Agricultura, San Salvador

Publication S. Handall and J. Woyke (1983) *Apicultura* (El Salvador: Ministerio de Agricultura y Ganadería) 167 pp. [B]

England *see under* United Kingdom

Ethiopia

Contact (Government) Animal and Fisheries Development Authority, POB 1052, Addis Ababa

Publications G. Mammo (1976) Practical aspects of bee management in Ethiopia (Pp. 69–79 in book by Crane (1976*a*), *see* Bibliography); BOTA 2 (13 refs)

Journal *Apiculture in Ethiopia*, Bee Research and Breeders, POB 7505, Addis Ababa

Fiji

Contact (Government) Ministry of Agriculture, Rodwell Rd, Suva
Publications A. G. Matheson (1987) *Apiculture in Fiji* (Nelson, New Zealand: Ministry of Agriculture) 82 pp. [B]
Journal *Beekeeping Newsletter*, Agriculture Dep., POB 44, Nausori

Finland

Contact (*BKA*) Suomen Mehiläishoitajain Liitto, Kasarminkatu 26 C 34, 00130 Helsinki 13
Supplier OY Hunajayhtymä, 32250 Kojonkulma
Publications E. Hämäläinen, S. Korpela and K. Långfors (1979) *Mehiläishoitajan käsikirja* (Keuruu: Otava) 201 pp. [B]
Journal *Mehiläinen*, pub. BKA above

France
see also French Guiana, Guadeloupe, Martinique, New Caledonia, Réunion

Contacts Office pour l'Information et la Documentation en Apiculture (OPIDA), Centre apicole d'Echauffour, 61370 Ste Gauburge
BKAs Union des Groupements Apicoles Français, 149 rue de Bercy, 75014 Paris; Union Nationale de l'Apiculture Française, 25 rue des Tournelles, 75004 Paris; Syndicat National d'Apiculture, 5 rue Copenhague, 75008 Paris
Supplier Ets Thomas Fils SA, 65 Rue Abbé George Thomas, BP 2, 45450 Fay-aux-Loges
Honey importer/packer Mellitag, 211-215 rue La Fontaine, 94120 Fontenay-sous-Bois
Publications J. Louveaux (1985) *Les abeilles et leur élevage* (Paris: OPIDA) 265 pp. [B]; J. Bonimond (ed.) (1975) *L'apiculture française. Les régions* (Paris: Union Nationale de l'Apiculture Française) 104 pp. [B]; bibliography by Casteljau (1983), *see* Bibliography of this book
Journals *Revue française d'Apiculture*, pub. 2nd BKA above; *L'abeille de France et l'Apiculteur*, pub. 3rd BKA above
Education Courses 4 to 20 weeks by Ministère de l'Agriculture, CFPPA, at: 341 route de Nantes, BP 740, 53017 Laval; route de Cambrai, Tilloy-les-Mofflaines, 62000 Arras; Quartier Les Gres, 83400 Hyères. Also 1-yr course at Centre de Promotion Sociale, 7 rue Colonel-Ferraci, 20250 Corte, Corsica; Diploma course on Apiculture Tropicale at Université de Paris VI, 24620 Les Eyzies
Libraries La Société Centrale d'Apiculture has an important library at 41 rue Pernety, 75014 Paris; the library at the Laboratoire de Neurobiologie Comparée des Invertébrés, INRA, 91440 Bures-sur-Yvette, near Paris, can also be visited
Museums Musée National d'Histoire Naturelle, 57 Rue Cuvier, 75231 Paris, and Musée des Arts et Traditions Populaires, route du Mahatma Gandhi, 75016 Paris, are believed to have bee material

French Guiana (France)

Contact (Government) Département d'Agriculture, Cité Rebard, BP746, Cayenne
Publication *see* 2nd ref. Guyana

Gabon

Contact (Government) Ministère d'Agriculture, Libreville
Publication R. Darchen (1979) L'ecologie apicole dans le forêt équatoriale du Gabon. *Gaz. apic.* 80(855): 99–102 [B]

Gambia

Contacts (Government) Ministry of Agriculture and Natural Resources, Forestry Dep., Banjul
BKA Gambia Beekeepers' Association, c/o Indigenous Business Advisory Service, Ministry of Economic Planning, 22 Anglesa St, Banjul
Publications O. Ängeby (1981) Biodling i Gambia. *Bitidningen* 80(6): 191–195 [B]; BOTA 2 (7 refs)

German Democratic Republic

Contact (*BKA*) Vereinigung Volkseigener Betriebe, Saat- und Pflanzgut der DDR, Clara-Zetkin-Str. 1, 43 Quedlinburg
Supplier M. Kempe, Ernst-Thälmann Str. 70, 936 Zschopau
Honey importer/packer AHB Nahrung Export-Import, Schicklerstrasse 5–7, Postfach 1503, 102 Berlin
Publications H. Westphal and 17 others (1975) *Imkerliche Fachkunde* (Berlin: VEB Deutscher Landwirtschaftsverlag) 543 pp. 3rd ed. [B]; bibliography by Droege (1962), *see* Bibliography of this book
Journal *Garten und Kleintierzucht, Ausgabe C (Imker)*, Reinhardstr. 14, POB 130, 1040 Berlin
Education (Library) Drory Bibliothek, Naturkundemuseum, Invalidenstr. 43, 104 Berlin
Museum Deutsches Bienen Museum Weimar; Thüringer Freilichtmuseum, Hohenfelden, Weimar

German Federal Republic

Contacts (Government) Each *Land* has a Beekeeping Institute, e.g. Bayerische Landesanstalt für Bienenzucht, Burgbergstr. 70, 8520 Erlangen
BKA Deutscher Imkerbund eV, Schollengasse 4a, 5307 Wachtberg-Villip 3
Suppliers Chr. Graze KG, Strümpfelbacherstr. 21, 7056 Weinstadt 2 (Endersbach); F. Wienold, Postfach 15, Dirlammerstr. 20, 6420 Lauterbach/Hessen
Honey importer/packer Rakemann & Co., 34 Bellevue, 2000 Hamburg 64
Publications E. Zander and F. K. Böttcher (1982) *Haltung und Zucht der Biene* (Stuttgart: Ulmer) 422 pp. 11th ed. [B]; K. Pfefferle (1982) *Imkern mit dem Magazin* (Münstertal/Schwarzwald: author) 250 pp. 6th ed. [B]; bibliography by Droege (1962), *see* Bibliography of this book
Journal *Allgemeine Deutsche Imkerzeitung*, Delta-Verlag KG, Liebfrauenstr. 43, 5205 Sankt Augustin 3
Education Courses 2–3 yrs at Berufsimkerschule, NLBBB, Wehlstr. 4a, 3100 Celle; Institut für Bienenkunde, Karl von Frisch Weg 2, 6370 Oberursel 1; 2 days, also 6 months at Bayerische Landesanstalt für Bienenzucht, Burgbergstr. 70, 8520 Erlangen
Library At above institutes
Museums More beekeeping museums than any other country, e.g. Armbruster Collection c/o Naturwissenschaftliche Sammlungen Berlin, Schlossstr. 69A, 1000 Berlin 19; Bienenkunde Museum, Altes Rathaus, 7816 Münstertal/Schwarzwald, open some afternoons; Karl-August-Forster-Bienenmuseum, Vöhlin-Schloss, 7918 Illertissen, Bayern, open several days weekly; also collections at 1st and 3rd addresses under Education

Ghana

Contacts (Government) Technology Consultancy Centre, University PO, University of Science of Technology, Kumasi
BKA Ghana Beekeepers' Association, POB 813, Accra
Publications S. O. Adjare (1984) *The golden insect: a handbook on beekeeping for beginners* (Kumasi: Technology Consultancy Centre) 104 pp. 2nd ed. [B]; BOTA 2 (13 refs)
Journal *Ghana Bee News*, pub. 1st Contact above
Education 1-week courses at Technology Consultancy Centre, 1st Contact

Greece
see also Crete, Rhodes

Contacts (Government) Direction for Livestock, Section Apiculture, Acharnon St 2, Athens
BKA Melissokomiki, Cooperative Union of Beekeepers' Associations, De Kellias 86, Nea Filadelfia, Athens 14341
Supplier BKA above
Publications G. B. Sellianakis (1978) [*The effects on Greek apiculture of entry into the Common Market*] (Athens: Melissokomiki) 122 pp. *In Greek* [B]; X. Thomaides (1979) Beekeeping in Ancient Greece. *Apiacta* 14(3): 97–108 [B]
Journal [*Melissokomiki Hellas*], 16 Anargyron St, Kallithea, Thessaloniki

Guadeloupe (France)

Contact (Government) Département d'Agriculture, Jardin Botanique, Basse-Terre
Publications BOTA 5 (2 refs)

Guam (USA)

Contact (Government) Agricultural Experiment Station, College of Agriculture, University of Guam, POB EK, Agana

Guatemala

Contact (Government) Ministerio de Agricultura, Guatemala
BKA Asociación Nacional de Apicultores, 6A Ave 19–15, Zona 1, Anapi
Publications J. D. Spence (1980) *La apicultura: guía práctica. Cómo trabajar con las abejas* (Nebaj: Proyecto Desarrollo Apícola) 222 pp. [B]; BOTA 5 (8 refs)

Guinea

Contact (Government) Ministry of Agriculture, Conakry
Publication B. Mitev, N. Stoilov and S. Kolev (1975) Study of the honey bee in Guinea Republic. *Proc. 25 Int. Apic. Congr.*: 355–356 [B]

Guinea-Bissau

Contact (Government) Ministère d'Agriculture, Bissau
Publication H. Couture and D. Guzzi (1985) Chronique africaine. *Abeille* 6(3): 17–18; (4): 17–18 [B]

Guyana

Contacts (Government) Ministry of Agriculture, POB 1001, Georgetown
BKA Guyana Beekeepers' Association, c/o address above
Publications J. Beetsma (1976) Improving honey production and disposal in Guyana and Surinam (Pp. 81–83 in book by Crane (1976*a*), *see* Bibliography); G. W. Otis and O. R. Taylor (1979) Beekeeping in the Guianas (Pp. 145–154 in book by Commonwealth Secretariat (1979), *see* Bibliography)

Haiti

Contact (Government) Ministère de l'Agriculture, des Ressources Naturelles, et du Développement Rural, Damiens, Port-au-Prince
Publications H. C. Mulzac (1978) Beekeeping for the good of Haiti. *Am. Bee J.* 118(5): 360–362 [B]; BOTA 5 (4 refs)

Hawaii (USA)

Contact (*BKA*) Hawaii Beekeepers' Association, POB 29675, Honolulu
Publications J. E. Eckert and H. A. Bess (1952) Fundamentals of beekeeping in Hawaii. *Ext. Bull. Univ. Hawaii* No. 55: 59 pp. [B]; BOTA 7 (28 refs)

Honduras

Contacts (Government) Ministerio de Agricultura, Tegucigalpa
BKA Asociación Nacional de Apicultores de Honduras, AP 1247, San Pedro Sula
Publication G. E. Patty (1980) Honey: honey output rising in Honduras. *USDA Foreign agric. Circ.* FHON 2–80: 8 pp. [B]

Hong Kong

Contact (Government) Agriculture and Fisheries Dep., Canton Road Government Offices, 393 Canton Road, 12/F Kowloon
Honey importer/packer Mitchell Cotts & Co. (Far East) Ltd, 15th floor, Kai Tak Commercial Bldg, 317–321 des Voeux Road, Central, Hong Kong
Publication Cheung Wai-Yen (1969) [*Methods of beekeeping*] (Hong Kong: Tak Lee Book Co.) [B]

Hungary

Contacts National Office of Apicultural Co-operatives, Garibaldi u.2, Budapest 1054; Méhészet, 2101 Gödöllő
BKA Szövetkezetek Országos Szövetsége, Szabadság-Tér 14, Budapest V
Supplier 1st contact above
Publications A. Nikovitz (ed.) (1983) *A méhészet kézikönyve* (Budapest: Hungaronektár) 826 pp., 2 vol. [B]; S. Kocsis (1976) Apiculture in Hungary and possibilities of its development. *Apiacta* 11(1): 42–47 [B]
Journal *Méhészet*, Blaha Lujza tér 3, 1959, Budapest VIII
Museum Kisállattenyésztési Kutatóintézet, Méhészet, 2101 Gödöllő, open to visitors

India
see also Nicobar and Andaman Islands

Contacts (Government) Central Bee Research Institute, Khadi and Village Industries Commission, 1153 Ganeshkhind Rd, Pune 411016
BKA All-India Beekeepers' Association, 1325 Sadashiv Peth, Pune 411030
Suppliers BKA above; also Rawat Apiaries (Himalayas), Ranikhet, Dist. Almore, UP
Publications B. S. Rawat (1982) *Bee farming in India* (Ranikhet, UP: Rawat Apiaries) 258 pp. [B]; F. A. Shah (1983) *Fundamentals of beekeeping* (Srinagar, Kashmir: Shah Beekeepers) 60 pp. [B]; R. C. Mishra and R. C. Sihag (1987) *Apicultural research in India* (Hisar: AICRP, Haryana Agricultural University) 120 pp; BOTA 3 (74 refs) + Satellite 28 (377 refs)
Journals *Indian Bee Journal*, pub. BKA above; *Indian Honey*, Indian Institute of Honey, Martandam Kuzhithuri 629163, Kanyakumari Dist., Tamil Nadu
Education Courses 15 days to 9 months at 1st Contact above; other courses at Kerala Agricultural Univ., Vellanikara, Trichura 680651; University of Pune, Dep. Zoology, Ganeshkhind, Pune 411007

(*continued*)

India *(continued)*

Library CBRI above has an important collection of bee books, and has Branch of the IBRA Library for Tropical Asia

Museum At CBRI above, open weekdays

Indonesia

Contacts National Beekeeping Development Board, Jakarta; Pusat Apiari Pramuka, Medan Merdeka, Timur 6, Jakarta; Apiary Research Centre, Cibubur, W. Java

Publications R. M. Sumoprastowo and R. A. Suprapto (1980) [*Modern beekeeping*] (Jakarta: Bhratara Karya Aksara) 217 pp. [B]; K. Patra (1988) *Beekeeping with* Apis cerana *in Indonesia* (Jakarta: Scout Movement Apiary Center, etc.) 24 pp.; BOTA 4 (20 refs)

Iran

Contact (Government) Animal Husbandry Organization, Sazeman Damparvari Keshvar, Floor 11, Ministry of Agriculture, Blvd Elizabeth II, Tehran

Publications F. Colson (1977) L'apiculture dans le développement agricole, un exemple iranien. *Bull. tech. apic.* 4(3): 13–21 [B]; BOTA 3 (22 refs)

Iraq

Contact (Government) Plant Protection Dep., Ministry of Agriculture, Abu-Ghraib, Baghdad

Publication Iraq Ministry of Agriculture (1973) *Promotion of modern methods of bee keeping in north Iraq* (Iraq: Ministry of Agriculture) 6pp. [B]

Ireland, Northern *see under* United Kingdom

Irish Republic

Contacts (Government) An Foras Talúntais, Beekeeping Research Unit, Clonroche, Co. Wexford

BKA Federation of Irish Beekeepers' Associations, Our Lady's Place, Naas, Co. Kildare

Supplier Tom Kehoe, 106 N Circular Rd, Dublin 7

Publications T. N. Hillyard and J. Markham (1968) *A survey of beekeeping in Ireland* (Dublin: An Foras Talúntais) 49 pp. [B]; J. K. Watson (1981) *Beekeeping in Ireland: a history* (Dublin: Glendale Press) 293 pp. [B]; bibliography by IBRA (1979*a*), *see* Bibliography of this book

Journal *Irish Beekeeper*, St Jude's, Mooncoin, Waterford

Education 5-day course in July at Franciscan College, Gormanston, Co. Meath

Israel

Contacts (Government) Ministry of Agriculture, Extension Service, Beekeeping Division, POB 7054, Hakirya, Tel Aviv, 61070

BKA Israel Beekeepers' Association, 8 Shaul Hamelekh Blvd, Tel Aviv

Publication G. Robinson (1975) Beekeeping in Israel. *Am. Bee J.* 115(4): 145–146 [B]

Education 3-month courses March-June at Foreign Training Dep., Ministry of Agriculture Extension Service, address above

Museum proposed at 1st Contact above

Italy

Contacts (Government) Universities and other institutes in individual provinces

BKAs Federazione Apicoltori Italiani, Corso Vittorio Emanuele 101, 00186 Rome; UNAPI, Viale Aldo Moro 16, 40127 Bologna

Suppliers Lega SDF, Via Armandi 19, 48018 Faenza; SAF s.n.c., Via Liguria 17, 36015 Schio (VI)

Honey importer/packer Ser-Fruits SA, Segrate, Milano

Publications L. Benedetti and L. Pieralli (1982) *Api e apicoltura* (Milan: Hoepli) 518 pp. [B]

Journals *Apitalia*, pub. BKA above; *Apicoltore moderno*, Osservatorio di Apicoltura dell'Università di Torino, Via Ormea 99, 10126 Torino

Museums Two can be visited by appointment: Osservatorio di Apicoltura 'Don Giacomo Angeleri', Strada del Cresto 2, 10132 Reaglie-Torino; Museo Apistico Didattico, via Memgardo 12, Bregnano, Como

Ivory Coast

Contact (Government) Institut de Recherches Agronomiques, BP 8035, Abidjan

Publication R. Borneck (1976) L'apiculture en Côte d'Ivoire. *Rev. fr. Apic.* (344): 334–335, 338–339 [B]

Jamaica

Contact (Government) Ministry of Agriculture, Plant Protection Div., Hope Gardens, Kingston 6
Publications J. K. Bianco (1985) *Beekeeping in Jamaica* (Peace Corps/Jamaica) 108 pp. [B]; BOTA 5 (17 refs)

Japan

Contacts Institute of Honeybee Science, Tamagawa University, Machida-shi, Tokyo 194
BKA [Japan Beekeepers' Association], 1–2 Surugadai Kanda Chiyoda-ku, Tokyo 101
Supplier Gifu Yoho Co. Ltd, Kano-Sakuradacho 1, Gifu-shi, Gifu 500-91
Honey importer/packer Sumimoto Shoji Kaisha Ltd, 1–2–2 Hitotsubashi, Chiyoda-ku, Tokyo
Publications K. Watanabe (1974) [*Modern beekeeping*] (Gifu: Nihon Yoho Shinkokai) 726 pp.; T. Sakai and M. Matsuka (1982) Beekeeping and honey resources in Japan. *Bee Wld* 63(2): 63–71 [B]
Journal [*Honeybee Science*], pub. 1st Contact above
Education 6-month courses at 1st contact above; 9-month courses at Rural Leaders Training Centre, 442–1 Tsukinokizowa, Nishinasuno, Tochigiken
Library 1st Contact above has collection of bee books, and has Branch of the IBRA Library for Oriental Asia
Museum Bee House, Livestock Center Park, Tsubakibora-Nakano, Gifu-shi, 502, open daily

Jordan

Contact (Government) Ministry of Agriculture, Amman
Publication W. S. Robinson (1981) Beekeeping in Jordan. *Bee Wld* 62(3): 91–97 [B]

Kampuchea (Cambodia)

Contact (Government) Dep. Agriculture, Phnom-Penh

Kenya

Contacts (Government) Beekeeping Branch, Ministry of Agriculture and Livestock Development, POB 274, 68228 Nairobi
BKA Kenya Beekeepers' Association, POB 40494, Nairobi
Supplier 1st Contact above
Publications K. I. Kigatiira (1984) Bees and beekeeping in Kenya. *Bee Wld* 65(2): 71–80 [B]; J Nightingale (1983) *A lifetime's recollections of Kenya tribal beekeeping* (London: IBRA) 37 pp. [B]; BOTA 2 (11 refs) + Satellite 26 (10 refs)
Education 1-yr and 3-yr courses at 1st Contact above
Library InterAfrican Bureau of Animal Research, Nairobi, has Branch of the IBRA Library for Africa

Kermadec Islands (New Zealand)

Publication New Zealand Beekeeper (1962) Bees for Raoul Island. *N. Z. Beekpr* 24(1): 28 [B]

Kiribati

Contact (Government) Division of Agriculture, POB 267, Bikenibeu, Tarawa
Publication G. M. Walton (1976) *Report on proposed beekeeping development for the Gilbert Islands* (Palmerston North: Ministry of Agriculture and Fisheries) 19 pp. [B]

Korea (Democratic People's Republic)

Contact (*BKA*) Korean Beekeepers' Association, Pyongyang
Publications BOTA 4 (2 refs)

Korea (Republic)

Contacts (Government) Institute of Korean Beekeeping Science, Suweon
BKA Korean Beekeeping Association, 201 Sambo Building, 173 Chongroku, Seoul 110
Publications Sang Yul Cho and Young Tai Cho (1955) [*Beekeeping*] (Seoul: National Federation of Korean Beekeepers' Associations) 642 pp. [B]; Seung Yoon Choi (1984) Brief report on the status of Korean beekeeping (Pp. 170–190 in book by FAO (1984), *see* Bibliography); BOTA 4 (6 refs)

Kuwait

Honey importer/packer Kuwait Catering Co., POB 1879, Salmiah

Lebanon

Contacts (Government) Ministry of Agriculture, Beirut
BKA Apiculture Protection and Improvement Society, POB 165908, Beirut
Publication R. P. H. Fleisch (1962/63) Apiculture libanaise. *Apiculteur* 106: 33–40, 123–127, 175–181, 199–202; 107: 40–45 [B]

Lesotho

Contact Tefobale Bee Research, POB 46, Mafeteng 900

Liberia

Contact (Government) Ministry of Agriculture, Tubman Blvd, Sinkor, Monrovia
Publication G. F. Clulow (1969) Afflictive aspects of African apiculture. *Glean. Bee Cult.* 97(9): 557–560 [B]

Libya

Contacts (Government) Ministry of Agriculture, Tripoli; also Agricultural Bank, POB 1100, Tripoli
BKA Beekeepers' Co-operation League, POB 14579, Tripoli
Publications Libyan Arab Republic (1973) [*Beekeeping*] (Tripoli: Ministry of Agriculture) 68 pp. *In Arabic* [B]; BOTA 1 (6 refs)

Luxembourg

Contacts (Government) Services Techniques de l'Agriculture, Service de l'Horticulture, 16 route d'Ésch, BP 1904, 1019 Luxembourg
BKA FUAL, 24 rue du Mulin, 8279 Holzem
Supplier Etienne Paulus, 21 rue du Stade, 3877 Schifflingen
Publications Luxemburgischer Bienenzeitung (1970) Bericht zur Lage. *Luxemb. Bienenztg* 85(3): 32–33 [B]
Journal *Letzebürger Beien-Zeitung*, pub. BKA above

Madagascar

Contact (Government) Ministère des Affaires Etrangères, Tananarive
Publications M. T. Chandler (1975) Apiculture in Madagascar. *Bee Wld* 56(4): 149–153 [B]; BOTA 2 (14 refs)

Madeira (Portugal)

Contact (Government) Dep. Agricultura, Funchal
Publication Abelhas (1980) Apicultura na Madeira e em Porto Santo. *Abelhas* 23(268): 39, 46 [B]

Malawi

Contact (Government) Chief Conservator of Forests, POB 30048, Capital City, Lilongwe 3
Publications M. N. Kawa (1987) *For better bee-farming in Malaŵi* (Maputo, Mozambique) 10 pp. [B]; BOTA 2 (8 refs)

Malaysia

Contact Faculty of Agriculture, Agricultural University of Malaysia, 43400 UPM, Serdang, Selangor
Publications A. Bakar et al. (1987) *Pemiliharaan lebah madu di kawasan getah* (Selangor: Rubber Research Institute of Malaysia) 84 pp.; M. Mardan (1984) Current status, problems, prospects and research needs of *Apis mellifera* in Malaysia (Pp. 191–197 in book by FAO (1984), *see* Bibliography)

Malta

Contact Dep. Agriculture and Fisheries (Horticulture Division), Government Farm, Marsa
Publication V. Farrugia (1979) Beekeeping in the Maltese Islands (Pp. 41–46, 190 in book by Commonwealth Secretariat (1979), *see* Bibliography)

Mariana Islands

Contact (Government) Agriculture Division, Dep. Resources and Development, Saipan
Publication Western Apicultural Society Journal (1981) Beekeeping cited as honey of a hobby. *W. apic. Soc. J.* 4(3): 134, 137–138 [B]

Marshall Islands

Contact (Government) as Mariana Islands

Publication A. P. Marshall (1981) Beekeeping in the Marshall Islands. *Glean. Bee Cult.* 109(8): 430, 433–434 [B]

Martinique (France)

Contact (Government) Département d'Agriculture, Av. Louis Georges Plissonneau, BP 861, Fort de France
Publications P. Bally (1976) La Martinique: sa géographie, son climat, sa flore, son apiculture. *Rev. fr. Apic.* (342): 230–233 [B]; BOTA 5 (3 refs)

Mauritius

Contacts (Government) Ministry of Agriculture, Réduit
BKAs Organisation of Mauritian Beekeepers, SSEC 1st floor, Galerie Réunies, 5 Bourbon St, Port Louis; Société Apicole de l'île Maurice, St Antoine, Goodlands
Publications BOTA 2 (4 refs)

Mexico

Contacts (Government) Departamento de Apicultura, Secretaría de Agricultura y Recursos Hidráulicos, Durango 138–503, Col. Roma, Mexico 7
BKA Union Nacional de Apicultores, Av. Uruguay 42–101, Mexico 1, DF
Supplier Miel Carlota S.A., AP 161–D, Queretaro III, Cuernavaca, Mor.
Publications J. F. Martínez Lopez (1979) *Apicultura* (Mérida: author) 218 pp. [B]; BOTA 5 (17 refs) + Satellite 29 (81 refs)

Morocco

Contact (*BKA*) Syndicat Professionnel des Apiculteurs du Maroc, BP 556 Rabat-Chellah
Supplier Agricola, 34 rue Beni, Amar, Casablanca
Publications E. Barbier (1974) Present and future of apiculture in Morocco. *Apiacta* 9(4): 188–191 [B]; BOTA 1 (27 refs)

Mozambique

Contact (Government) Tecnicas Basicas no Aproveitamento Racional de Natureza, Instituto de Investigação Cientifica Moçambique, Universidade Eduardo Mondlane, CP 1780, Maputo
Publications J. J. Ferreira Alcobia (1985) *Criação de abelhas* (Maputo: Ministério da Agricultura) 80 pp. [B]; BOTA 2 (5 refs)

Namibia

Contact (Government) Dep. Agriculture, Windhoek
Publication BOTA 2 (1 ref)

Nepal

Contacts (Government) Industrial Entomology Centre, Entomology Division, Khumaltar, Lalitpur, Kathmandu
BKA National Bee Development Council, 17/135 Lazimpat, POB 1601, Kathmandu
Publications K. Budhathoki (1982) [*Practical knowledge in beekeeping*] (Pokhara: Lumle Agricultural Centre) 88 pp. *In Nepali* [B]; BOTA 3 (7 refs)

Netherlands
see also Netherlands Antilles

Contacts (Government) Rijksbijenteeltconsulent, Proefbijenstand Ambrosiushoeve, Tilburgseweg 32, Hilvarenbeek
BKA Stichting Bedrijfsraad voor de Bijenhouderij in Nederland, Spoorlaan 50, Tilburg
Supplier Bijenhuis, Grintweg 273, 6704 AP Wageningen
Honey importer/packer NV Bijenstand Mellona-Adelshoeve, Bloemendaalsestraatweg 147, Santpoort
Publications J. G. de Roever (1977) *Bijen en bijenhouden* (Groningen: Erich Konstapel) 639 pp. [B]
Journals *Maandschrift voor Bijenteelt*, pub. Supplier above; *Bijenteelt*, Ravelijnstr. 2B, 4651 DT, Steenbergen
Museums Stichting Drents Bijenteeltmuseum, Hoeve Bekhof, de Hoek 5, Vledder, open some afternoons; Het Nederlands Openluchtmuseum, Schelmseweg 89, Arnhem, open daily, has small beekeeping exhibition; Royal Tropical Institute, Mauritskade 63, Amsterdam

Netherlands Antilles (Netherlands)

Contact (Government) Department van Landbouw, Bestuurschautoor Eilandgebeid Bonaire, Kralendijk, Bonaire
Publications BOTA 5 (2 refs)

New Caledonia (France)

Contact South Pacific Commission, BP D5, Noumea Cedex
Publications P. Dutant (1977) Quel type de ruche adopter? *Rev. agric. Nouv. Caléd.* (39): 7–12 [B]; *see also* Tonga

New Zealand
See also Chatham Islands, Kermadec Islands

Contacts (Government) Ministry of Agriculture and Fisheries, POB 2298, Wellington
BKA National Beekeepers' Association of New Zealand (Inc.), POB 4048, Wellington
Supplier Stuart Ecroyd Bee Supplies, POB 5056, 10 Sheffield Crescent, Burnside
Publications A. Matheson (1984) *Practical beekeeping in New Zealand* (Wellington: Government Printing Office) 185 pp. [B]; bibliography by Reid et al. (1988), *see* Bibliography of this book
Journal *New Zealand Beekeeper*, pub. BKA above
Education Bay of Plenty Community College, RD3, Tauranga, has 2-year course (correspondence + short residential periods); Telford Farm Training Institute, Balclutha, gives 4-day course in queen rearing
Library BKA above has a good library

Nicaragua

Contact (Government) Ministerio de Agricultura, Managua
Publications R. B. Kent (1979) Apicultural development in Central America and Panama: some historic and economic considerations (Pp. 167–181 in book by Commonwealth Secretariat (1979), *see* Bibliography); BOTA 5 (3 refs)

Nicobar and Andaman Islands (India)

Contact (Government) Department of Agriculture, Port Blair
Publications T. R. Dutta, R. Ahmed and S. R. Abbas (1983) The discovery of a plant in the Andaman Islands that tranquillizes *Apis dorsata*. *Bee Wld* 64(4): 158–163 [B]; BOTA 4 (3 refs)

Niger

Publication D. Moumouni and D. M. Caron (1985) Beekeeping potential of Niger. *Glean. Bee Cult.* 113(5): 271 [B]

Nigeria

Contact (Government) National Science and Technology Development Agency, Moor Plantation, PMB 5382, Ibadan
Publications J. A. A. Ayoade (1977) Beekeeping among the Tiv. *Niger. Fld* 42(1): 31–36 [B]; BOTA 2 (17 refs)

Niue Island

Contact (Government) Dep. Agriculture and Fisheries, POB 74, Niue
Publications J. B. Mackisack (1967) With bees to Niue Island. *Pacif. Isl. Mon.* 39(10): 143, 145, 147 [B]; *see also* Tonga

Norway

Contacts (Government) Institutt for Biavl, Norges Landbrukshøgskole, POB 233, 1370 Asker
BKA Norges Biavlsforening, Bergeveien 15, 1362 Billingstad
Supplier Honningcentralen A/L, Østensjøv 19, Oslo 6
Publications V. Hellern (ed.) (1984) *Hundre gyldne år* (Asker: Baerum) 167 pp. [B]
Journal *Birøkteren*, pub. BKA above
Education Courses at 1st Contact above
Museum Norges Birøkterlags Museum, at BKA above, can be visited

Oman

Contact (Government) Ministry of Agriculture and Fisheries, Ruwi
Publication R. W. Dutton and J. B. Free (1979) The present status of beekeeping in Oman. *Bee Wld* 60(4): 176–185 [B]

Pakistan

Contact (Government) Honeybee Research Programme, Pakistan Agricultural Research Council, L13 Almarcaz, F7 POB 1031, Islamabad
Supplier Syed Zafar Hasan Shah, Pak-beekeepers Society, 4–6 Dilkusha Chambers, Marston Rd, Karachi
Publications R. Ahmad (1981) *A guide to bee management in Pakistan* (Misc. Publ. Pakistan Agric. Res. Council) 38 pp. [B]; BOTA 3 (17 refs)

Panama

Contacts (Government) Programa Nacional de Apicultura, Ministerio de Desarrollo Agropecuário, Divisa (Herrera)
BKA Asociación Nacional de Apicultores de Panama, Apa. 1554, Panama 5
Publications *see* 1st refs, Costa Rica, Nicaragua; BOTA 5 (8 refs)

Papua New Guinea

Contact (Government) Dep. Primary Industry, Apiary Research Centre, POB 766, Goroka
Publications L. J. Kidd (1979) Introduction and spread of honeybees in mainland Papua New Guinea (Pp. 127–136 in book by Commonwealth Secretariat (1979), *see* Bibliography); BOTA 7 (12 refs)
Education Courses at Contact above

Paraguay

Contacts (Government) Servicio Nacional de Promoción Profesional, Calle Dr Molas Lopez e 5 y 6 (Villa Victoria), Asunción
BKA Asociación de Apicultores del Paraguay, c/o F. Schmidt, C.C 701, Asunción
Publication M. R. Quiñonez and J. R. W. Morinigo (1985) Antecedents and situation of apiculture in the Republic of Paraguay. *Proc. 30 Int. Apic. Congr.*: 410–412 [B]

Peru

Contacts (Government) Ministerio de Agricultura, 6to piso, Lima
BKA Asociación Peruana de Apicultores, Apa. 4578, Gral. Orbegozo 370, Brena, Lima
Publications R. B. Kent (1986) Beekeeping regions, technical assistance, and development policy in Peru. *Yb. Conf. Latin Am. Geogr.* 12: 22–23 [B]; BOTA 6 (3 refs)

Philippines

Contact Philippines Bee Research, Alaminos, Laguna 3724
BKA Philippine Beekeepers' Association, Manila
Supplier Imelda's Beekeeper Supplies, 1910 T. Benitez St, Malate, Manila
Publications A. S. Delima and E. J. Roberts (1987) *Beekeeping in the Philippines* (Philippines: Science and Technology Resource Agency ...) 87 pp. + annexes [B]; BOTA 4 (23 refs)

Poland

Contacts (Government) Institut Weterynarii, ul. Poznańska 35, Swarzedz k/Poznania
BKA Polski Zwiazek Pszczelarski, ul. Swietokrzyska 20, Warsaw
Publications J. Curyło et al. (1978) *Hodowla pszczól* (Warsaw: Państwowe Wydawnictwo Rolnicze i Leśne) 528 pp. rev. ed. [B]
Journals *Pszczelarstwo*, Al. Jerozolimskie 28, 00–024 Warsaw; *Pszczelnicze Zeszyty Naukowe*, Pszczelnictwa Instytutu Sadownictwa, ul. Kazimierska 2, 24–100, Puławy
Museums Skansen Pszczelarski, 1st Contact address, open daily; another at Radom; Muzeum im. Jana Dzierżona, ul. 15 Grudnia 12, 46–200 Kluczbork, open to the public

Portugal
see also Azores, Madeira

Contact (*BKA*) Sociedade dos Apicultores de Portugal, Rua S. Lazaro 130, 1 DTO Lisbon
Publications A. da Costa Moreira (1968) *Uma nova apicultura* (Porto: author) 2 vols. [Vol. 2, 264 pp., B]
Journal *As Abelhas*, Rua dos Carmelitas 5, 4000 Porto
Museum O Museu De Ovar, Ovar, has a beekeeping section

Puerto Rico (USA)

Contacts (Government) Ministerio de Agricultura, San Juan
BKA Asociación de Apicultores de Puerto Rico Inc., Apa 471, Lares, PR 00669
Publications D. M. Caron (1985) A look at beekeeping in Puerto Rico. *Am. Bee. J.* 125(4): 262–263; BOTA 5 (7 refs)
Journal *El Apicoltor*, pub. BKA above
Education 1-term courses at University of Puerto Rico, Bayamon Gardens Station, Bayamon 00620

Réunion (France)

Contact (Government) Ministère d'Agriculture, St Denis
Publications V. Kaczmarek (1988) Île de la Réunion: evolution chronique de l'apiculture. *Rev. fr. Apic.*: 348–349, 387–389 [B]

Rhodes (Greece)

Contact (Government) Department of Agriculture, Ródhos

Romania

Contact (*BKA*) Asociația Crescătorilor de Albine din Republica Socialistă România, Str. Iulius Fucik 17, Sector 2, Bucharest
Publications Asociația Crescătorilor de Albine din Republica Socialistă România (1986) *Manualul apicultorului* (Bucharest: pub. BKA above) 398 pp. 6th ed. [B]
Journal *Apicultura în România*, pub. BKA above
Education 10-day courses in June–July at Institutul International de Tecnologie si Economie Apicola, Str. Pitar Moș 20, Bucharest 70100
Museum Permanent beekeeping exhibition of BKA above, Bd. Fiscusului 42, sect. 1, Bucharest, open daily

Rwanda

Contact Regie des Centres Apicoles du Rwanda, BP 17, Kigali
Publications R. P. Tasse (1979) L'apiculture au Rwanda. *Gaz. apic.* 80(857): 147–149 [B]; BOTA 2 (4 refs)

Samoa, Eastern (USA)

Contact (Government) Ministry of Agriculture, Pago Pago
Publication *see* Tonga

Samoa, Western

Contact (Government) Ministry of Agriculture, Apia
Publication W. F. Moore (1984) Development of the honey and beekeeping industries in Western Samoa (Pp. 235–238 in book by FAO (1984), *see* Bibliography)

Saudi Arabia

Contact (Government) Ministry of Agriculture and Water, Riyadh
Honey importer/packer Ahmed M. Bamajally, POB 112, Jeddah

Scotland *see under* United Kingdom

Senegal

Contact (Government) Direction de la Santé et des Productions Animales, 37 Ave Pasteur, BP 67, Dakar
Publications M. N'Diaye (1976) Beekeeping in Senegal (Pp. 171–179 in book by Crane (1976*a*), *see* Bibliography); BOTA 2 (17 refs)

Seychelles

Contact (Government) Ministry of Agriculture, POB 54, Mahé
Publication G. Ratia (1984) *Rapport de mission sur l'apiculture seychelloise* (Mahé: Ministère du Developpement) 129 pp. [B]

Sierra Leone

Contact Dep. Agricultural Education, Njala University College, Freetown

Somalia

Contact Port Veterinary Officer, Ministry of Livestock, POB 924, Mogadiscio
Publications B. Monaco (1985) Appunti e considerazioni sull'apicoltura della Somalia. *Apic. mod.* 76(2): 35–42 [B]; BOTA 2 (3 refs)

South Africa

Contacts (Government) Plant Protection Research Institute, Private Bag 134, Pretoria
BKA South African Federation of Beekeepers' Associations, POB 4488, Pretoria 0001

Supplier Even-run Apiaries, 340 Boom St, Pietermaritzburg 3201
Publications R. H. Anderson, B. Buys and M. F. Johannsmeier (1983) *Beekeeping in South Africa* (Bull. Dep. Agric. Tech. Serv. No. 394) 207 pp. 2nd ed. [B]; BOTA Satellite 27 (177 refs)
Journal *South African Bee Journal*, pub. BKA above
Education 4-day courses at Wilgespruit Apiaries, POB 358, Rodepoort 1725

Spain
see also Balearic Islands, Canary Islands

Contact (*BKA*) Asociación Nacional de Apicultores de España, Poligono Industrial, San Anton S/L, Ayora, Valencia
Supplier Apicenter SA, Vizcaya 384, Barcelona 27
Publications J. M. Sepulveda Gil (1986) *Apicultura* (Barcelona: Biblioteca Agrícola AEDOS) 418 pp. 2nd ed. [B]; Primer Congreso Nacional de Apicultura, *see* J. B. Marquina Olmedo et al. (1983) in Bibliography
Journal *Vida Apícola*, Apdo 24.316, Barcelona

Sri Lanka

Contacts (Government) Bee Development Project, Bindunuwewa-Bandarawela
BKA Bee Farmers' Association, 41 Flower Rd, Colombo 7
Publications T. I. Szabo (1988) Sri Lanka: the bee paradise. *Am. Bee J.* 128(6): 405–406, 408–409 [B]; BOTA 3 (14 refs)

Sudan

Contact (Government) Ministry of Agriculture, Khartoum
Publications S. E. Rashad and M. S. El-Sarrag (1978) Beekeeping in the Sudan. *Bee Wld* 59(3): 105–111 [B]; BOTA 2 (9 refs)

Surinam

Contacts (Government) State Bee Officer, Ministry of Agriculture, Paramaribo
BKA Imkersvereniging Suriname, POB 459, Paramaribo
Publications *see* Guyana
Journal *Imker Koerier*, pub. BKA above

Swaziland

Contact Ministry of Agriculture and Co-operatives, Mbabane
Publication P. Bechtel and K. Gau (1988) *Introduction to beekeeping* (Mbabane: Ministry of Agriculture and Co-operatives & Near East Foundation) 92 pp.

Sweden

Contacts (Government) Bee Division, Dep. Animal Husbandry, Swedish University of Agriculture, Uppsala
BKA Sveriges Biodlares Riksförbund, POB 91, 590–20 Mantorp
Supplier Oscar Gustafsson & Co., Biredskapsfabrik AB, 4385 Tofta, 432 00 Varberg
Publications Å. Hansen (1980) *Bin och biodling* (Stockholm: LTs Förlag) 586 pp. [B]
Journal *Bitidningen*, pub. BKA above
Education Courses June–August at Bikonsult HB, POB 5034, Oja, 73300 Sala
Museum Biodlingsmuseet, Hembygdsparken, 29700 Degeberga, open some afternoons

Switzerland

Contacts (Government) Eidgenössische Milchwirtschaftliche Versuchsanstalt (Bienenabteilung), Leibefeld-Bern
BKA Fédération des Sociétés Suisses d'Apiculture, Via Regina 21, 6943 Vezia-Lugano
Supplier Bienen-Meier, 5444 Künten (AG)
Honey importer/packer Narimpex AG, Route de Reuchenette-Strasse 48, 2501 Biel
Publications G. Casaulta, J. Krieg and J. Spiess (eds) (1985) *Der Schweizerische Bienenvater* (Aarau: Sauerländer) 591 pp. 16th ed. [B]
Journals *Schweizerische Bienen-Zeitung* (German), Sauerländer AG, Postfach, 5001 Aarau; *Journal Suisse d'Apiculture* (French) and *L'Ape* (Italian), write c/o BKA above
Museum Naturhistorisches Museum Bern, 3005 Bern, Bernastr. 15, has small collection

Syria

Contact (Government) Ministry of Agriculture and Agrarian Reform, Damascus
Publication Mohamad Walid Kasso and Nabil Arkawi (eds) (1984) [*Beekeeping and honey production*] (Damascus: Al Taawonia Printing House) 232 pp. *In Arabic* [B]

Taiwan

Contact (*BKA*) Taiwan Beekeepers' Association, 201 Chung Hwa Rd, Section 2, Taichung
Publications Y. T. Kiang (1979) Beekeeping in Taiwan. *Am. Bee J.* 119(5): 363, 366–367 [B]; BOTA 4 (10 refs)

Tanzania

Contact (Government) Dep. Forests and Beekeeping, POB 65360, Dar-es-Salaam
Publications G. Ntenga (1976) Beekeeping development programmes in Tanzania (Pp. 147–154 in book by Crane (1976*a*), *see* Bibliography); BOTA 2 (23 refs) + Satellite 26 (39 refs)
Education 2-yr courses at Beekeeping Training Institute, POB 62, Tabora

Thailand

Contacts Bee Biology Research Unit, Faculty of Science, Chulalongkorn University, Bangkok
BKA Northern Thailand Beekeepers' Association, Chiangmai
Publications P. Akratanakul (1983) [*On bees and beekeeping*] (Bangkok: author) 182 pp. [B]; S. Wongsiri et al. (1985) [*Beekeeping in Thailand*] (Bangkok?: Funy Publishing) 160 pp.; BOTA 4 (7 refs)
Education Courses at 1st Contact above; (6 months) Khon Kaen University, Bee Division, Khon Kaen

Togo

Contact (Government) Ministère du Développement Rural, Lomé
Publication M. Petitjean (1975) L'apiculture au Togo. *Gaz. apic.* 76(814): 186 [B]

Tonga

Contact (Government) Ministry of Agriculture, Forests and Fisheries, POB 14, Nuku'alofa
Publication G. M. Walton (1976) Beekeeping development programmes in the tropical and subtropical Pacific (Pp. 159–164 in book by Crane (1976*a*), *see* Bibliography)

Trinidad and Tobago

Contacts (Government) Inspector of Apiaries, Ministry of Agriculture, Lands and Food Production, Harris St, Curepe
BKA Trinidad and Tobago Beekeepers' Association, c/o address above
Supplier Butcher's Glass Supplies, 85 Eastern Main Rd, Barataria
Publications G. A. Laurence and I. Mohammed (1976) Beekeeping in Trinidad and Tobago. *J. agric. Soc. Trin.* 76(4): 342–354 [B]; BOTA 5 (8 refs)

Tunisia

Contacts (Government) Ministère de l'Agriculture, Tunis
BKAs Association Tunisienne d'Apiculture, Office de l'Elevage et des Pâturages, La Rabta, Tunis; Fédération Nationale des Apiculteurs Tunisiens, 6 Ave Habib Thameur, Tunis
Publications A. Popa (1980) Beekeeping in Tunisia: its impact on other developing countries of the Mediterranean basin. *Wld Anim. Rev.* (34): 29–39 [B]; BOTA 1 (17 refs)

Turkey

Contact Gîda-Tarîm ve Hayvanchîlîk Tarîmsal, Arastîr-malar Genel Müdürlügü, Tavuculuk ve Arîkîlîk Entitüsü, PK 266, Ankara
Supplier Sedef Petek koll. şti., Hatice Aktuna ve Oğullari, Koreşehitleri Cad. No. 14 Zincirlikuyu, Levent-Istanbul
Publications K. Senocak (1956) [*Modern beekeeping*] (Ankara: Ankara Aricilik Kooperatifi) 192 pp. [B]; F. S. Bodenheimer (1942) *Studies on the honey bee and beekeeping in Turkey* (Istanbul: Numune Matbaasi) 179 pp. [B]; BOTA 1 (20 refs)

Tuvalu

Contact (Government) Dep. Agriculture, Vaiaku, Funafuti
Publication A. Thin (1984) Beekeeping on a coral atoll. *Bee Wld* 65(2): 57 [B]

Uganda

Contact (Government) Apiculture section, Ministry of Agriculture and Animal Industry, Kampala

Publications E. C. B. Nsbuga-Nvule (1979) The present development of beekeeping in Uganda (Pp. 75–77 in book by Commonwealth Secretariat (1979), *see* Bibliography); BOTA 2 (12 refs)

UK *see* United Kingdom

Union of Soviet Socialist Republics (USSR)

Contact National Committee of Apiculture, All-Union Ministry of Agriculture, Orlikov per 1/11, Moscow I-139
Publications G. A. Avetisyan (1978) *Apiculture* (Bucharest: Apimondia Publishing House) 270 pp. [B]
Journal [*Pchelovodstvo*], 107807, GSP, Sadovaya-Spasskaya 18, Moscow B-53
Education In *Pchelovodstvo* No. 4: 14–15 (1985) there is a list of about 100 addresses where courses are held, in 13 of the 15 Republics
Library Central Beekeeping Research Station, 391110 Rybnoe, Ryazan Province
Museum Central Beekeeping Research Station has important collection; visits by appointment

United Kingdom: general

Suppliers Robert Lee (Bee Supplies) Ltd, Willows Garden Nursery, Maidenhead Rd, Windsor, Berks; Steele & Brodie Ltd, Stevens Drove, Houghton, Stockbridge, Hants SO20 6LP, and Wormit, Fife DD6 8PG; Thorne (Beehives) Ltd, Beehive Works, Wragby, Lincoln LN3 5LA
Honey importer/packer Honey Importers & Packers Association, 6 Catherine Street, London WC2B 5JJ
Publications R. A. Morse and T. Hooper (1985) *The illustrated encyclopedia of beekeeping* (Sherborne: Alphabooks) 432 pp. [B]; bibliography by IBRA (1979*a*), *see* Bibliography of this book
Library IBRA headquarters and main library are at 16–18 North Road, Cardiff, CF1 3DY
Museum IBRA Collection of Historical and Contemporary Beekeeping Material, address above (enquire about visiting)

United Kingdom: England
see also United Kingdom: general

Contacts (Government) ADAS National Beekeeping Unit, Luddington, Stratford-upon-Avon, Warwicks. CV37 9SJ
BKA British Beekeepers' Association, National Agricultural Centre, Stoneleigh, Kenilworth, Warwicks. CV8 2LZ
Journal *Bee Craft*, pub. BKA above; *British Bee Journal*, 48 Queen St, Geddington, Northants. NN14 1AZ
Education 6-week course May–July at Minster Agriculture Ltd, 13 Upper High Street, Thame, Oxon. OX9 3HL. Also 1-week courses in summer at Hampshire College of Agriculture, Sparsholt, Winchester, Hants. SO21 2NF; Pershore College of Horticulture, Worcester WR10 3JP
Libraries Libraries with important collections of bee books include Ministry of Agriculture Library, 3 Whitehall Place, London SW1 2H8, and University of Reading Library, Whiteknights, Reading RG6 2AG; for IBRA library *see* UK, general

United Kingdom: Scotland
see also United Kingdom: general

Contacts (Government) Beekeeping Advisory Unit, North of Scotland College of Agriculture, Craibstone, Aberdeen AB2 9TR; also two other Colleges
BKA Scottish Beekeepers' Association, 9 Glenhome Ave, Dyce, Aberdeen AB2 0FF
Publications R. Couston (1972) *Principles of practical beekeeping* (Kilmarnock: Scottish & Universal Newspapers Ltd) 104 pp. [B]
Journal *Scottish Beekeeper*, pub. BKA above
Education 8-month courses at West of Scotland Agricultural College, Dep. Horticulture and Beekeeping, Auchincruive, Ayr KA6 5AE
Library The BKA's Moir Library, Central Public Library, George IV Bridge, Edinburgh EH1 1EG, has an important collection of bee books
Museum Scottish Agricultural Museum, Ingliston, Midlothian, has small collection

United Kingdom: Wales
see also United Kingdom: general

Contacts (Government) *see under* England above; also Welsh Beekeeping Centre, Brecon College of Further Education, Brecon, Powys LD3 9SR
BKA Welsh Beekeepers' Association, Derwen Fach, Llandygwydd, SA43 2QU
Publications M. R. Williams (1972) *Y fêl ynys* (Pontypridd: Cyhoeddiadan Modern Cymreis) 135 pp. [B]
Journal *Welsh Beekeepers' Association Quarterly Bulletin*, pub. BKA above

(*continued*)

United Kingdom: Wales *(continued)*

Education 1-yr course at University College Cardiff, Bee Research Unit, Cardiff CF1 1XL; 5-day beginners' courses at Green Meadow Apiaries, Glascoed, nr Pontypool, Gwent

Museum Welsh Folk Museum, St Fagans, Cardiff CF5 6XB, open weekdays, has small collection

United Kingdom: Northern Ireland

see also United Kingdom: general

Contacts (Government) Ministry of Agriculture, Belfast BT 35B

BKA Ulster Beekeepers' Association, Moyola Lodge, Castledawson, Co. Londonderry

Publications *see under* Irish Republic

United States of America (USA)

see also Guam, Hawaii, Puerto Rico, Samoa (Eastern), Wake Island

Contacts (Government) Bioenvironmental Bee Laboratory, US Dep. Agriculture, ARS, Beltsville, MD 20705

BKAs American Beekeeping Federation Inc., 13637 NW 39th Ave, Gainesville, FL 32601; Eastern Apicultural Society, 5 Rooney St, Northborough, MA 01532; Western Apicultural Society, 711 College, Woodland, CA 95695; April issue of 2nd journal below lists over 100 contact addresses

Suppliers Dadant & Sons Inc., Hamilton, IL 62341; Walter T. Kelley Co., Clarkson, KY 42726; A. I. Root Co., POB 706, 623 West Liberty St, Medina, OH 44258

Honey importer/packer Camerican International Inc., 260 Madison Avenue, New York, NY 10016

Publications United States Dep. Agriculture, Science & Education Administration (1980) *Beekeeping in the United States* (Agriculture Handbook, USDA No. 335) 193 pp. rev. ed. [B]; bibliography by T. S. K. and M. P. Johansson (1972*a*), *see* Bibliography of this book

Journals *American Bee Journal*, Hamilton, IL 62341; *Gleanings in Bee Culture*, pub. 3rd Supplier above

Education 2–5 yr postgraduate courses at Cornell University, Ithaca, NY 14853; Oregon State University, Corvallis, OR 97331; 3-week to 2-yr courses at Agricultural Technical Institute, Ohio State University, Wooster OH 44691; 10-week courses at Pennsylvania State University, University Park, PA 16802; shorter courses at University of Florida, Gainesville, FL 32611, and elsewhere

Libraries Special collection of bee books in Mann Library at Cornell University (above); Miller Library, University of Wisconsin, Madison, WI 53706; Jager Library, University of Minnesota, St Paul, MN 55701

Museum Hewitt Museum, Richards Road, Litchfield, CT, can be visited

Uruguay

Contact Centro de Estudios Apícolas, CC 701, Montevideo

BKA Sociedad Apícola Uruguaya, Av. Uruguay, 864, Montevideo

Publications H. Toscano (1979) Beekeeping in Uruguay. *Am. Bee J.* 119(3): 187 [B]; BOTA 6 (14 refs)

USA *see* United States of America

USSR *see* Union of Soviet Socialist Republics

Vanuatu

Contact (Government) Dep. Agriculture, Livestock and Forestry, POB 129, Port Vila

Venezuela

Contacts (Government) Ministerio de Agricultura y Cría in the different States, e.g. at Maturin (Estado Monagas), Barquisimeto (Estado Lara)

BKA Sociedad Venezolana de Apicultura, Ave Ricaurte 100–18, Valencia (Estado Carabobo)

Publications R. Gomez Rodriguez (1986) *Manejo de la abeja Africanizada* (Caracas: Apicultura Venezolana) 280 pp. [B]; BOTA 5 (26 refs)

Museum Museo Nacional de Apicultura Ignacio Herrera, Parque la Isla, Mérida

Vietnam

Contacts (Government) Ministry of Agriculture and Food Industries, Hanoi; Central Honeybee Company, Lang Ha St, Dong Da, Hanoi

BKA Vietnam Beekeepers' Association, 20 rue Nguyen-Bien, Ba-Dinh, Hanoi

Publications V. Mulder (1988) *Beekeeping in Vietnam* (Hanoi: Central Honeybee Company & KWT) 34 pp. [B]; BOTA 4 (5 refs)
Journal Ngành Ong, pub. 2nd Contact above

Virgin Islands (USA)

Publication A. and C. Krochmal (1985) Beekeeping in the Virgin Islands. *Glean. Bee Cult.* 113(3): 140, 157 [B]

Wake Island (USA)

Publications J. F. Larson (1972) *A. mellifera* discovers Pacific paradise. *Am. Bee J.* 112(10): 378–379 [B]; *see also* Tonga

Wales *see under* United Kingdom

Yemen (Arab Republic)

Publication E. R. Jaycox (1983) Starting a bee business in Yemen. *Speedy Bee* 12(5): 14, 16 [B]

Yemen People's Democratic Republic

Contact (Government) El-Kod Agricultural Research Centre, Abyan Governorate, El-Kod

Yugoslavia

Contacts (Government) Veterinarski fakultet, Bul. Jug. Narodne Armije 18, Belgrade
BKAs Savez pčelarskih Organizacija Jugoslavije, Molerova 13, 11000 Belgrade; Poslovno Udruzenje za Pčelarstvo Jugoslavije, 11 Bulevar 17A, Novi Belgrade
Publications J. Rihar (1981) *Pčelarenje nastavljacăma* (Ljubljana: Čebelarska Zadruge Ljubljana) 116 pp. 3rd ed. [B]
Journals *Pčela*, Pčelarski savez SR Hrvatske, 8 maja 26/III, 41000 Zagreb; *Slovenski Čebelar*, Cankarjeva c. 3/II, Ljubljana
Museum Muzeji radovlijiške občine, 64240 Radovljica, Linhartov trg 1, Slovenia, open daily in summer

Zaire

Contact (Government) Institut National pour l'Étude et la Recherche Agronomique, BP 1513, Kisangani
Publications L. Dubois and E. Collart (1950) *L'apiculture au Congo belge et au Ruanda-Urundi* (Bruxelles: Ministère des Colonies) 230 pp. [B]; BOTA 2 (14 refs)

Zambia

Contacts Forest Department Headquarters, Beekeeping Division, POB 228, Ndola, also Mwekera, POB Kitwe
Publications R. E. M. Silberrad (1976) *Bee-keeping in Zambia* (Bucharest: Apimondia) 76 pp. [B]; BOTA 2 (3 refs) + Satellite 26 (19 refs)

Zimbabwe

Contacts (Government) Senior Apiculturist, POB 8117, Causeway, Harare
BKA Zimbabwe Beekeepers' Council, POB 743, Harare
Supplier John Rau & Co. (Pvt.) Ltd, 2 Moffat Street, POB 2893, Harare
Publications P. Papadopoulo (1974/75) A guide for beginner bee-keepers. *Rhodesia agric. J.* 71: 27–28, 33, 89–94, 103–109; 72: 29–31, 63–66, 131–135 [B]; BOTA 2 (21 refs) + Satellite 26 (33 refs)
Journal *Beeline*, pub. BKA above
Education Courses at Dep. Conservation and Extension, POB 8117, Causeway, Harare

Bibliography

The Bibliography lists all publications referred to in the text of the book, including those in Further Reading and Reference, but not the many further publications included in the Gazetteer (Appendix 2).

References are listed below in alphabetical order of the first author, and then in year order. Publications are in the language of the title quoted except where otherwise indicated. A translated or transliterated title is in square brackets, as is a note describing an untitled publication.

The letter B at the end of a reference indicates that the publication is in the Library of the International Bee Research Association (16–18 North Road, Cardiff CF1 3DY, UK). A number such as 823/87 indicates an English summary in the IBRA journal *Apicultural Abstracts*; this example is Abstract number 823 in the 1987 Volume.

Aarhus, A. (1981) Virkninger av isolasjon i nettingbunn brett om våren. *Birøkteren* 97(4): 65–71 B, 1294/82

Abdellatif, M. A.; Abou-El-Naga, A. M. (1983) Brood rearing activity under high temperature and low relative humidity. *Proc. 2 Int. Conf. Apic. trop. Climates*: 412–414 B

Abeille de France et l'Apiculteur (1986) Les prêts bonifiés du Crédit agricole. *Abeille Fr. Apic.* (711): 549–553 B, 823/87

Adam, Brother (1968*a*) *In search of the best strains of bees* (Zell Weierbach, GFR: Walmar Verlag) 128 pp. Translated from *Auf der Suche nach den besten Bienenstämmen* (1966) B, 511/68

Adam, Brother (1968*b*) 'Isle of Wight' or acarine disease: its historical and practical aspects. *Bee Wld* 49(1): 6–18 B, 336/69

Adam, Brother (1987) *Breeding the honeybee: a contribution to the science of bee breeding* (Hebden Bridge, UK: Northern Bee Books) 118 pp. Translated from *Zuchtung der Honigbiene* (1982) B

Adams, J. (1984) *The long box hive and its operation in America* (Richmond, VA, USA: America-Kenya Publishing) 129 pp. B

Adams, J. (1985*a*) Waxless 'foundation' cheaper, easier, faster. *Speedy Bee* 14(1): 17 B

Adams, J. (1985*b*) Personal communication

Adams, J. (1986) *Catalog* (Richmond, VA, USA: American-Kenya Research and Development Corporation) 20 pp. B

Adey, M. (1985) Personal communication

Adey, M.; Walker, P.; Walker, P. T. (1986) *Pest control safe for bees: a manual and directory for the tropics and subtropics* (London: IBRA) 224 pp. B, 1318/86

Adjare, S. (1984) *The golden insect: a handbook on beekeeping for beginners* (Kumasi, Ghana: Technology Consultancy Centre) 104 pp. 2nd ed. B

Adjare, S. (1987) Ghana. *IBRA Newsletter* No. 11: 7 B

Adlakha, R. L.; Sharma, O. P. (1975) *Apis mellifera* vs. *Apis indica*. *Glean. Bee Cult.* 103(5): 160 B

Adlakha, R. L.; Sharma, O. P. (1976) *Stylops* (Strepsiptera) parasites of honey bees in India. *Am. Bee J.* 116(2): 66 B

Admad, R.; Camphor, E.; Ahmed, M. (1983) Factors affecting honey yield of the oriental bee, *Apis cerana* in Pakistan. *Pakistan J. agric. Res.* 4(3): 190–197 B, 499/86

Ahmad, R.; Camphor, E.; Ahmad, M. (1984*a*) Migration of the Eastern bee (*Apis cerana*) colonies and its effect on honey yield. *Pakistan J. agric. Res.* 5(1): 65–68 B, 500/86

Ahmad, R.; Jilani, G.; Muzaffar, N. (1984*b*) Domestication of *Apis dorsata* F. (Hymenoptera: Apidae). *Apiacta* 19(3): 69–70 B

Ahmad, R.; Muzaffar, N.; Munawar, M. S. (1985) Studies on biological control of hornet predators of honeybees in Pakistan. *Proc. 30 Int. Apic. Congr.*: 403–404 B

Ahmed, S. (1984) Nectar and pollen plants of Pakistan. *Pakistan J. For.* 34(2): 75–78 B, 911/87

Ahonen, K. (1983) Kennokansien pikasulatus nestekaasuliekillä. *Mehiläishoitaja* 17(1): 14–18. IBRA English translation E1619 B

Akratanakul, P. (1977) *The natural history of the dwarf honey bee,* Apis florea *F. in Thailand* (Cornell University: PhD Thesis) 91 pp. B, 419/78

Akratanakul, P. (1986*a*) *Beekeeping in Asia* (Rome: FAO) 112 pp. B

Akratanakul, P. (1986*b*, 1988) Personal communications

Akratanakul, P. (1987) *Honeybee diseases and enemies in Asia: a practical guide* (Rome: FAO) 51 pp. B

Alber, M. & 3 others (1955) Von der Paarung der Honigbiene. *Z. Bienenforsch.* 3(1): 1–28 B, 50/56

Albisetti, J. (1981) Propos sur les eucalyptus et l'apiculture dans le grand sud-ouest. *Rev. fr. Apic.* (397): 237–240 B, 167/82

Alfa-Laval (1988*a*) *Centri-Therm: ultra-short-time evaporator for heat-sensitive liquids* (Lund, Sweden: author) 8 pp. B

Alfa-Laval (1988*b*) *Contherm scraped-surface heat exchanger* (Lund, Sweden: author) 2 pp.

Alfa-Laval (1988*c*) *Convap scraped-surface evaporation plant* (Lund, Sweden: author) 2 pp.

Alford, D. V. (1975) *Bumblebees* (London: Davis-Poynter) 352 pp. B, 971/76

Allan, H. H. (1961) *Flora of New Zealand* Vol. 1 (Wellington: Government Printer) 1085 pp. B

Allen, M. D. (1957) Observations on honeybees examining and licking their queen. *Anim. Behav.* 5(3): 81–84 B, 410/59

Allen, M. D. (1958) Shaking of honeybees prior to flight. *Nature, Lond.* 181: 68 B, 116/59

Allen, M. D. (1960) The honeybee queen and her attendants. *Anim. Behav.* 8(3/4): 201–208 B, 236/62

Allen, M. D. (1965) The effect of a plentiful supply of drone comb on colonies of honeybees. *J. apic. Res.* 4(2): 109–119 B, 108/66

Alphandéry, R. (1981) *Un rucher nait* (Paris: Librairie de Vulgarisation Apicole) 192 pp. 5th ed. B

Alvarado Velasquez, H. A. (1975) *Estudio de rentabilidad sobre unas muestras tipo de apiarios en el oriente de Cundinamarca* (Bogotá: Universidad Santo Tomas de Aquino) 130 pp.

Amaral, E. (1957) Honey bee activities and honey plants in Brazil. *Am. Bee J.* 97(10): 394–395 B, 291/58

Ambrose, J. T. (1976) Swarms in transit. *Bee Wld* 57(3): 101–109 B

American Bee Journal (1985) USDA proposes revised grade standards for honey. *Am. Bee J.* 125(1): 12 B

American Bee Journal (1986) Hurricane Juan devastates Louisiana beeyards but not Cajun spirit. *Am. Bee J.* 126(1): 25–26 B

American Honey Producers' League (1924) *A treatise on the law pertaining to the honeybee* (Madison, USA: American Honey Producers' League) 88 pp. B

Anderson, B. N. (1966) The super system. *Am. Bee J.* 106(1): 6–7, 19 B

Anderson, E. M. (1979) Hive protection and bee disease eradication by heat sterilization. *Proc. 27 Int. Apic. Congr.*: 329–334 B, 1373/81

Anderson, R. H. (1963) The laying worker of the Cape honeybee, *Apis mellifera capensis. J. apic. Res.* 2(2): 85–92 B, 572/64

Anderson, R. H.; Buys, B.; Johannsmeier, M. F. (1983) *Beekeeping in South Africa* (Pretoria: Department of Agriculture) 207 pp. Bull. Dep. Agric. tech. Serv. No. 394, 2nd ed. B, 1st ed. 91/74

Andrich, G.; Fiorentini, R.; Consiglieri, A. (1987) Caratteristiche di alcuni tipi di propoli della Riviera Ligure. *Città Api* (28): 30–31, 34–35, 37–38 B, 321/88

Andruchow, L. (1983) *The economics of beekeeping in Alberta* (Edmonton: Alberta Agriculture) 58 pp. B, 920/86

Apiário (1977) Código Civil Brasileiro. *O Apiário* (26): 22–23 B

Apiculteur (1950) Le Cour de Paris juge un vol d'essaim. *Apiculteur* 94(1): 7–14 B, 76/50

Apimondia (1975) *Beekeeping in cold climate zones* (Bucharest: Apimondia Publishing House) 112 pp. B, 847/77

Apimondia (International Beekeeping Technology and Economy Institute) (1976) *Apitherapy today* (Bucharest: Apimondia Publishing House) 105 pp. B

Apimondia (1978) *A remarkable hive product: propolis* (Bucharest; Apimondia Publishing House) 250 pp. B, 1044/80

Apimondia (1981) *Primer simposio internacional de apicultura transhumante* (Bucharest: Apimondia Publishing House) 127 pp. B, 924/86

APV (1988) Personal communication

APV Crepaco (1988) Vertical scraped surface heat exchangers (Chicago: author) 4 pp. *Section 6 Bull. D-1-350*

Argyle, E. (1984) The Argyle method of beekeeping. *S. Afr. Bee J.* 56(6): 128–134 B, 1278/86

Armbruster, L. (1935) Litergewicht der einzelnen Bienenwohnungen. *Arch. Bienenk.* 16(1): 8–10 B

Armbruster, L. (1956) Veröffentlichungen von L. Armbruster. *Arch. Bienenk.* 33(1): 47–53 B

Armon, P. J. (1980) The use of honey in the treatment of infected wounds. *Trop. Doctor* 10: 91 B, 281/81

Asensio de la Sierra, E. (1984) Osmia (*Osmia cornuta* Latr.) pollinisateur potentiel des arbres fruitiers en Espagne (Hymenoptera, Megachilidae). *Proc. 5 Int. Symp. Poll.*: 461–466 B

Ashby, B. H.; Bailey, W. A.; Craig, W. L. (1982) Criteria for shipping bees in class 'D' aircraft cargo compartments. *Am. Bee J.* 122(6): 436–440 B, 587/83

Ashby, B. H. & 4 others (1986) A fiberboard tube package for shipping bees. *Am. Bee J.* 126(4): 264–267 B, 954/87

Ashcroft, W. J. C. (1986) The Hawke's Bay earthquake: its effect on an apiary. *N. Z. Beekpr* (189): 11 B

Asis, M. (1979) *El propóleo, un valioso producto apícola* (La Habana, Cuba: Ministerio de Agricultura) 124 pp. B, 291/81

Atkins, E. L.; Kellum, D.; Atkins, K. W. (1981) Reducing pesticide hazards to honey bees. *Leafl. Div. agric. Sci. Univ. Calif.* No. 2883: 24 pp. rev. ed. B

Atkins, E. L.; Kellum, D. (1986) Comparative morphogenic and toxicity studies on the effect of pesticides on honeybee brood. *J. apic. Res.* 25(4): 242–255 B, 274/88

Attar, Z. (1981) [Honey – an effective dental remedy.] *Stomatologia* 38: 261–268 *In Greek* B

Attri, B. S.; Singh, R. P. (1977) A note on the biological activity of the oil of *Lantana camara* L. *Indian J. Ent.* 39(4): 384–385 B, 159/81

Atwal, A. S.; Dhaliwal, G. S. (1969) Robbing between *Apis indica* F. and *Apis mellifera* L. *Am. Bee J.* 109(12): 462–463 B, 140/71

Atwal, A. S.; Goyal, N. P. (1973) Introduction of *Apis mellifera* in Punjab plains. *Indian Bee J.* 35(1/4): 1–9 B, 512/77

Aubert, S.; Gonnet, M. (1983) Mesure de la couleur des miels. *Apidologie* 14(2): 105–118 B, 963/84

Australasian Beekeeper (1978) Hive covers. *Australas. Beekpr* 80(5): 99–101 B

Australasian Beekeeper (1979) Moving beehives open entrance. *Australas. Beekpr* 81(6): 117–121 B, 588/81

Australasian Beekeeper (1986) Phostoxin (R) registered for wax moth control. *Australas. Beekpr* 87(8): 152–153 B

Australian Bee Congress, 1 (1973) *First Australian Bee Congress* (Bucharest: Apimondia Publishing House) 239 pp. B, 756/74

Autrum, H.; Metschl, N. (1963) Die Arbeitsweise der Ocellen der Insekten. *Z. vergl. Physiol.* 47: 256–273 B, 296/67

Avestisyan, G. A.; Gaidar, V. A. (1974) [Package bees used for honey production in Siberia.] *Pchelovodstvo* 94(5): 9–13 *In Russian* B, 510/77

Avetisyan, G. A. (1978) *Apiculture* (Bucharest: Apimondia Publishing House) 270 pp. B, 143/80

Avitabile, A.; Kasinskas, J. R. (1977) The drone population of natural honeybee swarms. *J. apic. Res.* 16(3): 145–149 B, 888/78

Avitabile, A. (1978) Brood rearing from late autumn to early spring. *J. apic. Res.* 17(2): 69–73 B, 949/77

Ayers, G. S.; Hoopingarner, R. A. (1987) Research imperatives for fixed-land honey production. *Am. Bee J.* 127(1): 39–41 B, 921/87

Ayers, G. S.; Hoopingarner, R. A.; Howitt, A. J. (1987) Testing potential bee forage for attractiveness to bees. *Am. Bee J.* 127(2): 91–98 B, 1232/87

Ayton, H. (1981) Bee keeping in Tasmania. *Bull. Dep. Agric. Tasmania* No. 58: 40 pp. B

Bailey, L. (1952) The action of the proventriculus of the worker honeybee, *Apis mellifera* L. *J. exp. Biol.* 29(2): 310–327 B, 264/53

Bailey, L. (1954) The respiratory currents in the tracheal system of the adult honeybee. *J. exp. Biol.* 31(4): 589–593 B, 108/56

Bailey, L.; Carlisle, E. (1956) Tests with acaricides on *Acarapis woodi* (Rennie). *Bee Wld* 37(5): 85–94 B, 357/57

Bailey, L. (1963) The 'Isle of Wight disease': the origin and significance of the myth. *Lect. cent. Ass. Beekprs*: 9 pp. B

Bailey, L. (1967) The world distribution of viruses of the honey bee. *Bull. apic. Doc. sci. tech. Inf.* 10(2): 121–129 B

Bailey, L. (1969) The signs of adult bee diseases. *Bee Wld* 50(2): 66–68 B, 450/70

Bailey, L. (1975) Recent research on honeybee viruses. *Bee Wld* 56(2): 55–64. IBRA Reprint M84 B, 806/75

Bailey, L. (1981) *Honey bee pathology* (London: Academic Press) 124 pp. B, 1314/82

Bailey, L. (1982) Viruses of honeybees. *Bee Wld* 63(4): 165–173. IBRA Reprint M111 B, 962/83

Bailey, L.; Carpenter, J. M.; Woods, R. D. (1982) A strain of sacbrood virus from *Apis cerana*. *J. invert. Path.* 39: 264–265 B, 1149/83

Bailey, L. (1985) *Acarapis woodi*: a modern appraisal. *Bee Wld* 66(3): 99–104. IBRA Reprint M116 B, 241/86

Bailey, L. (1988) Kashmir bee virus. *BBKA News* (68): 2 B

Bailey, N. E. (1970) Legal implications of beehive transporting on public highways. *Am. Bee J.* 110(3): 90–91 B

Bailey, R. (1969) Beekeepers bee ware on highways. *Am. Bee J.* 109(10): 391 B

Bailo, E. (1980) *Apicoltura pratica mediterranea* (Milan: Ottaviano) 217 pp. B

Bakar, A.; Napi, M.; Malik, A. (1986) Relationship of nectar flow in colony development and honey yield of *Apis cerana* under *Hevea brasiliensis* in Malaysia. *J. nat. Rubber Res.* 1(3): 176–186 B

Bakar, A. (1988) Personal communication

Banaszak, J. (1980) Badania nad fauną towarzyszącą w zasiedlonych ulach pszczelich. *Fragmenta Faunistica* 25(10): 127–177 B, 242/81

Bang, J.; Jørgensen, A. S. (1983) *Biavl for begyndere* (Fredericia: Tidsskrift for Biavl) 24 pp. B

Banker, R. (1975) Two-queen colony management. Pp. 404–410 from *The hive and the honey bee* ed. Dadant & Sons B

Bankova, V.; Marekov, N. (1984) [Propolis – chemical composition and standardization.] *Farmatsiya* 34(2): 8–18 *In Bulgarian* B

Banks, B. E. C.; Shipolini, R. A. (1986) Chemistry and pharmacology of honey bee venom. Chapter 7, pp. 329–416, from *Venoms of the Hymenoptera* ed. T. Piek (London: Academic Press) B, 1062/87

Barbosa da Silva, R. M. (1967) Influência do espaçamento entre caixilhos de melgueira, na produtividade da colméia móvel. *Bolm Ind. anim. (São Paulo)* 24: 209–217 B, 144/72

Batra, S. W. T. (1979) *Osmia cornifrons* and *Pithitis smaragdula*, two Asian bees introduced into the United States for crop pollination. *Proc. 4 Int. Symp. Poll.*: 307–312 B, 704/80

Battaglini, M.; Bosi, G. (1973) Caratterizzazione chimico-fisica dei mieli monoflora sulla base dello spettro glucidico e del potere rotatorio specifico. *Scienza Tecnol. Aliment.* 3(4): 217–221 B, 280/77

Bauer, M. (1985) Verbesserung der Trachtsituation für Bienenvölker in der Feldflur. *Bienenpflege* (1): 7–14 B, 1222/86

BBKA News (1981) Summary of the laws applying to the sale and supply of honey produced in the United Kingdom. *BBKA News* (March): 5–8 B

Bech, A. (1983*a*) Pesticidrester i honning. *Pointsopgave i Levnedsmiddeltoxikologi* No. 1: 19 pp. B, 1032/87

Bech, A. (1983*b*) Pesticidrester i honning – det er ikke noget problem. *Tiddskr. Biavl* 81(10): 251–252 B, 1033/87

Bechtel, P. H. (1989) Personal communication

Beck, B. F. (1935) *Bee venom therapy* (New York: D. Appleton-Century Co.) 238 pp. Reprinted 1981 B, 689/83

Beck, B. F. (1938) *Honey and health* (New York: Robert M. McBride & Co.) 272 pp. (2nd ed. with D. Smedley, 1944, omits some history) B

Bee Briefs NSW (1986) Bees and hives. *Bee Briefs* 3(2): 8 B

Bee Research Association, *see* BRA [became International Bee Research Association (IBRA) in 1976]

Bees and Honey (1980) Repelling ants. *Bees and Honey* (Jan.): 2 B

Bees and Honey (1987) Follower boards. *Bees and Honey* 4(1): 2–3 B

Beetsma, J. (1977) Places to visit: Bee markets and exhibitions in the Netherlands. *Bee Wld* 58(2): 92–93 B

Beetsma, J. (1979) The process of queen-worker differentiation in the honeybee. *Bee Wld* 60(1): 24–39. IBRA Reprint M98 B, 1284/79

Bee World (1933) The 'frosting' of granulated honey. *Bee Wld* 141(10): 119–120 B

Bee World (1936*a*) Do bees move eggs? *Bee Wld* 17(7): 74; also 26(2): 12 (1945), 30(5): 41–43 (1949) B

Bee World (1936*b*) [Notes.] 17(7): 84

Bee World (1939) Kearry v. Pattinson. Court of Appeal. *Bee Wld* 20(5): 50–51 B

Bee World (1947) [Waterglass for dissolving propolis.] *Bee Wld* 28(7): 58 B

Bee World (1962*a*) Propionic anhydride. *Bee Wld* 43(1): 4–5 B

Bee World (1962*b*) Bee products. *Bee Wld* 43(3): 64–65 B

Bee World (1964) Approaches to the problem of bee breeding. *Bee Wld* 45(4): 137–139 B

Bee World (1968) Clearing bees from honey supers. *Bee Wld* 49(2): 55–63 B

Bee World (1971) The effectiveness of queen mating stations. *Bee Wld* 52(1): 8–10 B

Bee World (1972) Attempts to obtain mating of honeybees in confinement. *Bee Wld* 53(2): 60–63 B

Bee World (1974) 'Isle of Wight' treasure trove. *Bee Wld* 55(1): 32–33 B

Bee World (1975) What we know about pollen. *Bee Wld* 56(4): 155–158. IBRA Reprint M86 B

Bee World (1976*a*) Harvesting pollen from hives. *Bee Wld* 57(1): 20–25. IBRA Reprint M86 B

Bee World (1976*b*) East, west – home's best. *Bee Wld* 57(4): 129–130 B

Bee World (1979) Foraging strategies. *Bee Wld* 60(3): 105–106 B

Bee World (1981) Correspondence and home study courses in beekeeping. *Bee Wld* 62(1): 31–33 B

Bee World (1982) Approval of varroa treatment. *Bee Wld* 63(3): 139 B

Bee World (1983) Honeybee pathology at Rothamsted since 1950. *Bee Wld* 64(1): 41–43 B

Bee World (1984) Using gamma radiation to control American foul brood (AFB). *Bee Wld* 65(2): 84 B

Beijerinck, W. (1940) *Calluna* – a monograph on the Scotch heather. *Verh. K. ned. Akad. Wet.* 38(4): 1–180 B

Beljavsky, A. G. (1933) Blister beetles and their relation to the honey bee. *Bee Wld* 14(3): 31–33 B

Bell, R. R. & 5 others (1983) Composition and protein quality of honeybee-collected pollen of *Eucalyptus marginata* and *Eucalyptus calophylla*. *J. Nutr.* 111(12): 2479–2484 B, 308/86

Bennett, C. F. (1964) Stingless-bee keeping in western Mexico. *Geogr. Rev.* 51(1): 85–92 B, 742/64

Bennett, H. (1975) *Industrial waxes* (New York: Chemical Publishing Co.) 413 + 323 pp. 2nd ed. B, 1345/78

Bentley, B. L.; Elias, T. S. (eds) (1983) *The biology of nectaries* (New York: Columbia University Press) 259 pp. B, 7/84

Benton, A. W.; Morse, R. A.; Stewart, J. D. (1963) Venom collection from honey bees. *Science, NY* 142(3589): 228–230 B, 431/64

Benton, A. W.; Morse, R. A. (1968) Venom toxicity and proteins of the genus *Apis*. *J. apic. Res.* 7(3): 113–118 B, 215/70

Berg, S. (1988) Paarungsflug der Drohnen (*Apis mellifera* L.) . . . (Universität Frankfurt: Thesis)

Bergner, K.-G. (1977) Honig, ein vorwiegend tierisches oder pflanzliches Produkt? *Probleme Ernäh.-Lebensm.-Wiss.* 4: 81–87 B

Berkhout, R. van (1987) The story of kiwicomb. *N. Z. Beekpr* (195): 38 B, 875/88

Berkin, A. (1987) *Banksia ornata* and bee keeping at Mt. Shaugh, S. A. *Aust. Bee J.* 68(8): 20 B, 863/88

Berry, R. (1987) The Arataki tube package: a major breakthrough in air-freighting bees. *N. Z. Beekpr* (195): 8–10 B

Berthold, R. (1987) Building your own beeswax processor. *Glean. Bee Cult.* 115(12): 687, 689 B

Berthold, R. (1988) A delicious way to increase sales: honey and fruit. *Glean. Bee Cult.* 116(7): 408–410 B

Bertram, B. C. R. (1986) Personal communication

Bessler, J. G. (1886) *Geschichte der Bienenzucht* (Stuttgart: W. Kohlhammer) 275 pp. B

Beutler, R. (1930) Biologisch-chemische Untersuchungen am Nectar von Immenblumen. *Z. vergl. Physiol.* 12(1): 72–176 B

Beutler, R. (1953) Nectar. *Bee Wld* 34: 106–116, 128–136, 156–162 B, 30/55

Bianco, J. K. (1985) *Beekeeping in Jamaica* (Kingston, Jamaica: Peace Corps) 108 pp. B

Bicchi, C.: Belliardo, F.; Frattini, C. (1983) Identification of the volatile components of some Piedmontese honeys. *J. apic. Res.* 22(2): 130–136 B, 301/84

Bielby, W. B. (1977) *Home honey production* (Wakefield, UK: E. P. Publications) 72 pp. B, 852/77

Bielby, W. B. (1986) Personal communication

Bienefeld, K. (1988) Dreissig Jahre Carnica-Reinzucht – Überblick und Ergebnisse. *Allg. dtsch. Imkerztg* 22(7): 221–226 B

Bienenvater (1986) [Radioactivity of honey.] *Bienenvater* 107(6): 209 B

Bijenteelt (1986) Imkeren in Indonesie. *Bijenteelt* 64(1): 4 *In Dutch* B

Bilash, G.D. (1987) Personal communication

Billingham, M. E. J. & 3 others (1973) An anti-inflammatory peptide from bee venom. *Nature, Lond.* 245(5421): 163–164 B, 892/74

Bindley, M. D. (1959) Cruelty to bees. *Bee Wld* 40(11): 284 B

Bisht, D. S.; Naim, M.; Mehrotra, K. N. (1982) Comparative efficiency of Jeolikote Villager and Newton beehives for maintaining *Apis cerana indica*. *Indian J. Ent.* 44(4): 368–372 B, 146/85

Blake, S. T.; Roff, C. (1972) *The honey flora of Queensland* (Brisbane: Department of Primary Industries) 234 pp. B, 630/73

Blaunac, Y. de (1987) *Plaisir du miel* (Paris: Robert Jauze) 318 pp. B

Blomfield, R. (1973) Honey for decubitus ulcers. *J. Am. med. Ass.* 224(6): 905 B

Bobrzecki, J. (1973) Intensywność powrotu pszczól przy przewozeniu rodzin pszczelich na małe odległości. *Zesz. nauk. Akad. roln.-tech. Olsztynie* (4): 235–247 B, 861/74

Bodenheimer, F. S. (1951) *Insects as human food: a chapter in the ecology of man* (The Hague: W. Junk) 352 pp. B

Bodnaryk, R. P. (1978) Bee venom peptides. *Adv. Insect Physiol.* 13: 106–115 B

Boetius, J. (1948) Über den Verlauf der Nektarabsonderung einiger Blütenpflanzen. *Beih. Schweiz.*

Bienenztg 2(17): 258–317. IBRA English translation E10 B

Bogdanov, S. (1984) Characterisation of antibacterial substances in honey. *Lebensm.-Wissensch. Technol.* 17(2): 74–76 B, 660/86

Bohart, G. E. (1972) Management of wild bees for the pollination of crops. *A. Rev. Ent.* 17: 287–312 B, 277/73

Böhny, W.; Müller, U.; Wüthrich, B. (1986) Berufsbedingte Inhalationsallergien bei Imkern. *Allergologie* 9(8): 337–340 B, 801/88

Bolton, J. L. (1962) *Alfalfa: botany, cultivation and utilization* (London: Leonard Hill) 474 pp. B, 306/63

Borchert, A. (1974) *Schädigungen der Bienenzucht durch Krankheiten, Vergiftungen und Schädlinge der Honigbiene* (Liepzig, DDR: S. Hirzel) 366 pp. B, 515/76

Borneck, R. (1981) L'Apiculture française et les abeilles domestiques. *Défense des Végétaux* (210): 251–255 B

Bornus, L. (1975) Hummelzuchtversuche haben eine reiche Tradition. *Proc. 3 Int. Symp. Poll.*: 279–290 B, 1221/76

Börtitz, S.; Reuter, F. (1977) Untersuchungen über den Fluorgehalt von Blüten in Gebieten mit bienengefährdenden Immissionen. *Arch. Gartenbau* 25(5): 247–255 B, 1052/79

Bosi, G.; Battaglini, M. (1978) Gas chromatographic analysis of free and protein amino acids in some unifloral honeys. *J. apic. Res.* 17(3): 152–166 B, 1075/79

BOTA, *see* Crane (1978*a*, 1978*b*)

Böttcher, F. K. (1941) 1/3 Zentner Pollen geerntet! – Mein neues Pollengerät. *Leipzig. Bienenztg* 56(2): 22–24 B

Böttcher, F. K. (1985) *Bienenzucht als Erwerb* (Munich: Ehrenwirth Verlag) 325 pp. 5th ed. B

BRA (1973) *World list of films on bees and beekeeping* (London: BRA) 67 pp. B, 81/74

Bradbear, N. (1983) *Laws and regulations (world-wide) relating to the importation of bees, the control and notification of bee diseases, and the registration and siting of hives* (London: IBRA) 27 pp. IBRA Bibliography No. 35 B

Bradbear, N. (1988) The world distribution of major honeybee diseases and pests. *Bee Wld* 69(1): 15–39 B. 1256/88

Brandeburgo, M. M.; Gonçalves, L. S.; Kerr, W. E. (1982) Effects of Brazilian climatic conditions upon the aggressiveness of Africanized colonies of honeybees. Pp. 255–280 from *Social insects in the tropics*, Vol. 1, ed. P. Jaisson (Paris: Presses de l'Université Paris-Nord) B, 532/84; see also 877/78

Brenzinger, M. (1987) East African beekeeping vocabularies: ki Zigua. *Afr. Arbeitspap.* 9: 113–123 B, 481/88

Brewer, J. W; Collyard, K. J.; Lott, C. E. (1974) Analysis of sugars in dwarf mistletoe nectar. *Can. J. Bot.* 52: 2533–2538 B

Brian, A. D.; Crane, E. (1959) Charles Darwin and bees. *Bee Wld* 40(12): 297–303 B, 313/60

British Columbia Department of Agriculture and Food (1986) Using hydrogen cyanide to kill bees. *Manitoba Beekpr* (Summer): 11–14 B

British Columbia Ministry of Agriculture (1984) Pollination and honey production, southern interior – estimated costs and returns. *Publ. Econ. Br. Min. Agric. BC* No. 267: 15 pp. B

British Standards Institution (1960) Beehives, frames and wax foundation. *British Standard* BS 1300: 1960, 28 pp. rev. B, 377/80

British Standards Institution (1964) Recommendations for the carriage of live animals by air. Fish, amphibians and invertebrates *British Standard* BS 3149: Part 9: 1964, 30 pp. B

Broadman, J. (1958) A review of the foreign literature on bee venom for the treatment of all types of rheumatism. *Gen. Pract.* 21(8): 13, 26, 28, 29 (also 21(5): 11, 12, 54, 55) B, 139/60

Broadman, J. (1962) *Bee venom: the natural curative for arthritis and rheumatism* (New York: Putnam) 220 pp. B

Brock, C. C. (1980) More on the moveable comb frameless hive. *Am. Bee J.* 120(10): 692–693 B

Bromenshenk, J. J. (1979) Investigation of the impact of coal-fired power plant emissions upon insects. *Rep. U.S. Envir. Prot. Agency* EPA 600/3-79-044: 215–239 B, 1048/81; see also 1363/80

Bromenshenk, J. J. & 3 others (1985) Pollution monitoring of Puget Sound with honey bees. *Science, NY* 227: 632–634 B, 650/86

Brooke, R. K. (1986) Personal communication

Brookes, A. (1967) Saving the back when moving hives. *Apic. W. Aust.* 2(5): 80 B

Brown, R. H. (1981) *Beeswax* (Burrowbridge, UK: Bee Books New & Old) 74 pp. B

Brückner, D. (1979) Effects of inbreeding on worker honeybees. *Bee Wld* 60(3): 137–140 B

Bryant, T. G. (1984) The ten frame brood nest. *Apiarist* (39): 8 B

Bryant, T. G. (1987*a*) Honey filtration unit. *N. Z. Beekpr* (196): 25–26 B, 974/88

Bryant, T. G. (1987*b*) Personal communication

Buchner, R. (1967) Über den Unterschied von Bienenhonig aus Blüten- und Honigtautracht. *Mitt. bad. Landesver. Naturk. Natursch.* 9(3): 589–593 B, 177/69

Buco, S. M. & 5 others (1987) Morphometric differences between Africanized and South African (*Apis mellifera scutellata*) honey bees. *Apidologie* 18(3): 217–222 B, 538/88

Budathoki, K.; Free, J. B. (1986) Comb support and attachment within transitional bee hives. *J. apic. Res.* 25(2): 87–99 B, 172/87

Budathoki, K.; Madge, D. S. (1987) Distribution of brood and food stores in combs of the honeybee, *Apis mellifera* L. *Apidologie* 18(1): 43–52 B549/88

Bühlmann, G.; Wille, M.; Imdorf, A. (1987) Messungen am Flugloch, Mai und Juni. *Schweiz. Bienenztg* 110(6): 245, 248–258 B, 1202/87

Bukharev, G. F. (1964) [Keeping bees in long hives.] *Trud. nauch.-issled. Inst. Pchelovodstva*: 58–72 *In Russian* B, 784/85

Büll, R. (1959–1970) *Vom Wachs: Hoechster Beiträge zur Kenntnis der Wachse* (Frankfurt am Main: Hoechst AG) Vol. I in 12 parts, 1097 pp. B

Bulman, M. W. (1955) Honey as a surgical dressing. *Middx Hosp. J.* 55(6): 188–189 B, 304/56

Bunney, M. H. (1968) Contact dermatitis in beekeepers due to propolis (bee glue). *Br. J. Derm.* 80: 17–23 B, 528/68

Burges, H. D. (1978) Control of wax moths: physical, chemical and biological methods. *Bee Wld* 59(4): 129–138. IBRA Reprint M96 B, 998/79

Burgett, D. M. (1973) *The occurrence of glucose oxidase in the honeys of social Hymenoptera* (Cornell University: PhD Thesis) 79 pp. B, 455/75

Burgett, D. M. (1974) Glucose oxidase: a food protective mechanism in social Hymenoptera. *Ann. ent. Soc. Am.* 67(4): 545–546 B

Burgett, D. M. (1980) The use of lemon balm (*Melissa officinalis*) for attracting honeybee swarms. *Bee Wld* 61(2): 44–46 B, 180/81

Burgett, D. M.; Akratanakul, P.; Morse, R. A. (1983) *Tropilaelaps clareae*: a parasite of honeybees in South-east Asia. *Bee Wld* 64(1): 25–28 B, 975/83

Burgett, D. M. & 3 others (1984) Evaluating honey bee colonies for pollination. *Publ. Ore. . . . Ext. Serv.* No. PN 245: 6 pp.

Burgett, D. M. (1986) Personal communication

Burking, R. C. (1986) The effects of wildfires on the Western Australian beekeeping industry. *Australas. Beekpr* 87(12): 255–258 B, 835/87

Burlando, F. (1978) Sull'azione terapeutica del miele nelle ustioni. *Minerva dermatolog.* 113: 699–706 B, 1030/80

Burtt, E. G. (1950) Personal communication

Busker, L. H. (1973) A mini-survey of hive stands. *Glean. Bee Cult.* 101(5): 144, 161 B

Butler, C. (1609) *The feminine monarchie* . . . (Oxford: printed Joseph Barnes) 120 leaves. Reprinted 1969, English Experience No. 81 B

Butler, C. G. (1954) *The world of the honeybee* rev. ed. 1974 (London: Collins) 226 pp. B

Butler, C. G.; Simpson, J. (1956) The introduction of virgin and mated queens, directly and in a simple cage. *Bee Wld* 37(6): 105–114, 124 B, 355/57

Butler, C. G.; Simpson, J. (1967) Pheromones of the queen honeybee (*Apis mellifera* L.) which enable her workers to follow her when swarming. *Proc. R. ent. Soc. Lond. A* 42(10/12): 149–154 B, 717/68

Cadapan, E. P. (1984) Beekeeping with *Apis mellifera* in the Philippines. Pp. 211–216 from *Proceedings of the expert consultation* . . . ed. FAO B

Caillas, A. (1960) *Manuel pratique du producteur de gelée royale* (Giens, Var: published by the author). Reprinted 1976 (Paris: Grancher) 63 pp. B, 268/61

Caillas, A. (1976) *Le pollen: sa récolte, ses propriétés et ses usages* (Paris: Grancher) 102 pp. 2nd ed. B

Cale, G. H.; Banker, R.; Powers, J. (1975) Management for honey production. Chapter 12, pp. 355–412, from *The hive and the honey bee* ed. Dadant & Sons B

Callow, R. K. (1963) Chemical and biochemical problems of beeswax. *Bee Wld* 44(3): 95–101 B, 657/64

Camargo, J. M. F. de (1970) Ninhos e biologia de algumas espécies de Meliponideos (Hymenoptera: Apidae) de região de Pôrto Velho, Território de Rondônia, Brasil. *Rev. Biol. trop.* 16(2): 207–239 B, 882/72

Camazine, S. (1986*a*) Queen rearing in São Paulo State, Brazil: a beekeeping experience of over 20 years. *Am. Bee J.* 126(6): 414–416 B, 951/87

Camazine, S. (1986*b*) Differential reproduction of the mite *Varroa jacobsoni* (Mesostigmata: Varroidae), on Africanized and European honey bees (Hymenoptera: Apidae). *Ann. ent. Soc. Am.* 79(5): 801–803 B, 238/88

Campbell, D. J. (1951) Hives of the world. *Bee Craft* 33(8): 186 B, 127/52

Campbell, R. W. (1961) Bag lifter as hive loader. *Aust. Bee J.* 42(10): 16–17 B

Canada National Research Council Associate Committee on Scientific Criteria for Environmental Quality (1981) *Pesticide-pollinator interactions* (Ottawa: NRCC) 190 pp. Publication No. 18471 B, 599/82

Canadian Beekeeping (1988) Closing the Canada–US border, a new concept. *Can. Beekeep.* 14(2): 27 B

Cantwell, G. E.; Shimanuki, H.; Retzer, H. J. (1971)

Containerized bees airdropped into cranberry bogs. *Am. Bee J.* 111(7): 272–273 B, 552/72

Cantwell, G. E.; Lehnert, T.; Travers, R. S. (1975) USDA research on ethylene oxide fumigation for control of diseases and pests of the honey bee. *Am. Bee J.* 115(3): 96–97 B, 826/76

Cantwell, G. E.; Shieh, T. R. (1981) Certan – a new bacterial insecticide against the greater wax moth, *Galleria mellonella* L. *Am. Bee J.* 121(6): 424–426, 430–431 B, 1293/82

Cárdenus, B. L. (1939) Flora melífera: monografía de la flora melífera de San Juan de la Costa. *Apicultura chil.* 2(9): 282–283

Carlson, D. A.; Bolten, A. B. (1984) Identification of Africanized and European honey bees, using extracted hydrocarbons. *Bull. Ent. soc. Am.* 30(2): 32–35 B, 897/87

Caron, D. M. (1978*a*) Other insects. Chapter 10, pp. 158–185, from *Honey bee pests, predators, and diseases* ed. R. A. Morse B

Caron, D. M. (1978*b*) Marsupials and mammals. Chapter 15, pp. 227–256, from *Honey bee pests, predators, and diseases* ed. R. A. Morse B

Caron, D. M. (1978*c*) Bears and beekeeping. *Bee Wld* 59(1): 18–24 B, 237/79

Caron, D. M. (1979*a*) Swarm emergence date and cluster location in honey bees. *Am. Bee J.* 119(1): 24–25 B, 515/80

Caron, D. M. (1979*b*) Observation bee colonies. *Ent. Leafl. Univ. Maryland* No. 103, 14 pp. B

Carpenter, M. (1978) Transporting palletized bees in refrigerator trailers. *Am. Bee J.* 118(1): 22–23 B

Casteel, D. B. (1912*a*) The behaviour of the honeybee in pollen collecting. *Bull. U.S. Bur. Ent.* No. 21: 36 pp. B

Casteel, D. B. (1912*b*) The manipulation of the wax scales of the honeybee. *Circ. U.S. Bur. Ent.* No. 161: 13 pp. B

Casteljau, C. de (1983) *Bibliographie d'apiculture de langue française* (Besançon: author) 160 pp. B, 96/84

Castro de Lepratti, M. de (1974) Honey flora in Uruguay. *Apiacta* 9(1): 45–47 B

Cavalloro, R. (ed.) (1983) Varroa jacobsoni *Oud. affecting honey bees: present status and needs* (Rotterdam: Balkema) 107 pp. B, 1266/84

Cavanagh, D.; Beazley, J.; Ostapowicz, F. (1970) Radical operation for carcinoma of the vulva. *J. Obstetr. Gynaecol. Br. Commonw.* 77(11): 1037–1040 B, 187/73

CCI (1987), *see* next entry

Centre du Commerce International CNUCED/GATT (1987) *Note sur les marchés pour les produits sélectionnés de la ruche au Royaume-Uni, en France et en Italie; perspectives et développements* (Geneva: CCI, UNCTAD/GATT) 35 pp. B

Chakrabarti, K.; Chaudhuri, A. B. (1972) Wild life biology of the Sundarbans forests: honey production and behaviour pattern of the honey bee. *Sci. Cult.* 38(6): 269–276 B, 1130/78

Chalmers, W. T. (1980) Fish meals as pollen-protein substitutes for honeybees. *Bee Wld* 61(3): 89–96 B, 228/81

Chambers, S. R. (1975) A unit for trapping clean pollen. *J. Agric. West Aust.* 16(1): 12–13 B, 674/77

Chambers, S. R. (1980) Harvest and transport of pollen. *Australas. Beekpr* 82(1): 17–19 B, 300/82

Chambers, S. R. (1981) Modified cavity lid built in inner cover. *Australas. Bkpr* 83(4): 78–80 B

Chambers, S. R. (1985, 1988) Personal communications

Chance, M. M. (1983) Honeybees observed feeding on the blood of a bear. *Bee Wld* 64(4): 177 B

Chandler, B. V. & 3 others (1974) *Composition of Australian honeys* (Canberra: CSIRO) 39 pp. Tech. Pap. Div. Fd Red. CSIRO Aust. No. 38 B, 845/75

Chandler, M. T.; Mdemu, E. (1975) Pollination research in East Africa: problems and promises. Pp. 57–65 from *Proc. 3 Int. Symp. Poll.* B, 1216/76

Chandler, M. T. (1980) Personal communication

Chandran, K.; Shah, F. A. (1974) Beekeeping in Kodai hills (Tamilnadu: India). *Indian Bee J.* 36(1/4): 1–8 B, 780/78

Chang, S. Y. (1988) Personal communication

Chaplin, L. M. (1969) [Hive entrance holes.] *Pchelovodstvo* 89(7): 19–21 *In Russian* B, 993/72

Chapman, M. (1985) Personal communication

Charles-Edwards, T.; Kelly, F. (eds) (1983) *Bechbretha: An old English bee-tract on beekeeping* (Dublin: Dublin Institute for Advanced Studies) 214 pp. B, 456/84

Chaubal, P. D.; Kotmire, S. Y. (1980) Floral calendar of bee forage plants at Sagarmal (India). *Indian Bee J.* 42(3): 65–68 B

Chaudhari, R. K. (1977) Bee forage in Punjab Plains (India). 1. Pathankot and adjacent villages. *Indian Bee J.* 39(1/4): 5–20 B

Chaud-Netto, J. (1978) Are diploid drones of *Apis mellifera* L. attracted by queen pheromone? *Ciência e Cultura* 30(5): 608–610 B, 1189/78

Chauvin R. (ed.) (1968*a*) *Traité de biologie de l'abeille* (Paris: Masson et Cie) 5 vols. 547, 566, 400, 434, 152 pp. B, 322/68

Chauvin R. (ed.) (1968*b*) *Biologie et physiologie générales*. Vol. 1 of Chauvin (1968*a*) B, 323/68

Chauvin R. (ed.) (1968*c*) *Système nerveux, comportement et régulations sociales*. Vol. 2 of Chauvin (1968*a*) B, 324/68

Chauvin R. (ed.) (1968*d*) *Les produits de la ruche*. Vol. 3 of Chauvin (1968*a*) B, 397/68

Chauvin R. (ed.) (1968*e*) *Biologie appliqueé*. Vol. 4 of Chauvin (1968*a*) B, 367/68

Chetaikin, N. V. (1982) [The contract system is beneficial to all.] *Pchelovodstvo* (5): 5–6 *In Russian* B, 357/84

Chevet, R.; Chevet, B. (1987) *L'arna aragonaise* (Bordeaux: author) 40 pp. B, 583/88

Chinese Standards Institution (1983) [The Chinese bee *Apis cerana* ten-frame hives.] *Chinese Standard* GB 3607–83, 10 pp. B

Christensen, F. (1984) *Biplanteflora* (Copenhagen: L. Launsø) 88 pp. B, 588/87

Chudakov, V. G. (1963) [The composition and properties of honey.] *Pchelovodstvo* 40(7): 18–19 *In Russian* IBRA English translation E700 B, 196/65

Chudakov, V. G.; Sorakina, V. Kh.; Vakhonina. T. V. (1976) [Standardization of beekeeping products: problems and perspectives.] Pp. 125-132 from [*Beekeeping – its industrial foundation*] ed. G. F. Taranov (Rybnoe, USSR: Beekeeping Research Institute) *In Russian* B

Cîrnu, I. V. (1980) *Flora meliferă* (Bucharest: Editura Ceres) 202 pp. B, 154/82

Cižmarik, J.; Matel, I. (1970) Examination of the chemical composition of propolis. I. *Experientia* 26: 713 B, 783/71

Clauss, B. (1982) *Bee keeping handbook* (Gaborone: Ministry of Agriculture) 76 pp. 2nd ed. B, 1st ed. 499/81

Clauss, B. (1983) *Bees and beekeeping in Botswana* (Gaborone, Botswana: Ministry of Agriculture) 122 pp. B

Clauss, B. (1984) Ex Africa: Botswana. Some observations on biology and behaviour. *S. Afr. Bee J.* 56(5): 113–116 B

Clauss, B. (1986) Personal communication

Clemson, A. A. (1980) The use of escape boards for clearing supers of bees. *Australas. Beekpr* 81(7): 139–145 B

Clydesdale, F. M. (1969) The measurement of color. *Fd Technol.* 23: 16–22

Cobey, S.; Lawrence, T. (1988) Commercial application and practical use of the Page–Laidlaw closed population breeding programme. *Am. Bee J.* 128(5): 341–344 B, 241/89

Codex Alimentarius Commission (1969) *Recommended European Regional Standard for Honey* (Rome: Joint FAO/WHO) Food Standards Programme); reprinted in *Bee Wld* 51(2): 97–91 (1970) B

Codex Alimentarius Commission (1983/84) *Proposed draft Codex Standard for Honey (World-wide Standard)* (Rome: FAO/WHO) CX/PFV 84/13, also Revisions ALINORM 85/20, Appendix IX, Appendix X B

Coggshall, W. L.; Morse, R. A. (1984) *Beeswax: production, harvesting, processing and products* (Ithaca, NY, USA: Wicwas Press) 192 pp. B, 1384/85

Colin, M. E. & 3 others (1986) La qualité des miels du commerce. *Cah. Nut. Diét.* 21(3): 219–222 B

Collins, A. M. & 4 others (1980) A model of honeybee defensive behaviour. *J. apic. Res.* 19(4): 224–231 B, 919/81

Collins, A. M. (1983, 1988) Personal communications

Collins, A. M. (1985) Africanized honeybees in Venezuela: defensive behaviour. *Proc. 3 Int. Conf. Apic. trop. Climates*: 117–122 B

Collins, A. M.; Rinderer, T. E. (1985) Effect of empty comb on defensive behavior of honeybees. *J. chem. Ecol.* 11(3): 333–338 B, 898/86

Collins, A. M. & 3 others (1987) Response to alarm pheromone by European and Africanized honeybees. *J. apic. Res.* 26(4): 217–223 B, 494/89

Combs, G. F. (1972*a*) The engorgement of swarming worker honeybees. *J. apic. Res.* 11(3):121–128 B, 616/73

Combs, G. F. (1972*b*) Distribution of food reserves in 'model' swarm clusters. *J. apic. Res.* 11(3): 129–134 B, 615/73

Commonwealth Secretariat; IBRA (eds) (1979) *Beekeeping in rural development: unexploited beekeeping potential in the tropics with special reference to the Commonwealth* (London: Commonwealth Secretariat) 196 pp. B, 1171/80

Congresso Brasiliero de Apicultura (1970–1988) [Proceedings] B

1	1970	[554/74	5	1980	[508/86
2	1972	[42/75	6	1984	
3	1974	[1144/77	7	1986	
4	1976	[1132/78	8	1988	

Congresso Internazionale Ceroplastica nella Scienza e nell'Arte I (1977) [Proceedings] 728 pp. B, 979/78

Connor, L. (1988) Personal communication

Conway, J. R. (1977) Analysis of clear and dark amber repletes of the honey ant, *Myrmecocystus mexicanus hortideorum*. *Ann. ent. Soc. Am.* 70(3): 367–369 B, 39/79

Cook, V. A. (1962) Drifting of bees: effect of hives placed in circle. *N. Z. J. Agric.* 105(5): 407–409, 411 B, 619/64

Cook, V. A. (1970) Puff ball control of bees. *N. Z. J. Agric.* 120(5): 72–73 B, 189/74

Cook, V. A. (1972) The carrying capacity of beekeeping areas. *N. Z. Beekpr* 34(4): 16–18 B

Cook, V. A. (1981) New Zealand honey from beech. *Bee Wld* 62(1): 20–22 B

Cook, V. A.; Bowman, C. E. (1986) *Mellitiphis alvearius*, a little-known mite of the honeybee colony, found on New Zealand bees imported into England. *Bee Wld* 64(2): 62–64 B

Cook, V. A.; Griffiths, D. A. (1985) Varroasis of bees: tobacco smoke detection. *Pamphl. Min. Agric.* No. 936: 2 pp. B

Cook, V. A. (1986) *Queen rearing simplified* (Geddington, UK: British Bee Journal) 63 pp. B, 1250/86

Cook, V. A. (1987) Personal communication

Cooke, M. J. (1986/1988) Queen-rearing – Ilse's way. *S. Afr. Bee J.* 58: 54–60, 78–84; 60: 6–9 B, 227/87

Cooks, W. M. (1984) Personal communication

Coon, C. S. (1972) *The hunting peoples* (London: Jonathan Cape)

Cooper, B. A. (1986) *The honeybees of the British Isles* (Derby, UK: British Isles Bee Breeders' Association) 158 pp. B, 568/87

Cornejo, L. G.; Rosi, C. O.; Dávila, M. (1971) Curva de aporte nectarífero mediante el uso de la colmena báscula. *Gac. Colmen.* 33(377): 260–261 B

Corner, J. (1976) *Beehive construction* (Ottawa: Canada Dep. Agriculture) 22 pp. Publication 1584

Corner, J. (1988) Personal communication

Cosenza, G. W. (1970) Estudo dos enxames de migração de abelhas africanas. *1 Congr. Bras. Apic.*: 128–129 B, 610/74

Cotton, W. C. (1842) *My bee book* (London: Rivington) 368 pp. B
Reprinted 1977 (Bath: Kingsmead Reprints) B

Couston, R. (1985) *Co-operative honey handling and extracting techniques* (East of Scotland College of Agriculture) 80 pp. Bulletin No. 30 B, 992/86

Cowan, T. W. (1908) *Wax craft* (London: Samson Low, Marston & Co.) 172 pp. B

Cramer, H. H. (1967) *Plant protection and world crop production* (Pflanzenschutznachrichten, Bayer) 524 pp.

Crane, E. (1949) Water and the honeybee colony. *Lect. cent. Ass. Bkprs*: 8 pp. B, 47/52

Crane, E. (ed.) (1951–88) *International Bee Research Association Dictionary of beekeeping terms with allied scientific terms.* Vols 1, 2, 4, London: Bee Research Association; Vol. 3, Warsaw: Pańстowe Wydawnictwo Rolnicze i Leśne; Vols 5–8, Bucharest: Apimondia Publishing House & IBRA; Vol. 9, Tokyo: Institute of Honeybee Science; Vol. 10, London: IBRA; each Vol. with Latin index. Vols 5–10 include new terms additional to those in Vols 1–4.
Vol. 1 (1951) English–French–German–Dutch, 74 pp. B, 74/52
Vol. 2 (1958) English–Italian–Spanish, 63 pp. B, 310/63
Vol. 3 (1964) English–French–German–Czech–Polish–Russian, 203 pp. B, 542/64
Vol. 4 (1971) English–Danish–Norwegian–Swedish, 108 pp. B, 337/72
Vol. 5 (1977) English–French–German–Russian–Spanish, 206 pp. B, 515/79
Vol. 6 (1978) English–Finnish–Hungarian, 104 pp. B, 458/80
Vol. 7 (1978) English–German–Dutch–Danish–Norwegian–Swedish, 238 pp. B, 259/80
Vol. 8 (1979) English–French–Italian–Spanish–Portuguese–Romanian, 252 pp. B, 444/82
Vol. 9 (1985) English–French–Japanese, 187 pp. B, 835/86
Vol. 10 (1988) English–French–Arabic, 196 pp. B

Crane, E. (1953) Acarine disease. *Bee Wld* 34(12): 246–247 B

Crane, E. (1968) Mites infesting honeybees in Asia. *Bee Wld* 49(3): 113–114 B, 187/71

Crane, E. (ed.) (1975*a*) *Honey: a comprehensive survey* (London: Heinemann in co-operation with IBRA) 608 pp. Reprinted 1976 with corrections B, 542/76

Crane, E. (1975*b*) The plants honey comes from. Chapter 1, pp. 3–76, from Crane (1975*a*) B

Crane, E. (1975*c*) The world's honey production. Chapter 4, pp. 115–153, from Crane (1975*a*) B

Crane, E. (1975*d*) Honey from other bees. Chapter 17, pp. 411–425, from Crane (1975*a*) B

Crane, E. (1975*e*) History of honey. Chapter 19, pp. 439–488, from Crane (1975*a*) B

Crane, E. (ed.) (1976*a*) *Apiculture in tropical climates* (London: IBRA) 208 pp. *Proceedings of 1st International Conference on Apiculture in Tropical Climates* B, 1155/77

Crane, E. (1976*b*) Unpublished report

Crane, E.; Townsend, G. F. (1976) *Index to Apicultural Abstracts 1950–1972* (Folkestone, UK: William Dawson for BRA) 2 vols. 1824 pp. B, 493/77

Crane, E. (1977) Dead bees under lime trees. *Bee Wld* 58(3): 129–130 B

Crane, E. (1978*a*) BOTA, i.e. *Bibliography of tropical apiculture* (London: IBRA) 24 Parts, as listed, 380 pp. B, 1181/80
1 Beekeeping in North Africa and the Middle East
2 Beekeeping in Africa south of the Sahara
3 Beekeeping in the Indian sub-continent (with Afghanistan and Iran)
4 Beekeeping in Asia east of India
5 Beekeeping in northern Latin America (with Brazil)

6 Beekeeping in southern Latin America
7 Beekeeping in the Pacific area
8 *Apis mellifera* of European and Asiatic origin in the tropics
9 *Apis mellifera* native to Africa
10 *Apis mellifera* hybrids, known as Africanized bees, in America
11 The Asiatic hive bee *Apis cerana*
12 The giant honeybee *Apis dorsata*
13 The little honeybee *Apis florea*
14 Beekeeping in the tropics with stingless bees
15 Bee forage in the tropics
16 Beekeeping management and equipment in the tropics
17 Indigenous materials, methods and knowledge relating to the exploitation of bees in the tropics
18 Bee diseases, enemies and poisoning in the tropics
19 Honey in the tropics
20 Beeswax and other hive products in the tropics
21 Descriptions of pollen grains in tropical honeys
22 Bees for pollination in the tropics
23 Apicultural development programmes (Information sheet)
24 Miscellaneous, indexes, acknowledgements, etc.

Crane, E. (1987*b*) *Bibliography of tropical apiculture. Satellite Bibliographies* (London: IBRA) 14 Parts, as listed, 238 pp. B, 142/81

S25 Beekeeping and bee research in Egypt
S26 Beekeeping and bee research in eastern Africa
S27 Beekeeping and bee research in South Africa
S28 Beekeeping and bee research in India
S29 Beekeeping and bee research in Mexico
S30 Beekeeping and bee research in Brazil
S31 Beekeeping and bee research in Argentina
S32 *Apis cerana*: laboratory studies
S33 Biology of stingless bees (Meliponinae)
S34 Bee forage in specific regions of the tropics
S35 Bee diseases and pests in specific regions of the tropics
S36 Honeys of specific regions of the tropics
S37 Descriptions of pollen grains of further tropical plants
S38 Bee pollination in specific regions of the tropics

Crane, E. (1978*c*) Places to visit: the British Museum's beeswax treasures. *Bee Wld* 59(1): 39–40 B

Crane, E. (1978*d*) Sugars poisonous to bees. *Bee Wld* 59(1): 37–38 B

Crane, E. (1979*a*) Directory of the world's beekeeping museums. *Bee Wld* 60(1): 9–23. IBRA Reprint M97 B, 1255/79

Crane, E. (1979*b*) Honey in relation to infant botulism. *Bee Wld* 60(4): 152–154 B

Crane, E. (1979*c*) Bees and beekeeping in the tropics, and trade in honey and beeswax with special reference to the Commonwealth. Pp. 1–19 from *Beekeeping in rural development* ed. Commonwealth Secretariat & IBRA B, 1182/80

Crane, E. (1980*a*) *A book of honey* (Oxford: Oxford University Press) 198 pp. B, 656/81

Crane, E. (1980*b*) Apiculture. Pp. 261–294 from *Perspectives in world agriculture* compiled by CAB (Farnham Royal, UK: Commonwealth Agricultural Bureaux) B, 140/81

Crane, E. (1981*a*) When important honey plants are invasive weeds. *Bee Wld* 62(1): 28–30 B

Crane, E. (1981*b*) Bee houses. *Bee Wld* 62(2): 43–45 B

Crane, E. (1982*a*) Introduction of non-native bees to new areas. *Bee Wld* 63(1): 50–53 B, 1184/82

Crane, E. (1982*b*) Learning about honey through fructose. *Bee Wld* 63(4): 174–177 B

Crane, E. (1983*a*) *The archaeology of beekeeping* (London: Duckworth) 360 pp. B, 104/84

Crane, E. (1983*b*) Honey and infant botulism: second report. *Bee Wld* 64(4): 148–149 B

Crane, E. (1983*c*) Surveying the world's honey plants. *Lect. cent. Ass. Bkprs*: 12 pp. B

Crane, E.; Walker, P. (1983) *The impact of pest management on bees and pollination* (London: Tropical Development and Research Institute) 232 pp. B, 347/84

Crane, E. (1984*a*) *Bibliographical tools for apiculture* (London: IBRA) 21 pp. IBRA Bibliography No. 20 B, 1160/85

Crane, E. (1984*b*) Bees, honey and pollen as indicators of metals in the environment. *Bee Wld* 65(1): 47–49 B

Crane, E. (1984*c*) Places to visit: Japan. *Bee Wld* 65(3): 139–141 B

Crane, E. (1984*d*) Honeybees. Chapter 65, pp. 403–415, from *Evolution of domesticated animals* ed. I. L. Mason (London: Longman Group) B, 1167/85

Crane, E.; Walker, P. (1984*a*) *Pollination directory for world crops* (London: IBRA) 183 pp. B, 664/85

Crane, E.; Walker, P. (1984*b*) Composition of honeys from some important honey sources. *Bee Wld* 65(4): 167–174 B, 990/85

Crane, E.; Walker, P.; Day, R. (1984) *Directory of important world honey sources* (London: IBRA) 384 pp. B, 1238/84

Crane, E. (1985*a*) Bees and honey in the exploitation of arid land resources. Chapter 12, pp. 164–175, from *Plants for arid lands* ed. G. E. Wickens, J. R. Goodin & D. V. Field (London: George Allen & Unwin) B, 906/86

Crane, E. (1985*b*) Some multipurpose trees that are important honey sources in the tropics and subtropics. *Proc. 3. Int. Conf. Apic. trop. Climates*: 192–197 B, 202/86

Crane, E. (1985*c*) *Selective annotated bibliography on the varroa mite and its control in honeybee colonies* (London: IBRA) 80 pp. IBRA Bibliography No. 37 B, 964/86

Crane, E. (1985*d*) Beekeeping. Chapter 12, pp. 213–230, from *Tools for agriculture: a buyer's guide to appropriate equipment* ed. Intermediate Technology, 3rd ed. Reprinted (1987) as *Beekeeping: some tools for agriculture* (London: Intermediate Technology Publications) 22 pp. B, 203/87

Crane, E.; Graham, A. J. (1985) Bee hives of the Ancient World. *Bee Wld* 66: 23–41, 148–170. IBRA Reprint M117 B, 917/85, 600/86

Crane, E.; Walker, P. (1985*a*) Important honeydew sources and their honeys. *Bee Wld* 66(3): 105–112 B, 214/86

Crane, E.; Walker, P. (1985*b*) Evidence on Welsh beekeeping in the past. *Folk Life* 23: 21–48 B, 1173/85

Crane, E.; Walker, P. (1986*a*) Some nectar characteristics of certain important world honey sources. *Pszczel. Zesz. Nauk.* 29: 29–45 B, 562/88

Crane, E.; Walker, P. (1986*b*) *Honey Sources Satellite 3. Chemical composition of some honeys* (London: IBRA) 42 pp. B, 312/87

Crane, E.; Walker, P. (1986*c*) *Honey Sources Satellite 4. Physical properties, flavour and aroma of some honeys* (London: IBRA) 56 pp. B, 308/87

Crane, E.; Walker, P. (1986*d*) *Honey Sources Satellite 5. Honeydew sources and their honeys* (London: IBRA) 33 pp. B, 194/87

Crane, E.; Walker, P. (1986*e*) *Honey Sources Satellite 6. Drought-tolerant and salt-tolerant honey sources* (London: IBRA) 93 pp. B, 187/87

Crane, E.; Walker, P.; Day, R. (1986) *Honey Sources Satellite 2. Plants listed alphabetically and by family; common name index: pollen grain information* (London: IBRA) 47 pp. B

Crane, E. (1988) Personal observation, in this or an earlier year

Crane, E. (1989*a*) History of beekeeping with *Apis cerana* in Asia. Chapter 1, from *Beekeeping with* Apis cerana *in tropical and subtropical Asia* ed. P. G. Kevan B

Crane, E. (1989*b*) Early honey production in St Kitts and Nevis. *Newsletter Nevis hist. Conserv. Soc.* 15 (March): 4–7 B

Crane, E. (1989*c*) Report on methods of removing water from honey. From *Beekeeping with* Apis cerana *in tropical and subtropical Asia* ed. P. G. Kevan B

Crawford, W. (1979) A new hive tool. *Glean. Bee Cult.* 107(7): 341 B

Crewe, R. M. (1985) Bees observed foraging on an impala carcass. *Bee Wld* 66(1): 8 B

Cronin, E. W.; Sherman, P. W. (1976) A resource-based mating system: the orange-rumped honey guide. *Living Bird* 15: 5–32 B, 787/78

Cross, D. J. (1983) Preservative treatments of wood used in hives. *Bee Wld* 64(4): 169–174 B, 574/84

Crosse-Upcott, A. R. W. (1956) Social aspects of Ngindo bee-keeping. *J. R. anthrop. Inst.* 86(2): 81–108 B, 283/61

Crundwell, J. (1985) An overall feeder. *Beekprs Ann. Suppl.* (1): 7 B

Cruz-Landim, C. de (1985) Avaliação fotográfica de digestão do pólen presente no intestino de operárias de *Apis mellifera* L. (Hymenoptera, Apidae). *Naturalia* 10: 27–36 B, 811/88

Currie, R. W. (1987) The biology and behaviour of drones. *Bee Wld* 68(3): 129–143 B

Cutts, J. M. (1961) The history of package bees. *Glean. Bee Cult.* 89(2): 87–89, 121 B

Dadant, C. (1874) *Petit cours d'apiculture pratique* (Chaumont: M. Dadant) 116 pp.

Dadant, J. C. (1977) From rag rolls to modern smoker. *Am. Bee J.* 117(6): 388–390 B

Dadant & Sons (ed.) (1975) *The hive and the honey bee* (Hamilton, IL, USA: Dadant & Sons) 740 pp. rev. ed.; new ed. to be published 1990 B, 859/77

Dadant & Sons (ed.) (1980) *The honey kitchen* (Hamilton, IL, USA: Dadant & Sons) 208 pp. B

Dadant & Sons (1987*a*) Personal communication

Dadant & Sons (1987*b*) [System for reducing the water content of honey.] *Am. Bee J.* 127(4): 272 B

Dade, H. A. (1962) *The anatomy and dissection of the honeybee* (London; BRA) 164 pp. Reprinted 1977 and 1985 B, 604/62

Daharu, P. A.; Sporns, P. (1984) Phenol residue levels in honey. *J. apic. Res.* 23(2): 110–113 B, 620/85

Daharu, P. A.; Sporns, P. (1985) Residue levels and sensory evaluation of the bee repellent, phenol, found in honey. *Can. Inst. Fd Sci. Tech. J.* 18(1): 63–66 B, 995/86

Dams, L. R. (1978) Bees and honey-hunting scenes in the Mesolithic rock art of eastern Spain. *Bee Wld* 59(2): 45–53. IBRA Reprint M93 B, 1276/78

Danka, R. G. (1986) Responses of Africanized bees to pollination management. *Am. Bee J.* 126(12): 828–829 B

Danka, R. G.; Rinderer, T. E. (1986) Africanized bees and pollination. *Am. Bee J.* 126(10): 680–682 B, 891/87

Dany, B. (1975) *Pollensammeln heute* (Munich: author) 32 pp. B

Darchen, R. (1960) Le polygynisme expérimental dans les sociétés d'*Apis mellifera*. *C. r. Acad. Sci., Paris* 250: 934–936 B, 128/61

Darchen, R. (1986) Personal communication

Dartington, R. (1985) *New beekeeping in a long deep hive* (Bridgwater, UK: Bee Books New and Old) 38 pp. B

Darwin, C. (1859) *On the origin of species by means of natural selection* . . . (London: Murray) 502 pp. B

David, R.; Brierley, J. E. C. (1978) *Major legal systems in the world today* (London: Stevens & Sons) 584 pp. 2nd ed.

Davies, A. M. C. (1975) Amino acid analysis of honeys from eleven countries. *J. apic. Res.* 14(1): 29–39 B, 459/75

Davies, A. M. C. (1976) The application of amino acid analysis to the determination of the geographical origin of honey. *J. Fd Technol.* 11(5):515–523 B, 1330/78

Davies, A. M. C. (1978) Proline in honey: an osmoregulatory hypothesis. *J. apic. Res.* 17(4): 227–233 B, 1296/79

Davies, A. M. C.; Harris, R. G. (1982) Free amino acid analysis of honeys from England and Wales: application to the determination of the geographical origin of honeys. *J. apic. Res.* 21(3): 168–173 B, 995/83

Davies, N. de (1944) *The tomb of Rekhmire at Thebes* (Salem, NH, USA: Ayer Co.)

Davis, A. R. & 3 others (1985) A portable incubator for transporting mature queen cells. *Can. Beekeep.* 12(5): 105–107 B, 940/86

Dawson, C. (1959) The beekeeper's other man. *N. Z. Beekpr* 21(2): 28–29 B, 160/60

Dawson, T. (1986) More on propolis. *Beeline* 23(4): 7 B

De Grandi-Hoffman, G. D; Hoopingarner, R.; Baker, K. (1984) Pollen transfer in apple orchards: tree-to-tree or bee-to-bee? *Bee Wld* 65(3): 126–133 B

De Grandi-Hoffman, G.; Hoopingarner, R.; Klomparens, K. (1986) Influence of honey bee (Hymenoptera: Apidae) in-hive pollen transfer on cross-pollination and fruit set in apple. *Envir. Ent.* 15(3): 723–725 B, 1199/88

De Jong, D. (1978) Insects: Hymenoptera (ants, wasps, and bees). Chapter 9, pp. 138–157, from *Honey bee pests, predators, and diseases* ed. R. A. Morse B

De Jong, D. (1982) Orientation of comb building by honeybees. *J. comp. Physiol. A* 146: 495–501 B, 834/84

De Jong, D.; Morse, R. A.; Eickwort, G. C. (1982) Mite pests of honey bees. *A. Rev. Ent.* 27: 229–252 B, 1338/82

De Jong, D. (1984) Africanized bees now preferred by Brazilian beekeepers. *Am. Bee J.* 124(2): 116–118 B, 529/85

De Jong, D. (1988) *Varroa jacobsoni* does reproduce in worker cells of *Apis cerana* in South Korea. *Apidologie* 19(3): 221–230 B

Delfinado-Baker, M.; Baker, E. W. (1982) Notes on honey bee mites of the genus *Acarapis* Hirst (Acari: Tarsonemidae). *Internat. J. Acarol.* 8(4): 211–226 B, 1284/83

Delfinado-Baker, M.; Styer, W. E. (1983) Mites of honey bees as seen by scanning electron microscope (SEM). *Am. Bee J.* 123(11): 812–813, 819 B, 247/85

Delfinado-Baker, M. (1985) An acarologist's view: the spread of the tracheal mite of honey bees in the United States. *Am. Bee J.* 125(10): 689–690 B

Delfinado-Baker, M. (1987) Personal communication

Delfinado-Baker, M.; Aggarwal, K. (1987*a*) Infestation of *Tropilaelaps clareae* and *Varroa jacobsoni* in *Apis mellifera ligustica* colonies in Papua New Guinea. *Am. Bee J.* 127(6): 443 B, 237/88

Delfinado-Baker, M.; Aggarwal, K. (1987*b*) A new *Varroa* (Acari: Varroidae) from the nest of *Apis cerana* (Apidae). *Int. J. Acarol.* 13(4): 233–237 B, 402/89

Delima, A. S.; Roberts, E. J. (1987) *Beekeeping in the Philippines: a feasibility study for small farmers* (Philippines: Science and Technology Resource Agency & Netherlands Development Organization) 87 pp. + annexes B

Demianowicz, Z. (1960/1963) [The honey yield of the main honey plants in Polish conditions.] *Pszczel. Zesz. Nauk.* 4(2): 87–104; 7(2): 95–111 *In Polish* B, 853, 854/64

Deodikar, G. B. (1965) Requirements for a co-operative honey processing house. *Indian Bee J.* 27(2): 77–80 B, 221/68

Deodikar, G. B.; Phadke, R. P.; Shende, S. G. (1966) A design for honey processing unit. *Indian Bee. J.* 28(1): 1–10 B, 225/68

Detroy, B. F.; Owens, C. D.; Whitefoot, L. O. (1975) Moving colonies of honey bees. *Am. Bee J.* 115(7): 268–271 B, 1104/76

Detroy, B. F.; Erickson, E. H. (1977) The use of plastic combs for brood rearing and honey storage by honeybees. *J. apic. Res.* 16(3): 154–160 B, 914/78

Detroy, B. F.; Whitefoot, L. O.; Hyde, G. M. (1979) Electrically heated uncapping knife and control. *Trans. Am. Soc. agric. Eng.* 22(4): 886–888, 893 B, 613/82

Dewan, S. M. A. L. (1984) Apiculture in Bangladesh. Pp. 131–141 from *Proceedings of the expert consultation . . .* ed. FAO B

Dewan, S. M. A. L. (1986) Personal communication

Diehnelt, B. (1966) The bee blower. *Am. Bee J.* 106(7): 246–247 B

Dietz, A.; Lambremont, E. N. (1970) Caste determination in honey bees II. *Ann. ent. Soc. Am.* 63(5): 1342–1345 B, 373/72

Dietz, A. (1975) Nutrition of the adult honey bee. Chapter 5, pp. 125–156, from *The hive and the honey bee* ed. Dadant & Sons B

Dietz, A.; Wilbanks, T. W.; Wilbanks, W. G. (1983) Investigations on long-term queen storage in a confined system. *Apiacta* 18(3): 67–70 B

Dietz, A.; Krell, R. (1986) Survey for honey bees at different altitudes in Kenya. *Am. Bee J.* 126(12): 829–830 B, 572/87

Dietz, A.; Hermann, H. R. (1988) *Biology, detection and control of* Varroa jacobsoni: *a parasitic mite on honey bees* (Commerce, GA, USA: Lei-Act Publishers) 80 pp. B

Dietz, W.; Martin, P. J.; Harcus, J. F. (1976) Bees as a supplement for pigs. *Can. J. Anim. Sci.* 56(4): 841 [abstract only] B, 743/79

Digbie, Sir Kenelme (1669) *The closet of the eminently learned Sir Kenelme Digbie, Kt, opened* (London: H. Brome) Transcript of 128 pp. containing receipts for mead, etc. (London: IBRA) 1983 B

Dinabandhoo, C. L.; Dogra, G. S. (1979) Dichlorvos for the control of acarine disease of Indian honey bee *Apis cerana* Fab. *Indian Bee J.* 41(1/2): 1–4 B, 813/81

Dines, A. M. (1978) For beginners. *Bee Craft* 60(2): 33–34 B

Dolgova, L. P. (1928) [Effect of some factors upon the visiting of honey plants.] *Opuit Pas.*: 176–179, 248–253 *In Russian* B

Dollin, L.; Dollin, A. (1985) Farming with Australian native bees. *Australas. Beekpr* 87(1): 14–16 B

Dollin, A.; Dollin,L. (1986) Tracing aboriginal apiculture of Australian native bees in the far north-west. *Australas. Beekpr* 88(6): 118–122 B, 813/87

Donadieu, Y. (1983) *Honey in natural therapeutics* (Paris: Maloine Editeur, S. A.) 28 pp. 2nd ed. B

Doner, L. W.; Chia, D.; White, J. W. (1979) Mass spectrometric $^{13}C/^{12}C$ determinations to distinguish honey and C_3 plant sirups from C_4 plant sirups (sugar cane and corn) in candied pineapple and papaya. *J. Ass. off. agric. Chem.* 62(4): 928–930 B, 1061/81

Donnelly, T. F. (1976) [Bee net for trucks]. *Aust. Bee J.* 57(3): 10 B, 946/77

Donovan, B. J.; Wier, S. S. (1978) Development of hives for field population increase, and studies on the life cycles of the four species of introduced bumble bees in New Zealand. *N. Z. Jl agric. Res.* 21: 733–756 B, 323/81

Dotimas, E. M.; Hider, R. C. (1987) Honeybee venom. *Bee Wld* 68(2): 51–70 B, 1386/87

Doull, K. M. (1974) Recent research on the use of pollen supplements. *Bee Wld* 55 *Suppl.*: 145–147. IBRA Reprint M81 B

Doutt, R. L. (1959) The case of the trespassing bees. *Bull. ent. Soc. Amer.* 5(3): 93–97 B, 361/60

Dreher, K. (1962) Die Erfahrungen mit der Carnica in Deutschland. *Bienenvater* 83(9): 235–240 B

Dreher, K. (1978) Werden die Bienen durch den Strassenverkehr gefährdet? *Biene* 114(10): 382–383 B

Drescher, W. (1969) Die Flugaktivität von Drohnen der Rasse *Apis mellifera carnica* L. and *A. mell. ligustica* L. in Abhängigkeit von Lebensalter und Witterung. *Z. Bienenforsch.* 9(9): 390–409 B, 919/70

Drescher, W.; Crane, E. (1982) *Technical cooperation activities: beekeeping. A directory and guide* (Eschborn, German Federal Republic: Deutsche Gesellschaft für Technische Zusammenarbeit GmbH) 166 pp. B, 1170/82

Droege, G. (1961) *Bibliographie der deutschen bienenwissenschaftlichen Literatur* (Berlin, DDR: Institut für Geflügel- und Pelztierzucht der Humboldt-Universität) 5 vols. 805 pp. + later supplements B, 54/64

Dubois, L.; Collart, E. (1950) *L'apiculture au Congo Belge et au Ruanda-Urundi* (Brussels: Ministère des Colonies, Direction de l'Agriculture, de l'Élevage et de la Colonisation) 230 pp. B, 9/58

Dudley, P. (1721) An account of a method lately found out in New England, for discovering where the bees hive in the woods, in order to get their honey. *Phil. Trans. R. Soc.* 31(367): 148–150

Dugat, M. (1947) *La ruche gratte-ciel à plusieurs reines* (Marlieux Ain: Abbaye N. D. des Dombes) 89 pp. 3rd ed. B

Dugat, M. (1948) *The skyscraper hive* (London: Faber & Faber) 78 pp. B

Duggan, M. (1984) Personal communication of tables giving records.

Dummelow, J. (1973) *The Wax Chandlers of London* (London: Phillimore & Co.) 204 pp. B, 683/74

Dumronglert, E. (1983) A follow-up study of chronic wound healing dressing with pure natural honey. *J. nat. Res. Council Thailand* 15(2): 39–66 B, 677/86

Dunham, M. (1988) Honeystix update. *WAS J.* 9(2): 387 B

Dustmann, J. H. (1971) Über die Katalaseaktivität in Bienenhonig aus der Tracht der Heidekraut-

gewächse (Ericaceae). *Z. Lebensmittelunters. u. Forsch.* 145(5): 294–295 B, 1055/72
Dutant, P. (1973) La récolte du miel sur la côte est en 1973. *Rev. agric. Nouv. Caléd.* (24): 14–15 B
Dutta, T. R.; Ahmed, R.; Abbas, S. R. (1983) The discovery of a plant in the Andaman Islands that tranquillizes *Apis dorsata*. *Bee Wld* 64(4): 158–163 B, 441/84
Dutta, T. R. & 3 others (1985) Plants used by Andaman aborigines in gathering rock-bee honey. *Econ. Bot.* 39(2): 130–138 B, 55/88
Dutton, R.; Simpson, J. (1977) Producing honey with *Apis florea* in Oman. *Bee Wld* 58(2): 71–76 B, 118/78
Dutton, R. W. & 3 others (1981) Observations on the morphology, relations and ecology of *Apis mellifera* of Oman. *J. apic. Res.* 20(4): 201–214 B, 498/82
Dyce, E. J. (1931) Fermentation and crystallisation of honey. *Bull. Cornell agric. Exp.* No. 528: 76 pp. B
Dyce, E. J. (1975) Producing finely granulated or creamed honey. Chapter 10, pp. 293–325, from *Honey: a comprehensive survey* ed. E. Crane (1975*a*) B
Dyer, F. C.; Gould, J. L. (1983) Honey bee navigation. *Am. Scient.* 71(6): 587–597 B
Dyer, F. C. (1987) New perspectives on the dance orientation of the Asian honeybees. Pp. 54–65 from *Neurobiology and behavior of honeybees* ed. R. Menzel and A. Mercer B
Dzierzon, J. (1848) *Theorie und Praxis des neuen Bienenfreundes* (Brieg, Germany [Brzeg, Poland]: Falch'sche Buchdrukerei) 208 pp.

Echeverry, R. (1984) *Flora apícola colombiana* (Bogotá: Biblioteca Científica de la Presidencia de la República) 238 pp. B, 1233/81
Echigo, T.; Takenaka, T.; Takahashi, K. (1982) [On chemical composition of carboxylic acids in royal jelly.] *Bull. Fac. Agric., Tamagawa Univ.* (22): 67–78 *In Japanese* B, 654/84
Echigo, T. (1986) Personal communication
Edlin, H. (1973) *Atlas of plant life* (London: Heinemann) 128 pp. B
Edrich, W. (1981) Night-time sun compass behaviour of honey bees at the equator. *Physiol. Ent.* 6(1): 7–13 B, 513/82
Egea Soares, A. E. (1985) Cardboard bait hives: a practicable alternative to capturing swarms. *IBRA Newsletter* No. 6: 3 B
Eischen, F. A.; Rinderer, T. E.; Dietz, A. (1986) Nocturnal defensive responses of Africanized and European honey bees to the greater wax moth (*Galleria mellonella* L.). *Anim. Behav.* 34(4): 1070–1077 B
Eischen, F. A. (1988) Overwintering performance of honey bee colonies heavily infested with *Acarapis woodi* (Rennie). *Apidologie* 18(4): 293–303 B
Eisenmann, H. (1982) Tannen- und Fichtensterben in Bayern. *Imkerfreund* 37(6): 226–227 B
Eisikowitch, D.; Masad, Y. (1980) Nectar-yielding plants during the dearth season in Israel. *Bee Wld* 61(1): 11–18 B, 1281/80
Eisikowitch, D.; Lupo, A. (1984) Simple method for controlling bee flight during day time. *Am. Bee J.* 124(10): 733–735 B, 979/85
Eisikowitch, D.; Dafni, A. (1988) The use and abuse of introducing honey plants. *Bee Wld* 69(1): 12–14 B
El-Badawy, A. (1969) *Fixation of Carniolan characteristics in the New Valley* (Cairo University: MSc Thesis) 116 pp. B, 942/79
El-Banby, M. A. (1963) Biological studies on the Egyptian honeybee race. *Abstr. 19 Int. Apic. Congr.*: 8–9 B
Elkon, B. (1955) The honey cookbook (New York: Alfred A. Knopf) 174 pp. B, 143/56
Ellis, M. (1987) Problem products and materials used by beekeepers. *Am. Bee J.* 127(5): 342–343 B, 1240/88
Elsharawi, M. O. (1988) Personal communication
Ely, J. (1985) Insurance necessary in lawsuit-conscious society. *Speedy Bee* 14(7): 14 B
Englert, E. (1986) Imkern im Blätterstock (Munich: Ehrenwirth) 90 pp. B
Erickson, E. H.; Thorp, R. W.; Briggs, D. L. (1977) The use of disposable pollination units in almond orchards. *J. apic. Res.* 16(1): 107–111 B, 648/78
Erickson, E. H. (1986) Personal communication
Erickson, E. H.; Carlson, S. D.; Garment, M. B. (1986) *A scanning electron microscope atlas of the honey bee* (Ames, IA, USA: Iowa State University Press) 292 pp. B, 1177/87
Espina, D.; Ordetx, G. S. (1983) *Flora apícola tropical* (Cartago, Costa Rica: Editorial Tecnológica de Costa Rica) 406 pp. B, 1234/84
Espina, D.; Ordetx, G. S. (1984) *Apicultura tropical* (Cartago, Costa Rica: Editorial Technológica de Costa Rica) 506 pp. 4th ed. B, *see* 455/83
Espina, D. (1986) *Beekeeping of the assassin bees* (Cartago: Editorial Technológica de Costa Rica) 170 pp. B
European Communities Council (1974) Directive of 22 July 1974 on the harmonisation of the laws of the Member States relating to honey (74/409/EEC). *Off. J. European Communities* No. L221: 10–14 B
Evans, H. E.; O'Neill, K. M. (1988) *The natural history and behavior of North American beewolves* (Ithaca, NY, USA: Cornell University Press) 278 pp. B

Evans, S. (1984) An economic survey of the honey industry in Victoria, 1980–81. *Res. Project Dep. Agric. Vic.* No. 186: 63 pp. B, 201/87

Fahd, T. (1968) L'Abeille en Islam. Chapter 6, pp. 61–83, from Chauvin (1986*f*) B

Fahn, A. (1979) *Secretory tissues in plants* (London: Academic Press) 302 pp. B

Fahn, A. (1986) Personal communication

Falanruw, M. C. (1983) Personal communication

Fang, W. A. (1981) Spring management for controlling swarming by Chinese honeybees (*Apis cerana*). *Zhongguo Yangfeng* (2): 5 B, 788/82

Fang Yue-zhen (1984) . . . European bees (*Apis mellifera*) in tropical and sub-tropical regions of China. Pp. 142–147 from *Proceedings of the expert consultation . . .* ed. FAO B

FAO (1978) *Pesticide residues in food. Index and summary 1965–1978* (Rome: FAO) FAO Plant Production and Protection Paper No. 11

FAO (1984) *Proceedings of the expert consultation on beekeeping with* Apis mellifera *in tropical and subtropical Asia* (Rome: FAO) 238 pp. B, 1178/85

FAO (1986) *Tropical and sub-tropical apiculture* (Rome: FAO) 283 pp. B, 69/88

FAO (annual) *FAO production yearbook* (Rome: FAO) *c.* 300 pp. B

FAO (annual) *FAO trade yearbook* (Rome: FAO) *c.* 400 pp. B

Farrar, C. L. (1946) Two-queen colony management. *Rep. U. S. Bur. Ent.* No. E-693 B

Farrar, C. L. (1953) Two-queen colony management. *Bee Wld* 34(10): 189–200 B, 33/55

Farrar, C. L. (1968) Productive management of honey-bee colonies. *Am. Bee J.* 108: 95–97, 141–143, 183–185, 228–230, 271–275, 316–317, 354–356, 392–393 B, 132/69

Farrugia, V. (1979) Beekeeping in the Maltese Islands. Pp. 41–46, 190 from *Beekeeping in rural development* ed. Commonwealth Secretariat & IBRA B

Faruga, A.; Puchajda, H.; Bobrzecki, J. (1975) [Utilization of hive waste in poultry feed.] *Zesz. nauk. Akad. roln.-tech. Olsztynie* No. 142: 151–158 *In Polish* B, 284/79

Fasler, A. (1975) Honey standards legislation. Chapter 13, pp. 329–354, from *Honey: a comprehensive survey* ed. E. Crane (1975*a*) B

Faure, R. (1979) *Atlas anatomique de l'abeille* (Paris: Union nationale de l'Apiculture) 52 pp. B, 171/81

Fedosov, N. F. (ed.) (1955) [*Beekeeper's encyclopaedia*] (Moscow, USSR: State Publishing House for Agricultural Literature) 419 pp. *In Russian* B, 79/58

Feeley, L. (1963) Queen fertility investigation. *Irish Beekpr* 17(2): 40–43 B, 87/64

Feinberg, W. (1983) *Lost-wax casting: a practitioner's manual* (London: Intermediate Technology Publications) 74 pp. B

Ferguson, F. (1987) Long term effects of systemic pesticides on honey bees. *Australas. Beekpr* 89(13): 49–54 B, 954/88

Fernando, E. F. W. (1978) Studies in apiculture in Sri Lanka. 1. Characteristics of some honeys. *J. apic. Res.* 17(1): 44–46 B, 1329/78

Fernando, E. F. W. (1979) The ecology of honey production in Sri Lanka. Pp. 115–125, 191–192 from *Beekeeping in rural development* ed. Commonwealth Secretariat & IBRA B, 878/88

Ferracane, M. (1987) Results of our 1987 plastic equipment survey. *Glean. Bee Cult.* 115(12): 715–717 B

Fierro, M. M. & 5 others (1988) Detection and control of the Africanized bee in coastal Chiapas, Mexico. *Am. Bee J.* 128(4): 272–275 B

Fini, M. A.; Sabatini, A. G. (1972) Indagine comparativa sul contenuto in idrossimetilfurfurolo dei mieli destinati al consumo diretto. *Scienza Technol. Aliment.* 2(6): 375–379 B, 1349/77

Fix, W. J.; Palmer-Jones, T. (1949) Control of fermentation in honey by indirect heating and drying. *N. Z. J. Sci. Tech. A* 31(1): 21–31 B, 146/51

Fletcher, D. J. C. (1975/76) New perspectives in the causes of absconding in the African bee (*Apis mellifera adansonii* L.) *S. Afr. Bee J.* 47(6): 11, 13–14; 48(1): 6–9 B, 1229/77

Fletcher, D. J. C.; Tribe, G. D. (1977) Natural emergency queen rearing by the African bee *A. m. adansonii* and its relevance for successful queen production by beekeepers. Pp. 132–140, 161–168 from *African bees: taxonomy, biology and economic use* ed. D. J. C. Fletcher B, 873, 874/78

Fletcher, D. J. C. (ed.) (1977) *African bees: taxonomy, biology and economic use* (Pretoria: Apimondia) 207 pp. B, 861/78

Fletcher, D. J. C. (1978) The African bee *Apis mellifera adansonii* in Africa. *A. Rev. Ent.* 23: 151–171 B, 1197/78

Fletcher, D. J. C. (1979) Management of *Apis mellifera adansonii* for honey production in Southern Africa. Pp. 86–89 from *Apicultura em clima quente* ed. V. Harnaj & H. Wiese B

Food and Agriculture Organization of the United Nations, *see* FAO

Forster, H. (1985) Personal communication

Forster, I. W. (1959) Concrete bottom boards for beehives. *N. Z. Jl Agric.* 98(3): 291–292 B, 369/59

Forster, I. W. (1972) Requeening honeybee colonies without dequeening. *N. Z. Jl. agric. Res.* 15(2): 413–419 B, 911/73

Forster, K. A. (1950) Chemie und medizinische Verwendung des Bienengiftes. Pp. 9–23 from *Aus chemisch-medizinischer Forschung* ed. H. Mack (Aulendorf, GFR: Cantor) B

Foti, N. & 3 others (1962) Untersuchung über die Überwinterung von Weiseln ausserhalb der Traube und ihre Verwendung in der Produktion. *Arch. Geflügelz. Kleintierk.* 11(5/6): 340–360 B, 615/63

Frankland, A. W. (1963) Treatment of bee sting reactions. *Bee Wld* 44(1): 9–12 B, 745/65

Franssen, C. J. H. (1931) Bijenteelt op Java en de biologie van *Apis indica* F. *Natuurhist. Maandbl.* 20(3/5): 15 pp. IBRA English Translation E 1546 B

Fraser, H. M. (1951) *Beekeeping in antiquity* (London: University Press) 145 pp. 2nd ed. (1st ed. 1931) B, 22/52

Fraser, H. M. (1958) *History of beekeeping in Britain* (London: Bee Research Association) 106 pp. B, 715/65

Free, J. B. (1958*a*) The behaviour of honeybees when their hive is moved to a new site. *Bee Wld* 39(5): 109–115 B, 617/63

Free, J. B. (1958*b*) The ability of honeybees (*Apis mellifera*) to learn a change in the location of their hives. *Anim. Behav.* 6(3/4): 219–223 B, 290/59

Free, J. B.; Butler, C. G. (1958) The size of apertures through which worker honeybees will feed one another. *Bee Wld* 39(2): 40–42 B, 336/59

Free, J. B. (1959) The transfer of food between adult members of a honeybee community. *Bee Wld* 40(8): 193–201 B, 268/60

Free, J. B.; Spencer-Booth, Y. (1961) Further experiments on the drifting of honey-bees. *J. agric. Sci.* 57: 153–158 B, 577/63

Free, J. B. (1967) Factors determining the collection of pollen by honeybee foragers. *Anim. Behav.* 15(1): 134–144 B, 508/68

Free, J. B. (1968) Engorging of honey by worker honeybees when their colony is smoked. *J. apic. Res.* 8(2): 135–138 B, 890/70

Free, J. B. (1970) *Insect pollination of crops* (London: Academic Press) 544 pp. Reprinted 1979 B, 478/71

Free, J. B.; Williams, I. H. (1970) Exposure of the Nasonov gland by honeybees (*Apis mellifera*) collecting water. *Behaviour* 37(3/4): 286–290 B, 947/72

Free, J. B.; Williams, I. H. (1976) The effect on the foraging behaviour of honeybees of the relative locations of the hive entrance and brood combs. *Appl. anim. Ethol.* 2: 141–154 B, 1207/77

Free, J. B. (1977) *The social organisation of honeybees.* (London: Edward Arnold) 68 pp. B, 140/79

Free, J. B. (1981) Biology and behaviour of the honeybee *Apis florea*, and possibilities for beekeeping. *Bee Wld* 62(2): 46–59. IBRA Reprint M105 B, 1241/81

Free, J. B.; Williams, I. H. (1981) Attachment of honeybee comb to sloping hive sides and side-bars of frames. *J. apic. Res.* 20(4): 239–242 B, 518/82

Free, J. B. (1982) *Bees and mankind* (London: George Allen & Unwin) 155 pp. B, 830/83

Free, J. B. & 5 others (1983) Using foraging honeybees to sample an area for trace metals. *Envir. Ent.* 9: 9–12 B, 325/84

Free, J. B.; Williams, I. H. (1983) Foraging behaviour of honeybees on oilseed rape. *Bee Wld* 64(1): 22–24 B

Free, J. B. & 4 others (1984) Honeybee Nasonov pheromone lure. *Bee Wld* 65(4): 175–181 B, 853/85

Free, J. B. (1987) *Pheromones of social bees* (London: Chapman & Hall) 218 pp. B, 540/88

Friedmann, H. (1955) *The honey-guides* (Washington: US National Museum) 292 pp. Bull. U.S. nat. Mus. No. 208 B, 181/56

Friedmann, H.; Kern, J. (1956) *Micrococcus cerolyticus*, nov. sp., an aerobic lipolytic organism isolated from the African honey-guide. *Canad. J. Microbiol.* 2: 515–517 B, 60/59

Fries, I. (1980) Vågkuperegistreringar vid Ultuna försöksbigård. *Bitidningen* 79 (7/8): 228–234 B

Frimston, J. D. (1966) Bee-keeping and the law. *Lect. cent. Ass. Beekprs:* 15 pp. B

Frisch, K. von (1919) Über den Geruchsinn der Bienen und seine blütenbiologische Bedeutung. *Zool. Jb.* Abt. 3 37: 1–238

Frisch, K. von (1965) *Tanzsprache und Orientierung der Bienen* (Berlin, GFR: Springer) 578 pp. B, 474/66

Frisch, K. von (1967) *The dance language and orientation of bees* (Cambridge, MA, USA: Belknap Press) 566 pp. B, 289/69

Frumhoff, P. C.; Baker, J. (1988) A genetic component to division of labour within honey bee colonies. *Nature, Lond.* 333(6171): 358–361 B

Fry, C. H. (1983) Honeybee predation by bee-eaters, with economic considerations. *Bee Wld* 64(2): 65–87. IBRA Reprint M112 B, 282/84

Fry, C. H. (1984) *The bee-eaters* (Calton, UK: Poyser) 304 pp. B, 951/85

Fukuda, H.; Sakagami, S. F. (1968) Worker brood survival in honeybees. *Researches Popul. Ecol. Kyoto Univ.* 10(1): 31–39 B, 900/70

Furness, C. (1984) *How to make beeswax candles* (Geddington, UK: British Bee Publications) 20 pp. rev. ed. B

Fyg, W. (1952) The process of natural mating in the honeybee. *Bee Wld* 33(8): 129–139 B, 133/53

Fyg, W. (1959) Normal and abnormal development in the honeybee. *Bee Wld* 40: 57–66, 85–96 B, 327/60

Gadbin, C. (1980) Les plantes utilisées par les abeilles au Tchad méridional. *Apidologie* 11(3): 217–254 B, 930/81

Galton, D. (1971) *Survey of a thousand years of beekeeping in Russia* (London: Bee Research Association) 90 pp. B, 868/71

Gałuska, H. (1970) Effect of the electric method of venom collection on the behaviour of the bee colony. *Zoologica Pol.* 20(2): 281–307 B, 608/73

Gałuska, H. (1972) The research on a most effective method of the collection of bee venom by means of electric current. *Zoologica Pol.* 22(12): 53–69 B, 685/77

Garoyan, L.; Taylor, S. (1980) Economic trends in the US honey industry. *Leafl. Coop. Ext. Univ. Calif.* No. 21219: 33 pp. B

Garside, M. (1987) Rapid, precise identification of Africanized honey bees. *Glean. Bee Cult.* 115(5): 270–271 B, 1214/87

Gary, N. E. (1961) Mass production of honey bee larvae. *Glean. Bee Cult.* 89(9): 550–552, 569 B, 675/63

Gary, N. E. (1963) Observations of mating behaviour in the honeybee. *J. apic. Res.* 2(1): 3–13 B, 792/64

Gary, N. E. (1975) Activities and behaviour of honey bees. Chapter 7, pp. 185–264, from *The hive and the honey bee* ed. Dadant & Sons B

Gary, N. E.; Witherell, P. C.; Lorenzen, K. (1979) Distribution of honeybees during water collection. *J. apic. Res.* 18(1): 26–29 B, 1313/79

Gary, N.; Lorenzen, K. (1980) How to construct and maintain an observation bee hive. *Leafl. Coop. Ext. Univ. Calif.* No. 2853: 20 pp.

Gautier, A. (1976) Ook apen lusten honing! *Maandbl. koninkl. vlaam. Imkersb.* 62(12): 443–446 B

Gayre, G. R. (1948) *Wassail! in mazers of mead* (London: Phillimore & Co.) 176 pp. B

Gayre, G. R.; Papazian, C. (1986) *Brewing mead – Wassail! in mazers of mead* (Boulder, CO, USA: Brewers Publications) 200 pp. B, 1351/87

Gedde, J. (1675) *A new discovery of an excellent method of bee-houses and colonies . . .* (London: D. Newman) 30 pp.

Gentry, C. & 3 others (1985) *A manual for trainers of small scale beekeeping development workers* (Washington, DC: Peace Corps) 407 pp. 2nd ed. B, 837/86

Gercke, A. (1985) *Das Bienenrecht* (Nordstemmen, GFR: author) 176 pp. B

Gerstung, F. (1926) *Der Bien und seine Zucht* (Berlin: Pfenningstorff) 525 pp. 7th ed.

Getz, W. M.; Brückner, D.; Parisian, T. R. (1982) Kin structure and the swarming behavior of the honey bee *Apis mellifera*. *Behavl Ecol. Sociobiol.* 10: 265–270 B, 574/86

Getz, W. M.; Smith, K. B. (1983) Genetic kin recognition: honey bees discriminate between full and half sisters. *Nature, Lond.* 302: 147–148 B, 166/84

Ghazali, H. M.; Hamidi, R. M.; Ming, T. C. (1987) Effects of microwave heat-treatment on the quality of starfruit honey. *A. Rep. Malaysian Beekeeping Research and Development Team:* 14–21 B

Ghisalberti, E. L. (1979) Propolis: a review. *Bee Wld* 60(2): 59–84 B, 1406/80

Ghotge, A. I. (1956) *A scientific approach to (wild) rock-bees.* (Ujjain, Madya Pradesh: All-India Khadi & Village Industries Board) 17 pp. B, 19/58

Gilbert, J. & 3 others (1981) Determination of the geographical origin of honeys by multivariate analysis of gas chromatographic data on their free amino acid content. *J. apic. Res.* 20(2): 125–135 B, 286/82

Giordani, G. (1953) Le api, le frutta e l'uva. *Apicolt. Ital.* 20(2): 27–29 B, 10/55

Giordani, G.; Vecchi, M. A.; Nardi, M. (1982) *Nozioni pratiche sulle malattie delle api* (Rome: Federazione Apicoltori Italiani) 208 pp. B, 593/84

Girotti, A. & 3 others (1977) Honey production and extraction. *Apiacta* 12(4): 49–53 B

Gleanings in Bee Culture (1917) Death of O. O. Poppleton. *Glean. Bee Cult.* 45(11): 878 B

Gleanings in Bee Culture (1986) Quick pesticide detector should prove valuable to beekeepers. *Glean. Bee Cult.* 114(4): 201 B, 1316/86

Gleim, K.-H. (1984) *Nahrungsquellen des Bienenvolkes 2. Die Honigtautracht* (St Augustin, German Federal Republic: Delta Verlag) 320 pp. B, 1264/85

Gleim, K.-H. (1985) *Nahrungsquellen des Bienenvolkes 1. Die Blütentracht* (St Augustin, German Federal Republic: Delta Verlag) 160 pp. 3rd. ed. B, *see* 895/78

Glock, J. P. (1891) *Die Symbolik der Bienen und ihrer Produkte* (Heidelberg, Germany: Weiss'scher Universitäts-Buchhandlung) 411 pp. B

Glukhov, M. M. (1955) [*Honey plants*] (Moscow, USSR: State Publishing House of Agricultural Literature) 512 pp. 6th ed. *In Russian* B, 415/57

Glushkov, N. M. (1956) [Experiments on combs with enlarged cells.] *Pchelovodstvo* 33(5): 32–40. *In Russian* B, 118/58

Gochnauer, T. A.; Hughes, S. J. (1976) Detection of *Ascosphaera apis* in honey bee larvae (Hymenoptera: Apidae) from eastern Canada. *Can. Ent.* 108: 985–988 B, 980/77

Goebel, R. L. (1984) Bee birds in Queensland. *Australas. Beekpr* 86(7): 137–138 B

Gojmerac, W. L. (1973) Building and operating an observation beehive. *Am. Bee J.* 113(9): 332–334 B

Gojmerac, W. L. (1980) *Bees, beekeeping, honey and pollination* (Westport, CT, USA: AVI Publishing Co.) 192 pp. B

Golding, R. (1847) *The shilling bee book* (London: Longman) 58 pp. B

Gomez Rodriguez, R. (1986) *Manejo de la abeja africanizada* (Caracas: Apicultura Venezolana) 280 pp. B, 1225/88

Gonçalves, L. S. (1974/75) The introduction of the African bees (*Apis mellifera adansonii*) into Brazil and some comments on their spread in South America. *Am. Bee J.* 114: 414–415, 419, 448–550; 115: 8–10, 24 B, 733/75

Gonnet, M.; Vache, G. (1984) *Le goût du miel* (Paris: UNAF) 146 pp. B, 1352/86

Gontarski, H. (1954) Fermentbiologische Studien an Bienen I. Das physiko-chemische Verhalten der kohlenhydratspaltenden Fermente. (*a*) Invertierende Enzyme. *Verh. dtsch. Ges. angew. Ent.*: 186–197 B, 49/56

Gontarski, H. (1955) Beitrag zur Analyse des Kittinstinktes der Honigbiene. *Z. Bienenforsch.* 3(3): 80–87 B, 66/62

Good, R. (1974) *The geography of the flowering plants* (London: Longman) 557 pp. 4th ed. B

Gooderham, C. B. (1938) Specific gravity of capped *vs* uncapped honey. *Rep. Canad. Dep. Agric. Dominion exp. Fms 1934–1936:* 17–18 B

Goodman, R. D. (1973) *Honey flora of Victoria* (Victoria, Australia: Department of Agriculture) 175 pp. B

Goodman, R.; Oldroyd, B. (1988) *Business opportunities in agriculture and horticulture: bees* (Leongatha: Victorian College of Agriculture and Horticulture) B

Gothe, F. (1914) Experimentelle Studien über Eigenschaften und Wirkungsweise der Honigdiastase sowie die Beurteilung des Honigs auf Grund seines Diastasegehaltes. *Z. Unters. Nahr. -u. Genussmittel* 28(6): 286–321

Gould, J. L. (1980) The case for magnetic sensitivity in birds and bees (such as it is). *Am. Scient.* 68(3): 256–267 B, 1269/81

Gould, J. L.; Dyer, F. C.; Towne, W. T. (1985) Recent progress in the study of dance language. *Fortschr. Zool.* 31: 141–161

Gould, J. L. (1987) Flower-shape, landmark, and locale memory in honeybees. Pp. 298–309 from *Neurobiology and behavior in honey bees* ed. R. Menzel & A. Mercer B, 499/89

Gould, J. L.; Gould, C. G. (1988) *The honey bee* (New York: Scientific American Library) 239 pp. B

Goyal, N. P. (1974) *Apis cerana indica* and *Apis mellifera* as complementary to each other for the development of apiculture. *Bee Wld* 55(3): 98–101. IBRA Reprint M76 B, 771/74

Goyal, N. P.; Atwal, A. S. (1977) Wing beat frequencies of *Apis cerana indica* and *Apis mellifera*. *J. apic. Res.* 16(1): 47–48 B, 115/78

Graze, C. (ed.) (1979) *Das Bienenhaus* (Weinstadt: author) 32 pp.

Greenaway, W.; Scaysbrook, T.; Whatley, F. R. (1987) The analysis of bud exudate of *Populus × euramericana* and of propolis, by gas chromatography–mass spectrometry. *Proc. R. Soc. Lond. B* 232: 249–272 B

Greenaway, W.; Scaysbrook, T.; Whatley, F. R. (1988) Composition of propolis in Oxfordshire, U.K. and its relation to poplar bud exudate. *Z. Naturforsch.* 43c: 301–305 B

Greenway, A. R. & 3 others (1975) Unusually severe granulation of winter stores caused by nectar from ivy, *Hedera helix*, in Ireland. *J. apic. Res.* 14(2): 63–68 B, 835/75

Greenway, A. R.; Simpson, J.; Smith, M. C. (1978) Granulation of ivy nectar and honey in the honey stomach of the honeybee. *J. apic. Res.* 17(2): 84–88 B, 1027/79

Griesohn, G. (1982) *Möglichkeiten einer Verbesserung der Trachtsituation in intensiv bewirtschafteten Feldkultur-Ökosystemen* (University of Tübingen: Diploma Dissertation) 92 pp. B, 556/83

Griffiths, D. A.; Bowman, C. E. (1981) World distribution of the mite *Varroa jacobsoni*, a parasite of honeybees. *Bee Wld* 62(4): 154–163. IBRA Reprint M107 B, 591/82

Griffiths, D. A. (1988*a*) Personal communication

Griffiths, D. A. (1988*b*) Functional morphology of the mouthparts of *Varroa jacobsoni* and *Tropilaelaps clareae* as a basis for the interpretation of their life-styles. Pp. 479–486 from *Africanized honey bees and bee mites* ed. G. R. Needham & 3 others B

Grobov, O. F. (1977) [The mite *Carpoglyphus lactis* and its role in the brood nest of honeybees.] Pp. 83–87 from *Arakhnozy i protozoinye bolezni sel'skokhozyaĭstvennykh zhivotnykh* ed. N. I. Stepanov (Moscow: Kolos) B, 1339/80

Gubin, V. A. (1953) [Why the cappings of northern and southern races of bees differ.] *Pchelovodstvo* (7): 14–15 *In Russian* B, 13/54

Guilfoyle, J. L. (1988) 100,000,000 West Australian honey bees destined for Canada. *Aust. Bee J.* 69(3): 29–30 B

Guinaudeau, C. (1984) Les haies brises-vent et leur contribution à la production du miel. *Bull. tech. Apic.* 11(1): 21–36 B, 521/85

Gulliford, R. B. (ed.) (1981) *Investing in commercial honey production (an outline and commentary on the 3rd Mansfied Report)* (Sydney: NSW Department of Agriculture) 9 pp.

Gunnison, A. F. (1966) An improved method for collecting the liquid fraction of bee venom. *J. apic. Res.* 5(1): 33–36 B, 760/66

Guy, R. (1971) A commercial beekeeper's approach to the use of primitive materials. *Bee Wld* 52(1): 18–24 B

Guy, R. D. (1975) A simple new hive and a beekeeping system for keeping the African bee. *S. Afr. Bee J.* 47(5): 8–11 B, 226/77

Guy, R. D. (1976) Personal communication

Haavi, O. (1973) Birøkt og driftsøkonomi. *Nord. Bitidskr.* 24(1): 14–18 B

Habermann, E. (1972) Bee and wasp venoms. *Science, NY* 177: 314–322 B

Hadorn, H.; Zürcher, K. (1974) Zuckerspektrum und Kristallisationstendenz von Honigen. *Mitt. Geb. Lebensmittelunters. Hyg.* 65(4): 407–420 B, 1007/77

Haffejee, I. E.; Moosa, A. (1985) Honey in the treatment of infantile gastroenteritis. *Brit. med. J.* 290: 1866–1867 B, 1006/86

Hall, R. (1977) Mobile extracting plant. *Australas. Beekpr* 79(1): 9–12 B

Hallim, M. K. I. (1986) Personal communication

Halmágyi, L.; Keresztesi, B. (1975) *A méhlegelő* (Budapest: Akadémiai Kiadó) 792 pp. B, 1078/76

Hameed, S. F.; Adlakha, R. L. (1973) Preliminary comparative studies on the brood rearing activities of *Apis mellifera* L. and *Apis cerana indica* F. in Kulu Valley. *Indian Bee J.* 35(1/4): 27–35 B, 479/77

Hamilton, R. (1960) Trucking packages from south to north. *Glean. Bee Cult.* 88(3): 142–145 B, 378/61

Hamman, E. (1958) Which takes the initiative in the virgin queen's flights, the queen or the workers? *Bee Wld* 39(3): 57–62 B, 260/59

Hänel, H.; Ruttner, F. (1985) The origin of the pore in the drone cell capping of *Apis cerana* Fabr. *J. apic. Res.* 16(2): 157–164 B, 68/87

Hansen, H. (1980) *Honey bee brood diseases* (Copenhagen: L. Launsø) 28 pp. Original in Danish (1977), published also in other languages B

Hansen, H. (1983) *Sygdomme og parasitter hos voksne bier* (Copenhagen: L. Launsø) 33 pp. B

Hansen, H. (1984) The incidence of the foulbrood bacterium *Bacillus larvae* in honeys retailed from Denmark. *Tidsskr. PlAvl* 88: 329–336 B, 1374/85

Hansmann, C.; Hansmann, L. (1959) *Viel köstlich Wachsgebild* (Munich: F. Bruckmann) 70 pp. B

Hanson, J. M.; Morley, J.; Soria-Herrera, C. (1974) Anti-inflammatory property of 401 (MCD-peptide), a peptide of the bee *Apis mellifera* (L). *Brit. J. Pharmacol.* 50: 383–392 B, 888/75

Hansson, Å. (1955) Fins honungsbiet kvar i vilt tillstånd i Norden? *Nord. Bitidskr.* 7(4): 123–126 B, 15/58

Hansson, Å. (1988) *Biväxter som alternativgröder* (Mantorp, Sweden: Sveriges Biodlares Riksförbund) 48 pp. B

Harbo, J. R. (1979) Storage of honeybee spermatozoa at −196°C. *J. apic. Res.* 18(1): 57–63 B, 1318/79

Harbo, J. R. & 3 others (1981) Development periods for eggs of Africanized and European honeybees. *J. apic. Res.* 20(3): 156–159 B, 477/82

Harnaj, V.; Wiese, H. (eds) (1979) *Apicultura em clima quente: Simpósio internacional* (Bucharest: Apimondia Publishing House) 255 pp. B, 831/83

Harrison, A. S. (1966) F-3H super extracting system progress report. *Am. Bee J.* 106(5): 168, 173 B

Hase, S. & 3 others (1973) [Changes in quality of honey caused by heating and storage.] *Nippon Shokuhin Kogyo Gakkai-Shi* 20: 248–256; 257–264 *In Japanese* B, 650, 651/77

Hassanein, M. H.; El-Banby, M. A. (1955) Studies on the brood-rearing activity, longevity and foraging behaviour of the Egyptian honeybee, *Apis mellifera* var. *fasciata*. *Ann. agric. Sci.*: 751–772 B, 582/64

Hauser, F. (1987) Wechselnde Temperaturen bestimmen Bienenleben. *Garten u. Kleintierz.* 26(8): 7 B

Haydak, M. H. (1943) Larval food and development of castes in the honey bee. *J. econ. Ent.* 36(5): 778–792 B

Haydak, M. H. (1957) The food of the drone larvae. *Ann. ent. Soc. Am.* 50(1): 73–75 B, 321/58

Haynie, J. D. (1973) *A demountable truck hive loader* (Gainesville, FL, USA: author) 3 pp. B, 633/74

Healtheries of New Zealand Ltd (1981) Wanted – pollen granules. *N. Z. Beekpr* 42(3): 24 *advertisement* B

Heath, L. A. F. (1982*a*) Development of chalk brood in a honeybee colony: a review. *Bee Wld* 63(3): 119–130. IBRA Reprint M110 B

Heath, L. A. F. (1982*b*) Chalk brood pathogens: a review. *Bee Wld* 63(3): 130–135. IBRA Reprint M110 B

Heath, L. A. F. (1985*a*) Occurrence and distribution of chalk brood disease of honeybees. *Bee Wld* 66(1): 9–15. IBRA Reprint M115 B, 947/85

Heath, L. A. F. (ed.) (1985*b*) *A case of hives* (Burrowbridge, Somerset, UK: BBNO) 73 pp. B, 601/87

Heinrich, B. (1979) Thermoregulation of African and European honeybees during foraging, attack and hive exits and returns. *J. exp. Biol.* 80: 217–229 B, 903/81

Heinrich, B. (1981*a*) The mechanisms and energetics of honeybee swarm temperature regulation. *J. exp. Biol.* 91: 25–55 B, 865/82

Heinrich, B. (1981*b*) The regulation of temperature in the honeybee swarm. *Scient. Am.* 244(6): 120–129 B

Heinrich, B. (ed.) (1981*c*) *Insect thermoregulation* (New York: John Wiley & Sons) 327 pp. B, 1051/82

Helmich, R. L. (1986) Mating European queens to European drones in areas with Africanized bees. *Am. Bee J.* 126(12): 830–831 B, 608/87

Helmich, R. L. & 3 others (1986) Comparison of Africanized and European queen-mating colonies in Venezuela. *Apidologie* 17(3): 217–226 B, 953/87

Hemptinne, J.; Hemptinne, J. L.; Desprets, A. (eds) (1985) *Une gestion de l'environnement pour une apiculture florissante* (Soignies, Belgium: Institut Provincial Supérieur Industriel du Hainaut) 74 pp. B, 917/87

Henderson, C. E. (1987) Preferred methods of removing bees from supers by New York State commercial beekeepers. *Glean. Bee Cult.* 115(9): 463–464 B

Hensels, L. G. M. (1981) *Drachtplantengids voor de bijenteelt* (Wageningen: Centre for Agricultural Publishing and Documentation) 118 pp. B, 155/82

Hepburn, E. J. (1982) A taste of honey. *S. Afr. Bee J.* 54(1): 3–5 B, 1369/82

Hepburn, H. R. (1980) How honeybees measure their pollen loads. *S. Afr. Bee J.* 52(3): 11–13 B

Hepburn, H. R.; Rigby, J. R. (1981) Strength of beeswax foundation in relation to cell orientation. *J. apic. Res.* 20(4): 234–239 B, 543/82

Hepburn, H. R.; Kurstjens, S. P. (1984) On the strength of propolis (bee glue). *Naturwissenschaften* 71: 591–592 B, 1014/86

Hepburn, H. R. & 4 others (1984) On the energetic costs of wax production by the African honeybee, *Apis mellifera adansonii*. *S. Afr. J. Sci.* 80: 363–368 B, 894/85

Hepburn, H. R. (1986) *Honeybees and wax* (Berlin, GFR: Springer-Verlag) 205 pp. B, 147/88

Hepburn, H. R.; Kurstjens, S. P. (1988) The combs of honeybees as composite materials. *Apidologie* 19(1): 25–36 B, 523/89

Heran, H. (1952) Untersuchungen über den Temperatursinn der Honigbiene (*Apis mellifica*) unter besonderer Berücksichtigung der Wahrnehmung strahlender Wärme. *Z. vergl. Physiol.* 34: 179–206. IBRA English translation E461 B, 159/53

Herbert, E. W.; Shimanuki, H. (1980) An evaluation of seven potential pollen substitutes for honey bees. *Am. Bee J.* 120(5): 349–350 B, 972/81

Herbert, E. W.; Argauer, R. J.; Shimanuki, H. (1986) The effect of an insect chitin synthesis inhibitor on honey bees. *Apidologie* 17(1): 73–76 B, 165/87

Hermann, H. R.; Blum, M. S. (1981) Defensive mechanisms in the social Hymenoptera. Chapter 2, pp. 77–197, from *Social insects* Vol. II ed. H. R. Hermann (London: Academic Press) B, 1067/82

Herron, K. M. (1988) High country beekeeping in two lessons (or one hard one). *N. Z. Beekpr* (197): 29–30 B

Hesse, M. (1981) The fine structure of the exine in relation to the stickiness of angiosperm pollen. *Rev. Palaeobot. Palynol.* 35: 81–92 B, 1047/82

Hetland, A. (1986) *Clostridium botulinum* sporer i norskprodusert honning? *Norsk. Vet. Årtidsskr.* 98(10): 725–727 B

Heywood, V. H. (ed.) (1978) *Flowering plants of the world* (Oxford: Oxford University Press) 336 pp. B, 960/79

Hillyard, T. N. (1958) Simple hive scales. *Bee Wld* 39(3): 102–103 B

Hillyard, T. N. (1960) Personal communication

Hirschfelder, H.; Sachs, H. (1952) Recent research on the acarine mite. *Bee Wld* 33(12): 201–209 B, 287/53

Hitchcock, L. E. (1986) Personal communication

Hobbs, G. A. (1967) Obtaining and protecting red-clover pollinating species of *Bombus* (Hymenoptera: Apidae). *Can. Ent.* 99(9): 943–951 B, 288/68

Hobson, J. V. (1983) Ferrocement as a material for hives. *Bee Wld* 64(3): 113–116 B, 572/84

Hocking, B.; Matsumura, F. (1960) Bee brood as food. *Bee Wld* 41(5): 113–120 B, 157/62

Hodges, D. (1952) *The pollen loads of the honeybee* (London: Bee Research Association) 120 pp. Reprinted 1985 (London: IBRA) B, 221/54

Hodgman, C. D. (1935) *Handbook of chemistry and physics* (Cleveland, OH, USA: Chemical Rubber Publishing Co.) 1951 pp. 20th ed.

Hodgson, N. B. (1955) Bee venom: its components and their properties. *Bee Wld* 36(12): 217–222 B, 173/58

Hodgson, N. B. (1973) *Children's books on bees and beekeeping* (London: BRA) 55 pp. IBRA Bibliography No. 14 B, 759/74
also (1973) Guide to children's literature on bees and beekeeping. *Bee Wld* 54(3): 121–138 B, 758/74

Hoffmann, J. A.; Hetru, C. (1983) Ecdysone. Pp. 65–88 from *Endocrinology of insects* ed. R. G. H. Downer & H. Laufer (New York: Alan R. Liss, Inc.)

Hogg, J.A. (1989) Comb honey in the halfcomb. *Am. Bee J.* 129(4): 230–234, 236, 238–240 B

Holm, E. (1986) *Artificial insemination of the queen bee – A manual for the use of Swienty's insemination apparatus* (Gedved, Denmark: Eigil Holm) 67 pp. B

Holm, S. N. (1973) *Osmia rufa* L. (Hym. Megachilidae) as a pollinator of plants in greenhouses. *Ent. scand.* 4(3): 217–224 B, 500/75

Hooker, T. (1983) Switching from comb to extracted honey production. *Am. Bee J.* 123(4): 311–313 B

Hooper, T.; Taylor, M. (1988) *The beekeeper's garden* (London: Alphabooks, A. & C. Black) 152 pp. B

Howe, S. R.; Dimick, P. S.; Benton, A. W. (1985) Composition of freshly harvested and commercial royal jelly. *J. apic. Res.* 24(1): 52–61 B, 1391/85

Howes, F. N. (1974) *A dictionary of useful and everyday plants and their common names* (Cambridge: University Press) 290 pp. B

Howes, F. N. (1979) *Plants and beekeeping* (London: Faber & Faber) 236 pp. 2nd ed. B

Hsiang, H. K.; Elliott, W. B. (1975) Differences in honey bee (*Apis mellifera*) venom obtained by venom sac extraction and electrical milking. *Toxicon* 13: 145–148 B, 679/77

Huber, F. (1792) *Nouvelles observations sur les abeilles* (Geneva: Barde, Manget & Co.) 369 pp. B

Huhlanen, C. N. (1987) [Honey not the main cause of infant botulism.] *Am. Bee J.* 127(5): 321 B

Hussain, S. A.; Ali, S. (1983) Some notes on the ecology and status of the Orangerumped Honeyguide *Indicator xanthonotus* in the Himalayas. *J. Bombay Nat. Hist. Soc.* 80(3): 564–574 B, 822/87

Huxley, A. (1984) *Green inheritance: the World Wildlife Fund book of plants* (London: Collins) 193 pp. B

Hyvönen, L. (1980) *Varying relative sweetness* (University of Helsinki EKT Series No. 546: Thesis) B

Hyvönen, L.; Espo, A. (1981) *Replacement of sucrose in bakery products: I Cakes and cookies; II Yeast leavened products* (University of Helsinki EKT Series No. 569, 570) B

Hyvönen, L.; Koivistoinen, P. (1981) Fructose in food systems. Pp. 133–177 from *Nutritive sweeteners* ed. G. G. Birch & K. J. Parker (London: Applied Science Publishers) B

IBRA (1977*a*) *Honey standards, laws and regulations (world-wide)* (London: IBRA) 17 pp. IBRA Bibliography No. 19 B

IBRA (1977*b*) *Drones: their behaviour and significance in beekeeping* (London: IBRA) 7 pp. IBRA Bibliography No. 21 B

IBRA (1977*c*) *Anatomy and physiology of drones and drone spermatozoa* (London: IBRA) 7 pp. IBRA Bibliography No. 22 B

IBRA (1977*d*) *The life cycle of normal (haploid) and of diploid drones* (London: IBRA) 5 pp. IBRA Bibliography No. 23 B

IBRA (1979*a*) *British bee books: a bibliography 1500–1976* (London: IBRA) 270 pp. B, 1184/80

IBRA (1979*b*) *World-wide standards for hive products except honey, and for equipment used in beekeeping and in processing hive products* (London: IBRA) 13 pp. IBRA Bibliography No. 24 B

IBRA (1981) *Garden plants valuable to bees* (London: IBRA) 52 pp. B, 1273/82

IBRA (1982) *Directory of suppliers of beekeeping equipment world-wide* (London: IBRA) 15 pp. B

IBRA (1983*a*) *World list of beekeeping journals, annual reports, and other serials currently received by IBRA* (London: IBRA) 30 pp. B

IBRA (1983*b*) *Laws and regulations (world-wide) to protect bees from pesticide poisoning* (London: IBRA) 11 pp. IBRA Bibliography No. 29 B

IBRA (1984) *Apiculture: Sources d'Information; Fuentes de Información en Apicultura.* French and Spanish editions of IBRA (1985*a*) B, 1164, 1165/85

IBRA (1985*a*) *Source materials for apiculture* (London: IBRA) 10 leaflets 2nd ed., as listed B

1. Suppliers of equipment for tropical and subtropical beekeeping. 6 pp.
2. Marketing bee products: addresses of importers and agents. 10 pp.
3. Planting for bees in developing countries. 10 pp.
4. Opportunities for training in apiculture worldwide. 10 pp.
5. Sources of voluntary workers for apicultural development. 12 pp.
6. Sources of grant-aid for apicultural development. 12 pp.
7. Obtaining apicultural information for use in developing countries. 6 pp.
8. Apicultural reference books for developing countries. 8 pp.
9. Educational aids on apiculture. 6 pp.
10. Writing about apiculture: guidelines for authors. 6 pp.

IBRA (1985*b*) Microfiche index to *Apicultural Abstracts*, 1973–1983 (London: IBRA) B

IBRA (1985*c*) Information for beekeepers in tropical and subtropical countries. *IBRA Leaflet*: 8 pp B

IBRA (1985/86) Audio-visual material obtainable from the International Bee Research Association. *IBRA List* No. 3: 8 pp. B

IBRA (1986*a*) *Honeybee mites and their control – a selected annotated bibliography* (Rome: FAO) 141 pp. Agricultural Services Bulletin No. 68/2 B, 1273/88

IBRA (1986*b*) Hive-aid. *IBRA Newsletter* No. 9: 6–7

IBRA (1988) *International Book Catalogue* (Cardiff: IBRA) 30 pp. B

Ifantidis, M. D. (1983) The movable-nest hive: a possible forerunner to the movable-comb hive. *Bee Wld* 64(2): 79–87 B, 246/84

Ikuse, M.; Sahashi, N.; Maeda, H. (1981) [Observations on pollen grains in Japanese honeys.] *Honeybee Sci.* 2(3): 97–104 *In Japanese* B, 970/82

Illinois Cooperative Extension Service (1983) *Evaluating honey bee colonies for pollination* (Urbana, IL, USA: author) 3 pp. B

Imdorf, A.; Bogdanov, S.; Kilchenmann, V. (1985) 'Zementhonig' im Honig- und Brutraum – was dann? *Schweiz. Bienenztg* 108: 534–544; 581–590 B, 1356, 1357/86

Indian Standards Institution (1961) Specification for hive stands. *Indian Standard* IS: 1735–1960, 4 pp. B

Indian Standards Institution (1970) Specification for beehives. *Indian Standard* IS: 1515–1969, 20 pp. B

Indian Standards Institution (1973) Code for conservation and maintenance of honey bees. *Indian Standard* IS: 6695–1972, 28 pp. B, 122/177

Indian Standards Institution (1975) Layout for honey processing plant. *Indian Standard* IS: 3891–1974, 8 pp. rev. ed. B
also Layout for a honey house. Part 1. Large scale honey handling units. IS: 7849 (Part 1)–1975, 8 pp. B

Indian Standards Institution (1976) Specification for comb foundation sheets. *Indian Standard* IS: 2072–1976, 3 pp. rev. ed. B

Indian Standards Institution (1977) Specification for squeezed honey. *Indian Standard* IS: 8464–1977, 8 pp. B, 1070/81

Indian Standards Institution (1979) Specification for comb foundation mill. *Indian Standard* IS: 3894–1978, 8 pp. rev. ed. B

Ingleton, J. F. (1976) La miel y su empleo en la industría del dulce. *Panadero Latinamericano* 31(1): 26–30 B, 975/78

Inoue, H. (1984) [Age-dependent changes in the composition of *Apis mellifera* venom.] *Honeybee Sci.* 5(2): 63–66 *In Japanese* B, 1411/85

Inoue, H.; Nakajima, T. (1985) A comparison of the venom compositions in *Apis cerana japonica* and *Apis mellifera*. *Proc. 30 Int. Apic. Congr.*: 441–443 B, 54/88

Inoue, H.; Nakajima, T.; Okada, I. (1987) The venomous components of worker and queen honey bees during their maturation and the seasonal generation. *Japan. J. Sanit. Zool.* 38(3): 211–217 B, 1191/188

Inoue, Tamiji (1985) Personal communication

Inoue, Tanji; Inoue, A. (1964) The world royal jelly industry: present status and future prospects. *Bee Wld* 45(2): 59–64, 69 B, 842/65

International Apicultural/Beekeeping Congresses (1897–1987) are listed in Section 16.82

International Bee Research Association, *see* IBRA

International Commission for Bee Botany (1982) *Second symposium on the harmonization of methods for testing the toxicity of pesticides to bees* (Hohenheim: International Commission for Bee Botany) 84 pp. B, 1287/84

International Commission for Bee Botany (1985) *Third symposium on the harmonization of methods for testing the toxicity of pesticides to bees* (Netherlands: Shell Internationale Petroleum Mij) 90 pp. B, 650/87

International Conference on Apiculture in Tropical Climates, 1st, *see* Crane (1976*a*)

International Conference on Apiculture in Tropical Climates, 2nd (1983) [*Proceedings of the*] *Second International Conference on Apiculture in Tropical Climates* (New Delhi: Indian Agricultural Research Institute) 728 pp. B, 821/85

International Conference on Apiculture in Tropical Climates, 3rd (1985) *Proceedings of the Third International Conference in Apiculture in Tropical Climates* (London: IBRA) 270 pp. B, 113/86

International Conference on Apiculture in Tropical Climates, 4th (1989) *Proceedings of the Fourth International Conference in Apiculture in Tropical Climates* (London: IBRA) 529 pp. B

International Symposium on Apitherapy, 1st (1975) *The hive products: food, health and beauty* (Bucharest: Apimondia Publishing House) 154 pp. B, 1175/76

International Symposium on Apitherapy, 2nd (1977) *New apitherapy research* (Bucharest; Apimondia Publishing House) 360 pp. B, 1035/80

International Symposium on Melliferous Flora (1977) *Honey plants – basis of apiculture. International Symposium on Melliferous Flora, Budapest*, (Bucharest: Apimondia Publishing House) 224 pp. B, 1217/78

International Symposium on Pollination:
1st, held in 1960, *Proceedings* (1962) ed. T. E. Mittler (Stockholm: Swedish Seed Growers Association) 224 pp. B, 434/63

International Symposium on Pollination: (*continued*)
2nd, held in 1964, *Proceedings* (1966) ed. E. Åkerberg & E. Crane *Bee Wld* 47(1) *Suppl.* 216 pp. B, 191/67
3rd, held in 1974, *Compte-rendu* (1975) *Bull. tech. apic.* 2 *Suppl.* 331 pp. B, 1197/76
4th, held in 1978, *Proceedings* (1979) ed. D. M. Caron (Beltsville, USA: Maryland Agricultural Experiment Station) 541 pp. B, 650/80
5th, held in 1983, *Compte-rendu* (1984) ed. J. N. Tasei (Paris: INRA) 493 pp. B, 661/85

International Trade Centre, UNCTAD/GATT (1977) *Major markets for honey* (Geneva: ITC, UNCTAD/GATT) 120 pp. B, 590/78

International Trade Centre, UNCTAD/GATT (1978) *The world market for beeswax: a high-value product requiring little investment* (Geneva: ITC, UNCTAD/GATT) 105 pp. B, 1348/78

International Trade Centre, UNCTAD/GATT (1986) *Honey: a study of major markets* (Geneva: ITC, UNCTAD/GATT) 167 pp. also in French, Spanish B, 1009/87

Ioirish, N. P., *see* Yoirish, N

Iparraguirre, F. (1967) *Economía apícola argentina* (Buenos Aires: FASA) 31 pp. B, 360/72

Iparraguirre, F. (1967) *Ordenamiento integral de la apicultura argentina* (Buenos Aires: FASA) 40 pp. B, 361/72

Isack, H. A., Reyer, H.-U. (1989) Honeyguides and honey gatherers: interspecific communication in a symbiotic relationship. *Science, N.Y.* 243: 1343–1346

Ismail, M. M.; Shamsudin, M. N. (1987) The economics of beekeeping industry in Malaysia. *A. Rep. Malaysian Beekeeping Research and Development Team*: 25–28 B

Istomina-Tzvetkova, K. P. (1953) [Reciprocal feeding between bees.] *Pchelovodstvo* (1): 25–29 *In Russian* English translation in M. Simpson (1954) B, 92/54

Iwanami, Y.; Sasakuma, T.; Yamada, Y. (1988) *Pollen: illustrations and scanning electronmicrographs* (Tokyo: Kodansha; Berlin: Springer-Verlag) 198 pp. B

Jacobs, M. R. (1981) *Eucalypts for planting* (Rome: FAO) 677 pp. 2nd ed. B

James, T. (1986) Personal communication in Chiang Mai, Thailand

Jander, R.; Jander, U. (1970) Über die Phylogenie der Geotaxis innerhalb der Bienen (Apoidea). *Z. vergl. Physiol.* 66: 355–368 B, 118/71

Jay, S. C. (1963) The longitudinal orientation of larval honey bees (*Apis mellifera*) in their cells. *Canad. J. Zool.*: 717–723 B, 551/64

Jay, S. C. (1964) The cocoon of the honey bee, *Apis mellifera* L. *Canad. Ent.* 96(5): 784–792 B, 733/65

Jay, S. C. (1965–69) Drifting of honeybees in commercial apiaries. *J. apic. Res.* 4: 167–175 (1965); 5: 103–112, 137–148 (1966); 7: 37–44 (1968); 8: 13–17 (1969) B, 478/66; 88, 702/67; 109/69; 393/70

Jay, S. C. (1969, 1970, 1980) Studies on hiving package bees. *J. apic. Res.* 8: 83–89, 91–97; 9: 71–78; 19: 242–247 B, 873, 874/70; 964/71; 991/81

Jay, S. C. (1971) How to prevent drifting. *Bee Wld* 52(2): 53–54 B

Jay, S. C.; Frankson, C. F. (1972) Effects of various hive covers, shading and ventilation on honeybees in the tropics. *J. apic. Res.* 11(2): 111–115 B, 362/73

Jay, S. C. (1979) The essence of beekeeping. *Bee Wld* 60(3): 140–142 B

Jay, S. C.; Dixon, D. (1982) Nosema disease in package honeybees, queens and attendant workers shipped to western Canada. *J. apic. Res.* 21(4): 216–221 B, 964/83

Jay, S. C. (1986) Spatial management of honey bees on crops. *A. Rev. Ent.* 31: 49–65; also *Bee Wld* 67(3): 98–113 B, 1032/86

Jay, S. C. (1987) Personal communication

Jaycox, E. R. (1963) Confinement of honeybee colonies to avoid pesticide losses. *J. apic. Res.* 2(1): 43–49 B, 645/64

Jaycox, E. R. (1980) Another way to get liquid honey from sealed combs. *Bees and Honey* (Nov.): 4 B

Jaycox, E. R. (1986) New 2-in-1 extractor. *Glean. Bee Cult.* 114(12): 617 B

Jean-Prost, P. (1957) Observations sur le vol nuptial des reines d'abeilles. *C. R. Acad. Sci., Paris* 245: 2107–2110 B, 115/59

Jean-Prost, P. (1987) *Apiculture* (Paris: Lavoisier) 579 pp. 6th ed. B

Jeffree, E. P. (1955) Observations on the decline and growth of honey bee colonies. *J. econ. Ent.* 48(6): 723-726 B, 190/56

Jeffree, E. P. (1959) The world distribution of acarine disease of honeybees and its probable dependence on meteorological factors. *Bee Wld* 40(1): 4–15 B, 65/61

Jeliński, M.; Wójtowski, F. (1984) *Melittobia acasta* Walker (Hym., Chalcidoidea, Eulophidae) mało znany pasożyt czerwia pszczelego. *Przegl. zool.* 28(4): 507–511 B, 635/86

Jensen, K. (1959) *Nye veier i birøkt: veiledning for 2-dronningskuber* (Oslo: Honningcentral) 131 pp. B, 204/60

Jeremiah (1932) How much honey do beekeepers eat? *Beekprs Item* 16(6): 252–254 B

Johannsmeier, M. F. (1983) Experiences with the Cape bee in the Transvaal. *S. Afr. Bee J.* 55(6): 130–138 B, 1225/85

Johannsmeier, M. F. (1984) Personal communication

Johansen, C. (1967) Encouraging the bumble bee pollinator of cranberries. *Publ. co-op. Ext. Serv. Wash. St. Univ.* EM2262: 3 pp. rev. B

Johansen, C. A.; Mayer, D. F.; Eves, J. D. (1978) Biology and management of the alkali bee, *Nomia melanderi* Cockerell (Hymenoptera: Halictidae). *Melanderia* 28(23): 25–46 B, 1523/79

Johansen, C. A. (1979) Honeybee poisoning by chemicals: signs, contributing factors, current problems and prevention. *Bee Wld* 60(3): 109–127 B, 271/80

Johansen, C. A. (1980) Mount Saint Helens blows the season for many Washington beekeepers. *Am. Bee J.* 120(7): 500–502 B

Johansen, C. A. & 7 others (1982) *Alfalfa seed integrated pest management – the Northwest Regional Program* (Pullman, WA, USA: Washington State University) 56 pp. B

Johansen, C. A.; Mayer, D. F (1985, updated annually) Protecting bees from pesticides. Pp. 18–23 from *Pacific North West insect control handbook* (Oregon State, Washington State, and Idaho Universities) B

Johansson, T. S. K.; Johansson, M. P. (1955) Royal jelly. *Bee Wld* 36: 3–13, 21–32 B, 211/56

Johansson, T. S. K.; Johansson, M. P. (1958) Royal jelly II. *Bee Wld* 39: 254–264, 277–286 B, 227/59

Johansson, T. S. K.; Johansson, M. (1967) Lorenzo L. Langstroth and the bee space. *Bee Wld* 48(4): 133–143 B, 754/69

Johansson, T. S. K.; Johansson, M. P. (1971*a*) Queen introduction. *Am. Bee J.* 111: 98–99, 146, 183–185, 226–227, 264–265, 306–307, 348–349, 384, 387 B, 392/73

Johansson, T. S. K.; Johansson, M. P. (1971*b*) Substitutes for beeswax in comb and comb foundation. *Bee Wld* 52(4): 146–156; also in (1978) below B, 381/73

Johansson, T. S. K.; Johansson, M. P. (1972*a*) *Apicultural literature published in Canada and the United States* (New York: author) 103 pp. B, 75/73

Johansson, T. S. K.; Johansson, M. P. (1972*b*) Bee-library of the late Rev. L. L. Langstroth. *Bee Wld* 53(1): 22–27 B

Johansson, T. S. K.; Johansson, M. P. (1973) Methods for rearing queens. *Bee Wld* 54(4): 149–175 B, 852/74

Johansson, T. S. K.; Johansson, M. P. (1976/1977) Feeding sugar to bees. *Bee Wld* 57: 137–143; 58: 11–18, 49–52. IBRA Reprint M92; also in (1978) below B, 231/78

Johansson, T. S. K.; Johansson, M. P. (1978) *Some important operations in bee management* (London: IBRA) 145 pp. B, 216/80

Johansson, T. S. K.; Johansson, M. P. (1979) The honeybee colony in winter. *Bee Wld* 60(4): 155–170. IBRA Reprint M101 B, 512/80

Johnson, E. (1976) *Bee keeping with the modern African hive* (Entebbe: Department of Agriculture) 26 pp. 2nd ed. B

Johnson, L. H. (1958) Asbestos cement hive covers and floor boards. *N. Z. Jl Agric.* 97(1): 63, 65 B, 301/59

Jolly, V. G. (1978) Propolis varnish for violins. *Bee Wld* 59(4): 158–161, 157 B

Jones, A.; Showers, R. E.; Moeller, F. E. (1964) Longevity of honey bees as affected by chemical gases and vapours commonly used in beekeeping. *Am. Bee J.* 104(9): 334–335 B, 162/65

Jones, B. (1983) *Gale's honey book* (London: Hamlyn) 128 pp. B

Jones, W. A. (1986) New bee blower developed. *Australas. Beekpr* 88(1): 7–9 B

Jordan, R.; Zecha, H. (1956) *Bienenkundliche Lehrtafeln* (Vienna: Bundeslehr- und Versuchsanstalt für Bienenkunde) 40 pp. B, 11/57

Joshi, M. A. (1982) Honey and beeswax trade in India. *Indian Bee J.* 44(1): 1–6 B, 219/84

Juliano, J. C. (1975) Montagem de um perfil tecnológico para o Estado do Rio Grande do Sul. *Anais 3 Congr. bras. Apic.*: 233–244 B, 1253/77

Kaatz, H. (1986) Heisser Honig? Nach Tschernobyl: auch der Honig ist radioaktiv belastet. *Allg. dtsch. Imkerztg* 20(7): 222, 224 B, 1029/87

Kafle, G. P. (1984) Personal communication

Kaiser, W.; Steiner-Kaiser, J. (1987) Sleep research on honeybees: neurophysiology and behavior. Pp. 112–120 from *Neurobiology and behavior of honeybees* ed. R. Menzel & A. Mercer B

Kalnins, M.; Detroy, B. (1984) Effect of wood preservative treatment of beehives on honey bees and hive products. *J. agric. Fd Chem.* 32: 1176–1180 B, 932/86

Kalnins, M. A.; Erickson, E. H. (1986) Extending the life of beehives with and without preservatives. *Am. Bee J.* 126(7): 488–491 B, 943/87

Kaloyereas, S. A.; Oertel, E. (1958) Crystallization of honey as affected by ultrasonic waves, freezing and inhibitors. *Am. Bee J.* 98(1): 442-443 B, 351/60

Kanematsu, H. & others (1982) [Amino acid analysis of honeys with different geographical and floral origin.] *J. Jap. Soc. Food Nutr.* 35(4): 297-303 *In Japanese* B, 269/85

Kannangara, A. W. (1940) Some bee plants of Ceylon. *Bee Wld* 21(10): 94–96 B

Kapil, R. P. (1959*a*) Anatomy and physiology of the alimentary canal of honeybee, *Apis indica* Fab. (Apidae, Hymenoptera). *Zool. Anz.* 163(9/10): 306–323 B, 738/63

Kapil, R. P. (1959*b*) Variation in the developmental period of the Indian bee. *Indian Bee J.* 21(1): 3–6, 26 B, 24/62

Kapil, R. P. (1960) Osservazioni sulla temperatura del glomere dell'ape indiana. *Apicolt. Ital.* 27(4/5): 79–83 B, 503/63

Kapil, R. P. (1962*a*) Anatomy and histology of the male reproductive system of *Apis indica* Fab. (Apidae, Hymenoptera). *Insectes soc.* 9(1): 73–90 B, 504/63

Kapil, R. P. (1962*b*) Anatomy and histology of the female reproductive system of *Apis indica* F. (Hymenoptera, Apidae). *Insectes soc.* 9(2): 145–163 B, 505/63

Kapil, R. P.; Dhaliwal, J. S. (1968/69) Biology of *Xylocopa* species. *J. Res. Ludhiana* 5(3): 406–419; 6(1) *Suppl.*: 262–271 B, 843/70, 314/71

Kapil, R. P. (1971) A hive for the Indian honeybees. *Apiacta* 6(3): 107–109 B

Karmo, E. (1984) Personal communication

Karmo, E. A.; Vickery, V. R. (1987) Pollen transfer in the hive. *Can. Beekeep.* 13(7): 163 B, 345/88

Kasianov, A. I. (1978) Standardization of hives in the Soviet Union. *Apiacta* 13(2): 76–78 B

Kawa, M. N. (1982) Personal communication

Kefuss, J. A. (1978) Influence of photoperiod on the behaviour and brood-rearing activities of honeybees in a flight room. *J. apic. Res.* 17(3): 137–151 B, 954/79

Kellogg, C. R. (1943) Colmenas rurales de tipo horizontal. *Fitofilo* 2(4): 18–28 B

Kemp, J. (1987) L'apiculture autrement. *Rev. fr. apic.* (460): 92–93; (461): 126–127; (462): 204–205 B, 588/88

Kempff Mercado, N. (1966) Abejas indígenas: su explotación racional. *Revta Univ. auton. G. R. Moreno* (23/24): 47–53 B, 59/68

Kennedy, M. J. (1986) Choosing paint for beehive timbers. *Australas. Beekpr* 88(4): 70–73 B, 946/87

Kent, R. B. (1979) Apicultural development in Central America and Panama: some historic and economic considerations. Pp. 165–181 from *Beekeeping in rural development* ed. Commonwealth Secretariat & IBRA B, 1213/80

Kenya. Ministry of Agriculture (1973) *Top bar hive* (Nairobi: Ministry of Agriculture) 4 pp. Leaflet No. 224 B

Kenya. Ministry of Agriculture; Canadian Apiculture Team (1974) *Bee keeping in Kenya* (Nairobi: Ministry of Agriculture) 38 pp. B

Keresztesi, B. (1977) *Robinia pseudoacacia*: the basis of commercial honey production in Hungary. *Bee Wld* 58(4): 144–150 B, 155/79

Kerr, H. T.; Buchanan, M. E. (1987) Acoustical identification of honey bees. *Amer. Bee J.* 127(12): 847 B

Kerr, W. E. (1957) Indrodução de abelhas africanas no Brasil. *Brasil apic.* 3(5): 211–213. IBRA English translation E1146 B, 184/58

Kerr, W. R.; Maule, V. (1964) Geographic distribution of stingless bees and its implications. *J. N. Y. ent. Soc.* 72(1): 2–18 B, 741/64

Kerr, W. E. & 4 others (1967) Observations on the nest architecture and behaviour of some species of stingless bees in the vicinity of Manaus, Amazon (Hymenoptera, Apoidea). *Atas Simp. Biota amazônica* 5 (Zool.): 255–309 B, 58/70

Kevan, P. G.; Bye, R. (1982) Comb honey production without foundation in Mexico. *Am. Bee J.* 122(5): 356–357 B

Kevan, P. G.; Lane, M. A. (1985) Flower petal microtexture is a tactile cue for bees. *Proc. Nat. Acad. Sci. USA* 82(14): 4750–4752 B, 853/87

Kevan, P. G. (ed.) (1989) *Beekeeping with* Apis cerana *in tropical and subtropical Asia* (London: IBRA) [not yet published]

Khalifman, I. A. (1960) [*The observation hive*] (Moskva: Izdatel'stvo 'Sovetskaya Rossiya') 55 pp. *In Russian* B, 267/61

Khan, M. R.; Razzaque, M. A. (1981) A microbiological approach to the causes of migration of honey bees from their hives. *Bangladesh J. Bot.* 10(1): 53–57 B, 112/83

Kidd, J. N. (1921) Relative to the two queen system. *Bee Wld* 3(7): 192 B

Kigatiira, K. I. (1974) Hive designs for beekeeping in Kenya. *Proc. Ent. Soc. Ont.* 105: 118–28 B, 224/78

Kigatiira, K. I. (1984) *The aspects of the ecology of the African honeybee* (University of Cambridge: PhD Thesis)

Kigatiira, I. K. (1985) The need for definitions of swarming, migration and absconding in honeybees. *Proc. 3 Int. Conf. Apic. trop. Climates*: 257 B

Kigatiira, I. K. & 3 others (1986) Using synthetic pheromone lures to attract honeybee colonies in Kenya. *J. apic. Res.* 25(2): 85–86 B, 123/87

Kigatiira, K. I.; Kahenya, W. A.; Townsend, G. F. (1988) *Prosopis* spp. as multipurpose trees in dry zones. *Bee Wld* 69(1): 6–11 B, 1218/88

Killion, C. E. (1975) Producing various forms of comb honey. Chapter 11, pp. 307–313, from *Honey: a comprehensive survey* ed. E. Crane (1975*a*) B

Killion, E. E. (1981) *Honey in the comb.* (Hamilton, IL, USA: Dadant & Sons) 148 pp. B, 266/83
Killion, E. E. (1986) Personal communication
Kilpi, K.; Hyvönen, L. (1982) Comparison of the browning reaction of different sugar-acid model systems due to sterilization. Pp. 127–149 from *Prkka koivistaisen 50-vuafisjuhlajjulkaisn* (University of Helsinki EKT Series No. 600) B
Kim, K. J.; Chung, Y. H. (1986) [A study on the distribution of genus *Tilia*.] *Korean J. Apic.* 1(1): 24–45 *In Korean* B
Kim, T. W.; Lee, Y. M. (1986) Flowering time of honey plants in Korea. *Korean J. Apic.* 1(1): 90–95 B, 591/87
Kingdon, J. (1977) Ratel, honey badger (*Mellivora capensis*). Pp. 86–103 from *East African mammals* (London: Academic Press)
Kirkor, S. (1959) Alergia u ludzi wywołana białkiem pszczelim. *Polsk. Tyg. lek.* (16): 748–749. IBRA English translation E703 B, 309/64
Kjaersgaard, H. (1985, 1988) Personal communications
Kleinschmidt, G. J. (1981) Standardization of hive equipment. *Apiacta* 16(1): 3–11, 14 B, 178/82
Kloft, W. J. (1963) Problems of practical importance in honeydew research. *Bee Wld* 44(1): 13–18, 24–29 B, 822/63
Kloft, W. J.; Kunkel, H. (eds) (1985) *Waldtracht und Waldhonig in der Imkerei* (Munich: Ehrenwirth Verlag) 329 pp. 2nd. ed. B
Knutson, B. M. (1964) A hive tipper. *Am. Bee J.* 104(10): 379 B
Koeniger, *see* Koeniger, N.
Koeniger, G.; Koeniger, N.; Fabritius, M. (1979) Some detailed observations of mating in the honeybee. *Bee Wld* 60(2): 53–57 B, 478/80
Koeniger, G. (1984) Funktionsmorphologische Befunde bei der Kopulation der Honigbiene (*Apis mellifera* L.). *Apidologie* 15(2): 189–204 B, 856/85
Koeniger, G. (1986*a*) Mating sign and multiple mating in the honeybee. *Bee Wld* 67(4): 141–150 B, 1208/87
Koeniger, G. (1986*b*) Reproduction and mating behavior. Ch. 10, pp. 255–280, from *Bee genetics and breeding* ed. T. Rinderer (1986) B, 106/88
Koeniger, G.; Ruttner, F. (1988) Paarungsverhalten und Anatomie der Fortplanzungsorgane. From *Die instrumentelle Besamung der Bienenkönigin* ed. R. F. A. Moritz
Koeniger, N. (1970) Factors determining the laying of drone and worker eggs by the queen honeybee. *Bee Wld* 51(4): 166–169 B, 378/72
Koeniger, N.; Weiss, J.; Ritter, W. (1975) Capture and management in cages of the colonies of the giant honey bees *Apis dorsata*. *Proc. 25 Int. Apic. Congr.*: 300–303 B, 418/78
Koeniger, N. (1976*a*) Neue Aspekte der Phylogenie innerhalb der Gattung *Apis*. *Apidologie* 7(4): 357–366 B, 114/78
Koeniger, N. (1976*b*) Interspecific competition between *Apis florea* and *Apis. mellifera* in the tropics. *Bee Wld* 57(3): 110–112, 100 B, 476/77
Koeniger, N.; Wijayagunasekera, H. N. P. (1976) Time of drone flights in the three Asiatic honeybee species (*Apis cerana, Apis florea, Apis dorsata*). *J. apic. Res.* 15(2): 67–71 B, 477/77
Koeniger, N. (1978) Das Wärmen der Brut bei der Honigbiene (*Apis mellifera* L.). *Apidologie* 9(4): 305–320 B, 1339/79
Koeniger, N.; Vorwohl, G. (1979) Competition for food among four sympatric species of Apini in Sri Lanka (*Apis dorsata, Apis cerana, Apis florea* and *Trigona iridipennis*). *J. apic. Res.* 18(2): 95–109 B, 124/80
Koeniger, N.; Koeniger, G. (1980) Observations and experiments on migration and dance communication of *Apis dorsata* in Sri Lanka. *J. apic. Res.* 19(1): 21–34 B, 476/81
Koeniger, N.; Koeniger, G.; Wijayagunasekara, H. N. P. (1981) Beobachtungen über die Anpassung von *Varroa jacobsoni* an ihren natürlichen Wirt *Apis cerana* in Sri Lanka. *Apidologie* 12(1): 37–40 B, 72/82
Koeniger, N. (1982) Interactions among the four species of *Apis*. Pp. 59–64 from *The biology of social insects* ed. M. D. Breed et al. (Boulder, CO, USA: Westview Press) B
Koeniger, N.; Veith, H. J. (1984) Spezifität eines Brutpheromons und Bruterkennung bei der Honigbiene (*Apis mellifera* L.). *Apidologie* 15(2): 205–210 B, 857/85
Koeniger, N. (1986) Personal communication
Koeniger, N. & 4 others (1988) Reproductive isolation by different time of drone flight between *Apis cerana* Fabricius, 1793 and *Apis vechti* (Maa, 1953). *Apidologie* 19(1): 103–106 B
Kolb, H. (1973) *Observation beehive* (Edmond, OK, USA: author) 4 pp. B
Kolmes, S. A. (1986) Age polyethism in worker honey bees. *Ethology* 71: 252–255 B, 906/87
Koltermann, R. (1973) Rassen- bzw. artspezifische Duftbewertung bei der Honigbiene und ökologische Adaptation. *J. comp. Physiol.* 85: 327–360 B, 181/76
König, B. (1985) Plant sources of propolis. *Bee Wld* 66(4): 136–139 B, 592/86
König, B.; Dustmann, J. H. (1985) Fortschritte der Celler Untersuchungen zur antivirotischen Activität von Propolis. *Apidologie* 16(3): 228–230 B

Koover, C. J. (1965) Clearing supers completely. *Glean. Bee Cult.* 93(2): 76–77 B, 590/65

Kortbech-Olesen, R. (1986) Honey exports: possibilities in a highly competitive market. *Int. Trade Forum* 22(3): 4–7, 30 B, 1322/87

Koster Keunen Inc. (1977) *Koster Keunen: specialists in beeswax since 1852* (Bladel, Netherlands: Koster Keunen Inc.) 48 pp. B

Kotze, W. A. G. (1949) Ton of honey from one colony in one year. *Glean. Bee Cult.* 77(8): 487–490 B

Kramer, F. E. (1985) A technique for protecting colonies during insecticide applications and moving bees in high temperatures. *Am. Bee J.* 125(3): 178–180 B, 1327/86

Krebs, J. R. (1975) The bee language controversy. *Nature, Lond.* 258: 109; *Bee Wld* 57(1): 25–26 (1976) B

Kreil, G. (1973) Structure of melittin isolated from two species of honey bees. *FEBS Letters* 33(2): 241–244 B, 346/74

Kreil, G. (1975) The structure of *Apis dorsata* melittin: phylogenetic relationships between honeybees as deduced from sequence data. *FEBS Letters* 54(1): 100–102 B, 113/77

Kreil, G. (1978) Biochemical surprise of a bee sting. *New Scient.*: 618–620 B

Krell, R.; Dietz, A.; Eischen, F. A. (1985) A preliminary study on winter survival of Africanized and European honey bees in Cordoba, Argentina. *Apidologie* 16(2): 109–117 B, 2/87

Krochmal, C. (1985) Hive cosmetics. *Glean. Bee Cult.* 113(10): 527–529 B

Kroes, R. C. A. (1951) The Amsterdam Bee Park. *Bee Wld* 32(2): 10–12 B

Krünitz, J. G. (1774) *Das Wesentlichste der Bienen-Geschichte und Bienen-Zucht* (Berlin: Joachim Paul) B

Kshirsagar, K. K. (1982) Current incidence of honeybee diseases and parasites in India. *Bee Wld* 63(4): 162–164 B, 958/83

Kuehl, L. J. (1988) Apparatus for removing moisture from honey. *U.S. Patent* 4,763,572: 7 pp.

Kumar, J.; Patyal, S. K.; Mishra, R. C. (1986) Evaluation of some essential oils as repellents to the Indian honeybee *Apis cerana indica* 1. Gustatory repellency and toxicity. *J. apic. Res.* 25(4): 256–261 B, 48/88

Kurennoi, N. M. (1953) [When are drones sexually mature?] *Pchelovodstvo* (11): 28–32 *In Russian*, English translation in M. Simpson (1954) B, 89/54

Kurennoi. N. M. (1954) [Flight activity and sexual maturity of drones.] *Pchelovodstvo* (12): 24–28 *In Russian* IBRA English translation E761 B, 189/56

Kuterbach, D. A.; Walcott, B. (1986) Iron-containing cells in the honeybee (*Apis mellifera*). *J. exp. Biol.* 126: 375–387, 389–401 B, 1192, 1193/88

Lacher, V. (1964) Elektrophysiologische Untersuchungen an einzelnen Rezeptoren für Geruch, Kohlendioxyd, Luftfeuchtigkeit und Temperatur auf den Antennen der Arbeitsbiene und der Drohne (*Apis mellifera* L.). *Z. vergl. Physiol.* 48: 587–623 B, 78/66

Laere, O. van.; Lagasse, A.; Mets, M. de (1969) Use of the scanning electron microscope for investigating pollen grains found in honey samples. *J. apic. Res.* 8(3): 139–145 B, 1011/70

Laere, O. van (1971) Construction de deux types de ruchettes experimentales. *Apidologie* 2(1): 111–116 B, 731/72

Laere, O. van.; Wael, L. de; Borneck, R. (1985) Expanded metal as foundation for comb construction in bee hives. *Proc. 30 Int. Apic. Congr.*: 335–337 B

Laffers, A. (1981) The mobile beehouse – a modern means of production. *Apiacta* 16(1): 12–14 B

Lagerman, B. (1985) Fribiodling. *Bitidningen* 84(4): 143–146 B

Lagerman, B. (1986) En ny spärrgallerbotten. *Bitidningen* 85(1): 14–17 B

Laidlaw, H. H. (1976) *Instrumental insemination of queen honey bees* (Hamilton, IL, USA: Dadant & Sons) set of 111 colour slides for use with book (next entry) B

Laidlaw, H. H. (1977) *Instrumental insemination of honey bee queens* (Hamilton, IL, USA: Dadant & Sons) 144 pp. B, 924/78

Laidlaw, H. H. (1979) *Contemporary queen rearing* (Hamilton, IL, USA: Dadant & Sons) 199 pp. B, 606/81

Laidlaw, H. H. (1981) Queen introduction. *Bee Wld* 62(3): 98–105 B

Laidlaw, H. H. (1986) Personal communication

Laidlaw, H. H.; Page, R. E. (1986) Mating designs. Ch. 13, pp. 323–344, from *Bee genetics and breeding* ed. T. E. Rinderer B, 193/88

Laidlaw, H. H. (1987) Instrumental insemination of honeybee queens; its origin and development. *Bee Wld* 68: 17–36, 71–88 B, 1271/87

Laidlaw, H. H.; Page, R. E. (1989) *Apiculture: introductory biology and husbandry* (Davis, CA, USA: University of California Press). In press

Langridge, D. F.; Ilton, C. D. (1981) *Beekeeping in Victoria* (Melbourne: Victoria Dep. Agric.) 139 pp. B

Langstroth, L. L. (1853) *Langstroth on the hive and the honey-bee, a bee keeper's manual* (Northampton, MA, USA: Hopkins, Bridgman & Co.) 384 pp. Reprinted

1914 (Medina, OH, USA: A. I. Root Co.) (2nd ed. 1957) B
Laperrouzaz, A. (1986) La fumée au bout des doigts. *Rev. fr. Apic.* (457): 504–505 B
Latif, A.; Qayyum, A.; Tufail, M. (1960*a*) Studies on Pakistan beeswax. *Bee Wld* 41(6): 153–156 B, 319/62
Latif, A.; Qayyum, A.; Manzoor-ul-Haq (1960*b*) Multiple- and two-queen systems in *Apis indica* F. *Bee Wld* 41(8): 201–209 B, 28/62
Latin America (1964) *Código Latinoamericano de alimentos* (honey, p. 160; artificial honey, p. 229; labelling, p. 61; containers, p. 53; trace metals, p. 33; health claims, p. 62)
Lavie, P. (1960) *Les substances antibactériennes dans la colonie d'abeilles* (Apis mellifica L.) (Université de Paris: Thèse doctoraux) 191 pp. IBRA English translation E801 B, 761/63
Lavie, P. (1967) Influence de l'utilization du piège á pollen sur le rendement en miel des colonies d'abeilles. *Ann. Abeille* 10(2): 83–95 B, 417/68
Lavie, P. (1968) Propriétés antibactériennes et action physiologique des produits de la ruche et des abeilles. Pp. 1–115 from Chauvin (1968*d*) B
Law Reports (1986) Tutton and others, v. A. D. Walter Ltd [report of 1984 case]. *The Law Reports 1 Q.B.*: 61–79
Layens, G. de; Bonnier, G. (1897) *Cours complet d'apiculture* (Paris: Librairie Générale de l'Enseignement) 446 pp. rev. ed. B
Lee, C. Y.; Kime, R. W. (1984) The use of honey for clarifying apple juice. *J. apic. Res.* 23(1): 45–49 B, 1323/84
Legge, A. P. (1976) Hive inserts and pollen dispensers for tree fruits. *Bee Wld* 57(4): 159–67 B, 1049/77
Lehnert, T.; Shimanuki, H. (1981) Oxytetracycline residues in honey following three different methods of administering the drug. *Apidologie* 12(2): 133–136 B, 618/82
le Maistre, W. G. (1936) The O. A. C. strainer. *Canad. Bee J.* 44(11): 279–280 B
Lensky, Y. (1958) Some factors affecting the temperature inside hives in hot climates. *Bee Wld* 39(8): 205–208 B, 412/59
Lensky, Y. (1963) [*Study on the physiology and ecology of the Italian honeybee in Israel*] (Hebrew University of Jerusalem: PhD Thesis) 103 pp. *In Hebrew, French summary* IBRA English translation E728
Lensky, Y. (1964*a*) Les régulations thermiques dans le ruche en été. *Ann. Abeille* 7(1): 23–45 B, 115/65
Lensky, Y. (1964*b*) L'économie de liquides chez les abeilles aux températures élevées. *Insectes soc.* 11(3): 207–222 B, 758/65
Lensky, Y. (1964*c*) Comportement d'une colonie d'abeilles à des températures extrêmes. *J. Insect Physiol.* 10(1): 1–12 B, 591/64
Lensky, Y.; Golan, Y. (1966) Honeybee populations and honey production during drought years in a suptropical climate. *Ser. hierosolymitana* 18: 27–42 B, 547/67
Lensky, Y. (1971) Rearing queen honeybee larvae in queenright colonies. *J. apic. Res.* 10(2): 99–101 B, 157/72
Lensky, Y.; Slabezki, Y. (1981) The inhibiting effect of the queen bee (*Apis mellifera* L.) foot-print pheromone on the construction of swarming queen cups. *J. Insect Physiol.* 27(5): 313–323 B, 509/82
Lensky, Y. & 6 others (1984) The tarsal gland of the honeybee (*Apis mellifera* L.) queens, workers and drones (Hymenoptera, Apidae) III. Chemical characterization. *Insect Biochem.* In press B
Lensky, Y. & 4 others (1985) Pheromonal activity and fine structure of the mandibular glands of honeybee drones (*Apis mellifera* L.) (Insecta, Hymenoptera, Apidae). *J. Insect Physiol.* 31(4): 265–276 B, 885/86
Lercker, G. & 4 others (1982) Components of royal jelly 2. The lipid fraction, hydrocarbons and sterols. *J. apic. Res.* 21(3): 179–184 B, 1003/83
Levin, M. D. (1983) Value of bee pollination to U. S. agriculture. *Bull. ent. Soc. Am.* 29(4): 50–51; also *Am. Bee J.* 124(3): 184–186 (1984) B
Levin, M. D.; Loper, G. M. (1984) Factors affecting pollen trap efficiency. *Am. Bee J.* 124(10): 721–723 B, 1030/85
Levinsohn, M.; Lensky, Y. (1981) Long-term storage of queen honeybees in reservoir colonies. *J. apic. Res.* 20(4): 226–233 B, 553/82
Lewis, W. (1985) Bee sampling. *UC Apiaries* 2(1): 2 B
Libert, M.; Rotthier, B. (1987) Een zeldzame bijeneter: de wespendief (*Pernis apivorus*). *Maandbl. vlaam. Imkerst.* 73(7): 281–283 B
Liebl, D. E. (1977) Method of preserving honey. *US Patent* 4,050,952: 1 p.
Lilley, W. (1983) Bee miners join British Columbia gold hunt. *Am. Bee J.* 123(9): 635–637 B
Lindauer, M. (1951/1953) Bienentänze in der Schwarmtraube. *Naturwissenschaften* 38: 509–513; 40: 379–385. IBRA English translations E91, E204 B, 103/52, 29/55
Lindauer, M. (1953) Division of labour in the honeybee colony. *Bee Wld* 34: 63–73, 85–90 B, 235/55
Lindauer, M. (1955*a*) The water economy and temperature regulation of the honeybee colony. *Bee Wld* 36: 62–72, 81–92, 105–111 B, 215/56
Lindauer, M. (1955*b*) Schwarmbienen auf Wohnungssuche. *Z. vergl. Physiol.* 37: 263–324. IBRA English translation E302 B, 202/57

Lindauer, M. (1957*a*) Communication among the honeybees and stingless bees of India. *Bee Wld* 38: 3–14, 34–39 B, 325/58

Lindauer, M. (1957*b*) Sonnenorientierung der Bienen unter der Aequatorsonne und zur Nachtzeit. *Naturwissenschaften* 44(1): 1–6. IBRA English translation E430 B, 267/60

Lindauer, M.; Kerr, W. E. (1960) Communication between the workers of stingless bees. *Bee Wld* 41: 29–41, 65–71 B, 19/62

Lindauer, M. (1987) Entfernungsweisung. *Nordwestdtsch. Imkerztg* 39(10): 274 B, 527/88

Linder, J. (1965) *Ruche Rivka (modèle intermédiare entre la ruche primitive et la ruche moderne – Langstroth)* (Dakar, Senegal: Ministère d'Economie Rural) 7 pp. B

Linder, J. (1966) *Ruche David (ruche grecque améliorée)* (Dakar, Senegal: Ministère d'Economie Rural) 4 pp. B

Liu, J. (1976) Personal communication

Lochhead, A. G. (1933) Factors concerned with the fermentation of honey. *Zentbl. Bakt. ParasitKde* II Abt. 88: 296–302 B

Loper, G. M.; Wolf, W. W. ; Taylor, O. R. (1987) Detection and monitoring of honeybee drone congregation areas by radar. *Apidologie* 18(2): 163–172 B, 132/88

Lopez, M.-C.; Nicotra, C. (1975) Investigations of organic air polluting agents fixed in the pollen collected by bees. *Proc. 25 Int. Apic. Congr.*: 473–476 B, 511/78

Lopez-Palacios, S. (1986) Catalogo para una flora apícola Venezolana (Mérida: CDCH) 211 pp. B, 1212/88

Lord, C. C. (1917) *An account of the restocking scheme inaugurated by the Crayford and District Beekeepers' Association 1915–1916* Unpublished report B

Lord, W. G.; Bambara, S. B.; Ambrose, J. T. (1985) Effect of Bee Calm device on honey bee aggressiveness. *Am. Bee J.* 125(4): 251–253 B, 1253/86

Lord, W. G.; Nagi, S. K. (1987) *Apis florea* discovered in Africa. *Bee Wld* 68(1): 39–40 B

Loring, M. (1978) Pesticide spraying – whose liability? *Indian Bee J.* 40(2): 40–41 B

Loring, M. (1981) *Bees and the law* (Hamilton, IL, USA: Dadant & Sons) 128 pp. B, 80/82

Loring, M. (1984) Bee droppings – a legal overview. *Am. Bee J.* 124(11): 773–774 B

Loring, M. (1985) Law for the beekeeper – a bee bailment. *Am. Bee J.* 115(6): 443–444 B

Louveaux, J. (1954) Études sur la récolte du pollen par les abeilles. *Apiculteur* 98(12) *Sect. sci.*: 43–50 B, 77/55

Louveaux, J. (1958/59) Recherches sur la récolte du pollen par les abeilles (*Apis mellifica* L.). *Ann. Abeille* 1: 113–188, 197–221; 2: 13–111 B, 330/61

Louveaux, J. (1962) Les caractères organoleptiques du miel. 2 pp. from *Vocabulaire technique des caractères organoleptiques et de la dégustation des produits alimentaires* ed. J. Le Magnen

Louveaux, J. (1966) Les modalités de l'adaptation des abeilles (*Apis mellifica* L.) au milieu naturel. *Ann. Abeille* 9(4): 323–350 B, 331/67

Louveaux, J. (1970) *Atlas photographique d'analyse pollinique des miels* (Paris: Service de la Répression des Fraudes et du Contrôle de la Qualité) 63 pp. + 53 plates. Annexes microphotographiques aux méthodes officielles d'analyse III B, 1026/71

Louveaux, J. (1973) The acclimatization of bees to a heather region. *Bee Wld* 54(3): 105–111 B

Louveaux, J.; Maurizio, A.; Vorwohl, G. (1978) Methods of melissopalynology. *Bee Wld* 59(4): 139–157. IBRA Reprint M95 B, 1078/79

Louveaux, J. (1985) *Les abeilles et leur élevage* (Paris: OPIDA) 265 pp. 2nd ed. B, 1st ed. 1249/81

Lovell, H. B. (1926) *Honey plants of North America* (Medina, OH, USA: A. I. Root Co.) 408 pp. B

Lovell, H. B. (1957) Let's talk about honey plants. *Glean. Bee Cult.* 85(2): 99, 123 B

Lovell, H. B. (1977) *Honey plants manual* (Medina, OH, USA: A. I. Root Co.) 96 pp. (rev. Goltz, L. R.) B

Low, W. H.; Kam Va Vong; Sporns, P. (1986) A new enzyme, β-glucosidase, in honey. *J. apic. Res.* 25(3): 178–181 B, 1333/87

Lowe, D. G. (1980) Propolis substitutes. *Bee Wld* 61(3): 120–121 B

Luchanska, E.; Kolev, I. D. (1984) [Nectar plants and forage in the Kyustendil district [of Bulgaria].] *Rast. Nauki, Sofia* 21(7): 123–128 B, 1240/87

Lunder, R. (1950) Kan biene drives tidlig fram om våren? *Nord. Bitidskr.* 2(2): 33–39 B, 55/54

Lüpke, J. (1980) Bienenhonig in Trockenform. *Kakao u. Zucker* 32(2): 43 B, 991/82

Maa, T. (1953) An inquiry into the systematics of the tribus Apidini or honeybees (Hym.). *Treubia* 21(3): 525–640 B, 146/59

Names beginning with Mc follow Mayer

Madsen, R. F. (1974) Membrane concentration. Pp. 251–301 from *Advances in preconcentration and dehydration of foods* ed. A. Spicer (London: Applied Science Publishers Ltd)

Mahindre, D. B. (1968) *Apis dorsata* – the giant bee of India. *Glean. Bee Cult.* 96(2): 102–108 B

Mahindre, D. B. & 7 others (1977) Nesting behaviour of *Apis dorsata*. *Proc. 26 Int. Apic. Congr.*: 314 B

Mahindre, D. B. (1983*a*) Handling rockbee colonies. *Indian Bee J.* 45(2/3): 72–73 B
Mahindre, D. B. (1983*b*) A suitable hive for *Apis cerana*. *Indian Bee J.* 45(2/3): 80–81 B
Mahindre, D. B. (1987) Personal communication
Maki, D. L. & 3 others (1987) Beekeeping in Mexico's Yucatan. *Am. Bee J.* 127(10): 708–711 B
Maki, D. L.; Wilson, W. T.; Cox, R. L. (1988) Influence of mite infestations (*Acarapis woodi*) and its effect on honey bee longevity. *Am. Bee J.* 126(12): 832 [abstract] B, 621/87
Mammo Gebreyesus (1986) Personal communication
Mangum, W. A. (1987) Building a regular or observation Kenya top bar hive. *Glean. Bee Cult.* 115(11): 646–648 B, 586/88
Manino, A. & 4 others (1985) Sequential carbohydrate variations from larch phloem sap to honeydew and to honeydew honey. *Apicoltura* 1: 93–103 B, 599/87
Manitoba Beekeeper (1985) 1983 Beekeeper Compensation Program. *Manitoba Beekpr* (Spring): 10 B
Manitoba Beekeeper (1986) Using hydrogen cyanide to kill bees. *Manitoba Beekpr* (Summer): 11–14 B, 928/87
Manjo, G. (1975) *The healing hand: man and wound in the Ancient World* (Cambridge, MA, USA: Harvard University Press) 571 pp. B, 999/77
Manley, R. O. B. (1933) Frosting. *Bee Wld* 14(12): 135–136 B
Manley, R. O. B. (1946) *Honey farming* (London: Faber & Faber) 293 pp. Facsimile reprint 1975 B
Mannheim, C. H.; Passy, N. (1974) Non-membrane concentration. Pp. 151–193 from *Advances in preconcentration and dehydration of foods* ed. A. Spicer (London: Applied Science Publishers Ltd)
Marchenay, P. (1979) *L'Homme et l'abeille* (Paris: Berger-Levrault) 2nd ed. 1984, lacks the illustrations B, 1187/80
Marchenay, P. (1981) Places of beekeeping interest in France. *Bee Wld* 62(3): 118–122 B
Marcovič, O.; Molnár, L. (1955) Príspevok k izolácii a stanoveniu vcelieho jedu. *Chem. Zvesti* 8(2/3): 80–90 B
Mardan, M.; Kiew, R. (1985) Flowering periods of plants visited by honeybees in two areas of Malaysia. *Proc. 3 Int. Conf. Apic. trop. Climates:* 209–216 B, 190/86
Mardan, M. (1987) Extruding liquid gobbets by Asian giant honeybees as thermoregulation mechanism. P. 243 from *Chemistry and biology of social insects* ed. J. Eder and H. Rembold (Munich: J. Peperny) B
Marekov, N. L.; Bankova, V. S.; Popov, S. S. (1984?) The practical value of the polyphenols taken from propolis. *Impact* (136): 369–374 B
Mariola, P. (1986) Beekeeper's backache, and backache relief. *Am. Bee J.* 126: 177–178; 255–256, 287 D
Markl, H. (1966) Schwerkraftdressuren an Honigbienen. *Z. vergl. Physiol.* 53: 328–371 B, 505, 506/68
Marletto, F.; Piton, P. (1976) Implanto per la disidratazione del miele mediante ventilazione. *Apic. mod.* 67(3): 81–84 B, 1010/77
Marletto, F.; Olivero, G. (1981) Ricerche su raccolta e utilizzazione della propoli da parta delle api. *Apicolt. mod.* 72(4): 131–140. IBRA English translation E1595 B, 170/83
Marletto, F. (1983) Caratteristiche della propoli in funzione dell'origine florale e dell'utilizzazione de parte delle api. *Apicolt. mod.* 74(5): 187–191 B, 525/85
Marletto, F. (1986) Personal communication
Marquina Olmedo, J. B.; Sáenz Barrio, C. C. (eds) (1983) *Primer Congreso Nacional de Apicultura 1983: ponencias y comunicaciones* (Madrid: Ministerio de Agricultura, Pesca y Alimentación) 246 pp. B, 844/86
Martin, B. (1977) *Batik for beginners* (London: Angus & Robertson) rev. ed.
Martin, E. C. (1958) Some aspects of hygroscopic properties and fermentation of honey. *Bee Wld* 39(7): 165–178 B, 350/59
Martin, P. (1963) Die Steuerung der Volksteilung beim Schwärmen der Bienen. Zugleich ein Beitrag zum Problem der Wanderschwärme. *Insectes soc.* 10(1): 13–42. IBRA English translation E942 B, 568/63
Martin, S. (1982) *Beekeeping: honey and other hive products* (Wellington, NZ: Development Finance Corporation) 60 pp. rev. ed. B, 853/86
Martin, U.; Martin, H.; Lindauer, M. (1978) Transplantation of a time-signal in honeybees. *J. comp. Physiol. A* 124: 193–201 B, 475/80
Martinovs, A. (1964) Om binas vattenbehov och nektardrag. *Nord. Bitidskr.* 15/16: 38–40. IBRA English translation E843 B, 99/66
Marz, R.; Mollay, C.; Kreil, G. (1981) Queen bee venom contains much less phospholipase than worker bee venom. *Biochemistry* 11(6): 685–690 B, 1187/83
Matheson, A. (1980) Easily-constructed paraffin-wax dipper. *N. Z. Beekpr* 41(4): 11–12 B
Matheson, A.; Reid, M. (1981) [Cappings spinners.] *N. Z. Beekpr* 42(3): 16–18 B
Matheson, A. (1984*a*) *Practical beekeeping in New Zealand* (Wellington: Government Printer) 185 pp. B
Matheson, A. G. (1984*b*) Current developments in greater wax moth control. *N. Z. Beekpr* (183): 24–26 B

Matheson, A. G. (1985*a*) Recent research into half-moon disorder in honey-bees. *N. Z. Beekpr* (188): 15 B

Matheson, A. G. (1985*b*) Options for beekeeping development: a Jamaican case study. *Proc. 3 Int. Conf. Apic. trop. Climates:* 104–108 B

Mathpal, Y. (1984) Newly discovered paintings in Central India showing honey collection. *Bee Wld* 65(3): 121–126 B, 809/85

Matsuka, M.; Watanabe, N.; Takeuchi, K. (1973) Analysis of the food of larval drone honeybees. *J. apic. Res.* 12(1): 3–7 B, 855/73

Matsuura, M. (1985) Group predation by the giant hornet *Vespa mandarina* on the colonies of honeybees and social wasps. *Proc. 30 Int. Apic. Congr.:* 244–247 B

Mattu, V. K.; Verma, L. R. (1985) Studies on the annual foraging cycle of *Apis cerana indica* F. in Simla Hills of northwest Himalayas. *Apidologie* 16(1): 1–18 B, 69/87

Mátyás, A. von (1932) *Fortschritte in der Bienenwohnungsfrage* (Budapest: Sajtóvállalat) 320 pp. B

Maung Maung Nyein (1984) *Study on traditional method of keeping Indian honey bees in Burma and keeping with modern method* (Rangoon: Burma Research Association) 9 pp. B, 67/87

Maurizio, A. (1950) The influence of pollen feeding and brood rearing on the length of life and physiological conditions of the honeybee. Preliminary report. *Bee Wld* 31(2): 9–12 B, 137/51

Maurizio, A. (1964) Mikroskopische und papierchromatographische Untersuchungen an Honig von Hummeln, Meliponinen und anderen, zuckerhaltige Säfte sammelnden Insekten. *Z. Bienenforsch.* 7(4): 98–110. IBRA English translation E838 B, 387/65

Maurizio, A. (1975*a*) How bees make honey. Chapter 2, pp. 77–105, from *Honey: a comprehensive survey* ed. E. Crane (1975*a*) B

Maurizio, A. (1975*b*) Microscopy of honey. Chapter 7, pp. 240–257, from *Honey: a comprehensive survey* ed. E. Crane (1975*a*) B

Maurizio, A. & 4 others (eds) (1975) *Der Honig: Herkunft, Gewinnung, Eigenschaften und Untersuchung des Honigs* (Stuttgart: Ulmer) 212 pp. 2nd ed. B, 543/76

Maurizio, A.; Grafl, I. (1982) *Das Trachtpflanzenbuch. Nektar und Pollen – die wichtigsten Nahrungsquellen der Honigbiene* (Munich: Ehrenwirth Verlag) 368 pp. 3rd ed. B, *see* 396/70

Maurizio, A. (1985) Honigtau – Honigtauhonig. Pp. 268–295 from *Waldtracht und Waldtrachthonig in der Imkerei* ed. W. J. Kloft & H. Kunkel B

Maxwell, H. (1987*a*) A small-scale honey drying system. *Am. Bee J.* 127(4): 284–286 B, 1341/87

Maxwell, H. (1987*b*) Martin Marietta Africanized honey bee detector discussed at Florida Bee Institute. *Am. Bee J.* 127(11): 791–792 B

Mayama, A.; Nakajima, A.; Echigo, T. (1982) [The relationship between consumer preference and chemical composition of sugar organic acid and free amino acid in honey.] *Honeybee Sci.* 3(3): 131–134 *In Japanese* IBRA English translation E1613 B, 645/83

Mayer, D. (1981) Tree fruit pollination and pollination contracts in Washington. *Am. Bee J.* 121(5): 365–366 B

McCormick, M. (1960) *The golden pollen* (Yakima, WA, USA: Yakima Bindery & Printing Co.) 160 pp. B

McCutcheon, D. M. (1984) Indoor wintering of hives. *Bee Wld* 65(1): 19–37. IBRA Reprint M113 B, 1248/84

McCutcheon, D. M. (1988) Bear damage in Canada. *Australas. Beekpr* 89(12): 155–156 B

McDonald, J. A.; Li, F. P.; Mehta, C. R. (1979) Cancer mortality among beekeepers. *J. occup. Med.* 21(12): 811–813 B, 1262/81

McDowell, R. (1984) The Africanized honey bee in the United States: what will happen to the U.S. beekeeping industry? *Econ. Rep. Nat. Resources Div. USDA No.* 519: 33 pp. B, 131/86

McGregor, S. E. (1952) Collection and utilization of propolis and pollen by caged honey bee colonies. *Am. Bee J.* 92(1): 20–21 B, 161/53

McGregor, S. E. (1976) *Insect pollination of cultivated crop plants* (Washington, DC: USDA) 411 pp. Agric. Handb. USDA No. 496 B, 1088/77

McLaughlin, T. (1973) *Candle making* (London, UK: Pelham Books) 94 pp. B

McLellan, A. R. (1977) Honeybee colony weight as an index of honey production and nectar flow: a critical evaluation. *J. appl. Ecol.* 14(2): 401–408 B, 627/79

Medler, J. T. (1962) Effectiveness of domiciles for bumblebees. *Proc. 1 Int. Symp. Poll.:* 126–133 B, 245/65

Mejia, G. & 4 others (1986) Acute renal failure due to multiple stings by Africanized bees. *Ann. intern. Med.* 104(2): 210–211 B

Melksham, K. J.; Jacobsen, N.; Rhodes, J. (1988) Compounds which affect the behaviour of the honeybee, *Apis mellifera* L.: a review. *Bee Wld* 69(3): 104–124 B

Mello, M. L. S. (1970) A qualitative analysis of the proteins in venoms from *Apis mellifera* (including *A. m. adansonii*) and *Bombus atratus*. *J. apic. Res.* 9(3): 113–120 B, 226/72

Méndez de Torres, L. (1586) *Tractado breve de la cultiva-*

ción y cura de las colmenas (Alcalá de Henares, Spain: author) 78 pp. Reprinted 1983 (Torrejón del Rey, Spain: Amurari) B

Menzel, R.; Freudel, H.; Rühl, V. (1973) Rassenspezifische Unterschiede im Lernverhalten der Honigbiene (*Apis mellifera* L.). *Apidologie* 4(1): 1–24 B, 107/75

Menzel, R.; Mercer, A. (eds) (1987) *Neurobiology and behavior of honeybees* (Berlin: Springer) 334 pp. B, 468/89

Merle, B. (1985) Étude technico-économique de quelques unités de production de miel, selon leur taille. *Bull. tech. apic.* 12(4): 169–208 B, 930/87

Merlin, R. (1978) Projet pour la normalisation des cadres. *Abeilles et Fleurs* (280): 8–10 B

Mertsin, I. (1980) [Transport of honeybees in combless packages.] *Doklady TSkhA* (266): 131–134 *In Russian* B, 913/84

Mertzig, J.-J. (1988) Monatsanweiser-Januar/Februar. *Lëtzebuerger Beien-Ztg* 99(2): 26–28 B

Merz, J. H. (1963) The objective assessment of honey aroma by gas chromatography. *J. apic. Res.* 2(1): 55–61 B, 896/63

Meyer, J. (1986) Gadgets and gizmos. *Am. Bee J.* 126(7): 508 B

Meyer, P. (1977) Nuclear energy and the environment. *Kerntechnik* 19(1): 9–13 B, 1084/79

Meyer, W. (1952) Die Kleinbauarbeiten unserer Bienen. *Dtsch. Bienenw.* 3(12): 237–240 B, 239/53

Meyer, W.; Ulrich, W. (1952) Zur Analyse der Bauinstinkte unserer Honigbiene. Untersuchungen über die 'Kleinbauarbeiten'. *Naturwissenschaften* 39(11): 264. IBRA English Translation E185 B, 159/54

Meyer, W. (1956*a*) 'Propolis bees' and their activities. *Bee Wld* 37(2): 25–36 B, 156/58

Meyer, W. (1956*b*) Arbeitsteilung im Bienenschwarm. *Insectes soc.* 3(2): 303–324. IBRA English translation E431 B, 16/59

Michael, A. S. (1980) Federal and State bee laws and regulations. *USDA Agric. Handb.* No. 335: 161–169 B

Michelbacher, A. E.; Hurd, P. D.; Linsley, E. G. (1968) The feasibility of introducing squash bees (*Peponapis* and *Xenoglossa*) into the Old World. *Bee Wld* 49(4): 159–167 B, 61/70

Michelsen, A.; Kirchner, W. H.; Lindauer, M. (1986) Sound and vibrational signals in the dance language of the honeybee, *Apis mellifera*. *Behav. Ecol. Sociobiol.* 18: 207–212 B, 904/87

Michener, C. D. (1974) *The social behavior of the bees* (Cambridge, MA, USA: Harvard University Press) 404 pp. B, 579/75

Michener, C. D.; Winston, M. L.; Jander, R. (1978) Pollen manipulation and related activities and structures in bees of the family Apidae. *Univ. Kansas Sci. Bull.* 51(19): 575–601 B, 705/80

Michener, C. D. (1979) Biogeography of the bees. *Ann. Mo. bot. Gdn* 66(3): 277–347 B, 69/81

Michener, C. D. (1987) Personal communication

Millard, S. (1987) The puff ball. *Brit. Bee J.* 115(4501): 208 B

Miller, W. R. (1976) *Let's build a bee hive* (Phoenix, AZ, USA: author) 92 pp. B

Ministerium für Ernährung, . . ., Baden-Württemberg (1985) *Landschaft als Lebensraum* (Stuttgart: Ministerium für Ernährung, . . ., Baden-Württemberg) 98 pp. B

Ministry of Agriculture, Fisheries and Food [UK] (1968) Beehives. *Bull. Minist. Agric. Lond.* No. 144: 27 pp. 4th ed. B

Ministry of Agriculture, Fisheries and Food [UK] (1985) *Pesticides: implementing Part III of the Food and Environment Protection Act 1985*, 62 pp. B

Mirnov, N. T. (1967) *The genus* Pinus (New York: Ronald Press) 602 pp.

Mitchener, A. V. (1940) The effect of color of hive covers upon the temperature within the hive. *J. econ. Ent.* 33(4): 649–650 B

Mitev, B. (1971) [Collection of bee venom using a weak electric current – its effect on the condition and the performance of the colony.] *Zhivot. Nauki* 8(1): 103–108 *In Bulgarian* B, 236/74

Mladenov, S. (1972) *Mierea şi therapia cu miere* (Bucharest: Editura Ceres) 264 pp. B

Moar, M. H. (1984) Personal communication

Möbus, B. (1985) Bees and the law. *Beekprs Ann. Suppl.* (1): 5–6 B

Mochida, S.; Haga, M.; Takino, Y. (1985) Chemical constituents and antimicrobial activity of Japanese propolis. *Proc. 30 Int. Apic. Congr.*: 455–456 B

Moeller, F. E. (1958) Relation between egg-laying capacity of queen bee and populations and honey production of their colonies. *Am. Bee J.* 98(10): 401–402 B, 177/61

Moeller, F. E. (1975) Effect of moving honeybee colonies on their subsequent production and consumption of honey. *J. apic. Res.* 14(3/4): 127–130 B, 460/76

Moeller, F. E. (1976) Two-queen system of honey bee colony management. *Prod. Res. ARS USDA* No. 161: 12 pp. B, 1262/77

Moeller, F. E.; Corley, C. (1977) Resmethrin. *Am. Bee J.* 117(11): 696 B

Moffett, J. O.; Morton, H. L. (1975) Repellency of surfactants to honey bees. *Envir. Ent.* 4(5): 780–782 B, 546/77

Moffett, J. O. (1985) Personal communication

Mogga, J. (1988) Personal communication

Mogga, J.; Ruttner, F. (1988) *Apis florea* in Africa: source of the founder population. *Bee Wld* 69(3): 100–103 B

Mohammad, A. (1984) Economic impact of beekeeping: a case study of Bangladesh. Pp. 111–122 from *Proceedings of the expert consultation* . . . ed. FAO B

Molan, P. C.; Russell, K. M. (1988) Non-peroxide antibacterial activity in some New Zealand honeys. *J. apic. Res.* 27(1): 62–67 B

Mollel, L. O. N. (1987) The use of puffballs *Langermannia wahlbergia* as bee repellents. *IBRA Newsletter* No. 11: 14 B

Moreaux, R. (1939) La connaissance de la maturité exacte du miel par l'abeille. *Bull. Soc. Sci. Nancy* (5): 110–114 B

Moreira, A. da C. (1968) *Uma nova apicultura. Vol. 2. Apicultura fixista, Apicultura mobilista, Apicultura simplificada* (Porto: author) 264 pp. B

Moret, C. & 4 others (1983) Un cas d'envenimation mortelle par piqûres multiples d'abeilles (*Apis mellifera* L). *Rev. med. Liège* 38(21): 815–822 B, 549/86

Morgan, P.; Percival, M. (1967) The rearing and management of bumble bees for students of biology. *Bee Wld* 48: 48–58, 100–109 B, 426/68

Morimoto, H. (1963) [On the feeding of drone honey bees] *Jap. Bee J.* 16(9): 258–263 *In Japanese* B, 578/64

Moritz, R. F. A. (1982) Präparatverteilung bei systemischer Therapie von Ektoparasitosen bei *Apis mellifera* L. *Apidologie* 13(2): 127–141 B, 615/83

Moritz, R. F. A.; Hänel, H. (1984) Restricted development of the parasitic mite *Varroa jacobsoni* Oud. in the Cape honeybee *Apis mellifera capensis* Esch. *Z. angew. Ent.* 97(1): 91–95 B, 243/85

Moritz, R. F. A. (ed.) (1988) *Die instrumentelle Besamung der Bienenkönigin* (Bucharest: Apimondia Publishing House)

Morland, D. (1931) Frosting. *Bee Wld* 12(4): 39–40 B

Morse, R. A. (1954) History of use of smoke for handling bees. *Glean. Bee Cult.* 82: 592–593, 637; 668–670 B

Morse, R. A.; Benton, A. W. (1964*a*) Notes on venom collection from honeybees. *Bee Wld* 45(4): 141–143 B, 339/66

Morse, R. A.; Benton, A. W. (1964*b*) Mass collection of bee venom. *Glean. Bee Cult.* 92(1): 42–45, 54 B

Morse, R. A. (1965) The effect of light on comb construction by honeybees. *J. apic. Res.* 4(1): 23–29 B, 112/66

Morse, R. A.; Benton, A. W. (1967) Venom collection from species of honeybees in south-east Asia. *Bee Wld* 48(1): 19–29 B, 67/68

Morse, R. A. & 3 others (1967) Observations on alarm substances in the genus *Apis*. *J. apic. Res.* 6(2): 113–118 B, 302/68

Morse, R. A.; Laigo, F. M. (1969*a*) *Apis dorsata* in the Philippines (including an annotated bibliography). *Monogr. Philipp. Ass. Ent., Inc.* No. 1, 96 pp. B, 586/71

Morse, R. A.; Laigo, F. M. (1969*b*) The Philippine spine-tailed swift, *Chaetura dubia* McGregor as a honey bee predator. *Philipp. Ent.* 1(2): 138–143 B, 994/70

Morse, R. A. (1970) Techniques used to move colonies of bees. *Glean. Bee Cult.* 98(10): 588–593, 635 B

Morse, R. A.; Boch, R. (1971) Pheromone concert in swarming honey bees (Hymenoptera: Apidae). *Ann. Ent. Soc. Am.* 64(6): 1414–1417 B, 614/73

Morse, R. A. (1972) *The complete guide to beekeeping* (New York: Dutton) 207 pp.; later editions include 4th, revised for UK market 1988 (London: Robert Hale) B

Morse, R. A.; Burgett, D. M.; Ambrose, J. T. (1973) Early introductions of African bees into Europe and the New World. *Bee Wld* 54(2): 57–60 B

Morse, R. A. (1974) How to make good combs. *Glean. Bee Cult.* 102(2): 38–39 B

Morse, R. A.; Steinkraus, K. H. (1975) Wines from the fermentation of honey. Chapter 16, pp. 392–407, from *Honey: a comprehensive survey* ed. E. Crane (1975*a*) B

Morse, R. A. (1976) Covers. *Glean. Bee Cult.* 104(7): 254–255 B

Morse, R. A. (ed.) *Honey bee pests, predators, and diseases* (Ithaca, NY, USA: Cornell University Press) 430 pp. B, 1016/79

Morse, R. A. (1979) *Rearing queen honey bees* (Ithaca, NY, USA: Wicwas Press) 128 pp. B, 982/81

Morse, R. A. (1982) Queen excluders. *Glean Bee Cult.* 110(3): 160–161

Morse, R. A. (1983) The economics of the beeswax market. *Am. Bee J.* 123(7): 514–515 B

Morse, R. A.; Nowogrodzki, R. (1983) Trends in American beekeeping. *Am. Bee J.* 123(5): 372–373, 375–378 B, 474/84

Morse, R. A. (1984) Honey production, harvesting, processing and marketing. Pp. 87–101 from *Proceedings of the expert consulation* . . . ed. FAO B

Morse, R. A.; Hooper, T. (eds) (1985) *The illustrated encyclopedia of beekeeping* (Sherborne, UK: Alphabooks) 432 pp. B, 1166/85

Mossedegh, M. S.; Komeili Birjandi, A. (1986) *Euvarroa sinhai* Delfinado & Baker (Acarina: Mesostigmata): a parasitic mite of *Apis florea* F. in Iran. *Am. Bee J.* 126(10): 684–685 B, 820/87

Motter, M. (1981) Alfalfa seed production in the San Joaquin Valley, California. *Bee Wld* 62(3): 111–114 B, 676/82

Mottl, J.; Štěrba, S.; Kodoň, S. (1980) *Vrby pro včelí pastvu*. (Prague: Český Svaz Včelařů) 128 pp. B, 530/82

Moulis, R. (1908) *Condition juridique des abeilles* (Thèse de doctorat)

Muid, M. (1988) Personal communication

Muir, N. (1984) A survey of the genus *Tilia*. *Plantsman* 5(4): 206–241 B

Mulder, V. (1986) *Beekeeping training in developing countries* (Wageningen, Netherlands: Agricultural University) 78 pp. B, 70/88

Mulder, V. (1988*a*) Personal communication

Mulder, V. (1988*b*) In press

Müller, E. (1938) Die Giftproduktion der Honigbiene. *Proc. VII Int. Congr. Ent.*: 1857–1864 B

Müller, E. (1950) Über Drohnensammelplätze. *Bienenvater* 75(9): 264–265 B, 180/50

Mulzac, H. C. (1979) Beekeeping in Belize. Pp. 155–163, 194 from *Beekeeping in rural development* ed. Commonwealth Secretariat & IBRA B

Murphy, A. P. (1980) Mechanical hive handling. *Australas. Beekpr* 81(9): 194–197 B

Murray, J. A. (1964) A case of multiple bee stings. *Centr. Afr. J. Med.* 10(7): 249–251 B, 502/67

Murrell, D.; Henley, B. (1988) Drying honey in a hot room. *Am. Bee J.* 128(5): 347–351 B

Muttoo, R. (1939) Editorial: [*Apis dorsata*]. *Indian Bee J.* 1(1): 1, 3, 43 B

Muttoo, R. N. (1952/53) Tree apiaries. *Indian Bee J.* 14: 87–89; 15: 25–30, 47 B, 202/58

Muttoo, R. N. (1957) Some so-called 'peculiarities' of behaviour of Indian honeybees as compared to the European bees. *Indian Bee J.* 19(3/4): 62–64 B, 317/59

Mwangi, R. W. (1985) Reasons for the low occupancy of hives in Kenya. *Proc. 3 Int. Conf. Apic. trop. Climates*: 61–63 B, 184/86

Myeong Lyeol Lee; Seung Yoon Choi (1986) Biometrical studies on the variation of some morphological characters in Korean honeybees, *Apis cerana* and *A. mellifera* L. *Korean J. Apic.* 1(1): 5–23 B

Naim, M.; Bisht, D. S. (1972) A simple method of disinfection of honeybee combs against wax moth. *Indian Bee J.* 34(3/4): 70–71 B, 507/76

Naim, M.; Phadke, K. G. (1972) Observations on the frequency of extraction and the honey yield. *Indian Bee J.* 34(1/2): 24–26 B, 71/74

Nair, K. S.; Chitre, R. G. (1980) Effect of moisture and temperature on the multiplication of honey fermenting yeasts in Indian honeys. *Indian Bee J.* 42(2): 39–47 B, 292/82

Nair, K. S. (1983) Physico-chemical characteristics of rubber honey in India. *Proc. 2 Int. Conf. Apic. trop. Climates*: 676–684 B, 1003/85

Nakamura, J. (1987) [Beekeeping in Nepal. Report on the Chepang tribe development programme.] *Honeybee Sci.* 8(3): 124–133 *In Japanese* B, 78/88

Nakamura, T. (1985) Quality standards of royal jelly for medical use. *Proc. 30 Int. Apic. Congr.*: 462–464 B, 313/88

Needham, G. R. & 3 others (eds) (1988) *Africanized honey bees and bee mites* (Chichester, UK: Ellis Horwood Ltd) 572 pp. B

Nelson, D. L. (1976) Western Canada's beekeeping industry. *Can. Agric.* 21(3): 23–24 B, 133/78

Nelson, D. (1982) Herring meal pollen substitute. *Can. Beekeep.* 9(9): 147 B

Nelson, D. L.; Jay, S. C. (1982) Producing honey in the Canadian Prairies using package bees. *Bee Wld* 61(3): 110–117 B

Netshiungani, E. N. (1981) Notes on the uses of indigenous trees in Venda. *J. Dendrology* 1(1/2): 12–17 B

Neukirch, A. (1982) Dependence of the life span of the honeybee (*Apis mellifica*) upon flight performance and energy consumption. *J. comp. Physiol. B* 146(1): 35–40 B, 481/83

New, D. A. T.; New, J. K. (1962) The dances of honeybees at small zenith distances of the sun. *J. exp. Biol.* 39(2): 271–291 B, 798/63

New South Wales Department of Agriculture (1983) *Inquiry by the Industries Assistance Commission into the honey industry* (Sydney: NSW Dep. Agric.) 26 pp. B, 984/86

Newton, D. C. (1968/69) Behavioural response of honeybees to colony disturbance by smoke. *J. apic. Res.* 7(1): 3–9; 8(2): 79–82 B, 85/69, 904/70

Newton, D. C. (1971) When bees are smoked. *Am. Bee J.* 111(5): 180–182 B

New Zealand Beekeeper (1987) How three million honey bees flew from Australia to Israel. *N. Z. Beekpr* (193): 30 B

New Zealand. Laws and Statutes (1969, amended to 1983) An Act to consolidate and amend the Apiaries Act 1927. *Act* No. 53, 1969 B

New Zealand Ministry of Agriculture and Fisheries (1981) *Honey export certification manual* (Wellington: author)

Nightingale, J. M. (1976) Traditional beekeeping among Kenya tribes, and methods proposed for improvement and modernisation. *Proc. 1 Int. Conf. Apic. trop. Climates*: 15–22 B, 1251/77

Nightingale, J. (1983) *A lifetime's recollections of Kenya tribal beekeeping* (London: IBRA) 37 pp. B, 467/84

Niijima, K. (1979) Further attempts to rear coccinellids on drone powder with field observations. *Bull. Fac. Agric. Tamagawa Univ.* (19): 7–12 B, 684/81

Niiranen, V. (1985) An efficient honey mixer to make creamed or spun honey. *Am. Bee J.* 125(12): 809–811 B, 321/87

Nilsson, S.; Praglowski, J.; Nilsson, L. (1977) *Atlas of airborne pollen grains and spores in Northern Europe* (Stockholm: Natur och Kultur) 160 pp. B, 715/80

Nixon, H. L.; Ribbands, C. R. (1952) Food transmission in the honeybee community. *Proc. R. Soc. B* 140: 43–50 B, 147/53

Nixon, M. (1982) Preliminary world maps of honeybee diseases and parasites. *Bee Wld* 63(1): 23–42 B, 592/84

Nixon, M. (1983) World maps of *Varroa jacobsoni* and *Tropilaelaps clareae* with additional records for honeybee diseases and parasites previously mapped. *Bee Wld* 64(3): 124–131 B, 1315/82

Nixon, M. (1985) Personal communication

Nogge, G. (1974) Die geographische Verbreitungsgrenze zwischen Westlicher und Östlicher Honigbiene. *Allg. dt. Imkerztg* 8(7): 163–165 B, 637/75

Nogueira-Neto, P. (1951) Stingless bees and their study. *Bee Wld* 32(10): 73–76 B, 155/53

Nogueira-Neto, P. (1953, 1970) *A criação de abelhas indigenas sem ferrão (Meliponinae)* (São Paulo: Chácaras & Quintais) 1st ed. 280 pp., 2nd ed. 365 pp. B, 86/54, 263/72

Nogueira-Neto, P. (1987) Personal communication

Nolan, W. J. (1925) The brood-rearing cycle of the honeybee. *Bull. U. S. Dep. Agric.* No. 1349: 56 pp. B

North American Apiotherapy Society (1978–1985) *Proceedings of the Society* Vols. 1–8 1st 3 are B, 645/82, 299/83, 1012/83

Norway. Laws and Statutes (1956) *Lovhefte for birøkt i Norge* (Oslo: Norges Birøkterlag) 44 pp.

Nowakowski, J.; Morse, R. A. (1982) The behavior of honey bees in sweet corn fields in New York State. *Am. Bee J.* 122(1): 13–16 B, 1249/82

Nowogrodzki, R. (1984) Division of labour in the honeybee colony: a review. *Bee Wld* 65(3): 109–116 B

Ntenga, G. (1972) Hive development in Tanzania. *Am. Bee J.* 112(1): 20–21 B

Nye, W. P. (1962) Extra supering and shading as factors in honey production in Northern Utah. *Glean. Bee Cult.* 90(7): 396–399, 436 B, 337/65

Ochi, T. (1981) [A new method to collect propolis.] *Honeybee Sci.* 2(1): 16 *In Japanese* B

O'Connor, R.; Peck, M. L. (1978) Venoms of Apidae. Pp. 613–659 from *Arthropod venoms* ed. S. Bettini (Berlin, GFR: Springer) B

O'Connor, R.; Peck, M. L. (1980) Bee sting: the chemistry of an insect venom. *J. chem. Educ.* 57(3): 206–209 B

Oertel, E. (1940) Mating flights of queen bees. *Glean. Bee Cult.* 68: 292–293, 333 B

Oertel, E. (1966) Our changing agriculture requires changes in beekeeping. *Am. Bee J.* 106(11): 406–408 B, 681/66

Oertel, E. (1973) Solar radiation and honey production. *Glean. Bee Cult.* 99(11): 397–398 B, 631/73

Oertel, T. (1984) Bee-sting anaphylaxis: the use of medical antishock trousers. *Ann. Emerg. Med.* 13: 459–460 B

Ohtani, T. (1989) [The working style of honey bee workers.] *Honeybee Sci.* 10(2): 49–58 *In Japanese* B

Okada, I. (1986) Personal communication

Okada, I. (1988) [Three species of wax moths in Japan.] *Honeybee Sci.* 9(4): 145–149 *In Japanese* B

Olbrich, H. (1982) Zuckerrohrstoppeln als Bienenweide. *Apidologie* 13(1): 92–95 B

Oldroyd, B. (1987) Detection of 'Africanization' in imported queens. *Australas. Bkpr* 89(1): 6–10 B, 502/88

Oliver-Bever, B. (1981) Medicinal plants in tropical West Africa 1. Plants acting on the cardiovascular system. *J. Ethnopharmacol.* 5: 1–71

Olsson, J. (1989) *Apis cerana*: management methods for prevention of absconding. *Proc. 4 Int. Conf. Apic trop. Climates*: 453–457 B

O'Neal, R. J.; Waller, G. D. (1984) On the pollen harvest of the honey bee (*Apis mellifera* L.) near Tucson, Arizona (1976–1981). *Desert Plants* 6(2): 81–94, 99–109 B, 558/86

Ono, M.; Okada, I.; Sasaki, M. (1987) Heat production by balling in the Japanese honeybee, *Apis cerana japonica* as a defensive behavior against the hornet, *Vespa simillima xanthoptera* (Hymenoptera: Vespidae). *Experientia* 43(9): 1031–1032 B, 118/89

Ono, M. (1989) Bionomics of *Apis cerana japonica*. From *Beekeeping with* Apis cerana *in tropical and subtropical Asia* ed. P. G. Kevan

Ordetx, G. S. (1952) *Flora apícola de la America tropical* (Havana: Editorial Lex) 334 pp. B, 240/52

Ordetx, G. S.; Zozaya Rubio, J. A.; Millan, W. F. (1972) *Estudio de la flora apícola nacional* (Chapingo: Dirección General de Extensión Agrícola) 95 pp. B, 804/76

Ormel, G. J. (1987) *Guide concis d'apiculture avec référence speciale à l'Afrique du Nord* (The Hague: Direction de l'Enseignement Agricole) 102 pp. B

Örösi-Pál, Z. (1976) Die Verteilung der Bienen im Stock nach Altersklassen während der Wanderung. *Apidologie* 7(2): 129–138 B, 947/77

Örösi-Pál, Z. (1980) Bee lice in the Americas. *Am. Bee J.* 120(6): 438–440 B

Ortega Sada, J. L. (1987) *Flora de interes apícola y polinización de cultivos* (Madrid: Ediciones Mundi-Prensa) 149 pp. B, 1205/88

Österlund, E. (1986) Looking inside a queen cell. *Am. Bee J.* 126(2): 110–111 B

Österlund, E.; Lagerman, B. (1986) Swedish 'FreeBee' system advocates frameless supers. *Am. Bee J.* 126(5): 338–342 B, 940/87

Österreichischer Imkerbund (1986) Strahlenlastung des Honigs: Festsetzung eines Grenzwertes. *Bienenvater* 107(6): 2-pp. insert B

Otis, G. W.; Taylor, O. R. (1979) Beekeeping in the Guianas. Pp. 145–154 from *Beekeeping in rural development* ed. Commonwealth Secretariat & IBRA B, 1217/80

Ouest Apiculture (1983) Un contrat type de pollinisation. *Ouest apic.* (37): 11 B

Owen, M. D.; Bridges, A. R. (1976) Aging in the venom glands of queen and worker honey bees (*Apis mellifera* L.): some morphological and chemical observations. *Toxicon* 14(1): 1–5 B, 165/78

Owen, M. D.; Braidwood, J. L.; Bridges, A. R. (1977) Age dependent changes in histamine content of venom of queen and worker honey bees. *J. Insect Physiol.* 23(8): 1031–1035 B, 926/79

Owen, M. D. (1978*a*) Venom plenishment, as indicated by histamine, in honey bee (*Apis mellifera*) venom. *J. Insect Physiol.* 24(5): 433–437 B, 1490/79

Owen, M. D. (1978*b*) Histamine in the venom of queen and worker honey bees (*Apis mellifera* L.). Pp. 589–598 from *Toxins: animal, plant and microbial* ed. P. Rosenberg (Oxford: Pergamon Press) B, 1484/79

Owens, C. D. (1962) A scale for weighing beehives. *USDA ARS* 42–72: 10 pp. B

Owens, C. D.; Benson, C. E. (1962) Confining honey bee colonies with burlap. *Am. Bee J.* 102(7): 260–262 B

Owens, C. D.; McGregor, S. E. (1964) Shade and water for the honey bee colony. *Leafl. U. S. Dep. Agric.* No. 530: 8 pp. B

Page, R. E.; Laidlaw, H. H. (1980) Egyptian beekeeping. *Am. Bee J.* 120(11): 776–779 B

Page, R. E. (1981) Protandrous reproduction in honey bees. *Envir. Ent.* 10(3): 359–362 B, 864/82

Page, R. E.; Laidlaw, H. H. (1982) Closed population honeybee breeding. *J. apic. Res.* 21(1): 30–37, 38–44 B, 1031, 1032/82

Pain, J. & 3 others (1962) Chemistry and biological activity of the secretions of queen and worker honey bees (*Apis mellifica* L.). *Comp. Biochem. Physiol.* 6: 233–241 B, 75/64

Palmer, D. J. (1961) Extraction of bee venom for research. *Bee Wld* 42(9): 225–226 B, 920/63

Paltrinieri, G. (1987) Personal communication

Pandey, R. S. (1977) Behavior of the Indian honey bee in double brood chamber hives. *Am. Bee J.* 117(10): 627 B, 1115/78

Pankiw, P. (1974) A comparison of queens from New Zealand and California for production of honey and package bees in Canada. *Bee Wld* 55(4): 141–145 B, 90/75

Papadopoulo, P. (1964) Bee houses. *Rhodesia agric. J.* 61(5): 106–107 B

Papadopoulo, P. (1965) Greek basket hives. *Rhodesia agric. J.* 62(2): 25–27 B

Papadopoulo, P. (1967) Apiary. *Rhodesia agric. J.* 61(4): 2 pp. B

Papadopoulo, P. (1987) Personal communication

Park, O. W. (1922) Time and labour factors in honey and pollen gathering. *Am. Bee J.* 62: 254–255 B

Park, O. W. (1925) The storing and ripening of honey by honeybees. *J. econ. Ent.* 18(2): 405–410 B

Park, O. W. (1927) Studies on the evaporation of nectar. *J. econ. Ent.* 20: 510–516 B

Park, O. W. (1928*a*) Further studies on the evaporation of nectar. *J. econ. Ent.* 21: 882–887 B

Park, O. W. (1928*b*) Time factors in relation to the acquisition of food by the honeybee. *Res. Bull. Iowa agric. Exp. Stn* No. 108 B

Park. O. W. (1946) Activities of honeybees. Chapter 5, pp. 125–206, from *The hive and the honey bee* ed. R. A. Grout (Hamilton, UL, USA: Dadant & Sons) B

Parker, C. (1986) A workforce that keeps profits sweet. *Beeline, Zimbabwe* 23(2): 3–8 B

Parker, F. D. (1985) A candidate legume pollinator, *Osmia sanrafaelae* Parker (Hymenoptera: Megachilidae). *J. apic. Res.* 24(2): 132–136 B, 342/86

Parker, F. D.; Frohlich, D. R. (1985) Studies on management of the sunflower leafcutter bee *Eumegachile pugnata* (Say) (Hymenoptera: Megachilidae). *J. apic. Res.* 24(2): 125–131 B, 341/86

Paterson, C. R.; Palmer-Jones, T. (1954) A vacuum plant for removing excess water from honey. *N. Z. J. Sci. Tech. Soc. A* 36(4): 386–400 B, 167/55

Paterson, C. R.; Palmer-Jones, T. (1955) Vacuum plant for removing excess water from honey. *N. Z. J. Agric.* 90(6): 571–578 B, 205/59

Paterson, P. D. (1986) Personal communication

Paterson, P. D. (1988) A Langstroth hive with top-bars instead of frames, for tropical African bees. *Bee Wld* 69(2): 63–68 B, 1238/88

Patetta, A.; Manino, A. (1978) Lavori gas-cromatografici sull'analisi glucidica dei mieli. *Cronache Chim.* (57): 9–13 B, 271/81

Patty, G. E. (1978) The honey industry of Mexico: situation and prospects. *USDA FAS* M-285 B

Paust, G. (1975) Income variation in the Western Australian beekeeping industry. *Farm Policy* 15(3): 85–89 B, 219/78

Paysen, J. (1977) JZs BZs honey harvest system. *Am. Bee J.* 117(2): 98–103 B

Paysen, J. (1987) A method of drying honey on a commercial scale. *Am. Bee J.* 127(4): 273–282 B, 1340/87

Pchelovodstvo (1988*a*) [Co-operation in beekeeping: the New Rules in full, with commentary by O. N. Kremeznoi.] *Pchelovodstvo* (6): 4–7 *In Russian* B

Pchelovodstvo (1988*b*) [How to make your own bee suit.] *Pchelovodstvo* (6): opp. 32 *In Russian* B

Pech, E. (1958) *Das Bienenrecht in der Deutschen Demokratischen Republik* (Berlin, DDR: Deutscher Bauernverlag) 116 pp. B, 7/60

Pechhacker, H.; Hüttinger, E. (1986) Die Gewinnung von Propolis mit Pressluft. *Bienenvater* 107(5): 160–161 B, 1043/87

Peer, D. (1984) Requeening with queen cells – a new techique. *Apiarist* (40): 3 B

Pelimon, C.; Baculinschi, H. (1955) Cercetări asupra compozitiei mierii din RPR. *Anal. Inst. Cerc. zoot.* 13: 621–637 B, 215/60

Pellett, F. C. (1938) *History of American beekeeping* (Ames, IA, USA: Collegiate Press) 213 pp. B

Pellett, F. C. (1947) *American honey plants* (Hamilton, IL, USA: Dadant & Sons) 467 pp. Reprinted 1976 B

Pence, R. J. (1981) Methods for procuring and bioassaying intact honey bee venom for medical use. *Am. Bee J.* 121(10): 726–731 B, 1399/82

Penfold, A. R.; Willis, J. L. (1961) *The eucalypts* (London: Leonard Hill) 550 pp. B, 539/62

Peng, Y.-S. & 3 others (1985) The digestion of dandelion pollen by adult worker honeybees. *Physiol. Ent.* 10(1): 75–82 B, 535/86

Peng, Y.-S.; Marston, J. M. (1986) Filtering mechanism of the honey bee proventriculus. *Physiol. Ent.* 11(4): 433–439 B, 551/87

Peng, Y.-S. & 3 others (1987) The resistance mechanism of the Asian honeybee, *Apis cerana* Fabr., to an ectoparasitic mite *Varroa jacobsoni* Oudemans. *J. Invert. Path.* 49: 54–60 B, 463/88

Percival, M. S. (1950) Pollen presentation and pollen collection. *New Phytol.* 49(1): 40–63 B, 62/51

Percival, M. S. (1961) Types of nectar in angiosperms. *New Phytol.* 60: 235–281 B, 844/64

Percival, M. S. (1965) *Floral biology* (Oxford: Pergamon Press) 243 pp. B, 650/65

Persano, A. L. (1980) *Apicultura práctica* (Buenos Aires: Editorial Hemisfero Sur) 300 pp. B

Pesante, D.; Rinderer, T. E.; Collins, A. M. (1987) Differential nectar foraging by Africanized and European honeybees in the neotropics. *J. apic. Res.* 26(4): 210–216 B, 501/89

Pesson, P.; Louveaux, J. (eds) (1984) *Pollinisation et productions végétales* (Paris: Institut National de la Recherche Agronomique) 663 pp. B, 668/85

Pesti, J. (1976) Daily fluctuations in the sugar content of nectar and periodicity of secretion in the Compositae. *Acta agron. hung.* 25(1/2): 5–17 B, 969/79

Peterson, M. (1985) African honeybees in East and West Africa, and Africanized honeybees in Venezuela: some observations on behaviour. *Proc. 3 Int. Conf. Apic. trop. Climates:* 109–111 B

Petkov, V. (1966) [Study on some legume-phacelia mixtures as forage plants and honey plants.] *Rastenievudni Nauki* 3(8): 127–133 *In Bulgarian* B, 492/76

Pflumm, W. (1968) Zum Verhalten nektarsammelnder Honigbienen an der Futterquelle. *Verh. dt. zool. Ges., Innsbruck* 37: 381–387 B, 115/72

Phadke, R. P. (1961) Some physico-chemical constants of Indian beeswaxes. *Bee Wld* 42(6): 149–153 B, 908/63

Phadke, R. P. (1962) Physico-chemical composition of major unifloral honeys from Mahabaleshwar (Western Ghats). *Indian Bee J.* 24(7/9): 59–65 B, 907/64

Phadke, R. P. (1967*a*) Studies on Indian honeys. 1. Proximate composition and physico-chemical characteristics of Indian multifloral apiary honeys from *Apis indica* bees. *Indian Bee J.* 29: 14–26 B, 478/72

Phadke, R. P. (1967*b*) Studies on Indian honeys. 2. Proximate composition and physico-chemical characteristics of unifloral honeys of Mahabaleshwar. *Indian Bee J.* 29: 33–46 B, 479/72

Phadke, R. P. (1968) Studies on Indian honeys. 3. Proximate composition and physico-chemical characterizations of honeys from the wild honey bees *Apis dorsata*, *Apis florea* and *Trigona*. *Indian Bee J.* 30(1): 3–8 B, 481/72

Phadke, R. P.; Nair, K. S.; Nandedkar, K. U. (1969) Indian beeswaxes. 1. Their physico-chemical constants. *Indian Bee J.* 31(2): 52–55 B, 812/72

Phadke, R. P.; Nair, K. S. (1970) Standards for Indian honeys and bee-waxes. *Indian Bee J.* 32(3/4): 68–74 B, 514/72

Phadke, R. P.; Nair, K. S.; Nandedkar, K. U. (1970) Studies on Indian honeys. IV. Minor constituents. *Indian Bee J.* 32(1/2): 28–35 B, 1056/72

Phadke, R. P.; Nair, K. S.; Nandedkar, K. U. (1971) Indian bees-waxes. 2. The nature of their chemical constituents. *Indian Bee J.* 33(1/2): 3–5 B, 438/73

Phadke, R. P.; Nair, K. S. (1973) Studies on Indian honeys. 5. Distinguishing characteristics of the apiary honey from the wild variety. *Indian Bee J.* 35(1–4): 36–39 B, 281/77

Phadke, R. P.; Phadke, R. S. (1975) Physico-chemical characteristics of waxes from Indian honey bees. II. Viscosity and refractive index. *Indian Bee J.* 37(1–4): 15–18 B, 1111/78

Phadke, R. P. (1982, 1987) Personal communications

Phillips, E. F. (1922) Beekeeping in the tulip-tree region. *Fmrs' Bull. US Dep. Agric.* No. 1222: 28 pp. B

Phiri, L. (1985) Personal communication

Phoon, A. C. G. (1983) Beekeeping in Malaysia. *Pertanika* 6 (Rev. Suppl.): 3–17 B

Pidek, A. (1977) Opłacalność produkcji pszczelarskiej w wyodrębnionych modelach pasiek. *Pszcz. Zesz. Nauk.* 21: 1–29; see also 31–43 B, 1362, 1363/79

Pinchinat, B.; Bilinski, M.; Ruszkowski, A. (1979) Possibilities of applying bumble bees as pollen vectors in tomato F1 hybrid seed production. *Proc. 4 Int. Symp. Poll.*: 73–90 B, 778/80

Pirker, H. J. (1978*a*) Package bee production in northern Canada. *Am. Bee J.* 118(1): 14–16 B

Pirker, H. J. (1978*b*) *First year* [guide to indoor wintering] (Debolt, Alberta: author)

Piyapidrat, S. (1986) Personal communication

Platt, J. L.; Ellis, J. R. B. (1985) Removing water from honey at ambient pressure. *US Patent* 4,536,973: 6 pp.

Platt, J. L. (1988) Personal communication

Plowright, R. C.; Jay, S. C. (1966) Rearing bumble bees in captivity. *J. apic. Res.* 5(3): 155–165 B, 827/67

Popescu, H.; Giurgea, R.; Polinicencu, C. (1985) *Extractul de propolis standardizat şi medicamentele Candiflor* (Bucharest: Centralá Industrială de Medicamente, Cosmetice, Coloranti, şi Lacuri) 226 pp. B, 1011/86

Popolizio, E. R.; Pailhe, L. A. (1973) Storing combs in 'wax-moth-safe' storage rooms. *Proc. 24 Int. Apic. Congr.*: 381–382 B

Poppleton, O. O. (1898) The advantages of single-storey hives over the double-deckers. *Glean. Bee Cult.* 26(9): 344–345 B

Popravko, S. A.; Gurevich, A. L.; Kolosov, M. N. (1969) [Flavonoid components of propolis.] *Khimiya Prir. Soedin.* 5(6): 476–482 *In Russian* B, 393/73

Popravko, S. A. (1976) [Plant sources of propolis.] *Pchelovodstvo* 96(7): 38–41 *In Russian* IBRA English translation E1612 B

Popravko, S. A. (1977) [Chemical organization of propolis and its standardization.] *Pchelovodstvo* 97(8): 21–23 *In Russian* IBRA English translation E1537 B, 1472/79

Portugal Araújo, V. de (1955*a*) Colmeias para 'abelhas sem ferrão' (Meliponini). *Bol. Inst. Angola* No. 7: 49 pp. B, 76/58

Portugal Araújo, V. de (1955*b*) Notas sôbre colônias de Meliponineos de Angola – Africa. *Dusenia* 6(3/4): 97–114 B, 45/65

Portugal Araújo, V. de (1960) A colmeia Dadant africana. *Ser. divulg. Agron. angolona* No. 17: 25 pp. B

Portugal Araújo, V. de (1974) *Apiários e instalações apícolas na extensão rural (tecnologia apícola) planalto central de Angola* (Nova Lisboa, Angola: Estado de Angola) 110 pp.

Powers, H. P. (1979) Plant pollination – a summary of state laws. *Proc. 4 Int. Symp. Poll.*: 446–474 B

Praagh, J. P. van (1972) Towards a controlled-environment room suitable for normal colony life of honeybees 1. Description and general observations. *J. apic. Res.* 11(2): 77–87 B, 100/73

Praagh, J. P. van (1975) Die Feuchtigkeit der Stockluft und die Bruttätigkeit der Bienen (*Apis mellifica* L.) in einem Flugraum. *Apidologie* 6(3): 283–293 B, 196/77

Praagh, J. P. van (1987) Personal communication

The Prairie beekeeping manual (1986) (Saskatchewan, Manitoba & Alberta Agriculture) 98 pp. rev. ed. B

Pratt, C. R.; Sikorski, R. S. (1982) Lead content of wild flowers and honey bees (*Apis mellifera*) along a roadway: possible contamination of a simple food chain. *Proc. Pa Acad. Sci.* 56(2): 151–152 B, 257/85

Prehn, D. (1986) Personal communication

Prime, F. J. (1958) [A bee sting remedy.] *Lancet* i: 378

Prokof'eva, L. V.; Savel'eva, G. B. (1984) [Factors affecting production costs and how to diminish them.] *Pchelovodstvo* (7): 3–5 *In Russian* B, 219/86

Pryce-Jones, J. (1944) Some problems associated with nectar, pollen and honey. *Proc. Linn. Soc. Lond.* 155: 129–174 B

Pryce-Jones, J. (1950) The composition and properties of honey. *Bee Wld* 31(1): 2–6 B, 219/52

Pryce-Jones, J. (1953) The rheology of honey. Pp. 148–176, 243–256 from *Foodstuffs: their plasticity and consistency* ed. G. W. Scott-Blair (Amsterdam: North-Holland Publishing Co.) B, 110/54

Prŷs-Jones, D. E.; Corbett, S. A. (1987) *Bumblebees* (Cambridge: University Press) 85 pp. B, 408/88

Punchihewa, W. & 3 others (1985) Observations on the dance communication and natural foraging ranges of *Apis cerana*, *Apis dorsata* and *Apis florea* in Sri Lanka. *J. apic. Res.* 24(3): 168–175 B, 495/86

Punchihewa, W. (1988) Personal communication

Purdie, J. D.; Doull, K. M. (1964) Beehive modifications for Australian conditions. *Australas. Beekpr* 66(5/6): 128–133, 159–162 B, 513/66

Pym, S. (1952) *A hive of suspects: an Irish detective story* (London: Hodder & Stoughton) 255 pp.

Pyramarn, K.; Wongsiri, S. (1986) Bee flora of four species of *Apis* in Thailand. *J. Sci. Res. Chula. Univ.* 11(2): 95–103 B

Quick, W. J. (1919) *Bee keeping* (Washington, DC: Federal Board for Vocational Education) 32 pp. Opportunity Monograph No. 37 B

Quiniones, A. C. (1988) Personal communication

Rabiet, E. (1984) *Choix et culture des plantes apicoles* (Jonzac, France: author) 422 pp. B, 903/86

Racal Safety Ltd (1986) Dustmaster is buzzing in UK apiary. *Australas. Beekpr* 87(12): 262–263 B

Rahman, K. A.; Singh, S. (1946) Size of the cell of the brood comb of the Indian honey bees. *Indian Bee J.* 8(11/12): 154–156 B

Rahman, K. A.; Singh, S. (1947) Preliminary studies on the bionomics of the Indian honeybee *Apis indica* F. *Indian Bee J.* 9(1/2): 6–8 B

Ramírez B., W. (1976) Temperaturas internas encajas de colmena de madera y asbesto-cemento. *Agroindustria* 5(10): 20–21 B, 912/78

Ramírez B., W.; Pontigo, M. (1979) Humedad relativa y temperatura interna en colmenas de madera y asbesto-cemento. *Agron. Costarricense* 3(1): 57–60 B, 225/81

Ramírez B., W. (1980) Producción anual de polen por una colmena en el bosque humedo premontano Costarricense. *Agron. Costarricense* 4(1): 111–113 B, 1462/81

Ramírez B., W.; Otis, G. W. (1986) Developmental phases in the life cycle of *Varroa jacobsoni*, an ectoparasitic mite on honeybees. *Bee Wld* 67(3): 92–97 B, 287/87

Ramsay, J. (1987) *Plants for beekeeping in Canada and the northern USA* (London: IBRA) 200 pp. B, 559/88

Rana, B. S. (1985) *Biological and economic characteristics of* Apis cerana indica *F. in relation to selection and breeding* (Himachal Pradesh University: Dissertation) 153 pp. B, 71/87

Rank, G. H. (ed.) (1982) *Proceedings of the First International Symposium on Alfalfa Leafcutting Bee Management* (Saskatoon, Canada: University of Saskatchewan) 281 pp. B, 1039/84

Ransome, H. M. (1937) *The sacred bee in ancient times and folklore* (London: George Allen & Unwin Ltd) 308 pp. Facsimile reprint 1986 (Burrowbridge: Bee Books New and Old) B, 92/88

Rashad, S. E. (1985) *Utilization of non-*Apis *bees as crop pollinators. Final Report* (Cairo: Cairo University) 144 pp. B, 21/87

Ratia, G. (1984) *Rapport de mission sur l'apiculture seychelloise* (Mahe: Ministère du Développement National) 129 pp. B, 797/88

Ratia, G. (1986) Personal communication

Ratnieks, F. L. W. (1986) *Effect of colony population size on the efficiency of nectar collection and honey production in honey bee* (Apis mellifera) *colonies* (Cornell University: MSc thesis) 88 pp. B, 582/87

Rawat, B. S. (1982) *Bee farming in India* (Ranikhet: Rawat Apiaries) 258 pp. B

Redma Consultants Ltd (1981) *The development of beekeeping and hive products in the Caribbean community* (Toronto: author) 228 pp. B

Reed, A. D.; Horel, L. A. (1976) Bee industry economic analysis for California. *Leafl. Div. agric. Sci. Univ. Calif.* No. 2345: 20 pp. B

Reich, J. J.; Reich, F. P. (1969) Method of repelling bees. *US Patent* 3,456,056 B, 774/69

Reid, M. (1975) Storage of queen honeybees. *Bee Wld* 56(1): 21–31. IBRA Reprint M83 B

Reid, M. (1976*a*) New hive cracker and lifter. *N. Z. Beekpr* 38(4): 9 B

Reid, M. (1976*b*) Palletised beehives speed site changing. *N. Z. Jl. Agric.* 133(4): 15, 17–18 B, 1257/78

Reid, M. (1981) Removing bees from honey houses. *N. Z. Beekpr* 42(2): 11 B

Reid, M.; Matheson, A.; Walton, G. (1988) *Bibliography of New Zealand apiculture* (Tauranga: Ministry of Agriculture & Fisheries) 140 pp. B

Rembold, H.; Lackner, B. (1978) Vergleichende Analyse von Weiselfuttersäften. *Mitt. dt. Ges. allg. angew. Ent.* 1(2/4): 299–301 B, 164/81

Rennie, J.; White, P. B.; Harvey, E. J. (1921) Isle of Wight disease in hive bees. *Trans. R. Soc. Edinb.* 52, IV(29): 737–779 B

Resch, K. (1974) Weiden (*Salix*): eine zusammenfassende Darstellung. *Bienenwelt* 16(10): 202–208 B

Ribbands, C. R. (1950) Autumn feeding of honeybee colonies. *Bee Wld* 31(10): 74–76 B, 109/52

Ribbands, C. R. (1953) *The behaviour and social life of honeybees* (London: Bee Research Association) 352 pp. B, 8/54
Ribi, W. A. (1987) The structural basis of information processing in the visual system of the bee. Pp. 130–140 from *Neurobiology and behavior of honeybees* ed. R. Menzel & A. Mercer B
Ricciardelli d'Albore, G.; Persano Oddo, L. (1978) *Flora apistica italiana* (Florence: Istituto Sperimentale per la Zoologia Agraria) 286 pp. B, 965/79
Ricciardelli d'Albore, G.; Tonini d'Ambrosio, M. (1981) Observations on the pollinators of the bay tree (*Laurus nobilis*). *Proc. 28 Int. Apic. Congr.*: 402–403 B
Ricciardelli d'Albore, G. C.; Battaglini Bernardini, M.; Isidoro, N. (1987) Sullo sviluppo delle ghiandole ipofaringee in api nutrite con pollini e sostituti. *Apicoltura* 3: 15–36 B
Richards, K. W. (1984) *Alfalfa leafcutter bee management in Western Canada* (Ottawa: Agriculture Canada) 54 pp. Publication 1495E, rev. ed. B
Richards, K. W. (1987) Alfalfa leafcutter bee management in Canada. *Bee Wld* 68(4): 168–178 B
Richards, O. W.; Davies, R. G. (1977) *Imm's general textbook on entomology* (London: Chapman & Hall) 418 + 936 pp. 10th ed. B, 711, 712/78
Richardson, J. A. (1977) High-performance plant species in reclamation. Pp. 148–172 from *Landscape reclamation practice* ed. B. Hackett (Guildford, UK: IPC Science & Technology Press) B, 974/79
Riches, H. R. C. (1982) Hypersensitivity to bee venom. *Bee Wld* 63(1): 7–22. IBRA Reprint M109 B, 1236/82
Riches, H. R. C. (1987) Honey and hay fever. *Bee Craft* 69(8): 12 B, 307/88
Riches, H. R. C. (1989) Bee venom sensitivity update. *Bee Wld* 70(1): 12–18 B
Rieth, J. P.; Levin, M. D. (1988) The repellent effect of two pyrethroid insecticides on the honey bee. *Physiol. Ent.* 13(2): 213–218 B, 1316/88
Rinderer, T. E.; Baxter, J. R. (1978) Effects of empty comb on hoarding behavior and honey production of the honey bee. *J. econ. Ent.* 71(5): 757–759 B, 958/79
Rinderer, T. E.; Baxter, J. R. (1980) Honey bee hoarding of high fructose corn syrup and cane sugar syrup. *Am. Bee J.* 120(12): 817–818 B, 151/82
Rinderer, T. E. (1981) Volatiles from empty comb increase hoarding by honey bees. *Anim. Behav.* 29(4): 1275–1276 B, 208/84
Rinderer, T. E. (1982) Regulated nectar harvesting by the honeybee. *J. apic. Res.* 21(2): 74–87 B, 499/83
Rinderer, T. E. & 3 others (1984) Nectar-foraging characteristics of Africanized and European honeybees in the neotropics. *J. apic. Res.* 23(2): 70–79 B, 499/85
Rinderer, T. E.; Collins, A. M.; Tucker, K. W. (1985*a*) Honey production and underlying nectar harvesting activities of Africanized and European honeybees. *J. apic. Res.* 24(3): 161–167 B, 569/86
Rinderer, T. E. & 3 others (1985*b*) Male reproductive parasitism: a factor in the Africanization of European honey-bee populations. *Science, N.Y.* 228: 1119–1121 B, 571/86
Rinderer, T. E. (ed.) (1986) *Bee genetics and breeding* (New York: Academic Press) 425 pp. B, 191/88
Rinderer, T. E. & 5 others (1986) Field and simplified techniques for identifying Africanized or European honey bees. *Apidologie* 17(1): 33–48 B, 160/87
Rinderer, T. E. (1987) Personal communication
Rinderer, T. E. & 6 others (1987*a*) Improved simple techniques for identifying Africanized and European honey bees. *Apidologie* 18(2): 179–197 B, 137/88
Rinderer, T. E. & 5 others (1987*b*) The proposed Honey-Bee Regulated Zone in Mexico. *Am. Bee J.* 127(3): 160–164 B, 1213/87
Rinderer, T. E. & 3 others (1987*c*) Differential drone production by Africanized and European honey bee colonies. *Apidologie* 18(1): 61–68 B, 543/88
Rinderer, T. E. (1988) The rediscovery of *Apis koschevnikovi*. *Am. Bee J.* 128(12): 807 B
Rindfleisch, J. (1980) A case for meliponiculture in pollination. *Am. Bee J.* 120(6): 468–470 B
Ritter, W.; Ruttner, F. (1980) Neue Wege in der Behandlung der Varroatose. Ameisensäure – Labor- und Freilandversuche. *Allg. dtsch. Imkerztg* 14(5): 151–153 B, 1033/81
Ritter, W. (1981) Varroa disease of the honeybee *Apis mellifera*. *Bee Wld* 62(4): 141–153. IBRA Reprint M106 B, 590/82
Ritter, W.; Delaître, N.; Ifantidis, M. (1984) Use of Folbex-VA in smoker to control varroa disease. *Apiacta* 19(2): 37–39 B, 969/85
Ritter, W. (1986) Versuche zur Entwicklung und Prüfung von Perizin, einem systemischen Medikament zur Bekämpfung der Varroatose der Honigbiene. *Tierärztl. Umsch.* 41(2): 105–108, 110, 112 B, 1296/86
Robb, D. (1951) The Robb cross-sectional observation hive. *Scot. Beekpr* 27(11): 216–217 B
Roberts, D. (1957) A plant for treating honey by the vacuum process. *N. Z. Beekpr* 19(3): 31–35 B
Roberts, D. (1958) A hive barrow aids handling of hives and supers. *N. Z. Jl. Agric.* 96(1): 82–84 B, 203/59
Robinson, F. A.; Nation, J. L. (1970) Long-chain fatty

acids in honeybees in relation to sex, caste, and food during development. *J. apic. Res.* 9(3): 121–127 B, 669/72

Robinson, F. A.; Oertel, E. (1975) Sources of nectar and pollen. Chapter 9, pp. 283–302, from *The hive and the honey bee* ed. Dadant & Sons B

Robinson, G. E.; Underwood, B. A.; Henderson, C. E. (1984) A highly specialized water-collecting honey bee. *Apidologie* 15(3): 355–358 B, 874/85

Robinson, G. E. (1986) The dance language of the honey bee: the controversy and its resolution. *Am. Bee J.* 126(3): 184–189 B

Robinson, G. E.; Page, R. E. (1988) Genetic determination of guarding and undertaking in honey-bee colonies. *Nature, Lond.* 333(6171): 356–358 B

Robinson, P. J.; French, J. R. J. (1986) Wood preservation in Victorian commercial apiaries; Beekeeping and wood preservation in Australia. *Aust. Bee J.* 67(1): 8–10; 11–14 B, 944, 945/87

Robson, C. H. (1986) Personal communication

Rocca, Abbot della (1790) *Traité complet sur les abeilles* . . . (Paris: Bleuet Père) 3 vols. 1494 pp. B

Rodgers, P. E. W. (1975) Honey quality control. Chapter 12, pp. 314–325, from *Honey: a comprehensive survey* ed. E. Crane (1975*a*) B

Rodrigues, L. (1986) Lightning strikes hive. *Am. Bee J.* 126(1): 47 B

Rodriguez López, C. (1985) Determinación espectrofotométrica del color de las mieles. *Vida apíc.* (16): 24–29 B, 1349/86

Roff, C. (1975) Honeybees, giant toads and hive stands. *Qd. agric. J.* 101(6): 689–691 B

Root, A. I.; Root, E. R. (1940) *The ABC and XYZ of bee culture* (Medina, OH, USA: A. I. Root Co.) 873 pp. B

Rósario-Nuñes, J. F. do (1973) Cortico melhorado com guias. Nova contribuição para o fomento da apicultura africana. Pp. 271–364 from *Livro de Homenagem ao Prof. Fernando Frade* (Lisbon: Junta de Investigações do Ultramar)

Rösch, G. A. (1925, 1930) Untersuchungen über die Arbeitsteilung im Bienenstaat. *Z. vergl. Physiol.* 2: 571–631; 12: 1–71 B

Rösch, G. A. (1927) Über die Bautätigkeit im Bienenvolk und das Alter der Baubienen. *Z. vergl. Physiol.* 6: 265–298 B

Rosin, R. (1988) Do honey bees still have a 'dance language'? *Am. Bee J.* 128(4): 267–268 B

Rossel, S.; Wehner, R. (1986) Polarization vision in honeybees. *Nature, Lond.* (323): 128–131 B, 491/88

Rossel, S.; Wehner, R. (1987) The bee's e-vector compass. Pp. 76–83 from *Neurobiology and behavior of honeybees* ed. R. Menzel & A. Mercer B

Rothenbuhler, W. C. (1964) Behaviour genetics of nest cleaning in honey bees IV. *Am. Zoologist* 4: 111–123; also papers I–III referred to there B, 166/66

Rothenbuhler, W. C.; Thompson, V. C.; McDermott, J. J. (1968) Control of the environment of honeybee observation colonies by the use of hive-shelters and flight-cages. *J. apic. Res.* 7(3): 151–155 B, 152/70

Rothenbuhler, W. C.; Kulinčević, J. M.; Thompson, V. C. (1979) Successful selection of honeybees for fast and slow hoarding of sugar syrup in the laboratory. *J. apic. Res.* 18(4): 272–278 B, 1318/80

Roubik, D. W.; Sakagami, S. F.; Kudo, I. (1985) A note on distribution and nesting of the Himalayan honey bee *Apis laboriosa* Smith (Hymenoptera: Apidae). *J. Kansas ent. Soc.* 58(4): 746–749 B, 1116/86

Roubik, D. W. (1986) Personal communication

Rousseau, M. (1972) Les abeilles domestiques et la pollution de l'environnement. *Bull. Off. int. Épizoot.* 77(9/10): 1473–1480 B

RSFSR (1977) [Standard for propolis] *RST RSFSR* 317–77 *In Russian*, published in Spanish in Asis (1979)

Ruddle, K. (1973) The human use of insects: examples from Yukpa. *Biotropica* 5(2): 94–101 B, 373/76

Rudnyanskaya, E. I. (1981) [Pollen yields from plants of the Compositae family.] *Pchelovodstvo* (4/5): 26–27 *In Russian* B, 169/82

Rüedi, J. P. (1955) *Rechtsfragen des Imkers* (Bern: Fachschriften Verlag des VDSB) 83 pp. B

Ruijter, A. de (1982) Tobacco smoke can kill varroa mites. *Bee Wld* 6(3): 138 B

Ruiz-Argueso, T.; Rodriguez-Navarro, A. (1975) Microbiology of ripening honey. *Appl. Microbiol.* 30(6): 893–896 B, 510/78

Ruttner, F. (1956) The mating of the honeybee. *Bee Wld* 57: 3–15, 23–24 B, 153/58

Ruttner, F. (1966) The life and flight activity of drones. *Bee Wld* 47(3): 93–100 B, 78/67

Ruttner, F. (ed.) (1975*a*) *Die instrumentelle Besamung der Bienenkönigin* (Bucharest: Apimondia International Institute for Bee Technology and Science) 122 pp. 2nd ed. B, 822/76

Ruttner, F. (1975*b*) African races of honeybees. *Proc. 25 Int. Apic. Congr.*: 325–344 B, 495/78

Ruttner, F.; Koeniger, N. (eds) (1977) Bienenhaltung in Flugräumen: Symposium über Flugraumtechnik. *Insectes soc.* 24(3): 279–302 B, 827/78

Ruttner, F. (1979) *Historische Entwicklung des Bienenstockes* (Bucharest: Apimondia Publishing House) 32 pp. B, 183/82

Ruttner, F. (ed.) (1983) *Queen rearing: biological basis and technical instruction* (Bucharest: Apimondia Publishing House) 358 pp. Translated from *Königinnen-*

zucht: biologische Grundlagen und technische Anleitungen (1980) B, 203/82

Ruttner, F.; Maul, V. (1983) Experimental analysis of reproductive interspecific isolation of *Apis mellifera* L. and *Apis cerana* Fabr. *Apidologie* 14(4): 309–327 B, 786/85

Ruttner, F. (1985) Characteristics and variability of *Apis cerana* (Fabr.) *Proc. 30 Int. Apic. Congr.*: 130–133 B, 52/88

Ruttner, F.; Pourasghar, D.; Kauhausen, D. (1985) Die Honigbienen des Iran. *Apidologie* 16(2): 119–137, 241–264 B, 78, 148/87

Ruttner, F. (1986) Geographical variability and classification. Chapter 2, pp. 23–56, from *Bee genetics and breeding* ed. T. E. Rinderer B, 44/48

Ruttner, F. (1988) *Biogeography and taxonomy of honeybees* (Berlin: Springer-Verlag) 284 pp. B, 1155/88

Ruttner, H. (1962) The percolation feeder. *Bee Wld* 43(1): 7–11 B, 839/63

Ruttner, H.; Ruttner, F. (1972) Untersuchungen über die Flugaktivität und das Paarungsverhalten der Drohnen V. Drohnensammelplätze und Paarungsdistanz. *Apidologie* 3(3): 202–232 B, 579/73

Ruttner, H. (1976) Untersuchungen über die Flugaktivität und das Paarungsverhalten der Drohnen VI. Flug auf und über Höhenrücken. *Apidologie* 7(4): 331–341 B, 191/78

Ryan, J. K.; Jelan, P.; Sauer, W. C. (1983) Alkaline extraction of protein from spent honey bees. *J. Food Sci.* 48(3): 886-888 B, 1328/84

Rye, B. (1986) Wasp control for beekeepers. *Aust. Bee J.* 67(5): 14–16 B

Sadovnikov, A. (1981) [How to obtain propolis.] *Pchelovodstvo* (1/2): 47–48 *In Russian* B, 640/82

Saito, Y. & 4 others (1980) [Detection of poisonous substances in honey which causes food poisoning.] *Bull. natn. Inst. hyg. Sci., Tokyo* (98): 32–35 *In Japanese* 1322/83

Sakagami, S. F. (1966) Techniques for the observation of behaviour and social organization of stingless bees by using a special hive. *Papers Dep. Zool. S. Paulo* 19: 151–162 B, 60/68

Sakagami, S. F.; Fukuda, H. (1968) Life tables for worker honeybees. *Researches Popul. Ecol. Kyoto Univ.* 10(2): 127–139 B, 121/71

Sakagami, S. F.; Matsumura, T.; Ito, K. (1980) *Apis laboriosa* in Himalaya, the little known world largest honeybee (Hymenoptera, Apidae). *Insecta Matsumurana* 19: 47–77 B, 469/81

Sakagami, S. F. (1982) Stingless bees. Pp. 362–423 from *Social insects* Vol. III, ed. H. R. Hermann (London: Academic Press) B, 787/83

Sakai, T.; Matsuka, M. (1982) Beekeeping and honey resources in Japan. *Bee Wld* 63(2): 63–71 B

Sakai, T.; Sasaki, M.; Tanaka, K. (1983) [Utilization of pollen loads from honeybees for artificial pollination of fruit trees, with special reference to the washing method used and long-term preservation.] *Honeybee Sci.* 4(2): 81–82 *In Japanese* B, 1021/84

Salem, S. N. (1981) Honey regimen in gastrointestinal disorders. *Bull. Islamic Med.* 1: 358–362 B, 276/85

Salmon, R. E.; Szabo, T. I. (1981) Dried meal as a feedstuff for growing turkeys. *Can. J. Anim. Sci.* 61: 965–968 B, 1383/82

Salvi, S. R. (1975) Seasonal management of the Indian hive bees at Pathankot, Punjab: a preliminary report. *Indian Bee J.* 37(1/4): 10–14 B, 1122/78

Sammataro, D.; Erickson, E. H.; Garment, M. B. (1985) Ultrastructure of the sunflower nectary. *J. apic. Res.* 24(3): 150–160 B, 370/86

Sammataro, D.; Avitabile, A. (1986) *The beekeeper's handbook* (New York: Macmillan) 148 pp. 2nd ed. B

Samsinak, K.; Vobrazkova, E.; Haragsim, O. (1978) *Melittiphis alvearius* Berlese, a little known bee mite. *J. apic. Res.* 17(1): 50–51 B, 1304/78

Sands, D. M. (1984) *The mixed subsistence-commercial production system in the peasant economy of Yucatan, Mexico: an anthropological study in commercial beekeeping* (Cornell University: PhD Thesis) 570 pp. B, 1187/85

Sanford, M. T. (1980) Pesticides and honeybees. *Coop. Ext. Serv. Ohio St. Univ.* No. 251: 6 pp. B

Sanford, M. T. (1983) Florida beekeeping almanac. *Circ. Fla Coop. Ext. Serv.* No. 537: 18 pp. B

Sanford, M. T. (1985) USDA now recognizes 'strained' extract honey. *Speedy Bee* 14(11): 5 B

Sanford, M. T. (1986*a*) A study in profitability for a mid-sized beekeeping operation. *Circ. Fla Coop. Ext. Serv.* No. 722: 39 pp. B, 581/88

Sanford, M. T. (1986*b*) A profitability model for a mid-sized beekeeping operation. *Circ. Fla Coop. Ext. Serv.* No. 692: 47 pp. B, 582/88

Santas, L. A. (1983) Insects producing honeydew exploited by bees in Greece. *Apidologie* 14(2): 93–103 B, 889/89

Saubolle, B. R.; Bachmann, A. (1979) *Beekeeping: an introduction to modern beekeeping in Nepal* (Kathmandu: Sahayogi Prakashan) 44 pp. B, 1166/80

Sawada, M. (1984) Personal communication

Sawyer, R. (1981) *Pollen identification for beekeepers* (Cardiff: University College Cardiff Press) 112 pp. + 50 punched cards B, 1271/82

Sawyer, R. (1988) *Honey identification* (Cardiff: Cardiff Academic Press) 115 pp. B

Schelotto, B.; Cornejo, L.; Llarias, A. R. (1973) Training blind people for beekeeping. *Proc. 24 Int. Apic. Congr.*: 593–594 B

Schepartz, A. I. (1966) Honey catalase: occurrence and some kinetic properties. *J. apic. Res.* 5(3): 167–176 B, 789/67

Scherhag, H. (1983) Betriebswirtschaftliche Analyse des Imkerbetriebes. *Allg. dtsch. Imkerztg* 17: 133–135, 173–175; 241–242; 273–274 B, 1230/86

Scheuermeier, U. (1988) Bienenhaltung in Nepal. *Schweiz. Bienenztg* 111: 25–29, 72–77, 123–127 B

Schley, P.; Büskes-Schultz, B. (1987) Die Kristallisation des Bienenhonigs. *Biene* 123: 5–10, 46–50, 114–118 B, 1011, 1342, 1343/87

Schmechel, M. (1981) Honeyman's heat tape. *Am. Bee J.* 121(6): 440 B

Schmid-Hempel, P.; Kacelnik, A.; Houston, A. I. (1985) Honeybees maximize efficiency by not filling their crop. *Behav. Ecol. Sociobiol.* 17: 61–66 B, 140/87

Schmid-Hempel, P. (1987) Efficient nectar-collecting by honeybees. I. Economic models. *J. Anim. Ecol.* 56: 209–218 B

Schmidt, J. O.; Thoenes, S. C. (1987) Honey bee swarm capture with pheromone-containing trap boxes. *Am. Bee J.* 127(6): 435–437 B, 206/88

Schmidt, J. O.; Thoenes, S. C.; Levin, M. D. (1987) Survival of honey bees, *Apis mellifera* (Hymenoptera: Apidae) fed various pollen sources. *Ann. ent. soc. Am.* 80(2): 176–183 B, 870/88

Schneider, G. (1908) Über eine Urwald-Biene (*Apis dorsata*) F. *Z. wiss. InsektBiol.* 4: 447–453 B

Schnepf, E. (1969) Sekretion und Exkretion bei Pflanzen. *Protoplasmatologia* 8(8): 1–181 B, 497/71

Schönitzer, K.; Renner, M. (1984) The function of the antenna cleaner of the honeybee (*Apis mellifica*). *Apidologie* 15(1): 23–32 B, 867/85

Schröder, O. (1954) Blühbeginn wichtiger Trachten und Zeit des Umhängens. *Leipzig Bienenztg (Allg.)* 68(2): 45–47 B, 210/55

Schuette, H. A.; Remy, K. (1932) Degree of pigmentation and its probable relationship to the mineral constituents of honey. *J. Am. chem. Soc.* 54: 2909–2913 B

Schuette, H. A.; Huenink, D. J. (1937) Mineral constituents of honey. II. Phosphorus, calcium, magnesium. *Fd Res.* 2: 529–528 B

Schuette, H. A.; Triller, R. E. (1938) Mineral constituents of honey. III. Sulfur and chlorine. *Fd Res.* 3(5): 543–547 B

Schuette, H. A.; Woessner, W. W. (1939) Mineral constituents of honey. IV. Sodium and potassium. *Fd Res.* 4(4): 349–353 B

Schwan, B. (1955) Fortsatta försök med artificiell uppwärmning av bikupor under våren. *Nord. Bitidskr.* 7(2): 40–42 B, 324/57

Schwarz, H. F. (1948) Stingless bees (Meliponidae) of the Western Hemisphere. *Bull. Am. Mus. nat. Hist.* No. 90: 546 pp. B, 3/52

Sechrist, E. L. (1922) Apiculture tropicale. *Proc. 6 Int. Apic. Congr.*: 64–73 B

Seeley, T. D.; Morse, R. A. (1982) *Bait hives for honey bees* (Ithaca, NY, USA: Cornell University) 8 pp. Inform. Bull. Cornell Co-op. Ext. No. 187 B

Seeley, T. D. (1983) Division of labor between scouts and recruits in honeybee foraging. *Behav. Ecol. Sociobiol.* 12: 253–259 B, 850/84

Seeley, T. D. (1985) *Honeybee ecology: a study of adaptation to social life* (Princeton, NJ, USA: Princeton University Press) 202 pp. B, 893/86

Seeley, T. D.; Levien, R. A. (1987) Social foraging by honeybees: how a colony tracks rich sources of nectar. Pp. 38–53 from *Neurobiology and behavior of honeybees* ed. R. Menzel & A. Mercer B

Segeren, P. (1988) *Beekeeping in the tropics* (Wageningen: Agromisa) 82 pp. Agrodok 32. French edition 1983 B

Severson, D. W. (1984) Swarming behavior of the honey bee. *Am. Bee J.* 124(3): 204–210, 230–232 B, 1198/86

Seyffert, C. (1930) *Biene und Honig im Volksleben der Afrikaner* (Leipzig: Voigtländer) 222 pp. B

Shah, A. M. (1984) Personal communication

Shah, F. A.; Shah, T. A. (1976) A note on the bee activity and bee flora of Kashmir. *Indian Bee J.* 38(1/4): 29–33 B, 128/80

Shah, F. A. (1980) Storing and mailing queen bees in a single cage. *Indian Bee J.* 42(1): 18–19 B, 70/82

Shah, F. A.; Shah, T. A. (1980) Early life, mating and egg laying of *Apis cerana* queens. *Bee Wld* 61(4): 137–40 B, 811/81

Shah, F. A.; Shah, T. A. (1982*a*) The role of Kashmir bees in exploiting beekeeping potential in India. *Indian Bee J.* 44(2): 37–42 B, 75/87

Shah, F. A.; Shah, T. A. (1982*b*) Multiple queen system in *Apis cerana* in Kashmir. *Indian Bee J.* 44(1): 11–12 B

Shah, F. A. (1983) *Fundamentals of beekeeping* (Srinagar: Shah Beekeepers) 64 pp. B

Shah, F. A. (1988) Personal communication

Shao-wen, L. & 5 others (1986) A comparative study of esterase isozymes in 6 species of *Apis* and 9 genera of Apoidea. *J. apic. Res.* 25(3): 129–133 B

Shaparew, V. (1979) Report on clustering bees on

front wall of hive in summer. *Can. Beekeep.* 8(3): 42–45 B

Shaparew, V. (1981) A conical bee escape board. *Glean. Bee Cult.* 109(9): 572–576 B

Shaparew, V. (1986) Beehive ventilation. *Glean. Bee Cult.* 114(9): 466–468, 470, 472 B

Sharma, K. G. (1984) Personal communication

Sharma, O. P.; Thakur, A. K. (1982) Beekeeping in Kangra Valley. *Indian Farming* 31(12): 33–35 B

Sharma, P. L. (1944) Queen introduction. *Indian Bee J.* 6(9/10): 152–154 B

Sharma, P. L. (1960) Observations on the swarming and mating habits of the Indian honeybee. *Bee Wld* 41(5): 121–125 B

Shearer, D. A. & 3 others (1970) Occurrence of 9-oxodec-*trans*-2-enoic acid in queens of *Apis dorsata, Apis cerana* and *Apis mellifera. J. Insect Physiol.* 16: 1437–1441 B, 67/72

Sheesley, W. R.; Atkins, E. L. (1986) Value of in-field honey bee watering sites in alfalfa seed pollination. *Appl. agric. Res.* 1(1): 6–7 B, 1399/87

Sheppard, W. S.; Huettel, M. D. (1987) Subspecific identification of honey bees using mitochondrial DNA analysis. *Am. Bee J.* 127(12): 851 B

Shipolini, R. A. (1984) Biochemistry of bee venom. Chapter 2, pp. 49–85, from *Handbook of natural toxins* Vol. 2, ed. A. T. Tu (New York: Marcel Dekker Inc.) 732 pp. B

Shipolini, R. A. (1986) Personal communication

Short, L. L.; Horne, J. F. M. (1985) Behavioral notes on the nest-parasitic Afrotropical honeyguides (Aves: Indicatoridae). *Am. Mus. Nov.* (2825): 1–46 B, 979/87

Short, L. L. (1986) Personal communication

Showler, K. (1985) *The observation hive* (Bridgwater, UK: Bee Books New and Old) 90 pp. 2nd ed. B

Shuel, R. W. (1955) Nectar secretion in relation to nitrogen supply, nutritional status, and growth of the plant. *Can. J. agric. Sci.* 35: 124–138 B, 83/56

Shuel, R. W. (1959) Studies of nectar secretion in excised flowers. II. The influence of certain growth regulators and enzyme inhibitors. *Can. J. Bot.* 37: 1167–1180 B, 307/65

Shuel, R. W.; Dixon, S. E. (1959) Studies on the mode of action of royal jelly in honeybee development II. Respiration of newly emerged larvae on various substrates. *Can. J. Zool.* 37: 803–813

Shuel, R. W.; Dixon, S. E. (1960) The early establishment of dimorphism in the female honey bee, *Apis mellifera* L. *Insectes soc.* 7(3): 265–282 B, 607/62

Shuel, R. W. (1975) The production of nectar. Chapter 8, pp. 264–282, from *The hive and the honey bee* ed. Dadant & Sons B

Shuel, R. W.; Dixon, S. E.; Kinoshita, G. B. (1978) Growth and development of honeybees in the laboratory on altered queen and worker diets. *J. apic. Res.* 17(2): 57–68 B, 916/79

Shuel, R. W. (1982) Personal communication

Shyu, S. T. (1983) [Economic study of the beekeeping industry in Taiwan.] *J. agric. Econ.* (33): 85–99 *In Chinese*

Sihag, R. C. (1982) Effect of competition with *Parkinsonia aculeata* L. on pollination and seed production in *Medicago sativa* L. *Indian Bee J.* 44(4): 89–90 B, 1044/86

Silberrad, R. E. M. (1969) The remotest colonies. *Br. Bee J.* 97(4192): 126–128 B

Silberrad, R. E. M. (1970) *Bee-keeping in Seychelles* (Republic of Seychelles: Dep. Agriculture) 21 pp. B

Silberrad, R. E. M. (1976) *Bee-keeping in Zambia* (Bucharest: Apimondia) 76 pp. B, 218/77

Simo, K.; Christensen, G. M. (1962) Quantitative analysis of sugars in royal jelly. *Nature, Lond.* 196: 1208–1209 365/69

Simonthomas, R. T.; Simonthomas, A. M. J. (1980) *Philanthus triangulum* and its recent eruption as a predator of honeybees in an Egyptian oasis. *Bee Wld* 61(3): 97–107 B, 251/81

Simpson, B. B. (ed.) (1977) *Mesquite: its biology in two desert scrub ecosystems* (Stroudsburg, PA, USA: Dowden Hutchinson & Ross) 250 pp. B, 280/80

Simpson, J. (1954) Effects of some anaesthetics on honeybees ... *Bee Wld* 35(8): 149–155 B, 204/55

Simpson, J. (1958*a*) The factors which cause colonies of *Apis mellifera* to swarm. *Insectes soc.* 5(1): 77–95 B, 187/59

Simpson, J. (1958*b*) The problems of swarming in beekeeping practice. *Bee Wld* 39(8): 193–202 B, 165/59

Simpson, J. (1960) The age of queen honeybees and the tendency of their colonies to swarm. *J. agric. Sci.* 54(2): 195 B, 71/62

Simpson, J. (1964) Dilution by honeybees of solid and liquid food containing sugar. *J. apic. Res.* 3(1): 37–40 B, 829/64

Simpson, J.; Fairey, E. M. (1964) How efficiently can wax be extracted from old brood combs by simple methods? *Bee Wld* 45(3): 99–103; see also 46(1): 9 (1965) B, 363/69

Simpson, J. (1972) Recent research on swarming behaviour including sound production. *Bee Wld* 53(2): 73–78, 86 B

Simpson, J.; Greenwood, S. P. (1974) Influence of artificial queen-piping sound on the tendency of honeybee, *Apis mellifera*, colonies to swarm. *Insectes soc.* 21(3): 283–287 B, 335/75

Simpson, M. (1954) *Some recent Russian researches on bees and beekeeping* (London: BRA) 36 pp. B, 232/57

Singh, G. D. (1957) Departmental collection of honey and wax, Bhinga Range, Baharal Uttar Pradesh. *Indian For*. 83(2): 113–117 B

Singh, S. (1940) ['Tearing down' of combs.] *Bee Wld* 21(6): 66 B

Singh, S. (1962) *Beekeeping in India* (New Delhi: Indian Council of Agricultural Research) 214 pp. Reprinted 1982 B, 291/64

Skillman, G. J. (1979) Queen bee imports – a trial run. *Anim. Quarant*. 6(2): 12–16 B, 614/81

Slessor, J. & 4 others (1988) Semiochemical basis of the retinue response to queen honey bees. *Nature, Lond*. 332(6162): 354–356 B

Smeltzer, G. G. (1988) Instructions for creaming honey. *Can. Beekeep*. 14(5): 110 B

Smirnov, A. M.; Luganskiĭ, S. N. (1987) [Parasitism of honeybees by *Senotainia*.] *Veterinariya* (6): 43–44 *In Russian* B, 1304/88

Smith, D. A. (1977) The first honeybees in America. *Bee Wld* 58(2): 56 B

Smith, E. (1971) A practical hive lifter. *Glean. Bee Cult*. 99(9): 329 B

Smith, F. G. (1951) Beekeeping observations in Tanganyika, 1950/51. *E. Afr. agric. J*. 17(2): 84–87 B, 207/52

Smith, F. G. (1953) Beekeeping in the tropics. *Bee Wld* 34(12): 233–248 B, 261/55

Smith, F. G. (1954) Notes on the biology and waxes of four species of African *Trigona* bees (Hymenoptera: Apidae). *Proc. R. ent. Soc. Lond. Ser. A* 29(4/6): 62–70 B, 185/56

Smith, F. G. (1958) Communication and foraging range of African bees compared with that of European and Asian bees. *Bee Wld* 39(10): 249–252 B, 69/62

Smith, F. G. (1960) *Beekeeping in the tropics* (London: Longmans) 265 pp. B, 362/60

Smith, F. G. (1961*a*) The races of honeybees in Africa. *Bee Wld* 42(10): 255–260 B, 634/62

Smith, F. G. (1961*b*) The African Dadant hive. *Bee Div. Pamphl. Tanganyika* No. 4: 15 pp. B, 471/62

Smith, F. G. (1964) The hive environment in hot climates. *J. apic. Res*. 3(2): 117–122 B, 116/65

Smith, F. G. (1966*a*) The honey super. *Glean. Bee Cult*. 94(7): 394–396 B

Smith, F. G. (1966*b*) The hive. *Bull. Dep. Agric. West. Aust*. No. 3464: 56 pp. B, 548/67

Smith, F. G. (1967) Deterioration of the colour of honey. *J. apic. Res*. 6(2): 95–98 B, 399/68

Smith, F. G. (1970) The gantry loader. *Apic. W. Aust*. 3(4): 57–63 B, 898/73

Smith, F. G. (1987) Personal communication

Smith, L. B.; Johnson, J. A. (1952) The use of honey in cake and sweet doughs. *Bakers' Dig*. 26(6): 113–118 B, 303/53

Smith, M. V. (1959) The production of royal jelly. *Bee Wld* 40(10): 250–254 B, 312/60

Smith, M. V. (1962) A portable incubator for transporting honeybee brood. *J. apic. Res*. 1: 33–34 B, 882/64

Smith, M. V. (1963) A new design in pollen traps. *Can. Bee J*. 74(4): 4–5, 8 B

Smith, M. V. (1967) Removal of bees from buildings and trees. *Publ. Ont. Dep. Agric*. No. 157: 5 pp. B

Smith, M. V. (1972) Marking bees and queens. *Bee Wld* 53(1): 9–13 B

Smith, R.-K. (1988) Europeanization of honey bees in South Africa. *Am. Bee J*. 128(5): 329–330 B, 184/89

Snelgrove, L. E. (1934) *Swarming: its control and prevention* (Bleadon, Somerset: author) 95 pp.; also many later editions B

Snelling, R. R. (1976) A revision of the honey ants, genus *Myrmecocystus* (Hymenoptera: Formicidae). *Bull. Nat. Hist. Mus. Los Angeles County* (24): 163 pp.

Snelling, R. R. (1987) Personal communication

Snodgrass, R. E. (1925) *Anatomy and physiology of the honeybee* (New York: McGraw-Hill) 327 pp. 2nd ed. (1956) *Anatomy of the honey bee* B, 90/57 Reprinted 1976, 1984 (Ithaca, NY: Cornell University Press)

Soares, A. E. E. (1981) Colméia de fibra de vidro: uma nova realidade na apictultura nacional. *Corr. Apic*. 2(7): 4 B

Søbstad, T. (1988) Økonomien i honning-produksjon. *Birøkteren* 104(2): 56–58 B

Soiffer, B. (1983) Floods destroy 8,000 hives in California. *Am. Bee J*. 123(5): 349–350 B

Solberg, Y.; Remedios, G. (1980) Chemical composition of pure and bee-collected pollen. *Meld. Norg. LandbrHøgsk*. 59(18): 2–12 B, 1463/81

Sommeijer, M. J. (1983) *Social mechanisms in stingless bees* (University of Utrecht: Thesis) 147 pp. B, 790/84

Southwick, E. E. (1987*a*) Cooperative metabolism in honey bees: an alternative to antifreeze and hibernation. *J. therm. Biol*. 12(2): 155–158 B

Southwick, E. E. (1987*b*) Personal communication

Southwick, E. E. (1988) Thermoregulation in honeybee colonies. Pp. 223–236 from *Africanized honey bees and bee mites* ed. G. R. Needham et al.

Spangler, H. G.; Owens, C. D. (1975) Why vibration from truck engines reduces bee flight. *Glean. Bee Cult*. 103(5): 151 B, 432/76

Spangler, H. G. (1984) Attraction of female lesser wax moths (Lepidoptera: Pyralidae) to male-produced and artificial sounds. *J. econ. Ent.* 77(2): 346–349 B, 718/85

Spangler, J. (1974) Zip-lock bag feeding. *Am. Bee J.* 114(9): 339 B, 784/75

Speedy Bee (1983) Miller Honey [Co.] develops dehydrated, pure honey. *Speedy Bee* 12(5): 11 B

Speedy Bee (1988) Dakota Gunness uncapper is revolutionarily simple. *Speedy Bee* 17(6):11 B

Spencer-Booth, Y. (1960) Feeding pollen, pollen substitutes and pollen supplements to honeybees. *Bee Wld* 41(10): 253–263 B, 108/62

Sperling, D.; Caron, D. M. (1980) The movable-comb frameless hive: 'appropriate technology' alternative to Langstroth hive? A pilot study. *Am. Bee J.* 120(4): 284–286, 288–289 B

Spiller, J. (1952) *The house apiary* (Taunton: author) 56 pp. B, 67/52

Sporek, K. F. (1982) Bee repellent for beekeepers. *Glean. Bee Cult.* 110(5): 286 B

Stace, P.; Bond, T. (1985?) *Honey houses: a guide to their design* (NSW, Australia: Dep. Agriculture) 32 pp. B, 993/86

Stanley, R. G.; Linskens, H. F. (1974) *Pollen: biology, biochemistry, management* (Berlin: Springer-Verlag) 307 pp. B, 585/76

Starostenko, E. V. (1968) [Propolization by bees of various races.] *Pchelovodstvo* 88(7): 30 *In Russian* B, 370/71

Steinhobel, F. (1972) The advantage of an 8-inch hive body. *S. Afr. Bee J.* 44(5): 4–5 B

Steinsholt, K. (1983) Honning som søtningsmiddel og smaksstoff i spisesis. *Meieriposten* 72(18): 341, 343, 348 B, 1382/85

Stephen, W. A. (1941) Removal of moisture from honey. *Sci. Agric.* 22(3): 157–169 B

Stephen, W. A. (1971) *An economic analysis of beekeeping operations* (Columbus: Ohio State University) 9 pp. B, 896/73

Stevenson, J. H.; Needham, P. H.; Walker, J. (1978) Poisoning of honeybees by pesticides: investigations of the changing pattern in Britain over 20 years. *Rep. Rothamsted exp. Stn for 1977* (2): 55–72 B, 584/80

Stirenko, V. V. (1983) [Method for processing honey.] *USSR Patent* SV 1,009,401 A *In Russian*

Stratford, C. A. (1987) Personal communication

Strickland, S. S. (1982) Honey hunting by the Gurungs of Nepal. *Bee Wld* 63(4): 153–161 B, 954/83

Stroempl, G. (1977) Distribution and use of basswood and lindens for honey production. *Am. Bee J.* 117(5): 208–301, 322 B, 892/78

Sugden, E. (1986) Research on honey bee environmental impact. *Aust. Bee J.* 67(8): 5–7 B

Svensson, B. (1989) Bee management and company management. From *Beekeeping with* Apis cerana *in tropical and subtropical Asia* ed. P. G. Kevan

Svoboda, J. (1962) [Arsenic poisoning of bees by industrial fumes.] *Sborn. čsl. Akad. zěměd Věd., Rostlinná Výroba*: 1499–1506 *In Czech* B, 186/65

Swan, K. (1956) *A receiver of stolen property* (London: Hon. Society of the Middle Temple) 21 pp. B

Sylvester, H. A.; Rinderer, T. E. (1986) Africanized bees: progress in identification procedures. *Am. Bee J.* 126(5): 330-333 B, 892/87

Szabo, T. I. (1977) Overwintering of honeybee queens 2. Maintenance of caged queens in queenless colonies. *J. apic. Res.* 16(1): 41–46 B, 237/78

Szabo, T. I.; Najda, H. G. (1985) Flowering, nectar secretion and pollen production of some legumes in the Peace River region of Alberta, Canada. *J. apic. Res.* 24(2): 102–106 B, 210/86

Szabo, T. I. (1986) Mating distance of the honeybee in north-western Alberta, Canada. *J. apic. Res.* 25(4): 227–233 B, 105/88

Szabo, T. I. (1988) Sri Lanka: the bee paradise. *Am. Bee J.* 128(6): 405–406, 408–409 B

Taber, S. (1961) Successful shipments of honeybee semen. *Bee Wld* 42(7): 173–176 B, 343/63

Taber, S. (1963) The effect of a disturbance on the social behavior of the honey bee colony. *Am. Bee J.* 103(8): 286–288 B, 759/65

Taber, S.; Owens, C. D. (1970) Colony founding and initial nest design of honey bees (*Apis mellifera* L.). *Anim. Behav.* 18(4): 631

Taber, S. (1986) Floods and mud slides in California destroy some colonies. *Am. Bee J.* 126(4): 248 B

Taber, S. (1987) *Breeding super bees* (Medina, OH, USA: A. I. Root Co.) 174 pp. B, 1248/88

Tabouret, T.; Mathlouthi, M. (1972) Essais de pasteurisation du miel. *Rev. fr. Apic.* (299, 300): 258–261, 301–304 B, 652/77

Tabouret, T. (1977) Vacuum drying of honey. *Apiacta* 12(4): 157–164 B, 718/79

Tabouret, T. (1979) Rôle de l'activité de l'eau dans la cristillisation du miel. *Apidologie* 10(4): 341–358 B, 1057/81

Tabouret, T. (1980) *Contribution à l'étude fondamentale de la pasteurisation du miel et des solutions aqueuses sursaturées de* D-*glucose* (Université de Dijon: Thèse pour. . . . Docteur) 165 pp. B, 615/82

Takenaka, T. (1982) [Chemical composition of royal jelly.] *Honeybee Sci.* 3(2): 69–74 *In Japanese* B, 290/83

Takenaka, T. (1984) [Studies on proteins and carboxylic acid in royal jelly.] *Bull. Fac. Agric. Tamagawa Univ.* No. 24: 101–149 *In Japanese* B, 337/87

Tamura, T. (1985) [Royal jelly from the standpoint of clinical pharmacology.] *Honeybee Sci.* 6(3): 117–124 *In Japanese* B

Tanda, A. S.; Goyal, N. P. (1979) Insect pollination in Asiatic cotton (*Gossypium arboreum*). *J. apic. Res.* 18(1): 64–72 B, 1518/79

Tanzania. Laws and Statutes (1957, amended to 1965) *The produce export (beeswax) rules, 1957* (Dar-es-Salaam) B

Taranov, G. F.; Ivanova, L. V. (1946) [Observations upon queen behaviour in bee colonies.] *Pchelovodstvo* (2/3): 33–39 *In Russian*, English translation in M. Simpson (1954) B, 65/50

Taranov, G. F. (1947) [A method of separating the swarming bees from the parent colony.] *Pchelovodstvo* (4): 22–28 *In Russian*, English translation in M. Simpson (1954) B, 109/50

Taranov, G. F. (1959) The production of wax in the honeybee colony. *Bee Wld* 40(5): 113–121 B, 156/60

Taranov, G. F. (1968) [Brood capping.] *Pchelovodstvo* 88(5): 20 *In Russian* B, 871/73

Taranov, G. F.; Azimov, T. N. (1972) [The length of life of honeybees.] *Pchelovodstvo* 92(3): 16–17 *In Russian* IBRA English translation E1294 B, 777/76

Tasei, J.-N. (1972) Observations préliminaires sur la biologie d'*Osmia* (*Chalcosmia*) *coerulescens* L. (Hymenoptera Megachilidae) pollinisatrice de la luzerne (*Medicago sativa* L.). *Apidologie* 3(2): 149–165 B, 711/74

Taylor, O. R. (1977) The past and possible future spread of Africanized honeybees in the Americas. *Bee Wld* 58(1): 19–30. IBRA Reprint M89 B, 926/77

Taylor, O. R.; Spivak, M. (1984) Climatic limits of tropical African honeybees in the Americas. *Bee Wld* 65(1): 38–47 B, 1224/84

Taylor, O. R. (1985) African bees: potential impact in the United States. *Bull. ent. Soc. Am.* 31(4): 15–24 B, 536/88

Taylor, S. R. (1959) Method and composition for pacifying bees. *US Patent* 2,900,300 B, 125/60

Temple, J. G. (1986) Personal communication

Tennant, K. (1956) *The honey flow* (London: Macmillan) 348 pp. B, 25/57

Terada, Y.; Garofalo, C. A.; Sakagami, S. F. (1975) Age-survival curves for workers of two eusocial bees (*Apis mellifera* and *Plebeia droryana*) in a subtropical climate, with notes on worker polyethism in *P. droryana*. *J. apic. Res.* 14(3/4): 161–170 B, 414/76

Tett, A. (1968) The beekeeper's delictual liability for bees. *Bee Line, Rhodesia* 5(1): 7–10; (2): 7–8 B

Tew, J. (1986) A literature review of the use of plastic bee hives and hive components. *Am. Bee J.* 126(8): 546–547 B

Tew, J. (1987) A single observation on the nest cleaning behavior of a honey bee colony. *Glean. Bee Cult.* 115(7): 411–412 B, 146/88

Tew, J. E.; Bare, C. H.; Villa, J. D. (1988) The Bee Regulated Zone in Mexico. *Am. Bee J.* 128(10): 673–675 B

Thakar, C. V. (1973) A preliminary note on hiving *Apis dorsata* colonies. *Bee Wld* 54(1): 24–27 B, 45/74

Thakar, C. V. (1976) The beekeeping development and research programme of the Khadi and Village Industries Commission, India. *Proc. 1 Int. Conf. Apic. trop. Climates*: 125–134 B

Thakur, A. K.; Garg, R.; Sharma, O. P. (1981) Vertebrate enemies of bees and their control in Dhauladhar mountains (Himachal Pradesh, India). *Indian Bee J.* 43(4): 112–113 B, 425/83

Thimann, R.; Aymard, G. (1982) *Flora apícola de Mesa de Cavacas y sus alrededores* (Guanare, Venezuela: Universidad Nacional Experimental de los Occidentales Ezequiel Zamora) 30 pp. B, 217/84

Thin, A. (1984) Beekeeping on a coral atoll. *Bee Wld* 65(2): 57 B

Thomas Fils SA (1987) Personal communication

Thompson, J. (1976) Beekeeping – the economics of migration. *Australas. Beekpr* 78(4): 78–83 B

Thorne, E. H. (1987) Personal communication

Thrasyvoulou, A. T. & 3 others (1985) Malathion residues in Greek honey. *Apidologie* 16(1): 89–94 B

Thurber, P. F. (1972) Tape your extractor. *Glean. Bee Cult.* 100(8): 247, 249 B

Thurber, P. F. (1983) How to load and move bees. *Glean. Bee Cult.* 111: 4–5; 82–83; 122, 124, 126; 192-193 B

Tikhonov, A. I. & 3 others (1977) [Standardization of propolis.] *Khim.-Farmatsevt. Zh.* 11(12): 113–118 *In Russian*, English translation in *Pharm. Chem. J.* 11(12): 1694–1699 (1978) B, 690/81

Tingek, S. (1986/87) *Life cycle ... in* Apis cerana (Universiti Pertanian Malaysia: Thesis)

Tingek, S. & 4 others (1988) Rediscovery of *Apis vechti* (Maa, 1953): the Saban honey bee. *Apidologie* 19(1): 97–102 B

Tirgari, S. & 4 others (1969) [On the occurrence and biology of little bee (*Micrapis florea* F.) and the first account on its migration (Hymenoptera, Apidae).] *II Natn. Congr. Ent. & Pest Control*: 4 pp. *In Persian* B, 859/71

Titěra & 3 others (1987) Metoda plošného odvčelení

krajiny. *Věd. Pr. výzk Úst. včelař. v Dole u Libčice* 9: 187–197 B, 1272/87

Todd, F. E.; Bretherick, O. (1942) The composition of pollens. *J. econ. Ent.* 35(3): 312–317 B

Toit, D. C. du; Toit, A. P. du (1987) Die winsgewendheid van byeboerdery. *S. Afr. Bee J.* 59(5): 98–106 B, 877/88

Tong, S. S. C. & 3 others (1975) Elemental analysis of honey as an indicator of pollution: forty-seven elements in honeys produced near highway, industrial and mining areas. *Archs envir. Hlth* 30(7): 329–332 B, 654/77

Torchio, P. F. (1985) Field experiments with the pollinator species, *Osmia lignaria propinque* Cresson, in apple orchards V. *J. Kans. ent. Soc.* 58(3): 448–464; also Parts I–IV (1981–1984) B, 702/86

Tosti, A. & 4 others (1985) Propolis contact dermatitis. *Contact Dermatitis* 12(4): 227–228 B, 1362/87

Tóth, G. (1986) [*The characteristics of honey composition, its microbiological evaluation and its importance in pharmacy.*] (Semmelweis Medical University, Hungary: Doctoral Dissertation) 120 pp. *In Hungarian* B, 682/87

Toumanoff, C. (1933) Documentation sur l'apiculture annamite. *Bull. écon. Indochin.*: 169–175 B

Toumanoff, C.; Nanta, J. (1933) Enquête sur l'apiculture au Tonkin. *Bull. écon. Indochin.*: 1015–1048 B

Toumanoff, C. (1939) *Les ennemis des abeilles* (Hanoi: Imprimerie d'Extrême-Orient) 178 pp. B

Towne, W. F. (1985) Acoustic and visual cues in the dances of four honey bee species. *Behav. Ecol. Sociobiol.* 16: 185–187 B, 113/89

Towne, W. F.; Kirchner, W. H. (1989) Hearing in honey bees: detection of air-particle oscillations. *Nature, Lond.* 244: 686–687

Townsend, G. F.; Burke, P. W. (1952) Beekeeping in Ontario for honey production and pollination. *Bull. Ont. agric. Coll.* No. 490: 53 pp. B, 47/54

Townsend, G. F.; Riddell, R. T.; Smith, M. V. (1958) The use of pollen inserts for tree fruit pollination. *Can. J. Pl. Sci.* 38: 39–44 B, 430/64

Townsend, G. F. (1963) Benzaldehyde: a new repellent for driving bees. *Bee Wld* 44(4): 146–149 B

Townsend, G. F.; Smith, M. V. (1969) Pollen storage for bee feed. *Am. Bee J.* 109(1): 14–15 B, 169/71

Townsend, G. F. (1974*a*) Beekeeping and agricultural development. *Wld Anim. Rev.* No. 12: 36–40 B

Townsend, G. F. (1974*b*) Absorption of colour by honey solutions from brood comb. *Bee Wld* 55(1): 26–28 B, 881/74

Townsend, G. F. (1975) Processing and storing liquid honey. Chapter 9, pp. 269–292, from *Honey: a comprehensive survey* ed. E. Crane (1975*a*) B

Townsend, G. F. (1976*a*) Transitional hives for use with the tropical African bee *Apis mellifera adansonii*. *Proc. 1 Int. Conf. Apic. trop. Climates*: 181–189 B, 228/78

Townsend, G. F. (1976*b*) Honey processing and collecting centres in East Africa. *Proc. 1 Int. Conf. Apic. trop. Climates*: 85–92 B

Townsend, G. F. (1978*a*) *Preparation of honey for market* (Ontario: Ministry of Agriculture and Food) 29 pp. B

Townsend, G. F. (1978*b*) A rational approach to standardization of beehive sizes. *Apiacta* 13(4): 145–148 B

Townsend, G. F. (1980) Standardization of bee hive sizes. *Proc. 27 Int. Apic. Congr.*: 477–479 B

Townsend, G. F. (1981) *Introductory apiculture* (Guelph, Canada: University of Guelph) text, filmstrips, taped commentary B, 445/82

Townsend, G. F. (1982) Standardization of beekeeping equipment. *Apiacta* 17(1): 3–4, 15 B

Townsend, G. F. (1983) Personal communication

Townsend, G. F. (1984) *Tropical apiculture* (Guelph, Canada: University of Guelph) text, filmstrips, taped commentary B, 838/86

Townsend, G. F. (1985) *Advanced apiculture* (Guelph, Canada: University of Guelph) text, filmstrips, taped commentary B, 67/88

Tredwell, E. J. (1964) *The Westley hive* (Winchester: Hampshire County Farm Institute) 14 pp. B

Tredwell, E. J. (1976) Personal communication

Trehen, K. N.; Longia, H. S. (1962) Some hints for the collection of *Apis dorsata* Fabr. *Indian Bee J.* 24(4/6): 53–54 B

Tribe, G. D.; Fletcher, D. J. C. (1977) A propolized nest in the open. *S. Afr. Bee J.* 49(4): 5–8 B, 1198/78

Tribe, G. D. (1982) Drone mating assemblies. *S. Afr. Bee J.* 54(5): 99–100, 103–112 B, 1207/83

Tribe, G. D. (1987) *Apis mellifera unicolor*: the honeybee of Madagascar. *S. Afr. Bee J.* 59(3): 50–52 B, 539/88

Tributsch, H. (1982) *When the snakes awake* (Cambridge, MA, USA: MIT Press)

Tryasko, V. V. (1951) [Signs indicating the mating of queens.] *Pchelovodstvo* (11): 25–31 *In Russian* B, 5/53

Tryasko, V. V. (1956) [Repeated and multiple mating of queens.] *Pchelovodstvo* (1): 43–50 *In Russian* B, 234/57

Tryasko, V. V. (1957*a*) [The mating sign of the queen and its characteristics.] *Pchelovodstvo* 34(4): 22–28 *In Russian* B, 186/58

Tryasko, V. V. (1957*b*) [Drones which have mated with the queen.] *Pchelovodstvo* 34(12): 29–31 *In Russian* B, 182/59

Tsuchiya, H. & 5 others (1977) [Studies on a poisonous honey originated from azalea, *Tripetaleia paniculata*.] *Kanagawa-ken Eisei Kenkyusho Kenkyu Hokoku* (7): 19–28 *In Japanese* 621/82

Tucker, K. W. (1978) Abnormalities and noninfectious diseases. Chapter 16, pp. 257–274, from *Honey bee pests, predators, and diseases* ed. R. A. Morse B

Tugwell, I. F. (1985) Personal communication

Tulloch, A. P. (1970) The composition of beeswax and other waxes secreted by insects. *Lipids* 5(2): 247–258 B, 202/73

Tulloch, A. P. (1973) Factors affecting analytical values of beeswax and detection of adulteration. *J. Am. Oil Chem. Soc.* 50(7): 269–272 B, 485/74

Tulloch, A. P. (1980) Beeswax – composition and analysis. *Bee Wld* 61(2): 47–62 B, 680/81

Tysset, C.; Rousseau, M. (1981) Le problème du microbisme et de l'hygiène des miels du commerce. *Rev. med. vet.* 132(8/9): 591–592, 595–600 B, 980/82

Ulrich, W. (1964) Geometrie und Entstehung der Bienenwabe. *Z. Bienenforsch.* 7(3): 62–71 B, 269/66

Underwood, B. A. (1986) *The natural history of* Apis laboriosa *Smith in Nepal* (Cornell University: MS Thesis) 88 pp. B, 62/88

United Kingdom. Laws and Statutes (1982) The Bee Diseases Control Order 1982. *Statutory Instruments* No. 107: 8 pp. B

United States Department of Agriculture, *see* USDA

University of California (1983*a*) Resmethrin and wax salvage. *Newsl. U. C. Apiaries* (February/March): 2–3 B, 911/85

University of California (1983*b*) ABF proposed uniform sanitation code for honey houses. *Newsl. U. C. Apiaries* (*June/July*): 3–8 B

Upadhyay, R. R.; Islampanah, S.; Davoodi, A. (1980) Presence of a tumor-promoting factor in honey. *Gann* 71(4): 557–559 B, 981/82

USA. Laws and Statutes (1980) Viable spores of the microorganism *Bacillus thuringiensis* Berliner: exemption from the requirement of a tolerance. *Fedl Register* 45(166): 56346–56347 B, 979/82

USDA (1980) *Beekeeping in the United States* (Washington, DC: USDA) 193 pp. Agric. Handb. USDA No. 335, rev. ed. B

USDA (1987) World honey situation. *USDA Foreign Agriculture Circular* FS 2-87 B

USSR. Laws and Statutes (1981) [*Veterinary legislation, Vol. 3*.] (Moscow: Kolas) *In Russian*

Vaillant, J. (1986) *Initiation à la génétique et à la sélection de l'abeille domestique* (Le Quesnoy, France: Maison de l'Apiculture) 374 pp. B, 195/88

Valle, O.; Aaltonen, M. (1969) Domestication trials on bumblebees. *Suom. maatal. Seur. Julk.* 113(2): 5–21 B, 510/74

Valli, E.; Summers, D. (1988) *Honey hunters of Nepal* (London: Thames & Hudson) 104 pp. B

Vandenberg, J. D. & 4 others (1985) Survival, behaviour and comb construction by honey bees, *Apis mellifera*, in zero gravity aboard NASA Shuttle Mission STS-13. *Apidologie* 16(4): 369–383 B, 173/87

Van Eaton, C. (1987) Commercial queen production in New Zealand. *Am. Bee J.* 127(11): 773–774, 785 B

Vanhaelen, M.; Vanhaelen-Fastré, R. (1979) Propolis I. Origine micrographie, composition chimique et activité thérapeutique. *J. Pharm. Belg.* 27(4): 889–890 B

Van Lawick-Goodall, J. (1968) The behaviour of free-living Chimpanzees in the Gombe Stream Reserve. *Anim. Behav. Mon.* 1(3): 159–311

Varela, F. G.; Wiitanen, W. (1970) The optics of the compound eye of the honeybee (*Apis mellifera*). *J. gen. Physiol.* 55(3): 336–358 B, 91/75

Velthuis, H. H. W.; Verheijen, F. J. (1963) Why the combination of sun and snow can be fatal to honeybees. *Bee Wld* 44(4): 158–162 B, 808/64

Velthuis, H. H. W. (1985) The honeybee queen and the social organization of her colony. Pp. 343–357 from *Experimental behavioral ecology* ed. B. Hölldobler & M. Lindauer (Stuttgart: G. Fischer Verlag) B, 1219/87

Verma, L. R. (1970) A comparative study of temperature regulation in *Apis mellifera* L. and *Apis indica* F. *Am. Bee J.* 110(10): 390–391 B, 68/72

Verma, L. R. (1984) Beekeeping in northern India: major constraints and potentials. Pp. 148–155 from *Proceedings of the expert consultation* . . . ed. FAO B

Verma, L. R. (1986) [Biological and economic characters of *Apis cerana indica* F.] *Honeybee Sci.* 7(4): 151–157 *In Japanese* B, 817/87

Verma, L. R. (1987) Biology of *Apis cerana* F. in relation to beekeeping development programme in Asia. *Paper prepared for 31 Int. Apic. Congr.*: 55 pp. B

Verma, S. (1983) Studies on the foraging behaviour of *Apis cerana indica* Fab. in Jeolikote (Nainital, India). *Indian Bee J.* 45(1): 5–7 B, 502/86

Vick, J. A.; Brooks, R. B. (1972) Pharmacological studies of the major fractions of bee venom. *Am. Bee J.* 112(8): 288–289 B, 961/73

Villa, J. D. (1986) Performance of Africanized colonies at high elevations in Colombia. *Am. Bee J.* 126(12): 835 B, 584/87

Villanueva, R. (1984) Plantas de importancia apícola en ejido de Plan del Rio, Veracruz, México. *Biotica* 9(3): 279–340 B, 184/87

Villeneuve, J.-L.; Houle, E.; Labonté, J. (1988) Pollen trapping versus honey production field report. *Am. Bee J.* 128(9): 612–613, 641 B

Villières, B. (1987) *L'apiculture en Afrique tropicale* (Paris: GRET; AFVP; ACCT) 220 pp. B

Vismatic Pty Ltd (1988) *Making honey flow* (POB 413, Clayfield, Qld 4011, Australia: author) 4 pp.

Visscher, P. K. (1983) The honey bee way of death: necrophoric behaviour in *Apis mellifera* colonies. *Anim. Behav.* 31(4): 1070–1076 B, 1206/86

Voges, K. (1981) The standard Langstroth beehive. *S. Afr. Bee J.* 53(4): 3–5, 7–8 B

Voges, K. (1983) A standard South African Langstroth beehive? *S. Afr. Bee J.* 55(3): 63–66 B

Vorwohl, G. (1976) Honeys from tropical Africa: microscopical analysis and quality problems. *Proc. 1 Conf. Apic. trop. Climates:* 93–101 B, 1345/77

Vries, R. de (1988) Bee keeping in the Konto River Project area: experiences with the Javanese honey bee (*Apis cerana Javanica*). *Working Paper* No. 17: 18 pp. + appendixes

Vuillaume, M. (1957) Contribution à la psychophysiologie de l'élevage des reines chez les abeilles I. *Insectes soc.* 4(2): 113–156 B, 332/59

Waddington, K. D. (1987) Perception of foraging costs and intakes, and foraging decisions. Pp. 66–75 from *Neurobiology and behavior of honeybees* ed. R. Menzel & A. Mercer B

Wafa, A. K.; El-Berry, A. A. (1971) Nesting behaviour of *Osmia latreillei* Spin. and *Osmia submicans* Mor. *Bull. Soc. ent. Egypte* 55: 363–372 B, 825/77

Wakhle, D. M.; Phadke, R. P.; Shakuntala Nair, K. (1983) Study on enzyme activities of pollen and honey bees (*Apis cerana* Fab.). *Indian Bee J.* 45(1): 3–5 B, 501/86

Wali-ur-Rahman, M. I. C. (1985) Bee-foraging plants at Peshawar, Pakistan. *Pakist. J. For.* 35(2): 71–76 B, 1230/87

Walker, P. (1976) *Annotated bibliography on propolis* (London: Bee Research Association) 14 pp. IBRA Bibliography No. 16 B

Walker, P. (1983*a*) *Beeswax: secretion and use by bees* (London: IBRA) 15 pp. IBRA Bibliography No. 30 B

Walker, P. (1983*b*) *Beeswax: processing* (London: IBRA) 11 pp. IBRA Bibliography No. 31 B

Walker, P. (1983*c*) *Beeswax: composition, properties, adulteration* (London: IBRA) 19 pp. IBRA Bibliography No. 32 B

Walker, P. (1983*d*) *Beeswax: uses and commercial aspects* (London: IBRA) 17 pp. IBRA Bibliography No. 33 B

Walker, P. (1983*e*) *Beeswax: publications of historical interest* (London: IBRA) 9 pp. IBRA Bibliography No. 34 B

Walker, P.; Crane, E. (1987) Constituents of propolis. *Apidologie* 18(4): 327–334 B, 1361/88

Waller, G. D. (1972) Evaluating responses of honey bees to sugar solutions using an artificial-flower feeder. *Ann. ent. soc. Am.* 65(4): 857–862 B, 126/74

Waller, G. D.; Torchio, P. F.; Stith, L. S. (1979) A comparison of *Apis mellifera* and *Osmia fulviventris* as pollinators of caged cotton in Arizona. *Proc. 4 Int. Symp. Poll.*: 299–305 B, 400/80

Walrecht, B. J. J. R. (1962) Over de biologische betekenis van de propolis. *Biol. Jb.* 30: 253–262 B, 576/63

Walrecht, B. J. J. R. (1963) Factors determining the capping of honey. *Bee Wld* 44(2): 77–80 B, 831/64

Walsh, C. (1985) 'Yes, those bee ordinances can be amended or repealed'. *Am. Bee J.* 115(5): 339–340 B

Walsh, F. (1983) *Standard beehives* (Worsley, UK: author) 24 pp. B, 931/86

Walsh, R. S. (1967) An experiment with queen banks. *N. Z. Beekpr* 29(4): 14–17 B, 752/68

Walsh, R. S. (1978) *Nectar and pollen sources of New Zealand* (Wellington: National Beekeepers' Association of New Zealand) 59 pp. rev. ed. B

Walton, G. M. (1974) *Proposals for the metric Langstroth hive* (Palmerston North, New Zealand: Ministry of Agriculture and Fisheries) 42 pp. B, 211/76

Walton, G. M. (1975) The metrication of beekeeping equipment. *Bee Wld* 56(3): 109–119 B, 210/76

Walton, G. M. (1979) Beech honeydew honey – a vast potential. *N. Z. Beekpr* 40(4): 6–9 B, 673/81

Warnke, U. (1976) Effects of electric charges on honeybees. *Bee Wld* 57(2): 50–56 B

Warren, C.; Warren, B. (1982) Plastic frames and foundation. *Bee Wld* 63(3): 104–105 B

Warren-Wren, S. C. (1972) *Willows* (Newton Abbot, UK: David & Charles) 180 pp. B

Warth, A. H. (1956) *The chemistry and technology of waxes* (New York: Reinhold) 602 pp.

WAS (1987) A few thoughts on queen introductions. *WAS Newsletter* 8(2): 345–346 B

Watmough, R. H. (1987) A leaf-cutter bee (Megachilidae) and a carpenter bee (Anthophoridae) as possible pollinators of lucerne (*Medicago sativa* L.) in the Oudtshoorn district. *S. Afr. Bee J.* 59(5): 114 B, 1061/88

Watson, J. K. (1981) *Bee-keeping in Ireland: a history* (Dublin: Glendale Press) 293 pp. B, 1245/81

Watson, L. R. (1927) *Controlled mating of queen bees* (Hamilton, IL, USA: American Bee Journal) 50 pp. B

Wayne, M. (1983) Clinical evaluation of the anti-shock trouser. *Ann. Emerg. Med.* 12: 342–346

Weatherhead, T. F. (1987) Preservation of hive equipment. *Aust. Bee J.* 68(5): 12–17 B, 185/88

Weaver, E.; Weaver, N. (1980) Physical domination of workers by young queen honeybees. (*Apis mellifera* L.: Hymenoptera: Apidae). *J. Kansas ent. Soc.* 53(4): 752–762 B, 142/82

Weaver, N. (1962) Control of dimorphism in the female honeybee. *Science, NY* 138(3544): 995 B, 778/64

Weaver, N. (1979) Possible recruitment of foraging honeybees to high-reward areas of the same plant species. *J. apic. Res.* 18(3): 179–183 B, 499/80

Weaver, N.; Weaver, E. C. (1981) Beekeeping with the stingless bee *Melipona beecheii*, by the Yucatecan Maya. *Bee Wld* 62(1): 7–19. IBRA Reprint M103 B, 1131/81

Weaver, N. (1987) Unpublished

Weaver, N. (1988) Personal communication

Weber, V. (1982) *Imkern leicht gemacht mit dem Alpentrogmagazin* (Munich: Ehrenwirth Verlag) 280 pp. B

Webster, T. C; Peng, Y.-S.; Duffey, S. S. (1987) Conservation of nutrients in larval tissue by cannibalizing honey bees. *Physiol. Ent.* 12(2): 225–231 B, 143/88

Wedmore, E. B. (1945) *A manual of bee-keeping for English-speaking beekeepers* (London: Arnold) 389 pp. 2nd ed. B
Facsimile reprint without plates 1975 (Warminster, Wilts: Bee Books New and Old) B

Weekly Law Reports (1985) Tutton and Others v. A. D. Walter Ltd. *Wkly Law Reps* (25 Oct.): 797–811 B

Wegner, A. M. R. (1949) A remarkable observation on the Indian honey-bee versus the yellow-throated marten from Java. *Treubia* 20(1): 31–33 B, 328/57

Weiss, K. (1983) Experiences with plastic combs and foundation. *Bee Wld* 64(2): 56–62 B

Weiss, K. (1984*a*) *Bienen-Pathologie* (Munich: Ehrenwirth Verlag) 252 pp. B, 239/87

Weiss, K. (1984*b*) Regulierung des Proteinhaushaltes im Bienenvolk (*Apis mellifica* L.) durch Brutkannibalismus. *Apidologie* 15(3): 339–353 B, 890/85

Weiss, K. (1986) *Zuchtpraxis des Imkers in Frage und Antwort* (Munich: Ehrenwirth) 232 pp. B, 190/88

Wellington, W. G.; Cmiralova, D. (1979) Communication of height by foraging honey bees, *Apis mellifera ligustica* (Hymenoptera, Apidae). *Ann. ent. Soc. Am.* 72(1): 167–170 B, 516/80

West Virginia Department of Agriculture (1975) The Kenya hive. *Leafl. W. Virginia Dep. Agric.*: 1 p. B

Wheler, G. (1682) *A journey into Greece* (London: T. Cademan), pp. 411–413 B

Whitcombe, H. J.; Douglas, J. S. (1955) *Bees are my business* (New York: G. P. Putnam's Sons) 245 pp. B, 267/55

Whitcombe, R. P. (1982) Experiments with a hive for little bees: some observations on manipulating colonies of *Apis florea* in Oman. *Indian Bee J.* 44: 57–63, 93–105 B, 814, 827/66

Whitcombe, R. P. (1984) *The biology of* Apis *spp. in Oman with special reference to* Apis florea *Fab.* (University of Durham: PhD Thesis) 621 pp. B, 107/86

White, *see* White, J. W.

White, B. (1987) Personal communication

White, G. (1986) Beekeepers' code of ethics. *N. Z. Beekpr* (189): 9 B

White, J. W.; Maher, J. (1951) Detection of incipient granulation in extracted honey. *Am. Bee J.* 91(9): 376–377 B, 160/52

White, J. W. & 3 others (1962) *Composition of American honeys* (Washington, DC: USDA) 124 pp. Tech, Bull. USDA No. 1261 B, 655/63

White, J. W. (1975*a*) Composition of honey. Chapter 5, pp. 157–206, from *Honey: a comprehensive survey* ed. E. Crane (1975*a*) B

White, J. W. (1975*b*) Physical characteristics of honey. Chapter 6, pp. 207–239, from *Honey: a comprehensive survey* ed. E. Crane (1975*a*) B

White, J. W. (1975*c*) Honey. Pp. 491–530 from *The hive and the honey bee* ed. Dadant & Sons B

White, J. W. (1978) Honey. Pp. 287–374 from *Advances in food research* ed. C. O. Chichester et al. (New York: Academic Press) B

White, J. W.; Doner, L. W. (1978) The $^{13}C/^{12}C$ ratio in honey. *J. apic. Res.* 17(2): 94–99 B, 1073/79

White, J. W.; Rudyj, O. N. (1978*a*) Proline content of United States honeys. *J. apic. Res.* 17(2): 89–93 B, 1064/79

White, J. W.; Rudyj, O. N. (1978*b*) The protein content of honey. *J. apic. Res.* 17(4): 234–238 B, 1446/79

White, J. W. (1979) Spectrophotometric method for hydroxymethylfurfural in honey. *J. Ass. off. analyt. Chem.* 62(3): 509–514 B, 663/81

White, J. W. (1980) Hydroxymethylfurfural content of honey as an indicator of its adulteration with invert sugars. *Bee Wld* 61(1): 29–37 B, 1398/80

White, J. W.; Siciliano, J. (1980) Hydroxymethylfurfural and honey adulteration. *J. Ass. off. analyt. Chem.* 63(1): 7–10 B, 1078/81

White, J. W. (1981) Natural honey toxicants. *Bee Wld* 62(1): 23–28 B

White, J. W. (1984) Instrumental color classification

of honey: collaborative study. *J. Ass. off. anal. chem.* 67(6): 1129–1131 B, 332/87

White, J. W. (1986) Personal communication

White, J. W. & 3 others (1988) Quality control for honey enterprises in less-developed areas: an Indonesian example. *Bee Wld* 69(2): 49–62 B, 1350/88

Whitefoot, L. O.; Detroy, B. F. (1968) Pollen – milling and storing. *Am. Bee J.* 108(4): 138, 140 B, 622/68

Whitehead, S. B. (1954) *Bees to the heather* (London: Faber & Faber) 153 pp. B, 251/54

Whyte, R. (1919) Hannemann – the inventor of the queen excluder. *Bee Wld* 1(7): 142 B

Wichman, F. W. (1976) Daytime move made easy. *Glean. Bee Cult.* 104(9): 356 B

Wickramasinghe, R. H. (1983) Bee keeping and its management on coconut, rubber, cashew, citrus, gingelly, fruit, timber and other plantations. *J. nat. Inst. Plantn Management* 2(1): 94–104 B

Wienands, A.; Madel, G. (eds) (1987) *Bibliographie* Varroa jacobsoni *Oudemans 1904* (Bonn: Univ. Bonn) 210 pp. 3rd ed. B, 229/88

Wiese, H. (ed.) (1985) *Nova apicultura* (Porto Alegre, Brazil; Livraria e Editora Agropecuária) 493 pp. 6th ed. B

Wille, A. (1964) Notes on a primitive stingless bee, *Trigona (Nogueirapis) mirandula. Rev. Biol. trop.* 12(1): 117–151 B, 236/66

Willett, L. S. (1988) Honey pricing: an economist's historical perspective. *Glean. Bee Cult.* 116(7): 400–401, 403 B

Williams, D. (1981) Small-scale cappings cleaner. *N. Z. Beekpr* 42(2): 25 B

Williams, I. H. (1972) Trap-nesting solitary bees for students of biology. *Bee Wld* 53(2): 123–135 B

Williams, I. H. (1985) The pollination of swede rape. *Bee Wld* 66(1):16–22 B, 1044/85

Williams, J. L. (1978) Insects: Lepidoptera (moths). Chapter 7, pp. 105–125 from *Honey bee pests, predators, and diseases* ed. R. A. Morse B

Willson, R. B. (1955) Meet the champions: Miel Carlota! *Glean. Bee Cult.* 83: 329–332; 408–410, 447; 473–476, 500 B, 69/56

Willson, R. B. (1975) World trading in honey. Chapter 14, pp. 355–377, from *Honey: a comprehensive survey* ed. E. Crane (1975*a*) B

Willson, R. B.; Crane, E. (1975) Uses and products of honey. Chapter 15, pp. 378–391, from *Honey: a comprehensive survey* ed. E. Crane (1975*a*) B

Wilson, E. O. (1971) *The insect societies* (Cambridge, MA, USA: Harvard University Press) 548 pp. B, 15/73

Wilson, H. F.; Milum, V. G. (1927) Winter protection for the honeybee colony. *Res. Bull. Wis. agric. Exp. Sta.* No. 75: 47 pp. B

Wilson, W. T.; Nunamaker, R. A. (1983) The incidence of *Nosema apis* in honeybees in Mexico. *Bee Wld* 64(4): 132–136 B, 598/84

Winston, M. L. (1979*a*) Intra-colony demography and reproductive rate of the Africanized honeybee in South America. *Behav. Ecol. Sociobiol.* 4(3): 279–292 B, 180/83

Winston, M. L. (1979*b*) The potential impact of the Africanized honey bee on apiculture in Mexico and Central America. *Am. Bee J.* 119: 584–586, 642–645 B, 906/81

Winston, M. L.; Otis, G. W.; Taylor, O. R. (1979) Absconding behaviour of the Africanized honeybee in South America. *J. apic. Res.* 18(2): 85–94 B, 510/80

Winston, M. L. (1980) Swarming, afterswarming, and reproductive rate of unmanaged honeybee colonies (*Apis mellifera*). *Insectes soc.* 27 (4): 391–398 B, 145/82

Winston, M. L.; Taylor, O. R.; Otis, G. W. (1980) Swarming, colony growth patterns, and bee management. *Am. Bee J.* 120(12): 826–830 B, 146/82

Winston, M. L.; Dropkin, J. A.; Taylor, O. R. (1981) Demography and life history characteristics of two honey bee races (*Apis mellifera*). *Oecologia* 48: 407–413 B, 137/82

Winston, M. L.; Taylor, O. R.; Otis, G. W. (1983) Some differences between temperate European and tropical African and South American honeybees. *Bee Wld* 64(1): 12–21 B, 896/83

Winston, M. L.; Fergusson, L. A. (1985) The effect of worker loss on temporal caste structure in colonies of the honeybee (*Apis mellifera* L.). *Can. J. Zool.* 63(4): 777–780 B, 1207/86

Winston, M. L. (1986) Can package bees and nuclei be produced commercially in British Columbia, Canada? *Am. Bee J.* 126(1): 36–38 B, 220/87

Winston, M. L. (1987) *The biology of the honey bee* (Cambridge, MA, USA: Harvard University Press) 281 pp. B

Wise, A. (1985) Wheelchair beekeeping. *S. Afr. Bee J.* 57(6): 135–137 B

Witherell, P. C. (1972) Flight activity and natural mortality of normal and mutant drone honeybees. *J. apic. Res.* 11(2): 65–75 B, 125/73

Witherell, P. C. (1975) Other products of the hive. Chapter 18, pp. 531–558, from *The hive and the honey bee* ed. Dadant & Sons B

Witherell, P. C. (1985) A review of the scientific literature relating to honey bee bait hives and swarm attractants. *Am. Bee J.* 125(112): 823–829 B

Witherell, P. C.; Lewis, J. E. (1986) Studies on the effectiveness of bait hives and lures to attract honey bee swarms – a possible survey tool for future use in Africanized honey bee eradication programs. *Am. Bee J.* 126(5): 353–354, 356–357, 359–361 B, 957/87

Witter, S. E. (1981) Removal without stings! *Glean. Bee Cult.* 109(5): 280 B

Witters, H. (1987) Het wegen van bijenkasten. *Maandbl. Vlaamse Imkersbond* 73(1): 10–13 B

Wittich, B. (1981) *A taste of honey* (Sherborne, Dorset: Alphabooks) 96 pp. B

Wójtowski, F. (1971) Bioekologiczne i techniczne problemy hod'owli i praktycznego użytkowania pszczół samotnic. *Wiad. ekol.* 17(1): 53–58 B, 516/73

Wollenweber, E. (1985) Chemistry and distribution of flavonoids in leaf resins of desert plants. *Proc. 3 Int. Conf. Chem. Biotechnol. biol. active nat. Prod., Bulg. Acad Sci.* 3: 99–112

Wollenweber, E. et al. (1987) *Z. Naturforsch.* 42*c*: 1030–1034

Wongsiri, S.; Tangkanasing, P. (1986) *Apis cerana* F. beekeeping in Thailand: problems and research needs. *J. sci. Res. Chula. Univ.*: 1–6 B, 1154/87

Wongsiri, S.; Lai, Y.-S.; Liu, Z.-S. (1986) Beekeeping in the Guangdong Province of China and some observations on the Chinese honey bee *Apis cerana cerana* and the European honey bee *Apis mellifera ligustica*. *Am. Bee J.* 126(11): 748–752 B, 815/87

Wongsiri, S. (1989) Queen production. From *Beekeeping with* Apis cerana *in tropical and subtropical Asia* ed. P. G. Kevan B

Wood, W. F. (1983) Anaesthesia of honeybees by smoke from the pyrolysis of puffballs and keratin. *J. apic. Res.* 22(2): 107–110 B, 260/84

Wootton, M. & 3 others (1976, 1978) Effect of accelerated storage conditions on the chemical composition and properties of Australian honeys *J. apic. Res.* 15: 23–28, 29–34; 17: 167–172 B, 1158, 1159/76; 1079/79

Wootton, M.; Edwards, R. A.; Rowse, A. (1978) Antibacterial properties of some Australian honeys. *Fd Technol. Aust.* 30(5): 175–176 B, 1002/80

Wootton, M.; Ryall, L. (1985) A comparison of Codex Alimentarius Commission and HPLC methods for 5-hydroxymethyl-2-furfuraldehyde determination in honey. *J. apic. Res.* 24(2): 120–124 B, 284/86

Woyke, J.; Ruttner, F. (1958) An anatomical study of the mating process in the honeybee. *Bee Wld* 39(1): 3–18 B, 50/62

Woyke, J. (1962) Natural and artificial insemination of queen honeybees. *Bee Wld* 43(1): 21–25, 91 B, 536/63

Woyke, J. (1964) Causes of repeated mating flights by queen honeybees. *J. apic. Res.* 3(1): 17–23 B, 795/64

Woyke, J. (1969) A method of rearing diploid drones in a honeybee colony. *J. apic. Res.* 8(2): 65–74 B, 917/70

Woyke, J. (1973*a*) Reproductive organs of haploid and diploid drone honeybees. *J. apic. Res.* 12(1): 35–51 B, 602/74

Woyke, J. (1973*b*) Instrumental insemination of *Apis cerana indica* queens. *J. apic. Res.* 12(3): 151–158 B, 344/74

Woyke, J. (1975) Natural and instrumental insemination of *Apis cerana indica* in India. *J. apic. Res.* 14(3/4): 153–159 B, 378/76

Woyke, J. (1976*a*) Population genetics studies on sex alleles in the honeybee using the example of the Kangaroo Island bee sanctuary. *J. apic. Res.* 15(3/4): 105–123 B, 924/77

Woyke, J. (1976*b*) Brood rearing efficiency and absconding in Indian honeybees *J. apic. Res.* 15(3/4): 133–143 B, 843/77

Woyke, J. (1977) Cannibalism and brood rearing efficiency in the honeybee. *J. apic. Res.* 16(2): 84–94 B, 503/78

Woyke, J. (1980*a*) International aspects of queen rearing around the world. *Bee Wld* 61(4): 132–137 B, 984/81

Woyke, J. (1980*b*) Evidence and action of cannibalism substance in *Apis cerana indica*. *J. apic. Res.* 19(1): 6–16 B, 472/81

Woyke, J. (1981*a*) *Flora apícola salvadoreña* (San Salvador: Ministerio de Agricultura y Ganadería) 14 pp. B

Woyke, J. (1981*b*) Comparative study of bees throughout the year, in temperate and tropical zones. *Apiacta* 16(3): 107–110 B

Woyke, J. (1984*a*) Survival and prophylactic control of *Tropilaelaps clareae* infesting *Apis mellifera* colonies in Afghanistan. *Apidologie* 15(4): 421–434 B

Woyke, J. (1984*b*) Personal communication

Woyke, J. (1988) Problems with queen banks. *Am. Bee J.* 128(4): 276–278 B, 246/89

Wulfrath, A.; Speck, J. J. (1955) *Enciclopedia apícola. I Tomo* (Mexico, DF: Editora Agricola Mexicana) 479 pp. B, 402/59

Wu Yan-ru; Kuang Bangyu (1987) Two species of small honeybee – a study of the genus *Micrapis*. *Bee Wld* 68(3): 153–155 B, 60/88

Xiao, H. L. (1982) [Techniques for keeping strong colonies of *Apis cerana sinensis*.] *Zhongguo Yangfeng* (2): 16–17 *In Chinese* B, 815/83

Yakovlev, A. S. (1975) Moving bees by daylight. *Proc. 25 Int. Apic. Congr.*: 527–529 B, 535/78

Yang, G. H. & 6 others (1981) [Designing a standard hive for the Chinese honeybee.] *Zhongguo Yangfeng* (6): 13–20 *In Chinese* B, 1166/82

Yeboah-Gyan (1988) Personal communication

Yener, E.; Ungan, Ş.; Özilgen, M. (1987) Drying behaviour of honey-starch mixtures. *J. Food. Sci.* 52(4): 1054–1058 B

Yerly, M. (1980) Hobbyists find strapping takes work, worry out of transporting bees. *Am. Bee J.* 120(11): 784–785 B

Yim, Y. J. (1986) [The effects of thermal climate on the flowering dates of plants in South Korea.] *Korean J. Apic.* 1(2): 67–84 B, 1229/87

Yoirish, N. [Ioirish, N. P.] (1959) *Curative properties of honey and bee venom* (Moscow: Foreign Languages Publishing House) 198 pp. B, 188/61
Reprinted 1977 (San Francisco: New Glide Publications)

York, H. F. (1975) Production of queens and package bees. Chapter 19, pp. 559–578, from *The hive and the honey bee* ed. Dadant & Sons B

Zalewski, W. (1965) Fosfatazy w miodach. *Pszczel. Zesz. Nauk.* 9(1/2): 1–34 B, 565/66

Zambia, Forest Department (1979?) *An introduction to frame hive beekeeping in Zambia* (Ndola, Zambia: Forest Department) 60 pp. B, 155/81

Zander, E.; Böttcher, F. K. (1984) *Krankheiten der Biene* (Stuttgart: Ulmer) 408 pp. B, 1258/86

Zeiler, C. (1984) *Ratschläge für den Freizeit-Imker* (Melsungen, GFR: J. Neumann-Neudamm) 132 pp. B

Zherebkin, M. V. (1964) [Rectal glands of honeybees.] *Pchelovodstvo* 41(3): 34–35 *In Russian* B, 89/66

Zimna, J. (1959) Facelia blekitna jako róslina miododajna. *Pszczel. Zesz. Nauk.* 3(2): 77–102 B, 848/64

Zmarlicki, C.; Morse, R. A. (1963) Drone congregation areas. *J. apic. Res.* 2(1): 64–66 B, 813/64

Zmarlicki, C. (1984*a*) Beekeeping with *Apis mellifera* and mite control in Burma. Pp. 45–49 from *Proceedings of the expert consultation* . . . ed. FAO B

Zmarlicki, C. (1984*b*) Evaluation of honey plants in Burma – a case study. Pp. 57–76 from *Proceedings of the expert consultation* . . . ed. FAO B, 1250/85

Zweigert, K.; Kötz, H. (1977) *An introduction to comparative law* (Amsterdam: North-Holland Publishing Co.) Vol. 1, 385 pp.

Plant index

Page numbers are cited under botanical (Latin) names of the plants, with cross-references from common names. One or more species names may be given on pages indexed under a genus (e.g. *Abies*). Plants in Tables 8.4A and 8.5A, and in Appendix 1, are indexed only if the name occurs elsewhere in the book. Algae, fungi and yeasts are included in the Subject Index, together with bacteria and other micro-organisms.

Geographical name index

Index entries are countries, territories, and continents and their regions.

Subject index

In this Index the words bee, honeybee and *Apis mellifera* are used as little as possible, and information on e.g. honeybee abdomen will be found under abdomen. There are many entries under *Apis*.